Sonja B. Grimm

Resilience and Reorganisation of Social Systems during the Weichselian Lateglacial in North-West Europe
An Evaluation of the Archaeological, Climatic, and Environmental Record

MONOGRAPHIEN

des Römisch-Germanischen Zentralmuseums

Band 128

Römisch-Germanisches Zentralmuseum
Leibniz-Forschungsinstitut für Archäologie

RGZM

Römisch-Germanisches Zentralmuseum
Leibniz-Forschungsinstitut für Archäologie

Sonja B. Grimm

RESILIENCE AND REORGANISATION OF SOCIAL SYSTEMS DURING THE WEICHSELIAN LATEGLACIAL IN NORTH-WEST EUROPE

AN EVALUATION OF THE ARCHAEOLOGICAL, CLIMATIC, AND ENVIRONMENTAL RECORD

Verlag des Römisch-Germanischen Zentralmuseums Mainz 2019

Die vorliegende Arbeit wurde vom Fachbereich 07 – Geschichts- und Kulturwissenschaften der Johannes Gutenberg-Universität Mainz im Jahr 2014 als Dissertation zur Erlangung des akademischen Grades eines Doktors der Philosophie (Dr. phil.) angenommen.

The present work was accepted in 2014 by the department 07 – Historical and Cultural Sciences of the Johannes Gutenberg University Mainz as dissertation to attain the academic degree of a Doctor of Philosophy (Dr. phil.).

Redaktion: Claudia Nickel, Marie Reiter (RGZM)
Satz: Claudia Nickel (RGZM)
Umschlaggestaltung: Reinhard Köster, Claudia Nickel (RGZM)

Bibliografische Information der Deutschen Nationalbibliothek

Die Deutsche Nationalbibliothek verzeichnet diese Publikation in der Deutschen Nationalbibliografie; detaillierte bibliografische Daten sind im Internet über **http://dnb.d-nb.de** abrufbar.

ISBN 978-3-88467-255-6
ISSN 0171-1474

Druck: Memminger MedienCentrum Druckerei und Verlags-AG
Printed in Germany.

CONTENTS

ACKNOWLEDGEMENTS

»It takes a village to raise a child.«
Proverb of unclear African origin

Comparable to a child born by a mother and raised by many, a thesis is usually written by one but accomplished only by the help of many. These »many« I would like to thank in the following part.

First of all my parents, Brigitte and Manfred Grimm, and my whole family have provided me with so much support – financially and emotionally – that I will always be grateful to them. Furthermore, the endurance and cheers of many friends (including some furry ones) also motivated and supported me during the whole process of this project and I feel blessed to have such good friends (and pets).

Scientifically, this Ph. D. project was advised by Prof. Dr. Sabine Gaudzinski-Windheuser who is a rich source of good advice I admittedly could not always take easily. However, one of her phrases served as guideline in this whole project and, presumably, continues in its effect:

> »I'm not interested in points or things, their moving and appearing nor the environment –
> I am interested in humans!«*

Moreover, without the freedom of study she allowed me to enjoy I would not have been able to develop my own grasp at the past. In the grand finale, she pushed me to my limits to write a better thesis than I had considered possible before myself. I'm truly thankful for this instance.

Additionally, Dr. Olaf Jöris accompanied the writing process with many helpful comments. Since my early days at university he introduced me, in particular, to the rich world of natural sciences as well as theoretical approaches in archaeology and as a result influenced my scientific interests significantly. Moreover, he set my records straight whenever necessary.

Dr. Martin Street invested an immense effort proof reading earlier versions. Moreover, from my first year working on this thesis, Martin pushed me gently into the scientific world by taking me to conferences, introducing me to colleagues, encouraging me to write articles, discussing my questions and ideas, and thereby sharing his enormous knowledge about Lateglacial archaeology with me. The present work is in large parts owed to his support and, certainly, his investment cannot be truly appreciated.

Dr. Mara-Julia Weber needs special emphasis because of our fruitful collaborative work on Lateglacial archaeology and, moreover, for her invitation to come to Schleswig for continuing this Ph. D. project at Schloss Gottorf during renovations in Neuwied. Furthermore, within the years since we met on a conference in Cologne 2006 some thousand e-mails were exchanged and hours of discussion on the phone or in person spent, often filled up with information, considerations, and ideas about Lateglacial archaeology of north-western Europe (and beyond). I am very glad to have met such a fantastic colleague and wish to continue the productive collaboration. Moreover, I hope to carry on the human discussion with the true friend I found in her.

A very special thanks is due to Prof. Dr. Berit Valentin Eriksen. Her publications were apparently shaping my own Lateglacial research by a general scientific guidance which compiled detailed studies of lithic artefacts with the results of further archaeological material to understand the behaviour of past hunter-gatherer societies. However, I am particularly grateful for her suggestion of my coming to Schleswig, for pushing me to come to an end, and advising me how to get there. Without her rigorous, sisterly tutoring and kind, midwife-like cheering up, this thesis would not be finished.

* Said in one of the first discussions on the subject of this Ph. D. project, translated from German by the present author.

An adequate thanks can also hardly be given to Prof. Dr. Michael Baales who I consider as one of my important scientific teachers who accompanied my studying life since my first day on an excavation (by chance this site was Niederbieber) and to whom I probably owe my persistent interest in the Lateglacial as a complex world of natural resources and human interactions.

Dr. Geoff Smith and Dr. Susan Harris boldly proof read what no man had read before! They provided useful comments on earlier versions of the text and converted it into a readable English. In the final phase, Dr. Elaine Turner and Dr. Jarod Hutson joined their English proof reading forces to fight my favour of using definite articles and howevers, my dislike of during, at, besides, and other prepositions or the creative use of them as well as my flexible placement of adverbs. Certainly, my English has improved by their comments. I wish to thank them very much for helping me in this ghastly task and, of course, mark all remaining mistakes as my own.

Several graphs and illustrations were improved significantly by good advice from Graeme Carruthers M.A. (Hons) MAAIS PifA for which I'm truly thankful.

Furthermore, for many years my work was embedded in Neuwied/Rhine at MONREPOS Archaeological Research Centre and Museum for Human Behavioural Evolution (MONREPOS) part of the Römisch-Germanisches Zentralmuseum, Leibniz-Forschungsinstitut für Archäologie (RGZM) that is a member of the Leibniz-Gemeinschaft. I was provided with a wonderful place to study and a library containing a rich corpus of Palaeolithic research. Thus, I appreciate these fundamental and essential aids for my project.

In addition, this large institution offered a wide range of fellow colleagues and by the plurality initiated diverse discussions. These discussions with many other colleagues also helped to broaden my views and at the same time sharpen them. Sometimes, these inputs affected more than only scientific matters.

Clearly, fellow archaeologists at MONREPOS consciously or sub-consciously belonged to these discussion partners and, therefore, I like to name the ones present at MONREPOS during my time there in alphabetic order: Anne-Katrin Ackermann M.A., Sylvie Bergmann M.A., Monica Brasser M.A., Enrico Brühl M.A., Dr. Alejandro García Moreno, Sabine Gayck M.A., Dr. Frank Gelhausen, Alexandra Güth M.A., Wolfgang Heuschen M.A., Johanna Hilpert M.A., Dr. Daniela Holst, Dr. Jörg Holzkämper, Dr. Radu Ioviță, Dr. Antje Justus, Dr. Jan Kegler, Dr. Lutz Kindler, Thomas Laurat M.A., Tim Matthies M.A., Dr. Ludovic Mevel, Dr. Luc Moreau, Frank Moseler M.A., Dr. Laura Niven, Heidi Panthermehl M.A., Matthias Probst M.A., Dr. Karen Rubens, Dr. Katsuhiro Sano, Dr. Martina Sensburg, Dr. Aritza Villaluenga Martinez, Monika Wagner, Dr. Stefan Wenzel, and Dr. Sibylle Wolf.

Moreover, only the staff at MONREPOS made a focused scientific atmosphere possible by overtaking some not so pleasant works and for which they deserve many thanks – here is one of them: DANKE, Gabi, Regina, Rosi, Ute! In particular, the »backbone« of the institution, Gabi Gräser, deserves true appreciation for managing all of us, our smaller and bigger problems, and running the whole house meanwhile.

Among the visiting colleagues I would like to mention Dr. Olivier Bignon, Dr. Julia M. Fahlke, Prof. Dr. Paul Pettitt, Dr. Marta Połtowicz-Bobaka, Dr. Felix Riede, and Dr. Tony Stewart. Their expertises gave further interesting »food for thought« in the process of this project.

In addition, many colleagues kindly provided me with literature and information and, furthermore, some of them were so kind to discuss drafts and ideas with me:

Prof. Dr. Nick Barton, Prof. emer. Dr. Dr. h. c. Gerhard Bosinski, Dr. Anne Bridault, Keri Brown M. Phil., Dr. Chantal Conneller, Dr. Dorothée Drucker, Prof. Dr. Dobrochna Jankowska, Dr. Kevan Edinborough, late Prof. Dr. Klaus Felix Kaiser, Dr. Tim Kerig, Hannes Napierala M.A., Drs. Marcel Niekus, PD Dr. Jörg Orschiedt, Dr. Valéry Sitlivy, Dr. Iwona Sobkowiak-Tabaka, Prof. Dr. Boris Valentin, and Dr. Jürgen Vollbrecht.

Moreover, late Dr. Roger Jacobi accumulated a great wealth of data about the (Lateglacial) archaeology of the British Isles and, therefore, will always be remembered. He transferred his database to Wessex Archae-

ology for further use. Wessex Archaeology, in the persons of Dr. Julie Gardiner and Jens Neuberger, made kindly parts of the PaMeLA (Palaeo-MesoLithic Artefacts) database available to me that I want to appreciate here although time did finally not allow to use this source in this project.
Special thanks are also due to Dr. Bernhard Weninger from Cologne and Dr. Tom Higham from Oxford for always answering kindly my questions about sometimes real basics of ^{14}C dating. Remaining mistakes in the present study are clearly owed to my ignorance.
In Schloss Gottorf, Schleswig, I was warmly welcome and had a good working atmosphere and schedule which I deeply appreciate. The kind and open help and constant encouragement from so many of the people in Schleswig shall be acknowledged here by mentioning some of them (in alphabetic order): Dr. Klaus Bokelmann, Dr. Aikaterini Glykou, Dipl. Karin Göbel, Dr. Julia Goldhammer, Dr. Daniel Groß, Dr. Sönke Hartz, Dipl. Charlotte Hegge, Dr. Vera Henkelmann, Ditte Skov Jensen, Wolfgang Lage, Dr. Harald Lübke, Dr. John Meadows, Dr. Elena A. Nikulina, Harm Paulsen, Thomas Poelmann, Dr. Ulrich Schmölcke, and Maike Tessars.
Some foundations and institutions provided financial or material support to this project:
The RGZM appreciably sponsored a license of the Microsoft Office package and made a licensed version of ArcGIS® 9.3 with a hardware key available for me. Moreover, application courses on GIS programs as well as a technical support not only for GIS questions was generously provided by the RGZM. Previously, ESRI Geoinformatik GmbH Kranzberg supplied me kindly with an ArcView® 9.0 Student Edition (EVA841553897) and 9.2 Student Edition (EVA862660258) within their graduate student programme. My thanks are also due to the Deutsche Forschungsgemeinschaft which founded travel expenses (KON 378/2006 GR 3276/1-1). The Prinz Maximilian zu Wied Foundation e.V. granted me a two year scholarship for which I am grateful. Finally, the Robert Bosch Foundation granted money to the former »Museum für die Archäologie des Eiszeitalters« for the »Denkwerk Projekt« and thereby financed me the opportunity to present findings of Pleistocene archaeology to the public.
To all of thee I hail:
»Hooray! The child has come of age!«

INTRODUCTION

Since the Club of Rome published its seminal study on »The limits of growth« (Meadows et al. 1972) in 1972, the interaction of the human population with its surrounding ecosystem and, in particular, the handling of finite resource supplies became a topic of major public debate. The publicity of this report fostered further studies on climate and environmental developments and the relation of these factors to human societies. These succeeding studies aimed to refine the data for simulation approaches to allow increasingly precise predictions of future developments. Subsequently, the simulations should provide maintainable solutions for political and social policy-makers concerning the most appropriate human behaviour in the future. Especially, the ability of past societies to adjust their policies of population growth and economic demands in response to climate and environmental change were investigated as examples of the ecological tolerance and resilience of social systems.

Many studies on past societies identified short-term shifts in the palaeoclimatic record as major causes of crisis which sometimes led to a collapse and/or reorganisation of societies during the Holocene (Weiss/Bradley 2001; Haug et al. 2003; Tainter 2008, 44-45). In contrast, Pleistocene social systems were frequently found to remain stable and unaltered across relatively long periods which could also encounter some marked shifts in the palaeoclimatic record (cf. Bettinger/Richerson/Boyd 2009). Perhaps, this inertia of Pleistocene societies is due to more flexible subsistence strategies of hunter-gatherer groups which were adapted to insecure environments and sustained stabilizing social security networks. Small alterations in the resource availability resulting from local environmental factors such as hydrological changes, vegetation development, variation in the behaviour of prey species, or natural disasters could be faced by these flexible, well-connected, and mobile groups with adaptation of single parts of the archaeologically observable record. These abilities formed the basis of the highly resilient social systems.

However, examples such as the changes in the subsistence strategies in northern German Mesolithic societies possibly related to the decline of hazelnut patches in response to a short cold episode in the Holocene (»8.2 kyrs event«; Holst 2010) suggested that the flexibility of such hunter-gatherer economies could lead to significant changes of the mobility and territorial patterns. Moreover, reorganisation of social systems was also observed in the Pleistocene record raising the question for the limits of resilience of these systems. For instance, the significant environmental changes provoked by abrupt climate shifts in the Weichselian Lateglacial (c. 16,200-11,700 years ago) and at the onset of the Holocene were discussed as a cause for the Neolithic revolution in the Levant (Hayden 1981; Bar-Yosef 1998; Maher/Banning/Chazan 2011; Simmons 2007; Rosen/Rivera-Collazoa 2012). This consecutive scenario suggested that confronted with regional to supra-regional changes flexible strategies of hunter-gatherers could fail to offer security and the necessity of a major reorganisation of the inert social system would arise. In addition, the stability and consequential predictability of the natural environment were suggested as important factors in the establishment of new behavioural patterns (Bettinger/Richerson/Boyd 2009).

These suggestions demonstrated the close connection to studies on strategies of hunter-gatherer groups to reduce risk and uncertainty. These strategies aimed to compensate for stress caused by minor variations of their environments (e.g. Halstead/O'Shea 1989a; Stein Mandryk 1993; Winterhalder/Leslie 2002). Natural disasters such as volcanic events form one example for these sometimes significant, but relatively short-termed variations of the environment. Although these short-lived events are useful as correlation points of the chronostratigraphic records, their impact on past social systems remained controversial (Grattan 2006; Weber/Grimm/Baales 2011; Cooper/Sheets 2012).

An important natural disaster affecting large parts of north-western Europe at the end of the Weichselian Lateglacial Interstadial was the Laacher See volcanic eruption (LSE; *c.* 13,000 years ago). Besides representing an important marker horizon, the Laacher See tephras (LST) covered and preserved mid-Weichselian Lateglacial horizons and numerous remains of hunter-gatherer groups on large areas in the vicinity of the volcano in the Central Rhineland. These remains were attributed to the *Federmesser-Gruppen* (FMG; see Material-Archaeology-Archaeological groups, p. 65-74) due to some characteristic artefact types. Thus far, near Bad Breisig a single concentration of archaeological remains still attributed to the mid-Weichselian Lateglacial represented the only assemblage recovered on top of the LST in this area (Baales/Grimm/Jöris 2001). These remains were still attributable to the FMG (see p. 163-168) suggesting that hunter-gatherer groups of this region could compensate this short-term natural disaster by relying on their widespread social system. In addition, a horizon which was stratigraphically lower than the mid-Weichselian Lateglacial horizon also yielded archaeological material. This material was well protected by a significant loess cover topped by the LST. These well preserved remains were classified as Late Magdalenian (see p. 56-65). Rich environmental records found in the two horizons below the LST allowed detailed reconstructions of the landscape inhabited by these hunter-gatherers and revealed very different habitats: A cold grass steppe landscape with horse, reindeer, arctic fox, and arctic hare in the loess covered period (Brunnacker 1978a; Street et al. 2006) and a temperate, light forest environment with red deer, aurochs, elk, and beaver immediately underneath the LST (Baales 2002; Baales 2006a). In addition to the landscape and differences in the archaeological material were apparent in the prey fauna, equipment, and settlement patterns as well as in exploitation strategies of lithic resources (Floss 1994; Baales/Street 1996; Street et al. 2006). According to the radiometric and chronostratigraphic results, the two different behavioural complexes prevailed during phases which lasted almost 1,000 years, whereas the intermediate period of change must be set from approximately 15,300 to 14,000 years ago. Significant climate and environmental changes of the Weichselian Lateglacial were recorded during this period in high-precision archives from Greenland and from north-western Europe (Steffensen et al. 2008; Litt et al. 2001). Occasionally, the terrestrial environment responded with some time lags to the climate changes (Jones et al. 2002; Litt/Schmincke/Kromer 2003). However, once a new stable ecosystem regime was established it lasted for several centuries. Thus, the hunter-gatherer communities inhabiting north-western Europe were confronted with important changes of their ecosystem during the Weichselian Lateglacial comparable to the groups in the Levant. In Northwest-Europe, long-term changes led to variations in the supply of essential resources. In consequence, climate and environment were assumed as triggers leading to the differences observed in the archaeological record (e.g. Bosinski 1989; Floss 2002).
However, although the chronological sequence is known in some detail for the Central Rhineland, only few comparative studies dealt with when and how exactly the differences between the Late Magdalenian and the FMG appeared. Yet, only comprehensive knowledge about tempo and mode of these changes makes causal assumptions of the precise triggers and a characterisation of the process possible. Probably, the small number of diagnostically conclusive assemblages which date to the intermediate period in Central Rhineland prohibited a more detailed study of the changing patterns. Nevertheless, comparable assemblages to the well stratified sites from the Central Rhineland were found from the middle-ranged mountain regions of Western and Central Europe up to the eastern border of Poland and in most parts of south-western Europe (e.g. Weniger 1989; Kozłowski/Kozłowski 1996, 81-84; Leesch/Cattin/Müller 2004; Ginter/Połtowicz 2007; Valentin 2008a). Thus, the occurrence of these patterns can be studied based on material from other regions such as northern France where more assemblages from the intermediate period were recovered (Fagnart 1997; Bridault/Bignon/Bemilli 2003; Valentin 2008a). In particular, the archaeological material from the adjacent western upland zone as well as the Paris Basin was considered to form a chronological and, perhaps, ethnic entity during the Late Magdalenian (Floss/Terberger 2002, 138;

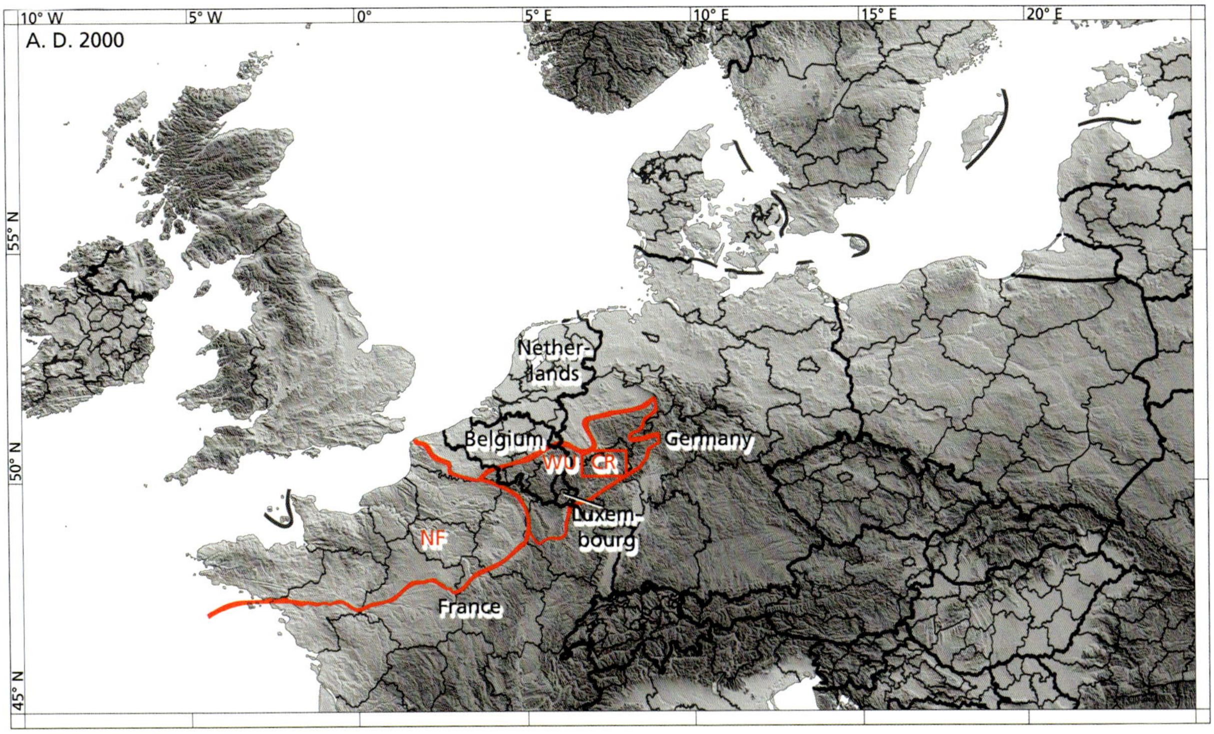

Fig. 1 Map of study areas (red) in north-western Europe with modern national (dark grey) and regional borders (medium grey; data source: Global Administrative Areas, http://www.gadm.org/country; freely available for academic and other non-commercial use). **CR** Central Rhineland. – **WU** western upland zone. – **NF** northern France. – For further details see Material-Environment-Landscapes , p. 31-40, and text.

cf. Arts/Deeben 1987). Furthermore, FMG assemblages comparable to those from the Central Rhineland were also found in northern France (Bodu/Valentin 1997; Bodu 2000a) suggesting the continuity of connections among these regions. In consequence, a common transformation process for these regions can be assumed. Thus far, no unambiguous FMG site was recovered in the upland zone between the Central Rhineland and northern France (De Bie/Vermeersch 1998; cf. Arts 1988) but numerous FMG sites are known from the adjacent lowlands (De Bie/Van Gils 2006). This difference in the distribution of sites is possibly explicable by the lack of open air sites, which were clearly preferred by the FMG, in the western uplands. Moreover, the Upper/Final Palaeolithic and Mesolithic levels in the cave sites were often found disturbed by natural agents or undocumented human interference. Nevertheless, based on the similarity of the FMG assemblages in the surrounding of the western upland zone, the same development as in northern France and the Central Rhineland is also assumed for this region. Therefore, in addition to the Central Rhineland, the framing western upland zone and northern France (**fig. 1**; for further details see p. 38-40) were selected to form the study area.

In France, the process of change between the Late Magdalenian and the FMG was termed Azilianisation (Bordes/Sonneville-Bordes 1979). Modern studies from northern France treated the Azilianisation as an evolutionary development (Bodu/Valentin 1997; Valentin/Fosse/Billard 2004), whereas the abbot Henri Breuil regarded the Azilian as a revolutionary process (Breuil 1913). The issue if changes occur in a revolutionary or in an evolutionary process is not restricted to human social systems. Comparable models on the »tempo and mode in evolution« (Simpson 1944) were discussed in palaeobiology (Eldredge/Gould 1972; Gould 1994; Eldredge et al. 2005). The observed scales and, in particular, the temporal frame of reference were thereby identified as important aspects for the interpretation of the examined process (Thomson 1992).

Consequently, the difference between the two views on the Azilianisation process as understood in the present study consists in the appearance of the changes: An evolutionary process is assumed as a constant adaptation. Thus, the differences between the Late Magdalenian and the FMG would arise from the accumulation of several, small-scale adaptations of hunter-gatherers over the more than 1,000 years between the two periods. These gradual effects on the various parts of life allow for a distinction of a succession of changes in the behaviour of the past hunter-gatherers. In contrast, a revolutionary process is suggested as a concentration of adaptations in a major shift. This major shift could be explained by a chain-reaction of closely related spheres of hunter-gatherers' life. For example, in mobile hunter-gatherer groups subsistence strategies and mobility patterns are closely related (see above). Moreover, changes in the mobility pattern could subsequently affect social networks. The chain-reaction could have been started when a threshold or limit was passed. The abrupt climate and environmental changes formed a strong external stimulus which could have pushed the flexible adaptive system of hunter-gatherers to its limits. However, if changes of social systems occurred in this threshold-like manner the factor causing the overstepping of the limit could be identifiable at the onset of the shift. Nevertheless, considering isolated lines of evidence cannot produce a reliable distinction between the two process types because single parameters also switch rapidly in an evolutionary process. Furthermore, both types affect probably the complete way of life of hunter-gatherers in a long temporal study. In the present study, the two types of process (revolution and evolution) are therefore assumed to be distinguished by the tempo in which they affect large parts of human life.

To reveal the tempo and mode of changes from the Late Magdalenian to the FMG as a case study and to examine these changes in regard to climate and environment as potential triggers, different lines of evidence (climate, environment, archaeology) need to be set in a common frame of reference. Thus, relevant material for the climate, environmental, and archaeological development from the relevant period are assembled in the present study. The relevant period spans the time from the Late Magdalenian to the FMG in Central Rhineland which approximates 16,200-12,800 years before 2000 A.D. (b2k). In the Lateglacial, the radiocarbon timescale is also commonly used due to the absolute dating of organic remains being essential for the construction of chronologies. On this timescale, the study period correlates to c. 13,200-10,800 ^{14}C years before present (^{14}C-BP) which refers to 1950 A.D. The material assembled in the present study includes Late Weichselian climate records (see Material-Climate, p. 7-30), environmental archives (see Material-Environment, p. 30-48), and radiocarbon databases (see Material-Databases, p. 49-53) as well as archaeological inventories (see Material-Archaeology, p. 53-244). In the present study, climate records form the chronostratigraphic baseline against which the environmental change and the process of human adaptation are examined to clarify the chronological succession. The environmental records as well as the archaeological assemblages are qualitatively examined concerning their precise chronostratigraphic position and the reliability as a high-resolution record. Then the material is correlated by the use of annually laminated archives and radiocarbon dating and set in the same high-precision time frame. Radiocarbon dates require a reliable calibration to be related to the chronostratigraphy of climate and environmental archives. Thus, the calibration records are also reviewed and a generally applied calibration curve (Weninger/Jöris 2008) is further refined for the Weichselian Lateglacial. By the use of pollen stratigraphies and directly dated macro-remains as well as directly dated faunal remains the progress of environmental change is modelled on regional scales. Furthermore, to identify the differences between the Late Magdalenian and the FMG, the archaeological material of the Central Rhineland is introduced per assemblage and with several standard values of the assemblages collected to equalise the data sets. The assemblages are grouped according to the resource exploitation strategies including subsistence but also indications of site function and mobility. The sorting into the groups is based on the attribution to predefined classes and the formation of indices. The material from northern France is presented in a comparable manner to locate changes in several lines of evidence. These lines of evidence and the changes therein are set in relation to the high-resolution chronostratigraphy by radiocarbon dating of the archaeological assemblages. This multi-linear approach is chosen because, as stated previously, small alterations are probable to occur in the behaviour of hunter-gatherers as buffering mechanism to perhaps local changes. Nevertheless, a major reorganisation is reflected by changes of numerous patterns in several lines of evidence occurring in a short time period. If this shift is observed a major trigger such as climate amelioration or forest development can be sought shortly before or at the onset of changing patterns. However, if indications for a major reorganisation are absent in the archaeological record small adjusting alterations were sufficient to cope the climate and environmental changes of the Weichselian Lateglacial. In this case, the resilience of the Late Magdalenian would have caused an adaptive process leading to the transformation into the FMG. Thus, the present study aims to characterise whether the change from the Late Magdalenian to the FMG reflected a revolutionary or an evolutionary process.

Studying this process of change in the Weichselian Lateglacial and, in particular, in north-western Europe is furthermore advantageous because adoption effects can be neglected. During the Last Glacial Maximum (LGM) most parts of Central Europe were de-populated (cf. Terberger/Street 2002; Verpoorte 2009) and only resettled by a major Magdalenian expansion coming from a south-western European refugium. This expansion direction was further supported by genetic data (Pereira et al. 2005). Thus far, no significant influence from people and/or ideas coming from the south-eastern refugium was observed in the archaeological record*. Hence, by the onset of the studied period the Late Magdalenian is the only archaeological group observable in Northwest-Europe. Furthermore, by the end of the studied period the known groups in north-western Europe descended directly from this Late Magdalenian. Changes occurred consequently within a single genetic inheritance and social transmission system. Therefore, the process of change without interfering adoption processes reflects an adaptive cycle of a Pleistocene social system. Increasing differentiation of groups occurred during the Weichselian Lateglacial and the more so during the Holocene. As a result, studying the process of change in the form of an undisturbed adaptive cycle is no longer possible afterwards.

Since further acculturation inputs are therefore excluded, understanding the process of change in the Weichselian Lateglacial represents a paragon for the behaviour of social systems under severe ecological stress. This knowledge contributes to the formation of general patterns of human behaviour under stress and is indispensable as an elementary component in creating future scenarios for a globalised world.

* In contrast, aDNA data meanwhile suggests a population turnover during the studied period by people coming from south- to south-eastern areas (Posth et al. 2016; Fu et al. 2016).

MATERIAL

To examine the process of change in human behaviour and how this process is motored by climatic and environmental change the present study is based on three types of archives: the climate, the environment, and the archaeology which comprises the remains of human behaviour.

In archaeological analyses, climate and environmental change were often taken as synonymous (Fort/Pujol/Cavalli-Sforza 2004; Gamble et al. 2004; Blockley et al. 2006; Riede/Edinborough/Thomas 2009), although high-resolution multi-proxy studies demonstrated that climate and environmental change were sometimes neither contemporaneous nor straightforwardly connected. Besides climatic parameters such as atmospheric and hydrological data, the environmental development is also dependent on factors such as geomorphological setting, pedogenesis, and/or migration rates of biotic populations (Lotter et al. 1992, 198; Hoek 2001; Hoek/Bohncke 2001, 1258. 1262; Litt et al. 2001, 1244; Jones et al. 2002; Price 2003). Therefore, an explicit difference is made between climate[1] as seen by proxies for temperature, moisture, or wind activity and environment[2] such as soil development, plant growth, or fauna presence. By this distinction, changes in human behaviour can be related more precisely in a chronological scale to specific exogenous forces.

CLIMATE AND CHRONOSTRATIGRAPHIC ARCHIVES

The climate archives (**fig. 2**) represent one of the three main types of archives in the present study combined with the environmental and the archaeological records. However, the climate is assumed as the main motor for environmental changes and, therefore, the process of climate change is used as basic chronostratigraphy in the present study.

The climate of north-western Europe is strongly governed by the North Atlantic and this climate regime is well recorded in the isotopic record of the Greenland ice-cores. Thus, the eventstratigraphy based on oxygen isotopes received from Greenland ice-cores (Björck et al. 1998; Walker et al. 1999; Jöris/Weninger 2000) is chosen as the basic chronostratigraphic terminology in the present study. Due to the rapid hemispheric ventilation in the atmosphere and, therefore, the relatively constant ratio of atmospheric oxygen isotopes across the hemisphere, the atmospheric oxygen isotopic record is well suited as a tool for correlating distant areas and various biospheres in a common chronostratigraphic framework. In contrast, the correlation and use of litho- and biostratigraphic nomenclature as chronostratigraphic units led to some confusion in the Late Weichselian terminology (see p. 41-45; cf. de Klerk 2004) and is therefore avoided in the present study.

1 According to the Intergovernmental Panel on Climate Change (IPCC), which was established by the United Nations Environment Programme (UNEP) and the World Meteorological Organization (WMO): »›Climate‹ refers to the average weather in terms of the mean and its variability over a certain time-span and a certain area.« (Baede et al. 2001, 87), whereas the weather »is the fluctuating state of the atmosphere around us, characterised by the temperature, wind, precipitation, clouds and other weather elements.« (Baede et al. 2001, 87). Hence, atmospheric values on temperature, precipitation, and wind activity are gathered to allow for statements on the change of Lateglacial climate.

2 By the term environment the natural environment defined as »all living and non-living things that occur naturally on Earth« (see http://en.wikipedia.org/wiki/Natural_environment [1/8/2011]) is meant and furthermore, the central concept of Frans Klijn and Helias A. Udo de Haes is followed regarding the environment »... as an ecosystem in order to incorporate all relevant interactions in the physical [i. e. natural] environment.« (Klijn/de Haes 1994, 90).

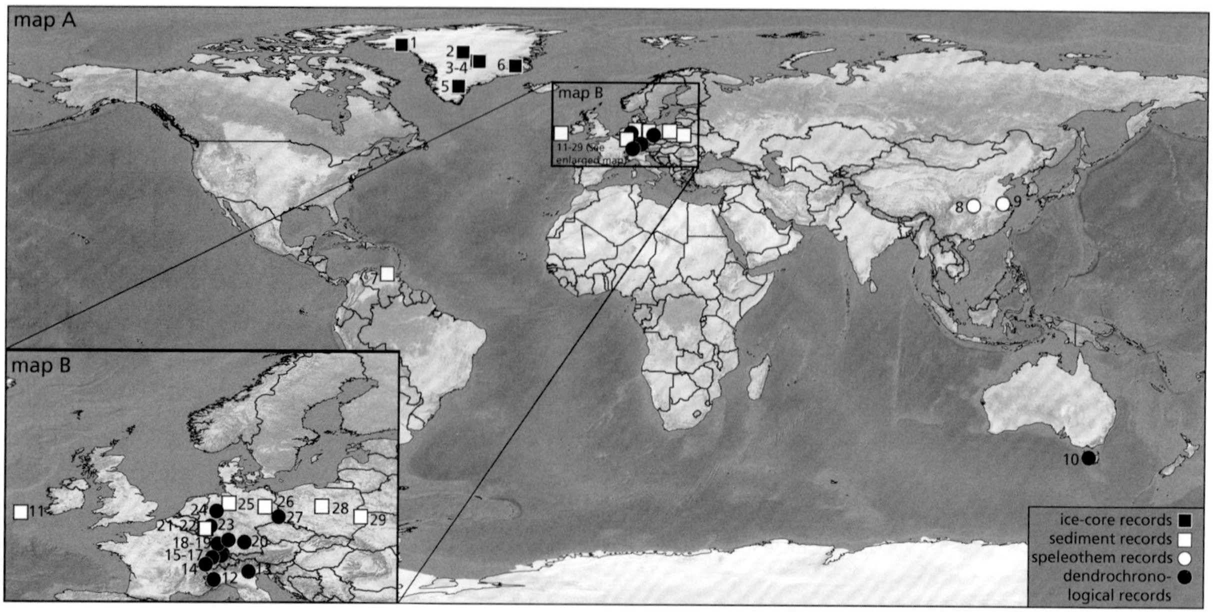

Fig. 2 **A** World map and **B** map of north-western Europe with sites yielding climate and chronostratigraphic relevant data. **1** Camp Century; **2** NGRIP; **3** GISP2; **4** GRIP; **5** DYE-3; **6** Renland; **7** Cariaco basin; **8** Qingtian cave; **9** Hulu cave; **10** Tasmania (Huon pines); **11** MD01-2461; **12** Avigliana; **13** Revine; **14** Ollon; **15** Avenches; **16** Zürich area (Landikon, Gänziloh, Birmensdorf, & Zürich-Wiedikon); **17** Dättnau; **18** Upper Rhine material; **19** Main material; **20** Danube, Isar, Iller, & Günz; **21** Meerfelder Maar; **22** Holzmaar; **23** Kruft; **24** Warendorf; **25** Hämelsee; **26** Rehwiese; **27** Lausitz (Cottbus, Reichwalde, Lohsa, Scheibe); **28** Lake Gościąż; **29** Lake Perespilno. – For references and further details see text.

However, terrestrial sequences with oxygen isotope records are limited and not present at any archaeological site. Therefore, further chronostratigraphic systems such as varve counting, correlation of marker horizons, and/or radiocarbon dating are necessary additions to build a solid chronostratigraphic sequence of the Weichselian Lateglacial in north-western Europe. However, radiocarbon dates require a calibration to solar years. The necessary data for this calibration curve were achieved from deep sea and terrestrial records which are also introduced and evaluated in the following. Usually, these records also produced climate data and, thus, by the use of these records, a direct correlation of climate with the radiocarbon calibration data is possible. This correlation facilitates the assembling of the different chronostratigraphic systems in a common time frame.

To create a common timescale for these different chronostratigraphic systems two points of reference are distinct in the present study: The year 2000 A.D. (b2k)[3] introduced by the construction of ice-core chronostratigraphy (Rasmussen et al. 2006) and the year 1950 A.D. (BP) used in the context of radiocarbon dating (Godwin 1962). The still very common use of the BP scale outside radiocarbon dating led to some confusion due to the incongruence of radiocarbon years with solar years (cf. van der Plicht/Hogg 2006). Thus, the convention of using an additional cal. (calendar, Stuiver/Kra 1986, Pearson 1987) set before the reference point is now commonly used to separate ^{14}C ages from calendar ages (including calibrated ^{14}C ages). In

3 The year 2000 A.D. (b2k, Vinther et al. 2006; Rasmussen et al. 2006) is preferred as a reference point in the current project when correlating stratigraphic sequences and, in addition, the point is applied to the calibrated ^{14}C ages instead of the recommended »cal. BC« because of easier comparability of the stratigraphic sequences. This reference year is chosen because the dating of many discussed sequences and assemblages neither originated from radiometric measurements as should be concluded from the use of the BP scale nor representing exact prehistorical dates as might be concluded when referring to the historic timescale (BC/A.D.). Instead, the dating of the episodes introduced in the present study represent correlations and scientific interpretations of various chronostratigraphic approaches.

the present work, ^{14}C and other radiometric measurements will be marked additionally (^{14}C-BP; TL-BP) to prevent any further misreading.
The studied time window comprises a period of approximately 3,500 calendar years (or 2,800 ^{14}C years) between the time characterised by the cold steppe horse hunters of the Late Magdalenian and the time of the FMG-assemblages originating from forest red deer hunters in the Central Rhineland (cf. Street/Terberger 2004, 294f.). The studied time period belongs completely to the Weichselian and, more precisely, to the Late Weichselian or Weichselian Lateglacial. The specification of the terms referring to the Weichselian period is therefore not further used in the following since it is regarded as unnecessary precision. The term Lateglacial is used in the present work according to Walker et al. 1999 and refers in the Greenland eventstratigraphy to the time from the onset of GS-2[4] to the onset of the Holocene (GH). In general, the Lateglacial can be subdivided into three main periods: The Upper/Late Pleniglacial (GS-2, furtheron: Late Pleniglacial), the Lateglacial Interstadial (GI-1, furtheron: the Lateglacial Interstadial) and the Lateglacial Stadial (GS-1, furtheron the Lateglacial Stadial).

Greenland ice-core records

The Greenland oxygen isotope eventstratigraphy (Björck et al. 1998) is well established as a reflection of climate on the North Atlantic seaboard. Therefore, this eventstratigraphy is frequently used as the basis for chronostratigraphic frameworks for north-western Europe. In this study, the Greenland oxygen isotope eventstratigraphy is used as the baseline to which the other climate and chronostratigraphic relevant records from the deep sea and terrestrial sites are correlated.
Thus far, several ice-core sequences from Greenland have been published (**fig. 2**; e.g. Johnsen et al. 2001). The European Greenland ice-core project (GRIP) drilled an ice-core (Dansgaard et al. 1993) at the summit of the Greenland ice sheet. This ice-core record was proposed as the stratotype for the eventstratigraphy (Björck et al. 1998; Walker et al. 1999). This proposal has been widely accepted and adopted due to the reliable, high-resolution chronostratigraphy related to the ice-core records. At the time of the proposal as stratotype, an age model (ss09) was preferred which was based mainly on the rate of surface snow accumulation. This age model was established by the correlation of the GRIP sequence with a further three Greenland ice-core records (DYE-3, Renland, and Camp Century). The correlation was based on reference horizons in the Holocene section, a multi-parameter identification of the annual layers in the Lateglacial and early Holocene part, and calculation with an ice flow model for the part beyond c. 14,550 years cal. b2k (Johnsen et al. 1992; Dansgaard et al. 1993).
However, instead of the GRIP record the ice-core records of the US-American Greenland ice-sheet project two (GISP2; Grootes et al. 1993) and the multi-national North Greenland ice-core project (NGRIP; Dahl-Jensen et al. 2002; Andersen et al. 2004) were occasionally used as comparatives (Hughen et al. 1998a; Wang et al. 2001; Sirocko et al. 2005). The GISP2 sequence was drilled 28 km west of the GRIP ice-core.

4 Following the eventstratigraphy based on oxygen isotopes as displayed in the Greenland ice-cores, the stadial (GS-2) following after the LGM (GS-3) was divided into three parts, of which the youngest/upper most part (GS-2a) referred to a severe cold (Björck et al. 1998, 288) and dry phase (Moreno et al. 2010). However, in a more recent correlation of various Greenland ice-core records the limit between GS-2a and GS-2b became blurred and could not be securely detected (Lowe et al. 2008). This boundary remained also indistinct in most of the preserved litho- and biostratigraphies except for the ELSA record (Sirocko et al. 2005). Stratigraphically a period requires detectable limits and, therefore, the whole GS-2 is considered in the present work as belonging to the Lateglacial, although the study focuses only on the unclear GS-2a which is regarded as arisen in research history (van der Hammen 1957; Mangerud et al. 1974; Björck et al. 1998) and is used in the present study according to the limit given in **fig. 3A**.

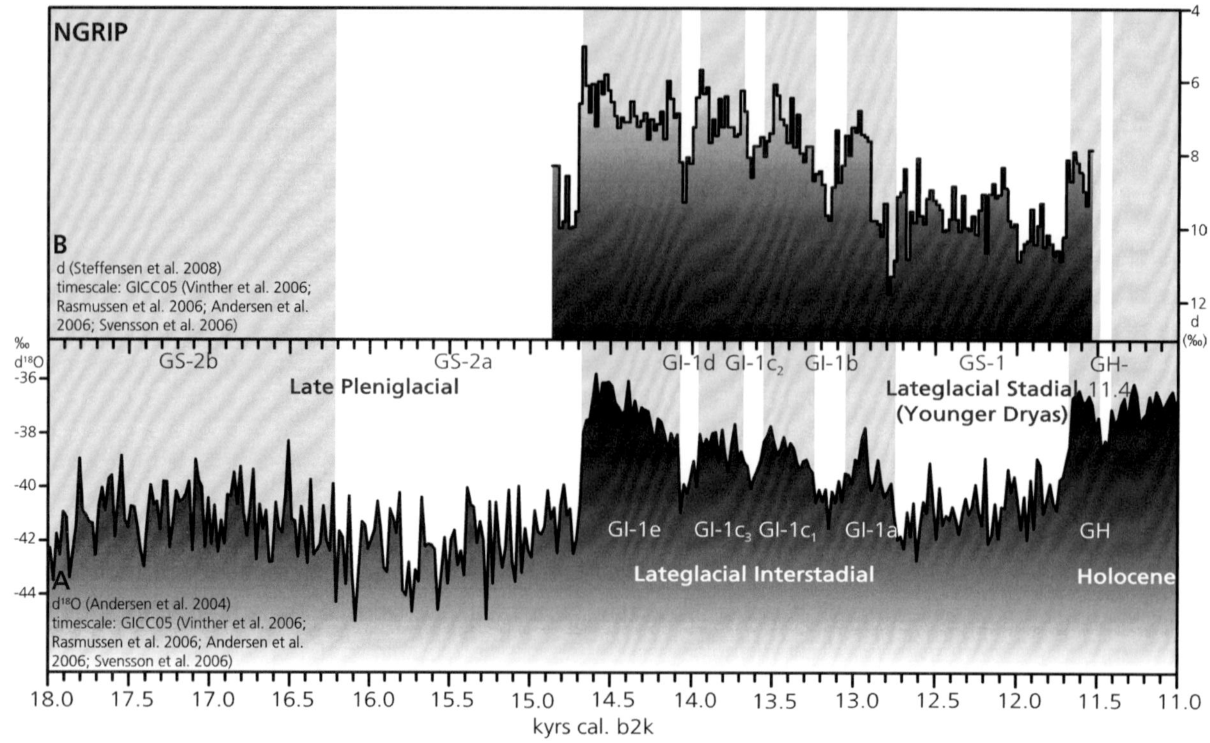

Fig. 3 Lateglacial isotope records from NGRIP in GICC05 (Vinther et al. 2006; Rasmussen et al. 2006; Andersen et al. 2006; Svensson et al. 2006; Rasmussen et al. 2008; Svensson et al. 2008). **A** oxygen isotope record (Andersen et al. 2004). Grey shaded areas represent periods of more interstadial oxygen isotope values than the surrounding values (for instance, the values in GS-2b are more interstadial than in GS-2a but still these values are as stadial as the values in GS-1). – **B** deuterium excess record (Steffensen et al. 2008). – For further details see text.

Visible layers from this record were counted in the Pleistocene part (Alley et al. 1997; Meese et al. 1997) and contributed regularly to the establishment of precise, high-resolution ice-core chronologies. The NGRIP ice-core came from a drilling north of the two summit ice-cores (Andersen et al. 2004).

Comparison to the GISP2 and NGRIP record and a correction of the isotope accumulation model with seawater-corrected isotope values led to an updated version of the generally preferred age model to the ss09sea age model (Johnsen et al. 2001). Although some incongruences existed in the details (Southon 2002), the Greenland ice-core sequences were usually used in this ss09sea age model until the presentation of the so-called Greenland ice-core Chronology 2005 (GICC05).

In 2006 this new chronology was presented (Vinther et al. 2006; Rasmussen et al. 2006; Andersen et al. 2006; Svensson et al. 2006). In general, this chronology was based on the detection and counting of annual layers in the records from GRIP, NGRIP, and DYE-3 (c. 65° northern latitude; Vinther et al. 2006). The annual layers were defined by a visible stratigraphy (greyscale refraction) and by the cyclic development of chemical parameters such as ammonium (NH_4^+) and calcium (Ca^{2+}) which were obtained in a continuous flow analysis (CFA; Rasmussen et al. 2006). The resolution of the single parameters differed between three and 50 years (Rasmussen et al. 2006, X-3 tab. 1). The three sequences were correlated by volcanic marker horizons such as documented in the electrical conductivity measurement (ECM) records of the ice-cores in the Holocene period. Below the Holocene/Pleistocene transition only data from the NGRIP record was used to construct the chronology (Rasmussen et al. 2006; Andersen et al. 2006). Below the beginning of the Lateglacial Interstadial, the sequence of NGRIP was synchronised with those of GRIP and GISP2 again by the ECM record and, furthermore, the NH_4^+ record (Rasmussen et al. 2008). Based on these correlations and the detection of annual layers, the chronology was counted back 60,200 years (Svensson et al. 2008).

onset of main isotope events	GICC05 (NGRIP)						
GH	11,703 ± 99 (d)	11,703 ± 99 (d)	11,681 ± 111 (^{18}O)	11,703 ± 101 (d)	11,686 ± 101 (dust)	11,691 ± 102 (Ca^{2+})	11,695 ± 102 (λ)
GS-1	12,896 ± 138 (d)	12,896 ± 138 (d)	12,819 ± 202 (^{18}O)	12,897 ± 140 (d)	12,805 ± 144 (dust)	12,804 ± 142 (Ca^{2+})	12,863 ± 160 (λ)
GI-1a	×	13,099 ± 143 (GRIP)	×	×	×	×	×
GI-1b	×	13,311 ± 149 (GRIP)	×	×	×	×	×
GI-1c	×	13,954 ± 165 (GRIP)	×	×	×	×	×
GI-1d	14,075 ± 169 (d)	14,075 ± 169 (d)	×	×	×	×	×
GI-1e	14,692 ± 186 (d); 14,680 ± 93 (^{18}O)*	14,692 ± 186 (d)	14,687 ± 187 (^{18}O)	14,693 ± 188 (d)	14,684 ± 188 (dust)	14,681 ± 188 (Ca^{2+})	14,683 ± 193 (λ)
GS-2a	×*	not clear	×	×	×	×	×
ref.	Rasmussen et al. 2006, tab. 4	Lowe et al. 2008, tab. 1; cf. Blockley et al. 2012, tab. 1	Steffensen et al. 2008, 683 tab. 1 (ramp points ± standard error of ramp fitting results, added with general counting error)				

Tab. 1 Comparison of dates (all given in years cal. b2k) for the onset of the main isotope events in the NGRIP ice-core. Parameters: **d** deuterium excess; **^{18}O** oxygen isotopes; **GRIP** onsets for $\delta^{18}O$ events and episodes in GRIP, depths given by Björck et al. 1998 and transferred to NGRIP depths using the volcanic markers of Rasmussen et al. 2006; **dust** dust content; **Ca^{2+}** calcium concentration; λ annual layer thickness. * date given in Andersen et al. 2006, 3254 tab. 1.

In addition, the ss09sea age model (Johnsen et al. 2001) was correlated to the GICC05 at 60,000 years and applied to the NGRIP record further back to 123,000 years (Wolff et al. 2010). Even though the counting error in the GICC05 approach was in general relatively low due to the cross-checking and the use of various parameters (Rasmussen et al. 2008, X-4), the error accumulated and increased with time depth. In the studied time period the counting error reached c. 130-230 years.

Nevertheless, depending on the publication date, some data correlated to the ice-core records as for example the ELSA sequence (Sirocko et al. 2005) are also set in an old age model and require an updated correlation if applied with records of the new chronology.

Due to the significance of the NGRIP data for the Lateglacial part of the GICC05, the NGRIP sequence was chosen as stratotype of the oxygen isotope eventstratigraphy in the present work (**fig. 3A; tab. 1**; cf. Lowe et al. 2008). More precisely, the Lateglacial data records originate generally from the NGRIP2 ice-core. The $\delta^{18}O$ record is given in a 20 year mean resolution (Rasmussen et al. 2006, X-14). The oxygen isotopes are considered in the ice-core records as proxy for the air temperature at the coring site and, thus, seasonal cycles (summer– winter) can be distinguished at sufficient high-resolution sampling. Furthermore, the oxygen isotope ratio is measured in relation to a standard of ocean water (Dansgaard et al. 1993) because the carrier substance originated from these ocean waters and, thus, the oxygen isotope ratio is also strongly dependent on the hydrological cycle (e. g. Keeling 1995; Johnsen et al. 2001; Landais et al. 2010). In Greenland the water cycle is mainly governed by the circulations of the North Atlantic which is the major factor of north-western European climate as well. Therefore, oxygen isotope records from limnic archives in north-western Europe which are also governed by the North Atlantic circulation were frequently correlated to the Greenland oxygen isotope record (e. g. Lotter et al. 1992; Grafenstein et al. 1999; Hoek/Bohncke 2001).

However, the deuterium excess was used to define the limits of stadial and interstadial conditions in the Greenland eventstratigraphy (see **tab. 1**) most recently due to an early and sharp reaction to changes in the North Atlantic circulation (Steffensen et al. 2008). The deuterium excess is an index calculated from oxygen and deuterium isotopes ($d = \delta^2H - (8x\delta^{18}O)$) and, therefore, a second-order parameter which is usually given in a 20 year mean resolution in the ice-core records (**fig. 3B**). In contrast to the oxygen isotopes, the deuterium excess is considered to reflect the ocean surface temperature in the region where the Greenland precipitation originated from (i.e. source region of moisture) or the seasonality of snow precipitation (Masson-Delmotte et al. 2005). However, changes in this parameter were assumed to precede changes in temperature or vegetation recorded in other archives (Lowe et al. 2008, 10 tab. 1). Therefore, the deuterium excess is not a suitable proxy for correlations with terrestrial archives from north-western Europe. In consequence, the limits of the eventstratigraphy have to be partially redefined in the present study according to the oxygen isotope sequence (see p. 245-247) to allow a correlation with the terrestrial archives. Yet, the simultaneity of the Greenland oxygen isotope records with deep sea or terrestrial records was discussed recurrently (Goślar/Arnold/Pazdur 1995; Blaauw et al. 2010). Nevertheless, some of the thereby described offsets were perhaps due to reaction time of the different proxies and/or the identification of the precise trigger (for example temperature or precipitation) causing changes in these proxy data. Moreover, correlations of the Greenland eventstratigraphy based on isotope records with other, also non-isotopic records from the deep sea (e.g. Koç/Jansen/Haflídason 1993) as well as from terrestrial archives (e.g. Wastegård 2005) can be accomplished by ECM records, visible volcanic ashes, and sulphate records. A detailed Lateglacial tephrostratigraphy from the NGRIP record was previously published (Mortensen et al. 2005). This analysis included the microscopically detected ash shards, the chemical composition of ashes, and the gypsum (Ca^{2+}) corrected sulphate content (or excess sulphate, cf. De Angelis et al. 2003) which is considered a good proxy for mainly volcanic sulphate input into the atmosphere. Nevertheless, this data set was presented in a previous age model and, thus, needs to be shifted to the GICC05 (see p. 247 f.) before using this indicator as a correlation factor.

Another mean of comparison is the record of beryllium isotopes (^{10}Be). These isotopes can be measured from the aerosols in Greenland ice. The production of these beryllium isotopes in the atmosphere is governed by the interaction with cosmic radiation comparable to the production of atmospheric radiocarbon isotopes. Thus, the beryllium record represents a tool for connecting the Greenland ice-cores to the ^{14}C calibration curve (Muscheler et al. 2000; Muscheler et al. 2008). However, a detailed beryllium record is thus far not available for the NGRIP sequence but is available for GRIP and GISP2 (Muscheler et al. 2005).

Deep sea records

In addition to the ice-core records, various records such as sediment or coral cores from the deep sea exist. Often these deep sea archives produced isotopic profiles (e.g. ^{14}C, ^{10}Be, ^{18}O), temperature curves, and/or sea-level sequences from foraminifera, diatoms, or corals (Fairbanks 1989; Bond et al. 1992; Koç/Jansen/Haflídason 1993; Hughen et al. 1998a; Dokken/Jansen 1999; Bard et al. 2000; Björck/Koç/Skog 2003; McManus et al. 2004). However, these sequences also provided insights in the development of the sea ice cover as well as the ice sheets in the adjacent lands, in particular on the northern Atlantic by layers of ice-rafted debris (IRD, Heinrich 1988; Bond et al. 1999; Dokken/Jansen 1999; McCabe/Clark/Clark 2005; Peck et al. 2007; Bigg et al. 2010; Stanford et al. 2011b). Some of the deep sea archives were dated by a series of radiometric measurements but generally the chronologies are only floating and were correlated with comparable proxy records from other archives, in particular from the Greenland ice-cores (McManus et al. 2004; Peck et al. 2007). The connection to a continuous chronostratigraphy is problematic as some of the atmospheric isotopes became deposited in the deep ocean with considerable time lags due to the global ocean-atmosphere exchange system (Dokken/Jansen 1999; Muscheler et al. 2000; Ikeda/Tajika 2002; Schrag et al. 2002; Fairbanks et al. 2005). Especially for the carbon cycle, these »lags« (reservoirs) are known to be inconsistent in time and space and, in fact, to vary considerably (Muscheler et al. 2000; Björck/Koç/Skog 2003; cf. Lowe/Walker 2000, 57f.; Schaub et al. 2008b, 77). Consequently, the time lags of the carbon isotope records from the deep sea needed to be estimated per time slice and per sequence. This estimate was usually also accomplished by correlation with other chronostratigraphies such as tephrochronology (Koç/Jansen/Haflídason 1993), ice-core chronologies (Bond et al. 1999), or partially dendrochronology (Hughen et al. 2004c) based on the assumption that isotopic events should correlate. Furthermore, for some sequences such as corals paired dating series (^{14}C and $^{230}Th/^{234}U$) with an additional stratigraphic control were another possibility (Fairbanks et al. 2005). In summary, these deep sea records can produce a supplementary confirmation of the established climatic developments but need to be evaluated with some caution.

Cariaco basin, Venezuela

Chronostratigraphies which are based on counts of annually laminated sediment couplets from the deep sea such as the floating varve chronology from the Cariaco basin, offshore Venezuela are very rare (**fig. 2**; Hughen et al. 1996b; cf. Ojala et al. 2012). In the Cariaco basin, several piston cores and additional ODP cores were drilled in the basin with four piston cores, in particular PL07-58PC and PL07-56PC, and one ODP core producing most of the results (Hughen et al. 2004c; Hughen et al. 2006). For instance, the PL07-58PC core provided a record of well preserved lipids (**fig. 4B**) which are considered as plant biomarkers distinguishing between arid and humid terrestrial plant communities (Hughen et al. 2004b). The varve chronology comprises c. 5,500 annual couplets spanning the period from the Holocene/Pleistocene transition to the transition from the Late Pleniglacial into the Lateglacial Interstadial (Hughen et al. 1998a; Hughen et al. 2000; Hughen et al. 2004c; Hughen et al. 2006). The lamination of the sediment were assumed to result from a generally bi-sected climatic cycle in the basin: The dry, windy season with coastal upwelling produced light-coloured, organic-rich plankton layers, whereas dark-coloured mineral grains from local river runoff formed the layers of the non-windy, rainy season (Hughen et al. 1996a; Hughen et al. 2000, 1951). Furthermore, almost 400 reliable ^{14}C dates were meanwhile made on samples for which the position in this laminated sequence is known with a 10-15 years exact precision (Hughen et al. 2000; Hughen et al. 2006). However, since the Cariaco basin sequence (**fig. 4**) is a floating chronology it also needs to be correlated, which can be accomplished by the ^{14}C record (Hughen et al. 2000) as well as by climatic changes which are displayed in the grey scale (relative reflectance) of the Cariaco basin sediment cores (Hughen et al. 1998a; Hughen et al. 2004a). The grey scale in combination with the laminae thicknesses are assumed to reflect the intensity of the local up-welling and the trade winds (Hughen et al. 2004b). These intensities are dependent on the position of the Inter-Tropical Convergence Zone (ITCZ) and, thus, the northern Atlantic thermohaline cycle. Hence, this record can also be used as confirmation for other records reflecting shifts in this oceanic exchange system such as the Central European dendrochronology (CEDC, Friedrich et al. 2004; see p. 25-30) to which the Cariaco basin record was wiggle-matched by the use of the $\Delta^{14}C$ variations (Hughen et al. 2000). The $\Delta^{14}C$ records of the CEDC and the Cariaco basin overlap currently in a 1,900 year long section between approximately 10,550 to 12,460 years cal. b2k. $\Delta^{14}C$ refers to the level of radiocarbon in the atmosphere (Stuiver/Polach 1977) which can be due to the global carbon cycle and

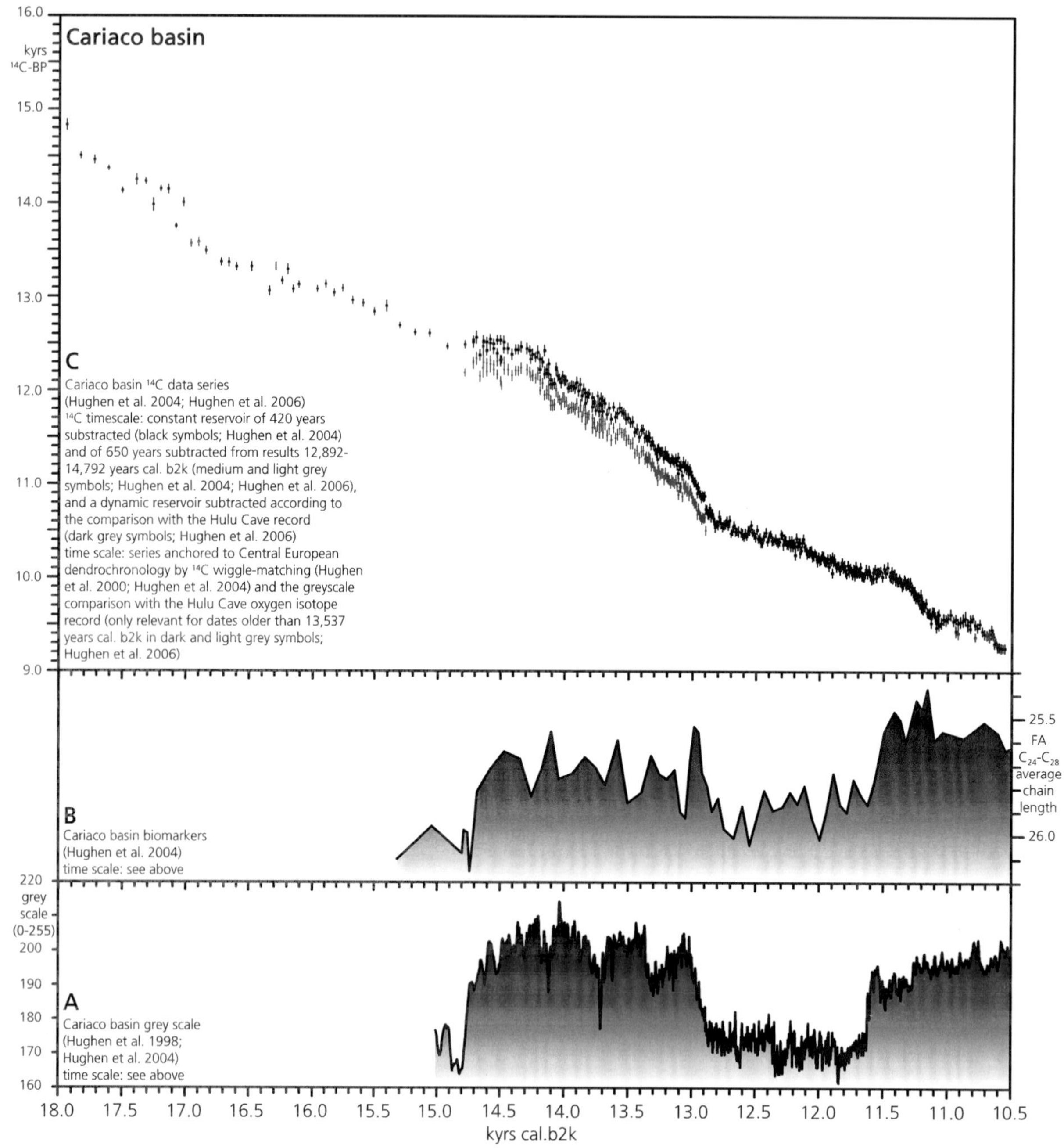

Fig. 4 Lateglacial records from the Cariaco basin with the varve timescale wiggle-matched to the Central European dendrochronology (Hughen et al. 2000, Hughen et al. 2004c) and correlated to the Hulu Cave record (Hughen et al. 2006). Timescale shifted from cal. BP to cal. b2k. **A** greyscale record (Hughen et al. 1998a; Hughen et al. 2004a). Grey shaded areas represent periods of more interstadial values than the surrounding values. – **B** average chain length index of C_{24}-C_{28} *n*-alkanoic homologs from the fatty acids (FA) of leaf waxes (Hughen et al. 2004b). – **C** series of ^{14}C dates in three different versions: one with constant reservoir of 420 years subtracted (black symbols); one with 650 years subtracted from the results between 12,892 and 14,792 years cal. b2k (medium and light grey symbols; Hughen et al. 2004a; Hughen et al. 2004c; Hughen et al. 2006), and one with a dynamic reservoir subtraction according to the correlation with the Hulu Cave record (for results older than 13,537 years cal. b2k; Hughen et al. 2006). – For further details see text.

the exchange of carbon isotopes between the atmosphere, the biosphere, and the oceans (Hughen et al. 2004a; Muscheler et al. 2004; Schuur et al. 2009) or due to the production rate of carbon isotopes in the atmosphere. The production rate of cosmogenic radionuclides such as ^{14}C or the beryllium-10 isotope (^{10}Be) in the atmosphere is motored by solar radiation, the geomagnetic field, and/or interstellar galactic cosmic ray flux (Laj et al. 2002; Muscheler et al. 2004; Hughen et al. 2004a; Muscheler et al. 2005). Since these

different reasons are globally effective, the changes should occur globally in a relatively short time period. However, due to the origin from the deep sea the ^{14}C dates from the Cariaco basin incorporated a reservoir effect (see p. 13) which is considered relatively constant at the Cariaco basin during the Lateglacial and the early Holocene (Hughen et al. 2000, 1951). Thus, 420 years were assigned to the marine reservoir effect and subtracted from the measurement results (Hughen et al. 2004c, 1183). Nevertheless, comparison with the floating German Lateglacial pine chronology (GLPC, Kromer et al. 2004; see p. 25-30) indicated a change in the reservoir ages from 420 years in the Holocene and the Lateglacial Stadial to 650 years in the Lateglacial Interstadial (**fig. 4C**; Hughen et al. 2004c, 1184f.; Kromer et al. 2004, 1205-1207). The GLPC is also still a floating chronology and at least two positions to anchor this chronology to the Cariaco basin record are plausible, therefore, a change in reservoir ages remains a preliminary assumption (Hughen et al. 2006, 3217). The Weichselian sequence of the Cariaco basin was again correlated to the archive from the Hulu Cave (see p. 17f.) also by wiggle-matching of the $\Delta^{14}C$ records (Hughen et al. 2006). This comparison indicated significant anomalies in the observed $\Delta^{14}C$ record between 15,050 and 45,050 years cal. b2k and based on simulation models were suggested to result from fluctuations in the reservoirs. Recently, the Cariaco basin sequence was fine-tuned to the oxygen isotope record of NGRIP using the variation of total reflectance or lightness of the sediment (Deplazes et al. 2013). This tuning was published too recently to be further integrated into the current project.
However, the chronology of this deep sea record was further described and evaluated in detail elsewhere because, in particular, the Cariaco basin record was frequently used in the construction of calibration curves passing beyond the current limit of dendrochronology because it bridges the Lateglacial Interstadial/Lateglacial Stadial transition (Lowe/Walker 2000; Reimer et al. 2004; Fairbanks et al. 2005; Weninger/Jöris 2008).

MD01-2461, offshore Ireland

MD01-2461 is a sediment core drilled on the north-western flank of the Porcupine Seabight, offshore south-west of Ireland (**fig. 2**; Peck et al. 2007). At this location, the sediment was subject to significant detrital carbonate-rich IRD (ice-rafted debris) deposition due to a route of iceberg drift which originated from the circum-North Atlantic ice sheets. Thus, the concentration of IRD is considered as a proxy for iceberg delivery to the site. Significant increase of this concentration reflects massive destabilisations of the contributing ice sheets and revealed periods of so-called Heinrich events (Heinrich 1988; Bond et al. 1992). During these events the formation of North Atlantic Deep Water is hindered and, thus, the transport of warm surface waters northwards is curtailed resulting in an exceptionally cold and dry climate in the northern Atlantic seaboard (Rahmstorf 2002; Stanford et al. 2011b). Based on a synchronised sea sediment stratigraphy, these Heinrich events were counted from top down beginning with Heinrich 0 which is related to the Younger Dryas (Rahmstorf 2002). Due to this connection with Heinrich events, the IRD records are as a proxy also of some relevance for the development of the north-western European climate. Furthermore, the end of Heinrich 1 occurred in the Late Pleniglacial and was related to the oldest part of the studied time period (cf. Stanford et al. 2011b).
The IRD concentration as well as its composition was analysed in a high-resolution approach on this sediment core (Peck et al. 2007). Furthermore, oxygen isotopes based on benthic foraminifera, magnetic susceptibility, the sea surface temperature (SST) based on the different presence of species in the planktonic assemblage, and some chemical parameters were also measured (**fig. 5**).
The chronostratigraphy of this record was formed by correlation of the SST record from the core to the GISP2 oxygen isotope record, ^{14}C dates, and the testing on the tephrostratigraphy (Peck et al. 2007).

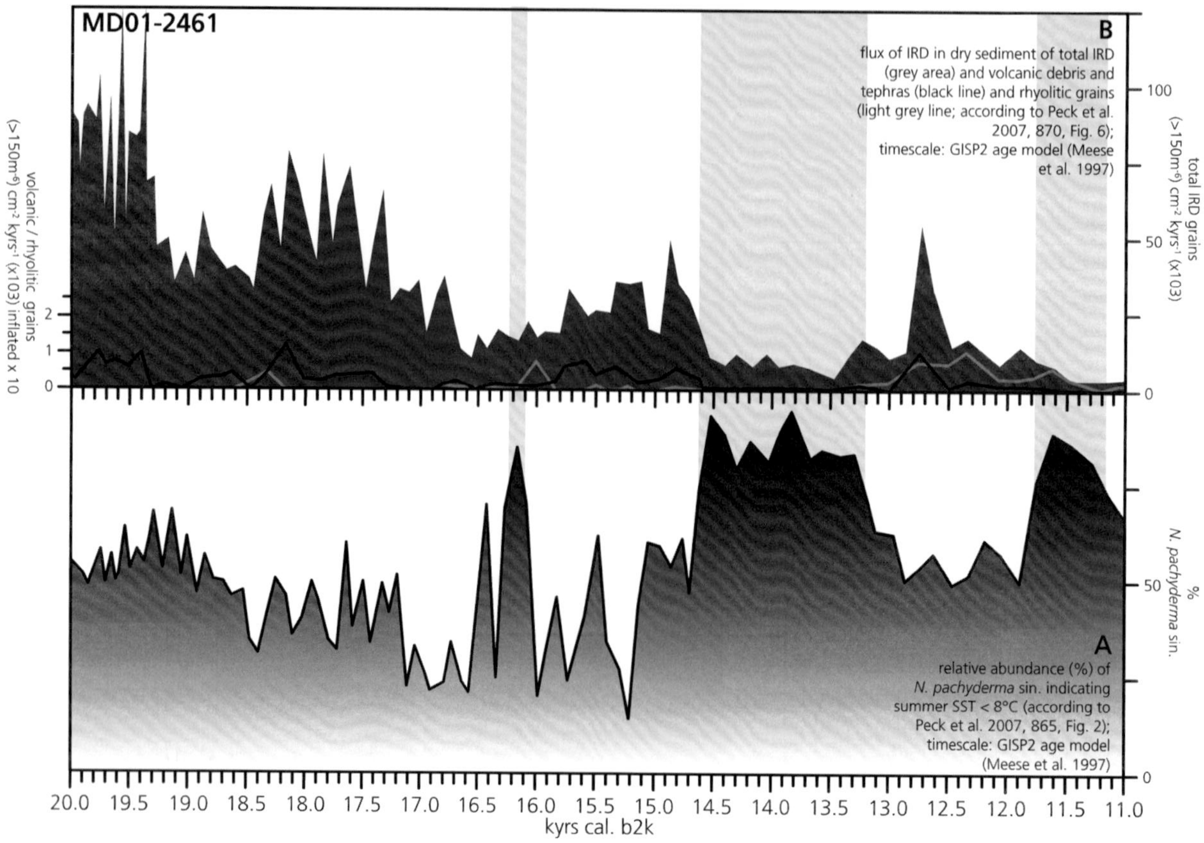

Fig. 5 Lateglacial records from the MD01-2461 with timescale according to GISP2 age model (Peck et al. 2007). Timescale shifted from cal. BP to cal. b2k. **A** relative abundance (%) of *Neogloboquadrina pachyderma* sinistral, an indicator for polar waters and dominant in planktonic assemblages in waters of summer SST < 8°C. Grey shaded areas represent periods of more interstadial values than the surrounding values. – **B** concentration of total IRD (grey shaded area) and volcanic debris and tephra (grey and black lines) in dry sediment. – For further details see text.

According to these correlations, the sediment core contained a sequence from approximately 10,050 to 60,050 years cal. b2k. However, due to the correlation with the GISP2 record in a previous age model this relative stratigraphy has to be correlated to the new GICC05.

Terrestrial records

Besides the sequences from Greenland and the deep sea, terrestrial records produced important information on the general climate development and, in particular, on the local impacts of these climatic regimes perceivable by Lateglacial hunter-gatherers. Some of these terrestrial records are as detailed as the glacial and oceanic archives. In contrast to deep sea records, an offset in the atmospheric values is either not as high or non-existent in continental sequences (including limnic ones). Therefore, these records are particularly important for the refinement of a Lateglacial chronostratigraphy. Furthermore, it is possible to directly connect the isotope events with the biotic stratigraphies in these records which help with describing the Lateglacial environment as well as relating these environmental developments to the global climate record. Additionally, the chronostratigraphy of these records can often be tested by radiometric dating as for instance in the Hulu cave speleothems (Wang et al. 2001) or the varve sequence of Holzmaar (Hajdas/Bonani/Zolitschka 2000).

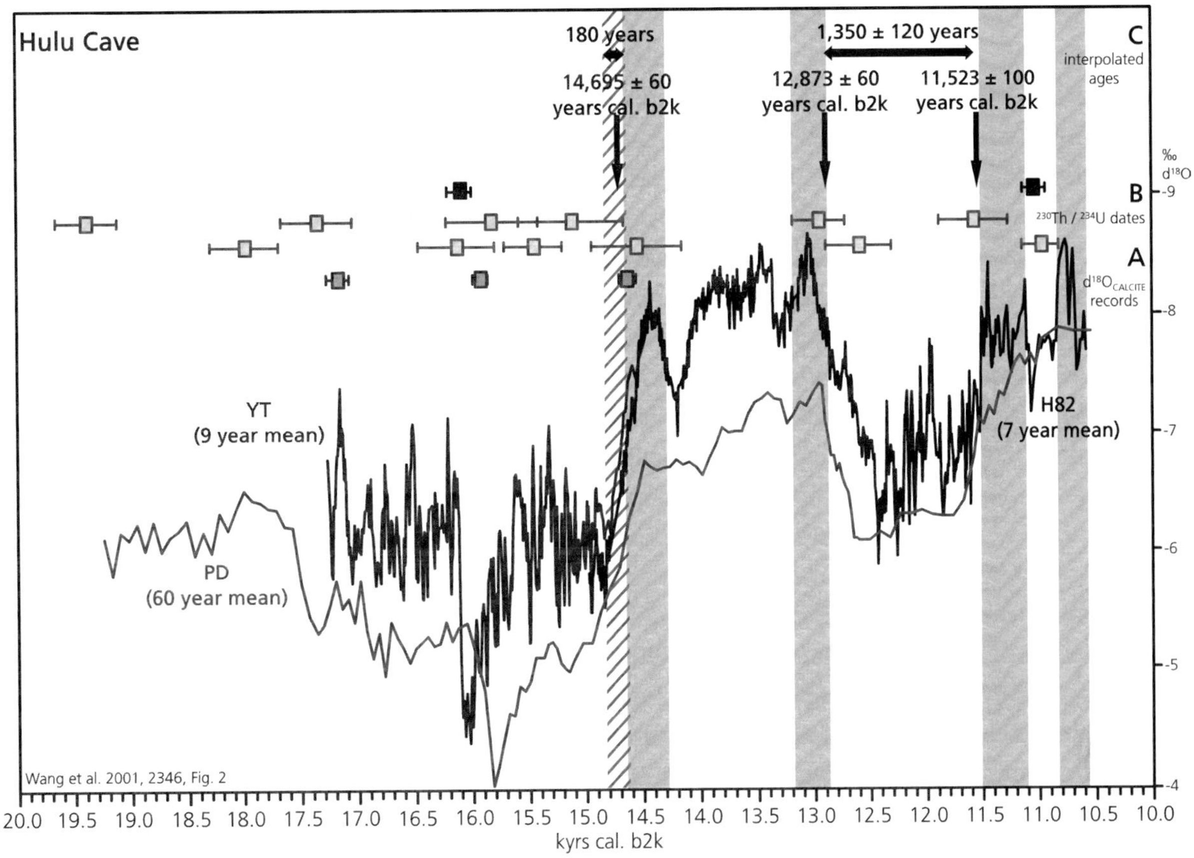

Fig. 6 Lateglacial records from the Hulu Cave stalagmites (reproduced fig. 2, Wang et al. 2001, 2346 but timescale shifted from cal. BP to cal. b2k). **A** oxygen isotope records from three colour coded stalagmites. – **B** $^{230}Th/^{234}U$ dates colour coded according to stalagmites in A. – **C** interpolated ages based on A, B, and laminae counting. Grey shaded areas represent more intense summer East Asian monsoon according to the oxygen isotope values. Hatched area represents the Late Pleniglacial to Lateglacial Interstadial transition. Duration for this transition and the Lateglacial Stadial are given on top. – For further details see text.

Speleothems

Continuously grown speleothems which cover large parts of the Lateglacial period are rare and only few are known from Europe (Wurth et al. 2000; Kempe et al. 2002; Moreno et al. 2010). The advantage of speleothems are possible seasonal bandings, a reliable output of $^{230}Th/^{234}U$ dates, and the combination with oxygen as well as carbon isotope records (Wang et al. 2001; Spötl/Mangini/Richards 2006).

Hulu Cave, China

The Hulu cave lies 28km east of Nanjing, China (**fig. 2**). Calcite samples for $\delta^{18}O$ analysis and a series of 59 $^{230}Th/^{234}U$ dates were taken along the growth axes of five speleothems from 35m deep inside the cave (Wang et al. 2001). A few TIMS-$^{230}Th/^{234}U$ dates in the late Pleistocene part of the oxygen isotope curve yielded standard deviations of ± 60-100 years (**fig. 6**; Wang et al. 2001, supplemental tab. 1). The origin of the isotope record from calcite might cause some offset though and it was therefore questioned (e.g. Hughen et al. 2006, 3221). However, the $\delta^{18}O$ record was supposed to mainly reflect the summer/winter precipitation ratio, which is influenced by the intensity of the Asian monsoon, and a close connection to the Greenland air temperatures could be identified (Wang et al. 2001, 2347). This assumption was also questioned recently suggesting that instead changes in the moisture source were reflected by the $\delta^{18}O$ record of the stalagmites (Maher/Thompson 2012). Nevertheless, the same study considered precessional forcing

of inter-hemispheric temperature gradients as possible drivers of the $\delta^{18}O$ variations in Chinese stalagmites. Yet, this forcing is again a global phenomenon and should affect climatic archives in a comparable manner and at comparable times. However, this forcing was assumed to change the position and intensity of the subtropical pressure cells which then resulted in the differences within the speleothem records. Thus, in this case, differences in the reaction time and the impact on the record are possible. For the construction of the Hulu Cave speleothem chronology the $\delta^{18}O$ curves of three speleothems were correlated (**fig. 6**) and the bands in the continuously grown speleothems were counted between 11,050 and 11,850 years cal. b2k (Holocene to GS-1) and between 13,050 and 14,650 years cal. b2k (GI-1 to GS-2a; Wang et al. 2001, 2346). These bands were assumed to represent annual cycles due to the congruency of the number of counted layer couplets with the results of the radiometric dating.
The high-resolution calcite $\delta^{18}O$ curve in combination with the low standard deviations of the $^{230}Th/^{234}U$ dates and the additional band counting made the Hulu sequence (**fig. 6**) a good terrestrial control stratigraphy in the Pleistocene chronology of the northern hemisphere and it was therefore used several times to adjust other sequences (Hughen et al. 2006; Weninger/Jöris 2008). An atmospheric ^{14}C record based on a Hulu Cave stalagmite (Southon et al. 2012) was published too recently to be integrate in the current project.

Qingtian Cave, China

The limestone cave is located in central China (**fig. 2**), near the southern edge of the Chinese Loess Plateau where the cave is strongly influenced by the seasonal cycle of the Asian monsoon (Liu et al. 2008). A stalagmite was recovered from 40 m inside the cave and produced a mostly undisturbed laminated section of 1,498 ± 21 couplets which were placed by seven $^{230}Th/^{234}U$ dates (**fig. 7**) to the time interval between 12,130 ± 80 years cal. b2k to 13,530 ± 90 years cal. b2k. Hence, this record covers the period from the mid-Lateglacial Stadial (GS-1) to the mid-Lateglacial Interstadial (GI-$1c_2$). Besides an oxygen isotope record, the ^{13}C isotopes and the layer thickness of the couplets were recorded for the laminated section. The layer thickness was assumed to reflect the meteorological changes outside the cave. Problems and restrictions of the record are comparable to the data from Hulu cave (see p. 17 f.). Nevertheless, the Qingtian record provided a detailed climate history for the mid-Lateglacial and the transitions therein.

Varve chronology

Seasonally laminated sediments occur not only in the deep sea but also in small and deep lakes which form sediment traps (Brauer et al. 1994). Some of these archives can reach an annual resolution which can only be disturbed by various errors concerning the counting of the couplets, taphonomic development, or changes in the deposition regime.
The currently longest counted sequence comes from the approximately 34 m deep Lake Suigetsu (Japan) where a floating chronology was established encompassing 29,100 couplets and was dated by correlation of the more than 330 ^{14}C dates on macrofossils from the lake with the CEDC to 8,880-37,980 years cal. b2k (Kitagawa/van der Plicht 1998). The sequence below this age was estimated to date down to c. 45,000 years due to the calculation of the sedimentation rate. However, the correlation with the calibration data indicated that parts of the sequence were miscounted or affected by hiatus (van der Plicht et al. 2004). These results warrant the exclusion of this dataset. Meanwhile, a new drilling program in the lake yielded more precise varve data and further radiocarbon dates which helped refining the age-depth-model of the sequence (Kossler et al. 2011; Bronk Ramsey et al. 2012). In the future, this new material promises to fill

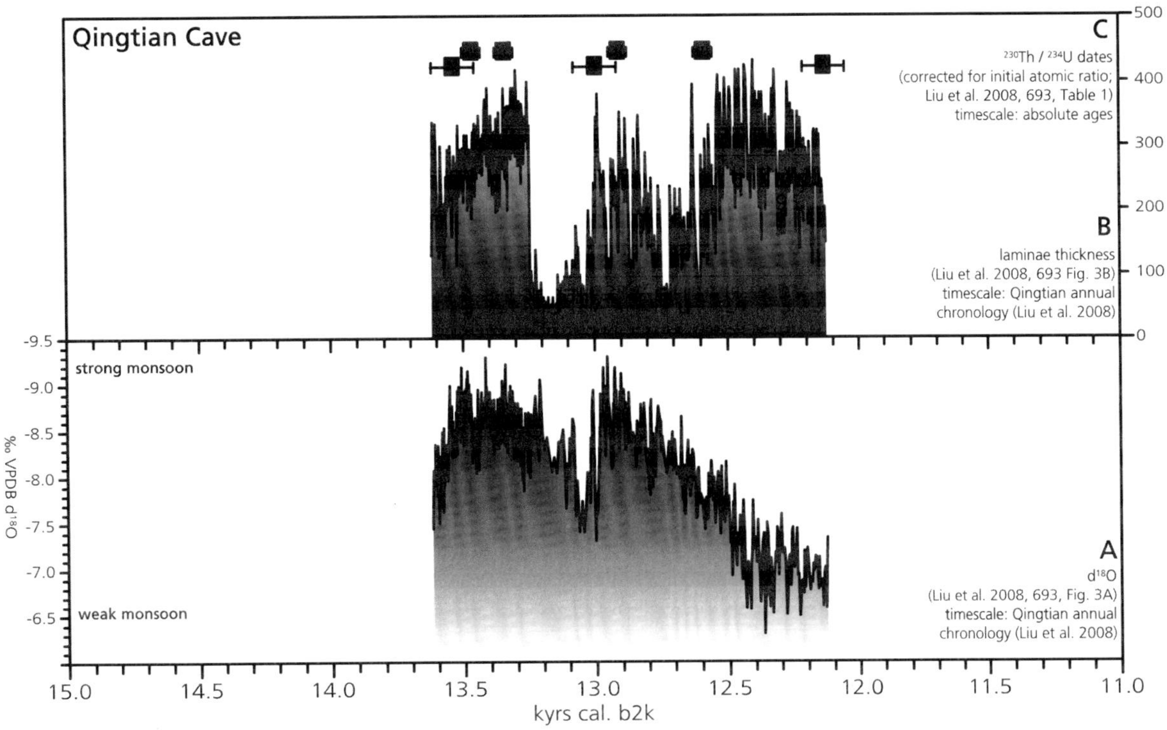

Fig. 7 Lateglacial records from the Qingtian Cave stalagmites in the Qingtian annually resolved chronology (Liu et al. 2008). Timescale is shifted from cal. BP to cal. b2k. **A** oxygen isotope record. – **B** layer thickness. – **C** corrected $^{230}Th/^{234}U$ dates. Grey shaded areas represent periods of strong monsoon according to the oxygen isotope values. – For further details see text.

some gaps and produced new insights in the Lateglacial chronology but the results are too recent for the incorporation in the present approach.

Another prominent Lateglacial varve sequence comes from the south-eastern Turkish Lake Van which covers the time period from the present day back to 13,750 ± 356 varve years cal. b2k (Landmann et al. 1996; Wick/Lemcke/Sturm 2003). Although the lake sequence yielded various additional analyses, no series of ^{14}C dates was published as an additional chronostratigraphic parameter. Thus, the evaluation of the chronostratigraphic reliability on the scale of the northern hemisphere was not possible. Therefore, this sequence was not further incorporated in the present study.

Holzmaar and Meerfelder Maar, Germany

In the middle-range mountains west to north-west of the Central Rhineland, lakes have formed in several volcanic maar depressions in the Quaternary Eifel volcanic field. These volcano lakes form natural sediment traps and laminated sediments were recovered from some of these locations such as Meerfelder Maar, Holzmaar, or Schalkenmehrener Maar (Brauer et al. 1994; Sirocko et al. 2005). Lamination, colour, and thickness of the couplets in these lakes are influenced by various climatic parameters controlling the water and sediment catchments of the lake as well as the developments of diatom population (Vos et al. 1997; Brauer/Endres/Negendank 1999; Zolitschka et al. 2000). The numerous sediment cores link laminated sediments to detailed geochemical analyses (Brauer et al. 2008; Sirocko et al. 2005; Lücke et al. 2003), pollen profiles (Litt/Stebich 1999), macrofossils (Hajdas et al. 1995), and diatom sequences (Brauer et al. 1999). Furthermore, ash layers from the surrounding Quaternary volcanoes served for chronostratigraphic correction and/or correlation of the various sequences (Zolitschka et al. 2000).

In a project on these Eifel Laminated Sediment Archives (ELSA), four of these sequences (Hoher List, Oberwinkler Maar, Dehner Maar, and Schalkenmehrener Maar) were correlated to form the »ELSA greyscale stack 2005« (ELSA greyscale) which can be considered as indicator for the organic carbon content and the introduction of silt (dust; Sirocko et al. 2005). Thus, in warm periods the record is mainly governed by the organic production of the lake catchment area and in cold periods by aeolian activity (Seelos/Sirocko/Dietrich 2009). Wind systems were reconstructed based on the geo-topography of the lakes and an analysis of the lithological composition of the introduced dust (Seelos/Sirocko/Dietrich 2009; Dietrich/Seelos 2009). The Lateglacial part of this greyscale was received from the Schalkenmehrener Maar where the Lateglacial Stadial section was not preserved sufficiently in the sediment cores drilled until 2005 and, therefore, data for this period is lacking in the record (cf. Sirocko et al. 2005, supplementary material). The chronostratigraphy is based on the varve chronologies of the sequences combined with radiometric age control (luminescence and ^{14}C dates) as well as the tuning of the ELSA greyscale to the NGRIP oxygen isotope record in the ss-09sea chronology (Sirocko et al. 2005; cf. Andersen et al. 2004). Consequently, if this record is applied it also requires tuning to the GICC05.

The most prominent lakes with laminated sequences spanning the Lateglacial are the Meerfelder Maar (MFM) and the Holzmaar (**fig. 2**), situated approximately 10km farther to the east than the former. In both sequences the stratigraphic marker horizons of the Lateglacial LST and the early Holocene Ulmener Maar Tephra (Zolitschka/Negendank/Lottermoser 1995) are present, allowing for a good correlation of the cores from the two localities (**fig. 8**). Additionally, a series of 41 ^{14}C dates was taken on terrestrial macrofossils from Holzmaar (Hajdas et al. 1995; Zolitschka et al. 2000) and from the MFM a further 51 samples were ^{14}C-dated (Brauer et al. 2000b) to allow further comparison with the ^{14}C calibration curves.

The Holzmaar sequence was laminated back from the present day and contained some 23,220 ± 810 couplets (Zolitschka et al. 2000). The upper c. 13,840 of these are of organic composition (Vos et al. 1997; Leroy et al. 2000), whereas the lower 9,380 couplets are clastic sediments. The latter are demonstrated to represent annual cycles by a sedimentation model and by comparison with solar cycles (Brauer et al. 1994; Vos et al. 1997). Missing or poorly developed varves in the Holocene part were detected by correlation with the ^{14}C dates of the CEDC record and the correction of 346 years between 3,550-4,550 years cal. b2k in the record was confirmed as correct by comparison with the same sequence in the MFM record (Zolitschka et al. 2000). However, due to this clear gap, the Lateglacial chronology has to be regarded as floating. The comparison with the MFM also indicated a hiatus of approximately 320 years within the Younger Dryas (**fig. 8**; Zolitschka et al. 2000) which had first been identified in the pollen record (Leroy et al. 2000).

In the MFM record the upper 1-2m (i.e. the last c. 1,500-2,000 years) were not analysed due to poor varve preservation (Brauer et al. 2008, supplementary information) but from there on 12,000 couplets were counted (Brauer/Endres/Negendank 1999, 19). This floating chronology was correlated with the Holzmaar record by the use of the Ulmener Maar Tephra (dated to 11,050 ± 215 varve years cal. b2k; Zolitschka 1998; Brauer/Endres/Negendank 1999). Above this correlation point some offsets might occur but the correlation of the Ulmener Maar Tephra at 11,050 years cal. b2k seemed reliable, based on comparison with both the ^{14}C record as well as the Holocene increase in tree-ring width of the CEDC (Brauer et al. 2000b), and therefore the offsets on top of this tephra are of no further interest here. Below this correlation layer some 3,020 couplets were counted in the MFM record (**fig. 8**) including the LST (Brauer/Endres/Negendank 1999) which occurred 1,880 couplets below the Ulmener Maar Tephra. In contrast, the Holzmaar record yielded only 1,560 varves between these two marker horizons. At the Holocene/Pleistocene transition the lamination in the MFM was weak and, thus, over 4cm (some 50 varve years) the exact counting becomes uncertain (Brauer/Endres/Negendank 1999). If these 50 varves were included, the hiatus in the Holzmaar record during the Lateglacial Stadial would also need to increase to 370 years and the onset of the Younger Dryas as well as the dating of the LST would also be pushed some 50 years older. Thus, this uncertainty was considered with the error estimate and not counted on top of the varve years. In the lower part of

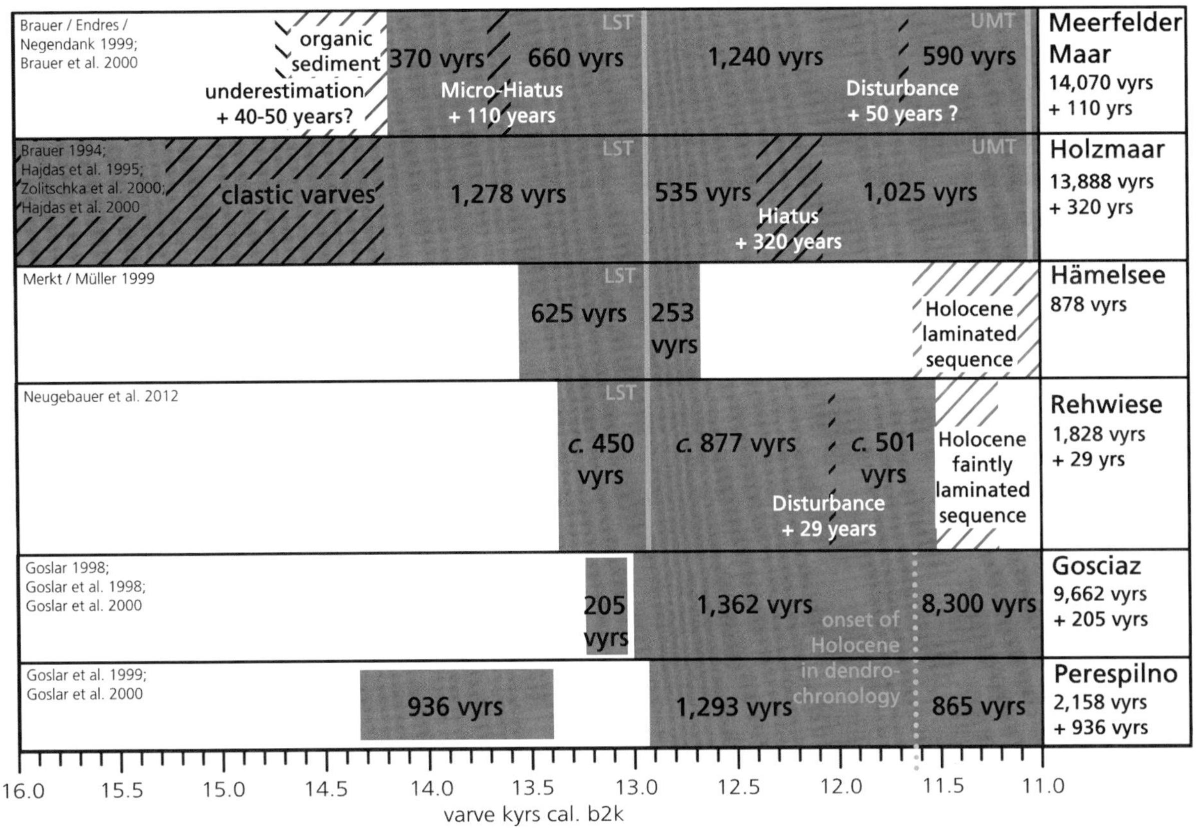

Fig. 8 Comparison of Lateglacial varve chronologies from Northern Europe. Timescales according to the varve chronologies and shifted from cal. BP to cal. b2k. Dotted lines: onset of the Holocene in oxygen isotope records. – Dashed line: sedimentological onset of the Holocene. – Solid line: tephra layer. – **LST** Laacher See Tephra; **UMT** Ulmener Maar Tephra. – For further details see text.

the laminated section, another c. 4 cm thick, disturbed section occurred which was estimated by correlation with the pollen zones in northern Germany to represent approximately 110 missing couplets (Brauer et al. 2000a, 231). However, if this part was compared to the Holzmaar record (Leroy et al. 2000) the most plausible correlation (psz 3-4/5) represented 340-380 couplets, i. e. 210-250 years more than were counted in MFM. According to the calculation of the sedimentation rate in this section of the MFM, 60-80 couplets could be missing (Brauer/Endres/Negendank 1999, 21). Thus, in the present study the »conservative« 110 varve counts were added. Additionally, the lowest 75 cm of the more organic sediment in the MFM sequence were not laminated. This section was estimated by extrapolation of depth and varve frequency in the 200 couplets following the onset of lamination (Brauer/Endres/Negendank 1999, 20) to have begun some 430 years earlier. However, an underestimation of up to 10 %, i. e. 40-50 couplets seemed possible (Brauer/Endres/Negendank 1999).

In combination, the two Eifel Maar records form a solid chronostratigraphy of the Lateglacial which can be connected to other records by the use of stratigraphic marker horizons as well as comparison of the carbon isotope record.

Hämelsee, Germany

The Hämelsee lies centrally on the southern North European Plain, near the transition towards the middle-range mountains (**fig. 2**). The lake presumably represents a sinkhole over Permian salt (Merkt/Müller 1999). Here partially laminated sediment formed as in several other lakes from northern Germany such as Lake Belau (Dörfler et al. 2012) and Lake Wollingst (Merkt/Kleinmann 1998; Müller/Kleinmann 1998).

These lakes were correlated using the Holocene event of the elm decline and various stratigraphic marker horizons such as the LST (**fig. 8**; Merkt/Müller 1999). For the Holocene the Saksunarvatn Ash was chosen as the correlation point. Sediments reaching back to the LST were found in the Lake Belau and the Hämelsee. Although the Holocene sequence of the Hämelsee reached back to the onset of the Holocene, the Preboreal section was not counted as being as reliable as the same sequence in the Plußsee (Merkt/Müller 1999, 44 fig. 3). On top of the Preboreal c. 1,600 couplets were counted in the Hämelsee, describing the Holocene sequence. Below the Preboreal a gap of inconsistent varve formation followed encompassing most parts of the Lateglacial Stadial. However, 253 laminated couplets were deposited on top of the LST. 625 couplets were counted below the LST. The unlaminated or poorly laminated sequences were correlated with well established chronostratigraphies from elsewhere (e.g. Lake Gościąż, MFM) because in northern Germany almost all lake sediments displayed a hiatus in sedimentation during the Lateglacial Stadial and sequences too short for the earlier parts of the Lateglacial Interstadial (Litt et al. 2001, 1239). However, the direct comparison of some parts of the MFM records with the segmented northern German records displayed some differences, for instance, in the duration of some events (Merkt/Müller 1999, 45) and led to the recalculation of some disturbed layers in the MFM (see p. 20; Brauer et al. 2000a). Nevertheless, these correlations need to be seen with some caution due to the uncertain difference in reaction time and the various positions in the succession of biozones. Microfacies analysis of the sediment revealed that a multitude of parameters had influenced the sedimentation in Hämelsee and that different processes led to the onset of the Allerød than those that led to the onset of the Younger Dryas (Merkt 1994, 60; Merkt/Müller 1999, 48).
Thus, the Hämelsee only provided a Holocene and a Lateglacial floating varve chronology which was partially confirmed and added by other northern German sequences such as the one from the Lake Belau, the Plußsee, the Lake Muggesfeld, and the Lake Wollingst. These floating northern German varve sequences can be correlated to other chronostratigraphies by stratigraphic marker horizons such as the LST or the Holocene Saksunarvatn Ash. In the Hämelsee record this chronostratigraphy is additionally combined with sedimentological, geochemical, pollen, spores, and rotifer analyses, as well as a series of some 15 ^{14}C dates on macrofossils. Therefore, the Hämelsee sequence can be used as a comparative record in the mid-Lateglacial.

Rehwiese, Germany

The palaeolake Rehwiese lies in the south-west of the Greater Berlin area within the North European Plain (**fig. 2**). The lake was part of a glacially formed water gap (*Grunewaldrinne*) which connected the Havel with the Spree and developed during the Weichselian maximum stage of the Scandinavian ice sheet in the area of modern Berlin.
The palaeolake Rehwiese consisted of three basins which were silted up by the mid-Holocene, some 6,000 years ago, and form part of a modern grassland park at the city limits of Berlin (Neugebauer et al. 2012). Four sediment cores were drilled at two locations in the centre of the still visible depression of the central basin of the palaeolake. At each location the two cores were drilled parallel with some overlap to produce two composite sequences (A/B and C/D) of laminated sediments and approximately 3.65 m thickness encompassing a period from the younger part of the Lateglacial Interstadial to the Early Holocene. The laminated sediments were formed by mainly calcite varves. Below the laminated sediments a silty layer was deposited on top of partially layered sands. On top of the Early Holocene varves a deposit with considerably poorer lamination followed which was overlain by some 6 m of a homogeneous calcareous gyttja. Finally, an approximately 1 m thick deposit of peat covered the lacustrine stratigraphy. Therefore, this laminated sequence represents another floating chronology.

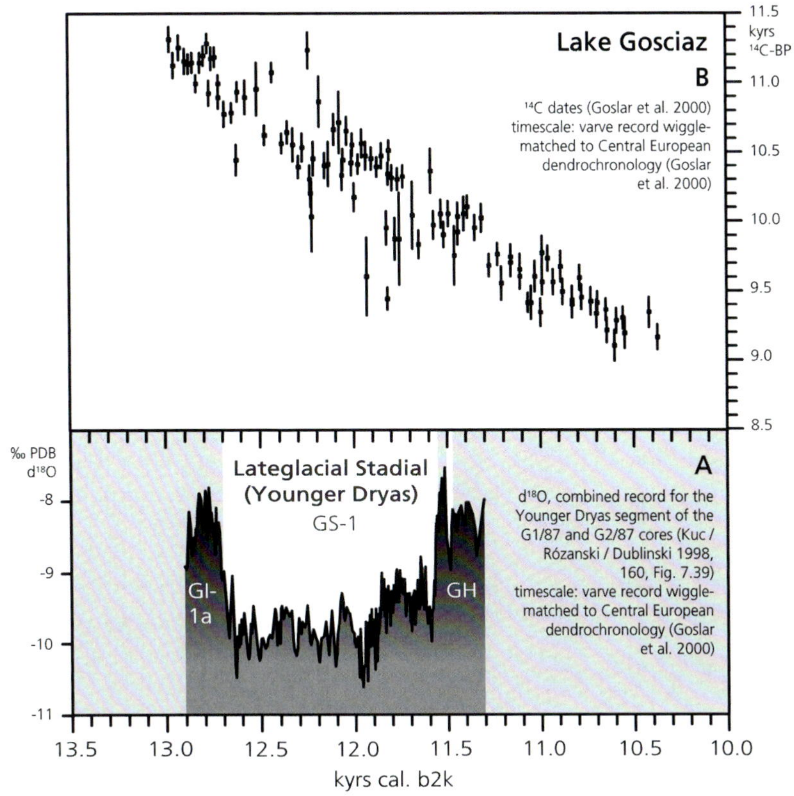

Fig. 9 Lateglacial record from Lake Gościąż in varve chronology wiggle-matched by ^{14}C isotopes to Central European dendrochronology (Goslar et al. 2000). Timescale is shifted from cal. BP to cal. b2k. **A** oxygen isotope record (Kuc/Różański/Dubliński 1998). Grey shaded areas represent periods of more interstadial oxygen isotope values than the surrounding values. – **B** series of ^{14}C dates (Goslar et al. 2000). – For further details see text.

However, in the lower part of the sequences the LST was found and used as the anchor point to connect the Rehwiese to the MFM record (**fig. 8**). On top of the LST 1,378 couplets and below some 450 couplets were counted with the confidence of c. 1 % counting error (Neugebauer et al. 2012). A disturbance of 7.4 cm in both composite sequences was interpolated by sedimentation rate to 29 couplets within the Younger Dryas section, around 12,040 years cal. b2k (Neugebauer et al. 2012). However, this profile represents the thus far most complete section of the Younger Dryas in the European varve records.

The material below the LST is not yet published. In addition to a detailed micro-facies analysis of the sediments from the LST upwards, geochemical analyses were performed and palynological results from a previously taken profile were correlated with a few high-precision pollen counts in short sub-sections (Neugebauer et al. 2012) and yielded one of the most reliable records for the Younger Dryas in northern Germany.

Lakes Gościąż and Perespilno, Poland

In the Polish part of the North European Plain two important Lateglacial stratigraphies with laminated sediments are from Lake Gościąż and Lake Perespilno (**fig. 8**). While Lake Gościąż is situated approximately in the centre of northern Poland, Lake Perespilno lies near the eastern border of the country and near the transition towards the middle-range mountains (**fig. 2**).

Lake Gościąż is one of the north-western lakes of the Gostynińskie Lake District, which is formed by over 60 lakes and is situated in the Płock Basin at the lower Vistula valley, north-west of Warsaw. The basin lies just inside the limits of the maximum extent of the Scandinavian ice sheet during the LGM. The lakes formed after the retreat of the periglacial conditions and the down cutting of the Vistula (Churski 1998). Similar to Gościąż, Lake Perespilno is located in the western part of a lake district with 67 lakes, which formed in a depression with restricted water runoff (Bałaga/Goslar/Kuc 1998). Some 300 km south of the maximum extent of the ice sheet the lakes were presumably formed due to karst and thermokarst processes in the underlying marl and chalk bedrock.

Laminated sediments from several cores taken at Gościąż cover a period from the Late Pleniglacial to the present day and produced intensively studied lithological and environmental sequences (Ralska-Jasiewiczowa et al. 1998). These were studied with geochemical and mineralogical analyses (Łącka/Starnawska/Kuźniarski 1998) and investigations of pollen (Ralska-Jasiewiczowa/Demske/van Geel 1998), diatoms (Marciniak 1998), cladocera (Szeroczyńska 1998), and isotopes (^{13}C, ^{18}O; Kuc/Różański/Dubliński 1998). Moreover, the cores provided samples for a series of 47 ^{14}C dates (**fig. 9**; Goslar/Arnold/Pazdur 1998; Goslar et al. 2000). In the seven cores from Perespilno, 51 ^{14}C dates were taken on macrofossils (Goslar et al. 2000); furthermore, isotopes (^{13}C, ^{18}O), geochemistry, and pollen were analysed (Bałaga/Goslar/Kuc 1998; Goslar et al. 1999).
In the upper part of the Gościąż stratigraphy, the poor development of the lamination prevented a reliable counting of couplets. Consequently, the 9,662 ± 90 couplets were not continuous to the present day and, hence, formed a floating chronology (Goslar 1998b). The organic sediments represent seasonal cycles and, thus, can be counted as varves. Significant gaps and uncertainties in the correlation generally occurred in the Holocene portion (8,300 ± 50 couplets). However, in the lower, Lateglacial part (1,362 ± 42 couplets + 205 uncorrelated couplets) some uncertainties arose at the onset of the Younger Dryas where a sand layer eroded some parts of the main counted cores. However, this gap was bridged by cores from the western and northern parts of the lake and, consequently, the gap was estimated to only represent 4 +6 / -2 varve years (Goslar 1998c). Generally, the cores were recovered in 1-2 m long segments and the correlation of these segments also provided a potential cause of miscounting. This problem applied in particular to the deepest segment of the deepest core which could not be correlated with certainty to the Lateglacial sequence.
These 205 couplets (**fig. 8**; Goslar 1998c) formed yet another independent floating, unconnected chronology and will not be used in the present study. Based on the lithological, palynological, and isotope studies, the Younger Dryas was clearly identified. Thus, the continuous counting reached back 222 ± 2 couplets beyond the onset of the isotopically determined Younger Dryas into the Lateglacial Interstadial. In the Younger Dryas section 1,140 ± 40 couplets were counted. In a monographic publication the Gościąż sequence was intentionally not compared to other records on a regional or global scale (Goslar et al. 1998b, 171-173), but wiggle-matching of ^{14}C dates on macrofossils with the CEDC record then in use allowed a fixation of the Holocene/Pleistocene limit and, furthermore, demonstrated the coincidental rise of the oxygen isotope record in Gościąż and the tree-ring width in the Central European pines (Goslar et al. 2000). Nevertheless, the shift of the CEDC since (Friedrich et al. 2004) implicated a comparable shift of the Gościąż record. The ^{14}C dates displayed a wide wiggling in the Lateglacial part. The $\delta^{18}O$ record in Gościąż was comparable to the tree-ring width mainly controlled by climatic parameters, in particular, the hydrological regime of the lake and the air temperature (Kuc/Różański/Dubliński 1998, 158). The varve thickness in Lake Gościąż seemed to be controlled by the organic productivity of lacustrine biota (Goslar 1998d, 105) and temperature during the flowering period (Goslar 1998d, 107).
At Perespilno only the lower 3 m of sediment was laminated (Bałaga/Goslar/Kuc 1998). The sequence comprises 3,105 ± 120 couplets (**fig. 8**; Goslar et al. 1999) which were not confirmed as representing annual cycles, but comparison with the Gościąż record suggested this assumption to be plausible (Bałaga/Goslar/Kuc 1998). The record was correlated mainly with Gościąż and comprised a period from the early Holocene to the mid-Lateglacial Interstadial (Goslar et al. 1999). Originally, two major gaps occurred within the sequence, one of which was due to coring and could be bridged by additional cores (cf. Bałaga/Goslar/Kuc 1998 and Goslar et al. 1999). The other one was a 10.8 cm long section of unlaminated sediment from the later part of the Allerød. According to the sedimentation rate in the surrounding sediments this part was first estimated to encompass 145 ± 20 couplets (Bałaga/Goslar/Kuc 1998) but later recalculated to 80 ± 60 couplets (Goslar et al. 1999). This gap bisected the core record into a part encompassing 2,158 ±

100 couplets, in which the transition from the Holocene to the Lateglacial Stadial and the transition to the Lateglacial Interstadial were present, and an older part containing 936 ± 35 couplets (Goslar et al. 2000). In the longer, younger section 168 ± 10 couplets represented the end of the Lateglacial Interstadial, 865 ± 20 years were assigned to the Holocene, and the remaining 1,125 ± 70 couplets were deposited during the Lateglacial Stadial. The older section was wiggle-matched to coral ^{14}C calibration data and, thus, dated to a period between 13,750 and 14,500 years cal. b2k (Goslar et al. 2000). However, calibration data sets spanning the early Lateglacial were refined since (e. g. Weninger/Jöris 2008) and the correlation should be reviewed in light of these new curves. No other isotopic measurements of the older part of the Perespilno sequence were published for more reliable comparison and so, for the moment, it needs to be considered as unconnected to the calendar age timescale and thus will not be used in the present study. Analyses of geochemistry and isotopes were connected with the upper part of the record as well as a palynological study and measurements of magnetic susceptibility (Goslar et al. 1999), which reflected the ratio of induced magnetisation to the magnetic field and could be used to identify volcanic activity in sediment records (Peters et al. 2010). However, a precondition for the latter analysis is a constant content of magnetic minerals in the sediment which seemed not to be the case in the record from Lake Gościąż (Nilsson 2006).

The main significance of the Polish records is the generally undisturbed lamination during the Younger Dryas as well as the extensive additional analyses. From Lake Gościąż so far the only continental high-resolution oxygen isotope record in combination with varve counting, a record of varve thickness, and a series of ^{14}C dates is known. The record could thus be correlated to the Greenland ice-cores by the use of the $\delta^{18}O$ data (Goslar/Arnold/Pazdur 1998) and also to the CEDC by wiggle-matching the carbon isotope series (Goslar et al. 2000) as well as comparing the varve thickness record (Goslar 1998a) which in addition allowed a correlation with other laminated sequences (Litt et al. 2001).

Dendrochronology

Tree-ring chronologies, with their annual resolution and content of carbon, provide a strong tool for constructing radiocarbon calibration curves (Stuiver 1982; Friedrich et al. 2004). Based on the assumption of comparable production rates of cosmogenic radionuclides due to the common dependence on solar radiation and geomagnetic field, the $\Delta^{14}C$ record can be correlated to the Beryllium (^{10}Be flux) record in Greenland ice-cores (Muscheler et al. 2008). Furthermore, by the means of tree-growth rate this terrestrial environmental record can partially be compared to the isotope eventstratigraphy (Friedrich et al. 2001b; Schaub et al. 2008a) because tree-ring width is controlled by the climatic conditions prevailing during the growth season and, thus, changes in the hydrological regime, in particular, can be inferred from this record (Friedrich et al. 1999). Additionally, the tree-rings were also sampled for further isotopes such as ^{2}H (deuterium) and ^{13}C, which in this record mainly reflect humidity and temperature during the summer months (Friedrich et al. 1999).

Unlike many other records, the CEDC extends back from modern times continuously (**fig. 10**). In general, the ^{14}C-dated samples comprised approximately 10 tree-rings and yielded ages with standard deviations of 15 to 35 radiocarbon years (Reimer et al. 2004). The previously described data sets showed clearly the significance of the CEDC for most of the other chronostratigraphic records. Besides representing the correlative sequence for various floating varve chronologies, the dendrochronological data set was also accepted by the major calibration curves as the most reliable ^{14}C calibration record (Fairbanks et al. 2005; Weninger/Jöris 2008; Reimer et al. 2009) and, consequently, the calibration results for dates from the time period covered by this data set are generally similar. However, if corrections are made to the tree-

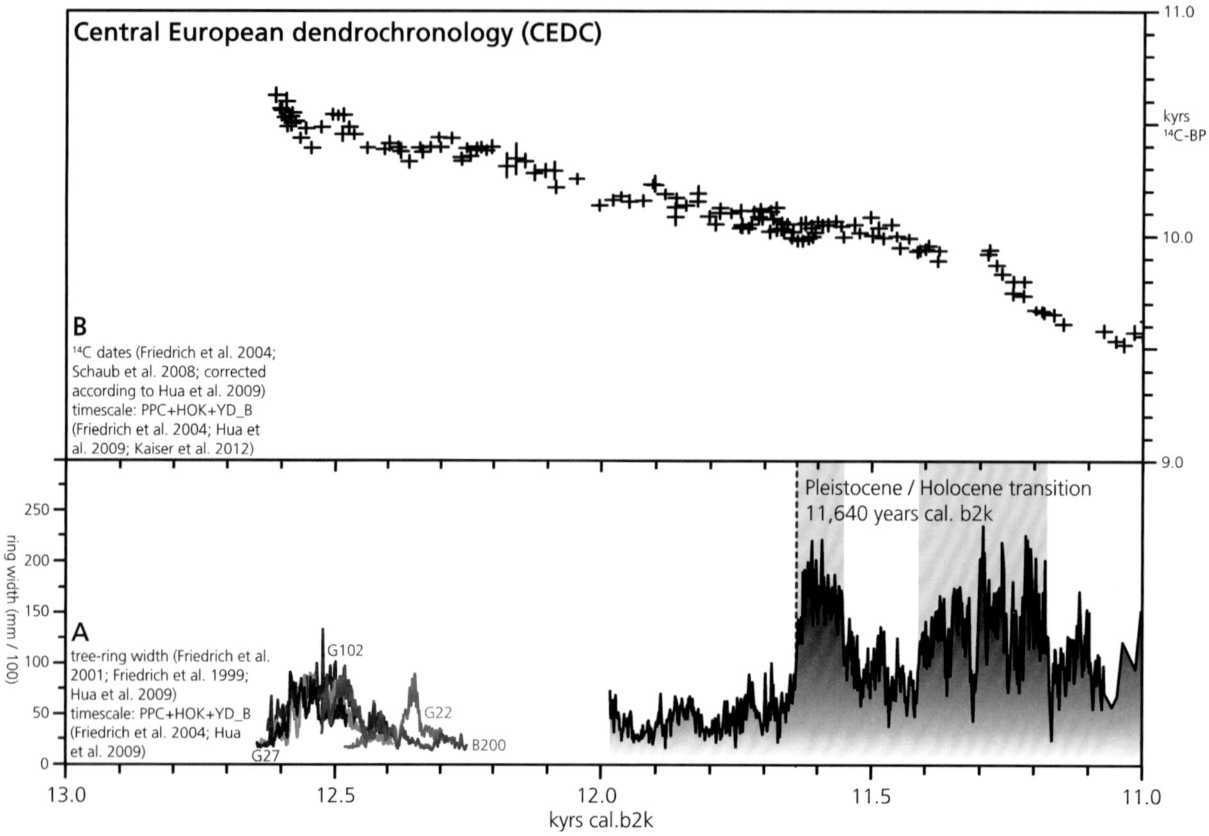

Fig. 10 Central European dendrochronology (CEDC; Friedrich et al. 2004; Schaub et al. 2008b). Timescale is shifted from cal. BP to cal. b2k. **A** tree-ring width record (Friedrich et al. 1999; Friedrich et al. 2001b) supplemented with tree-ring widths of five trees forming the YD_B chronology (Hua et al. 2009, 2983 fig. 1; scale is only approximated according to comparison with Schaub et al. 2008b, 79f. figs 4B and 5B). Grey shaded areas represent periods of intense tree-ring growth. – **B** series of ^{14}C dates (Friedrich et al. 2004; Schaub et al. 2008b). – For further details see **tab. 2** and text.

tree-ring sequences	sites of contributing sequences	ref.
CEDC	Upper Rhine, Main, Danube, & Lausitz material (HOC); Upper Rhine, Danube, Isar, Iller, & Günz material (PPC); Avenches & Ollon (PPC extension); Cottbus (YDPC); Birmensdorf (ZHW); Gänziloh & Birmensdorf (YD_B)	Friedrich et al. 1999; Friedrich et al. 2004; Schaub et al. 2008b
GLPC	Dättnau; Danube, Isar, Iller, & Günz material; Reichwalde; (the Kruft poplars were correlated to the GLPC but formed no part of it)	Kromer et al. 2004
LG pine ring widths	Avigliana; Revine; Dättnau; Danube & Isar material; Reichwalde, Lohsa, & Scheibe material; Warendorf	Friedrich et al. 2001b
CELM	Dättnau (Daeboech, Daeboeal, Daealch1-4); Landikon (Landboeal); Gänziloh (Gaenalch); Danube, Isar, & Günz material (Danube LG 2 & 3)	Schaub et al. 2008b; Kaiser et al. 2012
Huon pines	Stanley river/Tasmania	Hua et al. 2009
YD_A	Gänziloh	Schaub et al. 2008b

Tab. 2 Various tree-ring sequences and the sites yielding samples for these sequences (see **fig. 2**; cf. Kaiser et al. 2012). Abbreviations: **CEDC** Central European dendrochronology; **HOC** Hohenheim oak chronology; **PPC** Preboreal pine chronology; **YDPC** Younger Dryas pine chronology; **ZHW** Zürich-Wiedikon; **YD_B** Younger Dryas B sequence; **GLPC** German Lateglacial pine chronology; **LG** Lateglacial; **CELM** Central European Lateglacial master chronology; **YD_A** Younger Dryas A sequence.

ring data set (**tab. 2**), the chronologies in several other archives are also affected. In the Lateglacial this was especially apparent at the common correlation point of the Holocene/Pleistocene transition which was identified by a significant increase in the width of the tree-rings. This transition was last shifted from 11,620 years cal. b2k (Friedrich et al. 1999) to 11,640 years cal. b2k because of mis-correlation due to

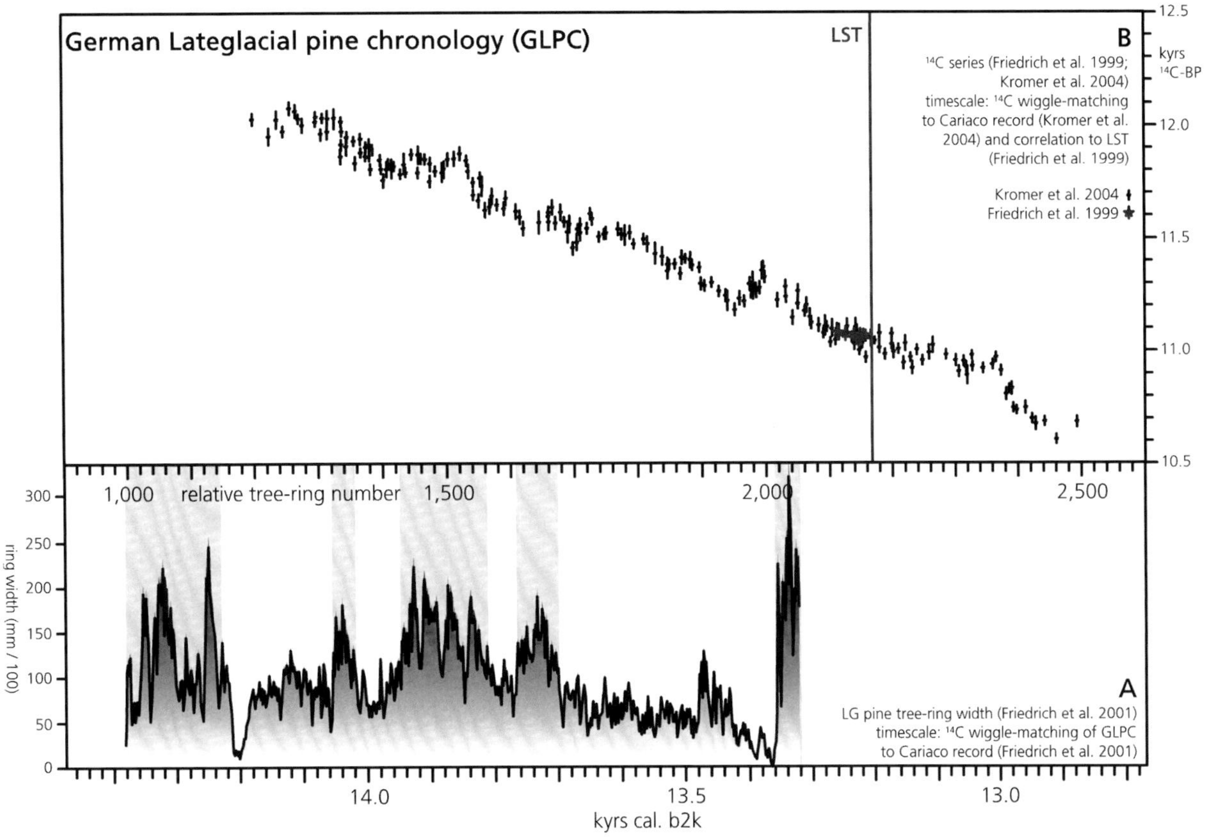

Fig. 11 German Lateglacial pine chronology (GLPC; Friedrich et al. 1999; Kromer et al. 2004) and Lateglacial (LG) pine tree-ring width sequence (Friedrich et al. 2001b). Timescale is shifted from cal. BP to cal. b2k. **A** tree-ring width record of LG pines (Friedrich et al. 2001b). Grey shaded areas represent periods of intense tree-ring growth. – **B** series of ^{14}C dates of GLPC (Friedrich et al. 1999; Kromer et al. 2004). – For further details see **tab. 2** and text.

cockchafer predation (Friedrich et al. 2004, 1120). Thus, correlations made with the dendrochronological record prior to this publication needed to be shifted accordingly. For a while the dendrochronological record extended reliably only into mid-GS-1 (12,460 years cal. b2k; Friedrich et al. 2004) but new data were presented lately pushing the CEDC further back into the early Lateglacial Stadial (12,644 years cal. b2k; Schaub et al. 2008b; cf. Hua et al. 2009; Reimer et al. 2009). Thus far, no detailed record of the tree-ring growth patterns including the extensions has been published (Schaub et al. 2008b) but the tree-ring growth sequences of the single trees forming the last extension (YD_B) were published in relation to each other (Hua et al. 2009, 2983 fig. 1). These sub-sections are used in the present study (**fig. 10**). However, even with this extension the continuous dendrochronology could not yet be connected unambiguously to the floating GLPC (**fig. 11**) or to the more recently developed Central European Lateglacial Master chronology (CELM; **fig. 12**). The former comprised a series of 1,382 tree-rings from 517 tree-ring sections (**tab. 2**). Furthermore, this set provided data about tree growth (Friedrich et al. 2001b) and produced also a series of 106 high-precision ^{14}C dates taken usually on samples encompassing ten tree-rings and giving ages with a standard deviation of 20 to 50 radiocarbon years (Kromer et al. 2004) and was thus used for calibration (e.g. Weninger/Jöris 2008). In addition to trees found in southern German gravel pits, Swiss loam pits, and eastern German bogs, trees buried in the deposits of the LSE contributed to this record (Friedrich et al. 1999, 36-38; Kromer et al. 2004, 1205). Approximately 1,040 rings were counted before and some 340 rings after the LSE. By this marker the GLPC set became correlative to other sequences which contained the LST.

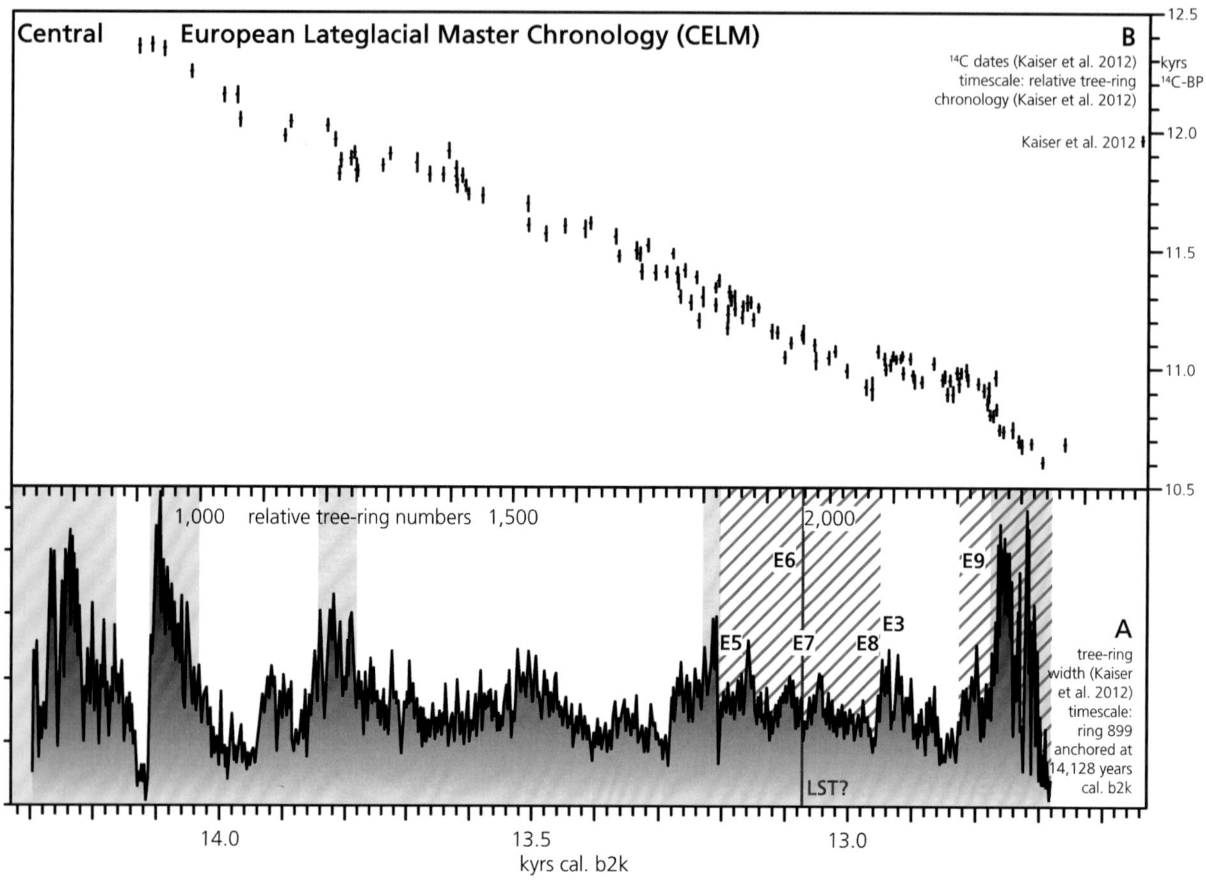

Fig. 12 Central European Lateglacial Master chronology (CELM; Kaiser et al. 2012). Timescale is shifted from cal. BP to cal. b2k. **A** tree-ring width record (Kaiser et al. 2012). Dark grey hatched areas events E6 and E9 according to Kaiser et al. 2012, 85f. fig. 5, where also the short-term events E5, E7 and E8 are given. E3 is set according to Schaub et al. 2008a, 36-38. Grey shaded areas represent periods of intense tree-ring growth. – **B** series of ^{14}C dates (Schaub et al. 2008b; Kaiser et al. 2012; cf. Kromer et al. 2004). – For further details see **tab. 2** and text.

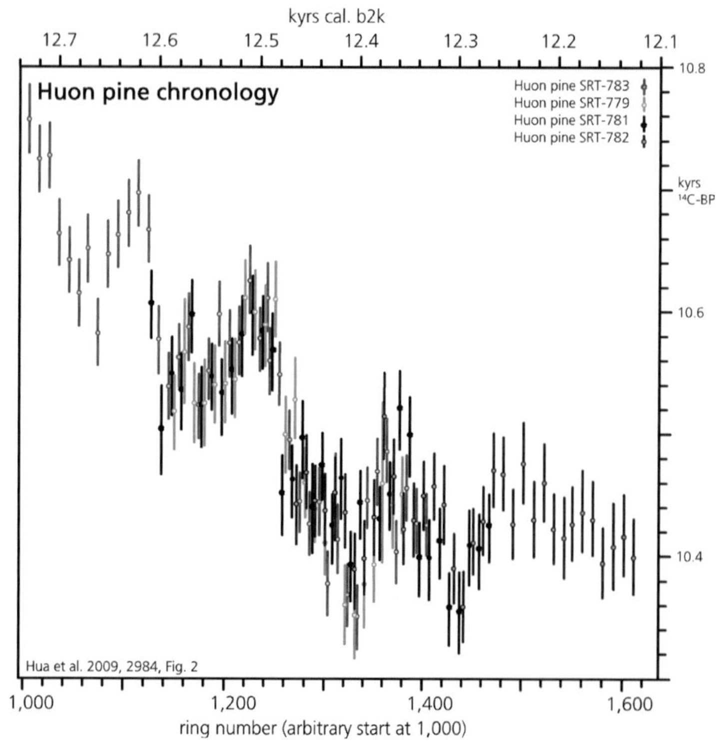

Fig. 13 ^{14}C dates from Huon pine record with relative tree-ring timescale (bottom) and results of wiggle matching with CEDC (top; Hua et al. 2009). Timescale is shifted from cal. BP to cal. b2k. – For further details see **tab. 2** and text.

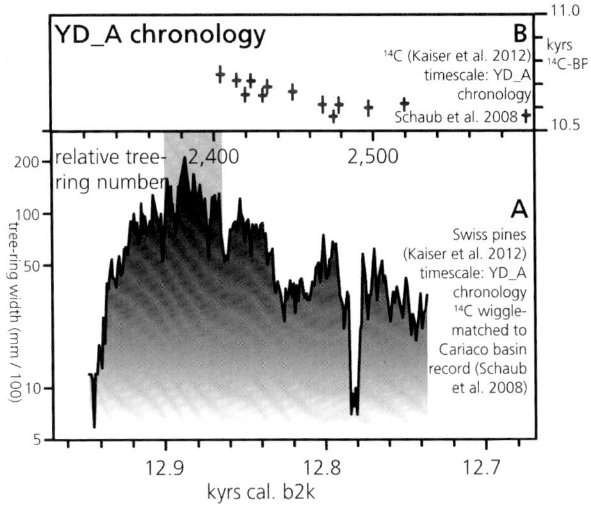

Fig. 14 Younger Dryas A chronology (YD_A). Relative tree-ring timescale wiggle matched with Cariaco basin and results of correlation to absolute timescale. Timescale is shifted from cal. BP to cal. b2k. **A** tree-ring width (Schaub et al. 2008b; Kaiser et al. 2012). Grey shaded area represents period of intense tree-ring growth. – **B** ^{14}C dates (Schaub et al. 2008b; Kaiser et al. 2012). – For further details see **tab. 2** and text.

In general, the CELM record comprises the same tree-ring sections but was further extended with younger sections and encompasses 1,606 tree-rings (Kaiser et al. 2012). In addition, a further 131 decadal samples produced ^{14}C ages for the Swiss material.

Two attempts to make the connection between the CEDC and the tree-ring data set from the Lateglacial Interstadial already exist:

One of these attempts connected the CEDC with the GLPC set by the aid of tree-ring sequences from Tasmanian Huon pines covering 617 years during GS-1 (Hua et al. 2009; **fig. 13**). The tree-ring width of these pines was measured couple-wise (i. e. annually) and samples for ^{14}C dating were taken in four to ten year intervals. However, due to the origin from Tasmania, i. e. on the southern hemisphere, an average hemispheric offset had to be subtracted. This offset was estimated to 40 years (Hua et al. 2009, 2985). Then the ^{14}C record was wiggle-matched to the revised extension of the CEDC and the GLPC. This data set was chosen for the construction of the IntCal09 calibration curve. However, in the light of small intra-hemispheric offsets in the ^{14}C content (Kromer et al. 2001) and inconstant inter-hemispheric offsets of uncertain dimension (Barbetti et al. 2004), a correlation of dendrochronologies from different hemispheres, in particular, remains problematic (cf. Hua et al. 2009, 2985).

In the other attempt, tree-ring data from the Swiss Gänziloh site were correlated to a 212-year long subset in the early Younger Dryas period (YD_A) combining tree-ring widths with ^{14}C dates (**fig. 14**). ^{14}C-dated samples from this subset covered usually ten tree-rings, although sometimes larger sections were necessary for ensuring reliable measuring results (Schaub et al. 2005, 13). The ten pines of the YD_A chronology set were used as an extension of the CEDC into the early Lateglacial Interstadial but the cross-dating process with the generally accepted dendrochronology was complicated by the generally short sectioned tree-ring sequences from the early Lateglacial Stadial (Kaiser et al. 2012). Therefore, the connection was established by wiggle-matching of the ^{14}C record with the Cariaco basin record (in the version of Hughen et al. 2000) resulting in a small and, thus, still insecure overlap of approximately ten tree-rings (Schaub et al. 2008b, 77. 82) and a more clear overlap of c. 50 tree-rings to the floating Zürich Lateglacial chronology 1 (ZH LG_1; (Schaub et al. 2008b, 82), i. e. the CELM data (cf. Kaiser et al. 2012, 83). A later revision of the Cariaco basin chronology resulted in a general shift 14 years older due to the revised onset of the Holocene in the CEDC (Hughen et al. 2004c, 1164) which should result in a gap of c. four years between the CEDC and the YD_A chronology. Moreover, the 14 years have to be added on the results given in Schaub et al. 2008b and Kaiser et al. 2012. In contrast, based on a correlation with the IntCal09 and, thus, the Australian Huon pine data, the YD_A chronology either overlaps almost entirely with the end of the CELM or should overlap

more clearly with the end of the CEDC record. At least 40-50 years overlap on both sides were assumed (Kaiser et al. 2012, 83) but neither has yet been confirmed dendrochronologically (cf. Schaub et al. 2008b, 79 fig. 4; 80 fig. 5; Kaiser et al. 2012, 82 fig. 3K-h). Short-term events recorded in the younger part of the CELM (**fig. 12**; cf. Schaub et al. 2008a, 36-38; Kaiser et al. 2012, 85f.) can be particularly useful in this comparison.
Nevertheless, in a numerical comparison the results of the two attempts are relatively similar. Although these results differed only by a few decades, this offset is of special significance for the high-resolution calibration curve of this period (pers. comm. Olaf Jöris, Neuwied) due to a strong increase in atmospheric ^{14}C isotopes at the transition to the Lateglacial Stadial. Therefore, the correlations are discussed in more detail elsewhere (p. 250-253 and p. 358-364).

Consequently, two major independent resources with annual resolution from modern times to the Pleistocene are available for the purpose of Lateglacial chronostratigraphy at the moment: the ice-core records from Greenland and the CEDC. Other records with annual resolution are correlated to these records and can help to further confirm the annual chronostratigraphy.
The CEDC has a higher resolution and, in addition, forms an essential part of the radiocarbon calibration curve. This correlation is important because of the connection of the climatic records to one another. However, as long as the connection across the early Lateglacial Stadial is not ascertained this record cannot be used for the Lateglacial Interstadial. Nevertheless, if the LST can be reliably attributed it can function as a solid correlation point.
The Greenland ice-core sequences are in contrast continuous records across the Lateglacial. The records are presented in the GICC05 based on NGRIP in the Lateglacial which is therefore used in the present study as a chronostratigraphic baseline to which the other records are related. The tephrochronological analysis excluded the presence of the LST particles in the ice-cores (Mortensen et al. 2005) and, thus, this important marker cannot be directly correlated. However, based on the ice-core depths the records can be transferred to the GICC05 chronology. This way it is possible to create a detailed insight into the climate development and precise chronological development. In particular, the correlation of the calibration records shall then be further used to relate the environmental as well as archaeological record (see Material-Databases, p. 49-53) to this chronostratigraphy.

ENVIRONMENTAL ARCHIVES

A fundamental presupposition of the present study is that the environmental change is of greater importance for the Lateglacial hunter-gatherers than climate change. This assumption is based on the fact that the natural environment provides the basic resources for the survival of hunter-gatherers such as food and raw material for clothing, equipment, or building shelters. These resources originate from geological deposits (minerals, fossils, stones, and lithic raw materials), the vegetation, and animals. Therefore, changes in the existence and/or accessibility of these natural resources have to have effects on the human activities in which these materials were involved. Moreover, the procurement strategies for raw materials influence the human mobility patterns and the supply of vital resources affects the human demography. Thus, the natural environment as providing the basic source for the daily life of hunter-gatherers also sets the framework in which humans can act.

Thus, in addition to the climate development and the changes of human behaviour, the natural environment is modelled by the use of different archives. In the present study, the landscapes form a fundamental archive on which the vegetation record is set. Landscapes refer to the physical geography of the studied areas, in particular, their geomorphological appearance, the geological and pedological deposits as well as the waterways. In general, changes in geomorphology and geology occur too gradually to be effective in the studied time period. Therefore, these factors are introduced in the following and assumed as mainly stable during the studied period. Nevertheless, the accessibility of resources could have been hindered by different natural circumstances such as pedological processes or changes in the water regime and/or the vegetation. Archives for the changes of water regimes are scarce in western Central Europe and fluctuations in lake levels which are frequently reconstructed elsewhere (Bohncke/Wijmstra 1988; Renssen/Isarin 2001; Magny 2001; Magny/Bégeot 2004) are particularly rare in the studied regions. Indications for the changes in the water regimes originate mostly from river bed constructions with a relatively low temporal resolution (Gibbard 1988; Antoine 1997; Pastre et al. 2003; Kasse et al. 2005). Thus, these general reconstructions are introduced in the following and adopted in the result chapter. For the reconstruction of the vegetation, pollen archives are used because these records deliver a general impression of a wider catchment area and, therefore, give a general cross-section of the available vegetation patches in the studied areas. Additionally, directly dated macro-fossils deliver information of the vegetation available near the findspots. The presence of selected fauna species can also be read from directly dated samples.

In the last decade, various genetic considerations were also introduced to the discussion about vegetation development (Hewitt 2000; Willis/van Andel 2004; Michalczyk et al. 2010; Habel/Assmann 2010). However, comparable to other genetic studies the chronological precision of these studies remained below the necessities of the present approach and, consequently, these studies are only mentioned anecdotally.

However, the reasons for changes in the above mentioned parts of the natural environment and the tempo of these changes are very heterogeneous.

Landscapes

The landscapes in Europe have changed significantly since the Lateglacial. To visualise some of these changes maps were made within the present study (see p. 253-259) based on modern satellite data. These data resulted from the shuttle radar topography mission (SRTM) mainly accomplished by the US-based National Geospatial-Intelligence Agency (NGA) and the National Aeronautics and Space Administration (NASA). The mission aimed to obtain elevation data on a near-global scale to generate a high-resolution digital topographic database of the planet. The mission was accomplished during eleven days in February 2000 with a specially modified radar system which flew on board of the Space Shuttle Endeavour. The resulting raw data (version 1) were reviewed by the NGA (version 2.1). These reviewed elevation data were made available online for public use (Farr et al. 2007; downloadable at http://dds.cr.usgs.gov/srtm/version2_1/SRTM30/). The SRTM 3×3 (arc-seconds) tiles reach an approximate 90 m resolution (cf. Smith/Sandwell 2003), whereas the SRTM30 data are of *c.* 900 m resolution (30×30 arc-seconds, cf. documentation file at http://dds.cr.usgs.gov/srtm/version2_1/SRTM30/) which is sufficient for the overview of north-western Europe. Furthermore, the data of the lower resolution were supplemented with measured and estimated bathymetric data (SRTM30_plus; Sandwell/Becker 2009) and published in 40 degree latitudinal by 50 degree longitudinal tiles. These tiles are used in the present study (downloaded as SRTM30_plus, version 6.0, at ftp://topex.ucsd.edu/pub/srtm30_plus/). Thus, the basis of the maps created for the present study are

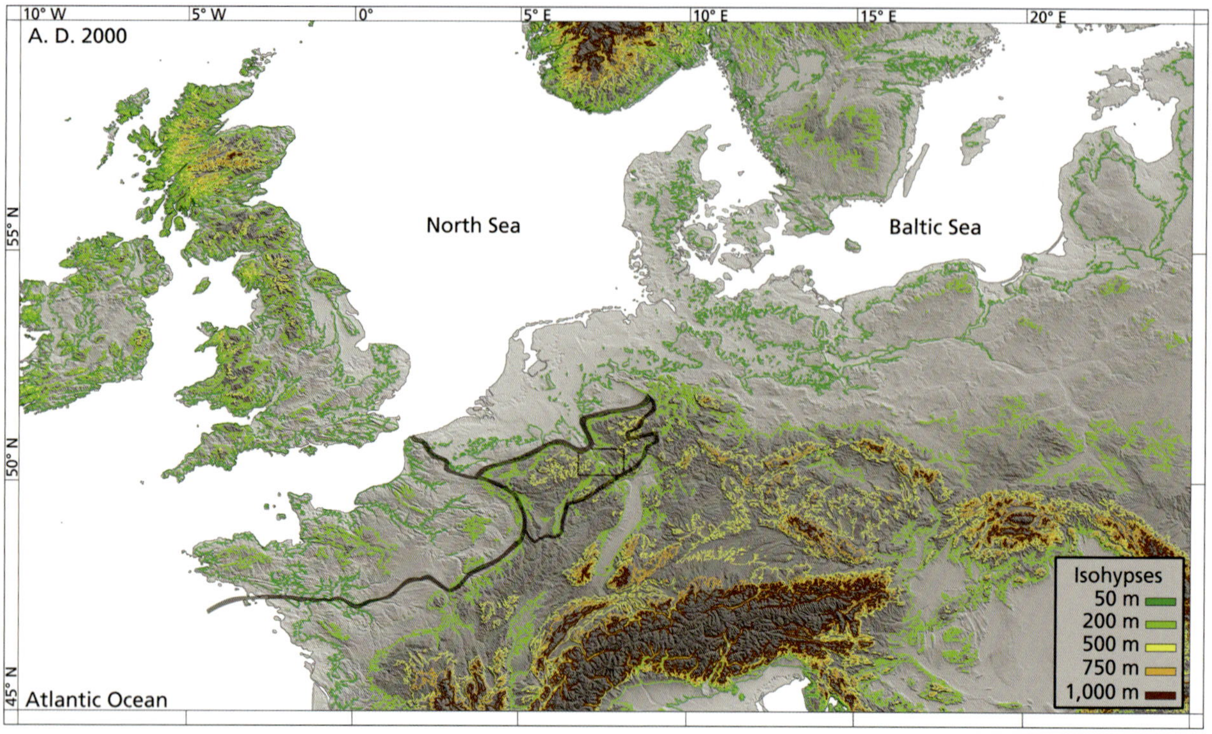

Fig. 15 Modern north-western Europe with selected modern isohypses (50 m – green, 200 m – green-yellow, 500 m – yellow, 750 m – orange-brown, 1,000 m – brown, all in m a. s. l.) and the sub-areas of the study (grey) indicated (see **fig. 1**). – For further details see text.

modern topographic data which in detail differ from the Lateglacial setting, for instance, due to isostasy, aeolian deposition, pedogenesis, and modern gravel pit quarrying.

However, some main geomorphological settings such as the presence of karst formations and geological deposits remained generally unaltered. These features are of interest because they can be assumed to have also influenced the perception of the landscape by the Lateglacial hunter-gatherers (cf. Rockman 2003a). In particular, for Lateglacial communities these features were presumably important criteria because they yielded natural shelter, micro-habitats, and/or lithic and mineral resources. In addition, the fresh water supply was certainly another substantial point. Besides the choice of the Lateglacial communities, the various geographical developments after the Lateglacial occupation further influenced the preservation of archaeological material in these areas. Therefore, evaluations of these large scale factors are necessary in large scale analyses to incorporate the bias factor.

In the following, the sub-areas delivering the archaeological material (i. e. the Central Rhineland and the western upland zone, northern France including the Paris Basin) are used to characterise the various landscapes within which Lateglacial humans had to adapt their way of life. This study area (**fig. 15**) is limited in the north by the North European Plain. However, to the south-west the Plain gradually passes into northern France. In the present study, the 50 m above sea level (a. s. l.) isohypse in combination with modern watershed limits (**fig. 16**) serves as an approximate distinction between the western North European Plain and northern France. The watersheds still belonging to the lowland region are the Scheldt and Aa which drain into the North Sea and on the northern French side the Liane, the Authie, and the Somme flow into the English Channel. Moreover, in the gradually rising southern part of the western North European Plain, the limit between the lowlands and northern France as well as the western uplands is chosen following the geology (**fig. 17**) with the Devonian uplands which frequently rise above 500 m a. s. l. and are also known as the Rhenish Slate Mountains or the Rhenish Shield marking the boundary. Still, the transition between the

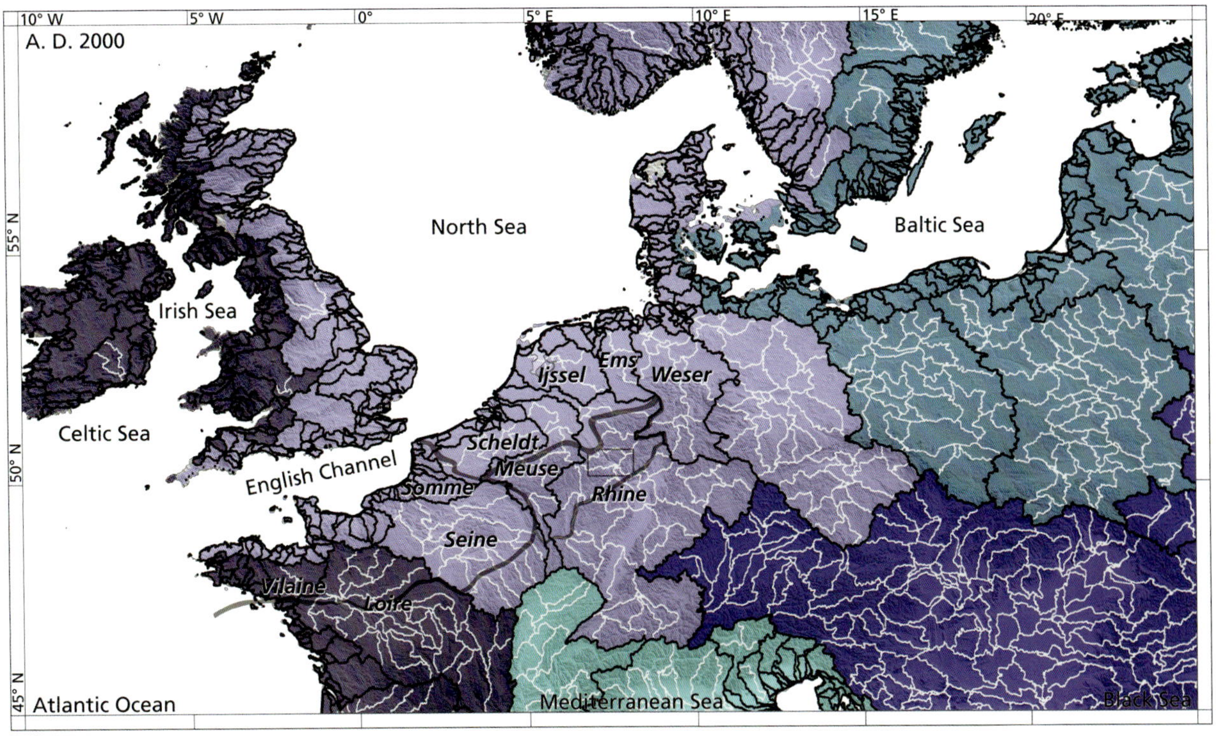

Fig. 16 Modern north-western European watersheds (data source: European Commission/Eurostat 2002; SADL [Spatial Applications Division Leuven] 2005; freely available for non-commercial use); major water catchment areas in the study area are named. Dark blue shaded: Atlantic draining systems; light blue shaded: North Sea draining systems; turquoise shaded: Baltic Sea draining systems; royal blue shaded: Black Sea draining systems; light turquoise shaded: Mediterranean Sea draining systems.

North European Plain and the adjacent uplands remains in some parts diffuse, hence, in this work a limit is chosen marked by a regular altitude of the modern terrain exceeding 200 m a. s. l. This isohypse is also taken as approximate limit between northern France and the western uplands.

To evaluate developments in the upland regions immediately adjacent to the North European Plain it is not necessary to consider the higher uplands farther to the south in this study. Indeed, in these southern regions the highly variable micro-climate and hence micro-ecotopes, which are probably due to the various effects of heterogeneous altitudes on regional climate and ecosystems (e. g. Bridault/Fontana 2003; Körner 2007), might obliterate some general developments within the Lateglacial environment which led to the changes observed in north-western Europe. The southern limit of north-western Europe is therefore determined at about 47° of northern latitude and/or at a recurrent altitude of the terrain above 750 m a. s. l. (**fig. 15**).

In the west the Atlantic Ocean forms a natural limit to the area.

The eastern limit of the study area is also partially defined on geomorphology but in part also due to geology. In particular, the Weser catchment area which generally consists of the Triassic Weser hills is chosen as the limitation. The western uplands are dominated by the Devonian Rhenish Shield which also contained the Central Rhineland region (**fig. 17**). This shield is surrounded by small Jurassic and Triassic formations and tends to reach higher altitudes than do the uplands adjacent to the east.

Plotting environmental data over the thus defined large areas cannot deliver a differentiated picture of each ecotope in this region (cf. Metzger et al. 2005) and for a more precise picture of the mosaic Lateglacial landscapes a more detailed analysis than is possible in a large scale meta-analysis such as the present study is necessary. Ideally, a detailed analysis would focus on only one of the sub-areas or on parts of it. However, a more general picture allows for a search for more universally valid patterns. The use of this type of approach

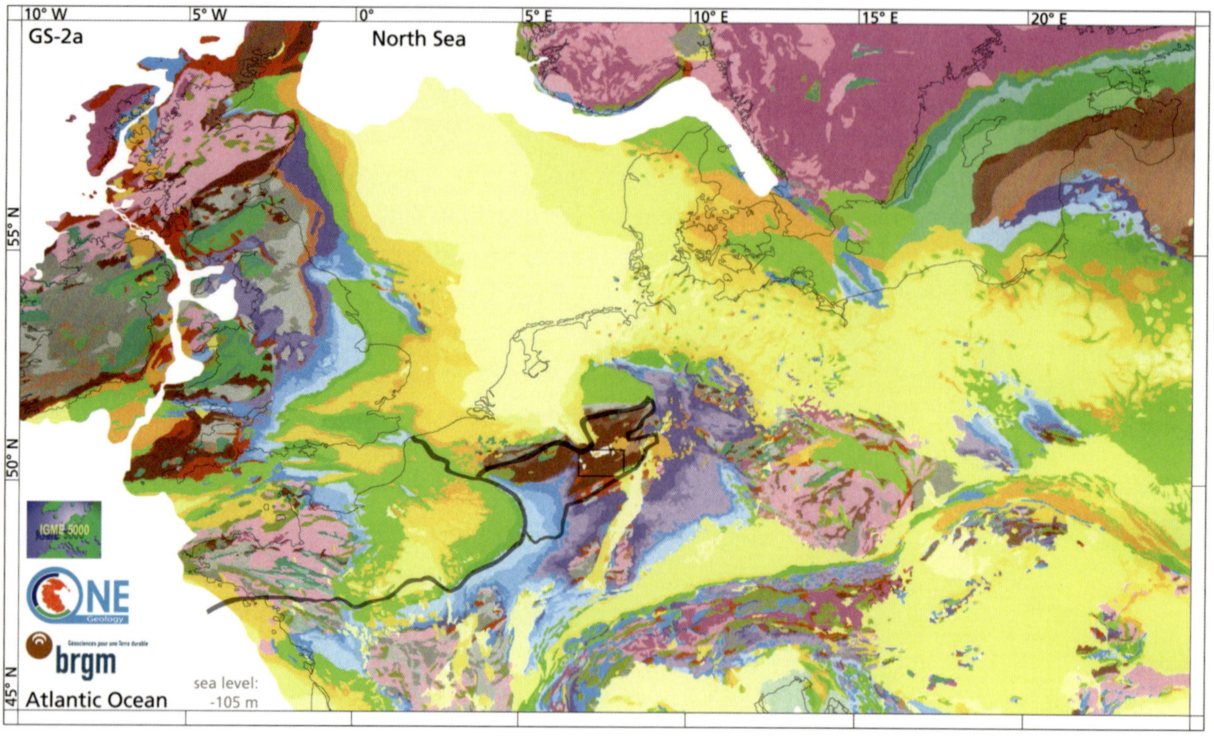

- Quaternary
- Pliocene
- Pliocene - Pleistocene
- Pliocene - Quaternary
- Miocene
- Neogene
- Neogene - Quaternary
- Oligocene
- Oligocene - Miocene
- Eocene
- Eocene - Oligocene
- Eocene - Miocene
- Eocene - Pliocene
- Palaeocene
- Palaeocene - Eocene
- Palaeogene
- Palaeogene - Neogene
- Cenozoic
- Late Cretaceous
- Late Cretaceous - Palaeocene
- Late Cretaceous - Eocene
- Late Cretaceous - Palaeogene
- Late Cretaceous - Neogene
- Early Cretaceous
- Cretaceous
- Cretaceous - Eocene
- Cretaceous - Miocene
- Cretaceous - Neogene
- Late Jurassic
- Late Jurassic - Early Cretaceous
- Late Jurassic - Cretaceous
- Middle Jurassic
- Middle Jurassic - Late Jurassic
- Early Jurassic
- Early Jurassic - Late Jurassic
- Jurassic
- Jurassic - Cretaceous
- Alpine
- Late Triassic
- Late Triassic - Jurassic
- Middle Triassic - Late Triassic
- Early Triassic
- Early Triassic - Middle Triassic
- Triassic
- Triassic - Middle Jurassic
- Triassic - Jurassic
- Triassic - Early Cretaceous
- Mesozoic
- Mesozoic - Neogene
- Late Permian
- Late Permian - Early Triassic
- Late Permian - Middle Triassic
- Early Permian - Middle Permian
- Permian
- Permian - Triassic
- Permian - Cretaceous
- Late Carboniferous
- Late Carboniferous - Middle Permian
- Late Carboniferous - Permian
- Early Carboniferous
- Carboniferous
- Carboniferous - Middle Permian
- Carboniferous - Permian
- Late Devonian
- Late Devonian - Early Carboniferous
- Late Devonian - Carboniferous
- Variscan
- Middle Devonian
- Middle Devonian - Late Devonian
- Early Devonian
- Early Devonian - Middle Devonian
- Devonian
- Devonian - Carboniferous
- Devonian - Permian
- Late Silurian
- Late Silurian - Devonian
- Wenlock - Early Devonian
- Early Silurian
- Silurian
- Silurian - Early Devonian
- Silurian - Devonian
- Late Ordovician
- Middle Ordovician
- Middle Ordovician - Late Ordovician
- Early Ordocvician
- Early Ordovician - Middle Ordovician
- Ordovician
- Ordovician - Silurian
- Ordovician - Devonian
- Late Cambrian - Early Ordovician
- Late Cambrian - Middle Ordovician
- Middle Cambrian - Late Cambrian
- Early Cambrian
- Cambrian
- Cambrian - Ordovician
- Cambrian - Silurian
- Palaeozoic
- Palaeozoic - Triassic
- Cadomian
- Caledonian
- Proterozoic III
- Proterozoic III - Middle Cambrian
- Proterozoic III - Cambrian
- Proterozoic III - Silurian
- Proterozoic III - Devonian
- Proterozoic III - Palaeozoic
- Proterozoic II
- Proterozoic II - Proterozoic III
- Proterozoic I
- Proterozoic I - Proterozoic II
- Proterozoic
- Proterozoic I - Proterozoic III
- Proterozoic - Ordovician
- Archean
- Archean - Proterozoic I
- Precambrian
- Archaen - Cambrian
- Precambrian - Palaeozoic
- Precambrian - Cenozoic
- undifferentiated
- unknown

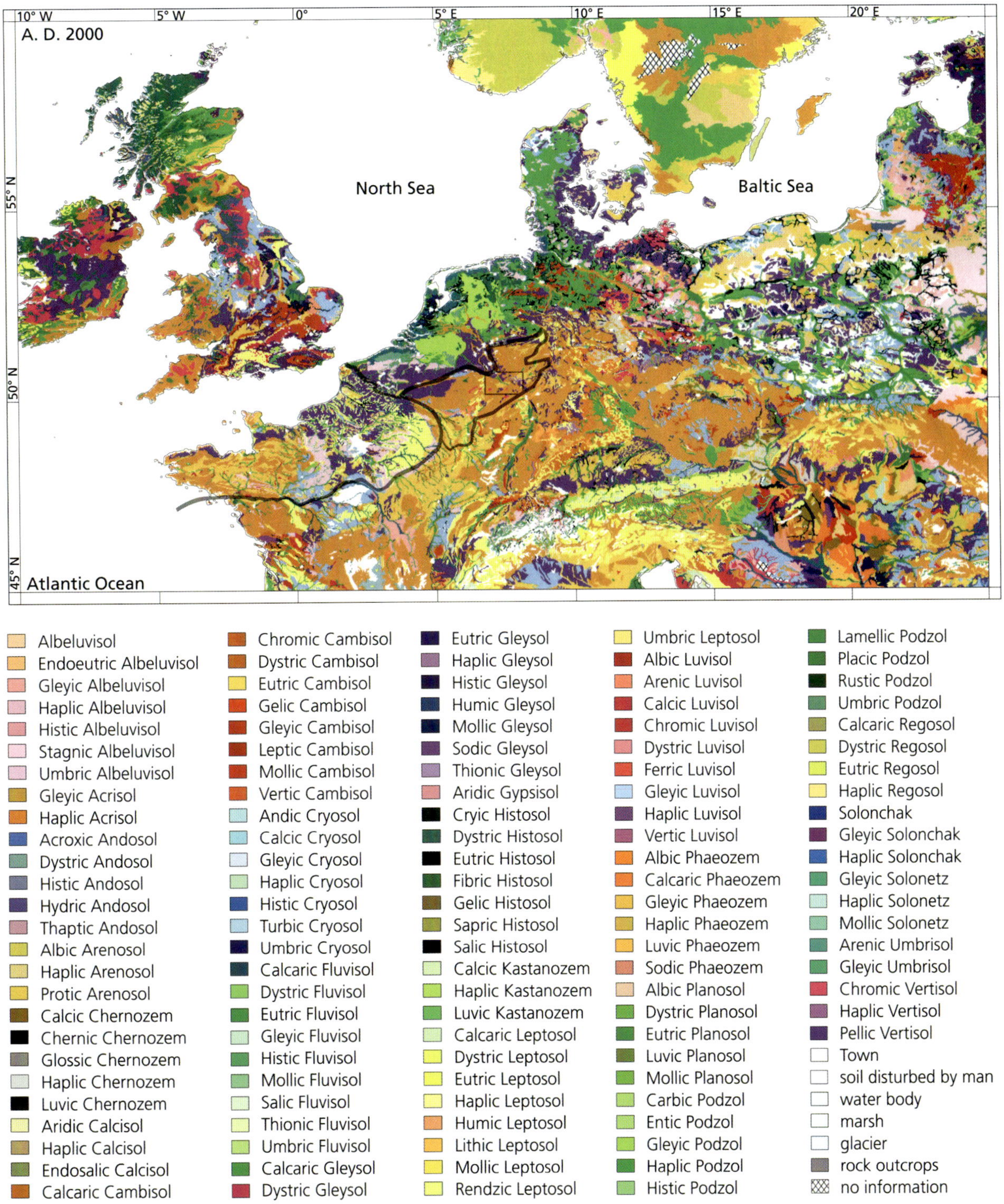

Fig. 18 Distribution of soils in modern north-western Europe with the sub-areas of the present study (WRB full soil code, data source: Soil Map Internet Server (SOMIS), version 2.0, van Liedekerke/Jones/Panagos 2006; http://eusoils.jrc.it/ESDB_Archive/ESDBv2/index.htm; cf. Jones/Montanarella/Jones 2005).

Fig. 17 Geology of north-western Europe with sea-level c. 105 m below modern sea level, the sub-areas of the present study, and modern coastlines indicated (data source for off- and onshore geology: IGME5000, (c) BGR Hannover, 2007; cf. Gradstein/Ogg/Smith 2004; with the permission of One Geology).

was shown to be of some value on the example of Lateglacial Britain (Rockman 2003a). North-western Europe is characterised today by various biomes but can be assigned in general to a northern European environment based on main climatic influences (Metzger et al. 2005).

Nevertheless, the varied preservation patterns of environmental data, especially on archaeological sites, leads to different scales of detailed information on the surrounding environment. Consequently, archaeological results can often only be compared on a very general level to the very detailed reconstructions of past environments. In particular, the deposition history and, thus, the temporal development were relatively well known for the environmental sites, whereas for the archaeological sites the gaps and over-representation of the environmentally defined parts of the stratigraphy were often difficult to estimate. The chronostratigraphic comparability of the two types of archives is therefore regularly restricted to a very general correlation.

Besides influencing the physical appearance of the landscapes, the geological units[5] (**fig. 17**) are also of further importance since they determine the possible existence of natural (rock) shelters as well as the availability of lithic raw materials for prehistoric hunter-gatherers. Formations from the Mesozoic era in particular can contain rich resources of fine-grained lithic raw materials and also various fossil materials such as amber or jet which were of some interest for Lateglacial hunter-gatherers. Hence, in some regions the rich variety of raw materials can explain decisions made by the prehistoric people. Other regions such as the Devonian uplands of the Rhenish Shield are poor in high quality lithic raw materials. In these regions especially studies of the origin of used lithic raw materials can often deliver precise patterns of spatial behaviour of the Lateglacial hunter-gatherers.

Besides the choice of Lateglacial humans, their archaeological visibility also depends on modern developments. For instance, the study area today comprises several countries and regions (**fig. 1**) with different policies of cultural heritage management and research strategies. These factors can also lead to divergent distribution patterns of Lateglacial sites, depending upon surveying, excavating, and reporting practices. In addition, various language areas are covered by this study.

Moreover, the Lateglacial archaeological horizons often immediately underlie the modern top soils[6] (**fig. 18**) or are even found incorporated within these soils. The soils and/or their development may therefore also have a considerable impact on the taphonomy of the archaeological assemblages. For example, in acidic environments the preservation of bone material is less probable (e.g. Nielsen-Marsh et al. 2006, 447; Turner-Walker 2008, 12), whereas pollen grains seem to be better preserved in these environments (e.g. Marshall 2007, 126). In addition to the development of the different soil types, water drainage appears also to play a major role in the preservation of organic material.

Besides the role in organic preservation, drainage systems have also contributed to the modelling of routes which were possibly taken by Lateglacial hunter-gatherers in their expansion into northern Europe assuming that river valleys indeed served as guiding influences for the mobility of Palaeolithic hunter-gatherers (e.g. Kobusiewicz 1999, 190f.; Conard/Bolus 2003, 333; Steele 2010, 2018). In general, north-western Europe is part of the north-western European water drainage system (**fig. 16**; cf. UNEP/DEWA~Europe 2004) meaning that the major river systems end in the Atlantic Ocean, partially, following a detour through the English Channel, the Celtic, the Irish, the North, or the Baltic Sea. Even though the geography of Europe was very different in the past, the mainland water catchment areas were presumably generally comparable. However,

5 Due to technical practicability, relatively detailed data for the geology as well as for the soils (see **fig. 18**) are presented in the maps of the present study even though only a sub-set of this data is needed for the description of the sub-areas.

6 In the following, terms and distribution are according to Jones/Montanarella/Jones 2005, 28-33. However, in the archaeological part the main stratigraphic description of the sites is followed.

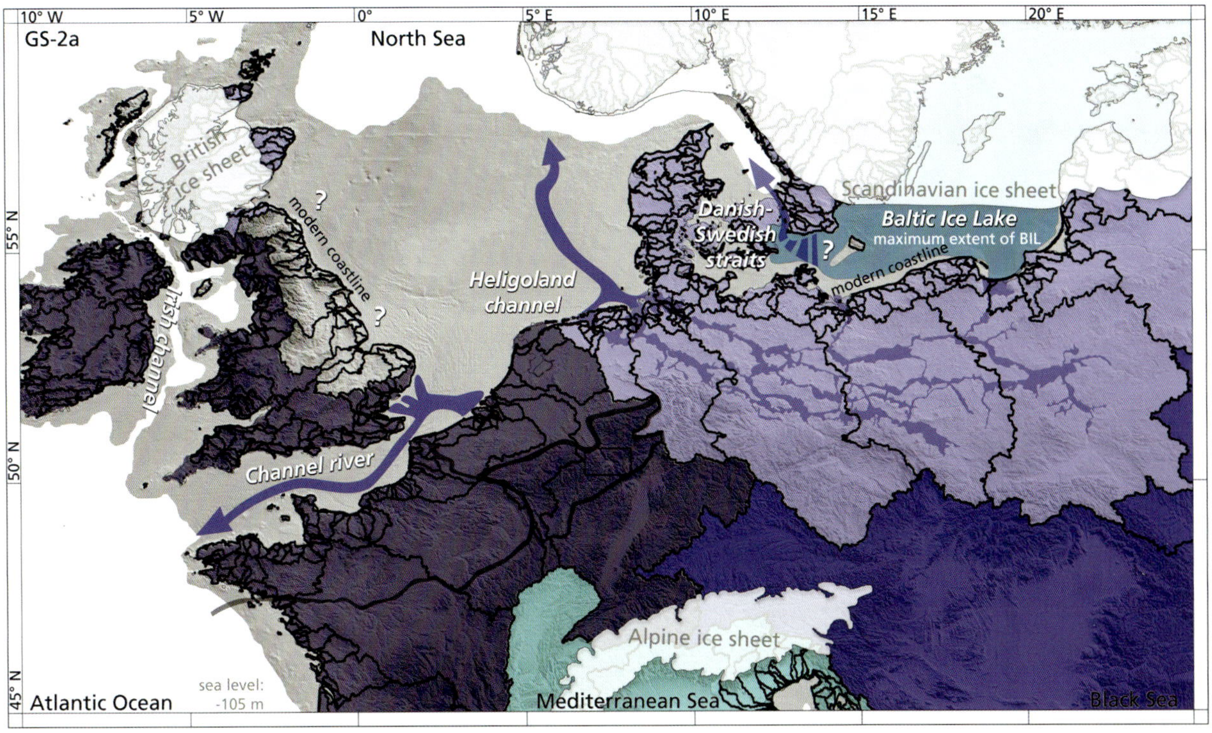

Fig. 19 North-western Europe at c. 16,000 years cal. b2k with major modern watersheds (data source: European Commission/Eurostat 2002; SADL [Spatial Applications Division Leuven] 2005; freely available for non-commercial use). Dark blue shaded: eastern Atlantic Ocean draining systems; light blue shaded: North Sea/northern Atlantic Ocean draining systems; light turquoise shaded: Mediterranean Sea draining systems; royal blue shaded: Black Sea draining systems; for most eastern British rivers the drainage remains uncertain thus far and therefore these areas are not shaded. Major large river systems in today submerged areas are indicated (cf. Björck 1995a; Konradi 2000; Streif 2004; Antoine et al. 2003a; Bourillet et al. 2003). Furthermore, the glacial valleys and water gaps are displayed in a medium blue shade in the North Sea draining system according to Wolstedt 1956, pl. I. – For further details see text.

during the Lateglacial the North Sea and English Channel system in particular were bisected more distinctly (**fig. 19**). Which rivers contributed to the northern and which to the Channel system depended not only on the existing topography of the landscape (Coles 1998) but also on the rising sea-level, i. e. water flowing into the northern North Sea basin from the North Atlantic Ocean and into the English Channel from the eastern Atlantic Ocean. With the inflowing water from the north a considerable area of land surface first became salt water marshes and was subsequently submerged in the salt water of the North Sea. This process was not accomplished before the early Holocene (e. g. Weninger et al. 2008). The water from the eastern Atlantic breaking through the English Channel also only reached the North Sea Basin in the Holocene and, subsequently, did not confluence with the waters coming from the northern Atlantic previously (cf. Uehara et al. 2006, 7; Shennan et al. 2000, 309 f.). Thus, during the studied period a land bridge connected the British Isles with continental Europe. Consequently, a considerable number of Lateglacial sites must be considered as lost due to the changed sea-level. Nevertheless, according to 3D seismic analyses in some sample areas a network of rivers, wetlands, and tunnel valleys existed on most parts of the North Sea basin during the Lateglacial (cf. Praeg 2003, Fitch/Thomson/Gaffney 2005, Lonergan/Maidment/Collier 2006, Gaffney/Thomson/Fitch 2007). Thus, how, from where, and when this landbridge was accessible and traversable depended presumably on the seasonality and the precipitation feeding the large river channels as well as the wetland networks.

Western upland zone

The western uplands form the higher elevated sub-area with altitudes regularly above 400 m a.s.l. and occasional altitudes as high as 750 m a.s.l. in the central part (**fig. 15**). This sub-area has not changed much geomorphologically since the Pleistocene. Only in some parts the appearance of the landscape changed moderately due to Late Pleistocene and early Holocene volcanic activity.

The main drainage systems (**fig. 16**) are Meuse, Moselle, and Rhine. The Weser which is oriented strictly northwards forms the limit to the east. The Rhine flows to the north-west and its tributaries, the Moselle and the Lahn, dissect the central part of the sub-area with their south-west to north-east directed valleys. However, most of the smaller river valleys are directed north to north-west and also the Moselle, in its upper reaches, flows from south to north along the German-Luxembourg border. In addition to this valley, the Meuse cuts a wide, south-east to north-west directed valley in the western limit towards northern France. The valleys of the Saar, the upper Moselle, and the Meuse approach each other south of Luxembourg (**fig. 1**) and traverse a wide north-south oriented depression. In this sub-area the drainage has not much changed since the Pleistocene with the only difference that then the rivers flowed into the large English channel river system instead of the North Sea (**fig. 19**).

Geologically, the south-western part is formed in general by the Jurassic uplands (**fig. 17**; cf. Asch 2005). Chert was occasionally found in a band of Triassic formations at the eastern limit of this part of the sub-area (Floss 1994). The Rhenish Shield which covers the north-eastern zone of the sub-area is a predominantly Devonian formation. This zone is rich in quartz and various types of quartzite as well as related metamorphic rocks (Floss 1994). The quartzite varieties are sometimes very fine-grained and homogeneous such as several varieties of Tertiary quartzite. Furthermore, chalcedony deposits were found in some places. However, this eastern part of the sub-area is generally poor in fine-grained siliceous raw materials compared to northern France and other adjacent regions. Hence, these fine-grained siliceous lithic resources were usually considered as imports. In consequence, the eastern part of this sub-area is particularly suited for the analysis of raw material origins and procurement distances, and/or the adaptation to less homogeneous raw materials. In addition, due to the major river valleys the area is a geographical pivot. In this context, the sub-area is also particularly interesting in the expansion process onto the North European Plain.

In fact, the initial area of the present study is located in the eastern part of this sub-area at the junction of the Rhine valley and its west-east running tributaries. Therefore, the archaeology from the Central Rhineland and its position within the pattern of north-western European Lateglacial developments is also of particular interest for the adjacent areas and their distant relations.

In contrast to the scarcity of fine-grained siliceous raw materials, haematite deposits are commonly known in the eastern part and occasionally in the western part of this sub-area. Pieces of this mineral can sometimes also be found in various river gravels in the eastern part. Jet, or at least cannel coal, deposits are possibly also present in the eastern sub-area but have not yet been unambiguously identified in the Rhineland (Allason-Jones/Jones 2001). Shell deposits (cf. Alvarez Fernández 2001, 548) were also possible but not yet localised in the Central Rhineland. However, Miocene shell deposits are well known from the southward adjacent Mainz Basin. Probably these deposits were also visited by Late Pleniglacial groups from the southern uplands (e.g. Pasda 1994) where rich chert and some jet deposits are known. Perhaps, investigations of sites from this area (Loew 2006; Serangeli/Terberger 2006) will help in the future to further analyse these north-south connections. However, raw materials originating from southern Central Europe farther south than the Mainz Basin were not common in the study area and a stable (exchange) connection farther to the south can in consequence not be established.

Soil formation is comparatively uniform with cambisols generally dominating the pedogenesis (**fig. 18**). This type is a young soil which developed under various environmental conditions (Jones/Montanarella/Jones 2005, 29). Preservation of organic material is relatively variable and in this region often influenced by the site type (preservation is generally good in caves; Terberger 1993) and/or cover by volcanic material (e.g. Baales et al. 2002).
Even though this sub-area is the highest elevated region with numerous river valleys and many caves, Lateglacial material most commonly originated from open air sites (see p. 52f.) which, as in the Paris Basin, are predominantly associated with river valleys. The good and extensive preservation of material below the LST indicated that this impression is partially biased by preservation. For example, in the Eifel region, well above the river valleys, single hearths survived in the Lateglacial Interstadial soil. However, the distribution of sites also revealed that the position in the river valleys reflected generally a topographic choice because the major sites were found only in the river valleys or on adjacent promontories. In southern Belgium the assemblages were recovered almost exclusively from rock shelters and caves (Dewez 1987; Otte 2000). In Luxembourg and the adjacent eastern French regions archaeological sites are rare and material was often found during surface surveys (Huet et al. 1995; Guillot/Guillot/Thévenin 2000; Spier 2000). However, most of the sites from this zone have a considerably recent research history.

Northern France

In the northern French sub-area, the relatively level terrain generally ranges between 50 and 150m a.s.l. and only in the western part are altitudes above 250m a.s.l. occasionally reached, and ones up to 400m a.s.l. are rare (**fig. 15**). In addition, large river systems characterise this sub-area (**fig. 16**). The south-east to north-west drainage direction of the river systems bisects northern France: In the west the Loire and the Vilaine system flow westwards into the eastern Atlantic Ocean and in the east the Seine and the Somme drain northwards into the English Channel. During the Pleistocene the tidal waters from the eastern Atlantic Ocean were not present due to the lower sea-level and the northern French rivers at that time formed tributaries to the large Channel river system draining into the eastern Atlantic Ocean (**fig. 19**). This system was further supplied by rivers in southern and eastern England along with Belgian, Dutch, and north-western German rivers. In contrast to the wide glacial valleys on the North European Plain, the northward draining rivers in northern France flow in comparably narrow meandering valleys more comparable to the western upland zone.
The region of the Seine, the Oise, and the Marne valleys is dominated by Palaeogene deposits (**fig. 17**; cf. Asch 2005) which are replaced towards the south by mainly Miocene formations. Geologically this sub-area is relatively uniform with a core region, which formed in the Cenozoic era, surrounded by Cretaceous deposits, which are again enclosed to the west and east by a band of Jurassic formations. In the east this band forms already a part of the western upland zone. Considerable amounts of fine grained lithic raw material (Senonian and Turonian flint) can also be found in the various river terraces of the core region (Valentin/Julien/Bodu 2002). In addition, some rare Tertiary, in particular Bartonian, chalcedonies occur in the archaeological inventories which, perhaps, originated from the Loire near Orléans or a no longer existing source on the plateau east of the Seine (Valentin 1995, 143. 165. 246. 350; Lang 1998, 91). The western part of the sub-area, in particular Brittany, was mainly formed by igneous (Palaeozoic and Proterozoic volcanic rocks) and metamorphic rocks (mainly Precambrian formations) which delivered lithic raw materials of considerably lower quality and, furthermore, restricted the organic preservation (Naudinot 2008). In consequence, lithic material analyses and technological studies in this part revealing import distances and adaptation to

less suited materials contrast the technological studies performed in the central Paris Basin where rich high-quality lithic raw materials where accessible almost everywhere in the river gravels.
In contrast to the geology, the soils in the western part are more homogeneous than in the rest of this sub-area. Cambisols clearly dominate the western part of this sub-area, whereas in the eastern part various soil formations (gleysols, luvisols, albeluvisols, leptosols, fluvisols) are common (**fig. 18**; cf. Jones/Montanarella/Jones 2005).
In the large river valleys of the Seine and the Somme some very rich and well excavated open air sites were extensively studied since the late 1960s (e.g. Fagnart 1997; Valentin 2008a). Rock shelter and cave sites are rare (see p. 52f.) and were often originally excavated at an early period in research history (e.g. Barois-Basquin/Charier/Lécolle 1996; Martin 2007a; Pigeaud et al. 2010). Numerous Lateglacial assemblages were identified comparatively recently in the west of this sub-area (Marchand et al. 2004; Naudinot 2008), however, the majority of these are surface collections. Furthermore, organic material was only recovered occasionally from any of the excavated sites there. Therefore, the making of a chronology in this part of the sub-area is mainly dependent on techno-typological analogies and the stratigraphy which again is often disturbed. Thus, in a study focusing on well dated and excavated archaeological assemblages the impression of human absence during the Lateglacial could arise, even though the material presented so far clearly demonstrated the opposite (Naudinot 2010).

Thus, the landscapes of the study area deliver rich resources which are distributed variously. Probably, this uneven distribution influenced the land-use strategies of mobile hunter-gatherer groups (cf. Butzer 1982, 211-278). For instance, the larger rivers in northern France provided not only water, food (fish), and orientation but delivered in their gravels also an important lithic raw material resource. In contrast, this resource needed to be imported to the Central Rhineland.
Furthermore, these landscapes form the premises upon which the living environment develops and, thus, influences the tempo and mode of this development.

Pollen stratigraphies

Along with the landscape, the vegetation needs to be considered as an important resource for the Lateglacial hunter-gatherer groups (cf. Gaudzinski-Windheuser/Jöris 2006, 46-50) if environmental factors are assumed to have affect human behaviour. Moreover, the vegetation forms as the primary producer of energy also the basic trophic level for the food chain. In addition, the presence of shrub and/or forest patches also changes the appearance of a landscape and the visibility of e.g. landmarks or prey therein. Pollen profiles are considered as important environmental archives in the present study because they give a general overview of these factors. Furthermore, the local presence of specific plants can be attested by directly dated macro-remains.
Preservation of biotic indicators is in general difficult and often leads to the preservation in specific surroundings such as aquatic environments or in assemblages biased by humans (e.g. burnt materials). However, burnt materials such as charcoal can also be preserved in small size classes, i.e. micro-charcoal and then human interference is not necessary causal to these preserved remains. Moreover, a quantitative analysis of non-human induced micro-charcoal, especially in stratigraphic sequences, can help deducing local as well as global fire regimes (Power et al. 2008). These regimes depend again very much on the local climatic conditions (Niemann/Behling 2008). For the studied area, visible charcoal analyses on a broad scale were partially recorded for the final part of the studied period in the so-called Usselo-soil (van der Hammen/van Geel 2008). However, systematic micro-charcoal analyses on a larger scale are not yet accomplished.

In fact, most of the well dated Lateglacial vegetation records were preserved in water-logged environments. Presumably, this overbalance of littoral and/or riparian communities influenced the picture of the Lateglacial landscapes significantly. For instance, the conditions at waterholes or in river valleys are in general more favourable for plants which is why littoral and/or riparian ecotopes are considered to respond quicker to climatic changes than their surrounding (Bennike/Seppänen 2004). Furthermore, vegetation communities with good water availability are usually more productive (Gouveia et al. 2008) and in sheltered areas such as river valleys vegetation is also more probable to develop steadily than in exposed areas (van Leeuwaarden/Janssen 1987).

However, in addition to hydrological conditions, in particular, water availability (Rodriguez-Iturbe 2000; Whittaker/Nogués-Bravo/Araújo 2007; Gouveia et al. 2008), vegetation development depends on multiple factors which range from extra-local to very local such as solar radiation (Whittaker/Nogués-Bravo/Araújo 2007), temperature (Hollister/Webber/Bay 2005), the development of the soil (Bennike/Seppänen 2004), or the distance to a refuge of plants (Whittington/Fallick/Edwards 1996). Moreover, these factors influence the gradual and migratory process (biome succession) after natural hazards and/or ice ages. For instance, Hartmut Usinger pointed out that the general succession of vegetation in the early Lateglacial sequences of northern Germany is very similar beginning with an initial maximum of *Hippophaë* sp., followed by subsequent maxima of *Helianthemum* sp. and *Juniperus* sp. (Usinger 1985, 10). The first increase of tree birches (*Betula pubescens*) is usually connected with the latter maximum. However, some studies on the distribution of tree pollen in the Lateglacial and early Holocene indicated that the development of deciduous forests (i.e. their refuge) is rather a western phenomenon, whereas in the east and the mountainous areas in the south the coniferous elements had a stronger influence (Cheddadi/Bar-Hen 2009, 375 fig. 5; 377). Thus, due to the migration distance of the various taxa the reforestation process and biotic succession is also not inevitably corresponding across Europe.

Moreover, preservation influences particularly the reliability of many quantitative analyses, such as pollen analysis, which are based on proportions and statistics and, hence, are in need of a sufficient amount of preserved material to deliver reliable results (Heiri/Lotter 2001). These quantitative pollen analyses were used for decades to date Lateglacial geological horizons within which archaeological remains were found (Schütrumpf 1937; Iversen 1942; Stampfuss/Schütrumpf 1970; Kaiser/de Klerk/Terberger 1999).

The general concept of pollen analysis is that pollen grains are counted per stratigraphic unit. These stratigraphic units represent usually a relative chronological sequence which in the best case can be on an annual resolution (see p. 18-25). The presentation of the counted pollen grains is made by calculating the pollen sums to a total of 100 %. This way a proportional development of the vegetation is combined with a relative chronostratigraphy.

However, the calculation of the pollen sums depends on the identification of the various pollen as well as the consideration about which pollen are incorporated in the calculation. The latter is connected to assumptions on which pollen are unrepresentative because they may be redeposited (Iversen 1942, 133 f.) or originate from an extra-local source and/or from species producing a particularly large number of pollen such as forest species as for instance *Pinus* sp. (Faegri/Iversen 1950). All these factors influence a percentage calculation of the pollen sums and, thus, the presentation in a pollen diagram. To manage some of these problems, some specific types of pollen are usually excluded from the pollen analysis. However, this extraction relies occasionally on assumptions that some studies have shown to be false for various reasons (e.g. Heikkilä/Fontana/Seppa 2009; Willis/van Andel 2004). Furthermore, for some of the chosen taxa presupposition on the presence in the Lateglacial based on the environmental development can be problematic due to the wide environmental variety of single species within a genus such as *Poaceae* (formerly: *Gramineae*), *Cyperaceae*, or *Betula* sp. Besides different preferences concerning their

pollen zone according to	Jessen 1935	Iversen 1942	Usinger 1985	Hoek 1997	van Geel/Coope/van der Hammen 1989	Pastre et al. 2000, added by Pastre et al. 2003	Litt/Stebich 1999	Goslar et al. 1999
Younger Dryas	Late Dryas (Zone III): Dryas flora; maximum of fir (*Pinus*) and willow (*Salix*) pollen	Younger Dryas (Zone III): tundra to light forests with occasional stands of birch (*Betula*) and spreading crowberry (*Empetrum*) heaths	Dryas-3 (DR-3): *Betula nana-Betula pubescens*-PAZ, *Empetrum* is typical, main expansion of *Pinus*, increasing values of *Betula nana*	Late Dryas (Zone 3): NAP-*Empetrum* PAZ, increasing NAP, decrease of *Pinus* and *Betula* (Zone 3a), increasing values of *Empetrum* (Zone 3b)	Late Dryas (Zone III): decline of *Pinus* and increase in *Betula* and NAP, e. g. *Artemisia*, *Helianthemum* but also *Empetrum*-type	Youngest Dryas (Zone 3): *Pinus* and *Artemisia* pollen zone, decreasing though still dominant *Pinus*, strongly increasing *Artemisia*, small increases of *Betula*, *Juniperus* and NAP	Younger Dryas biozone (III): NAP maximum between LST and UMT, increase also in *Salix* and *Juniperus*	Younger Dryas (zone c): *Artemisia*-Chenopodiaceae PAZ, increase of NAP especially *Artemisia* and Chenopodiaceae, but also *Ephedra distachya*, *Helianthemum*, *Pleurospermum austriacum*, *Gypsophila fastigiata* and *Juniperus*
Allerød	Allerød (Zone II): maximum of birch (*Betula*) pollen	Allerød oscillation (Zone II): forest with birch (*Betula*; zone IIa) and intruding pine (*Pinus*; zone IIb)	Alleröd (AL): *Juniperus-Betula nana-Betula pubescens*-PAZ, dominated by birch (*Betula pubescens* and *Betula nana*) and *Juniperus*; *Empetrum-Betula pubescens*-PAZ, cold episode within tree birch (*Betula pubescens*) forests with high values of *Empetrum*, increasing values of *Juniperus*, *Populus* and *Filipendula*; *Populus-Betula pubescens*-PAZ I and II: intermediate part (AL-b) with low birch, but high *Juniperus* and *Salix* pollen values, and a double *Populus* peak; intersected by *Artemisia-Betula pubescens*-PAZ with tree birch and *Artemisia* maxima; *Pinus-Betula pubescens*-PAZ, last part dominated by *Pinus* or tree birch depending of topography	Allerød (Zone 2): *Betula-Pinus* PAZ, increase in AP, *Betula* in particular, decrease of *Salix* (Zone 2a1), intruding *Pinus* with decreasing values of *Betula* and *Juniperus* (Zone 2a2), strong increase in *Pinus* (Zone 2b)	Allerød (Zones II) with *Betula* and *Pinus* phase, increase of *Filipendula*, whereas other NAP decrease	Allerød (Zone 2): *Betula* and *Pinus* pollen zone, sub-zone 2a: *Betula* increase with high values also of *Juniperus* and *Salix*, still high NAP values, especially of *Artemisia*, sub-zone 2b: *Pinus* increase and dominance over *Betula* and *Salix*, decrease of NAP, sporadic occurence of *Quercus* and *Corylus*, intermediate cold period with opening of vegetation by decrease in *Pinus* and *Betula,* an increase in *Juniperus*, *Artemisia*, Poaceae and Cyperaceae	Allerød biozone (II): 2nd Lateglacial tree birch (*Betula pubescens*) maximum in combination with the increase of pine (*Pinus*) and *Filipendula*	Alleröd (zone b): *Pinus-Betula* PAZ, *Pinus* and *Betula* dominant, *Artemisia*, *Thalictrum*, Chenopodiaceae regular, second part increase of *Populus* and *Filipendula*, decrease of *Thalictrum*, *Potentilla* and *Juniperus*
Older Dryas	Early Dryas (Zone I): maximum of fir (*Pinus*) and willow (*Salix*) pollen	Older Dryas (Zone Ic): tundra period, herb tundra with herbaceous plants and *Salix* pollen, decrease of *Betula*	Dryas-2 (DR-2): *Helianthemum-Betula nana*-PAZ, NAP dominant, *Betula nana* still dominant over increasing *Betula pubescens*, *Helinathemum* frequent, *Filipendula* occasionally	Early Dryas *sensu lato*, Earlier Dryas (Zone 1c): *Betula-Salix* PAZ, increase of *Salix*, *Juniperus* and NAP, decrease of *Betula*	Earlier Dryas (Zone Ic): *Artemisia* and *Helianthemum* maintain, whereas *Betula* and *Juniperus* decrease	Intermediate Dryas (sub-zone 1b): *Juniperus* and *Betula* pollen zone, increase in NAP	Older Dryas biozone (Ic): NAP maximum between Bølling and Allerød with *Artemisia* and Poaceae; *Betula* decreases	Older Dryas (zone a): *Betula-Salix* PAZ, NAP dominant, especially *Artemisia*, *Thalictrum*, Poaceae, Chenopodiaceae, Rubiaceae, high values of *Salix* and *Betula*

Tab. 3 Selected definitions of some main Lateglacial pollen zones. Abbreviations: **AP** arboreal pollen; **NAP** non-arboreal pollen; **PAZ** pollen assemblage zone.

pollen zone according to	Jessen 1935	Iversen 1942	Usinger 1985	Hoek 1997	van Geel/Coope/van der Hammen 1989	Pastre et al. 2000, added by Pastre et al. 2003	Litt/Stebich 1999	Goslar et al. 1999
Bølling	×	Bølling oscillation (Zone Ib): shrub tundra with pollen of Poaceae, Cyperaceae, *Artemisia*, *Salix* sp., and dwarf birch (*Betula nana*) pollen	Bölling (BÖ): first part of Alleröd	Early Dryas *sensu lato*, Bølling (Zone 1b): *Betula-Salix* PAZ, increase in AP, *Betula* in particular	Bølling *sensu lato*, Bølling *sensu stricto* (Zone Ib): increase *Betula* (*nana* and *pubescens*), *Juniperus*, *Hippophaë rhamnoides* regular	Bølling: first part with extending steppic formation with increasing values of Poaceae, *Artemisia* and *Juniperus*; second part is subzone 1a: *Juniperus* and *Betula* pollen zone, *Juniperus* and *Betula* dominant before *Salix*, high NAP with significant values of *Helianthemum*, *Thalictrum* and *Sanguisorba minor*	Bølling biozone (Ib): 1st remarkable *Betula* maximum attributed to tree birch (*Betula pubescens*)	×
Oldest Dryas	×	Oldest Dryas (Zone Ia): tundra period, treeless tundra with pollen of Poaceae, Cyperaceae, and *Salix*	×	Early Dryas *sensu lato*, Earliest Dryas (Zone 1a): *Betula-Salix* PAZ, rise of *Artemisia*, increase of species-rich herbaceous plant communities and dwarf bushes	Bølling *sensu lato*, Earliest Dryas (Zone Ia): Poaceae, Cyperaceae, *Artemisia* and *Salix*, *Helianthemum* and *Armeria* regular	×	Oldest Dryas biozone (Ia): declines in *Betula* and *Salix*; NAP maximum between Meiendorf and Bølling and before the expansion of tree birches (*Betula pubescens*)	×
Meiendorf	×	×	»Meiendorf-Intervall« (according to Menke 1968 and 1980; better not used): *Hippophaë-Betula nana*-PAZ, NAP dominant, *Betula nana*, *Hippophaë* and *Juniperus* numerous, increased *Helianthemum* values	×	×	×	Meiendorf biozone: weak *Betula* maximum with dwarf birch (*Betula nana*); decline in *Pinus* (redeposited), and increases in *Betula*, *Salix*, and *Juniperus;* first warming after the Pleniglacial	×
Late Pleniglacial	×	×	× (*Artemisia*-Poaceae-PAZ)	Zone Late Pleniglacial (LP): NAP PAZ, Poaceae, Cyperaceae, and some dwarf bushes	Upper Pleniglacial	Pleniglacial: open environments with sparse vegetation cover	Pleniglacial biozone: open, tree less vegetation with many grasses and sedges with a high amount of *Helianthemum* and the continuous occurrence of *Artemisia*, *Armeria*, *Ephedra*, and other heliophilous taxa	×

Tab. 3 (continued)

habitat, the different species within a genus have also different reaction times to climatic amelioration advising again to not use these general vegetation diagrams as correlatives with, for instance, the Greenland climate record.
Yet, since Johannes Iversen has introduced the calculation of a »total« pollen sum (Iversen 1942; Iversen 1947; de Klerk 2004, 270), it was generally applied to pollen records. He combined the tree pollen sum with non-arboreal pollen sums in an attempt to develop a method for describing specifically past habitats between forest and tundra environments so that the resulting periodisation could be used for chronostratigraphic considerations (Iversen 1942). In this total pollen sum the differentiation of arboreal pollen (AP) and non-arboreal pollen (NAP) is supposed to reflect the varying openness of the landscape (i.e. forest and tundra landscape; Iversen 1942, 139). However, the »openess« of the surrounding is not directly reflected by the distinction between AP and NAP because among the AP values not only trees are compiled but also shrub species such as *Juniperus communis* and dwarf shrub varieties such as *Salix polaris* or *Betula nana*. Therefore, macro-fossil analyses are of particular interest to distinguish, for instance, the rises of *Betula* sp. pollen due to *Betula nana*, a dwarf shrub of approximately 1 m height, from the ones due to *Betula pubescens*, a deciduous tree of usually 20 m height.
Prior to the introduction of Iversen's total pollen sum, other calculations (e.g. »tree pollen sum«) were used and in the last decade again other calculations were presented such as the upland pollen sum (de Klerk 2004). This latter calculation assumed that bogs were not present on the North European Plain during the early Lateglacial and, thus, wetland taxa were excluded from the calculation (de Klerk 2004, 277). Nevertheless, the differences resulting from the various calculation procedures and the implied assumptions reveal that also the mentioning of percentages of specific pollen to distinguish various sections of the vegetation development as displayed by pollen profiles (Usinger 1985; Bock et al. 1985) need careful consideration. In the present work only general tendencies such as the presence, absence, maxima and minima of some relevant taxa are therefore mentioned.
The definitions of the generally applied vegetation zones which are mainly based on pollen diagrams for Lateglacial north-western Europe are inconsistent and occasionally even conflicting (**tab. 3**). In addition, these pollen zones were sometimes used as chronostratigraphic markers based on the assumption of contemporary developments in north-western Europe (Firbas 1939). With increasingly high temporal resolutions of the pollen profiles the offsets of biome successions became visible in different areas and the various definitions then led to some confusion (**tab. 4**; cf. de Klerk 2004).
Meanwhile, the pollen zones are often replaced by pollen assemblage zones (PAZ) which describe the vegetation community more carefully without using arbitrary terms connected to chronostratigraphy (Usinger 1985, 15f.). However, these pollen zones describe comparable communities (e.g. pine-birch forests) which represent pure definitions of vegetation community types and should not be regarded as temporally equivalent across wide areas of north-western Europe.
To allow a general comparison and reveal the various developments of the Lateglacial environments across north-western Europe, the pollen profiles of the two sub-areas need to display the same types of pollen classes even though specific taxa may not play an important role in both sub-areas. Based on the classic environmental sequencing of Lateglacial north-western Europe (**tab. 3**; cf. Hoek 1997; Litt/Stebich 1999; Litt et al. 2001) some genera are chosen in the present work for comparison. These groups are among the AP birch (*Betula* sp.), pine (*Pinus* sp.), willow (*Salix* sp.), and juniper (*Juniperus* sp.) and among the NAP sage (*Artemisia* sp.) and the family of true grasses (*Poaceae*). Usually, filipendula (*Filipendula* sp.) and sunrose (*Helianthemum* sp.) would also be of interest but these species were not always recorded. As a pioneer species sea buckthorn (*Hippophaë* sp.) plays an important role in the recolonisation of northern (Usinger 1998) and higher mountainous habitats (Magny et al. 2006) where ice sheets shifted previous soils (including seeds

events	names given for correlative bio-, geo- and/or chronozones
GH	Holocene: Litt/Schmincke/Kromer 2003 Flandrian: Mangerud et al. 1974
GS-1	Younger Dryas: Mangerud et al. 1974; Lotter et al. 1992; Litt/Stebich 1999; Goslar et al. 1999 Late Dryas: Walker 1995; Hoek et al. 1999 Dryas 3: Lotter et al. 1992 Loch Lomond Stadial: Jones et al. 2002
GI-1a	Dryas 3: Lotter et al. 1992 Younger Allerød: Litt/Schmincke/Kromer 2003 Allerød c: Bokelmann/Heinrich/Menke 1983; Usinger 1981 Allerød: Mangerud et al. 1974; Litt/Stebich 1999; Hoek et al. 1999; Goslar et al. 1999
GI-1b	Allerød: Mangerud et al. 1974; Litt/Stebich 1999; Hoek et al. 1999; Goslar et al. 1999 Allerød b: Bokelmann/Heinrich/Menke 1983 Allerød c: Usinger 1981 Gerzensee fluctuation/oscillation: Lotter et al. 1992; Litt/Schmincke/Kromer 2003
GI-1c_1	Allerød: Mangerud et al. 1974; Lotter et al. 1992; Litt/Stebich 1999; Hoek et al. 1999; Goslar et al. 1999 Allerød a: Bokelmann/Heinrich/Menke 1983 Allerød c: Usinger 1981 Older Allerød: Litt/Schmincke/Kromer 2003
GI-1c_2	Older Dryas: Litt/Stebich 1999 Intermediate Dryas: Bokelmann/Heinrich/Menke 1983; Bock et al. 1985 Allerød b: Usinger 1981 Allerød: Mangerud et al. 1974; Lotter et al. 1992; Hoek et al. 1999; Goslar et al. 1999
GI-1c_3	Allerød: Lotter et al. 1992; Usinger 1998; Hoek et al. 1999 Allerød a: Usinger 1981 Older Dryas: Mangerud et al. 1974; Goslar et al. 1999 Bølling: Bokelmann/Heinrich/Menke 1983; Litt/Stebich 1999
GI-1d	Older Dryas: Bokelmann/Heinrich/Menke 1983; Clausen 1998 Earlier Dryas: Walker 1995; Hoek 1997; Hoek et al. 1999 Aegelsee fluctuation/oscillation: Lotter et al. 1992; Ammann et al. 1994 Oldest Dryas: Litt/Stebich 1999 Intermediate Dryas: Limondin-Lozouet et al. 2002 Dryas 2: Leroyer 1994 Bølling: Mangerud et al. 1974; Lotter et al. 1992
GI-1e	Meiendorf: Bokelmann/Heinrich/Menke 1983; Usinger 1998; Litt/Stebich 1999 Bølling: Mangerud et al. 1974; Lotter et al. 1992; Hoek et al. 1999 Old Dryas: Leroyer 1994 *Hippophaë* phase: de Klerk 2008
GS-2a	Bølling: Mangerud et al. 1974 Dryas 1: Lotter et al. 1992; Ammann et al. 1994; Leroyer 1994 earliest Dryas: van Geel/Coope/van der Hammen 1989; Hoek et al. 1999 Oldest Dryas: Magny et al. 2006; Clausen 1998 End Devensian: Jones et al. 2002 (Late) Pleniglacial: Hoek et al. 1999; Litt/Stebich 1999

Tab. 4 Selected nomenclature applied to the isotopic events in north-western Europe. Correlation to the Greenland eventstratigraphy was made by the use of oxygen isotope curves (Lotter et al. 1992; Hoek et al. 1999; Goslar et al. 1999; Jones et al. 2002; Magny et al. 2006), by varve counts (Goslar et al. 1999; Litt/Stebich 1999), by radiocarbon dating (Mangerud et al. 1974), or by analogies of the climatostratigraphy (van Geel/Coope/van der Hammen 1989; Bokelmann/Heinrich/Menke 1983; Leroyer 1994; Usinger 1998). However, the limits of these zones are not necessarily identical to the limits of the isotopic events (see p. 245-250).

and roots) away and left barren gravel moraines behind. Nevertheless, in the areas with a constant pedogenesis this pioneer is of minor importance and, therefore, not recorded in the present study.
As representatives for the complete sub-area, general pollen diagrams are chosen from each sub-area. Nevertheless, these general pollen diagrams were formed very differently. For the western upland zone the maar lake sequences are chosen (Litt/Stebich 1999) due to their reliable chronostratigraphy as well as their central position in the sub-area (**fig. 20**). In contrast, from northern France no comparably complete, high-resolution archive is known. Therefore many sequences which were spread across the sub-area (**fig. 20**) contributed to a synchronised pollenstratigraphy of northern France (Pastre et al. 2003).

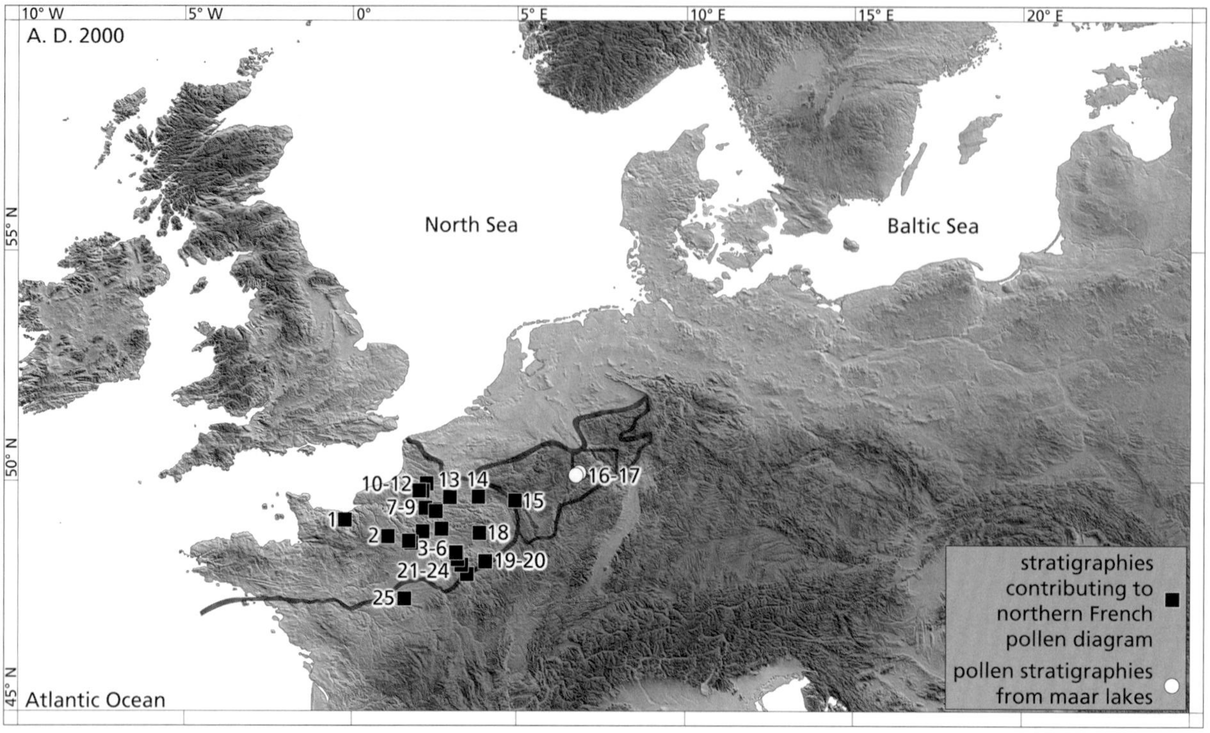

Fig. 20 Map of sites contributing stratigraphies to the main pollen diagrams used per sub-area. **1** Bellengreville (Leroyer 1994); **2** Acon (Limondin-Lozouet et al. 2002); **3** Angennes (Leroyer 1994); **4** L'Archet (Leroyer 1994); **5** Le Closeau (Pastre et al. 2000); **6** Fresnes-sur-Marne (Pastre et al. 2000); **7** Bresles (Leroyer 1994); **8** Sacy (Pastre et al. 2000); **9** Houdancourt (Ponel et al. 2005); **10** Famechon (Leroyer 1994); **11** Conty (Limondin-Lozouet et al. 2002); **12** Etouvie (Limondin-Lozouet et al. 2002); **13** Beaurains-lès-Noyons (Pastre et al. 2000); **14** Chivres (Leroyer 1994); **15** La Bar (Leroyer 1994); **16** Holzmaar (Litt/Stebich 1999; Leroy et al. 2000); **17** Meerfelder Maar (Litt/Stebich 1999); **18** Coizard-Joches I & II (Leroyer 1994); **19** Saint Pouange (Pastre et al. 2000); **20** Saint Lèger (Pastre et al. 2000); **21** Bazoches-lès-Brays (Pastre et al. 2000); **22** Pont-sur-Yonne (Pastre et al. 2000); **23** Étigny (Pastre et al. 2000); **24** Migennes (Limondin-Lozouet et al. 2002); **25** Mur-de-Solonge (Leroyer 1994).

Meerfelder Maar, Germany

The varved lake sediments in the western upland region delivered a precise sedimentologic stratigraphy (see p. 19-21; cf. Schettler/Rein/Negendank 1999; Lücke/Brauer 2004; Brauer et al. 2008) as well as a detailed Lateglacial pollen sequence for this area (Litt/Stebich 1999). Three cores from the Meerfelder Maar (MFM) were correlated to the MFM 6 sequence and used for the pollenstratigraphy as well as for the varve counting (Litt/Stebich 1999). Pollen zonation was made according to a cluster analysis which was based on a pollen calculation relying on a minimum of 500 grains per sample and excluding aquatic plants and pteridophytes (Litt/Stebich 1999, 7). Due to the correlation with the varve counting, exact onsets and ends for the clusters, i.e. biozones were given. For the studied time period six biozones were given (**fig. 21**): Pleniglacial, Meiendorf, Oldest Dryas, Bølling, Older Dryas, and Allerød. The latter is further sub-divided in Early, Middle (Gerzensee, cf. Lotter et al. 1992), and Late Allerød. The varve formation in the Meerfelder Maar began during the Meiendorf biozone and, thus, only for the four latter zones precise durations and dates for the onset can be given. The use of the classic palynological terms is combined with relatively precise definitions of how these terms are understood (cf. Litt/Stebich 1999, 10-14). However, these terms remain confusing and are therefore only used to refer to the described biozones (see **tab. 3**).

Fig. 21 Simplified Lateglacial pollen stratigraphy from Meerfelder Maar, MFM 6 (Litt et al. 2001, 1238 fig. 3). Timescale is according to varve chronology, i. e. interpolated below 14,070 years cal. b2k (see p. 20 f.) and is shifted from cal. BP to cal b2k. – For further details see text.

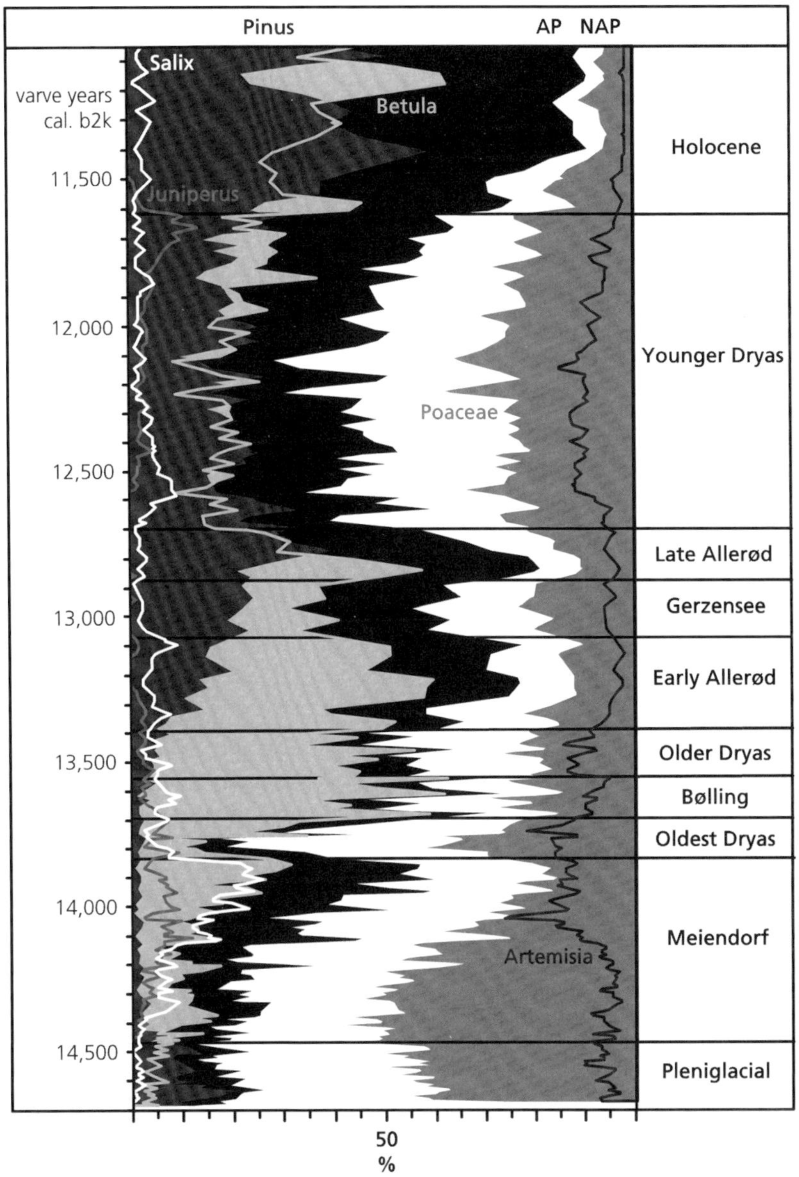

Pollenstratigraphy of northern France

In northern France no single comprehensive pollen record covering the Lateglacial sequence was found. Sequences comparable to the Meerfelder Maar were found only in the higher mountain regions of eastern France, for instance Grande Pile (Vosges; Beaulieu/Reille 1992; Guiter et al. 2003) and Lake Lautrey (Jura; Magny et al. 2006). However, for the former no detailed Lateglacial sequence is published yet and the latter, situated in the Jura mountains, is clearly too far away to be relevant for the Paris Basin or the Picardy.
Nevertheless, palaeoenvironmental analyses frequently accompanied the archaeological investigations of the last two decades in northern France (Leroyer 1994; Pastre et al. 1997; Limondin-Lozouet et al. 2002). Boreholes from some twenty sites scattered across the complete sub-area (**fig. 20**) provided material suitable for pollen analysis (Limondin-Lozouet et al. 2002). In general, palynological samples were taken every 5 cm from sequences containing organic and fine grained, slowly deposited sediments (Pastre et al. 2003, 2178). From the counted pollen the hygrophil species were excluded (cf. Leroyer/Allenet Ribemont 2011).

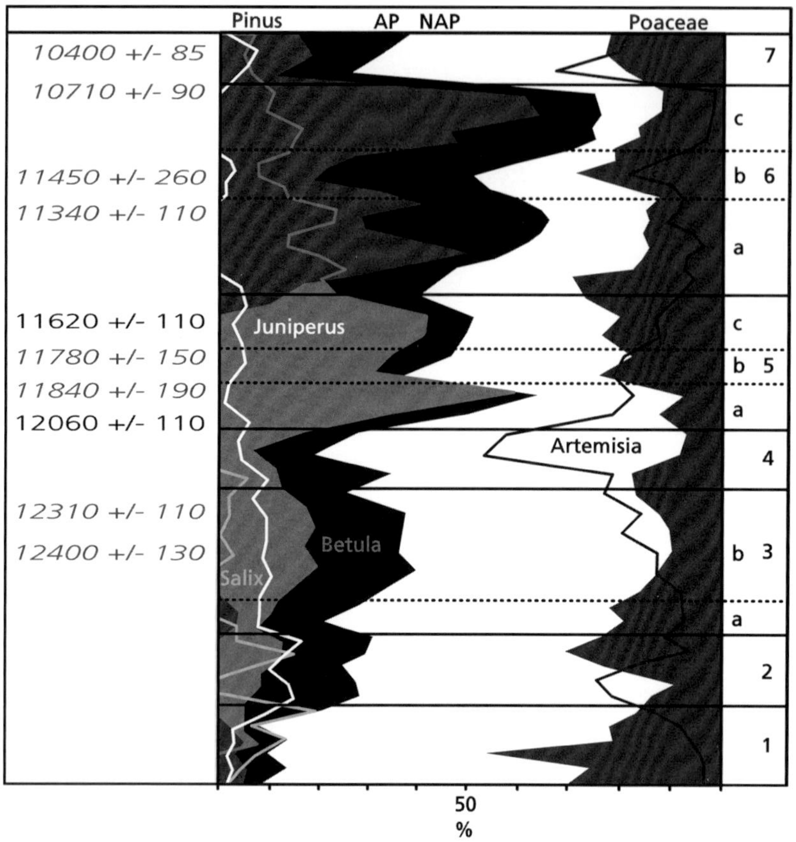

Fig. 22 Synchronised pollen stratigraphy of northern France (Pastre et al. 2003, fig. 3; cf. Ponel et al. 2005, fig. 5). Dates are ^{14}C ages in ^{14}C-BP. The ones set in grey and italic are made on sediment samples and the one set in black and italic is made on peat (cf. Limondin-Lozouet et al. 2002, tab. 1; Pastre et al. 2003, tab. 1). – For further details see text.

Frequently, these sequences covered only short sections of the Lateglacial and non-representative layers were also excluded (Limondin-Lozouet et al. 2002, 48). In addition, some ^{14}C dates were taken from these sequences but several of the samples were bulked sediment samples (Limondin-Lozouet et al. 2002, tab. 1; Pastre et al. 2003, tab. 1) which are not a reliable type of datable material (cf. Hiller et al. 2003). Thus, in contrast to the maar lakes from the western uplands, the temporal resolution of these sequences is less precise. Moreover, the sequences originated from different regions but usually from comparable limnic habitats. However, they were synthesised to deliver a standard pollenstratigraphy for northern France containing seven sub-zones which are numbered from bottom to top (**fig. 22**; Limondin-Lozouet et al. 2002; Pastre et al. 2003). Even though this compound sequence is based on the presuppositions of an instant and comparable vegetation development across the complete area, the constant stratigraphic evaluation should secure that only little gaps occurred in this succession. Furthermore, the wide distribution of the contributing sites sustains that the displayed developments affected most of the sub-area. Therefore, this record delivers a sufficiently precise record of the general vegetation development in northern France. In addition, this pollen stratigraphy was occasionally correlated directly with archaeological sites such as Conty (Limondin-Lozouet et al. 2002, 48; cf. Ponel et al. 2005).

Besides the pollen profiles, directly dated macro-fossils are useful material for reconstructing the local vegetation, and directly dated faunal remains also help to reconstruct the Lateglacial environment. However, this type of resource is presented in the next chapter because the evaluation of ^{14}C dates requires stratigraphic as well as chronostratigraphic considerations. Furthermore, the databases formed by the ^{14}C dates contribute in addition to information on the environment, also information on the dating of archaeological assemblages. Therefore, this material type is presented separately.

DATABASES

In general, the present study is structured according to the three main lines of research: climate, environment, and human behaviour. However, the databases are separated in this material presentation because this type of data is related to all three parts of the study. For example, the content of radiocarbon databases can be used to model faunal presence or to date single archaeological sites. In addition, a reliable radiocarbon calibration curve is largely dependent on climate archives (see p. 250-253).
The values in databases represent single episodes which are self-explanatory. Thus, single values can also be taken out of the database and be interpreted independently. In contrast, values in stratigraphies such as pollen profiles are only meaningful in relation to each other and, thus, the changes between and within the single values are significant. However, the single episodes recorded in the databases can also be compiled to firm up an event by overlapping results or create a sequence of accumulation and scarcity of dates. In the latter case, the independent data points are transferred into an eventstratigraphy.

Radiocarbon database

Since the introduction of radiocarbon measurements as an independent dating instrument in the late 1940s/early 1950s (Libby 1952), this method was used to date archaeological sites also in north-western Europe (e.g. Münnich 1957). Thus, a large corpus of ^{14}C dates from Lateglacial sites in north-western Europe has accumulated over the last 50-60 years. Besides the necessity of reviewing some of these results qualitatively (cf. Pettitt et al. 2003; Grimm/Weber 2008), the idea arose to use this increasing set of dates as a quantitatively analysable data set. This dates-as-data approach was introduced in the analysis of regional occupation patterns (Rick 1987). The main assumption was that an increase in human presence in a region also resulted in more activity and, thus, an increased production, deposition, and conservation of carbon material related to the human presence. Moreover, in a following step the results from adjacent regions were compared to indicate palaeodemographic movements (Housley et al. 1997; Gamble et al. 2005; Shennan/Edinborough 2007).
However, to make this quantitative approach possible a dataset consisting of as many reliable dates as possible is necessary. For the present study a database of almost 1,500 single radiocarbon dates (n=1,486) associated with the Lateglacial of north-western Europe was collected and reviewed (see p. 259-263 and p. 265-269). 322 of these dates originated from northern France and the western uplands including the Central Rhineland (**tab. 5**). Only some 240 of these dates fall into the relevant time slice from 13,200 to 10,800 years ^{14}C-BP.
The fundament for this database was formed by a large database of radiocarbon dates from Pleistocene and Holocene archaeological sites which was assembled at MONREPOS mainly by Martin Street and Olaf Jöris during the last decades and contained in mid-2009 some 8,600 single dates[7]. This fundament was supplemented by further sources (**tab. 6**).
For instance, a large database of radiocarbon dates (n=9,356) from Palaeolithic sites in western Europe was published online by the INQUA (PI: Pierre Vermeersch; http://ees.kuleuven.be/geography/projects/14c-palaeolithic). Julia Fahlke used this database, in particular, and collected further results in regard to the Lateglacial faunal succession in Central Europe creating another set of radiocarbon dates (n=934; Fahlke 2009).

7 Meanwhile the number has increased to some 12,000 single datasets (written comm., Martin Street).

attributes	number of ^{14}C dates in database	number of ^{14}C dates from the Central Rhineland	number of ^{14}C dates from the western uplands*	number of ^{14}C dates from northern France
all dates	1,489	123	185	137
dates from relevant time slice	894	101	136	101
dates on bone	769	85	142	79
dates on teeth	95	5	5	12
dates on antler	119	2	3	3
dates on charcoal	239	6	6	28
dates on wood	85	20	20	0
dates on reindeer (*Rangifer tarandus*)	208	5	20	14
dates on horse (*Equus* sp.)	195	24	40	21
dates on red deer (*Cervus elaphus*)	51	13	15	2
dates on elk (*Alces alces*)	26	6	6	0
dates on large bovids (Bos / Bison; *Bos primigenius; Bison priscus*)	55	3	3	13
dates on birch (*Betula* sp.)	19	4	4	0
dates on pine (*Pinus* sp.)	37	1	1	3
AMS dates	1,015	98	137	89
conventional dates	457	25	48	48

Tab. 5 Number of ^{14}C dates with selected attributes in present database. Note that these numbers are unaudited (cf. p. 259-263 and p. 265-269). Relevant time slice is 13,200-10,800 years ^{14}C-BP. * included are also the sites from the Central Rhineland database.

In addition, numerous radiocarbon dates (n=6,019) were published as PACEA geo-referenced database (d'Errico et al. 2011) which was mainly filled by southern European sites. In the supplemental databases the ^{14}C dates were given without comments on the technical and/or contextual reliability of these dates. In contrast, in the MONREPOS database some comments on a technical and/or contextual evaluation were already given (cf. Pettitt et al. 2003; Grimm/Weber 2008). A comparable evaluation of ^{14}C dates was accomplished for Lateglacial samples from the Netherlands and adjacent regions available in the mid-1990s (Lanting/van der Plicht 1996) and provided some useful information of some 106 dates incorporated in the present study.

Moreover, additional dates were searched from several datelists of various laboratories published in the journals Radiocarbon, Science, and Archaeometry. Unfortunately, with the commercialisation of the laboratories they often lost the right of publishing dates and, therefore, these truly helpful compilations of dates from a single laboratory were rarely published after the 1980s except for the Oxford laboratory (Archaeometry). Revisions of available dates for specific regions or periods partially fill up this gap (e. g. Street/Baales/Weninger 1994; Lanting/van der Plicht 1996; Bodu et al. 2009b; Jacobi/Higham 2009). Many such revisions were also consulted. Further information was looked up in the online accessible database of the Oxford radiocarbon accelerator unit (ORAU; online database requires registration, http://c14.arch.ox.ac.uk/embed.php?File=). Another group of dates on the Hamburgian settlement were made accessible by the Cologne laboratory (n=4), Jørgen Holm (n=9), and the Kiel laboratory (n=3). However, these dates were previously published (Grimm/Weber 2008). A further two dates from northern Germany were also discussed in a previous publication (Riede et al. 2010). A single date from Boppard made on a metapodial of red deer (*Cervus elaphus*; KIA-2644: 11,095 ± 55 years ^{14}C-BP; $\delta^{13}C$ = -25.1; see p. 161 f.) was kindly provided by Stefan Wenzel (VAT, RGZM/Mayen) and not published previously.

For each date the laboratory number and the standard deviation as well as the site location including name, region, country, latitude, longitude, and the approximate altitude were recorded to enable a geo-referencing of the location from which the date originated and the use of the dates in GIS programs. Moreover, the

database	MONREPOS 2009	INQUA	Fahlke dissertation	PACEA	present database
ref.	unpublished	webpage	Fahlke 2009, appendix 1.A	d'Errico et al. 2011	present study
n all ^{14}C dates	8,630	9,356	934	6,019	1,489
n ^{14}C dates attributed to relevant period	1,654	2,232	569	1,917	1,288
n ^{14}C dates from the relevant time slice	1,214	1,484	496	1,388	894
n ^{14}C dates from study area	500	599	218	282	322
n ^{14}C dates from relevant time slice & study area*	197	223	153	222	231

Tab. 6 Number of ^{14}C dates in main databases contributing to the present database. * Information from all these dates contributed to the present database but note that there is a significant overlap in dates and, therefore, an addition of these dates does not result in the number of dates in the database of the present study. Abbreviations: **Ref.** reference; **n** number of; **webpage** http://ees.kuleuven.be/geography/projects/14c-palaeolithic accessed 20th August 2007; **relevant period** Late Magdalenian to *Federmesser-Gruppen* and/or Late Pleniglacial to end of Lateglacial Interstadial. – Relevant time slice is 13,200-10,800 years ^{14}C-BP. – Study area see **fig. 1**.

position of the sample on the site in the horizontal as well as vertical stratigraphy was further specified if possible to evaluate the relation of the sample to the other material. Additional information from the site such as results of pollen analyses of the source layers were also recorded. Furthermore, the archaeological affiliation of the material from which the sample originated was noted to make a selection of dates for specific archaeological groups possible. Other contexts were described as palaeontological, palaeobotanical, or stratigraphic affiliation. In addition, the sample was described by its general type of material such as bone, antler, or charcoal (**tab. 5**) which was further defined by a more detailed description such as »phalanx, sin., cut-marks« or »bulked sample, calcined pieces«. Furthermore, if the species was determined this information was also recorded. As far as published data, further indicators of the composition of the sample were noted such as the gelatin yield, the percentage of modern carbon (pmC), the $\delta^{13}C$ value, the $\delta^{15}N$ value, and the C/N ratio. All Oxford dates (n=653) were checked for these values and, in general, the pretreatment procedure was also documented in the online database. Moreover, the dating method was distinct between counting of carbon ions by accelerator mass spectrometry (AMS; e.g. Fifield 1999) and the classic counting of radioactivity (β-counting or conventional dating; Libby 1952). As far as possible, a citation for the original publication was given. Otherwise, publications in which the date was used or which yielded further information on the sample were also recorded.

The pretreatment procedure as well as the additional information on the composition for the sample were used in a technical audit clarifying whether the given date produced a reliable age for the sample (see p. 259-263). The archaeological audit in contrast aims for the significance of the date for the age of the archaeological material (see p. 265-269). For example, a radiocarbon date can result in a reliable estimate for the death of an animal (»dated event«) such as a hare but the death of the hare might not have been related to the archaeological episode that was wished to be dated (»target event«, Dean 1978). In a pure archaeological audit the example date would be rejected, even though the date represents reliable evidence for the presence of the animal in this area. Thus, the distinction between technical and archaeological reliability is made to use the dates as indicators for the chronostratigraphic position of an archaeological assemblage as well as for the presence of vegetation or a fauna species in the area. The latter can help to model the environment and, in particular, indicate the availability of the specific resources in the studied region.

attributes	number of sites in database	number of sites from the Central Rhineland	number of sites from the western uplands*	number of sites from northern France
sites with selected directly data faunal material	135	9	21	11
sites with selected directly data plant remains	35	4	5	0
sites with pollenstratigraphies	94	1	9	22
sites with climate and calibration data	26	1	3	0

Tab. 7 Number of sites in the database of sites with climatic and environmental data. * included are also the sites from the Central Rhineland.

Database of Lateglacial sites from north-western Europe

In addition to the collection of radiocarbon dates, two geographic databases of locations of material associated with the Lateglacial of north-western Europe were created for the mapping of the material.

One set of databases consisted of findspots of climatic and/or environmental material (**tab. 7**). In the other one, archaeological sites and occasionally single concentrations on these sites (n=1,188) were compiled (**tab. 8**). In both types of databases, the locations were given by a specific name such as Gönnersdorf I or Miesenheim 4 and a region, a country, and their geographic coordinates (latitude, longitude, altitude) were assigned to this name. The coordniates are necessary to use information connected to these sites such as radiocarbon dates in GIS programs. The coordinates of each site/concentration were cross-checked using published maps of the sites and Google-Earth™. The coordinates are given in decimal degrees and the altitude in m a.s.l. However, depending on the resolution of the primary site information a precision code for the coordinates was given (0 – approximate area; 1 – the same parcel; 2 – findspot). Additionally, a regional code for the present study was created to facilitate the analysis of sub-sets across modern political borders. The environmental databases were distinct between sites yielding directly dated samples of selected faunal species (n=135) or plant species (n=35), sites yielding stratigraphic sequences (n=94), and sites yielding climate and calibration data (n=26). To be able to plot the sites on the relevant maps, the events were recorded for which these sites yielded data. Furthermore, in the table of the stratigraphies the type of material which was analysed as well as the type of dating and the type of correlation was noted. In the databases with directly dated material, the results of the technical evaluation were taken into account by the indication if temporal distributions might be false.

In the database of archaeological sites, for all concentrations the archaeological classification was recorded in three scales: The first scale was the epoch (i.e. A – Palaeolithic, B – Mesolithic, C – Neolithic etc.), second scale was a sub-period (A – Early/Lower, B – Middle/Intermediate, C – Late/Upper, D – Final/Top), and thirdly an archaeological group was given such as Late Magdalenian or FMG. Finally, a reference was given where the sites were presented and/or mentioned. However, some concentrations were introduced partially or only mentioned in publications thus far and, therefore, the following attributes were not always recordable for all concentrations. For instance, a code for the integrity of the archaeological material was currently given to only 769 concentrations. By this code problems such as stratigraphical disturbances, archaeological classification uncertainties, or both are identified. This distinction is of particular importance when studying assemblages from a transition period to distinguish possible palimpsests from assemblages with mixed characteristics of the two poles of the transition. However, to be able to judge this integrity, precise excavation standards are required. Therefore, the excavation standards were also considered (0 – collection, no excavation; 1 – collection with more detailed information on stratigraphy; 2 – excavations with rough excavation technique and/or imprecise documentation; 3 – modern standard excavation, i.e. with careful excavation techniques and good documentation including 3D-data for single artefacts).

attributes	number of sites in database	number of sites from the Central Rhineland	number of sites from the Western Uplands*	number of sites from northern France
all sites	1,188	53	133	237
all collections (with further information)	282 (57)	5 (2)	25 (10)	12 (3)
all excavated concentrations (modern standards)	602 (416)	47 (41)	79 (57)	145 (134)
material disturbed stratigraphically	37	3	5	0
archaeological attribution uncertain	301	4	35	27
material is archaeologically uncertain and stratigraphically disturbed	66	2	7	3
sites of Magdalenian-affiliation	100	11	27	16
sites of FMG-affiliation	158	32	34	49
sites of uncertain FMG- / Magdalenian-affiliation	86	2	14	18
open air sites	919	45	60	170
rock shelter sites	27	0	4	2
cave sites	112	4	35	3
sites with botanic material	142	23	30	37
sites with faunal material	260	46	65	69
sites with lithic material	572	42	74	139
sites with structures (evident)	106 (78)	13 (10)	19 (14)	24 (23)

Tab. 8 Number of concentrations for which selected attributes are recorded in the database of archaeological sites. * included are also the sites from the Central Rhineland.

Additionally, the site type was recorded for 1,061 concentrations to date indicating whether the concentration was found in a cave, a rock shelter, or in the open air. Moreover, the presence or absence of botanic evidence (n=503), of faunal material (n=631), of lithic material (n=622), and of settlement structures (n=367) from the sites was recorded. However, the number of sites where such material was present diverged considerably (**tab. 8**). For instance, even though some 260 concentrations out of 631 recordings yielded faunal material, this material was often poorly preserved and just represented by a few indeterminable bone fragments or only a malacological analysis was made. In addition, the lithic material was recorded in several sub-units such as number of cores, tools, or backed pieces. However, only the sites relevant to the present study are presented in more detail.

ARCHAEOLOGICAL SITES

Along with climate and environmental archives, archaeological assemblages form the fundament of the present study. The archaeological record yields the mainstay for the interpretation of changes in human behaviour because this record was created by human behaviour. The lithic artefacts were produced by humans in a usually conscious and largely intentional technical process. The faunal remains reflect a subsistence choice and/or hunting ability of the past humans. Besides those finds, the spatial organisation of these humanly created remains forms the archaeological record.

However, selection or modification by agents other than humans such as taphonomic processes have to be excluded from this record to extract the patterns created by past human groups. To prevent misinterpretation due to naturally formed patterns, the archaeological record has to be evaluated in regard to the site formation process. Furthermore, admixture of diachronic archaeological material needs to be reduced as far as possible to prevent mixing of unrelated behavioural expressions to a common development.

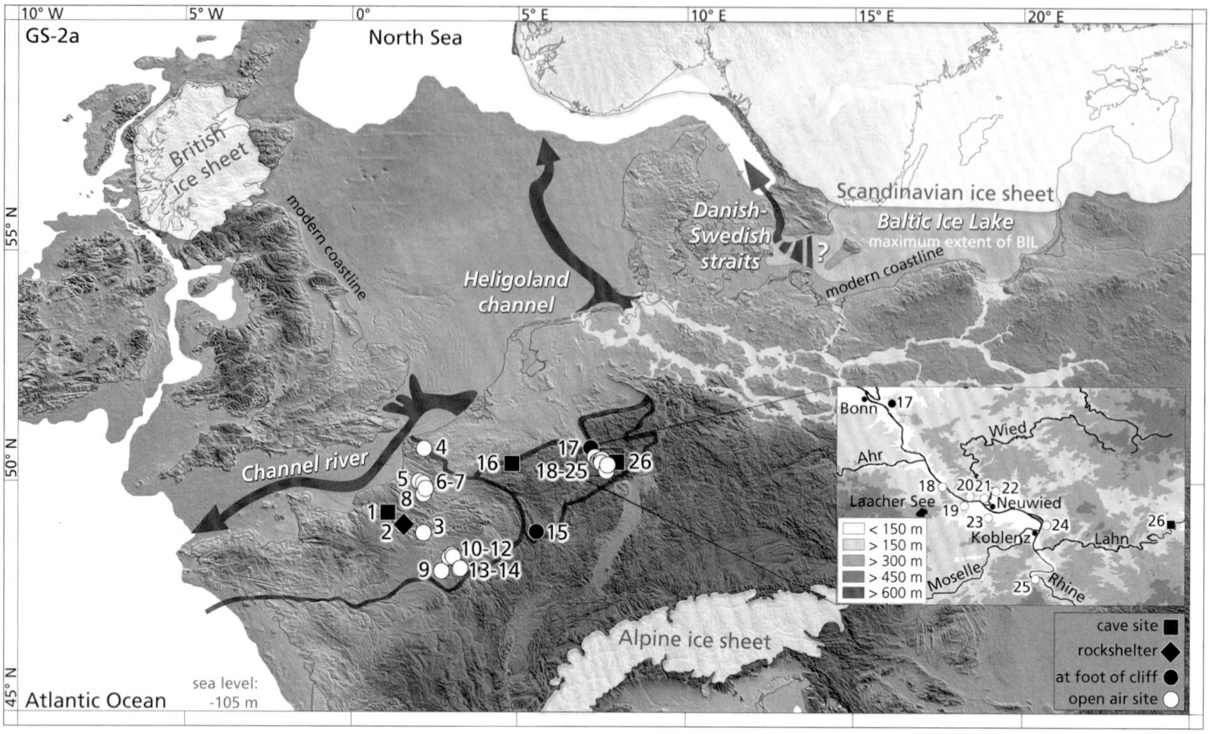

Fig. 23 Map of archaeological sites used in the present study. **1** Grotte de Gouy; **2** Bonnières-sur-Seine; **3** Le Closeau; **4** Hallines; **5** Hangest-sur-Somme; **6** Belloy-sur-Somme; **7** Saleux; **8** Conty; **9** Cepoy; **10** Pincevent; **11** Le Tureau des Gardes; **12** Le Grand Canton; **13** Étigny; **14** Marsangy; **15** Saint Mihiel; **16** Bois Laiterie; **17** Bonn-Oberkassel; **18** Bad Breisig; **19** Andernach-Martinsberg; **20** Gönnersdorf; **21** Irlich; **22** Niederbieber; **23** Kettig; **24** Urbar; **25** Boppard; **26** Wildweiberlei.

If uniform patterns in the choice of the material as well as their spatial deposition are observed in this reviewed record, a social agreement on the behaviour creating the patterns must have existed. Thus, these established patterns emerged from a social discourse and a communal decision-making process. Furthermore, a change of the social agreement meant to abandon the previously stipulated behaviour and to prefer an alternative behaviour which created a divergent pattern from the previously stipulated one in the archaeological record. In a diachronic comparison of archaeological assemblages, the establishment of alternative behaviours as new stipulated ones can be observed. Consequently, changes in the archaeological material and/or the position of this material in the spatial organisation reflect changes in human behaviour.

However, since humans are imperfect in the repetition of a specific behaviour (Eerkens 2000; Hamilton/Buchanan 2009), a general pattern usually consists of a range of socially acceptable variation. Moreover, the creation of the material also depends on availabilities and characteristics of resources such as lithic or organic raw materials. These conditions shape the variation additionally. Thus, the patterns as well as the typical tolerance ranges have to be identified to describe the stable states of behaviour reflected by the archaeological material. Furthermore, to specify the changes between a previous and a new state of agreement, these patterns and tolerance ranges have to be described for both states. In the example of Lateglacial north-western Europe, these two stable states are the Late Magdalenian and the FMG.

In the Central Rhineland, the archaeological remains of these two stable states are particularly well preserved due to the protective cover by the LST. This good preservation allowed for a relatively precise chronostratigraphic attribution of the assemblages as well as detailed spatial analyses of sometimes single concentrations of human activities. However, only a few assemblages were attributed to the transitional period between the Late Magdalenian and the FMG. The most important sites from the Lateglacial in Central Rhineland are introduced in the following (see Material-Archaeology-Central Rhineland, p. 75-170) to describe the stable

states and, thereby, specify the differences between them. In addition, the sparse information of the intermittent transition known from the Central Rhineland is outlined. Moreover, to fill this incomplete record of the transitional period, archaeological sites from northern France (see Material-Archaeology-Northern France, p. 182-244) comprising the Paris Basin and the Somme region as well as from the western upland zone (see Material-Archaeology-Western upland zone, p. 170-182) located between the Central Rhineland and northern France are included in the present study. These areas are chosen as supplements to the Central Rhineland because they were presumably inhabited by a closely related social network during the Late Magdalenian (Floss/Terberger 2002, 138; cf. Arts/Deeben 1987) and, probably, still in the period of the FMG (Bodu/Valentin 1997; De Bie/Vermeersch 1998; Bodu 2000a; De Bie/Van Gils 2006). In particular, the distribution of Late Magdalenian material in north-western Europe was assumed to indicate a large-scale common information network (cf. Schwendler 2012; Clarke 1968, 88-101). Thus, the progress of change from the Late Magdalenian to the FMG was probably influenced by the exchange of information within this network and/or by the possible secession from this network. In this case, archaeological assemblages dating to the transition period reflect individual decisions made in a large-scale social discourse on how to behave. Therefore, the social process of change from the Late Magdalenian to the FMG can be described more precisely by the use of reliable assemblages which consolidate the spatio-temporal web of observable decisions made in the process of changes. In addition, based on the similarities in the archaeological material from the Central Rhineland, northern France, and the western uplands during the Late Magdalenian and the FMG, a comparable process leading to the changes observable in the archaeological material between these two periods can be assumed for these three regions.
Since the archaeological assemblages form the mainstay for human behaviour, they are presented in detail (**fig. 23**). In particular, the research history, the setting, the stratigraphy, the archaeological material, the spatial patterns, and the chronology[8] of the archaeological sites are described in the following. Yet, a general description of the Late Magdalenian and the FMG is given previously to introduce the differences and commonalities between these two archaeologically defined groups.

Archaeological groups

The archaeological material found underneath the LST in Central Rhineland was generally attributed to two major Lateglacial groups: the Late Magdalenian and the FMG.

8 According to the Oxford Dictionary »chronology« refers to »the arrangement of events or dates in the order of their occurrence« (»chronology« Oxford Dictionaries, April 2010, Oxford University Press. Online accessed: 27 August 2012 <http://oxforddictionaries.com/definition/english/chronology>). Thus, the chronology of a site refers to the general chronostratigraphic setting of the site as well as to the relation of the different activities and processes forming the site. Therefore, considerations about the chronology should address the relation of different occupation episodes on the site to each other (internal chronology), the influence of natural processes other than humans (site formation process), and the relation of the dated material to the human activity (chronological evaluation). The latter is necessary to clarify the chronostratigraphic position of the human activities which shall be dated (target event) in contrast to the material which is dated (dated event, Dean 1978). For example, if a site is dated by the pollen spectrum of an overlaying deposit, the dated event represents only a *terminus ante quem* with an unknown gap to the time in which the archaeological assemblage was deposited. Equally, if an ivory point is ^{14}C-dated, the dated event is the death of the elephantidae with an unknown gap to the target event which could be the carving of the point as well as its use or its discard. For the Pleistocene, these gaps of the radiometric dated events to the target events were usually considered as attributable to the standard deviation and/or the calibration interval. However, the clarification of the relation of dated and target events becomes necessary with the increasing precision of radiocarbon dates as well as of the calibration curve because the gaps may no longer be covered by the methodological error margins. Furthermore, for the Lateglacial, this assumption is already challenged by the knowledge of the durability of some materials (Pétillon et al. 2011) and the probable use of fossil materials (Street et al. 2006).

These two entities reflected two chronologically successive sets of behaviour. Differences in some parts of these sets resulted probably from the adaptation to different surroundings. The different surroundings were formed by the variation of the climate regime and of the available resources such as vegetation and game. From these different surroundings, various necessities arose for the fulfilment of the physiological needs of humans (cf. Maslow 1943; Pittman/Zeigler 2007) such as protection against cold or moisture and repletion by vegetal and/or faunal resources. Thus, human groups had to adapt their behaviour to their surroundings to ensure their survival. How and when the Lateglacial groups adapted is the subject of the present study. Moreover, whether the adaptation occurred as a caesura or as a transformation period will be a result of this study. However, an assessment of the transition was often made -presumably in parts sub-consciously- when ambiguous assemblages from the transition period were attributed to one of the archaeological groups, to further sub-groups such as the Final Magdalenian or the Early Azilian, or to transitional industries. Hence, across Western Europe several archaeological assemblages were attributed to the Late Magdalenian, the FMG, and/or the transitional period between them. Although these previous attributions of the material are going to be mentioned in the presentation of the archaeological material, they are not considered as final assessments. However, to specify the similarity of the studied assemblages to the Late Magdalenian and/or the FMG, the general definitions and the main defining elements of the two archaeological groups are introduced in the following. Furthermore, these two entities set the beginning and the end point of the transition under study. Hence, the differences between the two groups are specifically emphasised. These differences allow for a description of the assemblages of the transitional period in terms of nearness to the one or the other archaeological group and focusing particularly on distinctive behaviours. Subsequently, a chronological order of these descriptions enables the reconstruction of the progress of changes in the Lateglacial revealing continuous and/or rapid developments.

The Late Magdalenian

The archaeological remains recovered from the lower or older horizon found in the loess deposit underneath the LST in the Central Rhineland were classified as Late Magdalenian. The Magdalenian forms one of the major archaeological entities in the European Upper Palaeolithic. This entity was named by Gabriel de de Mortillet based on the lithic and organic material from the site La Madeleine in the Vézère valley (de Mortillet 1869; de Mortillet 1872). He proposed the Magdalenian to cover the second part of the Late Quaternary.

The eponymous rock shelter, La Madeleine, was first excavated from 1863 to 1865 by Édouard Lartet and Henry Christy. The excavators published the material successively in the following decade (Lartet/Christy 1875). In the 19th century and the early 20th century, further excavations were conducted by Elie Massénat, Paul Girod, Émile Rivière, and Denis Peyrony (Capitan/Peyrony 1928). Finally, Jean-Marc Bouvier (Bouvier 1977) excavated with modern standards in this rock shelter. Successively, the archaeological record from the site comprised a rich inventory of organic artefacts and more than 12,000 lithic artefacts. Furthermore, a burial of an infant with presumably richly decorated clothes was found by Denis Peyrony in 1926 and ^{14}C-dated to the Younger Dryas period (Gambier et al. 2000). Besides La Madeleine, Denis Peyrony conducted further excavations in the western neighbouring rock shelter Villepin to receive a stratigraphic succession from the Magdalenian to the succeeding so-called Azilian (see p. 65-74; Peyrony 1936). Furthermore, the stratigraphic sequence from La Madeleine was combined with basically the stratigraphy from Le Placard by the abbot Henri Breuil in 1912 to construct what he considered a successive chronological development of the Magdalenian (Breuil 1913; cf. Breuil 1954).

Meanwhile, sites attributed to the Magdalenian were dated as early as approximately 17,500 years ^{14}C-BP (Ducasse 2012). However, the Badegoulian had been occasionally considered as the earliest stage of the Magdalenian or a Proto-Magdalenian (cf. Ducasse/Langlais 2007; Ducasse 2012, 151 fig. 1) and the earliest Badegoulian sites were dated to c. 19,000 years ^{14}C-BP. The upper temporal limit of the Magdalenian also depended on the definition of the Magdalenian and on its distinction from the Azilian (cf. Delporte 1966; Cheynier 1966). In consequence, this limit also ranged over a considerable period from approximately 12,800 to 11,500 years ^{14}C-BP (Straus/Leesch/Terberger 2012).

Independent of the precise definition, a continuous development of the Magdalenian was only established for a south-western European refuge thus far (Straus/Leesch/Terberger 2012). The Magdalenian was consequently considered a strictly south-western European innovation. It reached most parts of north-western Europe after the LGM in a demic expansion process which possibly included several waves (cf. Verpoorte 2009). This spread of people from a south-western refuge across the upland zones of north-western Europe during the Late Pleniglacial was further sustained by mtDNA evidence (Torroni et al. 1998; Pereira et al. 2005). Archaeological evidence for single incursions into northern France and the north-western European uplands were found for almost all periods of the Magdalenian (Terberger/Street 2002; Bodu/Chehmana/Debout 2007). However, these incursions became more frequent during a middle period of the Magdalenian and an established settlement of these areas was only associated with a late period of the Magdalenian (Housley et al. 1997; Verpoorte 2004; Gamble et al. 2005; Miller 2012; Połtowicz-Bobak 2012). Hence, the Magdalenian in general and the Late Magdalenian in particular were considered as the »ancestral« tradition to the succeeding Lateglacial archaeological groups of north-western Europe, both in a behavioural and a biological sense (cf. Schwabedissen 1954; Weber 2012; Pettitt/Rockman/Chenery 2012).

In general, the Magdalenian was bisected by Henri Breuil into a lower part and an upper part based on the stratigraphic occurrence of organic artefacts and pieces of art at Le Placard and at La Madeleine (Breuil 1913, 40-56). The lower, early part was mainly based on the sequence in Le Placard. In this part, bevelled points but no barbed points occurred and the art which mainly decorated the organic material was described as ornamental. In contrast, the organic inventory of the upper, late part was supplemented by barbed points and the art became figurative. The successive development of this part was found at La Madeleine. Breuil differentiated a further three sub-units within the two parts: the Magdalenian 1-3 for the Lower/Early Magdalenian and Magdalenian 4-6 for the Upper/Late Magdalenian. For the Upper/Late Magdalenian the distinction of the sub-groups was based on the development of the barbed points at the eponymous site of La Madeleine beginning with rudimentary barbed points (4), followed by barbed points with one row of barbs (5), and barbed points with two rows of barbs, i. e. harpoons (6). A final stage of this development was later also recovered at the neighbouring site Villepin in the succeeding Azilian horizons where some flat harpoons were found that were made of red deer (*Cervus elaphus*) antler and that were perforated at their base (Peyrony 1936). Comparable specimens were already known from the eponymous site of the Azilian, Mas d'Azil (Piette 1895). Although the general succession of the organic implements persisted, overlap led to occasional co-occurrence of some types (Langlais et al. 2012). Furthermore, barbed point types were meanwhile confirmed for an early period of the Magdalenian and, thus, these implements were only absent during an intermediate part of the Magdalenian, although some discussed proto-types also existed in this period (Pétillon 2008b). Furthermore, the variety of the barbed points was demonstrated to be related rather to function and use than chronostratigraphy (Julien 1982; Weniger 2000; Pétillon 2008b). The continuous co-existence of various types of barbed points into the Final Palaeolithic and Mesolithic (Smith/Bonsall 1991; Cziesla/Pettitt 2003; Cziesla 2007a; Cziesla 2007b) further sustained the use-oriented interpretation of the observed variability. In consequence, the classification of barbed points developed by Henri Breuil could not be confirmed as chronostratigraphically precise and, thus, relevant distinction.

Besides the classification according to organic types, Henri Breuil mentioned changes occurring in the lithic inventory between his stage Magdalenian 5 and 6 (Breuil 1913). In the Magdalenian 5, burins on truncations and small curve-backed points were widely spread, whereas the abbot observed a clear regionalisation of the assemblages in the Magdalenian 6. Nevertheless, in this latter stage the *bec-du-perroquet* (cf. Sonneville-Bordes 1960, 332 f.) and the generally curved and »hooked« burins were considered as a still common element. Based on his excavations at Laugerie-Haute, Denis Peyrony sustained three groups of lithic inventories in the Early Magdalenian (Peyrony/Peyrony 1938) but in contrast to Breuil he regarded the lithic material of the Late Magdalenian as extremely uniform (Capitan/Peyrony 1928; Peyrony 1936). According to Peyrony, several types of burins were usually dominant and, particularly, dihedral ones and those on truncation were frequent in all horizons of the Late Magdalenian. Furthermore, borers, end-scrapers on blades, and numerous composite tools occurred in all horizons of La Madeleine (Capitan/Peyrony 1928). Among the backed implements or laterally modified pieces (LMP)[9], backed bladelets and blades were the dominant type but also large *couteaux à dos* were found. Backed points occurred occasionally in Late Magdalenian assemblages but no uniform type of these points was observable. The numbers of points increased throughout the stratigraphic succession and a regular appearance of these implements marked the onset of the Azilianisation process. Thus, the lithic inventories used by Breuil to define the Magdalenian 5 and 6 were either assemblages which already fell into the transition to the Azilian or were results of admixtures. Consequently, the distinction of sub-groups by Henri Breuil based on the lithic inventories is obsolete. In contrast, the concept of a general uniformity of Late Magdalenian lithic inventories as observed by Denis Peyrony proved valid, in particular, in regard to the large distances of the distribution of Late Magdalenian assemblages.

Many further sub-divisions of the Magdalenian and the Late Magdalenian exist (e. g. Capitan/Peyrony 1928; Sonneville-Bordes 1960; Schmider 1982; Bosselin/Djindjian 1988; Le Tensorer 1998). In fact, the distinctions of various sub-groups were often similar to the system of Breuil and/or meant to refine or define this system more precisely. For instance, based on the refinement of typologies, a Middle Magdalenian comprising Breuil's sub-units 3 and 4 was distinct (e. g. Allain et al. 1985; Onoratini/Defleur/Joris 1996). The Middle Magdalenian was interpreted as the emergence of the typical Late Magdalenian (Langlais 2007). However, the lithic inventory of a Middle Magdalenian comprised also scalene triangles and microburins (Langlais 2007) which were distinctive for the Early Magdalenian sub-unit 2 (Onoratini/Defleur/Joris 1996). Comparably, the Magdalenian 1 was meanwhile renamed Badegoulian and considered an independent stage from the Magdalenian (Ducasse 2012). Furthermore, in northern Europe little or no organic tools were preserved in inventories with lithic material comparable to the Late Magdalenian. These assemblages usually comprised no backed bladelets but numerous backed points and, therefore, were attributed to a Final Magdalenian (Garrod 1926; Schwantes 1933) that was correlated to Breuil's sub-units Magdalenian 5

9 In the following the term »laterally modified pieces« (LMP; Caspar/De Bie 1996) refers to laminar pieces with abrupt modification along at least one lateral edge (cf. Jacobi 2004, 34-45). They are mainly assumed to represent lithic projectile implements (Rust 1943, 191; Allain 1979, 100-103; Leroi-Gourhan 1983; Moss 1986; Bratlund 1996b; Rots/Stapert/Johansen 2002; Holm 2003; Sano 2009) but micro-wear analysis also suggested that at least the larger pieces were used as cutting tools (»knives«; Rots/Stapert/Johansen 2002). Thus, under this term almost all Upper and Final Palaeolithic projectile implements and various »knives« as well as their fragments are subsumed. This collection in a single group allows the establishment of a main morphological class comparable to end-scapers and burins. By the use of this term, an easier general comparison of various stages within the process of change based on retouched artefact classes can be established. Moreover, the distinction between classic types of LMP is sometimes difficult (Grimm/Jensen/Weber 2012) and distinction in fragmented material during the Lateglacial is often impossible. The attribution especially of transitional types usually follows the find context (Jacobi 2004). This attribution practice also applies to fragments of LMP. Pointed distal fragments are often counted as points, whereas medial and basal fragments are often counted within the group of the backed blades and bladelets, although this attribution might be false. To prevent these presuppositions, these pieces are subsumed under the neutral term LMP and only complete or nearly complete types are further distinguished (see p. 275-282).

and 6 (see above). If these additional sub-divisions were considered and given a chronological meaning, the Magdalenian development would proceed from the Badegoulian to the Middle Magdalenian to the Final Magdalenian. Consequently, the question would arise what was meant with an Early and a Late Magdalenian. This exaggerated excursus in nomenclature, correlations, and sub-divisions is meant to indicate the ambiguity of this variously defined and used taxonomic system based on Breuil (Breuil 1913; cf. Breuil 1954) and partially mixed with Peyrony's distinction (Capitan/Peyrony 1928; Peyrony 1936) and other sub-divisions of the Magdalenian (Bordes 1958; Sonneville-Bordes 1966; cf. Ducasse 2012). In general, these early francophone approaches of using typologies were meant to establish chronologically relevant sub-divisions of the Magdalenian.

In contrast to the francophone approaches, early German archaeologists often followed the concepts of Gustaf Kossina (Kossinna 1911) and tried to distinguish mainly geographic sub-groups which were assumed to represent different ethnic sub-units (Schwabedissen 1954, 71-76). Perhaps, this focus on spatially distinct but potentially contemporaneous groups was also due to the presence of dominantly Late Magdalenian type inventories in the German upland areas. Moreover, the chronological control of these assemblages was often poor because many of them were surface collections. To increase the chronological control, archaeological assemblages were combined with analyses of the natural sciences such as pollen analysis to help construct the environments and allow some chronological differentiations to be made (e.g. Rust 1958). Therefore, excavated sites where a stratigraphic control was possible were increasingly preferred. Accordingly, Rudolf Feustel defined some groups within the Late Magdalenian (Lausnitz, Feustel/Teichert/Unger 1963; Oelknitz, Feustel 1956, 25) based on mainly excavated assemblages from the Thuringian Basin. Particularly in the Oelknitz group, the lithic assemblage was dominated by a variety of backed bladelets. These implements could be truncated on one or both ends, denticulated, or as well be retouched laterally. Yet, several pieces appear to be broken intentionally, perhaps, to be fitted in composite projectiles (cf. Leroi-Gourhan 1983). Furthermore, among the frequent burins the examples on truncation dominated. In the Oelknitz group, these truncations were often long and concave (*Lacan* type) and the borers were also rather elongated and sometimes with solidly retouched edges (i.e. *bec* type). In the following years, Feustel's Oelknitz group comprised most of the reliable Magdalenian assemblages in north-western Europe and, thus, could be taken as synonymous with the Late Magdalenian of Denis Peyrony (see above). Subsequently, Feustel further distinguished five variants: Oelknitz, Nebra, Moosseedorf, Ahlendorf, and Kniegrotte (Feustel 1979). These variants were particularly based on their lithic assemblages. In the Nebra group, the variety of backed bladelets was considerably smaller than in the other groups, with no denticulated or laterally retouched backed pieces at the eponymous site. However, except for this difference, the inventories of the Oelknitz group were basically identical to the Nebra group. In both groups, burins were made on truncations which were generally oblique. In the Mooseedorf and the Ahlendorf group of Feustel, the burins were also made on truncations but in contrast to the Nebra and Oelknitz group these truncations were mainly straight. The difference to the Kniegrotte was the presence of triangles in the Kniegrotte assemblage (cf. Höck 2000). Occasionally, these groupings are still used. For instance, Thomas Terberger discussed the attribution of the Central Rhineland Magdalenian to the *Nebraer Gruppe* (Nebra group, i.e. variant) based on the retouched lithic artefact inventory of Andernach as well as of Gönnersdorf (Floss/Terberger 2002, 135-138). He noted further similarities among the sites of the Nebra group in the presence of female figurines and spatial structures, in particular, in the evident structures such as pits and pavements (Floss/Terberger 2002, 135-138). Feustel had already suggested this relation of Nebra to the Gönnersdorf material due to the presence of reindeer among the faunal remains, the presence of *baguette demi-rondes* made of ivory, and the presence of female figurines (Feustel 1979). Except for the *baguette demi-rondes*, these characteristics also applied to the Oelknitz material

(Feustel 1979; cf. Gaudzinski-Windheuser 2011). Thus, a differentiation of the Oelknitz and the Nebra variant remained vague.

Besides organic and/or lithic typology, Rudolf Feustel considered prey choice to distinguish sub-groups in the Late Magdalenian. However, the exploitation of horse (*Equus* sp.) in contrast to reindeer (*Rangifer tarandus*) was meanwhile shown to be partially underestimated on Late Magdalenian sites (Bridault/Bignon/Bemilli 2003; cf. Gaudzinski/Street 2003). Furthermore, the preference for one of these species was probably seasonally influenced (Bridault/Bignon/Bemilli 2003; Bignon 2006; Bodu et al. 2006b). In consequence, sub-groups based on the faunal assemblages could represent environmentally and/or seasonally distinct but possibly contemporaneous groups with partially similar archaeological material and behavioural patterns (Gaudzinski-Windheuser 2012).

In northern France, the fundamental publications of André Leroi-Gourhan (Leroi-Gourhan 1964; Leroi-Gourhan 1965) introduced the concept of the *chaîne opératoire* (Leroi-Gourhan 1943; Leroi-Gourhan 1945; Leroi-Gourhan 1964; Leroi-Gourhan 1965) to the analysis of archaeological assemblages. This concept was first applied in the study of Pincevent (Leroi-Gourhan/Brézillon 1966; Leroi-Gourhan/Brézillon 1972). The *chaîne opératoire* sets the retouched artefacts into a sequence of technical decisions beginning with the choice of material and ending at the discard of the artefacts. In particular, this concept supplemented the archaeological analysis with technological studies and archaeological assemblages were hence studied in their entirety. This more complete collection of data made possible the definition of taxonomic groups of artefact combinations and their formation. For the Late Magdalenian, these techno-typological studies revealed a complex technical system aimed to produce blades and bladelets (Bodu 1993). A standardised preparation process preceded the production of these final products. Various knapping instruments were used during this process but organic hammers were assumed to be used particularly in the main production of the final products (Pelegrin 2000). With the additional refitting of the studied materials and the spatial analysis of their distribution, the studies became agent-based and in some cases suggested the instruction of apprentices by skilled master flintknappers (Pigeot 1990; Audouze/Cattin 2011).

In addition, the techno-typological and spatial analyses were further supplemented by analyses of the natural environment and the topographic setting of the sites comparable to the early German tradition. In this combination of the environmental context and the setting of the archaeological evidence into a regional context, the taxonomic groups could be tested on their emergence due to chronological successions, various environmental adaptations, different functions, and/or ethnic interactions. Based on these combined information, settlement systems could be identified in the archaeological record (cf. Binford 1980). Furthermore, considerations about a Lateglacial palaeoethnography were formulated (Schmider 1982; Pigeot 1987; Valentin 2008a).

Accordingly, Gerd-Christian Weniger established two regional sub-groups within the Middle/Late Magdalenian of the German and Swiss upland areas: a south-western and a northern group (Weniger 1989; cf. Eriksen 1991). His sub-division was based on a multi-characteristic approach including subsistence strategies, mobile special goods, and settlement patterns. The sites of the south-western group were located in southern Germany and northern Switzerland. For instance, the Lateglacial hunter-gatherers in this sub-group seemed to aggregate in larger sites in the lowlands during the autumn-winter season to hunt reindeer. During the summer season, these groups dispersed to smaller sites in the uplands and hunted mainly horse and ibex. The northern group included the sites from the Thuringian Basin and the Central Rhineland and, thus, the attribution of assemblages to a common sub-group equated the assemblages of the Nebra variant. For these assemblages, the seasonal data were scarce but in the Central Rhineland a reversed situation to the south-western group seemed possible with large autumn-winter sites focused on horse and small spring-summer sites focused on reindeer. However, in the Thuringian Basin no assemblages with significant proportions of reindeer were found.

Comparably, Terberger deduced from the presence of numerous evident structures on sites of the Nebra variant that the settlements should be interpreted as base camps (cf. Binford 1980). Based on this assumption, Terberger argued that the Nebra group was rather a chronological entity. Moreover, based on the raw material transports and ethnographic analogies, he considered an ethnic entity as possible in which Gönnersdorf and Andernach represented central places of local sub-groups which were presumably provided from camps such as Kanne or Orp located at the limits to the North European Plain. Together these sites formed part of an extended Nebra group (Floss/Terberger 2002, 138). Terberger further pointed out similarities with camp sites in the Paris Basin and set this extended Nebra group in the context of the Lateglacial expansion.

Furthermore, with the introduction of independent age measurements such as ^{14}C dating, the taxonomic groups could be tested for their chronological relevance (Ducasse/Langlais 2007; Langlais 2007; Langlais et al. 2012). Micro-wear analyses helped in addition to produce precise interpretations on the use of artefacts (Plisson 1985; Moss 1986; Rots 2005; Pétillon et al. 2011; Sano 2012b). These studies made testing of the typologies possible for differences in the functions of the assemblages.

In general, the distinction between an Early Magdalenian and a Late Magdalenian (*sensu lato*) remained. However, the Early Magdalenian developed in south-western France from the Badegoulian after or at the end of the LGM (Ducasse/Langlais 2007; Ducasse 2012). The adaptive flexibility of the Early Magdalenian is replaced by the increasing technical rigidity of the Middle Magdalenian during the Heinrich 1 event (Langlais 2007; Langlais 2011). In the sub-division based primarily on excavated sites in the Thuringian Basin, this Middle Magdalenian would equate to Rudolf Feustel's Kniegrotte variant (Feustel 1979; cf. Höck 2000) of his Oelknitz group (Feustel 1956, 25). Middle Magdalenian assemblages with narrow triangles disappeared presumably between approximately 13,200 and 13,100 years ^{14}C-BP (Thévenin 2000; Höck 2000; Ginter et al. 2005; Langlais et al. 2012). This period was paralleled by the end of the Heinrich 1 event and in this period the need to react on short-term needs was answered by some simplifications (Langlais 2008; Langlais et al. 2012). These changes resulted in highly standardised technical processes (Bodu 1993; Audouze/Cattin 2011) during the Late Magdalenian (*sensu stricto*) reflecting presumably a social system of strict conformism. This Late Magdalenian (*sensu stricto*) correlated to the other variants of Feustel's Oelknitz group (Feustel 1956; Feustel 1979). Moreover, the technology of this Late Magdalenian (*sensu stricto*), in particular, the lithic assemblage, »was greatly inspired by the Middle Magdalenian« (Langlais 2008, 231; cf. Langlais et al. 2012) which reflect the observation of Denis Peyrony that the Late Magdalenian (*sensu lato*) lithic assemblages are very similar (Capitan/Peyrony 1928). In Northern Europe, the independent character of the Final Magdalenian inventories was subsequently emphasised in terms such as Creswello-Hamburgian or shouldered point technocomplexes (e. g. Burdukiewicz 1981; Desbrosse/Kozłowski 1989). Furthermore, based on a correspondence analysis of retouched artefacts from various Lateglacial inventories, Bruno Bosselin and François Dijndijan clearly distinguish the Hamburgian and Creswellian from the Magdalenian by the absence of backed bladelets in the inventory (Bosselin/Djindjian 1988). Considering the reoccupation of Northern Europe as a social process sustained by comparative technological studies led to a reuse of the term Final Magdalenian for these northern inventories (Barton et al. 2003; Pettitt/Rockman/Chenery 2012; Weber 2012). Furthermore, comparable inventories were meanwhile found in the Paris Basin and attributed to a special variety of a Late or Final Magdalenian (*sensu lato*) named »*faciès Cepoy-Marsangy*« (Schmider 1982; Valentin 2008a). These similar developments in Northern Europe and northern France suggested some kind of connection between these regions and a related development (cf. Valentin 2008a; Weber 2012). Moreover, the presence of this sub-group in the Paris Basin makes a clear differentiation of the Final Magdalenian assemblages from the Late Magdalenian assemblages such as those from the unit IV

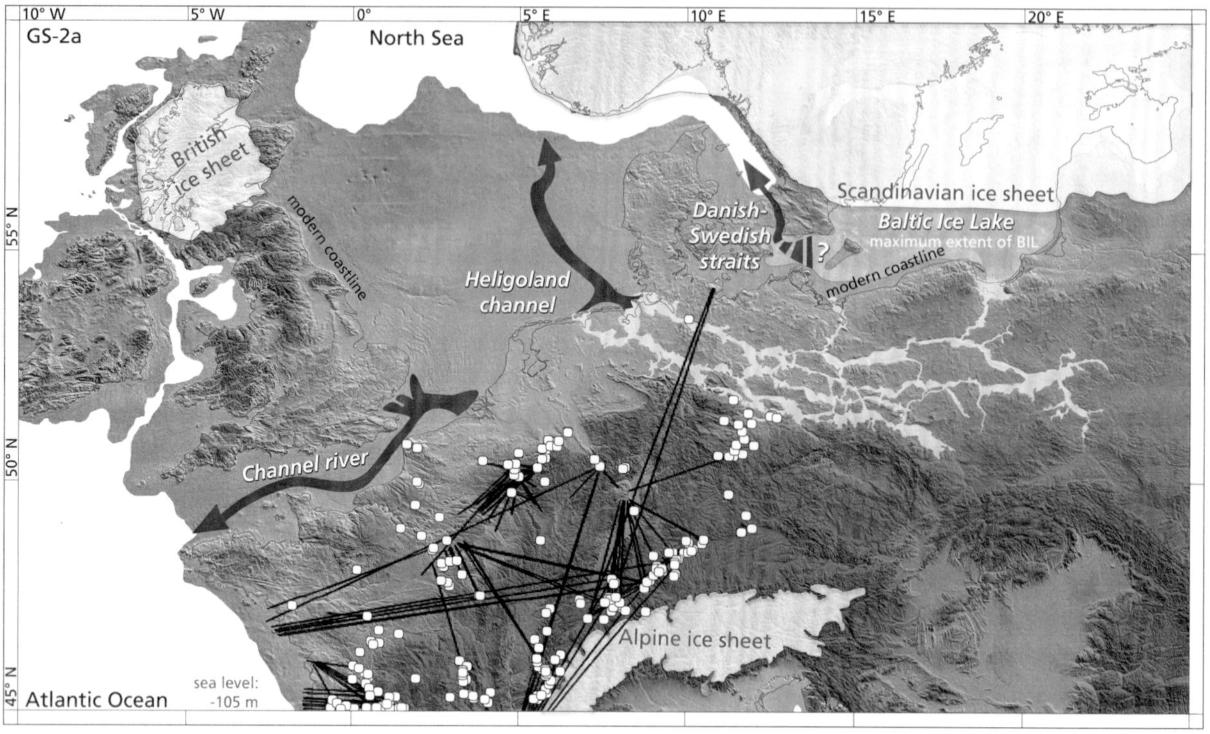

Fig. 24 Connections between Late Magdalenian (*sensu lato*) sites and source areas of raw material for personal ornamentation according to Schwendler 2012, 340 fig. 6 (cf. Weniger 1989, 361 fig. 8; Połtowicz 2007, 24 fig. 2; Widlok et al. 2012, fig. 8).

in Pincevent or several units in Étiolles possible (Pigeot 1987;Bodu 1993; Valentin 1995; Valentin/Pigeot 2000). However, the Northern European inventories were clearly dated to the transition period analysed in the present study (Barton et al. 2003; Grimm/Weber 2008) with some possible dates as early as 12,700 years ^{14}C-BP (Jacobi/Higham 2009). Consequently, the comparable inventories from northern France are incorporated in the present study (see p. 189-196, p. 204-210, and p. 220-226) and the Final Magdalenian is considered one of the earliest stages of the Azilianisation in the present study.

Thus, the Late Magdalenian as understood in the present study, is a Late Magdalenian *sensu stricto*. Sites of this Late Magdalenian are common in north-western Europe south of the English Channel and of the North European Plain suggesting this phase as a first phase of a permanent expansion into northern territories (cf. Housley et al. 1997; Gamble et al. 2005; Miller 2012). The connections of some raw materials discarded at these sites to the original source areas suggested that this large scale expansion process maintained a large network across north-western Europe (**fig. 24**; cf. Schwendler 2012). Independent of the interpretation of these materials as direct import or as trade goods, these long distance connections displayed information routes. In combination with the similarity of the archaeological material and behavioural patterns reflected by the settlement systems, these information routes indicated a shared social superstructure. Sites of this social network were found concentrated on the upland zones of north-western Europe and larger clusters of sites occurred particularly in relation to large river systems such as the Seine, the Meuse, the Rhine, the Rhône, the Danube, the Saale-Elbe, and the Vistula. However, watersheds (see **fig. 19**) were crossed to expand across the wide geographic distances from the Atlantic to eastern Poland (Połtowicz-Bobak 2009; Schwendler 2012). This Late Magdalenian occurred in a period of approximately 13,200 to 12,700 years ^{14}C-BP (cf. Street/Jöris/Turner 2012). These assemblages were related to the Late Pleniglacial cold and dry grass steppe. Late Magdalenian sites from the Paris Basin and Poland occurred clearly in the same type of

environment but the sampled material produced several younger ^{14}C dates (Débout et al. 2012; Połtowicz-Bobak 2012).
Some young dates from the Paris Basin (and Poland?) indicate perhaps that the dated material should be reviewed concerning its attribution to a Late Magdalenian *sensu stricto* or a later development. However, many of the young dates were probably unreliable due to various technical and contextual reasons. For instance, contaminations due to technical causes during the AMS dating procedure (e. g. Wohlfarth et al. 1998; Higham/Jacobi/Bronk Ramsey 2006) or due to contextual admixture such as bulked samples are common reasons for unreliable dates. A rigid evaluation of the existing databases is necessary to clear the record from these unreliable results (cf. Street/Terberger 2004; Grimm/Weber 2008; Jacobi/Higham/Lord 2009). Reviewed databases yielded increasingly precise and, hence, solid patterns of past developments (e. g. Charles 1996; Housley et al. 1997; Barton et al. 2003; Verpoorte 2004; Gamble et al. 2005; Grimm/Weber 2008; Jacobi/Higham 2009; Verpoorte/Šída 2009; Jacobi/Higham 2010).
Among the archaeological material, lithic inventories were the most commonly preserved pieces of evidence. Likewise the observations in south-western Europe, the assemblages attributed to the Late Magdalenian (*sensu stricto*) in north-western Europe were comparable concerning their inventories and their technical systems. In general, the lithic inventories were based on blades and bladelets as blanks. Technological analyses of Late Magdalenian sites in the Paris Basin demonstrated that a careful preparation procedure was an essential part of the blank production (Bodu 1993). This process was very standardised (Pigeot 1990; Audouze/Cattin 2011). Even though these standards were adapted to the material present in areas with fewer raw materials and/or material of poorer quality, the normative process was still detectable (Floss/Terberger 2002; Audouze/Cattin 2011).
As observed already by Denis Peyrony (Capitan/Peyrony 1928), Late Magdalenian inventories comprised various types of standardised tool types such as dihedral burins, burins on truncation, backed bladelets, end-scrapers on blades, borers, and composite tools. In general, the LMP class dominated among the retouched artefacts corresponding to the observation of Rudolf Feustel concerning the Oelknitz group (Feustel 1956; Feustel 1979). Some special sub-types such as *Lacan* type burins or elongated borers could indicate further possible sub-groups (or variants as Feustel termed them), perhaps, with some chronological relevance (cf. Street/Jöris/Turner 2012).
In general, the backed bladelets as well as the various organic points (bevelled points, barbed points) were assumed to represent projectile implements, even though no direct evidence for the use of projectiles in the Late Magdalenian was found. However, finds from the Late Magdalenian site Pincevent indicated a composite use of these implements (Leroi-Gourhan 1983). These composite projectiles were considered too heavy for the use as arrow heads (Pétillon et al. 2011). In addition, the use of spears and spear-throwers (atlatl) was also assumed for the Late Magdalenian due to comparable finds of organic and lithic projectile implements in Lower Magdalenian assemblages which also contained spear-throwers (Pétillon et al. 2011). Moreover, a spear-thrower from the south-western French site Isturitz was dated to the early Lateglacial Interstadial (Szmidt et al. 2009) suggesting at least the further occurrence of this technology in this area. Furthermore, no unambiguous evidence for bow-and-arrow technology has been found thus far dating to the time before the end of the Lateglacial Stadial. A pine fragment from Mannheim-Vogelstang was discussed as a fragment of a children's bow and dated to approximately the Middle Magdalenian/early Late Magdalenian (GrA-precise lab. no. not given: 14,680 ± 70 years ^{14}C-BP; Rosendahl et al. 2006). However, the classification as bow as well as the dating (stratigraphy; preservatives) are questionable and the use of old wood in later periods such as indicated by a pine shaft in Grieben (cf. Grünberg 2006) cannot be excluded. Nevertheless, the appearance of shaft smoothers was also considered as indirect evidence for the presence of arrows (Farmer 1994). These pieces were assumed to be used for straightening and smoothing wooden

shafts of projectiles (Rozoy 1978; Riede/Kristensen 2010) mainly based on ethnographic observations (Flenniken/Ozbun 1988). Even though the regular occurrence of these tools was noticed on sites in north-western Europe from the early Lateglacial Interstadial onwards (e.g. Luttenberg, Stapert 2005; Niederbieber, Loftus 1982; Le Closeau, *locus* R, Bodu 1998), first specimens were associated with early assemblages of the Late Magdalenian such as a fragment from the Polish site Dzierżysław 35 (Ginter/Połtowicz 2007). Moreover, this site was typologically attributed to the earlier phase of the Late Magdalenian or a Middle Magdalenian due to small triangles comparable to the pieces from the Kniegrotte. This typological attribution was further sustained chronologically by some radiocarbon results (Ginter et al. 2005; Wojtal 2007). In regard to the previous considerations, these microlithic triangles were possibly used as arrowheads. Perhaps, these rare evidences indicated that the knowledge about bow-and-arrow technology was present along with the concept of atlatl technology during the Late Magdalenian expansion.

On sites with a good preservation, organic tools or remains of their production were frequently found in Magdalenian assemblages. Therefore, these organic implements were also chosen as one of the first distinctive elements of Late Magdalenian assemblages. Commonly, organic remains in Late Magdalenian contexts comprised various types of hunting equipment such as barbed points, bevelled rods/points, baguettes demi-ronds as well as working tools such as awls, needles, or bâtons (Tinnes 1994; Lompré 2003; Pétillon 2006).

Major game species recovered at Late Magdalenian sites were horse (*Equus* sp.) and reindeer (*Rangifer tarandus*), often in large quantities on residential sites (Bridault/Bignon/Bemilli 2003; Bignon 2006; Street et al. 2006; Street/Turner 2013). These clearly dominant prey species were supplemented occasionally by bovids, mainly *Bison priscus*, and rarely saiga antelopes (*Saïga tatarica*) as well as mountainous species such as *Capra ibex* or *Marmota marmota* and, in addition, red deer (*Cervus elaphus*) in the southern regions. In northern regions red deer was usually only identified by teeth which probably reached the sites as ornamental items. Equally, mammoth (*Mammuthus primigenius*) and woolly rhinoceros (*Coelodonta antiquitatis*) remains often seemed to be collected and/or exchanged fossil items, but rarely could they also be attested as alimentary resource such as on the Polish site Wilczyze (Bratlund 2002). Additionally, the game frequently also included arctic hare (*Lepus timidus*), various fish species such as grayling (*Thymallus thymallus*), burbot (*Lota lota*), or salmon (*Salmo trutta/Salmo salar*), and birds, in particular, from the genera *Lagopus*, *Cygnus*, and *Anser*. Furthermore, carnivores were commonly identified in the Late Magdalenian assemblages, dominant among them is arctic fox (*Alopex lagopus*) but also canides (*Canis lupus* and/or *Canis familiaris*), bears (*Ursus spelaeus* and/or *Ursus arctos*) and sometimes hyena (*Crocuta* sp.).

Thus, a clear preference of prey and orientation on animals which occur in large herds in open landscapes such as horse and reindeer were observable. However, numerous supplemental fauna species revealed that the Late Magdalenian economy did not represent a specialised hunting community but in contrast showed generalised and flexible subsistence patterns (cf. Gaudzinski/Street 2003). Furthermore, some species were probably trapped specifically for further resources such as fur or jewellery (Álvarez Fernández 1999; Street/Turner 2013).

These flexible subsistence patterns and the considerations of Thomas Terberger about the sites of the extended Nebra group (Floss/Terberger 2002, 138) suggested that a single specific type of site alone cannot be attributed to the Late Magdalenian. Instead, various types of sites maintained a subsistence and settlement system. Moreover, the different functions of the single sites resulted in the absence of some settlement patterns such as pavements on various sites as well as to the absence of some categories of material such as art. For instance, portable art was lacking almost completely on the large French open air sites. At Étiolles, an engraved limestone was recovered as part of a stone ring around a hearth and represented rare evidence of Late Magdalenian art on open air sites of the Paris Basin. Other figuratively engraved stones such as the pieces from Pincevent or Cepoy (see p. 221 and p. 235) were clearly attributed to the Lateglacial Interstadial and archaeological assemblages which were distinct from the typical Late Magdalenian. By contrast, in the Central Rhineland, figurines and rich figurative art on schist plates were found on the Late Magdalenian sites (see p. 89 and p. 110f.). However, as mentioned, these sites were considered as base camps, whereas in the Paris Basin this simple model of base and special task camp was questioned (Bodu et al. 2011; Fougère 2011). Nevertheless, the large camp sites such as Pincevent or Étiolles were assumed to be regularly visited camps which were used flexibly for hunting as well as living (Julien/Karlin/Bodu 1987; Olive 2004; Bodu et al. 2009a; Fougère 2011). Based on Swiss Late Magdalenian sites, comparable suggestions were made that residential camps were moved to places of successful hunting episodes (Müller et al. 2006). However, the time spent at these sites appeared shorter than at the typical base camps. Nevertheless, combined techno-spatial analyses demonstrated the presence of skilled and unskilled, probably apprentice flintknappers on sites in the Paris Basin (Pigeot 1987; Bodu 1993; Valentin/Pigeot 2000) and suggested the accomplishment of further tasks such as transmitting flintknapping skills besides the hunting episode. Furthermore, pieces interpreted as jewellery were also found on these sites and indicated, as with the typical base camp sites, the particular long distance relations such as molluscs from the Atlantic in the Paris Basin (Alvarez Fernández 2001; Bodu et al. 2006b, 36), fossil molluscs from the Paris Basin in the Belgian Late Magdalenian sites (Dewez 1987; Álvarez Fernández 2009), Mediterranean molluscs in the Central Rhineland (Álvarez Fernández 2009), or Baltic amber pieces in Swiss assemblages (Schwab 1985; Leesch 1997; cf. Eriksen 2002). However, personal ornaments such as perforated animal teeth or molluscs and colourants are frequently present in considerable quantities, often of good quality, and also in large variety at Late Magdalenian sites (Álvarez Fernández 2009).

In conclusion, the Late Magdalenian (*sensu stricto*) can be distinguished by its comparable lithic inventory based on blade blanks and comprising various types of standardised tool types. Furthermore, organic material was used for various artefacts which again were formed in specific types such as bâtons, needles, or points. These generally very strictly defined types seem to reflect a very conformist behaviour. The procurement distances of raw materials as well as the varied degree of evident structures on sites suggested a complex settlement system. The origin of the various raw materials mark the Late Magdalenian as having a well connected social system.

The *Federmesser-Gruppen*

In the Central Rhineland, the material from the upper or younger horizon found directly underneath the LST was classified as *Federmesser-Gruppen* (FMG). At present the term FMG is often used to describe an agglomeration of archaeological groups in the European Final or Late Palaeolithic.

Originally, the FMG were defined by Hermann Schwabedissen based particularly on the material from several concentrations around Rissen near Hamburg (Schwabedissen 1939, 143) and Prandinge in the Dutch province of Friesland (Schwabedissen 1944b). Successively, he added some 40 lithic assemblages, mainly from the western part of the North European Plain, to the FMG (Schwabedissen 1944a) and published this group in a monographic volume (Schwabedissen 1954). According to him, these inventories were characterised by (Micro-) Gravette points and/or *Federmesser* (Schwabedissen 1944a, 116-119), burins of diverse but often crude types (Schwabedissen 1954, 61), and numerous end-scrapers that usually occurred as varied types and regularly as double end-scrapers (Schwabedissen 1944b, 51). Even though Hermann Schwabedissen thought the FMG were a regional sub-group of the Final Magdalenian and connected with the Magdalenian expansion into Central and Northern Europe (Schwabedissen 1944b), organic implements such as

name of archaeological group	eponymous/defining site	main reference(s)	defining elements
Azilian	Mas d'Azil (Ariège); Abri Villepin (Dordogne)	Piette 1889; Piette 1895; Peyrony 1936	lithic industry of Magdalenian habit but different tool composition: Azilian points, short & round end-scrapers and few burins; organic tools made of red deer antler, in particular, flat & perforated harpoons (i. e. points with two rows of barbs); painted or engraved gravels with geometric signs
Tjonger Group (previously: Kuinder culture)	sites alongside the Tjonger (Friesland) such as Prandinge, Donkerbroek, and Makkinga	Popping/Beijerinck 1933; Bohmers 1947; Bohmers 1961	various types of backed pieces with different dimensions; numerous end-scrapers, particularly short ones; various types of burins, often on truncation; flakes were the commonly used blank type
Federmesser-Gruppen	Rissen (Hamburg); Prandinge (Friesland); Wehlen (Niedersachsen)	Schwabedissen 1939; Schwabedissen 1944b; Schwabedissen 1954	Magdalenian-like lithic industry; numerous Gravette points, *Federmesser*, and backed blades; numerous end-scrapers on blades, often double end-scrapers but also short end-scrapers; many various types of burins
Tarnowian (including the Ostroměř group)	Tarnowa (Województwo wielkopolskie); Ostroměř (Hradec Králové)	Krukowski 1939; Vencl 1970a; Kozłowski/Kozłowski 1996, 83 f.	Magdalenian tradition; rare backed pieces of various types; numerous small end-scrapers; rare burins of small dimension & various types made on thick blanks; blades and elongated flakes were preferred blanks
Witowian	Witów (Województwo łódzkie)	Chmielewska/Chmielewski 1960; Chmielewska 1961; Chmielewski 1961; Kozłowski/Kozłowski 1996, 81-83	Epigravettian-like lithic industry; very small and variable backed pieces incl. bipointes in particular of a compact shape (Taucha type); small end-scrapers predominate; burins of various types usually made on thick blanks; flakes were the commonly used blanks
Penknife point phase	Mother Grundy's Parlour (Derbyshire); Robin Hood's Cave (Derbyshire)	Campbell 1977; Jacobi 1988; Barton/Roberts 1997; Barton/Dumont 2000	Creswellian tradition; more numerous but smaller assemblages than Creswellian; great variability of backed pieces but penknife points are characteristic; short end-scrapers, some round thumb-nail end-scrapers; burins on truncation; preferred blanks were small blades; soft hammerstone mode of percussion; varying overall standards and quality of lithic raw materials
Hengistbury Head type industries	Hengistbury Head (Dorset)	Barton 1992; Conneller/Ellis 2007; Barton et al. 2009; Pettitt/White 2012	straight-backed blades & bladelets; angle-backed points, shouldered points, and large tanged points; numerous end-scrapers which were often short; burins on truncations made on thick blanks; blades are preferred blanks; use of varied percussion instruments but dominantly soft hammerstone

Tab. 9 Characterisation for some selected FMG sub-groups.

bone or antler tools which were characteristic for the Magdalenian (Breuil 1913) were not considered in his definition due to the scarce preservation of such items in northern Germany, the Netherlands, and northern Belgium. However, several decades before Schwabedissen's publication, other groups of archaeological material which post-dated the Late Magdalenian were published. For instance, the Azilian was already defined in the late 19th century based on south-western French material (Piette 1889; Piette 1895; Peyrony 1936). The Azilian was regarded by Schwabedissen as another regional sub-group of the Final Magdalenian. According to Schwabedissen, the Azilian was distinct from the FMG by the presence of flat and perforated organic harpoons (i. e. biserially barbed points) and painted gravels (Schwabedissen 1954, 79).
The painted gravels are still unknown from unambiguous FMG contexts in north-western Europe. However, some painted stones and stone fragments were found in relation with Magdalenian horizons in southern Germany but several of these pieces represent presumable frost fragments of parietal art (Conard/Floss 1999; Conard/Malina 2011). In addition, although some single organic finds from the North European Plain and the adjacent uplands were meanwhile attributed chronostratigraphically to the FMG period (Veil et al. 1991; Terberger 1996; Clausen 2004; Grünberg 2006; Cziesla/Masojć 2007), organic artefacts are

thus far generally lacking from unambiguous FMG assemblages in Northern Europe. Only the sites in the Central Rhineland yielded organic artefacts which were clearly associated with FMG lithic assemblages. The organic items found in Northern Europe as well as in the Central Rhineland were very diverse but, in general, barbed points were found more frequently than other types. In contrast to the south-western French Azilian harpoons, the barbed points from north-western Europe were usually not perforated and of a greater variety. This variety referred to the general shape with uni- and biserially barbed points as well as to the morphology with denticulated barbs and finely carved barbs (cf. Baales 2002, 274-281; Tinnes in Baales 2002, 271-273; Cziesla/Pettitt 2003; Cziesla 2007b). Nevertheless, since barbed points were also a defining element for the Azilian, these implements can be regarded as one type of organic tool which was used across north-western Europe in the Lateglacial Interstadial and, thus, as one of the commonalities of these hunter-gatherer groups.

However, in contrast to the Late Magdalenian, the FMG material displayed a greater individuality and less standardisation in the shaping of these tools. This weakened conformism applied also to the lithic material, in particular, to the backed points. Perhaps, this decreased conformism after the highly standardised behaviour of the Late Magdalenian explains why the classification of the FMG/Azilian is still a matter of debate (Bodu/Valentin 1997; Baales 2002, 56-58; Heinen 2005, 102-109; Kegler 2007, 297-299).

Moreover, the observation of a decreased conformism among the FMG could also explain why so many regional and/or chronological sub-groups were defined, even though the general pattern of the archaeological material was similar (**tab. 9**). These various groups of Lateglacial assemblages were mainly defined during the late 19th and early 20th century when the influence of European nationalisms was high. In contrast, during the last decades of a unified Europe, the uniting aspects were more emphasised, also in archaeology, resulting in most of the assemblages from France to Poland and Bohemia (Vencl 1970b; Barton/Roberts 1996; Bodu/Valentin 1997; Kabaciński/Sobkowiak-Tabaka 2010; Riede/Laursen/Hertz 2011) and occasionally even to western Russia (Zhilin 1996) being attributed to the FMG as a common meta-group. Thus, the refusal or adoption of the term FMG across a wide geographical range was probably further influenced by tendencies of contemporary scientific policies to weight separating or uniting attributes in the various taxonomical classifications. The continuation of modern traditional preferences in the scientific community still finds its expression in the choice of using further names for the meta-group such as Azilian (Piette 1889; Piette 1895; Peyrony 1936; Kegler 2007; Valentin 2008a), Rückenspitzen-Kreis (Ikinger 1998), Arch-backed point technocomplex[10] (Schild 1984; Schild 1996), or Curve-Backed Point (CBP) groups (Kozłowski 1987).

Certainly, this wide adoption of a common term signalised an advanced expansion into Northern Europe in contrast to the Late Magdalenian. In addition to the areas occupied by the Late Magdalenian, territories settled by the meta-group in the Lateglacial Interstadial encompassed also central and southern Britain as well as the North European Plain. In general, these human groups were assumed to descend from a Magdalenian tradition which adapted to various landscapes during the Lateglacial expansion into north-western Europe and/or to a changing environment during the Lateglacial amelioration (cf. Schwabedissen 1954; Bodu/Valentin 1997; Coudret/Fagnart 1997; Baales 2002). In most of these groups an increasing importance of end-scrapers and the use of flexible knapping techniques on local materials were observed accompanied by a decreasing importance of burins and an elaborated blank production process. However, the definitions of the sub-groups greatly varied concerning the details used to define the group due to the material avail-

10 Arch-Backed Point Technocomplex were abbreviated as ABP (Schild 1984). This abbreviation could be easily confused with the term angle-backed points which refers to the Creswell and Cheddar type backed points. These angle-backed points were usually attributed to an older context (Creswellian, Barton et al. 2003; Jacobi 2004) and set in relation to the classic phase of the Hamburgian (shouldered point complex, Burdukiewicz 1986).

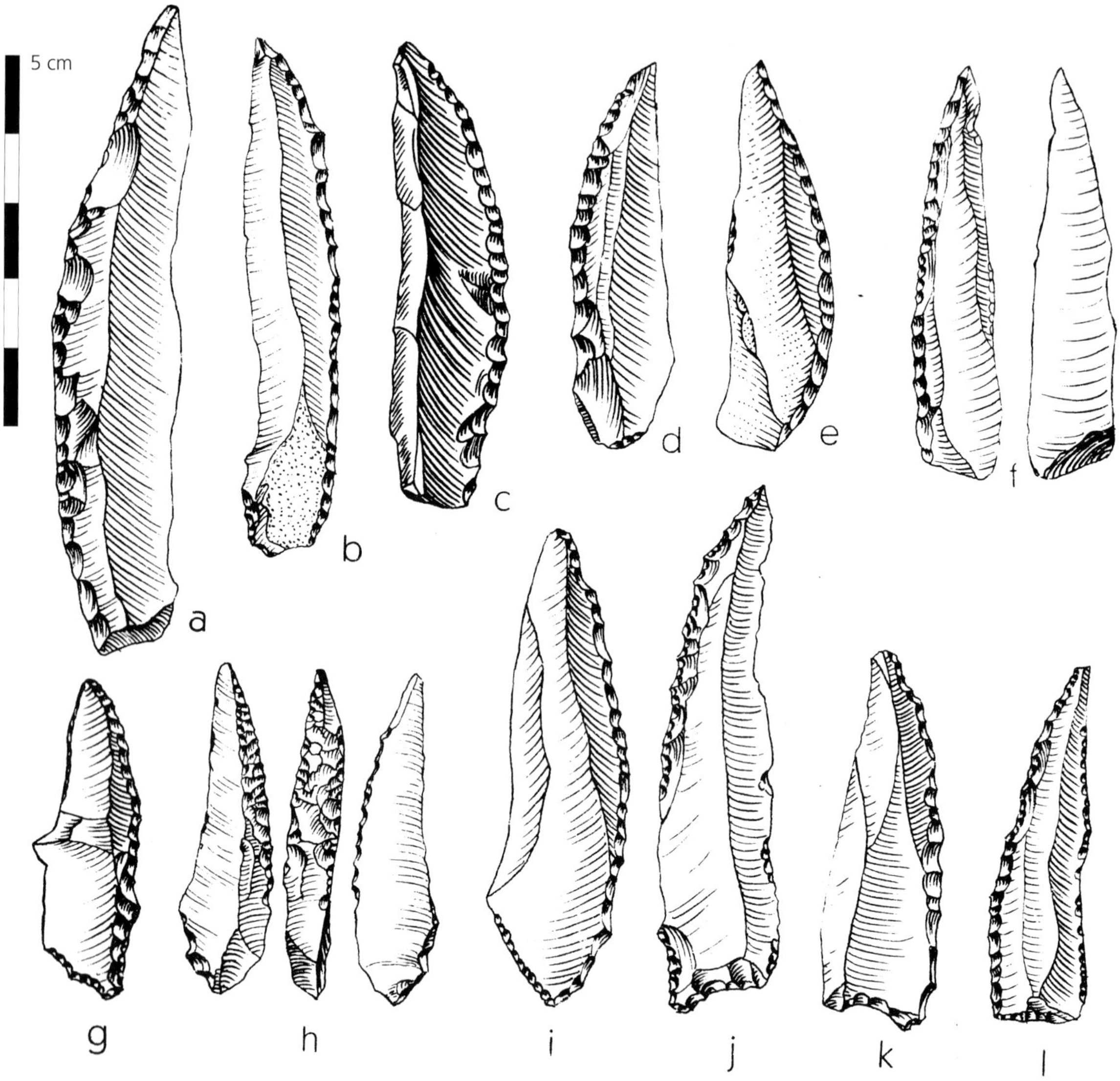

Fig. 25 *Federmesser* according to Hermann Schwabedissen (Schwabedissen 1954, 23 Abb. 11). **a** large backed blade (*couteau à dos*); **b**, **d**, **g-j** penknife point (*pointe à base rétrécie*); **c**, **f** *Federmesser* (*pointe à dos courbe et sans aménagement de la base*); **k-l** Malaurie point (*pointe à base tronquée*). – Point e displays a basal impact fracture and is therefore not classified. Perhaps, the same problem affects the points a-d and, thus, the classification of the pieces b-d remains provisional. However, the specimen a is clearly larger than usual and the attribution as a *couteau à dos* is clear. The point j falls in the transition from penknife point to Malaurie point.

able for classification and the focuses of research during the time when the definitions were formulated. In general, the distinction was based on the projectile implements assuming that these objects were subject to quicker changes due to fashionable reasons and the dependence on the changing Lateglacial environment. These assumed quicker changes would make a finer chronological and/or environmental distinction possible. Thus, precise and applicable definitions of sub-groups should be based on chronostratigraphically reliable material from a known environmental range.

The majority of the lithic assemblages used by Schwabedissen were surface collections which he regarded as reliable due to his expertise in lithic artefacts (cf. Schwabedissen 1955). However, these sites provided little or no environmental or chronostratigraphic information. In addition, the few excavated assemblages (n = 14) in his monograph were documented according to the standards of the time that, at least in the case of Andernach-Martinsberg, were proven to be insufficient by later works (Bosinski/Hahn 1972; Veil 1982). Particularly, these insufficient documentations usually prohibited a spatial analysis of the sites, including the

testing of palimpsest situations. However, this testing would be necessary, in particular, for the many sites which were found in sands because modern studies of analogous sites often showed the admixtures of various natural and anthropogenic events (cf. De Bie/Van Gils 2006). Thus, by modern standards the material from the standard volume about the FMG has to be rejected completely as a source of reliable comparisons. To define the FMG, other assemblages such as those sites from the Central Rhineland are required.
Furthermore, Hermann Schwabedissen yielded only a meagre catalogue of attributes to identify the FMG among which the *Federmesser* was the most distinctive one. However, his definition of the term *Federmesser* referred to any curve-backed monopoint (**fig. 25**; Schwabedissen 1954, 8. 23 Abb. 11; cf. Schwabedissen 1944a, 118f.). Schwabedissen adopted the term from Robert Rudolf Schmidt who had introduced the term *Federmesser* as a piece with one steeply retouched, arched formed edge (Schmidt 1912, 114) which is comparable to Gravette points but smaller than these mid-Upper Palaeolithic implements (Schmidt 1912, 136). However, Schwabedissen emphasised that the tip of the *Federmesser* in difference to a Gravette point was rather aligned with the (unretouched) lateral edge (Schwabedissen 1954, 8). Furthermore, the retouched edge should not contain an angle or a shoulder. Generally, these bladelets with curved lateral retouch had no basal modification even though various types of basal retouch even to the form of pointed tangs were acceptable for Schwabedissen (Schwabedissen 1954, 8).
Thus, many point types such as small Châtelperron or Micro-Gravette points (cf. Bohmers 1961, 23), Malaurie points, penknife points (cf. Fischer 1991, 102), as well as some special pieces of shouldered points (»tapered points«; Johansen/Stapert 2004, 35) were subsumable in the category *Federmesser sensu* Schwabedissen. He only differentiated the »*Halbmondmesser*« as a special type of *Federmesser*. These implements were equivalent to segments or bipoints.
Comparably, the Azilian point was previously very vaguely defined (Peyrony 1936; Bohmers 1956, 11; Sonneville-Bordes/Perrot 1956, 556) leading to the same uncertainty as the term *Federmesser*. Moreover, based on the finds from the rock shelter Villepin, bipoints were also included in the spectrum of the term Azilian point (Peyrony 1936). Therefore, the Azilian point also became widely synonymous with curve-backed point. Moreover, in south-western France a continuous development from the Magdalenian to the Mesolithic was assumed with the Azilian representing the Pleistocene part of this development (Breuil 1913). To distinguish the different stages of this development, various phases of the Azilian were described such as the *Azilien ancien*, *Azilien moyen*, *Azilien supérieur*, *Azilien récent*, *Azilien tardif*, *Azilien final*. However, mainly based on the stratigraphic results from eastern France (Girard/Bintz/Bocquet 1981; Pion 1990), a general bisection was suggested for the French Azilian with an older phase (*Azilien ancien*) dominated by bipoints and a younger phase dominated by monopoints (*Azilien récent*). In western Germany, Michael Baales comparably suggested to distinguish between the FMG and the Azilian by the definition of the two different types of points with the Azilian comprising the inventories with bipoints and the FMG the inventories with *Federmesser* (Baales 2002, 56-58). In contrast, Jan Kegler argued that in the eponymous inventory from Mas d'Azil no typical bipoints occurred (Kegler 2007, 298f.) and, consequently, the terms FMG and Azilian in a strict sense describe, in fact, the same type of inventories.
Furthermore, based on results from the Somme valley, Paule Coudret and Jean-Pierre Fagnart distinguished three phases within the FMG which they described as an initial phase with diverse armatures and two further phases with *Federmesser* dominant among the backed points (Coudret/Fagnart 1997). However, these *Federmesser* resembled again the definition of Schwabedissen containing also Malaurie points and *couteaux à dos*. Their intermediate phase was comparable to the former but the monopoints established as the more dominant LMP. Sites such as Saleux La vierge Catherine were attributed to the last phase of this FMG succession (Coudret/Fagnart 1997). In this phase assemblages yielded very comparable material to the FMG defined by Michael Baales based on the Central Rhineland sites (Baales 2002).

However, based on the stratigraphic sequence from the cave Bois Ragot at Gouex (Vienne) and the rock shelter Pont d'Ambon (Dordogne), a succession from the Late Magdalenian, the *Azilien ancien*, the *Azilien récent* to the Laborian characterised by Malaurie points was shown (Célérier/Chollet/Hantaï 1997). Moreover, in the sequence from Pont d'Ambon, the *Azilien récent* phases were characterised mainly by penknife points (*pointes à base rétrécie*; model 7) and the common Azilian point (model 1) which was described as »curve-backed without modification of the base« [11] (Célérier/Nisole/Beaune 1993, 89). Furthermore, the frequency distribution of points in Pont d'Ambon demonstrated the almost constant presence of this common Azilian point type or *Federmesser sensu stricto* from the Final Magdalenian to the Laborian (**fig. 26**; Célérier/Nisole/Beaune 1993).
This constant presence of *Federmesser s. s.* sustained on the one part the subsumption of the complete Lateglacial traditions with backed points as FMG [12]. However, on the other part this constant presence revealed that this type of implement was a poor choice as distinctive element in a chronological sequence. Moreover, as the wide geographical adoption of the term FMG had shown, this implement is also not well suited to identify geographic sub-groups (cf. Ikinger 1998, 45-54. 196-218).
The Pont d'Ambon sequence showed that for the classification of various backed points the number of pointed ends (mono- or bipoint) and the morphology of the basal part were particularly distinctive factors, besides the forming of the blunting lateral retouch (curve-, angle-, or straight-backed; see p. 275-282; cf. Brézillon 1968; Célérier/Nisole/Beaune 1993, 89-98; Ikinger 1998). In the context of projectile technology and hafting of projectile implements, these distinctions are of technical importance (Valentin 2008a, 144-160; cf. Grimm/Jensen/Weber 2012). Moreover, assuming that the breakage of the most valuable part was to be avoided, a validation of the various components of the projectile can be read from the various shapes of the lithic projectile implements and their potential predetermined breaking points.
However, the more precise distinction of curve-backed points (Célérier 1979; Célérier/Nisole/Beaune 1993, 89-92) and their stratigraphic significance (Célérier/Chollet/Hantaï 1997) could also explain the variously formulated impression that assemblages attributed to the FMG were relatively variable. In particular, the morphology of backed points, their frequency, and the proportion of backed bladelets were considered as variable factors (Taute 1963, 103-107; Barton et al. 2009). These factors were previously suggested to be functional, chronological, and/or traditional. For instance, Stefan Karol Kozłowski pointed out that smaller backed point types were more numerous in the mountain ranges of eastern Germany, the Czech Republic, and Poland, whereas larger specimens were predominant on the eastern German and Polish part of the North European Plain (Kozłowski 1987). Accordingly, the microlithic varieties of backed points on the distribution maps of Eva-Maria Ikinger were generally more frequent in the southern and mountainous areas than in the northern lowlands (cf. Ikinger 1998). Besides traditional preferences and/or environmental adaptations, the availability of high-quality lithic raw material and the applied blank production process could, in this case, have provoked the different proportions.
Thus, in addition to the composition of lithic tool inventories and particularly the LMP variety, the applied blank production process and the techniques used therein became a focus of studies examining the changes in the Lateglacial. For instance, characteristics found on cores and blanks were assumed to result from different knapping instruments and reflect various knapping concepts (cf. Bodu 1993, 43-49; Madsen 1996; Pelegrin 2000; Valentin 2000). In contrast to the shaping of projectile implements, these fundamental concepts

11 The original reads: »dos courbe sans aménagement de la base« and was translated by the present author.

12 Clearly, since the terms *Federmesser* and (common) Azilian point refer to the same implements, the same argumentation can be used to subsume the Lateglacial assemblages under the term Azilian.

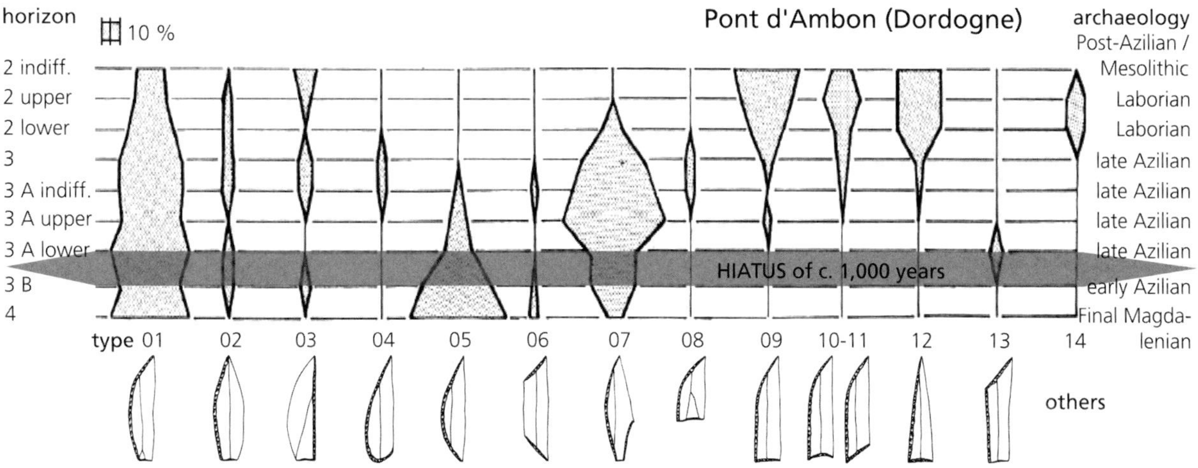

Fig. 26 Frequency distribution of various point types at the rock shelter Pont d'Ambon (Dordogne; modified after Célérier/Nisole/Beaune 1993, 92 fig. 32 and Célérier 1979, 461 fig. 1). Point types (*modèles*) are: **1** common Azilian point which is curve-backed without modification of the base; **2** symmetric point which has a curve-backed retouch symmetrical to the opposite, unretouched edge; **3** bulbous point which is a straight-backed point with a convex, opposite edge; **4** point with rounded base is a curve-backed point with a retouch forming a rounded or ogival base; **5** segment or bipoint which is formed by a regularly curved retouch affecting tip and base; **6** double-truncated point with a usually unretouched edge between the two oblique truncations; **7** point with narrowed base which is a curve-backed point with an oblique retouch on the base causing an almost tang-like impression; **8** short point is the shortened variant of type 1 meaning the length-width proportions of the point are <2:1; **9** typical Malaurie point with an almost straight-backed curved retouch and straight basal retouch; **10** Malaurie point with concave basal retouch; **11** Malaurie point with oblique basal retouch; **12** straight-backed point with various basal modifications (comparable to a Gravette point); **13** angle-backed point; **14** others.

were assumed as a generally more stable part of the *chaîne opératoire*[13], within a single archaeological group (Valentin/Pigeot 2000, 132) as well as in a diachronic perspective (Karlin/Julien 1994; Valentin 1995, 19-47). For the Paris Basin, Boris Valentin demonstrated a change of knapping instruments in the phase of producing the preferred blanks from the Late Magdalenian towards the Late Azilian by the variation of characteristics (Valentin 2008a, 128). This development began with stigmatas on the blanks and cores suggesting the use of mainly organic hammers in the Late Magdalenian (Valentin 2008a, 48), followed by an increasing number of artefacts displaying characteristics of a soft hammerstone (Valentin 2008a, 123-130) such as flint nodules with a calcareous cortex or various types of argilites, and the almost exclusive occurrence of indications for the use of soft hammerstones during the *Azilien récent* (Valentin 2008a, 48-51). This change led to an increased flexibility in the blank production process that resulted in a lower productivity of standardised blanks which were commonly selected for the projectile implements, whereas for the other categories of retouched artefacts also unstandardised by-products of the blank production were used (Valentin/Julien/Bodu 2002).

The uniformity of the various FMG sub-groups and, thus, of the FMG meta-group becomes apparent on a more abstract comparison with the Late Magdalenian. The quality of the lithic raw materials were of minor importance and the frequency of locally available raw materials increased (Coudret/Fagnart 1997;

13 The *chaîne opèratoire* (Leroi-Gourhan 1943; Leroi-Gourhan 1945; Leroi-Gourhan 1964; Leroi-Gourhan 1965; Audouze/Leroi-Gourhan 1981; Pigeot 1987; Audouze 1999) was translated variously into English as »action sequences or ›chains‹« (Leroi-Gourhan 1993, 233), »operational process« (Leroi-Gourhan 1993, 233), »operational sequence« (Lemonnier 1986, 181; White 1993, xviii), or »chain of technical operations« (Dobres/Hoffman 1994, 237). However, since the French term was variously described and was widely applied, no translation is used in the present study. The *chaîne opératoire* referred to the complete process of the material exploitation by humans beginning at the procurement of the raw material and ending with the discard of the used artefact. Moreover, the single stages of this process and the techniques used therein reflected normative behavioural patterns (Pigeot 1987) which were adapted to a specific economy (Bodu 1993, 38) and thereby to a specific environment. In this sense, the *chaîne opératoire* reflected complex cognitive constructs which were supposedly transmitted as a recipe-like concept (Pelegrin 2009; cf. Mesoudi/O'Brien 2008b).

Floss 2002; Conneller 2007). In general, the dimensions of the lithic artefacts tend to decrease towards the younger period. Based on characteristics on cores and blanks (cf. Bodu 1993, 43-49; Madsen 1996; Pelegrin 2000), mineral materials such as stones seemed to be preferred as knapping instruments, in particular, soft hammerstones (Valentin 2008a, 48-51). However, seen in a more complex relation of raw material and applied knapping techniques, the difference between the Late Magdalenian and the FMG can be described as follows: The Late Magdalenians chose their raw materials according to their technological concept, whereas the FMG adopted the technological concept to the available raw materials. Thus, the conforming to standards of the *chaîne opératoire* appeared higher during the Late Magdalenian period and more flexible strategies were applied during the period of the FMG.

Moreover, the inventories of retouched lithic artefacts in the FMG were clearly dominated by end-scrapers, burins, and backed implements. Among the LMP point types prevailed and, in general, *Federmesser s. s.* or common Azilian points were frequent. The end-scrapers were generally more numerous than the burins and occasionally more numerous than the LMP. Small thumb-nail end-scrapers were common. Further retouched pieces such as truncations, borers, or composite tools occurred in these assemblages but were infrequent. The organic material was rarely preserved and though a general idea was recognised, the single implement followed no standardised shape. Assuming that the equipment of the Lateglacial hunter-gatherers developed as adaptation to various environments and the available resources, the larger variety of material in the FMG than in the Late Magdalenian could reflect the increasingly diverse landscapes of north-western Europe.

In the spatial organisation, little evidence for constructions was found on FMG sites. However, the few constructions were neither as massive nor as frequent as in the Late Magdalenian. Regularly, these structures were related to hearths. Some evident hearths were identified by the concentration of charcoal and/or the alteration of the sediment. Moreover, latent hearths were frequently indicated by accumulations of burnt materials such as stones but also burnt bones or lithic artefacts. In contrast to the Late Magdalenian assemblages, these burnt materials occurred regularly on FMG sites, perhaps due to an increased focus of activities around open fires. Furthermore, no spatial distinction of skilled and apprentice flintknappers was observed in relation to these hearths, in contrast to the Late Magdalenian evidence from the Paris Basin (Pigeot 1987; Valentin/Pigeot 2000).

In addition, the distinction of various functions of sites such as was possible for the Late Magdalenian (hunting camp, residential site, procurement place) became less clear for the FMG. In north-western Europe, assemblages of the FMG meta-group were usually found in open air sites and, rarely, in caves or in rock shelters. Generally, the concentrations contained the complete set of the lithic *chaîne opératoire* with cores, blanks, and retouched artefacts. Furthermore, the common retouched artefact classes were usually present at all sites. Moreover, at sites with sufficient preservation faunal material was normally found, often in the form of calcined bones. Thus, the lines of evidence for interpreting the function of the sites were generally similar on the FMG sites.

Most sites in Northern Europe attributed to the FMG, the sites attributed to the Azilian in Western Europe, and the sites attributed to the penknife point phase as well as, probably, the sites attributed to the Hengistbury Head type industries in Britain belong to the FMG meta-group. However, the distinction of some of these assemblages to the so-called Final Magdalenian such as the Creswellian, the classic Hamburgian, or the Magdalenian faciès Cepoy-Marsangy (MfCM) remained partially unclear. Thus, this meta-group enclosed assemblages which formed part of the transition studied in the present work. Therefore, the use of FMG as a term for the meta-group would lead to misunderstandings. To prevent these misunderstandings in the present study, the term FMG is not further used for the meta-group but according to the French terminology Azilian is chosen as the term for the meta-group (cf. Valentin 2008a, 47 f.). This meta-group also enclosed younger developments such as the so-called Laborian (Le Tensorer 1981) and was contrasted by the Magdalenian. Moreover, by the use of this terminology, Azilianisation becomes the appropriate term for the transition from the Late Magdalenian to the type of assemblages found immediately below the LST in the Central Rhineland. However, the process of changes did not stop with the eruption of the Laacher See volcano but continued into the Mesolithic.

Nevertheless, the term FMG is used furtheron for a sub-group comparable to the proposals of Michael Baales (Baales 2002), the final phase of FMG proposed for the Somme valley (Coudret/Fagnart 1997), and/or the *Azilien récent* in the Paris Basin (Valentin 2008a). This sub-group was related by a common set of behaviours.

In the lithic *chaîne opératoire* not much attention was paid to raw material procurement. The material was of varying quality. The blank production appeared very flexible and was adapted to the varying quality of the raw material as well as to the anticipated results. Few regular blanks were selected for the transformation to projectile implements, whereas for the other tool classes no apparent blank selection was detectable. Sometimes, even decortication or preparation flakes were chosen as blanks for end-scrapers or burins. Usually, the dominant classes of retouched artefacts were LMP and end-scrapers. Among the LMP, backed bladelets were rare or absent. The point types were clearly dominated by *Federmesser s. s.* or the common Azilian point, although straight-backed points without further modifications were also common. The end-scrapers were usually small and regularly of the thumb-nail shape. The burins were usually on truncations but, in general, no standardised types of burins were observable. Truncations and borers were also unstandardised and occurred irregularly. Composite tools were rare and if present usually a combination of burins and end-scrapers. The preserved faunal material was often calcined. Due to generally poor preservation, butchering marks were infrequently evidenced, but the often highly fragmented state in combination with the traces of heating suggested a boiling and burning of some of the bones. Moreover, the spatial distribution often indicated a relation of the lithic and the faunal material. However, the majority of the faunal material remained undetermined due to the state of preservation. Among the determinable remains red deer (*Cervus elaphus*) usually dominated. Furthermore, beaver (*Castor fiber*), large bovids, usually determined as aurochs (*Bos primigenius*), and single horses (*Equus* sp.) were common. These species were infrequently supplemented by elk (*Alces alces*), wild boar (*Sus scrofa*), roe deer (*Capreolus capreolus*), caprids such as ibex (*Capra ibex*) or chamois (*Rupicapra rupicapra*), badger (*Meles meles*), and fish, in particular, northern pike (*Esox lucius*). In total, the fauna represented the community of light forest with sufficient open waters and meadows. Organic tools were rarely preserved, and those which were preserved were often made of antler. Pieces of art or ornament were also rare. Nevertheless, abstract engravings on a slate plate in Niederbieber or on a shaft smoother from the same site emphasised that this lack of ornaments was not only due to a poor organic preservation. In general, haematite was also absent from these FMG sites. Furthermore, evident structures were scarce. Evident hearths were sometimes identified by the concentration of charcoal and/or the alteration of the sediment. Besides these evident structures, latent hearths were often identified by frequently occurring burnt materials such as calcined bones or heat-altered flints. Perhaps, a focus on the works performed around the fire led to the increased burning of the archaeological material. In general, the hearths were surrounded by a dense cluster of mostly unburnt lithic artefacts. The analysis of the concentrations at Niederbieber showed that these concentration were generally structured in two opposing clusters of high artefact density (Gelhausen 2010; Gelhausen 2011a). However, these two clusters were not functionally distinct in general, for instance, between blank production and transformation into tools or between skilled and apprentice flintknappers (cf. Pigeot 1987). Furthermore, except for the clusters around a central hearth, hardly any other type of spatial organisation were found on FMG sites. Separated workshops for stone knapping or concentrations with special functions were rarely identified (De

Bie/Caspar 2000). Although some zones or clusters with blank production debris or specific tools were found, these accumulations only formed part of a larger concentration. Usually, the sites were formed by one or several of these typical concentrations focused on a central hearth and located on a small plateau. Furthermore, the sites with several concentrations could be formed by repetitive visits of a single small group as well as a single visit of several comparably behaving groups. Moreover, no assemblage attributed unambiguously to this sub-group was found in a rock shelter or a cave. In consequence, the distinction of various functions of sites became almost impossible for these FMG. According to Lewis R. Binford's classic differentiation of settlement behaviours among hunter-gatherer groups (Binford 1980), the differentiated sites of the Late Magdalenian seemed to represent the logistical system of collectors, whereas the evidence of these FMG sites appeared as a reflection of the general foraging system. Presumably, the Lateglacial reality was more complex than the model (cf. De Bie/Caspar 2000; Müller et al. 2006; Bodu 2010; Fougère 2011; Bodu et al. 2011; Gelhausen 2011b) but a change in the spatial organisations on the individual sites as well as in the variety of sites with different spatial organisations from the Late Magdalenian to the FMG is apparent. In the Central Rhineland, remains of this sub-group were recovered from the upper horizon of Andernach-Martinsberg, Urbar, Kettig, Niederbieber, Boppard, and Bad Breisig (see p. 93-101 and p. 143-168). The examples from the northern French sub-area were very similar and are therefore not further introduced in the present study. In the Somme Valley these sites include Saleux *La Vierge Catherine*, Saleux *Les Baquets* 234, the intermediate horizon at Conty, Hangest-sur-Somme I.3, the upper horizons of Hangest-sur-Somme III.1 as well as Belloy-sur-Somme (Coudret/Fagnart 1997; Fagnart 1997; Coudret/Fagnart 2006). In the Paris Basin the upper horizons of Le Closeau were attributed to this sub-group (Bodu 1998). Furthermore, the *locus* 33 at Le Cornet and in the Pays de la Loire some *loci* at Chaloignes represented this behavioural sub-group in the Haute-Normandie (Valentin/Fosse/Billard 2004; Marchand et al. 2008). In the western upland zone, no modern excavated site was unambiguously attributable to this sub-group. However, in the adjacent western part of the North European Plain several sites of this sub-group were recorded such as Meer (De Bie/Van Gils 2006), Rekem (De Bie/Caspar 2000), or Lommel (De Bie/Van Gils/Deforce 2009). Perhaps, comparable to the Paris Basin, the ephemeral character of the concentrations prevented a recognition of further *in situ* sites. Another reason for the lack of sites could be the choice of topographic settings that perhaps, resulted in a more severe destruction by down-slope erosion or wash outs. Nevertheless, this set of behaviours leading to the specific concentrations and assemblages described above as FMG was found on further sites on the western North European Plain such as Doetinchem (Niekus/Stapert/Johansen 1998; Johansen et al. 2000) and, perhaps, Horn-Haelen (Rensink 2002). Many other sites in this area such as the sites around Wierden (Deeben et al. 2006), Milheeze (Arts 2012), or Geldrop (Deeben 1988) as well as the classic site Usselo (Stapert/Veenstra 1988) clearly belonged to the Azilian tradition as well but their precise position in the Azilianisation process remains to be ascertained. Farther to the east, single concentrations such as Alt Duvenstedt 120a (Clausen 1999) and, possibly, Großlieskow and Kleinlieskow 120 (Pasda 2001; Pasda 2002) could also be attributed to this sub-group. This large geographical distribution and the relatively long chronostratigraphic occurrence of this sub-group made these FMG sub-group an equivalent to the Late Magdalenian. However, further assemblages or possibly groups of assemblages such as the MfCM, the Creswellian, the Early Azilian (*Azilien ancien*) bipoint phase, the Hengistbury Head industries, or the penknife point phase were found chronologically between and, perhaps, besides the Late Magdalenian and this FMG stage. The role of some of these assemblages and groups in the transition process remains to be clarified in the present study.

The Central Rhineland

Patterns of Late Magdalenian as well as FMG behaviour were well preserved in the archaeological record from the Central Rhineland. The major sites are introduced in the following. However, material attributed to the reorganisation period between these two entities was rarely found and, therefore, this record is supplemented by northern French sites and sites from the intermittent western upland zone.

Andernach, Rhineland-Palatinate

Research history

Herman Schaaffhausen was the first to conduct excavations on the Martinsberg in Andernach in 1883 (later named Andernach 1). In the context of the discovery of the site at Gönnersdorf, Gerhard Bosinski and Joachim Hahn re-analysed the material recovered by Schaaffhausen and stored in the Andernach Museum (partially with false labelling). The results of the re-analysis induced a test pit survey to find the excavation area of Schaaffhausen on the Martinsberg. This survey was organised by Stephan Veil in 1977 (Veil 1982, 394 f.). These works revealed the existence of two archaeological horizons explaining the FMG component in Schaaffhausen's Late Magdalenian material as admixture. However, by chance, the area excavated by Schaaffhausen was recovered in the course of a house construction in 1979, and a small rescue excavation was conducted under the supervision of Stephan Veil. From 1981 to 1983 Veil, accompanied by Martin Street, continued the excavation of the site (Andernach 2, c. 104 m²) that was attached south-westwards of the area examined by Schaaffhausen (Bolus/Street 1985). In total, three concentrations (I-III) of paved areas attributed to the Late Magdalenian were identified in these excavations as well as concentrations of FMG material (Andernach 2-FMG) around approximately three hearths. From 1994 to 1996 another area of the site (Andernach 3) some 15 m south-west of Andernach 2 was excavated with modern standards (81 m²) by Sylvie Bergmann, Jörg Holzkämper, and Jan Kegler. This area yielded a fourth concentration (IV) of large stone plates and pits from the Late Magdalenian (Holzkämper 2006) as well as another scatter of the FMG material (Andernach 3-FMG; Gelhausen/Kegler/Wenzel 2004). Due to time pressure inflicted by planned construction work, the southern area (32 m²) of Andernach 3 was only dug up in 12.5 cm × 12.5 cm × 5 cm blocks that were wet-sieved (Bergmann/Holzkämper 2002, 471). In total, some 245 m² (without some 130 m² uncovered by Schaaffhausen) were archaeologically investigated on the Martinsberg in Andernach.

Topography

Andernach is located at the north-western edge of the Neuwied Basin. The open air site is set at an altitude of c. 80 m a. s. l. on the south-eastward facing slope of the Martinsberg (**tab. 10**). The Martinsberg was formed by a Quaternary basalt lava stream alongside the Rhine which flows today approximately 1 km northwards of the site. During the Late Pleniglacial and Lateglacial Interstadial the Rhine bed was higher elevated and additionally braided. Thus, the river banks were closer to the site (c. 300 m; Veil 1982, 395-397). Northwards of the Martinsberg the hills forming the »Andernacher Pforte« (Andernach Gate) rise steeply forming a protective barrier for the settled area against northern winds. From this barrier to the south side of the Martinsberg only the gradually rising hills of the Eifel massif surround the slope. Southwards to north-eastwards of the site lies the wide Neuwied Basin. Thus, most times of a day the site is exposed to direct sunlight. Moreover, the brittle basalt underground induced erosive processes towards the lower lying floor of the Neuwied Basin. In fact, erosion channels disturbed the archaeological deposits in both areas and con-

site	approximate / average distance to water	type of water body	morphographic setting	site type	direction of cave opening / exposure
Andernach	close / near	stream / river	slope	open air	×
Gönnersdorf	close / near	stream / river	slope	open air	×
Wildweiberlei	near	river	slope	cave	W
Oberkassel	distant	river	valley bottom	*open air*	*SW*
Irlich	close / near	river	valley bottom	open air	×
Kettig	near	river	slope	open air	×
Urbar	close	spring	slope	open air	×
Niederbieber	close / near	stream / river	slope	open air	×
Boppard	close	river	valley bottom	open air	×
Bad Breisig	immediate / close	river	valley bottom	open air	×

Tab. 10 Topographic characteristics of sites from the Central Rhineland. **immediate** 0-29 m; **close** 30-149 m; **near** 150-500 m; **distant** 500 > m; **×** not restricted; **N** north; **S** south; **W** west; **E** east. Uncertain characteristics are set in italics. – For a discussion concerning the site type of Bonn-Oberkassel and for further details see text.

tained Late Magdalenian as well as FMG material. North-westwards and south-eastwards of Andernach 2 depressions isolated a »plateau« on which the settlement was set (Veil 1982). Presumably, this position further contributed to the erosive process. In addition, the erosion was possibly intensified by a small stream passing the site at a maximum distance of 100 m westwards (Holzkämper 2006, 5).

Stratigraphy

During the Andernach 2 excavations, a three-parted profile was recorded (Veil 1982, 397 f.) with the rutted basalt and its scree at the base overlain by a loess deposit and sealed by some 4 m of pumice of the LSE. The intermediate loess deposit was infiltrated by various fractions of basalt rubble depending on the proximity to the base but loess also filled in fissures in the basalt. In this lower part, the loess was light brown and calcareous, whereas in the upper half the loess was transformed into a brown loess loam with humus, iron, and manganese patches. The topmost centimetres are formed by a light brown loess loam. The general stratigraphic succession was confirmed by the excavations of Andernach 3 with the specification that the loess loam was in fact a stagnosol (»Pseudogley«) that was formed by waterlogging from a greyish brown loess (Kegler 2002, 501).

In this loess loam, two stratigraphically distinct archaeological horizons were observed (Veil 1982; Bergmann/Holzkämper 2002). The material of the lower horizon formed a continuous layer, whereas the archaeological remains of the upper horizon were scattered ephemerally mainly in the upper part of the loess loam, immediately underlying the LST. Based on the archaeological material, the lower horizon was attributed to a Late Magdalenian settlement and the upper horizon to a FMG occupation.

Charcoal was frequently found in small accumulations and attributed to this upper horizon. In Andernach 2, the pieces of charcoal were identified most commonly as willow (*Salix* sp.; 65.8 %) followed by pine (*Pinus* sp.), birch (*Betula* sp.), and a few pieces of elder (*Sambucus* sp.; Veil 1982, 400). In Andernach 3, birch and poplar (*Populus* sp.) as well as willow and Saliceae pieces, which were not further determinable, and singular specimens of *Prunus* sp., Maloideae (previously Pomoideae), and *Daphne* sp. were identified (Kegler 1999, 14). These determinations were consistent with an attribution to a Lateglacial Interstadial environment.

However, due to erosion the thickness of the loess deposit varies from a maximum of 50 cm to only 15 cm on the central part of the »plateau« (Veil 1982, 397). In addition, the archaeological material moved due to bio- and cryoturbation within the sediment. Moreover, in both modern excavation areas larger fissures in the basalt running NE-SW (Andernach 2) to N-S (Andernach 3) led to the formation of channels in the

site	excavated m²	total ≥ 10 mm (all)	cores and core fragments	retouched artefacts	% of burnt artefacts to total ≥ 10 mm (all)	ref.
Andernach, lower horizon	c. 245	6,518 (43,832)	55	1,831	0.5 (0.1)	1-4
Andernach I	c. 78	? (16,296)	30	530	×	1-3
Andernach II	c. 17	? (5,790)	3	503	×	1-3
Andernach III	c. 19	? (1,116)	5	181	×	1-3
Andernach IV	113	1,194 (19,817)	9	275	0.4 (0.0)	1; 4
Andernach, upper horizon	c. 245	3,794 (19,925)	66>	280	×	5-8
Andernach 2-FMG	104	1,377 (2,894)	25	156	2.25	5-6
Andernach 3-FMG	113	2,417 (17,031>)	41>	124	×	7-8
Gönnersdorf	687	33,000 (81,786)	309 / 311	4,855 / 4,320	< 1.0	5; 9-12
Gönnersdorf I	96	×	c. 35	1,657	×	9
Gönnersdorf II	215.5	? (29,742)	28	1,550	×	5; 9
Gönnersdorf III	c. 130	×	c. 145	928	×	9
Gönnersdorf IV	126	854 (2,758)	28	182	0.7	5; 10-11
Gönnersdorf SW	114.5	2,251 (2,804)	6	148	×	12
Wildweiberlei	c. 24	580 (662)	28	99	2.6 (2.3)	13
Oberkassel	×	1	0	0	0.0	14
Irlich	×	2	0	1	0.0	15
Kettig	242	3,834 (24,098)	61	352	8.1	15
Urbar	c. 17	516 (1,641)	8	120	10.8	16
Niederbieber	1,032.5	19,306 (119,954)	267	1,729	14.7 (2.4)	17-23
Niederbieber 1	48	2,583 (7,986)	33	272	31.1	17
Niederbieber 3	28	258 (1,402)	1	9	81.2	17
Niederbieber 4+17a	74	2,119 (17,201)	29	260	32.0	17-19
Niederbieber 5	40	1,572 (~7,361)	23	105	2.7	20
Niederbieber 6+10a	45	1,835 (12,941)	14	104	~30.1	18-19; 21
Niederbieber 7	36	1,040 (2,257)	16	125	4.6	22
Niederbieber 8	75	537 (11,336)	5	37	3.4	18-19
Niederbieber 9	123	2,088 (4,897)	31	174	4.4	18-19
Niederbieber 10	23	811 (2,314)	11	58	10.0	18-19
Niederbieber 11	56	499 (4,731)	11	66	3.6	18-19
Niederbieber 12	46	1,106 (6,804)	21	65	5.1	18-19
Niederbieber 13	39	697 (6,928)	9	50	8.6	18-19
Niederbieber 14	42	751 (6,595)	22	89	5.2	18-19
Niederbieber 15	30	284 (553)	7	9	9.6	18-19
Niederbieber 16	37	418 (5,383)	7	42	5.5	18-19
Niederbieber 17	51	162 (2,059)	2	31	6.3	18-19
Niederbieber 18	38	715 (13,991)	1	10	0.6	23
Niederbieber 19	28	159 (287)	7	9	2.5	23
Niederbieber 20	60	983 (3,404)	10	45	3.2	23
Bad Breisig	50	5,956 (45,480)	184	296	16.27 (2.15)	24

Tab. 11 Numbers of lithic artefacts recovered at sites from the Central Rhineland. Sub-assemblages are shaded grey. For Boppard these numbers are not available yet. – For further details see text. – References (ref.): **1** Bergmann/Holzkämper 2002; **2** Floss/Terberger 1987; **3** Floss/Terberger 2002; **4** Holzkämper 2006; **5** Floss 1994; **6** Bolus 1984; **7** Kegler 2002; **8** Gelhausen/Kegler/Wenzel 2004; **9** Franken/Veil 1983; **10** Sensburg/Moseler 2008; **11** Terberger 1997; **12** Buschkämper 1993; **13** Terberger 1993; **14** Schmitz/Thissen 1997; **15** Baales 2002; **16** Baales/Mewis/Street 1998; **17** Bolus 1992; **18** Gelhausen 2011a; **19** Gelhausen 2010; **20** Husmann 1989; **21** Thomas 1990; **22** Freericks 1989; **23** Gelhausen 2011c; **24** Grimm 2004.

sediment disturbing the archaeological deposits. In Andernach 2, such channels are particularly found in the eastern half. In Andernach 3, the channels cut through the centre of the excavation area. Partially, FMG material was found in these fissures.

site	distance classes of raw material exposure from the site					ref.
	0-5 km	6-15 km	16-60 km	61-250 km	251 km >	
Andernach, lower horizon	***Tertiary quartzite***; indurated slate (single); *local flint (single);* Devonian quartzite; quartz; (quartzitic) slate; (indurated) sandstone; basalt; *haematite; jet;* brown coal	***Tertiary quartzite*** (5 km > W)	***Tertiary quarzite*** *(20-50 km NNW); chalcedony (35-50 km NNW); local flint (single, NW); haematite (c. 20 km NW)*	*Triassic chert (single, 65 km > SW or 160 km > N);* **Western European flint** (95 km > NW); *Mesozoic quartzite (100 km > W);* Baltic flint (100 km > N); *Tertiary quartzite, type Ratingen (95 km > E or 105 km > N);* chalcedony (110 km > SEE); molluscs (70 km > SE)	*jet (255 km > SE);* Mediterranean molluscs (min. 680 km S, 800 km > SSW); Mediterranean/Atlantic whale bone (single, 800 km SSW, 900 km > SW)	1-5
Andernach, upper horizon	***Tertiary quartzite***; indurated slate (single); silicified limestone; Devonian quartzite; quartz	*Tertiary quartzite (5 km > W)*	**chalcedony** (30-35 km NNW); Eifel local gravel flints (15-50 km > W)	silicified limestone (60 km > S-SE); siliceous oolite (70 km > SE); Western European flint (95 km > NW); Baltic flint (100 km > N-NE)		1; 6-7
Gönnersdorf	indurated slate; rock crystal; *Jurassic chert*; slate; quartzitic slate; Devonian quartzite; quartz; *haematite; jet*	***Tertiary quartzite*** *(12 km > SE)*; basalt	***Tertiary quartzite*** *(16 km > SEE)*; Eifel local gravel flints (50 km > W); *haematite (c. 25 km > NW); opal (25 km > N)*	chalcedony (75 km > SE); siliceous oolite (75 km > SE); opal (75 km > SE); fossil curiosities (75 km > SE, 105-150 km W-SW, or 195 km > S); *Tertiary quartzite, type Ratingen (85 km > N);* **Baltic flint** (105 km > N); **Western European flint** (110 km > NW); *Palaeozoic quartzite (110 km > NW); fossil molluscs (110 km > NW)*	*jet (255 km > SE)*; jasper of Kleinkems type (305 km > S); Mediterranean molluscs (min. 680 km S, 800 km > SSW); *Atlantic molluscs (800 km > WSW)*	1; 8-9
Gönnersdorf SW	indurated slate (lydite); slate; quartzitic slate; Devonian quartzite; greywacke; Eisenschwarte; sandstone; quartz; haematite	Tertiary quartzite (12 km > SE); basalt	haematite (c. 20 km > NW)	chalcedony (75 km > SE); siliceous oolite (single; 75 km > SE); Tertiary quartzite, type Ratingen (85 km > N); **Baltic flint** (105 km > N); **Western European flint** (110 km > NW); Palaeozoic quartzite (single; 110 km > NW)	jasper of Kleinkems type (single; 305 km > S)	1; 10
Wildweiberlei	**indurated slate (lydite)**; basalt; slate; haematite	Tertiary quartzite (6 km > NE)		Baltic flint (120 km > N); *Western European flint (85 km > NW)*		1; 11

Tab. 12 Approximate distance classes of raw materials recovered at sites from the Central Rhineland. The most numerous lithic raw material(s) is/are set in bold. Raw materials of which only single artefacts were found are marked as (single). Sub-assemblages are shaded grey. Resources for which several places of origin are discussed are set in italics. For Gönnersdorf SW all materials found in the area are given, although several do probably not relate to the early Lateglacial Interstadial assemblage. – For further details see text. – References (ref.): **1** Floss 1994; **2** Floss/Terberger 2002; **3** Holzkämper 2006; **4** Alvarez Fernández 2001; **5** Langley/Street 2013; **6** Bolus 1984; **7** Kegler 2002; **8** Franken/Veil 1983; **9** Bosinski 1979; **10** Buschkämper 1993; **11** Terberger 1993; **12** Schmitz/Thissen 1997; **13** Baales 2002; **14** Baales/Mewis/Street 1998; **15** Gelhausen 2011a; **16** Wenzel 2004; **17** Grimm 2004.

site	distance classes of raw material exposure from the site					ref.
	0-5 km	6-15 km	16-60 km	61-250 km	251 km >	
Oberkassel	haematite		haematite (c. 20 km SW)	**flint** (Western European 35 km > NW; Baltic 80 km > N)		12
Irlich	haematite		**chalcedony** (30 km > NNW); haematite (c. 25 km NW)			13
Kettig	***Tertiary quartzite***; indurated slate; limonite; agate (single); indurated claystone (single); carnelian (single); quartz; Devonian quartzite; porphyry; quartzitic sandstone; argillaceous shale	Tertiary quartzite (8 km > W)	chalcedony (40 km > NW); Eifel local gravel flints (single; 55 km > W)	**Western European flint** (65 km > NW); *Tétange flint (90 km > SW);* Agate/Jasper of Weiselberg type (single; 95 km > SW); indurated claystone of Schaumberg type (115 km > SW); *Baltic flint (125 km > N)*		13
Urbar	**Tertiary quartzite**; indurated slate; Devonian quartzite; quartz; argillaceous shale; *quartzitic sandstone; haematite*		haematite (c. 35 km > NW)	Western European flint (single, 120 km > NW)		14
Niederbieber	***Tertiary quartzite***; indurated slate; Devonian quartzite; indurated limestone (single); quartzitic slate; basalt; argillaceous shale; *haematite*	**Tertiary quartzite** (8 km > SE)	**chalcedony** (35 km > NW); *haematite (c. 25 km NW)*	Western European flint (85 km > NW); Baltic flint (115 km > N); Triassic chert (170 km > SW); indurated claystone of Schaumberg type (115 km > SW)		1; 15
Boppard	indurated slate; quartz; Devonian quartzite; jet	**Tertiary quartzite** (10 km > SW or 20 km > N-NW)	chalcedony (35 km > NW or 45 km > SW); Eifel local gravel flints (55 km > NWW)	Western European flint (80 km > NW)		16
Bad Breisig	***Tertiary quartzite***; indurated slate; argillaceous shale (single); Devonian quartzite; sandstone; brown coal	**Tertiary quartzite** (7 km > S or 9 km > N)	chalcedony (single, 15 km > NNW); Eifel local gravel flints (45 km > W)	Triassic chert (single; 80 km > SW); Western European flint (85 km > NW)		17

Tab. 12 (continued)

Thus, in some parts of the excavated area the archaeological horizons are stratigraphically indistinguishable and the attribution of material to either the lower or the upper horizon became difficult. In consequence, the archaeological material was differentiated in these areas by tendencies in the choice of raw materials, techno-typological aspects, and varied horizontal distributions (Veil 1982, 398f.; Bergmann/Holzkämper 2002, 473; cf. Kegler 2002, 501-502). However, admixture of some pieces in the chronologically distinct archaeological material cannot be completely excluded. Hence, this constrain should be kept in mind when numbers are mentioned for this site because, in detail, they represent assumed attributions.
Nevertheless, in the following the material is presented according to the two distinct horizons.

Archaeological material of the lower horizon

In the lower horizon, a continuous band of settlement remains was observed (Veil 1982, 403). It produced in total 43,863 lithic artefacts (**tab. 11**) weighing c. 30 kg that were attributed to the Late Magdalenian occupation (Floss/Terberger 2002, 3; Holzkämper 2006, 89). However, if the splinters are excluded only some 6,500 larger pieces were recovered at this site. Nevertheless, due to the only partial excavation and the occasional desilification of Tertiary quartzite, Harald Floss supposed that the amount of material which reached the site was at least some 30 kg in addition combining for a total weight of approximately 60 kg (Floss/Terberger 2002, 7).
The main raw materials were Tertiary quartzite (c. 20 kg, n = 16,987) and Western European flint (max. 8 kg, n = 23,635). The latter occurred in some varieties at the site such as Rijckholt type or Simpelveld type, and clearly belongs to the distant raw material class with the nearest occurrences some 100 km to the north-west (**tab. 12**). In contrast, Tertiary quartzite is of local to regional origin but the exact source remained unidentified. Possible sources are situated at distances of 4 km south-westwards to some 30-40 km eastwards or north-westwards (Floss/Terberger 2002, 5-7; Holzkämper 2006, 95). The main type of Tertiary quartzite is a fine-grained material with occasional quartz inclusions of usually grey but also yellowish, greenish, or blackish colour. Besides these varieties, a piece of brownish colour with signs of river transport and 19 pieces of the coarse-grained Ratingen type occurred. The latter appeared to originate from a distant source some 100 km either northwards or eastwards (Floss/Terberger 2002, 17). Near the possible north-western regional source of Tertiary quartzite several outcrops of chalcedony are known (Floss/Terberger 2002, 8). This raw material was present in a small quantity in the Late Magdalenian assemblages of Andernach (n = 1,066) but originated possibly from sources in the Main region some 120 km to the south-east of the Martinsberg (Floss/Terberger 2002, 8f.). However, the characteristic cortex with wind abrasion was preserved only on some pieces, and also the number of pieces which were microscopically and chemically analysed was limited. Moreover, several pieces were patinated and, hence, Harald Floss could not exclude the region north-north-westwards as a possible origin of the Late Magdalenian chalcedonies (Floss/Terberger 2002, 9). Another distant source of lithic raw material was attested for the Baltic erratic flint (n = 1,291) that originated at a minimum of 100 km north to north-eastwards. The fifth raw material yielding more numerous artefacts was a Mesozoic quartzite (n ~ 880) which often was also referred to as Palaeozoic quartzite or Ardennes quartzite (Floss/Terberger 2002, 14-16). However, the exact source is thus far unknown but a proximity to the Western European flint was assumed (Floss/Terberger 2002, 14). Very low numbers of artefacts or single pieces (Floss/Terberger 2002, 19-22; Holzkämper 2006, 98) were found of local indurated slate/lydite (n > 6), Devonian quartzite (n = 1), a local flint (n = 1), and a blade of Triassic chert that might originate from some 65 km to the south-west or from 160 km north-eastwards (Floss/Terberger 2002, 21).
In addition to the lithic raw materials, several rock materials were brought to the site, some of them in considerable quantities and/or volumes. For instance, c. 414 kg of rocks were recovered only in Andernach IV (Holzkämper 2006, 57). These raw materials are basalt, quartz, some varieties of sandstone, various types of

site	blank production	cores total	1 platform	2 platforms, 1 preferred	2 platforms	3 and more platforms	others (e.g. fragments)	ref.
Andernach, lower horizon	blades	55	**15>**	8>	8>	×	9>	1-2
Andernach I	blades	38	**13**	8	8	0	9	1
Andernach IV	blades/bladelets	9	2	×	×	×	×	2
Andernach, upper horizon	flakes/bladelets	66>	×	×	×	×	×	3-4
Andernach 2-FMG	flakes/bladelets	26	6	1	9	4	6	3
Gönnersdorf	bladelets	309	**118**	? 2>	76	67	48	5
Gönnersdorf SW	bladelets	6>	**1**	**1**	×	×	×	6
Wildweiberlei	blades/bladelets	28	7	6	**11**	0	4	7
Kettig	flakes/short blades	60	**7**	5	5	5	1	8
Urbar	flakes/bladelets	8	1	×	×	**2**	**5**	9-10
Niederbieber	flakes/bladelets	267	11	**20**	**20**	14	3	11-13
Niederbieber 1	flakes/bladelets	33	1	**4**	3	3	×	13
Niederbieber 4	flakes/bladelets	29	2	7	**8**	4		13
Bad Breisig	bladelets/flakes	184	**50**	<46	32>	13	43	14

Tab. 13 General concept of blank production and core types recovered at sites from the Central Rhineland. For Boppard these numbers are not available yet. In Irlich no cores were found. Sub-assemblages are shaded grey. The most numerous class is set in bold. On some sites not all cores and core fragments were classified according to the number of platforms. In these cases the pieces displayed in the figures were classified. **×** this type is not mentioned or displayed in a publication. – For further details see text. – References (ref.): **1** Floss/Terberger 2002; **2** Holzkämper 2006; **3** Bolus 1984; **4** Kegler 2002; **5** Franken/Veil 1983, 83-148; **6** Buschkämper 1993; **7** Terberger 1993; **8** Baales 2002; **9** Mewis 1993; **10** Baales/Mewis/Street 1998; **11** Gelhausen 2011c; **12** Gelhausen 2011a; **13** Bolus 1992; **14** Grimm 2003.

quartzite, in particular, Devonian quartzite, and various slate types. These materials can be assumed as originating from local, nearby sources such as the basalt stream, the Rhine gravels, or primary slate resources such as the Krahnenberg at an approximate distance of 1 km north-westwards of the site (Holzkämper 2006, 57).

Among the lithic material, 55 cores were found in the modern excavations (**tab. 13**) of which 43 were made of Tertiary quartzite, five of Baltic erratic flint, four of Western European flint, two of chalcedony, and one of Mesozoic quartzite. Five Tertiary quartzite specimens can be classified as freshly prepared, i.e. complete cores of segmented shape measuring 10.7 to 15.7 cm (Floss/Terberger 2002, 26 f.). Further completely prepared cores were found in the Schaaffhausen material exceeding the pieces from the modern excavations by almost 5 cm in size (Floss/Terberger 2002, 26 f.). The cores of foreign materials are usually very small and exploited almost exhaustively. In general, the cores for blade and bladelet production clearly dominated. These cores frequently contained two prepared platforms but often only one was preferred. Hence, the cores were usually reduced from only one platform. The platforms were often additionally prepared and worked from only one knapping face (Floss/Terberger 2002, 33-37). Only approximately 8 % of the blanks from Andernach 2 wore traces of cortex, most frequently the cortex was on Mesozoic quartzite (22 %) and least frequently on Tertiary quartzite (5.4 %) suggesting that this raw material reached the site in a prepared shape (Floss/Terberger 2002, 47). In Andernach 3, the numbers in all raw materials were considerably lower with only a total of 78 specimens covered with cortex on some parts. In Andernach 2, almost a fifth to a quarter of the preserved butts of Tertiary quartzite and Western European flint blades and over a third of the Mesozoic quartzite blanks were made with a »talons en éperon«, whereas on the Baltic erratic flint and chalcedony blanks less than 5 % wore this spur (Floss/Terberger 2002, 40-60). In Andernach 3, this preparation was observed on some 4 % of the blanks (Holzkämper 2006, 101 f.).

1,831 retouched artefacts were attributed to the Late Magdalenian inventory in Andernach (**tab. 14**). The laterally retouched specimens are clearly the largest group (n = 607) but also a very heterogeneous group

site	total	LMP	end-scrapers	truncations	borers	burins	composite tools	others	ref.
Andernach, lower horizon	1,831	**367**	251	70	98	**395**	41	**613**	1-3
Andernach I	530	**122**	**186**	26	42	106	?	?	1-2
Andernach II	503	**151**	20	20	50	**191**	?	?	1-2
Andernach III	181	**56**	41	11	9	**51**	?	?	1-2
Andernach IV	275	**37**	23	9	10	**123**	6	**67**	3
Andernach, upper horizon	280	**128**	**41**	20	0	**41**	3	**49**	4-5
Andernach 2-FMG	156	**66**	**25**	11	0	22	1	**31**	4
Andernach 3-FMG	124	**62**	16	9	0	**19**	2	16	5
Gönnersdorf	4,855	**1,927**	229	98	480	**1,003**	260	858	6
Gönnersdorf I	1,657	**548**	71	40	180	**524**	121	173	6
Gönnersdorf II	1,550	**691**	139	92	168	**206**	25	**229**	7-8
Gönnersdorf III	931	**380**	46	17	152	**172**	23	138	9
Gönnersdorf IV	182	**86**	5	6	5	**22**	3	7	8
Gönnersdorf SW	148	**58**	7	2	12	**34**	3>	<30	10
Wildweiberlei	99	**20**	6	5	9	**12**	3	**44**	11
Irlich	1	**1**	0	0	0	0	0	0	12
Kettig	352	**100**	**117**	21	5	36	3	70	12
Urbar	120	**13**	**98**	2	0	2	0	5	13
Niederbieber	1,729	**551**	**279**	158	37	236	2>	**427**	14-15
Niederbieber 1	272	**69**	**58**	23	5	49	0	**68**	14
Niederbieber 3	9	**5**	0	0	0	**3**	0	1	14
Niederbieber 4+17a	260	**80**	**75**	25	2	34	0	44	14
Niederbieber 5	105	**34**	6	**9**	2	7	?	**39**	14
Niederbieber 6+10a	104	**53**	3	**9**	1	3	0	**30**	14
Niederbieber 7	125	23	14	**24**	5	**35**	1	**30**	14
Niederbieber 8	37	**12**	3	2	0	**13**	?	7	14
Niederbieber 9	174	**65**	**32**	15	4	14	?	**44**	14
Niederbieber 10	58	**25**	**10**	6	1	5	?	**11**	14
Niederbieber 11	66	**27**	**16**	5	1	7	1	10	14
Niederbieber 12	65	**26**	5	2	0	**13**	?	**19**	14
Niederbieber 13	50	**20**	**12**	1	1	4	?	**12**	14
Niederbieber 14	89	**25**	**21**	8	5	15	?	15	14
Niederbieber 15	9	**11**	**11**	5	0	4	?	**11**	14
Niederbieber 16	42	**19**	1	3	1	**4**	?	3	14
Niederbieber 17	31	**22**	7	**13**	0	12	?	9	14
Niederbieber 18	10	**6**	**3**	0	0	1	?	0	15
Niederbieber 19	9	**6**	1	1	0	1	?	0	15
Niederbieber 20	45	**23**	**10**	1	3	6	?	2	15
Bad Breisig	296	**114**	**98**	31	0	25	1	27	16

Tab. 14 Numbers of retouched lithic artefacts recovered at sites from the Central Rhineland. For Boppard reliable numbers are not available yet. Sub-assemblages are shaded grey. The two most numerous classes are set in bold (except for the »others« class which can result from various types is involved, then the three most numerous groups are given). ***** include double counting of working edges on composite tools but lack splintered and laterally retouched pieces. – For further details see text. – References (ref.): **1** Floss/Terberger 2002; **2** Floss/Terberger 1987; **3** Holzkämper 2006; **4** Bolus 1984; **5** Kegler 2002; **6** Franken/Veil 1983, tab. 18; **7** Sensburg 2007; **8** Sensburg/Moseler 2008; **9** Terberger 1997, Tabelle 12; **10** Buschkämper 1993; **11** Terberger 1993; **12** Baales 2002; **13** Baales/Mewis/Street 1998; **14** Gelhausen 2011a; **15** Gelhausen 2011c; **16** Grimm 2004.

and the retouch could occasionally result from use rather than intentional retouch (Floss/Terberger 2002, 128f.). Yet, in combination with the micro-wear analyses (Plisson 2002; Vaughan 2002), this group indicates intensive use of the lithic inventory. However, among the typical tool classes, burins (n=395) dominated the inventory followed by 367 LMP. On one dihedral specimen a basal thinning of the so called Kostienki end type was observed (Holzkämper 2006, 106). Among the burins, the ones on truncations are with 299 examples the most frequent type and 15.7 % were produced in the Lacan style (cf. Brézillon 1968, 179f.).

Dihedreal burins (n=87) and burins on natural edges or negatives (n=90[14]) occurred considerably less frequently. Based on the preserved lengths of some blanks, in particular of the broken specimens, Jörg Holzkämper considered the use of such burins in a hafted form (Holzkämper 2006, 107). Burin spalls were frequently used in the Andernach Magdalenian inventory as blanks for the preparation of borers or backed bladelets (Holzkämper 2006, 105). This efficient use hinted additionally at the intensive exploitation of the raw materials. Among the LMP, simple backed bladelets (n=281) are most common, although many were fragmented and, thus, a possible attribution to other types could not be excluded entirely. Sometimes the backed bladelets were also retouched on the opposite edge (n=88) and rarely a truncation (n=3) was added or the backed bladelet was supplemented with a truncation and a retouched opposite edge (n=2). No indication of a serial production such as notched ends were observed (Holzkämper 2006, 116). However, this absence was perhaps connected to the size of suitable raw material blanks and, thus, in combination with the frequent use of burin spalls, points to an alternative recipe for serial backed bladelet production in regions with poor supply of large nodules of fine grained, glassy raw materials. The 251 end-scrapers were generally made on blades and approximately 50 % were also laterally retouched and in one case basally thinned by a Kostienki end. 13 double end-scrapers were counted originating only from the excavations at Andernach 2. Among the 98 borers was one piece thinned with a Kostienki end. In Andernach 2 were seven double and one twin borer found. In general, these can be attributed to the perforator group but a few pieces (n=11) fell rather into the *Zinken* group (»Grobbohrer«, Floss/Terberger 2002, 116). The truncations form a small group (n=70). In Andernach 2, only one double truncation was found. Usually the truncations were straight (n=36), rarely concave (n=24), convex (n=11), or irregular (n=3; Floss/Terberger 2002, 119; Holzkämper 2006, 115f.). The intentionally retouched lithics were occasionally made with more than one edge and in 41 cases in a combination of various tool classes (combinations with lateral retouches [n=338] were excluded). Among these composite tools, the burin-end-scrapers formed the most numerous group (n=19). Further combinations were frequently with burins (n=10) but also the borer-backed bladelets (n=8) occurred often. In addition, on five artefacts the Kostienki end thinning was the only performed retouch and a presumably Middle Palaeolithic side-scraper was perhaps reused during the Magdalenian (Floss/Terberger 2002, 129). 62 of the retouched artefacts were subsequently used as some type of chisel resulting in a characteristic splintering (i.e. splintered pieces). In addition to these retouched specimens, a further 175 artefacts were transformed by use into such splintered pieces that are not considered among the retouched pieces in the present study since the defining »retouch« was not an intentional decision of the flintknapper on a normative shape or functional possibilities but resulted from chance during the use of the piece.

In Andernach 2, micro-wear analyses were conducted mainly on the Tertiary quartzite artefacts from concentrations I and III by Hugues Plisson (Plisson 2002) and on mainly flint artefacts of concentration II by Patrick C. Vaughan (Vaughan 2002). Both analyses pointed to a rather specialised use of end-scrapers on which occasionally haematite adhesives for hide and leather working were found (Plisson 2002, 152; Vaughan 2002, 169. 171). In contrast, burins were used more variedly. In general, they appeared to be resharpened and often combined with an edge that was used otherwise, for instance for cutting leather (Plisson 2002, 148; Vaughan 2002, 167-169). Several pieces show long-term and/or intensive use-wear patterns suggesting either an intensive work episode or the reuse of preferred pieces. The smaller LMP were presumably used glued onto a shaft of a projectile (Vaughan 2002, 169).

14 Here the »special types« are included but the combinations with backed bladelets (n=2) and borer (n=1) are excluded (Floss/Terberger 2002, 94).

site	*Coleodonta antiquitatis*	*Mammuthus primigenius*	*Rangifer tarandus*	*Equus* sp.	*Alces alces*	*Cervus elaphus*	*Capreolus capreolus*	*Capra ibex*	*Rupicapra rupicapra*	*Saiga tatarica*	*Bison* sp.
Andernach, lower horizon		+ (ivory)	+	++		+ (teeth)					?
Andernach, upper horizon				(+)	+	++	(+)		+		
Gönnersdorf	+	+	+	++		+ (teeth)			+	+	+
Gönnersdorf SW	(+)	(+)	(+)	++	+	+					
Wildweiberlei	+		+	+				+			
Oberkassel						+	(?)				
Irlich						+ (tooth)					
Kettig				+		++	+		+ (tooth)		
Urbar				?		++					
Niederbieber				+	+	++	(+)	+			
Niederbieber 1				(+)	++	++	?				
Niederbieber 3				+	+						
Niederbieber 4+17a				(+)		++	(+)	+			
Niederbieber 5					+	++					
Niederbieber 7						+					
Niederbieber 10						+					
Niederbieber 14				(+)		++ (teeth)					
Niederbieber 15						++ (teeth)					
Niederbieber 18				+	++	+					
Niederbieber 19				+	++	+					
Niederbieber 20					++	+					
Boppard						+					
Bad Breisig				+ (tooth)		++	+				

Tab. 15 Mammal species recovered at sites from the Central Rhineland. Sub-assemblages are shaded grey. Symbols are: **+** present; **( )** association is unclear; **?** classification/determination is uncertain; **++** major species making up more than 30 % of the faunal assemblage.– For further details see text. – References (ref.): **1** Street et al. 2006; **2** Street 1993; **3** Poplin 1976; **4** Poplin 1978; **5** Terberger 1993, 183 f.; **6** Street 2002; **7** Baales 2002; **8** Baales/Mewis/Street 1998; **9** Bolus 1992; **10** Gelhausen 2011a; **11** Husmann 1989; **12** Freericks 1989; **13** Gelhausen 2011c; **14** written comm. Stefan Wenzel, Mayen; **15** Grimm 2004.

In addition to the lithic inventory, at least 40 hammerstones and/or retoucheurs were recovered from the two excavation areas (Schulte-Dornberg 2002; Bolus 2002). Generally, these specimens were made of Devonian quartzite but in Andernach 2, three pieces were made of indurated sandstone and a single piece each made of quartzitic slate and of quartz were found. In all the concentrations, some 100 kg of mainly fragmented quartz were found. This material was presumably fractured and occasionally reddened and/or blackened by the use as heating stones (Eickhoff-Cziesla 1992, 129-134; Holzkämper 2006, 59). Gisela Schulte-Dornberg assumed in her Master thesis, in which she analysed 117 rock nodules and fragments from Andernach 2 and a further 206 pieces from Andernach 3, that the quartz fragments from the site

Bos primigenius	*large bovid*	*Sus scrofa*	*Canis lupus / familiaris*	*Alopex lagopus*	*Vulpes vulpes*	*Ursus arctos / speleaus*	*further carnivores*	*Lepus timidus*	*Lepus* sp.	*Castor fiber*	Aves	Pisces	ref.
			+	+				+			+	+	1
?										+		+	1-2
			+	++	+			+			+	+	1; 3-4
	(+)			(+)									1
				+		(+)		+			+		5
	+		++			+							6
													7
	+		(?)		(+)	(?)	(?) (*Martes martes*; *Mustela nivalis*)			+		(+)	1; 7
	+												8
+		+			(+) (tooth)		(+) (*Meles meles*)			+		+	1; 9-13
		?			(+) (tooth)		(+) (*Meles meles*)					+	1; 9
										++			9
(+)													1; 9; 10
+													11
		+ (teeth)								+ (teeth)			12
		+ (teeth)								++ (teeth)			10
													10
													10
													13
													13
										+ (teeth)			13
		+										++	14
			(?)		(+)		(?) (*Meles meles*)		(?)		(?)	(+)	15

Tab. 15 (continued)

could only be attributed to heating stones (Schulte-Dornberg 2000, 19). By contrast, Michael Bolus, who analysed 26 specimens from Andernach 2, thought that more pieces of quartz were used as hammerstones but that these pieces served possibly as heating stones afterwards and were thereby splintered, i.e. lost for analysis (Bolus 2002, 175). By comparison with ethonographic analogies and experimental studies, Schulte-Dornberg showed that the rock materials were probably used for a wide variety of activities such as anvil, grinding stones or weights for squeezing, perhaps, of vegetation material (Schulte-Dornberg 2000, 147f.).
Faunal remains were relatively well preserved in the lower horizon at Andernach (**tab. 15**), in particular in the sheltering pits (Street 1993; Holzkämper 2006, 144f.; Street et al. 2006). Dominant among the fauna remains was horse (*Equus* sp.) with a minimum of twelve individuals at Andernach 2 (Street et al. 2006, 762) and three individuals at Andernach 3 (Holzkämper 2006, 147). On-going reanalysis (Martin Street, in prep.) will perhaps reveal whether these numbers are additive or integrative. Nevertheless, the partially

site	season	indicator	ref.
Andernach, lower horizon	autumn/winter; spring/summer	eruptive stages of horse and reindeer teeth; salmon and geese migration	1-2
Andernach, upper horizon	late spring/summer?	eruptive stages of red deer and bovine teeth; absence of antler	1-2
Gönnersdorf	autumn/winter; spring/summer	foetal bones of horses; size of arctic fox teeth; only shed reindeer antler; foal hooves	1; 3-4
Wildweiberlei	late summer – early winter	high number of young animals; hunting of arctic fox for its pelt	5
Kettig	late summer/autumn	eruptive stages of red deer and roe deer teeth; red deer teeth cementum analysis	6
Urbar	late autumn/early winter	red deer teeth cementum analysis; red deer antler	7
Niederbieber	late autumn – spring	eruptive stages of red deer and horse teeth; red deer and elk antler; age of red deer	8-9
Bad Breisig	autumn?	size of red deer teeth	10

Tab. 16 Seasonality of sites from the Central Rhineland as indicated by faunal material. – References (ref.): **1** Street et al. 2006; **2** Street 1993; **3** Poplin 1976; **4** Poplin 1978; **5** Terberger 1993, 183f.; **6** Baales 2002; **7** Baales/Mewis/Street 1998; **8** Gelhausen 2011c; **9** Bolus 1992; **10** Grimm 2004, 16f.

fragmented character of the assemblage clearly indicated the intensive processing of these bones including the obtaining of grease or the use of bone spalls as raw material for needles, awls, and, perhaps, bone points. The horse material also formed part of stable isotope analyses (oxygen, carbon, nitrogen) to establish this material as climate and environmental proxy (Stephan 1999; Stevens/Hedges 2004) and, subsequently, to use the variations for chronological argumentation (Stevens et al. 2009b). Antler, teeth, and bones of reindeer (*Rangifer tarandus*) were also common, although many pieces had presumably reached the site as either ornament (cut incisors, cf. Álvarez Fernández 2009, 48f.) or as gathered raw material (shed antlers). Nevertheless, the bone material (Andernach 2 MNI = 3; Andernach 3 MNI ≥ 1) indicated also the processing as alimentary resource of this prey species. The eruptive stages of the reindeer and horse teeth suggested a hunting season of autumn/winter (**tab. 16**). Moreover, reindeer material was also sampled for carbon and nitrogen studies (Stevens et al. 2009b). A large bovid, probably bison (cf. *Bison* sp.), was represented by ten bones without observable modification, one molar or premolar, and eight incisors of which at least three were probably used as ornaments (Street et al. 2006, 762; Holzkämper 2006, 149). Material of large bovids was analysed for carbon and nitrogen isotopes and assumed to be attributable to the upper horizon (Stevens et al. 2009b). However, the results displayed two clearly distinct clusters, perhaps, suggesting an older and a younger phase that were attributed to two phases in the upper horizon (Stevens et al. 2009b, 144). According to the state of preservation and to the spatial distribution, one of the groups was related to a bovid bone that produced a relatively young date (OxA-998). Nevertheless, the two groups of stable isotope values among the large bovid samples from Andernach were also found in the bovid material from Gönnersdorf but with significantly lower nitrogen values. Generally, the material of the two sites showed high similarities and, therefore, Rhiannon Stevens and her colleagues considered this difference as a sign for the older age of the Gönnersdorf material (Stevens et al. 2009b, 144). However, then the question arises why the two distinct groups were also present at Gönnersdorf. Consequently, a possible association of some of the bovid material from Andernach to the Late Magdalenian occupation must not be neglected before further samples are directly ^{14}C-dated. The remains of at least two arctic hares (*Lepus timidus*) and five or seven arctic foxes (*Alopex lagopus*) were also found. The fauna was further supplemented by swans (*Cygnus* sp.) and geese (*Anser* sp.) as well as at least two individuals each of grouse (*Lagopus* sp.) and common raven (*Corvus corax*). Moreover, salmon (*Salmo* sp.) was frequently identified and one bone each of grayling (*Thymallus thymallus*) and of European perch (*Perca fluviatilis*) were found. The former as migratory fish indicated a fishing period during the spring/summer period (**tab. 16**; Street et al. 2006, 763). Moreover, the

site	points				fish hook	sewing instruments		knives	bâtons	axes	hammers	others	ref.
	bevelled	barbed, single row	barbed, double row	others (incl. tip fragments)		needles	awls						
Andernach, lower horizon	+ (a+i)	+ (a)	+ (a)	+ (i)		+ (b)	+ (b)		+ (a)		+ (a)	+ (b)	1-5
Andernach, upper horizon				+ (b)									6
Gönnersdorf	+ (a+i)	?		+ (i)	+ (b+a)	+ (b+a+i)	+ (b+a)		+ (a)			+ (b)	5
Wildweiberlei	+ (a)												7
Oberkassel												+ (b)	8
Irlich				? (b)									9
Kettig		+ (a)									+ (a)		9
Urbar												? (b)	10
Niederbieber				+ (b)									9; 11
Boppard												+ (b)	12-13

Tab. 17 Modified organic material from sites from the Central Rhineland. For symbols see **tab. 15**. Material abbreviations are: **(b)** bone; **(a)** antler; **(i)** ivory. – Within the brackets: **/** or; **+** and. – For further details see text. – References (ref.): **1** Veil 1982; **2** Street 1993; **3** Street et al. 2006; **4** Holzkämper 2006; **5** Tinnes 1994; **6** Baales/Street 1996; **7** Terberger 1993, 182; **8** Mollison 1928; **9** Baales 2002; **10** Baales/Mewis/Street 1998; **11** Bolus 1992; **12** Wenzel/Álvarez Fernández 2004; **13** Wenzel 2004.

remains of a carnivore represent probably a large canid, perhaps a small wolf (*Canis lupus*) or a dog (*Canis familiaris*, Holzkämper 2006, 150). In the upper horizon of Andernach (see p. 93-101) the presence of dogs was suggested by tooth marks on bones (pers. comm. Martin Street, Neuwied) but thus far the oldest direct evidence of dogs in Central Rhineland were the early Lateglacial remains from Bonn-Oberkassel (see p. 140) that were dated approximately 850 years (550 ^{14}C years) younger than the Late Magdalenian concentrations of Andernach. However, the few remains (seven postcranial fragments and a tooth) must be regarded as not further determinable than a large canid without further examination such as molecular analyses because morphologically significant regions were not preserved and the discussion of dog domestication remains highly controversial (e.g. Musil 2000; Raisor 2004; Germonpré et al. 2009; Napierala/Uerpmann 2010; Ardalan et al. 2011; Ovdov et al. 2011; Germonpré/Lázničková-Galetová/Sablin 2012).

In total, 253 organic artefacts made of 268 fragments were recovered at Andernach (**tab. 17**; Tinnes 1994, 5f.). The majority was made of antler (n = 163) followed by ivory (n = 64) and bone (n = 49). Besides the previously mentioned hammerstones, an organic hammer made of the proximal part of a reindeer antler was recovered at Andernach 2 (Tinnes 2002). Furthermore, ten bone retouchers were identified (Tinnes 1994, 118-122). Moreover, antler served as raw material for spalls (Veil 1982, 412; Holzkämper 2006, 154). Comparably, ivory and bone were also worked in such a manner at the site, producing spalls of various sizes (Veil 1982, 412-415; Tinnes 1994, 101; Holzkämper 2006, 154). However, the single and double bevelled rods found in Andernach were, in fact, made only of antler and ivory (Veil 1982, 415; Holzkämper 2006, 155). The two fragments of baguette demi-ronde were also made of ivory and the ten fragments of barbed points were made of antler (Tinnes 1994, 172). One of the latter might be a fragment with two rows of barbs. The only certain bone point was made of a cetacean bone, probably whale that was otherwise not identified in Andernach (Langley/Street 2013). These animals were hunted in Mediterranean Sea or the Atlantic Ocean and, thus far, comparable implements are only known from south-western France at a distance of approximately 1,000 km from Andernach (Langley/Street 2013). In addition, several needle and awl fragments made of bone as well as actual »needle cores« (Holzkämper 2006, 154) were found. Furthermore, two cut bone fragments showed polish on one end and, perhaps, belong into the context of hide-working

site	figurines	engravings			colour				jewellery					ref.
		»cut« cortex	figures / symbols on portables	figures / symbols on walls	spots	paintings on gravels	paintings on walls	colourants / powder	molluscs	amber / jet	drilled teeth	incised teeth	pen-dants	
Andernach, lower horizon	+ (i+b+s)	+	+ (s)		+			+	+	+ (j)	+ (h+c)		? (j)	1-5
Andernach, up-per horizon												+ (h)		6
Gönnersdorf	+ (i+a+s)	+	+ (s)					+	+	+ (j)	+ (h+c)	+ (h)	? (s+b)	3-5; 7
Gönnersdorf SW			+ (s)					+		(+ (j))			? (s)	8
Wildweiberlei		+	? (s)					+						9
Oberkassel	+ (b+a?)							+				+ (h)		10
Irlich								+			+ (h)	+ (h)		11
Urbar								(+)						12
Niederbieber			+ (s)					+						13
Boppard			+ (b)							(+ (j))				14

Tab. 18 »Special« goods recovered at sites from the Central Rhineland. Sub-assemblages are shaded grey. For symbols see **tab. 15**. Material abbreviations are: **(s)** stone material; **(l)** lithic material; **(am)** amber; **(j)** jet; **(b)** bone; **(a)** antler; **(i)** ivory; **(h)** herbivore; **(c)** carnivore. – Within the brackets: **/** or; **+** and. – For further details see text. – References (ref.): **1** Veil 1982; **2** Holzkämper 2006; **3** Höck 1995; **4** Álvarez Fernández 1999; **5** Tinnes 1994; **6** Baales/Street 1996; **7** Bosinski 1987; **8** Buschkämper 1993; **9** Terberger 1993, 158. 165 f.; **10** Henke/Schmitz/Street 2006; **11** Baales 2002; **12** Baales/Mewis/Street 1998; **13** Bolus 1992; **14** Wenzel 2004.

implements (Veil 1982, 416; Holzkämper 2006, 155). Moreover, in Andernach 2 a complete bâton percé made of antler and a further four fragmented antler specimens were recovered (Tinnes 1994, 143-145).
Along with the mammoth (*Mammuthus primigenius*) which is only represented by possible fossil ivory, red deer (*Cervus elaphus*) could be identified only by two perforated canine (Street et al. 2006, 762) which arrived presumably as ornaments at the site (**tab. 18**). Moreover, several animal teeth were perforated or cut and, thus, probably used as personal ornaments. Esteban Álvarez-Fernández (Álvarez Fernández 2009, 48) emphasised particularly the modified arctic fox premolars resembling female figurines of the so-called *Gönnersdorf* type that was focused on the presentation of a female torso (see below).
In Andernach, 22 of these female figurines were found. Two of these items were uncovered at Andernach 3 (Bergmann/Holzkämper 2002) and the other 20 originated from the areas Andernach 1 and 2 (Höck 1995, 268-277). These abstract figures were basically formed by a long shaft from which a wide bulge representing the bottom and sometimes a smaller bulge on the opposite side representing the breasts were carved out (cf. Bosinski/Fischer 1974; Bosinski/d'Errico/Schiller 2001). Thus, the *Gönnersdorf* type represented the torso of a woman with secondary sex characteristics and the indication of limbs but no feet, hands, heads, or ornaments were shaped. These figurines were generally made of ivory (n=20), one was possibly made of bone, and another one of quartzitic slate (Holzkämper 2006, 160f.). Three were fragmented under the pressure of the overlying sediment. Two pieces were probably unfinished (Höck 1995, 288) and one in the process of being reshaped (Höck 1995, 289). Moreover, in the material from the Schaaffhausen excavation a reindeer pedicle was found, that presumably was the waste of antler exploitation, but it was transformed by a few manipulations into a bird figurine (Bosinski/Hahn 1972).
Along with figurines, the slate was formed into small round plates with a central perforation and occasional engraved ornaments. The function of these pieces is unclear (e.g. Gaudzinski-Windheuser/Jöris 2006, 54; Álvarez Fernández 2009, 51).
Comparable to Gönnersdorf, the slate plates of Andernach were often engraved with detailed drawings of the Pleistocene fauna, in particular horses, as well as schematic humans such as the abstract female shapes of the *Gönnersdorf* type (Bosinski 1996).
Furthermore, various exotic items were found at Andernach such as molluscs from fossil deposits in the Mainz basin (approximately 70km south-eastwards) and from the Mediterranean Sea (at least some 800km in the south-west; Alvarez Fernández 2001; Holzkämper 2006, 159). Some pieces of a tabular brown coal and jet were also recovered from the site. The origin for both materials might be found in the local Rhine gravels (Holzkämper 2006, 158). However, the use of the materials, in particular of brown coal, remains uncertain (Álvarez Fernández 1999, 84; Holzkämper 2006, 158), but the transformation of jet into beads was attested at Gönnersdorf (Álvarez Fernández 1999) and, perhaps, the material was used comparably at Andernach.
In sheltered areas such as underneath the stone plates and inside pits, the horizon was still coloured by haematite that was also found in the form of some hundred smaller pieces at the site. The use as well as the origin of this material were discussed controversially (Holzkämper 2006, 86-88). Probably, the mineral used at Andernach came from the local gravels or a regional source based on the chemical similarity to material found some 20km north-westwards (Bosinski 1979, 138).

Spatial organisation of lower horizon

During the excavations several evident structures were documented (**tab. 19**; Eickhoff-Cziesla 1992) including almost 60 pits and a partially thick cover of slate, sandstone, quartzite, and basalt plates which could reach over 50cm in diameter (Veil 1982, 403; Holzkämper 2006, 64-85). These structures were used to define three large concentrations (I-III) in Andernach 2 and another one (IV) in Andernach 3.

site	evident structures					latent structures		grave	further	ref.
	pavement	stone setting	pit	hearth, stone packed	hearth, sediment alteration	hearth, latent	dwelling			
Andernach, lower horizon	+		+			+	+		? (cache?)	1-3
Andernach I	+		+						? (cache?)	1-2
Andernach II		+	+							1-2
Andernach III	+ (?)		+							1-2
Andernach IV		+	+			+	+			3
Andernach, upper horizon			?		+	+	+			4-5
Andernach 2-FMG						+	+			4
Andernach 3-FMG			?		+		+			5
Gönnersdorf	+	+	+	+	+	+	+			6-10
Gönnersdorf I	+	+	+		+	?	+			6-7
Gönnersdorf II	+	+	+			+	+			8-9
Gönnersdorf III	+		+			+	?			10
Gönnersdorf IV		+		+	?	+	+			9-10
Gönnersdorf V		+		+		+	?			7
Wildweiberlei	+	?	+		+		?			11
Oberkassel								+		12-13
Irlich								+		14
Kettig			?			+				14
Urbar				?		+				15
Niederbieber			+		+	+	+		? (cache?)	5; 16-19
Niederbieber 1					+		+			5; 16-17
Niederbieber 3						?				16
Niederbieber 4+17a					+		+			5; 16-17
Niederbieber 5						+				17
Niederbieber 6+10a						+				17
Niederbieber 7						+				17
Niederbieber 8						?				17
Niederbieber 9						+				17
Niederbieber 10						+				17
Niederbieber 11						+				17
Niederbieber 12						+	+			17
Niederbieber 13						+	+			17
Niederbieber 14						+				17
Niederbieber 15						+			? (cache?)	17; 18
Niederbieber 17						+	+			17
Niederbieber 19					+					19
Niederbieber 20			+		+					19
Boppard						+				20
Bad Breisig					+		?			21

Tab. 19 Structures on sites from the Central Rhineland. Sub-assemblages are shaded grey. For symbols see **tab. 15**; except **?** means possible but anthropogenic formation or relation to the archaeology is uncertain. – For further details see text. – References (ref.): **1** Eickhoff-Cziesla 1992; **2** Floss/Terberger 2002; **3** Holzkämper 2006; **4** Stapert/Street 1997; **5** Gelhausen/Kegler/Wenzel 2004; **6** Bosinski 1979; **7** Buschkämper 1993; **8** Sensburg 2007; **9** Sensburg/Moseler 2008; **10** Terberger 1997; **11** Terberger 1993, 161-166; **12** Verworn/Bonnet/Steinmann 1919; **13** Schmitz 2009; **14** Baales 2002; **15** Baales/Mewis/Street 1998; **16** Bolus 1992; **17** Gelhausen 2011a; **18** Baales 2003b; **19** Gelhausen 2011c; **20** Wenzel 2004; **21** Grimm 2004.

The plates used for the pavement were occasionally knapped to size and/or subsequently slurred by trampling (Veil 1982, 408). Between and below these plates the lithic and organic material was found that presumably fell through gaps and/or were deposited during an earlier occupation episode or phase. The fact that various occupation phases or even episodes formed the concentrations at Andernach was assumed from the various fillings of the pits that were also sealed occasionally by larger plates (Bergmann/Holzkämper 2002, 481-484). However, the time period passed between the various deposits in the pits is difficult to estimate and, thus, the concentrations might have been settled recurrently during specific seasons or for one longer time period.

The density of the stone pavement varied and only concentration I appeared almost completely paved. However, at least for concentration III this scarcity can be partially explained by the incomplete documentation because the larger part of this accumulation lay in the excavation area dug by Schaaffhausen in the 19th century. Moreover, only five pits were attributed to this concentration, whereas in the other concentrations between 13 and 22 pits were recovered. The pits clustered in sub-zones within the stone plate scatters. In these pit zones, the highest density of artefacts were recorded. In parts, this fact was surely due to the good preservation within the pits as a consequence of rapid deposition combined with reduced erosion. Presumably, the pits were dug into the ground for various reasons such as postholes, boiling pits, or refuse pits. Perhaps, some pits were used successively for various tasks. The pits were also considered as hearth pits, but no evidence of this type of use such as coloured and bricked sediment or charcoal deposits was observed. In particular, many exotic items were preserved in the pits and occasionally the accumulated material seemed as a deposition and storage for later use (cache) of these materials (Eickhoff-Cziesla 1992).

Although no evident hearth was detected in Andernach (Veil 1982, 408; Holzkämper 2006, 159), Jörg Holzkämper demonstrated the presence of a central hearth in the concentration IV as well as another hearth at the southern fringe of this concentration by the distribution patterns of various fire indicators such as charcoal or burnt quartz. However, charcoal was only preserved in sheltered areas of concentration IV. In contrast, no charcoal was associated with the lower layer in Andernach 2. Furthermore, burnt quartzes were found in the pits of concentration I and II (Eickhoff-Cziesla 1992, 135. 374) but assumed to rather point to a use of these structures as boiling or refuse pits than representing actual hearths. Some pieces of charcoal from concentration IV were identified as pine (*Pinus* sp., n=8), willow or poplar (Salicaceae, n=2), and daphne (*Daphne* sp., n=2).

Since the well excavated areas were subdivided by the extent of the pit zones and the concentration of stone plates, the majority of lithic artefacts was attributed to the almost completely excavated concentrations I and IV. Baltic flint cores were found exclusively in concentration I and the patinated side-scraper that was also made of Baltic flint was recovered in this concentration. Since the patinated surface was capped by some more recent negatives, Thomas Terberger assumed that the piece was gathered with the other raw material nodules of Baltic erratic flint at the source region (Floss/Terberger 2002, 129), i.e. the piece was probably taken for its suitable shape and raw material, not as a curiosity. However, the distribution of this generally dominant raw materials was varied. The Baltic erratic flint was found scattered across all of the site in varying quantities. In concentration II only some residual cores made of Tertiary quartzite that occurred generally in all concentrations were found in the periphery. The chalcedony cores were associated with concentration III (Floss/Terberger 2002, 26 Abb. 26), whereas the Western European flint cores only occurred in concentration IV. Furthermore, the number of cores associated with the concentrations also differs significantly (**tab. 13**) with a clear accumulation of cores in concentration I (n=38). The hammerstones were also most frequently connected to the pit area in concentration I, whereas retouchers were basically found in the periphery of concentration I. Moreover, further hammerstones were recovered in the remaining modern excavated pit zone of concentration III, but in concentration II few indications of these rock tools were found. This observation is concomitant with the rarity of cores from this area (n=3) and, thus, the assumption that blank production played no significant role in this concentration. A comparable picture was recovered in concentration IV where some hammerstones and further indications of an *in situ* blank production process were found but remained meagre (Holzkämper 2006, 100f. 140). Among the five occurrences where material from concentration II was fitted to material from concentration IV (Holzkämper 2006, 138) was one example of a serial blade production including a single preparation flake indicating that perhaps some blank production was performed in the south-western concentration IV, whereas in the north-eastern concentration II the blanks were further retouched (Holzkämper 2006, 101). In Andernach 2, the refitted material (n = 5,334) connected concentration I and III (i. e. E-W orientation) but left out concentration II. The lines running in the direction of concentration II from any of the other concentrations of Andernach 2 were generally associated with the fissures. Besides such inter-concentration connections, intra-concentration clusters were identified by spatial analysis. For example, in Andernach IV a second accumulation of material was found south of the pit zone and was interpreted in combination with the hearth at the fringe and an entrance situation of a suggested dwelling structure (Holzkämper 2006, 21. 32f.).

However, in all concentrations all types of retouched artefacts occurred, although sometimes in small numbers. Nevertheless, this presence implies that comparable activities were performed in all concentrations but in different intensity. In particular, the LMP, truncations, and borers were distributed relative evenly across the concentrations. In particular, no significant cluster of the LMP that could be identified as an accumulation, for instance, near a hearth was observed thus far in Andernach 2 and in Andernach 3. Although the LMP were generally associated with the suggested hearths, they also accumulated between these centres (Holzkämper 2006, 119f.). The end-scrapers were clearly concentrated in concentration I indicating some type of hide scraping in this area. The burins in concentration III were again mainly associated with the disturbance, and possibly originated from concentration II where the majority of burins (n = 154) was found (Floss/Terberger 2002, 98). In particular, the Lacan type burins occurred with a single exception in concentration II and suggested cutting activities.

The contradictory seasonal indications of the faunal remains can be explained by the spatial distribution of the indicative material: The reindeer and horse teeth suggesting the autumn/winter season were associated with the concentrations I and III, whereas the salmon remains were mainly found in concentration II. The carbon and nitrogen isotope analysis of the material from Andernach 2 could not establish significant differences between the three concentrations (Stevens et al. 2009b). The Andernach IV material provided no samples for isotope analyses nor were seasonal indications recognised yet.

The haematite colouring was not very intensive in Andernach and appeared concentrated in some locations where perhaps haematite lumps were ground to red powder (cf. Holzkämper 2006, 87f.). Whether these areas can be associated with the tanning of animal skins remains unclear, but the adhesive haematite on some end-scrapers, the micro-wear analysis of this type of lithic tools, and the generally congruent distribution of haematite and end-scraper concentrations suggests the use in a common process.

In the eastern part of Andernach 2, the figurines and organic artefacts were found associated with the disturbed channel (Höck 1995, 277-279) and further pieces came from concentration I, often from pits. Since at least the bird figurine originated from the excavation area of Schaaffhausen some art was connected with concentration III. Equally, the two pieces from Andernach 3 originated from such sheltered situations (Holzkämper 2006, 160).

The Magdalenian concentrations at Andernach were considered as habitation sites with (paved) dwelling constructions (Holzkämper 2006, 14-28). In the Central Rhineland the discussion of Magdalenian dwelling constructions and structuring of habitation sites was mainly based on the Gönnersdorf site but applied equally to the lower horizon of Andernach (Eickhoff-Cziesla 1992). However, this discussion is presented in more detail in the presentation of Gönnersdorf (see p. 116-122).

In general, the concentrations in Andernach were considered to represent a base camp due to the large variety of material, the often high number of single specimens within the various material groups, and the massive constructions. The recovered constructions seemed to be made for a longer duration, and for the creation of these structures, a group of people was needed.

Chronology of the lower horizon

Meanwhile, Andernach with its 35 ^{14}C dates is one of the most frequently ^{14}C-dated sites of the Lateglacial record in north-western Europe (**tab. 20**). However, several dates have to be rejected due to various reasons and some other dates referred to younger occupation events at the same site (see p. 265-269). Thus far, Late Magdalenian samples were only received from the concentrations I, II, and III. A temporal distinction of the dates from the various concentrations was not possible. They covered a period of 13,450-12,690 years ^{14}C-BP. After calibration this period fell completely into the Late Pleniglacial (16,920-15,000 years cal. b2k; see p. 466-470).

The stratigraphic position in a loess deposit as well as the presence of a fauna preferring cold and open habitat sustained the radiometric attribution.

Nevertheless, for the lower horizon at Andernach more considerations concerning the spatial structuring and, perhaps, chronological development can be further probed after the accomplishment of a currently prepared synthesis of the site that focuses particularly on the spatial patterns of this major Late Magdalenian site (pers. comm. Martin Street, Neuwied). This analysis will also help reveal the settlement dynamics between the concentrations by outlining various working areas as well as defining their precise temporal development.

However, according to the thus far known evidence, at least two quasi-contemporary occupation events (concentration I/III and concentration II/IV) were distinguished in the chronology of the lower horizon of Andernach. The temporal distinction of these events is beyond the possible resolution of radiocarbon dating. Furthermore, each occupation event was related to two concentrations that were formed by several occupation phases if not distinct episodes.

Archaeological material of the upper horizon

As explained previously, the 3,794 lithic artefacts mentioned in the various analyses for the upper horizon (**tab. 11**; Bolus 1984; Gelhausen/Kegler/Wenzel 2004) represent an approximate value. Besides the occasionally uncertain attribution of individual specimens, Michael Bolus did not specify whether his number of 2,894 lithic artefacts for Andernach 2 included splinters i.e. the value might be too high for artefacts ≥1 cm (Bolus 1984). However, Harald Floss mentioned only some 101 lithic artefacts less but clearly stated that only 1,377 pieces were larger than 1 cm (Floss 1994, 272). Yet, the size distribution of the additional 100 items remains unclear. Furthermore, Jan Kegler included only the chalcedony artefacts (Gelhausen/Kegler/Wenzel 2004, 71; cf. Kegler 2002, 501 f.), probably due to the problems of distinguishing pieces made of other raw materials from the Late Magdalenian artefacts i.e. the value for Andernach 3 might be too low. Moreover, refitted material proved that, in the material excavated by Schaaffhausen in 1883, further remains of the FMG could be found (Bolus/Street 1985). Thus, a reanalysis and/or a more comprehensive publication of the Andernach material with special reference to the lithic artefacts would be particularly appreciable for the FMG component. A technological reanalysis of the FMG material from Andernach by Ludovic Mevel is on-going. Preliminary observations by him indicated that, in general, the reduction of the materials was comparable for the FMG material in the two modern areas of Andernach, but that some raw material types were treated differently in two notable cases. In one case the difference is perhaps due to the quality of the raw material (Tertiary quartzite), and in the other case (Western European flint, type 11) a Late Magdalenian intrusion was considered possible (written comm. Ludovic Mevel, Nanterre).

site	lab. no.	years 14C-BP	± years	material	species	comment	years cal. b2k	ref.
Andernach, lower horizon III (?)	*OxA-10492***	*13,500*	*90*	*rib*	*Equus sp.*	*cut-marks;* REJECT: redating available	×	*1-4*
Andernach, lower horizon III (?)	OxA-10651**	13,270	180	phalanx	*Equus* sp.		17,100-15,500	1-4
Andernach, lower horizon II	OxA-1128	13,200	140	rib	*Equus* sp.	sample from pit	16,920-15,520	3-5
Andernach, lower horizon III (?)	OxA-10493**	13,185	80	rib	*Equus* sp.	cut-marks	16,720-15,640	1-4
Andernach, lower horizon I / upper horizon 2	GrA-16986	13,180	70	shaft fragment	*Cervus elaphus*	impact scar	16,670-15,670	6
Andernach, lower horizon I	OxA-V-2216-43	13,135	55	humerus	*Equus* sp.		16,540-15,620	4
Andernach, lower horizon I / upper horizon 2	GrA-16985	13,110	80	shaft fragment	*Cervus elaphus?*		16,550-15,550	6
Andernach, lower horizon II	OxA-V-2218-40	13,110	50	humerus	*Equus* sp.		16,390-15,630	4
Andernach, lower horizon II	OxA-1129	13,090	130	fragments	*Equus* sp.	sample from pit	16,680-15,400	3-5
Andernach, lower horizon III (?)	OxA-18409**	13,025	50	rib	*Equus* sp.	cut-marks; redating of OxA-10492	16,220-15,500	4; 7
Andernach, lower horizon I	OxA-V-2218-38	13,015	50	metatarsus	*Equus* sp.		16,210-15,490	4
Andernach, lower horizon III	OxA-1130	12,950	140	bone fragments	*Equus* sp.	sample from pit	16,410-15,170	3-5
Andernach, lower horizon I	OxA-1125	12,930	180	bone fragments	*Equus* sp.	sample from pit	16,530-15,010	3-5
Andernach, lower horizon I	OxA-1126	12,890	140	rib	*Equus* sp.	sample from pit	16,270-15,070	3-5
Andernach, lower horizon II	OxA-1127	12,820	130	bone fragments	*Equus* sp.	sample from pit	15,920-15,000	3-5
Andernach, lower horizon III	OxA-V-2223-37	12,675	55	humerus	*Equus* sp.	sample from pit	15,440-15,000	4
Andernach, upper horizon 2	*OxA-999*	*12,500*	*500*	shaft fragments	*Cervus elaphus*	REJECT: redating available	×	3-4; 8
Andernach, upper horizon 2	OxA-985	12,300	200	bone	*Rupicapra rupicapra?*	PROBLEMATIC: large standard deviation	15,340-13,620	3-4; 8
Andernach, upper horizon 2	OxA-V-2218-39	12,270	50	femur	*Equus* sp.		14,560-13,960	4
Andernach, upper horizon 2	GrA-16987	12,050	70	scapula fragments	*Castor fiber*		14,100-13,700	4; 6

Tab. 20 14C dates from sites from the Central Rhineland. If the sub-assemblage is known from which the sample originated, the sub-assemblage is given behind the site name. * dates which were pretreated by the use of ion-exchanged gelatin (Lab.code: AI) in the Oxford series (cf. Jacobi/Higham 2009, 1896); ** dates which might be contaminated due to the use of a method leaving traces of a humectant in the collagen (Lab.code: AF*) in the Oxford series (cf. Higham et al. 2007, S55 & S2). Doubtful dates are shaded grey. Rejected dates are shaded grey and set in italics and, in addition, the main reason for rejection is given in comment. For further details see p. 265-269 and text. The dates were calibrated with the calibration curve of the present study (see p. 358-364) and the calibration program CalPal (Weninger/Jöris/Danzeglocke 2007). The result range of 95 % confidence is given for the calibrated ages (years cal. b2k). – References (ref.): **1** Bronk Ramsey et al. 2002; **2** Stevens/Hedges 2004; **3** Street/Terberger 2004; **4** Stevens et al. 2009b; **5** Hedges et al. 1987; **6** Kegler 2002; **7** Higham et al. 2011; **8** Gowlett et al. 1987; **9** Street 1993; **10** Münnich 1957; **11** Schwabedissen 1957; **12** Higham et al. 2007; **13** Hedges et al. 1998b; **14** Brunnacker 1978d; **15** Evin/Marien/Pachiaudi 1975; **16** Evin/Marien/Pachiaudi 1978; **17** Street/Baales/Weninger 1994; **18** Baales/Street 1998; **19** Baales 2002, 11f. 40-45; **20** Baales/Mewis/Street 1998; **21** Lanting/Niekus/Stapert 2002; **22** kind permission of Stefan Wenzel; **23** Baales/Jöris 2001.

site	lab. no.	years ^{14}C-BP	± years	material	species	comment	years cal. b2k	ref.
Andernach, upper horizon 2	GrA-16991	12,040	70	shaft fragment	*Bos* sp./*Bison* sp.		14,090-13,690	4; 6
Andernach, upper horizon 2	GrA-16989	11,960	70	metatarsal fragment	*Cervus elaphus*		13,980-13,660	4; 6
Andernach, upper horizon 2	OxA-984	11,950	250	shaft fragments	*Cervus elaphus*	re-dating of OxA-999	14,670-13,190	3-4; 8
Andernach, upper horizon 2	OxA-1924	11,890	120	bone	*Cervus elaphus*		14,030-13,430	3-4; 9
Andernach, upper horizon 2	GrA-16990	11,820	70	bone	artiodactyl		13,800-13,480	4; 6
Andernach, upper horizon 2	OxA-997	11,800	160	bone	*Cervus elaphus*		13,980-13,300	3-4; 8
Andernach, upper horizon 3	GrA-16993	11,590	80	bone		northern concentration	13,590-13,230	4; 6
Andernach, upper horizon 2	OxA-998	11,370	160	bone	*Bos* sp./ *Bison* sp.		13,550-12,910	3-4; 8
Andernach, upper horizon 1	*H-85/91*	*11,300*	*220*	antler protein	Cervidae	REJECT: pretreatment, protein, material, uncertain association	×	4; 10-11
Andernach, upper horizon 3	GrA-16994	11,160	70	bone		southern concentration	13,190-12,870	4; 6
Andernach, upper horizon 2	*GrA-16521*	*10,970*	*60*	calcined bone		REJECT: calcined bone	×	4; 6
Andernach, upper horizon 2	*GrA-16613*	*9,490*	*45*	calcined bone		REJECT: calcined bone	×	4; 6
Andernach, upper horizon 3	*GrA-16618*	*7,550*	*40*	calcined bone		REJECT: calcined bone	×	4; 6
Andernach, upper horizon 2	*GrA-16616*	*7,360*	*40*	calcined bone		REJECT: calcined bone	×	4; 6
Andernach, upper horizon 3	*GrA-16621*	*5,775*	*40*	calcined bone		REJECT: calcined bone	×	4; 6
Andernach, upper horizon 1?	*H193-178*	*4,330*	*360*	bone		REJECT: carbonate fraction	×	4; 10
Gönnersdorf II	*OxA-10199***	*14,570*	*90*	ivory	*Mammuthus primigenius*	worked; REJECT: probably fossil material	18,030-17,590	3-4
Gönnersdorf I	*OxA-10239*	*14,380*	*100*	femur	*Mammuthus primigenius*	REJECT: probably fossil material	17,990-17,190	3-4
Gönnersdorf II	*OxA-10200***	*13,810*	*90*	molar fragment	*Coelodonta antiquitatis*	REJECT: probably fossil material	17,150-16,950	3-4
Gönnersdorf II	*OxA-10201***	*13,610*	*80*	molar fragment	*Coelodonta antiquitatis*	REJECT: probably fossil material	17,070-16,790	3-4
Gönnersdorf I	OxA-V-2223-39	13,270	55	metatarsal	*Equus* sp.		16,810-16,090	4
Gönnersdorf II	OxA-V-2223-40	13,165	55	metatarsal	*Equus* sp.		16,600-15,680	4
Gönnersdorf II	OxA-V-2223-41	13,095	55	scapula	*Bison* sp.		16,360-15,600	4
Gönnersdorf III	OxA-V-2223-43	13,075	55	metapodial	*Rangifer tarandus*		16,330-15,570	4
Gönnersdorf III	OxA-15295	13,060	60	metapodial	*Rangifer tarandus*	marrow fractured	16,310-15,550	4; 12
Gönnersdorf II	OxA-V-2223-31	13,010	55	metatarsal	*Rangifer tarandus*		16,220-15,460	4
Gönnersdorf I	OxA-V-2223-42	12,990	55	metatarsal	*Rangifer tarandus*		16,180-15,420	4

Tab. 20 (continued)

site	lab. no.	years ^{14}C-BP	± years	material	species	comment	years cal. b2k	ref.
Gönnersdorf I	OxA-5729*	12,910	130	rib fragments	*Equus* sp.		16,290-15,130	3-4; 13
Gönnersdorf	*KN-1980*	*12,910*	*105*	mollusc shell		REJECT: no association	×	4; 14
Gönnersdorf I	OxA-5730*	12,790	120	rib fragments	*Equus* sp.		15,810-14,970	3-4; 13
Gönnersdorf I	OxA-5728*	12,730	130	rib fragments	*Equus* sp.		15,760-14,720	3-4; 13
Gönnersdorf	Ly-1172	12,660	370	rib fragments	*Equus* sp.?	PROBLEMATIC: bulked sample	16,680-13,600	4; 15
Gönnersdorf	Ly-768	12,380	230	bones		PROBLEMATIC: bulked sample	15,520-13,680	4; 16
Gönnersdorf II	*OxA-2069*	*11,830*	*110*	ivory	*Mammuthus primigenius*	worked; REJECT: post-LST date	×	4; 17
Gönnersdorf	*Ly-1173*	*11,100*	*650*	rib fragments	*Equus* sp.?	REJECT: bulked sample; post-LST date	×	4; 16
Gönnersdorf	*KN-1979*	*10,540*	*210*	mollusc shell		REJECT: no association; post-LST date	×	4; 14
Gönnersdorf V	OxA-15296	12,385	65	radius	*Alces alces*	PROBLEMATIC: uncertain association	15,010-14,050	12
Wildweiberlei	OxA-18410**	12,835	55	modified antler	*Rangifer tarandus*	(dated after recognition of possible contamination)	15,660-15,300	7
Oberkassel	OxA-4793*	12,270	100	ulna	*Canis familiaris*		14,820-13,820	3; 13; 18
Oberkassel	KIA-4162	12,210	60	humerus	*Canis familiaris*		14,300-13,900	18
Oberkassel	OxA-4792*	12,180	100	humerus	*Homo sapiens*	female	14,470-13,750	3; 13; 18
Oberkassel	KIA-4161	12,110	45	maxilla, I2, I3	*Canis familiaris*		14,140-13,780	18
Oberkassel	OxA-4791*	11,780	90	os penis	*Ursus spelaeus* (?)		13,780-13,430	3; 13; 18
Oberkassel	KIA-4163	11,620	60	ulna	*Canis familiaris*		13,600-13,280	18
Oberkassel	OxA-4790*	11,570	100	humerus	*Homo sapiens*	male	13,600-13,200	3; 13; 18
Irlich	OxA-9736**	12,310	120	rib	*Homo sapiens*	adult; PROBLEMATIC: brownish bone	15,030-13,830	1; 19
Irlich	UtC-9221	12,110	90	long bone	*Homo sapiens*	red coloured; neonate	14,220-13,700	1; 19
Irlich	OxA-9848**	11,965	65	rib	*Homo sapiens*	neonate	13,980-13,660	1; 19
Irlich	OxA-9847**	11,910	70	femur	*Homo sapiens*	red coloured; adult	13,930-13,570	1; 19
Irlich	*OxA-9876***	*2,660*	*40*	skull cap	*Homo sapiens*	adult; REJECT: brownish bone; Holocene age	×	1; 19
Kettig	*GrA-12396*	*11,960*	*90*	bone		REJECT: calcined bone	×	1
Kettig	*GrA-14171*	*11,720*	*60*	bone		REJECT: calcined bone	×	19

Tab. 20 (continued)

site	lab. no.	years ^{14}C-BP	± years	material	species	comment	years cal. b2k	ref.
Kettig	*GrA-13389*	*11,710*	*50*	bone		REJECT: calcined bone	×	19
Kettig	Hd-18123	11,314	50	bones	cf. *Cervus elaphus*	PROBLEMATIC: bulked material	13,280-13,040	19
Kettig	*GrA-14762*	*11,210*	*60*	metacarpal	*Capreolus capreolus*	REJECT: calcined bone	×	19
Urbar	OxA-1137	11,350	120	bone	*Cervus elaphus*		13,440-12,960	3; 19-20
Niederbieber 9	*GrA-16622*	*11,290*	*40*	bone		REJECT: calcined bone	×	21
Niederbieber 3	OxA-1135	11,130	130	astralagus	*Equus* sp.	PROBLEMATIC: uncertain association	13,260-12,700	3; 19
Niederbieber 2 (19)	OxA-2066	11,110	110	bone	*Alces alces*		13,210-12,690	3; 19
Niederbieber 1	*OxA-1132*	*10,700*	*130*	bone	*Cervus elaphus*	REJECT: post-LSE date	×	3; 19
Niederbieber 4	*OxA-1136*	*10,480*	*130*	shaft fragment	*Cervus elaphus*	REJECT: post-LSE date	×	3; 19
Niederbieber 17	*GrA-18672*	*10,420*	*110*	bone		REJECT: poor carbon quality; post-LSE date	×	21
Niederbieber 7	*OxA-2067*	*10,390*	*100*	bone	*Cervus elaphus?*	REJECT: post-LSE date	×	3; 19
Niederbieber 17	*GrA-18881*	*10,390*	*80*	bone		REJECT: poor carbon quality; post-LSE date	×	21
Niederbieber 2 (19)	*OxA-1133*	*9,750*	*240*	bone	*Alces alces?*	REJECT: post-LSE date	×	3; 19
Niederbieber 2 (19)	*OxA-1134*	*6,250*	*130*	tooth	*Equus* sp.	REJECT: post-LSE date	×	3; 19
Boppard	KIA-26644	11,095	55	metapodial	*Cervus elaphus*		13,110-12,750	22
Bad Breisig	GrA-17493	10,840	60	charcoal	*Pinus* sp.		12,820-12,660	23
Bad Breisig	*GrA-17642*	*10,480*	*80*	charcoal		REJECT: uncertain association	×	23
Bad Breisig	*GrA-17716*	*10,220*	*60*	calcined bone		REJECT: calcined bone	×	23

Tab. 20 (continued)

In addition to the chalcedonies, which probably originated from Bonn-Muffendorf some 30 km north-north-westwards of Andernach (**tab. 12**), several pieces of various Western European flint varieties and Baltic erratic flint (cf. Floss 1994, 271-283) were singled out from the Late Magdalenian assemblage and attributed to the FMG inventory due to the lack of patination (Veil 1982, 400) and/or their typological classification (cf. Kegler 2002, 502). Furthermore, due to the latter reason, pieces of Tertiary quartzite and indurated slate were also attributed to the upper horizon of Andernach 3. Both materials occurred also in Andernach 2 where Tertiary quartzite was, in fact, the dominant raw material. Additionally, in Andernach 2 siliceous oolite was recorded as well as the silicified limestone. The former raw material originated presumably from the Mainz Basin, at least 70 km to the south-east, or possibly from the Rhine gravels. The silicified limestone was probably picked up from gravels, possibly local or farther upstream or from the Nahe, some 60 km south-east of Andernach. In total, approximately 20 nodules of 13 different varieties were assumed to have been brought to the site (Stapert/Street 1997, 179). The amount of pieces with cortex indicated the transport of small, unprepared nodules of the foreign raw materials to the site (Veil 1982, 400). Hence, the lithic raw material attributed to Andernach 2-FMG is very heterogeneous.

At least 66 core and core fragments were recovered during the two modern excavations (**tab. 13**). In Andernach 2 the blank production was described as an opportunistic strategy which produced particularly flakes (Bolus 1984, 60; Stapert/Street 1997, 179). However, this picture can already be revised by the preliminary observations of Ludovic Mevel: The various blanks indicated the flexible use of a soft hammerstone to receive various blank types depending, presumably, on the quality and, perhaps, on the exploitation stadium of the raw material (written comm. Ludovic Mevel, Nanterre; Mevel/Grimm 2019). The cores which were worked from two platforms were frequently not in a polar but transversal or angular orientation to one another (Bolus 1984, 49-54). Michael Bolus recognised a relatively large number of blanks with cortex and he suggested therefore the introduction of the material in raw nodules (Bolus 1984, 58f.). Thus far, only the retouched artefacts were analysed in Andernach 3 (Kegler 2002). Hence, the technology can only be partially read from the blanks chosen for further retouch which was mainly performed on bladelets for the LMP and dominantly flakes for the other tools, although blades and bladelets were also used (Kegler 1999). Consequently, the Andernach 3 material was presumably produced with a strategy similar to Andernach 2. This assumption was also supported by Ludovic Mevel's preliminary observations (written comm. Ludovic Mevel, Nanterre).

In total, 280 intentionally retouched pieces were found in the upper horizon (**tab. 14**). Furthermore, almost 200 pieces of tool production wastes such as end-scraper caps (n=134) and burin spalls (n ≥63) were identified. These specimens indicated the production and/or resharpening of the tools on the site. Thus far, no wastes of the LMP production were found, although the LMP (n=128) dominated clearly among the retouched artefacts. Almost 70 of these pieces were unambiguous points and according to Michael Bolus and Jan Kegler over 50 were *Federmesser*. Yet, the pieces displayed by Jan Kegler appear more variable in their morphology (Kegler 2002, 505 Abb. 5). Among the 39 backed blades and bladelets only one with two truncations and one laterally retouched piece occurred, all the others were simple ones. Four specimens from Andernach 2 were classified as pieces which were distally pointed by lateral retouch, but at least one of these specimens was used as a perforator (Bolus 1984; Plisson 1985). The truncations (n=20) were in fact dominantly oblique (n=15). End-scrapers and burins occurred in equal numbers (n=41). The former were dominantly of the small, thumb-nail variant on flakes (n=22). The latter group was diverse with those specimens on broken or natural edges (n=14) being most numerous followed by dihedral types (n=10) and burins on truncation (n=8). The three composite tools were all combinations of burin and end-scraper. Some 35 pieces were laterally retouched. A further six artefacts were identified as splintered pieces.

259 pieces were examined for micro-wear traces and on at least 65 traces of use were recognised. Besides the backed pieces, ten unmodified artefacts showed traces of use as parts of projectiles (Plisson 1985) and one backed piece yielded residues interpreted as hafting glue (Kegler 1999, 42).

In the upper horizon of each excavation area a retoucher was found (Bolus 1984; Kegler 2002). Moreover, hammerstones were probably associated with the upper horizon in Andernach 3 but these specimens were not further analysed thus far (Kegler 2002, 508).

Comparable to the other material, the fauna of the upper horizon from Andernach 2 (**tab. 15**; Street 1993; Street et al. 2006) was difficult to distinguish from the material of the lower horizon. However, 407 remains were determinable and clearly dominated by red deer (*Cervus elaphus*, n=209, MNI=5). Further large game species were present such as two large cattle (*Bos* sp., n=54), presumably aurochs (cf. *Bos primigenius*), and an elk (*Alces alces*, n=36). Furthermore, the remains of at least two chamois (*Rupicapra rupicapra*, n=26) and two beavers (*Castor fiber*, n=44) were identified. In addition, some 320 fish bones were found and attributed generally to fish of the Cyprinidae (n=89) family, usually Eurasian dace (*Leuciscus* sp.), in particular,

European chub (*Leuciscus cephalus*, n = 177). Moreover, remains of pike (*Esox lucius*, n = 47) were also frequent. Common minnow (*Phoxinus phoxinus*, n = 5) and European perch (*Perca fluviatilis*, n = 1) were rarely determined. The European bullhead (*Cottus gobio*, n = 1) is a small and bony fish which in the upper horizon of Andernach was probably a natural intrusion. The association of the few bones of horse (*Equus* sp., n = 4) and of the remains of roe deer (*Capreolus capreolus*, n = 5) with the upper horizon were also uncertain. Meanwhile a humerus of horse was dated slightly younger than the Late Magdalenian phase and a group of ^{14}C dates on horse, red deer, beaver, and bovid remains formed an intermediate period (see p. 468-470; Stevens et al. 2009b). A milk tooth of red deer and unworn teeth of large bovids as well as the absence of antlers could indicate a hunting season in late spring to summer (Street 1993, 135). Several of the bones displayed cut-marks and the spatial distribution suggested butchering, the extraction of marrow as well as discard activities on the site (Street 1993; Street et al. 2006, 776). A further 535 faunal remains larger than 3 cm were found in the upper horizon of Andernach 3 (Gelhausen/Kegler/Wenzel 2004, 71). The presence of red deer was also suggested for this concentration by a provisional analysis but the material was poorly preserved and a final analysis is not yet available (Street et al. 2006, 765).
Among the organic material of Andernach 2, a tooth of red deer was found which wore parallel incisions at its root (**tab. 18**). This piece is one of the rare items identified as special goods within the FMG inventories.

Spatial organisation of the upper horizon
FMG material was found in each modern area and was also present in the Schaaffhausen area according to the material from this collection (Veil 1982; Gelhausen/Kegler/Wenzel 2004; Street et al. 2006). Thus, a widespread scatter of FMG material on the Martinsberg is evident.
The distribution of burnt bone, quartz, burnt lithic material, and charcoal indicated the presence of three latent hearths in Andernach 2 and further supported the observation of two hearths in Andernach 3 (**tab. 19**). In Andernach 2, one hearth was situated in the eastern part near the infilled fissure lacking charcoal but surrounded by a lithic concentration. A second hearth dominated by charcoal of willow and situated in approximately the centre of the excavation area was associated with beaver remains (Stapert/Street 1997, 180). The third hearth of Andernach 2 was set only some metres westwards of the second, and contained birch, willow, and perhaps pine charcoal. It was associated with red deer and fish remains. The latter two hearths were located in the main lithic concentration of the upper horizon at Andernach 2. Moreover, the remains appeared directly overlaying the Late Magdalenian material, suggesting either a short time elapsed between these occupation phases, or a hearth which was dug in the sediment or the existence of structural post-holes (Stapert/Street 1997, 180) in which the material was transferred vertically within the stratigraphy. In fact, the faunal remains associated with the two hearths were generally dated to an early FMG period (Stevens et al. 2009b). The surroundings of these hearths were connected by refittings of lithic material and the faunal remains were conjoined across the complete excavated area (Stapert/Street 1997, 184). Density analysis with Dick Stapert's »ring and sector« method (Stapert 1992, cf. Wenzel 2009, 19) demonstrated some type of wall effect around these hearths (Stapert/Street 1997, 180-192). Yet, the analysis was based only on a small area with some overlap and generally few artefacts (Stapert/Street 1997, 180 f.). However, this analysis suggested areas of increased density (cf. Gelhausen 2010) for both hearths on the eastern side of the hearths and Dick Stapert and Martin Street assumed that the opposing distribution of various raw materials in this area indicated the presence of two flintknappers (Stapert/Street 1997, 190-192). Additionally, the western part of the Andernach 2 area was less disturbed, and the various raw material nodules showed some concentrated accumulations indicating varying knapping events (Stapert/Street 1997, 179). In the north-western area of Andernach 2, material of a black variety of a Western European flint (n = 146) was found which was of good quality and wore a fresh limestone cortex. This raw material

yielded some relatively long flakes (l=6.4cm) and could not be associated with any of the hearths of Andernach 2. Thus, this material might originate from an event which can be singled out (Stapert/Street 1997, 182-184; pers. comm. Martin Street, Neuwied). The chalcedonies occurred mainly in a southern concentration of Andernach 2. Perhaps this accumulation was associated with the Andernach 3 concentrations or another, unexcavated concentration. Material refitting might help prove an actual connection of the varied concentrations. In Andernach 2, two further concentrations were singled out east and west of the eastern fissure (Bolus 1984; Bolus 1991).
The two hearths of Andernach 3 were surrounded by archaeological material and also observable during the excavation by reddish and bricked sediment (Gelhausen/Kegler/Wenzel 2004, 71f.). In contrast to the Andernach 2, the highest density around the hearths from Andernach 3 occurred on the western to southern side of the hearth. Again, the backed implements were tendentially closer to the hearths then the other tool types. Due to density distributions of the isopach type (e.g. Gelhausen/Kegler/Wenzel 2004, 70; Gelhausen/Kegler/Wenzel 2005, 2-4; Wenzel 2009, 19f.) and the spread of refittings, Jan Kegler assumed the southern hearth to be encircled by some type of wall, probably a dwelling structure (Gelhausen/Kegler/Wenzel 2004). On the northern periphery of the southern accumulation, some pieces of charcoal, lithic debris, a bone fragment, and a small quartzitic slate plate were found in a 20cm deep pit of 10cm in diameter (Gelhausen/Kegler/Wenzel 2004, 72). However, this feature was located within one of the disturbing fissures as well as only some centimetres from the basalt distribution of the Magdalenian concentration and in addition only 1m west of the westernmost Magdalenian pit (cf. Gelhausen/Kegler/Wenzel 2004, 74 fig. 3d with Bergmann/Holzkämper 2002, 473 Abb. 2 and Holzkämper 2006, Abb. 3. 7). Thus, an association with the Magdalenian occupation and successive infilling also with younger material cannot be completely excluded. Nonetheless, Jan Kegler assumed the pit to represent a post-hole in the entrance area of the potential dwelling (Gelhausen/Kegler/Wenzel 2004, 74). The cores were dominantly recovered outside this supposed structure, several scattered clearly eastwards of the general artefact concentration but could partially be refitted to pieces in the southern concentration. Furthermore, Kegler considered the northern accumulation to represent an associated outside working area. Yet, the refittings from both areas were exclusive, but the densities appeared amalgamating in parts and the northern concentration was poor in cores and retouched specimens.

Chronology of the upper horizon

From the upper horizon in Andernach, 22 samples were taken for ^{14}C dating. Only two of these samples originated from Andernach 3 and further two were taken from the Schaaffhausen collection. The latter produced no reliable dates. The majority of dates (n=16) refers to Andernach 2 and a group of these dates (n=7) made in two different laboratories resulted in a consistent set (12,120-11,640 years ^{14}C-BP; 14,100-13,300 years cal. b2k) which dates the occupation to the mid-Lateglacial Interstadial (GI-1c_3; see p. 468-470). However, two dates (GrA-16986 and GrA-16985) produced Late Pleniglacial ages, although the preservation condition of the material and the determination suggested an attribution to the FMG assemblage (pers. comm. Martin Street, Neuwied). Another older date (OxA-985) made on chamois remains was not redated and could not be excluded. A more recently dated horse femur yielded a comparable age (OxA-V-2218-39) and indicated the possibility of an early Lateglacial phase at the site (cf. Stevens et al. 2009b). One date (OxA-998) was made on a bone from a large bovid and was considerably younger than the main group of dates from Andernach 2. Besides the possibility of contamination, the bone was found in the south of Andernach 2 related to a possibly younger chalcedony concentration. Furthermore, one set of calcined bones from Andernach 2 (n=3) and Andernach 3 (n=2) was part of a pilot study revealing the problems inflicted with calcined bones in the Lateglacial (Lanting/Aerts-Bijma/van der Plicht

2001; Lanting/Niekus/Stapert 2002) and, therefore, could be rejected. One sample with a relatively old result (OxA-999) was redated and produced a younger date (OxA-984). In Andernach 3, an older ^{14}C date (GrA-16993) was made on a bone recovered west of the northern hearth and a younger date (GrA-16994) was also made on a bone sample that was recovered from the material south-west of the southern hearth (Kegler 2002, 511-514). The two reliable dates are clearly younger than the main set of dates from Andernach 2 (**tab. 20**) and surround the younger bovid date of Andernach 2. In general, the latent hearths in the centre of the excavation area 2 were already considered as stratigraphically lower (Stapert/Street 1997, 180), and thus concomitant with an interpretation of a younger occupation in the southern concentration. However, the accumulations in Andernach 2 were formed dominantly by Tertiary quartzite and Western European flint and no other peculiarities for FMG assemblages concerning the technical, subsistence related, or spatial behaviour were observed thus far in this area. Nevertheless, the underlying basalt fissures and erosive process resulted in some disturbances of the site which can only be resolved in parts by a comprehensive spatial analysis which is currently accomplished by Martin Street (Neuwied, pers. comm.). Possibly, this examination allows further assumptions on the connections of the varied lithic and faunal material in the different episodes to be made. Clearly, an evaluation of the many ^{14}C dates from the upper horizon (see p. 265-269) can help to shape the hypothesis of the chronological development of the site in Andernach.

Chronology of the Andernach site

In general, the archaeological material was attributed to either the lower or the upper horizon thus far. However, the serial ^{14}C dating of faunal remains (**tab. 20**) revealed more than two clusters of reliable dates suggesting a possible intermediate event as well as a later event at the site.

In particular, two ^{14}C dates point to an early Lateglacial Interstadial dating. One date was on a sample of horse (*Equus* sp.; OxA-V-2218-39) and another one on bones of probably chamois (*Rupicapra rupicapra*; OxA-985). The possible chamois bones were dated early in the times of applying the AMS-procedure and, thus, the date encountered a relatively large standard deviation. Since another sample from the same series (OxA-999) resulted in a demonstrably too old date, the date for the chamois sample could also be questioned.

Nevertheless, the horse femur was dated and analysed very recently. This single piece of evidence was found in the south-western part of the central Andernach 2-FMG concentration and immediately outside the pit area of the Late Magdalenian Andernach I. Thus, either an unknown problem caused an erroneous dating of the horse bone or the date is correct. However, if the date is correct, the equivalent archaeological material is questionable. Perhaps, a good quality black flint could be associated with the more recent date because tools of the black variety were, in fact, found in the vicinity of the dated material (cf. Stapert/Street 1997, 183f.; Stevens et al. 2009b). If this association was correct, the evidence would hint to a transitional complex which technically and according to the dated fauna (horse) was still close to the Late Magdalenian but according to the stratigraphy and the tool types was already changing towards the FMG.

A later event at the site was suggested by three younger dates and appeared to be associated with the use of chalcedonies in the south of Andernach 2 and in Andernach 3.

Presumably, the spatial reanalysis that is currently in progress, supplemented by the on-going comprehensive technological studies of the material by Ludovic Mevel, can reveal the varied settlement dynamics forming the documented pattern and, thus, help associate the sample with lithic materials and/or spatial structures. Clearly, the results help to further evaluate the significance of the single date.

However, the various clusters of ^{14}C dates from the Late Magdalenian to the FMG showed perhaps that the Martinsberg in Andernach continued to represent a favourable settlement place from the Late Pleniglacial through the onset of the Lateglacial Interstadial into its younger part.

Gönnersdorf, Neuwied, Rhineland-Palatinate

Research history

In 1968 slate plates and bones were observed during a house construction which included a deep pit for a cellar at Gönnersdorf (Bosinski 1979, 14f.), a part of the village Feldkirchen that meanwhile became a district of the town Neuwied/Rhine. A small rescue excavation under the supervision of Gerhard Bosinski provided a profile and a first impression of the significance of this open-air site because along with lithic artefacts and a slate pavement, figurines, jet beads, drilled animal teeth, engraved plaquettes, and well preserved organic material was also recovered from haematite stained sediment. From 1970 to 1976 several field seasons uncovered 687 m² in an area of some 30 m (E-W) by 50 m (N-S) and in the north-east of this main area an additional 36 test pits of 1 m × 0.5 m width reached approximately 15 m farther to the north and almost 20 m farther to the east to locate the end of the site. These test pits yielded some higher densities of lithic remains immediately adjacent to the north-eastern corner of the main excavation area but the density ceased farther to the north and east. However, beyond this test pit area, north-east- and eastwards there was further space for settlement structures on the plateau, but existing development prohibited further investigations and, possibly, had destroyed evidence prior to the discovery in 1968. In the west of the central area a rescue excavation became necessary in 1970 due to the unplanned construction of a swimming pool (Sensburg/Moseler 2008, 2-4). According to the material yielded by the excavated ditches, further evident structures were destroyed in this area by the pool construction. Therefore, the excavated area of Gönnersdorf can be considered as a section of a possibly larger site.
The excavated square-metres were numbered consecutively according to the time of excavation. Based on previous excavation techniques applied in Russia and, moreover, the excavation techniques used in Pincevent (see p. 231 f.), the excavation of Gönnersdorf was based on the documentation and recovery of the archaeological material per square-metre which was further sub-divided into equal quarters. These quarter square-metres were uncovered in artificial strata down to the main archaeological horizon with the 2D documentation of finds and wet sieving of the loess sediment. Furthermore, in the main archaeological horizons several horizontal plana were prepared and photographed per square-metre (Bosinski 1979, 50-53). Moreover, below the main horizon the pits became visible and were numbered consecutively and documented in a horizontal as well as a profile drawing if possible. However, often the filling of these pits prohibited such a procedure and these structures were documented by multiple horizontal photos following the process of careful preparation of the single items filled into the pit (Bosinski 1979, 54f.). This combined documentation of artificial layers and horizontal photos allowed for the attribution of a third dimension to several of the finds. In addition, the general horizontal documentation of the site was supplemented by several N-S and E-W oriented profiles in the excavation area. The archaeological material was analysed frequently by researches with different research focuses, resulting in a large bundle of publications of varying detail and extent, complicating a compilation of the material in a common frame (Jöris/Street/Turner 2011, 66). For instance, occasionally the numbers given for a specific material group from the same part of the site were in detail divergent (e. g. Sensburg/Moseler 2008, 64 and Franken/Veil 1983, 83-148). Some basic numbers such as the total number of lithic artefacts from concentration III have not yet been collected or published enabling a separate description of this sub-set. However, syntheses of the Gönnersdorf material are in preparation (Jöris/Street/Turner 2011) or in the process of publication (Street/Turner 2013).

Topography

The Lateglacial site of Gönnersdorf is located at the northern end of the Neuwied Basin and lies -as the crow flies- approximately 2 km north-east of the Late Magdalenian site of Andernach-Martinsberg. The Gönnersdorf site is set on a promontory-like slope on the northern side of the Rhine, some 250 m northwards of the modern river (**tab. 10**). However, the modern Rhine flows at approximately 60 m a.s.l., whereas the site was located at an altitude of 100-105 m a.s.l. The terrain on the site sloped gradually down southwards but in the south-western area the inclination became slightly steeper. From the edge of the excavation area, the terrain slopes gradually southwards for almost 200 m before a steeper step of some 20 m cuts downwards to the river banks. In the east the terrain remains on approximately the same altitude as at the site and, thus, was elevated on average 30-40 m above the river valley. This plateau extends over an approximately 2 km × 2 km area which is limited by the Rhine and the Neuwied Basin in the south as well as the valley of the Wied in the east. However, 45 m to the west of the excavated area a small valley sloping downwards to the river was observed which levelled the steps towards the Rhine but also intensified the impression of a promontory-situation of the site. Perhaps, this small valley contained a small stream before the eruption of the Laacher See Volcano (Bosinski 1979, 29). Farther to the west the site is protected by the hills forming the Andernach gate which rise steeply some 80 m upwards and in the north the continuation of these ridges ascends more gradually to the Westerwald.

The massive evident structures on the site were one of the reasons for its discovery. Large slate plates formed pavements, partially multi-layered and often densely packed, and/or stone settings. In general, these structures led to the sub-division of the excavation area in five sub-areas. Two large accumulations of the dense pavements were recorded in the excavation area: one in the south-east (concentration I; Bosinski 1979) and another one some metres to the north-west, in the southern central part (concentration II; Sensburg 2007; Sensburg/Moseler 2008). To the north of the latter a smaller, but comparably dense accumulation was observed (concentration III; Terberger 1997) which almost merges into the northern part of concentration II. In the north-north-east of concentration III and separated by several metres without a structure, a large almost quadratic stone setting with a densely packed area of stones in its centre as well as another stone-filled structure to its north was named concentration IV (Terberger 1997; Jöris/Terberger 2001; Sensburg/Moseler 2008). In the excavated area south of concentration II and west of concentration I another area with an ephemeral distribution of smaller stone plates was recorded as concentration V (Buschkämper 1993). The latter was attributed to a chronologically younger phase due to stratigraphic, techno-typological, and radiometric reasons and, therefore, is discussed in greater detail elsewhere (see p. 127-133). However, the eastern part of this sub-area was clearly attributable to the large concentration I (Buschkämper 1993, 23) and, furthermore, the scatter from the northern part of the south-western sub-area originated probably from the large concentration II (Buschkämper 1993, 34; Franken/Veil 1983, 216-233).

Stratigraphy

In the south-eastern corner of the northern part of Gönnersdorf (concentration IV), a 7 m deep profile was taken. At this depth the bedrock was not reached but a reddish brown soil of presumably Eemian age. On top followed an over 1 m thick deposit of greyish brown, loamy loess. An early Weichselian age was proposed due to the presence of the so-called Metternich tuff which consisted of a 20 cm thick pumice deposit in the lower third of the loamy loess (Bosinski 1979, 29). The sediment within this greyish brown deposit was not homogeneous but interspersed with krotovinas and lime concretions. The upper part of this loess was capped by the up to 3 m thick deposit of the eruptive material called »Eltville tuff« which originated from an eruption in the East Eifel volcanic field dated to approximately 20,000 years b2k

(Pouclet/Juvigné 2009). This material contained several alternating layers of dark basaltic pumice and light fine-grained material which was deposited in a gully according to the different profiles of the pit. On top of this volcanic material several centimetres of another greyish loess deposit followed. In this loess deposit and spreading into the upper part of the basalt tuff deposit some bone and ivory fragments as well as lydite artefacts were found, suggesting the presence of an older settlement on the plateau near the Late Pleniglacial site (Bosinski 1979, 29). This possible previous use of this setting sustains the impression of a favourable topography of Gönnersdorf for hunter-gatherer occupations. Next in the stratigraphic column followed an approximately 1 m thick, yellowish, calcareous loess in the upper part of which the Late Magdalenian horizon was situated (Bosinski 1979, 27. 47). The top 10-12 cm were transformed during the Lateglacial Interstadial into a dark, humic loam which appeared so fatty that wet-sieving seemed impossible (Bosinski 1979, 47). This dark loam was correlated with an Allerød soil. In some areas of the site this fat sediment reached almost to the Late Magdalenian horizon, whereas in other areas some 20 cm of loess were between the first planum of the Late Magdalenian horizon and the base of this Lateglacial Interstadial soil. In the south-western area of the excavation, some archaeological material was scattered in this intermediate, untransformed loess and, possibly represented a separated archaeological horizon (see below; Buschkämper 1993, 22). Finally, the dark loam was overlain by the LST which was stripped along with the topsoil before the excavations.

Thomas Buschkämper suggested the younger horizon in the south-western part of the excavation due to several indications:

The material of the eastward adjacent concentration I was deposited on approximately the same level of the terrain as the remains in the south-western area. Thus, both sub-areas were probably affected comparably by erosion and erosive processes played presumably no role between these two areas. Erosion could consequently not explain stratigraphic differences between these areas. The intermediate, untransformed loess was on average 10 cm thick in the south-western area (Buschkämper 1993, 11) and the archaeological material of concentration V was found mainly in this untransformed loess, the upper archaeological planum, and punctually even at the base of the dark loam (Buschkämper 1993, 7. 17). The archaeological material of the eastern concentration I was found dominantly in the Late Magdalenian horizon in the loess. The evident structures from the western part of the south-western area which were attributed to concentration I were also on average some 15 cm deeper in the loess than the structures attributed to concentration V (Buschkämper 1993, 23). Thus, a second phase of deposition seemed indicated by the stratigraphic distribution. Furthermore, among the archaeological material, some fragments of possible backed points as well as some remains of an elk were found suggesting also a younger development. As an additional argument for a younger phase, Buschkämper pointed out that frost cracks were observed in the lithic material attributed to the Late Magdalenian but not in the lithic material attributed to concentration V (Buschkämper 1993, 16). Even though he noted that the identification of frost cracks is problematic, this difference in preservation could indicate that severe cold episodes had affected the Late Magdalenian material but not the material from the centre of the south-western area deposited stratigraphically higher. In consequence, the latter material was possibly deposited in a climatically ameliorated episode.

Thus, the material from the centre of the south-western part of the excavation area separated by Thomas Buschkämper represented presumably a distinct episode in the occupation history of the site. Therefore, the material from the south-western area (concentration V) is presented separately from the main concentrations of Gönnersdorf attributed to the Late Magdalenian (concentrations I-IV) in the following sub-chapters.

In addition to further geological analyses of the stratigraphy of Gönnersdorf (Brunnacker 1978b; Brunnacker 1978c), various scientific studies were performed to confirm the chronostratigraphic setting (Brunnacker 1978d; Koči 1978; Leroi-Gourhan 1978) as well as to model the natural environment during the time of occupation (Schweingruber 1978; Poplin 1978). However, the pollen analysis was based on only a few preserved specimens that were probably redeposited by the various ground movements such as bioturbation, sediment flows, or hydrological infiltration. This assumption also explained the difference to the malacological analysis suggesting covered ground vegetation with rare mesophile elements and only singular indications of semi-forested, hygrophile, or steppique species (Puisségur 1978). Along with the pollen, the occurrence of some small mammals (Malec 1978) such as the various Microtinae and mole (*Talpa europaea*) should be regarded critically concerning the possible intrusion in later periods.

Charcoal was frequently found at Gönnersdorf. In total, 612 samples were analysed and the majority were identified as fossil materials (flaky brown coal, n=240; Tertiary wood i. e. jet, n=174). This flaky brown coal was attested on other sites from the Central Rhineland as well and, perhaps, represented a source to feed the hearths. In addition, typical cold environment species such as pine (*Pinus* sp., n=106), willow (*Salix* sp., n=58), and, probably, juniper (cf. *Juniperus* sp., n=34) were found. However, only in one case Scots pine (*Pinus sylvestris*) was identified (Schweingruber 1978, 83).

Archaeological material attributed to the Late Magdalenian

In Gönnersdorf a large amount of material, judging by the weight as well as the numbers were deposited. The assumed distances from where the material was brought to this particular topographic position marked the site as a special and more permanent construction.

In total, 81,786 lithic artefacts (Floss 1994, 219) were uncovered and, although splinters and flakes smaller than 2 cm were dominant (n=48,786; **tab. 11**; Franken/Veil 1983, 93. 106. 124-126. 143-145; cf. Floss 1994, 224-235), the material weighted over 76 kg (Floss 1994, 219; cf. Franken/Veil 1983, 3. 6 Abb. 1).

The dominant raw materials were foreign flints (26.5 kg, n=49,321; Floss 1994, 219) which made up more than half of the assemblage (c. 57 %; Franken/Veil 1983; 60 %, Floss 1994, 219). 63 % of the flints could be attributed to the Western European flints (15.5 kg, n=31,024) that originated from sources of at least 110 km distance in the north-west and 18 % of the artefacts were made of Baltic erratic flint (~6 kg, n=9,093) originating from a minimum of 105 km northwards (**tab. 12**). The remaining 19 % were patinated pieces which could not be classified precisely. Furthermore, the local raw materials, lydite (28 kg, n=12,035) and Tertiary quartzite (19.5 kg, n=15,181), occurred frequently. Besides the local variety of the latter, the Ratingen type that possibly was gathered at a distant source some 85 km either northwards or eastwards (Floss/Terberger 2002, 17) was recovered rarely (n=212). Almost 3,000 artefacts were made of chalcedony which presumably did not originate from a local source but could possibly be associated with a source near Frankfurt some 75 km south-east of the site due to wind abrasion (Floss 1994, 224-226; cf. Franken/Veil 1983, 22). In addition, some 2,000 artefacts were made of siliceous oolite that probably originated from the same region as the chalcedony. Furthermore, Palaeozoic quartzite (n=146), Jurassic chert (n=30), rock crystal (n=23), opal (n=8), and a southern German jasper type (n=3) were found in small quantities. The latter raw material was found exclusively in the south-western part of the excavation area and, perhaps, was associated with a younger occupation event (see p. 127-133). The source of the jasper was presumably in the Freiburg region, some 300 km to the south. The opal could have originated from the same region as the chalcedonies but also possibly from the Siebengebirge approximately 25 km and more to the north. Even though the few pieces were clearly distinguishable from the chalcedony, Harald Floss considered them to probably originate from the same region as the chalcedonies which were found accumulated in the same part of the excavation area (Floss 1994, 234). The rock crystal pieces showed abrasions of river transport and were probably taken from the Rhine gravels near the site. According to the remaining cortex, the chert artefacts formed a single nodule which, perhaps, was also collected along the banks of the Rhine (Floss 1994, 231). The few pieces of Palaeozoic quartzite, that was also referred to as

Ardennes quartzite, were comparable to the same material from Andernach (Floss 1994, 232; see p. 80) and was identified as a Mesozoic quartzite which, possibly, originated from the same region as the Western European flint (Floss/Terberger 2002, 14-16).

Eduard Franken mentioned 309 cores (Floss mentions at least 311 specimens) from the site. These were dominantly made of lydite (n=224) but also pieces of Tertiary quartzite (n=47) and Western European flint (n=28) were recovered (**tab. 13**; Franken/Veil 1983, 83-87. 112-116. 130-134; Floss 1994, 224-235). The remaining cores were frequently of small dimension. The lydite pieces were predominantly exploited from one platform (n=106) and for the 32 pieces exploited from two platforms Franken mentioned several pieces where one platform was used dominantly (Franken/Veil 1983, 85). However, among the Tertiary quartzite cores the exploitation from two platforms was most frequent (n=22 to n=11) and among the flint cores this pattern was even clearer (n=24 to n=1). Some cores had three or more platforms and these pieces frequently were of an almost global shape and a small dimension suggesting that these raw material units were exhaustively used. Only four such cores were made of Tertiary quartzite, 13 of flint, and 50 of lydite. On some cores, crested preparations were still observable (e.g. Franken/Veil 1983, Taf. 7.3; 13.3), and some crested blades and bladelets made for example of Tertiary quartzite were found (Franken/Veil 1983, 123. 126). The remaining material with this preparation points to a blank production focused on lamellar blanks and, in particular, bladelets. The presence of butts of the »*en éperon*« type was not analysed but some cores displayed preparation which could have formed the necessary spur (e.g. Franken/Veil 1983, Taf. 14.2). In the Gönnersdorf assemblage the absence of cores and blank production debris was considered as significant concerning the way in which the raw material reached the site. For instance, the Tertiary quartzite of the Ratingen type seems to have reached the site only as ready made blanks (Floss 1994, 224) and, perhaps, the Palaeozoic quartzite as well (Floss 1994, 232 f.). The jasper and the opal items were presumably brought to the site as retouched artefacts (Floss 1994, 233 f.).

The reddened quartzite analysed by Eduard Franken was coloured by haematite, not heating, and in contrast to the denser lithic raw materials this red powder infiltrated into the more porous Tertiary quartzite material (Franken/Veil 1983, 53-55). However, Stephan Veil mentioned that only 0.9 % of the retouched artefacts and 0.02 % of the debris material of the retouching process showed traces of heating and noted that in concentration II the numbers were the lowest with only 0.5 % in contrast to around 1 % in the other areas (**tab. 12**; Franken/Veil 1983, 272. 296). In fact, Thomas Terberger counted only 20 artefacts with traces of heating in the northern area (Terberger 1997, 49-54), and a further 189 burnt pieces in the adjacent central area (Terberger 1997, 188). Based on observations of Hartwig Löhr and referring to the material of concentration IIa, Martina Sensburg emphasised the scarcity of burnt lithic material (Sensburg/Moseler 2008, 22). Finally, in his dissertation focused on the use of fire at Gönnersdorf, Frank Moseler found only few indications for burning in the lithic assemblage of Gönnersdorf based on comparisons with experimental heating of several stone varieties (pers. comm. Frank Moseler, Neuwied).

Stephan Veil found 4,855 pieces with 5,261 retouched working edges (Franken/Veil 1983, 259). This number was partially revised by later recounting (Terberger 1997; Sensburg/Moseler 2008, 64 tab. 3) resulting in some 4,320 identified retouched artefacts (**tab. 14**).

Among the LMP, the simple backed bladelets clearly dominated (Franken/Veil 1983, 288). Occasionally pieces with two retouched edges occurred, and truncated pieces were rarely found. Very scarce were pieces with notched backs. Refitted pieces revealed the production of small segments by intentional breakage of long ready retouched bladelets (Franken/Veil 1983, Taf. 33.12-13) and the metrics showed a dense clustering of the broken backed bladelets suggesting a high standardisation for these tools (Franken/Veil 1983, 290 f.).

In the northern most part of the south-western excavation area, three LMP made of southern German jasper were found which probably represented part of a basic equipment exchanged at the site (Floss 1994, 234). One piece was considered as a backed point (Floss 1994, 234) that was later questioned (Buschkämper 1993, 51). In the present study, the piece is attributed to the backed bladelets with partial lateral retouch of the opposite edge because in regard to the middle axis of the implement the »apex« is set on the same side as the straight-backed edge. Moreover, the vertical position in the level above the first planum as well as the horizontal position in the distributed material between the two major Late Magdalenian concentrations (Buschkämper 1993, 50-52) suggested a connection with the main, Late Magdalenian settlement event of the site.

The majority of the burins was made on truncation followed by dihedral burins and pieces on breaks (Franken/Veil 1983, 284). Burins of Lacan type occurred frequently among the pieces on truncation and occasionally transversal burins were recorded (Franken/Veil 1983, 286). The end-scrapers were usually made on blades and splintering of the edges was observable on several specimens (e.g. Franken/Veil 1983, Taf. 20.12 and 16; Taf. 21.9, 15-18 and 20). Truncations were commonly oblique but straight, concave, and convex truncations occurred also in notable quantities (Franken/Veil 1983, 286). The borers were frequently made on burin spalls. Perforators were the common type, some with an elongating retouch and some even with a *Zinken*-like design (Franken/Veil 1983, Taf. 19.22 and 27) but the Gönnersdorf pieces were never as massive as typical *bec* or *Zinken* and the small negative signaling the *Zinken* could originate from damages during the use of the pieces. Sometimes artefacts, particularly retouched ones were thinned by the so-called Kostienki ends.

A traceological analysis of 203 artefacts of Western European flint revealed the use of these implements on various materials as well as for various motions (Sano 2012b). The use-wear confirmed the proposed uses for backed bladelets as parts of projectiles, for perforators as drilling instruments, and for end-scrapers as generally hide-working tools. Moreover, the study also showed the heterogeneous and sometimes intensive use of burins, splintered pieces, or further artefacts which were not formally retouched. The variety of activities indicated by the different patterns were assumed to emphasise the function of Gönnersdorf as a residential occupation in contrast to short-lived settlements such as the »quarry site« Eyserheide or the »hunting camp« at Bois Laiterie (Sano 2012b).

Along with the lithic material, several hundred kilograms of stone material were brought to the site accumulating already in the barely structured northern part to more than 660 kg of various types of slate, quartzite, basalt, sandstone, and prophyr with an additional c. 41 kg of quartz (concentration IV, Terberger 1997, 69-102) and mounting to 1,551 kg of slate varieties, quartzite, and basalt with additional 252 kg of quartz in the northern central area of the excavation (concentration III, Terberger 1997, 273-296). These stones probably originated from the Rhine gravels and the arising cliff west of the site suggesting transport distances of 50-200 m (Bosinski 1979, 92-134). They were occasionally laid out as multi-layered pavement on the ground, as cover on pits, as enclosure of possible hearths, and, perhaps, as ballast stones for shelter covers on the ground. In contrast to the lithic material, these stones showed frequent indications of heating and probable contact to fire (pers. comm. Frank Moseler, Neuwied).

Moreover, some chopping tools (Bosinski 1979, 115-120; Terberger 1997, 76-79. 275-277) as well as several retouchers, usually of elongated shape (Bosinski 1979, 120-123. 136 f.; Terberger 1997, 275-278. 298 f.), were made of these materials.

The faunal analysis (Poplin 1976; Poplin 1978; Street/Turner 2013) determined more than a dozen species in the organic material (**tab. 15**). However, these were present in very different quantities and presumably brought to the site for various reasons.

The most common species were horse (*Equus* sp., MNI ≈ 55), arctic fox (*Alopex lagopus*, MNI ≈ 35), reindeer (*Rangifer tarandus*, MNI ≈ 5), and mountain hare (*Lepus timidus*, MNI > 7; Street et al. 2006; cf. Jöris/Street/Turner 2011; Street/Turner 2013). Of these four species all body parts were found at the site, proving the introduction of complete animals to the site. The remains of arctic fox showed cut-marks suggesting that these animals were also completely exploited and not only hunted for their pelt (Street/Turner 2013). Based on the size of the teeth, François Poplin identified arctic fox along with a few individuals of red fox (*Vulpes vulpes*; Poplin 1976, 46). According to the foetal bones of the horses, the size of the teeth of the arctic fox, and the exclusive presence of shed reindeer antler, most parts of the site were attributed to an autumn/winter occupations, whereas the southern central area (concentration IIa) produced larger foetal bones of horse and even some foal hooves suggesting an occupation during the spring and early summer months (**tab. 16**; Poplin 1976, 149; Terberger 1997, 160; Sensburg 2007, 155f.; Street/Turner 2013). Thus, the large numbers of animals, in particular, horses were probably accumulated by several hunting episodes which occurred over a period of several months, perhaps, several years.

Besides direct evidence from the faunal remains, Gönnersdorf also yielded indirect evidence by naturalistic engravings of different animals on the slate plates (**tab. 18**) providing detailed knowledge about these animals. Among the engravings, the proportion of horse to reindeer were comparable to the faunal material with numerous identified horses (Bosinski/Fischer 1980; Bosinski 2008, 17-22) in contrast to only occasionally observed reindeer images (Bosinski 2008, 49-55). However, the canon of engravings was not a one-to-one image of the faunal resource environment; further impressions and ideals governed the choice of motives. For example, foxes were not identified among the slate engravings. In the context of arctic fox clearly playing an important role in the faunal assemblage of Gönnersdorf, its absence in the canon of engraved animals is surprising. Hare was probably absent as well (cf. Bosinski 2008, 140; Jöris/Street/Turner 2011). In contrast, images of larger carnivores such as wolf (*Canis lupus*), bear (*Ursus* sp.), and lion (*Panthera* sp.) were found (Bosinski 2008, 105-113), although only the wolf was also present in the faunal assemblage. The presence of the latter was determined mainly by paw and cranial parts suggesting that some remains arrived in the form of a pelt at the site, whereas in other places such a pelt was perhaps made (Street et al. 2006, 761).

The megaherbivores of the classic mammoth steppe fauna were also identified at Gönnersdorf. Mammoth (*Mammuthus primigenius*) was determined by some pieces of ivory, some teeth fragments, a femur, and some ribs (Poplin 1976, 50). AMS-dating of the bones revealed that these pieces were some 1,000 ^{14}C years older than the Late Magdalenian occupation at Gönnersdorf (Stevens et al. 2009b). In particular, the femur was found in a position suggesting its use as a construction element (Bosinski 1979, 65). Thus, the mammoth material was presumably collected as raw material for construction and the production of artefacts. However, the existence of detailed knowledge about these animals was proven by the numerous engravings of mammoths on the slate plates (Bosinski/Fischer 1980).

Woolly rhinoceros (*Coelodonta antiquitatis*) was also only detected with a few teeth and remains of a mandible (Poplin 1978, 99; Street et al. 2006). Samples from these remains were dated to an age comparabe to the mammoth material (Stevens et al. 2009b). Again, on the slate plates detailed engravings of these animals were found (Bosinski 2008, 39-48). Nevertheless, both megaherbivores were probably not contributing to the alimentary provision, but their remains represented raw materials for the inhabitants of Gönnersdorf.

Large bovids were only rarely identified in the faunal assemblage and originated presumably from bison (*Bison priscus*), although the attribution to aurochs (*Bos primigenius*) could not be entirely excluded. According to the different habitats of these species and the indications of a grass steppe landscape in the surrounding of the site during the Late Magdalenian occupation, the attribution to bison was preferred. However, on the engravings attributed to large bovids the aurochs was clearly identified, whereas the bison was only partially observed (Bosinski 2008, 63-67 Taf. 77. 90-91). The two single ribs recovered from the south-eastern part were considered as tools for digging due to their smoothed ends (Poplin 1976, 53) and, hence, probably were also not hunting remains. Post-cranial remains of two large bovids (an adult and a young one) were found in the north of the site in a higher stratigraphic position and displayed marks of carnivore gnawing (Street et al. 2006, 762). Presumably, these individuals were remains of a carnivore attack some time after the abandonment of the site. However, in the central area a few post-cranial bones of a large bovid were recovered that perhaps represented prey remains but due to the small number these pieces were possibly introduced to the site as provision.

Saiga antelope (*Saiga tatarica*) had a presence comparable to the megaherbivores among the fauna remains with two bone fragments. Although these remains were extremely scarce, again this species occurred among the engravings (Bosinski 2008, 77-79). The occurrence of this dry desert inhabitant was of significance for the environmental reconstruction of the area during the Late Magdalenian period. However, the very scarce evidence could indicate that the hunting episode took place somewhere else and only the single remains reached the site as provision or for curiosity.

The mountainous species chamois (*Rupicapra rupicapra*) was only identified by teeth and mandible fragments. This sparse presence could also indicate the introduction in the form of a trophy or provision. On the slate plates this species was not observed but the related and equally mountainous ibex (*Capra ibex*) was engraved (Bosinski 2008, 71-75).

Red deer (*Cervus elaphus*) was only determined by some teeth which were usually drilled and, thus, probably arrived at the site in the form of adornments. On the slate plates some rare examples of red deer were identified (Bosinski 2008, 49-55). The post-cranial bones of at least two individuals were found in the south-western area and were presumably associated with a later occupation event (Street et al. 2006, 762).

Besides these major species, several bird species were determined: Mute swan (*Cygnus olor*), geese (*Anser* sp.), grouse (*Lagopus* sp.), probably rock ptarmigan (*Lagopus mutus*), snowy owl (*Nyctea scandiaca*), common raven (*Corvus corax*), and, perhaps, gulls (*Larus* sp.). This assemblage is almost identical with the evidences from Andernach. Various birds were also engraved on the slate plates at Gönnersdorf (Bosinski 2008, 115-128).

In contrast to Andernach, only few remains of fish were preserved (Poplin 1976, 63). However, the identified remains were with brown trout (*Salmo trutta*) and members of the Cyprinidae family comparable to the Andernach material but Gönnersdorf yielded additionally remains of burbot (*Lota lota*). Some engravings of fish were found (Bosinski 2008, 129-131).

In an attempt to explain the faunal co-occurrence of the various animals, François Poplin suggested five principal explanations (Poplin 1978, 101 f.): 1. the animals changed their behaviour, 2. the assemblage was formed over a considerable period of time, 3. the local topography allowed the co-existence of varied habitats, 4. the modern behaviour observed in the animals reflects only a range limited by the expansion of man, 5. the animals were present in the area at various seasons of the year. Certainly, all of these explanations as well as combinations of these are possible. However, excluding the probably fossil material (mammoth, woolly rhinoceros, bison?) and the pieces introduced as curiosity or, perhaps, from a distant source (bison?, saiga antelope?, red deer) the environment of Gönnersdorf was relatively similar to modern arctic or high mountain regions such as the Siberian tundra or the Carpathians and did not represent a typical Late Weichselian mammoth steppe habitat. Nevertheless, varied seasonality was suggested by the development of horse foetus bones, but the presence of varied habitats, in particular between the moist, protected valleys and the dry, exposed plateaus in the upland regions were variously sustained and, moreover, the vast amount of material at Gönnersdorf indicated the presence over a considerable period of time.

François Poplin found among the organic remains also at least five fragments of teeth which he distinguished from reindeer teeth and identified as human teeth (Poplin 1976, 3). In addition to a deciduous tooth, coronary fragments of adults with various stages of abrasion were found indicating the presence of at least three people (Poplin 1976, 68) of various age classes at the site. Presumably, these pieces were intra-vitam losses (Poplin 1976, 5).
Johann Tinnes described some 359 artefacts made of 368 fragments of bone (n=91), antler (n=222), and ivory (n=46) from Gönnersdorf (**tab. 17**; Tinnes 1994, 2). This rich organic inventory comprised debris such as cut pieces (n=7) or cores (n=31) as well as tools and some pieces of art. The majority of cores were made of antler (n=24), and only seven pieces were made of bone. The complete exploitation cycle of these raw materials was observable at Gönnersdorf (Tinnes 1994, 76) producing spalls of various sizes (Tinnes 1994, 80. 96. 98). However, the larger spalls were made of antler (n=39) and ivory (n=5), whereas the smaller needle blanks were made only of bone (n=5). Accordingly, the 71 needle and two awl fragments were generally made of bone (n=56) and less frequently of antler (n=13) and ivory (n=2; Tinnes 1994, 179). The two complete points and a further 41 fragments of single as well as double bevelled points (rods) were made only of antler (n=40) and ivory (n=3; Tinnes 1994, 146-153). Two fragments of baguette demi-rondes were also engraved with at least two identifiable animal heads (Tinnes 1994, 169f.). In addition, two fragments of fish hooks made of bone and antler were found (Tinnes 1994, 176f.). In contrast to Andernach, barbed points were not recovered except for a single barb (Tinnes 1994, 171f.) and no hammer instrument was found but also ten bone retouchers were identified (Tinnes 1994, 115f.). A single antler fragment of a bâton percé was found in the south-east of the excavation (Tinnes 1994, 142). Only two coin blanks were made of organic material (one of ivory, one of antler, Tinnes 1994, 188) in contrast to numerous specimens and fragments made of slate (Bosinski 1977; Terberger 1997, 93f. 292-295).
The previously mentioned engravings (**tab. 18**) encompassed not all hunted animals but instead yielded also species clearly not present at the site such as bears, lions, or seals (Hansen 2006; Bosinski 2008, 87-95). The details of many engravings sustained the assumption of comprehensive knowledge of these animals and, furthermore, the consideration that to gain this knowledge the engravers must have seen those animals. In contrast, images of humans (women, phantoms) were very poor in detail and almost abstract (Bosinski/Fischer 1974). Phantoms were assumed as frontal views of an abstract human portrait. These portraits indicate an upper body with shoulders a neck and an oval head with eyes and mouths depicted by circles. Often the oval forming the head ends in a jelly bag cap-like end giving the whole figure a ghost-shape outline. In contrast, the female silhouette displayed a profile from approximately the knees to the neck with accentuated breasts and bottoms and only occasional signs of arms. These images were so numerous and peculiar that they formed a specific type for which Gönnersdorf became eponymous (see p. 89; Bosinski 1992; Bosinski/d'Errico/Schiller 2001).
Moreover, 19 female figurines and figurine fragments were made in the same abstract style (Tinnes 1994, 195; Höck 1995, 256-266). The raw materials used to carve these specimens were ivory (n=10), antler (n=3), and Devonian slate (n=6). One of the antler pieces was classified as unfinished (Höck 1995, 288) suggesting that figurines were made at the site.
Among the faunal fragments were some teeth with drilling observed that fall into the category of jewellery. 38 examples of drilled arctic fox teeth were recorded and reached on average lengths of 1 cm. Esteban Álvarez-Fernández outlined that only nine of the 48 possible types of fox teeth (maxilla: I2, I3, C, P1, P2, P3; mandible: C, P2, P4) were chosen for this type of modification (Álvarez Fernández 1999, 88). Furthermore, he emphasised that the premolars (P2-P4) were often not completely preserved, which he assumed to be intentional to result in a similar, though smaller, shape to the female figurines (Álvarez Fernández 1999, 90-92). Besides fox teeth, six perforated canines of red deer were found (Álvarez Fernández 1999) as well

as three fragments of sawed incisors of the same species (Street et al. 2006, 761. 772). Also, reindeer teeth which were sawn from the maxilla were considered as pendants according to experimental examples (Alvarez Fernández 2001; Álvarez Fernández 2009).

58 jet beads were documented in Gönnersdorf of which 49 were finished and a further nine examples were of various finishing stages attesting the fabrication of these items on the site (Álvarez Fernández 1999). Among the generally very small (c. 1 cm long) finished beads, the discoid (n = 19) and the biconical types (n = 18) were most common. The origin of the raw material remains unknown. A famous modern source of jet is located some 250 km south-eastwards near Holzmaden (Wuerttemberg). However, during the Roman period a centre of jet carving independent from the British jet fabrication was located north of the Central Rhineland in the Bonn and Cologne area (Allason-Jones/Jones 2001) suggesting that these craftsmen had also access to a possibly local resource which is no longer known. Perhaps, the suggestion of the local Rhine gravels serving as a source for the jet (Álvarez Fernández 2009) that was also considered for the lower layer in Andernach (Holzkämper 2006, 158) could be a good explanation. Nevertheless, without chemical analysis of the objects, the origin of the jet cannot be further determined.

Along with these pieces of art and jewellery, some molluscs which were occasionally perforated were found and determined as *Dentalium (Antalis) vulgare* (n = 17), *Dentalium (Antalis) inaeqicostatum* (n = 5), and *Homalopoma sanguineum* (n = 5; (Strauch/Tembrock 1978; Alvarez Fernández 2001). Although some of these molluscs could originate from fossil sources in the southern Netherlands and Belgium (Strauch/Tembrock 1978), the latter two species were of no fossil origin and originated probably from the Mediterranean region (Alvarez Fernández 2001, 548) that, as the crow flies, lies 680 km to the south. However, the straight connection from Gönnersdorf to the Mediterranean would require connections across the Alpine ice sheet and, therefore, the Rhine-Rhône axis appeared a more probable route along which the molluscs had come with a minimum distance of 800 km. In addition, the *Dentalium vulgare* could originate from the Mediterranean, but this species was also known from the Atlantic coast (Alvarez Fernández 2001, 550).

Furthermore, fossil curiosities such as a Tertiary shark tooth which was presumably collected from fossil deposits in the Mainz Basin or a dinosaur vertebra from Luxembourg or Lothringia were brought to the site (Floss 1994, 244. 256 Abb. 159).

Haematite was used intensively in some parts of the site. Besides the red powder which coloured the sediment at some of the central areas, several lumps and pieces of haematite were discovered on the site. According to the chemical comparison, these pieces originated from the same source as material found some 25 km north-westwards of the site (Bosinski 1979, 138). Probably, the origin of the material was nearby this north-western occurrence.

Spatial organisation of the material attributed to the Late Magdalenian

The massive evident structures (**tab. 19**) on the site were one of the reasons for its discovery and allowed the sub-division of the excavation into five sub-areas (I-V). These large structures included pavements formed by stone settings and/or slate plates which were partially multi-layered and often densely packed.

Concentration I was significantly disturbed by the building of the house leading to the discovery of the site. The remaining structures were analysed by Gerhard Bosinski in the 1970s (Bosinski 1979) and supplemented by the work of Thomas Buschkämper (Buschkämper 1993, 23). They revealed some circular stone settings within the main pavement. Moreover, the stone settings in the south-western sub-area associated with the Late Magdalenian concentration I formed partially dense pavements (Buschkämper 1993, 23). Meanwhile, modern analytical tools require a combined reanalysis of these sub-areas to allow further interpretations of the position of the various parts in the formation of this south-eastern concentration as well as in the complete settlement process at the site. In the context of a comprehensive study of the complete site, this kind of reanalysis of the material from Gönnersdorf I is in progress but, thus far, the analysis was only partially accomplished (cf. Jöris/Street/Turner 2011; Street/Turner 2013). In constrast, concentration II was preserved almost completely. Only in the north-western part some imprecision of the excavation occurred due to a swimming pool construction (Sensburg/Moseler 2008, 3f.). The western part of concentration III was possibly also affected by this intrusion. The concentration IV appeared largely undisturbed.

Along with the large accumulations of material, denser groups of pits were observed. These structures were generally located in the central area of the large stone pavements. In concentration I 35 pits were identified (Bosinski 1979, 142-154) of which several were suggested to represent postholes (Bosinski 1979, 150). In concentration II two groups of pits were observed and justified another subdivision (Eickhoff 1989) into a concentration IIa where 21 pits were installed (Sensburg/Moseler 2008, 30) and a northwestward adjacent concentration IIb where a further nine pits were still preserved (Sensburg/Moseler 2008, 13-21). Spatial analysis of the lithic material further sustained the two distinct zones in concentration II (Sensburg 2007; Sensburg/Moseler 2008). Concentration IIa was the most completely preserved, whereas concentration IIb was significantly affected by the limits of the excavation area and the construction of the swimming pool (Sensburg/Moseler 2008, 51). Moreover, in the south-eastern part of the large stone pavement a circular structure was identified as IIc (Eickhoff 1988, 39). With this area, no pits were associated. In fact, Dick Stapert assumed that this part represented an intensively used entrance area to the other parts of the concentration II (Stapert 2003). Furthermore, in concentration III 21 depressions were documented of which Thomas Terberger described 17 structures as pits (Terberger 1997, 197-233). Spatial analysis of the lithic material revealed a minimum of three distinguishable clusters in concentration III with two distinct accumulations (zone B and C) between the centre of concentration III (A) and the limits of concentration IIa (Terberger 1997; Sensburg 2007). Moreover, Thomas Terberger considered another activity area in the east of concentration III. In concentration IV, as well as in the south-western area, no pits were recorded.

However, this distribution of pits could explain other spatial patterns such as the presence of molluscs or female figurines or other pieces of jewellery or art, because the preservation of fragile and organic specimens appeared to be associated with the pits and the intensive cover of stone plates (cf. Álvarez Fernández 1999, 96). Similarly, the intensity of red coloured sediment was mainly connected to these centres of the concentrations where also the original sediment and its additions were more protected against post-depositional outwash. Such post-depositional processes presumably affected further light elements such as lithic splinters or charcoal.

However, in concentration I a single and in concentration IV two hearths were observed during the excavation (Bosinski 1979, 64-68; Terberger 1997, 25-32; Sensburg/Moseler 2008, 39). In concentration I the hearth was dug into the ground and associated with a femur of a mammoth that was interpreted as part of a construction element around the hearth (Bosinski 1979, 65). Probably, the depression protected small material against down slope movements. In concentration IV one hearth which was set some metres to the north-east of the almost quadratic stone setting was filled with stones, and identified by darkened sediment (Sensburg/Moseler 2008, 68-70). The two dense accumulations of Gönnersdorf IV were connected by refitting, The connection lines between these hearths were oriented in NE-SW direction. These structures corresponded most closely to the hearth constructions known from the Late Magdalenian in the Paris Basin (Bodu et al. 2006b, 89-108) or northern Switzerland (Müller et al. 2006). However, charcoal material spread south-eastwards from the hearth, which was probably due to clearing out (Sensburg/Moseler 2008, 69) either by humans or by post-depositional horizontal movements or both. Nevertheless, the larger materials remained *in situ* after the human abandonment of the site. The stones were mainly fragments of quartzite, quartzitic slate, and sandstone but also quartz pieces occurred in this accumulation. Due to the latter and their attribution as cooking stones, the function of this hearth was associated with the preparation of food (Sensburg/Moseler 2008, 70). Based on the spatial distribution of different materials and indicators for hearths, Frank Moseler assumed repetitive clearing of material from the hearth (Sensburg/Moseler 2008, 69) which is consistent with Thomas Terberger's suggestion of several rearrangements of the material forming this structure (Terberger 1997, 29). Thus, for this comparatively small structure several episodes of use that were not performed quasi-simultaneously were reconstructed. A second hearth was observed in the centre of the almost quadratic stone setting of concentration IV and was also layed out with partially burnt stone material (Sensburg/Moseler 2008, 70). However, in contrast to the northern hearth, quartzes were rare, but near the hearth a large and heavy (15.5 kg) basalt block was found based on which Frank Moseler assumed that the function of this hearth was basically to produce light and warmth (Sensburg/Moseler 2008, 71).

Further hearths were located by spatial analyses. In these analyses, lithic artefacts with traces of heating were excluded as a potential indicator for hearths because these specimens were considered, for example, by Martina Sensburg in reference to an observation made by Hartwig Löhr, as too sparsely present on Magdalenian sites to result in significant accumulations (Sensburg/Moseler 2008, 22; cf. Sensburg 2007, 49). However, the scarce pieces were mainly found around stone rings reinforcing the suggestion that these settings represented limits of hearths. In contrast to the lithic material, charcoal and burnt quartz fragments were chosen as good indicators (Terberger 1997, 235-238; Sensburg/Moseler 2008, 25. 68).

By the mapping of these materials, Thomas Buschkämper located two hearths (F2 and F3) in the partially dense stone pavements found in the south-western sub-area and associated with the concentration I (Buschkämper 1993, 23). The suggested, short-lived hearth F2 was situated within the southern limit of a circular structure of 2-2.5 m in diameter (Buschkämper 1993, 7). The other hearth was situated farther to the south and surrounded by a ring of stones and dumps from cleaning of this hearth in the surrounding (Buschkämper 1993, 7). Both suggested hearths were not dug into the ground. Also based on the distribution of the hearth indicators, Martina Sensburg proposed the presence of five hearths in concentration IIa (Sensburg 2007), and a further three hearths in concentration IIb (Sensburg/Moseler 2008). In addition, Thomas Terberger reconstructed in the centre of concentration III a large hearth that was open and, perhaps, dug into the ground and later covered by the stone plates (Terberger 1997). A second possible hearth which was situated in the north of the central concentration III was assumed in a considerably smaller circular structure. Additionally, two smaller accumulations associated with hearths (zones B and C) were located south of the main concentration III merging almost into concentration II. Another area associated with fire was identified by Terberger in the eastern part of the sub-area but appeared more ephemeral (Terberger 1997, 239-247).

In his revision of the indications for the use of fire and their spatial distribution at the site, Frank Moseler considered the indications for *in situ* hearths in contrast to dumped or redeposited material more critically, and concluded that only six reliable hearths can be identified by various fire indicators such as charcoal and burnt bone at Gönnersdorf (pers. comm. Frank Moseler, Neuwied; Moseler 2014). In addition to the three structures already observed during the excavations, further spots were noticed in the central part of concentration II and one or two spots in the centre of concentration III (Moseler 2014). In concentration IIb the determined pieces of charcoal were spread without an identifiable concentration but clearly outside the hearths suggested by Martina Sensburg (Jöris/Street/Turner 2011, fig. 6B; Sensburg/Moseler 2008, 22), and therefore Frank Moseler could not locate an unambiguous hearth in this concentration (pers. comm. Frank Moseler, Neuwied). Based on the determination of the recorded charcoal pieces in concentration IIb as possible juniper (cf. *Junipers* sp.) wood charcoal and the further occurrence of such specimens in the centre of concentration IIa, the charcoal fragments in concentration IIb could represent a clearing of the hearth in concentration IIa and, thus, reflect at least two successive episodes in the use of the complex concentration II. However, the charcoal in concentration IIb could also be the remains of an undetected/not excavated hearth. Perhaps, another concentration of determined pieces of charcoal in concentration II reflected a dump area or a destroyed hearth associated with the stone circle IIc.

The distribution of lithic raw materials was, in general, helpful to distinguish various episodes in the settlement process and their position in the chronological development of the site (Terberger 1997, 166. 258-273; Sensburg 2007, 86-117. 142-150) as well as suggesting the presence of further structures such as dwelling structures (Sensburg 2007, 18-21. 26f.).

In Gönnersdorf, methods to analyse the spatial distribution of materials were tested particularly because the massive evident structures were considered as architectural elements (Jöris/Street/Turner 2011) and, therefore, as good indicators for the reliability of the latent spatial analysis. For example, Dick Stapert performed a ring-and-sector analysis on all Late Magdalenian concentrations of Gönnersdorf (Stapert 1990; Stapert 1991; Stapert/Terberger 1991; Boekschoten/Stapert 1993; Stapert 2003).

Thomas Buschkämper performed another ring-and-sector analysis for the two suggested hearths in the eastern part of the south-western sub-area (Buschkämper 1993, 106-110. 118-122): He concluded that a dwelling structure in the northern part, measuring approximately 2.5 m in diameter, existed with an opening oriented towards the south, and the northern hearth was located in this opening (Buschkämper 1993, 123). However, to prevent overlapping results, only eight 25 cm wide rings were chosen per structure i. e. in total 2 m around the assumed centre were examined. Nevertheless, some overlapping between the distribution around the southern hearth and the distribution within the circular structure were still observed (Buschkämper 1993, 108). Moreover, since Buschkämper did not choose a hearth as the centre of the rings, he assumed an unimodal frequency distribution to reveal the suggested barrier.

Clearly, this example showed the limits of the ring and sector method, but modern analytical tools such as density frequencies displayed in isolines (Gelhausen/Kegler/Wenzel 2004, 70; Wenzel 2009, 19f.) could help revealing overlapping areas. Moreover, the advantage of the latter method is the lower number of presuppositions such as the presence of a central hearth or a predefined barrier structure. In contrast, the fundamental calculations were more complex and, occasionally, led to misleading interpolations of the data (Wenzel 2009, 20). However, this problem affected, in particular, areas with low find densities which in Gönnersdorf were rather rare. Nevertheless, in the find rich areas, the analyst has to choose which lines are of significance (optimal isolines, Zimmermann et al. 2004; Zimmermann et al. 2009; cf. Holst 2014). Further distribution patterns such as refitting lines need to accompany these analyses to sustain the interpretations. Yet, the evidence of such refitting lines suggesting direct moves between two points should be considered critically due to possible transient stations (Sensburg 2007, 193f.).

Besides the circular structures identified by the ring and sector method (Stapert 1991; Stapert 1992; Buschkämper 1993), Gerhard Bosinski reconstructed a large and oval wall construction in Gönnersdorf I with a reverse lying main entrance towards the south-east based on the location of the evident structures such as postholes and large stone plates as well as the distribution of haematite stained sediment (Bosinski 1979). The circular wall structure reconstructed by Dick Stapert was set 1-2 m east of this entrance (Stapert 1991; Jöris/Street/Turner 2011), whereas the circular structure reconstructed by Thomas Buschkämper was located 1-2 m west of the back of the main structure (Buschkämper 1993; Jöris/Street/Turner 2011). Based on ethnographic analogies from the Inuit of Greenland, Dick Stapert and Lykke Johansen suggested that the main structure of concentration I served as a habitation place for a whole group, and that at some point a smaller group, perhaps mainly older people, stayed behind and moved into a smaller structure (Stapert/Johansen 1997, 60). In such a scenario, the chronology of this concentration would comprise at least two phases during a single occupation event.

Whether the smaller circular stone settings considered as dwelling structures by the ring and sector analyses were in fact teepee-like tents or other types of shelters cannot be ascertained, but future spatial analyses could sustain the presence of barriers. Thus far, these structures were neither observed to be intensively haematite stained, nor associated with depressions or pits, nor related to significant concentrations of the mainly processed animals (Street et al. 2006, fig. 12). However, ivory and mammoth molars were connected to the eastern-most circular structure in concentration I (Street et al. 2006, fig. 11a), suggesting the presence of a possible workshop. Moreover, these structures contained accumulations of mainly anorganic material and, therefore, represented intensely used zones of activities, probably lithic and other manufacturing. For concentration II, an oval shaped dwelling structure with an approximate 6 m extension in NW-SE direction and some 5 m in SW-NE direction was suggested by Dick Stapert with an intensively used and therefore also paved main entrance towards the south-east (Stapert 2003). This reconstruction incorporated the sub-concentrations IIa and IIb which were attributed to a dwelling of two families. This suggestion was based on the sector analysis and a classic gender attribution with men associated with the projectiles and stone knapping and women with »fine« tools such as borers (Stapert 2003, 11). In contrast, the spatial analyses of material densities performed by Martina Sensburg revealed a distinction of the two sub-concentrations IIa and IIb (Sensburg 2007; Sensburg/Moseler 2008). This distinction was already proposed by Sabine Eickhoff who, furthermore, distinguished the intensively used entrance in Dick Stapert's model as a circular structure (concentration IIc) in the south-east of the concentration (Eickhoff 1988, 39). In addition, Martina Sensburg reconstructed a heptagonal wall structure enclosing an approximately 40 m² large area with maximum extensions of 8 m × 7 m in concentration IIa (Sensburg 2007, 13-29), whereas the concentration IIb was considered as unlimited (Sensburg/Moseler 2008, 51). Due to the restrictions of the excavation area, Martina Sensburg concluded that the latter result had to remain hypothetical. Moreover, a red colouration of the sediment that was present in concentration I was observed only punctually in protected areas of concentration II (Sensburg 2007, 17 f.). Nevertheless, the occasional observation and the presence of the haematite powder in the pits attested to the intense use of this material also for concentration II. Comparable to the assumption of the ring analysis, a central zone of c. 3.5 m in diameter with a dense material concentration and a stone pavement surrounded by an approximately 1 m wide zone with poor finds was reconstructed in concentration IIa (Sensburg 2007, 7-11). All the pits were located within these zones. Around the find poor ring, an outer circle of lithic accumulations with an average width of 2 m was recorded that was also concomitant with the outer limits of the stone plate setting. Towards the south, a gap within the outer circle occurred which, perhaps, represented a frequently used passage and, thus, the main entrance to the supposed heptagonal structure (Sensburg 2007, 194). In the north-west, the find densities were partially blurred by the accumulations of concentration IIb (cf. Sensburg 2007, 8 Abb. 4; Sensburg/Moseler 2008, 26 Abb. 18; 29 Abb. 19). The distributions from concentration IIa seemed to superimpose and, thus, influence the patterns connected with concentration IIb. In consequence, the two accumulations either developed contemporaneously, or the material of concentration IIb was cleared by the later activities in concentration IIa. Towards the north-east, this outer circle was no longer present, probably due to the disturbances by the southern zones of concentration III. In the south-east, the outer ring was at least 1 m wider due to the adjacent activity area IIc. The adjacent concentration IIc was characterised by a large accumulation of faunal remains including cut red deer incisors, stone plates, charcoal, and burnt stones. Although the latter two could indicate a dump area, the former two materials suggested a workshop comparable to the eastern circular structure in concentration I, and the composition of the associated lithic inventory further sustained this interpretation (cf. Eickhoff 1988, 39; Sensburg 2007, 116 f.). This activity zone was probably separated from the main concentration IIa (Eickhoff 1988, 39) and a respect of the limits could indicate quasi-contemporary processes (cf. Sensburg 2007, 147-150).

In concentration III, the problem of various phases which disturbed the previous distributions became most apparent (Stapert/Terberger 1991; Terberger 1997). However, for the main concentration IIIa an oval to circular enclosure with an exit towards the south-east and activity areas in the south and east were suggested (Terberger 1997). Again, the reddening of the sediment by haematite powder was only documented in protected areas such as underneath stone plates and inside pits. Moreover, according to the various distributions and the filling of the pits, Thomas Terberger suggested two, possibly three phases in which concentration III was occupied (Terberger 1997, 313-315). Such a suggestion also implied that the large structures in Gönnersdorf were more permanent installations that, once constructed, were used again in following visits to the site. Thus far, for the scarce areas east and south, no spatial analysis was made focused on the questions of barriers or dwelling structures. These areas were rather considered as open air workshops or, perhaps, dump areas.

According to Dick Stapert's spatial analysis, the remains of the main concentration of Gönnersdorf IV represented an enclosed area of approximately 5 m in diameter with an exit towards the south-west (Stapert 1990). In general, this result was approved by following studies, only the shape of the structure was assumed to be rather trapezoid to almost quadratic (Jöris/Terberger 2001; Sensburg/Moseler 2008, 98). A detailed spatial analysis provided again indications for a previous use of the terrain and the cleaning of these previous remains by the constructors of the possible dwelling structure (Sensburg/Moseler 2008, 72-74). However, all analyses indicated that the north-eastern hearth was an installation in the open air and by some refitted material connected to the main concentration of Gönnersdorf IV.

The discussion on where to locate dwelling structures, either outside the main material concentrations or around these accumulations, as well as on the shape of these structures -round, polygonal, or trapezoid- is too broad a topic to be discussed in the present study (cf. Leesch/Bullinger 2012; Gaudzinski-Windheuser et al. 2011b). However, in the proposed large structures, static problems could have occurred (pers. comm. Olaf Jöris, Neuwied), in particular, if the presence of forests and, thus, the availability of sufficient building wood was limited. Therefore, the proposed latent barriers could have also resulted from other measures such as setting up of windbreakers or cleaning or partial outwash outside the paved areas. Whether the pavements were installed to prevent tripping into sharp material, to protect against the partially perhaps frozen floor, and/or functioned as a type of floor heating is also not thematised here. Nevertheless, these questions become important when considering the procurement of resources such as the immense amounts of slate plates and the substitute of these resources by, perhaps, brushwood or bark (Janes 1989; Grøn 1990). Certainly, the concentrations were places of intense and very heterogeneous activities.

Detailed studies of the single concentrations and the succession of the activities performed in these concentrations are numerous for Gönnersdorf (Poplin 1976; Bosinski 1979; Terberger 1997; Sensburg 2007; Sensburg/Moseler 2008). Coupled with analyses of the material from the complete site (Franken/Veil 1983; Tinnes 1994; Höck 1995; Álvarez Fernández 1999; Street/Turner 2013), the impression of an important base camp arose, where most types of activities in hunter-gatherer lives were performed. Hence, from the dressing of the game to consumption and curing as well as further processing into garment, tools, or art and final discard were evident at the site. The use of further resources was also proven such as lithic raw materials that were knapped and retouched for the use as tools on the site or as projectiles or blanks elsewhere, or haematite which was ground for the use as powder. Furthermore, the import of relatively numerous pieces described as jewellery and the manufacturing of jewellery (jet beads, animal teeth) as well as art (figurines, engravings) attested the satisfaction of growth needs (Maslow 1943) and suggested the performance and participation of the inhabitants in more complex social actions. In the following discussion, some main activities are mentioned for the general site to highlight the differences and similarities between the concentrations.

The numerous lithic raw materials were generally scattered across the complete site but with varying centres. Western European flint was clearly associated with concentration II, but in concentration III and concentration I also notable numbers were found (Franken/Veil 1983, 64-71 Abb. 43-49). However, Baltic flint, that occurred infrequently in concentration II, was more commonly used in the other two areas. The Tertiary quartzite was clearly dominant in concentration I but occurred regularly also in concentration III (Franken/Veil 1983, 33 Abb. 22). In particular, the haematite stained specimens were accumulated in these areas (Franken/Veil 1983, 54-55 Abb. 39-40). Lydite was massively concentrated in concentration III spreading into concentration IV and the northern part of concentration II (Franken/Veil 1983, 9 Abb. 2). In general, chalcedony was ephemerally scattered but with a slightly increased number in concentration III and IV (Franken/Veil 1983, 24 Abb. 14). The raw materials with small numbers such as opal or rock crystal were generally found in concentration III with some outliers in concentration IV and concentration II (Franken/Veil 1983, 57 Abb. 41). However, this short overview already suggested that concentration III represented the most diverse raw material composition, whereas concentration II was least diverse.

Cores of the common lithic raw materials occurred in all areas of the site and also crested blades and bladelets were found scattered across the site (Franken/Veil 1983). These distributions suggested some type of knapping and blank production was associated with all concentrations, only the processed raw materials varied. However, only concentration IV yielded low numbers of debris material of the blank production and, therefore, Thomas Terberger suggested that the blanks used there were taken from concentration III (Terberger 1997, 160). The introduction of blanks into concentration IV from elsewhere on the site was indicative for quasi-contemporary and interwoven activities on the site. However, the occurrence of various knapping spots on the site could be interpreted either as quasi-contemporary episodes within a single occupation event with, perhaps, different actors, or as different occupation events with the recurrent performance of the blank production. In addition, the different distributions of the lithic raw materials suggested the differentiated use of resources on the site.

However, Stephan Veil pointed out that various lithic raw materials were used for the production of various tool types in Gönnersdorf and, consequently, the different distribution of tool classes partially depended on the distribution of the raw materials (Franken/Veil 1983, 262-266). However, all tool classes occurred in all concentrations of the site suggesting that in general similar activities were performed in all concentrations of the site, only with different intensity and/or with materials of different quality for the task. In detail, LMP which on average made up some 40-45 % of the retouched artefacts were the most numerous retouched artefacts in all concentrations, only in concentration IV the values were the greatest. A traceological analysis made on some 200 flints from concentration II showed that the LMP were clearly associated with the use as projectile implements (Sano 2012b). Thus, their concentration could indicate places of repair of hunting equipment. According to the numbers of single pieces in the artefact class, burins, splintered pieces, and sometimes borers followed on the distribution proportion of LMP. Truncations, end-scrapers, and composite tools were also usually present but in significantly lower numbers. The traceological analysis of some Western European flint artefacts suggested a diverse use of these tools and emphasised the large spectrum of activities performed with these pieces.

Comparable to the various areas where blank production and tool modification occurred, the presence of various spots on the site where the same tools were used and/or discarded could not reveal whether these tasks were performed by the same actors in a relatively short succession, or by various actors quasi-contemporary within a longer occupation event or whether these patterns resulted from several successive occupation events.

However, the presence of all stages of the lithic production as well as all types of tools in the three large units of Gönnersdorf marked these centres as possible independent structures where at least the part of the livelihood of a hunter-gatherer group reflected by these implements was assured. Nevertheless, in concentration IV this assurance was not as completely evident as in the other concentrations due to the scarcity of evidence for the early preparation processes.

Similar to the lithic artefacts, horse was present in all concentrations, though in different numbers. Nevertheless, consumption of horse occurred probably in all concentrations, whereas the processing of the carcass was performed at various stations and with different intensities (Street/Turner 2013). For example, in concentration IV only two horses were securely determined, whereas concentration I produced indications for 14 and concentration II even for 28 individuals.

Considering that a modern adult Przewalski horse weighs a few hundred kilograms and assuming that prehistoric wild horses were of comparable or even larger size (Groves 1994), the transport of some 50 individuals to the site represented a considerable logistic effort. Several persons and possible transport equipment had to be involved in this effort, particularly, if the kill site was located at some distance to the site as was suggested for Late Magdalenian sites in Switzerland based on the analogy of a long fleeing distance of modern horses (Müller et al. 2006).

Along with horse, reindeer, arctic hare, fragments of mammoth, and various bird species were commonly spread species on the site, although they were not found in the scarce faunal assemblage from the northern concentration IV. In contrast, the remains of woolly rhino were only accumulated in concentration II, in particular IIb, and concentration III. Equally, the remains of bovids and caprids were mainly focused on these two central concentrations. In general, the fauna from concentration II and concentration III were comparable with the latter only containing less material and, thus, a connection of these main concentrations was suggested (Street et al. 2006, 761 f.). In this case the general organisation of the concentrations II and III with a large construction in the south-west and a smaller activity zone in the north-east were similar to the small ensemble in the area IV. However, the scarce seasonal evidence indicated a possibly successive use of the two large central concentrations. Moreover, the spatial analyses suggested several episodes on the two concentrations which, perhaps, were interrelated in a longer occupation event. Wolf remains were recovered in concentration I and II with the remains from concentration I representing perhaps a pelt (Street et al. 2006), whereas in concentration II rather the remains of producing a pelt were deposited. Arctic fox was dominantly associated with concentration I (MNI = 29), whereas on the other parts of the site only single specimens, in particular in the form of teeth pendants, were determined (Street et al. 2006; Street/Turner 2013). This focus on a single concentration and the high number of individuals in combination with the assumption of various hunting episodes forming the evidence would imply that the massive construction of this sub-area was perhaps occupied over a considerable period or recurrent visits. Furthermore, snowy owl and the majority of mammoth remains were very focused on concentration I.

Organic artefacts and debris of their production were unevenly distributed on the site and mainly associated with the large accumulations which can be due to preservation issues (Tinnes 1994). The most intense concentration of organic artefacts was with 181 fragments in concentration I followed by concentration II with 113 specimens (Tinnes 1994, 2). In concentration III and IV no organic artefacts were recovered.

The organic figurines were also mainly preserved underneath stone plates and in particular in pits. Consequently, the distribution of these pieces followed also the presence of these protective covers. The majority of female figurines made of organic raw materials were recovered from the northern part of concentration I, whereas the specimens made of slate were generally associated with concentration II (Höck 1995, 266 f.).

The production of pendants made of fox teeth was suggested for concentration I (Álvarez Fernández 1999) where also the main accumulation of this species was found. However, arctic fox teeth were also found in concentration II and III. In the former, the pieces without perforation were more common. Moreover, perforated red deer canines were also concentrated in concentration I (Álvarez Fernández 1999), whereas the red deer incisors were accumulated in concentration II (Street et al. 2006).

The molluscs were also mainly found in concentration I, in particular *Homalopoma sanguineum* was only recovered from this area. However, some *Denatlia* were found in all the three main areas (Alvarez Fernández 2001).

In the eastern part of concentration I some unfinished jet beads were found suggesting the location of a workshop in this part of the site, whereas in concentration II and III only single unfinished specimens occurred (Álvarez Fernández 1999). Additionally, many finished pieces were spread in the main concentrations, often found in pits suggesting loss or depot of these items (Álvarez Fernández 2009).

Engraved plates were found across the site but self-evident only in association with plate occurrences. Thus, the majority of engravings originated from the concentrations I and II, less frequently from concentration III, and only some examples were recovered in concentration IV and the south-western area. However, the number of single engraved individuals was significantly higher in concentration I (n = 377) than in concentration II (n = 178). For the most commonly represented types (females, n = 331; horses, n = 79; mammoths, n = 76), the distributions were very different (Bosinski/d'Errico/Schiller 2001, 291-298; Bosinski 2008): The engravings of horses were spread randomly on the site with single specimens found even in concentration IV and the south-western area. In contrast, female (n = 227) and mammoth engravings (n = 66) were significantly accumulated in concentration I where also most of the woolly rhino images (n = 13 out of n = 17) were found. The bird engravings (n = 22) were more frequent in concentration II (n = 15). Most of the less common motives were distributed evenly.

The resources used as firing material were different in the central concentrations[15]: In the western part of the centre of concentration III wood charcoal of willow (*Salix* sp.) was concentrated, whereas wood charcoal in the centre of concentration IIa was particularly determined as pine (*Pinus* sp.; cf. Street 2012). Occasional pieces of juniper (*Juniperus* sp.) were also found in this concentration, but these pieces were mainly scattered across concentration IIb. In all concentrations some type of brown coal supplemented presumably wood as fuel material (Schweingruber 1978). Equally, Tertiary wood was frequently used as supplementary in concentration III and was also found commonly in the eastern part of concentration IIa, only occasional pieces spread to concentration IV.

For concentration I the spatial analysis was thus far the least detailed one, but varied activities were still evidenced by the material from the concentration. In addition to fileting for possible alimentary use of horse and arctic fox, the preparation of pelts was suggested by the remains of arctic foxes and a wolf. Perhaps, the snowy owl remains were also indicative for the use of the feathers. In addition, the majority of organic artefacts were recovered from this area proving the particularly good preservation and also suggesting workshops for processing bone and dental material as well as reindeer antler. In this concentration many expressions and semifinished products of jewellery and art were found revealing the importance of artistic occupations in this area. Many pieces were found in pits and, particularly, one pit contained the probable remains of a complete necklace (Álvarez Fernández 2009, 53). The seasonal indicators attributed at least some of these activities to the cold and dark season where in the seasonal cycles of modern hunter-gatherers in arctic environments a reduction of the protein resources was observed (Halperin 1980) that could be matched by stored food and decreased caloric intake leading to more trapping instead of active hunting and the performance of other tasks inside a dwelling structure. In some arctic peoples such as the Copper Eskimo, aggregation camps were observed in this period of the year (Halperin 1980). The reanalysis of the

15 In concentration IV only a few pieces were determinable, whereas in concentration I the exact location remained unclear (cf. Peters 1969).

faunal material found also indications for the occupation during spring and early summer suggesting a longer period of use of this large structure (Street/Turner 2013).

In contrast, the activities of concentration II were usually considered to relate to the warmer period (Poplin 1978). These indications originated particularly from concentration IIa that influenced the distributions from concentration IIb. In a detailed spatial analysis of the large concentration IIa, Martina Sensburg distinguished various working areas based on the distribution of the various raw materials and the various classes of artefacts (Sensburg 2007, 112-117. 142-150). For example, she distinguished between inside and outside a proposed dwelling structure and emphasised the high amount of debris material outside the structure (Sensburg 2007, 60). In particular, in concentration IIc and in the scarce cluster south-east of this stone ring, cores of Western European flint were prepared and blanks were modified. Crested blades and bladelets made of this flint were also found basically in this south-eastern part where various raw materials were scattered. However, the main activity zone was located inside the supposed structure and organised around the central stone setting where all lithic raw materials were present and in addition preserved in almost all stages of exploitation. Nevertheless, Martina Sensburg emphasised that activities producing much debris or requiring a lot of space such as preparing of hides were performed outside the stone plate structure (Sensburg/Moseler 2008, 5). Moreover, according to the filling in the pits, she reconstructed an activity cycle within the centre of the structure beginning with a phase in which burins and backed bladelets were involved, perhaps accompanied by activities that made use of borers (Sensburg 2007, 142-145). In the next probable stage, blank production was one of the performed activities along with an intense production of burins and further renewal of backed bladelets. In this stage, activities were performed in which end-scrapers were used. Nevertheless, Sensburg recognised no substantial clearing in the distributions of the lithic material, leading her to assume a single occupation event (Sensburg/Moseler 2008, 5) according to a model of material diversification and the variability of spatial organisations established by Hartwig Löhr (Löhr 1979).

Besides the lithic production, processing of faunal material was also proven. All stages of processing of horses as well as arctic hare and reindeer were attested in this concentration. However, the reanalysis of the material showed that arctic hare was more dominant in concentration IIb than in concentration IIa which was clearly dominated by horse (Street/Turner 2013). Moreover, the spatial distribution of the faunal material yielded the south-eastern structure (IIc) as a third independent centre in this concentration. In contrast to concentration I, arctic fox played no important role. Perhaps, the absence was due to the warmer season when the pelt of the modern species is usually not as dense as in the colder season. Birds were comparably common as in concentration I, only snow owl was not determined among the species (Street/Turner 2013). However, rhino remains were more numerous, particularly in concentration IIb and the north-western part of concentration IIa but therefore mammoth was not as frequent as in concentration I.

Organic artefacts were frequent in this concentration (n = 113) with almost a third of the pieces (n = 40) originating from concentration IIb (Tinnes 1994). In contrast, the animal teeth interpreted as pendants (arctic fox and red deer incisors) were all found in the centre of concentration IIa and in concentration IIc where also a single unfinished jet bead was recovered (Álvarez Fernández 1999). Further, pieces of jewellery such as molluscs were found in this area (Alvarez Fernández 2001). Nevertheless, the indications for the production of these ornamental items were not as evident as in concentration I. Loss and discard seemed better explanations for the material in this area. In contrast to concentration I, the female figurines were usually made of slate in this concentration and the choice of engraved motives appeared reduced in comparison with concentration I. However, both activities were attested in the concentration. In general, the activities performed in concentration II were similar to the activities in concentration I, although often the numbers were smaller and the activities appeared not as intense as in concentration I. However, in this large

concentration a more detailed study of the faunal material supplementing Martina Sensburg's analyses could perhaps help to further differentiate various episodes and the development of this concentration, in particular in regard to the succession of the concentrations IIa, IIb, IIc, and in addition the influence of concentrations IIIb and IIIc on the spatial pattern.
Although almost all lithic raw materials were found dispersed across concentration III, the actually dominant raw material was different for each activity zone of this area: In the central area Tertiary quartzite dominated, whereas in the smaller northern part chalcedony was slightly more frequent, and in the eastern zone Baltic flint and Tertiary quartzite were slightly more common. In the zone IIIb, lydite artefacts were most numerous and in zone IIIc, the Western European flint occurred most frequently. However, the northern part of IIIa, the eastern zone, and the zone IIIc contained less artefacts than the central area IIIa and the adjacent zone IIIb. Since Western European flint and lydite were rarely found in the pits, they were considered as remains of a later phase (Terberger 1997, 235-248). Terberger suggested that the blanks found in concentration IV were taken from concentration III because of the low numbers of debris material of the blank production found in concentration IV (Terberger 1997, 160). Equally, he assumed concentration II and III to be interconnected in this way, whereas concentration I appeared rather unconnected according to the refitted material with concentration III, although the main raw materials from both concentrations were comparable. For instance, some crested blades of Tertiary quartzite that were common in concentration I were also found in concentration III and some cores of the same material were also found. In particular, in concentration IIIc a cluster of cores was found suggesting a working area there. In addition to the common raw materials, almost all exotic materials such as rock crystal were worked in this part of the site. In addition, fossil curiosities such as a shark tooth, a dinosaur vertebra, and a rhino bone were recovered here (Floss 1994, Abb. 159). Moreover, some finished and a few unfinished jet beads as well as a few perforated and not perforated arctic fox teeth were recovered from the central area III. Several pieces of each of these items as well as several more organic and anorganic artefacts including a perforated piece of lava were found together in a single pit in the south-west of the pit cluster (Álvarez Fernández 2009, 52 f.). Perhaps, these pieces represented the depot of a workshop. In summary, concentration III was the accumulation with the most heterogeneous spectrum of raw materials resulting from several activities mainly connected to lithic production. However, based on the technical, faunal, and ornamental remains the activities were again similar to those performed in concentration I and II but with the production of even less material than in concentration II.
In comparison with the other Late Magdalenian concentrations in Gönnersdorf, concentration IV yielded particularly small numbers of material. Nevertheless, according to the distribution of raw materials, Frank Moseler suggested that the majority of lydite artefacts were mainly cleared out before the main structure was set up (Sensburg/Moseler 2008, 99). However, the lydite distribution did not spread from the main centre in the southward located concentration III but originated from the north-west indicating further settlement activities west of the excavated area. The very scarce indications of a blank production attested that a significant amount of the blanks was prepared elsewhere. Also, the retouched tool inventory was very small. However, dependent on the analyst, half to almost two third of the retouched artefacts were LMP. Thus, these pieces were present more frequently than in other concentrations. Nevertheless, the other tool classes, in particular, burins, splintered pieces, and borers were comparably common as in other areas. Only some remains of horse were found in association with this concentration and, possibly, represent consumption debris. No organic artefacts, figurines, or jewellery were recovered from this area. This scarcity of organic material in general was partially attributed to the lack of pits and protecting stone pavement and, therefore, preservation could be an explanation for the rare finds. However, the presence of burnt quartzes led Frank Moseler to suggest boiling as activity around the north-eastern hearth (Sensburg/Moseler 2008,

99). Perhaps, this process further intensified the decomposition of the organic material used there. Nevertheless, the comparably small amount of lithic material, in particular knapping debris, indicated that some activities such as manufacturing processes were not performed intensively in this area. In regard to the refitted material, the evidence from the lithic inventory, and the general scarcity of faunal remains, concentration IV was probably a dependent concentration that was supplied from other areas of the site. However, some activities were performed around the north-eastern hearth and could have supplied, along with the main concentration, also other areas of the site.

Consequently, although some activities occurred at least once in all concentrations, some centres of for example blank production or horse exploitation appeared to have supplied other areas where for instance the blanks were used or horse was consumed. Further activities such as the manufacturing of mammoth ivory or jet appeared focused on only a few areas which, perhaps, reflected the lower necessity of such activities within the daily routine of hunter-gatherers. However, these focuses displayed a strict spatial distinction for some activities on the site.

This spatial organisation could result from either a single event in which different activities were performed quasi-contemporaneously in distinct spaces and possibly by different persons or from several temporally distinct events if these activities were for example performed in specific seasons such as perhaps the manufacturing of pendants from arctic fox teeth.

Although only a section of an even larger site was excavated in Gönnersdorf, a rich corpus of very diverse material, exceeding many Late Magdalenian sites, was uncovered within the remnants of massive structures. Comparable to other Late Magdalenian and, in particular, some cave sites, the material was spread by complex spatial dynamics within limited areas. In contrast to some other Late Magdalenian sites, these patterns resulted from numerous, diverse activities. In addition, the human material from the site suggested the presence of several age classes among the settlers of the site. Thus, Gönnersdorf represented one of the unambiguous examples of a base camp. Furthermore, the intense presence and, in particular, production of art, if understood as an expression of a growth motivation (Maslow 1955), could indicate a more important social component and support the conclusion that, perhaps, the site functioned as an aggregation camp. Nevertheless, to sustain this hypothesis the quasi-contemporary presence of different groups has to be proven in the chronological development of the site.

Chronology of the Late Magdalenian

Based on the Late Magdalenian material from Champrèyves and Monruz, the very complex spatial patterns of Late Magdalenian sites were recently highlighted by Clemens Pasda (Pasda 2012). Thereby, he illustrated the thesis that only repetitive visits of specific locations formed sites which become visible (and interpretable) in the archaeological record because only a considerable amount of material makes possible the recognition of the (secondary) activities performed at the site (Pasda 2012, 5-8). In fact, the complex settlement dynamics of Gönnersdorf have been prompting this discussion for some decades (cf. Jöris/Street/Turner 2011). The different possibilities were that the site was settled in one long-lasting occupation event by a single family group or by an aggregation of family groups or that the site was occupied several times by a traditionally successive group or even by several independent groups as a favourable location. Following from this discussion were questions such as how permanent the Late Magdalenian settlements were and how far the Late Magdalenians could be described as semi-sedentary hunter-gatherers.

Indications for answering this question with the help of spatial analysis were found on two levels of the site: On an upper level was the relation of the main concentrations to each other and on a lower level the relation of various episodes within the main concentrations. The latter is only mentioned occasionally in the following discussion because more detailed studies of some concentrations of the site were already published (Sensburg 2007; Sensburg/Moseler 2008) and further studies are still necessary for the other concentrations.

On the upper level, Thomas Buschkämper observed in the south-western area that the distribution of material attributed to concentration II ignored the structures attributed to concentration I and concluded that these two main concentrations resulted from two non-contemporary events (Buschkämper 1993, 23. 33 f.). However, as suggested in relation to the north-eastern hearth in concentration IV, down-slope erosion could have affected the lighter materials. Thus, if the remains of concentration I and II were deposited at the same time, erosion could have moved lighter material from concentration II downwards on top of the material from concentration I. Therefore, a detailed analysis of this part of the site including profile drawings could help resolve the stratigraphic relation of these two major concentrations.

Moreover, in the north-eastern part of concentration IIa the distributions were probably disturbed by material from the southern part of concentration III, i. e. concentrations IIIb and particularly IIIc, suggesting at least a successive use of these concentrations. In addition, the distribution of material from concentration IIb appeared limited or pushed aside by the construction of concentration IIa.

Thus, according to the disturbances of the material distributions, the western structures of concentration I existed perhaps previously to the southern extension of concentration II which probably was connected to the activities of concentration IIa. Although Martina Sensburg assessed the respecting of limits and the existing material exchange in concentration II as probable indicator for a quasi-contemporary existence (Sensburg/Moseler 2008, 50), she could not rule out that these patterns resulted from concentration IIa succeeding concentration IIb and, thus, clearing out the material. Moreover, the disturbances in the north-east of concentration IIa suggested that the latter was preceding at least concentration IIIc and, perhaps, also concentration IIIb. However, whether these disturbances occurred in a clearing within a single, long-term occupation event or in different occupation events remained unclear. Yet, according to the complex interconnections a relatively short period for the development of this part of the site was probable (Sensburg/Moseler 2008, 49 f.).

The different distribution of the lithic raw materials was applied as a reflection of the number of events or phases forming the site based on the assumption that these materials which originated from various, partially very distant, regions were introduced to the site either from different groups or at different times. Therefore, the interpretation of the raw material distributions was also coupled with the discussion of whether the site represented a revisited base camp or an aggregation camp. Another possibility for the presence of the diverse raw materials was the exchange of raw materials by some groups previous to the visit. In this case, the different raw materials could have been collected or otherwise received by a single group during a yearly round and then used and/or discarded in Gönnersdorf. Consequently, very diverse raw materials were perhaps used during a single occupation event. The least frequent material was, in this scenario, the material possibly collected early in the round and, therefore, reached the site only in small quantities, or a very distant material which was received in small quantities from another group. However, very frequent materials were conversely accessed shortly before the occupation of Gönnersdorf and/or easily accessible (and transportable) during the occupation. The Western European flint that dominated in concentration II came from a very distant source and, consequently, pointed to a long-distant movement comparable to the chalcedony in the concentration IV. If these distant materials were accumulated in such large quantities, transport must have been an issue resulting in the hypothesis of repetitive visits contributing to large quantity accumulation (Terberger 1997, 166). In contrast, the Tertiary quartzite reflected a regional movement prior to the settling of concentration I and the lydite from concentration III suggested very local moves. Thus, the latter could have been set up after the installation of one of the other concentrations and then supplied with the local material. In addition, the frequency of the deposited raw materials in the pits was used by

Thomas Terberger to distinguish two phases of raw material use: In one phase the pits were filled with four of the main raw materials, whereas two (Western European flint and lydite) were only found infrequently inside the pits and, thus resulted from another phase (Terberger 1997, 166). However, which of these phases occurred previously could not clearly be distinguished.

Moreover, assumptions based on the *chaîne opératoire* that required detailed analyses on the lower level to reveal various stages such as the scarcity of blank production debris in concentration IV was used as an indicator for temporal succession. In particular, some scarcer materials made possible detailed analyses on the succession of raw material use on the site (Sensburg 2007). However, following this line of evidence for the upper level of the complete site will become possible when all concentrations are presented in sufficient detail (cf. Jöris/Street/Turner 2011). For the two central, large units, this type of analysis was accomplished and revealed closely interwoven distributions of raw materials and possibly stages of production but the results led to very different interpretations: Thomas Terberger suggested at least two, possibly three distinct occupation phases, perhaps even occupation events in concentration III that were related to the possibly two-phased occupation of concentration II (Terberger 1997, 313-315), whereas Martina Sensburg considered a single occupation event for concentration II with various episodes (Sensburg 2007, 150-154. 202-204).

This discussion is of interest concerning the permanency of the Late Magdalenian settlement in Gönnersdorf and already arose from the massive structures and amounts of material found at the site as well as the assumption that such immense weights of stones accumulated successively. Nevertheless, whether this succession occurred during one very long or many stays could not be clarified.

Besides the distribution and overlapping of material, refitting lines were used as indications for spatial dynamics and, in particular, quasi-contemporaneity since the earliest analyses of Gönnersdorf (Bosinski 1979). However, Erwin Cziesla illustrated the general uncertainty of contemporaneity based on refitted material (Cziesla 1990, 184) and, in detail, Martina Sensburg clarified the uncertainty of stations traversed by the refitted material (Sensburg/Moseler 2008, 46-49). Nevertheless, frequent connections of material reflect a shared use of resources, and if such refits appear as directed mutually an exchange is a probable explanation and, in consequence, quasi-contemporaneous activities are a probable assumption. However, if these connections are only unidirectional one end served presumably as resource depot or dump but in consequence the distinction of such directions is based on the qualification of the refitted archaeological material (cf. Terberger 1997, 116-146). Moreover, besides relative synchronous activities a time lag between use and discard or deposition and reuse is possible, although in periods of strong aeolian activity the cover of the material limits its visibility and, thus, potential reuse.

Many refits were found between the western centre of concentration I and the south-western area (Franken/Veil 1983). Possibly, these represented flint workshops outside the centre of concentration I. Also from the south of this concentration, material was connected to the south-western area (Terberger 1997, 124) but in this case the former served perhaps as procurement area for the significantly younger concentration V. Single material from the peripheries of concentration I was refitted to material in the northern concentrations. In particular, refitted stone material connected the southern area with the centre of concentration III and concentration IV as well as the south-eastern part of concentration IIa. Along with the peripheries of concentration I, concentration IIa was also connected occasionally to the south-western area, but in these cases the south-western material appeared rather as dump remains. Moreover, a few refits to concentration IIb as well as to the concentrations IIIb and IV were found (Sensburg 2007). In contrast, the connections to the centre of concentration III and concentration IIIc as well as concentration IIc were numerous, displaying intensive exchange. Refitted material connected concentration IIb to all areas of the site except for concentration I (Sensburg/Moseler 2008, 46-51). In particular, the relations to concentration IIc and

IV were very frequent with concentration IIb serving as source for concentration IV. Some connections with the south-western area could also be interpreted as depot or dump. From the centre of concentration III some refits to concentration IIb were found and from concentration IIIb some pieces were connected to IIb, whereas from concentration IIIc only a single refit to concentration IIb was found. This pattern contrasted the intense connections between the areas in concentration III and concentration IIa. Hence, concentration III was frequently connected to some parts of concentration II. In particular, material from the centre of concentration III was frequently refitted to concentration IIa (Sensburg 2007) and to the northern transition between concentration IIa and IIb but only single items from the centre were refitted to concentration IIb (Sensburg/Moseler 2008) and very few to the accumulations in the adjacent areas IIIb and IIIc (Terberger 1997). Frequent refits of material from the centre were also achieved with the north-eastern hearth of concentration IV, whereas few connections to the main structure or from concentration IIIb and IIIc were found. Refits from concentration III to concentration I were very singular and made an accidental impression often connecting peripheral areas. Concentration IIIb was occasionally connected to concentration IIb, IIa, and IIIc. A few refits were achieved with the material from the eastern periphery between concentration III and IV. Material from concentration IIIb was rarely connected to other concentrations. In contrast, concentration IIIc was regularly connected to concentration IIa and IIc but only singularly to concentration IIb. In addition, some rare refits were found with the main structure of concentration IV and the south-eastern limit of the excavated area of concentration III. In concentration IV, the refitted material proved strong connections between the hearth in the north-east and the main structure. Moreover, both areas were frequently connected by quartzes, quartzite and slate plates as well as lithic implements to the various areas of concentration II. In contrast, the centre of concentration III served mainly as a source for the north-eastern hearth and was only rarely connected to the main structure. Furthermore, only single refits were made with concentration I (Terberger 1997, 109-160). The directions of all refits were suggested dominantly uni-directional connections supporting a hypothesis that the other concentrations served as sources for concentration IV. However, this pattern can be interpreted as quasi-synchronous concentrations »feeding« the northern most accumulation or as this northern concentration representing the latest occupation event exploiting the still visible remains from the previous events.

Certainly, frequent exchange of lithic materials occurred between the central concentrations suggesting an intense exchange of material among these accumulations. Many refitted pieces seemed to have served finally in the northern concentration IV. In contrast, the south-eastern concentration I was only connected by single refits of lithic and occasional refits of stone material. In particular, the relatively scarce examples to the centre of concentration I are explicable by unintentional movement of light material or use of still visible remains as a source or settling on no longer visible dumps. However, some connections to the southern, western, and eastern limits of concentration I could also indicate that the accumulation in the south-east was formed by several smaller and, perhaps, temporally distinct structures.

In a synopsis based on the seasonal indicators of the fauna, the filling of the pits, and the refitted material, in particular the Devonian quartzites, Thomas Terberger considered the first phase of concentration III, concentration IV, and possibly concentration I to have formed a winter settlement of a single group (Terberger 1997, 160). According to the refitted material, he further suggested that the concentration II existed previously to the second phase (Western European flint phase, perhaps, later or concomitant with lydite phase) of concentration III and concentration IV (Terberger 1997, 248) because he assumed that the blank production from concentration II had served as source for concentration IV (Terberger 1997, 160). However, the comparable raw material composition and the refitted material could also suggest that the blanks used in concentration IV were made in concentration III (Terberger 1997, 160). Nevertheless, the reanalysis of the distribution patterns in concentration IV by Frank Moseler suggested that the lydite phase was an early stage in concentration IV and the debris from this phase was partially moved aside by the later construction of the large stone structure (Sensburg/Moseler 2008, 99). Thus, if concentration II served as source depot for concentration IV, the lydite remains either represented a separated phase, or the main phase of concentration III occurred after the Western European flint and lydite phase.

Along with the general temporal distinction of the lithic raw material, faunal remains helped to refine the chronology of the site. In particular, the foetal and foal remains of horse were previously assumed as indicative for a seasonal difference in the use of concentration I and concentration II. According to this assumption, concentration I reflected one or several winter episodes (December – February). This seasonal classification was further sustained by the tooth development of reindeer and the significant amount of pelt production, in particular of arctic fox. Concentration II pointed rather to spring-summer hunting episodes (April – June; Poplin 1978; Street et al. 2006; Street/Turner 2013). However, the occupations could have been inhabited alternately but then the last stay would have been in the warm season, disturbing the winter material. In contrast, the very similar faunal composition of concentration II and III led Martin Street and his colleagues to consider these concentrations as interrelated (Street et al. 2006). The more recent reanalysis of the faunal material and its spatial distribution supplemented the previous seasonal indicators from the three main concentrations (Street/Turner 2013). Besides indicators for an occupation from January to April, horse remains indicative for June and September kills were also found in concentration I. The horse remains from concentration II yielded indications for hunting episodes during the whole year except for August, October, and December. Concentration III provided a very comparable pattern of horse kills as concentration I, only that there were indications for an October instead of a September hunting episode. Based on these results, an almost continuous settlement of all concentrations of Gönnersdorf appeared possible. The evidence from the faunal material was only scarce for the hottest summer months and the period of the longest day light which Martin Street and Elaine Turner suggested as a good point during the year for long range travels (Street/Turner 2013).

In contrast to the disturbances of the material distribution, the spatial distribution of reliable ^{14}C dates of samples determined to species revealed that a set of dates around 13,050 years ^{14}C-BP from the central excavation area, particularly concentrations IIb and III, were tendentially older than the ^{14}C results from the western centre of concentration I that ranged at 12,810 years ^{14}C-BP (Jöris/Street/Turner 2011, fig. 7a). However, single specimens in the south-east of concentration II and the eastern centre of concentration I produced older results. Another date obtained from an elk bone found in the south-western area resulted in a significantly younger age and, therefore, is discussed elsewhere (see p. 133). Moreover, a distinction between the episodes forming the central concentrations is thus far not possible based on the reliable ^{14}C dates. However, the long interval reflected by the reliable radiocarbon measurements (**tab. 20**; 13,325-12,600 years ^{14}C-BP) sustained the assumption of Gönnersdorf as a traditional occupation site that was settled repetitively over a considerable period but, equally, that some parts were possibly settled quasi-contemporaneously.

In summary, the evidences from the upper level of spatial analysis clearly suggested numerous occupation episodes at Gönnersdorf that reflected diverse activities and distinguished the site as a base camp. Although a relative succession of some of these episodes was possible, the attribution to various occupation events was not unambiguously possible due to the complex interrelations and the contrasting interpretations which were possible based on these connections. The impossibility to resolve the temporal relations also prohibits the assessment of Gönnersdorf as an aggregation camp thus far because the accumulations could also result from successive moves of a single group on the site and the various materials collected or received by exchange during a yearly round. In particular, this problem applied to the northern concentrations which formed a densely connected area and, perhaps, was also related to the peripheries of concentration I. In this connection of areas some activities were multiply observable such as the production of burins or the exploitation of horse, whereas other activities such as the manufacturing of jet beads or female figurines appeared focused to specific spaces. Nevertheless, the indications from stratigraphic observations, refitting, and some seasonal determinations suggested that the centre of concentration I with its strong focus on arctic fox could represent an independent event.

Archaeological material of the south-western area

Thomas Buschkämper counted 2,804 lithic artefacts among this younger assemblage (**tab. 11**; Buschkämper 1993, 30. 53). Due to the horizontally intermediate position between the two major Late Magdalenian concentrations and the additional bio- and cryoturbation processes in the sediment, the distinction of this small inventory from spoilt Late Magdalenian material was not unambiguously possible (Buschkämper 1993, 22). Moreover, the majority of artefacts were made of flint (n = 1,943). In general, the Western European flint (n = 1,040) from approximately 110 km to the north-west (**tab. 12**) dominated the south-western assemblage as well as the concentration II. In fact, the majority of Western European flint was found in the northern part of the south-western area and was connected to the concentration II by several material refits (Buschkämper 1993, Plan 12-13). The majority of Western European flints were primary Jurassic flints, only eleven pieces were identified as gravel flints, and one burin as Simpleveld variety. A further 752 pieces were made of Baltic flint which originated from a distance of at least 105 km northwards and were mainly distributed in the eastern part (Buschkämper 1993, Plan 14), partially associated with the part attributed to concentration I. In addition, local Tertiary quartzite (n = 386) as well as the foreign Tertiary quartzite of the Ratingen type that originated from some 85 km either northwards or eastwards (Floss/Terberger 2002, 17) were found in the south-western area. These materials were also common raw materials in the adjacent concentration I and, in fact, the majority of artefacts made of these raw materials were associated with the eastern most part which was attributed to concentration I (Buschkämper 1993, 48 Plan 19). Furthermore, the local lydite (n = 417) in different varieties including five pieces of radiolarite were probably brought to the site from the Rhine gravels in complete pebbles (Buschkämper 1993, 53). Chalcedony (n = 37), siliceous oolite (n = 8), Jasper (n = 3), and Palaeozoic quartzite (n = 1) were identified in small numbers. The latter originated from the north-west as well as the dominant Western European flint and was considered as intrusion from concentration II (Buschkämper 1993, 50). The jasper originated probably from the southern German Freiburg region, approximately 300 km to the south (Floss 1994, 233). The chalcedony and the siliceous oolite were probably collected from sources some 75 km south-eastwards in the Mainz Basin.

Generally, Thomas Buschkämper excluded the Western European flint material from concentration V because he regarded the raw material as scattered across the complete area, presumably from the northern concentration II (Buschkämper 1993, 33 f.). However, he also stated that some artefacts were possibly reused in concentration V (Buschkämper 1993, 110 f.). Based on this assumption, Buschkämper gave only few details on this raw material.

Thus, in addition to the six cores recorded by him for this area, further cores from Western European flint were not considered (**tab. 13**). Four cores were made of Tertiary quartzite and three of them were found in the structured area attributed to concentration I. The fourth core was found 1 m northwards of a residual core made of Baltic flint (Buschkämper 1993, 35) in the area westwards adjacent to the structures attributed to concentration I. The Tertiary quartzite was partially affected by desilification (Buschkämper 1993, 44-47) that also destroyed some surfaces on a Tertiary quartzite core (Buschkämper 1993, Taf. 2.1). However, the remaining negatives indicated a core with two platforms of which one was mainly used to exploit the piece. Another core was made of lydite and found in the north-western corner of the area, but the main lydite accumulation was in the vicinity of the other two cores. Furthermore, several pebbles with (n = 98) or with-

out single blows (n=22) as well as shattered pieces (n=212) of lydite were recovered from the sub-area. The unambiguous lydite core was exploited unidirectional along one lateral edge (Buschkämper 1993, 53f. Abb. 7) presumably aiming at elongated blanks, although the final knapping attempts ended in hinges. The majority of recovered blanks were bladelets as well as blades and burin spalls, indicating a blank production process focused on small elongated blanks.
Among the 148 retouched pieces recovered in the south-western area, LMP were most numerous with 58 specimens (**tab. 14**). The majority of the LMP was very fragmented but probably represented remains of backed bladelets (Buschkämper 1993, Abb. 6 Taf. 2-3) and resulted perhaps from a production of these implements by intentional breakage as suggested by refitted material on the other areas of Gönnersdorf (see p. 106). Moreover, some of the retouched material, in particular the LMP were associated with the structures attributed to the Late Magdalenian occupation (Buschkämper 1993, 33f. 48). An unambiguous classification was difficult due to the very fragmented preservation but five LMP were discussed as backed points or backed point fragments.
The proposed backed point specimens were diversely made. The single jasper point was questioned by Thomas Buschkämper (Buschkämper 1993, 51) because the small bladelet displayed a straight retouched back that was very comparable to the backed bladelets, whereas the pointed shape was due to the shape of the blank. Furthermore, an only partial, width reducing retouch was performed on the opposite lateral edge. Thus, this LMP is considered as laterally retouched backed bladelet in the present study. This piece was also recorded in the Late Magdalenian assemblage (see p. 107). In addition to this piece, two further backed bladelet fragments made of jasper were found in the northern part of the south-western area where the material was associated with scatters of the main Late Magdalenian concentrations, in particular concentration II (Buschkämper 1993, 50-52 Plan 11). Therefore, the jasper artefacts were probably not associated with concentration V. In addition, a point of siliceous oolite was mentioned but not further described (Floss 1994, 227) and, clearly, was not found in the concentration in the south-west (Buschkämper 1993, 49f.).
Consequently, only the four pieces made of Baltic flint are considered as possible points from the south-western area in the present study. The most complete backed point (1003/352) was already recognised by Stephan Veil (Franken/Veil 1983, 435 Taf. 29.4; Buschkämper 1993, 39). Another refitted specimen (a medial and a proximal piece) was also almost completely preserved and found approximately 4m north-westwards of the former artefact. A further two fragments were a proximal and a distal fragment. The two almost complete specimens fall in the range of curve-backed points, although the most complete specimen had no continuous lateral retouch leaving a little unretouched edge. Both pieces were retouched additionally on their proximal edges and along the basal part of the opposite edge. Moreover, two fragments were found, a proximal and a distal piece. The latter represented a retouched tip and yielded a shoulder angle. This piece was found near the two refitted fragments in the centre of the south-western area. Interestingly, a notch was set on the unretouched edge opposite to the tip part of this fragment. Such notches were variously observed in northern curve-backed fragments (Schwabedissen 1954, 64). However, whether this damage was intentional such as for fastening some hafting material or was due to use and represented, perhaps, a specific impact pattern remained unclear. A notch-like damage was observed on the retouched basal part of the most complete specimen and there clearly indicating an impact damage. The proximal fragment was pointedly shaped by two lateral retouches. Nevertheless, comparisons with prepared backed bladelets with a retouched opposite edge that were not broken into smaller fragments (e.g. Franken/Veil 1983, Taf. 33.8-11. 14 .19) advise to be cautious when judging broken pieces. Furthermore, two sided, pointed retouches could also represent transitions towards borers. Therefore, the possible pointed base fragment which was found in the south-eastern part of the area in the planum I of the Late Magdalenian

(Buschkämper 1993, 40) could also represent a worn out borer[16]. In contrast, the occasional mention of composite tools of backed bladelets and borers or broken borers in the Gönnersdorf assemblage could perhaps include further possible backed points or fragments of those (Franken/Veil 1983, 435 Taf. 29.9). In addition, the many medial fragments classified as backed bladelets could refit to tips and/or pointed bases and, thus, also represent backed points (Buschkämper 1993, 39). However, the complete specimen with the possible impact fracture on its base clearly prove the presence of backed points in the material of the south-western area.

Besides the LMP, burins (n = 34) were most frequently recovered. These tools were dominantly made on a truncation and were often shaped as the Lacan type. However, the latter were generally associated with the Late Magdalenian assemblage, in particular, with the eastern part of the south-western sub-area and, thus, concentration I. Furthermore, the dominance of Baltic flint and Tertiary quartzite among the burins reproduced the dominant raw materials of concentration I. Rarely dihedral pieces or burins on natural edges or fracture surfaces occurred. One such burin was made of Simpelveld flint that is a variety of Western European flint originating from the southern Netherlands and was found near the jasper pieces. Two further retouched artefacts of the Simpelveld flint were found in the northern concentrations III and IV (Buschkämper 1993, 32 f.). As with the jasper specimens, these three flint tools were considered as basic equipment. Nevertheless, this set of basic equipment was probably connected to the northern Late Magdalenian concentrations, not to concentration V. The seven end-scrapers were very worn out and the blanks were often broken. The remaining parts of the blanks were generally classifiable as wide blades. They were mainly concentrated in the eastern part of the south-western area. The twelve borers were clearly dominated by Baltic flint pieces (n = 11) which were occasionally made on burin spalls. Generally, they were piercers which were made often by two concave retouches from alternating surfaces. These specimens were found in the central area and the south-eastern corner representing the tool class that was most apparently connected to concentration V. Although this connection might further sustain the doubts about the classification of the backed point fragments, the morphometrics as well as the shaping of the different groups were significantly different with the suggested backed points being wider and thicker and retouched only from the ventral side. The two straight truncations of which one was also laterally retouched were made of Baltic flint (Buschkämper 1993, 155 Taf. 6.5 and .8). The three mentioned composite tools were all made of Baltic flint and all were combinations with burins; in two cases the burins were combined with end-scrapers and once with a borer (Buschkämper 1993, 37). They were generally found in or in close vicinity of the area attributed to concentration I. At least seven splintered pieces were identified in this area.

Besides lithic material, also rock stone material was recovered in the area. More than 225 kg of slate, Devonian quartzite, a singular basalt, and a few specimens of sandstone of different varieties (including a piece of greywacke and some pieces with a high iron ore content, so- called *Eisenschwarte*) as well as more than 40 kg of quartz were recovered from the south-western area. Many quartz pebbles and fragments (n = 312) were considered as cooking stones and hammerstones (Buschkämper 1993, 91-93). Some pebbles of Devonian quartzite were prepared as some type of chopping tool (n = 3) and a further seven pieces were probably used as some type of hammerstones or retouchers according to the fields of scratches on their cortex (Buschkämper 1993, 87-90). Moreover, two pieces of a coarser, almost sandstone-like variety showed signs of abrasion along one edge (Buschkämper 1993, 89 f.). The largest amount of stone material was introduced to the south-western area in the form of plates. The majority of the slate plates were

16 The complete backed point was found in the same square-metre, but a borer made of Baltic flint was also found in this quarter of the square-metre. Thus, the horizontal position does not favour any of those classifications. However, the measurements of the piece and the forming of the retouches as well as their direction were rather indicative for a backed implement (see p. 275-282).

modified along the edges, although the distinction between artificial retouch and depositional damage was not possible in all cases (Buschkämper 1993, 71). However, some plates wore unambiguous hitting marks indicating attempts of splitting or dividing larger plates into smaller ones (Buschkämper 1993, 64) or they wore notches of various sizes along their edges (Buschkämper 1993, 72). These pieces sustained an intentional shaping of these plates by the Pleistocene inhabitants. In addition, several of the many recovered plates, also those made of Devonian quartzite, showed traces of use such as heating by fire (Buschkämper 1993, 67-69. 85-87) or cut-marks (Buschkämper 1993, 77) or depressions that could indicate the use of these pieces as lamps (cf. Buschkämper 1993, 94-96). These used pieces accumulated in the part attributed to concentration I (Buschkämper 1993, 78) but the heated plates were also frequently found in the central part (Buschkämper 1993, 68f.). Moreover, 13 of the slate plates were intentionally engraved with mainly female shapes and mammoths (**tab. 18**). Occasionally, phantoms and horses were displayed and, furthermore, single examples of woolly rhino, and a possible seal were identified (Buschkämper 1993, 79). Nine of the engraved plates were associated with the remains of the Late Magdalenian concentration I in the east. However, four plates were recovered from the central part and displayed also females and a mammoth profile (Buschkämper 1993, 81f.).

14 coin blank pieces made of slate were also found in the south-western area but only one in the central part, whereas nine specimens were found in the areas attributed to the Late Magdalenian. Whether these items were considered as pendants or button-like tools or were used differently remains a matter of debate (cf. Bosinski 1977).

Furthermore, some 50 pieces of haematite were recovered from this sub-area with the majority (n=35) originating from the part attributed to concentration I (Buschkämper 1993, 26f.). In contrast to the large Late Magdalenian concentrations, red colouring of larger patches of the sediment were not observed in the south-western area.

The presence of four pieces of jet was discovered in the anthracological analysis and, probably, was not connected to bead production (Buschkämper 1993, 29). In addition to Tertiary wood, the charcoal pieces from the south-western area were, in fact, mainly identified as flaky brown coal that was concentrated in the area attributed to concentration II. In addition, five samples were determined as pine (*Pinus* sp.) and were scattered across the area but two pieces were found in the central part.

Comparable to the other archaeological materials, several pieces among the faunal material resulted probably from Late Magdalenian activities. For instance, only five bones of arctic fox (*Alopex lagopus*) were found scattered across the south-western area (**tab. 15**; Buschkämper 1993, 100) and, presumably, resulted from sprawl of concentration I (Street et al. 2006, 762). Reindeer (*Rangifer tarandus*) was also only present with a few bone and antler remains which were attributed to concentration I (Street et al. 2006, 762). Hare (*Lepus* sp.) was only determined in a single piece (Buschkämper 1993). Furthermore, a tooth of woolly rhino (*Coelodonta antiquitatis*) and ivory pieces of mammoth (*Mammuthus primigenius*) were probably also remains of the Late Magdalenian activities (Buschkämper 1993, 100; Street et al. 2006, 762).

Two bone fragments of bovids were identified in the south-western area. One was recovered in the eastern part of the site belonging to concentration I and the other one in the central part. The latter was not found in the more recent reanalysis of the faunal material (Street/Turner 2013). Four remains of two bovids (an adult and a young one) were found in a higher stratigraphic position in the northern part of Gönnersdorf (concentration IV, Bosinski 1979, 26) but carnivore gnawing indicated that those animals were probably killed by animal predators not humans (Street et al. 2006, 762; Street/Turner 2013). Further remains of large bovids were recovered from the concentrations I, II, and III (Street et al. 2006, 761). In concentration III, the piece was associated with a possibly disturbed structure (Buschkämper 1993, 100). Hence, the position of these bovid elements remains uncertain in the site context.

However, as in all areas of Gönnersdorf, the dominant species among the preserved faunal remains was horse (*Equus* sp., Buschkämper 1993, 99) of which at least 150 pieces were identified in this sub-area (Street et al. 2006, 762). The bones and teeth scattered across the complete area and refitted pieces suggested connections to the Late Magdalenian concentrations in the north and east. However, an accumulation was also observed in the central area (Buschkämper 1993, 99). The remains of at least five horses were deposited in this area (Street et al. 2006, 762).
Remains of elk (*Alces alces*) and red deer (*Cervus elaphus*) were found in a tendentially higher stratigraphic position (Buschkämper 1993, 102 f.). In contrast to the Late Magdalenian concentrations, the ten red deer elements of the south-western area were dominantly bone material and, probably, attributable to three individuals of which one was a calf that was younger than a year (Buschkämper 1993, 102; Street/Turner 2013). The two elk bones from the south-western area and the single carpal from the south-western part of concentration II were the only evidence of this animal in Gönnersdorf. A radius fragment was found in the central area of the south-western area, whereas the diaphyse fragment of the radius was found in the western corner of the area (Buschkämper 1993, 214 Plan 46).

Spatial organisation of the south-western area
The preserved structures in this part of Gönnersdorf contained neither a continuous, multi-layered pavement nor pits or postholes (Buschkämper 1993, 7). However, stone settings that partially formed dense pavements were observed, particularly in the eastern part, but these structures were associated with the Late Magdalenian concentration I (Buschkämper 1993, 23). Due to the presence of Baltic flint and Tertiary quartzite the south-western area was generally considered as an extension of concentration I during the excavation (Buschkämper 1993, 9). In addition, based on his analysis of all the retouched lithic artefacts from Gönnersdorf, Stephan Veil suggested that the south-western part belonged mainly to the larger area of concentration I according to the distances of refitted material and he recognised two partially overlapping activity areas (I. D and E) within the south-western area (Franken/Veil 1983, 216-233). However, in the northern part of the south-western area he assumed the influence of two activity areas (II. A and D) from concentration II. Hence, the episodes forming these large concentrations spread also on the south-western sub-area and blurred the picture of the spatial patterns in this area. Thomas Buschkämper observed a comparable pattern with remains attributable to concentration I in the east (Buschkämper 1993, 23) and material of concentration II in the north (Buschkämper 1993, 33 f.). Furthermore, he stated that the material from the northern concentration disturbed the eastern as well as the central scatter and, therefore, assumed that the northern concentration represented a non quasi-simultaneous event (Buschkämper 1993, 34), although the distribution in the central part were possibly formed by reuse of the material from the other concentrations (Buschkämper 1993, 111). However, due to the probable connection of the eastern as well as the northern parts of the south-western area to the Late Magdalenian, these structures were previously described (see p. 111-114).
Even though no continuous pavement except for the eastern part was observed, a light scatter of plates and plate fragments was documented across the south-western area but only in the central part another concentration occurred (**tab. 19**). This concentration formed a semi-circular ring of approximately 2 m in diameter and was found shortly underneath the pumice (Buschkämper 1993, 7 f. 17-19). At the south-western end of this structure another cluster of slate plates was considered as remains of a hearth which could not be related unambiguously to the stone circle, although the distinction of plates to the one or the other structure was often impossible at the contact zone (Buschkämper 1993, 8. 18). The two evident structures were surrounded by spaces with sparse scatters of stone fragments and were defined as concentration V. The plates forming concentration V were in general smaller than the plates from the Late Magda-

lenian pavement and in contrast to these dense scatters open spaces extended between groups of stones. A comparable structure was the almost quadrangular stone setting with a central hearth in concentration IV. However, this structure was more clearly defined and considerably larger than the small semi-circle.
Thomas Buschkämper performed a ring and sector analysis of the semi-circular structure setting the centre not to the hearth but in the centre of the semi-circle (Buschkämper 1993, 106-118. 216 Plan 48). From this analysis Buschkämper assumed an enclosed area due to the distribution of the lithic remains (Buschkämper 1993, 116) which were in general scarce inside the structure. According to ethnographic examples, he considered this area to represent the remains of a ridge tent of 2 m in diameter (Buschkämper 1993, 130). However, due to the overlapping of several structures and the given stone setting he did not include areas over 2 m from the geometric centre of the semi-circle. Thus, a reanalysis of this possible dwelling structure with modern spatial interpolations and in the context of the surrounding material would be of some interest. Inside the semi-circular structure three possible backed point fragments were found and immediately »outside« of this area the elk radius was deposited. However, the most complete backed point was found in the south-eastern corner of the sub-area where the cluster of plates from concentration I became sparser. All these pieces were made of Baltic flint which therefore was most clearly related to a typologically younger episode. However, this raw material was also applied in the Late Magdalenian concentration I. The Baltic flint and a Tertiary quartzite core from the south-western area were recovered in close vicinity outside the main Late Magdalenian scatter with small debris in the surrounding suggesting this assemblage does not represent the dumps of concentration I but rather an independent knapping episode. Thomas Buschkämper revealed, by mapping the Baltic flint material according to the various levels, that two distinct scatters were observable (Buschkämper 1993, 36 f.). However, since the upper scatter mainly fills the gaps in the lower scatter (cf. Buschkämper 1993, 184 f. Plan 16 f.) this distinction reflected possibly stratigraphical movements and inaccuracies in the excavation and documentation process (Buschkämper 1993, 10-15. 17-22).
Nevertheless, refitted material connected the semi-circular structure particularly with the northern Late Magdalenian concentrations, not the eastern one. Thomas Buschkämper suggested that the concentrations II and III were exploited by the inhabitants of concentration V and, therefore, he concluded that the semi-circular structure represented a later occupation phase than any of the other episodes from Gönnersdorf (Buschkämper 1993, 141 f.).
The indications for the possible hearth were the fragmentation of plates of a fine-grained slate variant, heat alteration of the quartzitic slate plates, and the presence of charcoal pieces (Buschkämper 1993, 134) as well as the occurrence of fragmented quartzes of which some were coloured by fire (Buschkämper 1993, 211 Plan 43). Probably, the structure was severely disturbed, but Thomas Buchkämper assumed the use for cooking and proposed the structure to represent a satellite hearth (Buschkämper 1993, 135). Moreover, he considered this hearth to have served as plate quarry for the semi-circular structure and concluded that these structures reflected two distinct occupation episodes (Buschkämper 1993, 135). In fact, neither faunal remains nor significant numbers of lithic material were found in the vicinity of this structure which could be due to cleaning out of the material or suggest another type of use of this concentration such as a dump zone.
Some of the plates from the circle as well as from the possible hearth were engraved with lines as well as mammoths and female outlines. Even though these engravings were comparable to the motifs in concentration I, only quartzites and no slate plates were refitted between these concentrations. Some refits were possible between the semi-circular structure and the possible hearth but the hearth was also not connected to concentration I but equally to the northern concentrations. Moreover, one of the female engravings found in the stone circle was depicted with special emphasis and, thus, was probably not brought to this

part without recognition but, perhaps, chosen for curiosity (Buschkämper 1993, 81). Nevertheless, the manufacture of these engravings by the creators of the stone circle cannot be excluded.
Contrary to the assumed hearth, a small cluster of mainly quartzitic slate plates at the other end of the semi-circle yielded many fire heated plates (Buschkämper 1993, 195 Plan 27). Partially, this difference could be due to the materials forming the clusters (Buschkämper 1993, 67). However, only a single pine charcoal was found in the vicinity and no observable accumulation of burnt quartzes was detected in this area. Nevertheless, faunal remains as well as lithic material, in particular the lydite and Baltic flint artefacts, appeared focused to this concentration in detail and the space between the semi-circular structure and the pavement of concentration I in general.

Chronology of the south-western area

The materials in the south-western area of Gönnersdorf were only partially distinguishable from the scatters between the main concentrations of the Late Magdalenian, and in particular from concentration I. In consequence, the numbers given here according to Thomas Buschkämper referred generally to the south-western area which was formed by various, probably non-simultaneous occupation episodes and only a very small sub-sample of these numbers was connected to concentration V.
The elk radius from the central area produced a considerably younger AMS date (Street in Higham et al. 2007, 16f.) than other reliable dates from Gönnersdorf (**tab. 20**) sustaining the stratigraphic tendency of a higher position in the sediment column. Thus, if the elk radius was related to the semi-circular structure of concentration V, these remains represented a presumably younger occupation episode. The presence of lithic point fragments in combination with elk and red deer bones resulted, perhaps, from a later occupation episode at Gönnersorf which was the last Pleistocene visit to the site according to the present evidence.

Chronology of the Gönnersdorf site

In summary, the chronological development of the site of Gönnersdorf comprised at least two different occupation events. A Late Magdalenian occupation and a significantly younger event with a thus far unknown impact in the south-western area which was presumably infrequently used during the Late Magdalenian. Furthermore, the Late Magdalenian occupation was clearly formed by various phases and episodes, perhaps, even two or more distinct occupation events. However, the distinction of these events was beyond the sensitivity of radiometric dating, and only a faint stratigraphic evidence suggested a chronological succession. Nevertheless, two Late Magdalenian occupation events would still suggest a semi-permanent settlement. Even though the Late Magdalenian settlement comprised a possible duration of some hundred years, the patterns of the material and its use remain relatively stable. Only in the south-western area were changes in the material and its distribution observed and, in fact, dated to a younger age.

Wildweiberlei, Diez, Rhineland-Palatinate

Research history

The Wildweiberlei was a small double cave within a limestone outcrop which was destroyed during the early 20th century quarrying along the Lahn river (Terberger 1993, 1). Some scarce remains of a prehistoric occupation in the Wildweiberlei cave were first revealed by a small excavation of Carl Rademacher in 1917 and led to further investigation in the cave conducted by him in 1919 (Terberger 1993, 150f.). Finally, in 1920 some 24 m² in six test areas (A-F) were excavated inside and outside the cave under the supervision of H. Heck and F. Kutsch due to the approaching lime quarry (Terberger 1993, 148f.). The financial and temporal pressure of these works as well as trouble with and among the participants caused imprecisions in the excavation and documentation process (Terberger 1993, 149-151. 154f.). In general, the position of the finds was recorded per 10cm deposit and the horizontal position was oriented to a fixed point. Refitting demonstrated that even artefacts from the topsoil were connected to the Late Magdalenian horizon and, thus, confirmed the material to represent a presumably single assemblage (Terberger 1993, 159). The main distribution of the archaeological material was recovered in the entrance area of the northern cave (area D) and the adjacent ledge leading north-westwards to the plateau (area F) and south-eastwards to the second entrance (area E). In this southern entrance (area B) as well as inside the cave (area C) only a few bones and lithic artefacts were found (Terberger 1993, 153). Perhaps this scarcity was due to the only thin remains of the loess deposit in this part of the cave system. Equally, at the back of the northern cave (area A), only 30cm of humic soil were excavated on top of the bedrock and only small bones, a mollusc, and some stones which were not considered as lithic artefacts were found. Nevertheless, refitted material connected the finds from the southern entrance with those of the northern cave suggesting that both galleries were used quasi-contemporaneously (Terberger 1993, 159f.).

Topography

The cliff in which the small Wildweiberlei double cave was located formed a promontory at the transition from a bottleneck situation to a widened part of the Lahn valley. The cave system was situated approximately 15m above the valley bottom and only accessible from the plateau (**tab. 10**; Terberger 1993, 148). The two main entrances of the cave were approximately 2.5m high and situated some 7m apart from one another. The two galleries were located in a west-east direction, thus opening towards the west where the valley widened. The ceiling became lower towards the eastern end where the caves were only some 1m high. A small passage connected the two approximately 9-10m long galleries. Further arms of the system ended in front of rubble fields.

Stratigraphy

Profiles recorded in the area of the northern cave were usually around 2m high but could also reach a height of up to 3.5m (area F), whereas the profiles in the southern cave did not exceed 1m. In those reduced stratigraphies the bedrock scree was overlain by only 15-35cm of loess on top of which some 30cm of topsoil followed (Terberger 1993,153). The archaeological remains were recovered from the complete stratigraphy. In the northern profiles the brownish bedrock scree gradually transformed to a dark or greyish brown layer in an approximately 30cm thick deposit (Terberger 1993, 155). Only in the entrance area D this layer was overlain by 7cm of small pebbles. On top of this alluvial deposit, 20cm of greyish brown sediment followed which were overlain by 15-20cm of a reddish yellow loessic deposit with some limestone scree. The greyish loess layer on top was found again in all northern profiles. This layer was approximately 20-25cm thick and contained in area D some limestone scree. The greyish loess was overlain by some 40-50cm of greyish yellow loess which in area D contained a significant amount of limestone scree. The archaeological material was generally associated with the upper part of this deposit. However, some artefacts spread vertically from the centre of this deposit up into the lower topsoil. Inside the cave, the loess was observed to be intensely to lightly red coloured by haematite powder at the transition to the next layer (Terberger 1993, 165). This layer consisted of some 20-30cm of light yellow loess which was at least partially transformed into a loess loam (Terberger 1993, 156). This loess loam was overlain by a burnt layer of supposed Latène Age on top of which the topsoil and finally limestone scree followed.

Archaeological material

In total, 662 lithic artefacts were recovered among which 82 pieces were classified as splinters (**tab. 11**; Terberger 1993, 167). Clearly, this number is too low especially in contrast to 28 cores, but probably this number reflects the variable quality of the excavation. Harald Floss observed that in contrast to the typical Late Magdalenian sites in the Neuwied Basin the lithic raw material of the Wildweiberlei yielded a more local character with lydite in four different varieties dominating the small inventory (Floss 1994, 263). Rolled pieces of this material occurred in the gravels of the Lahn and cores with still remaining gravel cortex were recorded in the Wildweiberlei assemblage (**tab. 12**). A fine-grained variety of Tertiary quartzite was also used and, according to a piece with cortex, this material was possibly collected in the Lahn gravels (Terberger 1993, 167). However, one end-scraper was made on a blade of a coarse-grained variety (Floss 1994, 263). Perhaps, this material originated from the Dornburg-Weilburg region which is located at a distance of at least some 15 km to the north-east (Floss 1994, 21-31). Furthermore, the locally available basalt was also exploited comparably to other lithic raw materials from a core (Terberger 1993, 167). Additionally, the LMP were generally made of Baltic flint which could be recovered in occurrences of at least 115 km northwards. Harald Floss assumed that besides Baltic flint also European flint was used (Floss 1994, 263). Yet, a verification of the used raw materials might, in particular for these foreign materials, be of interest concerning the social embedding of this site. However, cores (n = 2) and further debris of the flint material prove the introduction also of this foreign material in almost complete nodules.

Nevertheless, 23 of the 28 cores were made of the local lydite, a further two pieces were made of the local Tertiary quartzite, and another core was made of the local basalt. Furthermore, two basalt nodules and 16 lydite nodules display singular blank negatives indicating testing and rejection. In general, the cores were of small dimensions with a maximum length of 16 cm among the ready prepared but not exploited cores (n = 4, all lydite; Terberger 1993, 168). The 24 exploited cores were tendentially smaller than 6.5 cm and were formed in various shapes (**tab. 13**). Usually the cores were little prepared and yielded considerable amounts of cortex which regularly was found on the natural butt of the recovered blanks. However, 19 crested blades which were made of lydite except for two examples made of flint display some preparation on the cores. Nevertheless, only a single example was prepared completely and from both sites and a further two specimens partially from two sides. The majority were prepared from only one side (n = 13) and three blades were secondary crested blades (Terberger 1993, 171). The resulting blanks were, according to the negatives as well as the recovered material, rather blades than bladelets, but the bladelets were used more frequently as blanks for retouched tools (Terberger 1993, 170). Comparably to the majority of the cores, these blades surpassed lengths of 7 cm only occasionally (cf. Terberger 1993, 172 Abb. 58). Karin Terberger mentioned at least 15 burnt lithic artefacts (Terberger 1993, 161).

113 artefacts were identified as intentionally retouched pieces of which 20 were classified as LMP (**tab. 14**). Among the LMP only backed blades and bladelets were found but generally these tools were recovered in a fragmented state. One piece was completed by refitting (Terberger 1993, Taf. 72.6) and displayed that the modifying retouch did not extend along the entire lateral edge. Furthermore, the unretouched part was probably broken off intentionally from a shallow notch on the opposite edge. Few LMP had straight truncations (n = 5), of which one was additionally retouched on the opposite lateral edge. The burins (n = 12) were dominantly made on blades and of the dihedral type (n = 7). Only two burins were knapped from a truncation and only five oblique truncations without burin blow were found. In addition, nine borers were identified which were rather of the *Zinken* category with a blunt end but without massive retouches. The six end-scrapers were made on blades which were partially very long and also retouched along the lateral edges. Furthermore, the three composite tools were all combinations with end-scrapers, in two cases with a burin and in one case with a borer. The majority of retouched pieces were somehow laterally retouched

artefacts without further retouched working edges (n=44). Only one specimen was retouched continuously on both edges forming a pointed blade (Terberger 1993, Taf. 71.7), whereas the most common was a partial unilateral retouch (n=27) which was also the most common supplementary retouch to artefacts with other working edges (n=9; Terberger 1993, 179). Splintered pieces were not found among the inventory of the Wildweiberlei.

A block fragment of basalt revealed a roughly picked area forming a bowl-shaped depression in the small block. According to its position inside the northern cave and the settlement area as well as in comparison with the Gönnersdorf material the use as lamp was suggested (Terberger 1993, 183).

During the excavation limestone and slate plates were observed, of which some pieces were some 5 cm thick (Terberger 1993, 165). Some of the plates were described as quartzitic slate with edges which were trimmed and, thus, appear very comparable to the remains from Gönnersdorf. However, only 75 fragments of the slate plates were recovered (Terberger 1993, 158) and apparently no engravings were recognised on these pieces. Furthermore, quartzite and quartz pebbles which were black coloured, supposedly by heating, were documented (Terberger 1993, 164) but only two quartzite pieces were collected (Terberger 1993, 158). One was a hammerstone with a clearly defined field of scars (Terberger 1993, 166). Additionally, a quartzitic sandstone plate with modifications along the edges and traces of scraping was recovered from the ledge area (Terberger 1993, 166). Nevertheless, the rock material from the Wildweiberlei cannot be quantified anymore but according to the descriptions, at least comparable weights to those from Gönnersdorf IV can be expected.

Among the faunal remains, approximately two horses (*Equus* sp.) and two reindeer (*Rangifer tarandus*) were identified with an adult and a young animal in each species (**tab. 15**; Terberger 1993, 183). Moreover, five individuals of willow grouse (*Lagopus lagopus*), three of arctic hare (*Lepus timidus*), and one of arctic fox (*Alopex lagopus*) supplement the classic Late Pleniglacial cold steppe fauna (Terberger 1993, 184). The arctic fox was determined by four molars, a mandible, and a caudal vertebra. The presence of ibex (*Capra ibex*) was probably accounted for the mountainous surrounding, whereas the remains of a young cave bear (*Ursus spelaeus*) were perhaps part of the fossil deposit of the cave (Terberger 1993, 183 f.). The presence of material from a young woolly rhino (*Coleodonta antiquitatis*) could also represent fossil remains which were possibly collected to serve in the structuring of the cave. Yet, only ^{14}C dating could reveal whether some bones were actually fossil material.

In general, the seasonal indicators were scarce but Karin Terberger suggested a use of the cave in the summer and/or autumn months due to the high number of young animals, but she also considered that if arctic fox was hunted for its pelt the presence of this species would indicate a winter prey (**tab. 16**; Terberger 1993, 184). However, the fur of modern arctic foxes changes in October and again in April but can remain the same colour throughout the year in areas of stable climate (Nowak 2005, 80). Moreover, the evidence from Gönnersdorf suggested that the teeth of the animals were particularly chosen for jewellery production (Álvarez Fernández 1999) and, furthermore, that the animals were completely exploited (Street/Turner 2013). Thus, the pelt was not necessarily the most important reason for hunting this species. Nevertheless, if the Wildweiberlei was occupied for some weeks in autumn to early winter the various seasonal indicators were perhaps evidence of different episodes within a single occupation event at the cave.

A fragment of a possibly double bevelled point made of antler was recovered at the site (**tab. 17**; Terberger 1993, 182). The basal part was not well preserved and the second bevelled surface remained uncertain (Terberger 1993, 182). However, two prepared ridges were observed and had possibly an ornamental character (**tab. 18**). Moreover, two pieces of antler showed the fabrication of antler spalls by the application of groove technique. Both items were broken relatively recently and, therefore, could no longer be identified as whether they were representing shed or adjacent antler specimens.

Spatial organisation

During the excavation in 1920 a hearth was recognised by darkened sediment infiltrated with charcoal and burnt bone fragments in the northern cave (**tab. 19**). The presumable hearth was located near the northern wall in the entrance area of the cave. The hearth was 0.6 m × 0.7 m wide and deepened 30 cm into the sediment (Terberger 1993, 161). Furthermore, based on the diary of the excavators, Karin Terberger reconstructed four pits inside the cave floor (Terberger 1993, 161-165). These structures were on average 30 cm wide and 20 cm deep but in detail their shapes and measures were probably very different. A small pit was assumed south of the hearth. Probably it was filled with fragments of slate plates, lithic debris including cores, and hammerstones. East of this pit a limestone plate with several small depressions which was possibly used as an anvil was recorded. A further two pits were located east of the hearth towards the inside of the cave. The pit near the hearth was of an intermediate dimension and eastwards followed by the largest pit. The western pit was completely covered and the eastern pit was partially covered by a construction of three large stone plates including a limestone plate. Charcoal, haematite, fragments of slate plates, bone fragments, and lithic artefacts including cores and blanks were deposited inside the pits. In the pit closer to the hearth the upper 20 cm were filled by ashes which, in combination with the limestone plate, might have served as some type of heating, roasting, or drying construction (Terberger 1993, 163). A fourth pit of intermediate size was set at some distance from the hearth near the southern cave wall. Besides ashes and haematite, only blackened stones (presumably burnt quartzes and quartzites) were recovered in this pit. Moreover, at least some parts of the interior cave were probably paved with slate plates (Terberger 1993, 163-165). Additionally, the sediment within this structure was coloured to various intensities by haematite powder. Thus, according to the descriptions of the excavators, comparable efforts to those measures undertaken at Gönnersdorf and Andernach were invested to produce the evident structures at the Wildweiberlei. In the south-western part of the cave entrance, the intensity of the reddened sediment weakened and this approximately 2-3 m² wide area yielded relatively few archaeological remains. This area was considered as some type of dwelling structure and/or compartment partitioning for the living and sleeping area of a small group (Terberger 1993, 165).

Refitting of material indicated the connection of material from the two galleries and the forecourt. South of the hearth, the anvil and the inventory of the pit suggested a knapping place which was further sustained by the refittings. A comparable activity zone was described in area F where the cave transformed into a rock shelter with a sandstone plate with scratches and a quartzite hammerstone within some lithic artefacts. Based on the amount of material and the settlement patterns, Karin Terberger considered the Wildweiberlei assemblage as remains of a habitation site of a single small group of people which, in several working episodes, spread material in various activity areas across the cave system during a single occupation event of a few weeks (Terberger 1993, 166).

Chronology

According to the results of Karin Terberger, the activities in the Wildweiberlei can be assumed as quasi-contemporaneous. Three samples of humanly modified horse (*Equus* sp.) bones and reindeer (*Rangifer tarandus*) antler from the Wildweiberlei were submitted to the Oxford radiocarbon laboratory but first failed to produce reliably datable material (Street/Terberger 2004, 295). However, refinement of the dating procedure allowed the modified antler to be dated to 12,835 ± 50 years ^{14}C-BP (OxA-18410; **tab. 20**). Thus, by this date, the antler is placed to the younger end of the Neuwied Basin Late Magdalenian sites but still within the range of this classic Late Magdalenian period (Street in Higham et al. 2011, 1074). However, the collection and the use of the antler raw material could not be estimated precisely and, thus, in a general chronology of the Late Magdalenian in the Central Rhineland the occupation of the Wildweiberlei could be slightly younger than the Late Magdalenian occupations in the Neuwied Basin as considered previously (Terberger 1993, 180-182; cf. Street 1998b, 51). However, the behavioural patterns reflected by the material from the cave and its distribution within the cave system were still very comparable to the Late Magdalenian of the Neuwied Basin, only the more intense use of local materials and the setting inside a cave were distinct from Gönnersdorf and Andernach. The amount and diversity of the material were also smaller than in the two Neuwied Basin sites but this could be explained by the overall smaller size of the site.

Oberkassel, Bonn, North Rhine-Westphalia

Research history

In February 1914 track works within a basalt quarry revealed human remains in a sand deposit which was originally situated at the foot of a cliff which was facing south-westwards and already destroyed by the time of the discovery. Since the track workers collected the majority of bones the exact positions of the human material remained unknown. However, an inspection of the site by Max Verworn, Robert Bonnet, and Franz Heiderich a few days after the recovery in 1914 found an area of remaining red coloured sediment and occasional small bones where the workers had uncovered the human remains. Another two days later Gustav Steinmann, Charles Edgar Stehn, Hans Dragendorff, and Hans Lehner recorded a geological profile at the find spot and conducted a test pit survey in which they found human foot bones *in situ*. Furthermore, this survey revealed no finds outside the red stained sediments except for a small, patinated flint bladelet recovered from a sediment sample taken approximately 1 m away of the find area. Moreover, this focus on the limited red coloured area and the typological classification of the recovered artefacts led to the assumption of the remains representing a Late Magdalenian double burial. The Oberkassel evidence became particularly noticed because it was the only full-body double burial attributed to the Late Magdalenian (Schmitz/Thissen 1997, 202; Wüller 1999). The main results of the examinations in 1914 were published in 1919 by Verworn, Bonnet, and Steinmann in a monograph which for the time was very modern and comprehensive (Henke/Schmitz/Street 2006). According to the participants, the material was distributed to two collections (Rheinisches Landesmuseum and Geologisch-Paläontologisches Institut) and, perhaps, due to the high standard of the 1919 publication not further analysed until 1977. In this year Gerhard Bosinski initiated a unification of the material stored in the two archives (Street 2002). A redetermination of bones attributed to wolf (*Canis lupus*) as remains of a dog (*Canis familiaris*) by Erwin Cziesla, one of Bosinski's students at the time, resulted in several reanalyses of the whole assemblage (Henke 1986; Nobis 1986; Wüller 1993; Street 2002). Nevertheless, the site was considered lost to the basalt quarry until Ralf-W. Schmitz and Jürgen Thissen in cooperation with Renate Gerlach rediscovered the findspot by a field survey and geological test pit program in 1994 (Schmitz/Thissen 1997). Presumably, these reinvestigations at the site also led to the decision to propose organic remains from this site when the Oxford laboratory requested Upper Palaeolithic material from Germany for AMS dating. In the 2000s a prominent exhibition of important hominin remains in the context of the 150th anniversary of the discovery of the Neandertal human remains was held at the museum in Bonn where also the Oberkassel remains were stored (Rheinisches Landesmuseum Bonn 2006). The impulse from this project initiated again reinvestigations of the Oberkassel assemblage (Ralf W. Schmitz in Rheinisches Landesmuseum Bonn 2006, 350). Results of the reanalysis combined with modern methods such as molecular and stable isotope analyses are going to be presented in the context of the 100th anniversary of the recovery of the site in 2014 (Schmitz 2009).

site	sample	years IRSL-BP	± years	material	comment
Oberkassel	OD 1	11,870	1,030	sandy loess/sand	upper loess (c. find horizon)
Oberkassel	OD 2	11,420	1,240	sandy loess/sand	upper loess (below horizon)
Oberkassel	OD 3	12,180	1,750	sandy loess/sand	upper loess (below horizon)
Oberkassel	OD 4	11,780	1,910	sandy loess/sand	upper loess (below horizon)
Oberkassel	OD 5	11,170	1,780	sandy loess/sand	upper loess (below horizon)
Oberkassel	OD 6	11,380	920	sandy loess/sand	upper loess (below horizon)
Oberkassel	OD 7	11,210	1,400	sandy loess/sand	upper loess (below horizon)
Oberkassel	OD 8	11,430	1,540	sandy loess/sand	upper loess (below horizon)
Oberkassel	OD X	12,090	2,744	sand	upper loess (c. find horizon)

Tab. 21 IRSL dates from the stratigraphy at Oberkassel. Reference: Schmitz/Thissen 1997.

Topography

The site was set in the 19th century at the north-eastern foot of a basalt cliff at an altitude of 99m a.s.l. (Verworn/Bonnet/Steinmann 1919, 6). Before the basalt quarry, the cliff rose steeply c. 90m upwards east of the site. Nearby a small valley ascended gradually into the mountain ridges (**tab. 10**). The modern Rhine bed is located less than 1km to the west and approximately 40m lower than the site. On the opposite, western bank the river had widened the valley to some 5km extend.

Stratigraphy

The stratigraphy recorded by Gustav Steinmann was generally confirmed by the 1994 reinvestigations (Schmitz/Thissen 1997). In general, the human remains were found at the bottom of some 6m of basalt scree which was mingled with quartz gravels which originated from the terrace of the Rhine on top of the basalt cliff. Perhaps, the human bones were partially within 10cm thick of sandy loam which were underlying the basalt scree and on top of a 4m thick greyish yellow sand deposit which overlay the basalt bedrock (Verworn/Bonnet/Steinmann 1919, 6f.). The quarry workers reported that the skeletons were buried under larger flat basalt boulders supplemented by smaller and larger basalt gravels (Verworn/Bonnet/Steinmann 1919, 3. 191). Due to the good preservation of the skeletons Max Verworn assumed that these boulders were positioned intentionally and protected the human material subsequently. However, if the skeletons were embedded in intensely red coloured loam (Verworn/Bonnet/Steinmann 1919, 3) or sandy loams as considered possible by Gustav Steinmann (Verworn/Bonnet/Steinmann 1919, 6f.) the overlaying basalt boulders could represent a roof of shallow cave or rockshelter formed at the adjacent cliff and collapsed some time after the deposition of the dead (**tab. 10**). The hypothesis of a previously existing roof is further sustained by Max Verworn's description of the basalt scree in the immediate vicinity of the findspot stating that the deposits could only be moved by some effort including machines and strong man power (Verworn/Bonnet/Steinmann 1919, 4). Furthermore, Verworn considered the existence of a roof on the nearby cliff but he assumed this rockshelter as possible habitation site (Verworn/Bonnet/Steinmann 1919, 192). In this case the burial site of Oberkassel would equate the other evidences of intentional depositions of the dead from the early Lateglacial which were usually placed in caves or rockshelters (cf. Pettitt 2011, 226-231). Originally, Steinmann considered the lower sand deposit as an alluvial accumulation, precisely an upper terrace of the Rhine, whereas the modern reanalysis attributed the sands to an aeolian process. Of this loess no particles were found in the basalt scree but layers of basalt and/or quartz scree were found occasionally within the sands. The upper parts of the sands were dated by the infrared stimulated luminescence (IRSL) method (**tab. 21**) which belongs to the luminescence methods and is mainly applied to potassium feldspar-rich material (Wintle/Lancastert/Edwards 1994). This dating resulted in Lateglacial Interstadial to Early Holocene ages for the deposition of the sands and, thus, led to questioning of the typological attribution to the Late Pleniglacial (Schmitz/Thissen 1997, 201). However, meanwhile a tendency of IRSL ages to significantly underestimate the depositional age was considered (Kars/Wallinga/Cohen 2008) and the use of additional pretreatment (infrared stimulation) was suggested as producing more reliable results (Thomsen et al. 2008). Thus, these dates should be assumed as underestimates.

Archaeological material

The human remains represented two almost complete adult skeletons which were generally well preserved. The smaller one was determined as a young adult female, whereas the other remains originated from a mature-senile adult man who was estimated to have been at least older than 50 years. In the monograph from 1919 the remains were analysed morphologically in a comprehensive description (Henke/Schmitz/Street 2006, 244). This description stated that the bones were not manipulated post-mortem but coloured red with varying intensity sustaining the assumption of a burial of bodies and no secondary burial (Verworn/Bonnet/Steinmann 1919; Schmitz/Thissen 1997, 198). Generally, the female skeleton was more intensely covered by haematite than the male bones. The red stained body parts were also different: The female skeleton was particularly red coloured around the head, whereas the male skeleton was most intensely red coloured in the pelvis area. Nevertheless, both crania were damaged mainly on the right side and partially in the frontal part (Henke 1986). Some of these damages were presumably due to the recovery by the quarry workers (Verworn/Bonnet/Steinmann 1919, 2). Moreover, on the male skeleton fractures were observed on an elbow, a collar bone, and the cranium (Schmitz 2009) but the cause of these damages remained unmentioned. Possible causes include an accident, perhaps even a fatal one, during the lifetime of the man, as well as taphonomic processes such as sediment pressure or the fall of basalt boulders on the human remains. In the original publication the loss of several teeth was identified as *intra-vitam* process (Henke 1986).

Besides the human material, the red ochred sediment, and the single flint bladelet of uncertain association (**tab. 11**), almost 80 fauna remains were recovered at the site (**tab. 15**; Henke/Schmitz/Street 2006, 250 Tab. 1). Particularly, the presence of a possibly complete dog skeleton was frequently emphasised (Nobis 1986; Street 2002; Schmitz 2009). The importance of the Oberkassel dog was due to the conjecture as one of the earliest evidences of this domesticated animal for some time but meanwhile much older examples were attested in northern Eurasia (Germonpré et al. 2009; Germonpré/Lázničková-Galetová/Sablin 2012). Nevertheless, in the context of the consideration about local extinctions of dogs or the abandonment of the behaviour to keep dogs during the LGM and related to the Lateglacial expansion process into northern Europe the Oberkassel specimen retained a prominent position. In particular, the well documented morphological and chronological position of this specimen retained its position as important benchmark in the debate on use of dogs. Furthermore, the close connection to human behaviour in general and, in particular, the possible incorporation into ritual behaviour emphasised the importance of the Oberkassel dog. Among the 13 teeth and approximately 25 bones of dog no duplicates occurred and, furthermore, the morphological sizes were comparable suggesting the presence of a single, not fully grown individual (Street 2002, 275 f.).

In addition, brown bear (*Ursus arctos*), red deer (*Cervus elaphus*), and a bovid (*Bos primigenius/Bison priscus*) were determined by single teeth and bones (Street 2002). However, the red deer remains originated perhaps from another site (Henke/Schmitz/Street 2006, 251). Furthermore, a lumbar vertebra was considered as possible evidence of roe deer (*Capreolus capreolus*) and several bone fragments, in particular, those smaller than 1 cm could not be determined to species level.

A penis bone of bear which was ^{14}C-dated to the Lateglacial displayed cut-marks along one edge on which a thick adherence of haematite was observed and the root of an ochre stained red deer incisive was cut (**tab. 18**; Henke/Schmitz/Street 2006).

The immediately recognised organic artefacts led to the original attribution of the assemblage to the mid-Magdalenian. In particular, the flat figurine made of a long bone (or possibly antler) was often referred to as »contour découpé« comparable to examples from the Magdalenian IV (e.g. Bosinski 1989). However, a material, metric, and stylistic reexamination of the Oberkassel piece invalidated this hypothesis (Baales/Street 1998). The missing head prohibited an exact recognition of the modelled species but generally a larger cervid (roe deer, red deer, or elk, *Alces alces*) was perceived (Mollison 1928; Bosinski 1989). This interpretation as well as the piece itself showed clear parallels to the amber figurine of an elk from Weitsche (Heuschen et al. 2006, 25; Henke/Schmitz/Street 2006, 253f.; Schmitz 2009; Veil/Terberger 2009, 349f.).
The other artefact was also made of a long bone and resembles modern bone folders (Mollison 1928) and was attributed to the French Palaeolithic lissoirs (**tab. 17**; Verworn/Bonnet/Steinmann 1919, 186). Another comparison was a spatula such as the piece found at Boppard (Wenzel/Álvarez Fernández 2004) but in contrast to the Boppard piece the cross-section of the Oberkassel piece is rectangular with rounded edges (Verworn/Bonnet/Steinmann 1919, 186). The blunt edges were cut by angular lines. One end was pointed, polished, and yielded fields of scratches on both sides, whereas on the other end a schematic animal head was carved which according to Max Verworn represented possibly a rodent or marten. Another possible perception for the head is a female roe deer head. Perhaps, the piece was used as a hair pin (Mollison 1928) which was assumed from the observation of the workers who recovered the item underneath one of the skulls (Verworn/Bonnet/Steinmann 1919, 186). However, an application in activities comparable to the interpretations of spatulas is also possible (Wenzel/Álvarez Fernández 2004, 148).
Martin Street concluded that the fauna assemblage did not resemble the typical prey of a hunting episode because the remains served mainly no alimentary supply but could be classified as ornamental pieces (Street 2002, 285f.; Henke/Schmitz/Street 2006, 252).

Spatial organisation
According to the interrogated quarry workers the two skeletons were recovered in a not fully stretched position but not oriented in the same direction (Verworn/Bonnet/Steinmann 1919, 191). Furthermore, the previously mentioned stone settings as possible protection for the dead would be a unique observation for the Lateglacial in north-western Europe. However, since Irlich is also considered as burial at an open air site, this factor appeared not uncommon for the Central Rhineland, although generally the position in caves or rockshelters or on the forecourts of caves appeared the most common behavioural pattern in the Lateglacial (cf. Pettitt 2011, 226-231). In combination, these two unusual occurrences led the present author to consider a taphonomic explanation (e.g. roof collapse; see p. 139) for the formation of the Oberkassel evidence as probable. In this case, damages on mainly the right side of the faces and limbs of particularly the male remains suggested that the bodies were positioned on the left side and, therefore, the fall of the basalt boulders hit the right side first and harder. This part could probably be tested by the reanalysis of the human remains.
Nevertheless, the reported limited area of red ochre, the general presence of this material as well as of the special goods interpretable as items given into a grave or representing ornaments of special garments sustained the hypothesis of a double burial of an old man with a young woman and a dog.

Chronology
Seven subsequently taken ^{14}C dates (**tab. 20**) on the remains fall into the first part of the Lateglacial Interstadial. In contrast to the conclusion of the spatial arrangements, these dates are statistically not consistent with the hypothesis of a single event. Therefore, these dates have to be evaluated in greater detail elsewhere (see p. 265-269 and p. 470f.).

In summary, the Oberkassel evidence suggested that two people and a dog were buried in a possible rock-shelter, presumably in a crouched or semi-stretched position on their left side. Due to the lack of spatial documentation of the buried remains, the chronology has to rely on the radiocarbon dating which requires further evaluation.

Irlich, Neuwied, Rhineland-Palatinate

Research history

In 1957 some human remains with a few archaeological remains were discovered in sands which were used in a house construction. These sands were assumed to be transported there from a sand pit in Irlich, today part of Neuwied/Rhine (Baales 2002, 11). In late 1963 they were recorded in the district museum of Neuwied where they were forgotten in a box until 1998 when the Archaeological State Service overtook the box. In addition to the human bones, four artefacts and a note were stored in the box. The note written on December, 25th 1963 suggested the necessity to reinvestigate the site and, moreover, described that the sands at the site were in some places red coloured. However, if the remains were only discovered from a sand heap and, furthermore, recorded in the museum approximately six years later the question arises how the writer of the note received this information. Moreover, the human bones were after their rediscovery in 1998 AMS-dated to a Lateglacial age and, therefore, are mentioned here. Detailed anthropological and pathological studies of the material led by Jörg Orschiedt are on-going (pers. comm. Jörg Orschiedt, Mannheim).

Topography

Irlich is situated at the confluence of the rivers Wied and Rhine (**tab. 10**). The sand pit from which the human remains perhaps originated was situated c. 1 km west of this confluence at an approximate altitude of 60 m a.s.l. (Baales 2002, 11). On the other side of the Rhine and southwards of the site the Neuwied Basin spreads in a wide plain. The surrounding hills leave only a small bank before rising gradually some 35 m upwards to an almost 2 km wide plateau exposed to the south and protected towards the north by another hill step of more than 200 m altitudinal difference. On the north-western corner of this plateau the Late Magdalenian site of Gönnersdorf (see p. 103) was located.

Stratigraphy

The sand pits from the mid-20th century in the Neuwied Basin were known to be covered by the LST (Baales 2002, 11). In 2001, Michael Baales confirmed by a sequence from a small test pit in a wall of the former sand pit of Irlich that the pumice of the LSE had fallen on top of a sand deposit (written comm. Michael Baales, Olpe). Therefore, if these remains were shovelled from these sands they should predate the LSE.

Archaeological material

The human remains comprised some bone and teeth fragments including a completely preserved femur. Originally, some of the bones were identified as belonging to an adult individual and some to a juvenile child (neonate; Bronk Ramsey et al. 2002, 10f.; Baales 2002, 11). All children bones and several bones of the adult were red colour stained, whereas other specimens were uncoloured. Along with the possibility of indicating the presence of three individuals (Baales 2002, 11), the differently coloured bones were possibly caused by only partial cover of the humans by the reddening material (cf. Schmitz/Thissen 1997, 198).
However, in a first assessment of the on-going reanalysis Jörg Orschiedt and his colleagues described all bones as red colour stained (Orschiedt/Berg/Flohr 2011). Moreover, the bones were regrouped with the

left femur and a further five bones (from the left arm, cranial fragments and a lumbar vertebra) originating from one adult individual, presumably, a young adult female which according to porotic bone appositions suffered perhaps from a haemorrhagic disease, possibly scurvy (Orschiedt/Berg/Flohr 2011). The remaining bones were identified as three possible children of different ages: the youngest child was about 6-12 months old (cranial fragments, tooth fragment, rib fragments), one was approximately 4-8 years old (single fragment of left parietal), and another one was 5-10 years old (single right ischium).
The archaeological material stored with the human bones in the box comprised a LMP fragment (**tabs 11. 14**), a burin spall, an organic point or awl fragment (**tab. 17**), and a perforated and incised red deer incisor (**tab. 18**). The two lithic artefacts were made of a yellowish homogeneous raw material, perhaps a chalcedony variant (Baales 2002, 10f.) which could originate from the 30 km north-westwards located Bonn-Muffendorf or the Siebengebirge, east of Bonn (**tab. 12**). The tooth fragment was perforated at the end of the root and the root was incised with parallel lines. Furthermore, the grooves were slightly red coloured. The haematite i.e. red colouring material can be found in various resources within a distance of 50 km around Irlich. A not sufficiently large amount of collagen remained in the pointed artefact to date the piece reliably (Baales 2002, 12). Such singular artefacts and, in particular, the red colour appeared typically in funerary situations in the Upper and Final Palaeolithic as well as the Mesolithic. Therefore, the co-occurrence of these partially red coloured items with the red coloured human bones were considered as evidence of a possible burial situation (**tab. 19**; Baales 2002, 11).
All these pieces appeared typical for late Upper and/or Final Palaeolithic assemblages, although the pointed bone/antler fragment and the red colour rather indicated an association with the Late Magdalenian but the reliable ^{14}C dates are closer to the range of the main group of dates from the upper horizon of Andernach and, thus, a FMG context.

Chronology

Five AMS dates were taken on the human material but produced slightly heterogeneous results (**tab. 20**; cf. Bronk Ramsey et al. 2002) which require a more precise evaluation (see p. 265-269 and p. 471 f.). However, an uncoloured, adult skull fragment (OxA-9876) resulting in an age of the Hallstatt period either did not belong to the assemblage or was more severely contaminated. The other dates scatter in the first part of the Lateglacial Interstadial. Whether these remains were deposited in one or several events can only be clarified by an evaluation of the dates since further stratigraphic observations are lacking. However, the assemblage attributed to the sand pit of Irlich seemed to reflect the deposition of a young woman who probably died of a disease and three young children. These people were buried in a typical Upper Palaeolithic tradition with some grave goods and covered by red coloured material. Moreover, the ^{14}C dates of the human remains, the probable topographic position, the red colouration as well as the potential grave goods appeared also closely parallel to the finds of Bonn-Oberkassel which by ^{14}C dates was equally attributed to the first part of the Lateglacial Interstadial.

Kettig, Rhineland-Palatinate

Research history

After the removal of the pumice of the LSE in 1992 and the subsequent mechanical levelling of the uneven Allerød soil, bone splinters and a pre-core made of Tertiary quartzite were collected from a field east of the town Kettig (Baales 2002, 60). These finds led to the excavation of the site in 1993 under the supervision of Michael Baales. In total, an area of 242 m² including eight test pits and an approximately 20 m long test trench reaching several metres south-westwards from the site were uncovered during this campaign. However, some parts of the site were severely affected by the mechanical levelling because the Allerød soil was in these parts removed and refilled in other parts by this action. At least some 30 m² were totally capped and a further 30 m² were almost disturbed (cf. Baales 2002, 87 Abb. 40). Due to time pressure and the generally compact impression of the vertical find distribution (10-20 cm spread) the exact height of the artefacts was only taken in a small test area. This test revealed that the heights were relatively arbitrary and the artefacts originated presumably from a single horizon. Thus, the single artefacts in Kettig were only recorded two dimensionally, although an approximate height can be given due to the recorded altitudes of the excavated square-metre and the recording of artefacts laying considerably deeper than 20 cm underneath the surface (Baales 2002, 60-63). Furthermore, the sediment from the quarter-square-metres was sieved and the material was sorted out microscopically.

Topography

The Lateglacial open-air site on the eastern limit of the town Kettig is located some 1.6 km south of the modern banks of the river Rhine. In the Lateglacial, a branch of the river passed the site in a distance of only some 200 m in the north (**tab. 10**; Baales 2002, 65-69). Approximately 400 m westwards of the site the Kettig creek flowed into the Rhine branch during the Lateglacial forming a sediment fan which silted up the banks of the Rhine branch. The site is located almost in an angle formed by the Rhine branch and the sediment fan. Approximately 1.5 km eastwards another small stream (Lützelbach) ended in the Rhine branch in the Lateglacial. Furthermore, on the eastern limit of the site the terrain steeply falls some 2 m and in several steps over the next 20 m the terrain slopes down a further 4 m. This step was also observed at the northern limit. By these steps in the Allerød terrain a kind of small terrace was exposed above the floodplain of the Rhine and on this terrace the site was located. Some 300 m south of the site the terrain slopes upwards gradually over 20 m before the Kärlich hill rises steeply c. 100 m upwards. Presumably, the terrace was formed by erosion fans of this adjacent hill. In the south-west of the site the hill levels off into the valley of the Kettig creek. The hill continues in south-eastern direction for approximately 1.5 km.

Stratigraphy

In general, the stratigraphy was only recorded until 5 cm underneath the last finds. However, the main profile taken on the eastern limit of the site began with a 25 cm thick grey calcareous clay of Tertiary age (Ikinger in Baales 2002, 70). In a drilling, these Tertiary deposits were recorded over a thickness of c. 3.7 m containing up to 20 cm thick layers of brown coal (Baales 2002, 73). On top followed some 1.75 m of loam containing in the upper part the Lateglacial artefacts. The 25 cm of loam on top of the grey Tertiary clay was greyish, calcareous, and silty to sandy followed by 65 cm of less sandy loam which comprised single loesskindel. In the lower part of the overlaying deposit these loesskindel reached diameters of 2 cm. This deposit was 50 cm thick and greyish with light red flecks. The sandy component increased again. In the 15 cm on top the flecks were yellowish red and light greyish. The 20 cm on top were separated by a band of fine gravels. The loam was grey to blackish grey with a sandy component and slate component in the size fraction of sand. Partially humus formation was observable. In the upper half charcoal flitter and bands of iron oxide and manganese oxide were spread. Clearly, this part was affected by the Holocene pedogenesis. In this part of the profile the archaeological material was deposited with the heavier pieces laying in general deeper and the lighter pieces were presumably due to mechanisms moving these pieces upwards in the soil (Baales 2002, 63). This stratigraphic position several centimetres below the Allerød surface suggested a dating some time prior to the LSE. On top of the dark loam some 4 m of pumice from the LSE followed. One long West-East profile in the test trench eastward adjacent to the site recorded a first depression east of the site followed by gradual down slope to a gravel deposited dispersed with brown coal pieces. In the first depression a geological drilling was made which produced the three samples for macro-botanical analysis and the six samples for the pollen analysis. The palynological samples were taken from a depth of 6 to 28 cm below the surface but the pollen were relatively poorly preserved. In general, the pollen diagram was dominated by pine (*Pinus* sp.) pollen which could partially be due to the dominance of pine pollen during the Allerød or due to the introduction of foreign or redeposited pollen (Bittmann in Baales 2002, 74 f.). The redeposition of Tertiary pollen was observed occasionally in all samples. Birch (*Betula* sp.) pollen played a more important role in the upper samples. Possibly, the stratigraphy comprised the Late Pleniglacial to the end of the Lateglacial Interstadial. However, the chronostratigraphic attribution remained uncertain, although the upper part was presumably attributable to an interstadial period in the mid- to younger Lateglacial Interstadial. Comparable to the pollen, the few preserved macro-botanical remains were dominated by birch and Salicaceae. In addition to the pollen and macro-botanical samples, the determined pieces of charcoal (n = 20) were dominantly from poplar (*Populus* sp.; n = 15) or more generally Salicaceae (n = 1). In addition, two pieces were from a leave tree which could not be further determined. Furthermore, also presumably unburnt roots of poplar (n = 2) and Salicaceae (n = 1) were found. Only one piece of charcoal was attributed to birch (*Betlua* sp.). Thus, the determined macro-remains and pieces of charcoal indicated an alluvial forest environment. The mollusc samples which were taken on various spots on the site also suggested a light birch-pine forest environment (Mania in Baales 2002, 78-83). This light forested landscape with nearby open areas which were, perhaps, located in the upper parts of the adjacent hills were further sustained by the micro-faunal remains recovered from the sieved samples (Baales 2002, 83-86).

Archaeological material

The archaeological material from Kettig contained almost 4,000 lithic artefacts larger than 1 cm and including splinters some 24,000 pieces were recovered (**tab. 11**). In relation to the amount of material found at the site, only relatively few refits were found (n = 67; Baales 2002, 171-174). Furthermore, also the considerable stone material yielded only 164 refit lines of which many were related to thermally altered material (Baales 2002, 180 f. 183-185). Among the total number of lithic artefacts were some 1,952 burnt artefacts also indicating some intense contact to fire.

The raw material spectrum was very diverse on this site (**tab. 12**). The two most common raw materials making up some 38 % each were the local Tertiary quartzite and the Western European flint. According to the cortex, the Tertiary quartzite originated mainly from primary deposits and less frequently from river gravels (Baales 2002, 105 f.). Harald Floss recorded several deposits around the Laacher See at a distance of 8 km and more westwards of the site (Floss 1994, 21-33). Among the Western European flint, 15 varieties were observed and, in addition, three artefacts were attributed to the local Eifel variety which can for instance be found near Blankenheim in the Eifel. Some of the Western European flint varieties could be found already in some 60 km distance to the north-west in the gravels of the Rur and occasionally of the Erft (Floss 1994, 94-97). However, the majority of this raw material originated from Meuse gravels or primary deposits in at least 80 to 100 km north-westward direction (Baales 2002, 108-110). Along with these two main raw materials, indurated slate (18.9 %), Baltic erratic flint or flint of the Tétange type (3.6 %), and chalcedony (1.4 %) provided significant numbers of artefacts. The various types of indurated slate including a glassy, green variety originated probably from the Rhine gravel which were accessible near the site. The chalcedonies originated according to their colour spectrum and the inclusions from the Bonn-Muffendorf deposit (Baales 2002, 111). The use of glassy flint with bryozoa inclusions from the Baltic erratic moraines which were located at a minimum distance of 125 km northwards of Kettig could not be attested unambiguously due to the lack of the characteristic cortex. However, on other sites from the Central Rhineland this raw material was clearly attested. Nevertheless, the comparable Tétange type flint which can be found in the vicinity of Trier and in Luxembourg as well as in the gravels of the Moselle was considered as alternative classification. A further eight raw materials (indurated claystone of the Schaumberg type; limonite; reddish agate/jasper of the Weiselberg type; agate; greenish jasper; indurated claystone; carnelian; quartz) were represented by few or single pieces. In particular, the Schaumberg and Weiselberg type artefacts as well as the greenish jasper are of interest in the discussion concerning the origin of the glassy flint with bryozoa inclusions because these rare materials originated also from the area south to south-east of Trier. Thus, these raw materials came from some 100 km south-west of Kettig and, perhaps, flints of the Tétange type reached the site by the same connection. The limonite, agate, indurated claystone, carnelian, and quartz could be found in the gravels in the immediate vicinity (Baales 2002, 110).

The proportion of pieces with cortex was generally low suggesting that the raw material pieces were brought to the site in a partially prepared state. However, some of the foreign materials encompassed higher proportions of pieces with cortex indicating that these were brought merely tested to the site. In total, most of the materials seemed to be introduced to the site as raw material units and not as ready prepared blanks or retouched pieces (cf. Baales 2002, 107 Abb. 53). Nevertheless, some pieces were singled out as having reached the site as ready prepared tools which were discarded on the site (Baales 2002, 114-119). Based on the raw materials three end-scrapers, a burin, a *Federmesser*, and a blade were attributed to this primary working set on the site. In total, 60 core and core fragments were recovered in Kettig (**tab. 13**). The majority were made of Tertiary quartzite (n = 25) followed by indurated slate (n = 17), Western European flint (n = 13), chalcedony (n = 4), and Baltic/Tétange flint (n = 1). In general the cores were of very small dimensions. Some of these pieces were exploited exhaustively which was possible in a flexible exploitation strategy aiming not necessarily for blades or bladelets. Of the 22 cores and core fragments displayed by Michael Baales some specimens were exploited from platforms which were not opposing each other but which were located at an approximately right angle to one another. Although the general concept appeared very flexible, elongated blanks and in some examples blade sequences were observed (Baales 2002, 138; cf. Baales 1999). In particular, some short blades were identified among the greenish indurated slate material (Baales 2002, 138). A more detailed technological analysis of the assemblage from Kettig by Ludovic Mevel will be of particular interest concerning the differences of the applied concept depending on the raw materials. Furthermore, besides two claystone retouchers and a minimum of five hammerstones, a used hammer made of red deer antler (see p. 148) was found attesting the use of an organic knapping instrument on the site of Kettig.

The inventory of retouched artefacts comprised 350 pieces (**tab. 14**). End-scrapers were the most numerous class (n = 117) followed by LMP (n = 100). The end-scraper were generally small and some were very small. Micro-wear analysis revealed that these pieces wore only few alterations due to use or hafting but indicated resharpening (Pawlik in Baales 2002, 263-265). Five end-scrapers were double end-scrapers. Among the LMP four broken pieces were refitted and, in consequence, 98 single LMP were present at the site. However, these specimens were in general heavily fractured. However, many pieces were tip fragments allowing for the attribution to the backed point group. The blunting lateral retouches were generally shaped as a curved or straight edge. In addition, the blunting retouches were occasionally discontinuous and only partially applied to the blank. For instance, some microlithic points were found on which only the tip part was retouched in a type of an oblique truncation. Principally, these fall in the range of Zonhoven points but were made on significantly smaller bladelets than typical Zonhoven points. Including these implements, only twelve pieces can be considered as complete or almost complete LMP yielding a considerable variety of LMP types. Micro-wear analysis of five points (including two microliths) identified remains of some sort of impure birch resin on these implements (Pawlik in Baales 2002, 261-263; Baales 2002, 265-267). Based on these results, the small microliths were, in fact, glued with their basal part to the shaft indicating a use as a head of a projectile. In contrast, the larger *Federmesser* and straight-backed points were glued with their blunted lateral edge to the shaft comparable to evidences with backed bladelets (Allain 1979, 100-103; Leroi-Gourhan 1983). However, in difference to the regular backed bladelets, the wider part of a point would have protrude of the projectile in a barb-like way. Besides the LMP, four *Krukowski* micro-burins were found representing remains of the production of LMP on the site. The 36 burins were with few exceptions made on retouches and in one case classified as Lacan type. Only three dihedral pieces were identified. The size of the blanks as well as the forming of the retouch and the setting of the burin blow vary considerably. On five pieces a lateral retouch was added to the blank. Comparable to the end-scrapers the burins were often resharpened. A minimum of 35 burin spalls sustained this impression and indicated the preparation of these pieces on the site. The 21 truncations were in contrast to the burins made in general on regular blades or bladelets. Two pieces were double truncated. Five borers were also recorded. Two were made of Tertiary quartzite and only the blank was broken shortly underneath the working end. The other three borers were made of Western European flint. One of these was also only a fragment. The three fragments were heavily worn out and, perhaps, due to this intense use morphologically classifiable as Zinken. The two almost complete specimens were rather attributable to the piercer group. The three composite tools were all combinations with burins: twice with truncation and once with an end-scraper. In addition to these tools, 48 artefacts with lateral retouch were identified as well as a further ten pieces with lateral splinters and 22 splintered pieces.

Several gravels and quartzes were recovered from the site but some were possibly deposited naturally at the site. Five pieces showed indications of use, perhaps, as hammerstones (Baales 2002, 176) and two retouchers as well as a retoucher fragment made of argillaceous shale. A further 175 gravels or fragments of Devonian quartzite, quartzitic sandstone, and prophyry were recovered from the site weighing some 20.6 kg of which some 7 % were thermally altered. Some pieces were intentionally split to receive plates. In addition, 72 quartz pebbles and fragments were also found in the excavation area. Several of these specimens were reddened or fragmented, possibly due to the use as cooking stones.

In addition to the stone material, more than 57,000 fragments of bone were preserved at the site. The majority of this organic material (75.9 %) was burnt (Baales 2002, 97). In general the surface was not well preserved. Nevertheless, some cut-marks were observed on the 278 fragments of red deer (*Cervus elaphus*; **tab. 15**; Baales 2002, 202-206). The development of the teeth of the lower jaw as well as a teeth cementum analysis of the material suggested a hunting period from late summer to late autumn (**tab. 16**; Baales 2002, 226 f.). All body parts of this species were found and attributed to approximately eight individuals of red deer. More than 100 fish remains were detected and among them were pike (*Esox lucius*) and Cyprinidae identified (Krey in Baales 2002, 222-224). 34 remains were determined as beaver (*Castor fiber*). These were often burnt (Baales 2002, 219). A further 30 elements were attributed to approximately three roe deer (*Capreolus capreolus*). The eruptive and abrasive stages of teeth of this species suggested a hunting episode in the mid-summer to mid-autumn period, although a winter to autumn was also indicated by a not completely closed epiphyse. Probably, three horses (*Equus* sp.) were represented by 27 remains. Heavy butchering marks and fracturing for the probable marrow extraction were observable on these pieces. 25 fragments were attributed to two large bovids (*Bos* sp. / *Bison* sp.), presumably aurochs (*Bos primigenius*). Chopping marks indicated the opening of some bones, probably to receive the marrow (Baales 2002, 211). Approximately 18 remains of presumably fox (*Vulpes vulpes*) were discovered. Some teeth fragments were possibly attributable to chamois (*Rupicapra rupicapra*). Five pieces each were determined as brown bear (*Usrus arctos*) and marten (*Martes* sp.). A further two were determined as wolf (*Canis lupus*). The three latter species were attributed by the excavator to the background fauna (Baales 2002, 224). Further remains which presumably belong to the background fauna such as small rodents were also recorded at the site.

Besides the comparatively rich faunal remains, some pieces of organic artefacts were found (Baales 2002). One was a fragment of an antler spall (**tab. 17**). Furthermore, five pieces of an antler point with barbs along one row were uncovered. Three pieces were refitted to a small single barbed point. In addition, a basal piece of an antler was used as an organic hammer.

Spatial organisation

During the excavation no evident structures were observed but in the subsequent analysis a possible pit was detected by a small quartzite gravel broken in four pieces which lay in close vicinity considerably deeper than the other artefacts (**tab. 19**; Baales 2002, 86f.). However, sediment changes were not observed and, therefore, this position could have been caused by natural processes as well. Generally, horizontal moving due to natural processes such as erosion seemed negligible (Baales 2002, 86). Nevertheless, in the north-eastern corner of the excavation area the disturbances by mechanical levelling presumably affected the extend of the archaeological material. More importantly, the northern concentration associated with the indurated slate artefacts was located in a severely disturbed area.

The stone material was also partially affected by the moving of the sediment (Baales 2002, 179). Even though some accumulations of these materials were recorded, no structures created by these items were observed.

A light scatter of burnt lithics with smaller accumulations was observed on almost the complete site. However, a significant concentration was found in the north-eastern corner. In this area also the quartz, porphyry, Devonian quartzite, and burnt bone material accumulated and suggested the presence of a hearth. The burnt animal remains and the Devoninan quartzite formed a second accumulation approximately 2 m south-west of the possible hearth. Perhaps, an intense activity area was located there. The refitting of various materials suggested connections between the hearth and the southern activity zone. Moreover, another concentration in the south of the excavation area was also connected by refits and suggested as dump site by the excavator (Baales 2002). However, a small accumulation of quartz and Devonian quartzite was present in the northern area where also a light concentration of lithic artefacts, in particular, the greenish glassy lydite was observed. Single pieces of this type of indurated slate, Devonian quartzite, and quartz were also found in the concentration south-west of the hearth. This small northern accumulation was not connected by refits to the activity and hearth area. Furthermore, the tool as well as the raw material spectrum of the activity zone was much more diverse. Thus, if the northern concentration represented a dump area from this working area the dump would presumably have been brought at an early stage of use of the working area to the north when the material was less diverse also in the working area. Alternatively, the northern concentration represented a second, less intensely used knapping spot (cf. Baales 2001, 132). The latter interpretation is favoured by the excavator due to the presence of cores in the margins of the concentration. Furthermore, the almost exclusive occurrence of the few limonite pieces and the artefacts of indurated claystone of the Schaumburg type in this concentration sustained an independent use of this area. Perhaps, also the roe deer tooth which yielded another seasonal indication can be associated with this area. Clearly, the site would profit from a spatial analysis based on modern standards.

Chronology

The ^{14}C dates were generally unreliable (**tab. 20**). A more reliable date produced an age comparable to Urbar. This chronostratigraphic setting is in accordance with the stratigraphic and environmental indications. Based on the blank production process as well as the variety and the use of lithic tools two phases with several episodes were distinguished on the site (Baales 2002, 128f.). However, whether these phases were within a single occupation event, or in two distinct ones remains a matter of debate. Clearly, redating of this important material is necessary in general regarding its position in the transition process and in particular for clarifying the settlement history of the site.

Urbar, Rhineland-Palatinate

Research history

In Urbar, faunal and lithic material was first uncovered underneath the LST in 1966 by the teacher Günther Pausch who was constructing a sandbox in his private garden (Baales/Mewis/Street 1998). In this context 4.1 m² were dug without sufficient documentation. In 1972, a further 4.1 m² were excavated under the supervision of Hartwig Löhr. These square-meters were located adjacent to the north-east of the previous area and produced in particular a stratigraphic sequence as well as palaeoenvironmental samples. The sedimentological analysis provided a detailed stratigraphic sequence. Strongly burnt pieces could be determined by the anthracological analysis as some typical Lateglacial tree species (*Salix* sp., *Betula* sp., *Pinus* sp.) but neither the pollen nor the mollusc sample produced relevant results (Baales/Mewis/Street 1998). In 1979, further works in the garden added an uncovered area of 1.2 m² observed by the owner and in a test pit of 5.8 m² which was excavated east of the thus far known area in 1980 the limits of the archaeological concentration were reached. Finally, 7.2 m² were excavated in 1981 under the supervision of Hermann-Josef Fruth revealing the probable end of the find density towards the north. Due to partial overlapping of the campaigns, at Urbar a total of 16.7 m² was excavated with very heterogeneous documentation but often at least 2-dimensional recording of the finds (Baales/Mewis/Street 1998). However, the recovery of FMG material in Urbar started a number of further excavations related to the Lateglacial Interstadial in the Neuwied Basin.

Topography

The site in Urbar is set on a plateau on the eastern bank of the Rhine (**tab. 10**). Some 2 km south-west of the site is the confluence of the Moselle into the Rhine located which in the Lateglacial represented a large water delta. Towards the north-west of this confluence the large Neuwied Basin extends. The Rhine flows some 550 m west of the Urbar site and some 180 m west of the site the plateau falls steeply some 80 m to the valley floor. Towards the south and east the terrain of the plateau rises gradually a further 30 m upwards. Northwards the plateau gradually slopes 50 m over a distance of more than 1.1 km ending in a promontory position which is formed by the Rhine in the west and a deep valley of a small tributary (*Mallendarer Bach*). This small stream comes from the east and turns northwards approximately 600 m east of the site and flows around the promontory. On the plateau, some 40 m northwards of the site Hartwig Löhr registered a possible spring from which a light depression runs north-westwards towards the Rhine.

Stratigraphy

After the first observation in 1966 that the material was found underneath the pumice of the LSE, a stratigraphy was recorded and analysed in 1972 by Karl Brunnacker (Baales/Mewis/Street 1998). A drilling core made it possible to establish a 7.4 m deep sequence of which the lower c. 5.30 m were formed by varied loess deposits. These were intersected by layers of waterlogged material. The lower part was of sandier material which decreased towards the top. The upper 30 cm of the deposit were formed by a decalcified, loamy loess. It was overlain by a 20-25 cm thick deposit of yellowish to greyish brown loess loam which appeared mottled due to waterlogging processes. Intense bioturbation was observed in the form of many animal burrows which were partially filled with pumice from the deposits above. In the upper third of this layer, the majority of artefacts were found. However, they spread vertically over 20 cm in this layer, perhaps, due to the bioturbation (Brunnacker 1978e). On top followed a blackish brown loam layer of 15-25 cm thickness. This loam was considered as the upper part of the Allerød soil and was rutted with desiccation cracks of up to 5 mm width. This layer was sealed by a 45-90 cm thick deposit of pumice which was attributed to the LSE and which filled in the desiccation cracks of the underlying loam. The pumice deposit was bisected in a lower part of 35-70 cm of primary deposited material and an upper part of 10-20 cm formed by a secondarily deposited, greyish light brown pumice. Within the lower part some ferric precipitation was observed and this part was capped by an ash layer. On top of the pumice deposit, some 30 cm of greyish yellow weathered loam followed which were overlain by 15-25 cm of blackish brown topsoil.

Archaeological material

The lithic material comprised 1,641 pieces of which 516 were smaller than 1 cm (**tab. 11**). At least 15 refits were possible and further material was assumed to be clearly connected but could not be refitted. Some 10.8 % of the lithic artefacts wore traced of heating and, in particular, retouched pieces (22.5 %) were affected. Perhaps, this was due to the use near an open fire towards which unusable pieces were discarded. The majority of lithic artefacts was made of Tertiary quartzite (n_{all} = 1,463; $n_{\geq 1\,cm}$ = 979). In addition to these specimens, 146 attributed to five different varieties of indurated slate and a single retouch splinter of presumably Western European flint were found. Furthermore, 31 pieces made of Devonian quartzite were also considered as a part of lithic assemblage. Except for the flint, all raw materials occur locally. The Devonian quartzite and the indurated slate could be taken from the gravels of the Rhine (**tab. 12**). Known resources of Tertiary quartzite were recorded by Harald Floss 2-3 km north-north-east and some 4.5 km north-east of the site. Perhaps, the flint originated from some 120 km north-west.

Eleven pieces were identified as cores and core fragments (**tab. 13**). Ten of these were made of Tertiary quartzite and one of Devonian quartzite. The lack of cores made of indurated slate was considered by Harald Floss as result of the small excavation area because the origin in the near vicinity and the presence of cortex on some of the blanks clearly sustained knapping of this material on the site (Floss 1994, 270). In contrast, cortex was only partially present on pieces of Tertiary quartzite (Floss 1994, 268). Furthermore, the dimensions of the Tertiary quartzite cores were extremely small and the knapping directions were no longer identifiable on several negatives due to the too small size (Mewis 1993, 13). On the pieces with negatives indicating the knapping direction several platforms appeared to be used during the exploitation process. In addition, three of the fragments were assumed to represent a single core, although the pieces could not be refitted. Thus, the material was assumed to have been brought to the site in a final stage of exploitation (Floss 1994, 268). The core of Devonian quartzite was still partially covered by cortex and made of a thick flake. However, the ventral side was prepared and some blanks were knapped from a single platform. The lithic blanks showed indications for the use of various knapping instruments such as the forming of a lip (soft or indirect) as well as clearly distinct bulbs (hard direct).

The end-scrapers (n = 98) dominated clearly the 120 formally retouched artefacts (**tab. 14**). These end-scrapers were generally made on flakes, although some displayed regular ridges on their dorsal surface suggesting the origin of the blank from a series of regular, elongated blanks. However, many specimens were preserved only as fragments with an end-scraper cap. Furthermore, several end-scrapers were made on small blanks including several classic thumb-nail end-scrapers. Susanne Mewis performed a precise metrical analysis of the numerous end-scrapers revealing an almost static proportion of the pieces (Baales/Mewis/Street 1998, 256). The high number for the small area in combination with the observation of the low number of use-wear or resharpening of the pieces further sustained a short operating time for this tool class (Baales/Mewis/Street 1998, 260). Of the 13 LMP, seven were made of Tertiary quartzite and the remaining five of indurated slate. One complete backed point of Tertiary quartzite and three of indurated slate were found. A further four pieces of Tertiary quartzite were refitted to two complete backed points and the fourth backed point of indurated slate was only slightly damaged. The back of these LMP were usually retouched continuously. The blunting retouch was shaped curved or straight and exceptionally in a s-shape. Of the complete and almost complete specimens only one wears a basal retouch made from the ventral side. Only two burins were found. These tools were made on truncations which were set on small flakes of Tertiary quartzite. Furthermore, two truncations, four laterally retouched artefacts, and a splintered piece were found in the assemblage of Urbar.

Besides the lithic material, three retouchers with multiple scar fields were also recovered in the small area. These specimens were all made of argillaceous shale which can be collected from the various gravels of the Rhine. Furthermore, flat plates of Devonian quartzite and few pieces of slate were recovered from the site. In addition, some gravels of indurated slate, Devonian quartzite, and quartzitic sandstone were also recovered from the small area. In how far these pieces served as hammerstones or in the structuring of the settlement remained unclear due to a sometimes uncertain position on the site and more often due to the lack of unambiguous modification. Nevertheless, some Devonian quartzite plates wore impact points and cut-marks (Baales/Mewis/Street 1998, 266f.). Moreover, burnt quartzes were also found. All these materials were accessible in the nearby gravels of the Rhine.

In addition, few pieces of a coarse-crystalline haematite were excavated. However, these reveal no traces of use (Baales/Mewis/Street 1998, 267). Nevertheless, if the haematite originated from the same source as suggested for Gönnersdorf it was brought to the site from approximately 40km in the north-west. Although in regard to the much coarser structure, the material could also be recovered from the Rhine gravel.

Besides mineral material, some organic remains (n=505) were preserved at the site (**tab. 15**). 298 pieces were attributed to red deer (*Cervus elaphus*) including two pieces of antler and 96 teeth fragments (Baales/Mewis/Street 1998, 268). According to the teeth fragments a minimum number of seven individuals was suggested (Baales/Mewis/Street 1998, 274). Some pieces yielded clear cut-marks or chopping marks indicating the butchering of the animals on the site. One, possibly two bone fragments were heavier and determined as large bovid (*Bos* sp./*Bison* sp.), presumably aurochs (*Bos primigenius*). In addition, a rib fragment was attributed due to its dimensions to horse (*Equus* sp.). On this piece fine cut-marks were observed. Mainly based on the singularity of the material of horse and bovid the import as tool to the site was considered (Baales/Mewis/Street 1998, 276).

Spatial organisation

Due to the limited excavation area and the incomplete documentation, the spatial organisation was only partially reconstructible. Clearly, no evident structure was observed during the various campaigns (**tab. 19**). However, the numerous gravels and stone plates in the centre were perhaps remains of a stone packed hearth and a particularly large plate was considered as heating stone or working surface (Baales/Mewis/Street 1998, 266f.). The distribution of burnt quartzes and lithic implements suggested that a latent hearth was located in the area excavated in 1966. Thus, the hearth would have been situated in the main activity centre and in the surrounding various activities were performed such as the use of end-scrapers for possibly hide-working. In contrast, the organic remains were mainly found in the south-eastern corner a bit offside the main zone around the hearth. This observation is further sustained by the fact that no bone was mentioned to be burnt.

Chronology

According to the stratigraphy, the site at Urbar was not deposited immediately before the LSE but some time previously allowing several centimetres of soil to develop on top of the archaeological material. The

only ^{14}C date made on a red deer sample sustained this chronostraigraphic position (**tab. 20**). The anthracological analysis was partially questioned by Michael Baales and his colleagues (Baales/Mewis/Street 1998) due to the preservation status. However, the determined species were well documented in the Central Rhineland during the Lateglacial and the suggested alternative species (*Populus* sp.) was also recurrently documented underneath the LST. Nevertheless, the connection with the archaeological record remained uncertain and, thus, the proportion could reflect human choice as well as natural processes. Yet, a chronostratigraphic setting in the mid- to younger part of the Lateglacial Interstadial seems probable for the Urbar assemblage.
Furthermore, the small excavated area with the well defined concentration make a single occupation plausible, although the lack of heights, partial documentation, and the restricted area could also misguide the interpretation. The number of seven red deer appeared relatively high for a short occupation duration. However, the presence of stone material, in particular, assumed cooking quartzes and multiple tools suggested for hide-working could indicate an intensively used activity area, perhaps, in a longer period of occupation in a possible autumn/early winter camp. The relatively uniform inventory of retouched artefacts could be explained by a spatial organisation of working areas and the restricted excavation area not capturing all workspaces.

Niederbieber, Neuwied, Rhineland-Palatinate

Research history

The site Niederbieber was comparable to other sites such as Kettig discovered when in autumn 1980 the pumice of the LSE was removed for industrial purposes (Baales 2003a). The Allerød surface became partially disturbed or even capped (Baales 1998, 342). However, the exposed surface yielded first lithic artefacts, faunal remains, and charcoal accumulations which led to the first excavations. From 1981 to 1988 several areas (areas 1-7[17]) were excavated in general according to modern standards with 3D recording of the single finds, sieving of the sediment, and recording of the profiles. However, in some occasions time pressure prevented a 3D recording (Gelhausen 2011c, 5). The areas 1 and 4-7 were set in a near vicinity to each other on the plateau where the site was located. The area 2[18] was set some 30 m north-east of this main area and the area 3 was recovered some 90 m south-west of this main area in the valley below the plateau. Since 1982 a test pit programme setting two square-metre wide test pits every 10 m on the plateau had been conducted parallel to the excavation aiming for the stratigraphic and topographic development of the plateau (Bolus 1992, 4). Subsequently, the material from all areas on the plateau was presented in several M.A. theses of the University of Cologne. Moreover, Michael Bolus conducted in his dissertation a spatial analysis of two main concentrations recovered at the site (areas 1 and 4) as well as of the outlying area which probably represented a special task workshop (area 3) and the test pit programme (Bolus 1992). Equally, another M.A. thesis revised the material of the plateau concerning the presence of raw material units and hearths (Korn 1993). In a research priority programme of the German Research Foundation, the site was reinvestigated from 1995-1999 due to the prospect of good organic preservation and the almost undisturbed spatial patterns allowing more detailed information on settlement behaviour of this time period

17 Previously these concentrations were numbered with Roman numbers (I-VII) but the present study follows a suggestion to use the Roman numbers for the Late Magdalenian concentrations and Arabic numbers for FMG concentrations in the Central Rhineland (cf. Street et al. 2006).

18 This area 2 was meanwhile subdivided by Frank Gelhausen into the sub-areas 18, 19, and 20 (Gelhausen 2011c).

to be collected (Baales 1998; Baales 2003b). The archaeological material of these campaigns produced a further ten concentrations of lithic material. This material as well as the previously excavated areas from the main area of Niederbieber were analysed in a comprehensive spatial analysis of the site by Frank Gelhausen (Gelhausen 2010; Gelhausen 2011a; Gelhausen 2011b). Furthermore, he also reanalysed the material of area 2 with the same methods (Gelhausen 2011c). In total, some 1,020 m² were examined at Niederbieber of which almost 820 m² were located in the main area (areas 1, 4-17, test pits between 10 N / -5 W-E and 55 N / 61 W-E).

Topography

The site of Niederbieber is situated on a small promontory c. 25 m above the Wied valley (**tab. 10**). The river Wied passes the promontory 200-300 m in the north and west. Towards the north where the Wied valley narrows considerably the slopes are relatively steep, whereas in the west to south-west where the Wied widened the valley with a small meander the inclination becomes more gradual. The Wied flows a further 4 km from the site in south-south-west direction and then mouths into the Rhine. The Neuwied Basin extends south from this confluence. During the Allerød a small stream (*Herschbach*) ran down to the Wied valley some 30 metres westwards of the site forming the promontory position. In the *Herschbach* valley on the slopes to the Wied the excavation area 3 was located. However, the archaeological remains were mainly recovered from a small plateau with relatively little inclination and an extend of some 200 m in W-E direction and 100 m N-S. Some 10,000 m² of this spacious plateau were set under protection and partially excavated. Towards the north-east of the plateau hills rise up gradually to the low mountain ridges of the Westerwald. Eastwards the terrain rises gradually some 20 m above the site to a ridge which slopes farther to the east and south down to the wide valley of another small stream (*Aubach*). From this ridge the Neuwied Basin is partially visible today but this view is mainly possible due to the lack of tree cover. Approximately 1 km south of the site the Aubach flows into the Wied. At this confluence the valley is also widened and only some 2.2 km south-west of the site the Wied valley is again narrowed by hill flanks (*Heddesdorfer Berg*, Irlich) for approximately 1 km before the valley opens into the Neuwied Basin.

Stratigraphy

In general the stratigraphy of the plateau in Niederbieber was relatively simple. On top of the Wied gravels a loess loam had formed (Ikinger/Ikinger 1998). In the upper part of this loess loam the finds were recovered. However, in the south-western part of the promontory the stratigraphy was in some parts completely capped, perhaps, because the Allerød sediment was already reduced in the western parts due to erosive processes (Baales 1998, 342). Eastwards onto the plateau and, thus, the main excavation area the sediment deposit increased and reached 30-50 cm. In some parts with a thicker loess loam deposit the presence of two horizons was considered but refitting indicated that the artefacts belonged to one horizon (Gelhausen 2011c). However, the material was moved considerably by bioturbation and soil movement in the vertical axis. Nevertheless, the spatial analysis of Frank Gelhausen suggested that the movements did not substantially affect the horizontal position on most parts of the site (cf. Gelhausen 2010). Occasionally, small erosive channels and depressions affected the horizontal distribution of material (Baales 1998, 342; Gelhausen 2011c, 5). On top of the loess loam, approximately 1 m of pumice from the LSE was deposited and sealed the Allerød landscape. With the mechanical removal of this protective cover the preservation of the organic material was severely affected. In the first campaigns of areas 1 and 2 several faunal elements were still recovered, whereas in the late 1990s excavations occasionally only colour alterations in the sediment indicated the former presence of organic material. On top of the pumice, the Holocene topsoil had formed.

Archaeological material

The site yielded the largest FMG inventory in the Central Rhineland with over 19,000 lithic artefacts (> 1 cm) and in total almost 120,000 specimens (**tab. 11**). These remains originated from almost 20 distinct accumulations. 245 refit lines were detected on the site of which 29 connected distinct concentrations (Gelhausen 2011c; Gelhausen 2011a, 52-59). The proportion of burnt material varied considerably on the site. However, on average a sixth of the material showed traces of heat alteration.

The dominant raw materials varied between the concentrations (Street et al. 2006; Gelhausen 2011a, 14) and some previous classifications were meanwhile revised (cf. Husmann 1988; Floss 1994, 283-303; Gelhausen 2011a, 13-15). However, the majority of concentrations was dominated by the local Tertiary quartzite and the regional chalcedony (**tab. 12**; cf. Floss 1994, 283-303). Since rich Tertiary quartzite deposits exist at the upper Wied, this raw material can be found in the gravels of this river. However, some 7-8 km south-east of the site Harald Floss recorded a quartzite bank (Floss 1994, 21-26). According to inclusions the chalcedony originated probably from the deposit in Bonn-Muffendorf some 40 km north-west of the site. These main materials were accompanied by the equally local indurated slate, Baltic erratic flint, and regularly Western European flint. The Baltic erratic flint originated from the Baltic moraines some 115 km to the north of the site. The Western European flint was present in several varieties which were accessible at a distance of 85 km and more north-westwards. In addition to these common raw materials, some less frequent materials were found, some of which were attributed variously throughout the research history such as the indurated claystone of the Schaumberg type (cf. Gelhausen 2011a, 20f.). This material originated from the area south to south-east of Trier and, thus, at least 115 km towards the south-west of the site. The Triassic chert which was found occasionally on the site also originated presumably from a south-western area. Deposits of this material were found in the southern Eifel region and around Trier at a distance of some 95 km from the site but the type present at Niederbieber originated presumably from the region of the French-German border c. 170 km in the south-west (Baales 2003b). In singular cases the local quartz was also observed to be retouched.

267 artefacts were identified as cores or core fragments (**tab. 13**). Many of the pieces appeared exhaustively exploited based on their small dimensions. Although often a single preferred platform was observable, most pieces were exploited from two or more platforms which were often set angular to one another. Although the majority of blanks discarded on the site were flakes, several sequences of bladelet production as well as some elaborately prepared crested bladelets were found (Gelhausen 2011a, 25-27. 30). These finds evidenced that a complex blank production process was possible at Niederbieber and, in addition, that all stages of this process seemed to be present at the site. The 18 retouchers and five hammerstones which were recovered from the site (Gelhausen 2011a, 30-33; Gelhausen 2011c) further sustain the differentiated use of instruments in the *chaîne opératoire*. The high number of retouchers is significant but was partially due to a depot of five retouchers in area 3 (Bosinski et al. 1982). However, the intense use of these tools was sustained in area 4 where nine retouchers and retoucher fragments were found (Bolus 1992). These specimens were usually made of a local argillaceous shale and, thus, very soft stones in contrast to the hammerstones made of harder quartzite and quartzitic sandstone. More details on the various stages of the *chaîne opératoire* and the relation of these stages to the raw materials will probably be yielded by the on-going technological analysis of Ludovic Mevel.

The retouched artefacts (n = 1,671) were clearly dominated by LMP (n = 571; **tab. 14**). Some specimens were preserved relatively complete but many pieces were very fragmented. In general, the blunting retouch was continuously along one side. Besides curve-backed pieces, many backs were shaped straightly. Nevertheless, among the backed points the *Federmesser* is the most common type (Gelhausen 2011a, 36). Nevertheless, some specimens with basal retouch, representing formally Malaurie points, were also found (cf. Bolus

1992, 12 Abb. 9.4; Bolus 1992, 13 Abb. 10.1; 10.4; Bolus 1992, 108 Abb. 91.14). However, these often short specimen give the impression of broken of tips which were recycled by the additional basal modification. Some of the 279 end-scrapers were also occasionally recovered in a fragmented state. Usually these fragments seemed to be made on regular blades but refitting showed these to be either short blades or regular flakes (e. g. Bolus 1992, 54 Abb. 39.39; 39.43). However, the majority of end-scrapers were made on small flakes and several pieces were retouched on more than one edge forming the typical thumbnail end-scrapers (Gelhausen 2011a, 37). The burins formed a very heterogeneous group of varying dimensions and morphology. More commonly, the burin blows were set on a retouch; frequently a truncation but also breakage surfaces and natural surfaces as well as former burin blows were used as platform. The latter type (dihedral burins) were not very frequent. Nevertheless, often more than one burin blow was observable on the same artefact indicating resharpening activities (e. g. Bolus 1992, 57 Abb. 43.34-35; 43.44). In some cases the distinction towards bladelet cores was diffuse (Gelhausen 2011a, 38; cf. Bolus 1992, 61 Abb. 47.24-26). In Niederbieber, the truncations were comparably heterogeneous as the burins in their dimensions and morphology. In addition to some double truncated pieces, few truncations were retouched from the dorsal instead of the more common ventral face. In addition, over 30 borers were identified on the site. Even though these specimens were also variedly made, no examples of *bec* or *Zinken* types occurred. Composite tools were usually not mentioned in the presentation of the Niederbieber material but at least two pieces, a burin and end-scraper combination (Freericks 1989) and a possible combination of end-scraper and a truncation (Gelhausen 2011a, Taf. 15.9), were found. Furthermore, many laterally retouched (n = 237), partially retouched or unclassifiable pieces (n = 86) were found as well as splintered pieces (n = 34; Gelhausen 2011a, 40 f.).

Some 235 lithic artefacts from the areas 1, 4, and 5 were analysed for micro-wear patterns (Plisson 1985; Husmann 1988; Bolus 1992). Of the approximately 45 artefacts analysed in area 1 twelve wore traces of use in general on the unmodified edge and often this trace appeared related to cutting meat or hide. Furthermore, 52 pieces of the 160 analysed artefacts from area 4 wore traces of use. These use-wear patterns were also found most often on unmodified edges which were used for very short intervals and mainly on animal resources. Moreover, 13 artefacts of 28 analysed pieces from area 5 were also used frequently for cutting meat or for works on hides. Most of these implements were not exhaustively used which, perhaps, explains the occasionally large amounts of lithic artefacts for the processing of the mainly faunal material.

In addition to the previously described retouchers and hammerstones, numerous flat quartzitic slate pebbles were recovered of which some were split intentionally (Baales 1998, 353 f.). In area 15, heavy and large plates (2.2-4.5 kg) appeared deposited at the periphery of the lithic accumulation (Gelhausen 2011a, 31-33). Comparably, in area 2, concentration 19, broken slate plates were found near a hearth and considered as a type of anvil, perhaps for the smashing of animal bones (Gelhausen 2011c). However, this material can also be found in the gravels of the Wied and, thus, occurred also naturally at the site. Nevertheless, Frank Gelhausen doubted the origin of heavier items found within the loess loam deposit from the underlying gravels and, in particular, assumed pieces found within the artefact concentrations as a result of human activity (Gelhausen 2011a, 31). In addition to the quartzitic slate, Devonian quartzite and a type of basaltic lava was occasionally recorded. These materials were perhaps also present in the Wied gravels but certainly available in the region, for instance, in the Rhine gravels and the volcanic field of the Eifel (Bolus 1992, 76). Single stones wore scar fields or scratches sustaining their interpretation as anvils or supports (Baales 1998, 353 f.; Gelhausen 2011a, 31-33; cf. Schulte-Dornberg 2000). However, burnt quartz was only found in very small quantities (n = 14) in some test pits (Bolus 1992) and in area 2 (n > 18; Gelhausen 2011c). At both locations several of these pieces were refitted. Moreover, some pebbles wore traces of fire but did not appear to be used as cooking stones (Freericks 1989, 33).

Furthermore, in area 7 a small slate plate which was broken, perhaps, along two possible drill holes was found and displayed engraved crosshatch lines on both sides (Baales/Street 1996).
A comparable special find was a shaft smoother made of a relatively coarse, reddish sandstone found in area 2 (periphery of concentration 18; Gelhausen 2011c). The complete end of the lathy fragment (71 mm × 34 mm × 22 mm) was rounded and two surfaces were flattened naturally. One was the working surface which displayed a central groove. On one of the rounded lateral edges ten linear incisions were set transverse to the groove direction. These incisions were almost parallel and ran onto the unworked flattened surface where five lines ended in a triangular extension. These engravings were interpreted as female silhouettes of an even further abstracted *Gönnersdorf* type (Loftus 1982; Gelhausen 2011c). The use of this particular composition of abstract females silhouettes set in a row on the slate plates in the lower horizons of Gönnersdorf and Andernach as well as on the shaft smoother from the FMG site at Niederbieber was considered as evidence for a continuous tradition between the inhabitants of these sites. An alternative interpretation as use-wear was based on ethnographic observations of the pairwise use of shaft smoothers (Flenniken/Ozbun 1988) and consideration about a fixation of these pairs by entwining cords (pers. comm. Olaf Jöris, Neuwied). However, to create this type of use-wear in the hard sandstone considerable force or duration must be assumed which probably would affect all the corded areas not only some parts. Furthermore, if the grooves were incised for the better fixation of the cords the question remains why these had to have triangular extensions. Thus, in the case of the Niederbieber specimen an intentional engraving appeared more plausible.
The numerous red haematite pieces recorded in areas 1, 3, 4, 6, and 7 (Freericks 1989; Bolus 1992; pers. comm. Frank Gelhausen, Engelskirchen) could represent natural infiltration from the underlying Wied gravels. However, on one piece of haematite from area 1 rubbing marks were found (Bolus 1992, 79).
Among the approximately 817 determinable faunal elements, beaver (*Castor fiber*) was clearly dominant with 584 pieces. However, these pieces comprised 577 teeth fragments from the areas 2, 7, and 10 but only seven bone fragments from area 3. Furthermore, almost 500 fragments were found in a single square-metre and a further 73 within the same concentration (area 10; Gelhausen 2011a, 42). Thus, the many remains originated presumably from very few individuals. The second most numerous species was red deer (*Cervus elaphus)* to which some 164 pieces were attributed in the main area (Gelhausen 2011a, 40-43) and, in addition, a few bones were recovered in area 2 (Gelhausen 2011c). Again, of the specimens from the main area 92 elements were teeth fragments. According to the still determinable body parts that had reached the area 1, Michael Bolus considered complete carcasses being brought to the site (Bolus 1992, 180). Elk (*Alces alces*) was identified on 42 fragments of which one originated from area 3 (Gelhausen 2011a, 40) and 38 attributed to at least two animals were found in area 2 and the adjacent test pit from 1990 (Gelhausen 2011c). Thus, only three remains of elk were found in the main area of the site (areas 1 and 5). Furthermore, large bovids, probably, aurochs (*Bos primigenius*) were identified on four examples found singularly in areas 5, 13, 17, and 17a and a further two specimens per area were recovered in the areas 11 and 12 (Gelhausen 2011a, 40-42). Horse (*Equus* sp.) remains were also found rarely (n = 14) across the site in the areas 1, 2, 10a, 14, 17, and 17a. Comparable to beaver, wild boar (*Sus scrofa*) was mainly identified by teeth fragments (n = 25) of which 22 were found in area 7 (Freericks 1989) and three from area 10 (Gelhausen 2011a, 42 f.). A metapodial from area 1 was not unambiguously identifiable but could be from wild boar or roe deer (*Capreolus capreolus*; Bolus 1992) which was only once more determined on a bone from area 17a (Gelhausen 2011a, 41 f.). Only in area 4 were several elements of the ibex (*Capra ibex*) recovered (Bolus 1992). In addition, in area 1 a mandible and tooth attributed to badger (*Meles meles*) and a tooth of a fox (*Vulpes vulpes*) were found. Their association with the archaeological record remained uncertain. However, a tooth of a pike (*Esox* sp.) was also found in area 1. Even though no manipulation was found, the presence

of fish remains in this elevated position suggested the introduction by other agents, presumably, the Lateglacial human inhabitants.

In area 2, teeth of a young horse were found including milk incisors and premolars but the molars were not erupted. Assuming the last milk incisor breaking through at approximately 6-9 months of age in modern horses and the first molar between 9-12 months of age, the individual was approximately 6-12 months old. If the foal was born around April/May according to the modern most typical mating period occurring in June these faunal remains would indicate a time of death in the cold period (**tab. 16**). The absence of antler among the elk material could also indicate the cold period but therefore the presence of a male individual had to be proven and, moreover, the selection observable from the recovered material in this area suggested only partial transport to the site (Gelhausen 2011c, 14). Hence, these seasonal indicators are very vague because they incorporate many assumptions. Comparably, the indications from area 4 which were based on the remains of red deer encompassed a time window from November to April with the most probable phase between January and April (Bolus 1992, 134). This attribution was due to the eruptive stages of the molars and premolars on four mandible fragments and the presence of an antler fragment adjacent to the skull. The eruptive stages observed on two mandible fragments of red deer from area 17a suggested an age of 9-11 years and of 40 months (Gelhausen 2011a, 42). Assuming a birth period of early May to mid-June as is common for modern red deer in Central Europe, the age of the latter animal would also indicate autumn as the time of death.

Concerning the minimal number of individuals for red deer, a maximum of four individuals was assumed in area 4 and three individuals were recorded at area 1 (Bolus 1992). In addition, in area 17a which is considered a previously unexcavated part of area 4 and in area 2 a minimum of two individuals was assumed (Gelhausen 2011a, 42; Gelhausen 2011c). Furthermore, according to the remains at least one individual was found in the areas 5, 6, 7, 10, 10a, 11, 13, 14, and 15. Whether these remains attested 20 different animals or some parts from different areas were attributable to the same animals could only be answered by a comprehensive analysis of the faunal remains encompassing all areas of the site. However, this analysis was not yet accomplished for Niederbieber. As shown elsewhere (Street/Turner 2013; Leduc 2014a) such reanalyses of complete sites contribute to the understanding of the spatial organisation on the sites. In particular, differentiated functions of working spaces as well as the relation between those working places can be assumed. However, the generally restricted preservation favoured a selected preservation of durable parts such as teeth or mandibles clearly limiting the results of an analysis of the organic *chaîne opératoire*. In fact, the majority of still determinable fauna was found in the western part of the plateau where the pumice cover was thicker (Gelhausen 2011a, 41). Furthermore, the surfaces of the bones were often poorly preserved hindering the inspection for butchering marks. Nevertheless, on a few specimens in the areas 2, 4, 5, 10a, 11, and 13 cut-marks and/or intentional impact marks were observed (Gelhausen 2011a, 42). These manipulation were found on bones of red deer, elk, horse, and the large bovids indicating the processing of various species at the site. Furthermore, the majority of the faunal material was very fragmented and many bones were only preserved as very small, calcined pieces suggesting exhaustive exploitation of the material, perhaps, for marrow extraction (Bolus 1992, 180).

Nevertheless, the modified bovid bone from area 5 was also considered as a possible, roughly hewed tool (Baales 2002, 186). Moreover, two bone point fragments were recovered from the areas 1 and 4 (Bolus 1992, 75) clearly indicating the former presence of organic tools on the site.

A brachiopod and a trilobite fossil were found in a test pit without archaeological context (Bolus 1992, 16) and a third fossil was found in area 5 (Husmann 1988). Although these fossils were known from the Wied gravels and could perhaps be introduced to the site naturally, a human collection of these finds could not be excluded completely (Bolus 1992, 16).

Spatial organisation

Evident structures in Niederbieber were only confirmed in the form of hearths (Gelhausen 2011a, 44-48; Gelhausen 2011c, 5-11). These structures were identified by coloured and bricked sediment. This evidence was observed only in the areas 1, 2, and 4. In the areas 1 and 4 one hearth was found within the centre of the artefact accumulation. In area 4, the heat altered sediment surrounded by further burnt material was found 12 cm below the loess surface suggesting this as the archaeological horizon. The differentiated spatial distribution of charcoal determined as *Betula* sp. and as *Salix* sp. was suggested as indication for two firing episodes and the calcined bones were assumed as representing, perhaps, a third phase (Bolus 1992, 139). In area 2, John Loftus identified four structures with burnt material of which two were possible hearths (J2 in sub-area 19 and J4 in sub-area 20), whereas one accumulation (J1) located in sub-area 18 was assumed as a possible dump and the other one (J3) which was located in sub-area 20 was considered as a burnt wooden structure or multiple hearths (Gelhausen 2011c). Alternatively, the irregular structure could also represent a burnt tree. The burnt sediment of the two hearths reached some 5-10 cm below the surface. Within the southern hearth J4 a smaller structures was observed and another small structure was set less than a metre aside the hearth and was considered as a possible posthole with a stone plate at the bottom (Gelhausen 2011c). Furthermore, in area 2 five depressions were observed which were partially attributed to animal burrows or erosion but human interference could not be excluded completely (Gelhausen 2011c).

Furthermore, Frank Gelhausen identified further 14 latent hearths and a possible hearth in the main area (Gelhausen 2010, 54-58). He assumed the accumulation of burnt material such as lithic artefacts, calcined bones, and charcoal as well as the concentration of LMP as reliable indicators for hearths (Gelhausen 2011a; cf. Gelhausen/Kegler/Wenzel 2004). Only the areas 16, 17a, and, perhaps, 8 yielded no indication for a hearth. However, area 17a was assumed as extension of area 4 and in area 8 a hearth was considered probable due to the comparable pattern of the material concentrations (Gelhausen 2011c, 48). Hence, only area 16 was not associated with indications of fire. This area contained only ephemerally scattered material. Furthermore, in the areas 9 and 13 the existence of a second hearth was proposed. In addition, a possible hearth existed in the outlying area 3 (Bolus 1992). In combination with the limited inventory, this area was assumed to represent a workshop for hafting and retooling (Bolus 1992, 158-160. 183).

In area 15, Michael Baales identified a possible cache of two partially prepared, large flakes of chalcedony (Baales 2003b, 193-196). In combination with the depot of retouchers (Bosinski et al. 1982), these depositions suggested an intention to return to this site.

In total, Niederbieber yielded 19 distinct concentrations of lithic artefacts and mainly burnt bones (Gelhausen 2011a; Gelhausen 2011c). Frank Gelhausen demonstrated the highly similar organisation of the concentrations from the main area as well as the southern part of area 20 with usually two clusters of relatively high artefact density set opposed to each other with a hearth in the centre (Gelhausen 2011a, 48; Gelhausen 2011c, 33-36). In the areas 1, 4, 5, and 17 the pattern was observed that larger organic material, in particular, if unburnt was usually found in the periphery of the concentrations comparable to the cores (Gelhausen 2011a, 43). This distribution of larger pieces at the periphery of a concentration was suggested as possible indication for a barrier effect. A ring and sector analysis conducted for the concentration in the areas 1 and 4 was not indicative for a barrier effect (Bolus 1992, 181). However, based on density distributions of various find classes such as small lithic debris, LMP, or total retouched artefacts as well as larger pieces such as cores and, furthermore, the intra-concentration refits for some of these concentrations the presence of a dwelling structure was made probable (Gelhausen/Kegler/Wenzel 2004). These structures were proposed for the concentrations in the areas 1, 4, 12, 13, and 17 (Gelhausen/Kegler/Wenzel 2005; Gelhausen 2011a, 151-154. 169. 232-235).

In total, only 26 refitting complexes could connect different parts of the main excavation area (Gelhausen 2011a, 52-59. 257-261). Nine complexes were related to area 9, eight complexes were related to area 1, and six complexes to area 4. However, these areas also yielded dense artefacts clusters (see **tab. 11**). Areas 1 and 4 were in particular closely connected suggesting that perhaps these areas were used contemporaneously. Refits to the areas 2 and 3 were not attempted. However, within area 2 several pieces were refitted (Gelhausen 2011c). Moreover, based on the stratigraphic position of the majority of the material, Frank Gelhausen considered two phases possible in the concentration 20, an older phase with Baltic flint and a later one with Western European flint (Gelhausen 2011c). The chalcedony formed a discrete concentration in this area but the artefacts scattered as intensely as the Baltic flint material. The little vertical distribution of the Western European flints were assumed as indication for a deposition shortly before the LSE. However, the horizontal distribution was not much affected. This more diffuse spatial organisation, perhaps, more complex evident structures, and a deeper temporal use was only observed in area 2 not on the main area (Gelhausen 2011c). On the main area the undisturbed appearance of the horizontal distribution in combination with the highly similar organisation and comparable lithic inventories was assumed as indicative for little settlement dynamics and a short-termed creation of these discrete scatters on which recurrently the same activities were performed but which could not be further resolved temporally (Gelhausen 2011a, 248f. 261-266; Gelhausen 2011b).
In contrast, the analysis of Michael Bolus on details also of the faunal remains from the areas 1 and 4 suggested the introduction of complete carcasses to the site and assigned the lack of some body parts to the poor preservation and the high fragmentation rate which resulted presumably from intentional breakage for the extraction of marrow (Bolus 1992, 180). Thus, at least the butchering and exploitation process of the prey seemed to be accomplished at the site. Furthermore, the numerous end-scrapers which were frequently attributed to hide-working activity as well as the use-wear analyses sustain an interpretation of a comprehensive exploitation of the animals at the site. In addition, in the areas where faunal remains were preserved the spectrum appeared usually diverse which in a site of post-processing of a single hunting episode would be expected to be more specialised. Nevertheless, since the numbers were small, this impression could be blurred by the introduction of elements from a provision. Additionally, besides the lithic inventories which were comparably diverse as at other FMG sites in the single concentrations at Niederbieber, some organic tools were found as well as numerous stone tools which indicated intensive preparation works at the site. Furthermore, also some rare indication of lost personal ornaments were found. In summary, these various indications of differentiated activities at the site contradict the interpretation as simple, short hunting stops. According to a specialising index, that was proposed by Jürgen Richter (Richter 1990) who assumed longer occupation durations to produce more diverse retouched artefact assemblages, the calculated occupation time yielded relatively long occupation periods also for concentrations which appeared of short-lived, undisturbed character to Frank Gelhausen (Gelhausen 2011a, 251-254). He suggested that either the diversity calculation did not apply to the assemblage of Niederbieber or that the considered occupation time of the areas was false, although he assumed these to be more reliable as based on the archaeological record.

Chronology

The general, chronostratigraphic position of Niederbieber was attributed to the Lateglacial Interstadial due to the deposition of the finds in a loess loam covered by the pumice of the LSE. Moreover, the recovery of most artefacts in the 10cm immediately below the pumice and the preservation of most remains in a horizontally unaltered position suggested a deposition of these remains shortly before the LSE. In addition, the environmental indicators such as red deer and elk further sustained the attribution to the forested period in the Lateglacial Interstadial. A TL date from the site sustained this attribution in general (**tab. 22**;

site	lab. no.	years TL-BP	± years
Niederbieber	QTL51B	13,000	1,100

Tab. 22 TL date from Niederbieber. Reference: Bolus 1992, 19.

Bolus 1992). Furthermore, two technically reliable ^{14}C dates (OxA-2066 and OxA-1135) from the site (**tab. 20**) produced almost identical results from the 100-200 years before the LSE. However, the samples for these dates originated from area 2 (OxA-2066) and a test pit several metres west of area 3 (OxA-1135; Baales 2002, 41 f.; cf. Bolus 1992). Besides this spatial gap, the latter was also found in a stratigraphically higher position than the remains from the workshop in area 3. Thus, the relation of the sample for OxA-1135 to the archaeological material remained uncertain. The ^{14}C dates from the main excavation area produced unreliable results (see p. 265-269). The vague seasonal indications from area 2 as well as the areas 4 and 17a attributed the occupation of these concentrations very generally to a period from autumn to spring, more probable in the second half of the cold period. However, this comparable seasonal indications could sustain an actual co-existence of the concentrations as well as recurrent visits during more or less the same period of the year in a yearly round. This possibility of contrasting interpretations further sustained Frank Gelhausen's considerations that the connection of various episodes at Niederbieber to a common occupation event was almost impossible to prove by radiometric or stratigraphic analyses (Gelhausen 2011a, 254-257). Alternatively, he considered the various episodes at the site as quasi-contemporaneous and assumed that the episodes occurred within a relatively short period which allowed no further distinction between them. Furthermore, low settlement dynamics and/or respect of existing and still visible artefact heaps were suggested due to the little horizontal disturbance of the material combined with the few inter-concentration refits. The only exception were the more complex distributions in area 2 where different scatters appeared as a disturbance to other ones and the stratigraphic position showed different tendencies (Gelhausen 2011c). In addition, the two identified material depositions (Bosinski et al. 1982; Baales 2003b) suggested that further visits to the site were planned, perhaps, because Niederbieber was a habitually visited site in the seasonal cycle of these hunter-gatherers.

Thus, for the discussion of settlement behaviour on FMG sites, Niederbieber was particularly important due to the comparably good preservation. However, the major question was whether the concentrations, in particular of the main area, were created as diverse episodes during a single occupation event or created in successive, temporally distinct occupation events. Consequently, was Niederbieber an aggregation camp, a large base camp, or a recurrently visited place?

Comparable to Niederbieber, the Belgian FMG-sites Rekem (De Bie/Caspar 2000) and Lommel-Maatheide (De Bie/Van Gils/Deforce 2009) yielded several distinct concentrations of lithic artefacts and occasionally burnt bones. However, the FMG site at Niederbieber was in many characteristics distinct from the Belgian sites such as the topographic position on a promontory some 30 m above the river, whereas Rekem is situated on a sand bank in the Meuse valley and Lommel on the sand dune on the northern bank of a Lateglacial lake. This distinction may be due to the different geographic setting in the upland zone with steep valley walls in contrast to the south-western part of the North European Plain where the wide river plains or valleys with shallow water bodies are traversed by low sand dunes. Moreover, Rekem was considered as a large occupation site which developed by various episodes of workshops within one large occupation event (De Bie/Caspar 2000). The few undisturbed FMG concentrations at Lommel-Maatheide were accompanied by many Mesolithic accumulations and in contrast to Rekem assumed as settlement site which recurrently attracted Lateglacial and early Holocene hunter-gatherers. Frank Gelhausen also considered the similar patterns of the distinct concentrations at Niederbieber as a result from multiple, short episodes of hunting groups which prepared for a hunting episode and/or prepared the hunted prey for further transport (Gelhausen 2011b). Nevertheless, the refits as well as the diversity of the archaeological material marked the more complex nature of Niederbieber. Perhaps, the very different indications show the limits of classic distinctions of sites as hunting, base, or aggregation camps and suggest the presence of more flexible movements of residential camps, possibly, to nearby successful hunting grounds (cf. Müller et al. 2006).

Boppard, Rhineland-Palatinate

Research history

During construction works in December 2001 near the train station of Boppard, a small concentration of lithics and bones was found (Wenzel 2004) aside and partially disturbed by medieval latrines (Welker 2004). In approximately the following fortnight 10 m² were excavated according to modern standards, whereas due to the on-going construction work for a further 15 m² the coordinates were recorded and the sediment collected for wet-sieving. Since the foundation pit was already boarded at the discovery, the stratigraphy could only be documented in the underlying part. Moreover, the material was only published partially in a few articles thus far (Wenzel/Álvarez Fernández 2004; Wenzel 2004; Street et al. 2006) and, hence, not all lines of evidence can be quantified in the following.

Topography

Boppard is situated on the southern bank of a large Rhine meander (**tab. 10**). The site was excavated some 150 m south of the modern river bed and less than 10 m above the modern river. During the Lateglacial the site was probably also set in the floodplain of the river. Towards the south the terrain rises very gradually. Moreover, the adjacent plateaus which rise up to 200 metres above the valley floor are cut by several valleys. In particular, the *Fraubach* valley opens the plateau south of the site allowing access to the higher plateaus and, possibly, containing a stream which could have passed near the site in the Lateglacial.

Stratigraphy

The recorded stratigraphy was dominated by the position in the floodplains of the Rhine (Street et al. 2006, 758). Over sands and gravels an approximately 1 m thick sandy loam followed which was formed by high flood movements. On top of this unit a gravel band was deposited on top of which c. 10 cm of the brownish-grey sediment with the archaeological finds were preserved, whereas the upper part of this deposit was already capped by the construction works. Nevertheless, in some parts of the site remains of pumice from the LSE were observed above the sediment layer (Wenzel 2004, 13).

Archaeological material

The lithic inventory appeared comparable with other Central Rhineland FMG assemblages (**tab. 11**) and was dominated by the Tertiary quartzite. This raw material could be found in blocks some 10 km southwestwards of the site or in the wider surrounding of the Laacher See volcano some 20 km north to northwestwards of the site (**tab. 12**; cf. Floss 1994). Along with this raw material, indurated slate, quartz, and Devonian quartzite which could be collected in the local Rhine gravels were used at the site. In addition, chalcedony which occurred at a minimum distance of 35 km in the north-west or 45 km and more from the south-western Mainz Basin as well as a special local variety of the Western European flint named the Eifel local flint were found. The latter raw material represented residual deposits of the same formation epoch as the Western European flint. The Eifel local flint could be found some 60 km north-westwards as well as in various gravel deposits of the small streams in the Eifel some 45 km distance to the west of the site and in

the gravels of the Ahr river over 45 km in the north-west. Furthermore, the Western European flint from a minimum distance of 100 km north-westwards was also used in Boppard (Wenzel 2004, 14).
The retouched tools comprised small end-scrapers, burins on truncation, and backed pieces (**tab. 14**; cf. Wenzel 2004, 13 Abb. 1). Of the four published LMP specimens two can be classified as curve-backed point (Wenzel 2004, 13 Abb. 1.2-3), whereas another one appeared straight to angle-backed (Wenzel 2004, 13 Abb. 1.1). A thin LMP was retouched along both edges and, thus, resembled a double borer (Wenzel 2004, 13 Abb. 1.4). However, the lithic material from Boppard is not finally analysed and, therefore, not yet completely published.
Organic material was preserved on the site. Red deer (*Cervus elaphus*) is represented by some single teeth, a mandible with two teeth and a few bones, of which some were fractured for marrow extraction (**tab. 15**; Wenzel 2004, 14). A metapodial fragment of red deer was directly ^{14}C-dated (**tab. 20**; KIA-26644: 11,095 ± 55 years ^{14}C-BP, $\partial^{13}C$: -25.1, written comm. and kind permission Stefan Wenzel, Mayen) producing an age comparable to the dates of the LSE (Baales/Bittmann/Kromer 1998; Baales et al. 2002). Along with red deer, wild boar (*Sus scrofa*) was identified in the assemblage with seven teeth and a possible bone fragment (written comm. Stefan Wenzel, Mayen). The dental abrasion indicated the death of the animal(s) in the spring-summer season (Street et al. 2006, 765). However, due to the rare indications of this species in Central Rhineland, the origin of such indications remained a matter of debate on whether wild boar was imported from elsewhere or present in the Central Rhineland by the end of the Lateglacial Interstadial (Baales 2002, 28; Street et al. 2006, 765). Furthermore, a considerable number of fish remains was also recovered, of which some pieces were burnt and, therefore, probably associated with the human activity on the site.
Furthermore, a 17 cm long spall from a metapodial of red deer was transformed into a tool which was interpreted as smoother (**tab. 17**; Wenzel/Álvarez Fernández 2004). Alongside the edges of the spall several horizontal lines were cut which the excavator interpreted as decoration (Wenzel 2004, 14). This specimen was particularly comparable to Magdalenian examples from Cantabrian Spain and southern France, Final Palaeolithic spatulas from northern Italy as well as a Mesolithic piece from southern Germany (Wenzel/Álvarez Fernández 2004).
Two small pieces of jet were also excavated (Wenzel 2004, 14) but according to the size these were perhaps not used.

Spatial organisation

The Lateglacial remains of Boppard concentrated around a latent hearth (**tab. 19**; Street et al. 2006, 776) which was revealed by the accumulation of burnt bone fragments, heated lithic artefacts, and reddened quartz fragments (Wenzel 2004, 14). Assuming from the faunal and lithic remains in combination with the spatial distributions, the site equates for instance the concentrations of Niederbieber.

Chronology

According to the stratigraphy and the ^{14}C date the accumulation of Boppard can be placed chronostratigraphically shortly before the LSE (**tab. 20**). The high number of special goods, the topographic position on the river banks, and occurrence of wild boar single out the site in contrast to the other FMG sites, whereas the lithic raw materials and typological classifications, the dominance of red deer, and spatial organisation were similar to the other FMG sites in Central Rhineland. The publication of a more comprehensive analysis in the future will allow further considerations about function and/or preservation of the site.

Bad Breisig, Rhineland-Palatinate

Research history

The Lateglacial concentration near Bad Breisig was found by biologist and geologist Georg Waldmann in a profile on the southern limit of a gravel pit which he reinvestigated for plant imprints in the LST (Waldmann/Jöris/Baales 2001). He recognised lithics, burnt bone fragments, and a darker patch which indicated a hearth in the profile above the various deposits of the LSE. Due to quarry works half of the single concentration had already been lost by then. The collected finds and the description of the stratigraphy led to a first inspection by local archaeologists and consecutively an 18 days excavation under the supervision of Michael Baales and Olaf Jöris in autumn 2000. During this first campaign c. 9 m² around the hearth were excavated. All finds larger 1 cm were given 3D coordinates, drawn in a map, and described separately. Smaller finds were collected per 50 cm × 50 cm × 5 cm units equally as the sediment which was wet sieved later. The limits of the concentration were not reached in this campaign and thus, a second season of 20 days duration led by Michael Baales followed in spring 2001 and uncovered a further c. 37 m² according to the above mentioned standards. The limits of the archaeological scatter were reached in the second campaign (Baales/Grimm/Jöris 2001). However, the distribution also revealed that about half the concentration was lost in the gravel pit. Furthermore, five one square-metre wide test pits were laid out in approximately five metre distance to the excavation limits to look for further indications of scattered archaeological material but no further artefacts were found in the test pits. An inspection of the gravel pit profiles provided only Roman material. Furthermore, a surface survey on the surrounding field of the concentration yielded no finds, only on a neighbouring field few lithic pieces were collected.

Topography

The river Ahr flows into the Rhine some 3.5 km northwards of the Lateglacial site of Bad Breisig (**tab. 10**). The Ahr mouth had widened the Rhine valley on the western bank between Bad Breisig and Remagen (c. 8 km) to an approximately 1.5 km wide basin (the so-called Golden Mile; Schirmer 1990). The site is located on the western bank of the Rhine at the southern end of this basin where the valley is still about 1.3 km wide but almost half of the width is occupied by the modern river. A kilometre towards the south-east of the site steeply rising hills put an end to the Golden Mile. Southwards the Rhine valley is formed tube-like with a maximum width of 900 m and ends some 11 km south-westwards at the so-called *Andernacher Pforte* (Andernach Gate). Nevertheless, the walls of this tube are frequently cut by steep valleys of small streams on the east as well as on the west side (main ones: *Frankenbach*, *Vinxtbach* and *Brohlbach*) which are of some importance for the stratigraphic sequence (see below). South of this gate the wide Neuwied Basin opens. The modern river passes only a few hundred meters east of the site and some 10 m below the terrace where the site is located. However, the site is situated on the edge of lower Rhine terrace 2 and the incision into the modern terrace of the Rhine began presumably in the Lateglacial. Thus, depending on the accomplished process of the incision of the river, the site was probably set some 10-30 m away from the Rhine i. e. on the river bank during the time of occupation. On the eastern bank of the Rhine the hills rise steeply towards the Westerwald and, opposite of the site, these hills are cut by the valley of the Ariendorf creek. Towards the west of the site the terrain only rises very gradually for c. 400-500 m before the Eifel hills rise sharply some 140 m upwards. These hills are also cut in a west-west-south direction by a small valley named *Tiefpfad* (deep path) in which also a small stream flows. This stream ends in a pond at the foot of the hills but during the Lateglacial it flew perhaps into the Rhine not far south of the site.

Stratigraphy

In Bad Breisig the recorded stratigraphy began with the reddish brown Allerød soil which was sealed by a thin greenish grey tephra band (Baales/Grimm/Jöris 2001). Probably, this band can be equated with fallout ashes from an intermediate stage of the LSE. The Laacher See is located *c.* 11.5 km south of the site. The tephra band was overlain by up to 25 cm of grey fluvial sands. In the lower part of this deposit single, small pieces of pumice were found and in the middle part displaced ashes were intermingled. These sands represented presumably the sediment redeposited after a catastrophic dam burst (Park/Schmincke 1997). During the LSE the Rhine was dammed by various materials which were pushed through the incised valleys on the western walls from the volcanic centre into the tube-like valley part at the Andernach gate. Subsequently, an approximately 80 km² wide lake established in the Neuwied Basin. The water table raised up to 15 m and *c.* 0.9 km³ of water were stored. When the dam broke these waters were led out to the Rhine valley north of the dam in probably a single flood wave because the incisions of the flood can be traced up to Bonn, approximately *c.* 50 km north of the volcano. The flood wave washed away up to 4 m of pre-eruptive sediment from the valley floor and deposited secondary eruptive material such as the pumice-ash-sand deposit in Bad Breisig. On top of this deposit followed an approximately 10 cm thick layer of pumice on the site. The pieces were rounded and sorted by grain size. Therefore, this pumice layer swam probably on the lake waters and were spilled ashore when the waters of the lake drained through the broken dam. The pumice layer was sealed by a 7-8 cm thick, greenish dark grey tephra which was identified as a fallout ash from the late eruptive phase. Thus, the stratigraphy of Bad Breisig indicated that the dammed lake formed an episode of several days within the LSE (Baales et al. 2002). The greenish dark grey tephra was overlain by sands and silts with reworked volcanic material. Within these sands and silts a small band of obliquely deposited pumice can be found. On top of the sands and silts another layer of pumice followed. These units of sands, silts, and pumice were probably alluvial deposits at the river banks. If this deposition occurred within the eruptive process or shortly after remained unclear. The sealing layer of pumice was largely cut by the above formed high flood loams. This cutting was partially due to bioturbation but also due to erosion of some material from the unstable surface at the river banks after the eruption. The presence of single, heavily rounded pumice pieces could still be noticed in the lower half of the high flood loam deposit. This deposit was almost 1.1 m thick. 20-30 cm above the transition from the overlain, eruptive deposits the Lateglacial archaeological material was found. The lithic artefacts spread vertically over 35 cm within the high flood loam deposit but over 70 % of the material were found within 10 cm around the altitude of 66.35 m a.s.l. The vertical distribution can be explained by bioturbation, cryoturbation, and hydrological movements within the sediment. The dark brown Holocene topsoil formed the upper few centimetres of the profile.

Archaeological material

In the excavations 45,480 lithic remains were recovered which weigh more than 10 kg (**tab. 11**; Grimm 2004). 973 of the 5,956 lithic artefacts ≥ 1 cm displayed traces of heating. Cortex was still present on 478 of the pieces larger 1 cm. Only few artefacts were refitted (n = 69) but these refits connected all parts of the lithic concentration.

The main raw material was Tertiary quartzite (n_{all} = 38,059; $n_{\geq 1\,cm}$ = 5,183) which was presumably gathered at a nearby source (**tab. 12**). Around the Laacher See and the nearby Herchenberg some banks and weathered blocks of this quartzite were recorded as well as volcanic bombs of this material in basalt deposits (Floss 1994, 30). Thus, besides fluvial transport by the Brohl stream and the Rhine, the LSE could have directly deposited one or several blocks of Tertiary quartzite in the immediate vicinity of the site where the material possibly began weathering and thereby produced many of the flinders-type material. Western European flint (n_{all} = 7,145; $n_{\geq 1\,cm}$ = 708) was generally present in the form of gravel flints from presumably the Meuse but some pieces were identified as a possible local Eifel variety ($n_{all}=52$; $n_{\geq 1\,cm}=34$) and single pieces were classified as the sub-units of Vetschau flint, Lixhe-Lanaye flint, and Lousberg-/Schneeberg flint. In general, the sources of these varieties of Western European flints were located in north-western directions in a distance of 45 km and more from Bad Breisig. The third raw material which was more numerously present was indurated slate which originated from the local gravels. The few hundred pieces ($n_{all}=272$; $n_{\geq 1\,cm}=187$) were assigned to two nodules, one of the classic black lydite type, the other of a greenish grey, glassy variety. Furthermore, a single flake of argillaceous shale which was also available locally in the gravels, a flake of possible chalcedony, and two pieces of probable Triassic chert were recovered. The latter was not locally available. The next resource was located over 80 km south-west of Bad Breisig in the southern Eifel region. One of the pieces was transformed into an end-scraper. A well known chalcedony resource of the FMG in Central Rhineland was situated around 18 km north-north-west of the site at Bonn-Muffendorf. However, single occurrences of chalcedony were reported in the valleys south of Bad Breisig. Whether these single specimens reached the site in the form of tools which were only resharpened or discarded or if more pieces of these raw materials were dumped elsewhere can no longer be securely reconstructed due to the loss of half of the concentration to the gravel pit.

According to the number of blocks, cores, and core fragments from Tertiary quartzite (n = 144; **tab. 13**), the site of Bad Breisig was perhaps visited mainly due to a nearby source of Tertiary quartzite and the exploration of this source for the production of blanks. A further two core fragments were made of flint and another core fragment made of lydite was found in the concentration. 57 cores and core fragments were not prepared except for distal abrasion. Most commonly crested edges (on 31 pieces) and the rejuvenation of the platform by flaking (on 31 pieces) were observed. In a further 21 cases both types of preparations occurred together. Among the cores and core fragments of Tertiary quartzite, the ones with flaking in opposing directions dominated (n = 53), although one direction was often preferred. Unidirectionally knapped cores (n = 39) were more frequent than cores with two striking directions which were laterally set to one another (n = 26). Furthermore, polydirectionally exploited cores (n = 13) or pieces with a single negative (n = 11) occurred. The lydite core was only worked from one striking platform which displayed some negatives of platform rejuvenations. Two faces were exploited from this platform without any further preparations. The lower part of this core was still covered by the rolled cortex. In contrast, the flint core was exhaustively used with several striking platforms, several striking directions and faces. Often the cores were exploited alongside one or both thinner sides (lateral edges; n = 66) giving some of them a burin-like appearance. These laterally exploited cores indicated the aim of the blank production process to receive slim and long bladelets which were comparable to burin-spalls. This aim was also sustained by the length-width distributions of all complete blanks. Moreover, 221 blanks with the remains of cresting on their dorsal face were found indicating the wish of a controlled blank production. However, several of the crested blanks were secondary ones and the cresting was often made discontinuously, on one side of a natural surface, or by pressing rather than retouching. Thus, the effort to stabilise a crest for directing the knapping force was minimised where possible. Rarely pieces yielded indications of the use of an organic hammer or of soft hammerstones. The signals for the use of (hard) hammerstones were more frequently found on the blanks. Furthermore, on more than 45 artefacts the intentional or accidental removal of the bulb was noticed which in the latter case was related to a high force of the percussion. In fact, at least two hammerstones made of Devonian quartzite and a hard sandstone were also found on the site. Moreover, a retoucher of argillaceous shale with only few use-wear patterns was also recovered.

Even though the blank production process appeared as an important factor at the site of Bad Breisig, 296 retouched lithic artefacts were also recovered (**tab. 14**). 41 tool production or resharpening flakes (17 Krukowski »micro-burins«, 20 burin spalls, three end-scraper caps) suggested the knapping and/or resharpening of these tools on the site. In addition to the intentionally retouched specimens, splintered to bruised edges were observed on over 150 pieces.

Among the intentionally retouched artefacts the backed pieces were most numerous (n=114). The LMP were generally very fragmented. Consequently, only five pieces could be identified as backed blade or bladelet with some certainty. One of these was truncated; another very slim piece was truncated on both ends. One distal and two proximal fragments were only partially retouched appearing as backed bladelet production remains. However, they represent no typical micro-burins. Furthermore, the bulb was generally present on the ascertained backed bladelets and blades (except for the double truncated piece). Some eight pieces presumably represented point fragments and a further 38 specimens were reliably identified as backed points. Yet, of these 15 pieces were only tip fragments. Four backed points were basally retouched and classified as Malaurie points (Baales/Grimm/Jöris 2001). On seven pieces the blunted part was formed rather straight (Grimm 2004). On two particularly thick blanks the retouch appeared angular and reminded on the Andernach knives from Andernach 3-FMG. The remaining ten complete pieces were described as typical *Federmesser*.

Among the domestic tools the generally small end-scrapers (n=98) formed the largest group. 14 end-scrapers were made on a blade or a blade fragment. The truncations (n=31) and the burins (n=25) occurred in similar numbers. In both classes very diverse types were present. Among the burins those on truncation were the most common type and dihedral examples occurred only twice. A single composite tool was a burin on truncation combined with an end-scraper. 19 pieces were classified as laterally retouched. A further eight specimen were notched and/or only partially retouched.

Alongside the lithic artefacts, some stone material was discovered at the site and were at least partially associated with the concentration. The hammerstone which was made of a hard sandstone gravel was roughened along one long lateral edge. The Devonian quartzite hammerstone was split and knapped into an end-scraper-like shape. The inspection under a microscope produced no clear evidence for the use of this »retouched« edge. Possibly, this effort was expended for a better handling of the hammerstone. However, a comparable piece was recovered from the area 1 in Niederbieber where the use remained comparably uncertain (Bolus 1992, 77). In addition to these two hammerstones and the retoucheur mentioned previously, some small plates mainly of Devonian quartzite were found. At least in one case splitting of a quartzite plate into thinner pieces was observed.

Besides the stone material, a few, mainly burnt bone fragments (n=33) and teeth (n=13) were recovered from the sediment and determined by Michael Baales (**tab. 15**; Grimm 2004). Except for a tooth of horse (*Equus* sp.), all other teeth and teeth fragments were assigned to red deer (*Cervus elaphus*) which was the most dominant species with eleven bone fragments and at least three individual animals determined by the state of the teeth. One sample suggested an age of 3 ¼ to 3 ½ years of age at the time of death and, thus, suggested a possible killing episode between September and December (**tab. 16**). A further three diaphyse fragments can only be classified in a size class of red deer and/or roe deer (*Capreolus capreolus*). Six bone fragments were determined as roe deer. The other six determinable pieces represented possible admixtures, partially of recent age or from the background fauna: Two pieces were identified as European bullhead (*Cottus gobio*) which perhaps was deposited by a high flood event. A phalanx was classified as rabbit (*Oryctolagus cuniculus*) or hare (*Lepus* sp.) which could have dug itself into the archaeological horizon. A diaphyse fragment could also be classified as hare but due to the poor preservation the distinction to bird (*Aves*) or fox (*Vulpes vulpes*) was not possible. Another bone fragment was also identified as red fox but it was unburnt and made a particularly fresh impression. Therefore, this piece was probably of recent age. Finally, a caudal vertebra could originate from a small canid presumably a dog (*Canis* sp.) or a badger (*Meles* sp.). A calcined bone which was found in the profile but not further determined was dated to the late Younger

Dryas (**tab. 20**; GrA-17716) in the burnt bone project of the Groningen laboratory (Lanting/Niekus/Stapert 2002). The project demonstrated that the dating of burnt material from the Lateglacial produced generally no reliable results. Thus, this date was rejected.

Five larger pieces of charcoal were recovered from the blackened sediment and Julian Wiethold (Göttingen) could determine four of the charcoal pieces as pine (*Pinus* sp.), as conifer, as deciduous hardwood with a good water transporting vascular system (such as *Betula* sp., *Salix* sp., *Populus* sp., *Corylus* sp.), and as a »... brown coal-like piece, possibly the residue of a heavily coaled organic material« (Waldmann/Jöris/Baales 2001, 177, translated by the present author). Flaky brown coal was present in the lower terraces of the Rhine (Terberger 1997, 34) and frequently used as probably fuel material on the Late Magdalenian sites (see **tab. 12**, p. 89, p. 105, and p. 119). Besides the use for firing, another application of this resource was suggested by the evidence from the recently excavated FMG site of Wesseling north of Bonn where some pendants made of brown coal-like material were recovered (Heinen 2010). However, due to the intense firing of the Bad Breisig sample no further conclusions on the material and its use were possible. The last sample of charcoal was indeterminable.

Spatial organisation

In Bad Breisig a lithic concentration spread 2-3 m around an evident hearth (**tab. 19**). The hearth which was originally recognised in the profile became further apparent in the planum during the excavation. In addition to the concentration of burnt material such as calcined bone fragments, charcoal, and burnt flints, some reddish and blackening colour of the sediment and a few patches of bricked sediment were observed. The maximum length of the red coloured sediment was 0.4 m extending parallel to the profile and the maximum width of this colouring was 0.2 m into the excavated area. The brickened sediment was noticed slightly outside the reddish part in the blackened sediment. The blackened sediment extended in an east-west extension on c. 1.7 m and up to 0.7 m in north-south direction. Regarding the extension of the feature, probably half the hearth was lost to the gravel pit. The surrounding lithic concentration reached a maximum (preserved) extend of 4.5 m × 3 m. The dense concentration was directly south of the coloured sediment disrupted by a few centimetres wide strip with only few artefacts. This disturbance was presumably caused by ploughing which affected the southern limit of the concentration as well. If these disturbances were re-located to where the pieces were presumably deposited the already dense appearing concentration became even more solid. In the gravel pit profile two distinct concentrations at the same level separated by only a few centimetres were noticed. These centimetres were revealed in the horizontal display as a round to semi-circular spot in which the underlying pumice already emerged in the lower planum. Probably, a little tree or a bush had grown there sometime after the occupation and the roots partially disturbed the position of the Lateglacial material. The determination of a charcoal recovered from this area suggested the possible presence of hard wood during the Younger Dryas (see above). Nevertheless, the density of the lithic accumulation and the abruptness of its limits suggested that the site was only settled once because otherwise disturbance by further settlement dynamics could have occurred. Furthermore, the few refittings (n = 69) which connected all parts of the lithic concentration further substantiate the hypothesis of a single episode in which the archaeological material was deposited. Possibly, this generally remaining density was also due to the placing of the site in the high flood plain of the Lateglacial Rhine where the material could be quickly covered by sediment. However, the material appeared not much affected by water movements and, thus, the material remained generally *in situ*. Moreover, the abrupt decrease in material density could be explicable by some kind of restrictions of the concentration such as a wind shelter or the walls of a dwelling structure.

In general, the backed implements were found mainly in the immediate vicinity of the hearth, comparable to other Lateglacial sites. Moreover, around the hearth and in the eastern part of the concentration the re-

touched artefacts as well as the retouched debris were accumulated, whereas in the western part a majority of blank production debris was found. In particular, the fragmented cores and the little prepared cores were found in the western part and still exploitable cores were found tendentially in the east where also the hammerstones were recovered indicating that lithic production occurred around the hearth and in the eastern part of the concentration. The western part was either dump or storage area for lithic raw materials.
Thus, several occupation episodes formed the probably single occupation event. Along with the blank production, hafting and retooling took place as well as the consumption of red and roe deer. The compact accumulation of diverse materials in various activity areas suggested some type of barrier contributing to the dense concentration. The intense use of a local lithic resource reflected a good knowledge of the resources of the surrounding.

Chronology

The charcoal sample on deciduous hardwood and the pine sample were ^{14}C-dated. Surprisingly, the hardwood sample resulted in a mid-Younger Dryas age (**tab. 20**) but could be rejected due to its position in a more recent disturbance. In contrast, the pine sample was found in an area south of the intensely red coloured sediment. This sample produced a late Lateglacial Interstadial/early Lateglacial Stadial date (**tab. 20**) This date is regarded the most reliable radiometric date for the site because of the congruence with the temperate faunal material (in particular roe deer) and the stratigraphic position above the LST and in the lower part of the pre-Holocene high-flood loams. Thus, the Final Palaeolithic site at Bad Breisig was probably settled in the late Allerød, possibly at the transition to the Younger Dryas. The connections of the activities of the remaining site suggested that even though several episodes were involved, a single occupation event was reflected by this material.

Summary and context

In summary, the sites from the Central Rhineland were usually set near a watercourse, although only few sites were located immediately in the floodplain (**tab. 10**). The amounts of lithic material recovered from the sites were heterogeneous (**tab. 11**). This diverse numbers was not only a matter of the size of the excavated area as the varied concentrations from Niederbieber demonstrated.
In general, the resources for the raw materials were the same (**tab. 12**) but material from the very distant class was not found on the FMG sites. Moreover, the variety of lithic raw materials used at a single site decreased. Although the diversity of core types remained similar, the dominant type was less rigidly chosen on the FMG sites than on the Late Magdalenian sites (**tab. 13**) On all sites and in almost all concentrations the LMP were the most numerous artefact class or the second most numerous class (**tab. 14**). In the Late Magdalenian the burins were usually the second most numerous class, whereas in the FMG sites end-scrapers were generally this second most numerous class. The activities on the sites were consequently altered to produce a higher percentage of end-scrapers than burins. However, numerically the end-scrapers were more frequent on Late Magdalenian sites than on FMG sites and the burins were also regularly present in FMG assemblages. In fact, the FMG sites Niederbieber and Andernach displayed a larger variety in the most numerous classes on the level of concentrations.
Faunal remains were preserved regularly on the sites in the Central Rhineland (**tab. 15**) and yielded occasionally reliable information about the season of death for various animals (**tab. 16**). In contrast to the lithic retouched material, the faunal remains on FMG sites were homogeneously dominated by red deer (*Cervus elaphus*). The Late Magdalenian assemblages were predominated by horse (*Equus* sp.). Thus, the major spe-

cies horse (*Equus* sp.) was replaced by red deer (*Cervus elaphus*). Concerning the possible resource of furs, the arctic fox (*Alopex lagopus*) appeared rather replaced by badger (*Meles meles*) than by red fox (*Vulpes vulpes*). However, the connection of the latter species to the archaeological material was vague and, perhaps, the winter furs of arctic foxes were a resource which was discarded without substitution.
Organic artefacts were heterogeneously preserved (**tab. 17**). A diverse set of organic tools was usually found in the Late Magdalenian assemblages, whereas in the FMG assemblages rarely more than one type of organic artefacts was found.
Among the special goods were regularly engravings as well as jewellery (**tab. 18**). Although the presence of haematite was almost continuously attested, its use was more widely attested in the Late Magdalenian. Comparable to the organic artefacts, the types of special goods were more diverse in the Late Magdalenian than in the FMG assemblages. The occurrence of figurines ended between the Late Magdalenian and the FMG.
Hearths were the most commonly observed structures on the Central Rhineland sites (**tab. 19**). Evident structures were common on Late Magdalenian sites but rare on FMG sites resulting in the more ephemeral character of the FMG concentrations. However, in these concentrations latent structures were often reconstructed.
Finally, among the sites from the Central Rhineland were some of the most numerously ^{14}C-dated sites in the Lateglacial of north-western Europe (**tab. 20**). Nevertheless, many dates produced arbitrary results which required some evaluation.

Besides the material presented previously, further sites were attributed to the period from the Late Magdalenian to the FMG in the extended Rhineland region (e.g. Bosinski/Richter 1997, 25. 27-32) such as the small Late Magdalenian assemblage of the layer V in the Wildscheuer cave (Terberger 1993, 115-140; Floss 1994, 261-263). However, the attempt of dating fauna material from this horizon failed to meet the minimum requirements. In addition, the documentation of the finds and the stratigraphy at the Wildscheuer cave was occasionally poor prohibiting to test the mixed character of the material for its originality. Thus, the material could take a possible intermediate position and represent an industry of the transformation period. Alternatively, the observation of two Magdalenian horizons in the Wildscheuer cave by the first excavator Karl August von Cohausen in 1874 could have been correct and, thus material from two different events of which one might represent a considerably younger occupation (early CBP Group?) were admixed in the following century. Thus, the integrity of the material can be doubted but cannot be tested due to the lack of documentation and, therefore, the material is not further considered in the present study. Furthermore, a small piece of elk antler which was transformed into a fish figurine was collected in a valley (Leerschlucht) north of the Wildscheuer (Terberger 1993, 186f. Taf. 76). According to traces of fire on the piece, Karin Terberger assumed the figurine to originate from a nearby cave filling. She suggested a Holocene origin due to the raw material and the absence of clear stylistic parallels, although she could not exclude an Upper to Final Palaeolithic age completely (Terberger 1993, 187). In the context of elk occurring in Gönnersdorf V this piece could well represent an expression from the phase of transformation and, hence, a reexamination and, perhaps, a ^{14}C dating of this specimen could be of interest. In contrast, other previously supposed Late Magdalenian assemblages were meanwhile dated to other stages (Baales/Street 1998; Baales 2002, 10-12). In addition, many sites revealed singular activities such as a *Federmesser* shot into the ground at Miesenheim 2 (Street 1986), a possibly lost retoucher at Leutesdorf (Baales 2002, 13f.), or six hearths with no associated archaeological material or only few archaeologically relevant pieces covered by the LST (Berg 1994). These finds highlighted the roaming of people in the landscape, but due to the scarcity of this type of evidence patterns allowing the analysis of changes cannot be established from these sites. Therefore, they are not useful for the present study.

site	approximate / average distance to water	type of water body	morphographic setting	site type	direction of cave opening / exposure
Saint Mihiel	close	river	slope	rock shelter/ open air	NW-W-S
Bois Laiterie	close	stream	slope	cave	N

Tab. 23 Topographic characteristics of sites from the western uplands. **immediate** 0-29m; **close** 30-149m; **near** 150-500m; **distant** 500>m; **×** not restricted; **N** north; **S** south; **W** west; **E** east.

Further sites often considered in reviews of Rhineland material such as Alsdorf or Kartstein fall into the western upland zone of the present study (see below and p. 180-182). In addition, south of the study area the assemblages from Fußgönheim containing shouldered points (Stodiek 1987) were associated with the transitional period in the Rhineland. However, the material originated partially from collections and partially from test pits which revealed stratigraphic disturbances at the site affecting the position of the archaeological material. Thus, besides the location outside the study area, the material from Fußgönheim cannot be evaluated concerning its material integrity based on literature studies.

The western upland zone

In the archaeological record from the Central Rhineland only few concentrations were attributed to the transition period between the Late Magdalenian and the FMG. Thus, further sites are necessary to establish a chronologically dense web of archaeological assemblages across this period to characterise the progress of the transition. These archaeological assemblages have to originate from areas where a comparable development as in the Central Rhineland can be assumed to provide information on the same transition process. These areas can be identified by the comparison of the Late Magdalenian and the FMG material. Material from these periods which is particularly comparable to the material in the Central Rhineland was found in northern France meaning the Picardy and the Paris Basin as well as in the upland zone between the Central Rhineland and northern France. Substantial archaeological material from the period between the Late Magdalenian and the FMG was found in northern France (see Material-Archaeology-Northern France, p. 182-244) but also in the upland zone some sites were attributed to this period. However, only few of these sites appeared helpful in the description of the transition between the Late Magdalenian and the FMG.

Saint Mihiel, Meuse

Research history

At Saint Mihiel, a first excavation near the rock *La Roche Plat* was conducted in 1886 by Dr. Mitour. Besides traces of a hearth, some hundred lithic artefacts, animal bones, and reindeer antler were recorded. This excavation was situated in the northern part in front of a monolith. Another excavation by Dr. Lenez in the late 19th century uncovered some lithic flakes and reindeer antlers from the southern part. During the 1st World War some defensive works (trenches) were installed beside the monoliths and, thus, presumably in parts disturbed the thin stratigraphy (Stocker et al. 2006, 25).

From 1965 to 1970 a test trench of 12 m² on the southern end and an excavation on the northern side provided further stratigraphic information as well as archaeological material (Tixier 1968; Tixier 1973; Stocker et al. 2006, 25). In addition, the spoil heaps of the old excavations as well as of the defensive works were successfully examined for lithic material and bones (Tixier 1968). The presented material was not strictly dis-

site	excavated m^2	total ≥ 10 mm (all)	cores and core fragments	retouched artefacts	% of burnt artefacts to total ≥ 10 mm (all)	ref.
Saint Mihiel	12	602 (1,802)	6	16	×	1
Bois Laiterie	31.5	1,814 (3,369)	4	254	×	2-3

Tab. 24 Numbers of lithic artefacts recovered at sites from the western uplands. – For further details see text. – References (ref.): **1** Stocker et al. 2006; **2** Straus/Orphal 1997; **3** Sano/Maier/Heidenreich 2011.

tinguished between the old excavations, the collection from the spoil heap, and the more recent excavation. However, refitted material suggested the cohesion of the archaeological remains.

Topography

The rock named *La Roche Plat* is one of seven monolithic blocks, which are situated on the eastern side of the Meuse at the northern exit of the town Saint-Mihiel. Today the rock rises about 25 m high above the ground and offers no shelter (**tab. 23**; Stocker et al. 2006, 25). Probably, only a small overhanging roof had existed which collapsed in the Late Pleniglacial as the stratigraphy seems to indicate (layer IIa, cf. Stocker et al. 2006, 26). The terrace at the base of the rock was approximately 5 m wide and situated some 12 m above the Meuse bank which is situated at the end of the slope, some 80 m west of the terrace. There the Meuse flows from a comparably meandering part with a narrow valley in the south into a wide, less curved valley in the north. Due to this position the view from the rock towards the north-west is several kilometres wide. Towards the south-west a small basin is formed by a dry valley which entered the Meuse valley from the east.

Stratigraphy

The stratigraphy is comparably short. Above the calcareous rock a yellow sandy deposit followed which in the lower most 5-6 cm (layer III) was free of gravel and calcareous pieces but towards the rock wall formed into a breccia (Stocker et al. 2006, 26). On top followed some 15 cm of yellowish sands (layer II). Within these sands 15-20 cm large calcareous blocs occurred along with medium sized gravels. This deposit was rich in small mammal remains. The archaeological material was spread over the whole thickness of this layer. Towards the rock this layer was overlain by cryoclastic debris (layer IIa) which reached a maximum thickness of 45 cm. On top of this layer an up to 30 cm thick deposit of brown soil mixed with calcareous rubble formed. This layer thinned out to 3 or 4 cm towards the end of the terrace. There the soft modern topsoil directly overlaid the archaeological layer. Thus, admixture of fossil and modern bones, antlers, and also lithic artefacts occurred in this part (Stocker et al. 2006, 25).

Archaeological material

Although the site was relatively small, a diverse archaeological assemblage was recovered. However, the lithic assemblage was relatively small (**tab. 24**).

A Western European flint of good quality was presumably accessible at the opposite side of the Meuse (**tab. 25**; Stocker et al. 2006, 31). The near presence of the raw material resource was considered as an explanation to the lack of cortical flakes because the initial knapping stage would have taken place at the resource site (Stocker et al. 2006, 31). However, some single blanks were made of non-local material, generally of further Western European flint varieties but also of a indurated sandstone. Comparable material to the indurated sandstone was found some 18 km up the Meuse (Stocker et al. 2006, 31-34). For the foreign flint material an origin in the westward Marne region as well as in the Vosges, east of the site, was considered (Stocker et al. 2006, 31). In particular, the retouched artefacts were usually made of the foreign material (Stocker et al. 2006, 31).

site	distance classes of raw material exposure from the site					ref.
	0-5 km	6-15 km	16-60 km	61-250 km	251 km >	
Saint Mihiel	**Western European flint**; quartzite		indurated sandstone (18 km > S)	Western European flint (tools; 75 km > W or 110 km > E)		1
Bois Laiterie	psammite		***flint*** *(35 km > N-NE or 60 km > W); type 9 flint (25-30 km S)*	**flint** (60 km > W); type 9 flint (75 km > S-SW); fossil molluscs (e.g. Dameray, 160 km > SSW or Valmondois, 235 km > SW)		2-3

Tab. 25 Approximate distance classes of raw materials recovered at sites from the western uplands. The most numerous lithic raw material(s) is/are set in bold. Raw materials of which only single artefacts were found are marked as (single). Possible but not probable resources are set in italics. – For further details see text. – References (ref.): **1** Stocker et al. 2006; **2** Straus/Orphal 1997; **3** Miller/López Bayón 1997.

site	blank production	cores total	1 platform	2 platforms, 1 preferred	2 platforms	3 and more platforms	others (e.g. fragments)	ref.
Saint Mihiel	blades/bladelets	6	2	2	2	0	0	1
Bois Laiterie	blades	4	0	1	1	0	1	2-3

Tab. 26 General concept of blank production and core types recovered at sites from the western uplands. **x** – this type is not mentioned or displayed in a publication. – For further details see text. – References (ref.): **1** Stocker et al. 2006; **2** Straus/Orphal 1997; **3** Sano/Maier/Heidenreich 2011.

site	total	no. of retouched artefact groups	LMP	end-scrapers	truncations	borers	burins	composite tools	others	ref.
Saint Mihiel	16	4	**5**	**5**	0	3	3	0	0	1
Bois Laiterie	254	7	**114**	22	24	21	**31**	2	20	2-3

Tab. 27 Numbers of retouched lithic artefacts recovered at sites from the western uplands. The two most numerous classes are set in bold. – For further details see text. – References (ref.): **1** Stocker et al. 2006; **2** Straus/Orphal 1997; **3** Sano/Maier/Heidenreich 2011.

Although the resource was very close, only six cores were found at the site (**tab. 26**). Nevertheless, in regard to the total numbers in this small inventory these represent a significant number. However, the main preparations of the cores were presumably made elsewhere. The cores were not exhaustively used but had produced several larger flakes and blades. The shaping of these cores was relatively diverse (Stocker et al. 2006, 26-28) with two typical pyramid cores with one mainly used platform for the production of blades, two cores with a single platform for the production of bladelets, and a further two cores which were worked from two platforms to receive elongated flakes (cf. Tixier 1968, 347).

The retouched inventory comprised 16 retouched artefacts with the pieces from the previous excavations included (**tab. 27**; Stocker et al. 2006, 31). Furthermore, at least six blanks were used so intensely that they wear macroscopic use-wear patterns (Stocker et al. 2006, 31). The most numerous classes (n = 5) were LMP and end-scrapers. Among the LMP the fragments of backed points were clearly dominating. The blunted edge of these implements was made continuously and curved as well as discontinuously and angled. However, in one case, the piece seemed to display an impact fracture but it could also be classified as a burin on truncation (Stocker et al. 2006, 33 fig. 7.1). The end-scrapers were usually made on long blades. Furthermore, the three burins were made on truncation (n = 2) and on a breakage (n = 1). In addition, three borers were found of which one was identified as a perforator and two were attributed to the *bec* type. However, these were not very massively made, although one might represent a proper *Zinken*. Perhaps, this very limited inventory was substituted at the site, whereas the blanks of the local raw material were of greater

site	Saint Mihiel	Bois Laiterie*
Ovibos moschatus		+ (teeth)
Rangifer tarandus	++	+
Equus sp.	+	+
Alces alces		+
Cervus elaphus		+
Capreolus capreolus		?
Capra ibex	+	+
Rupicapra rupicapra		+
Saiga tatarica	+	?
Bison priscus	+	+
Canis lupus / familiaris		(+)
Alopex lagopus / Vulpes vulpes		+
Alopex lagopus	+	
Meles meles		+
Ursus arctos / speleaus	+	(+)
further carnivores	+	(+)
Lepus sp.	+	+
Aves	+	+
Pisces		+
ref.	1	2-4

Tab. 28 Mammal species recovered at sites from the western uplands. For symbols see **tab. 15**. ***** In this only the species attributed to YSS and BSC are included. – For further details see text. – References (ref.): **1** Stocker et al. 2006; **2** Gautier 1997; **3** Deville/Gautier 1997; **4** van Neer 1997.

site	season	indicator	ref.
Saint Mihiel	spring/autumn	reindeer foetus; reindeer antler	1
Bois Laiterie	summer – autumn	cementum increment analysis of reindeer teeth; migratory birds and fish	2-4

Tab. 29 Seasonality of sites from the western uplands as indicated by faunal material. – For further details see text. – References (ref.): **1** Stocker et al. 2006; **2** Stutz 1997; **3** Deville/Gautier 1997; **4** van Neer 1997.

importance and taken from the site. Nevertheless, a hammerstone made of quartzite was also found in the concentration and further confirmed the processing of lithic material on the site (Stocker et al. 2006, 28). Perhaps, the quartzite was taken from the gravels of the Meuse.

However, the main work on the site was probably related to bone and antler. Almost 500 antler remains of reindeer (*Rangifer tarandus*) were accumulated on the site (**tab. 28**). Five wore traces of use as a resource (Stocker et al. 2006, 34). Of the 926 bone fragments, 532 were determined (Stocker et al. 2006, 36). The majority was attributed to some seven reindeer (n = 224). According to the presence of a reindeer foetus as well as the presence of antlers still adjacent to the skulls two seasons were identified spring and autumn (**tab. 29**). However, the shed of antlers is also dependent on the nutrition, the climate, and the age of the reindeer. Therefore, a single occupation cannot be excluded completely. In addition, 41 remains were attributed to birds. For the other determined remains only the identified species were given: Horse (*Equus* sp.), bison (*Bison priscus*), saïga (*Saiga tatarica*), ibex (*Capra ibex*), arctic fox (*Alopex lagopus*), brown bear (*Ursus arctos*), hare (*Lepus* sp.), and polecats (*Mustela* sp.; Stocker et al. 2006, 36). This assemblage reflected a typical Late Magdalenian faunal assemblage. In addition, the analysis of small mammals also suggested a cold period environment (Stocker et al. 2006, 36f.).

The preservation of the surfaces was limited. Carnivore marks were only found on three pieces suggesting that these agents played only a minor role for the accumulation of the material. However, only a hare bone and on a rib of a medium-sized artiodactyl, presumably reindeer, cut-marks were observed (Stocker et al.

site	points				fish hook	sewing instruments		knives	bâtons	axes	hammers	others	ref.
	bevelled	barbed, single row	barbed, double row	others (incl. tip fragments)		needles	awls						
Saint Mihiel				? (b)					+				1
Bois Laiterie	+ (a)			+ (b)		+ (b)	+ (b+a)		+			+ (b+a)	2

Tab. 30 Modified organic material from sites from the western uplands. For symbols see **tab. 15**. Material abbreviations are: **(b)** bone; **(a)** antler; **(i)** ivory. – Within the brackets: **/** or; **+** and. – For further details see text. – References (ref.): **1** Stocker et al. 2006; **2** López Bayón et al. 1997.

site	figurines	engravings			colour				jewellery					ref.
		»cut« cortex	figures / symbols on portables	figures / symbols on walls	spots	paintings on gravels	paintings on walls	colourants / powder	molluscs	amber / jet	drilled teeth	incised teeth	pendants	
Saint Mihiel			+											1
Bois Laiterie		+	+		?	?		?	+				+ (s)	2

Tab. 31 »Special« goods recovered at sites from the western uplands. For symbols see **tab. 15**. Material abbreviations are: **(s)** coarse grained stone material; **(l)** lithic material; **(am)** amber; **(j)** jet, **(b)** bone; **(a)** antler; **(i)** ivory; **(h)** herbivore; **(c)** carnivore. – Within the brackets: **/** or; **+** and. – For further details see text. – References (ref.): **1** Stocker et al. 2006; **2** Lejeune 1997

2006, 36). Moreover, only a single fragment originated from a possible point with a groove made of antler (**tab. 30**). Nevertheless, few pieces wore grooves or engravings. Since several pieces wore figural engravings of two horse heads, a mammoth head, and an eye (**tab. 31**; Stocker et al. 2006, 34) the question remained whether these grooves were functional or ornamental. Nevertheless, the figural art was clearly comparable to the ones known from slate plates in the Central Rhineland (see p. 89 and p. 110). Thus, even though this large amount of material was gathered at the site, it displayed only few signs of use.

In summary, this assemblage appeared a typical Late Magdalenian one which was perhaps deposited at a special task camp where the remains of one or several reindeer hunting episodes were partially cached. However, the dominant presence of backed points among the LMP in this small assemblage was surprising.

Spatial organisation

During the 19th century excavation a hearth was observed which during the 1960s excavation was possibly sustained by a concentration of very small, burnt bone fragments found near the limit of the earlier excavation (**tab. 32**; Stocker et al. 2006, 26). Within the excavation two, possibly three discrete clusters were found. The one close to the rock yielded the majority of lithic material, whereas the second one formed a south-westward extension with a dense accumulation of bone and antler remains. This accumulation thinned out towards the west and was then followed by another small cluster. Comparable to the evidence from Niederbieber 12 (Gelhausen 2011b), an artefact free space was found in the northern part of the lithic accumulation, perhaps, also suggesting the seat of a flintknapper. However, this distribution indicated a relatively unaltered horizontal preservation. Presumably, the small site was used as post-processing a hunt in an area with sufficient high-quality material. Moreover, a cache-like installation for reindeer antler was also formed at the site.

site	evident structures					latent structures		grave	further	ref.
	pave-ment	stone set-ting	pit	hearth, stone packed	hearth, sediment alteration	hearth, latent	dwelling			
Saint Mihiel						?			? (cache?)	1
Bois Laiterie	+					+				2

Tab. 32 Structures on sites from the western uplands. For symbols see **tab. 15**; except **?** means possible but anthropogenic formation or relation to the archaeology is uncertain. – For further details see text. – References (ref.): **1** Stocker et al. 2006; **2** Straus/Martinez 1997.

site	lab. no.	years ^{14}C-BP	± years	material	species	comment	years cal. b2k	ref.
Saint Mihiel	Lv-2096	13,160	110	antler	*Rangifer tarandus*	bulked sample	16,740-15,540	1
Bois Laiterie	GX-20434	12,665	96	bone		YSS base	15,580-14,660	2
Bois Laiterie	OxA-4198	12,660	140	antler	*Rangifer tarandus*	double bevelled point, YSS	15,690-14,410	2-3
Bois Laiterie	GX-20433	12,625	117	bone		YSS top	15,590-14,430	2
Bois Laiterie	*GX-21380*	*9,235*	*85*	bone	*Homo sapiens*	REJECT: Mesolithic burial from Breccia base	×	4

Tab. 33 ^{14}C dates from sites from the western uplands. Rejected date is shaded grey and set in italics and, in addition, the main reason for rejection is given in comment. For further details see p. 265-269 and text. The dates were calibrated with the calibration curve of the present study (see p. 358-364) and the calibration program CalPal (Weninger/Jöris/Danzeglocke 2007). The result range of 95 % confidence is given for the calibrated ages (years cal. b2k). – References (ref.): **1** Stocker et al. 2006; **2** Charles 1996; **3** Hedges et al. 1994; **4** Krueger 1997.

Chronology

According to the environmental indicators and the ^{14}C date (**tab. 33**), the material was clearly attributed to the Late Pleniglacial. However, for how long the cache was collected remained unclear and the dating of a bulked sample of these antler fragment could overestimate the occupation period of the site.
Yet, the spatial distribution of the modern excavation suggested only a single workshop used presumably during a short period of time. Perhaps, further short-termed workshops were lost in the early excavations. Nevertheless, since the material recovered from these collections fitted in the excavated material, the lithic material seemed attributable to a single episode. Thus, the small lithic inventory sustained the hypothesis of a single event. However, the relatively large and diverse faunal assemblage could indicate several short visits to the site, possibly during the main migration period of reindeer in spring and autumn (Stocker et al. 2006, 36).

Bois Laiterie, Namur

Research history

The small cave Bois Laiterie was discovered in 1990 by Philippe Lacroix who in 1991 excavated some small test pits (Léotard/Lacroix 1997, 11). During this test excavations he realised that previous works by »pot hunters« had only caused superficial disturbance, whereas the Palaeolithic layers generally remained intact. During two field seasons in 1994 to 1995 an area of 23 m² was cleared from the remains of the former works in the cave and subsequently comprehensively excavated. This area encountered parts with remaining sediment deposits inside the cave as well as on the terrace in front of the cave. In addition, an area of 7 m² as well as a one-and-a-half square-metre wide test pit were examined in a passage towards an upper part of the cave and in the upslope part of the terrace (Straus 1997, 54). These additional areas produced few and no finds suggesting that the remains of the small camp were almost completely uncovered by these excavations. The supervision of these works were shared by Lawrence G. Straus and Philippe Lacroix with further assistance by Jean-Marc Léotard and Marcel Otte.

Topography

The small double cave is situated on the southern bank of a steep valley formed by the stream Burnot which is a small tributary to the Meuse (**tab. 23**). The confluence of the two watercourses is situated only 500 m westwards of the cave. The cave is cut into the Carboniferous limestone belonging to the Belgian Ardennes (Straus 1997, 25). The cave is situated c. 35 m above the present valley floor which was perhaps different in the Lateglacial (Straus 1997, 25) and above the entrance of the cave the hills rise approximately another 100 m. Two northward facing entrances lead into the cave system (Straus 1997, 28). Late Pleniglacial archaeological material was only found within the lower of the two entrances. The cave chamber opening behind the lower entrance sized at maximum 45 m² (Straus 1997, 30): It was 4-8 m wide, around 9.5 m deep, and approximately 2.5 m high during the Late Pleniglacial. Due to a connection to the upper chamber a constant draught was present inside this chamber (Straus 1997, 30). Thus, this small cave appeared as a rather uncomfortable place and, possibly, the ground was moist and required a pavement inside the cave. However, in comparison with the Lahn valley caves (see p. 133-138) a pavement inside a cave appeared common for a Late Magdalenian context. Moreover, the location at the cave entrance represented a strategically well chosen place for observation.

Stratigraphy

In Bois Laiterie as in most cave stratigraphies, the sequence was complex partially due to the protection inside and the erosion outside the cave (Straus 1997, 41-51). Inside the cave the bedrock was overlain by 7-15 cm of archaeologically sterile, grey sand (BGS). On top followed reddish clayey sands (RS) of up to 20 cm thickness containing almost no archaeological remains. In the area situated immediately inside the cave mouth grey sandy lenses were formed by disintegrated, decalcified scree (GL, YSS-grey lens or pseudo LGS). In the same area, these lenses were overlain by a pocket of silty clays (BSC) which contained artefacts. Inside the cave, 20-40 cm of coarse, grey sands (LGS) followed on top. This deposit was sterile of artefacts and was not documented outside the cave or in the cave mouth area. Overlaying was a yellowish-red to orange-brown sandy silt (YSS) which contained along with artefacts and faunal remains also some rock plates. This layer was 15 to 40 cm thick. On top followed various deposits depending on the topographic position: Inside the cave, 10-20 cm of grey sands (UGS) which contained few pieces of archaeological material were separated from YSS by some roof fall blocks. In a few profiles inside the cave, 15-25 cm of a grey-beige silt deposit (GBS) followed on top of UGS. This deposit was almost sterile of archaeological material. Along the cave wall GBS was partially covered by a 75-100 cm thick breccia containing pottery and bones, among which also human bones were identified. In the entrance area, the following light brown silts (LBS) showed already evidence of disturbance. On the terrace again large blocks delimited YSS and were overlain by reddish-brown colluvium (RC) which contained some amount of scree. Even though the archaeological remains were recovered from two stratigraphically distinct layers, the material appeared in a horizontal and vertical continuity, was typo-technologically indistinct, and could at least in parts be refitted. Hence, the remains from Bois Laiterie originated either from a single occupation event or more plausibly from several closely-spaces, nowadays indistinguishable episodes (Straus/Martinez 1997, 90). However, a projection of the finds in an approximate N-S profile indicated a significant downslope movement of a dense artefact band from inside to outside the cave (cf. Straus/Martinez 1997, 78 fig. 11). A micro-morphologicial analysis of the deposits from Bois Laiterie suggested that the material from YSS was deposited in a climatically variable period where severe cold episodes could still occur and, in fact, the layers following on top of the deposits containing the majority of archaeological remains were attributed to severe cold periods (Courty 1997). The palynological as well as the anthracological results from the site were rare and determined mainly as open forest species such as *Betula* sp. or *Corylus avellana* (Embry-Barbier 1997; Pernaud 1997). However, the intense sediment movements recommend caution about the relation of these remains and the archaeological unit. In fact, although the malacological analysis suggested the attribution of the YSS deposit to the onset of the Lateglacial Interstadial, the malacologist also considered a Preboreal disturbance of the material (López Bayón/Lacroix/Léotard 1997). This hypothesis was accompanied by the assumption based on the small mammals that the material was chronostratigraphically disturbed (Cordy/Lacroix 1997).

Archaeological material

Originally, the assemblage from Bois Laiterie was unambiguously attributed to the Late Magdalenian (Straus/Orphal 1997; López Bayón et al. 1997; Lejeune 1997). However, meanwhile some objections to this ascertained attribution occurred mainly based on the lithic inventory which yielded similarities to the Final Magdalenian faciès Cepoy-Marsangy (Sano/Maier/Heidenreich 2011).

In general, the lithic inventory (n = 3,369) was made of excellent quality material (**tab. 24**). The majority (91-96 %) is chalk flint, probably of Cretaceous age which originate perhaps from 35 km north to north-eastwards or some 60 km to the west (**tab. 25**). The only other raw material of considerable numbers (3-6 %) is a flint variety (type 9) which seemed to be secondary deposited and possibly originated from secondary deposits 25-30 km southwards or resources some 75 km south- south-westwards.

Only four cores were found (**tab. 26**). These cores were very small and exhaustively exploited. Still some very long blades were recovered at the site. Two of the cores were exploited from two platforms to receive blades or bladelets and display preparations of the platform (Sano/Maier/Heidenreich 2011). However, only few platform rejuvenation flakes and crested blades were found in the assemblage (Straus/Orphal 1997) indicating that the blank production was no important task at the site. The profiles of the blanks, the forming of their butts, and the shape of the bulb suggested the use of organic knapping instruments but predominantly of a soft hammerstone (Sano/Maier/Heidenreich 2011).

266 retouched ends were identified on 254 pieces (**tab. 27**). The biggest group was formed by LMP (n = 114) of which the majority were simple backed blades/bladelets (n = 91; Sano/Maier/Heidenreich 2011; cf. Straus/Orphal 1997). A truncated backed bladelet and three curve-backed points were already in the original publication observed, one point was identified as Microgravette (Straus/Orphal 1997, 237 fig. 12.21) and the other two as Azilian points (Straus/Orphal 1997, 255 Photo 1). In fact, one was a bipoint. The reanalysis of the lithic inventory with special emphasis on the traceological indicators revealed a total of 23 very heterogeneous points (Sano/Maier/Heidenreich 2011). Furthermore, the 35 burins were made more commonly on truncations (n = 14) than dihedral ones (n = 9). 24 blanks had at least one truncation. The 22 end-scrapers were various shapes but several (n = 10) were also retouched laterally. Some end-scrapers were made on very long blades (pers. comm. Katsuhiro Sano, Tokio; Straus/Orphal 1997, 235 fig. 10.5). Among the 21 borers several *becs* (n = 6 + 1 composite perforator-*bec* piece) occurred. Only two composite tools with various tool classes were found: a burin-end-scraper and a perforator-truncation. Among the other tools are laterally retouched, notched and denticulated pieces, as well as a raclette and two splintered pieces.

A first traceological analysis examined 129 artefacts of which eleven showed traces of use (Jardón Giner 1997). In a more recent study, 256 artefacts were analysed but 215 were too heavily patinated to produce reliable results (Sano 2009; Sano/Maier/Heidenreich 2011; Sano 2012b). Both study indicated that the

lithics were mainly used on hard material and that several pieces wore impact scars suggesting the use as projectile implements.
The 788 plates found in Bois Laiterie were made of a type of sandstone (psammite) which was available up the hills (Miller/López Bayón 1997). In total, the stone material weighted more than 120 kg. 26 of the plate specimens were engraved with lines and signs or wore traces of colour (Lejeune 1997).
The assemblage from Bois Laiterie yielded a rich faunal assemblage, of which many pieces could be identified (**tab. 28**). In addition to the large mammals, small mammals, frogs, and molluscs were analysed contributing to the environmental setting of site. However, some of the faunal remains spread over several stratigraphic units and, thus, their connection with the archaeological material remained often unclear. Therefore, the larger fauna from the YSS and BSC deposits were considered most informative for the Lateglacial human presence. The most numerous remains from the cave were determined as fox remains (*Alopex lagopus/Vulpes vulpes*; n = 184). Nevertheless, this small carnivore could have used the cave as shelter in winter and, thus, be a natural intrusion. Further small carnivores such as stoat (*Mustela erminea*; n = 6) as well as badger (*Meles meles*; n = 10), wolf (*Canis lupus*; n = 1), and lynx (*Lynx lynx*; n = 5) were recovered from the site (Gautier 1997). In how far further animals such as cave bear (*Ursus spaeleus*; n = 1) and cave hyena (*Crocuta crocuta*; n = 1) were older intrusions would require further AMS dates. In equal numbers as the foxes were the remains of arctic hare present (*Lepus timidus*; n = 143). However, horse (*Equus* sp.) remains were most numerous among the larger mammals in these deposits (n = 54) suggesting the presence of at least two horses (Gautier 1997). In total, three horses were assumed at Bois Laiterie and, perhaps, a European ass (*Equus hydruntinus*) was also found among the remains (Gautier 1997). Although reindeer (*Rangifer tarandus)* remains were slightly less numerous (n = 44), these specimens were attributed to a minimum of three animals. An ibex (*Capra ibex*) was indicated by 28 remains. Eleven elements were attributed to chamois (*Rupicapra rupicapra*), although these pieces might be over-represented. Additional small ruminant remains could be attributable to this species, to ibex, or less probable to saïga antelope (*Saiga tatarica*) or roe deer (*Capreolus capreolus*). A further eleven teeth and teeth fragments from musk ox (*Ovibos moschatus*) were attributed to a single animal. Furthermore, five remains were determined as originating from an elk (*Alces alces*), three remains from a bison (*Bison priscus*), and a further three pieces from two red deer (*Cervus elaphus*). Among the fauna attributed to the YSS and BSC deposits were also several bird species (Deville/Gautier 1997). These remains comprised willow grouse (*Lagopus lagopus*; n = 23), jackdaw (*Corvus monedula*; n = 9), partridge (*Perdix perdix*; n = 3), greylag goose (*Anser anser*; n = 2) as well as five elements of a large duck (*Anser* sp.), and buzzard (*Buteo* sp.; n = 2). Furthermore, single elements of falcon (*Falco* sp.), long-eared owl (*Asio otus*), eagle owl (*Bubo bubo*), whimbrel (*Numenius phaeopus*), swallow (*Hirundo rustica*), and also a domestic fowl (*Gallus gallus*) were found in these deposits. Greylag goose as well as whimbrel are migratory birds and were assumed as possibly nesting in Belgium during the Late Pleniglacial (Deville/Gautier 1997, 217). In addition, of 66 fish remains 37 pieces were identified as brown trout (*Salmo trutta fario*), a further 20 specimens were determined as burbot (*Lota lota*), and a further three remains were attributed to a graling (*Thymallus thymallus*; van Neer 1997). The distance of the cave to the streams as well as the dimensions of the fish was assumed as hindering the introduction of these remains by smaller carnivores and, therefore, a human introduction to the site was considered most plausible (van Neer 1997, 207). This faunal composition is clearly comparable to the assemblages from Gönnersdorf and the lower horizon of Andernach (see **tab. 15**, p. 85-87, and p. 107-109; cf. Street et al. 2006).
Moreover, some pieces yielded marks of carnivore gnawing (Gautier 1997, 183) which could indicate a natural introduction of these remains to the site. Furthermore, the calcination of a few bones was also considered as a possibly natural exposure to fire (Gautier 1997, 183). However, a conservative search for indications of human modification on the relatively fragmentary remain yielded an astralagus of reindeer,

a mandible of horse, and a humerus of wild boar with cut-marks (Gautier 1997, 183). The latter originated from a stratigraphically unclear situation but was attributed to a Holocene occupation due to its biogeographic history and its habitat preference (Gautier 1997, 184). Nevertheless, at least for the Paris Basin wild boar is meanwhile confirmed for the early Lateglacial Interstadial (Bignon/Bodu 2006).

Six mammal teeth including reindeer (n=2), ibex (n=2), musk ox (n=1), and elk (n=1) were analysed in a cementum increment analysis to provide indications for the death period of these animals (**tab. 29**; Stutz 1997). However, only the reindeer teeth indicated a summer or, perhaps, autumn kill which was in accordance with the presence of migratory birds such the greylag goose or possible migratory fish (cf. Deville/Gautier 1997).

Twelve pieces show traces of human modification (López Bayón et al. 1997). Five fragments were single-bevelled points which were all made of reindeer antler (**tab. 30**). Another point fragment was made of bone. Furthermore, several pieces made of bone as well as antler were identified as awls, needles, an assumed needle container, and remains of the antler spall production. Several of these specimens wore parallel cut-marks, perhaps for ornamental reasons.

Of the eight, fossil shells five were perforated (**tab. 31**). These pieces originated presumably from Tertiary deposits in the Paris Basin (Lejeune 1997).

Finally, a small perforated plate comparable to the ones made of slate in Gönnersdorf (Bosinski 1977) was made of a polished sandstone (Lejeune 1997).

Spatial organisation

A pavement was installed inside the cave and on the terrace by psammite plates of various sizes which were partially refitted (**tab. 32**; Miller/López Bayón 1997). This evident structure was a considerable investment. Burnt material was related to the grey lenses, immediately outside the cave and were considered as indications of a hearth (Straus/Martinez 1997, 96). Many activities were performed around this fire but a second activity area was assumed for the cave wall inside the cave. However, the refits suggested the connection of these areas and made investigators assume that the various clusters were created quasi-contemporary (Straus/Martinez 1997, 112). According to the traceological analysis, the presence of faunal remains, and the composition of the lithic inventory, several activities related to the processing of one or several successful hunting events can be assumed.

Chronology

A ^{14}C-dated human bone originated from a Mesolithic burial which was presumably dug into the Magdalenian horizon (Krueger 1997). However, the ^{14}C dates formed a highly consistent cluster (**tab. 33**). These three, statistically indistinguishable dates emphasise the hypothesis of a single occupation event or quasi-contemporaneous events. These dates were very reliable indicator for the human presence, in particular, because one of the samples was made on an organic artefact. The refits, the dense cluster of the profile projection, the size of the assemblage, the composition of the lithic inventory, the activities, and the horizontal distribution suggested a single occupation event which according to the seasonal indicators occurred in the warm period of the year. In regard to the massive structures and depending on which faunal remains were actually associated with the human presence, a longer occupation period than only a temporary hunting camp could be assumed.

Based on the diverse archaeological inventory, the evident structures, and the dating, the site cannot be unambiguously attributed to a Lateglacial archaeological unit. Bois Laiterie combines characteristics of the Late Magdalenian, the MfCM, and Azilian elements. This heterogeneity could indicate a palimpsest which developed perhaps during the transition period. However, a single inventory from a period of change cannot be excluded. Therefore, this assemblage is considered more generally as Final Magdalenian for the moment.

In the upland zone west of the Central Rhineland archaeological material was mainly found in caves or rock shelters. Open air material was often confined to surface collections. The valley bottoms were often traversed by water courses. Therefore, the morphographic setting of relevant sites was often in a position between the hill tops and the moist valley bottoms, i.e. on the slopes (**tab. 23**).
Moreover, due to the topographic conditions such as rock walls the extent of the excavation areas was often restricted (**tab. 24**). However, the numbers of the recovered lithic material were only partially influenced by this restriction. Presumably, the tasks performed in the excavated area had a greater influence. For instance, in Saint Mihiel where a more than 2.5 times smaller area than in Bois Laiterie was excavated 50 % more cores were found than in Bois Laiterie. This higher numbers as well as the higher proportion of material smaller than 1 cm suggest that the blank production took on a more important role in Saint Mihiel than in Bois Laiterie. This different focus of the two sites is also reflected in the raw material composition (**tab. 25**). Flints from the local gravels dominated in Saint Mihiel, whereas regional to distant flint varieties were generally preferred in Bois Laiterie. However, in both assemblages the blank production process was mainly focused on blades and bladelets (**tab. 26**).
The numbers as well as the composition of the retouched artefacts (**tab. 27**) further sustained the impression of different functions of the two assemblages.
In both assemblages faunal material was preserved (**tab. 28**). The composition of species at Saint Mihiel was clearly dominated by reindeer (*Rangifer tarandus*), whereas the composition at Bois Laiterie was much more diverse.
The seasonal indicators from both sites suggested human presence during the warmer period of the year from spring to autumn (**tab. 29**). Saint Mihiel was probably visited at the beginning and/or end of this period, whereas the occupation of Bois Laiterie occurred presumably within the warm period.
Comparable to the lithic tool inventory, the organic tools were more frequent and more diverse at Bois Laiterie than at Saint Mihiel (**tab. 30**). The materials with potential symbolic meaning (**tab. 31**) were also more common at Bois Laiterie than at Saint Mihiel.
Settlement structures were more evident at Bois Laiterie than at Saint Mihiel (**tab. 32**). Perhaps, the duration of the occupations were in addition to the function different between these two sites and, therefore, more effort was invested in the organisation of Bois Laiterie.
According to the ^{14}C dates (**tab. 33**), the assemblage at Saint Mihiel was deposited some centuries previous to the material from Bois Laiterie. However, the relation of the ^{14}C date from Saint Mihiel to the human activity seemed ambiguous and a closer temporal relation of the two assemblages is possible.

Besides the two presented sites, further inventories from the area west and east of the Rhineland were discussed in the context of the Late Magdalenian, the FMG, and the transition between the former two. Some important assemblages are mentioned briefly in the following. In the central Sieg valley, east of the Central Rhineland, an engraved retoucher was collected from the surface (Heuschen et al. 2006). On both sides of the piece a schematic animal silhouette was engraved. Both engravings were identified as probable elk. In reference to the engravings on slate plates in Gönnersdorf, the dating of the elk bone from the south-western area at Gönnersdorf, and the occurrence of elk in the art of the Lateglacial (cf. Veil et al. 2012), the retoucher was attributed to the Lateglacial Interstadial. However, this attribution relied only on typo-technological assumptions and lacked an independent and precise dating. Test pits and surface surveys near the findspot produced no conclusive results (Heuschen 2007).

In archaeological summaries of the extended Rhineland (e.g. Bosinski/Richter 1997, 25. 27-32) several sites from the Eifel region were frequently considered such as the Late Magdalenian site at Alsdorf (Löhr 1995a), possible Late Magdalenian and CBP Group artefacts in the Kartstein sequence (Baales 1996), or the small lithic CBP Group assemblage from the Katzensteine (Löhr 1995b).
The site at Alsdorf (North Rhine-Westfalia) was recovered in 1974 by an amateur archaeologists and, subsequently, excavated by Hartwig Löhr (Löhr 1995a). The excavation yielded 9,567 lithic artefacts among which 107 cores and 361 retouched pieces were found which were techno-typologically identical with the Late Magdalenian material from Central Rhineland (Löhr 1995a). Furthermore, structural similarities existed with Gönnersdorf and the lower horizon at Andernach. In addition to a possible pavement, pits were observed in which charcoal was preserved. However, the excavator assumed the charcoal to be a more recent intrusion and further organic material was not preserved. Hence, the site was not ^{14}C-dated but the stratigraphy and the structural and techno-typological comparison clearly indicated a chronological attribution to the Late Pleniglacial, similar to the Central Rhineland Late Magdalenian. The geographic setting of this site is of some interest since Alsdorf is situated at the transition from the upland to the lowland zone. Additionally, the setting in the resource area of good-quality Western European flint is in regard to the connections to the Central Rhineland notable. Moreover, based on the Alsdorf material Hartwig Löhr developed a model of the functional differentiation of Palaeolithic and Mesolithic assemblages depending on the duration of the settlement (Löhr 1979).
The lithic material from the Kartstein sequence was excavated in the early 20th century according to the standards of this time. Thus, the single finds lack most stratigraphic information (Baales 1996) and, therefore, the pieces could not be attributed reliably.
The material 1969-1970 excavated in the Katzensteine could not be unambiguously attributed to a chronostratigraphic position due to their occurrence in reworked sediment (Löhr 1995b).
In the Lower Rhineland, at the transition from the Eifel uplands to the North European Plain, large-scale surface surveys in relation to brown coal quarrying fields, produced several concentrations of possible Magdalenian material (Nehren 2001). In particular, the concentration at Altdorf-Güldenberg WW 95/79 yielded indications for fire use and an engraved schist plaquette along with some 300 lithic tools of non-local flint (Nehren 2001). The lithic inventory contained no backed pieces. However, burins, *becs,* and end-scrapers were present which in combination with an engraved small plate suggested an affiliation with the Late Magdalenian. Further thorough examinations of the probable Lateglacial concentrations at this site are going to be published and, perhaps, add on the knowledge about the use of the landscape by Lateglacial hunter-gatherers (e.g. Nehren 2001, 58). The currently available information on this material was too little for the present study and, therefore, the site was excluded. In general, a precise and independent dating of the concentrations in these survey areas remained difficult due to the often rather limited stratigraphic depth and the poor organic preservation. Hence, further material which might relate to the period between the Late Magdalenian and the FMG (Höpken 1995; Thissen/Krull/Weiner 1997) was also neglected in the present study.
In the Belgian and the Dutch part of the North European Plain some sites were recovered which were discussed in the context of some type of Creswellian (Charles 1999; Barton et al. 2003). The classic British Creswellian was attributed to a Final Magdalenian and, thus, the period between the Late Magdalenian and the FMG (see p. 56-74; Barton et al. 2003). Consequently, the Belgian and Dutch sites could additionally be considered in the present study. However, these assemblages lacked usually an independent and precise chronostratigraphic determination and, moreover, many of these sites were found outside the study area in a different geomorphological zone.
At the transitional zone from the Belgian and Dutch part of the North European Plain to the upland region some classic Late Magdalenian sites such as Kanne, Orp, Sweikhuizen, Mesch, and Eyserheide were found

(Rensink 1995; Rensink 2012). These sites were probably related to lithic procurement. Further sites recovered in this transitional area east of the German border were, for instance, the site Bois de St. Macaire at Obourg (Letocart 1970) or, perhaps, also the collection from Mont de l'Enclus at Orroir (Charles 1997a). Material from the uplands between the Central Rhineland and northern France such as the remains found in the caves of Engis (4e Grotte d'Engis, also: Caverne funaire; Charles 1997a) and those of Goyet (Toussaint/Becker/Lacroix 1998; Stevens et al. 2009a) was also discussed in the context of a Final Magdalenian. Due to the often poor documentation of excavations conducted early in the research history many assemblages represented rather collections with little additional information. These collections were occasionally shown to be the result of various archaeological events from very different periods (Stevens et al. 2009a). Therefore, the integrity of the material cannot be tested and the chronostratigraphic attribution remained vague. Further sites from the western upland zone such as Sy Verlaine were presumably disturbed (Dewez 1987; Charles 1996). In other sites such as the Trou de l'Ossuaire the chronostratigraphic indications were contradictory and/or the archaeological material was heavily disturbed by previous excavations (Leotard/Otte 1988; Charles 1997a). Some excavated sites such as Caverne de Bois de la Saute yielded material which suggested a position in the period between the Late Magdalenian and the FMG but the deposition of these assemblages were not well dated (Charles 1997a; Charles 1999). Furthermore, very small inventories such as the single block and its knapping debris at the Trou Jadot (Toussaint et al. 1993) allowed no detailed classification except for the suggestion of a very brief halt.
Farther to the south, the site Roc-la-Tour I was attributed to the Late Magdalenian and comparable to Alsdorf (Rozoy 1988). This site was also set in the Rhenish slate mountains and, comparable to Gönnersdorf or the lower horizon from Andernach, slate plates were recovered from this site. Moreover, these plates were also partially engraved. However, comparable to Alsdorf, the chronostratigraphic attribution relied only on typo-technological consideration because the acidic soil had destroyed all organic materials which could have helped with an independent attribution (Rozoy 1988, 139). Moreover, in this southern part of the upland zone between the Central Rhineland and northern France several sites yielded archaeological material which possibly originated from the period between the Late Magdalenian and the FMG (Huet/Thévenin 1995; Guillot/Guillot/Thévenin 1995; Huet et al. 1995; Spier 1997; Charles 1997a; Guillot/Guillot/Thévenin 2000; Spier 2000). These assemblages were frequently recovered in surface collections. This recovery made only typo-technological considerations about the inventories possible and lacked independent dating or spatial information. Nevertheless, the presence of these sites indicated a probably continuous settlement area from the Central Rhineland to northern France during the Late Magdalenian in particular and suggested in general the habitual use of this intermediate area by Lateglacial hunter-gatherers.

Northern France

The archaeological record is one of the main lines of evidence in the present study because it reflects the change in human behaviour. However, for the transition period between the Late Magdalenian and the FMG only few assemblages from the Central Rhineland were found and provided little evidence (see Material-Archaeology-The Central Rhineland, p. 75-170). Equally, in the western upland zone very few sites yielded reliable material from this transition period (see Material-Archaeology-Western uplands, p. 170-182). Therefore, assemblages from northern France meaning the Picardy and the Paris Basin which yielded material from this period were additionally considered in the present study. In this area also Late Magdalenian and FMG assemblages of comparable faciès as in the Central Rhineland were found suggesting a single social network during the Lateglacial.

site	approximate distance to water	type of water body	morphographic setting	site type	direction of cave opening / exposure
Hallines	immediate/close	stream	slope	open air	×
Le Grand Canton	immediate	river	valley bottom	open air	×
Bonnières-sur-Seine	near	river	foothills in valley bottom	rock shelter	N
Étigny-Le Brassot, south	immediate	river	valley bottom	open air	×
Le Tureau des Gardes	immediate	river	valley bottom	open air	×
Le Closeau	immediate	river	valley bottom	open air	×
Cepoy	immediate	river	valley bottom	open air	×
Marsangy	immediate	river	valley bottom	open air	×
Belloy-sur-Somme	immediate	river	valley bottom	open air	×
Gouy	close	river	slope	cave	NW
Pincevent III.2	immediate	river	valley bottom	open air	×
Conty	immediate	river	valley bottom	open air	×
Hangest-sur-Somme III.1	immediate	river	gravel heap in valley bottom	open air	×

Tab. 34 Topographic characteristics of sites in the Paris Basin. **immediate** 0-29 m; **close** 30-149 m; **near** 150-500 m; **distant** 500 > m; **×** not restricted; **N** north; **W** west. – For further details see text.

Hallines, Pas-de-Calais

Research history

Near Hallines, in the area where northern France gradually merges into the lowlands of the North European Plain, the collector J. Boutry found in 1968 a few lithic artefacts and bones which came up during construction works on the property of J. M. Levert (Fagnart 1997, 33). As a result a rescue excavation of 5 m² (Fagnart 1997, 35) under the supervision of Alain Tuffreau was conducted on the property. Test pits which were dug in 1969 on the other side of the road produced no further results. In 1977 canalisation work alongside the road also on the Levert property yielded only few further artefacts.

Topography

The rescue excavation area was located alongside the road from Hallines to Wisques on the northern slope of the valley formed by the Aa (**tab. 34**). The site was situated some 20-25 m above the valley floor at the foot of the Aa's middle terrace which formed a comparably steep slope to the north-east (Fagnart 1997, 33). From north-west to south-east the slope was cut immediately west of the site by a small thalweg within which the road was built.

Stratigraphy

Due to the topographic position the lithostratigraphy was not completely comparable on both sides of the road. On the excavated side of the valley a 1.5 m thick gravel deposit with various lenses of yellowish sandy silts was found on top of light yellow silts. According to the malacological analysis, the latter were deposited in pleniglacial conditions (Fagnart 1997, 33 f.). Within the gravel deposit several archaeological horizons were observed. In the present study, two levels which were found within the sandy lenses and which were separated by only 7 cm of small gravels are considered. In the 1990s, Jean-Pierre Fagnart analysed the material from the two horizons conjointly because he regarded the material as very homogeneous from a typological as well as a stylistic point of view (Fagnart 1997, 35). Nevertheless, the two distinct horizons suggested that this assemblage was deposited by at least two different, temporally succeeding episodes. The gravel deposit was overlain by a 10 cm thick layer of flint gravels and brown sandy clay on top of which a 1 m thick brown-

sites	excavated m² (p.r.n. relevant)	total ≥ 10mm (all)	cores & core fragments	retouched artefacts	% burnt artefacts of total ≥10mm (all)	ref.
Hallines	5	790	46	122	×	1
Le Grand Canton, lower horizon	c. 15	619	3	16	×	2
Le Grand Canton, upper horizon	1,020	24,556	1,015	1,029	×	2-3
Le Grand Canton, sector 1	60	1,021	188	113	×	2-3
Le Grand Canton, sector 2, upper horizon	504	23,535	827	916	×	2-3
Bonnières-sur-Seine*	20-30 (5-10)	1,316	7	13	×	4
Étigny-Le Brassot, south	201	8,252	present	c. 100	mentioned	5
Le Tureau des Gardes	463	60,912	942	3,442	×	6-7
Le Tureau des Gardes, *locus* 6	5	4,263	60	249	×	6
Le Tureau des Gardes, *locus* 7	140	7,551 (15,984)	84	784	× few	7
Le Tureau des Gardes, *locus* 10	85	14,091	276	1,188	×	6
Le Closeau, lower horizon	c. 1,000	5,650>	47>	495**	c. 3.4	8-9
Le Closeau, *locus* 4+50	182>	2,475	20	262**	3.6	8
Le Closeau, *locus* 46	193	2,309	23	201**	3.3	8
Le Closeau, greyish layers	25,568	41,397	827	823**	19.9	8
Le Closeau, *locus* 25 (top)	26	126	1	14	6.4	8
Cepoy	355	c. 30,000	355	c. 267	×	3; 10-11
Cepoy, sector 1	205	c. 14,500	252	142	mentioned	3; 10
Cepoy, sector 2	150	<15,000	103	125>	frequently	11
Marsangy<	213	21,618	379	642	mentioned	12-13
Marsangy N19	58	10,389	197	313	mentioned	12-13
Belloy-sur-Somme, lower horizon	c. 470	6,415	108	158	×	1
Gouy	?	116	1	16	×	14
Pincevent III.2	c. 280 (70)	489 (561)	10	31	18.4 (16.0)	15-16
Conty, lower horizon	100	(c. 2,000)	3	42	some	1
Hangest-sur-Somme III.1, lower horizon	150	839 (1,945)	21	100	some	1

Tab. 35 Numbers of lithic artefacts recovered at sites in the Paris Basin. Sub-assemblages of a site are shaded grey. **×** this type is not mentioned in a publication; ***** numbers refer to the complete material from the site including the old excavations, therefore, the complete excavation activities are given but the modernly excavated square-metres were also given in parentheses; ****** numbers do not include the pieces with diverse traces of modification and, therefore, can differ from the numbers in the publication. – For further details see text. – References (ref.): **1** Fagnart 1997; **2** Valentin et al. 1999b; **3** Valentin 1995; **4** Barois-Basquin/Charier/Lécolle 1996; **5** Lhomme et al. 2004, 733; **6** Lang 1998, 90; **7** Weber 2006; **8** Bodu 1998; **9** Bodu/Debout/Bignon 2006; **10** Allain et al. 1978; **11** Wenzel 2009; **12** Croisset/Schmider 1992; **13** Schmider 1992b; **14** Bordes et al. 1974; **15** Bodu/Orliac/Baffier 1996; **16** Orliac 1996b.

yellowish silt layer followed. This layer was separated by a 5cm thick band of chalk and flint fragments from a brownish sandy silt deposit of 1m thickness. Above this deposit 40 to 80cm of modern soil had formed.

Archaeological material

From the two horizons and including the various collections 995 lithic artefacts were found, of which 790 were larger than 2cm (**tab. 35**).

The major raw material was the local Senonian flint (**tab. 36**; Fagnart 1997, 35) but a few artefacts, especially tools, were made of a beige coloured silex which presumably came from the Tertiary plateau in the central Paris Basin (Fagnart 1997, 37). In this case, this chalcedony originated from at least about 130km south of the site.

Among the lithic artefacts, 46 cores were identified of which 29 were fragments or rather nodules with few knapped flakes (**tab. 37**). Only one prismatic core had two platforms, nine were single platform prismatic cores. Generally, blades were the aim of the blank production. The blanks often displayed an *en éperon* butt

site	distance classes of raw material exposure from the site					ref.
	0-5 km	6-15 km	16-60 km	61-250 km	251 km >	
Hallines	**Senonian flint**			Tertiary chalcedony (130 km > S)		1
Le Grand Canton, lower horizon	**Senonian gravel flint**		Tertiary chalcedony (45 km > WNW)	Tertiary chalcedony (85 km > SW)		2
Le Grand Canton, upper horizon	**Senonian gravel flint**; sandstone; granite; Nautilus sp. (c. 5 km SW)		gritstone (40 km > NW); *Tertiary chalcedony (45 km > WNW)*	Tertiary chalcedony (85 km > SW); Eocene mollusc (65 km > N or 65 km > NE)		2
Bonnières-sur-Seine	**local flint**; local chalcedony	local Cretaceous flint (?)				3-4
Étigny-Le Brassot, south	Senonian gravel flint					5
Le Tureau des Gardes	**local Cretaceous flint**; *Tertiary chalcedony;* sandstone		Tertiary chalcedony (20 km > WNW)			6
Le Closeau	**Campanian flint**; *Tertiary chalcedony* (not present in *locus* 25)		Tertiary chalcedony (25 km > WSW or 35-60 km > S)			7
Cepoy	**local Cretaceous gravel flint**; gritstone; sandstone; quartzite		Tertiary chalcedony (45 km > N or 60 km > WSW)			3; 8
Marsangy	**local Cretaceous flint**	local Campanian flint (6 km > NW)	quartzitic sandstone (35 km > SE); glossy sandstone (sinlge, 35 km > SE)	jasper (single, 65 km > SW); Tertiary chalcedony (80 km > NW); fossil molluscs (80 km > N; 90 km > NW; 130 km > W)		6-8
Belloy-sur-Somme	**local Turonian and Coniacian flints**					1
Gouy	local Senonian flint	high-quality Cretaceous flint (unknown origin)	**high-quality Cretaceous flint (unknown origin)**; Tertiary chalcedony (25 km > SE)			3; 9-10
Pincevent III.2	**local gravel flint**; *granite (single)*		Tertiary chalcedony (40 km > WNW)	Tertiary chalcedony (80 km > SW); granite (single, 150 km > SSE; 300 km > E)		11-12
Conty, lower horizon	**local Turonian flint**					1
Hangest-sur-Somme III.1 lower horizon	**local Turonian and Coniacian flints**					1

Tab. 36 Distance classes of raw materials recovered at sites in the Paris Basin. The most numerous lithic raw material(s) is/are set in bold. Raw materials of which only single artefacts were found are marked as (single). Resources for which several places of origin are discussed are set in italics. – For further details see text. – References (ref.): **1** Fagnart 1997; **2** Julien/Rieu 1999; **3** Valentin 1995; **4** Barois-Basquin/Charier/Lécolle 1996; **5** Lhomme et al. 2004, 737; **6** Lang 1998, 91; **7** Bodu 1998, 43-47. 335; **8** Wenzel 2009; **9** Bordes et al. 1974; **10** Fosse 1997; **11** Bodu/Orliac/Baffier 1996; **12** Orliac 1996b.

(Fagnart 1997, 37). The blades seemed generally to have been knapped with an organic hammer (Fagnart 1997, 37). However, the preparation of the cores was presumably made with a hammerstone.

Among the 122 tools were only three pieces attributable to the LMP (**tab. 38**). These three implements were pointed Magdalenian blades. The retouched pieces were dominated by 55 burins. Most of these were stroke on a retouch (n = 32), often on a lateral retouch or a notch, only once on a truncation. Several of the

site	blank production	cores total	1 platform	2 platforms, 1 preferred	2 platforms	3 and more platforms	others (e.g. fragments)	ref.
Hallines	blades	46	**9**	×	1	×	29	1
Le Grand Canton, lower horizon	bladelets/ blades	3	×	×	×	×	×	2-3
Le Grand Canton, upper horizon	bladelets/ blades	1,015	common	**dominant**	present	present	tested blocks	2-3
Bonnières-sur-Seine	blades	7	**1**	×	×	×	×	4
Étigny-Le Brassot, south	blades	×	**dominant**	present	present	×	×	5
Le Tureau des Gardes	blades/ bladelets	942	**common**	**common**	present	×	few abandoned fragments	6
Le Closeau, lower horizon	blades	47>	×	×	×	×	×	7
Le Closeau, greyish layers	blades/ flakes	827	×	×	×	×	×	7
Le Closeau, *locus* 25	blades	1	0	1	0	0	0	7
Cepoy	blades	254	14	10>	<21	×	9	2; 8
Cepoy, sector 1	blades	151	12	8>	**<20**	×	9	2
Cepoy, sector 2	blades	103	**2**	**2**	1	×	×	8
Marsangy	blades	379	**164**	×	<126	48	0	9
Belloy-sur-Somme, lower horizon	blades	108	19	**23**	18	1	47	1
Gouy	blades	1	0	1	0	0	0	10
Pincevent III.2	blades/ bladelets	10	3	**4**	2	0	1	11-12
Conty, lower horizon	blades	3	1	**2**	0	0	0	1
Hangest-sur-Somme III.1, lower horizon	blades	21	**6**	3 >	< 3	1	7 (c.)	1

Tab. 37 General concept of blank production and core types recovered at sites from the Paris Basin. Sub-assemblages are shaded grey. The most numerous class is set in bold. On some sites not all cores and core fragments were classified according to the number of platforms. In these cases the pieces displayed in the figures were classified. **×** this type is not mentioned or displayed in a publication. – For further details see text. – References (ref.): **1** Fagnart 1997; **2** Valentin 1995; **3** Valentin et al. 1999b; **4** Barois-Basquin/Charier/Lécolle 1996; **5** Lhomme et al. 2004, 733; **6** Lang 1998; **7** Bodu 1998; **8** Wenzel 2009; **9** Croisset/Schmider 1992; **10** Valentin 1995; **11** Bodu/Orliac/Baffier 1996; **12** Orliac 1996b.

burin blows on retouch ran transversally to the knapping direction of the blank (n=17). In addition to these specimens, ten dihedral burins, five burins on a broken edge, four burins on a natural edge, and two burins on the cutting edge (*Corbiac*) occurred. Furthermore, two burins were classified as mixed ones. Additionally, 21 burin spalls were found indicating the production of burins on the site. The only composite tool was a burin-end-scraper. End-scrapers formed the second largest group of tools (n=25). These specimens were usually made on long blades or blade fragments which were not laterally retouched except for one piece. The third major group of tools were borers (n=23) which were dominated by heavily retouched pieces, i.e. *bec/Zinken* (n=20). In addition, truncations (n=2), denticulated blades (n=3), laterally retouched (n=8), and splintered pieces (n=2) occurred in small numbers.

Furthermore, a small flint nodule was probably used as hammerstone.

In addition, fragments of heat altered sandstones were recovered during the rescue excavation.

Even though the faunal material from the site was comparably well preserved only some ten pieces were recovered (**tab. 39**). This assemblage was dominated by mammoth (*Mammuthus primigenius*) remains of which three vertebrae, two humeri, a part of a tusk, and some teeth fragments were identified. Besides these finds, only an equid tooth (presumably *Equus* sp.) and a rib fragment of an animal with the size of a red deer (cf. *Cervus elaphus*) were found. Nevertheless, the relation of these faunal remains with the archaeological material remained unclear (Fagnart 1997, 42).

sites	total	LMP	end-scrapers	trun-cations	borers	burins	composite tools	others	ref.
Hallines	122	3	**25**	2	23	**55**	1	13	1
Le Grand Canton, lower horizon	16	**8**	2	0	1	**5**	0	0	2
Le Grand Canton, upper horizon	1,029	174	**209**	32	123	**349**	70	72	2-3
Le Grand Canton, sector 1	113	**29**	22	4	7	**33**	10	8	2-3
Le Grand Canton, sector 2, upper horizon	916	145	**187**	28	116	**316**	60	64	2-3
Bonnières-sur-Seine	13	**4**	2	0	**4**	2	1	0	4
Étigny-Le Brassot, south	c. 100	**23**	**25**	×	×	15	×	×	5
Le Tureau des Gardes	3,442	**1,057**	509	184	442	**596**	104	528	6-7
Le Tureau des Gardes, *locus* 6	249	**122**	21	18	**29**	22	4	**33**	6
Le Tureau des Gardes, *locus* 7	784	**313**	**121**	47	80	86	16	**121**	6-7
Le Tureau des Gardes, *locus* 10	1,188	**336**	157	71	144	**231**	38	211	6
Le Closeau, lower horizon	495**	**81**	**98**	11	0	29	8	**268**	8
Le Closeau, *locus* 4 (+ 50)	252 (262)**	**40 (45)**	**46 (49)**	9 (10)	0 (0)	23 (23)	4 (4)	**130 (131)**	8
Le Closeau, *locus* 46	201**	**28**	**49**	0	0	4	3	**117**	8
Le Closeau, greyish layers	823**	**480**	**119**	41	24	31	2	66	8
Le Closeau, *locus* 25	14	**14**	0	0	0	0	0	0	8
Cepoy	c. 267	49	**68**	28	**50**	29	4	41	3; 9
Cepoy, sector 1	142	25	**39**	17	**34**	18	4	5	3
Cepoy, sector 2	125>	**22**	**29**	11	16	11	0	**36**	9
Marsangy	642	**168**	68	61	111	**179**	17	38	10
Marsangy N19	313	**86**	30	35	67	74	4	17	10
Belloy-sur-Somme, lower horizon	158	20	**23**	7	10	**21**	0	**77**	1
Gouy	116	**6**	1	2	**3**	**3**	1	0	11
Pincevent III.2	32 (35)	3 (5)	**8**	**7**	0	2	0	**11 (12)**	12-13
Conty, lower horizon	42	**24**	2	5	2	**6**	0	3	1
Hangest-sur-Somme III.1, lower horizon	100	**52**	6	7	0	**27**	0	9	1

Tab. 38 Numbers of retouched lithic artefacts recovered at sites in the Paris Basin. Sub-assemblages are shaded grey. The two most numerous classes are set in bold (except for the »others« class which can result from various types involved, then the three most numerous groups are given). **×** this type is not mentioned nor shown in a publication; ****** numbers do not include pieces with diverse traces of modification and, therefore, can differ from the numbers in the publication. – For further details see text. – References (ref.): **1** Fagnart 1997; **2** Valentin et al. 1999b; **3** Valentin 1995; **4** Barois-Basquin/Charier/Lécolle 1996; **5** Lhomme et al. 2004; **6** Lang 1998; **7** Weber 2006; **8** Bodu 1998, 169-173. 322-339; **9** Wenzel 2009; **10** Croisset/Schmider 1992; **11** Bordes et al. 1974; **12** Bodu/Orliac/Baffier 1996; **13** Orliac 1996b.

Spatial organisation

According to the observations during the small rescue excavation, the material appeared to be *in situ* (Fagnart 1997, 35). An accumulation of heat altered sandstone fragments, some pieces of charcoal, and calcined bone fragments in the upper horizon suggested the presence of a hearth immediately outside the excavated area (**tab. 43**; Fagnart 1997, 35). The composition of the lithic inventory suggested more specialised activities at the site related in particular to burins and end-scrapers and, perhaps, explained the low number of LMP. However, the LMP were frequently found concentrated around a hearth (Gelhausen 2011b) and, thus, the numbers would possibly be higher if the hearth was excavated.

site	*Mammuthus primigenius*	*Rangifer tarandus*	*Equus* sp.	Cervus elaphus	*Megaloceros giganteus*	*Capreolus capreolus*	bovid	*Bos primigenius*	*Sus scrofa*	*Canis lupus*	*Vulpes vulpes*	further carnivores	*Lepus* sp.	further small mammals	Aves	Pisces	ref.
Hallines	++		?	?													1
Le Grand Canton, upper horizon	(+)	+	++				+			+							2-3
Le Grand Canton, sector 1		+	++							+							2
Le Grand Canton, sector 2, upper horizon	(+)	+	++				+			+							3
Bonnières-sur-Seine		+	++	?	+				+								4
Étigny-Le Brassot, south		++		++													5
Le Tureau des Gardes	+	+	++				+			+				+ (Lagomorpha, Spermophilus)	+		6-7
Le Tureau des Gardes, *locus* 6			++				+							+ (Lagomorpha, Spermophilus)	+		6-7
Le Tureau des Gardes, *locus* 7			++				+										6
Le Tureau des Gardes, *locus* 10		+	++				+			+				+ (Lagomorpha)			6-7
Le Closeau, lower horizon			++	+			+		+	+		+ (*Panthera spelea*)	+		+		8-9
Le Closeau, *locus* 4			++	+					+				+		+		8-9
Le Closeau, *locus* 46			++	+					+	+		+ (*Panthera spelea*)	+				8-9
Le Closeau, greyish layers			+	++		?						(+) (*Meles meles*)					9
Marsangy		++	+	+													10
Gouy				+		+			+	+	+				+		7; 11
Pincevent III.2				+				+		+		+ (*Meles meles*)	+				12-13
Conty, lower horizon				+		+		++									1; 14
Hangest-sur-Somme III.1, lower horizon			++					++									1

Tab. 39 Mammal species recovered at sites in the Paris Basin. Sub-assemblages of a site are shaded grey. For symbols see **tab. 15**. – For further details see text. – References (ref.): **1** Fagnart 1997; **2** Alix et al. 1993; **3** Bridault/Bemilli 1999; **4** Barois-Basquin/Charier/Lécolle 1996; **5** Lhomme et al. 2004; **6** Lang 1998, 86-90; **7** Bignon/Bodu 2006; **8** Bemilli 1998; **9** Bodu 1998, 234. 285; **10** Poplin 1992, 40-44. 733; **11** Cordy 1990; **12** Bodu/Orliac/Baffier 1996; **13** Orliac 1996b; **14** Coudret/Fagnart 2006.

Chronology

A ^{14}C date was made on a mammoth vertebra and produced a comparably imprecise and old date falling calibrated into the Late Pleniglacial (**tab. 44**). If the fauna and the archaeological material were associated, the assemblage would be one of the oldest indication of Lateglacial re-settlement in northern France (Fagnart 1997, 42). However, besides the uncertain connection of the fauna with the lithic industry, the use of fossil material by Magdalenian hunter-gatherers was another possible explanation for the very old age, especially, since this type of behaviour was known from some Late Magdalenian sites such as Gönnersdorf (Street et al. 2006). Therefore, the dating of the rare additional bone material might be of some interest.
The lithic inventory and the fauna material as well as the stratigraphic position were consistent with a Late Pleniglacial to early Lateglacial Interstadial age of the assemblage. However, the observation of two distinct horizons could stratigraphically indicate the presence of two occupation events and, therefore, Jean-Pierre Fagnart's remark (Fagnart 1997, 42) is supported that further excavation might be helpful to gather more detailed data for a more precise classification of this site.

Le Grand Canton, Marolles-sur-Seine, Seine-et-Marne

Research history

The site Le Grand Canton near Marolles-sur-Seine was discovered in 1989 during archaeological surveys which were a preliminary stage for the construction of the motorway A5 (Alix/Rieu 1999). From 1990 to 1991 a total of 1,020 m² was excavated under the supervision of Jean-Luc Rieu and Philippe Alix. The archaeological material was found in depressions of the underlying river gravels. The excavation was located in an area of approximately 100 m × 100 m and was divided in sub-sections of 25 m × 25 m. Within this area three sectors (1-3) were excavated. Sector 1 was situated in the sections 1 and 2 and was the most southern sector. Sector 2 was located some 20 m northwards and was with almost 600 m² the largest excavated sector. The majority of this sector was located in section 18 but some uncovered square-metres lay already in the sections 9, 10, 17, 19, and 27. Some 25 m north-westwards of sector 2 followed sector 3. This sector was mainly found in section 25 but also parts of section 24, 33, and 34 were uncovered within the works on this sector.
Sector 1 was excavated carefully with 3D-measurements, whereas in the other two sector this procedure could not always be maintained and some parts of section 18 as well as sector 3 were uncovered in rescue works with collection of the material per square-metre or quarter-square-metre (Alix/Rieu 1999). Nevertheless, in sector 2 erosive displacements were observed in the layer above the main archaeological horizon including admixture with Neolithic material. However, in the bottom of the depression the material from sector 2 appeared *in situ* and in particular, the various indicators for hearths were examined in more detail. Furthermore, in section 9 at the limit of this large depression, a second horizon was observed underneath the main horizon with very well preserved organic remains. The extent of this lower horizon seemed limited but only some parts of the sector could be explored for this horizon also due to time pressure. In sector 3, a low number of material was found in an alluvial deposit in which organic material was not preserved well. However, the limits of the site were not reached. The archaeological excavations were accompanied by geological and various environmental studies.

Topography

The site Le Grand Canton was found in a wide basin formed by the Seine and the Yonne which confluence about 3-4 km west of the site (**tab. 34**; Alix et al. 1993). The site was located about 1 km south of the town Marolles-sur-Seine. The modern Seine passes approximately 1.5 km northwards and the modern Yonne flows at a distance of some 700 m south-west of the site in a large meander. However, in the alluvial plain several palaeochannels of both rivers were observed (Alix et al. 1993, 197). In particular, along the Seine this plain was morphologically varied by sand banks rising up to 5 m above the river (Lang 1998, 10), whereas the terrain along the Yonne remained relatively level (Alix et al. 1993, 198). Eastwards of the site the terrain remains relatively levelled for some 2-3 km before it rises gradually. More than 2 km north of the site, on the northern bank of the Seine, hills rise some 50 m upwards. This plateau is frequently cut by small valleys, often running parallel to the Seine. About 2.5 km south, on the southern bank of the Yonne, the terrain rises approximately 60 m upwards. Hence, the site was situated in approximately the centre of a wide alluvial plain. The site Le Grand Canton was set only a few metres above the modern Yonne on a low gravel terrace. In the Lateglacial the floodplain of various palaeochannels had formed an area of sand and gravel mounds on which the Lateglacial hunter-gatherers could settle between the various watercourses (Valentin 1995, 228).

Stratigraphy

Due to this landscape which was constantly changed by the channels of the two large rivers, the stratigraphy of the site is complex and the vertical stratigraphy depended on the position within this floodplain. Five main profiles were taken in the three sectors. One stratigraphy was approximately directed W-E and another one N-S in sector 2 as well as in sector 3. In sector 1 only a NNE-SSW directed stratigraphy was recorded.

The N-S stratigraphy in sector 2 produced the most comprehensive sequence. In general, the stratigraphy was formed by five stratigraphic units (Deloze et al. 1999, 27-32): The bedrock was not reached but the lowest unit was formed by calcareous scree (unit V). On top were alluvial gravels, sands, and clay deposited (unit IV). These deposits were cryoturbated and capped by erosion which led to the formation of the depressions. The alluvial material was overlain by yellowish to yellow-brown sandy silts which were partially affected by the development of a soil (unit III). This pedogenesis transformed the silts into a greyish compact calcite silt deposit. This approximately 20 cm thick deposit contained the lower archaeological horizon. The compact greyish calcite silts were cut by another erosive phase. This phase formed a limit on top of which the sediments were affected by pedogenic processes, whereas the underlying deposits were partially deformed by cryoturbation. The compact silt deposit was covered by a calcium carbonate deposit resulting from the outwashing of carbonates from the upper layers during the pedogenesis. This brownish soil was overlain by a brown clay layer on top of which 10-20 cm of yellowish silts followed. These silts were covered by a greyish green outwash horizon which blended into dark brown clayish silts. This clayish silt deposit was superimposed by almost 2 m of brown to brown-orange silts which were filling up the depressions in the terrain. In the upper part of these silts a blackish layer had formed in some of the depressions and yielded mainly Neolithic material. In the sector 1, these brown silts were more sandy and admixed with small gravels, presumably, as a result of colluvial processes. The unit II comprised all the layers from the erosive phase to the brown silts. Within this unit the material of the upper archaeological horizon was scattered. The sequence was covered by some 0.4 m thick, brown-grey silts which formed the topsoil.

In the pollen samples which were taken along the NNE-SSW transect of sector 1 and in the N-S profile of sector 2 only a low number of pollen was preserved (Deloze et al. 1999, 32-35). Moreover, among the few preserved pollen were modern as well as pre-Quaternary ones infiltrated in the Lateglacial deposits. In consequence, no reliable palynological attribution of the deposits was possible. The preserved molluscs were also rare (Deloze et al. 1999, 35-38). In addition, an admixture with post-glacial material could also not be excluded by the malacological analysis, although the hypothesis of an admixture was neglected due to the depth of the samples (Deloze et al. 1999, 37). Nevertheless, the temperate species were dominant among the preserved material from the archaeological horizons suggesting either an admixture or an attribution to an interstadial period (Deloze et al. 1999, 37). In fact, the preserved species were mainly indicating a steppe environment supplement by species of a covered and a semi-forested landscape. In a protected floodplain, this type of landscape could also occur in a not too dry and/or too cold stadial period.

Archaeological material of the lower horizon

This horizon was only examined on a small area and, therefore, the assemblage attributed to this horizon was not very numerous (**tab. 35**). Furthermore, the material was not presented in detail.

The majority of the raw material originated from the local Senonian flint gravels (**tab. 36**).

Only three cores were identified in the material (**tab. 37**). These pieces were not displayed and a technological analysis of them was not published. However, the 16 retouched artefacts were displayed (**tab. 38**). Based on the blanks used for these artefacts as well as the dorsal surface of these artefacts a production of regular blades and bladelets was deducible for this horizon.

The retouched artefacts were dominated by burins. They were made on fragments of regular blanks and exclusively on truncation (cf. Valentin et al. 1999b, 67 fig. 22). One of two end-scrapers was also made on a regular blade and the other one on a fragment of secondary crested blade. The LMP found in the lower horizon were short fragments of very thin and very slim bladelets. Some of these fragments were only partially retouched. In addition, a single *Zinken* was also found in this horizon.

Although the organic material was well preserved, neither numbers for these remains nor determinations were given. However, in comparison to the upper horizon, horse (*Equus* sp.) was presumably a major species. At least the ^{14}C-dated sample originated from this species.

Even though only a limited part of the archaeological material was excavated and published, an attribution to the Late Magdalenian seemed probable. However, in regard to the extract character of the material this attribution can only be provisional.

Spatial organisation of the lower horizon

This horizon was only examined on a small sample area and, thus far, nothing was published concerning the spatial distribution of the material from this sub-area.

Chronology of the lower horizon

Very little is known about the chronostratigraphy of the lower horizon. However, the horizon clearly preceded the upper horizon. Nevertheless, the single ^{14}C date on a horse sample (Gif-9606) produced a similar result to the reliable dates from the upper horizon (**tab. 44**). Thus, even though the horizons were stratigraphically distinct this difference was below the resolution of radiocarbon measurements. Calibrated this result dated the horizon into the early Lateglacial Interstadial (GI-1e/GI-1d). This attribution was in accordance with the stratigraphic position in a deposit affected by pedogenic processes as well as cryoturbation. Due to the limited information on this material an intra-site chronology could not be examined.

Archaeological material of the upper horizon

The upper horizon at Le Grand Canton yielded a very rich corpus of archaeological material (**tab. 35**). The remains from sector 1 and 2 weighted almost 350 kg (Valentin 1995). Thus far, the inventory from sector 3 was not published.

The raw materials originated mainly from the gravel deposits accessible at the site (**tab. 36**; Valentin et al. 1999a). The Senonian flint gravels were relatively small and of a poor knapping quality. 32 pieces were made of a foreign Tertiary chalcedony of uncertain origin. This material occurred in particular in the upper

horizon of sector 2 (n = 30). However, the scarcity of debris material suggested that this material was either mainly imported as ready made blanks and few prepared cores or exploited in an unexcavated part of the site (Valentin et al. 1999b, 80).
The availability of the raw material immediately at the site led to a relatively high number of cores, core fragments, and merely tested blocks ($n_{sector2}$ = 51; **tab. 37**). The exploitation strategy was clearly adapted to this local origin and the poor quality of the material. For instance, a choice towards elongated gravels was observed which made possible an effective use with a minimised preparation of the cores (Valentin et al. 1999b, 81). However, some preparation were identified on the blanks which occasionally had a facetted butt or one with an *en éperon* preparation. The cores were generally exploited from one platform or at least one preferred platform (Valentin et al. 1999b, 82). The aim of the blank production process were blades as well as bladelets. Although the number of bladelet cores was higher in the analysed sample (98 bladelet cores, 61 blade cores, nine flake cores), almost 30 of these examples served previously for the production of blades (Valentin et al. 1999b, 83). In total, some 400 cores were used for the blade production and some 550 cores were used for the bladelet production (Valentin et al. 1999b, 85). However, taking the change of use into account the numbers for the cores were relatively even. In a final stage, some cores were used for the production of small, elongated flakes which demonstrated the opportunistic concept of the flintknappers (Valentin et al. 1999b, 84). However, this almost complete exploitation also revealed an exhaustive use of suitable raw material even though the material was accessible in large quantities at the site. According to the indications on the blanks, the knapping instrument used dominantly in the blade production was an organic hammer, whereas in the preparation of the cores various knapping instruments from an organic hammer to a hard hammerstone were used (Valentin et al. 1999b, 81 f.).
The formally retouched material comprised 1,029 artefacts among which burins were clearly dominant (**tab. 38**). Approximately two third were dihedral examples and less than 7.5 % of the burins were on broken edges (Valentin et al. 1999b, 69). About a quarter of the burins were made on truncations of various shaping. The burins were commonly made on thick flakes and irregular blades. In contrast, the end-scrapers were usually made on long and regular blades. Among the composite tools the combination of burins and end-scrapers was the most frequent one. The vast majority of the LMP were made on very thin bladelets (1-3 mm thick) and were mainly fragmented. No example with a truncation was displayed and only few pieces were also retouched on the opposite lateral edge. These pieces belonged to another group of backed bladelets which were made on thicker blanks (3-5 mm) and which were usually also preserved as longer pieces. Within the upper limit of the latter group fell also seven backed points which were found in the upper horizon of sector 2 (n = 3) and sector 1 (n = 4). These points were generally angle-backed and rarely curve-backed (Valentin et al. 1999b, 73). Furthermore, two pointed bladelets were recovered from sector 1 which seemed to represent partially retouched angle- and curve-backed points.
In addition to the lithic material, some 1,300 kg of mainly burnt stone material were recovered at the site (Valentin 1995, 155) of which some 970 kg were recovered from sector 2 (Julien et al. 1999, 132). Among this material sandstone and granite accessible in the gravels of the Yonne were dominant. Only single gritstones were used which could originate from the Plateau Briard some 40 km north-westwards (Valentin et al. 1999a, 42).
The upper layer produced a large quantity of organic remains (**tab. 39**). The material from sector 1 and two samples from the sector 2 were analysed in detail (Alix et al. 1993; Bridault/Bemilli 1999). The larger sample from sector 2 weighted already more than 83 kg. More than 60 kg were determinable material (n = 3,328) of which some 32 kg were teeth material. Equal relations were assumed for the second sample from sector 2 (Bridault/Bemilli 1999, 51). Horses (*Equus* sp.) were in all sectors clearly dominant and in sector 3 the only determined species (Alix et al. 1993, 200). A minimum of 74 individuals of horse were identified in the

site	seasonality	indicator	ref.
Le Grand Canton, upper horizon	all year but probably not in winter	age structure of reindeer	1
Le Tureau des Gardes	all year but less frequent in winter	teeth eruption and bones of young horses	2
Le Closeau, lower horizon	all year but less frequent in summer	teeth eruption stages of horses	3-4
Conty, lower horizon	late winter season	not mentioned	5

Tab. 40 Seasonality of sites in the Paris Basin according to various indicators. – For further details see text. – References (ref.): **1** Bridault/Bemilli 1999; **2** Bignon 2006; **3** Bignon/Bodu 2006; **4** Bodu/Debout/Bignon 2006; **5** Coudret/Fagnart 2006.

larger sample from sector 2, a further 43 individuals in the smaller sample (Bridault/Bemilli 1999), and some 17 individuals in sector 1 (Alix et al. 1993). In addition to horse, only reindeer (*Rangifer tarandus*) occurred in larger quantities with a total of some 15 individuals. Furthermore, two teeth from sector 1 and a distal humerus fragment from sector 2 were attributed to wolf (*Canis lupus*). Moreover, a metacarpal fragment originated from a large bovid (*Bos* sp./*Bison* sp.) and a molar fragment was determined as mammoth (*Mammuthus primigenius*). The bovid bone was found in an area of low material frequency and the relation to the human presence remained unclear. The tooth fragments of mammoth and wolf showed also no traces of modification. According to the age structure of the horses and, in particular, of the reindeer as well as the varied presence of body parts on the site, repetitive stalking hunting events in the floodplain were suggested the best explanation for the accumulation (Bridault/Bemilli 1999, 64). Moreover, the removal of meat rich body parts was at least for the horses considered (Bridault/Bemilli 1999, 57). Although this representation at least for the reindeer remains seemed partially be related to preservation issues (Bridault/Bemilli 1999, 55), a selective preservation was probably not very influential in the preservation of horse (Bridault/Bemilli 1999, 62). In addition, the surfaces were affected by various processes during the deposition and, thus, no cut-marks were observed. Furthermore, only a few fractures were assumed as probably intentional and no burning of bones was observed (Bridault/Bemilli 1999, 55. 63). Based on the age structure of the reindeer only the winter period was excluded as hunting season which was sustained by the scarcity of antler on the site (**tab. 40**). However, Anne Bridault and Céline Bémilli considered these indicators as very vague because the age structure could also result from the differential preservation (Bridault/Bemilli 1999, 57).

Seven fragments of nautilus (*Nautilus* sp.) were found but displayed no traces of use. However, they were considered as possible containers due to their curved shape (Valentin et al. 1999a, 43). Along with the possibility of an alluvial origin, an *in situ* source with approved presence of large examples of *Nautilus* (*Hercoglossa*) *danicus* is known some 5 km south-westwards in the Bois d'Esmans. The nautilus fragments at Le Grand Canton were found in two groups: One group lay in section 9 (n=4) and one group was found in section 17 (n=3). The pieces were found in close relation to accumulations of burnt stone material. Based on this punctual spatial occurrence of the fragments, an alluvial transport seemed not probable.

Furthermore, a perforated Eocene molluscs (*Bayana lactea*) was also found in the surrounding of a hearth (**tab. 42**; Valentin et al. 1999a, 45). These fossils could originate from the region of Meaux (Seine-et-Marne) some 60 km northwards or the region of Montmirail (Marne) some 65 km to the north-east.

Spatial organisation of the upper horizon

In sector 1 a dense scatter of lithic material, faunal remains, and burnt stone material was found but no spatial organisation of the material was observed (Rieu/March/Soler-Mayor 1999, 97). In the sector 3, the material was in general less numerous and relatively even spread revealing only some concentration of burnt stone material accompanied by lithic material (Alix et al. 1993, 200). Thus, the focus of the spatial analysis was on the sector 2. In total, some 14 concentrations of mainly burnt stones, occasionally accompanied by charcoal were recognised in sector 2 and at least nine of them were interpreted as hearth

site	points				fish hook	sewing instruments		knives	bâtons	axes	hammers	others	ref.
	bevelled	barbed, single row	barbed, double row	others (incl. tip fragments)		needles	awls						
Le Closeau, lower horizon								? (pointed ribs)				+	1-2
Conty, lower horizon										? (a)			3

Tab. 41 Modified organic material from sites in the Paris Basin. For symbols see **tab. 15**. – For further details see text. – References (ref.): **1** Bemilli 1998; **2** Bodu 1998, 234. 285; **3** Fagnart 1997, 112. 117f.

site	figurines	engravings			colour				jewellery					ref.
		»cut« cortex	figures / symbols on portables	figures/ symbols on walls	spots	paintings on gravels	paintings on walls	colourants / powder	molluscs	amber/ jet	drilled teeth	incised teeth	pendants	
Le Grand Canton, upper horizon									+					1
Bonnières-sur-Seine					+									2
Étigny-Le Brassot, north		+												3
Le Closeau, lower horizon	?	+			+			+						4
Cepoy		+	+ (s)											5
Marsangy	?	+	? (l)						+				(+)	6
Gouy		+	+ (s)	+	+		+				+ (h)			8
Pincevent III.2			+ (l)											9
Conty, lower horizon			+ (a)											10
Hangest-sur-Somme III.1, lower horizon								+						1

Tab. 42 »Special« goods recovered at sites in the Paris Basin. For symbols see **tab. 15**. Abbreviations are: **(s)** coarse grained stone material; **(l)** lithic material; **(am)** amber; **(j)** jet; **(b)** bone; **(a)** antler; **(i)** ivory; **(h)** herbivore; **(c)** carnivore. – Within the brackets: **/** or; **+** and. – For further details see text. – References (ref.): **1** Valentin et al. 1999a; **2** Barois-Basquin/Charier/Lécolle 1996; **3** Lhomme et al. 2004, 732; **4** Bodu/Cary 1998; **5** Allain 1975; **6** Cremadès 1992; **7** Schmider 1992c; **8** Martin 2007a; **9** Bodu/Orliac/Baffier 1996; **10** Fagnart 1997.

constructions (**tab. 43**; Rieu/March/Soler-Mayor 1999). These structures were generally small but dense clusters and contained between 10kg and 45kg of stone material. However, two larger structures (1 and 6) with each yielding more than 70kg of stone material were found in the depression in the south of sector 2 (Rieu/March/Soler-Mayor 1999, 96 Tabl. 19; Julien et al. 1999, 133 fig. 66). These structures were also surrounded by the densest concentration of lithic artefacts, whereas a dense ring of faunal remains was found in 1-2m distance to the centres of these stone accumulations. However, in detail the proportions of blank production debris, various retouched artefacts classes, and the vicinity of faunal remains

site	evident structures					latent structures		grave	further	ref.
	pave-ment	stone set-ting	pit	hearth, stone packed	hearth, sediment alteration	hearth, latent	dwelling			
Hallines				?						1
Le Grand Canton, up-per horizon				+						2-3
Bonnières-sur-Seine				+	?	+				4
Étigny-Le Brassot, south				?		+				5
Le Tureau des Gardes				+						6
Le Tureau des Gardes, *locus* 7				+						6
Le Closeau, lower horizon		+			+	?	+			7-9
Le Closeau, greyish deposits						+				7
Cepoy		?		+						10-11
Marsangy			? (nests)	+			+			12
Belloy-sur-Somme, lower horizon				+						1
Pincevent III.2					+	+				13-14
Conty, lower horizon						+				1

Tab. 43 Structures on sites in the Paris Basin. Sub-assemblages of a site are shaded grey. For symbols see **tab. 15**; except **?** means possible but anthropogenic formation or relation to the archaeology is uncertain. – For further details see text. – References (ref.): **1** Fagnart 1997; **2** Rieu/March/Soler-Mayor 1999; **3** Julien et al. 1999; **4** Barois-Basquin/Charier/Lécolle 1996; **5** Lhomme et al. 2004; **6** Lang 1998; **7** Bodu 1998, 323-329; **8** Bodu/Debout/Bignon 2006; **9** Jöris/Terberger 2001; **10** Allain et al. 1978; **11** Wenzel 2009; **12** Schmider 1992b; **13** Bodu/Orliac/Baffier 1996; **14** Orliac 1996b.

were different suggesting some subtly different activities related to the structures (Julien et al. 1999, 152). Nevertheless, the large structure 1 was examined in more detailed and revealed a complex development of use. Furthermore, during the excavation no alteration of the sediment was observed in this structure and also a detailed analysis of the sediment revealed no indication for thermal alterations (Rieu/March/Soler-Mayor 1999, 119f.). According to an organic trace analysis, animal bones served perhaps also as fuel along with wood charcoal or the traces result from the roasting of meat in structure 1 (Rieu/March/Soler-Mayor 1999, 120-125).

In general, the lithic artefacts were scattered across the complete sector 2 in a relatively even density. This regular distribution could indicate significant disturbances by natural agents such as floods. Nevertheless, in the southern depression and also near the other hearths the densities usually increased. In particular, the differentiated distribution of cores and various retouched artefact classes indicated some recurrent clusters suggesting that non-human agents had only a minor influence on the distribution of the material (Bridault/Bemilli 1999, 60-63; Julien et al. 1999, 153). To clarify the relation of the different clusters to each other as well as to the hearths refitting of the material is the only possibility because stratigraphically as well as micro-stratigraphically only minor perturbations were found which prohibited attributions of the material to different events. Perhaps, the thus far unpublished spatial distribution of the small groups such as the backed points or the very few Tertiary chalcedony artefacts could already help to further precise the influences of the spatial dynamics on the site.

The variety of material clearly suggested that various tasks were performed at the site. Besides the production of blanks, these blanks were also transformed into tools. These tools were partially used on the site for the processing of the faunal resources but also in the repair of the hunting equipment. Parts of the faunal remains as well as some blanks were presumably taken away from the site. Nevertheless, based on the majority of retouched lithic artefacts serving the processing of faunal material, the character of the site seemed focused on the post-processing of successful hunting episodes. In addition, the blank production were a second focus in this vicinity of a material resource.

Chronology of the upper horizon

Three samples from the upper horizon in sector 2, section 18 and one sample from sector 1 were ^{14}C-dated (**tab. 44**). The sample from sector 1 produced two results which were very distinct. One date was the collagen fraction (OxA-3139) and the other were purified amino acids (OxA-3671). Both results were obtained with a pretreatment method including ion-exchanged gelatin which potentially caused some contamination (Higham/Jacobi/Bronk Ramsey 2006, 182). Usually, the purified amino acids represented the more reliable fraction but in this case the further treated fraction could have been subject of more intense contamination. However, the results from sector 2 scattered also over a considerable period. Nevertheless, the youngest date (Gif-9609) was probably due to the low amount of collagen in the sample and the consequently higher potential for contamination (Deloze et al. 1999, 38). Even though this very young date was rejected, a date on reindeer (Gif-9608) and the reliable date on horse (Gif-9607) were separated by 800 radiocarbon years. Were these results two representatives of two different phases? The date yielded by the lower horizon fell between those two dates of the upper horizon and, in fact, was almost identical with the younger date. However, the preservation of the sample material was described as generally poor (Deloze et al. 1999, 38). Therefore, only a small quantity of material was obtained for dating and produced no precise result. Moreover, the relation of the faunal material to the human activities remained often uncertain. Thus, several episodes were possible at the site but these episodes were not necessarily all related to human activity.
In most parts of the site, the dense scatter of archaeological material in the upper horizon was not useful in the reconstruction of an internal chronology. However, the development of structure 1 (Rieu/March/Soler-Mayor 1999) as well as the possible cleaning around hearth structures and the dumping of material in other areas (Julien et al. 1999, 153 f.) suggested that various episodes formed already the spatial patterns around single hearth structures. Nevertheless, the time between the various episodes could not be estimated and the relation of the various concentrations to one another remained also unclear (cf. Fougère 2011). The similarity of some lithic clusters in their composition as well as their spatial relation to a hearth indicated the recurrent performance of similar activities including processing and consumption of faunal material, blank production, and repair of equipment. In combination with the result of the faunal analysis that the large amount of horse remains was accumulated successively, repetitive visits to successful hunting grounds with a lithic resource supply appeared probable. In addition, the faunal analysis suggested that the organic remains were exposed in the open for some time. This exposed material would still be recognisable for preceding groups which, perhaps, also used this material remains of the former visits to the site. In consequence, an intense refitting program would be highly recommendable for this large palimpsest assemblage.

Bonnières-sur-Seine, Yvelines

Research history

A first excavation of 15-20 m² at the Abri de la Côte Masset at Bonnières-sur-Seine was conducted by Alphonse Georges Poulain in 1910. He located a hearth approximately 1 m from the back wall of the rockshelter. Around this feature Poulain found animal bones, lithic artefacts, and a sandstone which he interpreted as grinding stone. In 1991 the site was re-investigated by the archaeological service of Yvelines. Seven test

site	lab. no.	years ^{14}C-BP	± years	material	species	comment	years cal. b2k	ref.
Hallines	Gif-1712	16,000	300	vertebra	*Mammuthus primigenius*	PROBLEMATIC: fossil material?	19,840-18,680	1
Le Grand Canton, sector 2, upper horizon	Gif-9608	12,880	80	bone	*Rangifer tarandus*		15,810-15,290	2-4
Le Grand Canton, sector 1	OxA-3139*	12,650	130	phalange	*Equus* sp.	same sample as OxA-3671	15,660-14,420	2-5
Le Grand Canton, sector 2, lower horizon	Gif-9606	12,195	130	bone	*Equus* sp.	attributed to the Late Magdalenian	14,800-13,640	2-4
Le Grand Canton, sector 2, upper horizon	Gif-9607	12,080	115	bone	*Equus* sp.		14,250-13,650	2-4
Le Grand Canton, sector 2, upper horizon	*Gif-9609*	*11,420*	*100*	bone	Equus sp.	REJECT: low collagen content	×	2-4
Le Grand Canton, sector 1	*OxA-3671**	*11,030*	*105*	phalange	*Equus* sp.	REJECT: protein fraction; same sample as OxA-3139	×	2-5
Bonnières-sur-Seine, level 4a	GifA-93014	12,770	120	bone		PROBLEMATIC: association?	15,790-14,910	6
Étigny-Le Brassot, southern *locus*	OxA-10096 (Ly)	12,630	90	bone	*Rangifer tarandus*		15,540-14,540	3; 7-8
Étigny-Le Brassot, southern *locus*	Ly-9015	11,090	95	bone		PROBLEMATIC: bulked sample; association?	13,150-12,710	7
Le Tureau des Gardes, *locus* 10	AA-44216	12,520	130	radius	*Equus* sp.		15,510-14,070	3; 9
Le Tureau des Gardes, *locus* 6	Ly-6988	12,290	90	bone	*Equus* sp.	attributed to the Late Magdalenian	14,810-13,890	3-4; 9
Le Tureau des Gardes, *locus* 10	AA-44214	12,170	130	phalange	*Equus* sp.		14,680-13,640	3; 9
Le Tureau des Gardes, *locus* 10	AA-44215	12,160	120	humerus	*Equus* sp.		14,540-13,660	3; 9
Le Tureau des Gardes, *locus* 6	Ly-6989	11,560	100	bone	*Rangifer tarandus*	attributed to the Late Magdalenian; PROBLEMATIC: reindeer in France?	13,590-13,190	3-4
Le Closeau, test pit 17	*Ly-*	*12,575*	*75*	sediment		REJECT: sediment; no association	×	3
Le Closeau, lower horizon, *locus* 33	GrA-18860	12,510	80	diaphyse of longbone	*Equus* sp.		15,340-14,260	3; 10
Le Closeau, lower horizon, *locus* 33	GrA-18815	12,480	70	phalange	Bovinae		15,270-14,190	3; 10
Le Closeau, lower horizon, *locus* 46	AA-41881	12,423	67	bone	*Sus scrofa*		15,160-14,080	3; 10

Tab. 44 ^{14}C dates from sites in the Paris Basin. If the sub-assemblage is known from which the sample originated, the sub-assemblage is given behind the site name. * dates which were pretreated by the use of ion-exchanged gelatin (Lab.code: AI) in the Oxford series (cf. Jacobi/Higham 2009, 1896); ** dates which might be contaminated due to the use of a method leaving traces of a humectant in the collagen (Lab.code: AF*) in the Oxford series (cf. Higham et al. 2007, S55 & S2). Doubtful dates are shaded grey. Rejected dates are shaded grey and set in italics and, in addition, the main reason for rejection is given in comment. For further details see p. 265-269 and text. The dates were calibrated with the calibration curve of the present study (see p. 358-364) and the calibration program CalPal (Weninger/Jöris/Danzeglocke 2007). The result range of 95% confidence is given for the calibrated ages (years cal. b2k). – References (ref.): **1** Fagnart 1997; **2** Deloze et al. 1999, 38; **3** Bodu 2004; **4** Lang 1998; **5** Hedges et al. 1993b; **6** Débout et al. 2012; **7** Stevens/Hedges 2004; **8** Lhomme et al. 2004, 732; **9** Bignon 2006; **10** Bodu/Debout/Bignon 2006; **11** Bodu/Valentin 1997, 343; **12** Stevens/Hedges 2004; **13** Gowlett et al. 1986b; **14** Gowlett et al. 1986a; **15** Gilot 1997; **16** Fosse 1997, 242; **17** Bodu et al. 2009b; **18** Leroi-Gourhan/Brézillon 1966; **19** Delibrias et al. 1976; **20** Delibrias/Guillier/Labeyrie 1970; **21** Ponel et al. 2005.

site	lab. no.	years ^{14}C-BP	± years	material	species	comment	years cal. b2k	ref.
Le Closeau, lower horizon, *locus* 46	GrA-11665 (Ly-790)	12,360	60	femur	Cervidae		14,820-14,060	3; 10
Le Closeau, lower horizon, *locus* 46	GrA-11664 (Ly-789)	12,350	60	tibia	*Equus* sp.		14,830-14,030	3; 10
Le Closeau, lower horizon, *locus* 46	GrA-18816	12,350	70	femur	*Sus scrofa*		14,800-14,040	3; 10
Le Closeau, lower horizon, *locus* 56	GrA-18819	12,340	70	radius, dext.	Cervidae		14,810-14,010	3; 10
Le Closeau, lower horizon, *locus* 46	AA-41882	12,248	66	bone	*Panthera leo*		14,570-13,890	3; 10
Le Closeau, lower horizon, *locus* 4	OxA-5680* (Ly-166)	12,090	90	diaphyse	*Equus* sp.		14,180-13,700	3; 10-12
Le Closeau, lower horizon, *locus* 4	OxA-6338* (Ly-313)	12,050	100	diaphyse			14,150-13,670	3; 10-11
Le Closeau, light grey silts, test pit 17	*Ly-793*	*11,750*	*90*	sediment		REJECT: material; no association	×	3
Le Closeau, lower horizon, *locus* 4	*GrA-18762*	*11,640*	*70*	diaphyse of longbone	*Sus scrofa*	REJECT: association unclear	13,670-13,270	3; 10
Le Closeau, intermediate horizon, *locus* 51	Ly-570 (OxA)	11,275	85	charcoal			13,290-12,970	3
Le Closeau, intermediate horizon, *locus* 18	Ly-562 (OxA)	11,265	90	charcoal			13,280-12,960	3
Le Closeau, intermediate horizon, *locus* 34	Ly-566 (OxA)	11,240	90	charcoal			13,260-12,940	3
Le Closeau, intermediate horizon, *locus* 14	Ly-358 (AA-21677)	11,240	80	charcoal			13,240-12,960	3
Le Closeau, upper horizon, *locus* 32	Ly-565 (OxA)	11,205	100	charcoal			13,260-12,860	3
Le Closeau, undetermined horizon, *locus* 41	Ly-567	11,170	105	charcoal			13,270-12,750	3
Le Closeau, intermediate horizon, *locus* 19	Ly-561 (OxA)	11,165	90	charcoal			13,240-12,800	3
Le Closeau, undetermined horizon, *locus* 41	Ly-568	11,120	100	charcoal			13,200-12,720	3
Le Closeau, undetermined horizon, *locus* 48	Ly-569	11,105	95	charcoal			13,170-12,730	3
Le Closeau, top horizon, *locus* 25	Ly-564 (OxA)	10,885	85	charcoal	*Pinus sylvestris*?		12,990-12,630	3
Le Closeau, upper horizon, *locus* 8	OxA-6337* (Ly-312)	10,840	110	bone			13,010-12,570	3; 11
Le Closeau, top horizon, *locus* 25	Ly-563 (OxA)	10,755	90	charcoal	*Pinus sylvestris*?		12,800-12,560	3

Tab. 44 (continued)

site	lab. no.	years ^{14}C-BP	± years	material	species	comment	years cal. b2k	ref.
Le Closeau, upper horizon, *locus* 3	Ly-7189	10,670	110	charcoal			12,790-12,390	3; 11
Le Closeau, greyish deposits, St. IV	Ly-206 (OxA)	10,650	75	charcoal			12,760-12,440	3; 11
Le Closeau, greyish deposits, St. III	*Ly-7190*	*10,470*	*110*	charcoal		REJECT: no association	12,730-12,010	3; 11
Le Closeau, undetermined horizon, *locus* 45	GrA-11662	10,410	50	diaphyse			12,520-12,080	3
Le Closeau, lower horizon, *locus* 4	*GrA-18697*	*10,240*	*150*	phalange	*Sus scrofa*	REJECT: association?	×	3; 10
Le Closeau, lower horizon, *locus* 33	*GrA-10892 (Ly-814)*	*9,200*	*70*	diaphyse		REJECT: too young	×	3; 10
Le Closeau, upper horizon, *locus* 26	*GrA-10886*	*9,070*	*70*	jugale	*Equus* sp.	REJECT: too young	×	3
Le Closeau, light grey silts, I.F.P. 1	*GrA-10895*	*6,850*	*80*	bone		REJECT: too young	×	3
Le Closeau, upper horizon, *locus* 54	*GrA-18699*	*6,470*	*70*	diaphyse		REJECT: too young	×	3
Le Closeau, lower horizon, *locus* 46	*GrA-18763*	*6,420*	*110*	femur	*Sus scrofa*	REJECT: association?; too young	×	3, 10
Le Closeau, lower horizon, *locus* 4	*GrA-18701*	*5,380*	*100*	phalange	*Sus scrofa*	REJECT: association?; too young	×	3; 10
Le Closeau, upper horizon, *locus* 26	*GrA-10885*	*5,290*	*90*	jugale	*Equus* sp.	REJECT: too young	×	3
Le Closeau, intermediate horizon, *locus* 34	*GrA-10887*	*4,960*	*60*	bone		REJECT: too young	×	3
Le Closeau, undetermined horizon, *locus* 43	*GrA-10894*	*4,930*	*60*	tibia		REJECT: too young	×	3
Le Closeau, undetermined horizon, *locus* 48	*GrA-10675*	*4,520*	*80*	scapula		REJECT: too young	×	3
Le Closeau, intermediate horizon, *locus* 34	*GrA-10882*	*3,370*	*70*	bone		REJECT: too young	×	3
Marsangy, conc. N19	OxA-8453	12,140	75	tooth	*Equus* sp.	found in M16	14,220-13,780	3; 12
Marsangy, conc. D14	OxA-740	12,120	200	tooth	*Rangifer tarandus*	found in C14	14,920-13,440	3; 13
Marsangy, conc. N19	OxA-178	11,600	200	antler	*Rangifer tarandus*	found in P16	13,850-13,050	3; 14
Marsangy, conc. D14	*OxA-505*	*9,770*	*180*	bone	*Rangifer tarandus*	found in B12, REJECT: insufficient collagen	×	3; 13
Marsangy	*Lv-1215*	*5,000*	*350*	antler	*Rangifer tarandus*	REJECT: too young	×	15
Gouy	GifA-92346	12,050	130	bone		PROBLEMATIC: association?	14,250-13,610	3; 16

Tab. 44 (continued)

site	lab. no.	years ^{14}C-BP	± years	material	species	comment	years cal. b2k	ref.
Pincevent level IV.20, section 27	OxA-148	12,600	200	bone	*Rangifer tarandus*		15,800-14,000	3; 17
Pincevent level IV.30	ETH-37120	12,530	45	charcoal			15,260-14,540	17
Pincevent level IV.20	ETH-37119	12,450	45	charcoal			15,170-14,170	17
Pincevent level IV.21.3, section 25	OxA-149	12,400	200	bone	*Rangifer tarandus*		15,480-13,760	3; 17
Pincevent level IV, habitation 1	Gif-358	12,300	400	charcoal		PROBLEMATIC: contamination?	15,940-13,220	3; 17-18
Pincevent level IV.21.3, section 25	OxA-177	12,300	220	bone	*Rangifer tarandus*		15,390-13,590	3; 17
Pincevent level IV, habitation 1	Erl-6786	12,277	96	bone	*Rangifer tarandus*		14,810-13,850	3; 17-18
Pincevent level IV.20, section 37	OxA-467	12,250	160	charcoal		PROBLEMATIC: contamination?	15,080-13,640	3; 17
Pincevent level IV.20, section 27	Gif-6283	12,120	130	charcoal		PROBLEMATIC: contamination?	14,430-13,630	3; 17
Pincevent level IV.30, section 36	Gif-6310	12,100	130	charcoal		PROBLEMATIC: contamination?	14,360-13,640	3; 17
Pincevent level IV.40, section 36	Gif-5971	12,100	120	charcoal		PROBLEMATIC: contamination?	14,300-13,660	3; 17
Pincevent level IV.21.3, section 25	OxA-176	12,000	220	bone	*Rangifer tarandus*		14,640-13,320	3; 17
Pincevent level III.2, section 27	OxA-391	11,870	130	bone			14,010-13,410	3; 11; 13
Pincevent level IV.21.3, section 26	Gif-6284	11,800	130	charcoal		PROBLEMATIC: contamination?	×	3
Pincevent level IV, habitation 1	Lv-292	11,610	400	charcoal		PROBLEMATIC: contamination?	×	19
Pincevent level IV, habitation 1	Lv-293	11,310	330	charcoal		PROBLEMATIC: contamination?	×	19
Pincevent level IV, habitation 1	Lv-291	10,920	540	charcoal		PROBLEMATIC: contamination?	×	19
Pincevent level IV, habitation 1	GrN-4383	10,760	60	charcoal		PROBLEMATIC: contamination?	×	20
Pincevent level IV, section 9	Gif-349	9,840	350	charcoal		PROBLEMATIC: contamination?	×	20
Pincevent level IV.20, section 36	Gif-3480	9,460	170	charcoal		PROBLEMATIC: contamination?	×	3
Conty	*Ly-6998*	*12,370*	*70*	silty peat		same sample as Beta-86310; REJECT: no association; sediment	×	21
Conty	*OxA-6257* (Ly-286)*	*12,300*	*120*	antler	*Cervus elaphus*	REJECT: palaeontological sample	14,990-13,830	21
Conty	GifA-99527	12,220	90	charcoal			14,600-13,800	21
Conty, lower horizon	OxA-6151* (Ly-260)	11,890	90	metacarpal	*Bos primigenius*		13,960-13,480	1; 21
Conty	*Beta-86310*	*11,890*	*70*	silty peat		same sample as Ly-6998; REJECT: no association; sediment	×	21

Tab. 44 (continued)

site	lab. no.	years ^{14}C-BP	± years	material	species	comment	years cal. b2k	ref.
Conty, grey calcareous organic silts (6b3)	*GifA-99526*	*11,640*	*80*	charcoal		REJECT: only stratigraphic association	13,690-13,250	21
Conty, lower horizon	OxA-6148* (Ly-257)	11,620	90	diaphysis	*Bos primigenius*		13,670-13,230	1; 21
Conty, lower horizon	OxA-6149* (Ly-258)	11,560	90	diaphysis	*Bos primigenius*		13,570-13,210	1; 21
Conty, upper grey organic silts (6b1)	*Ly-284 (OxA)*	*11,540*	*80*	*bark*	*Pinus sylvestris*	REJECT: only stratigraphic association	13,560-13,200	21
Conty, lower horizon	OxA-6150* (Ly-259)	11,410	80	tibia	*Bos primigenius*		13,440-13,080	1; 21
Conty	*Ly-7407*	*11,130*	*80*	peat		REJECT: sediment	×	21
Conty	*Ly-7000*	*11,080*	*80*	peat		REJECT: sediment	×	21
Conty	*Ly-7408*	*10,960*	*85*	silt with organic horizon		REJECT: no association; sediment	×	21
Conty	*Beta-132166*	*10,790*	*80*	calcareous organic silts		REJECT: no association; sediment	×	21
Conty, lower horizon	*OxA-7653*	*9,815*	*60*	bone	*Equus* sp.	REJECT: too young; perhaps, top horizon?	×	12
Conty	*Ly-7409*	*9,310*	*60*	peat		REJECT: no association; sediment	×	21
Hangest-sur-Somme III.1, lower horizon	OxA-4432* (Ly-22)	11,660	110	molar	*Bos primigenius / Equus* sp.?		13,760-13,240	1
Hangest-sur-Somme III.1, lower horizon	OxA-4936* (Ly-86)	11,630	90	molar	*Bos primigenius / Equus* sp.?		13,700-13,220	1
Hangest-sur-Somme III.1, upper horizon	OxA-4935* (Ly-85)	10,920	90	vertebra	*Bos primigenius*		13,010-12,650	1

Tab. 44 (continued)

pits (A-G) and an area (H) alongside the rock wall was excavated under the supervision of Gilles Habasque. In the area H the limits of the excavations of Poulain were partially found. Four further concentrations of archaeological material were recovered in the 5-10 m² large modern excavations.

Topography

The rockshelter La Côte Masset is situated on the southern bank of the Seine, at the foot of Cretaceous rock formations which rise about 100 m above the site (**tab. 34**). The rock is partly made of dolomitic and Senonian flint layers. Around the site the formation is regularly cut by small valleys coming from the south or south-west. The rockshelter is situated at a meander of the Seine. Along the Seine, north of the rockshelter extends a wide basin which is c. 5 km long and approximately 1.5 km wide. The basin was formed by the confluence of the Epte into the Seine, some 4 km north of the site. Towards the north-east the Seine flows in an approximately 1 km wide valley for about 9 km until the next turning point occurs. This topographic setting makes the site a good point for observations (Barois-Basquin/Charier/Lécolle 1996, 36) and the rockshelter also opens towards this northern direction. The height of the rockshelter is around 4 m at the archaeological level (Barois-Basquin/Charier/Lécolle 1996, 33). The cavity is 7 m long and 1.6-2.6 m deep.

Stratigraphy

Above the dolomite bedrock a unit of brown clayey loess sands followed within which several pieces of chalk and cryofractured nodules of flint were found. This unit was overlain by brown and yellowish laminated sandy loess silts. They were superimposed by a unit of brown loess silts with compact chalk granules. These last two units were deposited under extreme dry conditions which was possibly associated with the Late Pleniglacial (Barois-Basquin/Charier/Lécolle 1996, 35). A light brown compact layer followed above which was composed of sandy loess clays and many chalk and flint fragments. The archaeological material was found in the lower part of this unit and the upper part of the underlain brown loess silts. Originally the material was probably deposited in the brown loess silt layer (Barois-Basquin/Charier/Lécolle 1996, 39) and then gradually moved upwards in the sediments into the light brown compact layer. Refitting of lithics from the two different layers indicated a single archaeological horizon. The compact unit was overlain by a brown layer of the same material. However, the upper deposit was much less compact and the chalk and flint fragments were smaller. This deposit was sealed by dark brown, loose, clayey loess sands and on top the Holocene topsoil had formed.

In the samples taken for palynological analysis from the site no pollen was preserved (Valentin 1995, 129).

Archaeological material

From both excavations a total of 1,316 lithics was recovered (**tab. 35**; Barois-Basquin/Charier/Lécolle 1996, 38). 86 of these pieces were still archived from the 1910 excavation. A further 318 pieces were found in the dumps of this excavation left at the site. 912 lithics were recorded *in situ* during the modern excavation. The material is partially patinated and many pieces are covered by a chalk crust. However, at least three different types of silex were identified as raw material (Barois-Basquin/Charier/Lécolle 1996, 38): A brown flint with grey inclusions, a fine-grained yellowish flint, and a translucent, light yellow flint or chalcedony. All of these are of local origin, possibly from the river deposits or the nearby, *in situ* flint layers which were presumably easy to access in the Lateglacial (**tab. 36**; Valentin 1995, 404). Only a few blades which were probably knapped elsewhere were made of a Cretaceous flint different from the local one (Valentin 1995, 408). A local origin of most of the raw materials was further emphasised by the presence of almost complete knapping sequences of the raw materials (Valentin 1995, 405). Moreover, the cortex parts and the found remains of the *chaîne opératoire* indicated an on-site testing of the raw material nodules.

The blank production was focused on elongated flakes or blades, even though the seven cores and core fragments were of an irregular shape and more flakes then blades were found (**tab. 37**). However, the blank indices strongly indicate that blades were the preferred blank type (Barois-Basquin/Charier/Lécolle 1996, 39). During the blank production process two types of hammer were in use: The preparation of the cores was performed with a hard stone or soft stone hammer, whereas the blades of the main sequence were struck with an organic hammer.

On the whole site only 13 lithic tools were found, of which ten were recovered during the modern excavations (**tab. 38**; Barois-Basquin/Charier/Lécolle 1996, 39). Among the four LMP were two shouldered points which were made on considerably shorter blades than the other tools. Two equally short backed fragments were retouched on narrower bladelets. In addition, four borers were found which were all attributed to the *bec* or *Zinken* group. Furthermore, there were two end-scrapers on blades and two burins on truncation as well as a composite tool of a dihedral burin and an end-scraper. All retouched artefacts wore also traces of use.

Along with these lithic artefacts, a sandstone which weighed over 1.5 kg was found during the excavation 1910. The sandstone originated probably from some kilometres away. The piece was recovered beside the observed hearth and its surface was polished and wore traces of red colour (**tab. 42**; Barois-Basquin/

Charier/Lécolle 1996, 39). Perhaps, the sandstone was used as a grinding stone, possibly, for minerals or something treated with colouring minerals.
A marcassite fragment which was found in a small concentration of lithics could not be proven to belong to the archaeological material due to lacking modification and the presence of this material in some geological layers of the site (Barois-Basquin/Charier/Lécolle 1996, 38). Nevertheless, a use of this mineral in an archaeological context and, in particular, in relation to a hearth is possible.
From the first excavation several bones were recovered. Subsequently, these remains were determined as coming from at least four horses (*Equus* sp.), a wild boar (*Sus scrofa*), and a *Megaloceros giganteus* (**tab. 39**; Barois-Basquin/Charier/Lécolle 1996, 36). In the modern excavation only some indeterminable bone fragments and a maxillary canine of a cervid were found. The canine might possibly be from a *Cervus elaphus* (Barois-Basquin/Charier/Lécolle 1996, 38). However, due to the presence of *Megaloceros giganteus* in the bone assemblage, the tooth could possibly also belong to the same individual.
Considering the early recovery of the material a ^{14}C date on the giant deer remains could be of particular interest to confirm the association with the Lateglacial material. Compared to the thus far confirmed distribution of this species during the Lateglacial, Bonnières-sur-Seine was situated considerably farther to the south where the appearance of giant deer was only dated to the LGM and earlier ages (Stuart et al. 2004). If the giant deer remains dated older but were still associated with a human presence, perhaps, associated with the Gravettian this finding could also be helpful in a discussion on human presence during this period in northern France (e. g. Pigeaud et al. 2010, 104-106). However, the presented archaeological material appeared congruent with only a Lateglacial human presence at the site.

Spatial organisation
Including the poorly documented concentration around a possible hearth from 1910 (**tab. 43**), the small assemblage was found in five different concentrations. Each concentration yielded a complete exploitation cycle of local material nodules including the preparation of cores, the blank production, and the transformation into formally retouched tools (Barois-Basquin/Charier/Lécolle 1996, 40). Thus far, material from the different concentrations could not be refitted which recommend caution about the integrity of the material with a single occupation episode.

Chronology
According to the stratigraphic evidence, the artefacts were deposited at the site shortly before or during the transition from the Late Pleniglacial to the early Lateglacial Interstadial. This chronostratigraphic position was further emphasised by the diverse fauna. A recently dated bone sample from the site produced a late Late Pleniglacial age (**tab. 44**). The piece was attributed to level 4a (Débout et al. 2012, 179 tab. 1) which were the light brown silts generally underlying the archaeological material underneath the rockshelter (Barois-Basquin/Charier/Lécolle 1996, 37). Thus, a bone from this level would predate the archaeological horizon. However, in a part excavated on the terrace in front of the abri (test pit A, area II) some strongly patinated lithic artefacts were found in the upper part of this level 4a. Moreover, refits related the material recovered from both levels and led Blandine Barois-Basquin and her colleagues to consider the level 4a as the actual living horizon (Barois-Basquin/Charier/Lécolle 1996, 39). Nevertheless, the relation of the bone sample to the archaeological material required further specification. Moreover, due to the lack of refits, multiple independent occupations of the site might be possible (Barois-Basquin/Charier/Lécolle 1996, 39). In this case, the exact relation of faunal material and lithic remains could be meaningful.
Nevertheless, the assemblages of each concentration appeared very small. These small numbers might be explicable with the site serving as an extremely short-term observation spot during hunting trips. However,

the various remains of the blank production process and the making of new tools, as well as the remaining of only some, fractured and heavily used tools on the site indicated the importance of the lithic preparation. On the one hand, this diverse remains could be in accordance with the hypothesis of an observation/hunting camp where hunters carried out other tasks while waiting for game to be spotted or while processing the carcasses after a successful hunt. On the other hand, the presence of bones from various animals in one concentration where also a grinding stone with possible traces of ochre were found in the combination with the existence of diverse tool types on the site rather suggested a non-specialised camp where various works were accomplished. Hence, the various concentrations could represent the dumps or remains from different workshops within the same occupation event (cf. Barois-Basquin/Charier/Lécolle 1996, 41).
To further evaluate the two different scenarios a spatial analysis of the site could be helpful, although the lacking spatial information of the main scatter would surely restrain the results (Barois-Basquin/Charier/Lécolle 1996, 40).

Étigny, Yonne

Research history

In 1998 archaeological material was excavated near the village Étigny at the site Le Brassot in two areas: a northern and a southern *locus* (Lhomme et al. 2004, 732). Although both *loci* were found in the same horizon, a considerable distance between them (c. 50 m) prohibited conclusions on the temporal connection of the two assemblages (Lhomme et al. 2004, 732). On the southern *locus* 201 m² were explored, whereas the northern *locus* was only tested on 4 m². The rest of this area which was estimated to encompass some 500 m² remained *in situ* for future research.

Topography

The site Le Brassot was situated on the western bank of the Yonne (**tab. 34**). The meandering of the river formed a wide valley at Étigny. On the western side of the modern river the hills rise very gradually. On the eastern bank, approximately a kilometre east of the site, the plateau rises abruptly 40 m upwards. Towards the north the Vanne river, an eastern contributory of the Yonne, further widened the valley at its confluence with the Yonne near Sens. Moreover, to the north the valley widens further to the confluence of the Yonne into the Seine. Towards the south the valley gradually narrows. Thus, Étigny is already set in the transition from the Paris Basin to the surrounding upland areas. In the large meander of Étigny and Gron a former river branch located only 250 m east of the site was considered as possibly active in the time of the occupation. Further palaeochannels of the Yonne were observed several hundred metres westwards of the site (Lhomme et al. 2004, 723 f.).

Stratigraphy

In Étigny a single artefact layer was found at the base of a grey soil which overlaid yellow silty sands. The soil was capped in its upper part by further yellow silty sands. These overlaying silty sands were in some parts laterally passed by sands which overlaid the deposits of the former river channel. Hence, these sands and the yellow silty sands above the grey soil might be contemporary (Chaussé 2003). However, the deposits of the river channel, peat and organic clay layers, had cut into the silty sand layer below the grey soil and it remained uncertain if the organic layers had completely capped the grey soil in the river bed (cf. Chaussé 2003, 42) or if the organic layers were contemporary to the grey soil (cf. Lhomme et al. 2004, 732). The biostratigraphical attribution based on a malacological and a palynological analyses were undertaken in the organic deposits of the river channel. In the pollen diagram pine pollen (*Pinus* sp.) dominated before birch (*Betula* sp.), juniper (*Juniperus* sp.), and willow pollen (*Salix* sp.; Chaussé 2003, 37). This result suggested an attribution to the forested period of the mid- to late Allerød which was further confirmed by the malacological analysis (Chaussé 2003, 37; Lhomme et al. 2004, 732). Hence, if the organic deposits and the archaeological horizon were contemporary, these results would clearly attribute the material of Étingy-Le Brassot to the younger part of the Lateglacial Interstadial. However, if the soil was capped, the down cutting of the river channel postdated the deposition of the archaeological material and this river branch did not exist contemporary with the Palaeolithic occupation and the biostratigaphical results would provide a *terminus ante quem*. Perhaps, in this scenario one or more of the palaeochannels westwards of the site were active at the time of occupation and, hence, could possibly produce biostratigraphic information on the occupation time.

Archaeological material of the northern *locus*

The northern *locus* yielded a very small assemblage of less than 200 pieces which were mainly determined as core preparation debris. Thus, the inventory was too small for a typological attribution but could well represent a supplementary knapping place to the southern *locus*. Yet, in this *locus*, the cortex of four flakes of which three were refitted was incised with many short, slightly curved lines before the stone knapping (**tab. 42**; Lhomme et al. 2004, 732).

No organic remains were recovered from the test pits in this area.

Archaeological material of the southern *locus*

On the southern *locus* 8,252 lithics of which more than half were smaller than 3 cm were excavated (**tab. 35**). These were generally made of local Senonian flint (**tab. 36**). The technology showed all characteristics of the so-called faciès Cepoy/Marsangy with differentiated use of percussion instruments and a blank production directed towards two sorts of blades (cf. Valentin 2008a, 123-125): long blades with curved profiles which were used for the production of the majority of tools and short blades with straight profiles served as blanks for the points (Lhomme et al. 2004, 733). Among the cores (**tab. 37**), the single platform specimens or these with a preferred platform were more frequent than the examples with two platforms and bidirectional exploitation (Lhomme et al. 2004, 733).

The approximately 23 LMP (**tab. 38**) which were found in the southern *locus* were mainly classified as various types of curve- and angle-backed points. The backed points were usually shaped from short blades. Typically for the so-called faciès Cepoy/Marsangy no specimen was identified as backed bladelet. In addition to these LMP, end-scrapers formed the most numerous tool class (n = 25) and were frequently made on long blades. Only 15 burins were found which were generally made on natural edges or as dihedral types (Lhomme et al. 2004, 733 f.). Some borers of which at least one belonged to the piercer group were also present at the *locus*.

Of the 102 recovered faunal elements only seven pieces were determined taxonomically (**tab. 39**). Along with several bones and teeth of reindeer (*Rangifer tarandus*), some elements of red deer (*Cervus elaphus*) were also identified (Lhomme et al. 2004, 733).

Spatial organisation

According to the spatial and vertical distribution, the material in both concentrations appeared relatively unaltered since its deposition and, hence, provided a high quality record.

Only one small accumulation of material was excavated in the northern *locus*. This area contained mainly debris of the preparation of cores including cortical material and preparation flakes. Hence, in this area a first step in the exploitation of the lithic raw material was performed and it could represent a knapping workshop.

On the southern *locus* the material was spatially distributed in two zones (Lhomme et al. 2004, 733). Due to the material found in each zone, the excavators suspected a functional differentiation. The northern zone contained mainly lithic debris and one third of the lithic tools indicating a workshop for the production and modification of blanks. In the southern zone, the majority of the preserved fauna and the tools were found. In the southern zone, heated stones indicated the presence of a hearth (**tab. 43**) and suggested a rather domestic function of this zone (Lhomme et al. 2004, 733). Between the two zones, no apparent limit was observed.

Chronology

The meander of Étigny yielded a complex chronostratigraphy with several palaeochannels and further archaeological sites such as Gron-Chemin de l'Évangile or Étigny-PLM revealing this part of the valley as interesting halt for different hunter-gatherer groups at various times.

Material from the southern *locus* of Étigny-Le Brassot produced two inconsistent ^{14}C dates (**tab. 44**). These dates have to be further evaluated in their biostratigraphic context (see p. 265-269).

The chronological relation of the two *loci* of Étigny-Le Brassot was not further specified except for a quasi-contemporaneity based on the stratigraphic position. In particular, for the northern *locus* a more precise chronostratigraphic position was not possible due to the lack of organic material. Moreover, refitting was not possible between the two *loci* but if they belonged to the same *chaîne opératoire* the first steps were performed presumably in the northern *locus* and further steps were accomplished in the southern *locus*. In total, the southern *locus* was considered as formed by peripheral activities suggesting the main settlement area in the not excavated west of the *locus*. Thus, at least several occupation episodes formed the assemblage but, perhaps, the site was even formed by some occupation events. Probably, further details of the spatial organisation and the sequence of the various processes on the site could be revealed by a publication of the results of two M. A. theses on the site (Soula 2000; Dumarçay 2001).

Le Tureau des Gardes, Marolles-sur-Seine, Seine-et-Marne

Research history

Quarry works at the site Le Tureau des Gardes in the Seine valley near Marolles-sur-Seine revealed Late Pleistocene material which was excavated between 1991 and 1998 under the supervision of Patrick Gouge and later Laurent Lang (Lang 1998, 82). Ten distinct *loci* (1-10) were observed in an excavation area of less than 500 m² (Lang 1998, 83). Of particular interest became *locus* 7 where a significant number of backed points were found. *Locus* 7 was found in 1996 during a test pit program conducted by Patrick Gouge and in 1997, an excavation of 140 m² by Laurent Lang and his team followed (Weber 2006, 163) which formed the largest of the ten areas (Lang 1998, 83). The second largest area was *locus* 10 with 95 m² (cf. Bignon 2006) followed by the *locus* 5 with 62 m² and *locus* 8 with 60 m². The *loci* 6 and 9 were only partially excavated on 5 and 14 m²; the rest of these concentrations remained *in situ* for future research.

Topography

The topography of the site Le Tureau des Gardes is similar to the site Le Grand Canton. Le Tureau des Gardes was also located about 1 km west of the town Marolles-sur-Seine in the wide basin formed by the Seine and the Yonne in the vicinity of their confluence (**tab. 34**; Alix et al. 1993). The confluence is situated about 3-4 km west of the site which was situated on the southern bank of the Seine and the northern bank of the Yonne. Several hundreds metres north of the site, on the northern bank of the Seine hills rise some 50 m upwards. This plateau is frequently cut by small valleys, often running parallel to the Seine. Towards the east the plateau gradually rises between the valleys of the Seine and the Yonne. About 4.5 km south, on the southern bank of the Yonne the terrain rises approximately 60 m upwards. Hence, the site was situated in a wide alluvial plain. The site Le Tureau des Gardes was set on the transition of the lower gravel terrace where a step of some 6-8 m was observable in the terrain. However, in the alluvial plain several palaeochannels of both rivers were observed (Alix et al. 1993, 197) and, in particular, along the Seine this plain was morphologically varied by sand banks rising up to 5 m above the river (Lang 1998, 10). The archaeological material was preserved in general only in hollow areas of the past surface (Lang 1998, 84). Also, in the first sector of Le Tureau des Gardes (later combined with sector II and re-named *locus* 7; Weber 2006, 163) a circular depression of some 60 cm depth and 15 m in diameter was observed (Lang 1998, 18).

Stratigraphy

The archaeological material was found in silty sands which overlaid sands and silts (Alix et al. 1993, 209). The latter deposit suffered from cryogenic movements. Thus, the site was settled after a period of severe cold such as the Late Pleniglacial. On top of the silty sands calcite sand layers followed. The samples which were taken for pollen analysis from these layers were too disturbed to produce a reliable result (Alix et al. 1993, 214). The malacological samples from a profile several 100 m north-west of *locus* 7 yielded a small, though statistically significant assemblage for the archaeological layer (Lang 1998, 85) which was composed of cold steppe as well as semi-forested species (Alix et al. 1993, 214). The latter indicated a temperate period where steppe elements were still present (Lang 1998, 85). The steppe component was further sustained by the presence of horse among the mammalian fauna. This transitional environment was probably congruent with the late Late Pleniglacial and the early Lateglacial Interstadial.

Archaeological material in general

In total, Le Tureau des Gardes yielded more than 60,000 lithic artefacts without splinters smaller than 2 cm (**tab. 35**; Lang 1998, 23. 90).

The main lithic raw material was a Cretaceous flint which originated from alluvial gravels (**tab. 36**). Only a minor portion (less than 1 %) was made of Tertiary (Bartonian) chalcedony. The origin of this material remained uncertain, although an available resource in the surrounding was considered due to the frequency of the material in the region (Lang 1998, 91). Nevertheless, the raw material reached the site only in the shape of blanks and retouched tools with the exception of *locus* 7 where a core of this material was exploited. Due to this selected import a transport over a wider distance was also considered possible for this Tertiary material.

942 cores were found on the site (**tab. 37**). Among the cores, the ones used for bladelet production were most common but typical blade cores were also found (Lang 1998, 25 f. 91 f.). The relation of these types was comparable to the presence of backed bladelets in the different concentrations. However, the original raw material nodules were not very large and, thus, the production of blades longer than 10 cm was rarely possible (Lang 1998, 91). The cores were prepared with one as well as two platforms, although the second platform was frequently used only to maintain the knapping surface for the first platform (Lang 1998, 91). Blanks with a butt indicating the *en éperon* preparation were rarely present.

The 3,442 retouched artefacts (**tab. 38**) were clearly dominated by the LMP (n = 1,057). Among the LMP were 558 backed bladelets and backed bladelet fragments and a further 499 LMP were identified as points and point fragments. The majority of the latter was found in *locus* 7 (n = 299). Besides this *locus*, only the *locus* 8 yielded more points (n = 67) than backed bladelets (n = 19). In contrast, in *locus* 6 not a single point was found among the 122 LMP (Lang 1998, 94). The burins were the second largest group (n = 596) and among them the dihedral types was most frequent (Lang 1998, 92). In addition, many of the composite tools (n = 104) were combinations with burins (cf. Weber 2006, 169). The 509 end-scrapers were almost exclusively made on long blades. The 442 borers were also frequently made on longer blades (Weber 2006, 169) and included *bec* and *Zinken* as well as perforators. The truncations (n = 184) were generally oblique ones.

Furthermore, more than 235 kg of sandstone and granite were recovered at the site (Alix et al. 1993).

Even though the preservation of faunal material (**tab. 39**) was poor due to the high acidity in the soil, the documented faunal material comprised 5,567 determinable elements in 1998 (Lang 1998, 86) and additional 2,866 fragments were identified in the material from *locus* 10 by Olivier Bignon in 2003 (Bignon 2006, 185 Tabl. II). Among the faunal remains determined in 1998, horse (*Equus* sp.; n = 4,197) clearly dominated with a minimum of 76 individual animals followed by reindeer (*Rangifer tarandus*; n = 1,323) with a minimum of 29 animals (Lang 1998, 87). The final analysis of the material from *locus* 10 further emphasised these numbers and added at least 69 horses and eleven reindeer to the previous numbers (Bignon 2006). Further species were only identified by rare numbers of animals such as three lagomorphs (Lagomorpha; n = 7) and two bovids (*Bos* sp./*Bison* sp.; n = 8) or even by singular remains such as wolf (*Canis lupus*; n = 1), mammoth (*Mammuthus primigenius*; n = 1), a spermophil (Spermophilus; n = 4; Lang 1998; 86-88), or birds (Aves; n = 16; Bignon 2006).

The various distribution of reindeer and horse was possibly due to taphonomy which affected the skeletal elements of reindeer stronger than the more resistant horse bones. However short- and intermediate-term chronological reasons such as the seasonal setting of the hunting episode or the general presence of larger herds of reindeer in the vicinity were also considered as reasons for this difference (Bridault 1996). The seasonal indications based on horse remains from *locus* 10 were more numerous for the warm season from spring to autumn but some indications for winter were also found (**tab. 40**; Bignon 2006, 193). These seasonal indicators allowed two hypotheses: Either *locus* 10 represented an occupation event encompassing a complete year or this *locus* was formed by various occupation events at various points in a complete yearly cycle (Bignon 2006, 193). Moreover, the age structure could be described as catastrophic, although the old and very old specimens were slightly over-represented for this model (Bignon 2006, 193).

One phalange from the site was possibly a human one (Bignon 2006, 193).

Spatial organisation in general

Thus far, only the evident structure from *locus* 7 was described in more detail (**tab. 43**). However, at least further three concentrations of burnt stone material were found at the site some might represent clearings of hearths (Alix et al. 1993).

Due to the fragmented preservation of the material in depressions in combination with partial rescue excavations for instance in *locus* 6, the connection between the ten sectors was generally difficult to establish (Lang 1998, 83 f.). For instance the successively excavated *loci* 6 and 7 were situated at a distance of 500 m to each other. Presumably, these scatters were as distant to one another as was the river for these places. Moreover, erosion within these depressions could not be excluded as important agent of the spatial distribution. However, a more comprehensive analysis of the refitted material and the spatial distribution of the various materials was not yet published (Lang 1998, 104).

Chronology in general

According to the two ^{14}C results from *locus* 6 (**tab. 44**), Laurent Lang considered a taphonomic pollution of the samples as possible (Lang 1998, 99). However, the additionally made AMS dates from *locus* 10 further

contributed to a heterogeneous impression of the results. In particular, since the assemblages with high numbers of backed points were supposed to be younger than the typical Late Magdalenian assemblages such as the inventory from *locus* 6 (Lang 1998, 100).
Nevertheless, based on the composition of the assemblages in these scattered patches and the indications of the seasonality, the site was probably visited repetitively for short episodes from one or perhaps some groups of horse and reindeer hunters (Lang 1998, 99-102) and, thus, the complete material represented a palimpsest of small installations and was accumulated over some decades or centuries alongside the river.

Archaeological material of *locus* 7

Over half of the 15,984 lithics found in *locus* 7 were splinters (n=8,433; **tab. 35**; Weber 2006, 165).
The lithic raw material was dominantly local Cretaceous gravel flint. Only some pieces were made of a Tertiary chalcedony including the only core made of this material found at the site. Presumably, the resource was at a more regional distance.
In contrast to other concentrations of the site, the bladelet production was of minor importance in this area and, clearly, the blank production process was focused on blades (Weber 2006, 169-171). Only five of the 84 cores and core fragments were related to the production of bladelet sequences (Weber 2006, 170). In general, the cores showed traces of two types of percussion instruments: organic hammers as well as of soft hammerstones (Weber 2006, 173). The blades which served as blanks for points reached considerable widths and were often received from the cores by direct knapping with a soft hammerstone (Weber 2006, 172 f.). A few cores exploited with a soft hammerstone showed traces of an unskilled use of the hammerstone (Weber 2006, 173).
The 784 retouched artefacts from *locus* 7 were clearly dominated by the 313 LMP (**tab. 38**) among which 299 points and point fragments were found. These points were mainly determined as shouldered points and angle-backed points (Lang 1998, 28-36). In contrast, only 14 backed blades were found. The 121 end-scrapers were almost exclusively made on long blades which were often finely splintered along the lateral edges (Weber 2006, 167). The 86 burins were dominantly dihedral ones which were made on shorter blades. The 16 composite tools were mainly combinations with burins. Among the 80 borers, only 21 perforator examples with finely retouched tips were identified (Weber 2006, 169), the others were classified as *becs* or *Zinken*. The truncations (n=47) were generally oblique.
Stone material, in particular, sandstone was also recorded with the emphasis on thermally altered material. Some 708 fragments weighing more than 82 kg were documented.
Several faunal elements (n=225) of which 84 were taxonomically determined were found in this area. These remains were representing at least seven horses (*Equus* sp.) and a single bone was identified as originating from a bovid, perhaps, *Bos* sp. or *Bison* sp. (**tab. 39**; Lang 1998, 87). Of particular interest in comparison with the total inventory, reindeer was completely lacking in this area.

Spatial organisation of *locus* 7

During the excavation a hearth was already recognised by a cluster of heat altered stone fragments, mainly sandstone (**tab. 43**; Weber 2006, 163). However, a depression in the excavated area led to some uncertainties in the interpretation of this distribution of the material and the burnt stones were suggested as a possible accumulation due to erosion (Lang 1998, 20). Nevertheless, the accumulation was found in the centre of the lithic concentration and this correlation led the excavator to rather assume these remains as *in situ* (Lang 1998, 20). In contrast, the bones were recovered outside the margins of the lithic accumulation and, thus, led to a discussion on the association of the bones with the lithic material (Lang 1998, 20; Weber 2006, 163 f.). However, the position of faunal remains at the outer margins or even outside the lithic concentration was also reported from other Lateglacial concentrations around a centralised hearth (Jöris/Terberger 2001; Gelhausen/Kegler/Wenzel 2005). Thus, a spatial distinction of the density of lithic and organic material reflected a possible functional differentiation of various areas within the organisation of the concentration and further supported the assumption of an (almost) *in situ* situation of this concentration.

However, the distribution of lithic artefacts was relatively dense over an area of 10 m × 10 m with various smaller accumulation within. A more detailed spatial analysis of the distribution of various types of artefacts and refitting of the material would be particularly helpful to further evaluate whether different activity areas or several uniform concentrations were reflected by these accumulations. The latter would indicate that probably several independent events formed this *locus*.

Chronology of *locus* 7

Radiometric measurements were not performed for this area. Thus, the chronostratigraphic setting relied on the general lithostratigraphy and techno-typological considerations. However, as Laurent Lang already demonstrated, the techno-typological differences could also result from a functional or economic difference of the various Magdalenian assemblages (Lang 1998, 98-102; Weber 2006, 174 f.). Moreover, according to the spatial distribution (Weber 2006, 163 Abb. 3), the possibility that the assemblage resulted from multiple episodes, perhaps, different occupation events could not be excluded (Weber 2006, 175).

Le Closeau, lower horizon, Rueil-Malmaison, Hauts-de-Seine

Research history

On the western periphery of Paris, at the site Le Closeau in the municipal of Reuil-Malmaison, Hauts-de-Seine, a total of 44,041 m² (25,568 m² yielded Palaeolithic material) was excavated between late 1994 and mid-1998 prior to construction of the motorway A86 (Bodu 1998, 20. 24). Further material was yielded by the on-going excavations until early 2000 (Bodu/Debout/Bignon 2006). The excavation area was divided in various larger sectors, areas, and trenches. Moreover, at least four Lateglacial archaeological horizons (inférieur, intermédiare, supérieur, extra-supérieur) were observed. Each horizon yielded several concentrations and, thus, 72 Lateglacial units (*locus*) were distinguished. The continued excavation until early 2000 produced at least a further four Lateglacial units (Bignon/Bodu 2006; Bodu/Debout/Bignon 2006). One area of c. 170 m × 185 m was located north-west of the Route Nationale (RN) 13 and included the sector 6, sector 27, trenches 1 & 4, area I. F. P. (Bodu/Debout/Bignon 2006; furtheron »North RN13«). In this area the *loci* 1-59 were found. Farther north-westwards only the sector 5 yielded single lithic finds from the late Allerød (Bodu 1998, 27). However, another area of 7,000 m² was situated south of the RN 13 (furtheron: South RN13). This area provided mainly material from the end of the Lateglacial and encompassed a further 19 units (*loci* A-S). Nevertheless, the *loci* were not equivalent with concentrations of archaeological material. Often a single concentration spread over two *loci* or a single *locus* comprised several accumulations. Furthermore, some accumulations seemed to blur into one another especially on the South RN13 area and, therefore, the distinction of single concentrations became difficult.

Le Closeau is, thus far, the largest excavated site with archaeological remains from the Lateglacial in north-western Europe. Besides Final Palaeolithic remains, Mesolithic, Neolithic, and other prehistoric as well as historic material was uncovered. Furthermore, various palaeoenvironmental analyses were conducted parallel to the archaeological investigations. The main responsible for the excavation and analysis of the Palaeolithic material was Pierre Bodu (Bodu 1998).

Topography

The areas with Lateglacial remains were situated some 300 m south-east of the modern Seine at altitudes between 21 and 24 m a. s. l. on the North RN13 area and 25 to almost 28 m a. s. l. in the South RN13 area (**tab. 34**; Bodu 1998, 36). The river meanders strongly in the region around Rueil-Malmaison and, thus, the site lies near a south-eastern vertex of meanders and within a wide river loop which at the height of the site is about 5-6 km wide (NW-SE) and 14-15 km long (SW-NE). North-westwards and south-eastwards further wide loops of the river formed a relatively levelled basin of over 10 km in N-S direction and 30 km in W-E direction with occasional elevations such as the Montmatre in Paris. The site Le Closeau is situated in the south-western corner of this basin in the flood plain of the Palaeo-Seine. During the Pleistocene the Seine flowed probably in a widely braided system. One branch passed possibly at the foot of the adjacent mountain range and formed the c. 500 m wide modern southern embankment (Bodu 1998, 23. 36). In general, the Pleistocene terrace gradually slopes towards the modern Seine. South and south-west of the excavated area the terrain quickly rises about 100 m to a hill range which follows the river on its southern and western bank. The range is intersected by a small, gradually ascending valley *c.* 250 m south-east of the excavation area and again some 1.3 km westwards of the site by the wider valley of Bougival. Due to the geological formation, rich lithic raw materials weathered out of these ridges (Bodu 1998, 43-48). During the Lateglacial another, approximately 25 m wide branch of the Seine or an oxbow lake was situated immediately north-west of the main excavation area (Bodu 1998, 36). The presence of this open water might be the reason for the rare finds from sector 5 (Bodu 1998, 27) which was situated inside this former water course. Another depression running through the North RN13 area represented perhaps another shallow branch which was possibly still active in the first phase of the settlement. However, this channel became landed successively during the time of the Lateglacial occupation but could be re-activated occasionally until the Younger Dryas (Folz et al. 2001, 927). By cutting into the gravel terrace, a gravel ridge accumulated in the north of the North RN13 area (Bodu/Debout/Bignon 2006, 714 fig. 3) and protected partially the area in the depression against higher water tides. The terrain in the area South RN13 was already situated 1-4 m higher than the highest level of the northern area including the gravel ridge in the north. The South RN 13 area was also intersected longitudinally by an artificial water course and, furthermore, disturbed by house constructions mainly on the northern part of this area.

The topographic developments on the site during the Lateglacial also led to partially different distribution of the archaeological remains in the different horizons.

The eight units associated with the lower horizon (*couche inférieur*) were situated at the northern limits of the North RN13 area. These units were the *locus* 4, the lower horizon of the *locus* 33, the *locus* 46 (Bodu 1998, 322-421) and, possibly, some artefacts in *locus* 50 which probably belong to the material from *locus* 4 (Bodu 1998, 170). In addition, the material from the later uncovered *loci* 56-59 (Bodu/Debout/Bignon 2006) were also attributed to the lower horizon. The material was grouped along the gravel ridge north-west of the shallow palaeochannel and clearly spared out this depression which was considered as an indication that this area was perhaps still water bearing by the time of the occupation of the lower horizon. Thus, these concentrations were probably placed on an island- or peninsula-like setting during the occupation period. Only the concentration of *locus* 4 was set at the bank of this possible palaeochannel which, perhaps, also caused the stronger patination of the material from this area (Bodu/Mevel 2008, 514).

The 13 *loci* attributed to the intermediate horizon (*couche intermediaire*) were found in the extreme south of the North RN13 area (Bodu 1998, 240 fig. 205). The stratigraphic development was comparably poor in this part of the North RN13 area. For instance, the Lateglacial horizon was generally overlain immediately by a layer containing Neolithic material. The distinction of the intermediate material from the upper horizon was therefore based on techno-typological considerations and assumed to be confirmed by ^{14}C dates

(**tab. 44**; Bodu 1998, 240). However, by a critical evaluation of the dates and the consideration about the standard deviation, this confirmation cannot be given unambiguously (see p. 265-269).
From 26 *loci* material was attributed to the upper horizon (*couche supérieur*). These *loci* were generally located in the south-eastern part of the North RN13 area, immediately alongside the edges of the palaeo-channel. Some material occurred even inside the depression at the southern end of the excavated area. Thus, the channel was presumably silted up at the time of these occupation episodes.
The very ephemeral material from *locus* 25 was attributed to the top horizon (*couche extra-supérieur*). The material originated from the uppermost part of the same stratigraphic position as the upper layer of Le Closeau (Bodu 2000a, 17). This *locus* was excavated on a surface of 25 m² in the central southern part of the North RN13 area during the summer 1996 (Bodu 1998, 236). Furthermore, most of the material from the South RN13 area was attributed to this uppermost horizon.

Stratigraphy
Almost 85 profiles were recorded on the two areas with Lateglacial remains (Bodu 1998, figs 35-36). In general, the site was set on the eastern bank of the Seine where a Weichselian gravel terrace reached a thickness of *c.* 10 m above the bedrock.
In the South RN 13 area, the gravels were interspersed by sandstone scree from the adjacent hills and were overlain by a deposit of brownish yellow silty gravels in the northern part (Gebhardt 1998, 68). These silty gravels were attributed to the Holocene but occasionally in the lower part some orange-brown sandy silts were preserved and regularly associated with some Palaeolithic material (Bodu 1998, figs 479-481). On top of the silty gravels followed a thin layer of Holocene brown silts which were overlain by brownish black modern rubble. Thus, in the South RN 13 area the stratigraphy was reduced significantly which was mainly due to erosive processes. Therefore, the distinction of the various archaeological horizons became difficult and, furthermore, in this part of the site only lithic material was preserved from the Lateglacial (Bodu 1998, 433).
In the part North RN 13 the stratigraphy was more complex, probably, due to representing the development of a braided flood plain. In this area yellowish fluvial sands followed generally on top of the Weichselian gravels. The lower parts of these sands were oxidised and of yellowish orange colour and contained only few gravels (Gebhardt 1998, 64). Light yellowish brown sands which were widely whitened and of a glossy structure followed on top. These two sand deposits were laminated. Furthermore, in the light yellowish brown sands the archaeological material of the lower horizon was found. In the area north-west of the gravel ridge, micro-morphological studies indicated pedogenic processes associated with the lower archaeological horizon (Chaussé 2005). However, the soil development in this layer was weak and Christine Chaussé concluded that the pedogenic process represented perhaps a first weak stabilisation of the grounds in the early Lateglacial Interstadial (Chaussé 2005, 93). The sands containing the lower horizon were overlain by a light yellowish orange sandy loam. In depressions on the site, this loam was covered by a gleyed sandy loam of yellowish brown colour on top of which clays followed. The two sandy loams were affected by processes of the pedogenesis resulting partially in a greyish colouring of the sediments. The archaeological material of the intermediate and upper horizon as well as the material of the top horizon (*locus* 25) originated generally from the upper part of these greyish deposits. Samples from the upper sandy loams failed to produce microscopic indications for particles of the LST (Pastre 1998). These upper loams were overlain by silts which were of light grey to yellow-orange colour but occasionally lenses of gravels in a sandy-silty matrix were interspersed. The silt deposit was partially sealed by bands of small calcareous gravels. The upper band formed already part of another yellowish brown silt deposit containing Neolithic artefacts and occasionally small gravels (Gebhardt 1998, 63). Sometimes, this deposit was slightly gleyed in the lower part. On top of the silt

deposit a dark organic silty soil formed. Further towards the Seine the latter two deposits were intersected by an up to 1 m thick deposit of whitish high flood silts. The sequence ended with the modern anthropogenic fillings of the flood plain formed by stone rubble in a context of brownish black organic silty soil.
The preservation of pollen in the sediments from the areas North and South RN13 were poor, prohibiting reliable classification (Leroyer/Allenet 1998, 100). However, in the area northwards of the gravel ridge (sector 3 which yielded mainly Mesolithic and Neolithic remains) deposits of a presumable Lateglacial age were reached and correlated to the layers on the area North RN13 (Leroyer/Allenet 1998, 106). According to this correlation, a four-phased Allerød section and a Younger Dryas section with three phases were identified palynologically in the greyish humic and the overlying clayish deposits. Both sections were dominated by pine (*Pinus* sp.) but in the younger section herbaceous pollen and birch (*Betula* sp.) became more numerous indicating an opening of the landscape. Furthermore, the lowest phase of the Allerød pollen deposit was characterised by a generally low number of arboreal pollen among which nevertheless pine dominated and a high number of pre-Quaternary pollen. These numbers indicated a poor and/or selective pollen preservation. Moreover, the assumed Younger Dryas section was found in sediments which were correlated to the upper Lateglacial Interstadial deposits on the North RN13 area (Leroyer/Allenet 1998, 106f.). The malacological analysis from this area suggested that the onset of the Younger Dryas might be found in the white silts on top of the correlated deposits (Limondin 1998, 114). Nevertheless, the preserved pollen in the stadial section and the lower Allerød phase (Clo a1) were only found in one test pit (no. 17). In regard to the complex stratigraphic development in the flood plain, a stratigraphic mis-correlation and/or erosion were further possible explanations as well as a disturbed palynological record due to a particular local vegetation development. Certainly, the pollen profiles indicated a generally light forested phase dominated by pine overlain by an even more open landscape with still existing light forests. Nevertheless, the exact correlation with the archaeological horizons remained unclear and, thus, the palynological analysis could not further refine the chronostratigraphic position. However, samples analysed from the various charcoal concentrations in the archaeological horizons were determined exclusively as Scots pine (*Pinus sylvestris*; Pernaud 1998). Since several of these charcoal concentrations were assumed to originate from natural fires, the combination of the correlated pollen profile and the anthracological analysis suggested an environment dominated by a stand of Scots pines in the river plain during the time from the end of the Lateglacial Interstadial to the onset of the Holocene. Additionally, the number of aquatic plant pollen typical for river plains decreased successively throughout the phases indicating an increasingly dry environment of this part of the plain during the Younger Dryas and suggesting a down cutting of the river into the modern bed (Leroyer/Allenet 1998, 108f.). In contrast, the malacological samples from the north-western part of North RN13 indicated increasing humidity from the early Lateglacial Interstadial to the Lateglacial Stadial (Limondin 1998, 115f.). Moreover, the samples taken from the palaeochannel deposits in the northern North RN13 area showed an increasing amount of aquatic molluscs species from base to top of the greyish deposits (Limondin 1998, 119). However, the preservation of shells in the sandy environment is usually poor (Limondin 1998, 118). Nevertheless, in comparison with malacological samples taken at other areas of Le Closeau the presence of freshwater species clearly increased in the palaeochannel part of North RN13 in the greyish deposits. However, of particular interest was the determination of the species *Avenionia brevis* which preferred underground water environments suggesting an at least regular upwelling of the ground water table in this area and, consequently, might explain the general absence of archaeological material in this possibly swampy channel band during the Lateglacial. Furthermore, a silting up of palaeochannels could explain the discrepancy between the decrease of water plants and the increase of molluscs inhabiting wet environments. However, these detailed analyses clearly indicated the specific local development of the Seine flood plain at Le Closeau and, therefore, restricted comparison with more regional or global developments.

With the lower archaeological horizon neither pollen nor charcoal could be associated, but molluscs and small mammal remains made possible some comments on the environment. However, the malacological zone summarised the whole Lateglacial Interstadial in one lower and the Lateglacial Stadial in another zone. The former was characterised by a dry and cool environment. Such a climate was assumed possible for GI-1d and, thus, in accordance with the calibrated ^{14}C dates from *locus* 4 (**tab. 44**) which was located adjacent to the profile. However, some of the cold adapted species found in this profile rather indicated special temperature ranges and were considered as pioneer species (Limondin 1998, 115). These species emphasised the presence of the on-going re-settlement process in the Lateglacial Paris Basin. Small mammal remains were only preserved in the lower layer at *locus* 4 but provided an environmentally indifferent picture for a flood plain with only remains of water vole (*Arvicola terrestris*) and common vole (*Microtus* cf. *arvalis*) as well as of an amphibian being determined (Mistrot 1998, 344).

Archaeological material in general
In total, some 48,000 Lateglacial lithic artefacts were found (**tab. 35**).
Among the lithic raw materials the Campanian flint which occurred at the site and in the immediate surrounding in secondary deposits (**tab. 36**) was clearly preferred in all horizons. Furthermore, a high-quality Tertiary chalcedony which might be of foreign origin (Bodu/Cary 1998, 335) but might as well be gathered at the foot of the hills west of Rueil-Malmaison (Bodu 1998, 44) was identified sporadically in the inventories. However, in the intermediate horizon this material was slightly more frequent, whereas it was absent from *locus* 25. Nevertheless, the retouched pieces of the latter *locus* were made of a higher quality material which perhaps originated from elsewhere (Bodu 1998, 237).
The blank production of the lower horizon was still very similar to the Late Magdalenian. In contrast, the concept of core reduction was more flexible in the upper horizons and appeared often unsystematic. Moreover, an increasing number of blanks yielded indications for the use of hard hammerstones in the main blank production process. However, the debris material from the top horizon was too small to produce significant indications.
Among the retouched lithic artefacts, the LMP dominated in all horizons accompanied by the end-scrapers (**tab. 38**).
In the greyish deposits some fragments of sandstone and, rarely, of quartzite were presumably used as hammerstones as well as a flint nodule with thick cortex.
In addition, inside a perhaps anthropogenic combustion structure in *locus* 38 a limestone was deposited which wore parallel ripple lines and traces of heating. Linear engravings in the cortex were also observed on five pieces from *locus* 20. In *locus* 28, a small bead was recovered but the association with the Lateglacial archaeology remained unclear. A pointed lump of ochre was found at the periphery of *locus* 14.
Furthermore, 5,000 fragments of faunal remains were attributed to the Lateglacial at the site. The majority of determinable remains were found in the lower horizon; in the upper horizon the preservation was limited, and in the intermediate horizon little faunal material was preserved. In the *locus* 25, no faunal material was found. However, in the upper horizon red deer (*Cervus elaphus*) superseded horse (*Equus* sp.) as dominant species. Nevertheless, both species were found in the two horizons. These species were the only commonalities. For the other species from the greyish deposits the determination was uncertain and/or the relation to the archaeological remains was unclear. Only for the lower horizon seasonal indicators were found (see p. 217).
In detail, the material was relatively diverse. The material from the lower horizon was attributed to the transition period from the Late Magdalenian to the FMG, whereas the remains from the intermediate, the upper, and the top horizon were clearly equivalents of the FMG material from the Central Rhineland. Therefore, only the lower horizon is described in more detail below.

site	sample	years OSL-BP	± years	material	comment	ref.
Le Closeau, C04, west of *locus* 43	A	19,200	1,300	sand size grains with few fine grain particles and clay	white sandy loam (attributed to Younger Dryas)	1-2
Le Closeau, C04, west of *locus* 43	B	18,400	1,300	sand size grains with few fine grain particles and clay	grey sand (attributed to the Lateglacial Interstadial)	1-2
Le Closeau, C04, west of *locus* 43	C	21,900	1,500	sand size grains with few fine grain particles and clay	yellowish sand (Weichselian fluvial terrace)	1-2

Tab. 45 OSL dates from the stratigraphy at Le Closeau. References (ref.): **1** Folz et al. 1998; **2** Folz et al. 2001.

Spatial organisation in general

Since the *loci* were not equivalent to concentrations, the distinction of single accumulations was difficult. However, counting from the lithic density map (Bodu 1998, fig. 2) the archaeological material was organised in some 90 concentrations (*c.* 60 in North RN13 and *c.* 30 in South RN13). Nevertheless, the archaeological material was published according to the *loci*. Consequently, in some inventories the impression of lacking parts could arise because this part was published as another *locus* or assemblages represented admixtures of spatially different episodes. These constraints should be kept in mind when reading numbers given for single *loci* from Le Closeau. Therefore, the major archaeological horizon (lower and greyish deposits) are given in the following. However, such large units are even more inevitable for admixing diachronic characteristics and, therefore, some single units are described as representative from the lower horizon. The material from South RN 13 yielded indicators for an even younger occupation of the site, perhaps, up into the Holocene but due to the lack of stratigraphic or radiometric indications these concentrations cannot be positioned reliably in a chronostratigraphic framework and, therefore, the South RN 13 is not included further in the present work.

On the site more than 350 »structures de combustion« were identified in the main archaeological horizons (Bodu 1998, 41; cf. Bodu/Debout/Bignon 2006, 718). These structures were filled with charcoal and, thus, yielding the main preserved organic material. However, some of these patches of charcoal were very small and referring to these structures as anthropogenic hearths was not always possible. In fact, some of these structures were probably results of natural events which were perhaps frequently a cause of large amounts of burnt lithic material (Bodu 1998, 49). Perhaps, the on-going analysis of the Lateglacial remains in the flood plain at Le Closeau (e.g. Bodu 2000b; Bodu/Debout/Bignon 2006; Bodu et al. 2009a) will provide more precise distinction between the natural and the anthropogenic patterns in the future.

Perhaps, a functional interpretation of the lithic assemblages from the different horizons – possibly, in combination with a temporal distinction – should be considered in the future research. Such results could possibly supplement the knowledge on complex spatial behaviour patterns in the FMG. Certainly, the on-going research such as refitting of material and spatial analyses on the rich inventory from the whole site will produce further insights in the settlement dynamics of this important site (Bodu/Debout/Bignon 2006, 718).

Chronology in general

Several ^{14}C dates taken on samples of charcoal and bone were helping in combination with the stratigraphy and the varied environmental analyses to reconstruct a reliable chronology for the development of the site. Nevertheless, the charcoal samples frequently originated from burning events which were not securely associated with the human occupation of the site and, thus, dated perhaps natural fires in the Lateglacial (Bodu 1998, 58). Thus, the ^{14}C dates from the site require a more comprehensive evaluation to become meaningful (see p. 265-269 and p. 474-481).

OSL measurements sampled at various heights in the stratigraphy of the intra-site palaeochannel produced clearly overestimated results (**tab. 45**). However, the reason for this false dating remained unclear (Folz et al.

1998; Folz et al. 2001), although the setting in fluvial sediments was a cause for a post-human displacement of the dated sediment (cf. Folz et al. 2001, 932).
In general, the sediment containing the lower archaeological horizon was attributed to the early Lateglacial Interstadial (GI-1e and, perhaps, GI-1d) based on the lithostratigraphy, the environmental indicators, and the radiometric results. The greyish deposits were generally attributed to the younger Lateglacial Interstadial (GI-1b and GI-1a) and the older part of the Lateglacial Stadial (GS-1) due to the stratigraphy and the environmental indicators. Although three archaeological horizons were distinguished in this deposit, the stratigraphic and chronological distinctions for the intermediate and the upper horizon were very thin. In fact, the stratigraphic positions and the radiometric results of the intermediate, upper, and top horizon were significantly overlapping (**tab. 44**). Therefore, the distinction of the archaeological horizon in the greyish deposits was based on typo-technological considerations. In general, the material of all three horizons in the greyish deposits was attributed to the FMG with an early phase (intermediate horizon), a typical phase (upper horizon) and, perhaps, a younger phase (top horizon) which was comparable to Bad Breisig in the Central Rhineland. However, the distinctions were gradual and other explanations such as functional differentiations of the inventories or the duration of use were possible alternative explanations. Probably, the on-going analysis of the settlement dynamics of these horizons from the greyish deposits will help explaining the differences in the material of the various *loci* in the future.
Nevertheless, a clear gap between the lower horizon (GI-1e – GI-1d) and the archaeological material from the greyish deposits (GI-1b – GS-1) became apparent by this short chronostratigraphic presentation. Whether this lack of material from GI-1c was due to the topography (concentrations were not excavated), the taphonomy (concentrations were not preserved), the chronostratigraphy (concentration were not identified as mid-Lateglacial Interstadial), a true settlement gap, or combinations of these possibilities cannot be answered by the present work. However, this observation of a gap in the data is of some significance in the consideration about the precise development from Late Magdalenian to the typical FMG.

Archaeological material of the lower horizon
5,650 lithic artefacts were found in the lower horizon (**tab. 35**). A further 337 lithic remains without traces of knapping were recorded (cf. Bodu 1998, 169f. tab. 12). In addition, approximately 1,500-2,000 knapped pieces were recovered from the *loci* 56-59 (cf. Bodu/Debout/Bignon 2006, 714 fig. 3) but exact numbers of these assemblages are not published yet. Therefore, the numbers in the following will only refer to the inventories from the *loci* 4 and 46 as well as the ones from the lower layer at the *loci* 33 and 50. The raw material is majorly Campanian flint which can be gathered on and around the site from secondary deposits (**tab. 36**). Only five blades, one of which was transformed into an end-scraper, were made of a high-quality Tertiary chalcedony which might be of foreign origin (Bodu/Cary 1998, 335) but might as well be gathered at the foot of the hills west of Rueil-Malmaison (Bodu 1998, 44).
47 cores were found in the four areas (**tab. 37**). The aim of the blank production process were clearly blades which is further supported by the almost exclusive use of this blank type for the tool production (Bodu/Cary 1998, 336). The characteristics on the blanks indicated the direct use of soft hammerstones in the production of the blanks. Such soft stones made of granite, limestone, sandstone, or flint with a thick calcareous cortex were also found in *locus* 46, partially with indications of use such as splintering, scratches, or picking. However, in *locus* 4 some blades with the »*en éperon*« butt type indicating the use of organic hammers were found, whereas this type was completely lacking at *locus* 46. In general, the butt was not retouched or wore only little modifications on the blanks in both assemblages.
In total, 861 artefacts were classified as tools in the lower horizon of Le Closeau (**tab. 38**). However, Pierre Bodu assumed that the modification on 366 pieces resulted from use rather than intended retouch (Bodu

1998, 170f.). Among this artefact class were also grouped specimens which otherwise would be classified as notched or laterally retouched pieces as well as burin spalls and other debris of the retouching process. Since these pieces were usually not displayed in figures, they cannot be exactly addressed. Furthermore, a conclusion on whether a retouch resulted from use or intention is often difficult when analysing the actual piece and even more difficult based on a drawing. Therefore, this group is generally not neglected in the classification of the retouched artefacts (**tab. 38**). Moreover, of the remaining 495 retouched artefacts were a further 267 blades identified as bruised blades. Again, judging from the drawings the bruised blades generally also form a heterogeneous group. They represented pieces with various kinds of modification along their edges such as notches or lateral retouch and are counted among the tool class »others« in the present study.

Thus, based on these restrictions, the most numerous formally retouched artefacts in the lower horizon were end-scrapers (n=98) followed by the LMP group (n=81). Judging from the relevant material displayed from the *loci* 4 and 33, the end-scrapers were made on long blades and the preferred point shape was the bipoint (Bodu 1998, figs 415. 418-419). The other tool classes were significantly less frequent with 29 burins, eleven truncations, and eight composite tools. The latter were mainly combinations of burin and end-scraper. Finally, one splintered piece was found among the material from *locus* 46.

Micro-wear analyses on almost 800 blanks from the lower horizon supported the hypothesis of intensive processing of meat and skin at this site (Christensen 1998; Bodu/Cary 1998; Mevel 2004). Nevertheless, the analysed lithics were used presumably also on bone, plants, and minerals (Christensen 1998) as well as on cortex which was also proven by the parallel linear engravings on remaining cortex parts of some flints (**tab. 42**; Jessen 1935).

In the *locus* 46, seven nodules with picking, scratching, and splintering marks were found indicating a use as hammerstones (cf. Bodu/Mevel 2008). On one of these nodules also some red colourant was preserved (**tab. 42**; Bodu/Cary 1998, fig. 434bis.2). Moreover, artificial grooves were observed on a sandstone block found in the same area. Another special find was an unmodified flint nodule found at the periphery of *locus* 46. The piece was heavily reddened by ochre. Possibly, this colouration is associated with the particular shape of the piece reminding of an anthropomorphic figure (Bodu/Cary 1998, fig. 437.2). However, the cylindrical form was perhaps simply handy and used in the grinding procedure of the red mineral.

In addition, in the *loci* 4 and 46 several boulder of sandstone, gritstone, and limestone were found. The lithic and faunal concentrations seemed to be limited by these stones weighing some 400kg (*locus* 4) and 600kg (*locus* 46) in total (Bodu/Debout/Bignon 2006).

In this horizon the preservation of faunal remains was significantly better than in the other horizons (**tab. 39**). The main represented species was horse (cf. *Equus arcelini*; Bignon/Bodu 2006) followed by red deer (*Cervus elaphus*) and wild boar (*Sus scrofa*; Bemilli 1998). Occasionally, also hare (*Lepus* sp.) remains were found. Rarely, other species were determined such as wolf (*Canis lupus*), cave lion (*Panthera spaelea*), a bovid (presumably *Bos* sp.), and a bird (Aves; Bemilli 1998, 405). Rarely carnivore gnawing was observed which, in general, was found rather on the material not affected by weathering (Bignon/Bodu 2006, 408). Also the traces of human processing of these bones was more commonly observed on unweathered material, although this material was usually more fragmented, probably due to human impact. In addition, this fragmented material was often burnt. Nevertheless, the traces of fire could have a natural cause in Le Closeau. According to the presence of body parts, the horses were processed completely at the site, whereas the meat parts of wild boar were probably taken elsewhere (Bignon/Bodu 2006). Based on the dental material of horse mainly from *locus* 46, these animals were hunted in all seasons (**tab. 40**; Bignon/Bodu 2006, 411f.). In particular, the late winter-early spring and the autumn period were indicated, whereas the indications for summer were least frequent. In contrast, the indications from *locus* 4 were very meagre and rather pointed to late spring and early summer as hunting season (Bodu/Debout/Bignon 2006, 723). Comparable to the considerations formulated about the material from Le Tureau des Gardes, these heterogeneous seasonal indicators could imply recurrently used areas or areas used over a considerably longer period. Three herbivore rib fragments were modified to a pointed shape which perhaps was used in the processing of animal skins (**tab. 41**; Bemilli 1998, 402). Furthermore, three bones of cave lion (n=2) and hare (n=1) show striated traces of human modification.

Spatial organisation of the lower horizon

Spatially, the remains of the lower horizon were found centred in two main areas:

In the south-west, the *locus* 4 was set immediately south-east of the gravel barrier and in the north-west, the *locus* 46 was located some metres north-west of this barrier and already some 2 m lower than the summit of this mound.

The main concentration of *locus* 4 was a dense accumulation of lithic artefacts with the vast majority of retouched artefacts deposited in the periphery of the main concentration. In addition, the few remains of the lower horizon of *locus* 50 represented a western appendix of the material from *locus* 4 rather than representing an independent working area (Bodu/Cary 1998, 322). Hence, these specimens were added onto the assemblage from *locus* 4. A large area of sediment inside the main concentration of *locus* 4 was blackened with charcoal. A few centimetres westwards of this blackened area, a zone of reddish colour was found as well as a very dense cluster of lithic material. On the opposite site of the charcoal patch, another but much smaller coloured zone was found. Presumably, these coloured zones represented a central hearth which was cleared occasionally (**tab. 43**; cf. Bodu/Debout/Bignon 2006, 721). Towards the north and east, the main concentration of *locus* 4 was limited by a small accumulation of large boulders (n=60; Bodu/Debout/Bignon 2006, 720). This limit was set some 20-50 cm away from the dense concentration. Some smaller stones scattered farther south-eastwards. This scatter was supplemented by lithic material forming a small concentration. East of this small concentration, a patch of reddish sediment and charcoal was found yielding only few lithic artefacts. This structure was either an external hearth or more probable an external dump area (cf. Bodu/Debout/Bignon 2006, 721). Some 4 m north-eastwards of the boulder accumulation another small lithic concentration was found. In the eastern periphery of this accumulation a small concentration of retouched artefacts was found.

Comparable to the *locus* 4, the *locus* 46 was also organised around a central hearth and limited by large boulders (almost 200; Bodu/Debout/Bignon 2006, 720f.). However, in *locus* 46 the stones were completely encircling the main concentration at a distance of some centimetres to the main lithic scatter. In addition, these boulders formed some type of an inner structure (cf. Jöris/Terberger 2001, 168-171). Again, the majority of retouched pieces was found in the periphery of the main concentration inside the boulder limits. Furthermore, outside the circle were again two further lithic concentrations, one towards the west and one towards the east of the boulder structure.

Moreover, the north-eastward adjacent *locus* 33 was connected to the main area of *locus* 46 by refitted lithic material. Probably, this additional concentration served as workshop for the retouching of tools, in particular hunting equipment, and/or as a butchering place (Bodu/Debout/Bignon 2006, 722). This setting again reflects the pattern of *locus* 4 where besides the retouched material in the periphery of the main concentration, another concentration of lithic tools was found in the north-eastern concentration.

Additionally, the *loci* 56 and 58 were probably satellite workshops as the refitting of lithic artefacts between these areas indicated (Bodu/Debout/Bignon 2006). Furthermore, the presence of cave lion in *locus* 56 and 46 suggested a connection of these satellites to the main concentration of *locus* 46. Hence, the concentrations in the *loci* 33, 56, and 58 were presumably supplementary working areas to the main concentration of *locus* 46.

Pierre Bodu stated that the lithic material was organised and not distributed randomly around the hearths (Bodu 2010). For instance, some of the blanks used for processing vegetal material were found in the periphery of the inner structure of *locus* 46 (Christensen 1998; cf. Jöris/Terberger 2001). Moreover, the faunal remains were clearly concentrated around the central hearths (Bodu/Mevel 2008). In *locus* 46, further faunal remains were found in the additional concentration in the west of the larger structure. In addition, blades with splintered edges by use were also found concentrated around the hearth areas (Bodu/Mevel 2008). Thus, some intense works related to the processing of faunal material seemed to have taken place in these areas.
From the *locus* 46 two blade fragments were connected by refitting to a blank production sequence from *locus* 4 which was set in a distance of some 80 m (Bodu/Debout/Bignon 2006, 722). Of some significance was the observation that the debris of the blank production was heavily patinated, whereas the blade fragments found at *locus* 46 were not patinated. This difference suggested that the blades reached *locus* 46 in a relatively fresh state and were not deposited in *locus* 4 earlier. However, various explanations could be found for this singular refittings and, therefore, Pierre Bodu and his colleagues remained sceptical whether this indication already sustained a contemporaneity of the two occupations (Bodu/Debout/Bignon 2006, 722). The composition of the processed faunal remains was almost identical in *locus* 4 and 46 (Bemilli 1998; Bignon/Bodu 2006). The faunal material could not be directly refitted even though it appeared very similar. Nevertheless, the horse material seemed to represent, at least, two independent cycles of animal processing centred on the two main concentrations (Bignon/Bodu 2006).

Chronology of the lower horizon
The samples of the faunal remains from the *loci* of the lower horizon were ^{14}C-dated and produced in general early Lateglacial Interstadial ages (**tab. 44**). In particular, the two reliable ^{14}C dates from *locus* 4 indicated after calibration a chronostratigraphic position in GI-1d which is in accordance with indications from the nearby sampled malacological analysis. However, this attribution to a severe cold episode seemed to contradict the faunal composition including wild boar. Perhaps, the occurrence of these forest inhabitants could be explained by sustaining forest stands in the protected river flood plains. Nevertheless, some samples of wild boar resulted in significantly younger dates proposing that either this animal was dispersed at a later period or that a systematic error occurred in these samples. The reliable ^{14}C dates from the *locus* 46 and its satellites produced older results falling to the transition from the Late Pleniglacial (GS-2a) and the onset of the Lateglacial Interstadial (GI-1e). This transition period was associated with rapid temperature rises and gradual afforestation. Nevertheless, in the early part of this transition still cold and dry environments as suggested by the malacological analysis still prevailed. Thus, the rare environmental data could not help to further refine the chronostratigraphic attribution of this horizon.
Based on the faunal material as well as the completeness of the lithic inventories at both areas, the two sub-areas of the lower horizon in Le Closeau seemed to represent two events which were temporally distinct. Whether the offset was only seasonally or, in fact, comprised some centuries could not be resolved, although the comparable typo-technological patterns, the similar spatial organisation, and the very equivalent faunal composition rather support the hypothesis of a short temporal offset. Furthermore, the singular refittings sustained a close chronological relation. However, the presence of various satellite concentrations and the effort in the construction of *locus* 4 and even more so *locus* 46 suggested a longer temporal use of these structures comparably to the Late Magdalenian installations at Gönnersdorf or the lower horizon of Andernach (see p. 89-93 and p. 111-122). Whether these central areas were used for one longer halt or repetitively could not be answered unambiguously. Nevertheless, if the settlement occurred repetitively, the various visits were not related to a single season.

Cepoy, Loiret

Research history

After the discovery of archaeological material during quarry works near Cepoy in 1972 a test pit program under the direction of Jacques Allain was conducted to decide which part of the site should be protected from quarry works for the future. Accordingly, the site *La Pierre aux Fées* was divided into two sectors. In the sector where the test pits yielded less material (sector 1) 205 m² were excavated in 1972 by François Guillon and Dominique Jagu with their team in a rescue project. The material was collected per square-metre and distinguished in two Palaeolithic horizons (horizon III and horizon IV). Sector 2 yielded a denser scatter of archaeological material in the test pits and was put under governmental protection. Between 1972 and 1977, another area of 150 m² was excavated by François Guillon and Dominique Jagu with their team in the protected sector, some 50 m north-east of the first excavation area. This second excavation was conducted according to modern standards with three-dimensional documentation and sieving of the sediment.

Topography

The site is situated on the eastern bank of the river Loing (**tab. 34**), some 500 m west of the town centre of Cepoy. In this part the river formed an approximately 1 km wide valley. Only at the site the valley is slightly narrowed to approximately 800 m and the valley turns from a SW-NE direction to a SSW-NNE direction. On the western side, the terrain rises steeply to a plateau which is elevated some 20 m above the valley floor. The hills on the eastern side rise also in a step of some 20-30 m but then the terrain rises further to a ridge some 50 m above the valley floor. Small valleys cut these uplands on both sides of the river valley only a few 100 m south of the site. Some 3 km south of the site the valley opens to a large basin around Montargis. Modern quarry works have exploited the floodplains around Cepoy and the resulting pits were subsequently filled with water leaving an artificial lake landscape. Thus, today the remaining part of the protected site is situated on an island within one of these lakes.

Stratigraphy

The stratigraphy began with alluvial gravel deposits which were overlain by up to 0.6 m of clay. In this deposit a lower archaeological horizon (V) was found yielding material which was attributed very generally to the Late Magdalenian. Only in this horizon some undetermined bones, charcoal, ashes, and ochre were found. However, these remains were only excavated in a small area of the protected part of the site and the rest remained covered *in situ* for future research. This material is not further considered in the present study. On top of the clay, yellowish white fine sands were deposited in a varying thickness of 0.1-0.8 m. Archaeological horizons were found at the base (horizon IV) and at the top (horizon III) of this sand deposit indicating that these occupations were placed on a sandy beach alongside the river. Most of the artefacts from horizon IV were found in a horizontal position (Valentin 1995, 314) suggesting little movement within the sediment and, hence, no recognizable disturbance of this archaeological material. This layer produced the majority of material and was used to describe the »faciès Cepoy/Marsangy«. The horizon III was mostly destroyed during quarry work and yielded too few diagnostic artefacts to further classify the industry. Comparable to the horizon V, this material is not further considered. The sands were overlain by some 0.3 m of topsoil in which the upper two archaeological horizons were found.

In some parts of the site, periglacial processes moved the alluvial gravels vent-like upwards into the fine sand deposit (Allain et al. 1978). Thus, stone material became naturally introduced into the archaeological horizons. In a maîtrise thesis, Claire Guillon reconstructed the presence of a clay-filled channel across sector 2 which she assumed to be younger than horizon IV (Wenzel 2009, 48 f.). However, Stefan Wenzel argued that this channel did not affect the archaeological horizon according to the horizontal position as well as the inclination of most artefacts in the vicinity of the channel (Wenzel 2009, 49). Nevertheless, some younger archaeological material such as pottery was found in the horizon IV clearly attesting some disturbances in the stratigraphy.

Archaeological material of horizon IV

The material of sector 1 from Cepoy comprised c. 14,500 lithics (**tab. 35**) and was used by Boris Valentin as basis for the definition of the Magdalenian »faciès Cepoy/Marsangy« (MfCM; Valentin 1995, 352).
The sector 2 provided more than 15,000 finds (Valentin 1995, 313) but among these finds were many unmodified gravels (Valentin 1995, 315). In contrast to the test pits, the find density in the excavated parts appeared poorer in sector 2 than in sector 1. A thorough analysis of the material from sector 2 was thus far not published (Wenzel 2009, 47). However, Stefan Wenzel included a short description of the excavated material from horizon IV in his chapter on the settlement structures of sector 2 (Wenzel 2009, 47-61).
Local Cretaceous flint which could be gathered from the river gravels was dominantly used as raw material (**tab. 36**). This local material was of a good knapping quality. In sector 1, only eight artefacts were made of different materials. The resources of these materials are unknown. However, three tools were made of a material which was comparable to Tertiary chalcedonies from the Île-d-France or material found near Orléans. An end-scraper and a point were made of materials which, perhaps, were also formed in the Tertiary. A further two points and two blades were made of foreign silices. In sector 2 only two artefacts made of different raw materials were found. One blade was made of a flint with rich microfossil inclusions and a blade fragment was made of a Senonian flint (Wenzel 2009, 51).
In total, 355 cores were recovered in Cepoy (**tab. 37**). For sector 1, Boris Valentin analysed the reduction sequence in detail (Valentin 1995, 343-370). The reduction sequence showed all the defining characteristics of the »faciès Cepoy/Marsangy« with the use of soft organic hammers for the production of long blades and the knapping with soft hammerstones for the production of short blades and flakes. Bladelets were no particularly required blanks. In sector 2, blades with a butt of the *en éperon* type were found (Wenzel 2009, 50) as well as cores indicating the production of short blades (Wenzel 2009, 53 Abb. 54.1). In addition, four flint nodules were found which were used as hammerstones according to the traces on their cortex (Wenzel 2009, 51. 54 Abb. 55).
Among the 142 retouched artefacts from sector 1, end-scrapers which were usually made on long blades were most numerous (n=39; **tab. 38**). In sector 2, the end-scrapers also formed the most numerous group (n=29) of the 125 retouched artefacts (Wenzel 2009, 51). Similar to sector 1, these tools were commonly made on long blades. In sector 1, the majority of borers (n=34) could be classified as *bec* (n=13) or *Zinken* (n=7). Although ten of the 16 borers were also classified as *bec/Zinken* in sector 2 (Wenzel 2009, 51. 57), this class was not as prominent as it was in sector 1. In sector 2, the LMP formed the second numerous group (n=22). Among the LMP were almost as many backed bladelets (n=8) found as there were shouldered points (n=9). In addition, five backed forms were found of which some were identified as bipoints, although the fragments might also be classified as shouldered points. In difference to sector 2, the 25 LMP of sector 1 were clearly dominated by 19 shouldered points and four curve-backed points but only two backed blades were identified. Among the 18 burins of sector 1 and the eleven burins of sector 2 was the dihedral type the most common one.
On several cores of sector 1 the cortex was engraved with lines. On two stone slabs (one fragmented into two pieces) figurative engravings were found. One shows the posterior legs of a quadruped, a further decoration shows a detailed horse head (Allain 1975, 468).
No organic material was preserved in this horizon.

Spatial organisation

Some hearths, clusters of heated stones, concentrations of lithic material, and possible stone settings were observed during the excavation (**tab. 43**; Allain et al. 1978, 10f.). However, thus far no detailed spatial analysis of sector 1 was published.

In contrast, the spatial distribution of the material of sector 2 was analysed in an unpublished maîtrise by Claire Guillon and a detailed study on the western part of this sector was published by Stefan Wenzel (Wenzel 2009, 47-61). A possible disturbance of the material by a fissure or channel in this area suggested by Guillon was ruled out by Wenzel according to the vertical and horizontal distribution of finds (Wenzel 2009, 48f.).

In the western part of sector 2, two evident structures were found. These structures were formed by dense clusters of partially burnt stone material and were interpreted as hearths (Wenzel 2009, 51-57). The northern one was mainly formed by gritstone plates and the other one was filled with gravels of flint and set almost centrally in the analysed area (Wenzel 2009, 51). However, the two areas were connected by a refitted plate (Wenzel 2009, 52). Burnt lithic material seemed to occur frequently and was spread over large parts of the analysed area. However, a concentration of burnt flint material was found east of the central hearth.

Moreover, several small concentrations of cores and blank production debris were found and considered as possible working spots. However, the cores were distributed relative randomly in the analysed area. Only in the area east of the hearth cores were significantly scarce. In contrast, a dense accumulation was found north-east of the southern hearth. In addition, the retouched artefacts were also spread on the complete area but a significant accumulation of LMP, in particular, shouldered points was found around the central hearth. Another concentration of various types of retouched pieces was located east of the central hearth. Perhaps, this area represented a place where blanks were transformed into tools. The hypothesis was further sustained by the spread of Krukowski micro-burins in this area (Wenzel 2009, 57). Between this concentration and the core concentration in the north-east, an area with relatively few artefacts was found. The dense concentration of burnt flint material was spread from the accumulation of retouched material into this artefact poor area. Refitting connected various concentrations but the core concentration in the north-east of the central hearth was spared out by these refits. In addition, Stefan Wenzel noted that the refitting lines seemed to create two clusters. These fan-like clusters were separated by the central hearth and ran into different directions: one fan towards the north-west and the northern hearth and the other fan spread to the south-east. According to Wenzel, this pattern seemed to show that activities were performed in strictly separated areas (Wenzel 2009, 58). Furthermore, he considered this strongly standardised occupation behaviour as only comparable to Late Magdalenian sites (Wenzel 2009, 61). Nevertheless, Stefan Wenzel also stated that some areas were affected by various settlement dynamics (Wenzel 2009, 57) and he concluded that repetitive but standardised activities formed the analysed area (Wenzel 2009, 61).

Even though no organic material was preserved, the comparison with other sites in river valleys such as Le Grand Canton and the dominance of end-scrapers and burins suggested the processing of faunal remains as an important task at the site. The number of LMP was relatively small but their distribution around hearth structures suggested some hafting and re-tooling activity there. In contrast to the retouched implements of the so-called *fond commun*, LMP were made at a site but used elsewhere. Thus, the scarcity of LMP at a site implied that the necessity of discarding used material was not very high and/or the rejects in the production were low. Furthermore, the blank production as well as the modification of lithic artefacts were also important activities at this site.

Chronology

The chronostratigraphic attribution of the assemblage was based only on the stratigraphic position which was near the change of a sediment regime. A comparable change of regimes in river valleys was often observed at the transition from the Late Pleniglacial to the Lateglacial Interstadial (Pastre et al. 2003). Therefore, the episodes which formed the assemblage from horizon IV in Cepoy dated probably to the early Lateglacial Interstadial.

Even though the material of sector 2 was thoroughly analysed, thus far, the number of occupation episodes and/or distinct occupation events remained uncertain. Consequently, an admixture of remains from various times could not be excluded. However, a separation of the material seemed not possible according to the horizontal distribution. Furthermore, Boris Valentin stated correctly that the coherence of the same technical and typological traits on various other sites was probably no accidental melange (Valentin 2008a, 125). Besides the technical comparability of the material, shouldered points, backed pieces, and backed bladelets were observed in close spatial vicinity on several spots of the site. In addition, the strictly standardised spatial behaviour suggested by Stefan Wenzel (Wenzel 2009, 61) also indicated that the performed activities were similar and, probably, the tools used in these activities as well. Thus, the repetitive accumulation of comparable material in a comparable manner should reveal the behavioural pattern at this site more clearly.

Marsangy, Yonne

Research history

The main excavation area of the site *Le Pré des Forges* in Marsangy was excavated from 1974 to 1981 on 220 m² under the supervision of Béatrice Schmider with modern standards. From 1972 until 1974, Henri Carré had already excavated two areas of test pits located north and south of the main excavation area.

Topography

The site is situated on the western bank of the modern river bed of the Yonne (**tab. 34**) and approximately 3 km south of Étigny. In this part, the Yonne valley is already flanked by a 90-100 m uprising plateau. The valley at Marsangy is about 2 km wide. This width is partly due to a meander of the Yonne and partly due to two tributary streams of which one (Montgerin) cut into the adjacent hills north-westwards and another one cut the hills south-west of the site. North of the site on the eastern bank of the Yonne another meander of the river widened the valley and, in addition, another valley coming from the east enlarged the Yonne valley. Hence, the site is located in an area with many east-west running valleys crossing the wide, north-south running Yonne valley. These valleys made the surrounding plateaus relatively easy accessible but they also caused considerable erosion of sediment towards the Yonne. Therefore, no further *in situ* concentrations were supposed to be found north of the excavated site.

Within the Late Pleistocene the Yonne began down cutting and regularising (Roblin-Jouve 1992, 29). In addition to a river branch immediately east of the site, another channel formed some 100 m farther to the east (Roblin-Jouve 1992, 26 fig. 12). This channel became the major stream in the Lateglacial but it was affected by erosion during the Holocene. In consequence, the river branch near the site was re-activated during the Holocene (Roblin-Jouve 1992, 27. 30) and the erosion connected with this process affected the archaeological horizon. Thus, some of the prehistoric material drifted downwards the river embankment (Schmider 1992d, 13-18). Nevertheless, the majority of the excavated material remained *in situ* (Schmider 1992d, 18).

Stratigraphy

On top of the river gravels two deposits of silts were recorded in Marsangy. These silts were overlain by loamy silts which became sandier in the upper part. In the lower part of these loamy silts the archaeological material from the excavation of Béatrice Schmider was found in a single horizon (Roblin-Jouve 1992, 27). The malacological analysis placed this part of the stratigraphy into a cold period in which the river banks were covered by regularly flooded grassland. On top of the sandy-loamy silts a very loamy silt layer with calcareous fragments and many molluscs was found.

In the northern area excavated by Henri Carré, a hearth construction was found on top of a lithic concentration indicating the presence of two distinct archaeological horizons. This evidence suggested a possible re-settlement of the site during the Lateglacial (Schmider 1992d, 13) and, thus, would imply at least two occupation events at the site. However, the structures in this northern area were more heavily affected by erosion as well as settlement dynamics than the structures in the main excavation area (Schmider 1992d, 13). In addition, structures of Hallstatt age further disturbed the Pleistocene concentrations of the site (Schmider 1992d, 11). Nevertheless, more recent analyses of small particles of charcoal from the hearths excavated by Béatrice Schmider suggested partial intrusion of Holocene material in these structures (Bodu et al. 2009b, 97-101), and, consequently, sustained a more complex stratigraphic development on the excavated area. In analogy, the distinction in the excavation of Carré should be considered with the necessary caution.

Archaeological material

The over 21,500 lithics from the main excavation (**tab. 35**) weighted some 405 kg (Croisset/Schmider 1992). The artefacts were dominantly made of local Cretaceous flint (**tab. 36**). Only four tools from the southern part of the main site were made of foreign material. This material was mainly identified as Tertiary chalcedony which possibly originated from some 80 km north-west of the site. One piece was made of a yellow jasper-like material of which the origin is unknown. Possible deposits are located some 100 km to the south (Schmider 1992a, 133). Some processed material from a single core in the northern concentration (N19) was made of quartzitic sandstone which originated from some 30 km south of Marsangy (Croisset/Schmider 1992, 89). A blade which was also found in the north of the main excavation was made of a glossy sandstone which, perhaps, originated from the same area as the quartzitic sandstone.

The inventory includes 379 cores and core fragments (**tab. 37**). Several of the fully exploited cores show knapping accidents which might be the result of inexperienced stone knappers (Pelegrin 1992, 111. 115). Blades were the main aim of the blank production process. Besides long blades, shorter ones were produced. The butt of the blanks was often well prepared (Croisset/Schmider 1992, 94 f.). A spur (*en éperon*) was regularly formed as preferred impact point (Pelegrin 1992, 111). The percussion was either direct hard or direct soft. In fact, some hammerstones were also found at the site (Bodu 1992).

642 tools are present within the material from the main excavation area (**tab. 38**). In general, the burins were the largest group of retouched artefacts (n = 179). These were frequently dihedral types or on broken edges. However, burins on truncations were also common, in particular, in the northern part of the site. In this part the LMP (n = 86) were also more frequent than the burins (n = 74). In total, the LMP (n = 168) were the second largest group of retouched artefacts. Among the LMP, simple backed blades (n = 116) occurred more than twice as often as backed points (n = 52). The points were mainly classified as shouldered points (n = 14 unbroken ones), although in some cases the shaping of the base even tended towards a tanged point (Schmider 1992a, 190 fig. 108.4; Schmider 1992a, 191 fig. 109.6). Furthermore, angle-backed points with a single angle (n = 2) and with two angles (Schmider 1992a, 191 fig. 109.5), *Federmesser* (n = 3), Malaurie points (n = 3), and a bipoint were found. Thus, the shaping of the points was very diverse. The third most numerous group were borers (n = 111). Among these implements, *becs* and *Zinken* were the dominant

types. Only in the northern part of the site, *becs* with very long formed tips (*Langbohrer*) were found. The end-scrapers (n = 68) were usually made on blades of which some were very long. The lithic inventory from the excavation of Henri Carré appeared similar with *becs* and shouldered points (Schmider 1992d, 11).
The poorly preserved fauna was dominantly determined as reindeer (*Rangifer tarandus*) but also bones and teeth of horse (*Equus* sp.) and red deer (*Cervus elaphus*) were found (**tab. 39**).
One perforated mollusc which originated from deposits either 80 km north, 90 km north-west, or 130 km west of Marsangy was found in the centre of the main excavation (**tab. 42**; Schmider 1992c, 231).
Furthermore, in the northern part of the main excavation area three stones with lines engraved on their cortex were found of which one might be an abstract figuration (Cremadès 1992). Another stone resembled a female figurine with an incision parting the upper part of the stone from the rest but this piece was found in a disturbed area in the south of the main excavation (Schmider 1992c).

Spatial organisation
In the excavation area of Béatrice Schmider four concentrations were distinguished. Henri Carré had excavated a further three concentrations: one several metres to the south and two farther north than Schmider's excavation area (Schmider 1992d, 11). The three concentrations found by Henri Carré were disturbed by prehistoric structures (Hallstatt age).
All four concentrations on the main excavation area were organised around a hearth which was filled with stones. 13 samples of charcoal material from around the presumable hearths were analysed. However, only samples from the southern most concentration were considered undisturbed (Bodu et al. 2009b). Within these samples, no wood charcoal was identified but many small, burnt bone fragments (Bodu et al. 2009b, 100 f.) indicating the possible use of bone material as fuel for the hearth. Within each concentration, several refits were found giving a relatively undisturbed impression. Béatrice Schmider stated that there was no significant difference in the lithic material from the four concentrations observable (Schmider 1992a, 223).
The three southern concentrations of the main excavation area were connected by refits and, hence, their material was not further distinguished during the analysis (Croisset/Schmider 1992). However, the connection of the southern most concentration (X18) to the intermediate ones (D14, H17) was considerably meagre based on three refitting lines. Since the concentrations were lying relatively close to one another and because the material was very homogeneous, the assignment of artefacts between the different concentrations to the one or to the other concentration was partially impossible.
The spatial organisation of the concentrations H17 and D14 were very comparable, whereas the concentration X18 appeared more ephemeral in material and the concentration N19 was much denser. Around N19 several nests of knapping debris were deposited, whereas around concentration H17 and D14 only two major zones of artefact accumulation were observed.
The three southern concentrations (X18, D14 and H17) yielded a comparable number of lithic artefacts (n = 11,229) and retouched artefacts (n = 329) as did the northern most concentration (n = 10,389 artefacts; n = 313 tools). Furthermore, 197 cores were found in concentration N19, whereas the other concentrations yielded only about a third each of these (H17: n = 64; D14: n = 56; X18: n = 62). Thus, Béatrice Schmider interpreted the differences as a functional variation of the concentrations (Schmider 1992a, 223; Schmider 1992e, 245) with the blank production process mainly occurring around concentration N19. She also noted that the concentration of material generally rises towards the north-western tributary (Schmider 1992d, 13).
The concentration N19 seemed to represent a single event and the thoroughly analysed material displayed all characteristics of the MfCM. Moreover, the spatial analysis based on the distribution of artefacts and the patterns of refitting lines indicated that this concentration was a workshop in the open air (Schmider 1984,

site	lab. no.	years TL-BP	± years	material	comment	ref.
Marsangy, conc. H17	Gif-M2	11,900	700	burnt sandstone	found in H16-17	1-2
Marsangy, conc. H17	Gif-M3	11,700	700	burnt sandstone	found in H16-17	1-2
Marsangy, conc. H17	Gif-M4	11,600	800	burnt sandstone	found in F16	1-2
Marsangy, conc. H17	Gif-M5	11,500	1,250	burnt sandstone	found in F18	1-2

Tab. 46 TL dates from Marsangy. References (ref.): **1** Valladas 1994; **2** Bodu 2004, 175.

176). However, the lack of refits suggested that this workshop did not serve as blank production area for the other concentrations.

The ring and sector method (Stapert 1992; cf. Gelhausen/Kegler/Wenzel 2004) which was created to detect barrier effects within accumulation patterns of material around centralised hearths was applied to the material from Marsangy. The result of the ring analysis indicated that the southern hearths were probably located inside limited spaces of approximately 4 m in diameter (Stapert 2003, 7).

In these small areas, various types of retouched pieces as well as the debris of their modification and debris of the blank production process were found. Thus, around these hearths similar activities were made in a repetitive pattern.

Chronology

For Marsangy the indications for a chronostratigraphic attribution are multiple. Four TL dates which were taken on burnt sandstones from the concentration H17 indicated that the occupation had taken place before the Younger Dryas (**tab. 46**; Valladas 1994). The archaeological horizon was situated in the lower part of loamy silts indicating a generally more temperate and humid regime than in the Late Pleniglacial. The presence of red deer combined with the remaining occurrence of reindeer pointed to an occupation at the transition from the Late Pleniglacial to the early Lateglacial Interstadial. Furthermore, the malacofauna placed the sediment of the archaeological horizon into a cold period which was consistent with a dating of the assemblages from Marsangy to GI-1d. Three ^{14}C dates were taken on reindeer material and a further date was taken on a tooth of horse. The results vary considerably (**tab. 44**). The incongruence of the dates resulted either from several occupation episodes or a thus far undetected contamination in the samples. However, the reliable ^{14}C dates sustained a chronostratigraphic position in GI-1e/d.

Even though the material cannot be fully distinguished for the three southern concentrations, their spatial organisation appeared unaltered. Nevertheless, the central concentrations D14 and H17 were connected by many refits suggesting either a contemporaneous existence or that one concentration served as resource for the other concentration. Nevertheless, these two concentrations were quasi-contemporaneous. The southern most concentration produced less material and represented perhaps another, short-termed occupation episodes or a special work place during the existence of the other concentrations. The relation to the northern concentration N19 remained unclear. However, since the same activities as in the other concentrations were indicated by the lithic material, this represented possibly another occupation event. Nevertheless, this area could also represent a knapping place which provided material for other areas along the river bank.

Belloy-sur-Somme, Somme

Research history

In the valley of the Somme near Belloy-sur-Somme, Victor Commont excavated a test pit in 1905 on the site La Plaisance which was discovered by surface survey in the late 19th century (Fagnart 1997, 43). In the following excavations between 1907 and 1910, some 916 m² were uncovered. Commont observed two archaeological horizons: In the lower horizon a small concentration yielded *bec* and bi-truncated implements and in the upper horizon many bruised blades were found. Some 50-200 m north-west of the limit of Commont's excavation, modern excavations were conducted between 1984 and 1990 after the promising results of a test pit in 1983. These works were supervised by Jean-Pierre Fagnart. In addition, between 1990 and 1992 another area was excavated in a dry valley west of the previous excavation area. The approximately 2,000 m² large area was divided in twelve sections. In particular, in the western part of these sections, three Lateglacial and a Mesolithic assemblage were found during the modern works. The Lateglacial assemblages were attributed to a Final Magdalenian (lower horizon), to the FMG (intermediate horizon), and to the Long Blade Technology (upper horizon; Fagnart 1997, 53-104). The Final Magdalenian material was found on an almost 500 m² large area in the very south-west on the lower slopes, whereas the FMG and the Long Blade Technology material was found farther uphill. The FMG were concentrated in the south-eastern part of the main excavation area and the Long Blade Technology was mainly recovered from the north of this area.

Topography

The site La Plaisance is situated on the eastern bank in the swampy valley of the Somme (**tab. 34**). At the site the valley is approximately 1 km wide. Towards the north and the north-east hills rise steadily 25-50 m above the valley floor. On the south-western bank of the Somme the adjacent plateau rises abruptly 50-65 m upwards. On both sides of the south-east to north-west flowing river the plateau is regularly cut by the valleys of small contributories, often running in a north-south axis. The concentration of archaeological material from the lower horizon was found on the south-eastern margins of a today dry valley within which several erosion events were documented (Fagnart 1997, 46 f. 53).

Stratigraphy

On top of the Cretaceous bedrock Victor Commont registered 2-3 m of gravels (Fagnart 1997, 44). These gravels were overlain by a yellowish white sandy clay deposit of approximately 40 cm thickness. This deposit was covered by a thin band of gravels on top of which about 1 m of whitish, calcareous silts followed. These silts were also reached in the modern excavations occasionally and were again covered by a thin gravel layer. These gravels were superimposed by greyish yellow silts which were identified as Late Pleniglacial loess. This loess deposit was about 0.9-1.6 m thick. In the profiles from the dry valley this loess was again covered by a very thin band of gravels which were partly ice-damaged. On top followed a blackish silt humus which was named Belloy-sur-Somme soil (Fagnart 1997, 46). This soil was 0.05-0.2 m thick. The pollen samples produced only a general attribution to the Lateglacial (Fagnart 1997, 48-51), presumably, due to the bioturbation which was often observed in this soil. The lower archaeological horizon was found at the base of this soil. The upper part of this deposit yielded artefacts assigned to the FMG. Moreover, the material classified as Long blade Technology (»*Belloisian*«, Fagnart 1997, 103 f.) was also found in the upper part of this soil. The Belloy-sur-Somme soil was overlain by a sandy silt layer which was attributed palynologically to the Boreal. The stratigraphic hiatus might be explained with erosive processes near the adjacent valley. Or the infiltration of modern pollen in a pre-existing sediment (Fagnart 1997, 51). This sandy silt was overlain by the modern topsoil. Only in the dry valley approximately 1-1.5 m of various peat and silt layers were formed between the sandy silts and the modern topsoil.

Archaeological material of the lower horizon

In the lower horizon, 6,415 lithics of which 1,880 pieces were larger than 2 cm were found (**tab. 35**).

The raw material (Turonian and Coniacian flints) was taken from local gravels, immediately accessible near the site (**tab. 36**).
The 108 cores and core fragments (Fagnart 1997, 55) were dominantly prismatic cores with two platforms (n=41) or one platform (n=19; **tab. 37**). A further seven cores were made on larger flakes and probably also with just one platform. Bladelets were removed alongside a lateral edge of these flakes. Only one multiple platform core occurred. The other pieces were either fragments (n=31), preforms (n=4), or merely tested blocs (n=5). On some cores traces were observed which demonstrated an unskilled handling of the hammerstone such as an apprentice use of the material (Valentin 2008a, 129f.). Furthermore, traces indicating the used of organic hammers were found on at least one core and demonstrated the use of organic hammers besides soft hammerstones (Valentin 2008a, 130). The traces were related to the production of some long blades. Moreover, the butt of several blades was facetted and some blades had a preparation of the »*en éperon*« type (Fagnart 1997, 55). In general, the blank production process in the lower horizon of Belloy-sur-Somme was clearly oriented towards the production of long blades (Fagnart 1997, 54f.). Bladelets were rare.
Among the 158 tools, 23 end-scrapers formed the most numerous tool class (**tab. 38**). The end-scrapers were generally made on long blades and only two examples were made on flakes. The 21 burins were also mainly made on blades. The dihedral types were slightly more numerous than the examples on truncation (Fagnart 1997, 59). Among the 20 LMP were six backed bladelets of which two were additionally truncated. Furthermore, one backed blade and a microlith of trapezoidal shape was found. The twelve backed points were of various shapes (Fagnart 1997, 63. 255) including shouldered points, angle-backed points, and curve-backed points. Furthermore, ten borers were found which were clearly dominated by *becs* and *Zinken* (n=8; Fagnart 1997, 59. 62f. 255). In general, the seven truncations were also made on blades. Only twice double truncated blanks were found. The majority of retouched artefacts were blades with various retouches. Combinations of different types of retouched working edges were not described.
Organic material was not preserved in this archaeological horizon.

Spatial organisation of the lower horizon

At least one hearth (M13) was identified by a concentration of burnt stone plates (**tab. 43**; Fagnart 1997, 54). The majority of lithic material was found around and northwards of this accumulation. In particular, in the north-east was a dense cluster of used blades indicating an intense working place there (Fagnart 1997, 54). In the north, several cores and core fragments were found. Further cores were deposited northwestwards and a dense cluster of retouched artefacts was found in this working area. In addition, material accumulations were found in two consecutive concentrations (G8/9 and E4) 6 and 10m westwards of the hearth. In these concentrations, further cores and retouched artefacts were found. A very small cluster containing two cores was also found in some 10m distance to the hearth but north-eastwards in the section 151. In section 151, some 13m eastwards of the hearth, a second more diffuse scatter of blanks was found.

Chronology of the lower horizon

The malacology in combination with the stratigraphy attributed the layer below the artefact horizon to the Late Pleniglacial. Consequently, the Belloy-sur-Somme soil in which the artefacts were found was probably formed in the Lateglacial Interstadial. Due to the position in the lower part of this soil, the artefacts of the lower horizon were presumably deposited in the first part of the Lateglacial Interstadial.
Jean-Pierre Fagnart assumed in analogy to Late Magdalenian sites in the Paris Basin that the lower horizon represented two patches with each being related to one or two knapping places in the periphery (Fagnart

1997, 54). In consequence, the lower horizon was clearly formed by several episodes, possibly two distinct events. Core and retouched artefacts were found in both patches. Nevertheless, except for a *couteau à dos* the LMP were found in the periphery of the concentrations related to the hearth. Thus, work related to these implements was either not performed in the north-western patch or the concentration related to this work was not excavated. Moreover, no indications for a hearth were connected to this area. Thus, either this patch represented a single occupation event which was possibly not completely excavated or very short-termed or this patch formed perhaps a satellite of the central area around the hearth. However, without refitted material the relation of the various concentrations remained unclear.

Gouy, Seine-Maritime

Research history

Presumably, the three successive chambers of the small cave *Grotte du Cheval* or *Gouy I* were just the remains of a larger system which was destroyed during the construction of a National road alongside the Seine in the 1930s (Graindor 1959, 87 f.). The small cave was discovered in 1956 by Pierre and Yves Martin (Breuil/Graindor 1959; Martin 2010). For preservation reasons the cave was put under protection in 1959. In the 1960s, some small, superficial excavations were conducted in the cave (Graindor 1965) and revealed a small inventory (Bordes et al. 1974).

Topography

The *Grotte du Cheval* is a small cave on the western bank of the Seine (**tab. 34**). The entrance is some 3 m above the modern road (Graindor 1965, 29 f.) and some 15 m above the modern level of the Seine. In regard to the partial destruction of the Cretaceous (Senonian) rocks and the inclination of the cave (Graindor 1959), the prehistoric entrance was probably further elevated in the cliff. The modern cave opens towards the north-west. The remaining part of the cave runs in a straight passage some 15 m into the rock with a downward inclination (Graindor 1959). In general, the passage is a very narrow but three successive chambers gradually widen up the cave.

Stratigraphy

The upper 2-2.5 m of the cave deposit was mainly composed of calcareous scree which eroded from the surrounding cliffs. However, the cave deposit was probably slumped into the cave until the material closed the entrance. Presumably, down slope erosion of the material after the road construction led to a more significant re-opening of the cave entrance (Graindor 1959). The archaeological remains were found during the excavations in the upper part of the deposit (Graindor/Martin 1972). However, some of the engravings of the wall seemed to continue below the cave deposit and were in consequence presumably older than the archaeological remains (Lorblanchet 1973). In addition, some pieces of local flints were classified as »pseudo-artefacts« and resulted from frost fracturing and compression in the deposit (Bordes et al. 1974). Probably, the sediment slumped into the cave was susceptible to the severe cryoturbation process outside the cave and, thus, prior to the relocation into the cave.

Archaeological material

Inside the cave a small lithic inventory (n = 116) was recovered (**tab. 35**). Even though the site was located in a Cretaceous rock formation with flint bands, the spectrum of raw materials was originally considered as extremely varied with more than three quarters of the assemblage being made of foreign material (**tab. 36**).

Less than a quarter of the material (n = 33) was assumed as the same material as the pseudo-artefacts and, thus, originated presumably from the immediate surrounding of the site. A further five materials were described mainly by colour, but prohibited further determination of the origin of the raw material (Bordes et al. 1974). However, in comparison with the descriptions of raw materials from other sites in the Seine valley at least some other pieces originated possibly from the fluvial deposits and in this case were also of local origin. Nevertheless, few pieces appeared as of a different, high-quality raw material (Fosse 1997, 242) which according to the cortex was a Cretaceous flint collected at a primary resource (Valentin 1995, 582). In addition, a blade resembled the Tertiary chalcedonies of the Île-de-France (Valentin 1995, 582).
The only core of the assemblages was also made of the very local raw material (**tab. 37**). It was described as a prismatic core with two platforms of which one was used dominantly (Bordes et al. 1974, 118). The blank production dominantly was aimed to create blades (Valentin 1995, 584f.). Among the various types of butt on the analysed blanks (n = 66) were few pieces (n = 3) with an *en éperon* type of preparation. François Bordes and his colleagues considered the material to be generally knapped directly with a hard hammerstone but they also found indications of an indirect knapping technique (Bordes et al. 1974, 118). Boris Valentin suggested the use of soft hammerstones (Valentin 1995, 583. 586). However, debris material is relatively rare in the assemblage of Gouy and either played no important role at the site or remained in unexcavated areas.
Among the heterogeneous finds were 16 retouched pieces (**tab. 38**). Six artefacts were classified as LMP, generally, as bipoints (n = 3). Another curve-backed point fragment was perhaps also a bipoint before fragmentation. Furthermore, another LMP fragment was an angle-backed remain. The sixth LMP was a large *couteau à dos* with the shape of a *Federmesser*. Boris Valentin noted that these points wore no traces of use (Valentin 1995, 569 note 10). However, in the retouches of a partially truncated *couteau à dos* traces of haematite were found (Bordes et al. 1974). Besides the LMP, three burins, three borers, two truncations, an end-scraper, and a composite tool were determined by François Bordes and his colleagues. In fact, the composite tool was an end-scraper which was modified on a burin. Thus, the piece did not represent a combination but a conversion. Of the three burins only one was made on truncation and the other two burin blows were set at a breakage negative. The end-scraper was made on an elongated flake. The opposite edge on the oblique truncation showed heavy traces of use and was possibly a broken borer. The borers were large, crude pieces with a *Zinken*-like working edge.
Hugues Plisson named in a comparative study of traceological analyses on material from caves also 113 pieces from the Gouy cave (Plisson 2007). They displayed the use for cutting or engraving of mineral material, butchering and cutting of hides, and working of bone material.
Among the fauna were only remains of large mammals preserved (**tab. 39**). In the assemblage red deer (*Cervus elaphus*), wild boar (*Sus scrofa*), and roe deer (*Capreolus capreolus*) were determined in addition to remains of wolf (*Canis lupus*), fox (*Vulpes vulpes*), a merlin (*Falco colombarius*), and a Western Jackdaw (*Corvus monedula*; Cordy 1990).
Furthermore, a drilled red deer canine was found (**tab. 42**; Lorblanchet 1973).
On the cave walls several engravings were documented (Martin 2007a) and some traces of colour were found (Martin 2004). Along with the eponymous horse, further horses, bovids, a possible bird, a female silhouette of Laline/Gönnersdorf type, several signs including vulvas, triangles, and barbed symbols were recognised thus far. The animal bodies were usually filled with parallel or hatched lines. Presumably, these various expressions of art were created in several visits to the cave reflected by different phases of rock art (Martin 2007b). Furthermore, some blocks with engravings were also recovered from the cave floor but whether these represent portable art or collapsed from the cave walls remained mainly uncertain (Martin 2007b).

Spatial organisation

The spatial distribution of the material within the calcareous scree was not testable.
Although the assemblage was relatively small, the composition was a relatively usual one with few remains of the blank production process, various retouched artefacts, various organic remains, and some ornamental pieces. Thus, the assemblage as well as the wear traces on the lithic artefacts suggested some type of engraving activities as well as processing of faunal remains. Nevertheless, the variability of raw material as well as the almost complete preservation of the LMP could also indicate some type of selection and/or bias.

Chronology

One of the bones was ^{14}C-dated (GifA-92346; **tab. 44**) and produced an age which calibrated fell to the transition from the first Lateglacial warming (GI-1e) to the first Lateglacial cooling period (GI-1d). According to the composition of fauna, the assemblage should rather originate from the latest GI-1e. However, neither was the contemporaneity of the faunal remains nor the relation to the archaeological material unambiguously proven (cf. Valentin 1995, 569).
Some of the engravings appeared to cover parts of the cave wall which seemed to continue below the horizon from which the lithic artefacts and bones were recovered. Thus, at least two phases seemed probable. This division in two events was also suggested by the composition of the cave art (Martin 2007b).
The significance of this cave was recurrently attributed to the most northern occurrence of cave art in France. However, meanwhile Lateglacial cave art was also found in the closer and wider surrounding of the site (Pigeaud et al. 2010) as well as in central England (Pettitt/Bahn/Ripoll 2007). For the present study, the archaeological material which presumably was related to the transition period between the Late Magdalenian and the FMG is of interest. Even though the analysis of Hugues Plisson indicted the use of some lithic artefacts for graving on mineral material (Plisson 2007, 129. 131), the question remained which engravings were made with this lithic material. Furthermore, Plisson could not exclude that some pieces were already used before they arrived at the site (Plisson 2007, 131). Thus, the relation of the archaeological material to the cave art remained as in most caves unclear.

Pincevent III.2, Seine-et-Marne

Research history

In the basin formed by the confluences of the Yonne and the Loing into the Seine the gravel pit of Pincevent had grown from 1926 to 1964 to an extent of approximately 12 ha. The pit was set on the southern bank of the Seine in the municipality La Grande Paroisse between Montereau-Fault-Yonne and Moret-sur-Loing. Since 1957 the gravel mining had been accompanied by archaeological investigations which focused on early Dark Ages, Gallo-Roman period, and Iron Ages. In 1963 first hearths related to Late Magdalenian material were recognised in the clays immediately above the quarried gravels (Leroi-Gourhan/Brézillon 1966, 263). However, only when in 1964 the gravel pit had cut again Palaeolithic layers the exploitation was stopped. The remaining land adjacent to the gravel pit, which was subsequently flooded, was declared site of national interest and several 1,000 m² were bought from the mining company.
From 1964 until 1985 the excavations were supervised by André Leroi-Gourhan and afterwards by members of his CNRS team, currently Pierre Bodu is in charge. The excavations which until the early 2000s explored only the western part of the protected site have been uncovering numerous concentrations in some 25 different archaeological horizons (Gaucher 1996; Bodu et al. 2006b, 8). To distinguish the various assemblages the large concentrations were named according to their vertical and horizontal position. The latter position

was defined by a square-metre grid which was plotted on the complete site and the adjacent part of the gravel pit with numbers 1 onwards for the west to east running axis beginning in the west and zones of 26 m length labelled alphabetically (A-Z) on the north-south axis beginning in the south. Habitation no. 1 (Leroi-Gourhan/Brézillon 1966) marked, thus far, the south-western end of the excavated area. After this first excavation, intersections were added at every 25 m from west to east. The thereby created 25 m (intersections) by 26 m (zones) wide rectangles (sections) were again numbered from south-west to north-east resulting in approximately 70 sections. Large concentrations were subsequently identifiable by the section in which they were found and further defined by the centre which were generally given the exact square coordinates such as horizon IV0, section 35, unit T125 (Bodu et al. 2006a) or horizon III.2, section 27, unit N91 (Bodu/Orliac/Baffier 1996). In fact, archaeological remains from horizon III.2 which is described in more detail below were recovered in the sections 17, possibly 18, and 27. In total, these areas comprise approximately 280 m² but faunal and lithic remains were recovered from only some 70 m² (Bodu/Orliac/Baffier 1996).

Pincevent was among the first sites to be excavated in a modern style with precise documentation of the position of the material. In addition, from the beginning of the investigations on the site the study of the setting, the surrounding, and the recovered material formed subject of a communal project of various specialists including besides archaeologists, for example malacologists and sedimentologists (cf. Leroi-Gourhan/Brézillon 1966, 267). Therefore, the site is presumably one of the most famous open air sites of the Late Upper Palaeolithic. Moreover, the excellent preservation of organic material along with lithic, mineral, and fossil assemblages in numerous concentrations, which were dominantly attributed to the Late Magdalenian, allowed for detailed analyses of subsistence, technical, and spatial behaviour (e.g. Leroi-Gourhan/Brézillon 1972; Enloe/David 1989; Bodu et al. 2006b). This good condition of the material was attributed to the favourable position in the floodplain of the Seine where the regular flooding covered the archaeological remains with sand and silt layers quickly after abandonment of the concentrations.

Topography

Approximately 7 km towards west-north-west of Pincevent the Loing flows into the Seine and some 5 km towards the east-north-east the Yonne joins the Seine (**tab. 34**). In this part the Seine flows from east to west before it directs towards the north-west shortly after the mouth of the Loing. The meandering Seine had formed a large valley of almost 2 km width in north-south direction in the surrounding of the site. Pincevent is set approximately in the centre of this valley with relatively levelled terrain for more than 1 km northwards and southwards. Around the Yonne mouth the width of the valley was further increased to c. 3 km. This widening reaches eastwards almost to Marolles-sur-Seine where the terrain rises more rapidly towards the east. Approximately 2.7 km westwards of Pincevent a ridge runs towards the Seine from the south and narrows the valley to only some 500 m width. Hence, the wide Seine-Loing-Yonne basin extends some 10-12 km with Pincevent set in its western part and the sites Le Tureau des Gardes and Le Grand Canton (see p. 189-196 and p. 206-210) located on the eastern end. Approximately 1 km north and south of Pincevent the terrain rises quickly some 40-60 m above the basin floor. However, directly south of the site the hills were cut by a small, dry valley which gradually ascended towards Ville-Saint-Jacques.

Stratigraphy

Situated in the floodplain, the stratigraphy at the site was dominantly influenced by the Seine. On top of the Cretaceous bedrock some 10 m of alluvial sediments beginning with light beige sands and gravels were deposited (Leroi-Gourhan/Brézillon 1966, 268; Roblin-Jouve 1996, 17-19). These sediments were partially exploited in the mining and termed horizon V in the general stratigraphic sequence of Pincevent. Analysed in more detail the lowest part (layer 6) is formed by obliquely deposited sands followed by an approximately 2 m thick sand and gravel deposit alternated with layers of clay (layer 5). This layer 5 is intersected by a c. 60 cm thick, yellowish brown clay band in which Middle Palaeolithic remains were observed (Roblin-Jouve 1996, 18).

The horizon V was overlain by the relatively homogeneous silts of the so-called horizon IV in which the Late Magdalenian remains were found. This deposit was divided in four subunits by two different classification systems. The sedimentological division was based on mainly pedological criteria (Roblin-Jouve 1996, 17-19), whereas the archaeological division was based on general depositional criteria (Orliac 1996a, 35-45):

According to the pedological criteria, the lower c. 50 cm of horizon IV which were formed by sands and clays (layer 4d) were affected by the overload of sediment as well as by microfaults. In consequence, these sediments were only preserved in protected areas. On top deposited beige silts were more clayish (layer 4c) and contained in the upper part the lower Late Magdalenian archaeological horizons such as level IV40. The level IV30 was situated at the base of the next subunit formed by sandy loam (layer 4b). In this layer 4b the stratification of the sediment turned from oblique to almost horizontal and wavy in the upper part where at the transition to the next subunit the various IV2 levels were deposited. On top followed a homogeneous, laminated sandy loam (layer 4a). In the upper part this sandy loam became increasingly sandy and at this stratigraphic position level IV1 was found. However, this part was often eroded. Late Magdalenian remains were not found above this sandy loam.

However, in the second division system the lower Late Magdalenian levels up to the horizon of habitation no. 1 belonged to the approximately 1.5 m thick phase of lower silts (PLI – *phase limoneuse inférieure*) which was followed by a transition phase of sandy loams (PTLS – *phase de transition limono-sableuse*) which ended at the level IV21. On top, between the horizons IV21 and IV201, an approximately 1 m thick sandy deposit (PS – *phase sableuse*) which was sterile of archaeological horizons was observed. The remaining sediment of horizon IV from level IV20 upwards was characterised as upper silts (PLS – *phase limoneuse supérieure*) which can be correlated approximately with the layer 4a in the pedological subdivision of horizon IV.

The pollen recovered in horizon IV were redeposited Tertiary material (Emery-Barbier/Rodriguez 1996, 53) and, hence, prohibited further environmental differentiation of this deposit. The malacofauna was also poorly preserved in these deposits (Emery-Barbier/Rodriguez 1996, 63). However, the few pieces which were found in horizon IV indicated a cold and dry steppe environment at the time of deposition.

These fine-grained layers were succeeded by horizon III which was formed by sands. In the lower section sands and gravels which were partially obliquely bedded were free of archaeological remains. Presumably, this lower section correlated with a formation of a new channel of the Seine (SLL) which eroded the silts of the horizon IV partially or completely in the north-western part of the area explored until the 1990s. The on average 20 cm thick layer 2b or horizon III.2 formed by greyish coarse sands with an important humic component (Roblin-Jouve 1996, 18) followed on top of the lower sands and gravels. Thus, the formation of the Seine channel and the connected erosive processes occurred before the deposition of the horizon III.2 (Orliac 1996a, 48 f.). However, this erosion represented a gap of unknown temporal dimension between the last Late Magdalenian and the archaeological remains found in the upper part of horizon III. Perhaps, the erosion was connected to an increase of occasional but significant flooding events suggested by the presence of alluvial molluscs (Emery-Barbier/Rodriguez 1996, 58. 64). The archaeological material from horizon III.2 is represented in more detail in the following. The 30 cm on top of layer 2b were of equally greyish humified coarse sands but these were very bioturbated and the upper part was eroded or altered by the overlying soils. The malacological analysis suggested that this upper horizon III represented the Younger Dryas to Preboreal transition (Emery-Barbier/Rodriguez 1996, 64). Moreover, at least one hiatus within horizon III and, perhaps, another one at the transition from horizon IV and III was proposed based on the sparse and quickly changing mollusc fauna. The malacological bisection within horizon III yielded a lower mollusc community which was comparable to horizon IV with relatively cold conditions but also with first indicators of shrubby and moist environments, whereas the upper community represented already a stronger forest component and a relatively moist ground (Emery-Barbier/Rodriguez 1996, 58-60). Only in the sediment above the archaeological horizon III.2 Late Quaternary and Holocene pollen were recovered (Emery-Barbier/Rodriguez 1996, 54-57). However, this observation confirmed that the material of III.2 was clearly deposited before the Holocene.

On top of the greyish humified coarse sands followed reddish, clayey sands which were already attributed to horizon II (layer 1c and b). These sands contained Neolithic to Gallo-Roman archaeology. Comparable to the upper part of horizon III, these deposits were bioturbated heavily, probably, because they were overlain by the topsoil (layer 1a) in which only Gallo-Roman remains were found.

Thus far, approximately 15 different Late Magdalenian horizons were identified of which at least seven formed living floors and were therefore frequently mentioned in the literature on the late Upper Palaeolithic (IV-0, IV-20, IV-21, IV-21-3, habitation no. 1, IV-30, IV-40; Orliac 1996a, 38; cf. Valentin 2008a; Bodu 2010). In addition, at least one horizon attributed to the Lateglacial Interstadial (III.2) was found (Bodu/Orliac/Baffier 1996; Orliac 1996b, 4). The vertical spread of the archaeological material encompassing approximately 10 cm within this Lateglacial Interstadial horizon was in contrast to the dense but thin Late Magdalenian horizons considerable and, probably, due to intense bioturbation in the horizon III (Bodu/Orliac/Baffier 1996, 69). However, the archaeological horizon III.2 was located stratigraphically 0.5-1 m higher than the last Late Magdalenian horizon and clearly fell to the transition period between Late Magdalenian and FMG. Therefore, the material from this horizon is considered in the present study. Although ^{14}C dates taken on samples of the Late Magdalenian material from horizon IV (Débout et al. 2012) also attributed these remains to the transition period, the archaeological material as well as the environmental data (e. g. Bodu et al. 2006b; Bignon 2006; cf. Limondin-Lozouet et al. 2002) clearly aligned these remains with the Late Magdalenian occupations of Gönnersdorf and the lower horizon of Andernach.

Archaeological remains of the horizon III.2

In total, 561 lithic artefacts were attributed to this horizon (**tab. 35**). Perhaps, a previously recovered blade which was found isolated in the same stratigraphic context and which wore splinters of use along the lateral edges could be added to this inventory. However, 72 pieces were only splinters. 119 artefacts and two stones were altered by heat.

The raw material was a Cretaceous flint which according to the rolled cortex was recovered from secondary deposits such as the river gravels (**tab. 36**; Bodu/Orliac/Baffier 1996, 69). Some pieces showed alterations due to frost (Orliac 1996b, 87). The resource of this material was presumably the same as for the Late Magdalenian at Pincevent but the quality for knapping of the chosen pieces was not as high as in the Late Magdalenian inventories (Bodu/Orliac/Baffier 1996, 72). Nevertheless, a blade from section 17 and, perhaps, a further six pieces (two backed points, four flakes) from the same section were made of higher quality material which was possibly a Tertiary chalcedony (Orliac 1996b, 87-89). Such material was described from other sites in Paris Basin and, possibly, originated from the Loire region or the hills west of the Seine. Thus, the material was brought from at least 40 km distance. Moreover, a granite cobble with traces of heating was recovered in section 17. Unless, a vein of granite was found by chance in the surrounding underlying bed rocks, the next resources of proper granite were recorded at least 150 km south-eastwards in the Morvan region or over 300 km eastwards in the Alsace-Lorraine region.

A total of ten cores were found (**tab. 37**) but additionally fragments and raw material nodules were also recovered suggesting the discard of inferior quality material. The artefacts yielding indications of the percussion technique indicated the use of direct hard hammerstone percussion throughout the complete *chaîne opératoire*. (Bodu/Orliac/Baffier 1996, 74f.). According to the technical observations on the blank production debris and the choice of the blanks for further retouching, the blank production of this horizon was focused on blades and bladelets (Bodu/Orliac/Baffier 1996, 72-74). Several of the blades and flakes found at the site showed clear traces of use such as splintering.

Furthermore, the largest group among the 32 retouched artefacts from horizon III.2 were flakes but mainly blades with unspecific retouches along one or more edges (n=11; **tab. 38**). In addition, a retouched blade from approximately the same stratigraphic level was attributed to this inventory. The most numerous class of typical tools were end-scrapers which made up a fourth of the retouched artefacts (n=8). However, seven pieces were identified as truncation mainly of the oblique type. Three LMP were recovered during the excavation and a further two pieces were previously collected on the surface. The two collected finds were comparable to an excavated point and, thus, identified as bipoint, although one of the pieces and the excavated specimen had a small oblique basal retouch which was set apart from the curved back by an angle. In addition, a medial fragment and a large, angle-backed *couteau à dos* were determined. Only two burins were found which were made either on a natural edge or breakage. However, borers and composite tools were absent.

Only 199 bone fragments were recovered from this horizon in section 27 (Bodu/Orliac/Baffier 1996, 81) and a further 25 fragments in section 17 (Orliac 1996b, 89 and 91). Some 15-20 elements were attributed to red deer (*Cervus elaphus*) and approximately a further ten pieces were attributed to aurochs (*Bos primigenius*; **tab. 39**). Although only one adult animal of each species was attested by these remains, the spatial distribution suggested that the faunal material was partially unconnected and, thus, result from various, independent hunting episodes. Moreover, single bones were determined as originating from wolf (*Canis lupus*), badger (*Meles meles*), and arctic hare (*Lepus timidus*). The later represented a typical member of the Late Pleniglacial cold steppe, whereas the other two were known from a wide range of habitats. Nevertheless, the singular occurrences of these species prohibited further connections to the human dynamics on the site. Further small mammals including five pieces of mole (*Talpa europaea*) were recovered. In general, there is no clear association of such animals to the archaeological remains and, in addition, many of them were usually considered as originating from Holocene burrowing which was in accordance with the fact that the greyish humic coarse sands on top of the archaeological level were heavily bioturbated.

The lack of horse remains needed special emphasis since an engraving represented such an animal (**tab. 42**). This engraving was in the calcareous cortex of a large flake. It represented a horse head which was comparable to several such images from the Late Upper Palaeolithic to the Mesolithic and was positioned stylistically at the transition between the realistic figurative and the expressive silhouette art (Bodu/Orliac/Baffier 1996, 85; cf. Pigeaud 2007). Although the piece was not found *in situ* but in the spoil heap of a ten year previously made mechanical detachment, refitting to material from horizon III.2 in section 27 was possible and proved the connection to this assemblage (Bodu/Orliac/Baffier 1996, 82).

Spatial organisation of the horizon III.2

The archaeological material originated from two concentrations, one in the north-eastern part of section 27 (Bodu/Orliac/Baffier 1996) and the other one in section 17 and, perhaps, spreading northwards into section 18 (Orliac 1996b, 87). In addition, an isolated blade and few bones were also recovered from the intermediate area, including the undisturbed parts of section 18 (Bodu/Orliac/Baffier 1996, 68). These finds suggested further episodes and, perhaps, a more complex settlement behaviour in this horizon. However, in section 27 the artefacts were associated with a brownish olive sandy soil (layer 2b), whereas in section 17 the horizon III was only formed by undifferentiated sands and the archaeological material occurred approximately from the middle of the deposit. Thus, the stratigraphic comparison did not make possible a decision between the two suggestions. Moreover, refitting attempts provided no result.

The only evident structures were two hearths from section 17. However, these structures were found some 3-4 m north of the small accumulation (Orliac 1996b, 87) where mainly blank production was performed. The western hearth was very small and observable by alteration of the sediment. In addition, some lithic material was associated with the western structure. The larger hearth was disturbed by bioturbation. Nevertheless, the sediment alteration allowed for the identification of a basin-set hearth. Within this basin a 10 cm large stone was found which was probably used as heating material. Presumably, this hearth was used in at least two phases. However, no unambiguous connection to the main lithic concentration could be established. 14 burnt flakes and six calcined fragments of bone were clustered in the eastern part of the main lithic accumulation that had a dimension of 4 m × 3 m. Possibly, a latent hearth or a dump was represented by these pieces. On the opposite side of the concentration, the core and some produced blades were found, whereas the retouched material was accumulated around the possible fire.

In contrast to section 17, a more significant accumulation of material was observed in the north-east of section 27 but no evident structure was documented in this area (Bodu/Orliac/Baffier 1996, 68 f.). Nevertheless, the distribution of fire indicators was spread across the accumulation with a diffuse cluster in the centre of the concentration (Bodu/Orliac/Baffier 1996, 69 f.). The fauna was mainly associated with this concentration which was only half preserved, whereas the second half was probably cut by mechanical diggers. Some of the raw material nodules were recorded in the periphery of the concentration. Thus, blank production was one of the main aims at this concentration. This interpretation was further supported by the rare number of retouched pieces. In fact, nine of the ten cores attributed to horizon III.2 originated from the concentration in section 27, indicating a significant blank production activity in this area. However, only a few of the resulting blanks were found in this accumulation (Bodu/Orliac/Baffier 1996, 74). Nevertheless, a bipoint was also recorded in this concentration. In addition, the low number of faunal material as well as of burins, borers, and end-scrapers implied that no significant processing of faunal material was accomplished in this area.

Chronology of the horizon III.2

From the section 27 an undetermined bone sample from the brownish olive soil was ^{14}C-dated (OxA-391) in the early days of AMS dating and produced a mid-Lateglacial age (**tab. 44**). Due to the lack of information on the dated material an unambiguous association with the human presence cannot be evaluated. However, faunal material was attested in relation to the archaeological material but it was also possible that the sampled material came from a natural intrusion. Nevertheless, the date was in accordance with the stratigraphic evidence of horizon III.2 which was separated from the Late Magdalenian horizons by a significant erosion deposit. If the date reflected human activity at the site the erosion encompassed the first part of the Lateglacial Interstadial. The molluscs on top of this erosion channel suggested still a cold though moister environment which could be correlated with the transition towards the mid-Lateglacial Interstadial (GI-1d/GI-1c_3). In addition, the development of an olive soil further sustained an attribution to the early Lateglacial Interstadial (GI-1c_3) where forest communities quickly spread forming the habitat for red deer and aurochs.

In contrast, many ^{14}C dates from the underlying horizon IV provided comparable results. However, the stratigraphic as well as the environmental data clearly attributed the Late Magdalenian occupation of Pincevent to a period comparable to the occupation period of Gönnersdorf and the lower horizon of Andernach. Thus, a systematic error or contamination needs to be considered. Many of these dates were taken on charcoal usually from hearths. However, a microscopic analysis revealed that the burnt material from hearths in Pincevent represented a conglomeration of materials (Bodu et al. 2009b). In samples taken for the same analysis in

Marsangy, modern intrusions were revealed (Bodu et al. 2009b, 99f.). Perhaps, the position in a floodplain intensified this partial exchange of material. Furthermore, this type of infiltration of more recent material could explain some of the very young results at Pincevent. In fact, two recently dated samples of charcoal material which was previously screened microscopically (ETH-37120, ETH-37119) resulted in considerably older dates (**tab. 44**). These results were still younger than the material from the Central Rhineland but if these results were calibrated they were clearly attributable to the Late Pleniglacial. Nevertheless, also dates on bone resulted in comparatively young result. However, these dates were made in the early use of AMS dating at Oxford and contained significantly high standard deviations. Perhaps, further dating on precisely chosen material could help clarifying the problems with the precise chronostratigraphic development of the site.
Certainly, the material from section 27 and the remains from section 17 represent two distinct occupation episodes. Moreover, the eastern hearth of section 17 was suggested as two-phased. However, whether the material originated from the same occupation event could no longer be tested. The technological analysis suggested that the lithic material from both sections was comparable and comparably different from the techniques applied to the material during the Late Magdalenian (Orliac 1996b, 92). Moreover, the sparse faunal evidence were also comparable. However, the relative stratigraphic position of these areas cannot be determined due to the considerable spatial distance of over 30 m and the reduced stratigraphic development in section 17. Perhaps, the two concentrations represented complementary activity areas within a single, larger occupation event. Nevertheless, in this scenario material should be refittable from the two concentrations but this was not possible thus far. Therefore, the two episodes could reflect recurrent visits to the site which would represent a continuation of the Late Magdalenian behaviour at the site.

Conty, Somme

Research history

The site *Le Marais* near Conty was found during surveys led by Thierry Ducrocq and Pascal Le Guen in advance of gravel exploitation in early 1994. Subsequently, Jean-Pierre Fagnart with his team conducted a rescue excavation of the material which was preserved *c.* 1.3 m underneath the water table (Fagnart 1997, 109). During this excavation a concentration of lithics and bones in the lower horizon was uncovered on an approximately 100 m² large area. However, due to the ingress of water, the spatial and stratigraphical positions were only observed in a very general way per square-metre.

Topography

The site was located north of Conty in the valley of the Selle (**tab. 34**) which confluences into the Somme almost 20 km north-east of the site. Around Conty the Selle flows approximately from south to north. The confluence of the tributary Evoissons into the Selle lies 200-300 m south of the site. Due to this confluence a small basin measuring approximately 1 km × 1 km was formed south-west of the site. Around the basin and on both sites of the approximately 500 m wide Selle valley the hills rise up gradually some 30 m. East of the site a small dry valley cuts the surrounding hills. The site was located centrally in the valley. According to the stratigraphic observations, the site was situated in the floodplain of the river between two stream beds during the Lateglacial (Fagnart 1997, 110f.).

Stratigraphy

Above river gravels a small band of sandy silts were deposited which were attributed by comparison with the development in the Somme valley to the Late Pleniglacial or early Lateglacial Interstadial (Fagnart 1997,

110). In some parts of the valley down cutting occurred in these deposits which were partially filled with peaty material (Antoine et al. 2003b). Palynologically, these peaty fillings revealed in their upper part a very high amount of arboreal pollen which was mainly due to a significant presence of willow (*Salix* sp.). The sandy silt deposits were overlain at the site by a 30-40 cm thick, very loamy, organic silt deposit which was assumed as correlative to the Belloy-sur-Somme soil (Fagnart 1997, 46). The artefacts of the lower archaeological horizon were found at the base of these silts, whereas the intermediate archaeological horizon was stratified in the upper part of this deposit. The intermediate horizon produced only a few lithic and faunal remains and was attributed to the end of the Lateglacial Interstadial. However, due to this chronostratigraphic attribution, the small number, and the publishing status, this material is not incorporated in the present study. In the palynological analysis the two parts of the deposit appeared very different (Antoine et al. 2003b): In the lower part the arboreal pollen were not as frequent as in the peaty deposits but at this level the pollen diagram was clearly dominated by birch (*Betula* sp.). In the upper part, the arboreal pollen were nearly as frequent as in the peaty deposits and at this level of the stratigraphy, the predominant pollen were pine (*Pinus* sp.). On top the loamy silt deposit 50-70 cm of very calcareous, sandy silts followed (Fagnart 1997, 110) which were attributed to the Younger Dryas by the malacological and the palynological analyses (Antoine et al. 2003b). In the upper part of these silts, the upper archaeological horizon was found and accordingly placed to the early Preboreal. This horizon provided material attributed to the Long Blade Technology with bruised blades. Loamy silts and the modern topsoil formed the upper part of the stratigraphy.

Archaeological material of the lower horizon

In the lithic concentration of Conty were approximately 2,000 artefacts found including the splinters of the blank production process (**tab. 35**). A more detailed study of the archaeological material (Coudret/Fagnart 2012; Fritz 2012; Auguste 2012) was not available to the present author in time to be incorporated in this study. Therefore, the following descriptions are based only on previous presentations of the material (Fagnart 1997; Coudret/Fagnart 2006).

They were made of the local Turonian flint (**tab. 36**) which due to its homogeneity is an excellent raw material. Presumably, the material was taken from a primary resource some 500 m west of the site.

At least three cores were recovered from the site (**tab. 37**). They were used to produce short blades. These blades were knapped with a soft hammerstone. The butt of these blanks was generally plain, rarely faceted, indicating little core preparation.

In the assemblage 42 retouched artefacts (**tab. 38**) and ten blades with heavy use-wear were detected. The LMP were clearly the largest tool category (n=24) which Jean-Pierre Fagnart divided in the couteaux à dos (n=7), the *pointes aziliennes* or *Federmesser* (n=16), and a single backed bladelet. As mentioned previously, the terms Azilian point and *Federmesser* were used variedly and according to the published drawings (Fagnart 1997, 114 fig. 89) several pieces were re-classified in the present study (see p. 275-282). Furthermore, six burins were found of which five were dihedral types and one was made on a truncation. However, five truncations were recovered. The end-scrapers were rare with two short examples. In addition, two borers were found and both classified as bec.

Furthermore, some 2,000 faunal fragments were recovered from the concentration. Due to the very good preservation the majority was determinable (**tab. 39**). Among the determined bones were those of aurochs (*Bos primigenius*) more frequent than the remains of red deer (*Cervus elaphus*; Fagnart 1997, 111). In addition, roe deer (*Capreolus capreolus*) was also attested (Coudret/Fagnart 2006, 731) which was, thus far, considered as a younger element in the Lateglacial fauna of North-western Europe (Sommer et al. 2009).

The good preservation also made it possible to observe various traces of human manipulation on the bones. An on-going analysis of these traces will provide more insights into the animal exploitation strategies of these mid-Lateglacial hunter-gatherers (Fagnart 1997, 111). Preliminary results of this archaeozoological analysis indicated an occupation of the site during the late winter season (**tab. 40**; Coudret/Fagnart 2006, 738).
An antler fragment of red deer was worked and decorated (Fagnart 1997, 112). This heavily fractured piece consisted of the basal shaft which was preserved over 20 cm long and the fractured lowest prong (Fagnart 1997, 118). The use of this object remained unclear but closest parallels were the so called Lyngby axes of northern Europe (**tab. 41**; cf. Clausen 2004). Even though the piece was very fractured, groups of engraved zig-zag lines were visible on several parts of the specimen (**tab. 42**; Fagnart 1997, 117). This type of geometric decoration was widely known in Lateglacial Europe (e.g. Rozoy 1990; d'Errico 1994; David/d'Errico/Thévenin 1998; d'Errico/Ucelli Gnesutta 1999; Veil/Terberger 2009).

Spatial organisation of the lower horizon

Due to the observed accumulation of burnt bones and lithic artefacts, a hearth was reconstructed in the east of the artefact scatter (**tab. 43**; Fagnart 1997, 111). A concentration of larger bone fragments was recorded some metres to the south of the hearth and the scatter of artefacts and, possibly, represented a dump site or a special working place.
The presence of the bones supplemented by the dominance of LMP and used blanks led to the interpretation of a specialised camp for hunting preparation and/or processing of the prey after a successful hunt. However, the diversity of the faunal material as well as the presence of a dump/special task area and the presence of the hearth could suggest a longer duration with more complex spatial organisation. Perhaps, the detailed archaeozoological study will help to explain this assemblage. For instance, if the accumulation of larger bones represented a workshop and some of the species were introduced as provision material, the hypothesis of a special task camp could be sustained.

Chronology of the lower horizon

Four bones of *Bos primigenius* were ^{14}C-dated (**tab. 44**) and produced a homogeneous scatter between 11,980 and 11,330 years ^{14}C-BP (Fagnart 1997, 112). Two dates from samples of charcoal (GifA-99526) and bark (Ly-284) taken for the stratigraphic analysis of the site yielded again comparable results. Further dates were taken on samples of sediment for the stratigraphic analysis (cf. Ponel et al. 2005). These dates were excluded due to the lacking association with the archaeological material as well as the general scepticism towards the reliability of sediment samples (see p. 265-269). However, the combination of all these dates clearly placed the human activity at the site of Conty to the mid-Lateglacial Interstadial. This chronological position was in accordance with the stratigraphy and could be further sustained by the palynological and malacological analysis.
The material from the lower horizon of Conty represented, perhaps, the remains of very few events in a temporally close succession or, more probable, of a single episode in the mid-Lateglacial Interstadial.

Hangest-sur-Somme III.1, Somme

Research history

The site III.1 from Hangest-sur-Somme was found in a test pit program conducted prior to gravel exploitation in the Somme valley. After quarrying work had already begun in the gravel pit, the site Hangest-sur-Somme III.1 was excavated in 1992 in a rescue project by Jean-Pierre Fagnart and his team. The work was particularly complicated by the elevated water table. Hence, the relevant sediment of 150 m² was mechanically recovered in 1-2 m³ and subsequently »excavated« (Fagnart 1997, 189). Thus, even though the excavation method was adapted to the circumstances of a high water table and the short time, the recovery of the material were according to modern standards.

Topography

The site was located only several 10 m away from the western banks of the modern Somme (**tab. 34**). In this area the river flows in a large, over 1 km wide valley. During the Lateglacial, the site was probably situated on a gravel heap which rose some 1-2 m above the floodplain (Fagnart 1997, 189). The terrain rises very gradually to the south-west but 300-400 m away from the site hills rise abruptly c. 60 m upwards. However, a small dry valley cuts into these hills south-west of the site. Approximately 800 m northwards another, larger valley cuts these hills and gradually ascends to the elevated plateau. On the eastern side of the floodplain the hills gradually rise some 40 m. The confluence of the Somme and its tributary the Nièvre is only a few hundred metres northwards of the site. This tributary had cut a wide valley into the hills on the eastern bank.

Stratigraphy

At the site the gravels of an old river terrace were overlain by a 50 cm thick deposit of Late Pleniglacial yellowish, sandy silts on top of which a small band of sand and gravels followed (5-6 cm). Yellowish calcareous, sandy silts of 35 cm thickness were deposited on top (Fagnart 1997, 189). These silts were attributed palynologically to the early Lateglacial Interstadial (Fagnart 1997, 190 f.). On top of these silts a blackish, organic silt horizon was formed which was assumed as correlative to the Belloy-sur-Somme soil (Fagnart 1997, 46. 190). The lower archaeological horizon was found at the transition from the sandy silts to this Lateglacial organic layer (Fagnart 1997, 189; Coudret/Fagnart 2006, 731). Palynologically the lower part of the organic silt horizon was correlated to the highest arboreal values and high numbers of birch (*Betula* sp.) pollen (Fagnart 1997, 190). However, the pollen profile was partially affected by the infiltration of Holocene material but this influence decreased with increasing depth (Fagnart 1997, 191). The increased diversity of the malacological assemblage indicated a developed vegetation cover (Fagnart 1997, 191). The upper archaeological horizon was located in the same deposit but some 10-15 cm higher in the stratigraphy. The arboreal pollen were considerably less frequent in this upper part than in the lower part. Pine (*Pinus* sp.) pollen increased but birch remained dominant throughout the pollen diagram. The occurrence of *Helicopsis striata* was considered as diagnostic for the Allerød (Fagnart 1997, 191). However, the upper archaeological horizon spread to the transition to the overlaying slightly sandy, loamy silts. This 20-30 cm thick deposit was affected by cryo-turbation and attributed malacologically to the Younger Dryas. The upper part of the stratigraphy was formed by Holocene brown and blackish organic silts and peat.

Archaeological material of the lower horizon

The 1,945 lithic artefacts of the lower horizon were mainly found in an area of 15 m × 15 m (**tab. 35**). The majority (n = 1,106) were smaller than 2 cm (Fagnart 1997, 193). Some artefacts were burnt.
The flint material used in the lower horizon of Hangest-sur-Somme III.1 was from the local Cretaceous deposits and of excellent quality (**tab. 36**; Fagnart 1997, 193).
21 cores and core fragments were recovered from the lower horizon (**tab. 37**). Blades were the main aim of the blank production process. Among the blades were some long ones but the majority were short blades. The indications of the knapping technique suggested the use of a soft hammerstone. Generally, the butt was

plain but facetted ones also occurred frequently. A considerable part (41 %) of the analysed butts showed traces of abrasion but no preparation of the *en éperon* type was observed (Fagnart 1997, 197).
Among the approximately 100 tools (**tab. 38**) the LMP formed the largest category (n=49; Fagnart 1997, 197. 255). In addition to 24 simple backed blades and bladelets, 13 truncated backed blades were identified. Three small truncated shouldered points were accompanied by three shouldered piece fragments. Furthermore, eight curve-backed points were recovered from the site among which a bipoint and a Malaurie point were found. Three large *Federmesser*-like shaped *couteaux à dos* were also found and one which resembled a large penknife point. Generally, the LMP were preserved comparatively completely. A long blade was finely retouched on its lateral edge and at the tip part an oblique abrupt retouch was made. In regard to abrading traces along the upper and lower part of the cutting edge, this piece could also be considered as *couteau à dos*. Nevertheless, formally it was regarded as truncation. A further six truncations were found in the assemblage. The 27 burins were dominantly made on truncation (n=20) and less frequently were dihedral types (n=6). The six end-scrapers were mainly made on flakes.
Numerous fragments of red ochre were also recovered from the site (**tab. 42**; Fagnart 1997, 193).
The fauna was poorly preserved. Determinable bones and teeth belonged to aurochs (*Bos primigenius*) and horse (*Equus* sp.; **tab. 39**; Fagnart 1997, 193). The co-occurrence of these species reflected perhaps the still mixed character of the mid-Lateglacial Interstadial landscape with forests in the floodplain where the bovid could roam but still enough grassland areas for the horses. A molar of aurochs as well as a tibia of a large herbivore, presumably also aurochs were ^{14}C-dated.

Spatial organisation of the lower horizon

Evident structures were not observed. The lithic artefacts were distributed in small concentrations of debris from the blank production process (Fagnart 1997, 193). Furthermore, the retouched artefacts formed distinct and differentiated clusters. Thus far, no spatial analysis of the material was published.

Chronology

Chronostratigraphically, the site was attributed to the mid-Lateglacial Interstadial based on the lithostratigraphy, the palynology, malacology, and composition of the mammal fauna. Furthermore, ^{14}C dating sustained this attribution.
According to the observations of the investigators of the site, the lower horizon from Hangest-sur-Somme III.1 represented a relatively undisturbed single occupation event (Coudret/Fagnart 2006, 738). However, without a detailed publication of the spatial information, the intra-site chronology cannot be further tested.

Summary and context

In general, the sites in northern France were usually located in valleys at the banks of rivers (**tab. 34**). The areas investigated around the sites vary considerably (**tab. 35**) mainly depending on the research history. For instance, the sites found in gravel pits were often analysed in large campaigns, whereas the extent of excavations in caves were naturally limited. However, the amount of lithic material seemed unrelated to the extent of the excavated area but in contrast appeared dependent on the number of concentrations in this area and the type of site. For example, the 5 m² from Le Tureau des Gardes, *locus* 6 yielded more than five times as many artefacts as the 5 m² from Hallines and almost as much material as the 75 times larger area from the *loci* 4, 46, and 50 of Le Closeau.

In the exploitation of raw materials a distinction between the Paris Basin and the Somme sites is observable (**tab. 36**). In the Paris Basin occasionally foreign lithic raw materials accompanied the local raw material of varying qualities, whereas in the Somme region only the local, usually high-quality material was used.
The general tendency in the blank production process is the disappearing importance of bladelets (**tab. 37**).
In contrast to the Central Rhineland, the general dominance of LMP was not observable in northern France, although some sites were also dominated by this tool class (**tab. 38**).
The preserved fauna showed diverse species compositions with horse (*Equus* sp.), reindeer (*Rangifer tarandus*), red deer (*Cervus elaphus*), and aurochs (*Bos primigenius*) being of some importance (**tab. 39**). However, the quality of the preserved material was usually not very good. Therefore, indications for human modification and the seasonal attribution of the sites are scarce (**tab. 40**). In consequence, organic artefacts were also very rarely preserved and/or recognisable (**tab. 41**).
In contrast, a group of diverse »special goods« including mineral and fossil material or engravings were usually better preserved (**tab. 42**). However, whether some of these materials such as ochre or the line engravings on cortex were of symbolic importance or utilitarian remained a matter of discussion (Barton 1992; Holzkämper 2006, 86-88).
Hearths represented also the typical structures on northern French sites (**tab. 43**). They were generally recognised by small and dense accumulation of gravels in the concentrations.
The corpus of ^{14}C dates from northern France is relatively rich (**tab. 44**) but many dates produced arbitrary results and, thus, the record has to be evaluated rigorously (p. 265-269 and p. 474-481).
Clearly, in northern France numerous sites from the Late Magdalenian were also found such as Pincevent (Bodu et al. 2006b; Débout et al. 2012) or Étiolles (Pigeot 1987; Olive 2004). Furthermore, sites of unambiguous FMG character such as Saleux *La Vierge Catherine* (Fagnart 1997, 131-143) were also found. These sites were clearly comparable to the record from the Central Rhineland and would not provide information on the transition between the Late Magdalenian and the FMG.
Therefore, these sites were not considered in the present study.

Besides the relevant assemblages presented previously, further assemblages from northern France provided potential information on the transitional period but were excluded for various reasons:
In the late 19th century the 17 m × 10 m wide Grotte de Clèves near Rinxent (Pas-de-Calais) was excavated and yielded Mousterian as well as supposed Magdalenian material along with three human mandibles (Fagnart 1997, 105 f.). The latter were considered as possibly intrusive burial remains of Chalcolithic age. Among the faunal remains were at least ten reindeer (*Rangifer tarandus*) identified which were accompanied by two red deer (*Cervus elaphus*), a bovid (*Bison* sp./*Bos* sp.), and a wild boar (*Sus scorfa*). This assemblage suggested either a transitional environmental period where wild boar and reindeer found suitable habitats in close vicinity or an admixture of the assemblage, perhaps, due to recovery. However, a piece of reindeer antler was dated to a Late Pleniglacial age (OxA-1343: 13,030 ± 120 years ^{14}C-BP; Hedges et al. 1988a) which is comparable to the oldest group of dates from Gönnersdorf and the lower horizon of Andernach (see **tab. 20**). In this case, the reindeer remains were possibly from a different episode than the other bones. In particular, if the reindeer antler was a shed piece it could also result from the collection of fossil material and in this case the radiometric date would date the shed of the antler but not the human collection of it. Indications of human modification were found on a rib of the bovid which was described as perforated (Fagnart 1997, 106). The lithic inventory comprised some 30 artefacts of which six were retouched. Besides a shouldered point, two end-scrapers made on large flakes were the most typical pieces. Moreover, a straight truncation, an oblique truncation with a retouch from the ventral on the shorter edge, and a further two pieces with lateral retouches were found. Thus, the lithic shapes equated rather material from the interme-

diate period and is therefore mentioned here. However, the lack of more detailed information due to the time of recovery prohibited detailed analysis of the assemblage. Comparable lithic material with additional faunal remains were reported from another cave and a rockshelter in the vicinity of the Grotte de Clèves but both collections from the 19th century are lost (Fagnart 1997, 106).

In Flixecourt (Somme), only a few kilometres northwards of Belloy-sur-Somme, a large assemblage attributed to the Long Blade Technology with bruised blades was recovered in the late 1980s and early 1990s (Fagnart 1997, 165). Besides this assemblage, a small lithic inventory comparable to the material from the lower horizon in Belloy-sur-Somme was recovered from this site. Moreover, some artefacts were of FMG character (Fagnart 1997, 165) and, perhaps, could originate from the later Lateglacial but these specimens were not well stratified. Thus far, these assemblages from Flixecourt were not thoroughly published and, therefore, this material is not included in the present study. Nevertheless, the close locational presence of possible chronologically distinct assemblages which, perhaps, were partially independent of a continuous tradition emphasised the persistent attraction of these places in the Somme valley for Lateglacial hunter-gatherers and advised caution in the interpretation of unstratified assemblages. For instance, the lithic inventories from the gravel pits in Dreuil-lès-Amiens (Somme) and Amiens-Étouvie (Somme) were recovered during various collections and excavations (Fagnart 1982; Fagnart 1997, 120-130). Jean-Pierre Fagnart and his crew finally excavated both sites successively from 1979 to 1982. However, no spatial organization was observed, due to the acidic ground no organic material was preserved, and the stratigraphy also did prohibited further narrowing of the dating of the geological layer. Nevertheless, due to the typo-technology of the material these assemblages were discussed in the context of the Hengistbury Head type industries (e.g. Barton 1992; Barton et al. 2009) and, hence, were attributed to the mid-Lateglacial Interstadial. However, since neither the chronostratigraphic position nor the integrity of the material can be ascertained, these inventories are not further considered in the present approach.

In contrast, the material from the *locus* 244 in Saleux *Les Baquets* produced along with an assemblage comparable to the Hengistbury Head type industries also a ^{14}C date (Coudret/Fagnart 2006; Barton et al. 2009). The site Saleux was excavated between 1993 and 2006 in three sectors. The southern one *La Vierge Catherine* yielded unambiguous FMG assemblages which were dated to the transition from the Lateglacial Interstadial to the Lateglacial Stadial (Fagnart 1997, 131-143). This chronostratigraphic position was in accordance with the stratigraphy and the palaeoenvironmental data. *Les Baquets* was the central sector and north of this a third, unnamed sector was excavated. All sectors were found along a palaeochannel in the Selle valley some kilometres north-east of Conty. In *Les Baquets* at least two concentrations were recognised. The northern one was *locus* 244. From this concentration, a metapodial of an aurochs (*Bos primigenius*) was ^{14}C-dated (GrA-18832 (Ly-1566): 11,640 ± 70 years ^{14}C-BP; Coudret/Fagnart 2004). The date is similar to the dates from Conty and Hangest-sur-Somme III.1. Besides aurochs, also red deer (*Cervus elaphus*) was determined in the faunal assemblage of *Les Baquets* (Coudret/Fagnart 2004). The lithic artefacts of *locus* 244 comprised very large curve-backed *couteaux à dos*, penknife points, *Federmesser* as well as some angle-backed pieces (Coudret/Fagnart 2004). Besides the LMP, end-scrapers and burins were common, whereas borers were rare. The lithic material scattered in several concentrations each presumably centred around a hearth (Coudret/Fagnart 2006). Thus, the spatial distribution suggested a complex spatial development of the site. However, the analysis of the material is on-going and too few details were published yet to incorporate this interesting assemblage in the present study. The 15 m south-eastwards located *locus* 234 was found in a stratigraphic higher position than the material from *locus* 244. This single concentration contained a human cranium (Coudret/Fagnart 2004; Coudret/Fagnart 2006). Furthermore, the lithic material was similar to the material from *locus* 244 only that among the LMP very large *couteaux à dos* occurred and also the *Federmesser* were of a considerable size (Coudret/Fagnart 2004). Of some interest was one piece which could be interpreted as Lyngby-Bromme type tanged point (Coudret/Fagnart 2004, 13 fig. 12.9). From this material two samples from aurochs were ^{14}C-dated (GrA-15945 (Ly-1141): 11,200 ± 70 years ^{14}C-BP; GrA-15946 (Ly-1142): 11,160 ± 70 years ^{14}C-BP; Coudret/Fagnart 2004) and sustained the younger chronostratigraphic position. However, for this material the analysis is also on-going.

In Normandy, the assemblage of the *locus* 33 at the site Le Cornet near the village Ambenay was techno-typologically related to the material from Conty, Saleux, *locus* 244, and the upper horizon at Hangest-sur-Somme III.1 as well as Pincevent III.2 (Valentin/Fosse/Billard 2004). This inventory yielded various types of monopoints, some very large *couteau à dos*, and a flexible exploitation strategy in which mainly soft hammerstones were used (Valentin/Fosse/Billard 2004). However, the remains were found in a poorly developed stratigraphy within a very acidic soil in which no organic remains were preserved. Thus, the chronostratigraphic attribution of the site was very generally possible to the Lateglacial but, furthermore, it was solely based on the comparisons of the lithic inventory. However, since the chronostratigraphic position cannot be ascertained independently, this assemblage is neglected in the present approach.

Faunal remains from the Abri du Mammouth near Saint-Pierre-d'Autils produced a ^{14}C date (GifA-92344: 11,040 ± 110 years ^{14}C-BP; Fosse 1997, 241) falling calibrated to the end of the Lateglacial Interstadial and, thus, in the chronological proximity to the LSE and the onset of GS-1. In general, the date was reliable but it remained unclear which sample was exactly dated. Furthermore, the association of the specimen to the human activity around the rockshelter was equally uncertain because the archaeological material appeared of Late Magdalenian character and was found in a possible colluvium (Fosse 1997, 241). Thus, the chronological indications were inconsistent and the specification unclear. Therefore, the date and the material were excluded from the present study. Nevertheless, due to the geographic position in northern France between the concentrations of sites in the Paris Basin and those in the Somme region a re-analysis of this assemblage could provide interesting new insights in the relations of these regions.

In the near future, the remains from a test pit near Decize (Nièvre) might yield further insights in the developments within the transition period, in particular, in the southern part of the Paris Basin. However, the analyses are, thus far, still in progress and, therefore, this material was only published in a very preliminary report (Nicoud et al. 2011). The 115 lithic remains and few burnt bone fragments from test pit 1044 were associated with two dark coloured pockets in an alluvial sand deposit. Some of the lithic remains were also burnt and, in consequence these dark spots might be considered as hearths. However, the sedimentological analysis is on-going and another organic origin of the dark colour such as soil formation cannot be excluded yet. The material was found in a small cluster within and around these pockets and was assumed as a small assemblage of a single occupation event. The LMP techno-typology could either be compared to assemblages such as Amiens-Étouvie (Fagnart 1982; Fagnart 1997, 120-130) and also to Conty and Hangest-sur-Somme III.1 but it could also be compared to material such as found in *locus* 25 of Le Closeau. Perhaps, the organic remains from this site allow for a more precise chronostratigraphic position which in combination with the spatial and techno-typological analysis of the material are of particular interest in this otherwise poorly known archaeological region.

METHODS

The present study aims to examine change in a human society within the context of climatic and environmental variation. Using an archaeological case study requires a detailed and independent chronology that can incorporate different sites in order to distinguish the different stages in this process of change.
Therefore, the material presented previously must be evaluated to facilitate the comparison of different archives in the same referential framework. For example, radiocarbon dating is the prevailing method for constructing chronological frameworks in Lateglacial archaeology. However, ^{14}C dates have to be calibrated because the radiocarbon timescale does not relate in a linear fashion to the calendar timescale. Consequently, a calibration of comparable radiocarbon ages does not necessarily result in comparable calender ages. The comparison of uncalibrated radiocarbon ages can lead to erroneous interpretations about the temporal relation of the dated materials (Blockley/Donahue/Pollard 2000b). Calibration of radiocarbon ages requires a calibration curve, which provides the calendar ages to a high degree of accuracy for the studied period. To construct a reliable calibration curve, high-resolution climate archives are vital. In addition, calibration results can, occasionally, be further distinguished through the use of the environmental data from the site. However, this environmental data is only useful if environmental development, at either site or broader scale, can be related to a calender timescale. This relation is usually established through events on a global scale such as volcanic eruptions or climate change. However, the response of the local environment to these global climate events depends on various factors relating, particularly, to the geographic and topographic site position. Thus, the site's spatial context is also important to create a reliable frame of reference.
In order to study changes in human behaviour within the context of climatic and environmental change requires a secure framework to produce reliable results. The current project uses intensive climate and environmental archives to provide background information necessary to create a spatio-temporal frame of reference. The methods used to create the referential framework and to evaluate changes in human behaviour are presented below.

CLIMATE AND CALIBRATION

Three main types of archives are used in the current study. The first type is the Lateglacial climate record (see Material-Climate, p. 7-30). These records formed the basis for the Lateglacial chronostratigraphy, helping to construct radiocarbon calibration curves (e.g. Jöris/Weninger 1998; Weninger/Jöris 2008; Reimer et al. 2009). Therefore, the climate and the calibration data are considered concurrently.

Defining the limits of Greenland oxygen isotope events

The Greenland eventstratigraphy (see p. 9-12) was selected as the base chronostratigraphy for this study because it combines global climate indicators with a high-resolution stratigraphy. Significant shifts in the oxygen isotope record were assumed to reflect both the onset and end of climate events. This type of shifts were defined as the main stratigraphic markers in the Greenland eventstratigraphy. This relation to climate

events allowed other oxygen isotope records to be correlated to the Greenland chronology using these stratigraphic markers (Björck et al. 1998; Jones et al. 2002; Magny et al. 2006; Weninger/Jöris 2008).
However, in the first publications of this eventstratigraphy it was stressed that the onset and end of the various climate events could register asynchronously (Björck et al. 1998; Walker et al. 1999). This asynchrony resulted from a variation in the exact response to these climate changes. The response to climate changes, which are equivalent to the onset of a climate event, depend on how sensitive the chosen proxies are to climate changes along with the geographic and topographic location of the sampling site. Thus, delimiting the same event in archives from different geographic locations and/or based on different proxy data could result in some chronological offset on an absolute timescale. In the construction of the eventstratigraphy this offset of exact age was considered less important than the general correlation of the archives. In fact, on a geological timescale the offsets were negligible. Consequently, other stratigraphic sequences were tuned to the limits given in the Greenland eventstratigraphy to produce a chronological sequence that could be related to the absolute timescale of the Greenland ice-cores. Therefore, comparisons of these sequences in regard to the question of leads and lags in relation to climate change became more difficult due to the tuned limits (Blaauw et al. 2010). Moreover, with the increasing number of high-precision records using the same proxies and constructing a detailed Lateglacial climate history, precise age estimates are important. For instance, the precise age for the limits of several events was compared in various ice-core records to construct the GICC05 (cf. Southon 2002; Vinther et al. 2006; Rasmussen et al. 2006). Furthermore, the precise age of the event limits also made the correlation with records possible that had no oxygen isotope curves but a comparably precise chronology, such as the varve records (Litt et al. 2001). The combination of these terrestrial records placed the climate history in a temporal framework related to environmental development. Thus, the exact definition of the Greenland isotopic events and, in particular, the shifts between these events is of interest for this project because these limits provide useful correlation points.
In the Greenland eventstratigraphy, the limits of the proposed events (GS-2a, GI-1e, GI-1d, GI-1c, GI-1b, GI-1a, GS-1, Björck et al. 1998; for further Pleistocene and Holocene events: Walker et al. 1999; Lowe et al. 2008) were set at the mid-point of the slope in the oxygen isotope curve (Lowe et al. 2008, 10 tab. 1, Note). However, for some shifts the development of the deuterium excess was used instead. A high-resolution analysis of the NGRIP ice-core provided the ages for the oxygen isotope changes (Steffensen et al. 2008), and these limits are used throughout the present study.
In contrast, for some episodes in the Lateglacial Interstadial (GI-1c_3, GI-1c_2, GI-1c_1) no limits were previously given in the literature, although these episodes seemed to be reflected in the environmental development in north-western Europe (Litt/Stebich 1999; Litt et al. 2001; cf. Hoek/Bohncke 2001; Magny et al. 2006). Therefore, these limit were defined for the current project based on the δ^{18}O record[19] of the NGRIP ice-core (see p. 9-12). Again the mid-point of the slope between the episodes was chosen as limit. However, the problem of high-resolution data is to identify the main slope (Lowe et al. 2008, 10 tab. 1, Note; Steffensen et al. 2008). In this study, a simple difference of one data point to the next was calculated to reveal steep changes between the points. However, in the fluctuating record of the Lateglacial Interstadial, several steep shifts, i.e. large differences occurred. Moreover, the fluctuations swung in two directions: positive and negative. Often a steep positive change was followed by a comparable steep negative change. Thus, these changes reflected the amplitude of the fluctuations within the event or episode. However, these amplitudes

19 This record was released September, 10th 2007 from the Centre for Ice and Climate, University of Copenhagen and is obtainable from www.iceandclimate.nbi.ku.dk/data/. It contained information on various parameters in the NGRIP ice-core given in 20 year means. These parameters were the δ^{18}O value, the age and the counting error according to the GICC05, and the depth in the ice-core.

of fluctuations were of no instant interest, instead, the general, long-term shifts were investigated. To recognise these long-term shifts, the amplitudes had to be neutralised and the resolution had to be lowered to reveal the more systematic shifts. This neutralization and lowering of the resolution was achieved by an addition of the previously calculated differences between the successive data points. In general, 21 years lay between two successive data points of the oxygen isotope record. By adding three to five of these differences to lower the resolution a time period of 63, 84, or 105 years was generally covered. If in this addition changes in the same direction (positive or negative) accumulated the value increased, whereas if the values were fluctuating the value became smoothed. Thus, in this lowered resolution the systematic changes became identifiable. Nevertheless, the limit was set at the mid-point between the two points yielding the largest difference in the recognised series of changes.
To test the usefulness of this approach, the data points in the NGRIP record from 18,000 to 10,000 years cal. b2k[20] were studied. The three comparable dates of the oxygen isotope record given by Jørgen Peder Steffensen and his colleagues (Steffensen et al. 2008) in the high precision analysis were equivalent within less than ten years to the results of the calculations used in this project.
An interesting by-product of this larger scale comparison of the fluctuations in the ice-core record was that the amplitudes were most marked during the Late Pleniglacial and diminished significantly with the beginning of the Holocene. These smaller amplitudes reflected a more stable climate regime that, probably, had an impact on the natural environment. For instance, a stabilised climate could lower the migration pressure and permit biotic communities to establish in a favourable place and adapt *in situ* to smaller fluctuations. Such a stabilization or destabilization was presumably recognisable by hunter-gatherer groups and would influence their adaptive strategies.

Correlation of tephrochonology with GICC05

Even though various high-precision dating methods have been developed, absolute simultaneity can rarely be demonstrated. For example, a tephra layer of a volcanic eruption is not deposited simultaneously but, depending on atmospheric movements and distance to the volcano, within hours, days, or months after the eruption (Rasmussen et al. 2008, 19). The exchange of atmospheric isotopic ratios to comparable values can, similar to the tephra deposit, take months or even a century across distant areas (Kromer et al. 2001, 2529; Hua et al. 2009, 2985; Rasmussen et al. 2008; 19; Magny et al. 2006, 426f.). The temporal resolution achieved by various dating methods for the Lateglacial included decadal to centennial deviations. Thus, in comparison to those deviations, volcanic tephras represent extremely high-precision, quasi-synchronous time markers.
Moreover, some tephra layers were deposited over large areas and in various archives such as ice-cores or varved lake sediments. Thus, the records from these various, tephra containing archives could be aligned along these markers. Therefore, a Lateglacial tephrochronology related directly to the Greenland ice-core chronology represents an important tool in combining various records within a common temporal frame.
A detailed Lateglacial tephrochronology of the NGRIP ice-core was published by Anette Mortensen and colleagues (Mortensen et al. 2005). However, this tephrochronology was presented without an age model. Instead the depth of the volcanic ash layers and single volcanic sherds in the NGRIP ice-core was provided. Some identified layers were correlated to ^{14}C-dated marine and continental records (Mortensen et al. 2005, 215 fig. 7; 217).

20 b2k refers to »before 2000 A.D.« (see p. 8 and p. 591).

However, for the present approach this succession of volcanic layers was correlated to the GICC05. This correlation was made using the depths given for the volcanic ash layers with the depths given in the 20 year mean record of the oxygen isotope curve set in GICC05 (see above). If no exact matches were found the distance between the two nearest depths given in the oxygen isotope record was scaled evenly by the number of years counted between these depths. The exact depth of the tephrochronology was rounded to the next of these even steps. Clearly, the record was not deposited evenly but the error is far beyond the counting error and, thus, negligible. The age of the Vedde Ash was given in the publication of the GICC05 as important marker (Rasmussen et al. 2006), and was adopted for this study.

In addition to the volcanic ash layers with exact depths, the calcium corrected sulphate content (SO^{2-}_4 – Ca^{2+}) for the Lateglacial and early Holocene was also provided by Mortensen and colleagues (Mortensen et al. 2005, fig. 2). Peaks in the calcium corrected sulphate content were assumed to reflect volcanic eruptions (Mortensen et al. 2005, 210). However, the increase of this content as well as peaks within this content during cold periods was apparent and attributed to lowered precipitation and, perhaps, increased wind activities leading to peaks from various sulphate sources (Mortensen et al. 2005, 216). Thus, a sulphate background signal that was not related to volcanic activity seemed to vary between the warmer and colder periods. However, the changes of this signal due to altered volcanic activity patterns could not be excluded entirely. The calcium corrected sulphate content was also given in relation to its ice-core depth and the development of the oxygen isotope record by Mortensen and colleagues (Mortensen et al. 2005, fig. 2). For the present project, this record was correlated to the GICC05 by depth and, in addition, the records were checked graphically for the relations of significant peaks in the corrected sulphate content to fluctuations in the oxygen isotope record. This record was considered important due to volcanic events that were not identified by tephra layers or single volcanic sherds in the ice-core record such as the LSE. The chronostratigraphic position of some eruptions that were related to a significant sulphate output such as the LSE can be discussed based on this record.

Correlation of oxygen isotope records with GICC05

The atmospheric oxygen isotope composition in the ice-core records depended mainly on the temperature at the coring site during formation (Dansgaard et al. 1993; Steffensen et al. 2008) and, as a consequence, on the climate regime. Therefore, this isotope composition behaved comparably in regard to its relative patterns across the globe and the isotopic records can be correlated to each other by matching of significant positive or negative excursions (i.e. wiggles). In the current study, peak values were used as correlation points for minor excursions, whereas the major shifts such as the onset of the GI-1e were correlated at the mid-point of the shift.

However, in comparison to the Greenland oxygen isotope records, terrestrial $\delta^{18}O$ values were influenced by further factors such as water temperature, isotopic composition of the water and the catchment runoff, precipitation, altitude, sedimentation rate, and the amount of allochtonous material in the sample (Siegenthaler/Oeschger 1980; Grafenstein et al. 1992; Hoek et al. 1999; Wang et al. 2001). These additional influences could cause some delay in the registration of changes and/or an admixture of different signals leading to arbitrary results. The possible sources of error such as admixture can be revealed from the records. The discussion of precise leads and lags of isotopic records (Blaauw et al. 2010) exceeds the subject of this study. A quasi-simultaneity of the isotopic signals has to be assumed to allow for the correlation according to the idea of an eventstratigraphy. However, the climate system of north-western Europe is generally governed by the North Atlantic circulation which also influences the Greenland record (e.g. Lotter et al. 1992; Moros et

al. 2004; Sirocko et al. 2005). This climate regime also influenced several additional signalling factors in terrestrial sequences such as precipitation and water temperature. Therefore, the signal of major shifts in this climate regime were intensified in the terrestrial records. Thus, isotopic compositions of authigenic deposits were strongly governed by the same major climatic signals as the Greenland isotope record and the major shifts should have occurred parallel in time. These time-parallel shifts are correlated.
To obliterate a major climate signal in the terrestrial records, several of the factors mentioned above as influencing the isotope signal need to co-occur. These obliterating situations occurred for instance in Gulickshof (Limburg, Netherlands) where an episode of ground water influx, presumably caused by the final melting of relict ground-ice, led to a disturbed oxygen isotope signal during the first part of GI-1c (Hoek/Bohncke 2001). In the southern German Ammersee a significant sub-decadal isotope excursion was registered during GS-1 and was also attributed to a high accumulation of detrital material caused by a warmer episode which destabilised the previously frozen soils (Grafenstein et al. 1999, 1657 fig. 4). In both cases, the introduction of larger proportions of allochtonous material caused the arbitrary values. This allochtonous component is usually considered within these sequences and can be identified. These periods of increased allochtonous impact have to be considered as problematic. However, if these disturbed episodes are relatively short they can be treated as gaps which can be spanned between two reliable correlation points. Nevertheless, the numerous additional factors related to the isotopic composition of terrestrial records influence the minor signals and as a consequence the correlation of single peaks was, and is, often not possible.
In summary, the correlation of the Greenland eventstratigraphy with terrestrial oxygen isotope records required archives formed mainly by authigenic deposition and for which the isotopic compositions of the contributing elements was, relatively, well known. The correlation was established by setting major shifts in the isotopic record as time parallel. Possible peaks of positive or negative excursions are only correlated in parts where this procedure seemed possible. Therefore, a correlation of the terrestrial oxygen isotope records to the Greenland eventstratigraphy should be regarded as an approximation, which should be confirmed by further indicators such as volcanic markers, counting of laminated sediments, and/or radiometric ages.

Correlation of laminated chronologies with GICC05

Since archives with no oxygen isotope signature cannot be correlated to the GICC05 using the eventstratigraphic wiggle-matching outlined above, other procedures are necessary.
In the case of varve records, a correlation to the ice-core records can be based on a numerical comparison of the couplets counted between marker horizons such as volcanic ash layers or between significant regime changes in the isotope record; changes in the sediment composition or other proxies such as pollen frequencies can also be used. If the counted ages for the identified events are statistically similar, a common factor ruling the shifts in the proxies can be assumed and, in consequence, a quasi-synchronous reaction is implied. Thus, the correlation of shifts in the different proxy records is again based on the assumption of time-parallel reactions to significant climatic events.
However, in both chronologies (ice-core and varve record) the counting errors are cumulative and increase to considerable numbers during the Pleistocene. Statistical comparison of the ages determined for various shifts during the Pleistocene record could result in similarity, even though a hundred years or more separated the detected shifts in the compared proxy records. In addition, if hiatuses occurred in the records and remain undetected, the numerical approach might be completely misleading. However, the identification of hiatuses remains difficult based on the numerical comparison alone. The development of further proxies in each record and between various records has to be compared to reveal the existence of a hiatus and to

number the missing couplets (cf. Leroy et al. 2000). In the sediment records hiatuses and/or disturbances of the sequence were usually identified in the lithological inspection.

Many of the terrestrial proxies are not as sensitive as the isotope record and are influenced by local factors. The assumption of recorded changes in different proxy values being time-parallel remains therefore an approximation and, in general, problematic. In particular, if vegetation changes such as the decrease or increase of arboreal pollen are correlated the influential factors of the local conditions are high. Moreover, by the setting of the identified changes as quasi-synchronous could potentially mean that leads and lags between the records can no longer be identified. Therefore, this type of correlation is not used solely to establish contemporaneous events but is supplemented by volcanic markers and/or radiometric ages if possible. Some caution is necessary in the choice of the radiometric ages because the varve chronologies usually originate from aquatic environments where a hard-water effect can potentially alter dates taken on samples of water plants (see p. 259-263). Thus, these samples need to be generally rejected.

Comparison of the Greenland oxygen isotope record with dendrochronological data can also follow a numerical approach. In some dendrochronological sequences, deuterium isotopes were also measured (Friedrich et al. 1999), which can reflect climatic stresses affecting the plant. However, climatic changes in the Lateglacial occasionally altered the isotopic composition of the dendrochronological material resulting in delayed or no reactions in this record (Friedrich et al. 1999). Therefore, the isotopic records from the two different archives were not compared further.

Instead, the Greenland ice-core record and the dendrochronologies are correlated by the growth patterns of tree-rings (Friedrich et al. 2001b). The growth of tree-rings is also governed strongly by locational factors, particularly during the growth period (Friedrich et al. 2001b; cf. Stokes/Smiley 1968, 9-11). Besides the nutritional composition of the soil, local climate conditions such as temperature or precipitation are an important factor influencing tree-ring growth. Even though these patterns could be affected by local topographic settings or micro-climatic conditions, they are in general also influenced by the more global climatic regime. Thus, major climatic shifts are also detectable in the tree-ring patterns (cf. Kaiser et al. 2012), particularly, if they originate from a similar region. Since climatic conditions are strong factors in the tree-ring growth, the lags to the oxygen isotope record should in general be smaller than in sedimentary records. Nevertheless, without additional marker horizons the correlation will remain vague. Volcanic deposits are usually not directly detectable in the dendrochronological record in contrast to the sedimentary record but can be assumed by their impact on growth patterns (Kaiser 1993). However, these short-term effects in the growth patterns can also be caused by other events such as floods (Kaiser et al. 2012). Therefore, the testing of correlations on marker horizons is problematic in these records except for the trees were directly related to the volcanic event (Baales/Bittmann/Kromer 1998; Kromer et al. 2004).

In summary, the numerical correlation of the Greenland oxygen isotope record with other terrestrial archives requires several points of cross-checking but they make relatively precise estimates of duration times for the single events possible.

Modifying the CalPal-2007$_{HULU}$ radiocarbon calibration curve

The necessity of calibrating ^{14}C dates to the solar or calendar timescale is meanwhile a well known issue in chronological considerations (e.g. Münnich 1957; Weninger 1986; Blockley/Donahue/Pollard 2000b; Jöris/Weninger 2000; Blaauw et al. 2007; Bronk Ramsey 2008; Grimm/Weber 2008; Weninger et al. 2009; Jöris et al. 2011). Archives combining a reliable calendric timescale with numerous ^{14}C dates are necessary to construct a reliable radiocarbon calibration curve. In general, these combined data sets are provided by climate archives. For the Holocene, the dendrochronological data is used, whereas for the Pleistocene portion of the radiocarbon calibration curve various, mainly marine climate data sets are used. This use of climate data is due to the relative solidity of these archives. Therefore, the construction of a reliable radiocarbon calibration curve for the Pleistocene is only possible in the context of climate archives. However, the construction of a uniform radiocarbon calibration curve for the Pleistocene is still under discussion (Weninger/Jöris 2008; Reimer et al. 2009; Bronk Ramsey et al. 2012). In particular, the part from the Lateglacial to the limits of the radiocarbon method remain a matter of controversy (van der Plicht et al. 2004; Bronk Ramsey 2008; Weninger/Jöris 2008; Reimer et al. 2009). Nevertheless, the Lateglacial calibration curves also vary in detail due to contrasting algorithms creating the curves (Bronk Ramsey/van der Plicht/Weninger 2001) and due to the incorporation of different data. So far, the generally accepted dendrochronological data end in the early Lateglacial Stadial at 12,644 years cal. b2k (cf. Schaub et al. 2008b; Hua et al. 2009; Reimer et al. 2009; Kaiser et al. 2012). Thus, the construction of a radiocarbon calibration curve relies on other, mainly marine data sets for the ages prior to this date (Schaub et al. 2008b). In general, the data sets used to construct the calibration curves are similar during the Lateglacial (Fairbanks et al. 2005; Weninger/Jöris 2008; Reimer et al. 2009). However, the evaluation of the data sets produced variable results because each were created for different reasons.

For instance, the IntCal radiocarbon calibration curve aimed to estimate the calendar age of a ^{14}C date or a series of ^{14}C dates in the statistically most reliable way (cf. Buck/Litton/Scott 1994; Heaton/Blackwell/Buck 2009). The mathematical reliability was consequently a major ambition of this curve. However, the increase of the statistical accuracy resulted in a decrease of the precision of the calibrated ages. This decrease was due to the rejection of possible uncertainties and the incorporation of partially large deviations combined with the application of a random walk prior model, this approach resulted in relatively accurate but not very precise calendar ages. By contrast, the CalPal radiocarbon calibration curve is aimed to produce a most precise calendar age to be able to relate the result exactly to the climatic and environmental development (Jöris/Weninger 1998; Jöris/Weninger 2000; Weninger/Jöris/Danzeglocke 2003). Therefore, further proxy data combined with the ^{14}C data sets such as the Cariaco greyscale were used for their comparability and consistency in the CalPal approach (Weninger/Jöris 2008). Thus, a second line of evidence was created. In accordance with the aim at precision, the decision on reliable data sets was made differently to the IntCal approach. The CalPal approach increased the precision by a careful selection of data sets and the single dates forming these data sets. The selection tried to eliminate potential dating errors and to synchronise the data sets with palaeoclimate signatures and, more generally, with calendric age models (Weninger/Jöris 2008). Thereby, large deviations and many uncertainties were often resolved. For instance, proposed modifications such as a change in the reservoir ages in the Cariaco data (see p. 13-15) were applied to this data set in the CalPal radiocarbon calibration (Weninger/Jöris 2008, 778f.). In contrast, the IntCal group also mentioned this proposal but banned the data due to the remaining uncertainty of the reservoir ages (Reimer et al. 2009, 1116). This ban resulted in a scarcity of precise data across an important cliff in the Lateglacial calibration record (cf. Reimer et al. 2009, 1120 fig. 2) and wider statistical error margins in this period. Nevertheless, the CalPal approach also required a statistically provable probability concept (Bronk Ramsey/van der Plicht/Weninger 2001). In an article, Bernhard Weninger and his colleagues demonstrated the non-commutative character of the radiocarbon calibration and identified their approach as a quantum probability concept (Weninger et al. 2011), which fits with their aim at precision.

The evaluation of archaeologically visible changes in human behaviour in the context of climatic and environmental developments requires precise dates. Furthermore, the relation of these dates to the high-precision chronostratigraphy of Lateglacial north-western Europe is an equally important requirement for this study. The current project aims to evaluate human behaviour within a tightly defined spatio-temporal framework. Thus, the approach used in the construction of the CalPal-2007$_{HULU}$ radiocarbon calibration curve and the CalPal program is preferred for the present study.

Thus far, the CalPal-2007$_{HULU}$ radiocarbon calibration curve is only stored and made readily applicable in the CalPal calibration program (Weninger/Jöris 2004). This program is furthermore preferred because it offers a non-dogmatic use of calibration data sets (cf. Reimer et al. 2009, 1112). Although the CalPal-2007$_{HULU}$ radiocarbon calibration curve is recommended and set as default, other calibration curves such as the IntCal04 calibration curve (Reimer et al. 2004) or single data sets such as the German Lateglacial pine chronology (see p. 25-30; Kromer et al. 2004) or the Cariaco dates (see p. 13-15; Hughen et al. 2006) can also be chosen as calibration data. Furthermore, the downloaded version of the CalPal program enables the user to review, change, and shift the single data sets. Thus, this calibration progam permits the construction of a new calibration curve by changing existing data and/or the inclusion of a new data sets.

Clearly, the »health warning«[21] (Pearson 1987, 103) of Gordon Pearson is still valid but the concern of people without detailed knowledge building their own calibration curves (Reimer et al. 2009, 1112) seems exaggerated. In fact, a short survey of archaeological papers of the past decade using calibrated data show that most archaeologists are aware of the complexity of calibration. Therefore, they took the advice of someone from a radiocarbon facility (e.g. Jacobi/Higham 2009; Weninger et al. 2009) and/or applied and quoted ready made calibration curves and instantly usable calibration programs (e.g. Barton et al. 2003; Gamble et al. 2005; Shennan/Edinborough 2007; Grimm/Weber 2008; Langlais et al. 2012). Furthermore, the wish for a consensual use of calibration data which »makes direct comparison between different studies easier« (Reimer et al. 2009, 1112) is also fulfilled by the use of the CalPal-2007$_{HULU}$ radiocarbon calibration curve and the CalPal program since the majority of archaeological studies of the last half-decade relating to the Pleistocene record used this set.

However, some alterations of the chosen CalPal-2007$_{HULU}$ radiocarbon calibration curve have to be performed for the present approach. Some data such as the early Younger Dryas tree-ring data from Switzerland (Schaub et al. 2008a) and Tasmania (Hua et al. 2009) became available only after the publication of the last update of this radiocarbon calibration curve (Weninger/Jöris 2008). These newly available data fall into the controversial period of the changing reservoir ages in the Cariaco record and help to identify the change of the reservoir ages in the Cariaco record more precisely. Since no new update is available thus far, the additional data series is correlated to the radiocarbon calibration curve in the present study.

Therefore, newly available data are graphically correlated to the existing calibration data sets based on the wiggles of the ^{14}C data series. Furthermore, the ^{14}C dates of the Gänziloo series were connected to the growth patterns of the tree-ring rings (see **fig. 13**). Thus, the position of these tree-rings widths based on the result of the previous wiggle-matching is tested against the relevant tree-ring data sets (Kaiser et al. 2012). In addition, a correlation with the tree-growth patterns allows for assumptions about the onset of the Lateglacial Stadial and, consequently, on its duration. Based on these assumptions, the data set can be compared numerically to the GICC05.

The resulting numerical calendar ages as well as the relating ^{14}C dates are used to construct a database using the Windows editor and the DAT-file of the Lateglacial pine chronology (Kromer et al. 2004) which was downloaded with the CalPal program as a blueprint. This resulting database is introduced into the CalPal composer

21 »Health warning! Proper calibration is not easy for the non-mathematician, but doing it incorrectly, wrongly interpreting the result, or even not understanding the potential of calibration may seriously damage your archaeology. Take advice from the experts – know what calendrical band-width is necessary for correct interpretation and discuss this with the dating laboratory, preferably before taking and certainly before submitting samples. Think first, not after you get the radiocarbon date.« (Pearson 1987, 103).

using the method suggested in the select_info file of the CalPal downloaded version. By the use of this database a modified version of the CalPal-2007$_{HULU}$ radiocarbon calibration curve is created in the CalPal composer. This modified version is used to calibrate the ^{14}C dates in this project. Thus, the underlying algorithms and probability concept of CalPal are used as well as the approach of constructing a solid calibration curve based on climate archive, only the data are supplemented in detail.

ENVIRONMENT

The second set of archives considered in this project are the environmental archives (see Material-Environment, p. 30-48) which are supplemented by directly ^{14}C-dated, vegetal and faunal samples (see p. 49-51). The natural environment formed the fundamental resource for the survival of hunter-gatherers. During the Lateglacial the natural environment underwent profound changes. In the present study, several approaches were used to illustrate these changes and relate them to the previously presented chronostratigraphy. In general, two aspects of the natural environment were considered: the physical geography and the living environment. Within the latter, the focus was set on the terrestrial vegetation and the mammal fauna.

Modelling of the physical geography

During Pleistocene climate changes the physical geography was fundamentally altered through rising and falling sea levels and advancing and retreating ice sheets alongside isostatic and eustatic changes in land relative to the sea level. To establish environmental change in relation to climate change and the environmental development, the physical geography must be known. Since current maps of the physical geography of Lateglacial north-western Europe were inadequate, the creation of adequate maps was the first step in modelling the Lateglacial environments.

Most changes in physical geography occur very gradually and, thus, changes only become detectable over long periods of time. Consequently, data on physical geography are usually displayed in maps with relatively low temporal resolution and the maps created for this study also encompass relatively long time intervals. In contrast, in maps with high temporal resolution changes to the physical geography become too gradual to be recognisable. Furthermore, some developments of the physical geography can only be identified by terminal points and not the gradual stages between these points. For example, the ice sheet development is mainly reconstructed by the position of end moraines. The development of glaciers is relatively complex (IUGG (CCS) – UNEP – UNESCO 2005) and was more complex in the past than originally assumed (Ivy-Ochs et al. 2009). Even with modern methods of dating exposure ages (Ivy-Ochs et al. 2008; Heyman et al. 2013), a high-precision decadal or centennial chronology for the developments of the European ice sheets during the Lateglacial is not possible. Thus, combining a specific ice sheet stage with for instance a sea level stage in a temporal high resolution map could easily lead to incorrect impressions because the changes in the physical geography progressed at varying paces. The resulting potential for combining asynchronous stages due to various, uncorrelated age models increases with the choice of increasingly high temporal resolution. Therefore, the accuracy of maps with a very high temporal resolution can be questioned for the Pleistocene.

However, the creation of maps with relatively long time slices requires that various mean values need to be applied, for example, for the ice sheet regression or the sea level. Some developments are coupled

map no.	events	limits of calendar ages (in years cal. b2k)	duration (in calendar years)	limits of the ^{14}C ages (in years ^{14}C-BP)
0	2000 A.D.	c. 5-0	5	×
1	GS-1	c. 12,750-11,700	1,050	10,750-10,000
2	GI-1c-a	c. 13,950-12,750	1,200	12,000-10,750
3	GI-1e-d	c. 14,700-13,950	750	12,350-12,000
4	GS-2a	c. 16,000-14,700	1,300	13,100-12,350
5	LGM (GS-3-GS-2c)	c. 26,550-19,050	7,500	22,000-16,000

Tab. 47 Approximate periods covered by the maps of Lateglacial north-western Europe. The ^{14}C ages were read from the CalPal2007$_{HULU}$ calibration curve and further compared to the ^{14}C dates and the eventstratigraphic attribution of their samples in the databases created for the present study. For the ^{14}C ages of the LGM cf. Street/Terberger 1999, 262 fig. 4.

altitude	RGB	shade	chroma	intensity
-250 m	148	160	0	139
-5 m	248	160	0	233
50 m	228	160	0	215
300 m	148	160	0	139
6000 m	28	160	0	26

Tab. 48 Settings of the greyscale shader used in Global Mapper.

such as the ice sheet regression and the sea-level. Furthermore, many of these developments were the result of the global climate system. Thus, to avoid incorrect correlations due to divergent age models, the long time slices were chosen according to the Greenland eventstratigraphy (Björck et al. 1998; Blockley et al. 2012).

For the present study, the Lateglacial was sub-divided into four time slices (**tab. 47**). These time slices were chosen as meaningful periods for this project due to their relation to the Greenland eventstratigraphy as well as the general climatic and environmental development of north-western Europe (cf. Litt et al. 2001; Lowe et al. 2008). Furthermore, with the same layouts as used for these four maps a modern map (A.D. 2000) and a map of the LGM (Clark et al. 2009; cf. Starnberger/Rodnight/Spötl 2011) were created.

In this project, north-western Europe was delimited at 45-60° northern latitude and 10° western to 25° eastern longitude.

A first set of these maps which extended 5° less in eastward direction was made available online for public use (Grimm 2007). The updated set is used in this study for the visualization of the setting of the various sites within Lateglacial Europe and to highlight the changing Lateglacial landscape. Meanwhile these maps are also available online in the project archive under »Change (Karten)«: http://monrepos-rgzm.de/forschung/projekte/projektarchiv.html.

Creating the base map

A single base map of north-western Europe and the surrounding sea beds was created as a basis for the four Lateglacial and the two additional maps. This base map was built from the SRTM30_plus record which included bathymetric data. This data set was read into the program Global Mapper v8.01 and displayed in a geographic projection (latitude/longitude). According to the SRTM documentation (cf. documentation file at http://dds.cr.usgs.gov/srtm/version2_1/SRTM30/), the 1984 revision of the World Geodetic System (WGS84) was chosen as reference datum.

Furthermore, a simple greyscale shader was specifically created in Global Mapper to display the data set. This shader had five fixed altitudes and relating grey values (**tab. 48**). The values between the fixed points were interpolated by the program. In addition, in the vertical options of the program the hill shading was enabled and the vertical exaggeration was turned to 8.1 times. Moreover, the light direction was set to 0 for

the altitude and 45 for the azimuth, and the ambient lighting was turned to 0.75. These settings resulted in a good visibility of the geomorphological relief of north-western Europe.
The resulting map was exported as 1200 dpi GeoTiff in the geographical limits named above.

Graphic revision of the base map

The resulting map was graphically modified by removing some perturbation especially in the bathymetric data. This perturbation were probably due to Holocene undulation movements of sea sediments and sediment fans from Lateglacial to Holocene river mouths (e.g. Bourillet et al. 2003; Busschers et al. 2007). Furthermore, in the cases where the physical geography was significantly altered such as in the case of open pit mining the relief was smoothed. In addition, gaps occurred occasionally in the maps and they were also filled according to their surrounding.

Compilation of the Lateglacial maps

To create maps of Lateglacial Europe from the base map various margins, for example, of the ice sheets, the sea, or of the Baltic Ice Lake had to be compiled. These margins were set as an additional layer on top of the base map. Mean values of the ice sheet regression, the sea level, or the development of the Baltic Ice Lake were created for the specified periods (**tab. 47**).
Mean values for the sea level changes were chosen according to the relative sea level development based on Barbados and Sunda Shelf data (Weaver et al. 2003, fig. 5B) and calibrated from this combined record at the mid-point of the time periods specified above. The sea level for the maps are 60 m b.s.l. for GS-1, 70 m b.s.l. for GI-1c-a, 90 m b.s.l. for GI-1d-e, and 105 m b.s.l. for GS-2a. In the LGM map the sea level was set to 120 m b.s.l. (Peltier/Fairbanks 2006), although meanwhile a minimum sea level of 123 m b.s.l. was identified for the LGM (Hanebuth/Stattegger/Bojanowski 2009). Maps with the different sea levels were created in Global Mapper by setting the water colour to opaque and the sea level to the specified values. These maps were exported as GeoTiffs in the geographical limits named above. From these GeoTiffs the area covered by the sea was chosen as extra sea-level layer. These five different layers were set on the base map creating the first stage of the LGM and the four Lateglacial maps.
Based on the studies compiled by Svante Björck (Björck 1995a), layers with the development of the Baltic Ice Lake were created for the maps 1-3 (map 1: Björck 1995a, fig. 5; map 2: Björck 1995a, fig. 4; map 3: Björck 1995a, fig. 2). In the periods of the maps 0 and 5 this lake was irrelevant. However, for the period of map 4 no conclusive data exist, although for the second half of this period indications were found that melt-water and icebergs passed through the Öresund to the Kattegatt which was gradually silted up (Björck 1995a, 21). Furthermore, sub-aquatic sediments from this period were only found at considerable modern water depths (Björck 1995a, 22). Nevertheless, some type of ice lake probably existed in this period and to take this in account the same extension as during GS-1 (Björck 1995a, fig. 5) is given for GS-2a. This stage is equivalent with the maximum southward extension of the Baltic Ice Lake.
The maximum extent of the British ice lakes was not ascertained to relate to the LGM and the dating for the various lake shore lines also remained uncertain (Clark et al. 2004, 368-370). Moreover, for the period encompassed by the Lateglacial maps no ice lake margins seemed relevant in Britain.
However, many smaller lakes and kettle-holes (Kaiser 2004; Clausen 2010) and, possibly, large wetland landscapes (Cziesla 2001, 392) as well as the major European rivers (Antoine et al. 2003a; Wallinga et al.

2004; Streif 2004; Ménot et al. 2006) changed the appearance of the landscape in detail (see **fig. 6**). For this type of information, no comprehensive compilation for the Lateglacial has been found thus far. In addition, the various stages of rivers or lakes were often not precisely dated or, in fact, datable. Therefore, this hydrological data was not included in the general Lateglacial maps.

The modern glacier margins are plotted onto map 0 (A.D. 2000) at a mean altitude of 2,490m a.s.l. according to the Simming Ferner glacier tongue in 1989 (Ivy-Ochs et al. 2006, 121 fig. 4) and a mean of the lowest elevation point of glaciers in the surrounding of Trins/Austria (IUGG (CCS) – UNEP – UNESCO 2005, 118-120 tab. A). Even though the various glacier margins occur at various altitudes a specific (mean) altitude was chosen because putting the contemporary altitude of each glacier onto the map would have meant an unnecessarily large data processing for the purpose of these maps. Certainly, in the near future such precise data of modern and, possibly, past moraines will be available as shape-files for GIS-programs. Thus far, the margin lines used in the present study represent rather schematic indication of the actual glaciers and were graphically smoothed in order to receive a graphic impression of the ice sheet. This graphic impression should indicate the imprecise reconstruction. This partial smoothing process was applied to the glaciers of the other maps as well, since they also rely on significant interpolations. As previously stated, correlations of moraines to the same glacier advance was occasionally difficult (Ivy-Ochs et al. 2009) and the ice sheets only retreated in some sub-units of the studied period. The position of the ice sheets in these periods can only be evaluated by the use of exposure dates and interpolated data. The various interpolations and choices are described in the following by ice sheet.

The Scandinavian ice sheet regression is well documented in Sweden (Lundqvist/Wohlfarth 2001, fig. 6) and the rest of Fennoscandia as well as in northern Germany and northern Poland (Boulton et al. 2001, fig. 12). The Swedish record was given priority due to the relevance of the area for the Lateglacial archaeology and, furthermore, the comparably fine-grained chronology which was mainly based on radiocarbon dating of the moraines. Therefore, the interpolated ^{14}C age ranges of the present maps were compared to the ^{14}C dates of the moraines (Lundqvist/Wohlfarth 2001, 1142-1145) and not the calibrated dates (Lundqvist/Wohlfarth 2001, fig. 6). The age ranges given for the moraines overlapping the period of the maps were selected. In the case of more than one moraine falling into the age range of the maps, the one dated closer to the mid-point of the period was chosen. Thus, the GS-1 ice margin limit was set at the Younger Dryas moraine (Lundqvist/Wohlfarth 2001, 1145), the GI-1c-a margin at the Levene moraine (Lundqvist/Wohlfarth 2001, 1144f.), the GI-1e-d margin at the Berghem moraine (Lundqvist/Wohlfarth 2001, 1144), and the GS-2a margin at the Göteborg moraine (Lundqvist/Wohlfarth 2001, 1143f.). The limits were followed through northern Scania and southern Kronoberg according to the interpolation in Boulton et al. 2001, fig. 12 and the margin lines given in Lundqvist/Wohlfarth 2001, fig. 6 south of Karlshamm. These margins could be connected through the Baltic Sea region to their margins in the Baltic States (Boulton et al. 2001, fig. 12). For the LGM the Brandenburg-Lezno margin was used as a maximum value (Boulton et al. 2001, fig. 12).

For the British ice sheet only the moraines of the Younger Dryas (Loch Lomond readvance ice limit) and the LGM (maximum extend of moraines) were given in a record of glacial landforms (Clark et al. 2004; cf. Sejrup et al. 2009). Thus, for maps 1 and 5 these available maximum values were chosen. For the present maps 2-4 the British ice margin lines were interpolated between the two known margin lines (Younger Dryas and LGM) according to the percentage of maximal regression between the corresponding marginal lines in the Scandinavian ice sheet (Younger Dryas to Brandeburg-Lezno margin; **tab. 49**). For instance, between the Brandenburg Lezno and the Younger Dryas margin, the Scandinavian ice sheet had shrunk by approximately 720km, whereas the British glaciers had changed between the maximum extend to the Loch Lomond readvance over some 285km. Between the Brandenburg-Leszno and the Göteborg moraine lay some 600km and, thus, the glacier had already shrunk some 85%. The same percentage for the British ice sheet would

map no.	distance (in km) from LGM margin (Scandinavian ice sheet)	% of regression between LGM and GS-1 margin	distance (in km) from LGM margin (British ice sheet)
5	0	0	0
4	c. 600	83.33	*237.49*
3	c. 650	90.28	*258.30*
2	c. 700	97.22	*277.08*
1	c. 720	100	c. 285

Tab. 49 Approximate regression distances of the Scandinavian ice sheet and calculated regression distances of the British ice sheet based on a percental relation to the Scandinavian ice sheet. Calculated values are set italics.

result in a regression of some 240km. At these calculated distances the relief was followed to create a potential ice sheet. Clearly, this linear interpolation of the British ice sheet according to the Scandinavian development for the specified periods neglects the closer position of the British ice sheet to the warming gulf stream and many other important issues contributing to the complex history of this ice sheet (cf. Hubbard et al. 2009). However, these approximations suffice to introduce the potential limitation which an ice sheet represents for the expansions of the living environment including human groups.
Moreover, at the time when the first set of maps were made, no reliable data from the central North Sea Basin was available and, thus, whether the British and the Scandinavian ice sheets have coalesced during the LGM was unclear (e.g. Sejrup et al. 2000; Sejrup et al. 2009; cf. Svendsen et al. 2004). A coalescence has meanwhile become probable for the maximum extend of the ice sheets (Carr et al. 2006). Presumably, this coalescence of the two ice sheets had considerable impact on the drainage systems of north-western Europe and, possibly, also on the whole climatic regime of the North Atlantic climate region (cf. Toucanne et al. 2009). However, by the beginning of the time period studied here the two ice sheets were already two separate formations (see **fig. 19**; Carr et al. 2006).
Although data on the development of the Alpine glaciers has been published before (e.g. Hantke 1980; Hantke 1983), these were not used in the present approach because these data sets were too detailed to be successfully and beneficially processed.
Further comprehensive data about the Alpine ice sheet has been published since the first set of maps was produced (Reitner 2007; Federici et al. 2008; Kerschner/Ivy-Ochs 2008; Ivy-Ochs et al. 2008; Ivy-Ochs et al. 2009). However, this data did not sufficiently change the general picture of the Lateglacial geography. Thus, to draw an Alpine ice sheet on the Lateglacial maps, specific altitudes were chosen for the margins of each stage comparable to the procedure with the modern glaciers. These altitudes were interpolated by adding or subtracting equilibrium line altitude (ELA) values from the values of the Gschnitz moraine at Trins/Austria. The upper end of this moraine was located at 1,410m a.s.l. with an ELA of -700m (Ivy-Ochs et al. 2006, 116-118, especially tab. 1). This moraine developed presumably during the Late Pleniglacial (cf. Kerschner/Ivy-Ochs 2008) and was chosen as mean altitude for map 4 (**tab. 50**). In the Younger Dryas the Egesen moraine developed which in the Gschnitz valley was found at 1,930m a.s.l. (Ivy-Ochs et al. 2006, 121 fig. 4) and the ELA was accordingly at -180m (cf. Ivy-Ochs et al. 2006, 117 tab. 1). Moraines reflect the transgression of glaciers and, thus, develop only in the periods of glacier advance. Due to the complex growth patterns of glaciers, moraines from different locations are difficult to combine to a single glacier front (cf. Egense moraine in Ivy-Ochs et al. 2009). Thus, the relatively stable and reliably determinable ELA was preferred as a calculation factor in contrast to the use of moraines. Even though the ELA and the altitude of the glacier front are not linearly related, the values from GS-1 and GS-2a in the Gschnitz valley give a linear frame in which altitudes can be estimated based on the ELA, also for regression stages (**tab. 50**). For instance, no moraines were associated with the Lateglacial Interstadial because the ice sheet was regressing during this event. In general, the ELA was considered to be above the Younger Dryas one, i.e. ELA

map no.	relative ELA (in m) in Gschnitz valley	altitude (in m a.s.l.)
0	×	2,490
1	-180	1,930
2	*-110*	*2,000*
3	*-325*	*1,735*
4	-700	1,410
5	-1,100	*1,010*

Tab. 50 Mean altitudes chosen for the Alpine ice sheet regression. Modern altitude based on the 1989 Simming Ferner glacier tongue (Gschnitz valley/Austria; Ivy-Ochs et al. 2006, 121 fig. 4). Lateglacial values are calculated based on relation of the relative equilibrium line altitude (ELA) and the altitude of the moraines in the Gschnitz valley/Austria (Ivy-Ochs et al. 2006). Calculated values are set in italics.

< -700 m (Ivy-Ochs et al. 2006, 118 tab. 1). However, the Daun stage seems to date only shortly before the onset of the Lateglacial warming (cf. Ivy-Ochs et al. 2008, 567) or may even represent the first cold readvance during the Lateglacial (Ivy-Ochs et al. 2006, 120). Therefore, an ELA for the first part of the Lateglacial Interstadial can be set according to the values of the Daun stage. For this stage an ELA of -400 to -250 is given and a mean of -325 m was chosen. Thus, the margins for limiting the early Lateglacial ice sheet were set to 1,735 m a.s.l. based on the ELA and glacier front relation in the Gschnitz valley. For the LGM, the ice sheet limits were also interpolated at an ELA of -1,100 m (Ivy-Ochs et al. 2006, 118 tab. 1). The second part of the Lateglacial Interstadial was given an ELA of -110 m assuming this ELA to be above the Egesen ELA of a minimum of -180 m (Ivy-Ochs et al. 2006, 118 tab. 1). Furthermore, this ELA of the second part of the Lateglacial Interstadial was assumed to possibly be above the ELA of the early Holocene Kromer/Kartell readvance which was at a minimum of -120 m. Finally, in the graphic implementation of the glaciers during the Lateglacial Interstadial, the major river valleys were considered to have been cleared from ice due to melt-water activity.

Alongside these changes, isostatic uplift influenced the physical geography especially in northern Europe (Kiden/Denys/Johnston 2002; Reicherter/Kaiser/Stackebrandt 2005; Shennan et al. 2006). However, the geomorphological relief in many regions that were strongly affected by isostatic uplift is, in general, very strong, particularly, in the transition areas of land and sea. Thus, the margins given in the maps for the sea-level were usually not affected by changes due to isostasy. Furthermore, isostatic uplift decreased significantly with the distance to the main glaciated areas and would only result in very slight differences in the shading of the maps. Nevertheless, if more accurate maps were planned these movements should be taken into account. However, the introduction of isostatic data would require a revision of the SRTM data before making the base map.

The recolonization of north-western Europe by a living environment was further restricted by the development of the grounds. For instance, the occurrence of permafrost and the pedogenic development of sediments alter the carrying capacity of these areas for vegetation species. However, pedogenesis is influenced by several local factors (see **fig. 18**) and the soil development in the Lateglacial was thus far only recorded in single profiles but not on a larger surface. The various zones of frozen grounds (seasonally frozen ground, sporadic permafrost, discontinuous permafrost, continuous permafrost) were probably very narrow in the Late Pleniglacial (Huijzer/Isarin 1997, 409) and permafrost was absent during the Lateglacial Interstadial (Vandenberghe 2001, 193). Only during the Lateglacial Stadial were significant parts of north-western Europe again under permafrost conditions (Isarin 1997).

In conclusion, the maps of Lateglacial north-western Europe give a static picture of gradually advancing processes and presumably do not reflect the landscape at any specific point in time. Perhaps, in the future more detailed and precise maps can be compiled by the use of data published in shapefiles or comparable formats. Nevertheless, the present maps represent good estimations of the landscape development at the various stages and provide an impression of the major constraints (water and ice cover) for the expansion of the human habitats.

Since neither orthomorphic nor equal-area nor equidistant projections are necessarily needed in the present study, the maps displayed in the present work were vertically stretched in a 2:3 proportion. The resulting maps appear a good mixture of equal-area and equidistant projections. However, if distances are given in the present work they were measured using the »ruler > line« tool on the free online program Google™ Earth.

Qualitative and quantitative assessment of data from the living environment

Besides the physical environment, the living environment also shaped the human habitat. Data from the Lateglacial living environment can be found in profiles (see p. 40-48; cf. Lotter et al. 1997; Lowe et al. 1999) and also in the form of directly dated samples (see p. 49-51). The pollen profiles are corrected for the chronostratigraphic constraints in their record but they are not further modified in the present study. In contrast, the directly dated samples require some evaluation and are then used to create frequency distributions to allow for assumptions on the presence of some species in the studied regions.

Evaluation of the radiocarbon database

^{14}C dating is one of the standard methods of dating in modern archaeology (Geyh 2005, 1. 69). In this project, ^{14}C dates form an integral part in contextualizing Lateglacial human activity and environmental developments. To accomplish and corroborate this contextualization, meaningful ^{14}C dates are necessary. To select these meaningful results, the existing record of ^{14}C dates has to be reviewed regarding the reliability of each sample. The reliability has to be questioned because »Errors in radiocarbon dating start at the moment of excavation and build through the process of dating« (Jacobi/Higham 2009, 1896). Generally, unreliable dates originate from inaccuracies in the apparatus and measuring procedure, contamination of the dated sample, and/or a mistaken association of the dated sample with the supposed significance of the result.

To account for the technical inaccuracies, the radiometric dating facilities give their results with a standard deviation. Therefore, this source of error is not further considered in the present study.

The necessity to evaluate ^{14}C dates for their significance has been emphasised since the early days of the technique (Johnson 1952; Dean 1978; Pettitt et al. 2003). The significance is dependent on the question that the ^{14}C date is intended to answer. In principle, the result of ^{14}C dating gives the age at which the incorporation of carbon into the sampled tissue stopped and the radiometric decay began. This age is equivalent with the death of the tissue. Thus, if a result is meant to date the death of the tissue the dated event is identical with the event that was supposed to be dated (target event, Dean 1978). In this case, concerns regarding potential contaminations form the main base for the test of reliability. However, the age of a sample is often considered as significant for other target events. For example, the age of a bone, which originated from an animal which was killed in a human hunting episode is assumed to date also the human activity. Often this association is not established as easily as in the given example and incorrect associations resulted in a divergence of the ^{14}C result and the target event. For instance, the heartwood is the dead tissue in the core of trees. If a sample from this part of the wood is dated the result can significantly predate the possible target event of the felling of the tree (Schiffer 1986). Therefore, the association of the two types of events (dated and target) has to be clarified and the period bridging the possible offset between the events has to be estimated (see p. 265-269).

The ^{14}C dates of environmental samples date the presence of selected environmental material relevant to Lateglacial hunter-gatherers in a particular region at a specific time. Thus, dated events mark the end of the target events, which are the life of the samples. In this case, old wood samples are well suited because they reflect the living years of the sample. In contrast, the results from faunal samples are more difficult to interpret because the place of death and the area of living are not necessarily identical. In particular, archaeological material such as bone and antler tools require a more cautious interpretation. These tools could potentially have been introduced to the region from elsewhere by Lateglacial hunter-gatherers and the sampled species were not in fact present in a region during the dated period (cf. Bratlund 1996a). Therefore, individual finds are evaluated critically in this study, particularly, if the sampled material originated from a human tool (cf. Riede et al. 2010, 307). The frequent occurrence of a species in a particular region during a specific time is considered a more reliable indicator for the presence of the species in the region.

The evaluation of contamination represent the major factor when considering which samples to use for modelling the Lateglacial environment. Contaminations can affect the dated samples during the lifespan of the dated material (Deevey et al. 1954; Olsen et al. 2010), during the deposition (Turney et al. 2000), during storage (Wohlfarth et al. 1998), and also during the dating procedure (Burky et al. 1998). In general, to evaluate the potential offsets various details of the ^{14}C-dated sample have to be accessible (Kra 1986, 766; Jacobi/Higham/Lord 2009, 7-9; cf. Lowe/Walker 2000; Pettitt et al. 2003).

Contaminations which occur prior to the dating procedure require a precise pretreatment chemistry (Hajdas 2008). However, samples that have incorporated »old« carbon isotopes (reservoir) during their lifespan must be rejected for dating past events. These samples are mainly water plants, shells, ostracods, fish, and other biota with strong dependence on hard water aquatic environments (Fischer/Heinemeier 2003; Rick/Vellanoweth/Erlandson Jon M. 2005; Olsen et al. 2010). In these environments, organisms absorb radiocarbon dissolved from the surrounding mineral throughout their lifespan. The contemporary atmospheric carbon signal is overprinted by these considerably older isotopes. Consequently, the concentration of carbon isotopes in the sampled tissue does not relate to the concentration of carbon isotopes in the atmosphere at the time of the death of the organism. Therefore, samples from these materials are in general rejected as unreliable in this project.

Furthermore, there are multiple ways of contamination during the deposition of the sample. Besides contextual contamination (see above, cf. p. 265-269), bio-geochemical processes in the soil (Hiller et al. 2003; Brock et al. 2011), particularly connected to (ground) water activity are a major source for contamination at this stage (e.g. Hedges/Millard 1995; Turney et al. 2000; Geyh et al. 1983; cf. Zazzo/Saliège 2011). In addition, contamination by micro-organisms can occur during storage. This type of contamination can be prevented by keeping the samples as sterile as possible, storing them in a dried state, and reducing storage times between recovery, analysis, and dating (Wohlfarth et al. 1998).

Radiocarbon dating facilities have tried to remove any type of contamination that occurred prior to the preparation of the sample for the dating (e.g. Münnich 1957; Gillespie/Hedges 1983; Lanting/Niekus/Stapert 2002; Higham/Jacobi/Bronk Ramsey 2006; Hajdas 2008; Olsen et al. 2008). Pretreatment protocols were developed, which are undergoing constant refinement and are regularly adapted to newly identified sources of error (Bronk Ramsey/Higham/Leach 2004, 18). These protocols increased the dating precision. Thus, contamination originating from deposition or storage are assumed to be sufficiently resolved by dating facilities for dates obtained since the late 1990s. For some of the earliest radiocarbon dates the incorporation of modern carbon during deposition or storage might represent a source of error. Therefore, these results should be checked for the $\delta^{13}C$ value to see if significant contamination is possible and/or for the contextual reliability of the resulting age. However, erroneous results were also due to the choice of sample material. In the »classic« method of radiocarbon dating (β-counting or conventional dating, Libby 1952) a significant amount of Pleistocene material was required to produce statistically reliable results. Due to the limited preservation of datable material in Northwest-European Lateglacial archaeology, bulked samples were often dated. This procedure is sufficient if the material is comparable and originated from the same event such as animals killed during a single hunting episode. In this case, this bulked sample also yielded a reliable date for the target event. More often, however, this relation of archaeological material remained unclear and was even proven false in some cases (e. g. Holocene admixtures in Stellmoor, Bratlund 1999). Consequently, bulked samples need to be more cautiously and rigorously assessed and compared to the overall site context. Nevertheless, results are not excluded due to the use of different methods for obtaining the ^{14}C date (β-counting or AMS, cf. Banks et al. 2008) because it is not the technique but the choice of sample that causes the contamination.

Further contamination is possible during the pretreatment and/or dating procedure. Generally, contamination during this phase can best be recognised and minimised by repetitive and transparent evaluation of the laboratories dating procedures (Bronk Ramsey/Higham/Leach 2004, 23) and inter-comparison of the results from various laboratories (Boaretto et al. 2002; Scott et al. 2004). For instance, over a decade ago problems with the ion-exchanged gelatin pretreatment (lab code: AI) were recognised in the Oxford Radiocarbon Accelerator Unit (Burky et al. 1998). This finding led to the abandonment of this method (Higham/Jacobi/Bronk Ramsey 2006, 182). Furthermore, in the initial phase of ultrafiltration pretreatment procedure at the same laboratory the collagen was insufficiently cleared from humectant (glycerine, Higham et al. 2007). This possible source of contamination needs to be mentioned because several Lateglacial dates are affected by these insecurities. Some of these measurements were redated suggesting that the impact on very small samples was more significant than on larger samples (Jacobi/Higham/Lord 2009, 9; Higham et al. 2007, S2). In general, the effect of small contaminations increases when the sampled material decreases in size (Wohlfarth et al. 1998, 144; Bronk Ramsey 2008, 259). Therefore, minimum amounts of datable carbon are required, in particular for AMS dating. Currently, this minimum is around 1 mg C for AMS facitilites (Higham/Jacobi/Bronk Ramsey 2006; Bronk Ramsey 2008). Samples with smaller amounts of carbon are considered unreliable in the present study. In general, the small sample sizes resulted from the wish to receive measurements with only minimal destruction of the dated sample. Perhaps in the future, for some materials, non-destructive sampling may become possible (cf. Steelman/Rowe 2004) and, presumably, will permit further important objects such as the Poggenwisch rod (Bosinski 1978) to be directly dated.

In addition, the dating procedure for various tissues and fractions requires specific consideration. For example, the collagen or amino acid fraction which represents approximately 30 % of the living tissue is normally dated in tooth, bone, and antler material (Burky et al. 1998). This fraction does not survive burning. However, a considerable portion of the Lateglacial archaeological fauna material has been only preserved in a charred state. In these samples the carbonate fraction can be dated (Lanting/Aerts-Bijma/van der Plicht 2001). A comparative dating program for cremated bone from Lateglacial and early Holocene sites produced concordant results for the Holocene but revealed some inconsistent results for the Lateglacial (Lanting/Niekus/Stapert 2002). Possibly, this inconsistency relates to an insufficient degree of bio-apatite recrystallization during the burning process leading to the still existing possibility of exchange processes with dissolved soil bicarbonate (cf. Olsen et al. 2008). Exchange processes with dissolved elements in the soil water occur in clay-containing sediments rather than sandy sediments. Moreover, these exchange processes are less of a concern in arid environments (Zazzo/Saliège 2011). Therefore, dates made on cremated bones, but without further structural analysis, have to be compared carefully to their context. The sediment structure and the drainage of the sediment as well as other independent dating results should be taken into account for this comparison. Often this type of additional information is only partially available and, thus, the majority of dates on or containing burnt bone material have to be rejected. In the future, additional analyses concerning the fire use and burning temperatures on archaeological sites (cf. Moseler 2014; Werts/Jahren 2007) can perhaps help to evaluate the preservation of reliably datable burnt bone material.

^{14}C-dated mammoth ivory occasionally produces dubious results such as the sample from a carved piece of ivory from the concentration II in Gönnersdorf (see **tab. 20**) yielding a mid-Lateglacial Stadial age (OxA-2069). Even though tusks grow continuously, the tusks are well vascularised throughout the animals life in contrast to wood and, thus, the tusks are living tissue. Consequently, ^{14}C dates on ivory should produce a date for the death of the animal not for the growth date of the tusk. In contrast to other dental material ivory is rather soft because it is largely formed of dentine (Heckel 2009, 76). Perhaps, contamination by infiltration occurs more easily in ivory than other dental material due to this soft consistency. However, the dense structure (Heckel 2009, 76) does not provide space for undetected infiltration of other substances on a molecular level. Therefore, ivory appears, in general, not to be a target of contamination and should not be rejected a *priori*.

To test technically reliable results against their context, the ^{14}C date has to be calibrated and the calibrated age can be compared to the chronostratigraphic indications of the context from which the sample originated.

The normal distribution of a ^{14}C measurement becomes more complex when plotted against a wiggling calibration curve and, in consequence, the result represents a function of the shape of the calibration curve (Blockley/Donahue/Pollard 2000a, 114; Weninger et al. 2011). The resulting probability is no longer evenly distributed on the calendar timescale and the calibrated age should therefore be given in time ranges (Pearson 1987, 100; Blockley/Donahue/Pollard 2000a, 114; cf. Geyh 2005, 70 f.). The mathematical probability for these calibration results is given by the various calibration programs, generally with 68 % or 95 % confidence ranges. Due to the unsteady calibration curve in the Lateglacial the 68 % must not be the best match (cf. Blockley/Donahue/Pollard 2000a) and, therefore, the 95 % version is chosen here. However, the higher confidence range which covers most possible readings from the calibration curve also blurs the result. This increase of imprecision is the case, especially, in strongly fluctuating parts or plateaus of the calibration curve. High-precision measurements may result in several, disconnected readings or a very long period on the calendar timescale. In general, the large period should be given as a calibration result. However, further considerations based on additional chronological information from a site can help to formulate the most probable interpretation. To tackle the interpretative problem of varied chronostratigraphic indicators such as pollen- or lithostratigraphy connected to the ^{14}C dates, statistical methods were developed (e.g. Buck/Litton/Scott 1994; Bronk Ramsey/van der Plicht/Weninger 2001; Blackwell/Buck 2003; Hamilton/Buchanan 2007; Jacobi/Higham 2009; Steele 2010). In particular, Bayesian approaches were used for this purpose (e.g. Rhodes et al. 2003; Wohlfarth et al. 2006; Blockley et al. 2008; Gearey/Marshall/Hamilton 2009; Blockley/Pinhasi 2011). However, many of these approaches neglected the quantum-theoretical aspects of radiocarbon calibration resulting in over-correction of the data by a frequency normalisation disregarding Heisenberg's uncertainty principle (Weninger et al. 2011). Moreover, for these approaches a considerable corpus of information and numerous radiometric dates in combination with the qualitative evaluation of the reliability of each date is necessary to produce reliable results. The majority of Lateglacial sites from north-western Europe does not, in fact, fulfil these requirements. Instead the tight connection with the eventstratigraphy based on the argumentative interpretation of the chronological information is attempted. Therefore, the relation of the context to the sample has to also be considered. Disturbances such as cryo- or bioturbation can cause displacements in the stratigraphic position of the sample to a higher or lower position. Furthermore, heavy samples such as bones can sink into softer ground or lighter sediment can be washed out and replaced by successive sediment, whereas heavier samples remain *in situ*. Therefore, the contextual confirmation of a ^{14}C date requires a good knowledge about the stratigraphic development at the site. If no reason for the rejection of a ^{14}C date is found it is regarded as reliable in this project.

In summary, a qualitative evaluation of ^{14}C dates requires a comprehensive understanding of the history of the sample in relation to the site formation process and its effects on the sample. The outcome of this evaluation is usually a significantly shrunken database (see **tab. 5**) which, nevertheless, produces preciser and more reliable results (cf. Grimm/Weber 2008).

Probability and frequency distributions of radiocarbon dates

To use reliable ^{14}C dates in an approach to model the Lateglacial environment required distinguishing the samples by species and from where the sample originated. Some species were selected (see below) and the reliable ^{14}C dates of these species were sub-divided according to the sub-areas of this study (see p. 31-40). These sub-sets were calibrated with the CalPal program and the reviewed Lateglacial calibration curve. Probability distributions of calibrated data sets have been considered as proxies reflecting frequencies of material remains created by human societies (dates-as-data; Rick 1987; Gamble et al. 2005). In this approach, the increase of ^{14}C dates during a specific period was assumed to be caused by increased human activity in this period (Gamble et al. 2004; Shennan/Edinborough 2007). Thus, a type of eventstratigraphy of human activity was created by the probability distributions. The recognition of divergent dates for the onset of a probability event in different areas was assumed as reflecting dispersal patterns (Housley et al. 1997; Hamilton/Buchanan 2007; Steele 2010).

In this study, the probability distributions of ^{14}C dates for the sub-sets are considered as reflecting the probable presence of the species in the sub-area. However, absence of a specific species cannot be proven by this proxy[22], only the presence can be shown as statistically probable. These data sets are partially biased by sampling strategies. Therefore, these probability curves are not taken as equivalent to demographic developments. Some further cautions are necessary in the interpretation of these distributions. In particular, calibration relics, the impact of the mathematical approaches creating the probability distributions (Blockley/Donahue/Pollard 2000a; Blackwell/Buck 2003; Shennan/Edinborough 2007), and preservation biases (cf. Surovell/Brantingham 2007) need to be excluded as factors influencing the distribution.

A comparable, event-like approach was used in palaeontology with a first and a last appearance date (FAD/LAD; e.g. Walsh 1998; Aaris-Sørensen 2009) setting the limits for the presence of a species in a specified area. For the period between these dates a continuous presence of the species is assumed. However, sometimes the evidence is sparse and then the continuity of the species presence is as questionable as the probability distributions. Additional information such as stratigraphic evidence, comparison to neighbouring areas, and modern analogies have to be integrated to sustain the conclusions (Aaris-Sørensen 2009, 7). Yet, if the database is sufficiently large, the combination of FAD, LAD and the probability distribution can provide a useful set in environmental studies to describe the statistically reliable presence of a specific species. This combined record contributes to the modelling of the regional environmental development.

However, ^{14}C dates on vegetal samples that were identified to species are rare in the Lateglacial record of north-western Europe. The typical species used as fuel for hearths on Lateglacial sites in north-western Europe were birch (*Betula* sp.), juniper (*Juniperus* sp.), pine (*Pinus* sp.), and willow (*Salix* sp.). However, only a few samples of birch and pine have been dated from the study area (see **tab. 5**). Their probability distribution produced no significant patterns. However, the presence of these samples can prove the presence of these resources at least in small numbers at specific periods.

22 Although the observation that »Absence of evidence is not evidence of absence...« (Sagan 1995) is a general scientific theorem and, therefore, appears trivial, it was neglected previously in the interpretation of ^{14}C date frequencies (Housley et al. 1997; Riede 2008; cf. Terberger/Street 2002).

Among the faunal samples the focus is set on the mammal species that were mainly hunted by Lateglacial humans such as reindeer (*Rangifer tarandus*), horse (*Equus* sp.), elk (*Alces alces*), red deer (*Cervus elaphus*), and large bovids such as bison (*Bison priscus*) and aurochs (*Bos primigenius*). Although the presence of specific faunal species could also provide insights into the Lateglacial climatic (Hernández Fernández/Peláez-Campomanes 2005) and the vegetation development (Fahlke 2009; Musil 2010), these species are only used as a general cross-checking parameter. Usually, the climate and the vegetation records are used to check the reliability of the faunal results.
To provide a more detailed picture of the possible presence or absence of selected species which in addition is more reliable than the probability distribution of only directly ^{14}C-dated specimens, tables of potential presence in the sub-areas were made containing columns for the different short-term isotope events and the selected fauna species in rows. The potential presence was indicated by simple colour coding (white: no indication; grey: possible indication; black: reliable indication). A comparison of this type of table with a pollen diagram is not possible in the same detail as a comparison of the probability distribution with the plant development. Nevertheless, in these tables the reliability of the data for a specific time is evaluated and therefore, these tables are an important supplement to the probability distributions.

Mapping the data from the environment

The finished maps of Lateglacial north-western Europe were read into a GIS program (previously ESRI ArcView® 9.3; since June 2012: Quantum GIS, version 1.8.0 »Lisboa«) by geo-referencing the four corner points of the maps. Within these programs the environmental data were plotted onto the relevant map by the use of the coordinates from the geo-databases (see p. 52 f.). For the mapping, the previously created sub-sets (see above) are used. The calibrated age range of each dated sample is compared to the limits of the Greenland eventstratigraphy and thereby the sample is attributed to the Lateglacial event to which it dated. With this attribution all samples of a selected species dating to a specified event could be plotted on the relevant maps.
Frequency distributions to interpolate areas of probable occurrences of a particular species during a selected time are technically possible based on the created point pattern. However, the numbers of samples per time slice are very small and, consequently, the extent of purely interpolated areas would be very large. As a result, the significance of frequency distributions based on these samples are doubted and, therefore, this technique is not used. Instead the single samples are plotted in a point pattern on the maps.
The maps of Lateglacial north-western Europe are also applied in a very conservative manner in the present study. Even though mathematical approaches to process large numbers of data as well as complex networks of spatial data sets are known in geography as well as in archaeology (Nakoinz 2009), the number of reliable pieces of evidence are too few in the present study to permit these statistical analyses to produce reliable results. In fact, sufficiently complex and detailed material databases to permit geostatistical analyses to produce reliable results for the Lateglacial on a large scale do not yet exist. Thus far, the first usable models of the Lateglacial landscape are only created on smaller scales (De Smedt et al. 2013). Although recommending the use and the consequent refinement of the geostatistical approaches for future research, the necessity of a qualitative evaluation of the existing record and a compilation of numerous data sets is regarded as a basic precondition to produce reliable results with the various types of geostatistical approaches. Therefore, this project tries to contribute to this future research aim by evaluating the present archaeological data and creating reviewed databases. Thus, the conservative use of the maps is, fundamentally, due to the material based orientation of the present study.

ARCHAEOLOGY

The third of the three main types of archives used in this study is the archaeological record. The archaeological material presented previously (see Material-Archaeology, p. 53-244) forms the basis for the analysis of past human behaviour. Systematic changes in this behaviour during the studied time frame are particularly relevant as they are considered to reflect the process of transition within a social system. The main aim of this project is to examine such a process on the example of the transition from the Late Magdalenian to the FMG way of life in the context of Lateglacial climatic and environmental developments. The character of the process can either be described as gradual or as rapid. In this study, the former is assumed to reflect an evolutionary process, whereas the latter is considered as a revolutionary process. The two types of processes (revolution and evolution) are assumed to be distinguished by the tempo in which they affect most parts of the human life (see Introduction, p. 1-5). Consequently, to distinguish between the two types of process, reliably dated assemblages are necessary. The formation of a reliable, chronological basis is therefore a key requirement to understand the transition process. This evaluation has to be supplemented by an analysis of the integrity of the archaeological assemblages. Furthermore, various lines of evidence have to be incorporated to identify various changing parameters such as lithic resource exploitation, changes in prey preference, or settlement systems and to prevent the overestimation of single rapid adaptation of one parameter. Therefore, the following methods were used to evaluate the chronological integrity of the assemblages as well as various approaches to systematically assess evidence for changing parameters within these assemblages.

Chronological evaluation of the integrity of the archaeological sites

A secure chronological attribution of the selected assemblages forms the base of the present study. The reliability of the chronological attribution of an archaeological assemblage depends on the integrity of the assemblage as well as on its position in the chronological development (cf. Weniger 1989; Crema/Bevan/Lake 2010; Fougère 2011). The position in the general chronological development is usually estimated by environmental chronostratigraphies and especially by ^{14}C dates. However, the latter require some evaluation regarding their reliability and significance (see p. 259-263 and below). Furthermore, the integrity of the assemblage can also be sustained by multiple ^{14}C dates. However, assemblage integrity is mainly constituted by a spatial analysis that is often combined with considerations about the *chaîne opératoire* of relevant parts of the lithic assemblage. Spatial analyses of Lateglacial sites have frequently been undertaken in the last decades (cf. Gaudzinski-Windheuser et al. 2011a). However, detailed analyses of each site exceed the possibilities of the present study but for many cases detailed analyses were already published (see Material-Archaeology, p. 53-244) and are used in the considerations about the internal chronology of the sites.

Evaluation of the radiocarbon database

Radiocarbon dating has become an integral part in contextualizing Lateglacial sites within the development of the surrounding environment as well as with other, often distant archaeological sites. Therefore, the verification of this type of date is an indispensable step in the chronological attribution of Lateglacial sites. Roger Jacobi and Tom Higham state that »Errors in radiocarbon dating start at the moment of excavation and build through the process of dating« (Jacobi/Higham 2009, 1896). Consequently, ^{14}C dates have to be evaluated in regard to all potential sources of error. However, the inaccuracies resulting from the apparatus and measuring procedure are accounted for in the standard deviation given by the radiometric dating facilities. In general, contaminations occurring prior to the dating procedure are removed by the dating facilities during sample pretreatment. However, some still important technical reasons for contamination and rejection of ^{14}C dates were presented previously (see p. 259-263) and apply also to the archaeological material.

Besides technical reasons for the rejection of radiocarbon results, many contextual reasons exist. These contextual reasons relate particularly to the significance of a ^{14}C date. The necessity to evaluate ^{14}C dates for their significance was already emphasised in the first substantial publication on radiocarbon dating (Johnson 1952). The evaluation of the significance of the dating result (dated event) for the questioned activity (target event) is a relevant part in the review process of the archaeological radiocarbon record (Dean 1978). Especially, questions concerning the site formation process and the position of the sampled material therein have to be challenged (cf. Pettitt et al. 2003).

Bulked samples that combine material of various ages are a common source of erroneous results and, occasionally, this combination can also occur unintentionally if the sampling was not performed carefully (cf. Bodu et al. 2009b).

Even though the combination of various events can be prevented by dating samples that consist of a single piece such as a bone, the contextual integrity and the archaeological significance of these samples can also be questioned (cf. Housley et al. 1997, 34). For example, the death of the dated tissue which started the radiometric clock can instantly relate to the human activity of interest if the dated tissue is a bone of an animal killed by humans. However, the association of the two types of events (dated and target) is not always so clear in archaeological assemblages.

In archaeological contexts, the dated event generally predates the target event, for example if the target event is an occupation episode in which »old« material was used for construction (animal skulls as trophies; mammoth femur from Gönnersdorf, Street/Terberger 2004, 296), as jewellery (fossil molluscs), for the carving of figurines and tools (fossil ivory, shed antler), or as fuel for hearths (jet and deadwood). In this case, the bridging period between the different events has to be established (cf. Dean 1978). To establish reliable estimates, other material that was more closely related to the human occupation should be dated. A good sample would be a bone with a probable deadly impact mark or a bone with cutmarks which suggested filleting of the meat. If the bone was of no use prehistoric humans would presumably not have made the effort to strip a bone of rotten meat which they could not consume. Consequently, the meat was probably relatively fresh[23] when processed by humans. The dated event (death of the animal) and the target event (filleting by humans) can be assumed as relatively close events in time.

The offset relating to the life history of the dated material such as the »old wood« effect were regularly considered to fall into the period covered by the standard deviations. However, based on the current accuracy and precision in ^{14}C dating, this reduction of all uncertainties to the period of standard deviation appears problematic. For example, in the case of charcoal which was burnt in a prehistoric hearth, a period shorter than the standard deviation is generally assumed for the bridging event between the

23 In fact, for this scenario some offset could also be made plausible in the presence of permafrost and cold water environments which could have served as natural refrigerator to store faunal material (Bokelmann 1979, cf. Grønnow 1987). However, the decay of soft tissues such as meat would began very quickly after the melting of permafrost. Therefore, this natural ice storage could not form a significant offset of ages in the Lateglacial (see p. 258). In cold water the meat is known to ferment (Grønnow 1987, 158) and, presumably, various organisms inhabiting this environment would began to decompose the meat. Thus, the storage of meat bearing parts in cold water environment is also temporally limited.

burning (target event) and the separation of the pieces of wood from the tree (dated event). In Lateglacial north-western Europe, the charcoal from archaeological sites was usually determined as pine (*Pinus* sp.), birch (*Betula* sp.), juniper (*Juniperus* sp.), willow (*Salix* sp.), or poplar (*Populus* sp.). Some of these shrubs and trees can live for several hundred years and, consequently, the heartwood would already produce a considerably older date than the sapwood which is contemporary with the separation. In this case, the dated event and the target event of wood diverge (Dean 1978, 228; Schiffer 1986, 16-19). Nevertheless, regularly occurring storms and forest fires keep the age structure, in general, comparatively young in areas comparable to the Lateglacial environment, such as Sweden. An analysis of age structure in pine (*Pinus sylvestris*) populations from eight localities in Sweden revealed that these trees rarely became older than 300 years (Agren/Zackrisson 1990, 1052). Some high-precision AMS-dates from periods with a relatively even part of the calibration curve such as the dates from Doetinchem in the Netherlands (Johansen et al. 2000) produce calibrated results for the Lateglacial which range between 200 and 300 calender years. Thus, an offset of more than 300 years, for example, due to the age of the tree would already fall outside the standard deviations. In addition, wood used for burning should preferably be dry because the moisture in fresh wood produces considerable amounts of smoke. Drier heartwood can be used but in forest environments also deadwood can be collected. Significant drying can occur in several months to a few years. However, the decay of the wood sets a natural limit to this offset. The natural decay resistance of the heartwood is, in general, considered relatively strong and, thus, the use of deadwood could result in an additional offset. Based on the reconstructed, generally moist, environments during the Lateglacial Interstadial and the modern environments in which pines grow, decreased decay times are possible for the specimens from the Lateglacial Interstadial (cf. Schiffer 1986, 17f.). Nevertheless, the survival of wood into later prehistoric periods, perhaps, in specific environments cannot be neglected as possible source of erroneous results (cf. Grieben, Grünberg 2006). Consequently, the use of heartwood from deadwood could result in some significant offset between the dated event (death of tree) and the target event (burning of wood). In total, this offset could surpass the standard deviations. In general, the history of the sample prior to becoming a resource for humans should always be kept in mind when relating dated and target event. Clearly, wood is an extreme example but this precaution also applies to faunal material. Although the majority of mammals does not become very old, the decay resistance of some hard tissues such as antler, bone, or dentine can be relatively high. For instance, modern experiences prove that in permafrost regions and, particularly, in melting permafrost regions significant offsets can occur between the death of the animal (mammoth) and the use of it as a resource (ivory trade, Haynes 1991, 47f.; Stuart et al. 2002;Kramer 2005). This use of fossil material was considered also probable during prehistoric times (Street et al. 2006, 761). Thus, the depositional history of the material prior to becoming a resource also requires consideration in the review process. However, to distinguish between the use of »old« material and an actual »old« occupation of a site is sometimes not possible (cf. Napierala 2008a).

Another offset could occur from the time of use of the organic material. In addition to the offset between the dated event and the target event, the question arises which is the target event of interest? The use of the implement could be of interest relating to the development of tool types, whereas the discard age could be interesting if dating the use of a particular location. For instance, hunters killed a reindeer and used the antler to carve an antler point. Modern experiments with organic implements revealed an astonishing durability of these pieces (cf. Pétillon 2006, 88-95). So, perhaps, the resistant tool was used for several years resharpened and, possibly, inherited to the next generation which finally lost or discarded the implement. Comparable considerations apply to special objects such as molluscs, figurines, or drilled animal teeth. However, this type of offset is also largely assumed to be covered by the calibrated age range for Lateglacial dates.

Even though the dated event generally predates the target event in archaeological contexts, sometimes the dating result post-dates the target event. In Lateglacial contexts, this type of offset can frequently be attributed to depositional displacements. For instance, a projectile can be shot into an otherwise unassociated context such as soft, limnic or fluvial sediments. If material from this sediment is dated, the target event (use or loss of implement) predates the dated event (deposition of the material). For example, the various single finds of Lateglacial organic points in north-western Europe (e. g. Stampfuss/Schütrumpf 1970; Czie-sla 2007b) can sometimes be younger than the context in which they were found. However, if the organic projectile is dated directly the target event post-dates the dated event again.

Finally, an incorrect attribution is also possible. For example, charcoal resulting from other events such as natural fires can be interchanged with charcoal from a hearth if the surrounding sediment is rich in charcoal and animal burrows. Furthermore, animal remains can enter a site through a multitude of natural causes and these remains are not always possible to distinguish from fauna introduced to the site by human activity. Besides humans, various kinds of predators such as wolves, foxes, or birds of prey can introduce bone material to a site. In cases of poor surface preservation of the bones, the cause for the damage on a bone cannot be detected with certainty and the distinction between human or others predators as the causal agent is unclear. Nevertheless, for some materials such as antler the introduction by other predators is less frequently attested. Furthermore, animals could have died naturally on the site (»background fauna«). However, if these »natural« remains are dated they produce an age unassociated with human activities.

The previous points illustrate that ^{14}C dates are only as reliable as the context of the sampled material and emphasise the importance of analysing the site formation process as well as the history of the sample for evaluating the reliability of radiocarbon dates. Each date of the radiocarbon database (see p. 49-51) was evaluated according to the technical reasons (see p. 259-263) and to the above described contextual reasons for rejection.

After this technical and basic contextual analysis, the reliable dates have to be calibrated to permit these dates to be compared with their climatic and environmental surrounding. Thus, calibration is a necessary step to result either in the confirmation of the ^{14}C date by its context or in a reconsideration of the previous assumptions regarding the reliability of the date or regarding the development of the context. Furthermore, this comparison with the context can also help to narrow the most probable reading from the date range or multiple readings resulting from the unsteady calibration curve.

Even though the strength of radiometric dating is the production of independent dates, the precision of the dating of a single archaeological assemblage (target event) can be increased by testing all dates from the same context against one another. For example, the sample history of reliable dates is clarified and the various dated events should relate closely to the same target event. Thus, by the overlapping age ranges the target event becomes more clearly defined. Therefore, several dates from a single assemblage are appreciable and, in addition, increase the precision of the radiometric age of the assemblage. To test whether the measured spread of several dates from a single site can be explained as a random process, a χ^2-test using the program Statave (Robinson 1988) is applied to the uncalibrated dates. The uncalibrated dates represent the raw measurements and their probability distributions can be described as Gaussian normal distributions. If the combined dates are consistent with the resulting weighted mean in the 2σ-range, they relate statistically to the same event. In contrast, the probability distributions of calibrated ages are altered further by the calibration curve and are no longer normal distributions which would complicate the test significantly. However, this constraint emphasises the importance of the previous technical and contextual evaluation to select already the reliable uncalibrated ^{14}C dates.

If by this time no reason for the rejection of a ^{14}C date is found, it is regarded as reliable in the present study.

Thus, ^{14}C dates help to clarify the chronological position of a single site and, moreover, compendia of regional ^{14}C dates can also be compared to show possible movements of dated characteristics (e.g. Rick 1987; Housley et al. 1997; Blockley/Donahue/Pollard 2000a; Blackwell/Buck 2003; Gamble et al. 2005; Shennan/Edinborough 2007; Hamilton/Buchanan 2007; Steele 2010). In fact, combined probability distributions of calibrated radiocarbon dates were themselves applied as a proxy for human activity in a specific region allowing for a conclusion on past population histories (dates-as-data; Rick 1987; Gamble et al. 2004; Gamble et al. 2005; Shennan/Edinborough 2007). However, several restrictions to this method have to be considered before interpreting population histories based on radiocarbon probabilities. In particular, preservation biases need to be excluded before discussing population dynamics. Only this revision allows for an interpretation of the results regarding the evidence of absence. Furthermore, over-representation of single, but numerously dated assemblages have to be levelled, which could be possible by using weighted means. Mathematical methods underlying the creation of probabilities from a calibration curve can cause some offsets (cf. Shennan/Edinborough 2007) and the strongly wiggling atmospheric carbon level can lead to the almost invisibility of some parts and over-representation of others in large series of ^{14}C dates (cf. Weninger et al. 2011). Since this is a systematic problem, it cannot be entirely solved mathematically but requires deductive argumentation. In this case, the positioning of the single archaeological units could help to evaluate the consistency of the observed probability distribution of ^{14}C dates. Furthermore, if the development of target events such as the use or discard of bone and antler points was aimed to be dated by these probabilities distributions, a bridging relation between the dated material and the target developments has to be established.

Therefore, these combined probability distributions are not used for the archaeological material in the present study. Since the age of the dated assemblages is established other graphs such as battleship diagrams or simple column graphs provide a clearer impression of the changes and continuities in the archaeological material.

Quantitative assessment of raw material sources

The decisions by hunter-gatherers concerning raw material procurement and exploitation provide insights into their economic behaviour within the landscape. Furthermore, the use of specific resources help to understand land use and, thus, mobility patterns (Floss 1994; Pettitt/Rockman/Chenery 2012). A diachronic perspective of these patterns could even reveal derived habits of landscape knowledge (cf. Rockman 2003b). In addition, if a tendency towards an increasingly careless and wasteful use of the raw material is detected, then a good knowledge and a reliable availability of the resources can be assumed (cf. Conneller 2007). However, due to the relation with the mobility pattern and the function of a site, the use of the raw material has to be considered within the character of the site and the function of the site within the settlement system. A first step towards classifying the sites according to this general handling of raw materials is attempted by a quantitative assessment of the raw material resources.

The exact origin of the resources other than mineral and fossil material is difficult to establish. Thus, in the present study only the fossil and the mineral, in particular, lithic raw material resources are considered further.

A general quantitative assessment of the lithic raw material resources is difficult to establish because often different values are given for the different raw materials. For example, often the numbers of artefacts made from a specific raw material are given but these numbers can relate to specific types or sizes of artefacts. However, the different raw materials are also given per weight or by proportion of the total artefacts. The

weight would be of particular interest because it allows for considerations about the transport efforts of the hunter-gatherer group. However, the weight is rarely given and, therefore, it is not useful in a comparative approach. Occasionally, only general tendencies of the raw material composition are given such as »the majority« and »a few«. This imprecision hampers a quantitative comparison of various assemblages.
To account for this variety a simplified differentiation of raw materials is used in which the various types of data are classified into three classes: major types, regular types, and minor types. The major types are the dominant raw materials which comprise 30 % of the lithic material or more. Regular types make up 5-29 % of the assemblage and minor types produced less than 5 % of the lithic artefacts. Thus, the classified raw materials are further differentiated according to the approximate minimum distance from which the material occurred (see **tabs 12. 25. 36**).
Furthermore, assuming an economically adapted use of the resources, the accessibility as well as the richness of a source influences the use of the raw material within an assemblage and, consequently, the position of the site within the settlement system. Thus, with the profile of the exploitation strategy used for a raw material, the availability of a source can be assumed (**tab. 51**) and tested against distances known from modern surveys. The influence of raw material availability on the settlement system can consequently be further considered.

Qualitative and quantitative assessment of the lithic assemblages

Lithic assemblages are the most common records on Pleistocene sites. To analyse these records a multitude of methods have been established. In the present study, the focus lies particularly on the composition and the size of the different assemblages to reveal changes in these general values.
The size and the composition of lithic assemblages were considered as dependent on the duration of a prehistoric occupation (Richter 1990) but also related to the function of the site (cf. Löhr 1979). The function is assumed from the activities accomplished at a site, which were partially based on the diversity of the material found at the site (butchering debris, hunting equipment, jewellery) and partially on the composition of the lithic tool inventories and the functions attributed to the tool classes. In fact, different activities influence the size of the assemblage in various ways and were accomplished over various periods of time. For instance, the blank production process produces more debris than the application of retouched tools. A skilled flintknapper requires several minutes to a few hours to complete a blank production process (Fischer 1990; cf. Olausson 2010, 45 f.), whereas a few retouched tools such as end-scrapers which were assumed to be used for hide work (Rots 2005; Sano 2012b) could reflect a considerably longer lasting activity. However, the size and composition of the assemblage is also influenced by what part of the site was excavated (complete, an activity zone etc.). To quantify the influence of the excavated area on the size of the assemblage a simple index of artefacts per excavated square-metre (density index = no. of artefacts / no. of excavated square-metres) was calculated. This index is named density index because it identifies the density of artefacts per square-metre. The density index is given as artefacts ≥ 10 mm per excavated square-metres and in parentheses the value for total artefacts per excavated square-metres is given.

Exploitation analysis

The exploitation analysis of archaeological assemblages should provide information on the handling of the resources. The handling of a resource, in particular, lithic raw material, has been identified as an important

source	rich	poor
distant	attentive use: some wastes, few heavily exploited cores, few discarded retouched artefacts	sparse, specialised use: single retouched artefacts
near	wasteful use: plenty wastes, hardly exploited cores, blanks, some retouched artefacts	specialised use: few wastes, few heavily exploited cores, few retouched artefacts

Tab. 51 Idealised model of raw material economy based on source availability.

difference between the Late Magdalenian and the FMG in Central Rhineland, whilst the origin of the resources remained almost unaltered (cf. Floss 2002). The Late Magdalenian exploitation was described as efficient relating to the transport of the raw materials. For instance, ready prepared cores or blanks were brought to the sites. Furthermore, the choice of usually high-quality raw material suggested that poorer qualities were discarded at the source of origin. In contrast, the FMG seemed to transport raw material nodules over considerable distances and only tested the material at the site where the raw material was used.

Consequently, the distance from where a raw material was brought to a site (see p. 269f.) and the way in which the material arrived at the site (raw nodule, ready made core, blank, ready made tool, or used tool) also shape the understanding of how past territorial ranges were used. This information allows for a consideration of mobility and the means of transporting material. Harald Floss had standardised a combined analysis of the *chaîne opératoire* with the raw material resources to distinguish the status of the material at the site (Floss 1994; Floss 2002). However, this analysis has been accomplished by him for the Central Rhineland and is only partially possible for the published material from the other parts of the study region.

In contrast, a more general exploitation analysis of the lithic material is used in the present study. Although more detailed studies such as the one described above are generally appreciated, for the more general character of the present study a more abstract and universal distinction of the handling of the raw material suffice. Therefore, the general index of the relation of cores to the complete lithic material including particularly debris (exploitation index = no. of artefacts [total] / no of. cores) is selected. Since the total number of artefacts is not always given, the same index is created with the no. of artefacts ≥10mm. This exploitation index is assumed to reflect the fragmentation of the assemblage and, thus, the handling of the raw material at the site. In an idealised model, in which the raw material was brought to the site only as raw nodules, and then the material was transformed, used, and discarded on the spot, this index would reflect the average number of artefacts which originate from a single core. In archaeological lithic assemblages, this idealised model is modified by the introduction of ready prepared cores, blanks, and/or tools to the site, the discard of material, which was made elsewhere, at the site, and the export of cores, blanks, and/or tools made at the site as well as taphonomic processes such as trampling (for the general problem of quantifying lithic assemblages cf. Hiscock 2002). Furthermore, this exploitation index is influenced by the function of the assemblage as well as the exploitation strategies of the raw material. Nonetheless, this exploitation index can provide an insight into how the various raw materials were differently reduced and whether differences can be found within assemblages with the same dominant raw materials. Due to the constraints, the results of the exploitation index are only considered in relation with further results from this study.

A further refined index of the ratio of blanks to cores would produce a value for the approximate number of blanks which were produced by a single core. This value could give an insight in the quality of the raw material, the technical abilities of the knappers, and/or the aim of the exploitation. Thereby, some assumptions about the choice of raw material can be made which further contribute to the understanding of the exploitation strategy and the resource economy. This refined index is not calculated in the present study because of inconsistently given data in the literature relating to the blanks and a general methodological question of which blanks should be selected for this index. Besides the blanks which the knapper aimed for, blanks were produced during the complete preparation process but many of these specimens were probably considered as waste products. For instance, cortical flakes were discarded by the Late Magdalenian, whereas in FMG assemblages these blanks were often transformed into end-scrapers. So the aimed blanks seemed very different ones. In this project questioning the procedure of change this different validation had to be considered for the creation of a blank-to-core-index but the archaeological distinction between this different validation of the blanks is difficult to establish, in particular during the transitional phase. Consequently, should all blanks including broken ones be taken into account for the exploitation potential of a core or only the complete blades and bladelets? In the future, a comprehensive study which has direct access to different assemblages and, thus, makes the comparison of various blank type indices possible is desirable. In particular, a study focused on this aspect could provide an interesting component to connect analyses relating to the technical thinking with those on the resource exploitation strategies of Lateglacial hunter-gatherers.

For the Paris Basin, Boris Valentin demonstrated that the changes in the exploitation strategy of cores were accompanied by the tendency of using two or more platforms on the cores (Valentin 1995, 567f.). Therefore, the cores were classified according to the numbers of platforms previously (see **tabs 13. 26. 37**). For those assemblages in which the numbers were not given, the figures of the cores were evaluated regarding the visible platforms and the directions of the blank negatives.

Furthermore, a second index which relates the retouched artefacts to the cores (function index = no. of retouched artefacts / no. of cores) is used to weight the influence of the two major functions of a lithic assemblage (blank production, tool use) at the site.

Functional analysis

The function of a site is of particular interest in the considerations about the position of the site in a hunter-gatherer settlement system (cf. Binford 1980). This settlement and mobility system were considered to change between the Late Magdalenian and the FMG. The Late Magdalenian record displayed characteristics of a forager system with a variety of settlement types, whereas the assemblages of the FMG seemed to indicate a collector system with a high mobility of mainly residential camps. However, the decrease of camps with special functions can only be revealed if the activities performed at the known sites were shown.

Indices such as the relation of retouched lithic material to the complete number of lithic artefacts (Löhr 1979; cf. Hiscock 1981) or cumulative diagrams of retouched tools (cf. Bohmers 1961) were regularly used to describe the function of a site and/or the intensity of its use. In this project, the percentage of retouched artefacts ≥10mm is given (% retouched artefacts = no. of retouched artefacts / [artefacts ≥10mm / 100]) and added in parentheses by the same index calculated with total artefacts instead of artefacts ≥10mm. Retouched artefacts reflected the importance of standardised tools for the activities at the site, whereas micro-wear analyses have proven the supplementary use of unretouched materials (e.g. Plisson 2002; Plisson 2007; Sano 2012b). Furthermore, the relation of retouched artefacts to the complete material depends on the type of raw materials used at the site and their splintering properties. Recalcitrant material produces more splinters and thereby increases the total amount of material. According to Hartwig Löhr (Löhr 1979), the size of an assemblage further depends on the duration and the intensity in which the activities were performed at a site. Therefore, Löhr set the percentage of the retouched artefacts in relation to the total number of the artefacts and compared the result to the diversity of the inventories (**fig. 27**). A very general

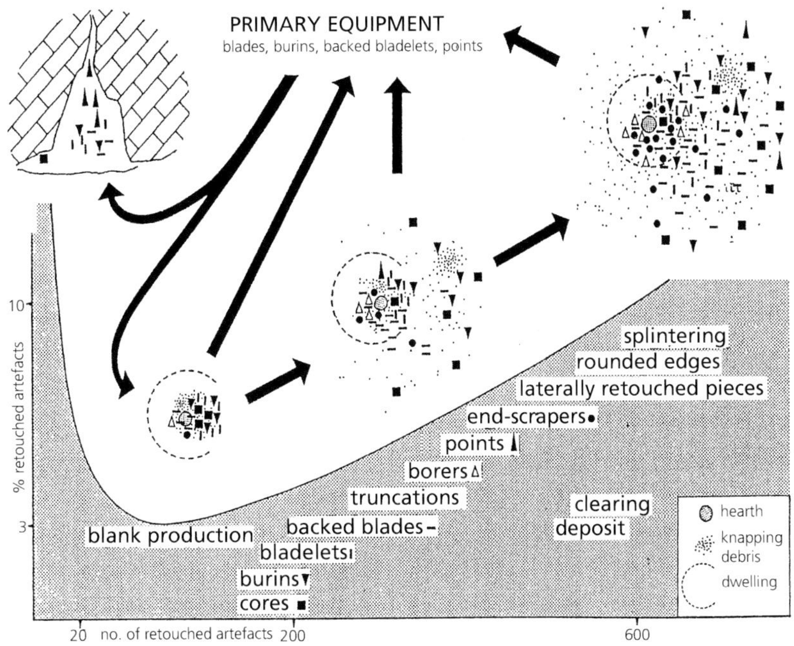

Fig. 27 Idealised scheme of the spatial development and the increasing diversity of lithic artefact assemblages on Upper Palaeolithic to Mesolithic sites depending on the duration of the occupation (Löhr 1979, 307 Abb. 33, translations by the present author).

classification was primarily used in the present study to distinguish assemblages of various sizes (**tab. 52**). However, these classes have to relate to the two different sets of data (total artefacts/artefacts ≥ 1 cm) given for the assemblages. The sub-division of the two sets of classes was made in a similar exponential way. Thus, if the two given classes vary significantly (more than one class), the material from the assemblage is fragmented into particularly small or particularly large pieces. Thus, a comparison of similarly sized assemblages compensates for the influence of raw material properties or duration on the proportion of retouched artefacts.

Based on the same suggestions of Hartwig Löhr and further observations of Claus-Joachim Kind (Kind 1985), Gerd-Christian Weninger developed a classification of assemblages to distinguish various types within the Magdalenian of north-western Europe (Weniger 1989). He based his classes on the number of cores (class A-E) and the number of retouched artefacts (class a-e). In the present study, the relation of cores to retouched materials (function index = no. of retouched artefacts / no. of cores) is also used as a functional indicator. In this functional index, the cores represent the blank production process, whereas tools reflect the activities performed with the material[24].

The composition of the retouched tool inventory was used to describe the function of sites and to distinguish between various types of assemblages. Therefore, the two most numerous formally retouched artefact groups were set in bold in the presentation (see **tabs 14. 27. 38**). Moreover, Laurent Lang distinguished three groups of Magdalenian inventories in the Paris Basin based on the amount of points among the LMP and the relation of end-scrapers to burins (Lang 1998, 96). His group 3 was correlative with the Late Magdalenian yielding no points and generally more burins than end-scrapers. Furthermore, Lang considered the burins being principally used for working antler material, in particular, to produce antler spalls for various types of points (Lang 1998, 99). Thus, changes in the relation of end-scrapers to burins were assumed as representing a functional change. However, he noted that *becs* (or *Zinken*) as well as other lithic

24 Even though the LMP were not used at the site, they are included in this index because they were produced at the site and the equipment which they represent was made or repaired at the site and, thus, their presence shows an activity of resource use other than preparation of the raw material.

class	numbers of lithic artefacts ≥1 cm	numbers of total lithic artefacts
0	1-500	1-1,500
1	501-1,000	1,501-3,000
2	1,001-2,000	3,001-6,000
3	2,001-4,000	6,001-12,000
4	4,001-8,000	12,001-24,000
5	8,000>	24,000>

Tab. 52 Numerical classes of lithic artefact assemblages chosen in the present study.

artefacts were possibly also used for working antler. An inference from these assumptions of Laurent Lang is that decreasing numbers of burins and (heavily retouched) borers could indicate a decrease of work with antler material and, perhaps, a diminishing importance of antler as a resource. Nevertheless, burins and borers could have also been replaced by other tools such as splintered pieces or unretouched blades (cf. Sano 2012b). Thus, along with the absence of worked antler, micro-wear analysis of several lithic artefacts including unretouched pieces should supplement the observations of the retouched inventory to conclude how important antler was as a material resource. Moreover, Lang also found a relation of the increasing number of lithic points with a decreasing importance of burins in his studied assemblages (Lang 1998, 96). He suggested that this relation reflected the presence of alternative projectile heads (bone/antler vs. lithic) but he could not decide whether the lithic points were used because antler was not available as material for projectile heads or if antler points were abandoned because lithic points were invented (Lang 1998, 99). This question cannot be answered by this study either. However, the relation of end-scrapers to burins (end-scraper-burin index = no. of end-scrapers / no. of burins) as well as the percental presence of borers (% borers = no. of borers / [no. of retouched artefacts / 100]) as alternative tools for working antler can be studied. Furthermore, the numerical presence of projectile points can also be compared to the numbers of burins (see **tabs 14. 27. 38**). The results of this comparison (point-burin index = no. of points / no. of burins) have to be seen critically in the present study because the points were only determined from the displayed LMP specimens. If the complete LMP inventories were analysed the numbers of points within the assemblages could alter and would, consequently, alter the results of the point-burin index. Nonetheless, the surveyed numbers used in the present study should be sufficient for a first assessment whether lithic points outnumbered the lithic artefacts supposed to be used in the work of bone and antler material. In this sense, the result of this assessment can be regarded as an indication for the function of an assemblage. In a comparison of the Lateglacial assemblages, this tendency can also be considered as an important parameter reflecting the chronological development of the hunting equipment.

The general diversity of the retouched artefact inventory is documented by counting the major tool classes which are common in the Lateglacial: LMP, burins, end-scrapers, truncations, borers, composite tools, and others[25]. If a class is not present, this class is recorded to see if a particular class of standardised tools becomes abandoned in the process of change (cf. Lang 1998). The abandonment of a typical retouched artefact class such as borers or burins can in general be regarded as a departure from a conforming behaviour which was common for the Late Magdalenian. Moreover, according to a model of Hartwig Löhr, the

25 Splintered pieces are not formally retouched tools but suitable lithics used as punches. Therefore, they are not included among the group of others. Retouched pieces such as end-scrapers were often secondarily used. In this case, the artefacts were counted among their primary use. According to the settlement model of Hartwig Löhr (Löhr 1979), splintered pieces appeared very late during the occupation of a site when the assemblages had become more diverse (cf. Richter 1990). Although pieces with splintering are occasionally described from FMG sites, »true« splintered pieces were hardly mentioned and usually not numbered in the lithic assemblage descriptions. The possible disappearance of these specimens is certainly an interesting fact for considerations about changes in the settlement system but due to this uncertainty in the literature the present project cannot evaluate when or whether these types disappeared.

	high number of cores	low number of cores
low diversity of retouched artefacts	special task camp: workshop	special task camp: hunting camp
high diversity of retouched artefacts	base camp	short-term habitation and/or far from source

Tab. 53 Idealised distinction of site types based on the diversity of retouched artefacts and the number of cores.

diversity of a lithic inventory, in particular, of the retouched artefacts increased with the number of activities performed at a site (few: special task/many: residential) and that the number of activities was related to the duration of the occupation of this site (Löhr 1979). Following this idea, Jürgen Richter introduced the diversity index of Simpson ($D = \sum(P_i 2)$)[26] to evaluate the degree of specialisation or diversity of lithic assemblages (Richter 1990). Furthermore, according to Löhr's and Richter's model, the presence of all the major tool classes indicates a relatively long duration and the accomplishment of several activities during the occupation. However, Löhr also emphasised that the production of further blanks and retouched artefacts is of some importance as a marker for longer durations. Consequently, the number of cores is an important value to put the diversity of retouched tools in relation to the duration and general function of the site (cf. Weniger 1989; Richter 1990). In the present study, the different numbers can be taken from previously presented tables (see **tabs 14. 27. 38**).

Based on these different values a simple model of the site function can be established (**tab. 53**). Sites with a high diversity of retouched tools suggest different activities at the site, whereas a low diversity indicates a special purpose for which the site was used. In assemblages with a high number of cores, several assumption are possible: either the exploitation of the raw material was a special task, many flintknappers were present, and/or the spot was used over a longer period. In contrast, low numbers of cores make a special task other than flintknapping, a small group of flintknappers, and/or a short duration of the camp more probable. Based on this simple model, changes in the settlement system due to changes in the function of sites should be identifiable.

Projectile analysis

Modern and historic hunter-gatherers of cold to temperate climate environments often obtain their nutritional needs from animal resources (Cordain et al. 2000). Therefore, hunting and fishing equipment such as projectile implements were considered as important tools in their subsistence strategies. Consequently, changes in this equipment could be a result of significant changes in the subsistence strategies as well as a trigger for alterations of these strategies. Moreover, these implements were altered more readily than tool classes such as burins, end-scrapers, or borers. This frequent change in the shape of the projectile implements make them better chronological markers than other tool types. Therefore, projectile implements were preferably used in the primary description and differentiation of lithic assemblages in an early classification process (see p. 55-74; Breuil 1913; Schwabedissen 1954; Taute 1968). The important relation of projectile implements to the subsistence strategies as well as the on-going discussions of differences in these implements relating to social information (Wiessner 1983; Mesoudi/O'Brien 2008a; Hamilton/Buchanan

26 In this formula, D is the diversity index, P is the minimal number of pieces of a given tool type divided by the total number of retouched artefacts, and i relates to the considered tool types. The latter precision is relevant because of the compilation of laterally and other artefacts without standardised retouches in the tool type »others« and the exclusion of splintered pieces as tool category in this project.

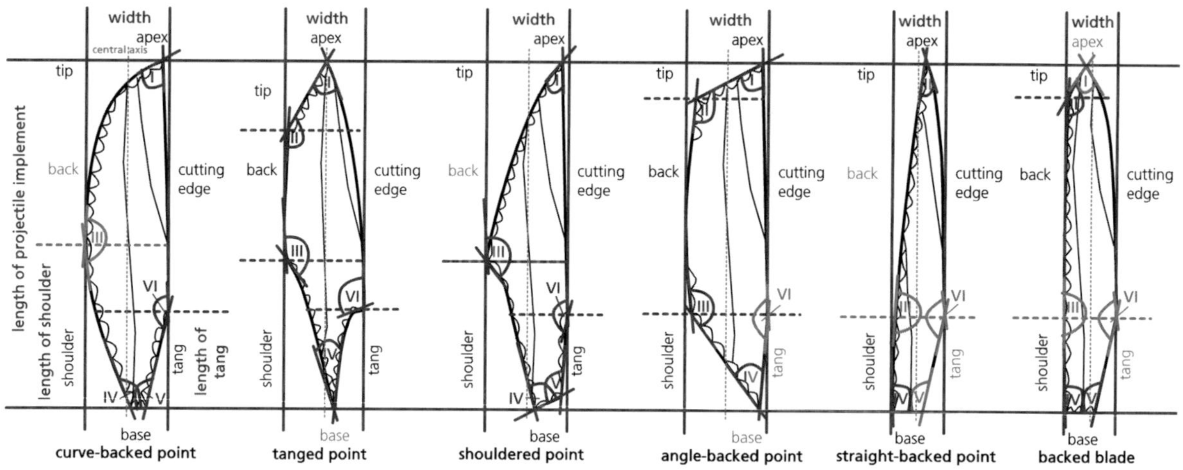

Fig. 28 Schematic outlines of various Lateglacial LMP. Terms and measured characters on various projectile implement types as applied in the present work (cf. Ikinger 1998, 41; Burdukiewicz/Schmider 2000, 98; Riede 2011). Characters, which are difficult to be distinguished and/or measured are given in light grey and were not measured. Angles are numbered: **I** tip-cutting edge angle; **II** tip-back angle; **III** back-shoulder angle; **IV** shoulder-base angle; **V** base-tang angle; **VI** tang-cutting edge angle. – For further details see text.

2009; cf. Barton 1997) and to their use (Cattelain 1997; Hughes 1998; Lyman/van Pool/O'Brien 2009; Riede 2010; Lombard/Haidle 2012) caused the continued importance of these implements in typological studies (Ikinger 1998).

In the Lateglacial, hunting equipment was generally associated with organic points and the lithic LMP. The former group was only rarely preserved in the Lateglacial Interstadial. However, the latter group frequently represented one of the most numerous classes of formally retouched pieces on archaeological sites (see **tabs 14. 27. 38**). In the literature, the different relation of pointed pieces to backed bladelets were considered as relevant (Lang 1998). This difference was related to different technical systems and is also of interest in the current project. The LMP were often very fragmented and the taxonomic classification of single fragments was often based on the surrounding context. Furthermore, within the FMG, the morphological distinction of the projectiles were considered as chronologically, functionally, and traditionally distinctive factors. However, the precise distinction in the morphology of the single point types often remained unclear. Therefore, in the present study a morphometric approach to classify the complete or almost complete implements was selected. Only with this type of quantification, can an evaluation of the impact of function or inventions and innovations on the LMP inventories be achieved.

Accordingly, the shape diversity of lithic projectile implements was established per selected assemblage. Therefore, all LMP which were identified in assemblages from the selected sites and illustrated in a readable figure were collected in a database, currently comprised of 1,201 entries. The figures were reviewed and classified according to a general concept of the Lateglacial projectile morphology and some major rules of attribution. Thus, a common denomination of the projectile parts is initially proposed (**fig. 28**).

Usually, two edges of the Lateglacial projectile blank were distinguished by a retouched and an unretouched edge. The latter remained sharp and is therefore termed the cutting edge. The apex is generally formed from the cutting and the retouched edge. However, in the case of backed blades this »apex« can be formed by two unretouched edges. The upper part of the LMP is either where the apex was located or for the backed blades/bladelets the distal part. The complete upper part of the LMP is named the tip. The intermediate part of the retouched edge was named the back. The lower part of the retouched edge was called the shoulder regardless of a retouched angle. In pieces without an obvious angle, the shoulder is regarded as the part between the maximum width and the base. In simple backed blades/bladelets, the proximal

class	primary criteria (general shape)	secondary criteria (forming of the retouched edge)
couteau à dos	size (length ≥80 mm and/or width ≥25 mm)	×
backed blade/bladelets	no point	×
simple point	point, no tang	angle or curvation, partial retouch (truncation)
straight-backed point	point, no tang	no angle, no curvation, apex beyond middle axis
curve-backed point	point, no tang	no angle but curvation
angle-backed point	point, no tang	angle/s, no curvation
tanged point	point, tang	×

Tab. 54 General classes of LMP used in the present study and the main criteria for their distinction. These general classes are used to distinguish the LMP diversity. – For further detail see text.

end is generally regarded as the base, except for pieces with a single truncation. These pieces are assumed to have been used as hafted as a part of a larger projectile (cf. Leroi-Gourhan 1983). In this case, the single truncation is assumed to have served as stabilizing retouch for the base of the piece against a recoil caused by the impact on the target. Therefore, the truncation is considered as the base. If the cutting edge was also retouched at the lower part, this retouch was classified as the tang. This denomination was comparable to the shoulder independent of the forming of this tang retouch. However, if the shoulder and the tang exhibited this retouch both had to form a concave for the distinction of a tanged point (see below). The limits of the three parts of the retouched edge were occasionally difficult to distinguish for some projectile implements. For example, the tip and the back merged in backed blades/bladelets and curve-backed points and the back part to the shoulder also merged in straight-backed points as well as backed blades/bladelets. Thus, some Lateglacial types cannot be distinguished by the morphology of the three parts of the retouched edge but require some further rules of classification as well as angles (including their presence and absence) as additional differing values.

Firstly, incomplete specimens were attributed in this secondary classification to an undifferentiated LMP class. Incomplete meant that the upper and/or lower part of the piece were significantly damaged by other modifications than human retouch. These other modifications were, for example, impact fractures or fractures due to sediment pressure. If pieces were refitted and the lower and the upper part of the LMP were preserved the refitted LMP pieces were regarded as one complete specimen. In the Late Magdalenian, backed bladelets were sometimes intentionally broken (**pl. 1, 28-30**). This behaviour produced a particular problem because a remaining piece was clearly preserved incompletely and had to be attributed to the undifferentiated LMP class. If the blank was completely refitted the numerous backed bladelets had to be counted as one piece. In this case, problems arise with the numerical and, thus, proportional differences for the comparison of the projectile implements. However, since the diversity of the projectile implements was the aim of the classification, the numerical problems could in general be neglected. Nevertheless, the most numerous class was identified to describe the inventory. Thus, to describe an inventory correctly according to the dominant projectile type some valid numbers had to be ascertained. Therefore, some almost complete pieces were also attributed to the backed blade/bladelet class in the Late Magdalenian assemblages following a contextual assumption.

Furthermore, if the size of a fragment already exceeded the metric limits of a *couteau à dos* (length ≥80 m and/or width ≥25 mm; cf. Le Tensorer 1981; Audouze/Beyries 2007) the piece was exceptionally attributed to the *couteau à dos* class. This class was not morphologically but metrically distinct (**tab. 54**) assuming that these exceptionally large specimens were used differently. According to use-wear analyses, these pieces were presumably used as knives or saws, perhaps, in the butchering process (Audouze/Beyries 2007, 190. 200; Sano/Maier/Heidenreich 2011, 1478). These large pieces could possibly have served as heads of

thrusting spears[27] (cf. Riede 2009), but they were assumed as too large and heavy to represent projectile implements (Valentin 2008a, 213f.). Nevertheless, the use of such heavy implements in heavier projectiles and/or hunting equipment cannot be excluded completely (Letourneux/Pétillon 2008; Sano 2009).
Secondly, implements were attributed according to their general shape. By this general shape, two further classes were added to the metrical class of the *couteaux à dos*: backed blades/bladelets and points. Within this general shape criteria, the tanged points could already be differentiated from the other Lateglacial points because of the intentional forming of the cutting edge. Intentional forming of the tang part was observed in various types of Lateglacial points which consequently fall into the transition such as some Hamburgian shouldered points or the penknife points (cf. Grimm/Jensen/Weber 2012). However, a proper tang is characterised by concave retouches along shoulder and tang part. Probably, this type of tang composed of the basal part related to the hafting properties of the implements and helped to fix the piece on top of a shaft. This hafting on top of a shaft is characteristic for implements which are traditionally understood as an arrow- or dart-head. In contrast to the other Lateglacial points, the tip part of the tanged pieces was occasionally not formally retouched, perhaps, because better cutting properties were more important than the stability of the lithic implement. Moreover, microscopic analyses on breakage patterns and adhesives on other Lateglacial points suggested that these implements were possibly used as barbs rather than heads of a projectile (Baales 2002; Rots/Stapert/Johansen 2002). Since the differentiation of use for these implements is considered as an important change, the tanged points are distinguished as a special class in this early step. The backed blades and bladelets are also assumed to have been used differently to pointed pieces in a composite projectile system.
They are differentiated from the other two types by neither displaying a tang nor an intentional point.
Among the points, a third rule was used to distinguish further groups. This rule was to distinguish the points according to the general form of the mainly retouched lateral edge. The forming could take a straight, an angled, or a curved shape and resulted from the relation of the tip part to the back and the back to the shouldered part of the implement (**fig. 28**). A denticulated retouch could partially obscure this general forming of the retouched edge. In this case, the drawing of mean lines through the denticulated parts helped to distinguish the underlying morphological tendency. For the British Creswellian, Roger Jacobi and Allison Roberts pointed to the existence of angle- and curve-backed points (Jacobi/Roberts 1992). This variant was also observed occasionally in the present study (**pl. 10, 4-5. 17**). In general, these pieces were attributed to either the angle-backed or the curve-backed group depending on the transition from the tip to the back part and the similarity to the other specimens of the same context. Furthermore, on occasions the shoulder part and sometimes additionally the back were not retouched. Thus, the retouch was applied only to the tip part and in a strict classification these pieces had to be attributed to the truncation class. However, several of these specimens were already singled out in the primary publications (e.g. Baales 2002, 140, projectile type E), usually, because they were significantly different from the typical truncations in the inventories. For example, many of these pieces were made on flakes rather than on blades or bladelets. In contrast to the retouches of typical truncations, the position of the retouches on the blanks of these pieces was often steeply oblique forming an acute angle of the retouched and the unretouched edge at the apex. These singled out specimens are named simple points in the present study. Perhaps, they already herald the move toward Mesolithic microliths. Straight backed point types are difficult to distinguish from backed blades/bladelets. Following the previously mentioned assumption that backed points and backed

27 In the present study, the simple distinction of terminology by the use of the weapon delivery system as proposed by Felix Riede is used. Thus, thrusted hunting instruments are named spears, thrown instruments (also by the use of spear-thrower/atlatl technology) are darts, and instruments fired from a bow are arrows (Riede 2009, 28; cf. Hughes 1998). Only the latter two types of instruments are subsumed as projectiles.

type	blank preference	pointed ends	forming of the blunted edge	continuity of retouch	forming of the basal retouch	forming of the base in relation to the blunted edge	micro-burin technique	modifications of the cutting edge
simple point	flake	monopoint	straight/curve	no	none	none	no	none
straight-backed point	none	monopoint	straight	yes	none	none	no	none
Micro-Gravette	none	monopoint	straight	yes	straight	straight (acute to approximate right-angle)	unknown	none
Federmesser	none	monopoint	curve	yes	none	none	no	occasionally, a notch shortly underneath the apex
penknife point	blade/bladelet	mono- / bipoint	curve	yes	convex/concave/straight	oblique (acute angle – bipoint)	unknown	occasionally, the base remained natural and instead the lower part was retouched (monopoint)
Malaurie point	blade/bladelet	monopoint	curve	yes	convex/concave/straight	approximate right angle	unknown	none
bipoint	regular blade	bipoint	curve	no	pointed	oblique (obtuse angle)	unknown	none
Cheddar point	regular blade	bipoint	angle	yes	straight	oblique (obtuse angle)	unknown	none
Creswell point	blade	monopoint	angle	yes	none	none	unknown	none
Hamburgian shouldered point	blade	monopoint	angle	no	none	none	yes	occasionally, the lower part was also retouched
Havelte point	regular blade	monopoint	angle	no	pointed/straight	oblique (acute angle)	yes	tang: oblique or concave retouch of the lower part forming a tang with the retouch of the principally retouched edge
Lyngy-Bromme type tang point	large flake	monopoint	angle/cavity	no	none	straight (approximate right-angle)	no	tang

Tab. 55 General characterisation of some Lateglacial LMP types according to the selected morphological factors. The micro-burin technique was usually associated with Mesolithic microliths but recent studies have revealed the presence of the technique also on Lateglacial material (Weber 2012) and emphasised the role in the removal of the bulb (Grimm/Jensen/Weber 2012) which shaped particularly the thickness and evenness of the implement.

blades/bladelets were used differently in the projectile installation, the difference of the two types must relate to the way the retouch transformed the complete blank and, in particular, the tip part. Consequently, the position of the two lateral edges in the tip part in relation to the complete blank was chosen as differentiation between backed blades/bladelets and straight-backed points. In the case of straight-backed points, the retouched edge crossed the middle axis of the blank and the apex was set in the half of the blanks unretouched edge. Furthermore, the unretouched edge and the retouched edge met in an acute angle. In backed blades/bladelets, the angle of the apex was variable with acute, right-, and obtuse angles occurring. In contrast to straight-backed points, the apex of backed blades/baldelets was always set on the side of the retouched edge in relation to the middle axis of the blank.

Thus, this classification differentiated seven main groups of projectile implements (**tab. 54**). The more of these classes were identified in an assemblage, the more diverse the composition of the projectile implements discarded at the site. The diversity of an inventory could indicate either a differentiated use of projectile types or an on-going transition of preferred LMP types. If different projectiles and/or projectile heads were used for hunting of different species (differentiated use) a diverse assemblage would represent the discard of varied hunting activities. However, in sites with organic preservation this hypothesis could be tested against the number and the diversity of the discarded faunal remains (see p. 282-285). If a change of preferred hunting equipment was reflected by the diversity of the LMP inventory a phase of trial-and-error in the technical system could be reflected. Furthermore, this process of variation in the hunting equipment could have been caused by changes in the subsistence strategies. Therefore, the comparison of this value to the preserved faunal remains seems particularly relevant with regard to the influence of the preferred prey on the preference of lithic projectile implements.

The weight of the projectiles could be an interesting, additional value for comparing the preference of the LMP. The weight of an implement probably correlates to its size. Both characteristics were possibly correlative with the use of the projectile in the different delivery systems such as arrow or dart (Riede 2009; Riede 2010; cf. Letourneux/Pétillon 2008) as well as the choice of prey such as whether small or large game, land mammal, or fish were hunted (Ellis 1997). However, the weight of the various implements could not be compared in the present study because this information was not usually available in the literature per piece. Occasionally, ranges of weight were recorded for the projectile points from a single assemblage. Assuming the correlation of size and weight, the sizes of the LMP were instead measured.

Therefore, the most complete LMP were chosen for this metric comparison which should demonstrate the diversity of the LMP inventories within each assemblage as well as between the various assemblages. This comparison should help to formulate suggestions on the influence of the available raw material as well as technical skills and operational habits of standardisation on the type and the size of the LMP. Various descriptions for the measuring of single types of Lateglacial points such as tanged points (Riede 2011), shouldered points (Burdukiewicz/Schmider 2000; Thévenin 2003), or curve-backed points (Ikinger 1998) exist. However, these attempts were, in general, designed for a specific type of LMP, whereas a combination of all these measuring systems supplemented with one for backed blades is needed in the present analysis. Thus, some major types of Lateglacial points were compiled and the major morphological criteria for the differentiation defined and systematised (**tab. 55**; cf. Schwabedissen 1954; Sonneville-Bordes/Perrot 1956; Bohmers 1961; Célérier 1979; Ikinger 1998; Burdukiewicz/Schmider 2000; Riede 2011; Grimm/Jensen/Weber 2012). Based on this system, which is adjusted to the variety of types present in the Lateglacial of northwestern Europe, the same values could be measured on various LMP (see **fig. 28**).

In addition, measurements of the maximum values for length, width, and thickness, the length of the shoulder and the length of the tang if it was present were taken. The length of the shoulder was only measured when an angle was observable.

This consideration reveals that measuring some angles is also important to describe the pieces morpho-metrically. Besides the angle between the back and the shouldered part, a further five angles were frequently observed on Lateglacial points: the apex angle between the cutting and the retouched edge, the angle between tip and back part, the angle between shoulder part and base, the angle between base and tang retouch, and the angle between the tang retouch and the cutting edge. Besides influencing the shape of the LMP significantly, the angles are assumed to be of technical importance which helped to position the LMP in the projectile shaft and to reinforce the edges against forces related to the impact. Furthermore, the angle between the ventral surface and the dorsal formed by the retouch of the tip could be related to the entering ability of the point. However, this part is often damaged and the profile of this region is not usually displayed. Therefore, this angle is not considered further.

Angles are only measured where retouched edges are involved because only these areas are obviously shaped by the decision of a human toolmaker. Some LMP types do not exhibit all six angles or fulfil the requirement of at least one retouched edge and, thus, these angles are considered as not present. For example, if the back and the shoulder part are not retouched the angle inbetween them is regarded as not present and, consequently, not measured. Moreover, the variety of LMP types poses further problems regarding the angle determination: In the case of curve-backed points the angle between the tip and the back and the one between the back and the shoulder are regarded as unmeasurable due to the curved outline. Principally, these angles are set to 180° in the case of straight-backed implements such as backed blades/bladelets. Another problem is the forming of the base which was sometimes reduced to a punctiform base, for example in some tanged points or in the bipoints (see **fig. 28**, tanged point). In this case the shoulder-base angle and the base-tang angle become identical. Furthermore, for some implements the tip-back and the back-shoulder angle are identical (see **fig. 28**, curve-backed point and shouldered point). In the database this factor is noted by using the Roman number of the identical angle.

In addition, further observations such as potential impact fractures were noted. Particular attention was paid to the occasionally observed notches on the unretouched edge, often in the apex region. These notches could represent intentional elements in the shaping of the pieces, for instance, to fasten a string or a special type of impact fracture (cf. Plisson 2009, 46). Clearly, these notches depend on the hafting and/or the use of the implements. If recorded more systematically this observation could help to further shape our understanding of past projectile technologies.

To receive the metrical values necessary for this comparison, drawings of complete or almost complete LMP were selected from each assemblage for measuring. These drawings were compiled in a graphic program (Adobe Photoshop) and scaled to a natural scale (M 1:1). The resulting 14 tables can be found in the supplementary figures in the appendix (**pls 1-14**). The measurements were taken with the measuring tool of the graphic program enabling to take the sizes as well as the angles. Thus, the potential error of the program applied to all specimens in a comparable manner and, therefore, can be neglected as a source of error in the comparison of the results. For the present project the measured data and morphological description of the various parts of the points had to rely on drawings of the implements. Clearly, the quality of these drawings could cause some imprecision of the measured data due to the different thickness of pens or the sometimes ambiguous character of drawn retouches. For the future, an analysis on the actual lithic implements would be more appropriate. For a first assessment of whether significant morpho-metric differences can be observed these drawings are sufficient.

In total, 335 implements were measured. Of these artefacts, the length and the width was preserved completely in 148 specimens which can be used in a metrical comparison of the material. This comparison is performed with the freely available data analysis package PAST (Palaeontological Statistics; Hammer/Harper/Ryan 2001). For the comparison of the data per assemblage, xy-graphs with convex hulls are

class	minimum number of individuals (MNI)
A	1-10
B	11-20
C	21-50
D	50>

Tab. 56 Classes of faunal assemblages based on minimum numbers of individuals (MNI) of larger herbivores (Weniger 1989, 346).

used. »The convex hull is the smallest convex polygon containing all data points« (Hammer/Harper/Ryan 2001, 24). The diagrams are interpreted with the assumption that the larger the areas of overlap emerge, the more similar are the data sets. Since standardisation and changing standardisations are of a particular interest in studies of transitions, the variance (Hammer/Harper/Ryan 2001, 37) of the different metrical values per assemblage is further compared in regard to the chronological development. Clearly, in future analyses more sophisticated statistical analyses are possible with sufficient morpho-metric data (cf. Iovițā 2009; Iovițā 2011). However, for the present study this simple comparison is sufficient for the presentation of diversity and similarities among the LMP from various sites.

Quantitative assessment of the faunal assemblages

Faunal assemblages from archaeological contexts represent the most instant reflection of Lateglacial subsistence strategies. Furthermore, these subsistence strategies allow for suggestions of site function and considerations about settlement behaviour. For example, if complete or almost complete carcasses of larger mammals were found at a site, a relatively close hunting ground was assumed (Müller et al. 2006). The transport of a horse which presumably weighted more than 250 kg (Groves 1994, 40) required considerable effort. Thus, the skinning and the first filleting of the animal was generally assumed to have taken place at the hunting ground from where the required material was transported to the residential camp. However, if the complete animals were valuable the residential camp must have been moved to the hunting ground (Müller et al. 2006) or the complete animals must have been brought to the residential site (cf. Street/Turner 2013). Both strategies represented considerable effort. In both cases, various factors such as the weight of the animal, the distance between the residential site and the hunting ground, and the size of the group contributed to the estimate of the effort. For instance, if the group size is small the moving of a residential camp is as difficult as with a big group but for the latter sufficient space is necessary at the new settling ground. In contrast, the transport of large mammals is easier with more people. Consequently, can the size of the group or the number of hunted animals decide whether to move the camp to the food or the food to the camp? Or are the classic distinctions of hunting grounds, special task camps, and residential sites (cf. Binford 1980) not applicable to the Lateglacial?

Thus, to quantify the assemblages a classification system based on a system developed by Gerd-Christian Weninger is used. He differentiated four classes of faunal assemblages based on the minimum numbers of individuals (MNI) of large herbivores found at Late Magdalenian sites to distinguish differences in the size of the sites (**tab. 56**; Weniger 1989, 346). Clearly, this classification is not correlative with economic gain of these resources such as leather/fur or meat (cf. Rozoy 1978). The different size within his large herbivore group such as ibex and horse and, consequently, the different output of, for instance, meat would make such a comparison very vague. In addition, even comparable resources were probably used differently, for example, horse hide clearly served a different purpose than the winter coat of an arctic fox. Therefore, a validation of the resources would be necessary but this validation system cannot be given for past societies. Hence, the MNI are considered to reflect a quantification of animals brought to the site or hunted at the site helping to characterise the function of the site. The additional information on the parts of the animals recovered at the site and the combination with the spatial analysis of the sites help to further evaluate the site function.

class	MNI of *Rangifer tarandus*	MNI of *Equus* sp.	MNI of *Bison priscus / Bos primigenius*	MNI of *Cervus elaphus*	MNI of *Alces alces*	MNI of *Capreolus capreolus*	MNI of *Sus scrofa*
1	1-15	1-10	1-10	1-5	1	1-3	1-5
2	16-30	11-20	11-20	6-12	2-3	4-6	6-12
3	31-75	21-50	21-50	13-30	4-6	7-16	13-30
4	75>	50>	50>	31>	7>	17>	31>

Tab. 57 Classes of faunal assemblages based on minimum numbers of individuals (MNI) for variously behaving larger herbivores. – For further details see text.

Besides the actual number of animals brought to the site, the conditions of preservation could change the classification significantly. In particular, preservation of bone material is probable to change with the changing climatic and environmental conditions, for instance, slower sedimentation due to the lack of aeolian deposition and the increasing attack by micro-organisms, humic acids, and roots within the sediment could result in a significantly higher destruction of the faunal material. Consequently, the changing environment of the Lateglacial could cause a classification into a lower class. For instance, the very low numbers of individuals (MNI = 5) found in the 25 *loci* of the upper horizon of Le Closeau were certainly due to the poor bone preservation in a temperate forest. In fact, faunal material was preserved only at eleven concentrations of the upper horizon and only four of those yielded determinable material (cf. Bodu 1998). Thus, a reliable MNI for the material brought to Le Closeau during the younger period could not be calculated, whereas the material from the lower horizon was well preserved and numerous (Bemilli 1998; Bignon/Bodu 2006) indicating the environmental changes as an important factor in the preservation of the faunal assemblages. Thus, a generalised classification of large herbivores according to the MNI is only useful as a comparison as long as the conditions of preservation are also considered as relevant agent in the formation of the assemblage.

Moreover, since the faunal composition in the landscape changed, the Lateglacial hunters had to adapt their behaviour to the new species if they remained in these landscapes. To compare the faunal assemblages similarly to Weniger (Weniger 1989) as an indicator for the formation of the assemblage, the considered species should reveal comparable ethologies in regard to their group sizes[28]. For example, a large number of reindeer (*Rangifer tarandus*) can be hunted in a single driving event when they congregate in herds of sometimes hundreds or thousands of animals (Baskin 1990; Reimers/Colman 2006). To receive a comparable number of elks (*Alces alces*) which are known to live a more solitary existence, many hunting events were necessary. Furthermore, in regard to the territorial ranges of elks (Ball/Nordengren/Wallin 2001), a large number of elks could only be accumulated over a considerable period of time or from a large territory. Thus, comparable numbers of these two species reflect very different human activities. The classes suggested by Weniger (Weniger 1989) are therefore adapted to the ethology of the species considered relevant in the present project (**tab. 57**). Horses (*Equus* sp.) were the most important prey species during the Late Magdalenian of the study area (Bridault/Bignon/Bemilli 2003; Street/Turner 2013). The data on group size in wild horses is rare because the main observation on horse groups come from animals raised in captivity (Houpt/Boyd 1994). However, a comparison with feral animals and wild living equids such as Zebras suggests that wild horses would have formed similar herds (Goodwin 1999). Also, the numerous depictions in Palaeolithic art (Pigeaud 2007), in particular, of horse groups such as the panel of horses in the Early Upper Palaeolithic Chauvet Cave (Clottes/Arnold 2001) suggest the presence of groups of horses. These groups could be of various compositions (stallion groups, harems, female and young groups). Nevertheless, with the changing environment of the Lateglacial Interstadial the behaviour of horses could also have adapted to

28 Special thanks is due to Sabine Gaudzinski-Windheuser, Monrepos, for suggesting this useful differentiation.

the more forested environment by decreasing group size (MacFadden 1992, 263-298). In total, the groups given by Weniger appear a good mixture of this species. Since steppe wisent (*Bison priscus*) and aurochs (*Bos primigenius*) are extinct, their group sizes are also unknown. For the aurochs some past observations exist and describe herds of a maximum of 30 animals in some parts of the year (Frisch 2010). For the European wisent (*Bison bonanus*) which were reintroduced into the wild in eastern Central Europe group sizes of 8-13 animals have been recorded (Pucek/Belousova 2004, 26). However, these animals inhabit mainly forest landscapes, whereas the American bison (*Bison bison*) were found in open grassland landscapes more comparable to the Late Pleniglacial landscape of Europe inhabited by the steppe bison. Large herds of American bison such as reported by early explorers visiting the Great Plains can no longer be attested (Knapp et al. 1999) but modern, free-ranging American bison are known to form groups of 3-26 animals but in general these groups were also around 16-17 animals (Fortin/Fortin 2009). Thus, the size classes given by Weniger are probably a good estimate for the large bovids. In contrast, reindeer groups appear on average larger with observed mean group sizes around 19 animals in the time of the largest diffusion of the animals and of more than 300 animals in the time of autumn migration (Baskin 1990). However, the groups of taiga reindeer are considerably smaller with occasionally only some 3-5 animals in a group (Baskin 1990). Due to the relatively large retreat areas during the Lateglacial, the occurrence of forest reindeer seems rather anecdotal in the Lateglacial record (cf. Riede et al. 2010). For this species a larger classification is chosen. As with most species used in the present study, red deer (*Cervus elaphus*) forms various group sizes depending on the seasons as well as on the habitat (Bonenfant et al. 2004). Even though large groups of up to 56 animals (Bonenfant et al. 2004, 885) were observed, this species usually forms smaller groups and solitary animals are also frequently observed (cf. Hebblewhite/Pletscher 2002). Hence, the classification for this species is selected smaller than for the previous species. Elk (*Alces alces*) is known to live mainly solitary in occasionally large territories (Ball/Nordengren/Wallin 2001). Therefore, this species is grouped in the smallest classes. In addition to the group of species compared by ^{14}C-dated material, the classification of faunal material found on archaeological sites is expanded by roe deer (*Capreolus capreolus*) and wild boar (*Sus scrofa*). Roe deer groups also vary in size but in general these groups are very small with around 4-5 animals (Pays et al. 2012). Male wild boars (*Sus scrofa*) are often solitary (cf. Ebert et al. 2010), whereas females and young ones can form groups of up to 42 observed animals (Hartley 2010). However, in general the groups are smaller and, thus, wild boar is treated as comparable to red deer in the present study.

The population density is another important factor for the group sizes as well as the impact of predators. These factors are relatively difficult to estimate for Lateglacial Europe and are therefore not further taken into consideration. However, since displacement processes are probable for some species in the Lateglacial, the five classes are supplement by a class 0 which indicates the absence of a species in an assemblage. Consequently, the variable classes can be compared within the archaeological assemblages as well as with the probability distributions for this species in the region (see p. 263f.).

The increasing diversity of species which in general lived in smaller groups could possibly result in a more generalised subsistence strategy. However, the few faunal assemblages from FMG sites were regularly described as dominated by either elk, red deer, or aurochs (Bokelmann/Heinrich/Menke 1983; Baales/Street 1996; Coudret/Fagnart 2006). Thus, a specialisation in an available resource could still be possible. To evaluate whether the specialisation patterns changed, the diversity index of Simpson ($D = \sum(P_i^2)$)[29] is calculated for the faunal assemblages analysed in this study.

29 In this formula, D is the diversity index, P is the minimal number of individuals of a given species divided by the total minimal number of individuals, and i relates to the calculated species. The latter precision is of some importance since in the present study amphibian and small mammals as well as other probable intrusions were excluded from the assemblages.

	small MNI	high MNI
low diversity	short episode	hunting camp
high diversity	provisioned/opportunistic episode	base or agglomeration camp

Tab. 58 Idealised distinction of site types based on the faunal diversity (Simpson index) and the total MNI (minimal number of individuals).

A previous study on the diversity of Magdalenian assemblages suggested that this index is useful in the distinction between specialised hunting camps and residential sites (Gaudzinski/Street 2003). Particularly in relation to the complete size of the assemblage (total MNI), differences of the site types should become evident (**tab. 58**). In this relation, very small MNI cannot produce high diversity indices due to the dependence of this index on the total MNI. However, the possible diversity values increase exponentially and with two animals an intermediate diversity can be reached and with three to five individuals values can be reached which are comparable to the anaylsed Late Magdalenian assemblages (Gaudzinski/Street 2003). Thus, relatively small MNI with high diversity are possible and would reflect a successful opportunistic hunting episode or a special task camp with provisioned food. However, more frequent are small assemblages with a low diversity which reflect a very short-term camp or a single hunting episode. High numbers of MNI combined with a high diversity reflect a generalised prey choice and speak in favour of a base or agglomeration camp, whereas sites with high MNI but with a low diversity indicate a hunting camp.
This simplified distinction works well for seasonally governed landscapes with large animal herds which were partially migratory and human hunters with a logistic settlement pattern (Binford 1980) such as assumed for the Late Magdalenian. However, several of these factors were changing in the transition to the FMG, for instance, the group size decreased and the animals were in general more faithful to their habitat. Thus, the number of animals which could be hunted during one episode decreased in comparison to driving events of large herds, whereas the increased diversity of species allowed for a higher diversity to be reached. Consequently, the differences between base/agglomeration camps and hunting camps could have been obliterated. However, these differences could have also shifted similarly towards smaller MNI and, thus, the difference between provisioned or opportunistic episodes and base/agglomeration camps could have been blurred. These possible obliterations have to be discussed before the question whether the settlement system changed from a logistic to a foraging one can be considered.
Hunting strategies as sensitive systems for the human survival, in particular in provisioning for otherwise meagre periods, are improbable to be changed easily but are more likely to be gradually adapted. Thus, if a sudden change was revealed by the comparison of the diversity index and the total MNI, this result would be a strong argument for a revolutionary change. However, a direct change from the one system to the next appears improbable and, thus, an intermediate stage could be possible. In fact, the relevance of the contribution of smaller mammals, particularly, in times of crisis was emphasised in the last decade (Bridault/Fontana 2003; Munro 2003; Napierala 2008b). Besides the migratory behaviour and the group size of the hunted species, the total MNI depends on further factors such as the size of the hunted animals. If the size of hunted animals decreases, the total numbers of animals have to increase to compensate for the size of the animals. Yet, smaller mammals such as hare are usually considered to be hunted or rather trapped more easily than larger mammals such as horse. Therefore, also the MNI of smaller mammals such as fox, hare, or beaver are recorded and given in a percentage relation to the MNI of the larger herbivores. If the transition from the Late Magdalenian to the FMG subsistence patterns caused a crisis, this value should increase temporarily before the FMG pattern was established.

	high diversity of retouched artefacts	low diversity of retouched artefacts
low diversity of fauna	short-term base camp/long-term hunting camp	short episode
high diversity of fauna	base or agglomeration camp	provisioned special task/opportunistic episode

Tab. 59 Idealised distinction of site types based on the faunal diversity (Simpson index) and the diversity of formally retouched artefacts (Simpson index).

Evaluation of the settlement behaviour

The general organisation (setting and hierarchy) of quasi-contemporary settlements within a defined area was described as settlement system (Binford 1980). Classically, Lewis Binford differentiated the forager and the collector settlement systems mainly based on the differences in their mobility (low residential mobility contrasted by high residential mobility) and the diversity of settlements with varied functions (few residential and many special task camps contrasted by many residential, few special task camps). Deriving from ethnography where the agents can still be observed and/or asked, the concept refers to a single social group and the area this group occupies. These groups and ranges can also be assumed in the Lateglacial but are considerably more problematic to be established from the archaeological record. In fact, such social structures in the past are aspired results and should not be prerequisite assumptions (cf. Thomas theorem, Merton 1949). However, these ethnographic examples helped to create models for the explanation of subsistence economies of the past and configuration of meso- and macro-scale predictions, for instance of the location of archaeological sites. However, both scales can also be based on a »bottom-up« approach of considering the already known archaeological sites without connecting them to a single group but a general habit (Bratlund 1996a; Gaudzinski/Street 2003; Banks et al. 2006; Banks et al. 2008). Furthermore, the geographical construct of a land-use system (Levy 1983; Amick 1996; Gaudzinski-Windheuser et al. 2011a) helps circuiting the territorial and social problem. This system refers to a limited study area in which quasi-contemporary settlements are set in relation to one another either based on their assumed function in human life cycles or their general distribution. In the latter, the sites can be distributed (cf. »compactness of settlement«, Murdock/Wilson 1972) either scattered, clustered, or concentrated allowing assumptions on environmental diversity, resource availability (Butzer 1982; Whallon 2006), and/or demographic stages (Bocquet-Appel et al. 2005). However, the study areas in these approaches need to be carefully chosen because they can reflect the exploitation area of more than one prehistoric group with perhaps complementary economies. Hence, for the analysis of a single group in a limited area the settlement system in the sense of referring to a single social unit and the area exploited by this group is still the best theoretical concept for a hierarchical ordering of human behavioural patterns.

Based on the idealised models from the previous two sub-chapters about the faunal and lithic composition of the assemblages (**tabs 53. 58**), the sites in the studied areas can be classified and their function evaluated. Furthermore, contrasting the two diversity indicators makes an additional confirmation of the use of the sites possible (**tab. 59**). However, some assumptions on what was to be found ideally on a site of a specific type must be formulated for a consistent classification and differentiation of the sites (**tab. 60**) but the validity of these characteristics must later be tested again using the archaeological record. By the diachronic composition of the assemblages, the development of the settlement systems within the sub-areas can be analysed by the possibly changing preference of site functions. These systems have to be balanced between several groups, for instance, to allow for communal hunting or the establishment of agglomeration camps of various types. This connection to foreign groups can also be reflected by exotic raw materials in the archaeological record.

type	used by	composition of archaeological evidence
agglomeration camp	several groups (contemporarily)	- larger structures - large assemblage with numerous and diverse material - high raw material diversity, often from distant sources - high number of cores - high number and diversity of retouched artefacts - numerous faunal remains with a high species diversity - single to several seasons - numerous exotic items and special goods - very diverse functions
base camp	one group	- larger structures - large assemblage with numerous and diverse material - diverse raw materials from local and distant sources (but during longer occupation, tendency towards one dominating, local raw material) - high number of cores - high number and diversity of retouched artefacts - numerous faunal remains with a high species diversity - several seasons possible - some exotic items and special goods - diverse functions
short-term residential camp	one group	- small structures - medium to small assemblage with diverse material - not very diverse raw material, mixture of local and distant sources - small number of cores - small number of diverse retouched artefacts - some faunal remains from some species - a single season - few if any exotic items and special goods - diverse functions
special task camp: hunting	team, one group, several teams or groups (contemporarily or at different times)	- small structures if any - small assemblage with a limited diversity - exclusive to very diverse raw materials from local to distant sources depending on the number of events and the origin of the participants - very small number of cores - medium to high number of specialised retouched artefacts - numerous faunal remains with very low diversity - usually a single season - no exotic items or special goods - specialised function
special task camp: workshop	team, one group, several teams or groups (contemporaneously or at different times)	- no structures - large assemblage with very limited diversity - almost exclusively the local raw material - very high number of cores - very small number of retouched artefacts - few if any faunal remains - various seasons - no exotic items or special goods - specialised function

Tab. 60 Idealised distinction of site types based on the archaeological evidence.

Consequently, the balance of these settlement systems forms the base for an inter-group communication and, thus, changes in the settlement system can be assumed to reflect major social transformations. Therefore, many of the values recorded in the present project contribute to the differentiation of how sites were used (**tab. 60**).

Mapping of archaeologically relevant variables

Finally, the maps of Lateglacial north-western Europe (see p. 254-259) were imported into a GIS-program (previously ESRI ArcView® 9.3; since June 2012: Quantum GIS, version 1.8.0 »Lisboa«) by geo-referencing the four corner points of the maps. The archaeological variables were also introduced into the GIS-program using the database of Lateglacial archaeological sites (see p. 52 f.). As with the environmental samples, the use of these maps is very conservative due to the relatively small numbers of reliable material and the large areas without data. However, since some variables have different frequencies of material found at a single site, this difference is displayed by the use of proportionally altered sizes of the symbols.

Systematisation of change and social systems in archaeology

The major focus of this project is the process of change within social systems. To achieve this aim, tempo and mode of change as well as social systems must be identified based on patterns identified in the archaeological material. A theoretical framework is required to create data that facilitates this identification. This framework should be based on quantitatively and/or qualitatively assessable observations in various parts and levels of the archaeological record.

Identifying tempo and mode of change based on archaeological material has a long tradition in archaeology (cf. O'Brien/Lyman 2002; Gamble et al. 2004; Valentin 2008a). In comparative analyses of archaeological assemblages, or parts of these assemblages, such as the retouched artefact inventory, similarities and differences in the analysed material were established. Quantification of the observed variations in a geographically and chronologically referenced dataset allowed the formation of quasi-contemporary groups and diachronic periods in the archaeological record. Thus, change between groups and between periods was identified. Moreover, analysing the chronological appearance and disappearance of these variations within a specific area and/or a specific group permitted estimations of the tempo of change in this area or within this group providing a basis to describe the mode of change between different periods (gradual, step-wise, or a mixture).

Besides identifying variation and establishing the tempo and mode of these changes, it is necessary to evaluate the observed patterns to recognise the agents. This valuation is relevant to establish social units and to understand the effectiveness of a change in past societies. For example, some observations of variation can relate to individual preferences and abilities reflected in the archaeological material such as alteration in retouch directions or the collection of peculiarities, whereas other variations relate to norms conformed to by several individuals such as those visible in the spatial organisation of a site or the use of a specific LMP type across wider areas; other changes reflect agreements of several groups as proposed for long distance transport of raw materials. The quantification of who performed a change makes a distinction possible between individual variation and changes of social norms. Thus, the relevance of this valuation becomes particularly apparent when comparing different definitions of archaeological units, such as those for the Late Magdalenian or the typical Azilian (see p. 55-74), which were often considered as normative reflections of past social units (Leroi-Gourhan 1993, 233; Barton 1997; Floss/Terberger 2002, 137 f.; Bradtmöller et al. 2012; Langlais et al. 2012).

Since endemic perspectives of prehistoric societies are not available, common patterns and characteristics in the archaeological record have to be taken as proxies for relatedness and communication between the creators of the record. Based on the assumed relatedness and communication, social units were established. However, a problem with using these proxies is the appearance of the convergence phenomena. Similarities

	equivalent	examples from Europe and the Near East
ATU 1	period	Palaeolithic, Mesolithic, Neolithic, LUP, Epipalaeolithic
	sub-period	Early Mesolithic, Late Epipalaeolithic, Pre-Pottery Neolithic
ATU 2	technocomplex and culture	Aurignacian, Arched backed Piece Complex, Magdalenian, Badegoulian, Natufian
	culture and industry	Upper Magdalenian, Late Natufian, PPNB, Final Creswellian
	industry and assemblage	Perigordian Vc, Magdalenian IV, Lower Epigravettian with shouldered points
ATU 3	artefacts and type fossils	*Zinken*, navettes, Mouilah points, lignite figurine, fox canine pendant
	attribute	scalar retouch, cut marks on animal bones, truncation, post-hole depth

Tab. 61 A provisional hierarchy of archaeological taxonomic units (ATU) for the study of population history (Gamble et al. 2005, 195 tab. 2; examples partially modified by present author).

probably occur in a comparison of universal principles which are limited in their variation. For instance, physical characteristics of lithic material limit the possibilities of successfully detaching elongated flakes (blades) from a core. Consequently, the appearance of similar conical blade cores such as in French Magdalenian and US-American Clovis contexts or similarities in essential hunting equipments such as bifacial points, known for instance from French Solutrean and US-American Clovis assemblages (cf. Bradley/Stanford 2004; Straus/Meltzer/Goebel 2005), are examples of convergence rather than the result of a transmission chain. Thus, to exclude convergence as reason for similarities in the archaeological record, a geographic and/or temporal relation between the analysed assemblages must, as a minimum, be established to argue for social transmission. Moreover, complex behavioural recipes such as the blank production or subsistence strategies were preferably taken as determinants for constructing social units. This preference was due to the assumption that complex behavioural recipes were formed by specific combinations of limited variables and these combinations could in detail only be acquired by social transmission (cf. Mesoudi/O'Brien 2008b).

If these complex behavioural recipes persisted over longer periods of time they could be assumed as traditional behaviour. Traditional behaviour is not necessarily an optimally adaptive behaviour but maladaptive behaviour becomes extinct or modified rather quickly. Karl W. Butzer already described human societies as an adaptive system which is formed by a web of parameters and which has to adjust within this interwoven web to internal and external changes (Butzer 1982, 286). He also suggested to borrow this adaptive system from biology to make an effective examination of the temporal dynamics of behavioural change possible in the form of scale analysis. In ecology, complex adaptive systems which were ordered hierarchically have been used since the 1950s (cf. Odum 1953). However, the connection with these systems, Butzer's adaptive systems, and the archaeological record still remains to be established. A hierarchical system was previously introduced in archaeology by David L. Clarke who provided a useful classification system of the archaeological record to form a theoretical basis for the analysis of this record (Clarke 1968).

Subsequently, an open and widely unbiased hierarchical system was borrowed in the form of biological taxonomy and combined with a classification similar to Clarke's system of the archaeological record to create archaeological taxonomic units (ATUs; Foley/Lahr 1997; Gamble et al. 2005). Clive Gamble and his colleagues structured this taxonomy of archaeological units for the study of population history (**tab. 61**) into a mega-scale (ATU 1), a meso-scale (ATU 2), and a micro-scale (ATU 3; Gamble et al. 2005, 195). They declared this hierarchy as a provisional one that can (and presumably has to) be further refined. This ongoing improvment process is in accordance with the concept of operational taxonomic units (Sokal/Rohlf 1962) that are clearly characterised as a temporary arrangement not a fixed set of attributes. Although the hierarchical concept was similar to the classification introduced by Clarke, it did not contain his attribute and artefact system which he had added as a sub-system to the material culture system (Clarke 1968, 134-145). In the present project, this sub-system is introduced as molecular scale (ATU 4) and relates to single

unit	level	equivalent	examples from NW-Europe	determinants
ATU 1 tradition / community	Level 1	epoch	Palaeolithic/Mesolithic	
	Level 2	period	Upper Palaeolithic/Early Mesolithic/spatial behaviour	
	Level 3	subperiod	Late Upper Palaeolithic/Final Palaeolithic/settlement behaviour	
ATU 2 alliance / population	Level 1	technocomplex	Magdalenian/Azilian/Tanged Point Complex/occupation behaviour	
	Level 2	industry	Late Magdalenian/*Federmesser-Gruppen*/settlement network	
	Level 3	faciès/group	*faciès Cepoy-Marsangy*/Nebra group/settlement system	
ATU 3 household / family group	Level 1	assemblage/settlement	animal processing/*chaîne opératoire*/residence camp/Gönnersdorf I-IV/Irlich	
	Level 2	artefact group/concentration	mammals/retouched artefacts/workshop	
	Level 3	artefact/structure	reindeer/curve-backed point/burin/hearth/accumulation	
ATU 4 individual	Level 1	type	cutmarked metatarsus/bipointe/burin on truncation/hearth pit	
	Level 2	attribute	filleting cutmark on animal/butt *en éperon*/oblique truncation/heat-coloured sediment	
	Level 3	trace	isotope analysis/micro-wear analysis/micromorphology	

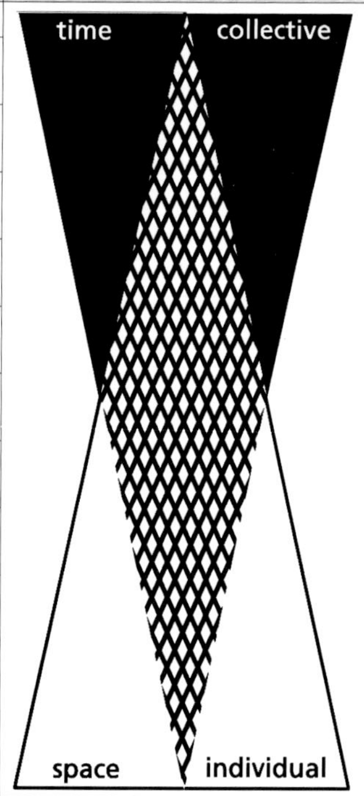

Tab. 62 A refined provisional hierarchy of archaeological taxonomic units (ATU) for the study of human behavioural patterns (modified after Gamble et al. 2005, 195 tab. 2).

artefacts and their attributes which can be analysed and compared in the archaeological record (**tab. 62**). These attributes are assumed to reflect a range of normative behaviours, which depend, at least partially, on the characteristics and composition of the piece (Clarke 1968, 140 f.).

Gamble and colleagues suggested that from the micro- to the mega-scale this provisional hierarchy was influenced increasingly by time, whereas the influence of space decreased (Gamble et al. 2005, 195). These two determinants are helpful when organising further data in this hierarchy: ATU 4 and 3 can be considered in a rather material and spatially oriented sense, whereas ATU 2 and 1 have a stronger systematic and temporal focus. Consequently, the approximate limits between the units in the present study are patterns based on single artefacts (ATU 4-ATU 3), on single sites (ATU 3-ATU 2), and on a relatively short period of time and/or a restricted geographical area (ATU 2-ATU 1). Thus, similar to single genes or individuals in biologic taxonomy, an application of this ATU system allows for a comparison of similarities and differences in archaeological assemblages based on single aspects of artefacts and single artefacts (cf. Riede 2011; Burdukiewicz/Schmider 2000). In addition, this system permits the ranking of research dependent on whether it was focused on single artefacts, assemblages, or groups of assemblages and the results can be arranged comparable to genera, families, clades, or trees (cf. Lewontin 1970). The flexible hierarchical order reflects the varying focus of research between reductionistic and increasingly holistic perspectives. Moreover, considering the previously highlighted valuation of similarities and differences, the number of people that conform to a norm also increases in this hierarchy. Hence, observations in ATU 4 usually reflects the creation of a single person; in ATU 3 a household or family group formed the observed structures and patterns; the similarities studied in ATU 2 can be assumed as collective agreements of a larger social unit or alliance, and in ATU 1 traditional behaviour of one or several groups can be proposed.

Various levels in the original provisional hierarchy were intentionally blurred by Clive Gamble and his colleagues »in order to emphasise that ATU 2s primarily assist the historical investigation of the spatial organization of human populations« (Gamble et al. 2005, 195). In this study, the units are further refined to facilitate the incorporation of further archaeological material analyses in the still provisional hierarchy of human behavioural patterns (**tab. 62**). Besides single archaeological objects, structural elements of settlement sites are introduced to allow considerations about mobility, settlement, and spatial behaviour within this system as a supplement to technological and subsistence focused analyses. Behavioural expressions as read from the archaeological material can through this approach be recorded on the molecular (ATU 4) or micro-scale (ATU 3) and become identifiable as patterns by the clustering on the meso-scale (ATU 2). In this way, a reliable fundament builds the meso-scale and based on this solid basis further classifications on a mega-scale (ATU 1) become possible, understandable, and/or interpretable. Archaeological records from various areas and periods can, in this way, be assembled in a common analytical set of behavioural patterns. The use of this system facilitates the distinction to a specific analytical level and, thus, assists the implementation levels of analysis as recommended in various evolutionary sciences (Lewontin 1970; Dean 1978; Thomson 1992). Furthermore, the concept of ATUs is deprived of most interpretative preconditions and incorrect preconditions made at an earlier stage of analysis can be modified more easily. Comparably, exact distinctions between the various levels and the units remain to be discussed and defined more precisely, if necessary. In combination, variations of single artefacts can be documented in ATU 4 as single phenomena and reflect individual reactions to an alteration of an internal or external force. In contrast, variations in ATU1 are diachronic developments and represent collective agreements, which can be applied in a variety of spaces. By using this system, contributions to various levels of this ATU system are possible and these contributions can supplement each other to enhance, systematically, the understanding of the development of human behaviour and its underlying mechanisms.

A connection of this hierarchy of archaeological material with observations of tempo and mode of changes in this material makes it possible to identify and distinguish large and slow changing variables from those that are small and fast in duration. The importance to differentiate between these variables was already described by ecologist Crawford S. Holling when formulating a theory of resilience and adaptive cycles of social-ecological systems (Holling 2001; Holling/Gunderson 2002; cf. Walker et al. 2006; Walker et al. 2012). Smaller variables are usually nested in larger variables and as a result stabilised by these slower changing variables, whereas innovation necessary to remain adaptive was provided by the faster changing, smaller variables. The balance of stability and variation between these variables can be disturbed by changes in external drivers. The application of this theory permits focusing on the impacts of climatic and environmental change as potential external drivers for human systems. The destabilisation of the analysed system by the external drivers can result in a process of reorganisation if the threshold of resilience of this system is passed. Consequently, considerations of system changes in this adaptive cycle model also reveal the effects of resilience on the development of a system.

The transformations within an adaptive cycle were also used as theoretical background to describe collapse and reorganisation in archaeology (Redman 2005; Bradtmöller et al. 2012). These studies were mainly focused on identifying adaptive cycles or different phases of these cycles but the process of change or the effects of resilience were usually not further considered in detail. Moreover, although different hierarchical levels were emphasised as important for the interpretation, this hierarchy was usually not applied to the archaeological record. Thus, the analogies to the archaeological record often remained at a general theoretical-interpretative level and were only occasionally filled with case studies (Redman/Kinzig 2003; Rosen/Rivera-Collazoa 2012).

In the present study, the provisional hierarchy of ATUs to study behavioural patterns is used to establish the different hierarchical levels within a social system and in combination with the tempo and mode of the observed changes help to reveal the resilience mechanisms of Lateglacial hunter-gatherer groups in

Northwest-Europe. Significant changes in human behaviour such as reactions to climate and environmental change, for instance by migration or variation of traditional behaviour, should be documented particularly on the indistinct meso-scale (ATU 2). Therefore, the present study is focused on the levels of ATU 3 and ATU 2. Characteristics of the archaeological material, observable attributes from archaeological sites, and patterns in both scales are documented and ordered into the hierarchical ATU system. Using this taxonomic system also helps in regard to questions of attribution of untypical assemblages such as the horizon III.1 of Pincevent, Bois Laiterie, and Gönnersdorf SW by hierarchically structuring the similarities and differences. Furthermore, this systematisation contributes to reveal how closely related groups are such as the Magdalenian faciès Cepoy-Marsangy (MfCM) and the Early Azilian or the Hangest-sur-Somme and Conty sites and the Early Azilian and/or the FMG.
In summary, using this hierarchisation of archaeological observations, impacts of changes found in the archaeological record can be studied on different levels of Lateglacial hunter-gatherer groups. Moreover, using such a hierarchical order for archaeological material allows for the human behavioural evolution to be integrated into the hierarchy of social-ecological systems.

RESULTS

Using the methods and material discussed previously, a well-founded discussion of changes in human social systems in the context of climatic and environmental developments was possible.

CLIMATE AND CHRONOLOGY

Limits of Lateglacial events in the Greenland oxygen isotope record

In the published oxygen isotope record from NGRIP (Andersen et al. 2004; Rasmussen et al. 2006), the data points were usually sampled every 20 years. In contrast to previous sub-divisions which were based either on the deuterium isotopes or on the GRIP instead of the NGRIP record (cf. Lowe et al. 2008; Blockley et al. 2012) and in order to establish a more detailed sub-division of the Lateglacial record, the amplitudes between two data points were used to identify the thus far undefined limits of isotope events within the Lateglacial Interstadial (**tab. 63**). Besides the difference between two data points, the limits were set based on the comparison of the actual oxygen isotope values and the difference accumulated in 100 years (see p. 245-247).

Assuming a relation of higher oxygen isotope values to a milder and more humid climate and of lower values to a cold and dry climate, a comparison of the actual oxygen isotope values makes considerations about the general climate regime possible. The lowest value (-44.97‰) occurred at 16,100 years cal. b2k[30] and the highest (-34,89‰) at 10,200 years cal. b2k. Thus, a mean oxygen isotope value lies around -40.00‰ and, therefore, this mean value is used as a distinctive level for the Lateglacial record. By this level, the oxygen isotope record of NGRIP between 18,000 and 10,000 years cal. b2k can be sub-divided in eight periods (18,000-16,240, 16,220-14,700, 14,680-13,260, 13,240-13,080, 13,060-12,780, 12,760-12,280, 12,260-11,760, 11,740-10,000 years cal. b2k).

Prior to 16,220 years cal. b2k the oxygen isotope values were very unsteady and also often below -40.00‰. However, between 18,000 and 16,240 this level was crossed 16 times. In contrast, during the 1,520 years between 16,220 years cal. b2k and the onset of the Lateglacial Interstadial after 14,700 years cal. b2k only two values lie slightly above this level (-39.98‰ at 15,080 years cal. b2k and -39.92‰ at 14,800 years cal. b2k). Thus, during the first period crossing of the level occurred eight times as often as in the period before the onset of the Lateglacial Interstadial[31]. After 14,700 years cal. b2k the values rise significantly and until 13,260 years cal. b2k the oxygen isotope values drop only three times below -40.00‰ (-40.94‰ at 14,080 years cal. b2k; -40.21‰ at 14,040 years cal. b2k; -40.05‰ at 13,640 years cal. b2k). The former two low points mark a cold phase in the early Lateglacial Interstadial (GI-1d) and the third relates to a short cold reversal in the mid-Lateglacial Interstadial (GI-1c_2). Between 13,240 and 13,080 years cal. b2k, only the value at 13,100 years cal. b2k (-39.77‰) is above -40.00‰. This short cold period can be related to the terminal Lateglacial Interstadial cold event GI-1b. This event was probably equivalent to the so-called

30 b2k refers to »before 2000 A.D.« (see p. 8 and 5).

31 If an identical number of years were chosen in the first part, relating to the period between 17,760 and 16,240 years cal. b2k, the numbers would only slightly decrease to 15 points above -40.00‰ and 7.5 times more crossings than in the following period.

Gerzensee oscillation (Lotter et al. 1992) or the Inner/Intra-Allerød Cold Period (IACP; Lehman/Keigwin 1992) which were observed in several marine and continental records from the north-western European region (e.g. Ammann et al. 2000; Litt/Schmincke/Kromer 2003; Wennrich et al. 2005, 280; Huber et al. 2010, 144). From 13,060 to 12,780 years cal. b2k, the values are generally above -40.00‰. From 12,760 to 12,280 years cal. b2k, the values are usually below this level, only at 12,540 years cal. b2k (-39.09‰) it is crossed again. Between 12,260 years cal. b2k and the onset of the Holocene, the values are still mainly below -40.00‰ but this level is crossed eight times. This observation sustains previous findings of a milder and more unsteady late part of the Lateglacial Stadial (Weber/Grimm/Baales 2011; cf. Bakke et al. 2009). From 11,740 years cal. b2k to the end of the survey at 10,000 years cal b2k no value lies below -40,00‰ anymore. Based on this general comparison of the actual values, the onset of the Holocene can be proposed around 11,740 years cal. b2k, the onset of GI-1 around 14,680 years cal. b2k, and the onset of the GS-2a around 16,220 years cal. b2k.

In the comparison of the differences which accumulated in 100 years between the oxygen isotope values, a uni-directional and substantial change of the climate regime for an intermediate duration can be assumed if the accumulated value increases. The clearest of these changes in the surveyed record occurred between14,740 years cal. b2k (-42.47‰) and 14,600 years cal. b2k (-35.81‰) where three successive accumulation values are higher than 4‰ with the intermediate value even crossing the 5‰ limit. Furthermore, the two prior values (14,780-14,680 years cal. b2k and 14,760 and 14,660 years cal. b2k) lie also above 3‰. These extreme values as well as their concentration illustrate the onset of the Lateglacial Interstadial as the most significant change in the climate regime as observed in the oxygen isotope record between 18,000 and 10,000 years cal. b2k. After the onset of the Lateglacial Interstadial, the 4‰ level is only crossed once more in the surveyed period, between 11,760 and 11,660 years cal. b2k. This change indicates the onset of the Holocene. Between 18,000 and 14,740 years cal. b2k, accumulated values of over 4‰ occurred at three positions: between 16,620 and 16,520 years cal. b2k, between 15,680 and 15.580 years cal. b2k, and between 15,380 and 15,280 years cal. b2k. In the latter two cases, the values decrease further reflecting the continued deterioration of the values during GS-2a or, perhaps, alternative limits for the onset of GS-2a. At 16,520 years cal. b2k the highest oxygen isotope value (-38.29‰) prior to the onset of the Lateglacial Interstadial occurred in the surveyed record.

Generally, the limits were set according to the mean of the steepest part of the record. Therefore, the simple difference of the oxygen isotope values between two data points is considered. This difference reflects the short-term fluctuations in the climate record. Very large differences of over 4‰ between two data points, thus, within 20 years, occurred only twice in the surveyed Pleistocene record: between 16,240 to 16,220 years cal. b2k the value decreased by 4.38‰ from -39.89‰ to -44.27‰ and between 15,280 and 15,260 years cal. b2k, the oxygen isotope value increased from -44.91‰ to -40.55‰ (+4.36‰). The former, very large difference, correlates with the assumed onset of GS-2a based on the comparison of the actual oxygen isotope values. Therefore, the onset of GS-2a is set to this limit. At the latter position, a comparison to the previous as well as the two following data points reveals a significant back and forth fluctuation of the values for approximately 100 years between 15,300 and 15,200 years cal. b2k. This flickering overlaps partially with the accumulated decrease between 15,380 and 15,280 years cal. b2k. At this depth in the NGRIP ice-core was also a bimodal ash deposited (cf. Mortensen et al. 2005). Perhaps, this combination of fluctuating oxygen isotope values coupled with increased volcanic activity suggests an unstable climate (cf. Zielinski 2000; Hall/Meiklejohn/Bumby 2011; Evan 2012; Kutterolf et al. 2013).

The simple difference between two data points surpassed 3‰ twelve times in the record (**tab. 63**). Only one of these rapid changes occurred after the onset of the Lateglacial Interstadial between 12,200 to 12,180 years cal. b2k where the oxygen isotope value decreased by 3.18‰ from -38.91‰ to -42.09‰.

NGRIP years b2k	NGRIP depth (m)	NGRIP $\delta^{18}O$ (‰)	max. counting error (years)	years to data point above	difference of $\delta^{18}O$ to data point above	c. 40 years accumulated amplitude	c. 60 years accumulated amplitude	c. 80 years accumulated amplitude	c. 100 years accumulated amplitude	»event«	volcanic ashes
10,000	1,384.34	-35.10	83	20	0.45	-0.36	0.25	-0.16	0.04		
10,020	1,385.88	-34.94	83	20	-0.16	0.29	-0.52	0.09	-0.32	2nd highest point of the record	
10,040	1,387.37	-35.11	84	20	0.17	0.01	0.46	-0.35	0.26		
10,060	1,388.82	-35.79	84	20	0.68	0.85	0.69	**1.14**	0.33		
10,080	1,390.23	-35.21	85	20	-0.58	0.10	0.27	0.11	0.56		
10,100	1,391.72	-35.32	85	20	0.11	-0.47	0.21	0.38	0.22		
10,120	1,393.18	-35.29	85	20	-0.03	0.08	-0.50	0.18	0.35		
10,140	1,394.61	-35.45	86	20	0.16	0.13	0.24	-0.34	0.34		
10,160	1,396.05	-34.99	86	20	-0.46	-0.30	-0.33	-0.22	-0.80	3rd highest point of the record	
10,180	1,397.61	-35.02	87	20	0.03	-0.43	-0.27	-0.30	-0.19		
10,200	1,399.13	-34.89	87	20	-0.13	-0.10	-0.56	-0.40	-0.43	highest point of the record	
10,220	1,400.55	-35.93	87	20	**1.04**	0.91	0.94	0.48	0.64		
10,240	1,402.07	-35.54	88	20	-0.39	0.65	0.52	0.55	0.09		
10,260	1,403.52	-36.33	88	20	0.79	0.40	**1.44**	**1.31**	**1.34**		
10,280	1,405.04	-36.39	89	20	0.06	0.85	0.46	**1.50**	**1.37**		
10,300	1,406.44	-36.01	89	20	-0.38	-0.32	0.47	0.08	**1.12**		
10,320	1,407.91	-36.01	89	20	0.00	-0.38	-0.32	0.47	0.08		
10,340	1,409.42	-35.20	89	20	-0.81	-0.81	**-1.19**	**-1.13**	-0.34		
10,360	1,410.77	-36.12	89	20	0.92	0.11	0.11	-0.27	-0.21		Saksurnavatn Ash at 1409.89 m (10,347 ± 89 b2k*)
10,380	1,412.14	-36.18	89	20	0.06	0.98	0.17	0.17	-0.21		
10,400	1,413.48	-36.36	89	20	0.18	0.24	1.16	0.35	0.35		
10,420	1,414.80	-35.96	90	20	-0.40	-0.22	-0.16	0.76	-0.05		
10,420	1,414.80	-35.96	90	20	-0.40	-0.22	-0.16	0.76	-0.05		
10,440	1,416.18	-36.12	90	20	0.16	-0.24	-0.06	0.00	0.92		
10,460	1,417.50	-36.04	90	20	-0.08	0.08	-0.32	-0.14	-0.08		
10,480	1,418.84	-35.53	90	20	-0.51	-0.59	-0.43	-0.83	-0.65		
10,500	1,420.22	-36.34	90	20	0.81	0.30	0.22	0.38	-0.02		

Tab. 63 NGRIP oxygen isotope record with calculated amplitudes and volcanic ashes. First four columns taken from the file »GICC05_NGRIP_20y_10sep2007« at www.iceandclimate.nbi.ku.dk/data/ (see p. 247-245). Accumulated amplitudes were calculated with values in the lines above in the previous column. To facilitate the reading of the table, NGRIP $\delta^{18}O$ values are set in bold if they are below -40.0‰. Very low values (< -42.49) are shaded in light grey and very high values (-36.0 >) are shaded in dark grey. In addition, if the amplitude values surpassed +/- 1.0 they were also set in bold. Furthermore, increasing values were shaded in increasingly dark grey from +/- 2.0 to 5.22. Volcanic ashes are given according to Mortensen et al. 2005. * given ages are according to Rasmussen et al. 2006, X-12 tab. 4. All other ages of the volcanic ashes are calculated based on the first two columns assuming constant deposition between the value before and after the depth given for the ash layer.

NGRIP years b2k	NGRIP depth (m)	NGRIP $\delta^{18}O$ (‰)	max. counting error (years)	years to data point above	difference of $\delta^{18}O$ to data point above	c. 40 years accumulated amplitude	c. 60 years accumulated amplitude	c. 80 years accumulated amplitude	c. 100 years accumulated amplitude	»event«	volcanic ashes
10,520	1,421.55	-35.46	90	20	-0.88	-0.07	-0.58	-0.66	-0.50		
10,540	1,422.87	-35.73	90	20	0.27	-0.61	0.20	-0.31	-0.39		
10,560	1,424.23	-35.86	91	20	0.13	0.40	-0.48	0.33	-0.18		
10,580	1,425.59	-35.99	91	20	0.13	0.26	0.53	-0.35	0.46		
10,600	1,426.93	-35.89	91	20	-0.10	0.03	0.16	0.43	-0.45		
10,620	1,428.22	-36.12	91	20	0.23	0.13	0.26	0.39	0.66		
10,640	1,429.54	-36.13	91	20	0.01	0.24	0.14	0.27	0.40		
10,660	1,430.85	-36.16	91	20	0.03	0.04	0.27	0.17	0.30		
10,680	1,432.22	-35.50	91	20	-0.66	-0.63	-0.62	-0.39	-0.49		
10,700	1,433.50	-35.77	92	20	0.27	-0.39	-0.36	-0.35	-0.12		
10,720	1,434.84	-36.05	92	20	0.28	0.55	-0.11	-0.08	-0.07		
10,740	1,436.11	-36.45	92	20	0.40	0.68	0.95	0.29	0.32		
10,760	1,437.43	-35.88	92	20	-0.57	-0.17	0.11	0.38	-0.28		
10,780	1,438.71	-36.22	92	20	0.34	-0.23	0.17	0.45	0.72		
10,800	1,439.92	-36.83	92	20	0.61	0.95	0.38	0.78	**1.06**		
10,820	1,441.16	-35.84	92	20	-0.99	-0.38	-0.04	-0.61	-0.21		
10,840	1,442.44	-36.59	93	20	0.75	-0.24	0.37	0.71	0.14		
10,860	1,443.80	-36.09	93	20	-0.50	0.25	-0.74	-0.13	0.21	peak	
10,880	1,445.12	-37.23	93	20	**1.14**	0.64	**1.39**	0.40	**1.01**		
10,900	1,446.33	-37.60	93	20	0.37	**1.51**	**1.01**	**1.76**	0.77		
10,920	1,447.53	-37.25	93	20	-0.35	0.02	**1.16**	0.66	**1.41**		
10,940	1,448.64	-37.15	93	20	-0.10	-0.45	-0.08	**1.06**	0.56		
10,960	1,449.92	-37.00	93	20	-0.15	-0.25	-0.60	-0.23	0.91		
10,980	1,451.03	-38.01	94	20	**1.01**	0.86	0.76	0.41	0.78	low point	
11,000	1,452.29	-36.53	94	20	**-1.48**	-0.47	-0.62	-0.72	**-1.07**		
11,020	1,453.54	-37.05	94	20	0.52	-0.96	0.05	-0.10	-0.20		
11,040	1,454.75	-36.44	94	20	-0.61	-0.09	**-1.57**	-0.56	-0.71		
11,060	1,455.98	-36.64	94	20	0.20	-0.41	0.11	**-1.37**	-0.36		
11,080	1,457.18	-36.89	94	20	0.25	0.45	-0.16	0.36	**-1.12**		
11,100	1,458.38	-37.11	94	20	0.22	0.47	0.67	0.06	0.58		
11,120	1,459.59	-36.71	95	20	-0.40	-0.18	0.07	0.27	-0.34		
11,140	1,460.85	-36.43	95	20	-0.28	-0.68	-0.46	-0.21	-0.01		
11,160	1,461.97	-37.13	95	20	0.70	0.42	0.02	0.24	0.49		
11,180	1,463.16	-36.62	95	20	-0.51	0.19	-0.09	-0.49	-0.27		
11,200	1,464.37	-37.25	95	20	0.63	0.12	0.82	0.54	0.14		

Tab. 63 (continued)

NGRIP years b2k	NGRIP depth (m)	NGRIP $\delta^{18}O$ (‰)	max. counting error (years)	years to data point above	difference of $\delta^{18}O$ to data point above	*c.* 40 years accumulated amplitude	*c.* 60 years accumulated amplitude	*c.* 80 years accumulated amplitude	*c.* 100 years accumulated amplitude	»event«	volcanic ashes
11,220	1,465.52	-37.39	95	20	0.14	0.77	0.26	0.96	0.68		
11,240	1,466.67	-37.60	95	20	0.21	0.35	0.98	0.47	**1.17**		
11,260	1,467.90	-36.57	95	20	**-1.03**	-0.82	-0.68	-0.05	-0.56		
11,280	1,469.13	-36.16	96	20	-0.41	**-1.44**	**-1.23**	**-1.09**	-0.46	peak	
11,300	1,470.36	-36.58	96	20	0.42	0.01	**-1.02**	-0.81	-0.67		
11,320	1,471.51	-37.54	96	20	0.96	**1.38**	0.97	-0.06	0.15		
11,340	1,472.66	-36.67	96	20	-0.87	0.09	0.51	0.10	-0.93		
11,360	1,473.88	-36.91	96	20	0.24	-0.63	0.33	0.75	0.34		
11,380	1,475.02	-37.45	96	20	0.54	0.78	-0.09	0.87	**1.29**		
11,400	1,476.16	-37.63	96	20	0.18	0.72	0.96	0.09	**1.05**	low point	
11,420	1,477.23	-37.11	97	20	-0.52	-0.34	0.20	0.44	-0.43	peak	
11,440	1,478.16	-38.46	97	20	**1.35**	0.83	**1.01**	**1.55**	**1.79**	steepest part mean: 11,430 b2k	
11,460	1,479.15	-38.26	97	20	-0.20	**1.15**	0.63	0.81	**1.35**		
11,480	1,480.13	-38.72	97	20	0.46	0.26	**1.61**	**1.09**	**1.27**	low point	
11,500	1,481.24	-37.41	97	20	**-1.31**	-0.85	**-1.05**	0.30	-0.22	steepest part mean: 11,490 b2k	
11,520	1,482.32	-37.84	97	20	0.43	-0.88	-0.42	-0.62	0.73		
11,540	1,483.41	-36.90	97	20	-0.94	-0.51	**-1.82**	**-1.36**	**-1.56**		
11,560	1,484.59	-36.60	98	20	-0.30	**-1.24**	-0.81	**-2.12**	**-1.66**		
11,580	1,485.69	-36.87	98	20	0.27	-0.03	-0.97	-0.54	**-1.85**		
11,600	1,486.95	-36.78	98	20	-0.09	0.18	-0.12	**-1.06**	-0.63		
11,620	1,488.10	-36.41	98	20	-0.37	-0.46	-0.19	-0.49	**-1.43**	first peak	
11,640	1,489.26	-36.57	98	20	0.16	-0.21	-0.30	-0.03	-0.33		
11,660	1,490.44	-36.68	98	20	0.11	0.27	-0.10	-0.19	0.08		
11,680	1,491.45	-38.49	98	20	**1.81**	**1.92**	**2.08**	**1.71**	**1.62**	steepest part mean: 11,670 b2k	rhyolitic ash at 1491.48 m (11,681 b2k)
11,700	1,492.33	-39.04	99	20	0.55	**2.36**	**2.47**	**2.63**	**2.26**	mean between low point and peak: 11,690 b2k; Steffensen et al. 2008, mean of multi-proxy values: 11,698 ± 99 b2k	
11,720	1,492.94	-39.60	99	20	0.56	**1.11**	**2.92**	**3.03**	**3.19**		
11,740	1,493.62	-39.72	100	20	0.12	0.68	**1.23**	**3.04**	**3.15**		
11,760	1,494.26	**-40.85**	100	20	**1.13**	**1.25**	**1.81**	**2.36**	**4.17**	lowest point before GH	
11,780	1,494.85	**-40.55**	101	20	-0.30	0.83	0.95	**1.51**	**2.06**		

Tab. 63 (continued)

NGRIP years b2k	NGRIP depth (m)	NGRIP $\delta^{18}O$ (‰)	max. counting error (years)	years to data point above	difference of $\delta^{18}O$ to data point above	c. 40 years accumulated amplitude	c. 60 years accumulated amplitude	c. 80 years accumulated amplitude	c. 100 years accumulated amplitude	»event«	volcanic ashes
11,800	1,495.43	-39.87	102	20	-0.68	-0.98	0.15	0.27	0.83		
11,820	1,496.06	**-40.45**	102	20	0.58	-0.10	-0.40	0.73	0.85		
11,840	1,496.68	**-40.97**	103	20	0.52	**1.10**	0.42	0.12	**1.25**		
11,860	1,497.25	-39.17	104	20	**-1.80**	**-1.28**	-0.70	**-1.38**	**-1.68**		
11,880	1,497.81	-38.93	104	20	-0.24	**-2.04**	**-1.52**	-0.94	**-1.62**		
11,900	1,498.35	**-41.52**	105	20	**2.59**	**2.35**	0.55	**1.07**	**1.65**		
11,920	1,498.93	-39.73	106	20	**-1.79**	0.80	0.56	**-1.24**	-0.72		
11,940	1,499.53	**-41.94**	106	20	**2.21**	0.42	**3.01**	**2.77**	0.97		rhyolitic ash at 1499.14 m (11,927 b2k)
11,960	1,500.15	**-40.34**	107	20	**-1.60**	0.61	**-1.18**	**1.41**	**1.17**		
11,980	1,500.74	**-41.54**	108	20	**1.20**	-0.40	**1.81**	0.02	**2.61**		
12,000	1,501.29	**-40.54**	108	20	**-1.00**	0.20	**-1.40**	0.81	-0.98		
12,020	1,501.85	-39.60	109	20	-0.94	**-1.94**	-0.74	**-2.34**	-0.13		
12,040	1,502.51	**-40.17**	110	20	0.57	-0.37	**-1.37**	-0.17	**-1.77**		
12,060	1,503.09	-39.32	110	20	-0.85	-0.28	**-1.22**	**-2.22**	**-1.02**		
12,080	1,503.65	**-40.77**	111	20	**1.45**	0.60	**1.17**	0.23	-0.77		
12,100	1,504.25	**-40.74**	112	20	-0.03	**1.42**	0.57	**1.14**	0.20		
12,120	1,504.78	**-41.89**	112	20	**1.15**	**1.12**	**2.57**	**1.72**	**2.29**		
12,140	1,505.35	**-40.61**	113	20	**-1.28**	-0.13	-0.16	**1.29**	0.44		
12,160	1,505.87	**-41.76**	114	20	**1.15**	-0.13	**1.02**	0.99	**2.44**		
12,180	1,506.37	**-42.09**	114	20	0.33	**1.48**	0.20	**1.35**	**1.32**		Vedde Ash at 1506.18 m (12,171 ± 114 b2k*)
12,200	1,506.94	-38.91	115	20	**-3.18**	**-2.85**	**-1.70**	**-2.98**	**-1.83**	inner GS-1 peak; end of inner GS-1 warming	
12,220	1,507.49	**-41.40**	115	20	**2.49**	-0.69	-0.36	0.79	-0.49		
12,240	1,508.03	**-40.90**	116	20	-0.50	**1.99**	**-1.19**	-0.86	0.29		
12,260	1,508.58	-39.90	117	20	**-1.00**	**-1.50**	0.99	**-2.19**	**-1.86**		rhyolitic ash at 1508.18 m and 1508.26 m (12,246 b2k and 12,248 b2k)
12,280	1,509.20	**-40.71**	117	20	0.81	-0.19	-0.69	**1.80**	**-1.38**		
12,300	1,509.74	**-41.01**	118	20	0.30	**1.11**	0.11	-0.39	**2.10**	onset of inner GS-1 warming	
12,320	1,510.29	**-40.41**	119	20	-0.60	-0.30	0.51	-0.49	-0.99		
12,340	1,510.78	**-40.95**	119	20	0.54	-0.06	0.24	**1.05**	0.05		
12,360	1,511.34	**-41.73**	120	20	0.78	**1.32**	0.72	**1.02**	**1.83**		rhyolitic ash at 1511.34 m (12,360 b2k)

Tab. 63 (continued)

NGRIP years b2k	NGRIP depth (m)	NGRIP $\delta^{18}O$ (‰)	max. counting error (years)	years to data point above	difference of $\delta^{18}O$ to data point above	*c.* 40 years accumulated amplitude	*c.* 60 years accumulated amplitude	*c.* 80 years accumulated amplitude	*c.* 100 years accumulated amplitude	»event«	volcanic ashes
12,380	1,511.83	**-41.17**	121	20	-0.56	0.22	0.76	0.16	0.46		
12,400	1,512.41	**-40.73**	121	20	-0.44	**-1.00**	-0.22	0.32	-0.28		
12,420	1,512.95	**-41.21**	122	20	0.48	0.04	-0.52	0.26	0.80		
12,440	1,513.56	**-41.05**	123	20	-0.16	0.32	-0.12	-0.68	0.10		
12,460	1,514.20	**-41.39**	123	20	0.34	0.18	0.66	0.22	-0.34		
12,480	1,514.75	**-40.02**	124	20	**-1.37**	**-1.03**	**-1.19**	-0.71	**-1.15**		
12,500	1,515.27	**-41.91**	125	20	**1.89**	0.52	0.86	0.70	**1.18**		
12,520	1,515.84	**-40.62**	125	20	**-1.29**	0.60	-0.77	-0.43	-0.59		
12,540	1,516.38	-39.09	126	20	**-1.53**	**-2.82**	-0.93	**-2.30**	**-1.96**		
12,560	1,516.89	**-40.95**	127	20	**1.86**	0.33	-0.96	0.93	-0.44		
12,580	1,517.46	**-40.54**	127	20	-0.41	**1.45**	-0.08	**-1.37**	0.52		
12,600	1,517.94	**-41.43**	128	20	0.89	0.48	**2.34**	0.81	-0.48		
12,620	1,518.39	**-42.71**	129	20	**1.28**	**2.17**	**1.76**	**3.62**	**2.09**	lowest point in GS-1	
12,640	1,518.90	**-40.59**	129	20	**-2.12**	-0.84	0.05	-0.36	1.50		
12,660	1,519.50	**-41.82**	130	20	**1.23**	-0.89	0.39	**1.28**	0.87		basaltic ash (Tv-1 / I-THOL-2) at 1519.1 m (12,647 b2k)
12,680	1,520.04	**-40.92**	131	20	-0.90	0.33	**-1.79**	-0.51	0.38		
12,700	1,520.59	**-42.27**	131	20	**1.35**	0.45	**1.68**	-0.44	0.84	lowest point after GI-1	
12,720	1,521.21	**-41.82**	132	20	-0.45	0.90	0.00	**1.23**	-0.89		
12,740	1,521.82	**-41.95**	132	20	0.13	-0.32	**1.03**	0.13	**1.36**	low point a	
12,760	1,522.36	**-41.06**	133	20	-0.89	-0.76	**-1.21**	0.14	-0.76		
12,780	1,522.91	-39.86	134	20	**-1.20**	**-2.09**	**-1.96**	**-2.41**	**-1.06**	peak a; steepest part mean: 12,770 b2k	intermediate ash at 1522.27 m (12,757 b2k)
12,800	1,523.44	**-40.17**	134	20	0.31	-0.89	**-1.78**	**-1.65**	**-2.10**		
12,820	1,524.02	**-40.36**	135	20	0.19	0.50	-0.70	**-1.59**	**-1.46**	low point b; mean between last peak and lowest point after GI-1: 12,820 b2k	
12,840	1,524.70	-39.88	136	20	-0.48	-0.29	0.02	**-1.18**	**-2.07**		
12,860	1,525.30	-38.98	136	20	-0.90	**-1.38**	**-1.19**	-0.88	**-2.08**	peak b	
12,880	1,525.94	-39.81	137	20	0.83	-0.07	-0.55	-0.36	-0.05	Steffensen et al. 2008, mean of multi-proxy values: 12,884 ± 139 b2k	rhyolitic ash at 1525.88 m (12,878 b2k)
12,900	1,526.67	**-40.11**	138	20	0.30	**1.13**	0.23	-0.25	-0.06	low point c	
12,920	1,527.42	-39.42	138	20	-0.69	-0.39	0.44	-0.46	-0.94		
12,940	1,528.24	-37.79	139	20	**-1.63**	**-2.32**	**-2.02**	**-1.19**	**-2.09**	last peak	

Tab. 63 (continued)

NGRIP years b2k	NGRIP depth (m)	NGRIP δ18O (‰)	max. counting error (years)	years to data point above	difference of δ18O to data point above	c. 40 years accumulated amplitude	c. 60 years accumulated amplitude	c. 80 years accumulated amplitude	c. 100 years accumulated amplitude	»event«	volcanic ashes
12,960	1,528.98	-38.25	139	20	0.46	**-1.17**	**-1.86**	**-1.56**	-0.73		
12,980	1,529.76	-39.11	140	20	0.86	**1.32**	-0.31	**-1.00**	-0.70		
13,000	1,530.73	-38.96	140	20	-0.15	0.71	**1.17**	-0.46	**-1.15**		
13,020	1,531.50	-39.87	141	20	0.91	0.76	**1.62**	**2.08**	0.45		
13,040	1,532.26	-39.55	141	20	-0.32	0.59	0.44	**1.30**	**1.76**		intermediate ash at 1531.93 (13,031 b2k)
13,060	1,533.12	-39.49	142	20	-0.06	-0.38	0.53	0.38	**1.24**	peak	
13,080	1,533.86	**-40.52**	142	20	**1.03**	0.97	0.65	**1.56**	**1.41**	low point; steepest part mean: 13,070 b2k	
13,100	1,534.55	-39.77	143	20	-0.75	0.28	0.22	-0.10	0.81	peak	
13,120	1,535.21	**-40.57**	143	20	0.80	0.05	**1.08**	**1.02**	0.70		
13,140	1,535.89	**-40.17**	144	20	-0.40	0.40	-0.35	0.68	0.62		
13,160	1,536.56	**-41.51**	145	20	**1.34**	0.94	**1.74**	0.99	**2.02**		
13,180	1,537.23	**-40.08**	145	20	**-1.43**	-0.09	-0.49	0.31	-0.44		
13,200	1,537.94	**-40.53**	146	20	0.45	-0.98	0.36	-0.04	0.76		
13,220	1,538.64	**-40.10**	146	20	-0.43	0.02	**-1.41**	-0.07	-0.47		
13,240	1,539.37	**-40.58**	147	20	0.48	0.05	0.50	-0.93	0.41	low point	
13,260	1,540.00	-39.45	147	20	**-1.13**	-0.65	**-1.08**	-0.63	**-2.06**	steepest part mean: 13,250 b2k	
13,280	1,540.80	-39.30	148	20	-0.15	**-1.28**	-0.80	**-1.23**	-0.78	mean between peak and low point: 13,270 b2k	rhyolitic ash at 1540.3 m (13,277 b2k)
13,300	1,541.65	-38.97	148	20	-0.33	-0.48	**-1.61**	**-1.13**	**-1.56**	peak	
13,320	1,542.48	-39.03	149	20	0.06	-0.27	-0.42	**-1.55**	**-1.07**		
13,340	1,543.37	-39.35	149	20	0.32	0.38	0.05	-0.10	**-1.23**		
13,360	1,544.20	-39.07	150	20	-0.28	0.04	0.10	-0.23	-0.38		
13,380	1,545.02	-38.56	150	20	-0.51	-0.79	-0.47	-0.41	-0.74		
13,400	1,545.94	-38.36	151	20	-0.20	-0.71	-0.99	-0.67	-0.61		
13,420	1,546.78	-38.76	151	20	0.40	0.20	-0.31	-0.59	-0.27		
13,440	1,547.75	-38.14	152	20	-0.62	-0.22	-0.42	-0.93	**-1.21**		
13,460	1,548.55	-38.56	152	20	0.42	-0.20	0.20	0.00	-0.51		
13,480	1,549.46	-38.79	153	20	0.23	0.65	0.03	0.43	0.23		
13,500	1,550.33	-38.33	153	20	-0.46	-0.23	0.19	-0.43	-0.03		
13,520	1,551.27	-37.74	154	20	-0.59	**-1.05**	-0.82	-0.40	**-1.02**	peak	
13,540	1,552.17	-38.08	155	20	0.34	-0.25	-0.71	-0.48	-0.06		
13,560	1,553.16	-38.29	155	20	0.21	0.55	-0.04	-0.50	-0.27		

Tab. 63 (continued)

NGRIP years b2k	NGRIP depth (m)	NGRIP $\delta^{18}O$ (‰)	max. counting error (years)	years to data point above	difference of $\delta^{18}O$ to data point above	c. 40 years accumulated amplitude	c. 60 years accumulated amplitude	c. 80 years accumulated amplitude	c. 100 years accumulated amplitude	»event«	volcanic ashes
13,580	1,553.95	-38.99	156	20	0.70	0.91	**1.25**	0.66	0.20	mean between low point and peak+steepest part: 13,580 b2k	rhyolitic ash at 1553.85 m (13,578 b2k)
13,600	1,554.75	-39.36	156	20	0.37	**1.07**	**1.28**	**1.62**	**1.03**		
13,620	1,555.50	-39.64	157	20	0.28	0.65	**1.35**	**1.56**	**1.90**		
13,640	1,556.26	**-40.05**	157	20	0.41	0.69	**1.06**	**1.76**	**1.97**	low point intermediate GI-1	
13,660	1,557.08	-39.21	158	20	-0.84	-0.43	-0.15	0.22	0.92		
13,680	1,557.89	-39.09	158	20	-0.12	-0.96	-0.55	-0.27	0.10	mean between peak2 and low point: 13,670 b2k	
13,700	1,558.84	-38.67	159	20	-0.42	-0.54	**-1.38**	-0.97	-0.69	peak2	
13,720	1,559.72	-38.93	159	20	0.26	-0.16	-0.28	**-1.12**	-0.71	mean between peak1 and low point: 13,710 b2k	
13,740	1,560.62	-37.66	160	20	**-1.27**	**-1.01**	**-1.43**	**-1.55**	**-2.39**	peak1, steepest part mean: 13,730 b2k	
13,760	1,561.51	-38.45	160	20	0.79	-0.48	-0.22	-0.64	-0.76		
13,780	1,562.43	-39.42	161	20	0.97	**1.76**	0.49	0.75	0.33		
13,800	1,563.36	-38.81	161	20	-0.61	0.36	**1.15**	-0.12	0.14		
13,820	1,564.34	-37.86	162	20	-0.95	**-1.56**	-0.59	0.20	**-1.07**		
13,840	1,565.20	-39.05	162	20	**1.19**	0.24	-0.37	0.60	**1.39**		
13,860	1,566.16	-37.96	163	20	**-1.09**	0.10	-0.85	**-1.46**	-0.49		
13,880	1,567.16	-38.00	163	20	0.04	**-1.05**	0.14	-0.81	**-1.42**		
13,900	1,568.06	-38.43	164	20	0.43	0.47	-0.62	0.57	-0.38		
13,920	1,568.98	-38.21	165	20	-0.22	0.21	0.25	-0.84	0.35		
13,940	1,569.95	-38.61	165	20	0.40	0.18	0.61	0.65	-0.44		
13,960	1,570.75	-38.03	166	20	-0.58	-0.18	-0.40	0.03	0.07	peak	
13,980	1,571.51	-39.99	166	20	**1.96**	**1.38**	**1.78**	**1.56**	**1.99**	steepest part mean: 13,970 b2k	
14,000	1,572.26	-39.03	167	20	-0.96	1.00	0.42	0.82	0.60	mean between low point and peak: 14,000 b2k	
14,020	1,572.97	-39.73	167	20	0.70	-0.26	**1.70**	**1.12**	**1.52**		
14,040	1,573.59	**-40.21**	168	20	0.48	**1.18**	0.22	**2.18**	**1.60**	low point	basaltic ash at 1573.0 m (14,021 b2k)
14,060	1,574.28	-39.90	168	20	-0.31	0.17	0.87	-0.09	**1.87**		rhyolitic ash at 1573.95 m (14,050 b2k)

Tab. 63 (continued)

NGRIP years b2k	NGRIP depth (m)	NGRIP $\delta^{18}O$ (‰)	max. counting error (years)	years to data point above	difference of $\delta^{18}O$ to data point above	c. 40 years accumulated amplitude	c. 60 years accumulated amplitude	c. 80 years accumulated amplitude	c. 100 years accumulated amplitude	»event«	volcanic ashes
14,080	1,575.03	**-40.94**	169	20	**1.04**	0.73	**1.21**	**1.91**	0.95	low point early GI-1	
14,100	1,575.90	-38.05	169	20	**-2.89**	**-1.85**	**-2.16**	**-1.68**	-0.98	steepest part mean: 14,090 b2k	
14,120	1,576.77	-38.34	170	20	0.29	**-2.60**	**-1.56**	**-1.87**	**-1.39**		
14,140	1,577.59	-38.01	170	20	-0.33	-0.04	**-2.93**	**-1.89**	**-2.20**	mean between peak and low point: 14,150 b2k	
14,160	1,578.54	-38.68	171	20	0.67	0.34	0.63	**-2.26**	**-1.22**		rhyolitic ash at 1577.61 m (14,140 b2k)
14,180	1,579.30	-38.04	171	20	-0.64	0.03	-0.30	-0.01	**-2.90**		intermediate ash at 1579.15 m (14,176 b2k)
14,200	1,580.19	-37.59	172	20	-0.45	**-1.09**	-0.42	-0.75	-0.46		
14,220	1,581.18	-37.46	173	20	-0.13	-0.58	**-1.22**	-0.55	-0.88	peak	
14,240	1,582.16	-38.42	173	20	0.96	0.83	0.38	-0.26	0.41		
14,260	1,583.16	-37.76	174	20	-0.66	0.30	0.17	-0.28	-0.92		
14,280	1,584.06	-37.19	174	20	-0.57	**-1.23**	-0.27	-0.40	-0.85		
14,300	1,585.16	-37.13	175	20	-0.06	-0.63	**-1.29**	-0.33	-0.46		
14,320	1,586.22	-37.05	175	20	-0.08	-0.14	-0.71	**-1.37**	-0.41		
14,340	1,587.24	-37.54	176	20	0.49	0.41	0.35	-0.22	-0.88		
14,360	1,588.24	-36.81	176	20	-0.73	-0.24	-0.32	-0.38	-0.95		
14,380	1,589.20	-37.49	177	20	0.68	-0.05	0.44	0.36	0.30		
14,400	1,590.16	-36.07	178	20	**-1.42**	-0.74	**-1.47**	-0.98	**-1.06**		
14,420	1,591.12	-37.17	178	20	**1.10**	-0.32	0.36	-0.37	0.12		
14,440	1,592.08	-37.29	179	20	0.12	**1.22**	-0.20	0.48	-0.25		
14,460	1,593.14	-37.04	179	20	-0.25	-0.13	0.97	-0.45	0.23		
14,480	1,594.21	-36.86	180	20	-0.18	-0.43	-0.31	0.79	-0.63		
14,500	1,595.05	-36.26	180	20	-0.60	-0.78	**-1.03**	-0.91	0.19		
14,520	1,596.12	-36.11	181	20	-0.15	-0.75	-0.93	**-1.18**	**-1.06**		basaltic ash at 1595.1 m (14,501 b2k)
14,540	1,597.21	-36.14	181	20	0.03	-0.12	-0.72	-0.90	**-1.15**		
14,560	1,598.32	-36.14	182	20	0.00	0.03	-0.12	-0.72	-0.90		
14,580	1,599.47	-36.98	182	20	0.84	0.84	0.87	0.72	0.12		
14,600	1,600.52	-35.81	183	20	**-1.17**	-0.33	-0.33	-0.30	-0.45	peak; highest value of the record prior to the Holocene	
14,620	1,601.40	-36.74	184	20	0.93	-0.24	0.60	0.60	0.63		
14,640	1,602.32	-37.48	184	20	0.74	**1.67**	0.50	**1.34**	**1.34**		

Tab. 63 (continued)

NGRIP years b2k	NGRIP depth (m)	NGRIP δ18O (‰)	max. counting error (years)	years to data point above	difference of δ18O to data point above	c. 40 years accumulated amplitude	c. 60 years accumulated amplitude	c. 80 years accumulated amplitude	c. 100 years accumulated amplitude	»event«	volcanic ashes
14,660	1,603.31	-37.52	185	20	0.04	0.78	**1.71**	0.54	**1.38**		
14,680	1,604.17	-37.97	185	20	0.45	0.49	**1.23**	**2.16**	0.99	mean between low point and peak: 14,670 b2k; Steffensen et al. 2008, mean of multi-proxy values: 14,688 ± 186 b2k	
14,700	1,604.89	**-40.64**	186	20	**2.67**	**3.12**	**3.16**	**3.90**	**4.83**	steepest part mean: 14,690 b2k	
14,720	1,605.38	**-41.96**	187	20	**1.32**	**3.99**	**4.44**	**4.48**	**5.22**		
14,740	1,605.96	**-42.47**	189	20	0.51	**1.83**	**4.50**	**4.95**	**4.99**	low point before GI-1	
14,760	1,606.49	**-40.81**	189	20	**-1.66**	**-1.15**	0.17	**2.84**	**3.29**		
14,780	1,607.08	**-41.03**	191	20	0.22	**-1.44**	-0.93	0.39	**3.06**		
14,800	1,607.68	-39.92	191	20	**-1.11**	-0.89	**-2.55**	**-2.04**	-0.72	peak	
14,820	1,608.28	**-40.68**	191	20	0.76	-0.35	-0.13	**-1.79**	**-1.28**		
14,840	1,608.80	**-42.44**	192	20	**1.76**	**2.52**	**1.41**	**1.63**	-0.03	low point	
14,860	1,609.36	**-40.78**	193	20	**-1.66**	0.10	0.86	-0.25	-0.03		
14,880	1,609.88	**-41.18**	194	20	0.40	**-1.26**	0.50	1.26	0.15		
14,900	1,610.39	**-40.14**	194	20	**-1.04**	-0.64	**-2.30**	-0.54	0.22	peak	
14,920	1,610.90	**-41.47**	195	20	**1.33**	0.29	0.69	-0.97	0.79		
14,940	1,611.37	**-41.76**	195	20	0.29	**1.62**	0.58	0.98	-0.68		
14,960	1,611.83	**-41.65**	196	20	-0.11	0.18	**1.51**	0.47	0.87		
14,980	1,612.30	**-40.88**	196	20	-0.77	-0.88	-0.59	0.74	-0.30		
15,000	1,612.83	**-42.61**	197	20	**1.73**	0.96	0.85	**1.14**	**2.47**		
15,020	1,613.32	**-42.37**	198	20	-0.24	**1.49**	0.72	0.61	0.90		
15,040	1,613.72	**-41.54**	198	20	-0.83	**-1.07**	0.66	-0.11	-0.22		
15,060	1,614.18	**-43.17**	198	20	**1.63**	0.80	0.56	**2.29**	**1.52**	low point	
15,080	1,614.66	-39.98	199	20	**-3.19**	**-1.56**	**-2.39**	**-2.63**	-0.90	peak	
15,100	1,615.13	**-43.51**	200	20	**3.53**	0.34	**1.97**	**1.14**	0.90	low point	
15,120	1,615.55	**-42.76**	202	20	-0.75	**2.78**	-0.41	**1.22**	0.39		
15,140	1,615.98	**-40.12**	204	20	**-2.64**	**-3.39**	0.14	**-3.05**	**-1.42**		
15,160	1,616.39	**-42.26**	204	20	**2.14**	-0.50	**-1.25**	**2.28**	-0.91		
15,180	1,616.84	**-42.99**	205	20	0.73	**2.87**	0.23	-0.52	**3.01**		
15,200	1,617.32	**-42.08**	205	20	-0.91	-0.18	**1.96**	-0.68	**-1.43**		
15,220	1,617.73	**-40.57**	206	20	**-1.51**	**-2.42**	**-1.69**	0.45	**-2.19**	peak	
15,240	1,618.18	**-43.57**	207	20	**3.00**	**1.49**	0.58	**1.31**	**3.45**	low point	

Tab. 63 (continued)

NGRIP years b2k	NGRIP depth (m)	NGRIP $\delta^{18}O$ (‰)	max. counting error (years)	years to data point above	difference of $\delta^{18}O$ to data point above	c. 40 years accumulated amplitude	c. 60 years accumulated amplitude	c. 80 years accumulated amplitude	c. 100 years accumulated amplitude	»event«	volcanic ashes
15,260	1,618.67	**-40.55**	208	20	**-3.02**	-0.02	**-1.53**	**-2.44**	**-1.71**	peak	
15,280	1,619.12	**-44.91**	209	20	**4.36**	**1.34**	**4.34**	**2.83**	**1.92**	2nd lowest point of the record	
15,300	1,619.63	**-41.99**	209	20	**-2.92**	**1.44**	**-1.58**	**1.42**	-0.09		bimodal ash at 1619.58 m (15,298 b2k)
15,320	1,620.11	**-41.29**	211	20	-0.70	**-3.62**	0.74	**-2.28**	0.72		
15,340	1,620.56	**-42.08**	211	20	0.79	0.09	**-2.83**	**1.53**	**-1.49**		
15,360	1,621.02	**-40.75**	213	20	**-1.33**	-0.54	**-1.24**	**-4.16**	0.20		
15,380	1,621.45	**-40.66**	214	20	-0.09	**-1.42**	-0.63	**-1.33**	**-4.25**		
15,400	1,621.92	**-40.04**	214	20	-0.62	-0.71	**-2.04**	**-1.25**	**-1.95**	peak	
15,420	1,622.38	**-43.26**	215	20	**3.22**	**2.60**	**2.51**	**1.18**	**1.97**		
15,440	1,622.83	**-42.38**	216	20	-0.88	**2.34**	1.72	**1.63**	0.30		
15,460	1,623.27	**-42.89**	217	20	0.51	-0.37	**2.85**	**2.23**	**2.14**		
15,480	1,623.77	**-41.61**	218	20	**-1.28**	-0.77	**-1.65**	**1.57**	0.95		
15,500	1,624.25	**-41.95**	219	20	0.34	-0.94	-0.43	**-1.31**	**1.91**		
15,520	1,624.71	**-43.57**	219	20	**1.62**	**1.96**	0.68	**1.19**	0.31		
15,540	1,625.17	**-41.88**	220	20	**-1.69**	-0.07	0.27	**-1.01**	-0.50		
15,560	1,625.60	**-43.21**	221	20	**1.33**	-0.36	**1.26**	**1.60**	0.32		
15,580	1,626.03	**-44.56**	223	20	**1.35**	**2.68**	0.99	**2.61**	**2.95**	low point	
15,600	1,626.44	**-42.31**	224	20	**-2.25**	-0.90	0.43	**-1.26**	0.36		
15,620	1,626.87	**-42.15**	224	20	-0.16	**-2.41**	**-1.06**	0.27	**-1.42**		
15,640	1,627.34	**-42.12**	226	20	-0.03	-0.19	**-2.44**	**-1.09**	0.24		
15,660	1,627.74	**-42.20**	226	20	0.08	0.05	-0.11	**-2.36**	**-1.01**		
15,680	1,628.16	**-40.39**	226	20	**-1.81**	**-1.73**	**-1.76**	**-1.92**	**-4.17**	peak	
15,700	1,628.56	**-43.50**	226	20	**3.11**	**1.30**	**1.38**	**1.35**	**1.19**		rhyolitic ash at 1628.25 m (15,685 b2k)
15,720	1,628.98	**-43.08**	227	20	-0.42	**2.69**	0.88	0.96	0.93		
15,740	1,629.41	**-44.63**	228	20	**1.55**	**1.13**	**4.24**	**2.43**	**2.51**	3rd lowest point of the record	
15,760	1,629.81	**-42.79**	230	20	**-1.84**	-0.29	-0.71	**2.40**	0.59		
15,780	1,630.21	**-44.02**	231	20	**1.23**	-0.61	0.94	0.52	**3.63**		
15,800	1,630.60	**-43.85**	231	20	-0.17	**1.06**	-0.78	0.77	0.35		
15,820	1,631.06	**-40.21**	231	20	**-3.64**	**-3.81**	**-2.58**	**-4.42**	**-2.87**	peak	
15,840	1,631.51	**-41.70**	233	20	**1.49**	**-2.15**	**-2.32**	-1.09	**-2.93**		
15,860	1,631.99	**-41.15**	234	20	-0.55	0.94	**-2.70**	-2.87	**-1.64**		
15,880	1,632.52	**-40.81**	234	20	-0.34	-0.89	0.60	**-3.04**	**-3.21**		

Tab. 63 (continued)

NGRIP years b2k	NGRIP depth (m)	NGRIP $\delta^{18}O$ (‰)	max. counting error (years)	years to data point above	difference of $\delta^{18}O$ to data point above	c. 40 years accumulated amplitude	c. 60 years accumulated amplitude	c. 80 years accumulated amplitude	c. 100 years accumulated amplitude	»event«	volcanic ashes
15,900	1,632.96	**-43.15**	235	20	**2.34**	**2.00**	**1.45**	**2.94**	-0.70	low point	
15,920	1,633.38	**-42.98**	235	20	-0.17	**2.17**	**1.83**	**1.28**	**2.77**		
15,940	1,633.83	**-41.55**	237	20	**-1.43**	**-1.60**	0.74	0.40	-0.15		
15,960	1,634.32	**-40.53**	237	20	**-1.02**	**-2.45**	**-2.62**	-0.28	-0.62	peak	
15,980	1,634.76	**-41.60**	240	20	**1.07**	0.05	**-1.38**	**-1.55**	0.79		
16,000	1,635.26	**-41.94**	241	20	0.34	**1.41**	0.39	**-1.04**	**-1.21**		
16,020	1,635.70	**-42.30**	243	20	0.36	0.70	**1.77**	0.75	-0.68		
16,040	1,636.21	**-41.42**	245	20	-0.88	-0.52	-0.18	0.89	-0.13		
16,060	1,636.66	**-41.29**	246	20	-0.13	**-1.01**	-0.65	-0.31	0.76		
16,080	1,637.12	**-42.99**	247	20	**1.70**	**1.57**	0.69	**1.05**	**1.39**		
16,100	1,637.54	**-44.97**	248	20	**1.98**	**3.68**	**3.55**	**2.67**	**3.03**	lowest point of the record	
16,120	1,637.99	**-43.25**	249	20	**-1.72**	0.26	**1.96**	**1.83**	0.95		
16,140	1,638.43	**-40.35**	250	20	**-2.90**	**-4.62**	**-2.64**	-0.94	**-1.07**	peak	
16,160	1,638.85	**-43.59**	252	20	**3.24**	0.34	**-1.38**	0.60	**2.30**	low point	
16,180	1,639.30	**-41.86**	254	20	**-1.73**	**1.51**	**-1.39**	**-3.11**	**-1.13**		
16,200	1,639.76	**-41.64**	254	20	-0.22	**-1.95**	**1.29**	**-1.61**	**-3.33**		
16,220	1,640.23	**-44.27**	255	20	**2.63**	**2.41**	0.68	**3.92**	**1.02**	low point	
16,240	1,640.66	-39.89	255	20	**-4.38**	**-1.75**	**-1.97**	**-3.70**	-0.46	peak; steepest part mean: 16,230 b2k	
16,260	1,641.06	**-42.36**	257	20	**2.47**	**-1.91**	0.72	0.50	**-1.23**	low point	
16,280	1,641.51	**-41.99**	257	20	-0.37	**2.10**	**-2.28**	0.35	0.13		
16,300	1,641.94	**-40.71**	259	20	**-1.28**	**-1.65**	0.82	-3.56	-0.93		
16,320	1,642.37	**-42.10**	260	20	**1.39**	0.11	-0.26	2.21	**-2.17**		
16,340	1,642.79	**-42.33**	261	20	0.23	**1.62**	0.34	-0.03	**2.44**		
16,360	1,643.27	**-42.50**	262	20	0.17	0.40	**1.79**	0.51	0.14	low point	
16,380	1,643.76	-39.74	264	20	**-2.76**	**-2.59**	**-2.36**	-0.97	**-2.25**	peak	
16,400	1,644.22	**-42.81**	265	20	**3.07**	0.31	0.48	0.71	**2.10**	low point	
16,420	1,644.71	**-41.62**	265	20	**-1.19**	**1.88**	-0.88	-0.71	-0.48		
16,440	1,645.15	**-40.74**	266	20	-0.88	**-2.07**	**1.00**	**-1.76**	**-1.59**		
16,460	1,645.64	**-42.58**	267	20	**1.84**	0.96	-0.23	**2.84**	0.08	low point	
16,480	1,646.11	**-41.26**	268	20	**-1.32**	0.52	-0.36	**-1.55**	**1.52**		
16,500	1,646.55	**-41.18**	269	20	-0.08	**-1.40**	0.44	-0.44	**-1.63**		
16,520	1,646.99	-38.29	269	20	**-2.89**	**-2.97**	**-4.29**	**-2.45**	**-3.33**	peak	
16,540	1,647.48	**-40.38**	272	20	**2.09**	-0.80	-0.88	**-2.20**	-0.36		
16,560	1,647.90	**-41.68**	272	20	**1.30**	**3.39**	0.50	0.42	-0.90		
16,580	1,648.30	**-41.52**	273	20	-0.16	**1.14**	**3.23**	0.34	0.26		

Tab. 63 (continued)

NGRIP years b2k	NGRIP depth (m)	NGRIP $\delta^{18}O$ (‰)	max. counting error (years)	years to data point above	difference of $\delta^{18}O$ to data point above	c. 40 years accumulated amplitude	c. 60 years accumulated amplitude	c. 80 years accumulated amplitude	c. 100 years accumulated amplitude	»event«	volcanic ashes
16,600	1,648.75	**-40.58**	273	20	-0.94	-1.10	0.20	**2.29**	-0.60		
16,620	1,649.20	**-42.77**	275	20	**2.19**	**1.25**	**1.09**	**2.39**	**4.48**		
16,640	1,649.73	**-42.78**	277	20	0.01	**2.20**	**1.26**	**1.10**	**2.40**	low point	
16,660	1,650.24	**-40.88**	278	20	**-1.90**	**-1.89**	0.30	-0.64	-0.80		
16,680	1,650.75	**-41.55**	279	20	0.67	**-1.23**	**-1.22**	0.97	0.03		
16,700	1,651.20	**-40.65**	280	20	-0.90	-0.23	**-2.13**	**-2.12**	0.07		
16,720	1,651.66	**-42.69**	281	20	**2.04**	**1.14**	**1.81**	-0.09	-0.08	low point	
16,740	1,652.10	-39.36	282	20	**-3.33**	**-1.29**	**-2.19**	**-1.52**	**-3.42**	peak	
16,760	1,652.56	**-40.72**	283	20	**1.36**	**-1.97**	0.07	-0.83	-0.16		
16,780	1,653.03	**-41.80**	284	20	**1.08**	**2.44**	-0.89	**1.15**	0.25	low point	
16,800	1,653.52	**-40.92**	285	20	-0.88	0.20	**1.56**	**-1.77**	0.27		
16,820	1,653.96	-39.26	287	20	**-1.66**	**-2.54**	**-1.46**	-0.10	**-3.43**	peak	
16,840	1,654.49	**-40.64**	288	20	**1.38**	-0.28	**-1.16**	-0.08	**1.28**		
16,860	1,654.97	-39.46	290	20	**-1.18**	0.20	**-1.46**	**-2.34**	**-1.26**		
16,880	1,655.41	**-40.35**	292	20	0.89	-0.29	**1.09**	-0.57	**-1.45**		
16,900	1,655.84	**-40.30**	293	20	-0.05	0.84	-0.34	**1.04**	-0.62		
16,920	1,656.28	**-42.10**	294	20	**1.80**	**1.75**	**2.64**	**1.46**	**2.84**		
16,940	1,656.66	**-41.25**	295	20	-0.85	0.95	0.90	**1.79**	0.61		
16,960	1,657.11	**-42.41**	296	20	**1.16**	0.31	**2.11**	**2.06**	**2.95**	low point	
16,980	1,657.54	**-40.69**	298	20	**-1.72**	-0.56	**-1.41**	0.39	0.34		
17,000	1,658.02	**-42.12**	299	20	**1.43**	-0.29	0.87	0.02	**1.82**		
17,020	1,658.50	**-40.42**	300	20	**-1.70**	-0.27	**-1.99**	-0.83	**-1.68**		
17,040	1,658.94	**-42.23**	301	20	**1.81**	0.11	**1.54**	-0.18	0.98		
17,060	1,659.44	**-40.00**	303	20	**-2.23**	-0.42	**-2.12**	-0.69	**-2.41**		
17,080	1,659.92	-39.76	306	20	-0.24	**-2.47**	-0.66	**-2.36**	-0.93		
17,100	1,660.37	-38.99	306	20	-0.77	**-1.01**	**-3.24**	**-1.43**	**-3.13**	peak	
17,120	1,660.89	**-42.35**	308	20	**3.36**	**2.59**	**2.35**	0.12	**1.93**	low point	
17,140	1,661.36	-39.80	309	20	**-2.55**	0.81	0.04	-0.20	**-2.43**	peak	
17,160	1,661.88	**-40.26**	310	20	0.46	**-2.09**	**1.27**	0.50	0.26		
17,180	1,662.31	**-40.48**	310	20	0.22	0.68	**-1.87**	**1.49**	0.72		
17,200	1,662.80	**-40.56**	312	20	0.08	0.30	0.76	**-1.79**	**1.57**		
17,220	1,663.28	**-40.17**	313	20	-0.39	-0.31	-0.09	0.37	**-2.18**		
17,240	1,663.78	**-41.24**	313	20	**1.07**	0.68	0.76	0.98	**1.44**		
17,260	1,664.18	**-42.24**	315	20	**1.00**	**2.07**	**1.68**	**1.76**	**1.98**	low point	
17,280	1,664.59	-39.81	316	20	**-2.43**	**-1.43**	-0.36	-0.75	-0.67		

Tab. 63 (continued)

NGRIP years b2k	NGRIP depth (m)	NGRIP $\delta^{18}O$ (‰)	max. counting error (years)	years to data point above	difference of $\delta^{18}O$ to data point above	*c.* 40 years accumulated amplitude	*c.* 60 years accumulated amplitude	*c.* 80 years accumulated amplitude	*c.* 100 years accumulated amplitude	»event«	volcanic ashes
17,300	1,665.04	-39.80	317	20	-0.01	**-2.44**	**-1.44**	-0.37	-0.76	peak	
17,320	1,665.49	**-41.58**	318	20	**1.78**	**1.77**	-0.66	0.34	**1.41**	low point	
17,340	1,665.89	**-40.55**	321	20	**-1.03**	0.75	0.74	**-1.69**	-0.69		
17,360	1,666.33	**-40.64**	322	20	0.09	-0.94	0.84	0.83	**-1.60**		
17,380	1,666.86	-39.94	323	20	-0.70	-0.61	**-1.64**	0.14	0.13	peak	
17,400	1,667.31	**-41.25**	325	20	**1.31**	0.61	0.70	-0.33	**1.45**		
17,420	1,667.78	**-42.94**	328	20	**1.69**	**3.00**	**2.30**	**2.39**	**1.36**	low point	
17,440	1,668.19	**-42.38**	328	20	-0.56	**1.13**	**2.44**	**1.74**	**1.83**		
17,460	1,668.65	**-41.52**	329	20	-0.86	**-1.42**	0.27	**1.58**	0.88		
17,480	1,669.09	**-40.75**	330	20	-0.77	**-1.63**	**-2.19**	-0.50	0.81		
17,500	1,669.55	**-40.72**	331	20	-0.03	-0.80	**-1.66**	**-2.22**	-0.53		
17,520	1,669.95	**-41.49**	332	20	0.77	0.74	-0.03	-0.89	**-1.45**		
17,540	1,670.38	**-41.22**	333	20	-0.27	0.50	0.47	-0.30	**-1.16**		
17,560	1,670.80	-38.85	333	20	**-2.37**	**-2.64**	**-1.87**	**-1.90**	**-2.67**	peak	
17,580	1,671.20	**-40.46**	333	20	**1.61**	-0.76	**-1.03**	-0.26	-0.29		
17,600	1,671.65	**-41.85**	334	20	**1.39**	**3.00**	0.63	0.36	**1.13**	low point	
17,620	1,672.12	-39.58	335	20	**-2.27**	-0.88	0.73	**-1.64**	**-1.91**	peak	
17640	1,672.61	-39.70	336	20	0.12	**-2.15**	-0.76	0.85	**-1.52**		
17,660	1,673.08	**-40.89**	337	20	**1.19**	**1.31**	-0.96	0.43	**2.04**		
17,680	1,673.63	**-40.04**	338	20	-0.85	0.34	0.46	**-1.81**	-0.42		
17,700	1,674.11	**-40.70**	338	20	0.66	-0.19	**1.00**	**1.12**	**-1.15**		
17,720	1,674.49	**-42.53**	340	20	**1.83**	**2.49**	**1.64**	**2.83**	**2.95**	low point	
17,740	1,674.92	**-41.35**	341	20	**-1.18**	0.65	**1.31**	0.46	**1.65**		
17,760	1,675.35	**-41.31**	342	20	-0.04	**-1.22**	0.61	**1.27**	0.42		
17,780	1,675.80	**-41.04**	343	20	-0.27	-0.31	**-1.49**	0.34	**1.00**		
17,800	1,676.27	**-40.90**	343	20	-0.14	-0.41	-0.45	**-1.63**	0.20		
17,820	1,676.80	-38.95	345	20	**-1.95**	**-2.09**	**-2.36**	**-2.40**	**-3.58**	peak	
17,840	1,677.31	**-41.45**	346	20	**2.50**	0.55	0.41	0.14	0.10		
17,860	1,677.78	**-42.40**	346	20	0.95	**3.45**	**1.50**	**1.36**	1.09		
17,880	1,678.20	**-43.36**	347	20	0.96	**1.91**	**4.41**	**2.46**	**2.32**	low point	
17,900	1,678.67	**-41.14**	347	20	**-2.22**	**-1.26**	-0.31	**2.19**	0.24		
17,920	1,679.16	**-40.93**	348	20	-0.21	**-2.43**	**-1.47**	-0.52	**1.98**		
17,940	1,679.64	**-42.52**	349	20	**1.59**	**1.38**	-0.84	0.12	**1.07**		
17,960	1,680.05	**-41.74**	350	20	-0.78	0.81	0.60	**-1.62**	-0.66		
17,980	1,680.47	**-43.47**	351	20	**1.73**	0.95	**2.54**	**2.33**	0.11		
18,000	1,680.96	**-42.27**	351	20	**-1.20**	0.53	-0.25	**1.34**	**1.13**		

Tab. 63 (continued)

main isotope events	onset in NGRIP (GICC05)		
GH-11.4 (PBO)	×	×	**11,490 ± 97** (^{18}O)
GH	11,703 ± 99 (d)	**11,681 ± 111** (^{18}O)	11,670 ± 99 (^{18}O)
GS-1	12,896 ± 138 (d)	12,819 ± 202 (^{18}O)	**12,770 ± 133** (^{18}O)
GI-1a	13,099 ± 143 (GRIP)	×	**13,070 ± 142** (^{18}O)
GI-1b	13,311 ± 149 (GRIP)	×	**13,250 ± 147** (^{18}O)
GI-1c_1	×	×	**13,580 ± 156** (^{18}O)
GI-1c_2	×	×	**13,730 ± 160** (^{18}O)
GI-1c_3	13,954 ± 165 (GRIP)	×	**13,970 ± 166** (^{18}O)
GI-1d	14,075 ± 169 (d)	×	**14,090 ± 169** (^{18}O)
GI-1e	14,692 ± 186 (d)	**14,687 ± 187** (^{18}O)	14,690 ± 187 (^{18}O)
GS-2a	not clear	×	**16,230 ± 255** (^{18}O)
ref.	Lowe et al. 2008, 10 tab. 1	Steffensen et al. 2008, 683 tab. 1	present study

Tab. 64 Dates (given in years cal. b2k) for the onsets of the main Lateglacial isotope events in the NGRIP ice-core record (GICC05). These dates are the calculated results of the present study (right; see text and **tab. 63**). In addition, the dates based on the weighted mean dates calculated from the $\delta^{18}O$ (^{18}O) record of NGRIP given in the high-resolution approach of Steffensen et al. 2008 (see **tab. 1**) and the onsets of the events proposed by the INTIMATE group are given for comparative reasons. The latter are based on data of the deuterium excess record (d) of NGRIP and the onsets for $\delta^{18}O$ events in GRIP (GRIP) which were transferred by depths (given by Björck et al. 1998) and volcanic markers (Rasmussen et al. 2006) to NGRIP (Lowe et al. 2008; cf. Blockley et al. 2012). The GH-11.4 event (previously 11.2, cf. Walker et al. 1999, 1147f.) or Preboreal Oscillation (PBO) is regarded by these authors as difficult to define, whereas this fluctuation was considered clearly visible in the three ice-cores (NGRIP, GRIP, DYE-3) used for the construction of the GICC05 in the Holocene (Vinther et al. 2006, X-7). Set in bold are the limits used in this study.

In the following sample, significant volcanic activity related to the Vedde Ash (see p. 310-318) was again observed, sustaining a possible interrelation of volcanic activity and the climatic development as observed in the oxygen isotopes in Greenland.

The different frequency of rapid changes between two data points indicate the greater instability of the Pleistocene, in particular, the Late Pleniglacial climate (twice above 4‰, eleven times above 3‰). In contrast, the climate from the Lateglacial Interstadial onwards is characterised by greater stability in the intermediate- and short-term means.

Even though these intermediate- and short-term differences were not as extreme as in the Late Pleniglacial, the decisions where to position the exact limits of the sub-events within GI-1 were made as outlined above (**tab. 64**). The mid-point of the highest difference value was also used for the onset of GS-1. The transition from GI-1a to GS-1 stretched over almost 250 years and was relatively diffuse due to several flickering episodes (**fig. 29**). Therefore, the value given for this limit in different approaches varies between different proxies (cf. Steffensen et al. 2008) as well as between the decisions taken by researchers to set the limit to the onset, the end, the mean between them, or the steepest part of the transition (see **tab. 64**). In a high-resolution study of the oxygen isotope values, the highest value before the transition is given at 12,925 years cal. b2k and the lowest value after the transition is given at 12,712 years cal. b2k (Steffensen et al. 2008). These ages allow the calculation of the mean value between them (**tab. 64**). However, since the current project works with the steepest part this date for the onset of GS-1 is neglected. The last value above -40.00‰ occurred at 12,780 years cal. b2k and for the following 500 years this value was only once reached again. Before 12,780 years cal. b2k values below -40.00‰ were reached three times: at 12,800, 12,820, and 12,900 years cal. b2k. However, since between 12,900 and 12,800 years cal. b2k no value is below -40.00‰, this part is rather attributed to the interstadial part. The accumulated values were also highest between 12,800 and 12,700 years cal. b2k (-2.10‰) but comparably large accumulations occurred between 12,940 and 12,840 years cal. b2k (-2.09‰) and between 12,860 and 12,740 years cal. b2k (-2.08‰; -2.07‰). This short survey marks the period between 12,940 and 12,700 years cal. b2k as the relevant deterioration phase. The steepest gradient between two data points (-1.63‰) occurs between

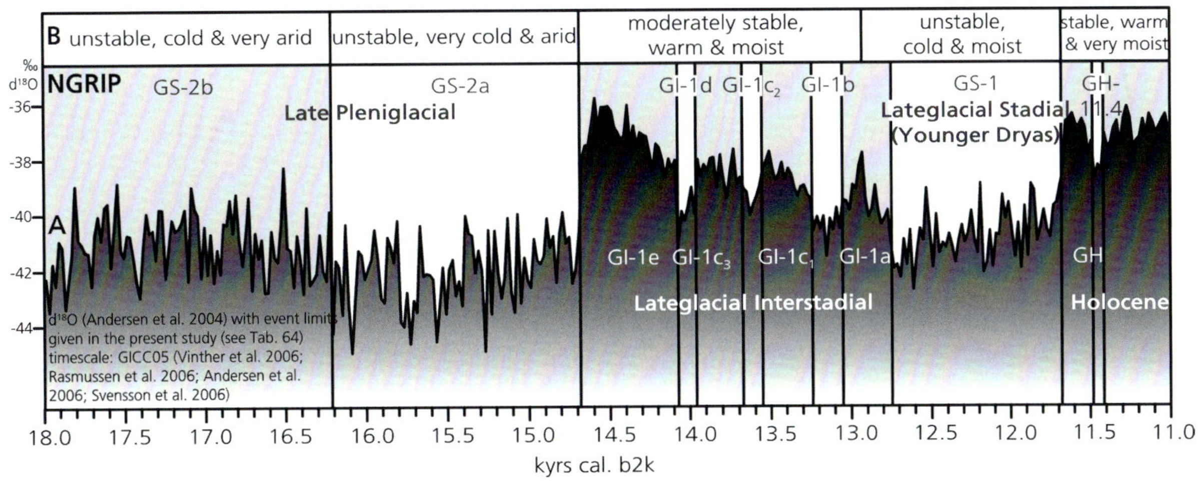

Fig. 29 Oxygen isotope record from NGRIP. **A** Limits (black bars) of the eventstratigraphy (see **tabs 63-64**). Grey shaded areas represent periods of more interstadial values than the surrounding values (for instance, the values in GS-2b are more interstadial than in GS-2a but still these values are as stadial as the values in GS-1). – **B** General climatic interpretation of the isotope record (supplemented with indications of aridity connected to the Heinrich 1 event and the continental precipiation; Hatté/Guiot 2005; Stanford et al. 2011b).

12,940 and 12,920 years cal. b2k which becomes even steeper if the last peak is set to 12,925 years cal. b2k in the high-resolution record. This decline continues to 12,900 years cal. b2k after which the last, short of higher values follows. The steepest decline (-2.12‰) in the vicinity of the transition phase occurred between 12,640 and 12,620 years cal. b2k where the lowest value of GS-1 was reached. Consequently, the values afterwards increased again and, therefore, no accumulated change towards a deterioration was observable. Prior to this steep gradient a basaltic ash layer was identified as Tv-1/I-THOL-2 in the ice-core records (Mortensen et al. 2005). In the biostratigraphic analysis of the northern Icelandic Lake Torfadalsvatn, where the Tv-1 ash was found, the ash layer occurred shortly before the onset of the terrestrial Younger Dryas as defined by the sediment and the pollen composition (Rundgren 1995). Thus, according to this comparison and a correlation of the Younger Dryas with GS-1, this major gradient would mark the onset of GS-1. Yet, this gradient marks rather the end of the extreme deterioration but not the transition. The only other gradient surpassing 1‰ between two data points in this part of the record laid between 12,780 and 12,760 years cal. b2k (-1.20‰). Since the three indicators (-40.00‰ limit, accumulated amplitude, and gradient between two data points) concentrate around this period, the onset of GS-1 is also set to 12,770 years cal. b2k. However, the comparison with the northern Icelandic biostratigraphy already advise to be prudent with the comparison of terrestrial sequences.

Besides the limit for the onset of GS-1, a further two high-resolution ages for the oxygen isotope events were available (Steffensen et al. 2008). Since these dates relate closely with the limits found in this study (**tab. 63**), these ages were chosen for the limits of the oxygen isotope record as the more accurate ones (**tab. 64**).

With the limits of the events also the duration of the events can be determined which can help in the numerical comparison with other annually resolved archives.

In the following sub-chapters, the onset of the Holocene (GH) or the end of the Lateglacial Stadial (GS-1) is occasionally emphasised because this limit is relatively well defined and represents an important correlation point for the Lateglacial records (cf. Material-Climate, p. 12-30). In contrast, the onset of GS-1 or end of GI-1 is not so well suited as correlation point because the transition between warm climate and cold climate values is a relatively long interval with several fluctuations. Furthermore, the marked onset of the Lateglacial Interstadial that appeared very pronounced in the oxygen isotope record is also used infrequently because in many records the lower part of the Lateglacial Interstadial was not reliably preserved. Thus, the number of onsets of oxygen isotope event as chronological marker is rather limited. Therefore, volcanic markers become an important additional help in a correlation of Lateglacial records.

The Lateglacial chronostratigraphy of north-western Europe

For the development of a reliable, high-resolution chronostratigraphy of the Lateglacial only a few suitable records exist. Some of these few records have to be compiled to approach a comprehensive result which can be used for the north-western European record. In this study, the suitable records are compared to the isotope record from the Greenland ice-cores. Some of the suitable records are considered to make further precision of the ice-core chronologies possible but others are correlated to the NGRIP ice-core in the new Greenland ice-core chronology (GICC05) for a higher precision. However, since some of these records are based on proxy data other than isotopes, a correlation creates new standard chronologies for these different proxy records based on the GICC05.
The correlation of these variable records and archives in the same high-precision chronology makes considerations about the interrelation of various climatic and environmental developments necessary. Therefore, climatically and environmentally independent correlation markers are relevant in the construction of a generally valid chronostratigraphy.

Tephrochronology

Besides the limits of the oxygen isotope events, the volcanic record related to NGRIP represents an important set of correlation points. Therefore, volcanic events have to be identified and dated. The identification is usually in the form of ash particles and their chemical composition but also other volcanic indicators such as sulphate peaks can be used. However, the latter does not identify a particular volcanic eruption but only that a volcanic eruption occurred. Dating an eruption is possible in several ways: Material found in the ashes can be dated radiometrically or the ash is set into a chronostratigraphic framework such as a standard pollen profile or a laminated sequence. To use these volcanic events as relative chronostratigraphy, they are best set in relation to other volcanic events. However, many records contain only a single volcanic indicator and therefore a relatively precise dating of some important events such as the Vedde Ash or the Laacher See eruption (LSE) is necessary.
The Ca^{2+} corrected sulphate content in ice-cores is assumed to reflect mainly the volcanic sulphate input into the atmosphere (De Angelis et al. 2003). Sometimes direct evidence of volcanic eruptions is preserved in the ice-cores in the form of tephra layers or single ash shards which allow for a determination of the geochemical composition of the ash and, thus, assumptions about its origin. To use these important markers in this project, the ashes and ash shards as well as the gypsum (Ca^{2+}) corrected sulphate content recorded in NGRIP (Mortensen et al. 2005) were transferred to the GICC05 timescale by the use of depths (**tab. 63**) and a continuous accumulation model (see p. 247 f.).
The resulting graph reveals a tendency of more frequent volcanic activity relating to cold events (**fig. 30**). Moreover, five of the seven prominent peaks in the Ca^{2+} corrected sulphate content occurred during cold periods. Thus, besides the increase of frequency also the intensity of the volcanic activities increased during the cold periods. A connection of glaciation and volcanism was considered previously (Hall/Meiklejohn/Bumby 2011; Kutterolf et al. 2013) as well as the impact of volcanism on the climate system (Zielinski 2000) but these relations are of no substantial relevance in the current study.

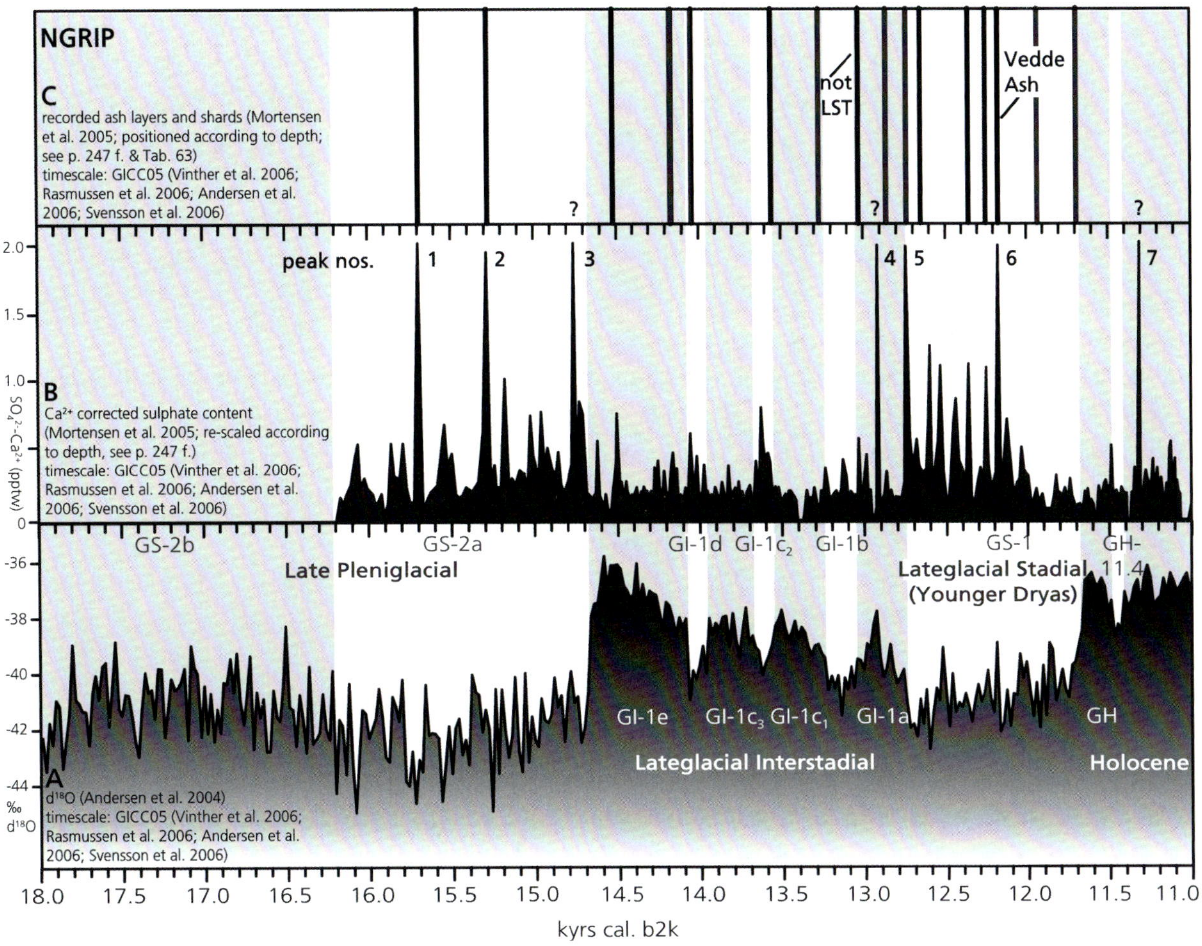

Fig. 30 Volcanic indicators in NGRIP. **A** Oxygen isotope record in NGRIP with event limits (shaded grey and white; see **fig. 29**). – **B** Ca^{2+} corrected sulphate content recorded in NGRIP (Mortensen et al. 2005) was rescaled and positioned in the GICC05 according to depth. The peaks with significantly high values are numbered. – **C** Ash layers (black bars) and single ash shards (grey bars) identified in NGRIP (Mortensen et al. 2005) were positioned in the GICC05 according to depth. Significant peaks of the Ca^{2+} corrected sulphate content (see B) without a correlating ash in NGRIP were marked by a ?. – For further details see p. 247f. and text.

Many of the identified volcanic ashes found in NGRIP originated from Icelandic volcanoes but their ashes were seldom dispersed into Central Europe as visible tephra layers (cf. Blockley et al. 2007). However, three of the seven significant peaks in the Ca^{2+} corrected sulphate content in NGRIP were not related to ash shards: peak 3 at 14,766 ± 191 years cal. b2k, peak 4 at 12,918 ± 138 years cal. b2k, and peak 7 at 11,308 ± 96 years cal. b2k. This absence of volcanic deposits indicate the dependency of the volcanic ash dispersal to the meteorological conditions and atmospheric circulations. Thus, identifiable particles of volcanic eruptions were not deposited evenly. Furthermore, site formation processes could relocate particularly micro-particles within the stratigraphic column or remove them completely (cf. Weber et al. 2010; Brock et al. 2011; Housley et al. 2012). Therefore, the quest for micro-particles, the so-called micro- or cryptotephras, not only in ice-cores but also in marine and terrestrial sequences requires a wide-spread and fine-meshed sampling program supplemented by a comprehensive site formation analyses. However, due to the quasi-synchrony and the independence of these marker horizons, the construction of a Lateglacial tephrostratigraphy as well as a database on the geographic distribution of the tephras are important chronostratigraphic endeavours. Environmental and archaeological stratigraphies in northern Europe are currently studied to approach this endeavour (Lowe 2001; Turney et al. 2006; Lowe 2011; Housley et al. 2012; cf. http://c14.arch.ox.ac.uk/reset/embed.php?File=WP6.html).

A known prominent volcanic marker horizon in the Lateglacial is the Vedde Ash (Blockley et al. 2007; Housley et al. 2012; Lane et al. 2012b). This ash was unambiguously identified in NGRIP at 12,171 ± 114 years cal. b2k (Mortensen et al. 2005; Rasmussen et al. 2006) and correlated to a significant peak (no. 6) in the Ca^{2+} corrected sulphate content (**fig. 30**). The position of the Vedde Ash was already used as marker horizon in the construction of the GICC05 (Vinther et al. 2006; Rasmussen et al. 2006). However, this marker was thus far only identified in non-laminated parts of the continental sequences (Blockley et al. 2007; Lane et al. 2012a) making a numerical comparison, for example of the duration between the Vedde Ash and the end of the Lateglacial Stadial or between the LSE and the Vedde Ash, impossible. However, this tephra is searched for in on-going cryptotephra analysis programs. A possible impact of the Vedde Ash was detected in the Gościąż sequence and dated to 12,181 ± 53 years cal. b2k. This date is in accordance with the date from NGRIP. In the future, an anchoring of stratigraphies on this marker horizon can probably help to confirm other correlations.

In the final Lateglacial Interstadial, the LSE and its tephras (LST) form an important correlation marker. Thus far, LST were not identified in the Greenland records. Ash found in NGRIP a few centimetres below peak 4 were previously considered to represent the LST but due to the chemical composition this layer was attributed to the Icelandic Hekla volcano (Mortensen et al. 2005, 214). Above the significant peak 4 another peak followed but with lower values. This peak was related to rhyolitic ash found in NGRIP and again improbable to relate to the mainly phonolitic LST. In GISP2, the LSE was assumed to be represented by peak values in the total volcanic sulphate (Zielinski et al. 1996) and the electric conductivity measurements (ECM; Taylor et al. 1993a; Baales et al. 2002) because the LSE is usually related to a significant sulphate input into the atmosphere based on a petrological method (Schmincke/Park/Harms 1999). If peak 4 in the Ca^{2+} corrected sulphate content of NGRIP was interpreted comparably to the sulphate content in GISP2, the age difference between the Vedde Ash and the possible LSE signal comprised 747 ± 24 years in NGRIP. Furthermore, the difference between the onset of the Lateglacial Stadial as defined in this study and the assumed LSE would be 148 years, and if the decline in the deuterium excess was chosen as limit for the Lateglacial Stadial only 22 years passed between the possible LSE signal and the onset of GS-1 (**tab. 30**). The age difference to the onset of the Holocene (see **tab. 29**) would comprise 1,237 ± 27 years.

In contrast to the Vedde Ash, the LST was found in the laminated segments of several varve records such as in Meerfelder Maar (MFM), Holzmaar, Hämelsee, and Rehwiese as well as in the stratigraphies of Lake Siethen (Kleinmann/Merkt/Müller 2002), Lake Steisslingen (Eusterhues et al. 2002), and Soppensee (**tab. 30**; Lotter 1991a; Blockley et al. 2008; Lane et al. 2011). Moreover, trees buried by this natural hazard were incorporated in the GLPC (Kromer et al. 2004) which also forms an important Lateglacial radiocarbon calibration data set. Consequently, the position of the LSE is also relevant for the radiocarbon calibration.

Besides the evidence of some laminated archives, a previous age estimate of the LSE considered several radiometric dates from samples found within the LST near the volcano and concluded that the LSE occurred in late spring/early summer of 12,966 years cal. b2k (Baales et al. 2002). In this previous estimate, the period between the LSE and the onset of the Lateglacial Stadial[32] and the duration of the Lateglacial Stadial were taken from the laminated archives and the weighted mean was chosen as bridging period to the onset of the Holocene which was defined by the CEDC (Baales et al. 2002). This onset was corrected since this publication and the ice-core chronologies used to bridge the duration of the Lateglacial Stadial were also refined further. Therefore, this date cannot simply be adopted in this study.

32 In the terrestrial archives the Younger Dryas is considered the equivalent of the Lateglacial Stadial (Lotter et al. 1992; Hoek/Bohncke 2001; Litt et al. 2001).

However, the approach of estimating the age of the LSE remains valid. The onset of the Holocene was a relatively short-term transition, which occurred in several chronostratigraphic archives within only a matter of decades. Therefore, the Pleistocene/Holocene transition represents a better marker horizon than the often diffuse onset of the Lateglacial Stadial. The transition from the Lateglacial Interstadial to the Lateglacial Stadial stretched frequently over several decades or some centuries such as in the NGRIP oxygen isotopes (see p. 293-310).
However, the precise onset of the Holocene is dependent on the defining parameters, in particular to which main factor this parameter responded (e. g. air temperature, precipitation, wind intensity) and in which time. Thus, the defining parameters of the Holocene have to be named to attempt a reliable correlation along this marker (**tab. 65**). The onset of the Holocene in the CEDC was defined by an increase of tree-ring width which doubled within less than 30 years around 11,640 years cal. b2k (Friedrich et al. 1999, 29; Friedrich et al. 2004). This increase was quasi-simultaneously recorded in trees from several sites in the Danube basin and, therefore, interpreted as an indicator of changes in climatic factors influencing tree-ring growth such as a better water supply, higher temperatures, or a longer growing season (Friedrich et al. 1999, 29-31). However, the development of the associated ^{14}C age scale in the younger portion of the Younger Dryas and the older portion of the Holocene revealed an earlier onset of changes in this proxy than the increase in tree-ring growth. A change in the carbon isotope exchange between the atmosphere and the ocean reservoir was considered as a possible reason for this earlier change and this explanation would indicate »a gradual increase of ocean transport already several centuries before the end of YD [present author: Younger Dryas, end defined by the increase of tree-ring width]« (Friedrich et al. 1999, 32). Thus, proxies which are more dependent on the ocean dynamics could have reacted earlier on the changes towards the Holocene amelioration than the tree-ring growth. Another problematic issue in this way of estimating the age of the LSE is that the Lateglacial Stadial was often documented incompletely in annually resolved archives. The duration of the Younger Dryas is, therefore, difficult to establish reliably, in particular in terrestrial archives where the LST was also found. In north-western Europe, only the MFM, the Rehwiese, and the Polish varve records seemed to contain a complete varve sequence of the Younger Dryas but the Polish records did not contain the LST (**fig. 31**). Furthermore, the deep sea record from the Cariaco basin was assumed to contain a complete record of the Lateglacial Stadial (Hughen et al. 1998a; Hughen et al. 2004c). However, indications for the LSE were also absent from this record and, therefore, the Cariaco basin record is not further considered in this matter. Moreover, problems within the chronostratigraphy of this record during the Lateglacial and particularly the Lateglacial Stadial are discussed in a following sub-chapter.
To estimate the precise duration of the Lateglacial Stadial, the onset must be defined. However, several proxy values changed gradually to stadial conditions at this transition and the response of some terrestrial proxies depended on very local factors. For example, Holzmaar is located only a few kilometres north-east of MFM. Even though the varve chronologies from both maars were correlated along the LST and the early Holocene Ulmener Maar tephra, the period between the LST and the onset of the sedimentologically (260 years) as well as the palynologically defined Younger Dryas (274 years) in the Holzmaar record was considerably longer than in the MFM record (201 and 200 years respectively; **tab. 65**). Moreover, the calculated high-precision ages for the beginning of the transition from GI-1 to GS-1 in NGRIP varied in different proxies between 12,939 (layer thickness) and 12,870 years cal. b2k (calcium ions) and for the end of the transition between 12,896 (deuterium excess) and 12,712 years cal. b2k (oxygen isotopes; Steffensen et al. 2008, 683 tab. 1) with mean ages for the onset between 12,897 (deuterium excess) and 12,804 years cal. b2k (calcium ions). Since approximately 200 years were counted between the LSE and the onset of the Younger Dryas in the terrestrial laminated archives (**tab. 65**), an estimated age difference for the onset of the Lateglacial Stadial of more than 90 years depending on the defining parameter is an important problem

archive	years between LSE and the onset of the Lateglacial Stadial	defining characteristics of the Lateglacial Stadial	duration of the Lateglacial Stadial	defining characteristics of the Holocene	age for onset of the Holocene (in years cal. b2k)	age of LSE (in years cal. b2k)	ref.
NGRIP (sulphate peak)	148 / 22	oxygen isotope decrease / deuterium excess decrease	1,089 ± 22 / 1,193 ± 39	oxygen isotope increase / deuterium excess increase	11,681 ± 111 / 11,703 ± 99	12,918 ± 138	Mortensen et al. 2005; see text and **tab. 64**
Meerfelder Maar (MFM)	200	increase of varve thickness, westerly winds, and NAP	1,039 ± 22 / 1,089 ± 39	increase of *Juniperus* sp. and *Filipendula* sp., biogenic opal, total organic carbon, decrease of sediment accumulation rate / regular organic varves	11,690 ± 40 / 11,640 ± 40	12,930 ± 40	Brauer/Endres/Negendank 1999; Litt/Stebich 1999; Brauer et al. 2008
Holzmaar	205 / 260-274	end of *Stephanodiscus parvus* dominance / increase of minerogenic sediment, Poaceae and *Artemisia* sp., decrease of *Betula* sp.	715 ± 43 / 692 ± 41 / 654 ± 39 (c. 320 varves missing)	increase of planktonic diatoms / increase of organic detritus and *Filipendula* sp. / increase of *Pinus* sp. and decrease of NAP	11,690 ± 129 / 11,658 ± 128 / 11,682 ± 129	c. 12,930	Lotter/Birks/Zolitschka 1995; Leroy et al. 2000; Zolitschka et al. 2000
Hämelsee	197 (70-200)	increase of clay/silt laminae, layer thickness, grain size, and NAP (reworking of calcite, deposition of sand)	× (56 varves into the Younger Dryas)	increase of AP	11,610 ± 118	c. 12,950	Merkt/Müller 1999
Rehwiese	205 (200-223)	decrease of *Betula* sp. and varve thickness; increase of NAP and *Pinus* sp.	982 ± 10 / 1,035 ± 10	increase of calcite precipitation / decrease in *Artemisia* sp. and *Juniperus* sp. and increase in *Betula* sp.	11,743 ± 40 / 11,690 ± 40	c. 12,930	Neugebauer et al. 2012
Lake Siethen	200	increase of NAP and cold water rotifer; decrease of calcareous content and *Betula* sp.	× (some varves into the Younger Dryas)	establishment of calcite lamination, decrease of NAP and cold water rotifer	×	c. 12,950	Kleinmann/Merkt/Müller 2002
Lake Steisslingen	190 (240)	sediment composition; increase of *Pinus* sp. and NAP (end of varve formation)	× (c. 50 varves into the Younger Dryas)	predominance of *Betula* sp. and decrease of *Pinus* sp.	*c. 11,630*	c. 12,930	Eusterhues et al. 2002
Soppensee	185-235	increase of calcite varves and of NAP	c. 400 varves (several unlaminated segments)	high values of *Pinus* sp. and decrease of NAP; increased pollen accumulation rates; oxygen isotope increase	c. 11,036	c. 12,900	Lotter 1991b; Hajdas et al. 1993; Hajdas/Bonani/Zolitschka 2000
Dättnau (tree-ring growth decrease)	*192 (214-314)*	*disturbance in growth patterns; decrease of mean segment length*	×	× *(in CEDC: increased width of tree-rings)*	*11,640 (CEDC)*	*13,250-13,060 ± 70; 12,900*	Friedrich et al. 1999; Kaiser et al. 2012
GLPC (Kruft trees)	c. 200	increase in $\Delta^{14}C$	× (140 tree-rings into the Younger Dryas)	× *(in CEDC: increased width of tree-rings)*	*11,640 (CEDC)*	*c. 13,200*	Kromer et al. 2004

Tab. 65 Dating of the eruption of the Laacher See volcano in relation to the onset of the Lateglacial Stadial and the Holocene. Problematic correlations are set in italic. **×** not given; **AP** arboreal pollen; **NAP** non-arboreal pollen. – For further details see text.

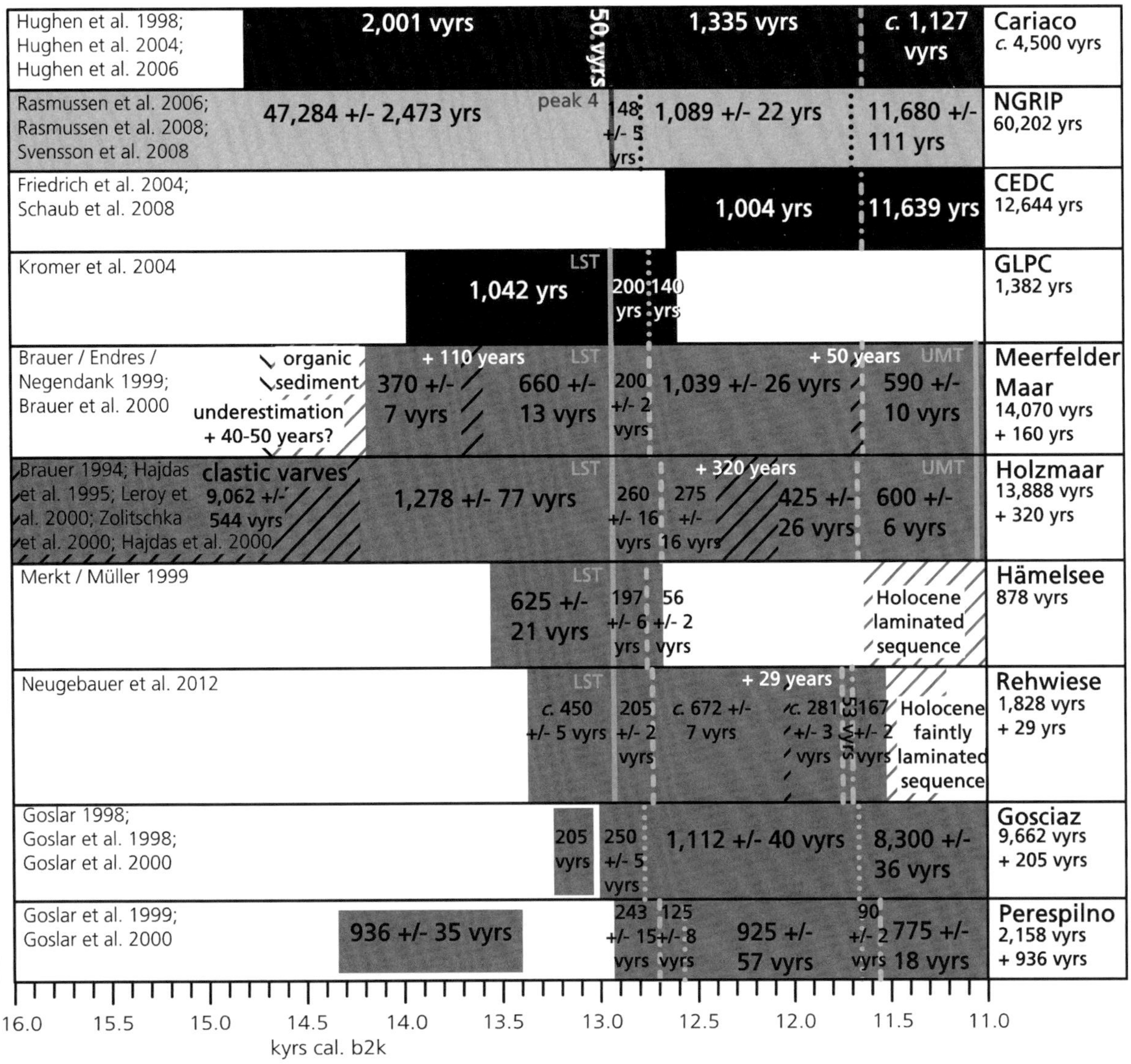

Fig. 31 Archives with laminated sequences in the Lateglacial (see **fig. 10**). Solid lines: position of tephra layers; dashed lines: sedimentologically defined limits; dotted lines: isotopically defined limits; dashed and dotted lines: palynologically or dendrochronologically defined limits.

when dating the LSE. Thus, comparable to the onset of the Holocene the defining parameters for the onset of the Younger Dryas have to be mentioned to be able to correlate quasi-simultaneously reacting proxy data sets.

The most reliable age estimate for the LSE was established in the MFM record. According to the varve chronology from the MFM, the LSE was dated to 12,930 ± 40 years cal. b2k (Brauer et al. 1999, 324; cf. Brauer et al. 2000b, 363). The transition towards the Holocene began around 11,690 years cal. b2k with an increase of juniper (*Juniperus* sp.) and *Filipendula* sp. in the pollen profile and a poor preservation of the varves (Brauer et al. 1999, 325 f.). From 11,640 years cal. b2k onwards, well defined organic varves were formed again and juniper and the NAP decreased in the pollen profile at this point, whereas birch (*Betula* sp.) rose significantly (Litt/Stebich 1999). Thus, the date of 11,640 years cal. b2k marks the establishment of interstadial conditions, whereas the results of climate change towards the Holocene were identifiable some

50 years earlier. Consequently, the period between the LST and the onset of the Holocene encompassed 1,240-1,290 ± 26[33] varve years in the MFM record depending on the correlation.

The date for the onset of the climate change towards the Holocene is comparable to the NGRIP record (11,681 years cal. b2k, see **tab. 64**), whereas the date for the establishment of interstadial conditions is concomitant with the growth increase in tree-ring width in the CEDC (11,640 years cal. b2k; Friedrich et al. 2004). In combination with the considerations about the carbon isotopes of the CEDC, the development in the MFM record suggests different response times of the proxy data depending on the influence of different climatic factors. The oxygen isotope record in NGRIP as well as the air temperature and general hydrology near the MFM appear more instantly governed by the Northern Atlantic circulation (cf. Sirocko et al. 2005; Brauer et al. 2008). If the two dates from NGRIP and MFM are tested for coherence in a χ^2-test, they are absolutely consistent ($p = 100\,\%$) in a weighted mean age of 11,689 ± 38 years cal. b2k. In contrast, the tree-ring growth pattern of the CEDC and the organic productivity in the MFM seem to lag some decades behind this amelioration. If the periods between the Holocene amelioration and the volcanic indicator in NGRIP and MFM are statistically tested, they are consistent in a statistical weighted mean of 1,238 ± 17 varve years ($p = 100\,\%$).

In the Rehwiese record, the LST was set according to the MFM record to 12,930 years cal. b2k. In contrast to many other chronological archives including the MFM, the biostratigraphic transition towards the Holocene was considered a more gradual process in the Rehwiese record. The first indications of a change in the sediment composition towards interstadial conditions occurred with an increase of calcite precipitation after 11,744 years cal. b2k. If this change was selected as indicator as in the MFM record, the period between the LST and the onset of the Holocene were only 1,193 ± 12[34] varve years. Since the varve chronologies of Rehwiese and MFM were correlated along the LST, this difference indicates that the sedimentological onset of the Holocene was not correlative in the two records. Concomitant with the change in the sediment composition at 11,744 years cal. b2k, an increase in juniper, *Artemisia* sp., and birch pollen occurred (Neugebauer et al. 2012, fig. 5). However, the first signals assumed to relate to climatic change in the pollen profile were decreasing values of *Artemisia* sp. and juniper and a small increase of birch pollen at 11,690 years cal. b2k (Neugebauer et al. 2012, 95). However, other species followed gradually in the Rehwiese pollen profile and full interstadial values were reached according to the palynological analysis around 11,500 years cal. b2k when *Filipendula* sp. increased in this record (Neugebauer et al. 2012, 95). Thus, the palynological onset of the Holocene proceeded very differently in the Rehwiese and the MFM records. The decrease of juniper and the concomitant increase of birch occurred some 50 years earlier at Rehwiese than at the MFM, whereas the increase of *Filipendula* sp. was recorded some 190 years earlier in the MFM record. Moreover, the *Filipendula* sp. increase occurred before the juniper decrease in MFM, whereas it occurred after the juniper decrease at Rehwiese.

Thus, the palynological and the sedimentological onset of the Holocene are different in the two records. In contrast, the biostratigraphic and sedimentological onset of the Younger Dryas is similar in both records (**tab. 65**). The biostratigraphic onset was observed between 12,730 and 12,707 years cal. b2k in the Rehwiese record and a sharp decrease in varve thickness occurred 205 varves after the LST (Neugebauer et al. 2012, 96). A possibility of explaining the different ages in the two records during the Pleistocene/Holocene transition is an underestimation of missing varves in a 7.4 cm thick, disturbed section in the Rehwiese cores within the younger half of the Younger Dryas (Neugebauer et al. 2012, 94). However, a bipartition was observed in the varve composition of the Rehwiese and the MFM record during the Younger Dryas. The limit between the two sub-portions of the Younger Dryas was dated to 12,320 years cal. b2k and to 12,290

33 Assuming a counting error of 1-2 % (Brauer/Endres/Negendank 1999, 21).

34 Assuming a counting error of 1 % (Neugebauer et al. 2012, 94).

years cal. b2k respectively (Neugebauer et al. 2012, 100). Thus, this limit fell in the Rehwiese record into the portion between the LST and the disturbed section and still the response of the sediment composition in the Rehwiese record was preceding the MFM record by approximately 30 years. Consequently, the difference in the response time of the sediment composition in the two records seem to gradually increase from the correlation along the LST to the onset of the Holocene. Based on the palynological comparison of the two records, it appears improbable that this difference resulted from a systematic underestimation in the Rehwiese record. Presumably, different factors were the main influences in the sediment composition in these records. For instance, the more continental position and the closer vicinity of the Rehwiese archive to the Scandinavian ice-sheet were suggested as possible differentiating factors (Neugebauer et al. 2012, 100). However, a final decrease of total varve thickness, thickness of the calcite sub-layers, and potassium coincided with a minimum of total organic carbon (TOC) falls in the Rehwiese record to approximately 11,620 years cal. b2k (Neugebauer et al. 2012, fig. 8). In the pollen profile, birch values slowly decrease, whereas pine (*Pinus* sp.) begins to increase at this point. Juniper and artemisia were still present but remained at small values (Neugebauer et al. 2012, fig. 5). After this position several parameters stabilise on interstadial values and, therefore, this position can be considered as the establishment of the Holocene. This establishment is comparable with the onset of the Holocene in the CEDC and to the organic varves in the MFM. Statistically, these three dates are also consistent (mean of χ^2-test: 11,626 ± 11 years cal. b2k; p=78.33 %). The period from the LST to this point encompasses 1,312 ± 13 varve years which is also statistically consistent with the duration of this period in the MFM (mean of χ^2-test: 1,308 ± 12 varve years; p=75.0 %).

If the various mean values were compared and added the age of the LSE would range between 12,934[35] and 12,927[36] years cal. b2k. This age estimate falls around the counted age in the MFM which is partially due to the strong influence of this record on this estimated age.

As previously mentioned, the LST was not detected in the two Polish lake records. In Lake Gościąż the varve formation began between 13,116 and 12,932[37] years cal. b2k and in Lake Perespilno the varve formation began at approximately 12,870 years cal b2k (**fig. 31**). Thus, if the LSE is dated to approximately 12,930 years cal. b2k, the LST should be present within the laminated part of Lake Gościąż. According to the mean duration of the period between the LST and the onset of the Holocene in the MFM and the Rehwiese record, the LST could possibly be also found at the onset of the varve formation in Lake Perespilno. The lack of a clear indicator for this event in the Polish lakes could suggest an older age for the LSE. However, it might also be possible that only small amounts of air-borne ashes reached the Polish lakes and resulted in no or an ephemeral or invisible deposit and an undetectable impact on the lake environments. Cryptotephra analyses from these two sites are therefore of some importance for the Lateglacial chronostratigraphy.

An oxygen isotope analysis was conducted on both Polish records. To receive another age estimate for the period between the Holocene and the LSE, the duration of the Younger Dryas in the two Polish records could be taken and added by the years between the LSE and the onset of the Younger Dryas. Both records were assumed to be directly correlative to the NGRIP oxygen isotope records and, therefore, the 148 years between the peak 4 and the onset of GS-1 in NGRIP could be added to the duration. Alternatively, if the terrestrial influence on these records was considered stronger, then the approximately 200 years between the LST and the onset of the Younger Dryas had to be added to the duration of the Younger Dryas. If the limits of the Younger Dryas are set according to the steepest part of the oxygen isotope records (**fig. 31**),

35 This is the result of adding 1,308 ± 12 varve years to 11,626 ± 11 years cal. b2k based on the CEDC, MFM terrestrial stabilisation and Rehwiese terrestrial stabilisation.

36 This is the result of adding 1,238 ± 17 varve years to 11,689 ± 38 years cal. b2k based on NGRIP and MFM climatic response.

37 This age range depended on the varve counting error and the error of the Pleistocene/Holocene limit (see p. 329-333).

the duration of the Lateglacial Stadial in Lake Gościąż ranged between 1,260 and 1,312 varve years and between 1,073 and 1,125 varve years in Lake Perespilno. The latter seems too short and an increase in the water levels preceding the Pleistocene/Holocene transition was possibly the reason for an early reaction of the oxygen isotopes in this record (Goslar et al. 1999, 908). The calculated duration in the former record is consistent with the NGRIP and climatic indicators from MFM as well as with the organic varve formations in the MFM and the establishment in the Rehwiese record. Thus, this calculation further sustains the age estimate of the LSE along the MFM.
Besides the varve records, the position of the LSE was also located in the GLPC (**fig. 10**). ^{14}C dates from poplars buried by the LST near the village of Kruft were integrated in the data set of the GLPC (Kromer et al. 2004, 1205). The sampled poplars still contained their bark and a date was taken from the outermost rings of poplar no. 8 (Hd-18438, 11,065 ± 22 years ^{14}C-BP; Baales/Bittmann/Kromer 1998) allowing only a minimal temporal difference between the dated growing phase and the LSE. Further ^{14}C dates from the region were also made on barks and produced comparable results (Baales et al. 2002). The onset of the Younger Dryas was determined in the GLPC by an increase of $\Delta^{14}C$ relating to a significant decline in the calibration curve. According to the integration of the poplars in the GLPC, the last rings of the poplars were formed some 200 tree-ring years before the onset of this significant decline. Thus, the period between the LSE to the onset of the Younger Dryas was in this record consistent with the European varve records (**tab. 65**). Based on this relation, the position of the LSE also became an important issue for the radiocarbon calibration curve (see p. 358-364). Originally, the GLPC was ^{14}C wiggle-matched to the Cariaco basin record though the significant decline in the calibration data set and the LSE was thereby estimated to date to approximately 13,200 years cal. b2k (Kromer et al. 2004). In this comparison, the younger portion of the Cariaco basin record was correlated to CEDC (Hughen et al. 2000) and most of the Pleistocene portion was compared to an age-depth model based on the snow accumulation in GISP2 (Hughen et al. 1998a; cf. Meese et al. 1997), which was later also used in the construction of the ss09sea age model (Johnsen et al. 2001). However, due to the subsequent shift in the older part of the Preboreal pines which are a part of the CEDC, this result must be shifted some 70 years older (Friedrich et al. 2004, 1116). The Pleistocene part of the Cariaco basin record has been correlated since to the chronology of the Hulu Cave record (Hughen et al. 2006) and recently to the GICC05 (Deplazes et al. 2013). These correlations resulted in minor shifts of the Cariaco basin calibration data set at the transition of the Lateglacial Interstadial to the Lateglacial Stadial. Nevertheless, these shifts indicated the insecurity of the Cariaco basin chronology in this part. Therefore, a correlation of the GLPC along the LST in the European varve records could provide a more reliable and precise age estimate.

In summary, this compilation of the partially different definitions of the Younger Dryas and the age difference between the LSE and the onset of the Younger Dryas as well as the Holocene demonstrate that the various climatic and environmental reactions are time transgressive and, thus, an onset for these events is often not well suited as a correlation point in a chronostratigraphic framework. Another volcanic episode such as the Vedde Ash or the Ulmener Maar tephra are much better suited for this type of comparisons.
Finally, based on the above correlations and dates, the most reliable dating of the LSE remains the age estimated from the MFM at 12,930 ± 30 years cal. b2k. Peak 4 of the Ca^{2+} corrected sulphate content in NGRIP is absolutely consistent with this date.

Oxygen isotope records

Besides the Greenland ice-cores, four archives with oxygen isotope records were presented in the climate and chronostratigraphic archives (see Material-Climate, p. 17 f. and p. 23-25): the Hulu Cave, the Qingtian Cave, Lake Gościąż, and Lake Perespilno. These four archives are terrestrial sequences and only the latter two came from north-western Europe. The formation and composition of these terrestrial isotope records were more complex than in the Greenland ice (see p. 248 f.). The oxygen isotope records from the Greenland ice-cores were considered to reflect, in particular, the past air temperature at the coring site (Steffensen et al. 2008). In contrast, the oxygen isotope records from the Chinese speleothems were assumed as proxy for the intensity of the Asian monsoon at the sampled site (Wang et al. 2001; Liu et al. 2008). The Lake Gościąż oxygen isotopes probably reflected the air temperature at the coring site as well as the local hydrological regime (Kuc/Różański/Dubliński 1998). Due to the comparable development of Lake Gościąż and Lake Perespilno, the formation of the oxygen isotopes in both records were probably influenced by the same mechanisms. Based on the more complex formation, offsets between the terrestrial and the ice-core isotope records need to be discussed in regard to whether they represent deficiencies in the chronology of the records or a difference in reaction time for the oxygen isotope signal due to different influences. In the former, the records should be shifted to the reliable and/or more precise age estimate. In the latter case, this procedure would lead to incorrect correlations of the complete record (cf. Blaauw et al. 2010). The Polish lakes yielded varve chronologies which make further numerical comparisons possible. The records from the Chinese speleothems were combined with band counting, radiometric and, thus, absolute ages giving a second way of verifying the reliable position of the chronology. Moreover, the cumulative counting error of laminated stratigraphies was counter-checked by the radiometric ages and thereby the age estimate became more precise, particularly in the older Pleistocene record.
Of the four terrestrial oxygen isotope records used in this study, the long archive of the Hulu Cave (**fig. 6**) was used, particularly in the recent decade, to define the Pleistocene chronostratigraphy more precisely (e. g. Hughen et al. 2006; Weninger/Jöris 2008). This archive comprises the complete Lateglacial, whereas the other three records yielded only sub-periods from the younger half of the Lateglacial.
In the Hulu Cave, absolute ages were given for the three major shifts in the Lateglacial record: the onset of the Lateglacial Interstadial, the end of the Lateglacial Interstadial, and the onset of the Holocene (**tab. 66**). In addition, the onset of a last severe period in the Late Pleniglacial which is probably comparable to the onset of GS-2a in the ice-core records was observable as a sharp gradient (Wang et al. 2001). Furthermore, the limits for an episode of higher oxygen isotope values which perhaps correlates to GI-1b were also given and age estimates made for the onset of GI-1b and GI-1a were possible (Liu et al. 2008). The Pleistocene-Holocene transition was estimated to date around 11,523 ± 100 years cal. b2k, the transition of GI-1 to GS-1 to c. 12,873 ± 60 years cal. b2k, the onset of GI-1a to 13,160 ± 100 years cal. b2k, the onset of GI-1b to 13,320 ± 90 years cal. b2k, and the onset of the Lateglacial Interstadial to 14,695 ± 60 years cal. b2k (Wang et al. 2001; Liu et al. 2008, 695). Clearly, the numerical comparison with the NGRIP record (see **tab. 66**) already reveals a difference of 70 to 160 years for the four youngest limits, whereas the onset of the Lateglacial Interstadial is very similar in both records. The sharp gradient marking the onset of the final severe period in the Late Pleniglacial was recorded in the oxygen isotope record of the high-resolution stalagmite YT and dated to 16,123 ± 60 years cal. b2k. Due to the large counting errors in the ice-core records at this depth, the date suggested for the onset of GS-2a in NGRIP and the date from the Hulu Cave are, despite over 100 years difference, consistent.
In the original publication, the Hulu Cave age for the onset of GS-2a is based on $^{230}Th/^{234}U$-dating in the three stalagmites and on band counting in the YT stalagmite (Wang et al. 2001). Between 16,141 and

main isotope events	NGRIP	Hulu Cave		Qingtian Cave		Lake Gościąż	Lake Perespilno
GH-11.4 (PBO)	11,490 ± 97	×	11,118	×	×	×	×
GH	11,681 ± 111	11,523 ± 100	11,524	×	×	11,662 ± 50	11,650 (^{18}O); 11,560 (sediment)
GS-1	12,770 ± 133	12,873 ± 60	12,873	12,630 / 12,820	12,680 (^{18}O); 12,635 / 12,713 (lt)	12,774 ± 90[2]	12,550-12,600 (^{18}O); 12700 (sediment)
GI-1a	13,070 ± 142	13,160 ± 100	13,144	13,040 ± 100	12,988 (^{18}O); 13,007 (lt)	×	×
GI-1b	13,250 ± 147	13,320 ± 90	13,370	13,220 ± 100	13,211 (^{18}O); 13,248 (lt)	×	×
GI-1c$_1$	13,580 ± 156	×	×	×	×	×	×
GI-1c$_2$	13,730 ± 160	×	×	×	×	×	×
GI-1c$_3$	13,970 ± 166	×	14,152	×	×	×	×
GI-1d	14,090 ± 169	×	14,313	×	×	×	×
GI-1e	14,687 ± 187	14,695 ± 60	14,658	×	×	×	×
GS-2a	16,230 ± 255	16,123 ± 60	16,123	×	×	×	×
ref.	see **tab. 64**	Wang et al. 2001; Liu et al. 2008, 695	Wang et al. 2001, supplemental tab. 2*	Liu et al. 2008	Liu et al. 2008, supplemental tab. 1-2*	Kuc/Różański/Dubliński 1998; Ralska-Jasiewiczowa/Demske/van Geel 1998; Goslar/Arnold/Pazdur 1998	Goslar et al. 1999

Tab. 66 Comparison of the dates (given in years cal. b2k) for the onsets of the main Lateglacial oxygen isotope events in various records. **(lt)** layer thickness; **(^{18}O)** oxygen isotope; ***** reading and calculation are made by the present author from supplemental tables (see p. 245-249). – For further details see text.

16,105 years cal. b2k approximately every 9th band was sampled. In this period of 36 years the values increased for 2.01 ‰ and remained relatively high afterwards for more than a century (Wang et al. 2001, supplemental tab. 2). If the oxygen isotope record of the Hulu Cave is treated comparable to the NGRIP record, the sharpest increase between two data points (1.21 ‰) occurred between 16,128 and 16,118 years cal. b2k. One ^{230}Th/^{234}U date came from some 220 bands after the sharp gradient in this high-resolution stalagmite (YT-02: 15,908 ± 59 years cal. b2k, Wang et al. 2001, supplemental tab. 1). In the low resolution PD stalagmite two ^{230}Th/^{234}U dates were before the sharp gradient (PD-490: 16,139 ± 333 years cal. b2k, Wang et al. 2001, supplemental tab. 1) and from after the sharp gradient (PD-470: 15,831 ± 402 years cal. b2k, Wang et al. 2001, supplemental tab. 1). The high-resolution H82 stalagmite was not sampled for oxygen isotopes in this region and, thus, the relation of the ^{230}Th/^{234}U date (H82-2: 16,115 ± 105 years cal. b2k, Wang et al. 2001, supplemental tab. 1) to the gradient is uncertain. Nonetheless, the comparison of the band counting and the radiometric dating in the YT stalagmite are very consistent and the additional information from the other stalagmites are also consistent with this high-precision date. The lowest oxygen isotope value in the surveyed period in the NGRIP record is comparably dated to 16,100 years cal. b2k (**tab. 63**). However, no sharp decrease was recognised in the ice-core record immediately before this point (**fig. 32**). Although a higher value (-40.35 ‰) occurred at 16,140 years cal. b2k, the decrease was more gradual in the ice-core record and lay within the generally high values of the fluctuating amplitude of the

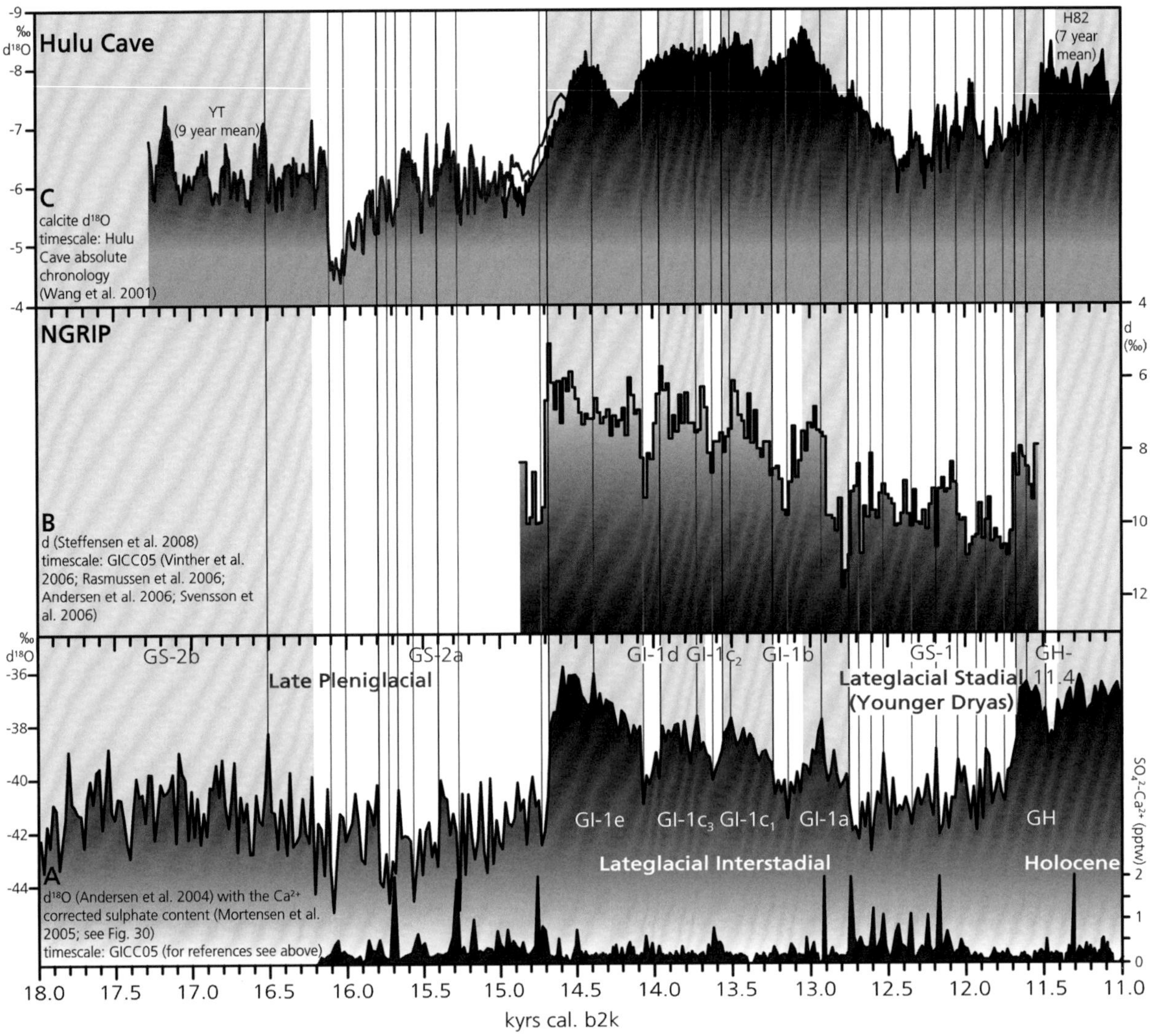

Fig. 32 Comparison of the NGRIP record with the Hulu Cave record. Thin bars given as guide lines (light grey: low; medium grey: high; black: transition). **A** Oxygen isotope record from NGRIP with the Ca^{2+} corrected sulphate content but without associated ash layers (see **fig. 30**). Grey shaded areas represent events of more interstadial values than the surrounding values. – **B** Deuterium excess record from NGRIP. – **C** Oxygen isotope record of the Hulu Cave stalagmites YT (grey) and H82 (black; Wang et al. 2001, fig. 2). – For further details see p. 248 f. and text.

Late Pleniglacial. A shift of the NGRIP record to the more precise date for the onset from the Hulu Cave can be neglected for several reasons: One reason is that the difference of the two records is well within the counting errors of NGRIP. Another reason is the comparatively gradual development of the NGRIP oxygen isotope record and, thus, some chronological offset could be explained by different reaction times and amplitudes between the two records. If further high and low values in the Late Pleniglacial part of the records are compared (**fig. 32**), a very similar behaviour of the two records is apparent. Thus, the onset of GS-2a might have begun with a deterioration in the Atlantic region reflected in the Greenland record and the Asian monsoon record only reacted as this development continued for a longer period.

The transition from the Late Pleniglacial to the Lateglacial Interstadial was estimated to have taken some 180 years centred around 14,695 ± 60 years cal. b2k in the Hulu Cave record (Wang et al. 2001). The estimates for the dating in the Hulu Cave were based on band counting in the H82 and the YT stalagmites, a TIMS-$^{230}Th/^{234}U$ date some 200 bands before the maximum value (14,439 years cal. b2k) in the YT

stalagmite, and two $^{230}Th/^{234}U$ dates from the PD stalagmite taken before and after the transition. The $^{230}Th/^{234}U$ date from the YT stalagmite produced an age of 14,613 ± 71 years cal. b2k (YT-01; Wang et al. 2001, supplemental tab. 1), which was within 30 years in relation to the band counting. In this part of the record a sample from the YT stalagmite contained approximately 20 layers. Thus, the $^{230}Th/^{234}U$ date and the band counting appear very consistent. In comparison with the accumulated counting error of the NGRIP record, the standard deviation of the TIMS-date is considerably smaller and helps to make a Lateglacial chronostratigraphy more accurate. In contrast, the samples from the PD stalagmite taken from before the transition (PD-348: 15,073 ± 464 years cal. b2k) and from after the transition (PD-387: 14,505 ± 391 years cal. b2k; Wang et al. 2001, supplemental tab. 1) were much less precise. In the YT record, the amplitude that accumulated in 100 years usually exceeded 0.5‰ during the period from 14,806 to 14,599 years cal. b2k, marking these 207 years as the main transition. The oxygen isotope values shifted in this period from -6.14‰ at 14,793 years cal. b2k to -7.56‰ at 14,599 years cal. b2k with the sharpest gradient at the beginning between 14,793 and 14,777 years cal. b2k (0.32‰; Wang et al. 2001, supplemental tab. 2). The H82 stalagmite is sampled in closer spaces in this part (as low as two bands per sample) and, therefore, the sequence appeared less stable. The accumulated amplitudes shifted generally more than 0.5‰ between 14,711 and 14,455 years cal. b2k. Thus, this period began a bit later and was more gradual than in the YT stalagmite. However, if the oxygen isotope values are compared, the last high value (-5.87‰) was also in the H82 sequence at 14,793 years cal. b2k but the first low (-7.29‰) was already reached at 14,604 years cal. b2k (Wang et al. 2001, supplemental tab. 2). This was followed by a short and small set back before values continued to decrease to the maximum of -8.27‰ at 14,441 years cal. b2k. The YT record was only sampled until 14,439 years cal. b2k where it also reached its lowest value (-8.05‰). Also the steepest gradient between two data points occurred in a comparable position in both records: in YT record, this shift occurred between 14,660 and 14,637 years cal. b2k (0.25‰) and record a comparable shift (0.24‰) occurred between 14,660 and 14,656 years cal. b2k in the H82. Since, in this project the mid-point of this steepest shift is usually selected for the onset the limit between GS-2a and GI-1 would fall to 14,658 years cal. b2k in the Hulu Cave record (**tab. 66**). The mid-points between the first peak and the last minimum were equivalent to the NGRIP record (YT: 14,696 years cal.b2k; H82: 14,699 years cal. b2k). The dates for the limits and for the mid-points of the transition as well as the estimate for the duration of the transition are almost identical to the observations in the NGRIP record in which the transition was recognised between 14,780 and 14,600 years cal. b2k (see p. 294). Even though this transition is numerically almost identical in both records, the exact development of the proxy record appeared rather inconsistent as was demonstrated by significant shifts (**fig. 32**). The intensity of the reactions were different: some short-term fluctuations observed in NGRIP were not reflected in the Hulu Cave record, and the latter seemed in general to progress more gradually. Opposite reactions of marine and speleothem $\delta^{18}O$ on glacial melting were discussed as partially causing this more gradual development at the limits of the Lateglacial Interstadial in the Hulu Cave record in comparison to the Greenland ice-cores (Wang et al. 2001, 2347). Therefore, the offset of these limits is attributed to the different response times and, therefore, the two records are assumed as chronologically correlative and no record is shifted.

Furthermore, the transition from the Lateglacial Interstadial to the Lateglacial Stadial was also a very gradual process which stretched over several hundred years in the oxygen isotope record from the Hulu Cave stalagmites (Wang et al. 2001, fig. 2). Besides a possibly suppressed response time in the Chinese record, the development from GI-1a to GS-1 was also very gradual in the NGRIP record characterising this transition as a slow-progressing, long-term change. The date of 12,873 ± 60 years cal. b2k for the limit in the Hulu Cave record was based on band counting in the high-resolution H82 stalagmite and two $^{230}Th/^{234}U$ dates in the low-resolution PD stalagmite. One of these $^{230}Th/^{234}U$ dates was taken at the last interstadial peak (at

17.9 mm; PD-210: 12,962 ± 227 years cal. b2k) and the second date was taken at the first stadial minimum in this record (at 12.0 mm; PD-140: 12,607 ± 294 years cal. b2k). In this part of the high-resolution H82 stalagmite, a sample comprised usually seven bands. This closely sampled oxygen isotope record is again more unstable and, thus, several fluctuations become visible in this transition. An interstadial peak occurred already at 13,062 years cal. b2k (-8.67‰) and the stadial minimum is reached at 12,442 years cal. b2k (-5.88‰). The mean between these two points is 12,752 years cal. b2k. The first value below 8.0‰ was recorded at 12,952 years cal. b2k and the last value above 8.0‰ occurred at 12,917 years cal. b2k. The first value below 7.0‰ occurred at 12,614 years cal. b2k and the last value above 7.0‰ was recorded at 12,573 years cal. b2k. The only value below 6.0‰ was the stadial minimum. Thus, in the 620 bands between the two main peaks, values changed only gradually. The highest accumulated amplitude occurred at the end of the transition between 12,545 and 12,442 years cal. b2k. In this period the oxygen isotope values increased by 1.03‰ in 103 years and also the next value from 12,538 to 12,435 years cal. b2k was higher (0.80%) than any other accumulated amplitude in this transition. In the 620 year long transition, further periods where several values of higher amplitudes accumulated over approximately 100 years followed closely on one another can be found. One of these periods occurred between 12,731 and 12,607 years cal. b2k (0.61‰, 0.47‰, 0.55‰, 0.58‰). A second period occurred at the onset of the transition between 13,062 and 12,931 years cal. b2k (0.61‰, 0.76‰, 0.53‰, 0.60‰, 0.81‰) and continued with single values crossing a 0.5‰ amplitude until the accumulated amplitude for the time from 12,917 to 12,814 years cal. b2k. Towards the end of this period of accumulated amplitudes, the steepest gradient between two data points (0.40‰) occurred between 12,876 and 12,869 years cal. b2k. This gradient was used to determine the onset of GS-1 (**tab. 66**; cf. Wang et al. 2001, supplemental tab. 2). The second steepest gradient occurred at the end of the transition between 12,456 and 12,449 years cal. b2k (0.37‰) and the third steepest at the onset of the transition between 13,014 and 13,007 years cal. b2k (0.30‰). Furthermore, at five positions the values increased for more than 0.25‰: 13,034-13,028 years cal. b2k, 12,731-12,724 years cal. b2k, 12,676-12,669 years cal. b2k, 12,559-12,552 years cal. b2k, and 12,476-12,470 years cal. b2k.

Thus, detailed inspection of the oxygen isotope record from the Hulu Cave displays a comparable, stepwise deterioration at the end of the Lateglacial Interstadial as the Greenland ice-cores (cf. p. 293-310). The first indications for a deterioration in the Hulu Cave preceded those in the NGRIP record by some 130 years, whereas the minimum values of the deterioration seemed to succeed those from the Greenland ice-core by some 200 years. However, the accumulated amplitude values and a relatively steep gradient occurred around 12,731 years cal. b2k and, thus, highlighting a reaction to climatic changes around the same period when oxygen isotope values in the NGRIP record also changed significantly. Furthermore, the extremely low values of GS-1 in NGRIP appeared to not be registered in the Hulu Cave record but the minimum in GS-1 at 12,620 years cal. b2k in NGRIP was followed by a transgression of the oxygen isotope values below 7.0‰ in the Hulu Cave record. By 12,607 years cal. b2k, the values in the Hulu Cave record stabilised. Subsequently, the final phase of the deterioration in the Chinese stalagmite corresponded with a short increase of values in the NGRIP record around 12,540 years cal. b2k (-39.09‰) and the subsequent deterioration. In this case, the reactions of the Hulu Cave and the NGRIP record were quasi-synchronous during this time period and, perhaps, the climatic factors influencing the Hulu Cave record also buffer the registration of stadial limits by contrasting signals. Another possibility is that the Hulu Cave record and the NGRIP record were not synchronous in this part and, consequently, one of the records had to be shifted. However, these remarks serve to illustrate the difficulty of defining the onset of the Lateglacial Stadial.

The original age estimate for the onset of the Holocene in the Hulu Cave record was based on band counting in the high-resolution stalagmite H82, on a $^{230}Th/^{234}U$ date from the same stalagmite, and a $^{230}Th/^{234}U$ date for the transition in the PD stalagmite (low resolution; PD-85: 11,584 ± 310 years cal. b2k, Wang et al.

2001, supplemental tab. 1). This $^{230}Th/^{234}U$ date was taken between a low point interpolated to date to 11,770 years cal. b2k and a peak dating to 11,532 years cal. b2k (mean: 11,651 years cal. b2k). In H82, the end of the first Holocene warming was dated by a TIMS-$^{230}Th/^{234}U$ measurement to 11,052 ± 97 years cal. b2k (H82-1; Wang et al. 2001, supplemental tab. 1). In this record, the first full interstadial values were reached after a sharp increase of the $\delta^{18}O$ values between 11,536 (-6.99‰) and 11,515 years cal. b2k (-8.04‰; mean: 11,526 years cal. b2k; Wang et al. 2001, supplemental tab. 2). The latter date correlated with the transition towards the GH-11.4 event in the NGRIP record (cf. Walker et al. 1999, 1147f.; Vinther et al. 2006, X-7). The increase of the oxygen isotope values in the Hulu Cave was preceded by several fluctuations. These fluctuations between two data points began around 11,695 years cal. b2k in the H82 stalagmite. The sharpest gradient between two data points occurred between 11,626 and 11,620 years cal. b2k where the values deteriorated by 1.12‰ to the lowest value before the establishment of interstadial conditions. Very high values of accumulated amplitudes were reached only after 11,620 years cal. b2k with the highest values (0.94‰) falling between 11,559 and 11,447 years cal. b2k. Before this period, accumulated amplitudes above 0.5‰ were only reached twice: between 11,717 and 11,620 years cal. b2k (0.60‰) and in the period between 11,667 and 11,559 years cal. b2k (0.83‰; 0.64‰; 0.67‰;0.50‰). Thus, although the full Holocene mode of the Asian monsoon, which is probably reflected by the Hulu Cave oxygen isotope record, was reached around 11,525 years cal. b2k, a transition period from the Lateglacial Stadial to the Holocene mode began already around 11,720 years cal. b2k (Wang et al. 2001, supplemental tab. 2) with a period of various changes concentrated around 11,623 years cal. b2k. This limit succeeded the limit for the Pleistocene-Holocene transition in the North Atlantic by several decades but the first indicators of a changing climate regime were again comparable to the onset of the Holocene as indicated by the deuterium excess records in NGRIP. Perhaps, the notably delayed reaction of the Hulu Cave oxygen isotope record to Holocene amelioration could be explained as the onset of the Lateglacial Interstadial with opposite reactions of marine and speleothem $\delta^{18}O$ on glacial melting in the two types of records (Wang et al. 2001, 2347). In fact, a significant meltwater flux into the oceans (meltwater pulse 1b [MWP-1b][38]) was discussed for the early Holocene (Fairbanks 1989, 639) and this input might have caused again some delayed reactions in the Hulu Cave record. However, a chronological offset cannot be excluded thus far.

In summary, the detailed comparison of the oxygen isotope records from the Hulu Cave and NGRIP across the three main transitions in the Lateglacial record indicates that the two records react quasi-synchronously to some major climatic impulses. The remaining offsets can partially result from counting/dating errors and partially from various response times to the climatic triggers. Stalagmites respond slower and/or less intense, and often only on long-term impulses, whereas the ice-core records react rapidly and, therefore, these records appear to have a greater fluctuation. It is possible that the dependence of the oxygen isotopes in the stalagmites on various sources resulted occasionally in a neutralization of the climatic signals. Due to these different responses of the speleothem records and the ice-core records at the onsets of some isotopic events, the limits given by the Hulu Cave record should not be simply correlated to the limits of the NGRIP record or vice-versa.

To further evaluate whether the offset between the Hulu Cave record and the NGRIP record is due to different response times or due to problems in the chronologies, the Hulu Cave oxygen isotope record can be compared to the same proxy record from the Central Chinese Qingtian Cave (**fig. 33**; Liu et al. 2008)

38 The existence and the significance of the Holocene MWP-1b was questioned (Bassett et al. 2005) but influx of meltwater and a possibly accelerated rate of sea-level rise is undoubted (Bard/Hamelin/Delanghe-Sabatier 2010), 1237). Furthermore, the various freshwater outbursts from the North American ice-dammed Lake Agassiz are well documented (Teller/Leverington 2004). In fact, an outflow of comparable dimensions to the massive mid-Lateglacial outburst was recorded for the early Holocene. This Holocene outflow was discussed to have had an impact on the thermohaline circulation of (at least) the North Atlantic Ocean (Teller/Leverington/Mann 2002).

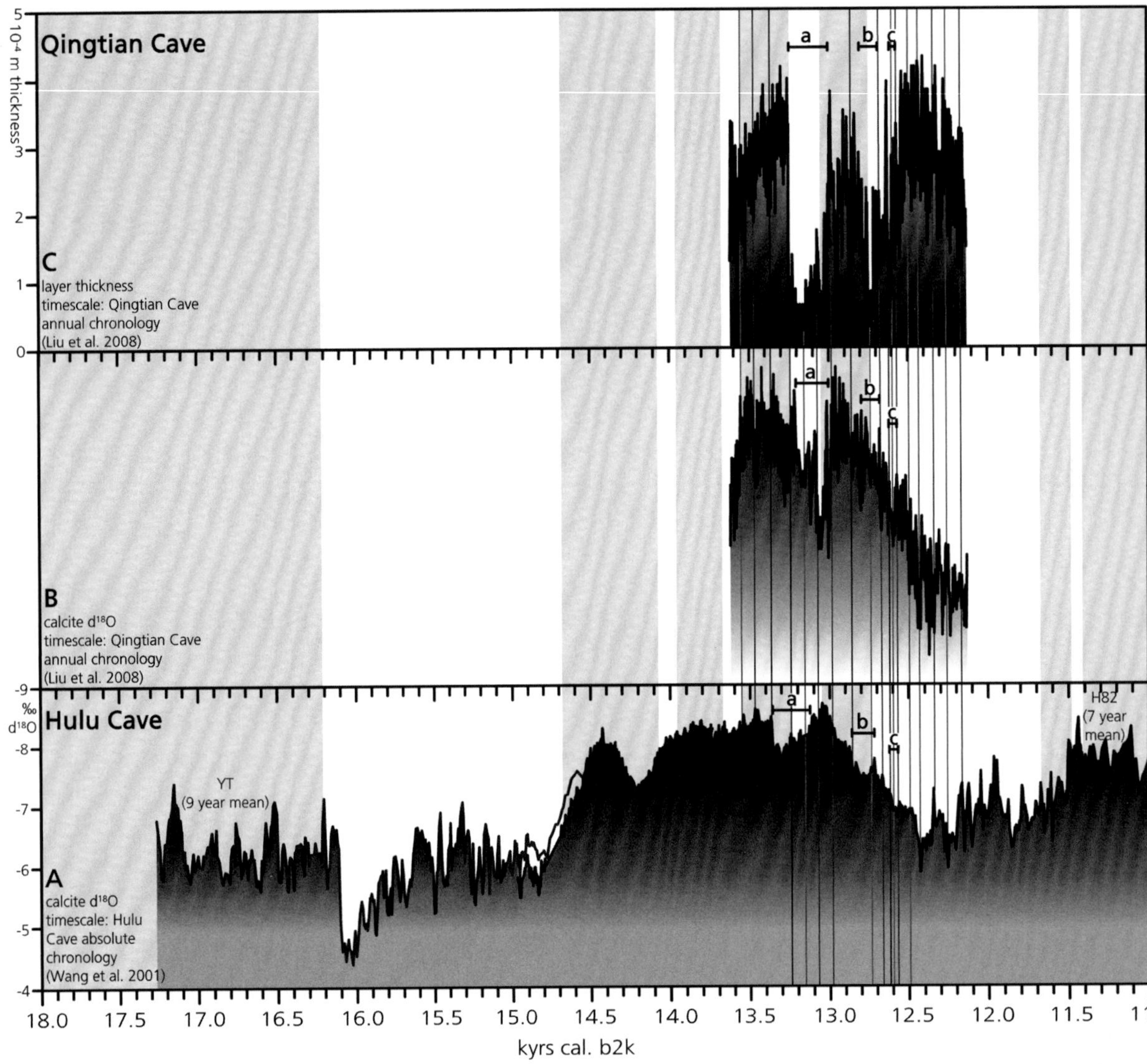

Fig. 33 Comparison of the Hulu Cave record with the Qingtian Cave record. Thin bars given as guide lines (light grey: low; medium grey: high; black: transition). Grey shaded areas represent events of more interstadial values than the surrounding values in NGRIP (see **fig. 29**). Compared episodes a, b, and c are indicated by crossbars. **A** Oxygen isotope record of the Hulu Cave stalagmites YT (grey) and H82 (black; Wang et al. 2001, fig. 2). – **B** Oxygen isotope record of the Qingtian Cave (Liu et al. 2008, fig. 3). – **C** Layer thickness record of the Qingtian Cave (Liu et al. 2008, fig. 3). – For further details see text.

which subsequently can be compared with the NGRIP record (**fig. 34**). The Qingtian record is another high-precision speleothem record with seven high-precision $^{230}Th/^{234}U$ dates that confirmed the band counting and dated the sequence from the mid-Lateglacial Interstadial to the early Lateglacial Stadial. A comparison of the Hulu Cave record helped in particular to define various episodes within the younger half of the Lateglacial Interstadial in both records. Both Chinese oxygen isotope records were assumed to respond to the same control mechanism and revealed similar reactions (Liu et al. 2008, 695). For example, the onset of the Lateglacial Stadial is also a very gradual process in the Qingtian Cave record (**fig. 7**). This transition encompassed the period from approximately 12,820 to 12,440 years cal. b2k (mean: 12,630 years b2k; Liu et al. 2008). In detail, the oxygen isotope value changed over several hundred years between a maximum in the terminal Lateglacial Interstadial (12,908 years cal. b2k, -9.14‰) and the first low point in the Lateglacial Stadial (12,495 years cal. b2k; -6.93‰; Liu et al. 2008, supplemental tab. 1). The lowest value is not reached before 12,372 years cal. b2k (-6.29‰; mean between the two maxima: 12,640 years cal. b2k).

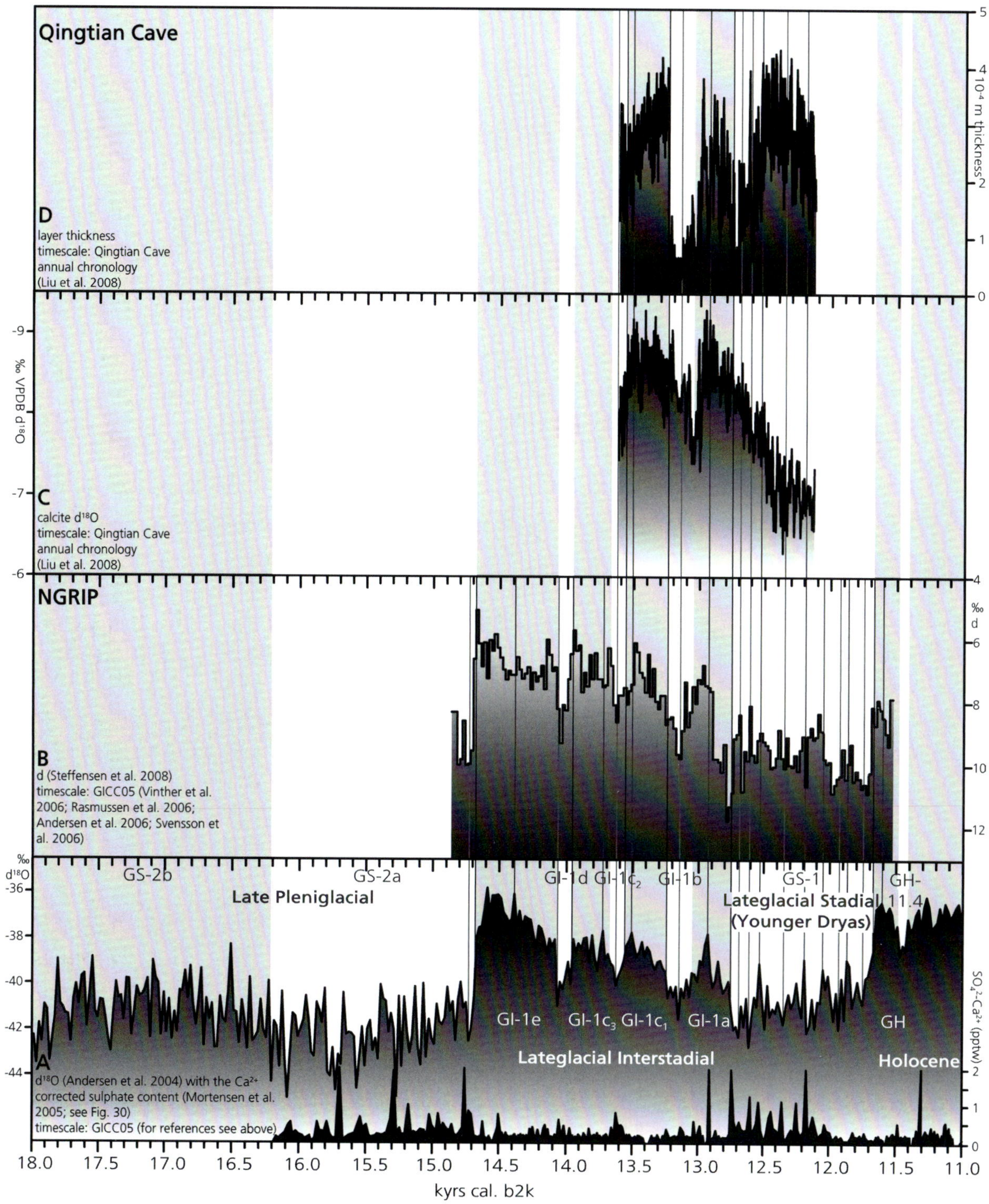

Fig. 34 Comparison of the NGRIP with the Qingtian Cave record. Thin bars given as guide lines (light grey: low; medium grey: high; black: transition). **A** Oxygen isotope record from NGRIP with the Ca^{2+} corrected sulphate content but without associated ash layers (see **fig. 30**). Grey shaded areas represent events of more interstadial values than the surrounding values. – **B** Deuterium excess record from NGRIP. – **C** Oxygen isotope record of the Qingtian Cave (Liu et al. 2008, fig. 3). – **D** Layer thickness record of the Qingtian Cave (Liu et al. 2008, fig. 3). – For further details see text.

These values were several decades younger than the maximum values in the Hulu Cave record and, thus, more comparable to the NGRIP record. In this transition, the most significant decrease between two data points occurred between 12,794 and 12,791 years cal. b2k (-0.87 ‰) in the Qingtian Cave record. This decrease followed an even larger increase between 12,798 and 12,796 years cal. b2k (0.94 ‰). After this fluctuation, the values again increase gradually. Seven smaller fluctuations with amplitudes ranging between 0.44 and 0.72 ‰ followed in the next 300 years. A high value in the accumulated amplitudes (-1.12 ‰) occurred between 12,766 and 12,668 years cal. b2k. However, the largest accumulated amplitudes occurred between 12,535 to 12,425 years cal. b2k (-1.12 ‰; -1.56 ‰). This accumulation is almost identical with the highest accumulated amplitude in the Hulu Cave record between 12,545 and 12,442 years cal. b2k. From a total of seven larger fluctuations between two data points in the Qingtian Cave record five occurred in this period of the highest accumulated amplitude. The largest (-0.72 ‰) fluctuation was between 12,485 and 12,482 years cal. b2k which is possibly correlative with the shift between 12,456 and 12,449 years cal. b2k or 12,476 and 12,470 years cal. b2k in the Hulu Cave record. Even though the two records were similar, the transition in the Qingtian Cave oxygen isotope record yielded some chronological offsets to the developments in the Hulu Cave record. Besides the oxygen isotope record, the development of the layer thickness could be compared in the Qingtian Cave record. Variations in layer thickness were assumed to depend primarily on soil CO_2 production, cave temperature, and hydrological conditions (Liu et al. 2008). Between 12,820 to 12,804 years cal. b2k the layer thickness was of medium (100-250 µm) to very thick (250 µm >) dimensions. Subsequently, the layer thickness began decreasing (Liu et al. 2008, supplemental tab. 2) and in the period between 12,759 and 12,713 years cal. b2k the layers became very thin (generally below 100 µm thickness). After this low, a period of significant increase of more than 175 µm between two data points followed (12,713 and 12712 years cal. b2k). The values gradually increased and began a short period of very thick values between 12,634 and 12,626 years cal. b2k, which ended again with a significant shift between 12,626 (373.87 µm) and 12,625 years cal. b2k (194.84 µm). The values decreased relatively steeply and a short period of very thin layers followed between 12,611 and 12,600 years cal. b2k. Between 12,599 (100.05 µm) and 12,598 years cal. b2k (284.35 µm) the last significant shift towards higher values occurred and from 12,567 years cal. b2k onwards very thick layers prevailed in general (Liu et al. 2008, supplemental tab. 2).

Thus, although the development of the Hulu Cave and the Qingtian Cave record was observable as similar in the general course of interstadial and stadial values, the intensity of these values were sometimes very different between the two caves. The differences in the amplitude were considered as a result of the different altitudes and the different geographical locations (Liu et al. 2008, 695).

Furthermore, if the two records are compared in detail, chronological offsets occurred in the onset of the Lateglacial isotope events (**tab. 66**). The offset between the two records increased significantly towards the onset of the Lateglacial Stadial (159 years for onset GI-1b, 153 years for the onset GI-1a, and 193 years for the onset of GS-1; **tabs 66-67**). To make the comparison independent of the Greenland developments three short-term fluctuations (episodes a-c) in the stalagmite records were compared in the transition from the Lateglacial Interstadial to the Lateglacial Stadial (**fig. 33; tab. 67**). For these episodes, the offset between the Hulu Cave and the Qingtian Cave records almost vanished towards the Lateglacial Stadial (**fig. 33; tab. 67**). Episode a is probably correlative with the Greenland isotope event GI-1b. This episode was also dated by a $^{230}Th/^{234}U$ date on the Qingtian Cave stalagmites which produced an age of 12,998 ± 82 years cal. b2k (QT-198; Liu et al. 2008, tab. 1) and was situated some 25 layers before the transition to the last warm period of the interstadial part as given for the Qingtian Cave (Liu et al. 2008, 695). Episode a had approximately the same duration in the oxygen isotopes of the Hulu Cave (13,370-13,144 years cal. b2k, 226) as in the same proxy record from the Qingtian Cave (13,211-12,988 years cal. b2k, 223 years) but began

record	Hulu Cave, oxygen isotopes	offset	Qingtian Cave, oxygen isotopes	offset	Qingtian Cave, layer thickness	offset to Hulu Cave
end of episode c	12,584	10 years	12,594	5 years	12,599	-15 years
onset of episode c	12,628	1 year	12,629	-3 years	12,626	2 years
duration of episode c	44 years	-9 years	35 years	-8 years	27 years	17 years
end of episode b	12,735	-52 years	12,683	13 years	12,696	39 years
onset of episode b	12,873	-80 years	12,793	-9 years	12,784	89 years
duration of episode b	138 years	-28 years	110 years	-22 years	88 years	50 years
end of episode a	13,144	-156 years	12,988	19 years	13,007	137 years
onset of episode a	13,370	-159 years	13,211	37 years	13,248	122 years
duration of episode a	226 years	-3 years	223 years	18 years	241 years	-15 years
ref.	Wang et al. 2001, supplemental tab. 2		Liu et al. 2008, supplemental tab. 1		Liu et al. 2008, supplemental tab. 2	

Tab. 67 Comparison of the dates (given in years cal. b2k) and the durations (given in stalagmite band years) for some episodes in the Chinese stalagmite records (see **fig. 33**). Episode a is considered as the equivalent of GI-1b and the onset of episode b is taken as limit between GI-1 and GS-1. Durations are shaded light grey. – For further details see text.

almost 160 years earlier in the Hulu Cave (**tab. 67**). The reaction of the layer thickness in the Qingtian Cave had a longer duration and was comparable to the development of the oxygen isotopes in the Hulu Cave. However, the response of the layer thickness in the Qingtian Cave also began over 100 years later than the oxygen isotopes in the Hulu Cave (13,248-13,007 years cal. b2k, 241 years). Furthermore, the highest values of episode a occurred early in the Hulu Cave, whereas in the Qingtian Cave the oxygen isotope values were highest in the late part of this episode. In the beginning of this episode, the Qingtian values were similar to the Hulu Cave record but around 13,020 years cal. b2k an extreme decline in the monsoon intensity was recorded in the Qingtian Cave lasting for some 40 years. The layer thickness was comparable to the Hulu Cave record but a short episode of thicker layers occurred between 13,029 and 13,015 years cal. b2k similar to the oxygen isotopes from the same stalagmites. The offset between the Qingtian Cave and the Hulu Cave record was either a result of chronological reasons or different geographical location. In the latter case, the more maritime climate at the Hulu Cave could have concealed an extreme reaction of the monsoon intensity in the second part of episode a and/or the more continental position of Qingtian Cave had enhanced it. A possible reason for varied changes in the monsoon patterns are the aerosol dispersal of volcanic eruptions as was shown by the comparison of multi-proxy data and historical eruptions as well as modern general circulation models (Anchukaitis et al. 2010; Turner/Harris/Highwood 2012). The dispersal of aerosols could partially depend on the geographical position of the volcano and also on the season of the eruption (Timmreck/Graf 2006). Simulations of the distribution of the aerosols from the super-volcanic LSE, which occurred during this period, suggested no significant differences in the monsoon intensity between the positions of the Qingtian Cave and the Hulu Cave (Graf/Timmreck 2001). Furthermore, the Ca^{2+} corrected sulphate content of NGRIP was relatively steady in this period but the comparison with the NGRIP record correlated this reaction with an almost concomitant decline in the deuterium excess and the oxygen isotope record from NGRIP. Perhaps, this strong signal from the North Atlantic had a higher impact on the more western Qingtian Cave than the eastern Hulu Cave which was probably more strongly influenced by the western Pacific.

In direct comparison the layer thickness of the Qingtian Cave is very similar to the development of the deuterium excess in the NGRIP record (**fig. 34**). The deuterium excess is assumed to be related to the ocean surface temperature in the source region of the Greenland precipitation, i.e. a region near and around the equator during the glacial periods (Masson-Delmotte et al. 2005) and the layer thickness of these relatively southern stalagmites is assumed to be controlled by the hydrological conditions, among other local fac-

tors (Liu et al. 2008). Thus, a closer relation of the layer thickness and the deuterium excess is not very surprising. However, the oxygen isotope records of the speleothems are considered to be influenced by the Asian monsoon and its seasonality (cf. Wang et al. 2001). The seasonality of the Asian monsoon depends on the atmospheric circulation over the Asian landmass and the Indian Ocean and the Chinese Sea and the intensity of the monsoon is thought to be linked to various atmospheric and oceanic factors such as the thermohaline circulation of the oceans or the atmospheric moisture transfer as well as to solar activity (Wang et al. 2005). Thus, the Asian monsoon system is particularly governed by the difference of the Asian landmass to the mostly equatorial watermasses. Thus, the oxygen isotope record from Qingtian Cave can also be comparable to the development of the deuterium excess but this relation is only partially observable in the Qingtian Cave record. The general development of the oxygen isotope record is very comparable to the development of the deuterium excess values in NGRIP but the precise timing seems more comparable to the oxygen isotope record from NGRIP. In contrast, the Hulu Cave record appears generally a bit offset to the oxygen isotope record from NGRIP within the Lateglacial Interstadial, whereas the comparison during the Late Pleniglacial revealed a very similar development (**fig. 32**). Thus, these offsets in the Hulu Cave fall into the range analysed in detail for the H82 stalagmite. This record was correlated to the YT stalagmite record at the onset of the Lateglacial Interstadial (see p. 321 f.) and a $^{230}Th/^{234}U$ date in the early Holocene. Since the offsets seems to occur only during the Lateglacial Interstadial and the transition to the Lateglacial Stadial, the radiometric dating seems to be in accordance with the band counting again. Moreover, the Qingtian Cave record which is sustained with seven $^{230}Th/^{234}U$ dates falls into this period and again shows a high chronological similarity to NGRIP (**tab. 66**). Thus, with these two records (Qingtian Cave and NGRIP) share high concordence, it seems more plausible that the reason for the offsets originated in the Hulu Cave record. A possible reason for the offsets of the Hulu Cave could be that the formation of the bands in H82 altered during the Lateglacial Interstadial. Another reason could be shifts in the seasonality of the monsoon signals for the two Chinese records in this period (Liu et al. 2008). Based on the comparison of the Qingtian Cave and the Hulu Cave records, the NGRIP and Qingtian Cave correlation is given more priority in the present project for the precision of the Lateglacial chronostratigraphy. However, the highly comparable development of the Qingtian Cave and the NGRIP record and/or the explanations for the offsets due to opposing signals make no further shifts in the Lateglacial chronostratigraphy necessary.

In contrast to the Chinese oxygen isotope records, the record from Gościąż provided very sharp signals for the onset and the end of the Lateglacial Stadial with 150 varve years and 80 varve years between the peak values (**tab. 66; fig. 9**; Kuc/Różański/Dubliński 1998, 160). Probably, this closer resemblance to the NGRIP record was due to the more comparable influences on the oxygen isotope record in Lake Gościąż and NGRIP (**fig. 35**). The limits were set to the mid-point between the two peak values, not the sharpest gradient between them. If the sharpest gradient was chosen the limits would shift *c.* 20 years younger for the onset and about eight years older for the end of the Lateglacial Stadial. Even though this difference is well within the counting error of the estimated duration of 1,140 ± 40 varve years (Kuc/Różański/Dubliński 1998), the duration with these years subtracted (1,112 years) emphasises the almost identical duration of GS-1 in Lake Gościąż and NGRIP (1,089 years; **tab. 68**).

However, the dating of the onset of the Holocene in the oxygen isotope record (mean value between two peak values) from Lake Gościąż was established by varve counting (Kuc/Różański/Dubliński 1998; Goslar 1998c) and the correlation of the varve thickness to the width of tree-rings in the CEDC (see p. 23-30; Goslar 1998a; Goslar/Arnold/Pazdur 1998). Thereby the onset of the Holocene was dated to 11,560 ± 50 years cal. b2k and the onset of the Lateglacial Stadial to 12,700 ± 90 years cal. b2k (Goslar/Arnold/Pazdur 1998, 166). A supplementary age estimation was based on wiggle-matching of a ^{14}C data series to the calibration record from the CEDC (Goslar et al. 1998a) and resulted in ages of 11,490 ± 120 years cal. b2k and

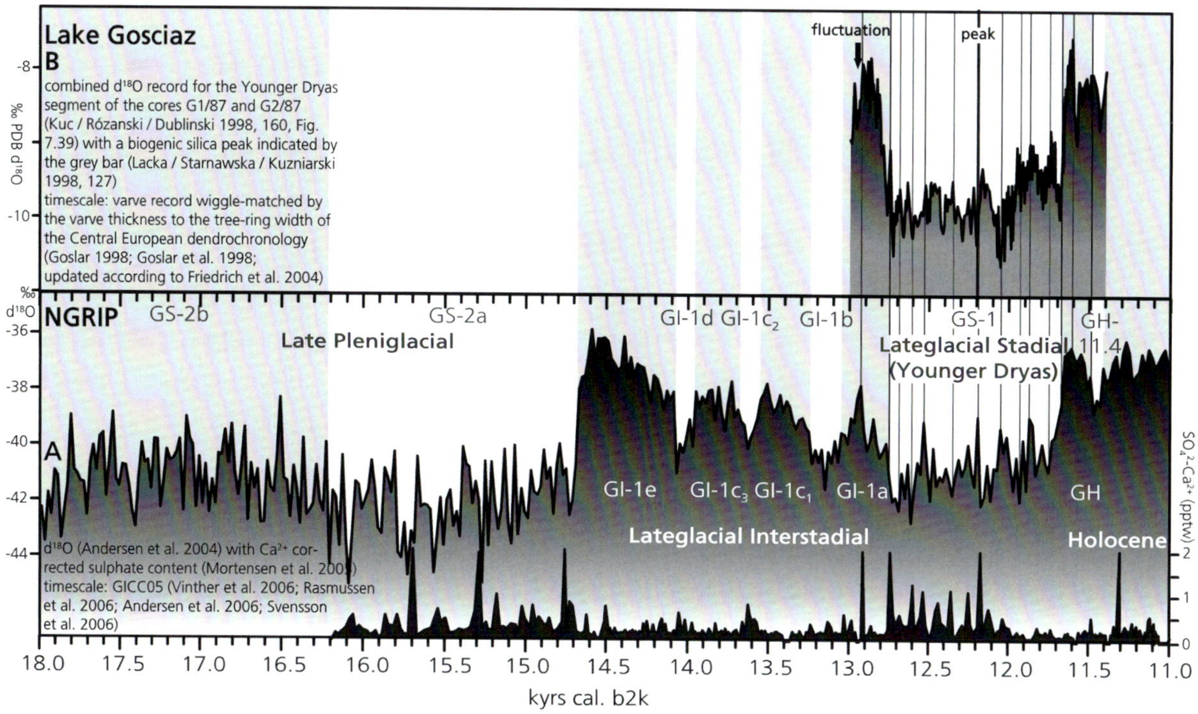

Fig. 35 Comparison of NGRIP and Lake Gościąż. Thin bars given as guide lines (light grey: low; medium grey: high; black: transition). **A** Oxygen isotope record from NGRIP with the Ca^{2+} corrected sulphate content but without associated ash layers (see **fig. 30**). Grey shaded areas represent events of more interstadial values than the surrounding values. – **B** Oxygen isotope record from Lake Gościąż (Kuc/Różański/Dubliński 1998; Goslar 1998c; shifted according to Friedrich et al. 2004) and biogenic silica peak indicated as dark grey bar (Łącka/Starnawska/Kuźniarski 1998). – For further details see text.

12,630 ± 130 years cal. b2k for the same limits (Goslar/Arnold/Pazdur 1998). A revision of this correlation suggested a shift of a further 14 years older (Spurk et al. 1998) but the necessity of this shift was questioned because the difference was well within the error estimate of the dendrochronology (Goslar et al. 2000, 336). In this revision, the onset of the Holocene was determined at varve no. 1072 which was dated to 11,546 ± 36 years cal. b2k according to the ^{14}C wiggle-matching (Goslar et al. 2000). The Holocene part of the dendrochronological record has since been shifted another 80 years older (Friedrich et al. 2004) resulting in a shift of the calculated age for the onset of the Holocene to 11,626 ± 36 years cal. b2k. However, the ^{14}C series from Lake Gościąż (**fig. 9**) displayed some uncertainties in comparison with the CEDC, especially in the Lateglacial period (Goslar et al. 2000, 341), and the record from Lake Perespilno behaved comparably due to the correlation with the Gościąż series (Goslar et al. 1999; Goslar et al. 2000). A constant offset between the ^{14}C data series from Lake Gościąż and the dendrochronological calibration record remained after the additional shifts were performed. Reservoir effects were not usually considered in data series from freshwater environments but more recent studies pointed to contamination by freshwater reservoirs which could occasionally result in offsets of some hundred years (Olsen et al. 2010). Therefore, these dates are rejected pending a revision of geochemical processes which could explain the constant offset of the ^{14}C dates. Nonetheless, the shifts of the dendrochronological data equally affected the ages given by the correlation of the varve thickness from the Gościąż record with the dendrochronological data. A shift of these ages to 94 years older resulted in and age of 11,654 ± 50 years cal. b2k for the onset of the Holocene and 12,794 ± 90 years cal. b2k for the onset of the Lateglacial Stadial. If the record is shifted accordingly and, in addition, the limits of the Lateglacial Stadial are shifted to the steepest part of the transitions in the record, the resulting dates of 11,662 ± 50 years cal. b2k and 12,774 ± 90 years cal. b2k are almost identical to those

main isotope events	NGRIP	(GICC05)	Hulu Cave		Qingtian Cave	Lake Gościąż	Lake Perespilno
GS-1	1,089 ± 22 yrs		1,350 (*1,349*) yrs		×	1,140 (*1,112*) ± 40 yrs	950-900 yrs (*1,125 ± 70 yrs*)
GI-1a	300 ± 9 yrs	1,917 ± 54 yrs	1,822 (*1,785*) yrs	287 (*271*) yrs	410 / 220 (*308*) yrs	×	×
GI-1b	180 ± 5 yrs			160 (*226*) yrs	180 (*223*) yrs	×	×
GI-1c$_1$	330 ± 9 yrs			×	×	×	×
GI-1c$_2$	150 ± 4 yrs			×	×	×	×
GI-1c$_3$	240 ± 6 yrs			×	×	×	×
GI-1d	120 ± 3 yrs			(*161*) yrs	×	×	×
GI-1e	597 ± 18 yrs			(*345*) yrs	×	×	×
GS-2a	1,543 ± 68 yrs		1,428 (*1,465*) yrs		×	×	×
ref.	see **tab. 64**		Wang et al. 2001; Liu et al. 2008, 695		Liu et al. 2008	Kuc/Różański/Dubliński 1998	Goslar et al. 1999

Tab. 68 Duration of the main Lateglacial isotope events in the oxygen isotope records. The limits of the NGRIP ice-core record (GICC05) are set according to the onsets given in **tab. 64**. The duration is given in ice-core years (yrs) and with the increase of the counting error within the event, not the accumulated total counting error. For the Chinese stalagmite records the limits are set according to the published values and the durations are given in stalagmite band years (yrs). For Lake Gościąż the estimated duration is given according to Kuc/Różański/Dubliński 1998 and in varve years (yrs). For Lake Perespilno also varve years (yrs) are given. The value in parentheses is the change in the sediment record of this lake. Correlations to comparable limits as used for the isotope records from the Greenland ice-core (see **tab. 64**) are given in italic and in parentheses. – For further details see text.

from the ice-core records (**tab. 66**). Some minor inconsistencies still remained between the oxygen isotope records from Lake Gościąż and NGRIP but the general pattern is very similar (**fig. 35**). In regard to the onset of the Holocene, the offset between the two records can be explained by the more complex formation and resulting composition of terrestrial isotope values. For instance, fluctuations in the hydrological system of the lake and its surrounding (e. g. Starkel 2002) could mask the climatic signal of the air temperature for the onset of the Holocene such as discussed for the lowest part of the Gościąż record (Kuc/Różański/Dubliński 1998, 159). Thus, a delayed reaction time of terrestrial climate signals compared to the ice-core records is an explainable lag. Consequently, a correlation of terrestrial records with the Greenland ice-core eventstratigraphy on the steepest gradient of the Holocene/Pleistocene transition is not advisable for approaches utilising high resolution records (cf. Friedrich et al. 1999, 33). This lag of terrestrial proxies becomes even more apparent through the comparison for the palynological onset of the Younger Dryas which followed some 20 years after the isotopic onset (cf. Ralska-Jasiewiczowa/Demske/van Geel 1998, 138-140; Goslar/Arnold/Pazdur 1998, 166f.). The onset of the Holocene in the palynological record was almost concomitant with the increase in the varve thickness and the oxygen isotopes but the first indications of a changing environment were recorded some 30 years before this limit and continued some 10-20 years afterwards (Ralska-Jasiewiczowa/Demske/van Geel 1998, 140). However, the establishment of light shrub forest environments succeeded the isotope limit by almost 50 years and the expansion of trees was recorded 200-300 years later in this area (Ralska-Jasiewiczowa/Demske/van Geel 1998, 143).

In comparison with the MFM, this record contained enough varves between the onset of the Holocene and the onset of lamination to also contain the LST. However, visible ashes of the north-eastern fan of the LST were not found in the two Polish lakes. Either the LST was not deposited in the Polish lake sediments due to wind directions or the LST is present in the stratigraphy but not recognised yet, perhaps due to size (cryptotephra) or taphonomy (re-deposition). In fact, a small fluctuation in the lowest part of the oxygen isotope record (**fig. 35**) could correlate to a short-termed climatic deterioration such as caused by a volcanic eruption (cf. Friedrich et al. 1999).

A comparable short-termed, single episode was formed by a concomitant peak in biogenic silica (**fig. 35**) and phosphorus (sample size: six varve years) defining an eutrophication stage of the lake (Łącka/Starnaw-

ska/Kuźniarski 1998, 127). According to the difference of 527 ± 3 years between the onset of the Holocene and the biogenic silica peak in the Gościąż record, this peak would date to 12,181 ± 53 years cal. b2k. Eutrophication can occur due to nutrient surplus which may occur as a result of changes in the depositional environment (Pennington 1981, 216f.). In fact, the maximum varve thickness was reached in this part of the record (Goslar 1998d, 106). Possibly, this increased deposition in Lake Gościąż was connected to a short amelioration period which was also recorded in the southern German Ammersee (Grafenstein et al. 1999, fig. 4) and caused presumably by thawing of soils which were frozen during the first severely cold part of the Lateglacial Stadial (Grafenstein et al. 1999). In the MFM sequence, a sharp change in the sediment record occurred 440 years after the onset of the Younger Dryas, i.e. 650 years before the onset of the Holocene around 12,290 years cal. b2k (Brauer et al. 1999, 325; Lücke/Brauer 2004). In contrast to the Ammersee record, varves became thinner mainly due to the absence of diatom bloom but, comparable to the Ammersee record, also due to a characteristic »graded silt layer [developed] at the base indicating surface runoff processes during the snow-melt in spring or early summer« (Brauer et al. 1999, 325). This amelioration in combination with the more open vegetation (Ralska-Jasiewiczowa/Demske/van Geel 1998), constant or increased precipitation (Goslar et al. 1998b), and higher wind actions (cf. Brauer et al. 2008; Neugebauer et al. 2012) would have dissolved more minerals and resulted in a higher sediment flow which resulted in the nutrient surplus and eutrophication of the lake. Perhaps, the biogenic silica peak in the varve record of Lake Gościąż also correlated with this accumulation of surface runoff processes following a climatic amelioration phase which according to the MFM chronology proceeded the Vedde Ash by some 120 years.

In the eastern German Lake Siethen, where besides the LST also the Icelandic Saksunarvatn Ash was documented, a peak of the ash concentration was recorded concomitant with peaks of silicon (Si), titanium (Ti), and aluminium (Al) in the mid-Younger Dryas (Kleinmann/Merkt/Müller 2002, 60-62). In this record, a greater than 10 cm thick influx of clastics was observed and assumed to come from the littoral zone due to the general increase of algae, in particular, the (rather eutrophic) *Cosmarium* sp. However, the very common littoral cladocera species *Bosmina longirostris* (Gr.) which also prefers eutrophic habitats sharply decreased at this point similar to the pollen of *Pinus* sp. and *Juniperus* sp., whereas a short and sharp increase of *Betula* sp. and *Salix* sp. pollen was recorded (cf. Kleinmann/Merkt/Müller 2002, 60 Abb.1). Stratigraphically below this disruption, laminated sediments of a better quality than the early Younger Dryas were deposited for a short period (Kleinmann/Merkt/Müller 2002, 61) and reflected a possible amelioration phase. Whether the disruption in the Lake Siethen sequence was solely associated with the preceding indications of amelioration and subsequent sediment flows or was additionally connected to the deposition of a volcanic ash such as the Icelandic Vedde Ash cannot be decided without a cryptotephra analysis of the profile.

However, varve thickness was generally high during the Lateglacial Stadial in Lake Gościąż and the short-termed and not recurrent peak in phosphorus remains unexplained. Furthermore, an indicator diatom taxa inhabiting eutrophic lakes (*Asterionella formosa*; Marciniak 1998, 147) had a minimum value at the corresponding depth (Marciniak 1998, 145f.). A comparable pattern (increased input of silica, but no clear eutrophication indicators among the diatoms) was reported from north-eastern German sequences as a reaction to the deposition of the LST (de Klerk et al. 2008). Nevertheless, in the cladocera community which was sampled at other intervals, a small increase of the eutrophic littoral species *Bosmina longirostris* was observable in the record from the shallow northern bay around the depth of the biogenic silica peak, but this increase of eutrophic indicator species was not found in the cores from the deeper parts of the lake (Szeroczyńska 1998) from where the records usually originated.

If the short-lived episode in the Gościąż record was interpreted in analogy to the north-eastern German sequences as a reaction to volcanic activity, the Vedde Ash would be a very probable candidate. This ash

was identified in the NGRIP record at 12,171 ± 114 years cal. b2k concomitant with the end of a mid-GS-1 amelioration phase. The date of 12,181 ± 53 years cal. b2k for the biogenic silica peak is consistent with the date for the Vedde Ash. Thus, if this association was correct, it would further confirm the chronology of the Gościąż record.
However, the Vedde Ash marks the end of an intense period in the Ca^{2+} corrected sulphate content in NGRIP beginning with peak 5 and containing several Icelandic ash deposits (**fig. 30**). Thus, a correlation with an older volcanic event in the NGRIP record is also possible and rather probable than a correlation with a younger eruption than the Vedde Ash (**fig. 35**). If the biogenic silica peak correlated to an older volcanic event, some error in the chronology of Lake Gościąż exceeding the counting errors had to be taken into account.
Between this peak and the onset of the Younger Dryas, as defined by the oxygen isotope values, some 585 varves were deposited. The period between this peak and the onset of the Younger Dryas is very similar to the period between the Vedde Ash and the onset of GS-1 in the NGRIP record (see above and p. 312).
Furthermore, the period between the Vedde Ash and the Ca^{2+} corrected sulphate peak 4 in NGRIP comprised 747 years and the period between the ice-core date of the Vedde Ash and the LST as dated by the MFM would comprise 759 years. Consequently, the number of varves counted below the biogenic silica peak in Lake Gościąż (n = 827 varves) indicates again that the varve formation in Lake Gościąż had begun before the LSE and, thus, the LST could be found within the laminated part of this record. In fact, if the Gościąż and the NGRIP records are correlated along the biogenic silica peak and the Vedde Ash a small fluctuation in the lowest part of the oxygen isotope record (**fig. 35**) correlates with the sulphate peak in the NGRIP record.
Since the association of the biogenic silica peak with a cryptotephra, particulary the Icelandic Vedde Ash, remains to be confirmed, the ice-core record is not shifted 18 years older towards the more precise Gościąż date to achieve a more precise Lateglacial chronostratigraphy. Furthermore, the chronology of Lake Gościąż relies on more analogies and correlations than the continuous NGRIP record and the difference of the biogenic silica peak and the Vedde Ash falls within the error estimates of both records.
In the other Polish stratigraphy, Lake Perespilno, an extreme peak or low in several minerals in the mid- to late Younger Dryas part of the sequence is concomitant with a small fluctuation in the magnetic susceptibility around varve no. 1923 (cf. Goslar et al. 1999). However, this indication of a volcanic eruption occurred some 300 years before the onset of the Holocene in this record and in comparison with the position of the Vedde Ash in the NGRIP record as well as with the biogenic silica peak in the Gościąż sequence, the Perespilno event might be some 100-200 years younger.
The short chronology from Lake Perespilno was synchronised palynologically with the record from Lake Gościąż (Bałaga/Goslar/Kuc 1998; Goslar et al. 1999) and, therefore, the close resemblance seems unsurprising. However, the oxygen isotopes from Lake Perespilno reacted with some offset to the NGRIP and the Lake Gościąż record, in particular during the Lateglacial Interstadial to Lateglacial Stadial transition (**tab. 66**). This offset was attributed to the changing budget of water in the lake environment (Goslar et al. 1999, 908). Perhaps, the more continental climate influenced absorption of the strong signals from the North Atlantic already in the Perespilno record and, therefore, the onset of the Lateglacial Stadial of this record is more comparable to the developments in the Chinese speleothems, particularly the Qingtian Cave, than the NGRIP record. However, at the Pleistocene/Holocene transition the onset of the oxygen isotopes in Perespilno is again very comparable to the Lake Gościąż and the NGRIP record suggesting that the hydrological impact was not very strong during this transition in Perespilno.
As a result for the Lateglacial chronology, the terrestrial oxygen isotope records confirmed the NGRIP limits in general and only partially helped to test and minimise the counting error. However, this detailed comparison provided some interesting suggestions in regard to the climate system and/or the registration in non-

ice-core records: The components additional to the air temperature in the terrestrial oxygen isotope records not only seemed to make the climatic system more complex but also made it appear more inert to climatic impulses of the North Atlantic. In particular, the Asian monsoon was clearly affected by the impact of the oceanic signal but only continued developments in the oceans seemed to result in changes in the monsoon system. This decreasing sensitivity of terrestrial regimes resulted in more stable climates in the terrestrial sphere. A more continuous stability provided the possibility for adaptation and the establishment of stable biotic communities. However, some question relating to the influence of the hydrological components remained to be further evaluated. For example, how significant were the contrasting climatic signals of the air temperature and the hydrological system on the oxygen isotope records of terrestrial archives? How did the influence change, particularly during transitional periods? As long as the formation and composition of the considered terrestrial records is not understood in more detail, correlations of the terrestrial records with the Greenland ice-core records need to consider possible that offsets due to various reaction times between these records occurred and not due to incorrect chronologies.

Chronologies from laminated archives

Besides the terrestrial oxygen isotope curves, the dendrochronological archives and the varve records provided important climatic records combined with a high-precision chronology. These records give insights into the reaction of the ocean (Cariaco basin) and to the responses of the terrestrial sphere to the changes in the atmosphere and oceans.

The CEDC is the only climate and chronostratigraphic record which yielded a continuous chronology from modern day into the Lateglacial. In this archive, the growth patterns of the tree-rings served as climate proxy (Friedrich et al. 1999; Hua et al. 2009). A sudden increase of tree-ring widths was identified at 11,640 years cal. b2k (Friedrich et al. 2004) when tree-rings became on average more than one millimetre thicker within some 20 years (cf. Friedrich et al. 1999). The onset of the Holocene in the CEDC was often correlated with the isotopically defined onset of the Holocene event (GH) in the ice-core records. The period of thick tree-ring values lasted only c. 100 years after which the tree-ring width decreased again in two steps to values almost as low as previously. Around 11,410 years cal. b2k widths increased again but the tree-rings remained relatively unstable with large fluctuations. In the Preboreal pine record the isotopes ^{2}H and ^{13}C were measured (Friedrich et al. 1999, fig. 8). These records displayed constant increase since the Lateglacial Interstadial. A small decline was observable in the isotope values around the second increase in the tree-ring width at 11,410 years cal. b2k followed by an accelerated increase of the isotope values. Thus, the isotope values in the trees reacted to the Holocene amelioration with a delay of approximately 250 years in comparison to the tree-ring width. Possibly, water supply for the trees increased due to massive river discharges induced by melting glaciers and this surplus of water counteracted the temperature amelioration at the onset of the Holocene (Friedrich et al. 1999, 33). Hence, the isotopic record from the dendrochronological data set was not correlative to the extremely rapid onset in the isotope record from the Greenland ice-cores but displayed a delay comparable to the oxygen isotope record at the Hulu Cave.

However, the onset of the Holocene in the CEDC post-dated the signal by 41-63 years for the onset of GH in the five proxy records (oxygen isotopes, deuterium excess, insoluble dust content, calcium content, and annual layer thickness) analysed in a high-resolution study of the NGRIP ice-core (Steffensen et al. 2008). Clearly, this offset falls into the counting errors of NGRIP which has reached 98 years at this point (cf. Vinther et al. 2006; Rasmussen et al. 2006). In this case, the NGRIP record should be shifted towards the

main isotope events	NGRIP	CEDC	GLPC in Cariaco correlation (Australian)	Swiss correlation (Schaub et al. 2008a data)	CELM in IntCal09 corrected Swiss correlation
GH-11.4 (PBO)	11,490 ± 97	11,549 / 11,473	×	×	×
GH	11,681 ± 111	11,640	×	×	×
GS-1	12,770 ± 133	×	12,960 (12,810)	12,870 (12,900)	12,841-12,813
GI-1a	13,070 ± 142	×	×	13,170 (13,015)	13,013-12,985
GI-1b	13,250 ± 147	×	*13,440*	13,470 (13,280)	13,358-13,330
GI-1c_1	13,580 ± 156	×	*13,770*	×	×
GI-1c_2	13,730 ± 160	×	*13,810*	×	×
GI-1c_3	13,970 ± 166	×	*14,195*	×	14,220-14,192
GI-1d	14,090 ± 169	×	*14,215*	×	14,242-14,214
GI-1e	14,687 ± 187	×	×	×	×
GS-2a	16,230 ± 255	×	×	×	×
ref.	see **tab. 64**	Friedrich et al. 2004	Friedrich et al. 2001b; Kromer et al. 2004; Hua et al. 2009	Schaub et al. 2008a; Schaub et al. 2008b	Kaiser et al. 2012

Tab. 69 Comparison of the dates (given in years cal. b2k) for the onsets of the correlatives of the main Lateglacial isotope events in the European dendrochronological records. In the GLPC in Cariaco correlation the relative tree-ring no. 1,000 was correlated with 14,370 years cal. b2k (Kromer et al. 2004, 1206) and, thus, shifted the relative tree-ring chronology c. 50 yrs older than in Friedrich et al. 2001b. Italic values are according to the correlations given in Friedrich et al. 2001b but the ages were interpolated according to the tree-ring numbers in Kromer et al. 2004. Grey values were read from the published tree-ring growth patterns by the present author. – For further details see text.

tree-ring record but the question would arise: which of the proxy data should be considered correlative with the tree-ring width? The three main factors influencing the tree-ring growth are air temperature, precipitation, and duration of the growing season (Friedrich et al. 1999, 31; Friedrich et al. 2001b, 1226; cf. Schaub et al. 2008a, 30). Thus, higher temperatures, better water supply, and/or a longer growing season would have caused thicker tree-rings. Michael Friedrich and his colleagues assumed that the great similarity of the tree-ring growth patterns in various regions was mainly controlled by air temperature because precipitation was influenced by more local factors (Friedrich et al. 2001b, 1226). However, the combination of hydrological and thermal influences on the terrestrial oxygen isotope records was previously considered as reasons for delayed reactions (cf. p. 319-334). Perhaps, Friedrich and his colleagues underestimated the magnitude of the water stress influencing these terrestrial records and, consequently, the tree growth patterns. Another possible explanation for the offset of the two data sets, besides the counting errors, are changes in the influencing factors of the Greenland oxygen isotopes during this transitional period (cf. Landais et al. 2010; Thornalley et al. 2011). Influencing factors of the Greenland oxygen isotopes different from the air temperature are, for instance, evaporative origin or seasonality of the precipitation (Jouzel et al. 1997). These factors are related to the thermohaline circulation of the North Atlantic and are considered as major influences of the deuterium excess record (cf. Masson-Delmotte et al. 2005). In fact, the deuterium excess is the first high-resolution proxy record to change towards Holocene conditions (Steffensen et al. 2008). Consequently, a greater influence of these factors on the oxygen isotopes in Greenland could have led to a response of this proxy preceding the atmospheric thermal signal. Thus, the offset must not necessarily have a chronological reason but could reflect different response times. Therefore, both records remain unshifted and due to the counting error of the NGRIP record the limits for the onset of the Holocene are considered consistent.

The CEDC extends, thus far, only into the early mid-Younger Dryas and ends at 12,643 years cal. b2k (**tab. 69**; Friedrich et al. 2004; Schaub et al. 2008b; cf. Kaiser et al. 2012). The gap into the early Lateglacial Stadial is not yet unambiguously bridged. Thus, the ice-core records as a continuous data set across this

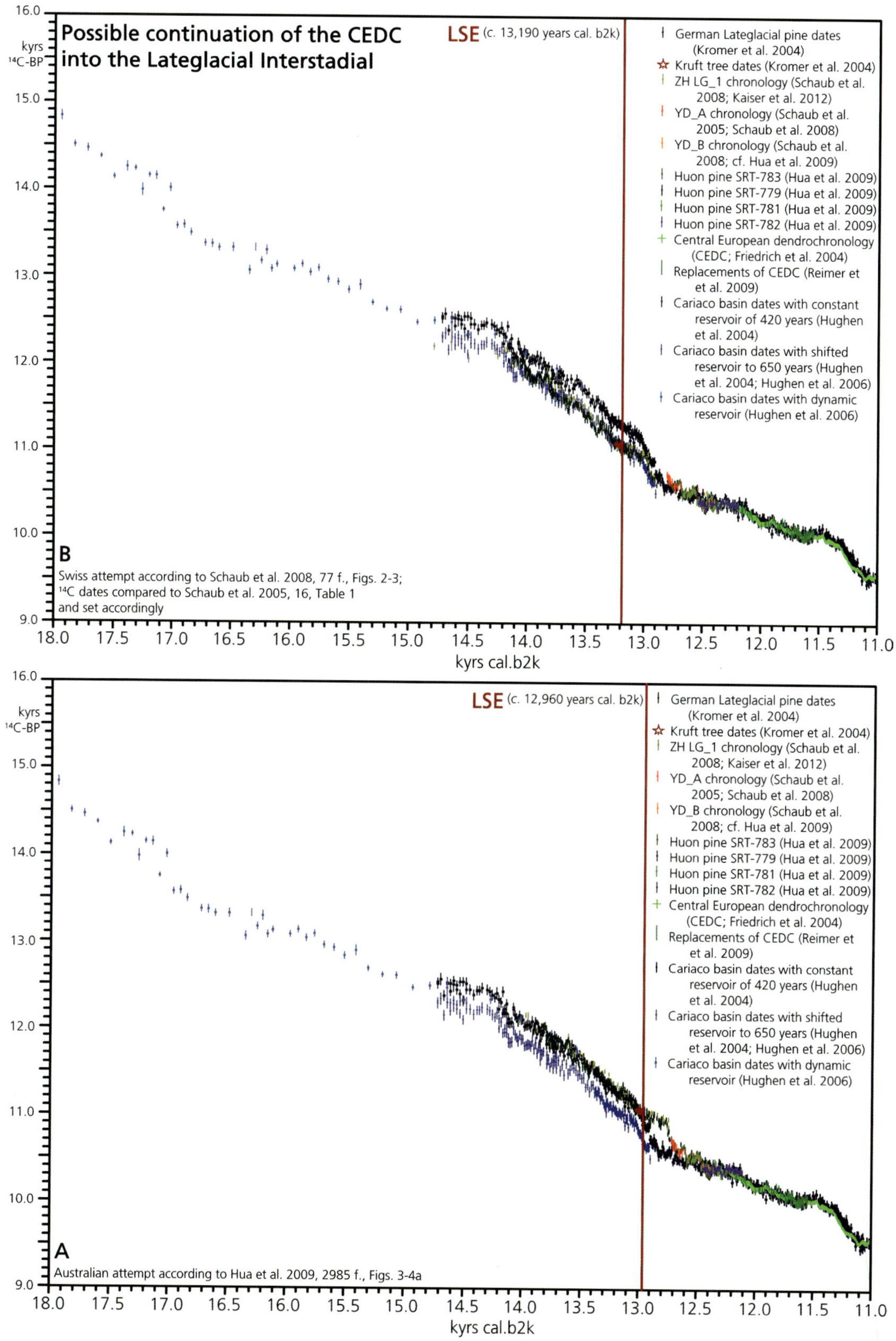

Fig. 36 Comparison of the ^{14}C calibration series according to **A** the Australian attempt (Hua et al. 2009) and **B** the Swiss attempt (Schaub et al. 2008b). Timescales shifted from cal. BP to cal. b2k. – For further details see text.

gap provided the more robust chronology. Nevertheless, the correlation of the tree-ring data can provide interesting insights in the reaction of the tree communities to climate changes, sustain the numerical chronostratigraphy, and are very valuable in the creation of a reliable ^{14}C calibration. The two recently proposed extensions of the CEDC and connections to the Lateglacial tree-ring chronologies (Kromer et al. 2004; Schaub et al. 2008a; Kaiser et al. 2012) produced comparable result for the duration of the Younger Dryas (Schaub et al. 2008b; Hua et al. 2009). Thus far, both records were wiggle-matched by their ^{14}C series to the existing calibration data but no dendrochronological confirmation was given for these correlations yet (cf. Kaiser et al. 2012).

The Australian Huon pines were correlated to the GLPC (**fig. 36A**). The onset of the Younger Dryas in the GLPC is defined by a sharp change in $\Delta^{14}C$ values (Kromer et al. 2004) which according to the new correlation began instantly after the correlation around 12,810 years cal. b2k and continued over 130 years (Hua et al. 2009, 2986). Consequently, the GS-1 would last 1,170 years (**tab. 70**). This rapid change was dated some 240 years older in the original correlation of the GLPC to the Cariaco basin data set (Kromer et al. 2004; Hua et al. 2009, 2985).

A comparable shift of the GLPC some 300 years younger than in the original correlation and also 60 years younger than suggested by the Australian Huon pines was previously suggested based on the best fit in a correlation of $\Delta^{14}C$ values with a climate corrected ^{10}Be flux record (Muscheler et al. 2008). The correlated $\Delta^{14}C$ record was based on the GLPC as well as CEDC data sets and the climate corrected ^{10}Be flux record was taken from GISP2 with a chronology synchronised on the GICC05 timescale. However, by this correlation the end of the GLPC would date to 12,550 years cal. b2k and would clearly have overlapped with the extension of the CEDC into the Lateglacial Stadial (Schaub et al. 2008b; Hua et al. 2009). The age for the onset of the Younger Dryas would shift to c. 12,690 years cal. b2k. This correlation was rejected by the Australian as well as the Swiss attempt. Besides inaccuracies of the correlation arising from the precision of the ^{10}Be measurements, further uncertainties can result from an uncorrected possible climatic influence on the deposition of the ^{10}Be isotopes, from unknown reservoirs of the ^{14}C ages, and from the chronologies of both data sets in general (Muscheler et al. 2008, 2). Thus, further improvements of these sectors are necessary, for example a closely sampled ^{10}Be record from NGRIP. With these improvements a more direct comparison of the ^{14}C records and the ice-core records could help to further enhance the precision and accuracy of the Lateglacial chronostratigraphy.

The Swiss attempt at bridging the early Younger Dryas gap was principally through correlation to the deep sea record from the Cariaco basin (**fig. 36B**; Schaub et al. 2008b). Even though the reservoir ages in the Cariaco basin record were considered relatively constant, significant fluctuations were suggested for the last glacial period (Hughen et al. 2006). For the Lateglacial this type of shift from reservoir ages of approximately 420 years during the Lateglacial Stadial to 650 years for the period between 13,050 and 14,050 years cal. b2k was signified by the comparison with the floating GLPC chronology (Kromer et al. 2004; Hughen et al. 2004c). The Swiss attempt applied this shift from approximately 12,900 years cal. b2k backwards, i. e. from the end of the transition from the Lateglacial Interstadial to the Lateglacial Stadial as defined by the ^{14}C isotope curve (Schaub et al. 2008a). Additionally, the Swiss data set was correlated to the Dättnau sequence and, consequently, to the CELM (Kaiser et al. 2012) as well as to the GLPC itself (Schaub et al. 2008b). The onset of the Younger Dryas was again correlated with the sharp gradient in the ^{14}C data series and an increase in the average growth rate of the tree-rings (Schaub et al. 2008b, 82 f.). In addition, a reduction of the segment lengths covered by the single trees was observed concomitantly. This transition occurred according to the data of Matthias Schaub and his colleagues some 250 years after the end of the extended Holocene dendrochronology (Schaub et al. 2008b, 83 f.) and, thus, around 12,893 years cal. b2k. According to the correlation with the Cariaco basin record, the onset of the Younger Dryas was set to tree-ring

main isotope events	NGRIP (GICC05)		GLPC (Australian attempt)	Swiss data (Schaub et al. 2008a) data)	CELM	MFM		Holzmaar	Rehwiese	Hämelsee	Cariaco basin
GS-1	1,089 ± 22 yrs		1,320 (1,170) yrs	1,230 (1,260) yrs	1,201-1,173 yrs	1,089 (1,039) yrs (sediment)		1,012 yrs (sediment); 974 (pollen)	982 yrs (microfacies); 1,230-1,017 yrs (pollen)	×	1,335 (1,235) yrs (grey scale)
GI-1a	300 ± 9 yrs	1,917 ± 54 yrs	×	300 (115) yrs	200-144 yrs	671 yrs (pollen)	1,341 yrs (preserved; interpolated: 1,851 ± 180 yrs [sediment]; 1,771 ± 180 yrs [pollen])	229 yrs (pollen)	660 yrs (pollen+ sediment)	625 yrs (pollen+ sediment)	400 yrs
GI-1b	180 ± 5 yrs		×	300 (265) yrs	373-317 yrs			274 yrs (pollen)			
GI-1c₁	330 ± 9 yrs		330 yrs	×	890-834 yrs			40 / 380 yrs (pollen)			400 yrs
GI-1c₂	150 ± 4 yrs		40 yrs	×		190 yrs (pollen)		×	×	×	
GI-1c₃	240 ± 6 yrs		385 yrs	×		130 yrs (pollen)		×	×	×	400 yrs
GI-1d	120 ± 3 yrs		20 yrs	×	50-1* yrs	130 yrs (pollen)		×	×	×	
GI-1e	597 ± 18 yrs		×	×	×	220 yrs (preserved; interpolated [sediment+pollen]: 650 yrs)		×	×	×	c. 608 yrs
GS-2a	1,543 ± 68 yrs		×	×	×	×		1,820 yrs	×	×	
ref.	see **tab. 64**		Friedrich et al. 2004; Kromer et al. 2004; Hua et al. 2009	Friedrich et al. 2004; Schaub et al. 2008b; Schaub et al. 2008a	Friedrich et al. 2004; Kaiser et al. 2012	Brauer/Endres/Negendank 1999; Litt/Stebich 1999; Brauer et al. 2008		Leroy et al. 2000; Zolitschka et al. 2000	Neugebauer et al. 2012	Merkt/Müller 1999	Hughen et al. 2000; Hughen et al. 2004b

Tab. 70 Durations of the main Lateglacial isotope events and the correlatives in the various laminated sequences. The limits of the NGRIP ice-core record (GICC05) are set according to the onsets given in **tab. 64**. The duration is given in ice-core years (yrs) and with the increase of the counting error within the event, not the accumulated total counting error. * one year is given here because negative durations are not possible for events but if the oldest possible onset for GI-1c is subtracted from the youngest possible onset of GI-1d the result would be -6 years. The limits in the Meerfelder Maar record (MFM) are set according to changes in the varve composition (sediment) for the main changes in the Lateglacial (Brauer/Endres/Negendank 1999; Brauer et al. 2008) but only differences in the vegetation record (pollen) are recognised during the Lateglacial Interstadial and the limits are set accordingly (Litt/Stebich 1999). In the Holzmaar record 320 varves are added in the GS-1 section (Leroy et al. 2000; Zolitschka et al. 2000).

no. 1,350 which dated approximately to 12,870 years cal. b2k (**fig. 36**; Schaub et al. 2008b, 82[39]). In this correlation, the Younger Dryas would have encompassed 1,230 years (**tab. 69**).

In a numerical comparison, the Australian results more closely resembled the oxygen isotope record from NGRIP and the $^{230}Th/^{234}U$ date from the Hulu Cave than the Swiss attempt. In fact, in a more recent presentation (Kaiser et al. 2012), the Swiss YD_A chronology was compared to the IntCal09 calibration which relied on the Australian attempt (Reimer et al. 2009). This suggestion would shift the YD_A chronology some 70 years younger on the calendar timescale than was suggested by Matthias Schaub and his colleagues (Schaub et al. 2008b). In this case, an overlap of the YD_B chronology with the YD_A record should be observable over some 70 years in the dendrochronological record. In the CELM a sharp decline in ^{14}C ages was recorded between the maximum values dated to c. 2,215 and c. 2,330 years relative tree-ring ages with the steepest part around the relative tree-ring no. 2,246 (Kaiser et al. 2012, fig. 6). Besides the ^{14}C ages, the number of available trees significantly decreased whereas the tree-ring width generally increased but began to wiggle strongly (Kaiser et al. 2012, fig. 5). However, a smaller excursion in the tree-ring width began some 50 tree-rings earlier around the ring no. 2,200 and the onset of the Younger Dryas as an environmental event displayed by tree growth patterns (event 9) was set to this 50 years earlier increase (Kaiser et al. 2012, 86 f.). In this comparison, ring no. 899 of the CELM was anchored at 14,142-14,114 years cal. b2k (Kaiser et al. 2012). Consequently, the onset of event 9 dated to 12,841-12,813 years cal. b2k. Nevertheless, uncertainties in the IntCal09 data, in particular, at the onset of the Lateglacial Stadial caused Klaus Felix Kaiser and his colleagues to consider the anchorage of the CELM with the IntCal09 calibration record only as an approximation (Kaiser et al. 2012, 87).

Since the onset of the Lateglacial Stadial was a very gradual transition in climatic and environmental parameters, a better correlation point should be used for the construction of a Lateglacial chronostratigraphy. The widely recognised volcanic LSE event marks a good correlation point in this period. Some poplars (*Populus* sp.) preserved near the volcano within the LST were matched to the GLPC (**fig. 10**) and, thus, related the dating of the LSE exactly to this still floating dendrochronological record (Baales/Bittmann/Kromer 1998; Friedrich et al. 1999, 34; Kromer et al. 2004, 1205). The sharp decline of ^{14}C ages at the onset of the Younger Dryas post-dated the LSE in this record by some 200 years (Kromer et al. 2004, 1205 f.).

For a better understanding the floating GLPC (Kromer et al. 2004) and the equally floating CELM (Kaiser et al. 2012) are correlated based on some shared tree-ring sequences and ^{14}C dates (see **tab. 2; fig. 37**). In the dendrochronological comparison, the first 200 years of both records were assumed to be mainly influenced by Dättnau sequence (cf. Kaiser et al. 2012, 86; Friedrich et al. 2001b, 1226). The comparison reveals that the two records correlate either in the tree-ring growth patterns or in the ^{14}C dates. The ^{14}C dates and tree-ring growth patterns were correlated in both records according to the respective relative tree-ring numbers[40]. However, in the two possible correlations, different positions of the LSE became apparent. This clarification is of some importance if the LSE is used as a correlation point. The assumed impact of the LSE which was identified by a disturbance in the tree-ring growth patterns (E7) in the Dättnau record (Friedrich

39 For the onset of the Younger Dryas various dates were given such as 12,900 years cal. b2k (i.e. ring no. 1,320) or ring no. 1,400 (i.e. 12,820 years cal. b2k; Schaub et al. 2008b, 83) but ring no. 1,350 corresponds to a significant decrease in the Allerød to Younger Dryas transition (Schaub et al. 2008b, 83) and is selected as onset according to the convention in the current study.

40 The comparison of the ^{14}C dates given in Kromer et al. 2004, 1204 fig. 1 & supplementary data and the ones given in Friedrich et al. 2001b, 1227 fig. 7 indicate that the same relative ring numbers/positions were used and, consequently, the connection of ^{14}C dates and tree-ring growth patterns seems correct for the GLPC. The CELM data were given in the same article, the tree-ring growth patterns in the relative Zürich scale (Kaiser et al. 2012, 86 fig. 5) and the ^{14}C dates in the SWILM relative scale (Kaiser et al. 2012, 87 fig. 6). These two scales were assumed to be identical and Klaus Kaiser did not correct this assumption as false in a written communication in May 2011. Still, this assumption remains a possible source of erroneous correlations.

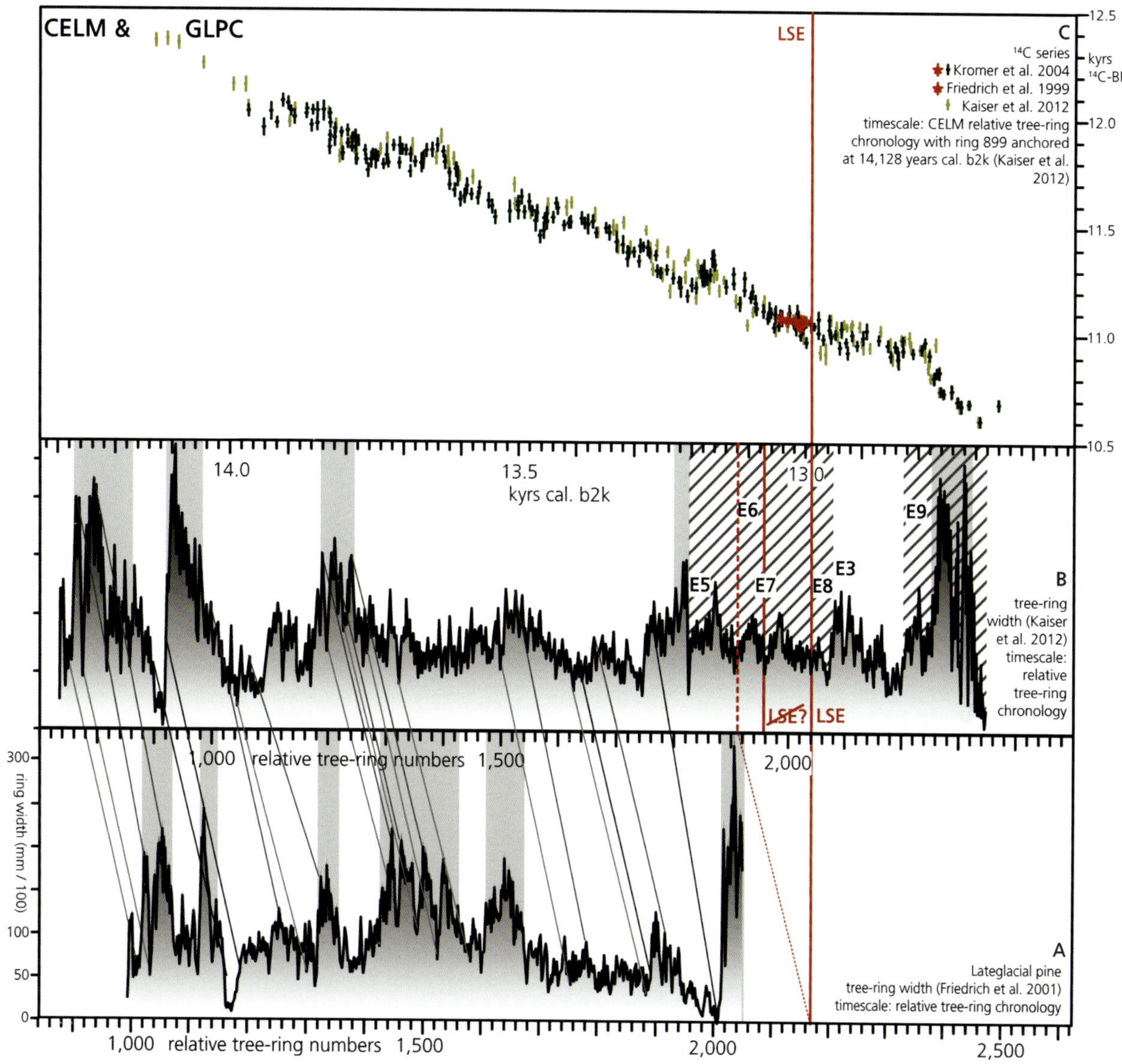

Fig. 37 Correlation of Central European Lateglacial Master chronology (CELM; Kaiser et al. 2012; see **fig. 11**) with the German Lateglacial pine chronology (GLPC; Kromer et al. 2004; see **fig. 10**) along the shared ^{14}C dates. Grey shaded areas relate to periods with an average tree-ring growth of more than 1.25 mm/yr. Thin bars are given as guide lines (light grey: low; medium grey: high; black: transition. – Dotted line: LST correlation) for the correlation along the tree-ring growth patterns. **A** Tree-ring width record of the GLPC given with relative tree-ring number as timescale (Kromer et al. 2004). – **B** Tree-ring width record of the CELM given with relative tree-ring number as timescale (Kaiser et al. 2012). Dark grey hatched areas are events E6 and E9 according to Kaiser et al. 2012, fig. 5, where also the short-term events E5, E7 and E8 are given. E3 is set according to Schaub et al. 2008a, 36-38. – **C** ^{14}C series (some dates were set slightly apart for better readability). Timescale according to the correlation of CELM with IntCal09 (Kaiser et al. 2012) and shifted from cal. BP to cal. b2k.

et al. 1999, 34) was not correlative with the position of the LSE as assumed from the poplars in the GLPC (Friedrich et al. 1999, 36-38). Besides the growth disturbance E7 which occurred in what was assumed to be GI-1b, further short-termed disturbances in the Swiss records were documented, for example, a some 100 years younger event (E8) fell to the transition from the assumed GI-1b to the assumed early GI-1a (Kaiser et al. 2012, 85 f. fig. 5). Another short-termed but sharp event (E3) occurred another 20-25 years later than E8 within early GI-1a (Schaub et al. 2008a). The latter two positions were in accordance with the position of the LST in laminated sequences of Central Europe based on the vegetation development (Lotter et al. 1992; Litt et al. 2001, 1247 tab. 3; Magny et al. 2006). However, the correlation of the CELM with the GLPC revealed that none of these events is equivalent with the LSE in the GLPC which according to the correlation

of ^{14}C dates post-dated the E7 in the CELM by some 80 years (**fig. 37**). If the two records were correlated along the tree-ring growth patterns the LSE in the GLPC would predate the E7 in the CELM record by some 40 years. Thus, the two different correlations are offset by approximately 125 years with the ^{14}C wiggle matching resulting in the younger ages.

According to the classic correlation of the ^{14}C dates of the floating GLPC with the Cariaco basin calibration data set, the LSE was dated to approximately 13,200 years cal. b2k (Kromer et al. 2004, fig. 2). Based on the shift suggested by the Swiss attempt, the LSE would date to approximately 13,070 years cal. b2k, whereas the Australian Huon pine attempt would result in an age of approximately 12,965 years cal. b2k. The IntCal09 corrected Swiss attempt resulted in an age of approximately 12,961-12,913 years cal. b2k. These latter dates are consistent with the date for the LSE in the MFM varve record (12,930 ± 30 years cal. b2k) as well as the Ca^{2+} corrected sulphate peak 4 in NGRIP at 12,918 ± 138 years cal. b2k (see p. 9-12 and p. 310-312. 317 f.; cf. Mortensen et al. 2005). Since all dates relate on the correlation of the ^{14}C ages, the possible offset of 125 years in the tree-ring growth patterns make no difference in these ages. In general, the Swiss attempt would be given priority in the present analysis due to the intra-hemispheric and also Central European character of the underlying data but the offset between the date for the LSE as given by the Swiss attempt to date the MFM as well as assumed by the sulphate content of NGRIP is too significant to be neglected.

In addition, the correlation of the tree-ring growth patterns is difficult due to partially intense reactions of the tree-rings to very local factors (cf. Schaub et al. 2008a). The position to which the YD_A chronology was placed according to the suggestion of the IntCal09 corrected Swiss attempt appears to correlate with the onset of the Younger Dryas (event 9) in the CELM (Kaiser et al. 2012) and the onset of the extended CEDC record. The exact position in the current project was chosen according to the correlation of the GLPC ^{14}C data set along the LSE in the MFM record. In this correlation, a significant overlap of the YD_A chronology with the YD_B chronology would have also resulted in a small overlap of the CELM with the YD_B dendrochronological record (**fig. 38**). The very low values at the younger end of the CELM as well as the very low values at the onset of the YD_B chronology correlate with a short but significantly low value in the YD_A chronology. This low point correlates to the lowest point in the NGRIP oxygen isotope record in GS-1. Furthermore, a small peak in the YD_A chronology followed by another low point immediately before the end of the YD_A chronology are correlative with the developments at the onset of the YD_B chronology. In this correlation the assumed onset of the Younger Dryas in the CELM dendrochronological record correlates with the onset of the Lateglacial Stadial in the oxygen isotopes in NGRIP (**fig. 38**). This onset of the Younger Dryas in the dendrochronological records was marked in the CELM and also in the YD_A chronology by an increase of the tree-ring width which quickly declined again in the YD_A chronology but further increased in the CELM dendrochronological record. The number of trees for the CELM is very low in this part and the tree-ring growth patterns are influenced intensely by age-related factors (Kaiser et al. 2012, 86) making the pattern of the YD_A chronology more reliable in this part. Furthermore, if the 125 years offset between the two different correlations of the GLPC and the CELM were based on a miscorrelation in the CELM record (see above) the dendrochronological record of the CELM would have to be shifted 125 years older on the timescale. In this case, the CELM and the YD_A chronology would only overlap for a few years or decades. Moreover, the increasing tree-ring width values at the end of the CELM would correlate with the end of GI-1a and the final decline in these values would be concomitant with the onset of the GS-1 in the NGRIP record. Even though these uncertainties about the position of the CELM record remain, the position of the YD_A according to the revised Swiss attempt appears reliable based on the dendrochronological comparison but the very short overlap of the involved records and the reduction in number of trees makes this correlation still insecure.

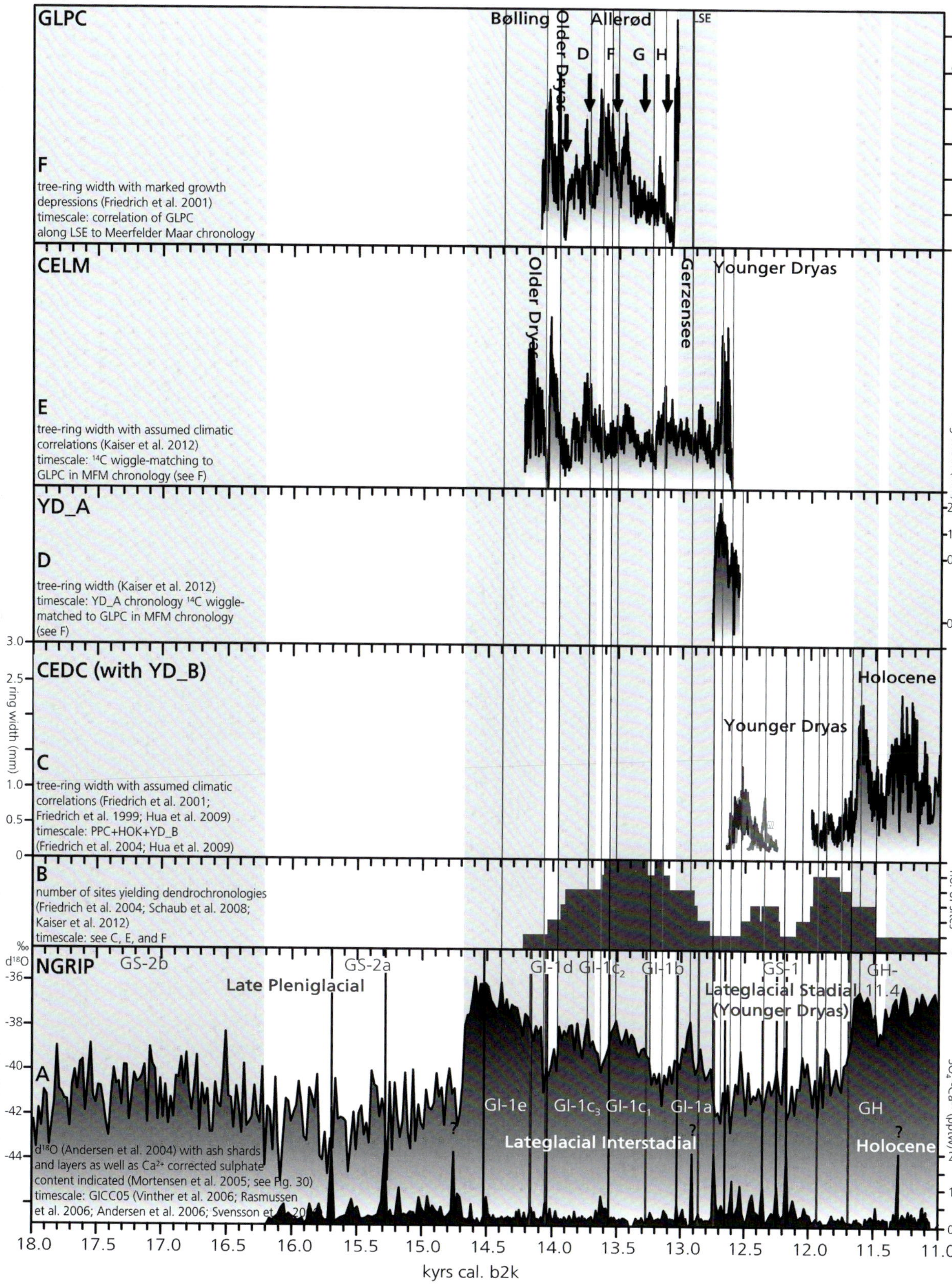

Fig. 38 Comparison of NGRIP with the European dendrochronologies. Thin bars given as guide lines (light grey: low; medium grey: high; black: transition). **A** Oxygen isotope record from NGRIP with the Ca^{2+} corrected sulphate content but without associated ash layers (see **fig. 30**). Grey shaded areas represent events of more interstadial values than the surrounding values. – **B** Number of sites from Central Europe yielding dendrochronologies (Friedrich et al. 2004; Schaub et al. 2008b; Kaiser et al. 2012). Timescale is according to C and E. – **C** Central European dendrochronology (Friedrich et al. 2004; Schaub et al. 2008b; Hua et al. 2009). – **D** YD_A chronology (Schaub et al. 2008b; Kaiser et al. 2012). – **E** Central European Lateglacial Master chronology (Kaiser et al. 2012). – **F** German Lateglacial pine-chronology (Friedrich et al. 2001b; Kromer et al. 2004). – Correlations of tree-ring growth patterns with climatic bio-chronozones in C-F are given according to the respective authors. For further details see text.

Without a secure connection between the CEDC and the Lateglacial tree-ring chronologies (YD_A, GLPC, CELM) the full chronostratigraphic potential of the latter cannot be explored. Thus far, the chronostratigraphic value of the dendrochronological record in the Lateglacial Interstadial depends partially on how good the tree-ring growth patterns reflect the climate signal (cf. Friedrich et al. 2001b; Schaub et al. 2008a). However, a comparison of the various attempts at identifying the isotope events in the Lateglacial Interstadial section of the tree-ring growth patterns illustrates the importance of very local *stimuli* and also the apparently varied response times of the tree-rings (**fig. 38**; cf. Friedrich et al. 2001b; Kaiser et al. 2012). These local factors were also revealed by the number of trees per site which is usually given in the original publications and indicates besides the reliability of the chronology also the density of tree cover at this spot and the local preservation quality. In contrast, the number of sites containing preserved trees reflects the general conditions of preservation as well as the spread of forest cover in Central Europe (**fig. 38B**). However, to evaluate the exact number of sites is partially difficult because some chronologies such as the Danube one were made of trees from several spots in the gravels. These trees were often driftwood which stranded on the river banks and got covered by the gravels. Their origin cannot be further specified but the Danube or the Upper Rhine valley and are therefore counted as a single site in this study. The gravels in the Upper Rhine valley also produced several pine remains (cf. Friedrich et al. 2004, fig. 1) of which at least some can be attributed to the Late Pleniglacial (cf. Rosendahl et al. 2006). However, this material did not contribute to a dendrochronology and cannot be exactly positioned on a calender time scale. Therefore, this region is not counted in the graph.

Besides the CEDC, the Eifel maars provided an almost continuous chronology up to present day. However, due to discontinuous varve formation in the upper part of the MFM record and a hiatus in the early Holocene of this record as well as disturbances in the mid-Holocene part of the Holzmaar record (Brauer et al. 2000b; cf. Lücke et al. 2003), these records cannot be regarded as completely continuous. Based on a comparison of both records an almost continuous varve chronology could be established that was correlated along the Ulmener Maar tephra at 11,050 years cal. b2k and along the LST in the final Lateglacial Interstadial. The micro-hiatus of approximately 240 varves in the early Holocene of the MFM occurred some 1,000 varves on top of the Ulmener Maar tephra and, therefore, did not affect the Lateglacial.

In the MFM record the transition between the Younger Dryas and the Pleistocene was defined by a decrease in varve thickness and a change from a minerogenic to organic dominated sedimentation at 11,640 ± 40 years cal. b2k (**tab. 71**; Brauer/Endres/Negendank 1999; Brauer et al. 2000b). This change was concomitant with an increase of pine (*Pinus* sp.) in the pollen profile marking the palynological onset of the Holocene (Litt/Stebich 1999). However, the decrease of the varve thickness was gradual and lasted until 11,580 ± 40 years cal. b2k (cf. Brauer/Endres/Negendank 1999, fig. 4). In the 4 cm below the transition, the varve preservation was poor and as a result the exact point of change could not be set unambiguously based on the varve thickness. According to the varve chronology, this section of poorly preserved varves began at 11,690 ± 40 years cal. b2k if a maximum error of 20 % for the recognition of varves was taken into account (Brauer et al. 1999, 324). In the Holzmaar sequence, also a change from organic to rather organic-minerogenic sediment was observed and dated to 11,658 ± 128[41] years cal. b2k (Leroy et al. 2000, 54-57). In the pollen composition, an instant response of *Filipendula* sp. which indicate mean temperatures in July of a minimum of 10°C was used to characterise the onset of the Holocene which preceded the change in the sediment by some decades and was set to 11,682 ± 129 years cal. b2k (Leroy et al. 2000, 64). The increase of pine occurred some decades later around 11,625 years cal. b2k and was succeeded by an

41 The error estimate is based on a counting error of 1.1 % which adds up constantly in the Holocene and Lateglacial part of the record (Brauer et al. 1994, 329; Zolitschka 1998).

main isotope events	NGRIP	MFM	Holzmaar	Hämelsee	Rehwiese	Cariaco Basin
GH-11.4 (PBO)	11,490 ± 97	×	×	×	×	×
GH	11,681 ± 111	11,640 (11,690) ± 40 (varve thickness)	11,690 ± 129 (diatom); 11,682 ± 129 (pollen); 11,658 ± 128 (sediment composition); 11,640 ± 128 (AP increase)	11,660 ± 119 (amelioration); 11,628 ± 118 (sediment composition; algae; caldocera); 11,610 ± 118 (pollen)	11,690-11,410 (pollen); 11,743 (micro-facies)	11,630 ± 16 (grey scale)
GS-1	12,770 ± 133	12,729 ± 40 (varve thickness + pollen)	12,670 (sediment)[1]; 12,656 (pollen, PAZ 9)[1]	c. 12,750	12,725 (micro-facies); 12,730-12,707 (pollen)	12,965 ± 16 (grey scale); 13,015± 10 (^{14}C)
GI-1a	13,070 ± 142	c. 12,900 (pollen)	12,885 (pollen, PAZ 7)[1]	c. 12,950	×	×
GI-1b	13,250 ± 147	c. 13,100 (pollen)	13,159 (pollen, PAZ 6)[1]	c. 13,175	×	c. 13,365
GI-1c_1	13,580 ± 156	13,400 ± 40 (pollen)	13,199 (pollen, PAZ 5)[1]; 13,539 (pollen, PAZ 3)1	c. 13,320	×	×
GI-1c_2	13,730 ± 160	13,590 ± 40 (pollen)	13,355 (pollen, PAZ 4)[1]; 13,635 (pollen, PAZ 2, Older Dryas)[1]	×	×	c. 13,765
GI-1c_3	13,970 ± 166	13,730 (*13,840*) ± 40 (pollen) 13,750 (*13,860*) ± 40 (sediment)	13,723 (pollen, PAZ 2, Bølling)[1]	×	×	×
GI-1d	14,090 ± 169	13,850 (*13,960*) ± 40 (pollen + sediment)	×	×	×	c. 14,165
GI-1e	14,687 ± 187	14,070 (*14,180*; onset of varve preservation; interpolated onset [sediment+pollen]: 14,500 (*14,610*) ± 220)	14,730 ± 300 (peak of sedimentation rate; other onsets [increased lacustrine productivity]: 14,240 ± 270; [pollen preservation, onset allochtonous varves]: 14,208[1])	×	×	c. 14,773
GS-2a	16,230 ± 255	×	16,550	×	×	×
ref.	see **tab. 64**	Brauer/Endres/Negendank 1999; Litt/Stebich 1999; Litt et al. 2001; Brauer et al. 2008	Leroy et al. 2000; Zolitschka et al. 2000	Merkt/Müller 1999; Litt et al. 2001	Neugebauer et al. 2012	Hughen et al. 2000; Hughen et al. 2004b; Hughen et al. 2006

Tab. 71 Comparison of the dates (given in years cal. b2k) for the onsets of the correlatives of the main Lateglacial isotope events in the varve records. Ages in grey are according to the reading of the present author. Italic ages set in parentheses in the MFM column contain additional 110 years due to a comparison of a 4 cm slump and turbidite section with the Lake Wollingst (Brauer et al. 2000a). [1] Varve chronology according to Leroy et al. 2000 with additional 320 years for the ages before the mid-Younger Dryas. – For further details see text.

increase of birch (*Betula* sp.) around 11,600 years cal. b2k. Furthermore, an increased occurrence of planktonic diatoms were already observed at 11,690 ± 129 years cal. b2k in the Holzmaar record (Leroy et al. 2000, 66f.). Hence, the Holocene/Younger Dryas transition in the Holzmaar record dated to the same period as in the MFM record (Brauer et al. 2000b) but in detail the response times were different. The sedimentological response was some 30 years slower in the Holzmaar than in the MFM, whereas the first palynological indicator reacted some 40 years previous to the MFM record. If in both records the increase of pine is compared, the Holzmaar record again lagged behind the MFM record by some 15 years. Even though the localities are situated very close to each other, the topography might have caused the different response times of the vegetation and as a result the establishment of the new sedimentological regime. Nevertheless, the development of the varves in the MFM and the Holzmaar record documented exactly the offset between the limits of this important transition in the Greenland oxygen isotope record and the CEDC (see **tab. 69**) with an onset of changing climates around 11,690 years cal. b2k and an establishment of the new climatic and environmental conditions around 11,640 years cal. b2k.

The onset of the Younger Dryas was previously determined by the general composition and formation of the sediment as well as the palynological development from 12,730 years cal. b2k (Brauer et al. 1999; Litt/Stebich 1999). More recently, a high-resolution analysis of the MFM sediments identified the precise onset at 12,729 varve years cal. b2k (Brauer et al. 2008). This date marked a significant shift in wind track patterns in north-western Europe. These patterns are influenced by fluctuations in the polar front (i.e. drifting oceanic ice), which is also an important factor influencing in the deuterium excess (Guiter et al. 2003, 70). However, the change in deuterium excess marking the onset of the Lateglacial Stadial was recorded almost 170 years earlier in the NGRIP record (Steffensen et al. 2008). In the MFM record, an approximately 30 year prelude to this shift was detected with short intervals of varying wind and moisture climate and the varve thickness gradually began decreasing some 50 years before the shift (Brauer et al. 2008, 521 fig. 2). Although this period clearly post-dates the change in the deuterium excess (12,896 ± 138 years cal. b2k), it coincided almost perfectly with the onset of GS-1 in the oxygen isotope record (**tab. 64**). In the deuterium excess record, a small recovery of the values can be observed after the significant low values at the onset of this new deuterium excess regime (**fig. 3**). Thus, the coincidence of the changed wind track patterns in the MFM record with the decreased air temperatures in Greenland supportes the suggestion of different response times to the onset of the Lateglacial Stadial in the high latitudes atmosphere and in the low latitudes ocean. If this process of the Lateglacial Stadial cooling was related to some type of Heinrich event (Andrews et al. 1995; Dokken/Jansen 1999) various phases of this event with different effects at different latitudes could be a possible explanation for these offsets (cf. Stanford et al. 2011b). However, a problem in the varve chronology of the Eifel maars cannot be excluded completely. For instance, the Younger Dryas section of the nearby Holzmaar only encompasses 654 varve years and the limit from the Lateglacial Interstadial to Lateglacial Stadial would date to 12,350 varve years cal. b2k (Leroy et al. 2000, 57). Since the two maar records were correlated along the Ulmener Maar tephra and the LST, the amount of varves between those markers were relevant for the interpolation of the missing varves in the Holzmaar record. According to this comparison some 320 varves were missing in the Holzmaar record (Leroy et al. 2000; Brauer et al. 2000b). Based on sharp changes in the pollen record, Suzanne A. G. Leroy and her colleagues assumed that the majority of varves were missing in the mid-Younger Dryas section around 12,075 varve years cal. b2k (Leroy et al. 2000, 67). This could explain the considerably short duration of the Younger Dryas. However, bioturbation and underestimation due to poor varve quality were also considerable as alternative explanations for missing varves. Nevertheless, the onset of the Younger Dryas as revealed by the sediment as well as by the pollen is considerably younger than in the nearby MFM also with the 320 additional varve years (Leroy et al. 2000). Since the records were correlated tephrochronologically, different response times or pos-

sible inconsistent definitions appear the most probable explanation for this discrepancy. Moreover, in the MFM record, no disturbed sediment was observed within the Lateglacial Stadial except for the 4 cm below the Pleistocene/Holocene transition and, thus, only an unrecognised hiatus in this record could explain a chronological offset. In contrast to the Holzmaar record, no sharp change in the pollen profile was observed (cf. Litt/Stebich 1999).

In the MFM, a change from organic to clastic varves was observed at 12,290 varve years cal. b2k reflecting a significant change in the depositional system of the lake (Brauer et al. 1999, 325). This change reflects reduced wind intensities (cf. Brauer et al. 2008) that could be correlated with a warming episode recognised during the mid-GS-1 in the NGRIP oxygen isotope record (see p. 293 f. and **tab. 63**; cf. p. 332). If this correlation is correct the MFM record and the NGRIP record were still quasi-simultaneous in the mid-Lateglacial Interstadial. Even though the LST has not, to date, been indentified in the NGRIP record (cf. Mortensen et al. 2005), peak 4 in the Ca^{2+} corrected sulphate content (**fig. 30**) could be correlated with the dating of the LST in the MFM record. If this correlation is substantiated, the MFM and the NGRIP record were almost synchronous from the onset of the Holocene to the LSE.

The limits within the Lateglacial Interstadial were mainly based on the development of the pollen record in both maars (Litt/Stebich 1999; Leroy et al. 2000) because the sediment showed only small and gradual lithological differences in this part (Brauer/Endres/Negendank 1999; Brauer et al. 2000b; Brauer et al. 2000a). However, in the MFM record some 4 cm of slump and turbidite were found below 13,590 years cal. b2k. Based on a palynological correlation with the northern German Lake Wollingst, a further 110 varve years were added to the varve chronology (Brauer et al. 2000a). In the Holzmaar record, the comparable palynological zone of an increase of birch ended by an increase of sagebrush (pollen zone 2, Bølling part) was clearly much shorter than in the MFM (Leroy et al. 2000, 62 f.). Thus, either the additional 110 years overestimate this period in the Eifel region or 162 varves had to be added to the Holzmaar record below this point. The end of this period dated to 13,635 years cal. b2k and, thus, c. 45 years older than in the MFM. Besides a chronological gap in the MFM record, a delayed response in the MFM record could be a possible explanation for this difference. For both records there are no further marker horizons which could tie them reliably to the calendar chronology and the limits set according to the palynological analysis are very arbitrary in relation to the onsets of the isotope events.

In general, the lithological sub-divisions are more comparable to the main isotopes events than the pollen record but the only remaining lithological limits are the onset of varve formation and organic sediment in the MFM record and the onset of biogenic varve formation in the Holzmaar record which were both related to the onset of the Lateglacial Interstadial (Brauer/Endres/Negendank 1999; Leroy et al. 2000; Zolitschka et al. 2000). In the MFM, the varve formation began only around 14,070 years cal. b2k (with added 110 years: 14,180 years cal. b2k). At this point the Lateglacial Interstadial lake environment stabilised but more organic than clastic sediments were already deposited 75 cm below this onset (Brauer/Endres/Negendank 1999). Assuming a constant sedimentation in comparison to the first 200 continuous varves, the 75 cm between the onset of the varve formation and the onset of the Lateglacial Interstadial were calculated to have encompassed at least a further 430 varves (Brauer/Endres/Negendank 1999; Brauer et al. 2000a). Thus, according to this calculation the onset of the Lateglacial Interstadial in the MFM occurred by least at 14,500 years cal. b2k (with added 110 years: 14,610 years cal. b2k). If the 50 varves of assumed underestimation (Litt et al. 2001) were added to this calculated age the onset of the Lateglacial Interstadial in the MFM record is again almost identical with the onset of the Lateglacial Interstadial in NGRIP (**tab. 71**). A sedimentological transition reflecting an increased lake productivity occurred around 14,250 years cal. b2k in the Holzmaar record (Zolitschka et al. 2000; cf. Lücke et al. 2003). The preservation of sufficient pollen began around 14,208 years cal. b2k (Leroy et al. 2000). A significant peak in the sedimentation rate dated

to 14,730 years cal. b2k was related to thawing in the drainage area and subsequent meltwater input into the lake (Zolitschka et al. 2000). This episode was possibly correlative with the climatic amelioration at the onset of the Lateglacial Interstadial. This limit predated the onset of the Lateglacial Interstadial in the oxygen isotope records by only a few decades. Furthermore, in the clastic varves a peak of the total inorganic carbon around 16,550 years cal. b2k reflected probably a peak of aeolian input (Zolitschka et al. 2000). Possibly, this aeolian input correlated to the onset of the dry GS-2a event as observed in the Hulu Cave record (see p. 319-321). However, if these episodes were correlative the ages in the Late Pleniglacial section are overestimated in the Holzmaar record.

In addition to the Eifel maars, several lakes and kettleholes in northern and north-eastern Germany produced laminated sequences. The sequences of the northern German lakes were correlated along palynologically steep gradients from the elm decline and the chronology was based on a composite of well laminated parts from various lakes (Plußsee, Hämelsee, and Belauer See; Merkt/Müller 1999, 44 fig. 3) especially from the well preserved layers in the Hämelsee record (Merkt/Müller 1999, 49). The onset of the Holocene was set according to the rapid spread of birches at 11,610 ± 118[42] years cal. b2k (Merkt/Müller 1999, 49), even though increasing birch values indicate a possible climatic amelioration some 50 years earlier (Merkt/Müller 1999, 55). An abrupt change in the sedimentation (grain size and concentration of clastics) accompanied by changes in the cladocera and algae communities occurred 15-20 years earlier than the palynological onset (Merkt/Müller 1999, 58). The change from clastic to organic sedimentation in the Hämelsee record was consequently placed at approximately 11,628 ± 118 years cal. b2k. The geochemical composition of the sediment took a further 15 years after the palynological onset of the Holocene until interstadial values were established. Thus, the transition from the Pleistocene to the Holocene was recorded in the northern German records some decades later than in the Eifel maars. This transition still revealed a very comparable pattern with first climatic indications preceding the significant changes in the environmental parameters which were almost concomitant to developments further south. However, below the transition, the lamination in the northern German lakes was generally poor and, consequently, exact age estimates were hard to establish.

Only the north-eastern German Rehwiese record yielded a reliable varve formation throughout this period. This record was correlated to the MFM record using the LST. Comparable to the northern German records, a first small increase of birch dated to 11,690 years cal. b2k occurred alongside a decrease of sagebrushes (*Artemisia* sp.) and juniper (*Juniperus* sp.). These changes marked the onset of the Pleistocene to Holocene transition in the Rehwiese record. In contrast to the Holzmaar record, the increase of *Filipendula* sp. began considerably later around 11,500 years cal. b2k. In the Rehwiese record, the end of the transition was set to the expansion of bulrush (*Typha latifolia*; Neugebauer et al. 2012), which indicates a mean July temperature of 13°C (Isarin/Bohncke 1999). This expansion occurred in the non-laminated part of the sediment and was interpolated to date to 11,410 years cal. b2k. Thus, the palynological transition was very gradual in this record but the first climatic changes were also recorded at 11,690 years cal. b2k. In comparison with the Eifel maars, the Rehwiese pollen suggested a temperature gradient which shifted from south-west to north-east and mean July temperatures which were common for the onset of the transition in the Eifel were established only some 180 years later in the Berlin area. In the Rehwiese record, the sedimentological and the geochemical change preceded again the palynological changes by some 50 years (Neugebauer et al. 2012, 100). In particular, the varve formation and the calcite precipitation shifted already at this early point around 11,743 years cal. b2k. This early onset suggested a response of the lake environment to the climate

42 The overall counting error for the Lateglacial part of the Hämelsee is given as 3 % with control counts providing counting errors of less than 2 % (Merkt/Müller 1999, 47). The counting error was calculated as adding up from the elm decline downwards with 2 %.

change some 30 years before the high-resolution proxy records in Greenland began reacting (Steffensen et al. 2008). If this early response of the lake environment is considered improbable some 50 varves must be missing between the LST and this onset of the Holocene in the Rehwiese record. In regard to the development of the pollen, the chronology of the Rehwiese record seems comparable to the Eifel maars and the northern German lakes and, thus, this early response of the lake appears rather valid. Moreover, the early reacting proxies reflect decreasing allochtonous input into the lake and an increasing primary productivity as well as, indirectly, an elongated warm period (Neugebauer et al. 2012, 97). Due to the generally low allochtonous input the presence of a local vegetation cover throughout the Lateglacial stadial was suggested for the Rehwiese record (Neugebauer et al. 2012, 101) and this presence of an established vegetation cover could explain the almost instant response in this record.

In the sediment record of Rehwiese, a second part of the Younger Dryas began at 12,326 varve years cal. b2k and was considered correlative to a warm episode which was recorded for example in the varve record of the MFM and in the oxygen isotope record of NGRIP. However, Ina Neugebauer and her colleagues advised some caution with this correlation because the more lateral position of Rehwiese in relation to the North Atlantic climate system might have caused some differences in detail (Neugebauer et al. 2012, 100). In the comparison of the chronology of the Rehwiese and of the MFM record, this consistent onset of the intermediate phase in the Younger Dryas indicates that the two records are still correlative at this part of the Lateglacial chronostratigraphy.

The onset of the Younger Dryas was in contrast to most other records, a very rapid transition in the Rehwiese record. It occurred in less than 25 years around 12,719 varve years cal. b2k in the pollen profile and within a year at 12,725 varve years cal. b2k in the sedimentological record (Neugebauer et al. 2012). This onset correlates with the onset of the Younger Dryas in the MFM. Since the Rehwiese record was correlated to the MFM chronology along the LST only a 210 varves below this transition, this correlation seems unsurprising. In the northern German lake sequences, the onset of the Younger Dryas was correlated with the duration of the Younger Dryas in Lake Gościąż (Merkt/Müller 1999). In the Hämelsee record a total of c. 675 varves were preserved and can be sub-divided in three sections: some 50 varves on top of the Younger Dryas/Allerød transition, 197 varves between this transition and the LST, and 428 varves below the LST. This number of varves below the LST correlates with the approximately 450 varves in the Rehwiese record (Neugebauer et al. 2012, 96). Further, uncorrelated sections provided estimates for the duration of pollen zones such as the Lake Wollingst (Brauer/Endres/Negendank 1999). In regard to the significant offsets between the palynologically defined zones (**tab. 69**) and the additional uncertainty arising in the chronology from uncorrelated sequences and untested chronostraigraphies, these parts are not further considered in this study.

The European lake varves have been chronostratigraphically correlated to the NGRIP record. However, the accumulated counting error of these records was comparable to the counting error of NGRIP and, therefore, produced no more precise dating of the limits. In addition, these records documented climatic shifts and combined them with biotic reactions. In particular, wind, air temperature, and water availability had significant impacts on the vegetation record. This direct combination of climate indicators and vegetation development indicated that occasionally vegetation zones are not concomitant with oxygen isotope and climatic developments. In contrast, vegetation records sometimes reacted to more local developments and are consequently not well suited as chronostratigraphic markers.

Besides the various European varve chronologies, the deep sea record from the Cariaco basin yielded a floating laminated chronology (**fig. 4**). The $\Delta^{14}C$ values from the laminated parts of the Cariaco basin grey scale were correlated with those of the CEDC. For the parts older than the end of the dendrochronology, the laminated parts and the grey scale were previously correlated to the accumulation rate in GISP2 and its chronology (Hughen et al. 1998a; cf. Hughen et al. 2004c). Subsequently, the grey scale was correlated

with the Hulu Cave record (Hughen et al. 2006). Three of the 46 tie points of this correlation fall into the Lateglacial: the Pleistocene/Holocene transition, the Interstadial transition/Lateglacial Stadial, and the Late Pleniglacial/Lateglacial Interstadial transition. The anchorage of the Lateglacial chronology in the Cariaco basin record to the CEDC and, thus, the onset of the Holocene remained widely unaffected since the comparison to the Hulu Cave record was only used for the period below 14,750 years cal. b2k. In the section between 12,450 and 14,750 years cal. b2k $^{230}Th/^{234}U$-dated fossil corals supplement the correlation. The onset of the Holocene was set according to a correlation of a steep gradient in the grey scale record to the CEDC to 11,630 years cal. b2k (Hughen et al. 2004c).

In the Cariaco basin, a gradual change from the Lateglacial Interstadial to the Lateglacial Stadial was recorded by the means of the grey scale, the $\Delta^{14}C$, and the biomarker record (Hughen et al. 2004b; Hughen et al. 2004c, 1164). A last peak occurred around 13,080 years cal. b2k in the grey scale indicating weak trade-winds and a concomitant minimum occurred in the $\Delta^{14}C$ record (with values of c. 11,200 years ^{14}C-BP[43]) following on a small wiggle in both records (Hughen et al. 2004b, 1958 fig. 3). This onset of declining/increasing values was discussed as a result of a shut down of the North Atlantic deep water formation and, thus, the onset of the Younger Dryas was placed to the last peak of the sediment and isotope record (Broecker 2003, 1519). At this point the values for the arid grassland vegetation were almost as strong as at the onset of the Lateglacial Stadial but the wet forest vegetation increased significantly again afterwards. A last peak of prevailing forest vegetation was signalled by the biomarkers some 50 years after the last peak of the other data. Around 12,940 years cal. b2k (c. 10,890 years ^{14}C-BP) a short but sharp decline occurred in the grey scale which is followed by increased values in the $\Delta^{14}C$ record. However, both curves recovered quickly again to almost pre-decline values. The steepest part of the $\Delta^{14}C$ incline followed some 25 years later (c. 10,880 years ^{14}C-BP). In the grey scale a prolonged depression formed around this steep incline beginning c. 12,910 years cal. b2k and lasting some 20 years. Afterwards the reflectance of the Cariaco sediments shortly recovered before a last steep decline occurred. Subsequently, minimal values for the sediment reflectance and the forest vegetation and maximum values of $\Delta^{14}C$ occurred around 12,885 years cal. b2k (c. 10,700 years ^{14}C-BP), some 200 calender years after the last $\Delta^{14}C$ and grey scale peak (Hughen et al. 2004b, 1958 fig. 3). In the same period some 500 years pass on the ^{14}C timescale forming a steep decline or »cliff« in the calibration curve. After this lowpoint the values in the grey scale stabilise for some 120 years before some extreme minima occurred after c. 12,760 years cal. b2k (c. 10,570 years ^{14}C-BP). During these 120 years, the $\Delta^{14}C$ data wiggled considerably with a general increase of the values before the absolute maximum of this time window was reached at c. 12,830 years cal. b2k (c. 10,600 years ^{14}C-BP). This maximum was followed by decreasing values with a minimum at 12,710 years cal. b2k (c. 10,570 years uncorrected ^{14}C-BP). Also, the vegetation development signalled a slight recovery of the wet forest environments until c. 12,830 years cal. b2k but was followed by peak values of arid grasslands around 12,710 years cal. b2k. If only the very low values in the Cariaco basin grey scale were counted the Younger Dryas comprised only 1,235 varve years (Hughen et al. 2004c, 1164). Moreover, the pattern displayed by the marine data from Cariaco basin resemble closely the Greenland ice-core data and showed comparably that the gradual trend towards stadial conditions was a long process (cf. Fiedel 2011) with vegetation proxies lagging behind the climatic parameters by some decades. Furthermore, if the correlation of the Cariaco record is correct it might indicate that the long lasting process began some 150 years earlier in the low latitude marine (last peak: 13,080 years cal. b2k) than in the high latitude ice-core record (last peak: 12,925 years cal. b2k).

43 The ^{14}C ages are given with a constant reservoir correction of 420 years because the exact position of a switch in the reservoir ages from 420 to 650 years at the Lateglacial Interstadial to the Lateglacial Stadial transition identified by comparison of the Cariaco basin record with the floating GLPC (e. g. Hughen et al. 2004c; Kromer et al. 2004) is not known.

Nevertheless, in such a case the stabilisation of the grey scale values concomitant with the changed deuterium excess data clearly suggested a shift in the low latitudinal North Atlantic regime as a source of this stadial event. Furthermore, the position of this shift explains the offset in the atmospheric parameters tied to the higher latitudes. Yet, an offset in the chronology cannot be completely excluded.

According to the comparison of the period between the onset of the Holocene and the LSE in the MFM and the duration of the Lateglacial Stadial in the Cariaco basin, the LSE would have occurred in the early Younger Dryas of the Cariaco basin. Besides the possibility that the Younger Dryas began considerably early in the subtropical Cariaco basin, this age difference can be explained by an overestimation of the duration of the Lateglacial Stadial in the Cariaco basin or a lack of at least 45 varve years in the MFM record. The possibility of overestimation was rejected by Konrad Hughen and colleagues (Hughen et al. 2004c, 1164), although a relatively high proportion of the record was formed by weakly laminated (c. 35 %) or unlaminated parts (< 1 %; Hughen et al. 2004c, 1163). This composition made the counting of varves problematic and required the correlation of the Cariaco basin varve chronology with other records. However, without an indication for the position of the LSE in the Cariaco record, a conclusion between the possible underestimation of the MFM record and a potentially different climatic regime in the Cariaco basin based on this single evidence should not be made. However, further indications for an overestimation in the chronology of the Cariaco record were provided by the radiocarbon calibration data set (see p. 359-364).

In addition, comparison of the limits within the Lateglacial Interstadial in the oxygen isotope and deuterium record of NGRIP and the grey scale record of the Cariaco basin several tie points in the Cariaco basin predate the reactions in the NGRIP record (**fig. 39**). An update of the record resulted in a lengthening of several parts of the record including the Younger Dryas (Hughen et al. 2004c). However, a change in the sedimentation regime resulting in an overestimate of the varves, for instance due to the deposition of four layers per year, was excluded for the Younger Dryas (Hughen et al. 2004c, 1164). However, if the wiggles in the Younger Dryas are compared, the records of the deuterium excess in NGRIP and the Cariaco basin grey scale are very similar until the early GI-1b event. Only after this point the records vary considerably. At the onset of GI-1b, the age difference between the oxygen isotope record from NGRIP and the Cariaco basin grey scale ranged around 110 years, whereas at the onset of GI-1c_2 the difference decreased to 35 years and it increased again to 75 years at the onset of GI-1d (**tab. 71**).

Since the onset of the Lateglacial Interstadial was correlated to the Hulu Cave record (Hughen et al. 2006), this part is again very comparable to the NGRIP record. The last minimum values occurred around 14,790 years cal. b2k and the peak in the grey scale is reached around 14,755 years cal. b2k. In contrast, the maximum values in the fatty acids and the ^{13}C record occurred some decades or centuries later (Hughen et al. 2004b). The onset is considerably sharper than the onset of the Younger Dryas but it predated the NGRIP records by some 100 years (**tab. 71**). This offset of approximately 100 years at the beginning of the Lateglacial Interstadial was previously recognised in the comparison of the Cariaco basin record with the GISP2 chronology and demonstrated a gradual development during the latter half of the Lateglacial Interstadial (Hughen et al. 1998b). The observed difference is either due to an underestimate of ice-core layers in NGRIP

→

Fig. 39 Comparison of NGRIP with the Cariaco basin record. Thin bars given as guide lines (light grey: low; medium grey: high; black: transition). **A** Oxygen isotope record from NGRIP with the Ca^{2+} corrected sulphate content but without associated ash layers (see **fig. 30**). Grey shaded areas represent events of more interstadial values than the surrounding values. – **B** Deuterium excess record from NGRIP. – **C** greyscale record (Hughen et al. 1998a). – **D** average chain length index of C_{24}-C_{28} *n*-alkanoic homologs from the fatty acids (FA) of leaf waxes (Hughen et al. 2004b). – **E** series of ^{14}C dates in three different versions: one with constant reservoir of 420 years subtracted (black symbols); one with 650 years subtracted from the results between 12,892 and 14,792 years cal. b2k (medium and light grey symbols; Hughen et al. 2004c; Hughen et al. 2006), and one with a dynamic reservoir subtraction according to the correlation with the Hulu Cave record (for results older than 13,537 years cal. b2k; Hughen et al. 2006). – For further details see text.

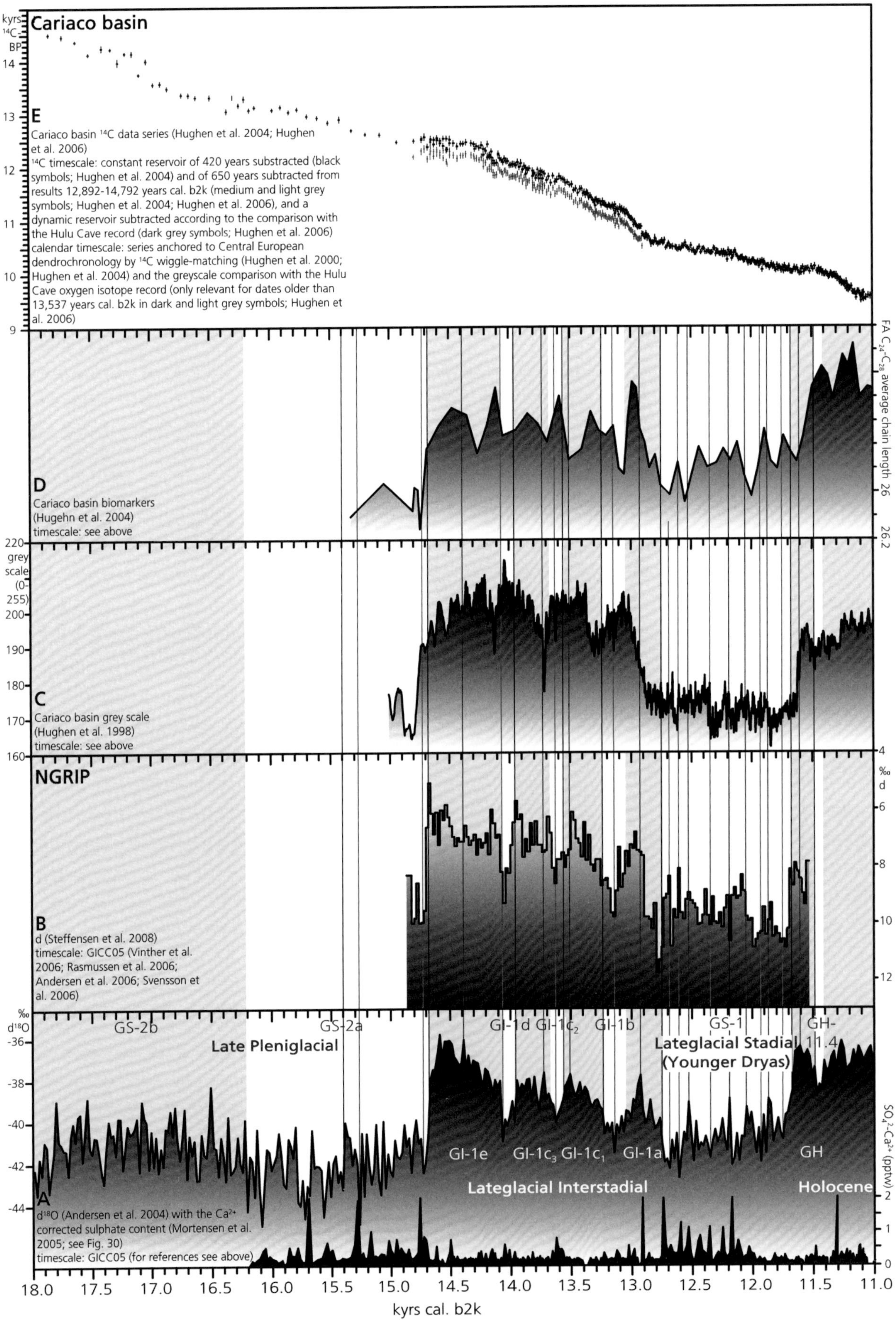
Cariaco basin
kyrs 14C-BP
E
Cariaco basin 14C data series (Hughen et al. 2004; Hughen et al. 2006)
14C timescale: constant reservoir of 420 years substracted (black symbols; Hughen et al. 2004) and of 650 years subtracted from results 12,892-14,792 years cal. b2k (medium and light grey symbols; Hughen et al. 2004; Hughen et al. 2006), and a dynamic reservoir subtracted according to the comparison with the Hulu Cave record (dark grey symbols; Hughen et al. 2006)
calendar timescale: series anchored to Central European dendrochronology by 14C wiggle-matching (Hughen et al. 2000; Hughen et al. 2004) and the greyscale comparison with the Hulu Cave oxygen isotope record (only relevant for dates older than 13,537 years cal. b2k in dark and light grey symbols; Hughen et al. 2006)
D
Cariaco basin biomarkers
(Hugehn et al. 2004)
timescale: see above
FA C24-C28 average chain length
C
Cariaco basin grey scale
(Hughen et al. 1998)
timescale: see above
grey scale (0-255)
NGRIP
B
d (Steffensen et al. 2008)
timescale: GICC05 (Vinther et al. 2006; Rasmussen et al. 2006; Andersen et al. 2006; Svensson et al. 2006)
‰ d
‰ d18O
GS-2b
GS-2a
Late Pleniglacial
GI-1d GI-1c2 GI-1b
GS-1
Lateglacial Stadial
(Younger Dryas)
GH-11.4
GI-1e GI-1c3 GI-1c1 GI-1a
GH
Lateglacial Interstadial
Holocene
A
d18O (Andersen et al. 2004) with the Ca2+ corrected sulphate content (Mortensen et al. 2005; see Fig. 30)
timescale: GICC05 (for references see above)
SO42--Ca2+ (pptw)
kyrs cal. b2k

or an overestimate in the Cariaco basin. The similarity of the ages in the NGRIP and the Hulu Cave record at the onset of the Lateglacial Interstadial suggests that rather an overestimation occurred in the Cariaco basin record. Another possible reasons for the offset are various response times and an earlier response in the Cariaco basin. However, the deuterium excess data relating Greenland to the lower latitudes during glacial periods (cf. Masson-Delmotte et al. 2005) reacted in a comparable fashion to the other proxies in NGRIP. Thus, an overestimation of 100 years in the Cariaco basin record appears as a probable reason but an exact onset for this age difference cannot be identified unambiguously. Nevertheless, the comparison of the ^{14}C calibration data set from the Cariaco Basin with the European dendrochronological records (see p. 359-363) sustains the necessity of adjusting the chronology of Cariaco Basin record. Due to the connection with the radiocarbon calibration, this adjustment is further specified later.

In conclusion, the Holocene amelioration appears to have begun around 11,680 years cal. b2k or, more precisely, the climate system influencing the North Atlantic climate zone switched probably to an interstadial mode after 11,703 ± 99 years cal. b2k (cf. Rasmussen et al. 2006; Steffensen et al. 2008; Walker et al. 2009). This shift in the climate regime caused general changes in the Northwest-European climate system and, in particular, in the glacio-hydrological cylce resulting in processes such as thawing of previously frozen grounds and/or melting of glaciers. To some degree, these processes probably masked the global climatic signals in the terrestrial records. The continental system restabilised some decades later (cf. Muscheler et al. 2008) and the terrestrial sequences of north-western Europe produced a signal of amelioration and increased organic productivity almost simultaneously around 11,640 years cal. b2k.
Consequently, for the purpose of correlating north-western European sequences in a general chronostratigraphy the type of record and of the signal needs to be considered. Since these data sets mainly represent terrestrial sequences which besides temperature also react on further factors such as the hydrological surrounding, a correlation at 11,640 years cal. b2k (dendrochronology, Lake Gościąż, Lake Perespilno, MFM) is used in the current project. Nevertheless, some processes such as the succession of biomes is controlled by various factors ranging from global to local scale. Thus, also parameters such as the concentration of clastics in sediments may react later in some regions, for example, in more northern latitudes or areas in the vicinity of an ice sheet (cf. Levesque/Cwynar/Walker 1997), whereas other areas such as places in more southern latitudes or protected valleys (cf. van Leeuwaarden/Janssen 1987) could react more instantly on an amelioration. However, the correlation of the various important Northwest-European records suggested that this delay fell within the standard deviation or counting errors of most records. In consequence, the climatic and environmental change from the Late Pleistocene to a fully established Holocene occurred across north-western Europe within a century or less.
For the purpose of considering human reaction on climatic change, it needs to be kept in mind that the climatic as well as first environmental changes (decreasing dust intensity, temperature amelioration, thawing of permafrost, and melting of glaciers) began presumably some 30-70 years earlier than the definition of the Holocene onset. Consequently, the Lateglacial/Early Holocene hunter-gatherer groups probably faced first signs of (severely) changing environments one or more generations before Holocene conditions were established in the terrestrial data. Adaptation process of these groups to their changing surrounding could have begun from the first signs onwards, i.e. even before the Holocene was established in the terrestrial environmental records.
Comparable to the onset of the Holocene, the onset of the Lateglacial Stadial in north-western Europe is not a sudden event, although it occurred abruptly in some proxy records. In general, the onset of the Lateglacial Stadial was a continuous process which took several decades to a few centuries to be accomplished. This gradual transition was a general pattern of this climatic event. The here described development of a

gradual cooling terminated by an abrupt, more intense cooling is a typical onset in a stadial cycle (Wolff et al. 2010). Various parameters were affected during the transition. For instance, the $\Delta^{14}C$ as well as the grey scale record from the Cariaco basin display a comparable pattern of a peak following on the short-term episode in the records and then a gradual decline followed over the next c. 200 years before stadial conditions were established. In several records, this transition lasted approximately 200 years. After this transition, the human surrounding transformed gradually into a cold adapted landscape in some parts of Europe, possibly accelerated by increased wind intensities and storm tracks (Brauer et al. 2008). In contrast, in protected areas the environment seemed partially unaffected (Price 2003) or almost unaltered (cf. Lotter et al. 1992, 195 f.; Eusterhues et al. 2002, 356). Moreover, the event encompassed more than 1,000 years in which the climate as well as the environmental record changed. The impact of this event on human behaviour therefore needs to be considered in comparison to the actual changes affecting the immediate surrounding (cf. Weber/Grimm/Baales 2011).

The onset of the Lateglacial Interstadial was a very comparable process as the onset of the Holocene. The few climatic records which yielded a signal of this major Lateglacial event recorded a very sharp change towards warm/moist climate conditions at 14,690 years cal. b2k. First indications of the amelioration began a few decades earlier, around 14,730 years cal. b2k. In contrast to the Holocene, the pedogenic development slowed the environmental development in many areas (e. g. MFM, Holzmaar) and, thus, a full interstadial mode was reached some centuries later.

The sub-division within the Lateglacial Interstadial was only comparable in the oxygen isotope records. Due to the additional influences, in particular of the hydrological system on the sites, the dates for the limits were also partially offset in the oxygen isotope records. The onset of the various sub-events as read from the vegetation proxies displayed significant discrepancies between the records and disqualify as reliable correlation markers. For example, the onset of a colder period within the younger Lateglacial Interstadial, the so-called Inner or Intra Allerød Cold Period (IACP, Lehman/Keigwin 1992, cf. Obbink/Carlson/Klinkhammer 2010), was considered to be correlative with GI-1b as well as the Gerzensee Oscillation (Lotter et al. 1992). The duration of GI-1b in the oxygen isotope records ranges between 160 and 180 years (see **tab. 68**), whereas the duration for this period in the other proxy records such as the tree-ring growth patterns or the pollen spectra exceeded this duration by 94 to 193 years (see **tab. 70**). In addition, the onset of this period in the European varve lakes post-dated the onset of GI-1b in NGRIP by 75 to 150 years (see **tab. 71**). The Cariaco basin greyscale recorded the onset of a colder period approximately 115 years before the onset of GI-1b and with the assumed shift of this record , the signal in the Cariaco basin would still precede the NGRIP record by 15 years and the terrestrial signals by some 110 to 210 years. Since the records appear relatively well correlated at the major transitions, these significant offsets within the Lateglacial Interstadial suggest that the »minor« climate signals were not strong enough to surpass rapidly the influence of local factors on the various terrestrial proxy records. Thus, the use of these »minor« limits as correlation markers can lead to significant confusion in the comparison of the terrestrial records. In contrast, volcanic tephras represent reliable markers. For instance, interstadial environments were recorded for a further c. 200 years after the LSE in the final Lateglacial Interstadial in the terrestrial records (e. g. Brauer et al. 2008; Merkt/Müller 1999; Eusterhues et al. 2002) and also the sharp decline in $\Delta^{14}C$ succeeded the LSE by some 200 years (Kromer et al. 2004). Thus, the use of this marker related the terrestrial tephrostratigraphies to various other proxy records. Even though some uncertainties about the exact position of the LSE remained, the quasi-contemporaneity of the direct evidences for the LSE was unquestioned. Consequently, these reliable markers should be used in particular for the chronological comparison of terrestrial records within the major isotope events.

The Late Pleniglacial was a period with high amplitudes in the oxygen isotope record of the NGRIP ice-core. This period was rarely recorded in the terrestrial records. In the records where the Late Pleniglacial

was documented, the numbers of the proxies were usually limited and the few values preceding the onset of the Lateglacial Interstadial were in general reflecting a very cold and dry environment. The onset of this particularly cold and dry phase in the Late Pleniglacial was rather diffuse and, possibly, non-synchronous in the various records.
A rather general observation in this comparison is the sharp signals of ameliorations in contrast to the more diffuse onsets of climatic deteriorations. Thus, the onset of ameliorations are better suited as correlation gradients. Nevertheless, short and sharp cold peaks are also good correlation indicators but these short-lived events are sometimes recorded poorly in the vegetation records.

Further stratigraphies

Stratigraphies from the deep sea usually have to be correlated to the Greenland ice-core records along tephrochronological and/or lithostratigraphic correlation points as well as according to radiometric dating, temperature developments, and/or isotope sequences. These deep sea records reflect climatic developments in the ocean and, thus, indicate changes in a major driving force of the global climate. Therefore, these archives are relevant indicators for the development of the climate in north-western Europe. In the present study, the deep sea record from MD01-2461 (**fig. 5**), offshore Ireland, was chosen as a representative for these records because it provides additional information which were considered relevant for the possible impact of the ocean on the north-western European climate. The timescale of this sediment core was originally set according to the GISP2 age model (Meese et al. 1997) and, thus, must be partially changed to correlate to the NGRIP record in GICC05.
In the stratigraphy of MD01-2461, volcanic ash layers were found of which one was assumed to represent the Vedde Ash (Peck et al. 2007, 864 f.). Furthermore, a zone of increased volcanic debris including a peak of rhyolitic material post-dated the Heinrich 1 event in this record (**fig. 40**). The MD01-2461 record was left in the GISP2 age model and simply shifted according to the peak in the rhyolitic ash grains assumed to reflect the Vedde Ash to match with the marker horizon of the Vedde Ash in the NGRIP record (**fig. 40**). When the record was shifted this way the zone of increased volcanic activity post-dating the Heinrich 1 event became correlative with an increase of volcanic activity after the rhyolitic ash layer at 1628.25 m depth in NGRIP (**fig. 40D**; Mortensen et al. 2005, 215 f.). Shifted into this position, the development of the sea surface temperature (**fig. 40C**) is also very similar to the development of the oxygen isotope (**fig. 40A**) as well as the deuterium excess record from NGRIP (**fig. 40B**). Therefore, no further alterations of the deep sea record were performed to correlate it to the GICC05. Although some stretching, compressing, and/or shifting might be necessary to fit this record exactly to the GICC05, the lack of further reliable correlation marker prohibits further precision in this project. Thus, the chronology of this record is a tuned timescale and not an independent one. This dependence on a correlation to NGRIP influenced the temporal relation between this record and other records and inhibits further specifications of the durations of various events and major conclusions on the numerical chronostratigraphy based on this record. Nevertheless, some observations can be made based on the relation of the different proxies in this record.
For instance, during the Late Pleniglacial, very low sea surface temperatures as indicated by the relative abundance of planktonic species were concomitant with an almost stagnant phase of IRD accumulation after a period of extreme fluctuations. Perhaps, this low in the sea surface temperature was correlative with the deterioration of the climate towards GS-2a. If this record was compared to the Hulu Cave oxygen isotopes, the deterioration in the deep sea record would have to be shifted 500 years younger. However, in the MD01-2461 record this stagnant period was correlated with the Heinrich 1 event (Peck et al. 2007).

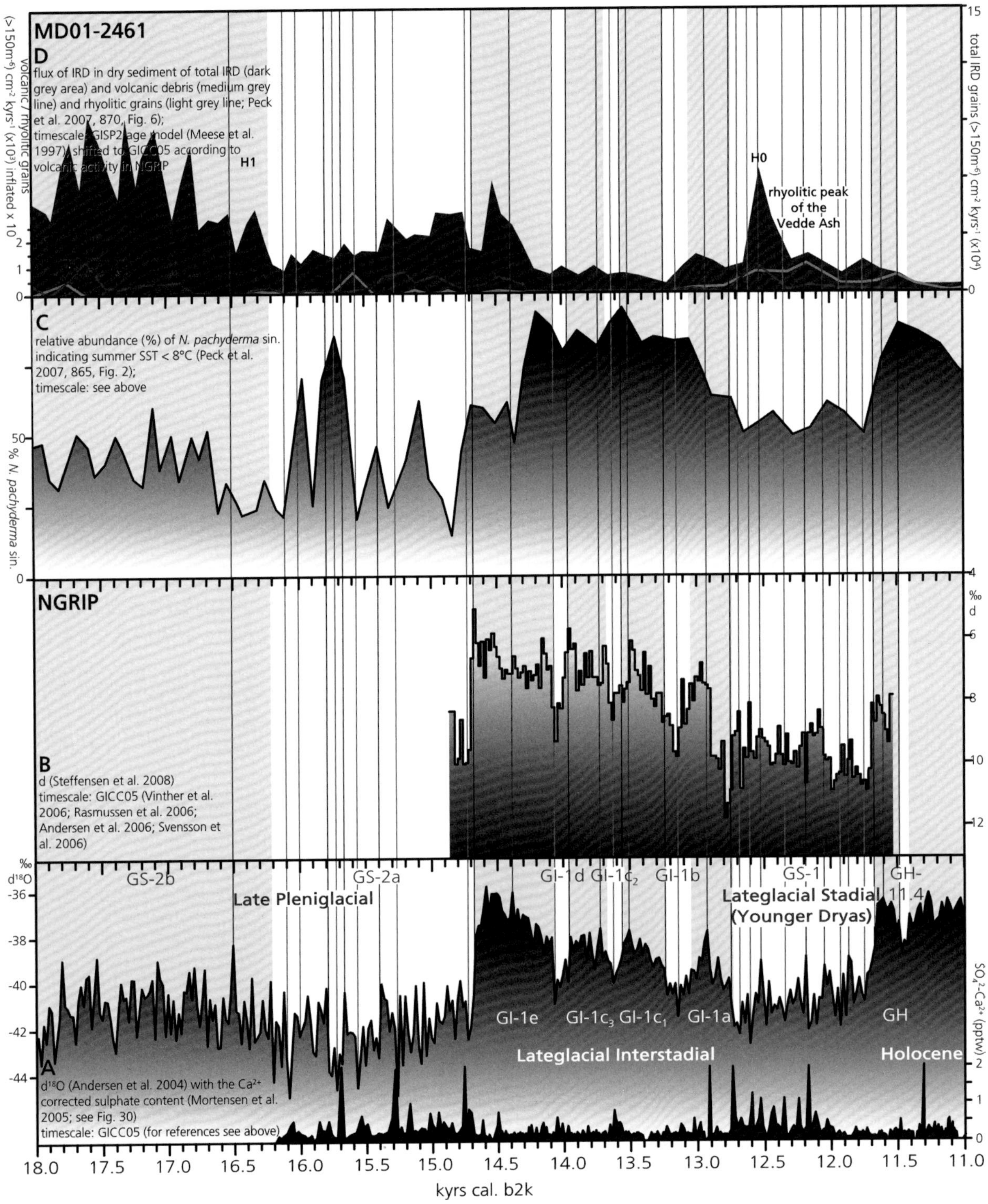

Fig. 40 Comparison of NGRIP and MD01-2461. Thin bars given as guide lines (light grey: low; medium grey: high; black: transition). **A** Oxygen isotope record from NGRIP with the Ca^{2+} corrected sulphate content but without associated ash layers (see **fig. 30**). Grey shaded areas represent events of more interstadial values than the surrounding values. – **B** Deuterium excess record from NGRIP. – **C** relative abundance (%) of *Neogloboquadrina pachyderma* sinistral, an indicator for polar waters and dominant in planktonic assemblages in waters of summer SST < 8°C, in MD01-2461 (Peck et al. 2007) correlated to NGRIP in GICC05 by tephrochronology. – **D** concentration of total IRD (black) and volcanic debris and tephra (grey line) in dry sediment of MD01-2461 (Peck et al. 2007) correlated to NGRIP in GICC05 by tephrochronology. **H1** Heinrich 1 event (Peck et al. 2007; cf. Phase 3 in Stanford et al. 2011b); **H0** Heinrich 0 event (Andrews et al. 1995; cf. Clark et al. 2002). – For further details see text.

In a more recent meta-study of the development of the Heinrich 1 event in the North Altantic, the Heinrich 1 event was sub-divided in several phases beginning as early as 19,050 years cal. b2k (Stanford et al. 2011b). The second phase occurred between approximately 17,550 to 16,750 years cal. b2k and was characterised by »a large-scale iceberg release and melting in the open North Atlantic IRD belt« (Stanford et al. 2011b, 1062) which is a very comparable pattern to the development of IRD before the Heinrich 1 event as specified in MD01-2461 (**fig. 40D**). In the early third phase of the Heinrich 1 event according to Jennifer Stanford and her colleagues, IRD deposition decreased in combination with increased freshening of the surface waters which was perhaps caused by the admixture of surface waters from the Nordic Seas (Stanford et al. 2011b, 1062).

This freshening reached its maximum around 15,150 years cal. b2k and decreased afterwards again. This maximum was clearly not related to a significant IRD deposition in their records. If the MD01-2461 record was shifted 500 years younger to match the deterioration in the Hulu and NGRIP records, the IRD flux of Heinrich 1 (as defined by Peck et al. 2007) would not cease before 15,750 years cal. b2k and the lowest value of IRD in the Late Pleniglacial would occur around 15,650 years cal. b2k.

This low of IRD was followed by an episode of two significant warm peaks in the sea surface temperature during and after which the IRD deposition began to increase again and the sea surface temperatures had some significant lows. The first and most significant low in the sea surface temperature would correlate approximately with 15,150 years cal. b2k in the shifted version. The warm episode clearly preceded the warming of the Lateglacial Interstadial and encompassed possibly several hundred years. Such an amelioration prior to the onset of the Lateglacial Interstadial was mentioned for the British records around 15,550 years cal. b2k by Jennifer Standford and her colleagues (Stanford et al. 2011b, 1062), and the peaks in the sea surface temperature proxy would clearly coincide with this amelioration if the record was shifted 500 years younger. However, the radiocarbon dates which dated the mentioned peak value in the British coleopteran record (Atkinson/Briffa/Coope 1987) can be considered as unreliable by modern standards and the warming identified by this record was meanwhile qualified as correlating with the early Lateglacial Interstadial (Coope/Lemdahl 1995; Coope et al. 1998). Thus far, a warm episode before the Lateglacial Interstadial with the intensity suggested by the planktonic record from MD01-2461 was not observed elsewhere.

Nevertheless, a warming of this intensity in the sea surface waters offshore Ireland would clearly have affected the climate of north-western Europe. One possibility is that the onset of preservation in various environmental records correlated with this warm episode. In the Eifel maars, this onset occurred only some centuries before the onset of the Lateglacial Interstadial and in the Swiss records this onset of the preservation occurred approximately around 15,200 years cal. b2k (Lotter et al. 2012). However, in the Austrian Längsee a sequence comprising material from approximately 19,100 to 13,200 years cal. b2k was analysed in regard to the vegetation change based on pollen and lake water temperature development based on the diatom communities (Huber et al. 2010). Approximately 13 cm below the sharp increase of up to 6°C marking the early Lateglacial Interstadial in the diatom-based water temperature, an up to 3°C warmer episode occurred between 390-380 cm profile depth. In the pollen profile the arboreal pollen, in particular those of *Betula* sp., *Pinus* sp., and *Juniperus* sp., began rising in this depth. The pine pollen reached a first small maximum in the lower part of this episode, whereas juniper had its maximum in the upper part of this episode. Non-arboreal pollen such as the Poaceae and *Artemisia* sp. remained relatively high. A ^{14}C date (KIA-30226[44]) made of three wood fragments from immediately above this episode (377-375 cm) produced an age of 12,467 ± 48 years ^{14}C-BP and another date (KIA-30227) from the younger part of this episode (383-380 cm) was made on a bulked sample of wood fragments, macro-remains of *Betula* sp., and seeds of *Carex* sp. and yielded an

44 This date series was made several years before the technical problems in the Kiel laboratory occurred (cf. http://www.uni-kiel.de/leibniz/).

age of 13,059 ± 74 years ^{14}C-BP (Huber et al. 2010, 136 tab. 3). The two radiocarbon dates following stratigraphically below this episode contained too small amounts of carbon to produce reliable results. Thus, the onset of this amelioration episode cannot be further determined in this record. If the date for the end of this episode is calibrated it ranges around 15,600 years cal. b2k. This age is correlative with the end of the warm episode in the sea surface temperatures if the tuned MD01-2461 record was not shifted 500 years younger. Furthermore, following on this warm episode the sea surface temperatures of the deep sea record had some significant lows which were also observable in the diatom-based water temperature of the Längsee (Huber et al. 2010, 137 fig. 2). If the tuned MD01-2461 record was shifted to match the deterioration of GS-2a, the end of the second peak would only shortly precede 15,150 years cal. b2k and 15,600 years cal. b2k would correlated to the onset of the warm episode in MD01-2461. In addition, after the very strong, second peak in MD01-2461, the period of the more intense deposition of volcanic material followed which according to the comparison with NGRIP is correlative to the increased Ca^{2+} corrected sulphate content before the onset of the Lateglacial Interstadial.

Although the warm episode which followed on the IRD minimum in the Late Pleniglacial was not observed in the records used by Stanford and her colleagues (Stanford et al. 2011b), the ages given by them for the sub-phases 2 and 3 of the Heinrich 1 event are generally in good agreement with the tuning of the MD01-2461 record as performed in this study without the additional shift towards the onset of the GS-2a deterioration. According to this tuning, the decrease in sea surface temperatures (**fig. 40**) appears to rather predate the decline of GS-2a in the NGRIP and the Hulu Cave record and, in fact, if the correlation is sustainable the warming episode would be concomitant with the significant decrease in the Asian monsoon intensity recorded at the Hulu Cave (see **fig. 32**).

Another observation in this record is that the onset of the Lateglacial Interstadial as well as the Lateglacial Stadial were related to peaks of ice rafting debris (IRD) and, thus, iceberg melting events. These melting events could perhaps also explain the delayed reaction in the Chinese oxygen isotope records (see p. 322-329).

With the rapid onset of the amelioration in the sea surface temperatures at the onset of the Lateglacial Interstadial, the deposition of IRD decreased. The Lateglacial Interstadial optimum recorded in the Greenland ice-core and the terrestrial records (Coope et al. 1998) was perhaps associated with this decrease of iceberg melting. However, this continued amelioration led to another important melting event which slowed the warming of the sea surface temperature. The sea surface temperature development formed a moderate plateau during the iceberg release. After this melting event had ceased, the sea surface temperatures established at relatively high levels where they approximately remained until the end of the Lateglacial Interstadial (**fig. 40**). In general, the IRD content decreased very gradually and only minor increases of the IRD deposition suggested further short-termed cooling events after the last peak and following sharp decline in the early Lateglacial Interstadial melting event.

Shortly after the minimum of IRD deposition was reached in the final Lateglacial Interstadial, a first increase of the record led to a first sharp decline in the sea surface temperatures and when the IRD deposition ceased again the sea surface temperatures stopped for a plateau comparable to the values of the onset of the Lateglacial Interstadial. With the very rapid increase to a peak of IRD deposition which was occasionally referred to as Heinrich 0 event (Andrews et al. 1995; Dokken/Jansen 1999; Clark et al. 2001; Broecker et al. 2010) the sea surface temperature decreased further and remained approximately on this level which was a bit lower than the moderate first plateau in the Lateglacial Interstadial.

Another important information which is possible in the MD01-2461 record is that the peak of this Heinrich 0 event predates the Vedde Ash and by the present correlation the offset encompassed more than 300 years (**fig. 40**). Perhaps, the decreasing intensity of this Heinrich 0 event caused the amelioration phase which

was observed in some European records and assumed to have caused some significant melting and surface runoff processes (Brauer et al. 1999; Grafenstein et al. 1999; Kleinmann/Merkt/Müller 2002).
Thus, even though the deep sea record of MD01-2461 was tuned, it provided interesting considerations about the relation of climatic developments and also of environmental developments in north-western Europe. The latter can become particularly useful information in the comparison of the environmental development with the archaeological record (see Discussion-Change, p. 565-568; cf. Bradtmöller et al. 2012).

The Lateglacial radiocarbon calibration curve

The necessity to calibrate radiocarbon ages has often been emphasised (see p. 250-253, p. 259-263, p. 265-269, and references there). Therefore, a large body of high-resolution calibration data sets was collected over the last decades for the northern and for the southern hemisphere (McCormac et al. 2004; Reimer et al. 2004; Hua et al. 2009).
The most reliable of these data sets were dendrochronological sequences which provide an annually resolved calender timescale supplemented by a series of high-precision ^{14}C measurements. For the data sets from the northern hemisphere, the dendrochronological material sampled for the ^{14}C measurements comprised usually approximately 10 tree-rings and produced ages with standard deviations ranging between 15 and 50 ^{14}C years (Reimer et al. 2004). With every new supplementary data, the dendrochronological sequence and, thus, the calender timescale was monitored. Thereby, disturbances caused by cockshafer infestation were detected in the Holocene part resulting in a minor revision of the timescale in 2004 shifting the onset of the Holocene in the CEDC some 20 years older (Friedrich et al. 2004). This shift also affected data sets which were correlated to the previous versions of the CEDC along the Pleistocene/Holocene transition. In addition, the CEDC was extended further into the Lateglacial Stadial and currently constitutes a continuous data set from modern times to 12,644 years cal. b2k (Schaub et al. 2008b).
For the Lateglacial Interstadial the floating, 1,382 years long GLPC (Kromer et al. 2004) is available. The onset of the Lateglacial Stadial is recorded as a steep decline of the ^{14}C ages in this record and the Laacher See eruption (LSE) is documented by a series of dates from poplars which were buried during the eruption (Baales/Bittmann/Kromer 1998; Friedrich et al. 1999; Kromer et al. 2004). The GLPC was thus far not connected to the extended CEDC (Kaiser et al. 2012) but once this connection is achieved, the dendrochronological calibration data set extends into the early Lateglacial Interstadial.
To bridge the gap between the CEDC and the GLPC, other calibration data sets such as the deep sea records from the Cariaco basin were used in the past (Kromer et al. 2004). However, two tree-ring sequences were published in the last five years and assumed to bridge this older part of the Lateglacial Stadial: The Tasmanian Huon pines (Hua et al. 2009) and pines from the Swiss Gänziloo site forming the Younger Dryas A (YD_A) chronology (Schaub et al. 2008b). Based on wiggle-matching of the two new ^{14}C data series with the existing dendrochronological data and/or the deep sea record from the Cariaco basin, the two data sets were considered to fill the gap between the CEDC and the GLPC (see p. 358-364). However, the overlap of the YD_A chronology with the CEDC and the GLPC respectively seemed not sufficient to allow a dendrochronological anchoring of the three data sets (Kaiser et al. 2012). The ^{14}C dates from the YD_A chronology were wiggle-matched to the Cariaco basin record which was itself variously tuned in this portion (Hughen et al. 1998b; Hughen et al. 2004c; Hughen et al. 2006; Deplazes et al. 2013).
The Tasmanian Huon pines originated from the southern hemisphere and, therefore, the associated ^{14}C ages had to be corrected for an inter-hemispheric offset in ^{14}C ages (Hua et al. 2009). The inter-hemispheric offset resulted generally from the larger ocean surface on the southern hemisphere and, therefore, the

greater interaction of the atmospheric and oceanic reservoirs. On average, this increased exchange resulted in older ages on the southern hemisphere. Comparable to the ocean reservoir in ^{14}C data series from deep sea records, the inter-hemispheric offset was not constant and changed its behaviour particularly in periods of different ocean ventilations between almost no offset to almost 100 years. However, an average value of 40 years were substracted from the Huon pine ^{14}C ages (Hua et al. 2009, 2985). This possible fluctuation of the dates some 40-60 years on the ^{14}C age scale have to be kept in mind when comparing this data set to the calibration data sets from the northern hemisphere. Moreover, using this data set in the construction of a radiocarbon calibration curve implicates for each date the possibility of influencing the calibration curve towards too young or too old ages. Therefore, other data sets without inter-hemispheric offset should be preferred in the northern hemisphere. However, in the dendrochronological data from the northern hemisphere the ^{14}C samples are not yet as numerous and densely spaced in this period as in the Huon pine sequence.

To be independent of the tuning of the Cariaco basin record and the assumptions on ocean reservoirs as well as the inter-hemispheric offset, this study firstly positions the GLPC in relation to the CEDC along the onset of the Holocene and the LSE as known from the Lateglacial chronology in the Central European terrestrial sequences (**fig. 31**). The onset of the Holocene is known in the CEDC, whereas the LSE is clearly located in the GLPC. The varve records from MFM and Rehwiese bridge the Lateglacial Stadial and the LST as well as the establishment of the Holocene were identified in these records. The temporal distance of the two marker horizons in these records is further sustained by the findings in NGRIP and Lake Gościąż (see p. 310-318). Thus, the onset of the Holocene is marked by the increase of tree-ring growth in the CEDC (11,640 years cal. b2k) and the GLPC is anchored along the position of the LSE (ring no. 2160) to 12,930 years cal. b2k. Numerically, the two records overlap already some 55 calendar years in this correlation. However, neither closely sampled ^{14}C ages nor a standardised tree-ring growth curve have become available for the older end of the CEDC thus far. Nevertheless, the few available dates preset the general pattern of the radiocarbon calibration curve and make an anchoring of the Swiss pine and the Huon pine calibration data sets along this preset possible (**fig. 41**). In this correlation, the Huon pine sequence overlapped with the last 500 years of the CEDC and the first 100 years of the GLPC and the Swiss Lateglacial Master chronology (SWILM, Kaiser et al. 2012). The Swiss YD_A chronology is also correlative with these first 100 years of the GLPC/SWILM and the Huon pines. However, the Huon pine chronology and the YD_A chronology end immediately before the steepest part of the sudden decrease of ^{14}C ages marking the onset of the Younger Dryas in the GLPC record (Kromer et al. 2004). The GLPC and SWILM calibration data sets continue the dendrochronological calibration data set to almost 13,900 years cal.b2k and the dendrochronological record to almost 14,000 years cal. b2k in this position (**figs 38. 41**).

For the periods older than 13,900 years cal. b2k, dendrochronological calibration data sets from the northern hemisphere have not been established (cf. Reimer et al. 2009; Kaiser et al. 2012). Instead other records such as deep sea records have to be used for the construction of a calibration curve in these older periods (Hughen et al. 1998b; Jöris/Weninger 1998; Fairbanks et al. 2005; Weninger/Jöris 2008; Reimer et al. 2009).

The Cariaco basin record is one of the rare calibration data sets which yielded results for the complete Lateglacial Stadial and Lateglacial Interstadial. However, the calendar age model of this record was only partially confirmed by the laminated sediments because the lamination was often disrupted and/or incomplete (Hughen et al. 2004c). Therefore, the calender ages given for the ^{14}C dates were estimated due to sediment accumulation and tuning with other chronologies such as the one from the GISP2 ice core (Hughen et al. 1998a), from the Hulu Cave (Hughen et al. 2006), or the NGRIP ice core (Deplazes et al. 2013). These estimates could partially be incorrect and, thus, observing differences between radiocarbon calibration data sets from the European dendrochronologies and from the Cariaco basin could be due to incorrect calender

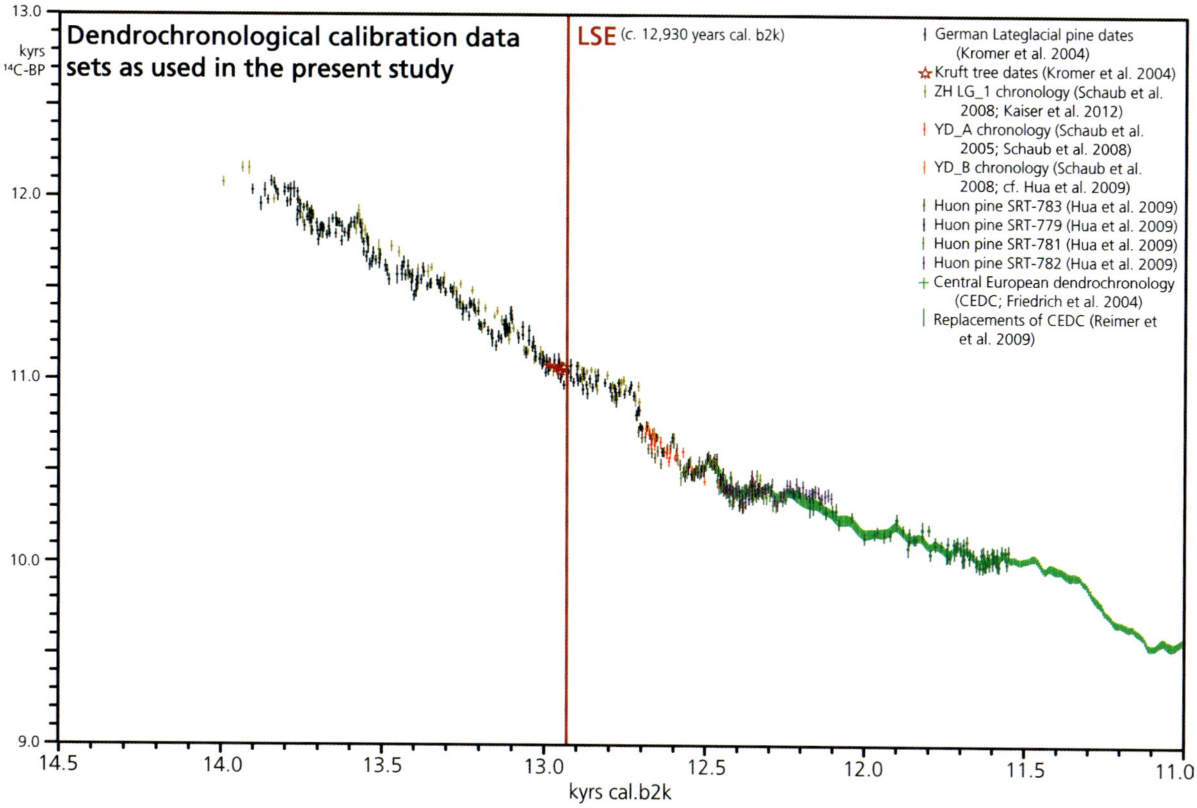

Fig. 41 Radiocarbon calibration data sets from Lateglacial dendrochronological records and location of the LSE in the GLPC. Records correlated as described in the text.

age estimates in the Cariaco basin record as well as incorrect reservoir estimates in the ^{14}C ages of this record. A comparison of the ^{14}C data sets from the CEDC and from the 2004 Lateglacial dates from the Cariaco basin (Hughen et al. 2004c) revealed that these sets were consistent across the Pleistocene/Holocene transition and into the Lateglacial Stadial. However, the steep decline at the onset of the Lateglacial Stadial occurred considerably earlier in the Cariaco basin record. Thus, the two data sets appeared to have develop at a different pace within the older part of the Lateglacial Stadial. The duration of the Lateglacial Stadial as documented in the greyscale record from the Cariaco basin was almost 250 years longer than the same period in the NGRIP oxygen isotope record or the European varve records (**tab. 70**).Consequently, the different comparisons suggested an overestimation of the duration of the Lateglacial Stadial in the Cariaco basin record. A possible explanation for this longer duration could be that the laminated sediments in the Cariaco basin had formed during some periods more than two sub-layers per year. However, this increased formation of sub-annual layers was previously suggested for the Younger Dryas part of the Cariaco basin record but was rejected by Konrad Hugen and his team (Hughen et al. 2004c).

Nevertheless, the Cariaco basin records were compressed by 210 years between 12,901 and 12,320 years cal. b2k which consequently relates to 12,691 to 12,320 years cal. b2k (**fig. 42**). This compression was based on a comparison of the ^{14}C data sets with dendrochronological calibration record and a correlation with this record along the steep decline at the onset of the Lateglacial Stadial. The remaining GI-1 part of the 2004 ^{14}C data set from the Cariaco basin (original data between 14,723 and 12,909 years cal. b2k) was shifted accordingly by 210 years with the youngest date correlating to a calender age of 12,699 years cal. b2k.

Based on a comparison of the Cariaco basin greyscale with the isotope records from NGRIP and the Hulu Cave, a shift of 100 years towards the younger was previously suggested for the Cariaco basin record (see p. 350-352). This shift seemed necessary because in the early Lateglacial Interstadial the calender ages in

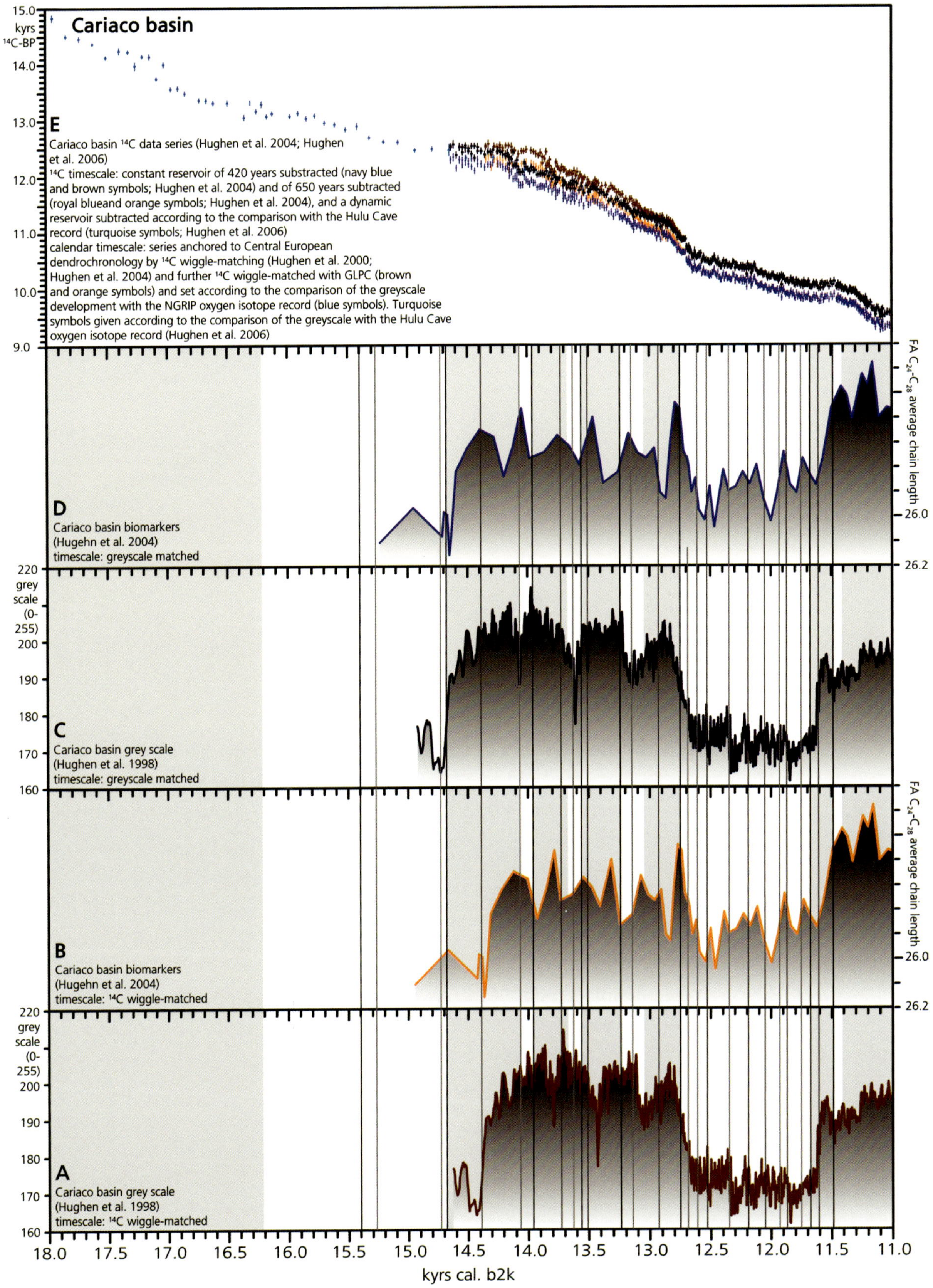

Fig. 42 Comparison of two different age models for the Cariaco basin record. **A**, **B**, **E** (orange and brown symbols) the record according to the best match with the dendrochronological radiocarbon calibration data sets as used in the current approach; **C**, **D**, **E** (blue symbols) the record according to a mixture of ^{14}C wiggle-matching and comparison of the Cariaco basin greyscale to the NGRIP record (see **fig. 39**). Thin bars relate to the NGRIP record and are given as guide lines (light grey: low; medium grey: high; black: transition). – For further details see text.

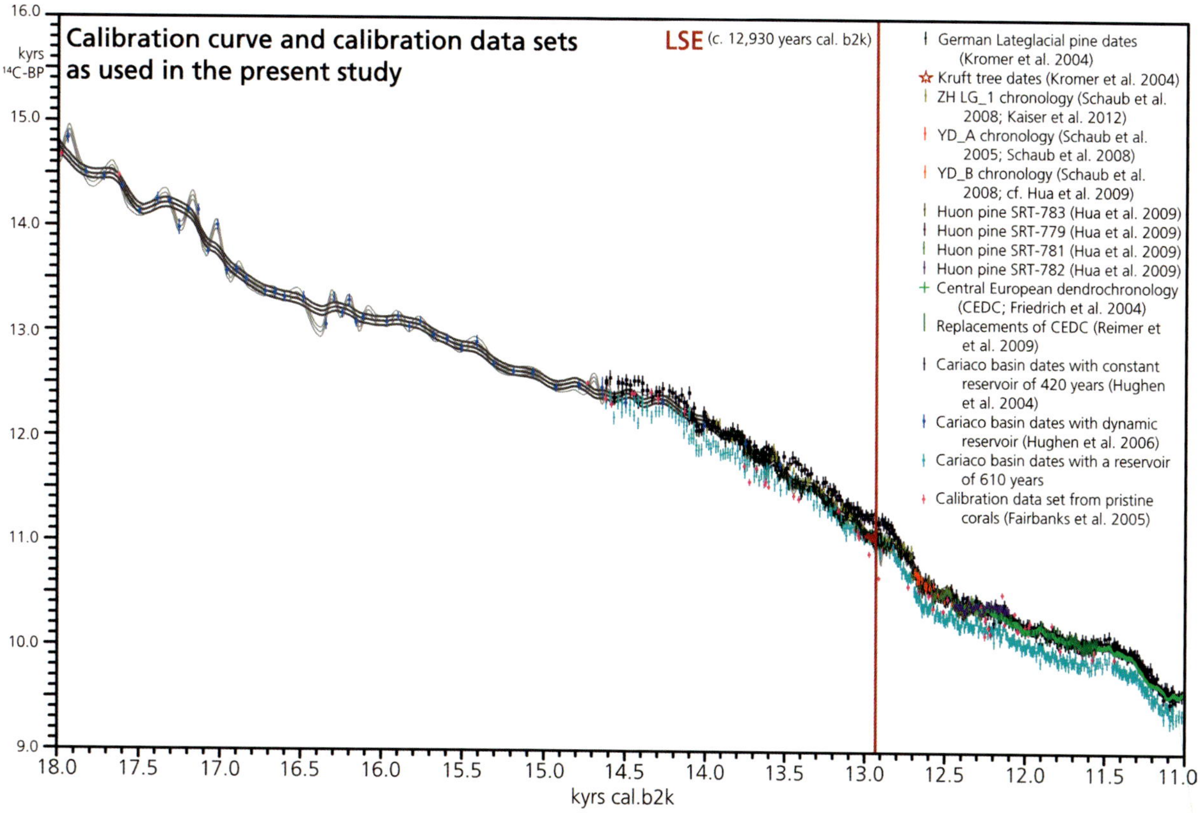

Fig. 43 Calibration curve (with 2σ spline dispersion lines; thick dark grey lines) constructed for the present study and calibration data sets as used in this project. In light grey the calibration curve (with 2σ spline dispersion lines) before smoothing is given. – For further details see text.

the Cariaco basin record seemed to encompass again longer durations than in the NGRIP or the Hulu Cave record. However, after the above described shift to 12,699 years cal. b2k, the onset of the Lateglacial Interstadial was some 100 years younger in the Cariaco basin than in the NGRIP isotope records or the Hulu Cave oxygen isotope curve. Nevertheless, comparing the GI-1 ^{14}C data segment of the Cariaco basin record further to the European dendrochronologies, a compression of additional 174 years could be suggested resulting in the GLPC to begin at 14,339 years cal. b2k. In this case, the ^{14}C ages were consistent and the reservoir ages switched along the Lateglacial Interstadial to Lateglacial Stadial transition from 420 ^{14}C years to 650 ^{14}C years and remained relatively constant on this level in the Lateglacial Interstadial. Only at one point (original data between 14,089 and 14,076 years cal. b2k; shifted and compressed: 13,805 and 13,754 years cal. b2k) the reservoir ages seem to shortly switch to the lower values (420 ^{14}C years) again. However, if the greyscale was correlated accordingly, this proxy record reflected climate changes very differently to the oxygen isotope record from Greenland (**fig. 42A-B**).

In contrast, if the data set older than 12,699 years cal. b2k was shifted according to a comparison of the greyscale with the NGRIP oxygen isotopes, the Cariaco basin record should be stretched again resulting in the earliest ^{14}C date being set at 14,735 years cal. b2k. Then the sub-segment between 14,056 to 14,735 years cal. b2k is compressed by 113 years to an end age of 14,622 years cal. b2k. In this correlation, the steep decrease of ^{14}C ages began in the Cariaco basin record around 12,830 years cal. b2k (**figs 42-43**). In comparison with the GLPC/SWILM data sets, the previously suggested sudden increase of reservoir ages in the Cariaco basin occurred between 12,818 and 12,776 years cal. b2k (**fig. 43**; original values: 13,028 and 12,986 years cal. b2k; cf. Hughen et al. 2004c; Kromer et al. 2004). In contrast to the original suggestion, this correlation suggested a smaller switch in the reservoir ages to only 610 ^{14}C years. Moreover,

several, short-termed switches in the reservoir ages during the younger half of the Lateglacial Interstadial, in particular during the equivalents of GI-1c_1 and GI-1b, seemed possible according to this comparison. Between approximately 13,500 and 14,150 years cal. b2k, the reservoir ages switch again to reservoir ages around 420 ^{14}C years (**fig. 43**). A comparison of the Cariaco basin record to the Hulu Cave record already suggested more instable reservoir ages for the Cariaco basin area during the Weichselian Pleniglacial than during the Holocene (Hughen et al. 2006). Furthermore, a simulation study based on a circulation model analysed the development of Atlantic reservoir ages during the Lateglacial Stadial in the context of a shut down of the Atlantic meridional overturning circulation (Ritz/Stocker/Müller 2008). Possibly, the changes in the thermohaline circulation of the North Atlantic indicated by the Heinrich 0 event in MD01-2461 could be correlated with this anomaly. This simulation study suggested a smaller shift in the reservoir ages at the onset of the decline in Δ^{14}C in the transition from the Lateglacial Interstadial to the Lateglacial Stadial followed by a gradual increase of the reservoir ages to higher values than the usual ones and an abrupt, short-termed increase at the recovery of the Atlantic meridional overturning circulation (Ritz/Stocker/Müller 2008, 207). This development was probably indicated by the comparatively steep development of the ^{14}C ages of the Lateglacial dendrochronologies in comparison to the more oblique development of the Cariaco basin dates. Thus, instable episodes of the reservoir ages in the Cariaco basin occurred possibly already during the Lateglacial.

Since the dates given in the Hulu Cave revision of the Cariaco basin record (Hughen et al. 2006) were already changed to a calendar timescale according to a comparison with the oxygen isotope record from the Hulu Cave, the changes of the calendar timescale (shifting, stretching, and compressing) were not applied to these dates. Furthermore, these dates were given with dynamic reservoirs and therefore no additional change on the ^{14}C timescale was considered necessary.

For the period older than 14,622 years cal. b2k, only these Hulu Cave revised dates from the Cariaco basin were available. Further calibration data to supplement these dates could come from pristine corals. In fact, calibration data sets from corals are considered as more accurate and precise than the greyscale data (Fairbanks et al. 2005; Weninger/Jöris 2008). However, Fairbanks and colleagues excluded coral data which could contain more than 0.2 % calcite from their calibration data set due to the possible contamination of the sample and, consequently, of the radiocarbon results (Fairbanks et al. 2005, 1782). Furthermore, they redated some coral samples to achieve more accurate results with higher precisions. Thus, this record compiles the most reliable calibration data sets from pristine corals. The reservoir ages were calculated based on the comparison with the Holocene CEDC data. Comparable to the reservoir ages in the Cariaco basin, a change in these reservoir ages cannot be excluded and seems in parts probable based on the comparison with the Lateglacial Stadial part of the CEDC (**fig. 43**). Moreover, this compilation provided no dates between 14,735 and 17,630 years cal. b2k. Thus, this part of the calibration curve is solely based on the Cariaco basin dates.

These data sets were all compiling in the above described chronologies in the CalCurve composer of the CalPal program and interpolated by the use of the »variable« method[45] resulting in a new calibration curve. The difference between the spline of this curve and the data reached on average 28 ^{14}C years (spline shape of 15.0) but could manually be shifted to only 24 ^{14}C years (spline shape of 8.0 to 11.0; **fig. 43**, light grey curve). However, for the current calibration curve the spline was created with a spline shape of 20.0 (average age difference of 35 ^{14}C years) to smooth some extreme wiggles (**fig. 43**). These wiggles occurred in the GS-2a and GS-2b part of the curve where only the Cariaco basin dates formed the curve. Since these

45 This method uses IMSL (International mathematics and statistics library) in Fortran and the subroutine CSSMH to calculate a smooth cubic spline approximation from noisy data followed by the subroutine CSVAL to evaluate the cubic spline. The result is a cubic spline with variable shape due to a variable SMPAR (smoothing parameter).

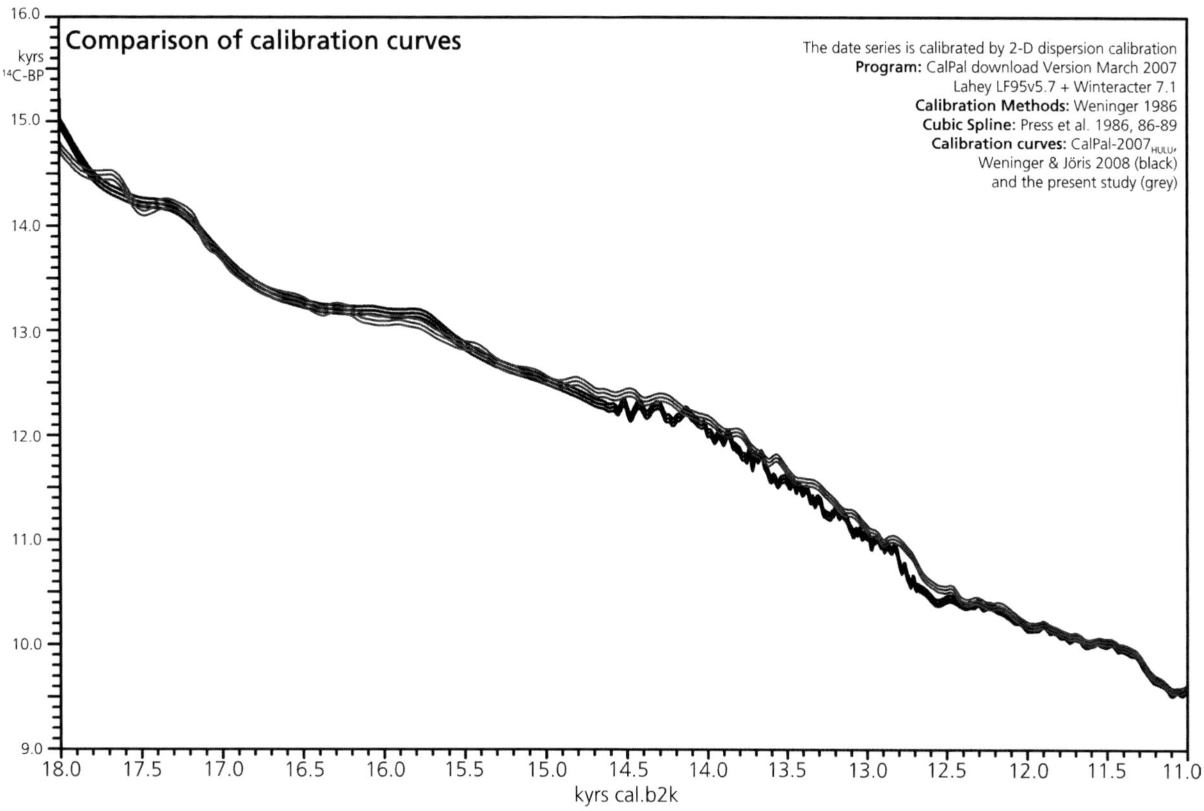

Fig. 44 Comparison of CalPal-2007$_{HULU}$ calibration curve (black) with the calibration of the current project (grey). – For further details see text.

dates could not be further verified in regard to their reservoir ages and a swinging back-and-forth of some 300 ^{14}C years within approximately 200 calender years seemed unusual, a smoothed spline was chosen assuming that some of the Cariaco basin dates might still incorporate inaccuracies.

The resulting calibration curve differs in detail from the CalPal-2007$_{HULU}$ radiocarbon calibration curve, particularly during the Lateglacial Interstadial (**fig. 44**). The Holocene and the younger part of the Lateglacial Stadial are identical in the two calibration curves and the Late Pleniglacial part is also very similar. Thus, the differences are mainly based on the approximately 40 years younger position of the LSE in the curve of this study. Consequently, the calibration with this curve will generally produce the same results as a calibration with the CalPal-2007$_{HULU}$ radiocarbon calibration curve; only in the Lateglacial Interstadial younger results than a calibration with the CalPal-2007$_{HULU}$ radiocarbon calibration curve have to be expected.

Since the calibration curve of this study is closely oriented on the European varve chronologies and the NGRIP record, the calibration results probably relate closely with the events and episodes within the Lateglacial Interstadial. This close correlation should help to relate the environmental and archaeological changes within this period more precisely.

ENVIRONMENT AND CHRONOLOGY

The environmental development of Lateglacial north-western Europe was established in the present study using various lines of evidence: the physical geography, the vegetation development, and the faunal successions in the studied areas.

Physical development of Lateglacial north-western Europe

General remarks

The created maps of Lateglacial north-western Europe display the changes in the sea-level as well as the approximate development of the major ice-sheets. In addition, the Lateglacial landscapes were influenced by isostasy, rivers, and permafrost.

The former was caused by the regressing of the European ice sheets and the subsequent decreasing ice load. For the initiation of ice sheet growths the mean summer temperature is a limiting factor, whereas the volume of ice sheets depends on additional factors such as mean annual temperature or the annual snowfall (Kageyama et al. 2004). Thus, the ice sheet volume can also decrease in cold periods such as GS-2a due to the lack of moisture and snowfall, whereas summer cooling, caused for instance by dammed glacial lakes, can initiate ice sheet growth in periods of temperature amelioration (Krinner et al. 2004). However, visually displayed over long periods, the general trend in the Lateglacial was a regression of the ice sheets causing uplifting of the grounds freed from the ice load. Due to this isostasy as well as other tectonic movements, the exact offset between the modern altitudes and altitudes during the Lateglacial are imperfectly known (cf. Kiden/Denys/Johnston 2002; Reicherter/Kaiser/Stackebrandt 2005; Shennan et al. 2006; Amantov/Fjeldskaar/Cathles 2011). Even though the maps reflect mean values for relatively long periods, the influence of the isostatic uplift on these Lateglacial maps is negligible because of the large scale of the maps. The total difference of the altitudes due to isostatic and/or tectonic movements along, for example, the Belgian coast was probably below 5 m in comparison to the Dutch coast in the studied time period (cf. Kiden/Denys/Johnston 2002). Partially, the uplift was superseded by the accumulating sedimentation load (cf. Amantov/Fjeldskaar/Cathles 2011). Thus, in many parts of the map the difference would be hardly observable due to the large steps in the colour interpolation (see p. 254).

However, the significant movements of the earth's crust caused earthquakes (Mörner 1999; cf. Reicherter/Kaiser/Stackebrandt 2005) and tsunamis (Mörner 1999; Bondevik et al. 2005; Weninger et al. 2008) in northern Europe during the Lateglacial and the early Holocene. Thus far, only very few tsunamis were documented for the Lateglacial (Mörner 1999; Mörner 2008). Due to the geomorphological relief and the position of the coastlines, these natural hazards affected mainly northern Europe in this period and they seemed to have had no direct impact on the human occupation in these presumably sparsely settled areas in this period. Nevertheless, the tsunamis, perhaps in combination with the earthquakes, probably had some impact on the hydrographic development of the Baltic Sea region in general and the development and clearance of the outflow of the Baltic Ice Lake in particular (Mörner 1999; Jensen et al. 2002). In this context, these natural hazards could have also affected the roaming patterns of various mammals and possibly also the human occupation patterns in southern Scandinavia by influencing the natural barrier between Scania, Zealand, Funen, and Jutland (cf. Björck 1995b; Eriksen 1996; Larsson 1996a; Larsson 1999; Schmölcke et al. 2006; Larsson 2009).

Besides the difficulties already inflicted on the reconstruction of the onshore landscapes such as the tectonic movements, the cover by massive sediment deposits, and/or the precise dating of the geomorphological features, the exact development of the Baltic Sea basin is further complicated by the restricted accessibility of the submerged areas, for instance due to weather, currents, and marine traffic. However, based mainly on offshore borehole surveys and onshore stratigraphies, a general history of the development of the Baltic Sea region was established and revealed in part a very different landscape from the modern circum-Baltic region (Björck 1995b; Harff/Björck/Hoth 2011).

The drainage of the European rivers was also in some parts very different from the drainage of modern times and their valleys underwent several changes in the Lateglacial. However, in general the river systems were too small to be acknowledged in the maps of Lateglacial north-western Europe. Moreover, changes in the fluvial system are difficult to date exactly, for various reasons. For example, geo-chemical processes can contaminate the dated samples (cf. Hiller et al. 2003) or later developments led occasionally to the partial or complete destruction of the Lateglacial channels (Chaussé 2003; Badura/Jary/Smalley 2013; cf. Park/Schmincke 1997). However, a general development of most large European river systems is known. In general, the discharge regime and thereby the incision patterns and branching of the rivers changed during the Lateglacial. These changes were influenced by various factors such as the water absorbing capacity of the underground, the melting of glaciers, and the precipitation. For the latter, no direct proxy data are available but based on a geochemical alteration of the sediment due to vegetation growth a simulation model was applied to the Nussloch sequences (Antoine et al. 2001) and yielded a reconstruction of the Lateglacial precipitation patterns (Hatté/Guiot 2005). Even though this analysis refers to a site south of the study are in the Upper Rhine Rift, the general tendency is assumed to be comparable to the study areas. Moreover, the Rhine in the Central Rhineland received its waters from this region.

The changing river courses had some impact on the geomorphology of the valleys, their accessibility, and their ability to serve as guide routes or barriers. For the latter, the seasonality and the according flooding or drying up of the valleys also played an important role.

Besides the large outflow of the Baltic Ice Lake, two further very large drainage rivers had formed during the Lateglacial: one in the area of the English Channel (Antoine et al. 2003a) and the other one west of the Cimbrian Peninsula (Streif 2004; cf. Badura/Jary/Smalley 2013). In addition to the English and northern French rivers which drain into the English Channel until modern times, the Channel river was supplied by the large western European river systems such as the Scheldt, the Meuse, and the Rhine as well as several (south-)eastern English rivers such as the Thames (Busschers et al. 2005; Busschers et al. 2007; Antoine et al. 2003a; Bourillet et al. 2003; Wallinga et al. 2004; Ménot et al. 2006, Makaske/Maas/van Smeerdijk 2008). However, this river was probably formed by several meandering river channels intersected by sand dune formations which were usually set on the northern side of the banks (Hijma et al. 2012). Furthermore, many of the large rivers behaved, presumably, in a comparable way before the confluence. Thus, the area where the Thames and the Rhine coalesced, probably joined by several other smaller rivers, was a large water landscape with a network of river channels. In general, this large water landscape was probably comparable to modern river deltas (Gilg et al. 2000), which can be environmentally productive and therefore favourable landscapes (Bianchi/Allison 2009) but occasionally also migration barriers.

The large river west of the Cimbrian Peninsula was mainly formed by the large rivers systems of the Ems, the Weser, and the Elbe which formed a 30-40 km wide channel (Streif 2004, 9) flowing around Helgioland (Konradi 2000; cf. Houmark-Nielsen/Henrik Kjær 2003) towards the central North Sea basin and from there presumably northwards into the Norwegian Trench. In addition, many western Jutland and Schleswig-Holstein rivers, which flowed in partially deeply incised valleys, formed tributaries to the Heligoland channel, whereas other Danish rivers flowed directly northwards to the Norwegian Trench, the Skagerrak, or the Kattegat. Some of the eastern rivers could have drained into one of the Danish straits and the Baltic Ice Lake. Further large river systems in the eastern North European Plain such as the Oder or the Vistula mainly drained into the Baltic Ice Lake area already during GS-2a (Starkel/Gębica/Superson 2007). During the maximum glaciations in the Late Pleistocene, the northwards drainage of these rivers was temporarily blocked by the Scandinavian ice sheet and glacial valleys and water gaps were established which connected these river systems westwards with the Elbe, the Weser, and the Ems drainage system (Wolstedt 1956; Badura/Jary/Smalley 2013). However, the northwards drainage of these rivers was reestablished soon after

the retreat of the Scandinavian ice sheet (Starkel/Gębica/Superson 2007; cf. Badura/Jary/Smalley 2013). Nevertheless, some basins in the glacial valleys and water gaps remained water-filled into modern times (Neugebauer et al. 2012, 92), whereas others fell dry and were partially filled with sediment and served probably as source for aeolian sediment removal (Badura/Jary/Smalley 2013). Consequently, when exactly the connections ceased and/or until when these connections were perhaps occasionally reactivated, for example when ice-bergs in the Baltic Ice Lake dammed the outflow of meltwaters through the river mouths, remained uncertain. Moreover, the exact topographic development in the area of the modern Baltic Sea is difficult to reconstruct (cf. Lemke et al. 2002; Meyer/Harff 2005; Reicherter/Kaiser/Stackebrandt 2005; Harff et al. 2007; Bellec/Diesing/Schwarzer 2010; Hoffmann/Reicherter 2012). The difficulty is in particular due to the cover by modern sediments, the partial destruction of old topographies by more recent tidal and neotectonic movements, and the problem of dating the submerged structures more precisely.

A comparably difficult area to reconstruct is the basin of the modern North Sea. A land-bridge emerged between the Cimbrian Peninsula and the British Isles during the periods of low sea-level in the Pleistocene such as the LGM. This land which was occasionally referred to as Doggerland (Coles 1998, 47; Coles 2000; cf. Gaffney/Thomson/Fitch 2007) submerged underneath the North Sea again in the Holocene rise of the global sea-level (cf. Weninger et al. 2008). This flooding ended the Lateglacial drainage system.

The general difficulties of reconstructing this Lateglacial drainage system as well as the landscape of Doggerland are the same as in the river systems and the Baltic Sea basin (see above; cf. Ward/Larcombe 2008). Moreover, the majority of new studies on the North Sea basin is based on 3D seismic analyses which can date the developments only relative to each other. Additional borehole samples can help if the samples contain datable material or provide indications to differentiate marine, coastal, freshwater, or terrestrial environments. However, the precise temporal development of the North Sea basin often remains a matter of debate thus far (Moreau et al. 2012). Therefore, the possible notes on the Lateglacial development of this area remain vague. According to 3D seismic analyses which were conducted in some sample areas, a network of rivers, wetlands, and tunnel valleys existed on most parts of the North Sea basin during the Lateglacial (Praeg 2003; Fitch/Thomson/Gaffney 2005; Lonergan/Maidment/Collier 2006; Gaffney/Thomson/Fitch 2007; Stewart/Lonergan/Hampson 2012). However, some rivers from eastern England must have ended in this area but whether they formed dammed lakes in trough-like situations such as the Outer Silver Pit and/or supplied the wetland complex of Doggerland, drained towards the north, or supplied the Channel river (cf. Toucanne et al. 2010), remains unclear thus far. However, with the gradual flooding of the North Sea basin, the drainage areas supplying the Doggerland wetlands, the Channel river, or the Heligoland channel constantly decreased and at least in the case of Doggerland this suspension of the water supply could have caused droughts prior to the final flooding.

Besides water straits, permafrost development played an important role in limiting the expansion of biomes in north-western Europe during the Late Pleistocene (cf. Huijzer/Vandenberghe 1998, 408f.). In permafrosts, only a small part of the subsurface thaws in the warmer seasons making the development of organic life only possible for a limited period and at a limited planting depth. In addition, water also becomes frozen in these grounds and the availability of water consequently decreases. Even though permafrost features were relatively well defined and, therefore, regularly observed in the geological record (Huijzer/Isarin 1997; Renssen/Vandenberghe 2003), a precise dating of the processes creating these features is difficult (Isarin 1997). Permafrost develops under severe cold conditions depending on the soil surface temperature for which air temperatures above the ground, type of vegetation, snow cover, surface slope, and the sediment composition of the ground are controlling factors (Delisle et al. 2009; cf. Haeberli et al. 2010; Vandenberghe et al. 2012). Moreover, a relation of the latitudinal extend of permafrost areas in Late Pleistocene Europe to the sea-ice cover in the North-Atlantic was observed (Renssen/Vandenberghe 2003; Vandenberghe et

al. 2012). As a result, the general distribution of permafrost fluctuated with the episodic variations of the North Atlantic sea-ice cover such as during Heinrich events. The sea-ice cover influenced the patterns of the isotherms of the air masses reaching western Europe (Renssen/Vandenberghe 2003) and, generally, a mean annual air temperature of -1°C appears as the limit for the development of permafrost (Delisle et al. 2009). Continuous permafrost, which relates to the presence of permafrost in all places, can develop in regions with a fine-grained sediment texture and a mean annual air temperature of -4°C, whereas in regions with sandy and gravely sediments a mean annual air temperature of about -8°C is necessary to develop continuous permafrost (Renssen/Vandenberghe 2003; cf. Vandenberghe et al. 2012). Thus, in most areas with a mean annual air temperature of -8°C permafrost can develop but sporadic places with permafrost can also persist for a relatively long duration during the process of amelioration (cf. Busschers et al. 2007, 3242), especially in particularly cold places such as northern slopes or deep valleys. Furthermore, the freezing of the ground reaches considerable depths in areas of continuous permafrost and the process of thawing in these great depths is a relatively gradual and slow process (cf. Vandenberghe et al. 2012, 17f.). In consequence, a relatively quick reestablishment of full permafrost conditions becomes possible in periods when permafrost are thawed only superficially and rapid returns of colder climates occur. Thus, the final disappearance of permafrost conditions is also difficult to estimate exactly.

The accumulation of the loess and coversands (**fig. 45**) is comparably difficult to date. The development and dispersal of these aeolian deposits was generally associated with cold and dry climates (Koster 1988; Kasse 2002; Antoine et al. 2003c) and, thus, their sedimentological composition can be partially compared with more precisely dated climate archives (Antoine et al. 2013; Jary/Ciszek 2013). In addition, the sequence of biotic remains preserved in the wind borne material, in particular pollen, was used for a more precise temporal attribution (Hoek 1997; Kolstrup 2007; Antoine et al. 2013). However, the acidic aeolian sediments are not well suited for pollen preservation (cf. Marshall 2007, 8-10. 122-124). Furthermore, the number of most organic indicators decreases in cold and dry climates and, consequently, the number of relevant material for the creation of a biostratigraphy was often reduced already at the time of the sedimentation. Thus, the stratigraphies with preserved biotic indicators is often biased towards locations with wetter conditions (Kolstrup 2007) in which the preservation was enhanced. The silty and sandy structure of these sediments occasionally allowed for post-depositional exchange of micro-remains such as pollen or cryptotephra particles. These exchanges are again enhanced in areas with wetter conditions and, thus, could pass undetected if they did not produce noticeably arbitrary results and/or if no detailed micro-morphological study of the sediment was conducted (cf. Weber et al. 2010; Housley et al. 2012). Therefore, a chronology of these aeolian sediment sequences made by the use of cryptotephras and micro-fossils becomes more reliable if it is supplemented by micro-morphological and/or pedological studies.

The loess deposits accumulated throughout the Pleistocene but the largest loess deposits originated from the Late Pleistocene, in particular from a period between approximately 30,000 to 15,000 years ago in western Europe (GS-3 and GS-2; Antoine et al. 2003c; cf. Frechen/Oches/Kohfeld 2003). This tendency was also observed in the Eastern European loess sequences (cf. Molodkov/Boliskhovskaya 2006), although thicker deposits of older material were also well preserved in some eastern Central European profiles (Madeyska 2002; Antoine et al. 2013). The formation of loess continued to the onset of the Lateglacial Interstadial and ceased then. The dispersal began again in some parts of Europe during the Lateglacial Stadial (Hilgers et al. 2001).

In contrast to the loess formation, the coversands were generally related to the North European Plain (**fig. 45**) and were mainly assumed to be deposited during the Lateglacial (Kasse 2002; Kolstrup 2007). In the western North European Plain, two phases of coversand deposition (Older and Younger Coversand) were generally distinguished and considered as important marker sections to compare lithostratigraphies

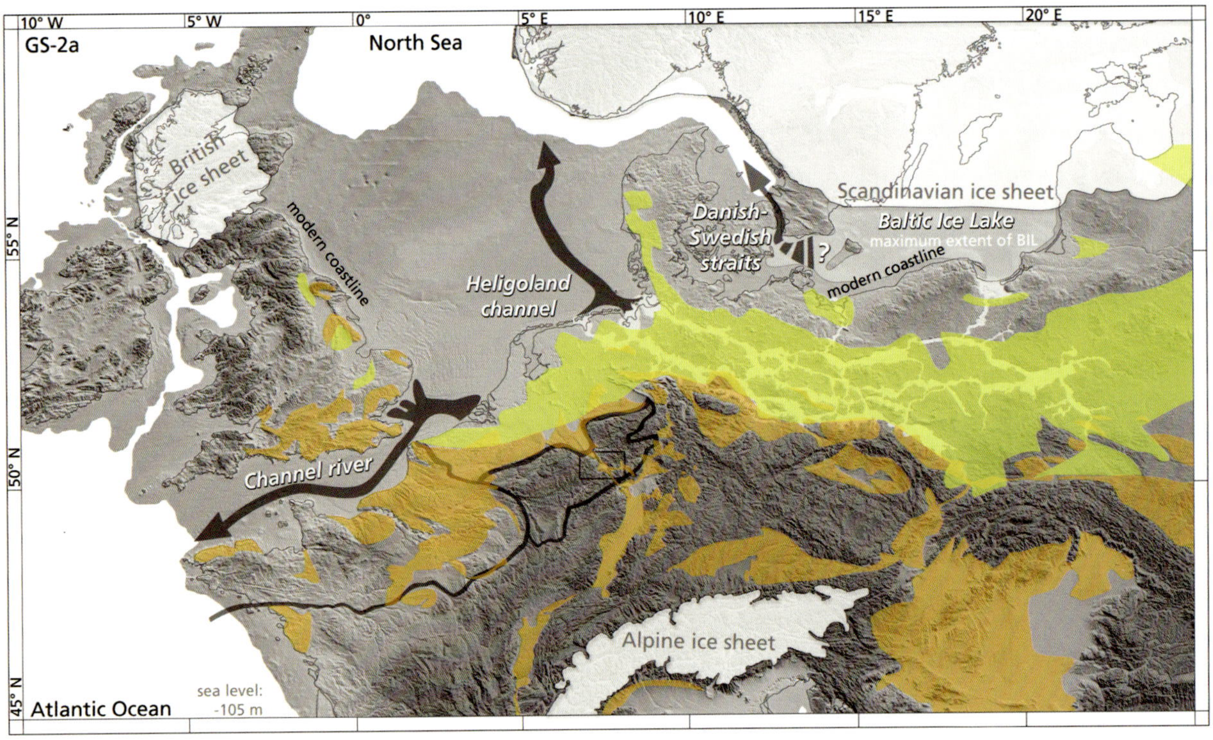

Fig. 45 Map of north-western Europe during GS-2a with approximate limits of significant aeolian deposits and the study areas (see **fig. 1**). Sea-level is set according to Weaver et al. 2003, 1712 fig. 5. An Irish Channel is supported by data from McCabe/Clark/Clark 2005; McCabe/Cooper/Kelley 2007; Edwards/Brooks 2008. The glacially formed valleys and water gaps on the North European Plain are highlighted in light grey (according to Wolstedt 1956; cf. Badura/Jary/Smalley 2013). The limits of the ice sheets are based on data from Lundqvist/Wohlfarth 2001; Boulton et al. 2001; Clark et al. 2004 (cf. Clark et al. 2012); Ivy-Ochs et al. 2006 (see p. 253-259). The loess distribution (orange shaded area) includes besides loess also sandy loess, loess derivates, and loess-like deposits but no aeolian sands or alluvial loess (Haase et al. 2007, fig. 9; Antoine et al. 2013, 18 fig. 1). The sand belt (yellow shaded area) is sketched according to Kasse 2002, 509 fig. 1 and Zeeberg 1998, 128 fig. 1.

(van Geel/Coope/van der Hammen 1989; Hoek 1997; Kolstrup 2007). Based on refined chronologies and more precise characterisations of the deposited material as well as concerns relating to the terminology, Kees Kasse differentiated between sedimentary facies in the coversands and phases of aeolian activity (Kasse 2002). By this distinction, he revealed that the aeolian activity occurred approximately in the same periods and climate regimes as the loess dispersal. However, studies of modern inland drift sands and coversand areas showed that comparable sand sheets also formed in a generally temperate climate regime (Koster 2009). In fact, Kasse also recognised that the activity phases continued for some time into the interstadial conditions Kasse 2002, 523). He explained the increased and widespread formation of sand sheets with four factors: (i) the increasing size of proglacial deflation areas with unconsolidated material which became available after the regression of the British ice sheet and, in particular, the Scandinavian ice sheet (cf. deflation areas indicated by Antoine et al. 2013, 18 fig. 1), (ii) the sparseness of vegetation to bind the material, (iii) the high ratio of wind energy and sediment availability (cf. Brauer et al. 2008), and (iv) the absence of major barriers on the North European Plain (Kasse 2002, 515f. 521-523). In fact, these factors are generally the same as for the formation of loess (Pye 1995). Therefore, the largely overlapping deposition periods are unsurprising.

The loess deposits, which are often several metres thick, have significantly influenced the European geomorphology and have still changed the topography after Lateglacial Interstadial (Hilgers et al. 2001). These deposition affected, in particular, the study area. The sometimes comparably massive aeolian sand sheets

also covered the Lateglacial topography and these coversands have thereby also changed the geomorphology in the affected regions significantly, for instance when forming sand dunes or filling depressions in the landscape (Hoek 2000, 500; cf. Sandersen/Jørgensen 2003). Loess and coversands were, in general, related to rivers (Kasse 2002; Smalley et al. 2009; Badura/Jary/Smalley 2013; cf. Haase et al. 2007), which provide the substrate and wind patterns which spread the substrate (Pye 1995; Koster 1988; Antoine et al. 2009). The general grain-size of the transported substrate makes the difference between loess (silt) and coversands (sand). However, in some areas transitional deposits were observed and described for example as sandy loess (Pye 1995; cf. Haase et al. 2007). Thus, in these areas the distribution maps of coversands and loess partially overlap. Furthermore, distribution maps of loess deposits usually include besides loess deposits also loess derived deposits as well as loess-like sediments (Haase et al. 2007) due to the aeolian origin and the post-depositional processes (Pye 1995). Mapping the various types of loess and evaluating the period of deposition of the loess to map only the Late Weichselian loess deposits exceeds the possibilities of the present study (cf. Antoine et al. 2009; Antoine et al. 2013). A distinct mapping of the coversands which formed during different periods is also beyond the scope of the present study. Nevertheless, an approximate distribution of the European loess belt and the European sand belt is a reminder that modern topographic measurements such as the SRTM data can hardly provide reliable information on the geomorphology of the Lateglacial landscape in these regions.
Besides these aeolian deposits, river dunes and drift sands also altered the landscape in areas of the large river systems (cf. Kasse et al. 2005; Koster 2009; Hijma et al. 2012).
Thus, many areas of north-western Europe were constantly changed by wind borne material, rivers and larger water bodies as well as by tectonic movements during the Lateglacial. The following maps can only produce a static picture of these processes.

Development within four sub-periods

The development of the physical geography in Lateglacial north-western Europe is displayed by maps that reflect four major sub-periods of the Lateglacial: the Late Pleniglacial (GS-2a), the early Lateglacial Interstadial (GI-1e-d), the mid- and late Lateglacial Interstadial (GI-1c-a), and the Lateglacial Stadial (GS-1).
The map relating to north-western Europe during GS-2a (**fig. 46**) reflects approximately the physical geography during the occupation of the Late Magdalenian sites such as Gönnersdorf, Andernach (lower level), Verberie, or Étiolles.
In this period the large European ice sheets had already retreated significantly in comparison to the maximum extension during the LGM. Perhaps, this retreat was related to the generally high aridity during the period following the LGM (cf. Hatté/Guiot 2005). The northern ice sheets which probably had coalesced during the LGM (Sejrup et al. 2009) were clearly separated by the onset of GS-2a (Carr et al. 2006).
The glacial front of the Scandinavian ice sheet had retreated to a latitude in southern Sweden uncovering most of Scania (Boulton et al. 2001; Lundqvist/Wohlfarth 2001). Approximately at the onset of GS-2a, a short-termed readvance of the glacial front was observed (cf. Lundqvist/Wohlfarth 2001, 1131; Sejrup et al. 2009). Probably, this readvance related to the so-called Halland coastal moraine (Lundqvist/Wohlfarth 2001). The well defined Göteborg moraines were already found some 10 km and more further inland (Lundqvist/Wohlfarth 2001). These moraines indicated a second, intense readvance in the early GS-2a. Thus, the velocity of the movements of the glacial front was relatively high in the Late Pleniglacial. Around the mid-Late Pleniglacial, the ice sheet had already retreated over the distance covered by the short-termed readvances, presumably due to the extreme dryness.

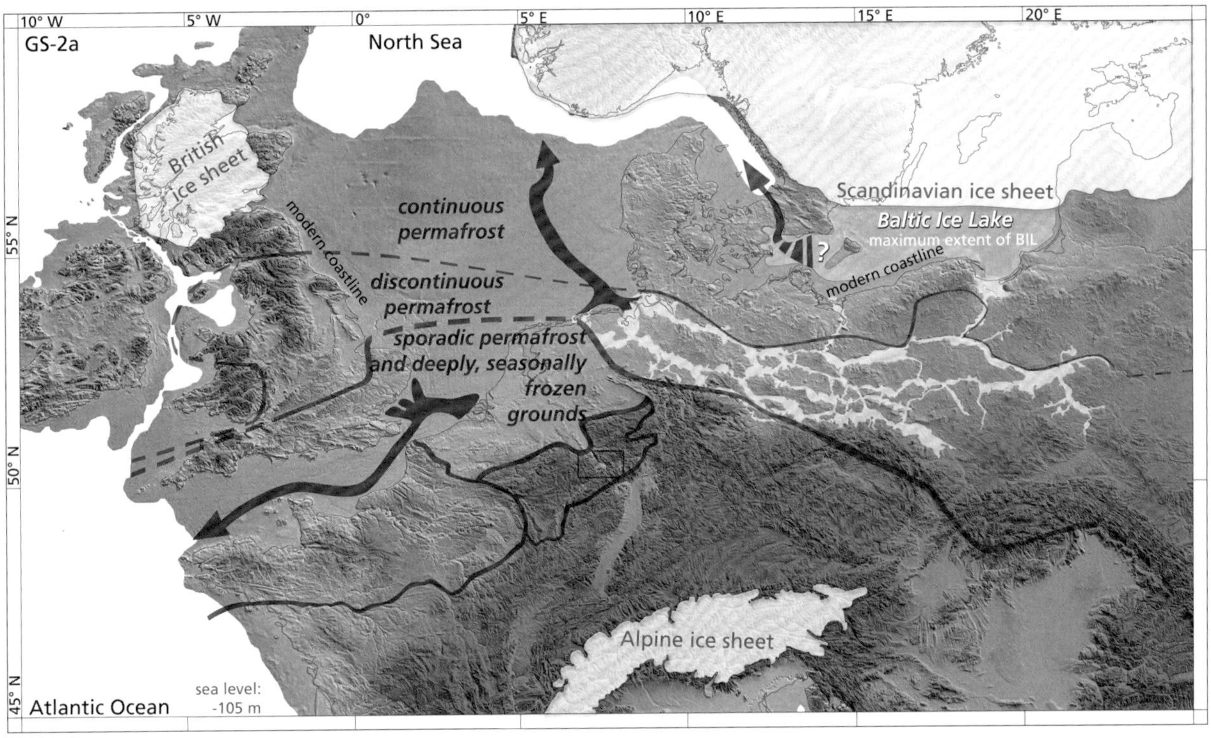

Fig. 46 Map of north-western Europe during GS-2a with approximate limits of permafrost zones and the study areas indicated (see figs 1. 45). Permafrost limits are approximations according to the data in Huijzer/Vandenberghe 1998 and the ice wedge presence in Wilczyce (Fiedorczuk/Schild 2002; Irish et al. 2008) as well as assumptions based on the glacial observations in the Giant Mountains (Nývlt/Engel/Tyráèek 2011) and the western Carpathians (Makos/Nitychoruk/Zreda 2013). For the areas where the permafrost lines are dashed no data from which the limits can be deduced was available.

If the proportion and speed of the regression of the British ice sheet was comparable to the Scandinavian one the British ice sheet had reached approximately the Scottish Central Belt during GS-2a. The Grampians, the north-western Highlands, and the Inner Hebrides were probably still covered by glaciers. Based on model calculations, Anthony Brooks and colleagues assumed that the British ice sheet had melted completely by c. 15,000 years b2k (Brooks et al. 2011), whereas a model based on the ice sheet related landforms considered relic ice sheets to have still existed in Scotland and Ireland at this time (Clark et al. 2012). However, in comparison to these reconstructions, the British ice sheet of the present study is overestimated indicating that the retreat patterns of the British ice sheet were probably faster than the retreat of the Scandinavian ice sheet. Perhaps, this lower velocity of the Scandinavian ice sheet can be explained by the slowing of the purging ice once the sea estuary through the Norwegian trench was established (cf. Clark et al. 2012). The disintegrating British-Irish ice sheet formed a glacial melt-water channel between Ireland and Great Britain approximately 16,000 years b2k (Edwards/Brooks 2008). Through this channel the meltwaters of the remaining British ice-sheet could partially be drained (cf. Clark et al. 2012). This ice channel disconnected Ireland from the European continent and established its island history. However, situated between the ice-channel and the slowly retreating sea-ice cover of the North Atlantic, Ireland remained probably a cold steppe to icy desert environment at least until the early to mid-Lateglacial Interstadial (cf. Huijzer/Vandenberghe 1998; Wickham-Jones/Woodman 1998).

The retreat of the Alpine ice sheet into higher altitudes freed less landmass than the retreat of the northern ice sheets. In these often short distances between the different Alpine moraines, a more complex history of retreat and readvance could be established (Ivy-Ochs et al. 2008). A widespread readvance of Alpine glaciers termed »Gschnitz-Stadial« was a several centuries long process which according to the so-called

exposure dates (based on ^{10}Be isotopes) was considered to be related to the ice-rafting event of Heinrich 1 (Ivy-Ochs et al. 2006). According to these dates, the maximum margins of this »Gschnitz-Stadial« were reached approximately at the onset of GS-2a. At its maximum extent, the ice front stagnated for at least several decades (Kerschner/Ivy-Ochs 2008, 61). Subsequently, the ice sheet continued to retreat into higher altitudes during GS-2a but two further, poorly dated readvances (Clavadel/Senders and Daun) occurred before the onset of the Lateglacial Interstadial. The relation of these readvances to the development of the Scandinavian ice sheet in this period remain unclear due to the uncertain chronology. How much such Alpine readvances can be related to tectonic processes (Norton/Hampel 2010) and/or changing patterns of moisture and precipitation (cf. Turk 2012), particularly during the Heinrich 1 event in the North Atlantic (Stanford et al. 2011b), remains to be analysed.

During the LGM further ice sheets developed on some middle range and higher mountain ranges such as the Jura Mountains (Buoncristiani/Campy 2011), the Vosges Mountains (Andreoli et al. 2006), or southern German Black Forest (Fiebig/Ellwanger/Doppler 2011) alongside the three major European ice sheets. However, by the onset of GS-2a, the majority of these smaller ice sheets had already shrunken to single glaciers of which some persisted into the Holocene, for instance in favourable locations in the Vosges Mountains (Andreoli et al. 2006) as well as in the Giant Mountains (Nývlt/Engel/Tyráèek 2011). In general, these glaciers retreated further during the Late Pleniglacial and Lateglacial Interstadial, for example the relic glaciers of the larger glacial ice sheets in the southern Carpathians (Urdea/Reuther 2009). The LGM ice sheet of the Polish High Tatra, part of the Carpathians, had also shrunk to a single glacier, but comparable to the Scandinavian and the Alpine ice sheets a last major readvance occurred at the onset of GS-2a (Makos/Nitychoruk/Zreda 2013). This readvance was followed by a stepwise melting accelerated by the onset of the Lateglacial Interstadial.

Thus, higher mountain ranges such as the Alps or the Carpathians as well as possibly some smaller areas in the lower mountain ranges such as in the Vosges Mountains or in the Giant Mountains still formed glacial barriers to migration during GS-2a.

Moreover, permafrost could also limit expansion processes in this period (**fig. 46**). The zone of continuous permafrost had probably retreated into northern Europe leaving only some sporadic permafrost relics in north-western Europe (Huijzer/Vandenberghe 1998, 409). Patches of permafrost were still recorded in the early Lateglacial Interstadial in northern Rhineland (Busschers et al. 2007, 3242) suggesting at least sporadic permafrost conditions survived in north-western Europe during GS-2a. However, in neither northern France nor in Central Rhineland were indications of permafrost found that could be related to the Late Magdalenian occupation of these regions. According to the proxy data for temperature (Huijzer/Vandenberghe 1998; cf. Coope et al. 1998), the ground in these regions were probably still deeply frozen for some time throughout the year. These severe cold seasons are confirmed by the archaeological material disturbed by cryoturbation. Furthermore, some lithic material exhibited frost cracks and/or were heavily patinated suggesting the influence of a cold climate. The faunal remains from the archaeological sites comprised animal species such as reindeer (*Rangifer tarandus*), arctic hare (*Lepus timidus*), and arctic fox (*Alopex lagopus*) all of which indicate a cold tundra/steppe environment.

In contrast, the Late Magdalenian material from the south-eastern Polish site of Wilczyce was mainly preserved in a large ice-wedge (Fiedorczuk/Schild 2002; Fiedorczuk et al. 2007) that was still active and successively filled during the period when the site was visited by the Late Pleniglacial hunter-gatherers (Irish et al. 2008). Perhaps, the warm period in the North Atlantic region suggested by the deep sea record of MD01-2461 affected western Europe more intensely and accelerated the withdrawal of the permafrost conditions in the areas adjacent to the North Atlantic (cf. Coope et al. 1998; Renssen/Vandenberghe 2003). In contrast, the more continental climate preserved, at least discontinuous permafrost conditions, further to

the south in eastern Europe. Another possibility is that the severe aridity during this period (cf. Hatté/Guiot 2005; Fletcher et al. 2010), prohibited the formation of frost within the dry sediment of western Europe. Since the active ice wedge of Wilczyce was recorded in the vicinity of the relatively protected basin formed by the San confluence into the Vistula, the adjacent higher mountain ranges of the Carpathians were probably also influenced by these cold climate conditions during GS-2a. This review shows that well dated records are not very exhaustive for illustrating and delimiting biome expansions.

Besides the cryosphere, the hydrosphere shaped major changes in the appearance of north-western Europe. The retreating ice sheets in northern Europe gave space for the rising northern Atlantic waters to intrude into the open lands from the north. By the onset of GS-2a, the sea-level had risen approximately 15-20 m above the LGM minimum of -123 m (Hanebuth/Stattegger/Bojanowski 2009). In particular, the deep Norwegian trench and the adjacent Skagerrak and Kattegat straits were filled by sea water soon after the deglaciation of these areas. Thus, this fjord-like estuary was presumably already formed before the onset of GS-2a (Houmark-Nielsen/Henrik Kjær 2003). The readvance of the Scandinavian ice sheet at the onset of GS-2a also reached coastal and marine parts of offshore Norway and partially filled the Norwegian trench with ice streams and ice bergs (Sejrup et al. 2009). The intrusion of sea water into the estuary was at least partially blocked by this glacial front and the formation of pack ice is very probable during this episode. The ice sheet retreated again during the Late Pleniglacial but at some points ice bergs were still launched into the Skagerrak and the Norwegian Trench at the onset of the Lateglacial Interstadial (Houmark-Nielsen/Henrik Kjær 2003, 781).

In the Baltic region a land bridge possibly existed in this period. With increasing deglaciation during GS-2a, the amounts of melt-waters combined with ice-bergs increased in the region south of the Scandinavian ice sheet and created the Baltic Ice Lake with an outflow through the Öresund area (**fig. 47**; Björck 1995a, 21). Through this outflow further ice-bergs were injected into the Kattegat, Skagerrak, and finally the Norwegian trench. To what extent these various sources of drifting ice-bergs resulted in the formation of pack ice or in some areas, such as the Öresund, fast ice for at least some of the year remains to be studied.

Sample studies on the Late Weichselian geomorphology of the North Sea basin, particularly offshore Denmark, highlights deep tunnel valleys which were largely filled shortly after their development (Kristensen et al. 2007; Kristensen et al. 2008). Some of the tunnel valleys were dead end valleys which depending on the individual geomorphology could serve well as traps for groups of migrating animals. Based on the sedimentary record of the North Sea basin, the area offshore from northern Germany appeared relatively level in the Late Weichselian but also contained some steep sided valleys (Streif 2004, 25; cf. Passchier et al. 2010). Comparable to the tunnel valleys in northern Germany (Clausen 2010), Lateglacial deposits of relic glacial lakes and/or kettleholes were found in several of these North Sea valleys (Kristensen et al. 2007). Thus, the landscape of the eastern North Sea landmass was characterised by a water rich landscape cut by tunnel valleys comparable to Lateglacial northern Germany.

Through this landscape the large Heligoland channel river flowed northwards (see p. 366 f.). This drainage was still supplied by the majority of the Danish rivers as well as the large rivers from the North European Plain (cf. Houmark-Nielsen/Henrik Kjær 2003). Even though a wide channel bed was identified in the Heligoland channel (Konradi 2000), the exact filling is unknown. An analogy for the general fluvial style of the large rivers in the Late Pleniglacial (van Huissteden/Kasse 2001; Kasse et al. 2005; Starkel/Gębica/Superson 2007; cf. Antoine et al. 2003a), would most probably be a large braided system rather than a single wide channel. Since the channels in braided river system are usually shallow, this braided river landscape was probably crossable outside melt-water and/or rainy seasons. In regard to the recent morphology of the sea floor and the reconstruction of tunnel valleys, the Heligoland channel river was flanked by a wide but limited floodplain. In times of highwaters, the space for the additional water was consequently restricted and the

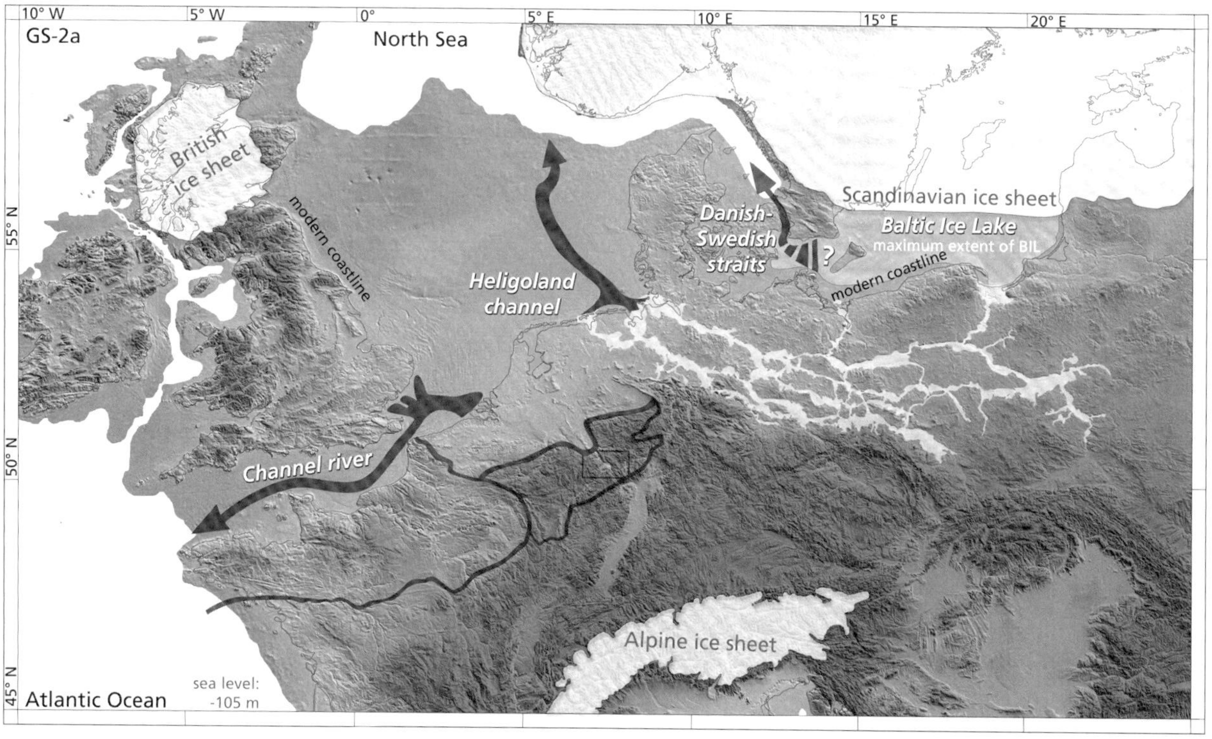

Fig. 47 Map of north-western Europe during GS-2a with the drainage system and the study areas indicated (see **figs 1. 45**). For the Baltic Ice Lake and possible outflows see Björck 1995a. The white question marks indicate the uncertain drainage of the rivers in the basins of the Baltic Sea and the North Sea. On the North European Plain the glacially formed valleys and water gaps are highlighted in light grey (according to Wolstedt 1956; cf. Badura/Jary/Smalley 2013). The Heligoland channel is sketched according to Konradi 2000 and Streif 2004 (cf. Badura/Jary/Smalley 2013). For the North Sea basin small-scale data gradually increases but still remains difficult to date (cf. Fitch/Thomson/Gaffney 2005; Ward/Larcombe/Lillie 2006; Gaffney/Thomson/Fitch 2007; Ward/Larcombe 2008; Stewart/Lonergan/Hampson 2012). The English Channel river is sketched according to Bourillet et al. 2003; Lericolais/Auffret/Bourillet 2003; Gupta et al. 2007; cf. Gibbard 1988; Antoine et al. 2003a; Ménot et al. 2006; Hijma et al. 2012.

Heligoland channel river probably developed into a wide and deep river during these water-rich seasons. This river presumably had some areas with strong currents and could have formed a barrier to, for instance, migrating reindeer herds.

Borehole and 3D seismic analyses from the western and southern North Sea basin are relatively numerous (Praeg 2003; Fitch/Thomson/Gaffney 2005; Passchier et al. 2010; Hijma et al. 2012; Moreau et al. 2012; Stewart/Lonergan/Hampson 2012). Comparable to the eastern sector of the North Sea basin, many tunnel valleys formed in the south-western North Sea basin and also some large sand ridges such as the Dogger Bank or the Brown Bank (cf. Ward/Larcombe 2008). Even though the exact dating of the observed structures and deposits in in this region are still a matter of debate (Moreau et al. 2012), the tunnel valleys are assumed to have formed during a glaciation prior to the onset of GS-2a. To what extent these tunnel valleys were already filled with sediment or were still recognisable as valleys and whether they contained open waters during GS-2a remains, to date, uncertain. In the south sector of the North Sea basin, the major western European rivers coalesced into the large English Channel river which formed a large braided river system during this period (see p. 366; Antoine et al. 2003a; Gupta et al. 2007; Hijma et al. 2012). In regard to the barrier effect this river system could have had on migratory and expansion processes, the same considerations as to the river system in the Heligoland channel apply. In contrast to this northern system, the English Channel river flowed through areas with only sporadic permafrost and emptied into an Atlantic region without significant sea-ice cover. Furthermore, in many parts of the English Channel the morphology allowed the water courses to spread across the complete width of the modern channel. Thus, in periods of

higher water volume, this area would probably have changed into an even wider water landscape. In the context of the drier conditions during GS-2a, this type of water landscape would probably have formed a fertile and therefore attractive area for various species.
The rivers of north-western Europe were generally flowing in high-energy braided systems (Mol/Vandenberghe/Kasse 2000; Lewis/Maddy/Scaife 2001; Antoine et al. 2003b; Kasse et al. 2005; Starkel/Gębica/Superson 2007; cf. Vandenberghe 2003) but some developed locally into meandering river types (Pastre et al. 2003; Turner et al. 2013).
Even though several barriers such as glaciers or large water straits existed for expansion and migratory processes in north-western Europe, these blocked pathways could often be bypassed or crossed during more favourable seasons.
Alongside the conditions related to cold climate and hydrology, the aeolian distribution of sediments shaped the Late Pleniglacial landscape. Based on the position of some Late Magdalenian assemblages within loess deposits, two observation can be made: (i) the majority of loess was already deposited at the beginning of the Late Magdalenian and (ii) loess was still deposited after the Late Magdalenian occupations. The rapid deposition of the archaeological material in aeolian sediments was occasionally assumed as a factor of relatively good preservation. However, at the Late Magdalenian sites of Gönnersdorf and Andernach, this enhanced preservation applied mainly to material deposited in pits and underneath stone plates, whereas faunal material deposited on the surface was strongly weathered, poorly preserved, or probably completely unpreserved (Street/Turner 2013). Thus, the dispersal of loess appeared reduced concomitant with the Late Magdalenian occupation of the Central Rhineland sites.
A possible explanation for the reduction of loess dispersal could be a change in the North Atlantic sea-ice cover. The sea-ice distribution in the North Atlantic influenced the temperatures over the ocean and was thereby connected with the western European air masses resulting in the influence on the North Atlantic climate on the distribution of permafrost (Renssen/Vandenberghe 2003) as well as the loess dispersal in western Europe (Antoine et al. 2013). If the correlation of MD01-2461 is correct, the Late Magdalenian settlement in northern Central Europe can be related to a more favourable period following the Heinrich 1 event. In this period, the sea-ice melted and milder temperatures prevailed in north-western Europe resulting in a regression of permafrost and changes in the wind tracks and intensities yielding a possible reduction in loess accumulation. However, loess deposition did not cease completely and as the warm episode ended the loess dispersal probably increased again. Consequently, the hunter-gatherers of the Late Pleniglacial and the early transition period had to cope with loess bearing winds which periodically varied in their intensity during this time and potentially also with the seasons.
Thus, north-western Europe was still affected by significant geomorphological changes during the Late Pleniglacial and the landscape was presumably exposed to significant seasonal changes driven by the climate system.

The second map (**fig. 48**) provides an outline of Europe during the early Lateglacial Interstadial (GI-1e and GI-1d), which approximates to the time when concentrations such as the south-western area of Gönnersdorf, the concentrations of Marsangy, or the lower horizon of Le Closeau were occupied.
In this period, the retreat of the northern European ice sheets was accelerated by the quickly rising temperature. However, the melting of the global glaciers increased the global sea-level and the northern European estuary expanded. Perhaps this expansion led to an increased area affected by purging ice which again could have slowed the retreat of the Scandinavian ice sheet (cf. Clark et al. 2012). Furthermore, the precipitation and consequently the snowfall also increased in this period which could explain the dating of the Berghem-Moslätt moraines to the early Lateglacial Interstadial (Lundqvist/Wohlfarth 2001, 1131f.).

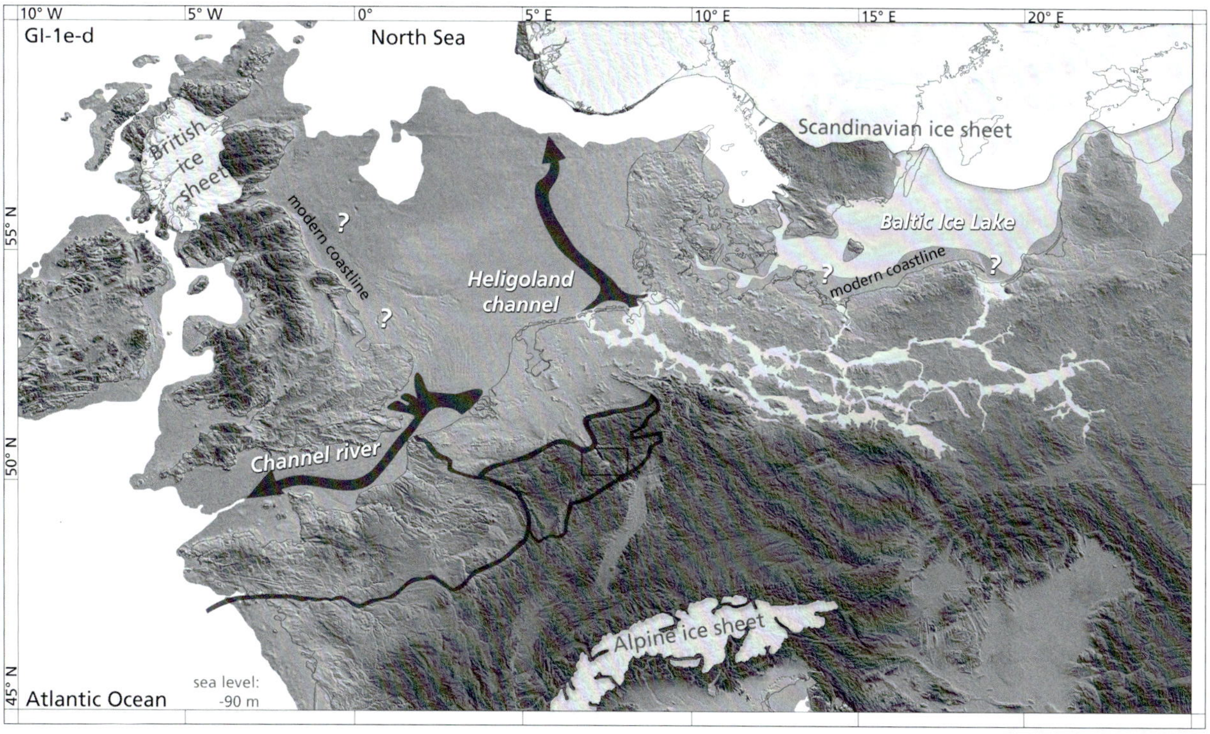

Fig. 48 Map of north-western Europe during GI-1e-d with the drainage system and the study areas indicated (see **figs 1. 45. 47**).

These moraines are smaller than the Göteborg moraines but they were, in many places, characterised by two series of ridges. Perhaps, this formation reflected the reactions to the climatic conditions of the early Lateglacial Interstadial with a first advance at the onset due to increased precipitation, a short recovery after the temperature maximum, and a short-termed readvance due to the cooling of GI-1d. Comparison with the general regression of the Scandinavian ice sheet suggests that the British ice sheet had retreated further inland and probably lost the connection to the Irish channel suring this time. The Central Belt area in Scotland was largely free of glaciers and only the western Grampians and the Highlands were ice covered. Landforms from this period were not identified yet (Clark et al. 2004, cf. Clark et al. 2012) but according to a model calculation, a British ice sheet had already melted completely by this time (Brooks et al. 2011). The only archaeological site in this northern range attributed to approximately this period was discovered at Howburn (Ballin et al. 2010). The chronological attribution was based on typo-technological comparisons with northern German material because no organic material was found in the thin stratigraphic sequence. The site was located south of the Central Belt of Scotland and according to the regression pattern, used in the present study, along with two more recent retreat scenarios (Brooks et al. 2011; Clark et al. 2012), the area around Howburn was already deglaciated during GS-2a. Thus, the location of the site and position of a possible glacial front are very remote from one another, providing no counterindications for the chronological attribution of the site nor the existence of a relic ice sheet.

In the Alps, the Daun readvance was generally dated to the period before the onset of the Lateglacial Interstadial (Ivy-Ochs et al. 2006; Ivy-Ochs et al. 2008). In comparison with the Scandinavian ice sheet and the climatic conditions, a dating of this readvance to the onset and earliest Lateglacial Interstadial might be possible. However, the increasing temperatures in the early Lateglacial Interstadial led to a quick retreat of the ice sheet above the moraines of the Lateglacial Stadial (Ivy-Ochs et al. 2008). The Lateglacial Stadial moraines overran possible glacial structures situated in higher altitudes and thereby destroyed potential evidence for the ice sheet development during the Lateglacial Interstadial (Ivy-Ochs et al. 2006, 117 tab. 1).

In general, the permafrost conditions were assumed to have retreated to a thin band around the major ice sheets (Huijzer/Vandenberghe 1998) but some evidences of sporadic permafrost were still detected in the northern Rhineland (Busschers et al. 2007, 3242). A possible explanation for the continued presence of permafrost is can be found through reference to modern analogies (Osterkamp/Romanovsky 1999). For example, deeply frozen permafrost thaws in a very gradual process and cold periods, such as GI-1d, could lead to the quick reestablishment of permafrost conditions, in particular in permafrost-friendly places such as northern slopes where the thawing process was not previously completed. In some protected parts of north-western Europe, the first soil horizons started to develop during this period (Schwark/Zink/Lechterbeck 2002).

The temperature rise at the onset of the Lateglacial Interstadial resulted, besides the rapid melting of ice sheets, into a significant rise of the global sea-level, termed melt-water pulse 1a (Weaver et al. 2003). This pulse comprised a few hundred years in which the global sea-level rose approximately 20 m, whereas the global sea-level had previously risen 20 m in approximately 5,000 years and after the pulse further c. 1,200 years passed before the global sea-level had risen another 20 m. The exact timing of this pulse remains a matter of debate but is clearly associated with the early Lateglacial Interstadial (Weaver et al. 2003; Stanford et al. 2006; Deschamps et al. 2012). During this pulse, the previously stable coastlines of northern and western Europe were significantly shifted inland and submerged thereby possible long established places for hunting and/or gathering of marine resources (cf. Hansen 2006; Langley/Street 2013). However, depending on which timing is chosen the yearly sea-level increase ranged between 4 and 6 cm (Stanford et al. 2006; Deschamps et al. 2012). Perhaps, this increase was detectable for the hunter-gatherers during yearly migrations but in comparison with a human lifetime, this slowness also allowed the gradual adaptation of human groups to this rising sea-level.

In the Baltic region an ice lake was well established in this period, although the continuous uplift of the surrounding areas placed the shorelines significantly below the modern coastlines of the Baltic Sea (Björck 1995a, 21 f.). The more western Storebælt and Lillebælt straits were not significant outlets but possibly formed minor drainage channels towards the Kattegatt (Lemke et al. 2002; Krienke 2002; Jensen et al. 2005). At present the limited availability of environmental analyses suggests that this lake remained a freshwater reservoir (Björck 1995a, 23). The outflow of this ice lake was through the Öresund area, which was probably continuously cut down in this time as indicated by the eroded material deposited in the Kattegat. In the area between the Öresund strait and the Kattegat, delta-like deposits were found suggesting a water landscape with a braided channel network during the early Lateglacial Interstadial (Björck 1995a, 22). In analogy to modern river deltas (e. g. Gilg et al. 2000), this area probably formed a favourable habitat during some parts of the year as well as a natural barrier, due to flooding, during a few months. In some seasons, however, such as late summer the area was probably passable. Thus, excursions into Scania seemed to still be possible during this time.

The chronological development of the North Sea basin is not known in detail. Besides the intrusion of sea water from the north and the presence of the major river systems in the Heligoland channel and the English channel, seismic 3D sample studies revealed the presence of a dense net of water channels and tunnel valleys on the basin floor (Praeg 2003; Fitch/Thomson/Gaffney 2005; Kristensen et al. 2007; Passchier et al. 2010; Hijma et al. 2012; Moreau et al. 2012; Stewart/Lonergan/Hampson 2012). During the Lateglacial Interstadial, this water-rich landscape was probably still very similar to the landscape that had formed at the same time in northern Germany (Clausen 2010).

The precise chronological development of the two major river systems in the North Sea basin is not well known for the Lateglacial. However, it can be assumed to have behaved comparably with the rivers that formed them. In general, the river type changed from braided to meandering systems in this phase (Mol/Van-

denberghe/Kasse 2000; Lewis/Maddy/Scaife 2001; Antoine et al. 2003b; Pastre et al. 2003; cf. Vandenberghe 2003). Therefore, the dominant river channel incised into a deep channel (Starkel/Gębica/Superson 2007). In some areas, such as the Elbe valley, the transition process predated the onset of the Lateglacial Interstadial (Turner et al. 2013), whereas in other areas such as the northern Rhineland the transformation was a more continuous transformation process lasting into the mid-Lateglacial Interstadial (Kasse et al. 2005). The local vegetation cover and the presence of frozen grounds mainly determined the river type (Mol/Vandenberghe/Kasse 2000). These determinants changed during the early Lateglacial Interstadial but their change depended, among others, on the geographic position and the distance to the vegetation refugia. Furthermore, the different response times of the river systems to the climatic changes was also related to the grain size of the transported sediments, and/or geomorphological features such as the basin configuration and the size of the catchment area (Mol/Vandenberghe/Kasse 2000; Vandenberghe 2003). The transported sediment was generally related to the energy of the river but partially also associated to the aeolian deposits. In a detailed study of the sedimentary development in the northern Rhineland, a continuation of the aeolian deposition during this early part of the Lateglacial Interstadial were attested (Busschers et al. 2007, 3242). Moreover, patches of permafrost were also identified in this study. Thus, these observations of frozen grounds and a continued sediment input could explain the delayed response in the northern Rhineland.
In total, this early Lateglacial Interstadial forms the transitional period in which the climatic and geographic conditions changed from the Late Pleniglacial to the Lateglacial Interstadial mode.

The third map (**fig. 49**) encompasses the mid- to late Lateglacial Interstadial (GI-1c_3 to GI-1a). In this period, humans settled at sites such as Andernach (upper level), Kettig, Niederbieber, Conty, or Le Closeau (upper horizons).
Concurrently, the Scandinavian ice sheet retreated relatively rapidly into middle Sweden and exposed Scania and Blekinge along with most areas of Västergötland and Småland. According to the related ^{14}C dates and varve chronologies, the diffuse Trollhättan moraines were possibly correlative with a readvance during GI-1c_3 and/or GI-1c_2 and the well defined Levene moraines were possibly formed in a short and coherent readvance in GI-1b (Lundqvist/Wohlfarth 2001). In this period, the sea waters had entered the lowlands of middle Sweden from the west and the Scandinavian glaciers still functioned as barrier for the intrusion of the sea-waters further to the east than Lake Vättern (Lundqvist/Wohlfarth 2001, 1134).
Whether a relic British ice sheet still existed in this period remains unclear (cf. Clark et al. 2004; Brooks et al. 2011). According to model calculations, the British ice sheet would have vanished before GS-2a (Brooks et al. 2011). Landforms relating to this stage have not been identified thus far (Clark et al. 2004; cf. Clark et al. 2012). In comparison with the regression velocity of the Scandinavian ice sheet, a small, inland ice sheet would have still existed at the onset of this period covering the western Grampians and the southern Highlands. However, this ice sheet could have retreated significantly during this relatively stable warm and moist period and, perhaps, fallen into several large glaciers or vanished completely. The moraines of the following Lateglacial Stadial readvance that represent one of the few evident stages for the development of the British ice sheet covered a smaller area in the same region. The presence of moraines indicates that the ice cover had retreated to an area smaller than the GS-1 ice sheet before the readvance. How far back the ice sheet retreated prior to the readvance of the Lateglacial Stadial cannot be distinguished. However, the only Scottish material attributed to this period originated from Kilmelfort Cave (Saville/Ballin 2009) which was located west of the possible relic ice sheet. Thus, the archaeological record is in accordance with the possible reconstruction of a relic ice sheet.
The Alpine ice sheet had retreated above the moraines of the Lateglacial Stadial and was partially disintegrating into single glacier fields (Ivy-Ochs et al. 2006). Although the short-termed cold phases within the

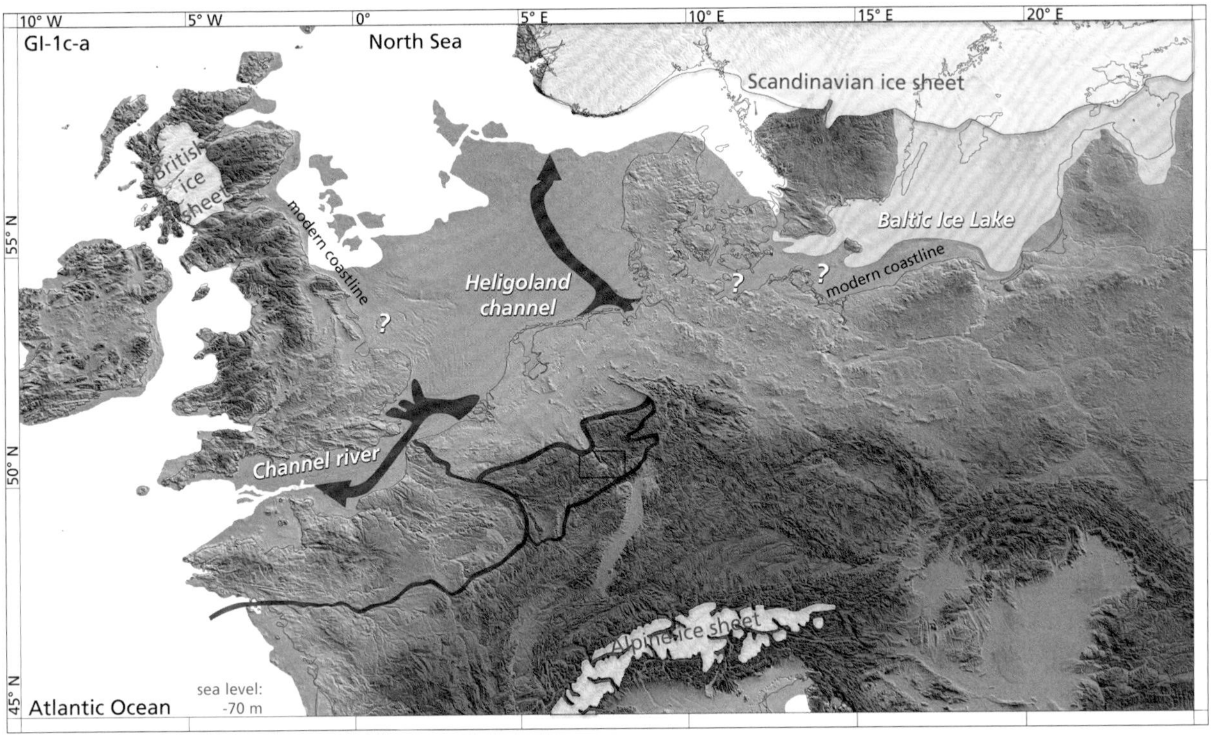

Fig. 49 Map of north-western Europe during GI-1c-a with the drainage system and the study areas indicated (see **figs 1. 45. 47**).

Lateglacial Interstadial probably resulted in readvances, the possible evidence for these readvances was depleted by the readvance of the Lateglacial Stadial which overran these older moraines (Ivy-Ochs et al. 2008, 567). The relatively high temperatures and precipitation probably strengthened the melt-water discharge and, thus, the river valleys were uncovered from the glaciers.

For this period no indications of permafrost were found in north-western Europe. Thus, the various types of permafrost had probably retreated to thin zones around the ice sheets. The increasing vegetation cover in many parts of north-western Europe led to stabilisation of the sediments and decreasing erosion processes (Tolksdorf et al. 2013). In addition, pedogenic processes transformed the top levels of the often aeolian or alluvial sediments in most parts of north-western Europe into various soils during this period.

The rate of the global sea-level rise further slowed down to a rise of approximately 2 cm per year.

The Baltic Ice Lake and the sea level in the Kattegat had reached an equilibrium at approximately 7 m b.s.l. during the mid-Lateglacial Interstadial (Björck 1995a, 22). However, the Baltic area was still affected by uplifting and the Öresund strait was probably eroded to the bedrock at this time hindering the flow of waters out of the ice lake. Thus, the sea level of the Baltic Ice Lake gradually increased above the sea level in the Kattegat. These raising water levels increased the outline of the Baltic Ice Lake and also enabled a continuous outflow through the Öresund strait and, in fact, an overflowing of the bedrock sill (Björck 1995a, 22 f.). This overflowing further enlarged the water strait in this area and, probably, made the area more impassable. Around the Inner Allerød Cold Period the lake level sank rapidly but the reasons for this drainage remained unclear (Björck 1995a, 23-25). Possibly, a connection to the Skagerrak was opened across the Swedish lowlands around Mount Billingen or the Store Belt Strait was opened as additional outflow. The latter case appeared improbable to Svante Björck because the Baltic Ice Lake was dammed up again afterwards (Björck 1995a, 23). However, in the former case, at least for the eastern part, a possibly subglacial connection was suggested and that the subsequent blocking of this strait by ice-bergs prohibited the introduction of saline waters into the freshwater lake (Björck 1995a, 23). The Kattegatt area was affected by

significant tectonic activities (Jensen et al. 2002) including an earthquake accompanied by a large tsunami during this time (Mörner 2005; Mörner 2008). Perhaps, this short-termed event allowed additional straits to be opened and large amounts of water to be drained by these straits (cf. Jensen et al. 2002). However, as a result, the lake level fell below the threshold of the Öresund strait and a landbridge between Zealand and Scania was established. Probably, the other Danish straits also became dry connecting Zealand further to Funen and Jutland (cf. Bennike/Jensen 2011). Thus, a wide connection between northern Germany and southern Scandinavia existed towards the end of the Lateglacial Interstadial.
Towards the west, a large landbridge still connected Great Britain with the continent. Since the chronological development of the North Sea basin is not known in detail, the development of the dense net of water channels and tunnel valleys revealed by seismic 3D sample studies on the North Sea basin floor cannot be further distinguished during the Lateglacial (Praeg 2003; Fitch/Thomson/Gaffney 2005; Kristensen et al. 2007; Passchier et al. 2010; Hijma et al. 2012; Moreau et al. 2012; Stewart/Lonergan/Hampson 2012). The development of the two major river systems in this area, the English Channel river and the river west of the Jutland peninsula, is not distinct for the various periods of the Lateglacial. However, in analogy to the general development of fluvial systems of the north-western European rivers during the Lateglacial, some observations on the development of these wide river systems can be made. The increasing vegetation cover and soil formation led to a stabilisation of the landscape and, consequently, a decrease in grain size of the sediments transported in the rivers. Thus, during this phase most European rivers were meandering systems (Mol/Vandenberghe/Kasse 2000; Antoine et al. 2003b; Pastre et al. 2003; Kasse et al. 2005; Starkel/Gębica/Superson 2007; Turner et al. 2013). Presumably, the two major river systems also incised into meandering beds.
During this period, the full Lateglacial Interstadial mode of the climatic and physical conditions had been established. In contrast to the Late Pleniglacial, the aeolian dispersals ceased, soils developed, the ice sheets and permafrost conditions had retreated, temperatures and precipitation were generally high and the wind intensity was low.

The fourth map of Lateglacial north-western Europe reflects the period of the Lateglacial Stadial (**fig. 50**). Reliable archaeological evidence from this period are scarce (Weber/Grimm/Baales 2011) and absent in the Central Rhineland and almost absent in the central Paris Basin. In the western uplands, some sites could be attributed to this period (Baales 1996) and some sites in the Somme basin might be related with the late Lateglacial Interstadial based on techno-typological analysis (Fagnart 1997).
Even though in all three major ice sheet regions distinctive moraines were related to the Younger Dryas, the glacial front of the Scandinavian ice sheet had further retreated in comparison to the mid-Lateglacial Interstadial. The moraines attributed to the Younger Dryas were detectable across the Scandinavian peninsula and reflect in some parts a stillstand of the glaciers and in other parts a significant readvance (Lundqvist/Wohlfarth 2001, 1136). The glaciers continued to retreat, possibly related to the warmer episode within GS-1 or the onset of the Holocene. With the gradual uncovering of the Central Swedish lowlands, sea waters could expand further eastwards, whereas the Baltic Ice Lake pushed into the lowlands from the east. Finally, the dam between these two large open waters broke and the higher Baltic Ice Lake drained into the sea waters. At this point the Baltic Sea basin became a part of the northern European estuary. Nils-Axel Mörner suggested that this possible catastrophic event was related to a tsunami dated to 10,430 Swedish varve years BP (Mörner 1999). This Swedish varve age correlated approximately with the onset of the Preboreal oscillation in the GRIP ice-core (Andrén/Björck/Johnsen 1999) and is in accordance with a previous attribution to the transition from the Lateglacial Stadial to the Holocene (Björck 1995a).
Consequently, the stillstand and/or readvance of the Scandinavian ice sheet around the latitudes of the Central Swedish lowlands encompassed the complete Younger Dryas period.

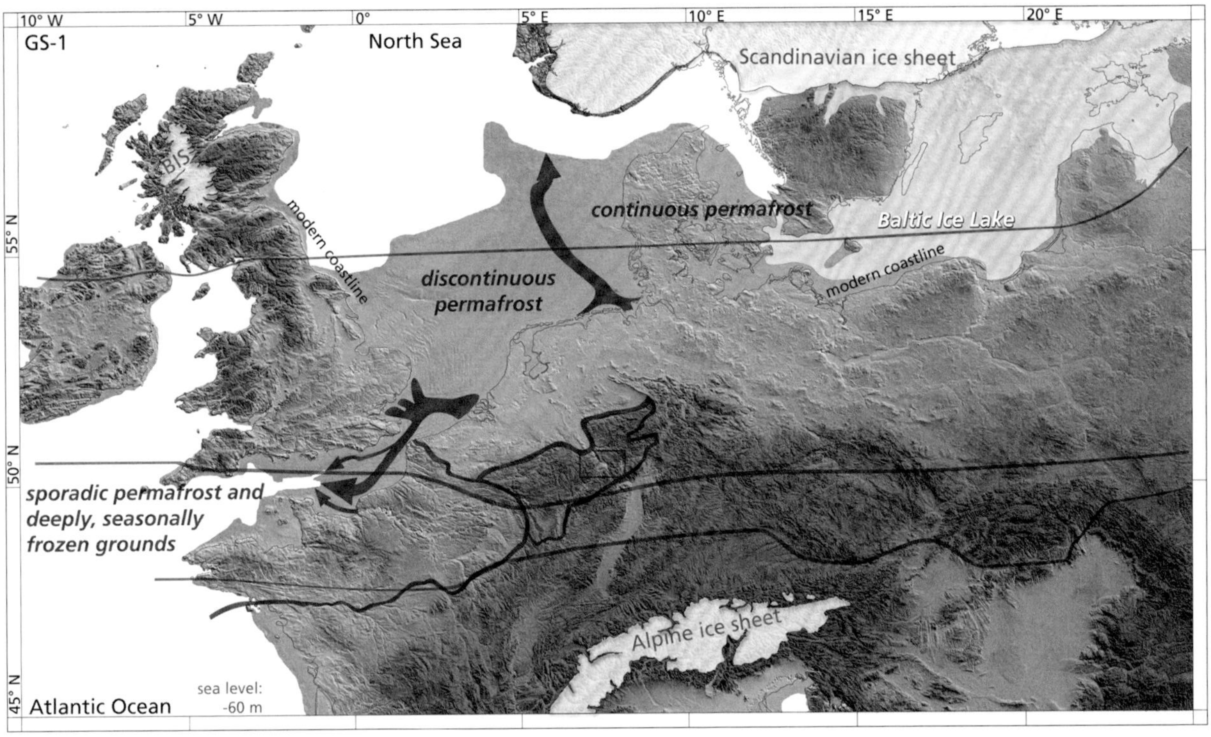

Fig. 50 Map of north-western Europe during GS-1 with approximate limits of permafrost zones and the study areas indicated (see **figs 1. 45. 47**). The permafrost zones are set according to Isarin 1997, 324 fig. 7. The eastern continuation is set according to the temperature gradients in Velichko et al. 2002, fig. 5b.

The moraines of the so-called Loch Lomond readvance are unambiguously attributed to the history of the British ice sheet (Clark et al. 2004). The small inland ice sheet covered the western central Grampians and the southern Highlands reaching across the Great Glen fault. Again, archaeological material from this time is scarce from this area. For example, organic material recovered from the Reindeer Cave (Creag nan Uamh Caves) was dated to the late Lateglacial Stadial (Lawson/Bonsall 1986). This cave is located north-eastwards of the ice sheet. Thus, the location of Scottish sites yielding archaeological material attributed to the Lateglacial is never in conflict with the reconstructions of the British ice sheet.

In the Alps, the glacial front readvanced to lower altitudes and formed the so-called Egesen moraines (Ivy-Ochs et al. 2009). Since summer temperatures were considered to be lower in the Younger Dryas (Isarin/Bohncke 1999), the strength of the spring and summer melt-water streams probably decreased. With the weaker melt-water impulses, the main rivers were less able to keep their channels free of ice. Therefore, the Alpine ice-sheet became more compact again. The maximum expansion of the Egesen moraines was presumably related to the first cold part of the Lateglacial Stadial (Ivy-Ochs et al. 2009). From this maximum extension the ice sheet retreated relatively quickly again with the onset of the warm phase within the Lateglacial Stadial. Two further readvances seem to be related to the onset of the Holocene and these readvances were possibly again due to the increase of precipitation (Ivy-Ochs et al. 2009).

Periglacial conditions returned to north-western Europe with the intense cold reversal in early Lateglacial Stadial (Isarin 1997). Evidence for continuous permafrost was found in southern Scandinavia, Scotland, and northern Ireland. The zone of discontinuous permafrost encompassed most of Poland, northern Bohemia, northern and eastern Germany, the Netherlands, most of Belgium, and the remaining parts of the British Isles. In western Germany the southern limit of the discontinuous permafrost coincided approximately with the southern limit of the Eifel, the Taunus, and the Vogelberg. According to the compilation of René F. Isarin, the southern limit of the zone of deep seasonal frost ran through Brittany, the Paris basin, along the Alpine

foothils in southern Germany and northern Austria, and through southern Slovakia (Isarin 1997, 324 fig. 7). After the first period with severe climate conditions, a period of milder temperatures occurred within the Lateglacial Stadial.

Besides the possible retreat of the major ice sheets and potential increase of melt-water discharge, permafrost and deeply frozen ground probably also thawed in some areas during this period resulting in a massive rinsing out of sediments (cf. Grafenstein et al. 1999). Perhaps, this combination of amelioration with high water discharges is an explanation for the development of several eroded layers or colluvia (Guiter et al. 2003, 70), landslides (Vardy et al. 2010), and ceasing of continuous varve formation in this period (Merkt/Müller 1999; Leroy et al. 2000). To what degree the following, final phase was cold enough to reestablish the permafrost conditions remains to be examined. Erosion and landslide events occurred frequently into the transition to the Holocene indicating a continuous instability of the grounds (van Vliet-Lanoë et al. 1992; Bos 2001; Magny/Bégeot 2004).

The continuous rise in the global sea-level slowed down marginally during the Lateglacial Stadial and in the approximately 1,100 years of this stadial, the sea-level rose some 15 m. The sea-level rise was accelerated again at the onset of the Holocene but whether this resulted in another meltwater pulse (1B) and to which period this possible meltwater pulse is dated to is a matter of on-going debate (Bard/Hamelin/Delanghe-Sabatier 2010; Rodrigues et al. 2010; Stanford et al. 2011a).

At the beginning of this period, the lake level of the Baltic Ice Lake recovered from the late Lateglacial Interstadial drainage and was increased in volume again (**fig. 51**). Perhaps the reason for this increase was that the Mount Billingen drainage was blocked by ice again (Björck 1995a, 24). The raising water level overflowed the Öresund strait again and ended the landbridge. Towards the end of this period, the final drainage of the Baltic Ice Lake occurred and lowered the water level in some areas for approximately 25 m (Björck 1995a, 26 f.). In this period, a several kilometres wide connection to the northern oceans was established through the Central Sweden lowlands. This wide connection transformed the freshwater lake into a brackish water basin, the so-called Yoldia Sea. The opening of this connection was related by Nils-Axel Mörner to a major earthquake followed by a tsunami which cleared the strait of ice-bergs (Mörner 1999). However, a hiatus in the stratigraphic sequences from the Lillebælt area suggested also that the Danish straits were involved in the final drainage which cut and eroded some older sediments (cf. Bennike/Jensen 2011). Subsequently, a landbridge was established between Scania, Zealand, Funen, and Jutland due to the uplifting of these areas above the water level. The abrupt lowering of the water level established, on the one hand, significant areas of land, which were initially rather unstable due to the silty deposits of the former ice lake and, on the other hand, drained previous littoral habitats significantly. Consequently, the environments around the Baltic Ice Lake were influenced by these new hydrographic conditions that probably occurred concomitant with the Holocene amelioration and the subsequent melting of the remaining ice-sheets and thawing of the permafrost soils.

The filling of the deeper channel along the British east coast with sea waters from the north did not probably occur before the Holocene (cf. Ward/Larcombe 2008). Thus, the North Sea basin presumably formed a large landbridge which connected England, France, Belgium, the Netherlands, northern Germany, and Denmark. However, the excess from England to France was limited by the wide English Channel river and the west of the Cimbrian Peninsula flowed another large river. The low resolution of the chronological development of the North Sea basin makes an attribution to a specific period in the Late Weichselian impossible. Based on the general development of fluvial styles of some north-western European rivers during the Lateglacial Stadial, these two major river systems in the North Sea area could be further characterised. The response of the European rivers varied. Some remained in a meandering system and further incised into their beds (Antoine et al. 2003b; Pastre et al. 2003; Turner et al. 2013), whereas others switched back to

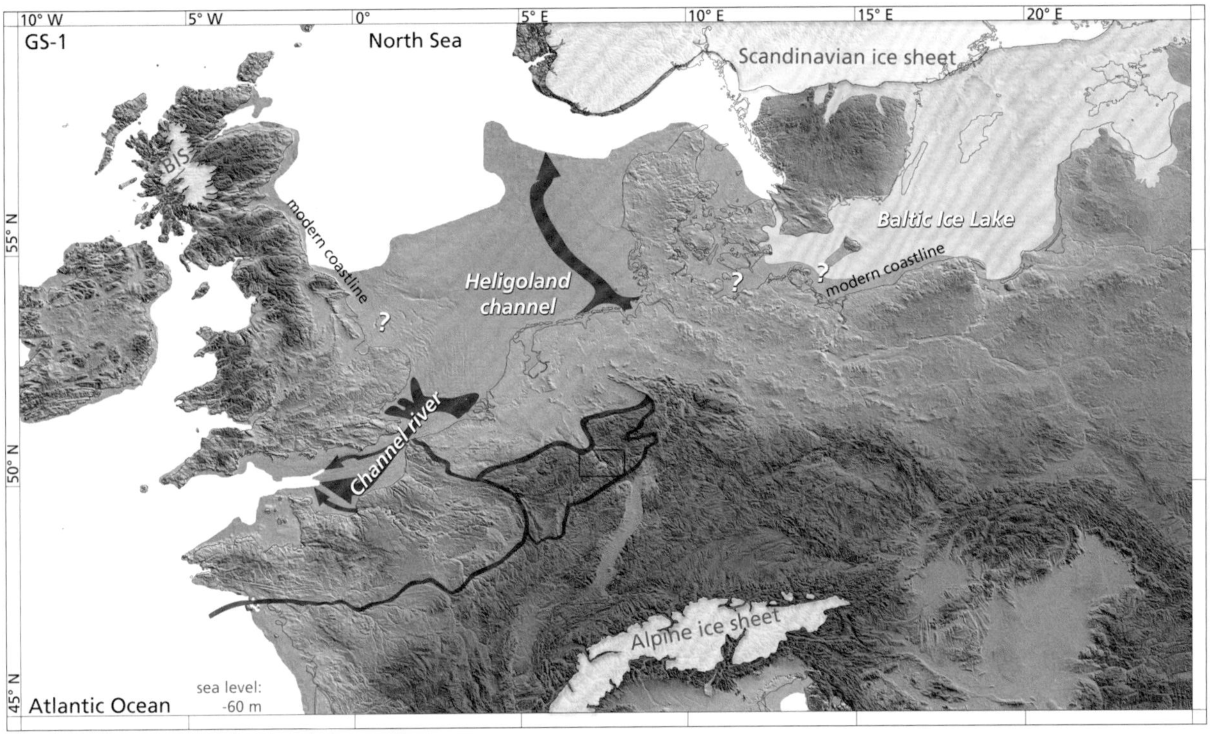

Fig. 51 Map of north-western Europe during GS-1 with the drainage system and the study areas indicated (see **figs 1. 45. 47**).

a braided system (Mol/Vandenberghe/Kasse 2000; Lewis/Maddy/Scaife 2001; Starkel/Gębica/Superson 2007). This response was partially dependent on the geographic position of the rivers, for instance, the Rhine and the Meuse further incised into their beds in the Netherlands, whereas they temporarily reestablished braided systems in western Germany (Busschers et al. 2007). Furthermore, the geomorphological setting seemed also of importance because the stream in the Niers valley was further incised, whereas the Rhine in this area switched to a braided system (Kasse et al. 2005). Besides the relation to the presence of permafrost, the storm track pattern and the dispersal of aeolian sediment partially influenced the behaviour of the rivers due to the creation of unstable bank sediments which could increase the amount of sediment transported by the river. The aeolian dispersal restarted in some places during the Lateglacial Stadial and deposited significant amounts of sediments which changed the local geomorphology (Hilgers et al. 2001; Kaiser/Clausen 2005).

In the western Cimbrian Peninsula the coversands filled several valleys resulting partially in a block and/or redirection of the rivers running into the Heligoland channel (cf. Kaiser/Clausen 2005). Moreover, some tsunamis from this period (Mörner 2008) could have changed the land in the northern estuary and lead to redirections of rivers directly northwards. Thus, the catchment area of the Heligoland channel possibly decreased temporarily during the Lateglacial Stadial. In regard to the periglacial conditions, the increased sediment input, and decreased water amount, the Heligoland channel river probably also returned to a braided system during the Lateglacial Stadial.

Even though the climate signal of the Lateglacial Stadial is comparable to that of the Late Pleniglacial, the development of the landscape during the Lateglacial Interstadial resulted in partially different reactions. In fact, the climatic and geographic indications of the Lateglacial Stadial appeared, in general, more comparable to the conditions during the transition period in the early Lateglacial Interstadial when aeolian deposition still occurred and rivers changed between meandering and the braided system.

Lateglacial vegetation development in the study areas

Besides oxygen isotope records, vegetation profiles have provided the most prominent and distinct sub-division of the Lateglacial Interstadial. However, these vegetation sub-periods have not necessarily correlated across a wider geographic areas due to the partial dependence on regional developments. Moreover, they were also not reacting instantaneously to the developments that occurred in the isotope eventstratigraphy of the Greenland records. Thus, the development of vegetation should be checked against an independent chronology.

Chronology of the pollen profiles

In the present study, two pollen profiles were used as the main source of information for the vegetation development in the studied sub-areas. The Meerfelder Maar (MFM) record provided insights into the changes in the Central Rhineland and the western uplands in general (Litt/Stebich 1999). Furthermore, the MFM yielded a well established varve chronology (Brauer/Endres/Negendank 1999), which is correlated to the oxygen isotope record from NGRIP (see p. 293-310). The LST was an important marker in this record and was dated to 12,930 years cal. b2k in the varve chronology.

In contrast, the record from the Paris Basin is a synchronised pollen diagram (Pastre et al. 2003) that was made of several pollen profiles from the Paris Basin (see **fig. 20**; cf. Leroyer 1994). Since the pollen diagram were checked against various pollen- and lithostratigraphies in the region, the resulting synchronised sequence represents a reliable relative chronology. To relate this succession to the calender timescale, some pollen profiles were supplemented with ^{14}C dates, which were usually made on bulked sediment samples (Pastre et al. 2003, tab. 1; cf. Limondin-Lozouet et al. 2002, tab. 1). In this study, sediment samples are assumed to produce in general no reliable age estimates. This unreliability is due to material of different ^{14}C ages being mixed as a result of the often complex depositional and geochemical processes within sediment deposits and, in particular, within freshwater environments (Hiller et al. 2003; Brock et al. 2011).

However, the synchronised pollen diagram has also been supplemented by two dates made on wood samples and a further two dates made on other plant remains (Pastre et al. 2003, tab. 1; cf. Ponel et al. 2005, figs 3 and 5). All four samples were recovered at Houdancourt in a palaeochannel deposit that was analysed in a multi-proxy project that, unfortunately, yielded a fragmented pollen profile. The calibration results of the four ^{14}C dates are checked in the following against the results of this multi-proxy analysis and the isotope record of NGRIP.

The coleopteran composition of the lowest palaentomological sample from Houdancourt suggested that these remains originated from the time of the thermal optimum in the Lateglacial (Ponel et al. 2005, 2458). The plant remains for the lowest ^{14}C date were found below this palaeontomological sample but already within an organic silt deposit which formed on top of the Pleistocene sands and gravels (Pastre et al. 2003). This stratigraphic position suggested that the most reliable attribution of this date was between the onset of the Lateglacial Interstadial and the end of the calibrated age range (15,480-14,200 years cal. b2k).

The second oldest date was produced from a wood sample and originated from a transition of organic silt deposits to fibrous peat. The calibrated age range of the ^{14}C date (14,210-13,650 years cal.b2k) overlapped the date range of the cold reversal of GI-1d and also partially touched the date range of the cold GI-1c_2 episode (see **tab. 64**). Within the top part of the organic silt, the palaeoentomological sample 7 was taken (Ponel et al. 2005, fig. 3). The upper part of this sample contained coleoptera species indicating a cold climate (Pastre et al. 2003, 2180). On top of the position of the wood sample, birch (*Betula* sp.) and juniper

(*Juniperus* sp.) pollen increased in the Houdancourt profile before a small decrease affected the arboreal pollen in general. Alongside this small decrease in arboreal pollen, birch and juniper increased again and steppic herbs and the true grasses (Poaceae) decreased significantly. Assuming some relation of the climate and the environment as reflected by the pollen composition, this smaller decrease could be related to the short-termed cold episode of GI-1c_2 and the colder species in the coleopteran sample 7 could be attributed to the GI-1d event. The wood originated from the top part of this sample and well before the possible onset of GI-1c_2 in the pollen record. Thus, the most reliable attribution is the first part of GI-1c_3 for this date.
The second youngest date was also made on a wood sample. The calibrated age ranges from 13,720 to 13,200 years cal. b2k and could relate to almost the complete GI-1c_2 and the complete GI-1c_1, as well as the onset of GI-1b (see **tab. 64**). The sample originated from the middle of the fibrous peat where the material became siltier. This position relates approximately to the transition of the palaeoentomological samples 4 to 3 (Ponel et al. 2005, fig. 3). According to the coleoptera composition, the summer temperatures decreased slightly between these two samples. The molluscs suggest still water conditions during this period. On top of the sample in the pollen profile, birch pollen decreased to minimal values, whereas juniper, willow (*Salix* sp.), and pine (*Pinus* sp.) values illustrated a small increase. In addition, the values for *Artemisia* sp. pollen increased suggesting an opening of the landscape. Thus, in general, the sample seemed to originate from the onset of a moist episode with colder summers that resulted in an opening of the vegetation cover. If GI-1c_2 was recorded in the lower part of the pollen diagram, this episode could represent a first indication of the deterioration towards GI-1b and the most reliable position for this calibrated age would be the transition from GI-1c_1 to GI-1b. In contrast, if this deterioration is related to GI-1c_2, the episode in the lower part of the pollen profile could be related to another event. This correlation could further sustain the attribution of the second oldest date to the onset of GI-1c_3 because of the increasing distance of the older wood sample to the position of GI-1c_2 in the profile. According to the correlation of the Houdancourt profile to the synchronised pollen diagram of the Paris basin, two further cold episodes occurred on top of the second oldest date before the onset of the Younger Dryas (Ponel et al. 2005, fig. 5). Within the deterioration on top of the position of the sample the pine pollen began to increase (Pastre et al. 2003, fig. 3). The second deterioration appears of longer duration and more significant and could correlate with GI-1b. Therefore, an attribution of the second youngest date to GI-1c_2 appeared more probable. However, since on top of the sample the palaeoentomological and the palynological data indicate increasing deterioration towards a colder climate, the calibrated age referred either to the onset of GI-1c_2 which corresponded to the lower calibrated age range or to the younger part of GI-1c_1 in which the isotope values from NGRIP indicate a continuous deterioration of the North Atlantic climate towards GI-1b. This position corresponds to the middle of the calibrated age range.
The calibrated age range of the youngest date (13,280-12,960 years cal. b2k) overlapped GI-1b. This sample consisted of plant material and was found in the upper part of the fibrous peat. The paleoentomological sample 2 was taken at approximately the same height in the profile (Ponel et al. 2005, fig. 3). According to the coleopteran data, the summer temperatures increase again in this period, whereas the winter temperatures appeared to decrease but this impression is due to the increasing data range for the winter temperatures (Ponel et al. 2005, fig. 4). In the pollen profile, arboreal pollen are more dominant than in the previous sections. The pine pollen are more abundant than the birch pollen and further increased in this section. Even though the arboreal pollen are abundant, the general number of pollen is low and the higher numbers of steppic herbs than in the previous sections as well as the significantly increased juniper values indicate a colder and more open episode within a forested phase below the stratigraphic position of the sample. Probably, this position related to the end of GI-b and the onset of GI-1a. Moreover, this position for the youngest sample further suggests that the position of the second youngest sample to a position before

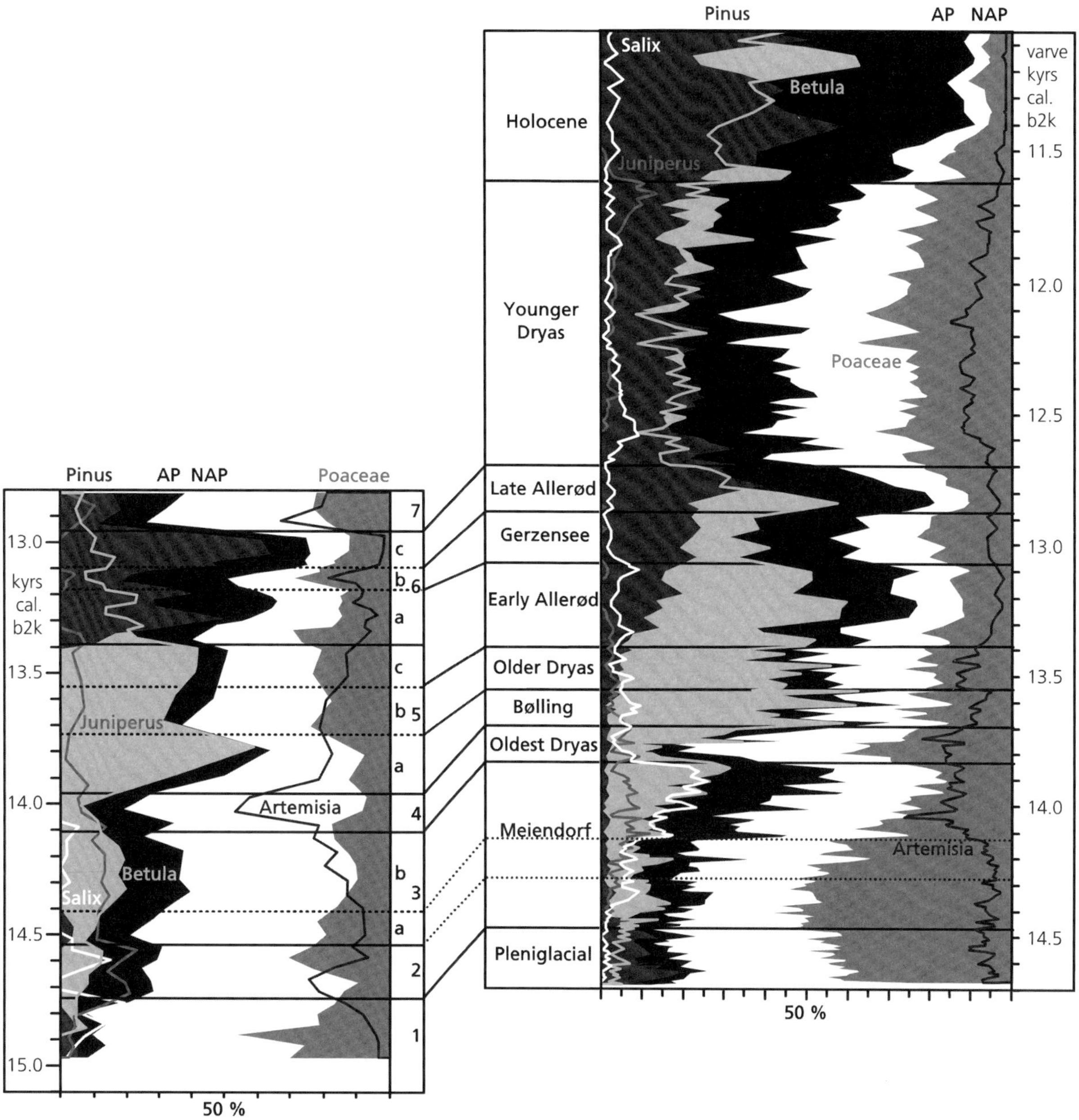

Fig. 52 Synchronised pollen diagram of the Paris Basin fitted to the calender timescale by the calibration of the associated and reliable ^{14}C dates (left) and the simplified Lateglacial pollen diagram from Meerfelder Maar, MFM 6, according to the varve chronology (right; see p. 40-48). The grey lines serve as guidelines for comparable palynological zones. Grey dotted lines: possible sub-zones in the Meiendorf stage of the MFM 6 diagram which are comparable to the Paris Basin record. **AP** arboreal pollen; **NAP** – non-arboreal pollen. – For further details see text.

GI-1b. On top of the youngest sample, the fibrous peat continued for another few centimetres before it was overlain by the calcareous silts of the Younger Dryas.

The synchronised pollen diagram was fitted to the calender timescale according to the oldest and the youngest date. The diagram between these two points was stretched and clinched according to the two intermediate calibrated ages (**fig. 52**).

However, the MFM pollen profile also requires some modification. In the MFM Bølling zone, a 4.2 cm thick deposit of slump and turbidite was recorded which due to the lack of lamination could not be included in the varve chronology (Brauer et al. 2000a). Thomas Litt and Martina Stebich counted only 130 varve

couplets in their Bølling pollen zone (Litt/Stebich 1999, 8f. 13), whereas 140 varve years were attributed to this period according the varve thickness measurements (Brauer et al. 2000a). Based on a palynological comparison with northern German lakes, 110 varves were suggested to be missing due to the 4 cm of slump and turbidite. Clearly, this palynostratigraphic correlation across several hundred kilometres and two very different geomorphological regions during the mid-Lateglacial Interstadial forms an additional source of error in the precision of the chronology. However, the varve thickness above and below this disturbance varied significantly between approximately 0.25 and 2.0 mm per year and up to c. 3.0 mm per year some 100 years below the disturbance. Thus, based on the varve thickness some 100 years before and after the deposit, an age estimate for the unlaminated 42 mm ranges between 14 and 168 years. Since the comparison with the northern German lakes produced more precise age estimates within this range, they were accepted as probable substitution. In regard to the different duration of the Bølling zone in the pollen and the varve thickness study of the MFM, a mean of additional 120 years was added to the pollen profile of the MFM at the position of the disturbance. In addition, a comparison of the pollen diagrams from Holzmaar and Meerfelder Maar indicated a further 35 years difference between these two Eifel records after the addition of the 120 varve years in the MFM (see p. 346). However, this age offset could be explained by different response times. An impact of the deposition of the LST on the varve preservation as observed in other laminated stratigraphies such as the Soppensee (Hajdas et al. 1993) was not documented or considered unimportant for the MFM chronology.

If the northern French diagram was correlated to the calender timescale and the additional varve years were added in the MFM pollen profile as described above, the two diagrams can be compared in more detail (**fig. 52**).

The onset of comparable pollen assemblage zones (PAZ) occurred in general some 50-200 years earlier in northern France than in the Eifel region. Thus, the onset of the uppermost PAZ 7 (Younger Dryas) in the French diagram correlated to approximately the mid-Gerzensee zone in the MFM record. In regard to the short-lived climate events in the Lateglacial Interstadial, a lagged response of more than 100 years to the same climatic event appears improbable. Moreover, parallel analyses of pollen and oxygen isotopes from Swiss lakes (Lotter et al. 1992) and Polish lakes (Ralska-Jasiewiczowa et al. 1998; Goslar et al. 1999) indicated much shorter response times with only annual to decadal lag of the vegetation to climatic changes (cf. Litt et al. 2001).

Several explanations can be found for the lagging onset of the PAZ in the MFM record in comparison to the synchronised Paris Basin diagram:

The development of the vegetation in the western upland zone can be offset due to the generally higher altitudes and the more northern and eastern location of this zone resulting in a later beginning of the process of the Lateglacial biome succession. Further possible explanations are that the correlation of the zones is incorrect or the chronologies and/or their correlation is imprecise or in the composite diagram from the Paris Basin a hiatus remained unrecognised.

The latter cannot be fully excluded but the French pollen diagram was synchronised from several profiles in the Paris Basin. Consequently, a hiatus must have affect all relevant profiles from northern France. A hiatus of this extent would have been recognised also in the archaeological records but at Le Closeau a relatively continuous record was found throughout the Lateglacial Interstadial (see p. 210-219). Therefore, an undetected hiatus seems not very probable as an explanation for the offset between the Paris Basin and the western upland zone.

The chronologies of both records could be imprecise. Although some varves in the MFM record could have been lost due to the deposition of the LST and would require further varves to be added in the chronology of the Lateglacial Interstadial, the comparison of the MFM sediment record with the isotope record of NGRIP indicated a relatively high similarity (see p. 345). The difference between the onsets in the MFM sediment

record to the onsets of the NGRIP isotope events ranged between 77 and 130 years (**tab. 71**) which was within the accumulated standard deviations of the records. Accordingly, only a few decades could have been lost by the deposition of the LST in the MFM record. In contrast to the sediment record, the onset of the PAZ in MFM were generally delayed in comparison to the onset of the NGRIP isotope events by some 130 to 180 years (**tab. 71**). This direct comparison within the MFM record sustains that the response to changes in the North Atlantic climate system was quicker in the sediment system than in the vegetation. Thus, an offset between the MFM pollen diagram to the NGRIP isotope events are probable.
In contrast, the calibration results from the Paris Basin were tuned along the NGRIP record. Furthermore, the chronology of the Paris Basin diagram is based only on four calibrated ^{14}C dates from one of the contributing profiles and, thus, an imprecise correlation is a possible source of error. In particular in regard of the onset of the PAZ 7 because this zone remained in the original chronology due to the first reliable age estimate being located in the PAZ 6b and the PAZ 6c (Pastre et al. 2003). Therefore, additional, reliable dates from various profiles are desirable for future comparisons with this synchronised French pollen diagram.
In addition, the resolution of the French diagrams did not reach the very high resolution of the laminated MFM diagram and further resolution of the French diagrams was lost by the synchronisation. Consequently, the MFM record could be segmented more precisely along single wiggles in the pollen profile, whereas the northern French diagram reflects the general tendency of the pollen composition in the Lateglacial. Therefore, some sharp but short-term wiggles in the MFM record could remain unrecorded in the French diagram, whereas other, well preserved events could also be overestimated by the stretching of the Paris Basin diagram. This resolution difference hindered the correlation of the records and made probable that the pollen assemblage zones are not exactly correlative. Additionally, the pollen zones were defined differently in detail and, therefore, some almost contemporary developments could be overseen by the comparison along the original PAZ.
Thus, the PAZ of the original publications are neglected in the following comparison. However, the correlation of the two pollen profiles along a calendar timescale shows that an equation of palynologically defined zones and chronozones across far distances is problematic and, occasionally, misguiding.

Comparison of the pollen profiles

Comparing the synchronised pollen diagram from the Paris Basin and the pollen diagram from the varve counted MFM, some comparable developments but also some differences become apparent.
Some of the differences are independent of the chronological correlation. For instance, in the Paris Basin, juniper pollen are more important than willow, whereas in the MFM record willow is usually more important. Furthermore, birch was the most important arboreal pollen (AP) type in the MFM record, whereas in northern France pine became significantly more important during the upper part of the diagram. Among the non-arboreal pollen (NAP), Artemisia pollen are occasionally more important than the pollen of the true grasses (Poacecae) in the Paris Basin. In the Lateglacial section of the MFM record, Poacecae are always significantly more important than Artemisia.
Perhaps, these differences resulted from climatic influences due to the longitudinal and topographic setting of the profiles. For example, the various types of willow and juniper are relatively comparable in their general climatic and soil requirements[46] but willows usually need more moisture than juniper. Thus, the greater

46 These requirements are considered according to statistical indicator values created for the use in analyses of local conditions (Ellenberg et al. 2001).

importance of willow in the MFM could indicate that this middle range mountain region received more precipitation and/or that the soils could store more water than the sediment at the sites from the relatively level northern France. This impression of a higher moisture is further supported by the supersession of birch by pine in the upper part of the French diagram. Pine is generally more indifferent towards moisture than several birch types which require relatively wet grounds. Thus, with increasing aridity these trees become less frequent. The generally high frequency of Artemisia pollen further sustains the impression of drier grounds in the Paris Basin. Artemisia types require in general more light and less moisture than many true grasses such as *Agrostis* sp., *Poa* sp., or several species of *Festuca* sp. Thus, the catchment area of the northern French pollen diagram appeared to be generally drier than the MFM catchment area. Modern precipitation data also documented a difference in the annual average precipitation between the western Eifel and Ardennes region, the Central Rhineland, and the Paris Basin (Nixon et al. 2003, Map 5.1). The latter two areas receive approximately the same amount of annual rainfall, whereas the western Eifel and the Ardennes region receives clearly more precipitation. This increased precipitation is due to the general atmospheric circulation patterns in Europe. The moist air is transported from the Atlantic onto the continent where the clouds usually pass across the more level regions closer to the Atlantic and along the English Channel but precipitate in the higher altitudes of the Eifel and Ardennes region. Based on the pollen profiles, this climatic sub-area with increased precipitation seemed to have already existed in the Lateglacial.

Besides these general differences, some chronologically different developments can be observed in the two pollen diagrams. Compared to the NGRIP record (**fig. 53**), the Paris Basin profile appears to reflect the climatic changes as revealed in the NGRIP isotope record more instantly than the MFM record. However, this impression is partially due to the correlation of the calibrated ^{14}C ages from the northern French record along the NGRIP isotope record and partially due to the higher resolution of the MFM diagram.

Assuming that the chronologies and the correlations of both records are reliable, pollen preservation in the Paris Basin began approximately 300 years before the onset of the Lateglacial Interstadial, whereas in the MFM only some decades of the Late Pleniglacial were recorded. Thus, pollen were preserved approximately 200 years earlier in northern France than in the MFM profile. Perhaps, increasing moisture and increasing vegetation cover leading to a changing geochemical composition of the sediment as well as decreasing redeposition processes contributed to a better preservation of these microfossils (Bennett/Willis 2002; Lebreton et al. 2010; Tweddle/Edwards 2010). The western upland zone seemed to have a generally moister climate regime than the Paris Basin but the longer duration of frozen grounds and the stronger impact of aeolian activities (see p. 367-370) could have contributed to the later onset of pollen preservation in this area.

In both regions, the relatively short preserved pollen zone of the Late Pleniglacial is dominated by the NAP. In particular, high proportional values of true grasses suggested the presence of widespread grasslands in north-western Europe at this time. However, the pollen concentration remained generally low which further indicated a sparse vegetation cover (Litt 1999, 7). The low AP comprised mainly willow in the Paris Basin and pine in the MFM record. The often high amount of pine pollen but also the occurrence of pollen from thermophilous trees such as oak (*Quercus* sp.) were frequently assumed to represent redeposited material or material from long-distance transport (Litt/Stebich 1999, 7; Cheddadi et al. 2006). Nevertheless, macrobotanical remains have started a discussion about the presence of cryptic northern refugia for some tree species such as pine (Stewart/Lister 2001; Willis/van Andel 2004; Birks/Willis 2008). However, in both pollen diagrams the proportion of pine pollen vanished to insignificance after the first 150-250 years when vegetation covers and conditions of preservation had probably stabilised. This pattern further sustains the impression that the pine pollen originated mainly from redeposited sediments and that with the ceasing redeposition of sediment the values of pine pollen decreased significantly in both regions.

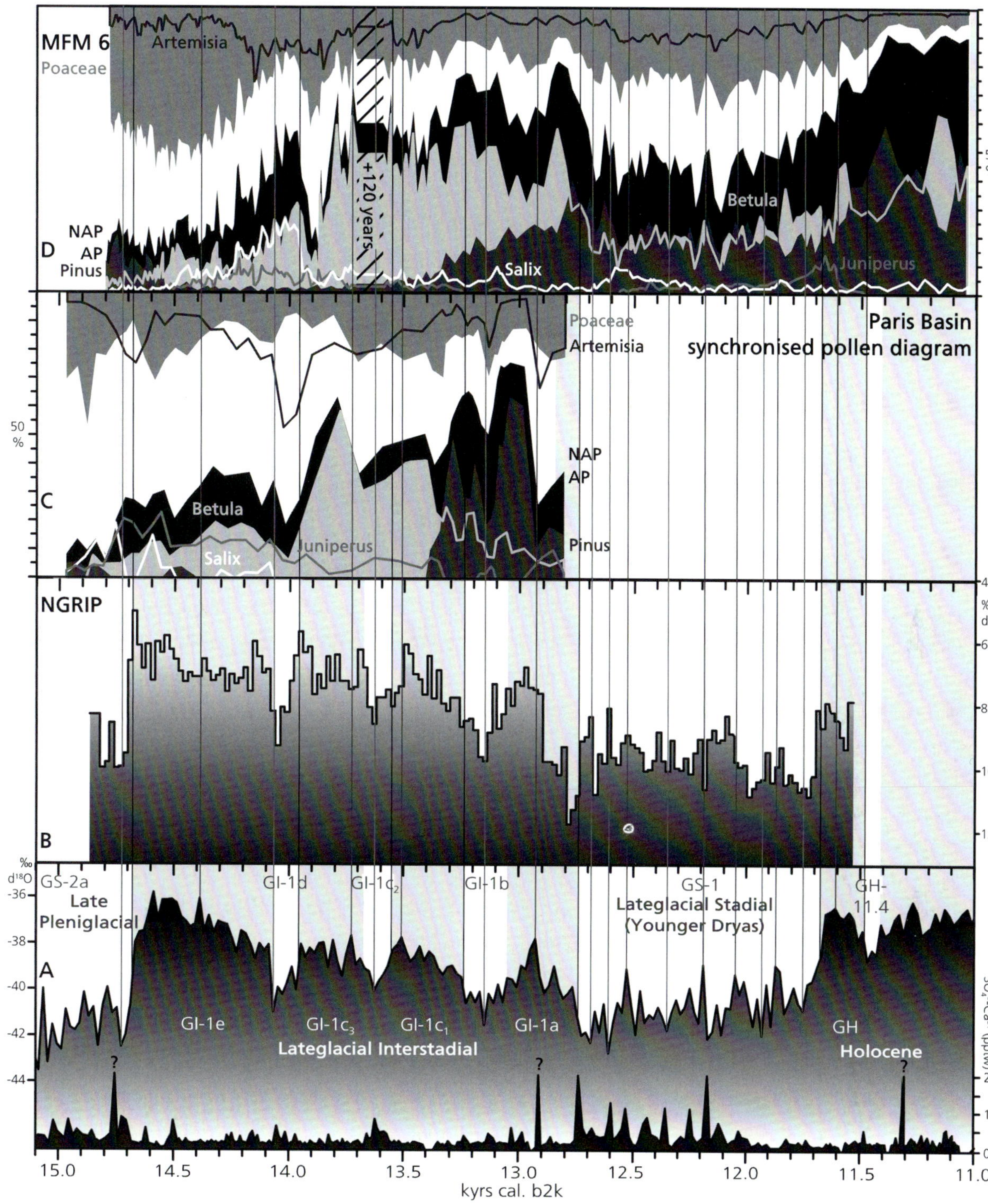

Fig. 53 Comparison of the isotope record from NGRIP and the pollen diagrams used in this project. Grey shaded areas represent events of more interstadial values than the surrounding values in the oxygen isotope record (A). Thin bars given as guide lines (light grey: low; medium grey: high; black: transition). **A** Oxygen isotope record from NGRIP with the Ca^{2+} corrected sulphate content but without associated ash layers (see **fig. 30**). – **B** Deuterium excess record from NGRIP. – **C** synchronised pollen diagram from the Paris Basin, fitted to the calender timescale by the ^{14}C calibration (see **fig. 52**). – **D** simplified Lateglacial pollen diagram from Meerfelder Maar, MFM 6, according to the varve chronology. – **AP** – arboreal pollen. – **NAP** non-arboreal pollen. – For further details see p. 40-48 and text.

Shortly before the onset of the Lateglacial Interstadial, a first peak of AP occurred in both records. In the Paris basin, pine pollen had already vanished from the diagram and the peak was due to a peak of willow (*Salix*. sp.) followed by a peak of juniper (*Juniperus* sp.). In the MFM record, the peak was still formed by pine pollen. The AP pollen in both records decreased around the onset of the Lateglacial Interstadial, only the values of birch (*Betula* sp.) began to increase at this point. When the birch pollen reached a first peak in the French record, juniper and willow pollen formed equally peaks resulting in a first peak of AP in the early Lateglacial Interstadial. In contrast to the French record, the AP remained low at the onset of the Lateglacial Interstadial. Among the AP, pine pollen remained the most abundant species for another 100 years and were then suppressed by birch pollen. From thereon, birch was usually predominated above pine in this archive, only in the transition from the Lateglacial Interstadial to the Lateglacial Stadial and in the early Lateglacial Stadial pine pollen occurred again more abundant than birch pollen.
In the Paris Basin diagram, the first higher AP values again decreased after the GI-1e maximum in the oxygen isotope record of NGRIP and are synchronous with a decline in the deuterium excess record (**fig. 53**). Furthermore, a basaltic ash layer was recorded at this point in the NGRIP record (**fig. 30**) and attributed to the Icelandic Laki volcanic system (Mortensen et al. 2005). Historic eruptions from this volcanic system significantly affected the climate and environment of Central Europe (and beyond) for the following years (Grattan 2006). Since modern atmospheric circulation patterns were in all likelihood already established at the time of the Lateglacial Interstadial, the impact of an eruption in the Laki fissure on Europe was possibly comparable to these historic events. In the French pollen record, the low values of AP were accompanied by higher proportions of pine pollen which probably reflected another short episode of increased introduction of allochtonous material into the stratigraphies. Perhaps, this introduction was related to extreme hot and dry summers followed by spring flooding events which were also quoted as a direct result of a historic Laki eruption (Thordarson/Self 2003). However, if this eruption had affected north-western Europe, crypotephras should be detectable. Therefore, cryptotephra analyses, searching particularly for basaltic ash, could be a considerable help to ascertain chronological correlations within this period and, perhaps, to position archaeological assemblages during this period. This type of findings is particularly important because a longer plateau in the calibration curve makes a precise chronological attribution more difficult. In the MFM diagram a small increase of pine and a decrease of birch was documented around this period but concomitantly a peak of willow occurred which kept the AP values high. However, with the decrease in willow pollen the AP values also decreased so that a low of all AP values, comparable to the French pollen record, occurred some 100 years later in the MFM diagram. Nevertheless, the pollen concentration slowly increased after this point. In the northern French diagram the values of AP began to increase again at this time, in particular due to birch reestablishment, whereas juniper values remained relatively constant. Willow remained undetected after the possible introduction of allochtonous material and reappeared only for a short period towards the end of GI-1e. After a last peak at the onset of GI-1d, willow pollen disappear from the French pollen sequence. In addition, juniper pollen began to gradually decrease at this point.
In the MFM, a significant increase of willow is also accompanied by an increase in AP and birch pollen increase towards the end of GI-1e. However, the steep decline in AP, birch, and willow pollen values in the MFM at the beginning of the Oldest Dryas biozone (Litt/Stebich 1999) correlates, in the present chronology, with the end of GI-1d and occurred again 100 years later than these low values in the French diagram. Without the additional 120 years, the extreme behaviour of the MFM record in the Oldest Dryas zone corresponds to the mid GI-1c_3. Even though a correlation with the cold event GI-1d was usually assumed, the existing chronology cannot confirm this correlation and rather indicates an extreme reaction of the regional vegetation on an event after the onset of the interstadial episode GI-1c_3. A comparable pattern was recently reported from the Weerterbos region, some 150 km north-westwards of the MFM in the southern Netherlands (van

Asch et al. 2013). Previously, combined records of oxygen isotopes and pollen from this region suggested a delayed reaction of the vegetation record to the isotopic cold events GI-1d and GI-1b (Hoek/Bohncke 2001, 1262). Thus, the close correlation of the Paris Basin diagram might be due to the correlation of the chronology with NGRIP (see p. 384-386) and the asynchronic reaction in the MFM could be the more reliable development in the Lateglacial environment. However, in the French diagram, a significant peak of *Artemisia* sp. pollen occurred during the period of low AP values, whereas the true grass values also decreased. This pattern possibly suggests a widespread opening of the landscape because most species of the Artemisia genus require full-light conditions, whereas many species of the various genera within the Poaceae family grow in partially shaded areas. The temperature tolerance of the various genera varies considerably. Thus, the development in the pollen diagram could indicate in analogy to the onset of GS-1 that the impact of a temperature decrease in GI-1d did not affect the environment as severely as aridity and wind breakage (cf. Brauer et al. 2008). However, this combination of low precipitation and high wind activity would also result in an increase of allochtonous material in the records. Indicators for this allochtonous intrusions as suggested by the previous increases of pine pollen were not found in this part of the diagram but in the MFM varve record an increase of varve thickness was documented for this biozone (Brauer/Endres/Negendank 1999). Furthermore, the pollen concentration dropped in this zone to almost Late Pleniglacial values (Litt/Stebich 1999, fig. 4). Moreover, the Artemisia pollen formed only a short peak towards the end of the Oldest Dryas biozone but remained less significant than the Poaceae in this region. Comparable to the Paris Basin, willow and juniper only occurred in small proportions after the onset of this biozone in the MFM.
According to the present correlation, the following steep increase of birch pollen and AP in the MFM occurred quasi-contemporaneously with the same increase in the Paris Basin. In the former, this increase is assumed to correlate with a spread of birch trees and, thus, the spread of first light forests. This interpretation is further sustained by the increase of pollen concentrations to generally higher values. In addition, this peak was accompanied by lower values of juniper, Artemisia, and Poaceae pollen in the French sub-area indicating a decline of an open shrub vegetation. However, the increase in the MFM is not gradual but characterised by extreme fluctuations between high and low AP and birch pollen proportions. This wiggling record is, however, obscured in the synchronised Paris Basin diagram.
After this first increase, a gap due to the slump and turbite deposit followed in the MFM. and In the northern French records, a steep decline of AP and birch values followed and it was accompanied by a slight rise in values of Artemisia and juniper and a peak of Poaceae. This episode in the French diagram correlates with the end of GI-1c_3 and the transition towards the cold event GI-1c_2 in NGRIP. These significant reactions in the French diagrams and the destruction in the MFM within GI-1c_3 can be aligned with the disturbed results from oxygen isotope records in the southern Netherlands (Hoek/Bohncke 2001). This disturbance was assumed to be related to increased groundwater flux due to the final melting of relict ground-ice (Hoek/Bohncke 2001, 1262). However, GI-1c_2 is characterised by a period of higher values in the gypsum corrected sulphate record from NGRIP. Only in the later part of this period a rhyolitic ash layer of possible northern Icelandic origin was detected. The preceding high values of the sulphate could be best explained by globally high volcanic activity, possibly related to the cold but still wet climate event and glaciological developments.
Consequently, the cold event of GI-1d seems to indicate a beginning for approximately 500 years of unstable environmental conditions during which the first shrub vegetation of late GI-1e was replaced for a short period by an open herb steppe followed by the establishment of first light forests in the sub-areas. However, these general developments were accompanied by extreme climatic conditions relating in particular to the water balance and resulting in fluctuating values in the pollen diagrams but also to very differential conditions of preservation.

In the northern French pollen profiles, the AP, birch, and juniper values began increasing again within GI-1c_2, whereas Artemisia continuously decreased and was surpassed by Poaceae. Even though the values did not reach the peak value of GI-1c_3, a widespread light forest with some grass meadows appears to have been established by the onset of GI-1c_1 in the Paris Basin. In the MFM record, the values were still fluctuating, also at a lower level than during GI-1c_3, but the amplitude of the peaks and lows as well as the frequency seemed to decrease indicating a slowly stabilising environment. The Artemisia values remained relatively high in the Eifel, possibly suggesting still more open meadows on higher elevated plateaus of the middle range mountains.

In mid-GI-1c_1 the gypsum corrected sulphate content of NGRIP had its lowest values before the Holocene. In the French record, this low was followed by another sharp decrease of AP, juniper and birch pollen. In contrast, the percentage of pine pollen began increasing. In the MFM diagram, a short peak of AP influenced by an increase in birch followed by willow pollen occurred after the low in the sulphate content. This short AP peak is also followed by a steep decline in AP and birch pollen. The low in AP and birch pollen was accompanied by a peak of Poaceae and a first peak value of pine pollen which surpassed the percentages of willow in this episode. In the Paris Basin, pine became the predominant arboreal genus after this episode, whereas the relative birch concentration decreased gradually. This supersession marked the end of a pioneering phase and the establishment of denser and darker forest covers. This interpretations is further supported by the rather low frequency of Artemisia pollen in the Paris Basin diagram. The importance of pine also increased in the MFM during this period but, as previously mentioned, birch remained the most important arboreal pollen type in this sequence. Thus, at the onset of GI-1b, birch was clearly dominant in the MFM where the AP already presented approximately 75 % of the pollen. In contrast, pine pollen was already dominant among the AP that represented approximately two thirds of all pollen in the synchronised Paris Basin sequence. The values of pine and birch were, in some areas, still close. With the onset of GI-1b the AP values in both diagrams decreased again, whereas Poaceae increased. In the Paris Basin, this increase was accompanied by increasing Artemisia values, although the values of this genus still remained below the Poaceae values. In the second half of GI-1b, willow and juniper pollen had a peak value in the MFM sequence. The juniper peak can also be found in the Paris Basin diagram but began and ended a few decades before the peak in the MFM.

Following the juniper peak, the AP again increased significantly due to a pine peak and accompanied by a steep decrease of Artemisia pollen in the Paris Basin. In contrast, AP and birch pollen began decreasing in the MFM record but comparable to the Paris Basin pine pollen also continued to increase in the MFM. The lowest value of AP and birch occurred before the highest value in GI-1a and corresponds to the onset of PAZ 7 (Younger Dryas) in the Paris Basin diagram. However, this correlation might be due to the lack of reliable correlation points in this part of the French diagram. The sharp decline of pine pollen and AP accompanied by the steep increase of Artemisia pollen which surpass again the Poaceae pollen preceded the decline of oxygen isotopes and the sulphate peak 4 in the NGRIP record and, consequently, the deposit of the LST in the MFM. However, a comparison with the development of the deuterium excess record in NGRIP (**fig. 53**) possibly suggests that the significant alterations of the climate system reflected by this proxy (temperature at source of Greenland precipitation) influenced the vegetation in northern France. Based on the insecure chronology of this part, this hypothesis still requires testing against future stratigraphies. In the MFM region, the AP and birch pollen increased again after the LST to a last peak within the second half of GI-1a. Before the end of GI-1a, the birch values drop sharply and only at this point the pine values increased and surpassed the birch pollen in the MFM. With the first severely low values in NGRIP, the AP record of MFM also decline reflecting the opening of the landscape towards the Younger Dryas. This impression is further sustained by the decline of pollen concentrations. However, the AP values remained on a level comparable to the later

part of GI-1e and the pine as well as birch pollen remained relatively high. This pattern suggests that forests might have retreated to more sheltered areas but remained present in north-western Europe during this Lateglacial Stadial.

Evidence from botanical macro-remains

To accomplish a construction of the local environments, directly dated botanical macro-remains (including pieces of charcoal) that were previously determined to species level should supplement the reconstructions based on pollen profiles (Birks 2003; Bennike/Seppänen 2004; Mortensen et al. 2011). This supplement is necessary because of general sources of error with sole reliance on pollen assemblage for past vegetation reconstruction. General sources of error are, for example, the proportional presentation of these diagrams and the disproportional production of pollen as well as the differential effects of the dispersal and of diverse taphonomic processes such as redeposition on the pollen species (Odgaard 1999; Bennett/Willis 2002; de Klerk 2004, 271; Lebreton et al. 2010). These potential errors affect, in particular, environments producing only small sedimentary pollen assemblages (cf. Mortensen et al. 2011, 2535).

In contrast to pollen, macro-remains often make a more precise species determination possible. Moreover, if these specimens are found in environmental archives, a relatively close vicinity of the plant and the deposited remain can be assumed after the exclusion of natural redeposition processes. In contrast, remains from archaeological sites do not necessarily originate from a close vicinity but could be introduced by humans. Yet, these specimen answer, instantly, the archaeological question relating to the general availability and use of a resource. Furthermore, depending on the weight of the macro-remain and the consistency of the sediment, sinking of the remains into softer ground can cause some chronological offset between the sediment and the plant remain. In general, later intrusions due to bioturbation and other post-depositonal processes are possible and, therefore, directly dated plant remains are very useful in the reconstrution of past environments. Directly dated macro-remains can be considered in the form of probability distributions or first and last appearance dates to test the results of the pollen analyses. However, the directly dated and determined macro-remains from the sub-areas of this study are in too low number to be useful in a probability distribution. For instance, from the Central Rhineland only one sample of pine (*Pinus* sp.) was directly dated (Bad Breisig, GrA-17493: 10,840 ± 60 years ^{14}C-BP; see **tab. 75**).

Nevertheless, the presence of determined botanical macro-remains can help in the discussion about the local environment. For example, in the on-going debate about whether small stands of trees or small forests survived in sheltered topographies during the LGM and the Late Pleniglacial, the evidence of macro-remains is one of the essential arguments in favour of their survival (Stewart/Lister 2001; Willis/van Andel 2004; Birks/Willis 2008). The reconstruction of a patchy or mosaic environment during the Late Pleniglacial is of some interest to the palaeontological discussion of non-analogue or disharmonious assemblages (cf. Fahlke 2009; Polly/Eronen 2010) as well as for archaeological considerations about available resources.

For the Late Pleniglacial period, pine and willow charcoal found at Gönnersdorf (Schweingruber 1978), Andernach IV (Holzkämper 2006), and Pincevent (horizon IV; Thiébault 1994; **fig. 54; tab. 72**) should be considered in the reconstruction of the environment. In a more recent microscopic analysis of hearth remains from Pincevent, no pine material was found and only remains attributed to birch (*Betula* sp.) and willow (*Salix* sp.) were determined (Bodu et al. 2009b). At Gönnersdorf, birch was also frequently found and these three typical species (pine, willow, birch) were further supplemented by juniper (*Juniperus* sp.). In Andernach IV, no birch remains were identified but the spectrum was supplemented by a single remain of daphne (*Daphne* sp.). Among the poorly preserved charcoal remains of Étiolles, birch was determined and

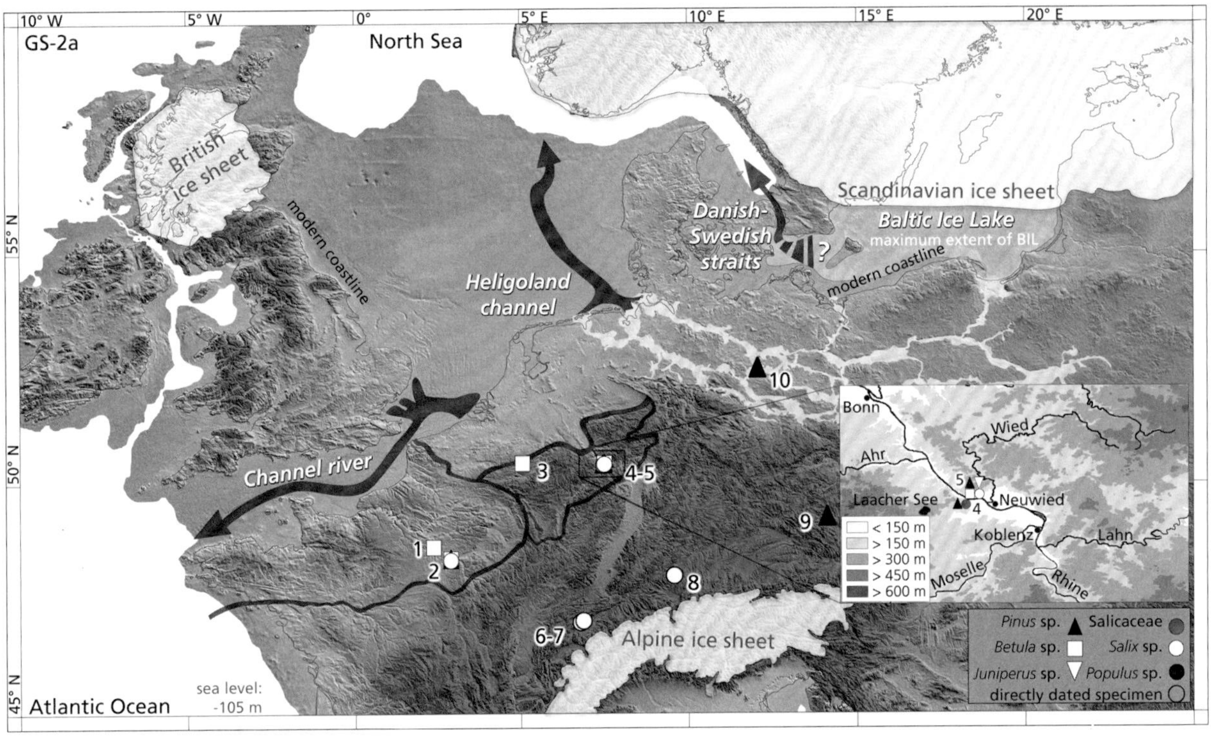

Fig. 54 Map of north-western Europe during GS-2a with sites yielding determined plant macro-remains (within the sub-areas) supplemented by directly dated plant macro-remains (outside the sub-areas). **1** Étiolles; **2** Pincevent; **3** Bois Laiterie; **4** Andernach-Martinsberg; **5** Gönnersdorf; **6** Monruz; **7** Champréveyres; **8** Schussenquelle; **9** Putim; **10** Grieben. – For further details see text.

another piece of the Betulaceae family showed characteristics of the European hornbeam (*Carpinus betulus*; Thiébault 1994). If the determination of the latter is correct, it appeared to represent a younger intrusion and, thus, also the attribution of the birch sample to the Late Pleniglacial can be questioned.

In anthracological analyses of Magdalenian material from the Swiss Lake Neuchâtel, willow (*Salix* sp.) was the dominant species and assumed to originate from the dwarf shrub *Salix retusa* (Leesch et al. 2012). Several directly dated samples confirmed the Late Pleniglacial attribution of this material (Leesch 1997; Bullinger/Leesch/Plumettaz 2006). Besides willow, birch (*Betula* sp.) was again determined. These arboreal species were supplemented with numerous remains of flowers and herbs in the well preserved lower horizon from Monruz (Bullinger/Leesch/Plumettaz 2006). Furthermore, charcoal from the Late Magdalenian Moosbühl site also indicate the typical composition of pine, willow, and birch (Bullinger/Lämmli/Leuzinger-Piccand 1998). However, a convential date made on a sample of birch bark that was found beside a hearth produced an early Lateglacial Interstadial age (B-2316: 12,060 ± 150 years ^{14}C-BP, 14,370-13,570 years cal. b2k; Feustel 1980, 120). Further directly dated charcoal from this site frequently resulted in Holocene ages (Bullinger/Lämmli/Leuzinger-Piccand 1998) suggesting chemical and/or botanical contamination. Further to the north-east, at the Late Magdalenian site of Schussenquelle, a piece of charcoal was determined as willow and also directly dated to the Late Pleniglacial (ETH-6154: 12,630 ± 120 years ^{14}C-BP, 15,600-14,440 years cal. b2k; Schuler 1994). Thus, the presence of willow in north-western Europe during the Late Pleniglacial appears evident. Furthermore, the reliable remains show that besides willow, also pine wood and occasionally birch was available during the Late Magdalenian occupation of northern Europe.

However, these remains may not have originated from tree species because these genera also include small (dwarf) shrub species such as *Pinus mugo*, *Betula nana*, *Salix polaris*, or *Salix reticulata*. These species would sustain the reconstruction of a steppe-like environment comparable to modern analogues in higher altitudes. In pollen diagrams from north-western Europe, the Late Pleniglacial and early Lateglacial Interstadial

species	Andernach lower horizon IV	Gönnersdorf	Pincevent, horizon IV	Étiolles	Bois Laiterie	Le Closeau	Andernach, upper horizon		Conty	Urbar	Kettig	Niederbieber	Bad Breisig
							2-FMG	3-FMG					
Pinus sp.	**×**	**×**	**×**			**×**	×	**×**	**×**	×		×	×
Saliceae / *Salix* sp. / *Populus* sp.	×	×	**×**				**×**	×		**×**	**×**	**×**	?
Betula sp.		×	×	×	×		×	**×**		×	×	**×**	?
Daphne sp.	×							×					
Juniperus sp.		×											
Carpinus betulus				(?)									
Corylus avellana					(**×**)								
Sambucus sp.							×						
Pomoideae								×					
Prunus sp.								×				?	

Tab. 72 Presence of determined botanical remains (mostly charcoal) on the archaeological sites used in this project (for further details and references see Material-Archaeology, p. 75-133, p. 143-161, p. 163-168, p. 175-179, p. 210-219, and p. 231-237). The most common species are set in bold. Symbols: **×** present; **?** uncertain determination; **()** doubtful association with the archaeological horizon.

presence of birch was usually attributed to the dwarf-shrub *Betula nana* (Lotter 1999; Merkt/Müller 1999; Jones et al. 2002). However, macrofossils of tree birch (*Betula pubescens*) occurred in the MFM and in northern Germany during the early Lateglacial Interstadial (Litt/Stebich 1999; Merkt/Müller 1999) suggesting a rapid expansion at the onset of the Lateglacial Interstadial and/or a survival of small patches in sheltered areas during the Late Pleniglacial. The usually small dimensions of the recovered charcoal material and the high destruction by the firing process did often prohibit further determination but one piece in Gönnersdorf showed characteristics typical for the tree species Scots pine (*Pinus sylvestris*). A review of the evidence for the presence of Scots pine in northern Europe during the Late Weichselian in combination with modern genetic variation among Scots pines suggests that evidence for its presence in northern Europe during GS-2a was almost absent and that this pine dispersed mainly from south-eastern refugia into northern Europe (Cheddadi et al. 2006).

Thus, although a presence of some stands of pine, willow, and/or birch trees or larger shrubs in the flood plains or in sheltered brook valleys cannot be excluded for the Central Rhineland, driftwood from more favourable southern regions as the Upper Rhine rift or possible old wood from gravel deposits washed out during seasonal flooding could have also served as fuel. In the Rhine gravels, remains of Weichselian pines are still being documented (cf. Friedrich et al. 1999; Rosendahl et al. 2006) and sources for fossil material were certainly known among the Magdalenian inhabitants of Gönnersdorf and Andernach as proven by fossil materials identified regularly in the hearth remains from these sites. Therefore, only a direct dating of some determined pieces could help evaluate the presence of Scots pines during the Late Pleniglacial.

A piece of Scots pine from Grieben (Saxony-Anhalt) was dated to the second half of GS-2a (OxA-13284: 12,620 ± 50 years ^{14}C-BP, 15,390-14,870 years cal. b2k; Grünberg 2006). The pine wood served as shaft in a Mesolithic composite tool. If this date is correct, it proves the availability as well as the use of old wood by Mesolithic hunter-gatherers and suggests the availability of larger pines during the Late Pleniglacial in Central Germany. Possibly, this piece originated from a nearby south-western glacial refugium (Cheddadi et al. 2006) and reached this area by river transport. This interpretation was further sustained by a directly dated piece of pine charcoal (GrA-36010: 13,010 ± 60 years ^{14}C-BP, 16,240-15,440 years cal. b2k) from the Magdalenian site at Putim (Verpoorte/Šída 2009) which is located near one of the potential refugia in south-eastern Europe. In addition, a small pine charcoal fragment from the ice wedge cast at Wilcyzce was directly dated and produced a Late Pleniglacial age (Poz-14892: 12,770 ± 120 years ^{14}C-BP, 15,790-14,910 years cal. b2k) which was considered a minimal age due to the amount of material (Fiedorczuk et al. 2007, 120). Besides the pine charcoal, Magdalenian material and a human foetus were found which dated to a comparable age range as the Late Magdalenian remains from Gönnersdorf and Andernach, lower horizon (see p. 466-468; Fiedorczuk et al. 2007; Ginter/Połtowicz 2007; Irish et al. 2008). However, the position near a wide river valley which drained one of the potential Scots pine refugia again made the use of non-primary material possible. Thus far, these pieces are the only determined and directly-dated macro-fossil of pine found in this period.

Comparable with Gönnersdorf and Andernach some macro-botanical remains were dated to GS-2a from their context that must be considered cautiously. For example, the site of Bois Laiterie produced dates comparable to the Grieben specimen and the charcoal remains from this site were determined as birch and hazel (*Corylus avellana*; Pernaud 1997). This latter species is unknown from Lateglacial pollen assemblages but it is well known from Holocene stratigraphies. Therefore, this charcoal can be assumed as a younger intrusion, although the presence of small pioneering patches of hazel and birch during moister and milder episodes might have been possible (cf. Pernaud 1997, 144).

This outline shows how sparse and questionable the direct evidence is for the reconstruction of the Late Pleniglacial vegetation. Additional information can be gathered from few profiles with almost continuously

preserved pollen from this period such as La Grande Pile (Beaulieu/Reille 1992). The resolution of this long stratigraphy is relatively low and, thus, the exact chronological attribution of the vegetation to GS-2a was based only on the stratigraphic succession and correlations with Central French biostratigraphies. In general, the impression of the Late Weichselian vegetation developments can be documented from this sequence. The phase approximately correlaties to the GS-2a period and has been described as corresponding »to a crisis of aridity« (Beaulieu/Reille 1992, 435). This impression is further confirmed by pollen records from marine boreholes offshore Europe which registered »higher relative abundance of steppic taxa than during the Last Glacial Maximum« (Fletcher et al. 2010, 2849) in this period and thereby suggests severe cold and dry conditions relating to the Heinrich event 1 (Fletcher et al. 2010).
In contrast to the vegetation records, faunal remains are often better preserved and some faunal species allow for further assumptions on the past vegetation (see p. 412-432). In general, this faunal evidence confirms the picture of a dry and cold steppe environment.
In conclusion, the general picture of a cold and dry grass steppe landscape in the study area during the Late Pleniglacial remains valid but it might be supplemented by grove communities in small refuge areas, particularly in moister low- and middle-range mountain areas and/or sheltered floodplains (cf. Birks/Willis 2008). Even though the presence of these small stands of trees cannot be excluded based on the macro-botanical evidence, it is also not yet proven because the exact source of the macro-botanical material remains uncertain. Besides more sheltered regions, some remains could have originated from dwarf species and/or fossil sources. Therefore, the direct dating of pieces determined to species level from this period is desirable. Nevertheless, the infrequent presence of larger plant material in occasionally good preserved conditions on archaeological sites for this period indicates that wood was a rather scarce resource in north-western Europe during the Late Pleniglacial. However, the presence of wood charcoal in hearths on archaeological sites proves the availability of this material as a resource for fueling fires and suggested the probable availability of this resource for making equipment such as projectile shafts or building constructions such as frames for drying fish or hides. For dwelling constructions or hunting equipment some longer and more solid pieces of wood were required. In the case that no substitute material was available, unsustainable handling of this resource appears very improbable. Wood could have been replaced by faunal material such as prompted by the considerably older, Eastern European mammoth houses (Pidoplichko 1998) or the mammoth femur in the hearth construction of Gönnersdorf (Street et al. 2006). Furthermore, wood could have originated from other arboreal species such as *Juniperus communis*. Although juniper is usually only of a small shrub size, it can occasionally become larger. However, these longer woody parts are usually not straight but rather distorted, which is useless for hunting equipment but could serve in dwelling constructions. Thus, for the hunting equipment, the scarcer straight offshoots of shrubs and trees must have been used. Based on the assumed scarcity of this material, some questions about the handling of this resource arise such as: Were the composite techniques, which are known for lithic and (faunal) organic artefacts, extended to a wooden shaft part? Or were the constructions designed with predetermined breaking points allowing rather a loss or fracturing of the lithic and/or faunal material? The composite technique of wooden shafts was observed in the Ahrensburgian arrows from Stellmoor (Rust 1943) and possibly the result from a longer tradition of composite projectiles intended for a sustainable handling of scarce resources (Bokelmann 1991).

For the Lateglacial Interstadial, pollen profiles are better preserved and the number of sites producing determinable wood charcoal also increases (**tab. 72; fig. 55**). However, a plateau in the calibration record spreads the calibrated age ranges across the final Late Pleniglacial period and the first half of the Lateglacial Interstadial and makes a precise attribution to this early part, based only on radiocarbon dating difficult, in particular for dates with larger standard deviations. Moreover, the developments observed in the pollen

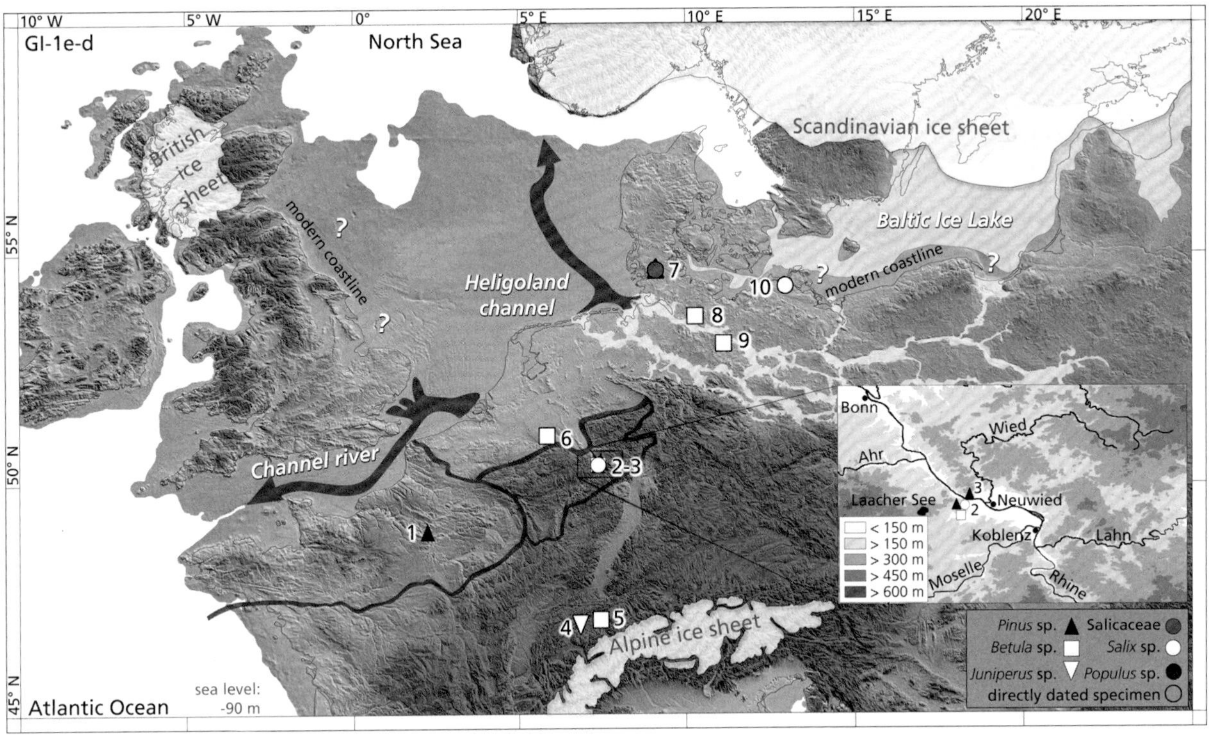

Fig. 55 Map of north-western Europe during GI-1e-d with sites yielding determined plant macro-remains (within the sub-areas) supplemented by directly dated plant macro-remains (outside the sub-areas). **1** Le Closeau; **2** Andernach-Martinsberg; **3** Gönnersdorf; **4** Monruz; **5** Moosbühl; **6** Gulickshof; **7** Ahrenshöft LA 73; **8** Poggenwisch; **9** Grabow 15; **10** Endingen Horst VI. – For further details see text.

records for the period corresponding to GI-1e to GI-1c_3 cannot be evaluated in greater detail based on the macro-remains due to this lack of precision. In fact, some material was attributed to this early period based on sedimentology, approximate vegetation development in the region (from pollen profiles), and the associated archaeological material. Thus, vegetation reconstruction for this period occasionally suffered from problems related to chronology and circular arguments.

For example, the calibrated age range of the material from Pincevent, horizon IV and Étiolles spread into the Lateglacial Interstadial but due to the assumed relation with the archaeological material, the material including the wood charcoal should be attributed to the Late Pleniglacial. However, younger intrusions of charcoal and/or a continued occupation into the early Lateglacial Interstadial cannot be excluded. Thus, the age range of the calibrated ^{14}C dates could be reliable and, thus, also date the presence of the botanical macro-remains. This consideration applies equally to the remains from Bois Laiterie. These possible early Lateglacial Interstadial remains were supplemented by the pine charcoal from Le Closeau. The range of directly dated samples attributed this species only to the late Lateglacial Interstadial (GI-1b and GI-1a). However, pine charcoal found in hearths associated with the lower horizon could be, indirectly, attributed to the early Lateglacial Interstadial. In fact, the dates on bones also spread into the later part of the Late Pleniglacial but the complex stratigraphy suggested that the horizon from which the material originated post-dated the Late Pleniglacial. As long as the dated remains and/or the associated macro-fossils were not moved stratigraphically upwards, this part of the age range can, consequently, be neglected. However, the presence of pine in the Paris Basin must still be questioned for the early Lateglacial Interstadial because of the unclear associations and the remaining uncertainty about the origin of the botanical material.

The same uncertainty about the association of the macro-botanical material with dated material applies to the pine charcoal found in the south-western area of Gönnersdorf and the pine, willow, and birch charcoal

associated with the FMG concentrations in Andernach 2. The macro-botanical remains on archaeological sites from the sub-areas are usually only indirectly attributed to GI-1e and GI-1d or falls into the date range due to the calibration plateau (Bois Laiterie).
However, some material outside the sub-areas could be attributed to this early part of the Lateglacial Interstadial with some certainty.
Four directly dated samples of juniper from Monruz (sector 1 and sector 2, south) were associated with the younger, Azilian horizon and yielded calibrated ages which scattered between the final Late Pleniglacial and the early Lateglacial Interstadial (Leesch/Cattin/Müller 2004, 167. 183). One of these samples came from a disturbed area (hearth B11) and produced an almost entirely Late Pleniglacial age (ETH-23271: 12,570 ± 90 years ^{14}C-BP, 15,450-14,450 years cal. b2k; Leesch/Cattin/Müller 2004, 183). Thus, these age ranges suggest the use of juniper in a very early Lateglacial Interstadial context. Besides juniper, willow and pine were associated with the Azilian hearths from Monruz, sector 2-South (Leesch/Cattin/Müller 2004) but due to some disturbances the association of these species with a younger occupation phase remains uncertain. However, pine from an organic layer was directly dated and the calibrated age range encompasses the late GI-1e and GI-1d (Leesch/Cattin/Müller 2004). Based on the stratigraphic position, above the upper (Azilian) horizon, and due to the organic composition of the layer, an attribution to the mid-Lateglacial Interstadial appears more plausible. Nevertheless, the Central European pine chronologies reached into this period and, consequently, some groves of Scots pines were already growing at the end of GI-1e in Switzerland (Kaiser et al. 2012). Thus, the organic deposit could also originate from the latest part of GI-1e. This attribution would further emphasise that the Azilian occupation occurred in the very early Lateglacial Interstadial.
Even though charcoal samples were not yet dated directly to this period in Champréveyres, the Azilian material consisting of willow, juniper, and birch probably of comparable dates to the material from Monruz. A previously mentioned date was made on birch bark from the Moosbühl site. This date, perhaps, suggests a short reoccupation of this site comparable to the south-western area of Gönnersdorf. This indication could be further sustained by the occurrence of possible *Federmesser* and triangles (Bullinger/Lämmli/Leuzinger-Piccand 1998) as well as the presence of Baltic amber material (Schwab 1985) comparable to the Champréveyres site (Leesch 1997). However, the sampled material could also represent a natural intrusion that only indicates the presence of birch wood at this site during the early Lateglacial Interstadial but was not related to the archaeology. Furthermore, juniper seemed of some importance in the Alpine foothills region (cf. Lotter et al. 1992; Magny et al. 2006).
The increasing importance of birch, as indicated by pollen profiles, is also documented by directly dated macro-fossils (**fig. 55**). Evidence for this increasing importance usually comes from sites in the vicinity of major river systems such as Gulickshof (Hoek et al. 1999), Poggenwisch (Tromnau 1992; Lanting/van der Plicht 1996; cf. Grimm/Weber 2008), and Grabow 15 (Tolksdorf et al. 2013). As well as providing expansion routes along the drainage channels, this pattern indicates the preference of this genus to grow on well-watered grounds.
At the onset of the Lateglacial Interstadial, the increasing importance of birch was probably related to dwarf birch (*Betula nana*). Directly dated leaves of this species found in Gulickshof, southern Netherlands, produced a date at the transition between the Late Pleniglacial and the early Lateglacial Interstadial (Hoek et al. 1999). Dwarf birch was also observed in early GI-1e in the north-western English stratigraphy from Hawes Water where it was replaced by European white birch (*Betula pubescens*) around the end of GI-1e and within GI-1d according to comparisons with the oxygen isotope record (Jones et al. 2002). In addition, the first macro-fossils of European white birches occurred in the MFM sequence at the end of the Meiendorf biozone which corresponded approximately to the end of GI-1d (Litt/Stebich 1999). Thus, the expansion of birch trees across north-western Europe had occurred during the very early part of the Lateglacial Intersta-

dial with first communities established before the first significant increase recorded in the pollen diagrams. However, evidence from the Danish Slotseng site indicated a much later transition from open tundra-like environments reflected by the presence of *Betula nana* and *Dryas octopetala* to light forests with European white birches (Mortensen et al. 2011). Presumably, this delayed reaction showed the persistent influence of the Scandinavian ice sheet and the glacial meltwaters in the region around the modern Baltic Sea.
The numerous indications of Saliceae are possibly related to the return of colder conditions. A piece of charcoal from a hearth at Ahrenshöft LA 73 (classic Hamburgian horizon) was attributed to *Salix* sp. or *Populus* sp. and another fragment from an upper horizon (Havelte Group) was determined as pine (Clausen 1998). These specimens were directly dated and produced statistically identical ^{14}C ages which attributed the occupation of the site towards the end of GI-1e, GI-1d, and the early GI-1c_3 (Grimm/Weber 2008). A charcoal fragment determined as *Salix* sp. or *Populus* sp. found in Reichwalde resulted in a comparable date (KIA-13412: 12,193 ± 58 years ^{14}C-BP, 14,250-13,890 years cal. b2k; Vollbrecht 2005). This date was younger than dates made on calcined bones from the same archaeological feature. However, the latter material was shown elsewhere to be affected by some type of contamination in the Lateglacial (Lanting/Niekus/Stapert 2002). Besides the dated material, wood remains of birch, willow, and pine were preserved in this gravel pit reflecting a Lateglacial forest community (Friedrich et al. 2001a). However, the precise relation of the archaeological charcoal material and these environmental finds remains uncertain due to the large area and the complex stratigraphy. A piece of willow wood found in a test pit near the Final Palaeolithic site Endingen (Terberger 1996; Street 1996) was stratigraphically attributed to the first cold phase in the Lateglacial Interstadial (GI-1d, early Dryas; Kaiser/de Klerk/Terberger 1999). Due to a large standard deviation and the calibration plateau, the conventional date made of this sample was absolutely consistent with this attribution (Hv-20987: 12,360 ± 245 years ^{14}C-BP, 15,540-13,620 years cal. b2k; Kaiser/de Klerk/Terberger 1999).
This compilation indicates that with the return of colder conditions (GI-1d) birch material was presumably substituted with more readily available material such as willow.

After GI-1d, the pollen diagrams document a significant increase in arboreal pollen that was assumed to reflect the spread of light forest environments. However, some disturbances were also observed during the following mid-Lateglacial Interstadial period (GI-1$c_{3\ \text{and}\ 2}$; see p. 378-380). Based on the directly dated botanical macro-fossils this period, in particular GI-1c_2, appears as a transition between the early Lateglacial Interstadial and the late Lateglacial Interstadial. The early Lateglacial Interstadial was still characterised by a relatively open environment with shrub communities, light forests, and a significant amount of meadows, whereas in the late Lateglacial Interstadial the forest communities predominate the reconstruction. Depending on the locality, the tempo of the development varied. For instance, more open communities persisted into this intermediate phase in some more exposed areas, whereas in sheltered areas such as the Seine valley a denser forest developed.
Partially, this impression is due to the calibration curve that becomes more reliable for this period due to the increasing evidence but also due the plateau encompassing the late Late Pleniglacial and the early Lateglacial Interstadial ending within GI-1d. Thus, more precise chronological attributions became possible after the end of the plateau. Consequently, this period, particularly the early part (GI-1c_3), provides the end point for the calibrated age range of many of the previously mentioned dated material.
In addition, some material that produced a longer dated age range towards the late Lateglacial Interstadial could not often be attributed more precisely. This imprecision was because many of the associated biostratigraphies distinguished only between a first pioneer episode and a last forested episode separated by a cold interval in the Lateglacial Interstadial. Usually, the first phase was equivalent to the early Lateglacial Interstadial (GI-1e-d), whereas for the later phase the attribution was occasionally difficult to distinguish

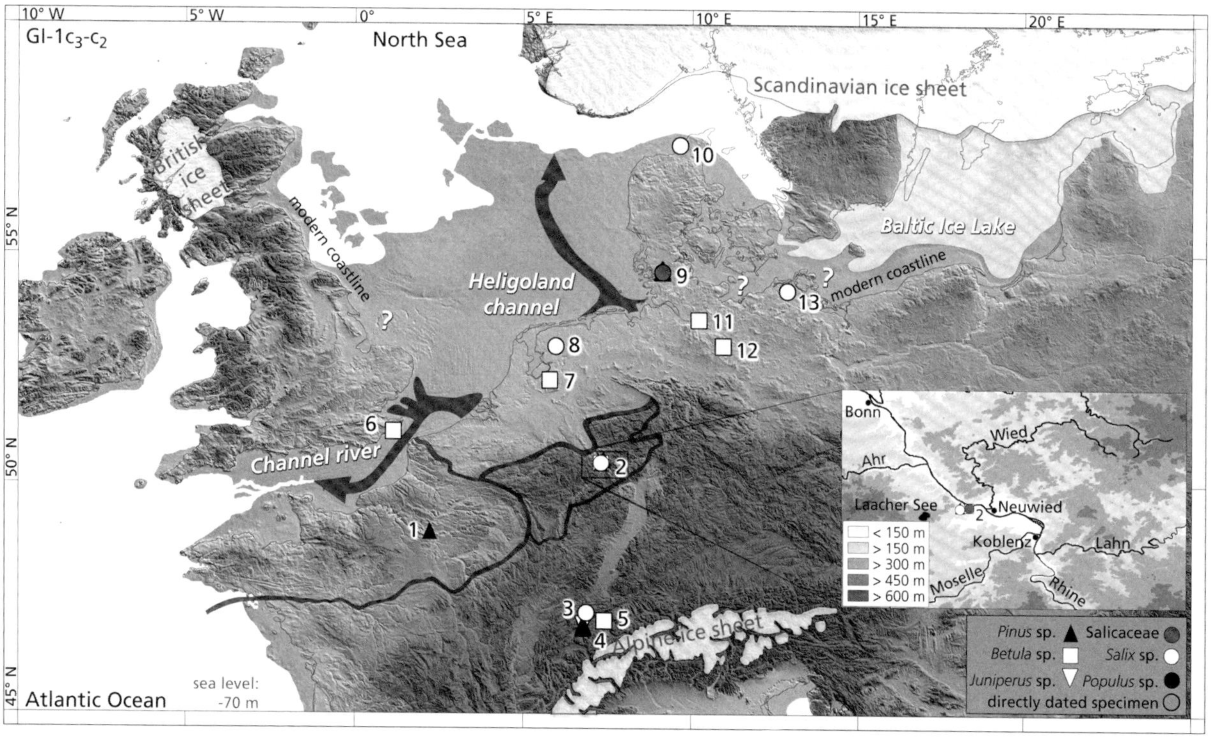

Fig. 56 Map of north-western Europe during GI-1c$_{3-2}$ with sites yielding determined plant macro-remains (within the sub-areas) supplemented by directly dated plant macro-remains (outside the sub-areas). **1** Le Closeau; **2** Andernach-Martinsberg; **3** Grotte du Bichon; **4** Monruz; **5** Moosbühl; **6** Holywell Coombe; **7** Usselo; **8** Oldeholtwolde; **9** Ahrenshöft LA 73; **10** Nørre Lyngby; **11** Poggenwisch; **12** Grabow 15; **13** Endingen Horst VI. – For further details see text.

whether it represented a preserved mid-Lateglacial Interstadial (GI-1c) or formed only the onset of this last forested episode (GI-1c-a). Consequently, the attribution of dated macro-fossils from the onset of this type of later forested period could relate to the mid-Lateglacial Interstadial as well as to the onset of the late Lateglacial Interstadial.

In the sub-areas, ^{14}C dates made on bones from the lower horizon of Le Closeau helped to date the pine charcoal on this site. These dates ended during the early part of this period (**fig. 56**). Furthermore, the FMG material from Andernach 2 could also originate from this period. Both attributions are vague and, consequently, the vegetation development of this period is poorly documented in the sub-areas.

Outside the sub-areas, most of the previously mentioned directly dated material from northern and eastern Germany but also the juniper material from Monruz and the birch bark from Moosbühl ranged up to this period and ended in it. Only the pine charcoal from the organic layer at Lake Neuchâtel probably dated to this period. In addition, the sites contributing material to the various dendrochronologies became more numerous around this time (Kaiser et al. 2012) indicating the presence of small groves of pines, at least in favourable places in Switzerland and eastern France, as well as in the Danube valley and eastern Germany. However, birch and willow seem, in general, to remain the predominant species. For instance, the calibrated age range of some directly dated willow remains from the new investigations at Nørre Lyngby (Aaris-Sørensen 1995) began in this period and spread into the late Lateglacial Interstadial. This range could suggest a continuous patch of willows in this area in the second half of the Lateglacial Interstadial. This onset of a shrub vegetation in northern Denmark is consistent with the first indications of light forests in Central Denmark (cf. Mortensen et al. 2011).

In addition, a fragment of willow charcoal that was recovered in the Swiss cave Le Bichon also produced a date which ranged from this period into the late Lateglacial Interstadial (ETH-4246: 11,680 ± 120 years ^{14}C-

BP, 13,790-13,230 years cal. b2k; Morel 1993). The cave bear and the human remains which were related with the charcoal yielded comparable results. Only a second fragment of charcoal produced a younger date, possibly due to contamination.

The calibrated age range for several pieces of willow charcoal found at the northern Dutch site of Oldeholtwolde (Johansen/Stapert 2004) also began in this period. Besides the several pieces of willow, a piece of pine charcoal was directly dated and produced a younger date suggesting a late Lateglacial Interstadial presence. According to the general succession known from the pollen profiles (Hoek 1997), these attributions seemed possible but the stratigraphic position and the archaeological context of the material were contradictory (Grimm/Weber 2008). If the dates were reliable, they would indicate the presence of willow in the northern Netherlands during the mid-Lateglacial Interstadial. Nevertheless, sample contamination is a possible reason for the relatively young results. Therefore, the remains from Oldeholtwolde have to be considered cautiously when discussing the development of the Lateglacial vegetation.

In addition, a birch sample recovered in a peat during excavations of the well-known Usselo site (Stapert/Veenstra 1988; van Geel/Coope/van der Hammen 1989) produced a vague age which after calibration spread across almost the entire Lateglacial Interstadial (R-106: 11,800 ± 280 years ^{14}C-BP, 14,390-13,030 years cal. b2k; Alessio/Bella/Cortesi 1964). However, according to the correlation of the peat with the pollen stratigraphy of the site, the sample originated probably from a period equivalent to GI-1c_3.

Directly dated birch samples from the biostratigraphy at Holywell Coombe yielded some results of mid- to late Lateglacial Interstadial age (Hedges et al. 1993a). One fragment from a lower section was considerably older stretching into the Late Pleniglacial. This attribution could be explained by the setting of the site near the English Channel river and an early expansion of dwarf birches along these favourable watercourses but the date remained doubtful and was therefore not used in this study.

Even though the calibration possibilities are relatively good, only a few directly dated macro-fossils can be attributed, unambiguously, to this period. Perhaps, the disturbance observed in the pollen diagrams during this period also had a negative effect on the preservation of macro-fossil material.

This poverty is set in clear contrast to the late Lateglacial Interstadial from which considerably more material was preserved. In particular, in the Central Rhineland, this better preservation was due to the cover of the LST that created an exceptional environmental archive.

In particular, the directly dated results of charcoal of poplars (*Populus* sp.) from Miesenheim 2 and the Brohl valley ranged from the onset of this period to the LSE but also the birch remains found at Thür encompassed the complete period until the LSE (**fig. 57**; Street 1986; Hedges et al. 1993b; Baales et al. 2002). Besides individual pieces of occupation evidence from Andernach 2, the Andernach 3 location was attributed to this period by the ^{14}C dates (Kegler 2002). The charcoal from this site comprised birch, willow, poplar, as well as Saliceae and singular specimens of a drupe tree (*Prunus* sp.), an apple tree (Maloideae), and again daphne (Kegler 1999, 14). According to the dated material, the other FMG sites in the Central Rhineland were also settled during this period (see **tab. 20**, p. 143-162, and p. 466-473). The species determined from these sites consisted of pine, willow, poplar, and birch. In Niederbieber, a possible drupe tree (*Prunus* sp.) was identified. Thus, temperate, light forest communities had certainly established in the Central Rhineland by this period.

Furthermore, imprints of leaves, flowers, and fruits of various taxa were preserved in the LST. Among many other plants, these imprints proved the presence of bird cherry (*Prunus padus*), European white birch (*Betula pubescens*), which was also present in a smaller arctic variety (subsp. *tortuosa*), silver birch (*Betula pendula*), European common aspen (*Populus tremula*), the shrub-sized *Salix caprea* and *Salix pentandra* as well as elder (*Sambucus nigra*, Baales et al. 2002). In comparison with the pollen profiles, this exceptional archive proved the variety within single genera and further showed that genera which were recorded sparsely in

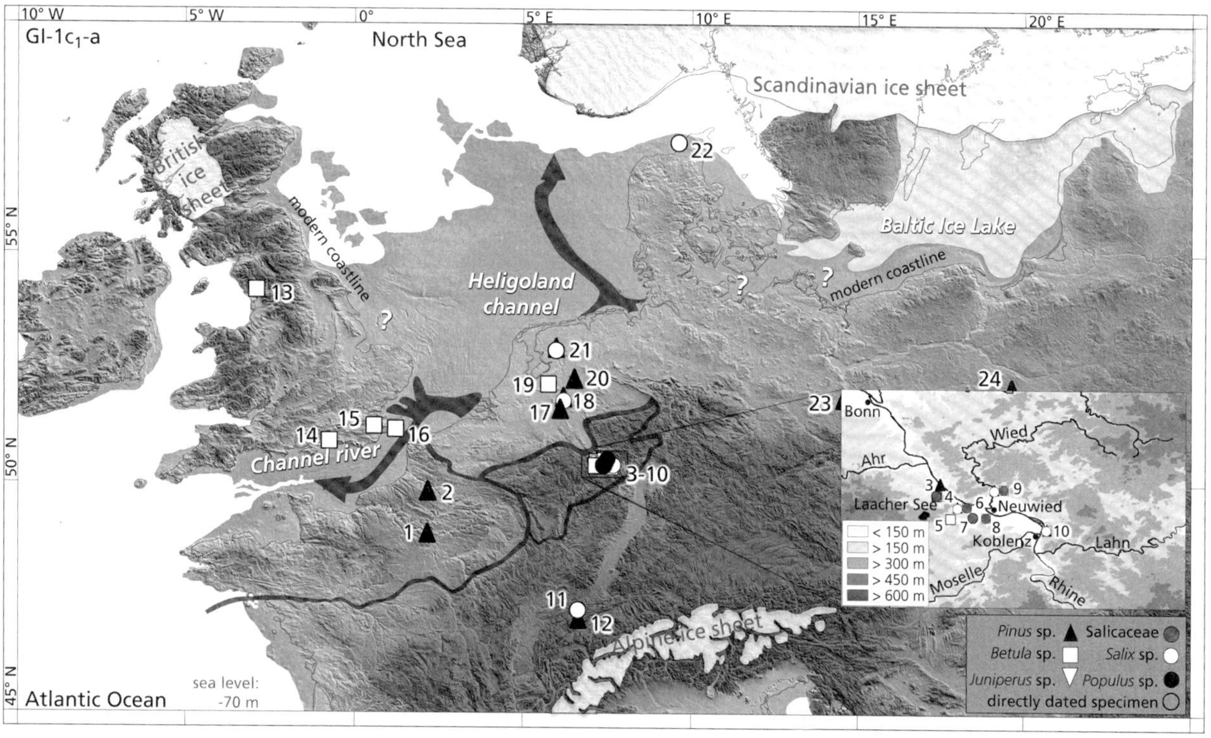

Fig. 57 Map of north-western Europe during GI-1c_1-a with sites yielding determined plant macro-remains (within the sub-areas) supplemented by directly dated plant macro-remains (outside the sub-areas). **1** Le Closeau; **2** Conty; **3** Bad Breisig; **4** Brohl Valley; **5** Thür; **6** Andernach-Martinsberg; **7** Miesenheim 2; **8** Kettig; **9** Niederbieber; **10** Urbar; **11** Grotte du Bichon; **12** Monruz; **13** Hawes Water; **14** Westhampnett; **15** Cherry Garden Hill; **16** Holywell Coombe; **17** Hülm; **18** Doetinchem; **19** Usselo; **20** Wierden Enterse-Akkers HS; **21** Oldeholtwolde; **22** Nørre Lyngby; **23** Groß Lieskow; **24** Witów (1-P, layer 4a). – For further details see text.

the pollen record such as willow could still occur numerously in favourable locations. Furthermore, the scarcity of pine charcoal as well as the rarely determined imprints of pine are consistent with the MFM pollen diagram where deciduous trees, in particular birch, remained more important than pine until the very end of the Lateglacial Interstadial.

For the remaining centuries of the Lateglacial Interstadial, the record from the Central Rhineland is almost absent which was tempting to interpret as a significant environmental hazard in this region. However, the pollen diagram from the MFM but also from other sites in Central Europe proved that the vegetation cover reestablished shortly after the LSE and prior to the onset of the Lateglacial Stadial (Merkt/Müller 1999; Litt/Schmincke/Kromer 2003). Directly dated macro-fossils are again very sparse and due to a small plateau and/or a very gradual decline in the calibration curve around the LSE, the attribution of the material to before or after the LSE remains a matter of speculation without a tephra layer in the stratigraphy. In the Central Rhineland the material from the Bad Breisig sites was found on top of the volcanic deposits and suggested that pines began growing again in the floodplains before the onset of the Lateglacial Stadial. This finding is consistent with the short dominance of pine pollen in the MFM sequence at the end of the Lateglacial Interstadial.

In the northern French sub-area, pine bark found in the stratigraphy of Conty is dated to the first part of this period (Ly-284: 11,540 ± 80 years ^{14}C-BP, 13,560-13,200 years cal. b2k; Ponel et al. 2005). This spread of pine in northern France is further sustained by the evidence from the intermediate and upper horizon of Le Closeau which produced also remains of pine charcoal (see **tab. 44** and p. 215). Presumably, some of this charcoal originated from natural fires (cf. Bodu 1998, 49). Nevertheless, these remains indicate widespread pine groves in the northern French river valleys during the late Lateglacial Interstadial. According to the directly

dated specimens from Le Closeau, some pines remained growing in this location until the early Lateglacial Stadial when this type of floodplain forest first seemed to spread into the Central Rhineland.
On the British Isles, directly dated evidence for birch becomes more numerous suggesting the spread of more deciduous light forests in this region (Hedges et al. 1993a; Bronk Ramsey et al. 2000; Jones et al. 2002). Besides the macro-fossils, the pollen diagrams from the British Isles also indicated that pine was generally of minor importance on the British Isles during the Lateglacial (e.g. Walker/Coope/Lowe 1993; Brooks/Mayle/Lowe 1997; Jones et al. 2002; Walker et al. 2003).
In contrast, the observation of a spread of denser, pine dominated forests into sheltered valleys of mainland Europe during the early part of this period was further sustained by the available dendrochronological material which became more abundant in regard to the number of sites as well as the number of trees on a single site during this period (Kaiser et al. 2012). In addition, directly dated pine charcoal from Oldeholtwolde which originated from the so-called Usselo soil dated to this period (Johansen/Stapert 2004) and, thus, dated comparable to other material from this marker horizon (van der Hammen/van Geel 2008; Kaiser et al. 2009). Besides confirming the presence of wild fires, the numerous charcoal particles from this horizon further sustain the picture of widespread forests during this period. However, the pine material from Monruz is possibly dated to the onset of this period but the attribution to the later part of this period (GI-1b-a) was increasingly unreliable.
Nevertheless, numerous further directly dated pine specimens from northern sites further contributed to the picture of the previously observed expansion of pine forests in the floodplains after the LST. For instance, a pine cone from Hülm produced an age that fell into the plateau around the LST and the onset of GS-1 (GrA-23744: 11,080 ± 50 years ^{14}C-BP, 13,110-12,740 years cal. b2k; Kasse et al. 2005). Another directly dated charcoal fragment determined as Scots pine from the Dutch site Doetinchem was also attributed to the very end of the Lateglacial Interstadial and the transition to the early Lateglacial Stadial (GrA-13686: 10,870 ± 50 years ^{14}C-BP, 12,860-12,660 years cal. b2k; Niekus/Stapert/Johansen 1998; Johansen et al. 2000). In the eastern part of the North European Plain, this increasing predominance of pine became apparent a bit early within the cold event of GI-1b as suggested by the material from Groß Lieskow (Bittmann/Pasda 1999) and Witów (Tauber 1966).
Moreover, these pine forests did not presumably retreat completely during the Lateglacial Stadial, even though a clear decline occurred which resulted in the gap in the dendrochronological material. Nevertheless, a pine charcoal fragment from a hearth at the site Geldrop Mie Peels resulted in an early Lateglacial Stadial age (OxA-2563: 10,610 ± 100 years ^{14}C-BP, 12,760-12,320 years cal. b2k; Hedges et al. 1992) and indicated a continued presence of pines in the Netherlands (**fig. 58**). Comparably, a fragment of Scots pine charcoal was recovered at the northern German site of Melbeck and the date (Hv-17306: 10,515 ± 95 years ^{14}C-BP, 12,730-12,130 years cal. b2k; Richter 1992) also proved the availability of pine material during the early to mid-Lateglacial Stadial in this northern area. Furthermore, in a fluvial context at Wustermark 22 a pine branch from the mid-Lateglacial Stadial was preserved (unpublished lab. no.: 10,389 ± 57 years ^{14}C-BP, 12,520-12,040 years cal. b2k; Hanik 2009).
Consequently, even though pollen profiles, in particular, from northern Europe, document a significant decline in arboreal pollen during the Lateglacial Stadial, the sparse macro-botanical evidence suggest the continued presence of pine groves and, thus, indicated a more taiga-like environment in these areas during the Lateglacial Stadial. In the north-west, pines remained absent but single birch and willow remains were found (**fig. 58**). The limited contribution of sandy and loessic areas to fluvial sediments in northern France suggested in addition to the macro-remains that the ground cover was sufficient in this period to prevent erosional processes (Pastre et al. 2003, 2184). However, further to the north where permafrost conditions temporarily returned vegetation changed again to a more tundra-like habitat (cf. Mortensen et al. 2011).

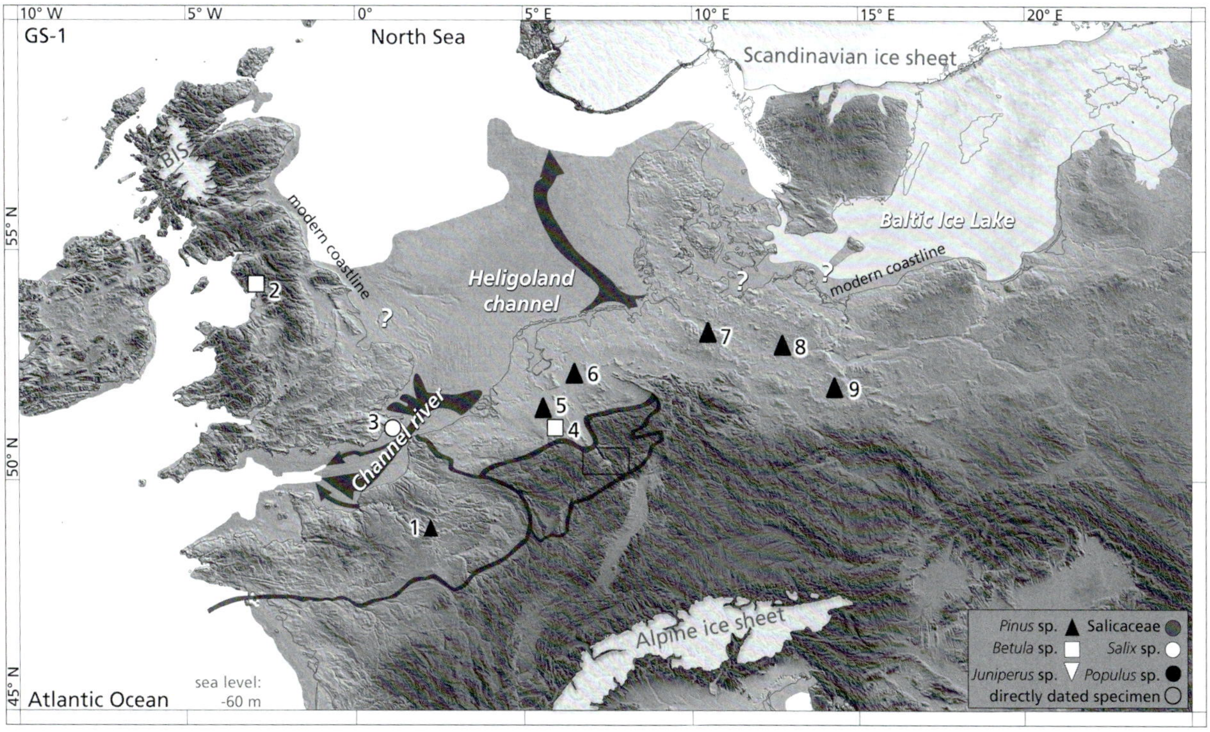

Fig. 58 Map of north-western Europe during GS-1 with sites yielding determined plant macro-remains (within the sub-areas) supplemented by directly dated plant macro-remains (outside the sub-areas; see **tab. 73**). **1** Le Closeau; **2** Hawes Water; **3** Holywell Coombe; **4** Gulickshof; **5** Geldrop Mie Peels; **6** Wierden Enterse-Akkers HS; **7** Melbeck; **8** Wustermark 22; **9** Groß Lieskow. – For further details see text.

Moreover, the number of sites producing macro-remains decreased significantly which was possibly due to a decline in conditions of preservation such as due to increasing deposit of acidic coversands, increasing taphonomic disturbances such as intensified high flood events in areas of frozen grounds, opening of vegetation covers, or a combination of these possibilities (cf. Weber/Grimm/Baales 2011).

In summary, the pollen profiles and the botanical macro-fossils reveal the following vegetation development in the Lateglacial (**fig. 59**):

During the Late Pleniglacial period, the landscape was characterised by a sparse grass steppe vegetation that was possibly supplemented by some stands of shrubs and trees in sheltered and moist areas. This light shrub vegetation expanded at the onset of the Lateglacial Interstadial. The first expansion of light forest communities into sheltered areas could only be proven for the latter part of GI-1e in north-western Europe. Pioneer forests began to spread in the mid-Lateglacial Interstadial (GI-1c_3). However, this period appeared characterised by inconstant climatic conditions resulting in unstable grounds and, thus, the expansion of the light forests communities appeared as a more vibrant development. Furthermore, this instability seemed to have affected the conditions of preservation at palaobotanical as well as archaeological sites and, consequently, the reliable evidence for the vegetation development from this time is particularly sparse. At the end of this period, the environment had shifted from the still relatively open landscapes to a landscape dominated by light forests. In some drier and more sheltered areas, these forests were already darker and dominated by groves of pines, in particular Scots pine. These darker forests increasingly became more dominant and by the end of the Lateglacial Interstadial pine dominated forests formed the main vegetation community in north-western Europe. These communities were so well established that they seemed to persist as groves in favourable areas throughout the Lateglacial Stadial. With the onset of the Holocene the conifer dominated forests established quickly over wide parts of the study area.

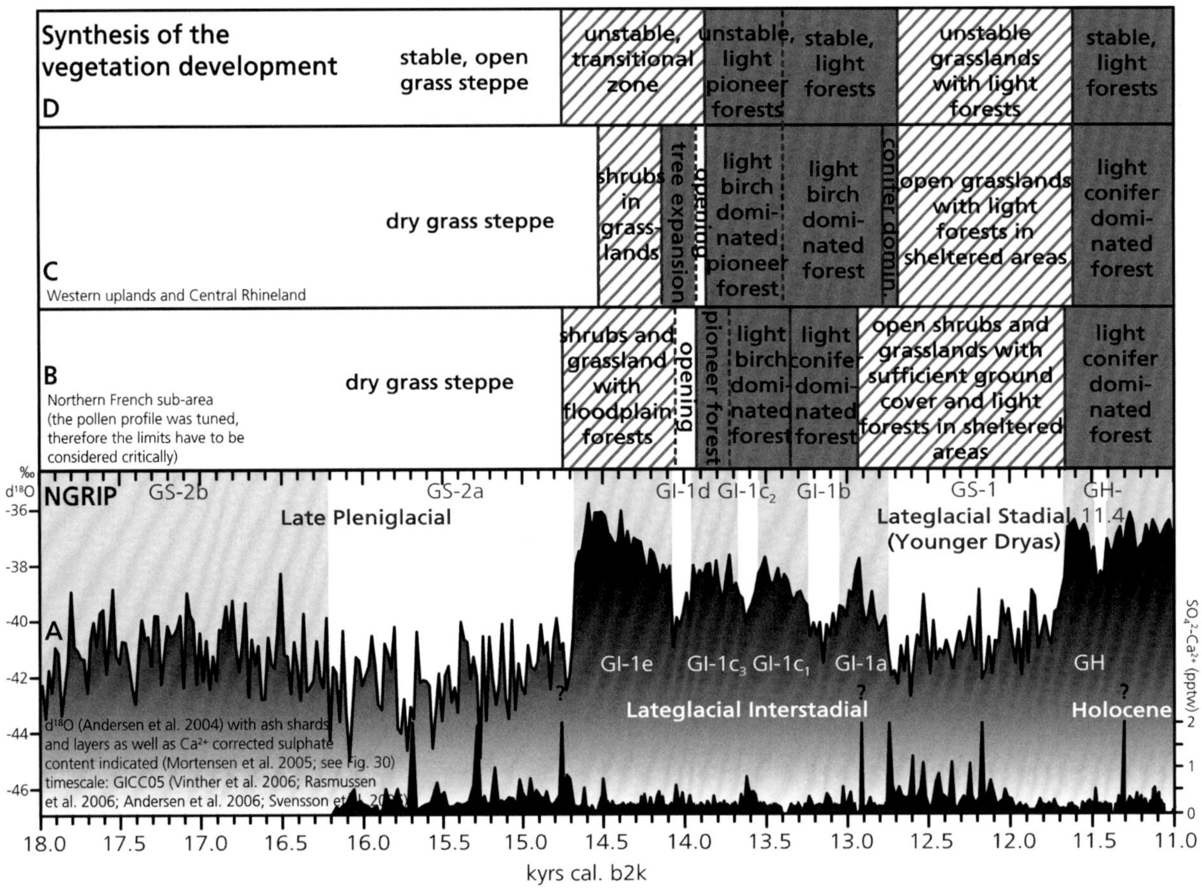

Fig. 59 Synthesis of the vegetation development in the sub-areas of this study (see **figs 53-58**). White backgrounds: cold and open environments; hatched background: transitional environments; grey shaded background: temperate forest environments.

site	lab. no.	years ^{14}C-BP	± years	material	species	comment	years cal. b2k	ref.
pine (*Pinus* sp.), n=29								
Groß Lieskow, Brandenburg (D)	*LZ-1346*	*18,960*	*180*	charcoal/ sediment	*Pinus* sp.	archaeological horizon; PROBLEMATIC: big pieces of charcoal mixed with sand; REJECT: too high concentration of sand?	×	1
Mannheim-Vogelstang, Baden-Württemberg (D)	*GrA-*	*14,680*	*70*	wood	*Pinus sylvestris*	modified; bow fragment? REJECT: long storage, too low amount of material	×	2
Putim, Jihočeský (CZ)	GrA-36010	13,010	60	charcoal	*Pinus* sp.		16,240-15,440	3
Wilczyce, Świętokrzyskie (PL)	Poz-14892	12,770	120	charcoal	*Pinus* sp.	from ice wedge cast; PROBLEMATIC: »minuscule fragment…and, therefore, is considered … as of minimal age.« (Fiedorczuk et al. 2007, fig. 5)	15,790-14,910	4
Grieben, Saxony-Anhalt (D)	OxA-13284	12,620	50	wood	*Pinus sylvestris*	haft of coarse Mesolithic stone hatchet; PROBLEMATIC: long storage, use of old wood	15,390-14,870	5-6
Ahrenshöft LA 73 south, Schleswig-Holstein (D)	KIA-3605	12,200	60	charcoal	*Pinus* sp.	horizon 1	14,260-15,440	7
Monruz, sect. 2, Neuchâtel (CH)	ETH-17108	11,945	95	charcoal	*Pinus* sp.	possibly a cone was within the sample, above archaeological horizons	14,020-13,580	8
Monruz, sect. 1, Neuchâtel (CH)	ETH-20726	11,610	80	charcoal	*Pinus* sp.	above archaeological horizons	13,610-13,250	8
Oldeholtwolde, Friesland (NL)	OxA-2560	11,300	110	charcoal	*Pinus* sp.	from Usselo soil; no association to archaeology	13,350-12,950	9
Hülm, Nordrhein-Westfalen (D)	GrA-23744	11,080	50	cone	*Pinus* sp.		13,100-12,740	10
Wierden-Enterse Akkers HS, Overijssel (NL)	GrA-25906	11,070	60	charcoal	*Pinus* sp.		13,090-12,730	11
Witów (1-P), horizon 4a, Łódzkie (PL)	K-952	11,020	170	charcoal	*Pinus* sp.		13,190-12,630	12
Groß Lieskow, Brandenburg (D)	LZ-1350	11,000	110	charcoal	*Pinus* sp.	archaeological horizon; PROBLEMATIC: big pieces of charcoal mixed with sand	13,080-12,680	1
Groß Lieskow, Brandenburg (D)	LZ-1349 (or -1348?)	10,970	80	charcoal	*Pinus* sp.	archaeological horizon; PROBLEMATIC: big pieces of charcoal mixed with sand	13,040-12,680	1
Groß Lieskow, Brandenburg (D)	LZ-1352	10,960	80	charcoal	*Pinus* sp.	archaeological horizon; PROBLEMATIC: big pieces of charcoal mixed with sand	13,030-12,670	1
Groß Lieskow, Brandenburg (D)	LZ-1347	10,870	105	charcoal	*Pinus* sp.	archaeological horizon; PROBLEMATIC: big pieces of charcoal mixed with sand	13,010-12,610	1
Usselo, Overijssel (NL)	*Y-139-2*	*10,880*	*160*	sediment/ charcoal?	*Pinus* sp.	REJECT: material, mainly sandy peat?	×	13

Tab. 73 ^{14}C dates from selected plant samples in north-western Europe outside the study area. If the sub-assemblage is known from which the sample originated, the sub-assemblage is given behind the site name. Otherwise the province and the country is given. **?** (behind the species) uncertain species determination. Doubtful dates are shaded grey. Rejected dates are shaded grey and set in italics and, in addition, the main reason for rejection is given in comment. For further details see p. 259-263, p. 265-269, and p. 412-429. The dates were calibrated with the calibration curve of the present study (see p. 358-364) and the calibration program CalPal (Weninger/Jöris/Danzeglocke 2007). The result range of 95 % confidence is given for the calibrated ages (years cal. b2k). – References (ref.): **1** Bittmann/Pasda 1999; **2** Rosendahl et al. 2006; **3** Verpoorte/Šída 2009; **4** Fiedorczuk et al. 2007; **5** Grünberg 2006; **6** Higham et al. 2007; **7** Clausen 1998; **8** Leesch/Cattin/Müller 2004, 167. 183; **9** Johansen/Stapert 2004; **10** Kasse et al. 2005; **11** Deeben et al. 2006; **12** Tauber 1966; **13** Barendsen/Deevey/Gralenski 1957; **14** Johansen et al. 2000; **15** Hedges et al. 1992; **16** Lanting/van der Plicht 1996; **17** Hanik 2009; **18** Gob 1990; **19** Leesch 1997, 21. 199; **20** Bullinger/Leesch/Plumettaz 2006; **21** Schuler 1994; **22** Kaiser/de Klerk/Terberger 1999; **23** Vollbrecht 2005; **24** Otlet/Slade 1974; **25** Morel 1993; **26** Aaris-Sørensen 1995; **27** Kabaciński/Schild 2005; **28** Hedges et al. 1993a; **29** Hoek et al. 1999; **30** Tromnau 1992; **31** Tolksdorf et al. 2013; **32** Barr 1975; **33** Hedges et al. 1995; **34** Alessio/Bella/Cortesi 1964; **35** Jones et al. 2002; **36** Bronk Ramsey et al. 2000; **37** Niekus 2006.

site	lab. no.	years ^{14}C-BP	± years	material	species	comment	years cal. b2k	ref.
Doetinchem, Gelderland (NL)	GrA-13686	10,870	50	charcoal	*Pinus* sp.		12,860-12,660	14
Monruz, sect. 1, Neuchâtel (CH)	ETH-20725	10,800	75	charcoal	*Pinus* sp.	above archaeological horizons	12,800-12,640	8
Groß Lieskow, Brandenburg (D)	LZ-1349 (or -1348?)	10,660	80	charcoal	*Pinus* sp.	archaeological horizon; PROBLEMATIC: big pieces of charcoal mixed with sand; below aeolian sands – intrusion of these sands?	12,760-12,440	1
Geldrop Mie Peels, Noord Brabant (NL)	OxA-2563	10,610	100	charcoal	*Pinus* sp.	hearth	12,760-12,320	15; 16
Wierden-Enterse Akkers HS, Overijssel (NL)	GrA-24580	10,610	60	charcoal	*Pinus* sp.		12,720-12,440	11
Groß Lieskow, Brandenburg (D)	LZ-1351	10,520	100	charcoal	*Pinus* sp.	archaeological horizon; PROBLEMATIC: big pieces of charcoal mixed with sand; below aeolian sands – intrusion of these sands?	12750-12,110	1
Groß Lieskow, Brandenburg (D)	LZ-1353	10,420	100	charcoal	*Pinus* sp.	archaeological horizon; PROBLEMATIC: big pieces of charcoal mixed with sand; below aeolian sands – intrusion of these sands?	12,660-11,980	1
Wustermark 22, Brandenburg (D)	?	10,389	57	branch	*Pinus* sp.	fluvial accumulation	12,520-12,040	17
Wierden-Enterse Akkers HS, Overijssel (NL)	*GrA-23937*	*10,350*	*60*	charcoal	*Pinus* sp.	REJECT: too small amount of material	×	11
Groß Lieskow, Brandenburg (D)	LZ-1345	9,780	75	charcoal	*Pinus* sp.	PROBLEMATIC: big pieces of charcoal mixed with sand; REJECT: Holocene age but found below aeolian sands – intrusion of these sands?	×	1
Elsloo-Tronde, Friesland (NL)	*GrN-4869*	*7,790*	*95*	charcoal	*Pinus* sp.	from hearth; REJECT: Holocene intrusion / Mesolithic feature?	×	18
Meer IV, Antwerp (B)	*GrA-10290*	*5,940*	*180*	charcoal	*Pinus sylvestris*	from hearth; REJECT: Holocene intrusion?	×	16
willow family (*Saliceae*) / willow (*Salix* sp.) / poplar (*Populus* sp.), n=42								
Champréveyres, sect. 1, Neuchâtel (CH)	UCLA-2760	17,695	210	charcoal	*Salix* sp.	from hearth in lower horizon; PROBLEMATIC: too old for stratigraphy but dated twice with comparable results; old wood?	21,800-20,600	19
Monruz, sect. 1, Neuchâtel (CH)	ETH-6413	13,330	110	charcoal	*Salix* sp.	from hearth in lower horizon	16,730-15,490	19-20
Monruz, sect. 2, north, Neuchâtel (CH)	ETH-6421	13,140	120	charcoal	*Salix* sp.	from hearth	17,000-16,000	8; 19
Monruz, sect. 1, Neuchâtel (CH)	ETH-6420	13,120	120	charcoal	*Salix* sp.	from hearth in lower horizon	16,700-15,460	19-20
Monruz, sect. 1, Neuchâtel (CH)	ETH-6416	13,070	130	charcoal	*Salix* sp.	from hearth lower horizon	16,640-15,360	19-20
Champréveyres, sect. 1, Neuchâtel (CH)	UZ-2285	13,050	155	charcoal	*Salix* sp.	from hearth in lower horizon	16,680-15,280	19
Monruz, sect. 1, Neuchâtel (CH)	ETH-6417	13,030	120	charcoal	*Salix* sp.	from hearth in lower horizon	16,500-15,340	19-20
Monruz, sect. 1, Neuchâtel (CH)	ETH-6412	12,970	110	charcoal	*Salix* sp.	from hearth in lower horizon; PROBLEMATIC: the $\delta^{13}C$ value is surprisingly high	16,340-15,260	19-20
Champréveyres, sect. 1, Neuchâtel (CH)	UZ-2283	12,950	155	charcoal	*Salix* sp.	from hearth in lower horizon	16,470-15,110	19
Monruz, sect. 1, Neuchâtel (CH)	ETH-6415	12,900	120	charcoal	*Salix* sp.	from hearth in lower horizon	16,190-15,150	19-20
Monruz, sect. 1, Neuchâtel (CH)	ETH-6419	12,880	120	charcoal	*Salix* sp.	from hearth in lower horizon	16,100-15,140	19-20
Champréveyres, sect. 1, Neuchâtel (CH)	UZ-2286	12,870	135	charcoal	*Salix* sp.	from hearth in lower horizon	16,160-15,080	19

Tab. 73 (continued)

site	lab. no.	years ^{14}C-BP	± years	material	species	comment	years cal. b2k	ref.
Monruz, sect. 1, Neuchâtel (CH)	ETH-6414	12,840	120	charcoal	*Salix* sp.	from hearth in lower horizon; PROBLEMATIC: the $\delta^{13}C$ value is surprisingly low	15,910-15,110	19-20
Champréveyres, sect. 1, Neuchâtel (CH)	UZ-2171	12,730	135	charcoal	*Salix* sp.	from hearth in lower horizon	15,790-14,670	19
Champréveyres, sect. 1, Neuchâtel (CH)	UZ-2175	12,630	130	charcoal	*Salix* sp.	from hearth in lower horizon	15,630-14,390	19
Schussenquelle I, Baden-Württemberg (D)	ETH-6154	12,630	120	charcoal	*Salix* sp.	lower horizon	15,600-14,440	21
Champréveyres, sect. 1, Neuchâtel (CH)	UZ-2172	12,620	145	charcoal	*Salix* sp.	from hearth in lower horizon	15,660-14,300	19
Champréveyres, sect. 1, Neuchâtel (CH)	UZ-2177	12,600	145	charcoal	*Salix* sp.	from hearth in lower horizon	15,630-14,230	19
Ahrenshöft LA 73 south, Schleswig-Holstein (D)	*KIA-3606*	*12,550*	*+1,170 / -1,020*	charcoal	*Salix* sp./ *Populus* sp.	horizon I; REJECT: too low amount of carbon	×	7
Champréveyres, sect. 1, Neuchâtel (CH)	UZ-2173	12,540	140	charcoal	*Salix* sp.	from hearth in lower horizon	15,560-14,080	19
Champréveyres, sect. 1, Neuchâtel (CH)	UZ-2174	12,510	130	charcoal	*Salix* sp.	from hearth in lower horizon	15,470-14,070	19
Champréveyres, sect. 2, Neuchâtel (CH)	UZ-2287	12,500	145	charcoal	*Salix* sp.		15,500-14,020	8; 19
Endingen, Horst VI, Mecklenburg-Vorpommern (D)	Hv-20987	12,360	245	wood	*Salix* sp.		15,540-13,620	22
Reichwalde 5055, Sachsen (D)	KIA-13412	12,193	58	charcoal	*Salix* sp./ *Populus* sp.		14,250-13,890	23
Ahrenshöft LA 73 north, Schleswig-Holstein (D)	KIA-3833	12,130	60	charcoal	*Salix* sp./ *Populus* sp.	horizon II	14,180-13,780	7
Sproughton, Devil's Wood Pit, Suffolk/England (GB)	HAR-260	11,940	180	branch	*Salix* sp.	PROBLEMATIC: not associated with archaeology; uncertain sedimentation	14,260-13,380	24
Oldeholtwolde, Friesland (NL)	OxA-2558	11,810	110	charcoal	*Salix* sp.	PROBLEMATIC: stratigraphic position and archaeology in contrast to date	13,860-13,420	9; 15
Ahrenshöft LA 73 south, Schleswig-Holstein (D)	KIA-3606	11,750	60	charcoal	*Salix* sp./ *Populus* sp.	horizon I	13,740-13,460	7
Oldeholtwolde, Friesland (NL)	OxA-2561	11,680	120	charcoal	*Salix* sp.	PROBLEMATIC: stratigraphic position and archaeology in contrast to date	13,790-13,230	9; 15
Grotte du Bichon, Neuchâtel (CH)	ETH-4246 (UZ-2423)	11,680	120	charcoal	*Salix* sp.		13,790-13,230	25
Nørre Lyngby, Nordjylland (DK)	AAR-1509	11,590	130	wood	*Salix* sp.		13,710-13,150	26
Oldeholtwolde, Friesland (NL)	GrN-10274	11,540	270	charcoal	*Salix* sp.	PROBLEMATIC: stratigraphic position and archaeology in contrast to date	13,920-12,880	9; 16
Oldeholtwolde, Friesland (NL)	OxA-2559	11,470	110	charcoal	*Salix* sp.	PROBLEMATIC: stratigraphic position and archaeology in contrast to date	13,540-13,100	15
Nørre Lyngby, Nordjylland (DK)	AAR-1507	11,340	120	wood	*Salix* sp.		13,410-12,970	26
Nørre Lyngby, Nordjylland (DK)	AAR-1508	11,260	120	wood	*Salix* sp.		13,330-12,890	26
Nørre Lyngby, Nordjylland (DK)	AAR-1510	11,230	150	wood	*Salix* sp.		13,370-12,770	26
Oldeholtwolde, Friesland (NL)	GrN-12280	11,080	280	charcoal	*Salix* sp.	no relation with archaeology	13,450-12,490	9; 16
Grotte du Bichon, Neuchâtel (CH)	ETH-4245 (UZ-2422)	10,950	180	charcoal	*Salix* sp.	PROBLEMATIC: age significantly younger than other ages from the site	13,150-12,590	25
Mirkowice 33, Wielkopolskie (PL)	UtC-8617	10,210	120	leaf	*Salix* cf. *polaris*	PROBLEMATIC: association uncertain	12,420-11,420	27

Tab. 73 (continued)

site	lab. no.	years ^{14}C-BP	± years	material	species	comment	years cal. b2k	ref.
Holywell Coombe, Kent/England (GB)	OxA-2608	10,160	110	wood	*Salix* sp.		12,290-11,330	28
Holywell Coombe, Kent/England (GB)	OxA-2606	9,900	100	wood	*Salix* sp.		11,800-11,120	28
Sproughton, Devil's Wood Pit, Suffolk/England (GB)	*HAR-259*	*9,880*	*120*	macro remains	*Salix* sp.	REJECT: not associated with archaeology; bulked sample; uncertain sedimentation	×	24
birch (*Betula* sp.), n = 17								
Holywell Coombe, Kent/England (GB)	OxA-1751	13,160	400	fruits, cone scales	*Betula* sp.		17,360-14,600	28
Gulickshof, Limburg (NL)	UtC-3196	12,480	90	leaves	*Betula nana*		15,320-14,120	29
Poggenwisch, Schleswig-Holstein (D)	GrN-11254	12,460	60	wood	*Betula* sp.		15,210-14,170	30
Grabow 15, Niedersachsen (D)	KIA-41862	12,125	50	charcoal	*Betula* sp.	from archaeological feature	14,160-13,800	31
Moosbühl, Bern (CH)	B-2316	12,060	150	bark	*Betula* sp.	found *in situ* beside hearth	14,370-13,570	32
Holywell Coombe, Kent/England (GB)	OxA-1974	11,830	140	fruits, cone scales	*Betula* sp.		13,970-13,370	28
Krumpa, Sachsen-Anhalt (D)	OxA-4499	11,810	100	fruits, cone scales	*Betula pendula / pubescens* (previously: *alba*)		13,830-13,430	33
Usselo, Overijssel (NL)	R-106	11,800	280	wood	*Betula* sp.	PROBLEMATIC: found in peat; association to archaeology uncertain	14,390-13,030	34
Krumpa, Sachsen-Anhalt (D)	OxA-4498	11,660	100	fruits, cone scales	*Betula pendula / pubescens* (previously: *alba*)		13,760-13,240	33
Holywell Coombe, Kent/England (GB)	OxA-2352	11,600	100	charcoal	*Betula* sp.	PROBLEMATIC: the $\delta^{13}C$ value is surprisingly low	13,650-13,210	28
Cherry Garden Hill, Kent/England (GB)	OxA-2242	11,580	100	charcoal	*Betula* sp.		13,610-13,210	28
Holywell Coombe, Kent/England (GB)	OxA-2353	11,520	90	charcoal	*Betula* sp.		13,540-13,180	28
Hawes Water, Lancashire/England (GB)	NERC-32069	10,980	60	fruit	*Betula* sp.		13,020-12,700	35
Westhampnett, West Sussex/England (GB)	OxA-4166	10,880	110	charcoal	*Betula* sp.		13,020-12,620	36
Gulickshof, Limburg (NL)	GrA-4309	10,800	90	leaves, fruits	*Betula* sp.		12,850-12,610	29
Hawes Water, Lancashire/England (GB)	CAMS-45852	10,160	220	fruit	*Betula* sp.		12,620-11,100	35
Vledder, Drenthe (NL)	*GrA-10938*	*6,150*	*60*	charcoal	*Betula* sp.	REJECT: Holocene intrusion in sand?	×	37
juniper (*Juniperus* sp.), n = 5								
Monruz, sect. 2 south, Neuchâtel (CH)	ETH-23271	12,570	90	charcoal	*Juniperus* sp.	from hearth	15,450-14,450	8
Monruz, sect. 1, Neuchâtel (CH)	ETH-17974	12,370	110	charcoal	*Juniperus* sp.	from hearth in upper horizon	15,190-13,910	8
Monruz, sect. 2 south, Neuchâtel (CH)	ETH-23272	12,355	85	charcoal	*Juniperus* sp.	from hearth	14,980-13,980	8
Monruz, sect. 1, Neuchâtel (CH)	ETH-17973	12,165	130	charcoal	*Juniperus* sp.	from hearth in upper horizon	14,660-13,620	8
Holywell Coombe, Kent/England (GB)	OxA-2346	9,820	90	fruits	*Juniperus communis*		11,530-11,090	28

Tab. 73 (continued)

Lateglacial faunal successions in the study areas

The presence of selected faunal species is established by the frequency of directly ^{14}C-dated samples of these species and by the geographic distribution of the faunal remains, in particular on the archaeological sites.

The probability distributions

For the sub-areas, the total numbers of ^{14}C dates from the relevant species are in general small (see **tabs 74-76**) and during the Late Pleniglacial clearly biased by the preference for hunting horse which consequently provided a large quantity of datable material. Thus, the probability distributions of ^{14}C dates from the sub-areas (**figs 60-61**) indicate several problems previously discussed regarding the interpretation of these distributions (see p. 263f.).

According to the probability distribution of the Central Rhineland and the western uplands (**fig. 60**), a major change in the selected fauna composition occurred at the onset of GI-1d. In contrast, a comparable change occurred in the Paris Basin considerably later, during GI-1c_3 (**fig. 61**). In both areas, this change seemed to correlate with a significant increase of arboreal and particularly birch tree pollen.

The probability distribution of the Lateglacial fauna from the Central Rhineland and the western uplands can be subdivided in a total of four sections numbered stratigraphically from top to base. The second and the fourth can be further divided into two sub-units. The sections 4 and 3 occurred before the major change in the fauna composition and section 1 and 2 were identified afterwards. The oldest section (4) relates to the Late Pleniglacial and ended at the onset of the Lateglacial Interstadial. The following section 3 comprised the early Lateglacial Interstadial. After the major change, section 2 related to the mid- and late Lateglacial Interstadial and, finally, the section 1 occurred in the LateglacialStadial.

Section 4 was dominated by numerous directly dated horse (*Equus* sp.) remains. Furthermore, this section can be bisected in an earlier part in which reindeer (*Rangifer tarandus*) dates were also frequent and a later part in which more horse remains were dated but no more reindeer material. In the latter, directly dated musk ox (*Ovibos moschatus*) remains appeared in some Belgian sites (see **tab. 74**), whereas the steppe wisent (*Bison priscus*) from Gönnersdorf was dated to the earlier part of this sub-unit (see **tab. 75**).

Reindeer and musk ox are considered as a typical representative of cold, arctic conditions, whereas steppe wisent seemed in general more common in arid steppe environments (von Koenigswald 1999) and horse usually inhabits grasslands. Thus, to interpret this distribution in a successive climatic and environmental way is tempting but this interpretation is only possible within a more complete faunal context. For instance, musk ox seemed to be avoided by Late Weichselian hunters (Pushkina/Raia 2008). Thus, material was only found in the Belgian Late Magdalenian sites but not in comparable sites in the Paris Basin or the Central Rhineland. Perhaps, this distribution was due to the range area of the presumably small remaining population of this species in Europe after the LGM (Campos et al. 2010). Consequently, the presence of musk ox in this area could indicate the presence of more intense arctic conditions in the north-western part of the western uplands and a regression of these conditions in the Central Rhineland as suggested by the decreasing number of reindeer dates. Moreover, the probability distribution of reindeer in Central Rhineland could suggest that the presence of these animals was more probable in a younger phase of the Magdalenian occupation of Gönnersdorf. In additon, a retreat of steppe wisent during this younger sub-unit could be interpreted as reflecting increasingly moist conditions.

However, this temptation reveals the problems of interpreting probability distributions without the context of the dated specimens because the distribution of reindeer remains at Gönnersdorf was not noticeably differ-

site	lab. no.	years ^{14}C-BP	± years	material	species	comment	years cal. b2k	ref.
pine (*Pinus* sp.), n=0								
willow family (*Saliceae*) / willow (*Salix* sp.) / poplar (*Populus* sp.), n=0								
birch (*Betula* sp.), n=0								
reindeer (*Rangifer tarandus*), n=21								
Trou des Blaireaux (Vaucelles)	Lv-1385	16,270	230	antler	*Rangifer tarandus*	FAD reindeer in western uplands; PROBLEMATIC: bulked sample	20,160-18,880	1
Trou des Blaireaux (Vaucelles)	Lv-1558	16,130	250	antler	*Rangifer tarandus*	PROBLEMATIC: bulked sample	19,940-18,780	1
Schlaederbach valley/Plateau Haed (Oetrange, LUX)	Lv-466	16,070	450	antler	*Rangifer tarandus*		20,370-18,450	2
Trou des Blaireaux (Vaucelles)	Lv-1433	13,930	120	antler	*Rangifer tarandus*	PROBLEMATIC: bulked sample of female antlers	17,540-16,900	1
Trou des Blaireaux (Vaucelles)	Lv-1309D	13,850	335	antler	*Rangifer tarandus*	PROBLEMATIC: bulked sample	17,920-16,160	1
Trou des Blaireaux (Vaucelles)	Lv-1314	13,790	150	antler	*Rangifer tarandus*	PROBLEMATIC: bulked sample	17,210-16,850	1
Trou des Blaireaux (Vaucelles)	Lv-1434D	13,730	400	antler	*Rangifer tarandus*	PROBLEMATIC: bulked sample of female antlers	17,970-15,570	1
Saint Mihiel	Lv-2096	13,160	110	antler	*Rangifer tarandus*	PROBLEMATIC: bulked sample	16,740-15,540	3
Bois Laiterie	OxA-4198*	12,660	140	antler	*Rangifer tarandus*	double bevelled point, YSS	15,690-14,410	1; 4
Coléoptère (Bomal-sur-Ourthe), lower horizon	Lv-717	12,400	110	bone	*Rangifer tarandus*		15,230-13,950	5
Trou da Somme (Waulsort)	OxA-4199*	12,240	130	antler point	*Rangifer tarandus*	artefact	14,880-13,720	4
Remouchamps	OxA-4191*	10,800	110	metacarpus	*Rangifer tarandus*	cut-marks	12,940-12,540	5
Remouchamps	Lv-535	10,380	170	bone fragments	*Rangifer tarandus*	PROBLEMATIC: bulked sample	12,770-11,610	6
Remouchamps	OxA-3634*	10,320	80	maxilla	*Rangifer tarandus*	cut-marks; PROBLEMATIC: the $\delta^{13}C$ value is surprisingly high (cf. OxA-9031)	12,510-11,870	5
Kartstein (Mechernich)	OxA-9031	10,220	75	femur	*Rangifer tarandus*	PROBLEMATIC: »the $\delta^{13}C$ value is surprisingly high« (M. Baales / O. Jöris / B. Weninger in Bronk Ramsey et al. 2002, 31)	12,260-11,660	7

Tab. 74 ^{14}C dates from selected plant and faunal samples from the western upland areas. ***** dates which were pretreated by the use of ion-exchanged gelatin (Lab.code: AI) in the Oxford series (cf. Jacobi/Higham 2009, 1896). Rejected date is shaded grey and set in italics and, in addition, the main reason for rejection is given in comment. For further details see p. 259-263, p. 265-269, and p. 412-429. The dates were calibrated with the calibration curve of the present study (see p. 358-364) and the calibration program CalPal (Weninger/Jöris/Danzeglocke 2007). The result range of 95 % confidence is given for the calibrated ages (years cal. b2k). – References (ref.): **1** Charles 1996; **2** Gilot 1970; **3** Stocker et al. 2006; **4** Hedges et al. 1994; **5** Hedges et al. 1993b; **6** Gob 1990; **7** Bronk Ramsey et al. 2002; **8** Baales 1996, 42; **9** Léotard 1993; **10** Stevens et al. 2009a; **11** Hedges et al. 1993a; **12** Stevens/Hedges 2004; **13** Germonpré 1997; **14** Hedges et al. 1988b; **15** Charles 1997a.

site	lab. no.	years ^{14}C-BP	± years	material	species	comment	years cal. b2k	ref.
Kartstein (Mechernich)	KN-4254A	10,030	60	shaft fragments	cf. *Rangifer tarandus*	PROBLEMATIC: bulked sample	11,880-11,280	7
Kartstein (Mechernich)	KN-4252	10,000	50	shaft fragments	cf. *Rangifer tarandus*	PROBLEMATIC: bulked sample	11,790-11,270	7
Kartstein (Mechernich)	*KN-4254B*	*9,900*	*45*	shaft fragments	cf. *Rangifer tarandus*	PROBLEMATIC: bulked sample; REJECT: probably different fraction of KN-4254	11,480-11,240	7
Kartstein (Mechernich)	*KN-4254C*	*9,685*	*50*	shaft fragments	cf. *Rangifer tarandus*	PROBLEMATIC: bulked sample; REJECT: probably same as KN-4254 but other fraction	11,360-10,800	7
Kartstein (Mechernich)	KN-4072	9,550	90	bone fragments	cf. *Rangifer tarandus*	PROBLEMATIC: bulked sample	11,270-10,630	7-8
Kartstein (Mechernich)	KN-4073	9,530	90	shaft fragments	cf. *Rangifer tarandus*	PROBLEMATIC: bulked sample	11,270-10,590	7-8
horse (*Equus* sp.), n=14								
Trou des Blaireaux (Vaucelles)	OxA-4200*	13,330	160	ulna	*Equus* sp.	possible cut-marks	17,110-15,710	4
Trou du Frontal (Furfooz)	*Lv-1750*	*13,130*	*170*	»cut bone splinters«	*Equus* sp. (and/or *Canis lupus*)	PROBLEMATIC: bulked sample; REJECT: species attribution unclear	16,870-15,350	1; 9
Trou du Frontal (Furfooz)	Lv-1749	12,950	170	»cut bone splinters«	*Equus* sp.	PROBLEMATIC: bulked sample	16,520-15,080	1; 9
Trou de Chaleux (Hulsonniaux)	OxA-3633*	12,880	100	bone	*Equus* sp.	cut-marks	15,920-15,240	10-12
Coléoptère (Bomal-sur-Ourthe), lower horizon	OxA-3635*	12,870	95	phalange	*Equus* sp.	cut-marks; PROBLEMATIC: the δ^{13}C value is surprisingly high	15,850-15,250	5
Sy Verlaine-sur-Ourthe	OxA-4014*	12,870	110	pisiform	*Equus* sp.	possible cut-marks	15,950-15,190	4
Trou de Frontal (Furfooz)	OxA-4197*	12,800	130	bone	*Equus* sp.		15,870-14,950	12
Trou de Chaleux (Hulsonniaux)	OxA-3632*	12,790	100	bone	*Equus* sp.	cut-marks	15,720-15,080	10-11
Goyet, 3rd cave, horizon 1	OxA-V-2223-48	12,775	55	bone	*Equus* sp.		15,580-15,220	10
Goyet, 3rd cave, horizon 1	GrA-3237	12,770	90	bone	*Equus* sp.	cut-marks, coloured by ochre	15,670-15,070	10; 13
Trou des Nutons (Furfooz)	OxA-4195*	12,630	140	phalange	*Equus* sp.	cut-marks	15,660-14,340	4
Trou de Chaleux (Hulsonniaux)	OxA-V-2216-45	12,630	55	bone	*Equus* sp.	cut-marks	15,400-14,880	10
Goyet, 3rd cave, horizon 1	Utc-8957	12,560	50	bone	*Equus* sp.		15,380-14,540	10
Trou de Chaleux (Hulsonniaux)	OxA-V-2216-44	12,375	50	bone	*Equus* sp.	cut-marks	14,810-14,090	10

Tab. 74 (continued)

site	lab. no.	years ^{14}C-BP	± years	material	species	comment	years cal. b2k	ref.
elk (*Alces alces*), n=0								
wild boar (*Sus scrofa*), n=1								
Trou de Chaleux (Hulsonniaux)	*OxA-4193**	*3,060*	*85*	humerus	*Sus scrofa*	cut-marks; REJECT: too young; intrusion	×	4
red deer (*Cervus elaphus*), n=2								
Presles, uncertain cave	OxA-1344	10,950	200	mandible	*Cervus elaphus*	unmodified	13,190-12,550	14-15
Trou des Nutons (Furfooz)	*OxA-4194**	*2,210*	*80*	naviculo-cuboid	*Cervus elaphus*	cut-marks; REJECT: too young / younger intrusion?	×	4
large bovids (*Bison priscus / Bos* sp. / *Bos primigenius*), n=0								
others, n=5								
Trou de Chaleux (Hulsonniaux)	OxA-4192*	12,860	140	bone	*Ovibos moschatus*	cut-marks	16,150-15,030	4; 10
Trou da Somme (Waulsort)	OxA-8308	12,815	75	bone	*Ovibos moschatus*		15,680-15,240	7; 10
Goyet, 3rd cave, horizon 1	OxA-12121	12,775	50	bone	*Ovibos moschatus*		15,580-15,220	10
Goyet, 3rd cave, horizon 1	GrA-3238	12,620	90	bone	*Ovibos moschatus*	cut-marks	15,520-14,520	10; 13
Goyet, 3rd cave, horizon 2	KIA-22275	12,380	60	bone	*Alopex lagopus*		15,520-14,520	10

Tab. 74 (continued)

site	lab. no.	years ^{14}C-BP	± years	material	species	comment	years cal. b2k	ref.
pine (*Pinus* sp.), n=1								
Bad Breisig	GrA-17493	10,840	60	charcoal	*Pinus* sp.	from the hearth	12,820-12,660	1
willow family (*Saliceae*) / willow (*Salix* sp.) / poplar (*Populus* sp.), n=17								
Brohl valley 1	KN-3803	11,510	90	wood	*Populus* sp.	within LST	13,530-13170	2-3
Miesenheim 2	KN-3533	11,460	90	wood, root	*Populus* sp.	PROBLEMATIC: root material	13,510-13,110	4
Miesenheim 2	KN-3531	11,460	100	wood, trunk	*Populus* sp.		13,530-13,090	4
Miesenheim 2	KN-3532	11,390	90	wood, root	*Populus* sp.	PROBLEMATIC: root material	13,410-13,050	4
Miesenheim 2	KN-3534	11,360	110	wood, root	*Populus* sp.	PROBLEMATIC: root material	13,430-12,990	4
Miesenheim 2	KN-3517	11,290	95	wood, root	*Populus* sp.	PROBLEMATIC: root material	13,300-12,980	4
Brohl valley 1	KN-3802	11,280	100	charcoal	*Populus* sp.	within LST	13,310-12,950	2-3
Brohl valley 1	KN-3801	11,260	95	wood	*Populus* sp.	within LST	13,280-12,960	2-3
Brohl valley 1	KN-3800	11,240	110	wood	*Populus* sp.	within LST	13,300-12,900	2-3
Miesenheim 2	KN-3516	11,230	95	wood, root	*Populus* sp.	PROBLEMATIC: root material	13,270-12,910	4
Miesenheim 2	KN-3518	11,080	220	wood, bark	*Populus* sp.		13,350-12,590	4
Miesenheim 2	KN-3520	11,070	100	wood	*Populus* sp.		13,140-12,700	4
Miesenheim 2	KN-3519	11,040	220	wood, bark	*Populus* sp.		13,300-12,580	4
Miesenheim 2	OxA-2611	11,030	110	wood	*Populus* sp.		13,120-12,680	5
Miesenheim 2	OxA-2609	10,960	110	wood	*Populus* sp.		13,060-12,660	5
Miesenheim 2	OxA-2610	10,960	110	wood	*Populus* sp.		13,060-12,660	5
Miesenheim 2	*ETH-*	*10,840*	*195*	wood	*Populus* sp.	REJECT: post-LSE date	×	2
birch (*Betula* sp.), n=4								
Thür	KN-2870	11,250	95	wood	*Betula* sp.		13,270-12,950	2
Thür	KN-2869	11,110	90	wood	*Betula* sp.		13,170-12,730	2
Brohl valley	unknown	11,085	90	wood	*Betula* sp.	within LST	13,150-12,710	2-3
Thür	KN-2868	11,050	120	wood	*Betula* sp.		13,150-12,670	2
reindeer (*Rangifer tarandus*), n=5								
Gönnersdorf III	OxA-V-2223-43	13,075	55	metapodial	*Rangifer tarandus*		16,330-15,570	6
Gönnersdorf III	OxA-15295	13,060	60	metapodial	*Rangifer tarandus*	marrow fractured	16,310-15,550	6-7

Tab. 75 ^{14}C dates of selected plant and faunal samples from the Central Rhineland. If the sub-assemblage is known from which the sample originated, the sub-assemblage is given behind the site name. * dates which were pretreated by the use of ion-exchanged gelatin (Lab.code: AI) in the Oxford series (cf. Jacobi/Higham 2009, 1896); ** dates which might be contaminated due to the use of a method leaving traces of a humectant in the collagen (Lab.code: AF*) in the Oxford series (cf. Higham et al. 2007, S55 & S2); **LSE** Laacher See (Volcanic) Eruption; **LST** Laacher See tephra. Doubtful dates are shaded grey. Rejected dates are shaded grey and set in italics and, in addition, the main reason for rejection is given in comment. For further details see p. 259-263, p. 265-269, and p. 412-429. The dates were calibrated with the calibration curve of the present study (see p. 358-364) and the calibration program CalPal (Weninger/Jöris/Danzeglocke 2007). The result range of 95 % confidence is given for the calibrated ages (years cal. b2k). – References (ref.): **1** Baales/Grimm/Jöris 2001; **2** Baales et al. 2002; **3** Street/Baales/Weninger 1994; **4** Street 1986; **5** Hedges et al. 1993b; **6** Stevens et al. 2009b; **7** Higham et al. 2007; **8** Higham et al. 2011; **9** Bronk Ramsey et al. 2002; **10** Stevens/Hedges 2004; **11** Street/Terberger 2004; **12** Hedges et al. 1987; **13** Hedges et al. 1998b; **14** Evin/Marien/Pachiaudi 1975; **15** Evin/Marien/Pachiaudi 1978; **16** Baales 2002, 11 f. 40-45; **17** Kegler 2002; **18** Gowlett et al. 1987; **19** Street 1993; **20** Baales/Mewis/Street 1998; **21** kind permission of Stefan Wenzel.

site	lab. no.	years ^{14}C-BP	± years	material	species	comment	years cal. b2k	ref.
Gönnersdorf II	OxA-V-2223-31	13,010	55	metatarsal	*Rangifer tarandus*		16,220-15,460	6
Gönnersdorf I	OxA-V-2223-42	12,990	55	metatarsal	*Rangifer tarandus*		16,180-15,420	6
Wildweiberlei	OxA-18410**	12,835	55	antler	*Rangifer tarandus*	modified; (Lab.code: AF*, but dated after recognition of possible contamination)	15,660-15,300	8
horse (*Equus* sp.), n=24								
Andernach, lower horizon III (?)	*OxA-10492***	*13,500*	*90*	rib	*Equus* sp.	cut-marks; REJECT: redating available	×	6; 9-11
Andernach, lower horizon III (?)	*OxA-10651***	*13,270*	*180*	phalanx	*Equus* sp.	REJECT: relatively high C/N ratio and relatively high $\delta^{15}N$	×	6; 9-11
Gönnersdorf I	OxA-V-2223-39	13,270	55	metatarsal	*Equus* sp.		16,810-16,090	6
Andernach, lower horizon II	OxA-1128	13,200	140	rib	*Equus* sp.	sample from pit	16,920-15,520	6; 11-12
Andernach, lower horizon III (?)	OxA-10493**	13,185	80	rib	*Equus* sp.	cut-marks	16,720-15,640	6; 9-11
Gönnersdorf II	OxA-V-2223-40	13,165	55	metatarsal	*Equus* sp.		16,600-15,680	6
Andernach, lower horizon I	OxA-V-2216-43	13,135	55	humerus	*Equus* sp.		16,540-15,620	6
Andernach, lower horizon II	OxA-V-2218-40	13,110	50	humerus	*Equus* sp.		16,390-15,630	6
Andernach, lower horizon II	OxA-1129	13,090	130	fragments	*Equus* sp.	sample from pit	16,680-15,400	6; 11-12
Andernach, lower horizon III (?)	OxA-18409**	13,025	50	rib	*Equus* sp.	cut-marks; redating of OxA-10492; (Lab.code: AF*, but dated after recognition of possible contamination)	16,220-15,500	6; 8
Andernach, lower horizon I	OxA-V-2218-38	13,015	50	metatarsus	*Equus* sp.		16,210-15,490	6
Andernach, lower horizon III	OxA-1130	12,950	140	bone fragments	*Equus* sp.	sample from pit	16,410-15,170	6; 11-12
Andernach, lower horizon I	OxA-1125	12,930	180	bone fragments	*Equus* sp.	sample from pit	16,530-15,010	6; 11-12
Gönnersdorf I	OxA-5729*	12,910	130	rib fragments	*Equus* sp.		16,290-15,130	6; 11; 13
Andernach, lower horizon I	OxA-1126	12,890	140	rib	*Equus* sp.	sample from pit	16,270-15,070	6; 11-12
Andernach, lower horizon II	OxA-1127	12,820	130	bone fragments	*Equus* sp.	sample from pit	15,920-15,000	6; 11-12

Tab. 75 (continued)

site	lab. no.	years ^{14}C-BP	± years	material	species	comment	years cal. b2k	ref.
Gönnersdorf I	OxA-5730*	12,790	120	rib fragments	*Equus* sp.		15,810-14,970	6; 11; 13
Gönnersdorf I	OxA-5728*	12,730	130	rib fragments	*Equus* sp.		15,760-14,720	6; 11; 13
Andernach, lower horizon III	OxA-V-2223-37	12,675	55	humerus	*Equus* sp.	sample from pit	15,440-15,000	6
Gönnersdorf	Ly-1172	12,660	370	mainly rib fragments	*Equus* sp.?	PROBLEMATIC: bulked sample	16,680-13,600	6; 14
Andernach, upper horizon 2	OxA-V-2218-39	12,270	50	femur	*Equus* sp.		14,560-13,960	6
Gönnersdorf	Ly-1173	11,100	650	mainly rib fragments	*Equus* sp.?	PROBLEMATIC: bulked sample	14,520-11,280	6; 15
Niederbieber 3	OxA-1135	11,130	130	astralagus	*Equus* sp.	PROBLEMATIC: uncertain association	13,260-12,700	11; 16
Niederbieber 2 (19)	*OxA-1134*	*6,250*	*130*	tooth	*Equus* sp.	REJECT: post-LSE date	×	11; 16
elk (*Alces alces*), n=6								
Gönnersdorf V	OxA-15296	12,385	65	radius	*Alces alces*		15,010-14,050	7
Miesenheim 4	OxA-3585*	11,310	95	rib	*Alces alces*		13,340-12,980	5; 11
Miesenheim 4	OxA-3584*	11,190	90	rib	*Alces alces*		13,230-12,870	5; 11
Miesenheim 4	OxA-3586*	11,190	100	rib	*Alces alces*		13270-12,830	5; 11
Niederbieber 2 (19)	OxA-2066	11,110	110	bone	*Alces alces*		13,210-12,690	11; 16
Niederbieber 2 (19)	*OxA-1133*	*9,750*	*240*	bone	*Alces alces*?	REJECT: post-LSE date	×	11; 16
wild boar (*Sus scrofa*), n=0								
red deer (*Cervus elaphus*), n=13								
Andernach, lower horizon I / upper horizon 2	GrA-16986	13,180	70	shaft fragment	*Cervus elaphus*	impact scar	16,670-15,670	17
Andernach, lower horizon I / upper horizon 2	GrA-16985	13,110	80	shaft fragment	*Cervus elaphus?*		16,550-15,550	17
Andernach, upper horizon 2	*OxA-999*	*12,500*	*500*	shaft fragments	*Cervus elaphus*	REJECT: redating available	×	6; 11; 18
Andernach, upper horizon 2	GrA-16989	11,960	70	metatarsal fragment	*Cervus elaphus*		13,980-13,660	6; 17
Andernach, upper horizon 2	OxA-984	11,950	250	shaft fragments	*Cervus elaphus*	re-dating of OxA-999	14,670-13,190	6; 11; 18
Andernach, upper horizon 2	OxA-1924	11,890	120	bone	*Cervus elaphus*		14,030-13,430	6; 11; 19

Tab. 75 (continued)

site	lab. no.	years ^{14}C-BP	± years	material	species	comment	years cal. b2k	ref.
Andernach, upper horizon 2	OxA-997	11,800	160	bone	*Cervus elaphus*		13,980-13,300	6; 11; 18
Urbar	OxA-1137	11,350	120	bone	*Cervus elaphus*		13,440-12,960	11; 15; 20
Kettig	Hd-18123	11,314	50	bone	cf. *Cervus elaphus*	PROBLEMATIC: bulked material	13,280-13,040	16
Boppard	KIA-26644	11,095	55	metapodial	*Cervus elaphus*		13,110-12,750	21
Niederbieber 1	*OxA-1132*	*10,700*	*130*	bone	*Cervus elaphus*	REJECT: post-LSE date	×	11; 16
Niederbieber 4	*OxA-1136*	*10,480*	*130*	shaft fragment	*Cervus elaphus*	REJECT: post-LSE date	×	11; 16
Niederbieber 7	*OxA-2067*	*10,390*	*100*	bone	*Cervus elaphus?*	REJECT: post-LSE date	×	11; 16
large bovids (*Bison priscus* / *Bos* sp. / *Bos primigenius*), n = 3								
Gönnersdorf II	OxA-V-2223-41	13,095	55	scapula	*Bison* sp.		16,360-15,600	6
Andernach, upper horizon 2	GrA-16991	12,040	70	shaft fragment	*Bos* sp./ *Bison* sp.		14,090-13,690	6; 18
Andernach, upper horizon 2	OxA-998	11,370	160	bone	*Bos* sp./ *Bison* sp.		13,550-12,910	6; 11; 18
others, n = 3								
Kettig	*GrA-14762*	*11,210*	*60*	metacarpal	*Capreolus capreolus*	REJECT: calcined bone	×	16
Miesenheim 2	*OxA-2608*	*10,820*	*110*	bone	*Capreolus capreolus*	REJECT: post-LSE date	×	11
Andernach, upper horizon 2	GrA-16987	12,050	70	scapula	*Castor fiber*		14,100-13,700	6; 17

Tab. 75 (continued)

site	lab. no.	years ^{14}C-BP	± years	material	species	comment	years cal. b2k	ref.
pine (*Pinus* sp.), n=3								
Conty	Ly-284	11,540	80	bark	*Pinus* sp.	PROBLEMATIC: reworked material?	13,560-13,200	1
Le Closeau, top horizon, *locus* 25	Ly-564	10,885	85	charcoal	*Pinus* sp.	PROBLEMATIC: uncertain association	12,990-12,630	2
Le Closeau, top horizon, *locus* 25	Ly-563	10,755	90	charcoal	*Pinus* sp.	PROBLEMATIC: uncertain association	12,800-12,560	2
willow family (*Saliceae*) / willow (*Salix* sp.) / poplar (*Populus* sp.), n=0								
birch (*Betula* sp.), n=0								
reindeer (*Rangifer tarandus*), n=14								
Thèmes-Ferme de la Bouvière, Yonne/Bourgogne	OxA-7342	14,340	130	tooth	*Rangifer tarandus*	PROBLEMATIC: palimpsest or contaminated samples?	17,970-17,090	3
Rinxent	OxA-1343	13,030	120	antler	*Rangifer tarandus*		16,500-15,340	4-5
Le Grand Canton, sector 2, upper horizon	Gif-9608	12,880	80	bone	*Rangifer tarandus*		15,810-15,290	5-7
Étigny-Le Brassot, southern *locus*	OxA-10096	12,630	90	bone	*Rangifer tarandus*		15,540-14,540	5; 8-9
Pincevent level IV.20, section 27	OxA-148	12,600	200	bone	*Rangifer tarandus*		15,800-14,000	5; 10
Pincevent level IV.21.3, section 25	OxA-149	12,400	200	bone	*Rangifer tarandus*		15,480-13,760	5; 10
Pincevent level IV.21.3, section 25	OxA-177	12,300	220	bone	*Rangifer tarandus*		15,390-13,590	5; 10
Pincevent level IV, habitation 1	Erl-6786	12,277	96	bone	*Rangifer tarandus*		14,810-13,850	5; 10-11
Marsangy, conc. D14	OxA-740	12,120	200	tooth	*Rangifer tarandus*	found in C14	14,920-13,440	5; 12
Pincevent level IV.21.3, section 25	OxA-176	12,000	220	bone	*Rangifer tarandus*		14,640-13,320	5; 10
Marsangy, conc. N19	OxA-178	11,600	200	antler	*Rangifer tarandus*	found in P16; PROBLEMATIC: reindeer in France?	13,850-13,050	5; 13
Le Tureau des Gardes, *locus* 6	Ly-6989	11,560	100	bone	*Rangifer tarandus*	attributed to the Late Magdalenian; PROBLEMATIC: reindeer in France?	13,590-13,190	5-6
Marsangy, conc. D14	*OxA-505*	*9,770*	*180*	bone	*Rangifer tarandus*	found in B12, REJECT: insufficient collagen	×	5; 12

Tab. 76 ^{14}C dates from selected plant and faunal samples from northern France. If the sub-assemblage is known from which the sample originated, the sub-assemblage is given behind the site name. * dates which were pretreated by the use of ion-exchanged gelatin (Lab.code: AI) in the Oxford series (cf. Jacobi/Higham 2009, 1896); ** dates which might be contaminated due to the use of a method leaving traces of a humectant in the collagen (Lab.code: AF*) in the Oxford series (cf. Higham et al. 2007, S55 & S2). Doubtful dates are shaded grey. Rejected dates are shaded grey and set in italics and, in addition, the main reason for rejection is given in comment. For further details see p. 259-263, p. 265-269, and p. 412-429. The dates were calibrated with the calibration curve of the present study (see p. 358-364) and the calibration program CalPal (Weninger/Jöris/Danzeglocke 2007). The result range of 95 % confidence is given for the calibrated ages (years cal. b2k). – References (ref.): **1** Ponel et al. 2005; **2** Bodu 2004; **3** Higham et al. 2007; **4** Hedges et al. 1988a; **5** Fagnart 1997; **6** Lang 1998; **7** Hedges et al. 1993b; **8** Stevens/Hedges 2004; **9** Lhomme et al. 2004, 732; **10** Bodu et al. 2009b; **11** Leroi-Gourhan/Brézillon 1966; **12** Gowlett et al. 1986b; **13** Gowlett et al. 1986a; **14** Gilot 1997; **15** Deloze et al. 1999, 38; **16** Bignon 2006; **17** Bodu/Debout/Bignon 2006; **18** Hedges et al. 1997; **19** Stevens/Hedges 2004; **20** Bodu/Valentin 1997, 343; **21** Coudret/Fagnart 2004; **22** Fagnart/Coudret 2001.

site	lab. no.	years ^{14}C-BP	± years	material	species	comment	years cal. b2k	ref.
Marsangy	*Lv-1215*	*5,000*	*350*	antler	*Rangifer tarandus*	REJECT: contamination?	×	14
horse (*Equus* sp.), n=22								
Le Grand Canton, sector 1	OxA-3139*	12,650	130	phalange	*Equus* sp.	same sample as OxA-3671	15,660-14,420	5-7; 15
Le Tureau des Gardes, *locus* 10	AA-44216	12,520	130	radius	*Equus* sp.		15,510-14,070	5; 16
Le Closeau, lower horizon, *locus* 33	GrA-18860	12,510	80	diaphyse of longbone	*Equus* sp.		15,340-14,260	5; 17
Le Closeau, lower horizon, *locus* 46	GrA-11664 (Ly-789)	12,350	60	tibia	*Equus* sp.		14,800-14,040	5; 17
Le Tureau des Gardes, *locus* 6	Ly-6988	12,290	90	bone	*Equus* sp.	attributed to the Late Magdalenian	14,810-13,890	5-6; 16
Étiolles	OxA-5995*	12,250	100	bone	*Equus* sp.	from horse concentration A17?	14,790-13,790	5; 18
Le Grand Canton, sector 2, lower horizon	Gif-9606	12,195	130	bone	*Equus* sp.	attributed to the Late Magdalenian	14,800-13,640	5-6; 15
Le Tureau des Gardes, *locus* 10	AA-44214	12,170	130	phalange	*Equus* sp.		14,680-13,640	5; 16
Le Tureau des Gardes, *locus* 10	AA-44215	12,160	120	humerus	*Equus* sp.		14,540-13,660	5; 16
Marsangy, conc. N19	OxA-8453	12,140	75	tooth	*Equus* sp.	found in M16	14,220-13,780	5; 19
Le Closeau, lower horizon, *locus* 4	OxA-5680* (Ly-166)	12,090	90	diaphyse	*Equus* sp.		14,180-13,700	5; 17; 19-20
Le Grand Canton, sector 2, upper horizon	Gif-9607	12,080	115	bone	*Equus* sp.		14,250-13,650	5-6; 15
Le Grand Canton, sector 2, upper horizon	*Gif-9609*	*11,420*	*100*	bone	*Equus* sp.	REJECT: low collagen content	×	5-6; 15
Le Grand Canton, sector 1	*OxA-3671**	*11,030*	*105*	phalange, protein	*Equus* sp.	REJECT: fraction; same sample as OxA-3139	×	5-7; 15
Belloy-sur-Somme, upper horizon, north	OxA-724	10,260	160	tooth	*Equus* sp.		12,650-11,370	12
Belloy-sur-Somme, upper horizon, north	OxA-722	10,110	130	tooth	*Equus* sp.	re-dating of OxA-461	12,260-11,220	12
Belloy-sur-Somme, upper horizon, north	OxA-723	9,890	150	tooth	*Equus* sp.		11,960-11,000	12
Conty, lower horizon	OxA-7653	9,815	60	bone	*Equus* sp.	PROBLEMATIC: origin – top horizon?	11,380-11,180	19
Belloy-sur-Somme, upper horizon, north	OxA-462	9,720	130	tooth	*Equus* sp.		11,470-10,710	12
Le Closeau, upper horizon, *locus* 26	*GrA-10886*	*9,070*	*70*	jugale	*Equus* sp.	REJECT: younger intrusion?	×	5
Belloy-sur-Somme, upper horizon, north	*OxA-461*	*8,010*	*110*	tooth	*Equus* sp.	REJECT: low collagen content; redating available (OxA-722)	×	12
Le Closeau, upper horizon, *locus* 26	*GrA-10885*	*5,290*	*90*	jugale	*Equus* sp.	REJECT: younger intrusion?	×	5
elk (*Alces alces*), n=0								
wild boar (*Sus scrofa*), n=6								
Le Closeau, lower horizon, *locus* 46	AA-41881	12,423	67		*Sus scrofa*		15,160-14,080	5; 16; 20
Le Closeau, lower horizon, *locus* 46	GrA-18816	12,350	70	femur	*Sus scrofa*		14,830-14,030	5; 16-17; 20
Le Closeau, lower horizon, *locus* 4	GrA-18762	11,640	70	diaphyse	*Sus scrofa*	PROBLEMATIC: association doubtful	13,670-13,270	5; 17
Le Closeau, lower horizon, *locus* 4	GrA-18697	10,240	150	phalange	*Sus scrofa*	PROBLEMATIC: association doubtful	12,600-11,360	5; 17
Le Closeau, lower horizon, *locus* 46	*GrA-18763*	*6,420*	*110*	femur	*Sus scrofa*	REJECT: younger intrusion?	×	5; 17

Tab. 76 (continued)

site	lab. no.	years ^{14}C-BP	± years	material	species	comment	years cal. b2k	ref.
Le Closeau, lower horizon, *locus* 4	*GrA-18701*	*5,380*	*100*	phalange	*Sus scrofa*	REJECT: association doubtful; younger intrusion?	×	5; 17
red deer (*Cervus elaphus*), n=4								
Le Closeau, lower horizon, *locus* 46	*GrA-11665 (Ly-790)*	*12,360*	*60*	femur	Cervidae	REJECT: species determination unclear	14,820-14,060	2; 17
Le Closeau, lower horizon, *locus* 56	*GrA-18819*	*12,340*	*70*	radius	Cervidae	REJECT: species determination unclear	14,810-14,010	2; 17
Conty, palaeontological sample	OxA-6257* (Ly-286)	12,300	120	antler	*Cervus elaphus*		14,990-13,830	1
Saleux, trench A17	Beta-170494	11,180	50	diaphyse	*Cervus elaphus*		13,180-12,940	21
large bovids (*Bison priscus* / *Bos* sp. / *Bos primigenius*), n=14								
Thèmes-Ferme de la Bouvière	OxA-8049	13,580	180		bovid	PROBLEMATIC: palimpsest or contaminated samples?	17,210-16,290	3
Conty, lower horizon	OxA-6151* (Ly-260)	11,890	90	metacarpal	*Bos primigenius*		13,960-13,480	1; 5
Hangest-sur-Somme III.1, lower horizon	*OxA-4432* (Ly-22)*	*11,660*	*110*	molars	*Bos primigenius* / *Equus* sp.?	REJECT: species attribution unclear	13,760-13,240	5
Saleux-Les Baquets (244)	GrA-18832 (Ly-1566)	11,640	70	metapodial	*Bos primigenius*		13,670-13,270	22
Hangest-sur-Somme III.1, lower horizon	*OxA-4936* (Ly-86)*	*11,630*	*90*	molars	*Bos primigenius* / *Equus* sp.?	REJECT: species attribution unclear	13,700-13,220	5
Conty, lower horizon	OxA-6148* (Ly-257)	11,620	90	diaphysis	*Bos primigenius*		13,670-13,230	1; 5
Conty, lower horizon	OxA-6149* (Ly-258)	11,560	90	diaphysis	*Bos primigenius*		13,570-13,210	1; 5
Conty, lower horizon	OxA-6150* (Ly-259)	11,410	80	tibia	*Bos primigenius*		13,440-13,080	1; 5
Saleux-Les Baquets (234)	GrA-15945 (Ly-1141)	11,200	70	femur	*Bos primigenius*		13,210-12,930	21
Saleux-Les Baquets (234)	GrA-15946 (Ly-1142)	11,160	70	M2 inf.	*Bos primigenius*		13,190-12,870	21
Saleux-La Vierge Catherine (114)	OxA-4932* (Ly-81)	11,010	80	diaphyse	*Bos primigenius*		13,060-12,700	5; 21
Hangest-sur-Somme III.1, upper horizon	OxA-4935* (Ly-85)	10,920	90	vertebra	*Bos primigenius*	PROBLEMATIC: the δ^{13}C value is surprisingly low	13,010-12,650	5
Saleux-La Vierge Catherine (114)	OxA-4933* (Ly-82)	10,800	140	diaphyse	*Bos primigenius*		13,030-12,470	5; 21
Hangest-sur-Somme II.1, lower horizon	Gif-9355	10,140	110	bone fragments	*Bos primigenius*		12,260-11,300	5
others, n=0								

Tab. 76 (continued)

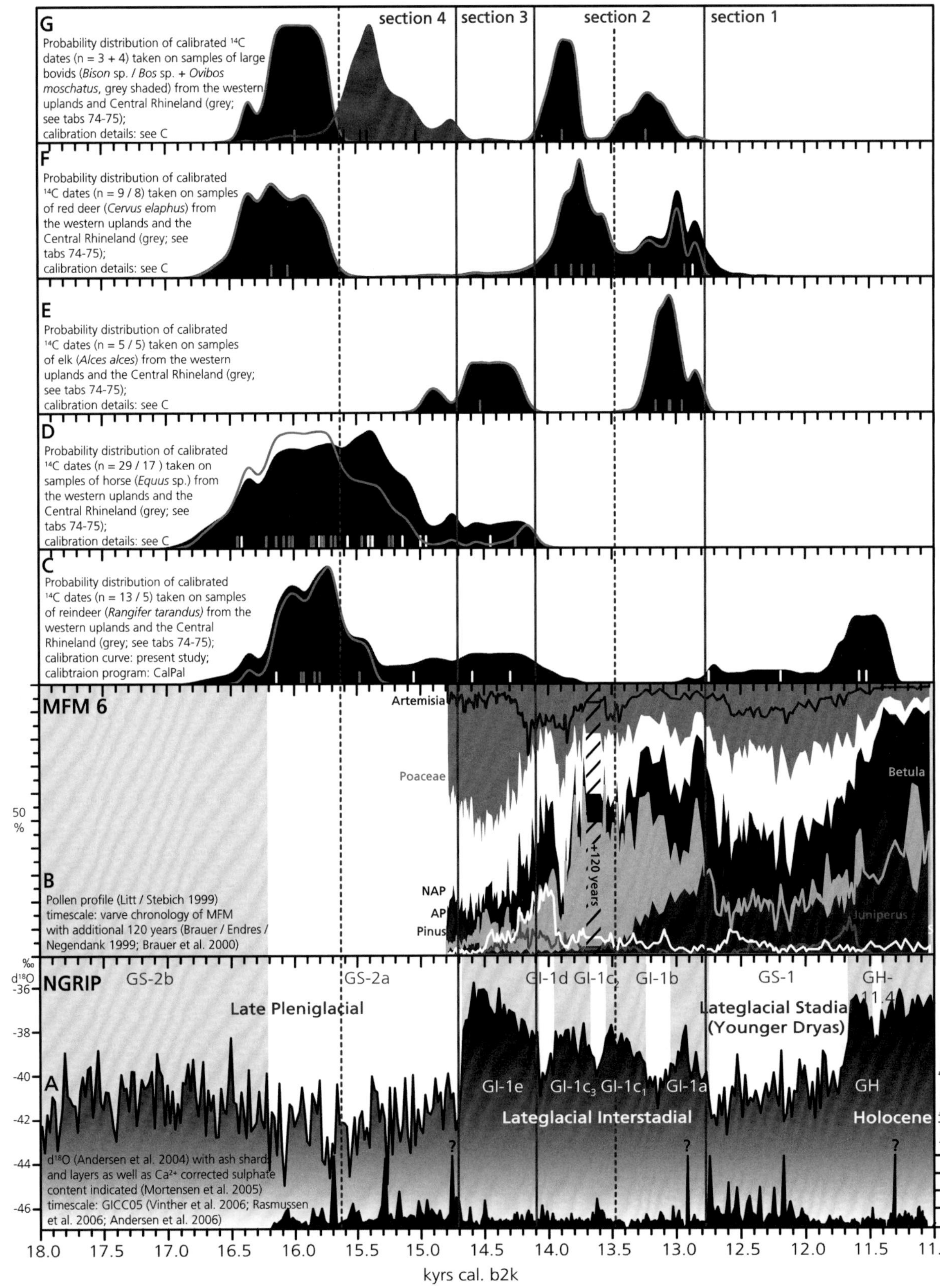

Fig. 60 Probability distribution of ^{14}C dates made on samples of the selected fauna species (**C-G**) from the western uplands (see **tab. 74**) and, in particular, the Central Rhineland (grey lines; see **tab. 75**) in comparison to the pollen diagram from the Meerfelder Maar (**B**) and the oxygen isotope record as well as the gypsum corrected sulphate content of NGRIP (**A**; see **fig. 53**). **G** grey shaded area represents the probability distribution for the arctic species musk ox (*Ovibos moschatus*).

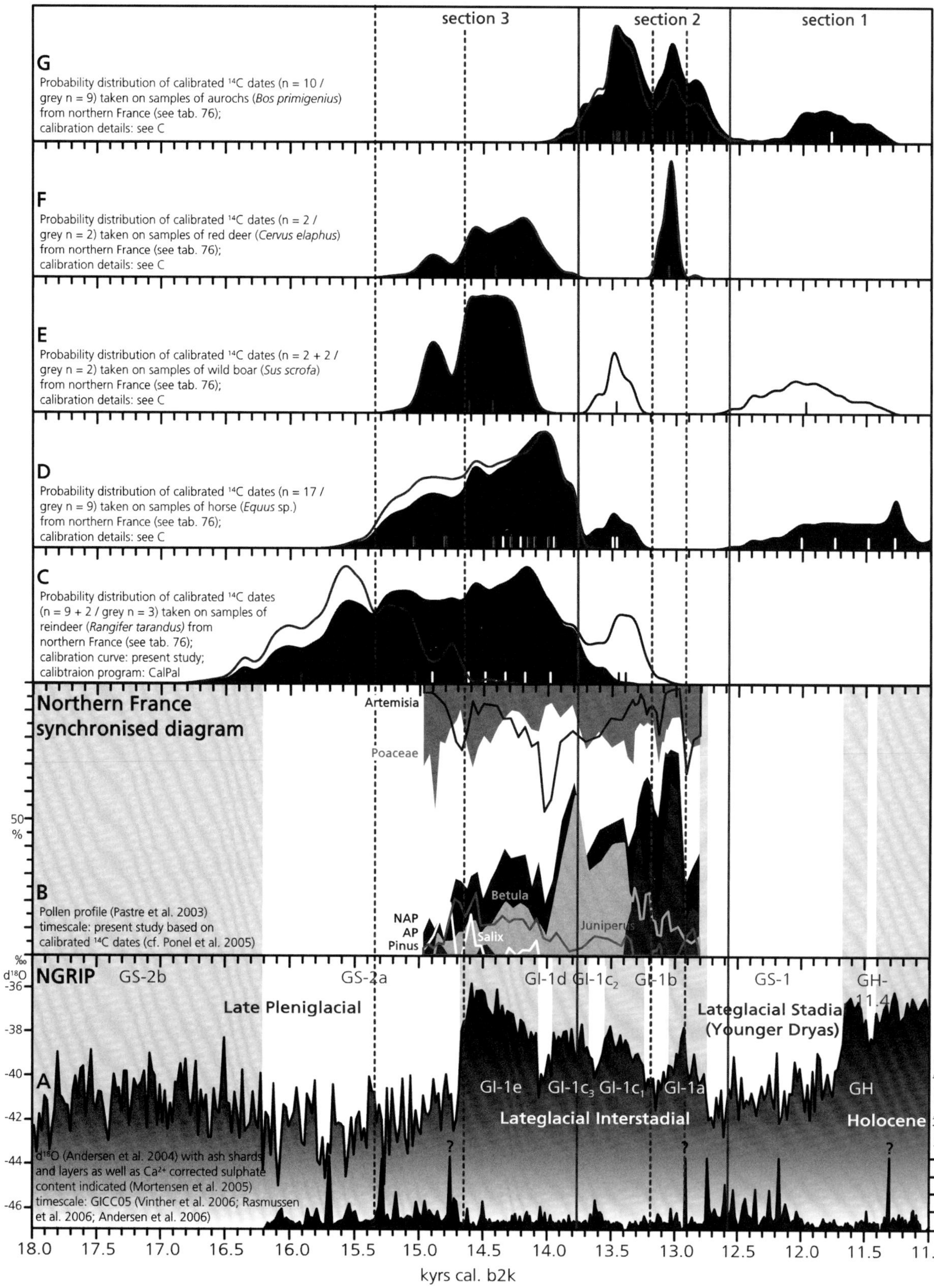

Fig. 61 Probability distribution of ^{14}C dates made on samples of the selected fauna species (**C-G**) from the Paris Basin (see **tab. 76**) in comparison to the pollen diagram from the Paris Basin (**B**) and the oxygen isotope record as well as the gypsum corrected sulphate content of NGRIP (**A**; see **fig. 53**). In **C-G** black lines with white background refer to problematic material and grey lines refer to more recently made, reliable dates (see text).

ent from the distribution of horses. This spatial context suggested no chronologically younger phase related to the presence of reindeer. Moreover, the Belgian faunal assemblages yielded, besides musk ox, remains from steppe wisent and aurochs (*Bos primigenius*; Germonpré/Sablin 2002) which prefers more forested habitats. The remains of steppe wisent and aurochs from the Belgian assemblages were not directly dated but according to the stratigraphic evidence these species occurred in the western uplands quasi-contemporaneously with musk ox. However, the directly dated examples from the Central Rhineland yielded distinct results for steppe wisent and aurochs and both species were never found together in a single assemblage. Therefore, direct dates from the western uplands on these large bovid species could help to determine further the reliability of these collections and/or the possibility of more varied environment in the western upland zone. Thus, these direct dates could alter the probability distribution of large bovids from the western uplands and, consequently, change this criterion for a bisection of the Late Pleniglacial unit.
Furthermore, the bisection of section 4 was also established based on the distribution of red deer (*Cervus elaphus*) remains because material from Andernach was dated to the early part of this section. However, a presence of red deer was not necessarily proven by the dated remains because the relation of the dated bones with the Late Magdalenian occupation remained uncertain and, therefore, a contamination of the samples could not be excluded. Furthermore, the ascertained Late Magdalenian material consisted mainly of dental material which was considered as jewellery material and, thus, during the Late Pleniglacial red deer was possibly not present in the Central Rhineland (Sommer et al. 2008, 720-723).These problems arose when the probability distributions are interpreted without considering the context of the sampled material. In contrast, at the transition to section 3, problems relating to calibration relics occurred in the probability distributions. Extra-fluctuations suggested a possible earlier appearance of elk (*Alces alces*) and/or a continued presence of large bovids and horses. These additional peaks were due to the plateau in the calibration curve. Thus, these deflections demonstrate methodological problems which are due to the dependence on the calibration curve.
Section 3 represented a transitional faunal composition in which some dates of reindeer and horse occurred, supplemented by elk in the Central Rhineland. Directly dated material from red deer and larger bovids is unknown from this period. However, the total number of directly dated specimens as well as of ascertained archaeological sites decreased significantly in this period making a reliable discussion about the fauna of this period difficult. Besides the elk remains from the south-western area of Gönnersdorf, single horse remains from the Trou de Chaleux (Stevens et al. 2009a) and Andernach (Stevens et al. 2009b) as well as single reindeer remains from the Trou da Somme (Hedges et al. 1994; Lanting/van der Plicht 1996) and the Grotte du Coléoptère (Hedges et al. 1993b; Lanting/van der Plicht 1996) were dated to this transitional section. The material from the Trou da Somme including an antler point was dated at the Oxford laboratory during the use ion-exchanged gelatin (lab. code: AI) which could cause some contamintation (Burky et al. 1998; Higham/Jacobi/Bronk Ramsey 2006). Thus, besides a potentially contamintated date, the material could have been introduced to this region from elsewhere due to the artefact character of the piece. Moreover, post-depositional disturbances at this site make a reliable stratigraphic attribution impossible (Miller/Noiret 2009). The remaining reindeer date was made again on antler and, thus, the presence of the species near the site can be questioned. The date fell onto the calibration plateau at the transition from GS-2a to GI-1e and, consequently, this reindeer material is dated probably to the end of section 4. Two further dates from this horizon proded an older and a younger result. The younger date was produced from bone splinters that could indicate that either material was relocated in the cave sediments and/or disturbed during excavations in the early 20th century. The older date was from a horse bone that bore cut marks and, thus, represented a reliable indicator of the Late Magdalenian occupation. Thus, the presence of reindeer in the western uplands during the period of section 3 still remains to be ascertained. The dates on

the horse bones were made during a project by Rhiannon Stevens about the nitrogen and carbon isotopic signatures of horse bones in order to reconstruct the local environment (Stevens et al. 2009b; Stevens et al. 2009a). Thus, they were thoroughly analysed and both samples were of a different isotopic signature than the Late Magdalenian samples. The isotopic composition of the piece in Chaleux suggested that this horse relied on a more mixed diet than the other horse remains in this assemblage. Consequently, the authors assumed that the horses which yielded the younger samples lived in a different environment and considered, for both sites, a younger episode as a possible explanation (Stevens et al. 2009b, 143f.; Stevens et al. 2009a, 661).
However, these later episodes have remained, thus far, undetected in both archaeological assemblages. Therefore, another possible explanation involves undetected contamination of the samples. Without further samples of comparable age and/or comparable isotopic signature, these intermediate phases cannot be verified. Consequently, the elk from Gönnersdorf remains the only reliably dated species for this section in the Central Rhineland and the western uplands. Thus, this section appears more like a gap of data than an actual transitional period. This gap in the record is not clearly shown by the probability distribution but no assemblage from the Central Rhineland or the western uplands was attributed unambiguously to GI-1d. Only the calibrated age range of the single date on horse from Andernach 2 could indicate human presence in the Central Rhineland across this period. However, this single date remains doubtful. In the western uplands, the calibrated age range of the Trou da Somme specimen encompassed this period but apart from this material, no other assemblage was attributed to this or the immediately following period. A directly dated red deer mandible from Presle (Hedges et al. 1988b) which was calibrated to the transition towards the Lateglacial Stadial is the further evidence of faunal development in this area. Hence, the temporal classification of the directly dated faunal remains from the western upland zone revealed a gap of directly dated specimens during the mid-to late Lateglacial Interstadial (GI-1c_2, GI-1c_1, and GI-1b) and the dates for an occupation during the early Lateglacial Interstadial (GI-1e and GI-1d) were also sparse. Consequently, the process of faunal change remains poorly understood for this area.
However, the major change in the probability distributions is observable in the following section 2 which can be further sub-divided in an earlier and a later part. Already during the earlier phase, no dates for the previously dominant species, reindeer and horse, were recorded. Instead, dates of red deer and large bovids, usually determined as aurochs, become frequent. These species indicate a light forest environment. Thus, the cold event GI-1d marked the end of a steppe landscape in the eastern sub-area and section 2 represented a fully established Lateglacial Interstadial environment. In general, the number of dated specimens increased in this section but the number of archaeological sites remained low in this earlier part. Only from the upper horizon of Andernach 2 were directly dated samples obtained, proving human occupation in this early phase after the faunal exchange.
In the later part of section 2, the total number of ^{14}C dates increased only gradually but the number of assemblages attributed to this period increased significantly. This younger sub-unit of the Lateglacial Interstadial section was related to the stabilisation of the climate and a forested environment in the western upland zone (**fig. 60B**). The red deer and aurochs dates were supplemented by directly dated elk remains. However, elk was also determined in the material of the upper horizon from Andernach 2 but not dated. Thus, even though a continued presence of elk in the Central Rhineland and the western uplands since the onset of the Lateglacial Interstadial remains a matter of discussion, due to the scarcity of data attributed to the early Lateglacial Interstadial, it had been regularly present in the Central Rhineland since the onset of the mid-Lateglacial Interstadial.
Section 2 of the probability distributions ended at the onset of the Lateglacial Stadial but in the Central Rhineland the preservation of organic remains decreased significantly with the absence of the LST. Only at

Bad Breisig were calcined remains of red deer, roe deer, and horse recovered. No Lateglacial faunal material younger than this assemblage, from the transition of the Lateglacial Interstadial to the Lateglacial Stadial, has been recovered in the Central Rhineland. Also, in the western uplands, reliably dated archaeological assemblages vanished almost completely at the transition to the Lateglacial Stadial but several Ahrensburgian assemblages such as Remouchamps and Kartstein were attributed to the younger half of Lateglacial Stadial. However, some of the directly dated reindeer remains from Remouchamps yielded some early Lateglacial Stadial ages. Besides reindeer, horse was also identified in this assemblage and also in the other, unambiguously later Ahrensburgian assemblages. These remains suggested the presence of some open grasslands. Furthermore, in some of these assemblages arctic fox (*Alopex lagopus*) was present indicating cold climatic conditions. From the Hohlen Stein in Kallenhardt a mixed assemblage was recovered that contained, alongside these three species also aurochs, wild boar, elk, beaver, red deer, and roe deer (Baales 1996). These species indicate that moist and forested areas were also found, occasionally, in the Belgian Ahrensburgian assemblages such as the beaver in the horizon 6B of the Grotte du Coléoptère or the wild boar in Fonds de Forêt (Dewez 1987). André Thévenin considered the presence of these species as an indication of the late Lateglacial Stadial to early Holocene age of these assemblages (Thévenin 1990), whereas Michael Baales assumed these species as Holocene intrusions (Baales 1996). A Holocene age for the wild boar from the Magdalenian collection of the Trou de Chaleux indicates that Holocene admixtures in these cave collections were possible (Hedges et al. 1994). Nevertheless, the detailed environmental study from the Kartstein and also the small mammal fauna from Coléoptère demonstrates the variability of the landscapes in the middle-range mountain areas during the Younger Dryas. This variability, due to topography, was also suggested for the same period in Britain (Price 2003). Thus, without direct dates of samples from the questionable species, their presence cannot be excluded, in particular due to the increasing indications for a return of moist and forested conditions to the western uplands during the younger half of the Lateglacial Stadial. The Holocene faunal development was not considered in the present study.

In the Paris Basin, three sections were established in the probability distribution.
The oldest section 3 and the intermediate section 2 were further sub-divided into three sub-sections.
Section 3 is dominated by horse and reindeer dates. Only a few direct dates of reindeer were attributed to the oldest sub-section. In the intermediate sub-section, horse supplemented the reindeer material, and the youngest sub-section correlates approximately with the section 3 in the western uplands. In contrast to the eastern sub-area, dates of wild boar and red deer supplemented horse and reindeer material during this period in the Paris Basin. The onset of this sub-section corresponds to the onset of the Lateglacial Interstadial and follows on an increase in the arboreal pollen that was due to increases in juniper, willow, and, later, birch pollen. Direct dates of the supplementary fauna species vanished with a decrease in the arboreal pollen and an increase in Artemisia, whereas the probability of horse dates increased. The end of this section correlated to a significant increase of the arboreal pollen and particularly the birch pollen and, thus, this environmental position is comparable to the onset of section 2 in the Central Rhineland.
Section 2 corresponds to the forested phase and was dominated by dated aurochs material in the northern French records. Thus, the major change in the faunal composition also occurred between the sections 3 and 2 in the Paris Basin. However, in northern France this transition occurred in the mid-Lateglacial Interstadial and, thus, 350 years later than in the Central Rhineland. In the oldest sub-section, the dated aurochs material was supplemented by some dates on horse and in the intermediate sub-section by red deer. In the youngest sub-section of section 2, direct dates were only made on aurochs remains.
Section 1 again corresponds to the fauna of the Lateglacial Stadial which consisted of horse and aurochs in the Paris Basin.

Nevertheless, possible problems with the quality of the ^{14}C dates on bones from the Paris Basin and in particular from the area near Marolles-sur-Seine makes this probability distribution problematic. Results were often younger than considered possible based on the stratigraphic and environmental evidence but thus far no consistent problem has been identified. However, direct comparison with the synchronised pollen diagram from the Paris Basin (**fig. 61B**) suggests that the probability distributions could be explained by the development of a denser plant environment. Since the synchronised pollen diagram from the Paris Basin was correlated to the NGRIP record also by using calibrated age ranges, the probability distributions and the pollen diagram are subject to the same methodological problems. Consequently, potential offsets resulting from methodological reasons are negligible. Thus, if contamination of the material is excluded, the comparable developments in the plant and the faunal environment indicate probable interrelations.
Nevertheless, the botanical macro-remains suggested a denser environment in most parts of northern France than in the Central Rhineland. Thus, even though correlation with the pollen diagram could explain the long persistence of a cold grass steppe adapted fauna, the previously outlined climatic and environmental development of the western sub-area clearly contradicts this longer presence. Furthermore, reliable dates on the mainly forest adapted fauna from the lower horizon in Le Closeau were older than many faunal remains from unambiguously Late Magdalenian assemblages. Considering the general vicinity, for example, of Étiolles and Le Closeau as well as the environmental indications from these sites (Rodriguez 1994; Olive 2004; Bignon 2009), a quasi-contemporaneity is improbable because of the numerous differences. Therefore, contamination of several ^{14}C-dated samples seems worth considering.
Possibly, the deposition in alluvial plains and, thus, the interaction with groundwater resulted in undetected diagenetic alterations of faunal material (Hedges/Millard 1995; Hedges/Millard/Pike 1995; Hedges 2002). These possible alterations have resulted in the younger ages. However, if dates made on potentially contaminated bulked samples or made in the early days of AMS, possibly with an insufficient pretreatment protocol, were rejected in general from the list, very few dates remained to develop the probability distributions (**fig. 61**, grey lines). However, the general sub-division would remain the same except for numerous dates on reindeer being rejected. According to this reduced list of dates, reindeer seemed to have vanished from the Paris Basin already with the first increase of arboreal pollen (**fig. 61C**). This final decrease of the probability distribution corresponded approximately to the decrease of reindeer dates in the Central Rhineland. According to this observation, reindeer might have shifted its range areas further northwards in the final phase of the Late Pleniglacial. Moreover, with the disappearance of reindeer herds, horse became more important but with the spread of light forests probably also on the higher plateaus during the mid-Lateglacial Interstadial. The subsequent importance in the probability distributions speaks in favour of a diachronic model of the occupation in the Paris Basin with horse dominated assemblages following on sites where mainly reindeer was hunted (Enloe 2000). However, the stratigraphic, environmental, and archaeological comparison of several sites such as Verberie, Étiolles, and various concentrations of Pincevent IV (Roblin-Jouve 1994; Rodriguez 1994; Bignon 2006) suggested that several of the assemblages dominated by reindeer or by horse as well as sites dominated by both species were deposited in the same period. Thus, a synchronic settlement with a complementary subsistence system appears probable during the Late Plenigacial in the Paris Basin (Bignon 2006). In addition, this contemporaneity suggests at least the occasional presence of reindeer also in the final part of the Late Pleniglacial. Moreover, the various comparisons singled out Marsangy as a clearly younger settlement episode. In the small faunal inventory from this site, reindeer was found besides horse and red deer remains indicating a more continued presence of single reindeer in this region into the early Lateglacial Interstadial. This scenario for the Paris Basin is comparable to the more continued, though sporadic indication of reindeer in the western uplands. In contrast to the Belgian cave sites, reindeer bones were found besides antler sustaining the presence of these animals as potential prey in northern France.

In summary, this detailed comparison of the probability distributions of the selected fauna species from the sub-areas with the contextual evidence of the species and the pollen diagram from the same areas helped to visualise relations between the faunal and the plant environment but a cautious interpretation of the distributions as well as the relations. In particular, small numbers of dated specimens from a few selected assemblages and/or the few selected species bias the outcome of these probability distributions significantly. Differential conditions of preservation due to the bone structure of the species, the taphonomic processes at the sites as well as in different climates and environments further influence the outcome of the distributions. Therefore, probability distributions of ^{14}C dates should not be interpreted without additional information such as the first and last appearance dates and/or supplementary analysis such as maps showing the distribution of the species. These supplementary information make an evaluation of the absence or presence of a species in the studied areas based on the development in surrounding areas possible. A combination of these data allows for a well rounded interpretation of the presence of selected fauna species in the study area.

Supplementary information on the presence of selected faunal species

In general, the ^{14}C dates used in this study represent the first appearance dates (FAD) of the selected species in the sub-areas after the LGM (GS-3, *c.* 27,550-23,350 years cal. b2k, *c.* 23,000-19,600 years ^{14}C-BP). The LGM was followed by a short-termed interstadial (GI-2, *c.* 23,350-22,900 years cal. b2k; *c.* 19,600-18,800 years ^{14}C-BP) and the tri-parted GS-2 (GS-2c: 22,900-20,900 years cal. b2k and *c.* 18,800-17,800 years ^{14}C-BP; GS-2b: 20,900-16,230 years cal. b2k and *c.* 17,800-13,200 years ^{14}C-BP; GS-2a: 16,230-14,687 years cal. b2k and *c.* 13,200-12,400 years ^{14}C-BP; Lowe et al. 2008; Rasmussen et al. 2008; Weninger/Jöris 2008). For horse and reindeer were older dates than those given for the presented sites known in or close to the sub-areas. Besides the FAD, the collection of dates (**tabs 74-76**) also provided last appearance dates (LAD) in the sub-areas for some of the glacial species such as reindeer, musk ox, or steppe wisent. In contrast, many of the species indicate temperate forest environments such as wild boar, beaver, aurochs, red deer, and roe deer, and inhabited the forests of the sub-areas until historic times (Ziswiler 1965; von Koenigswald 2004). Consequently, no LAD for these species are given by the present project.
In contrast to the probability distributions, the use of FADs and LADs does not allow for detecting short phases of absence for a species in the analysed areas (cf. Aaris-Sørensen 2009, 7). Therefore, tables of potential presence were made per the sub-area also including the indirectly dated material (**tabs 77-79**). These tables revealed a clear gap in the data from the western uplands during the mid- to late Lateglacial Interstadial. Furthermore, in the French record several species were not found suggesting biased preservation either due to the absence of the species, the different preservation properties of faunal remains, a selection of the prey and a better preservation or more probable determination in an archaeological assemblage, or a combination of these factors. In particular, species indicating very arctic conditions such as musk ox as well as indicators of very moist environments such as elk were missing. The latter is consistent with the assumptions based on the general comparison of the pollen profiles.

Supplementary information by further faunal indicators

Besides the selected fauna, further animal species can help to reconstruct the past environments. In particular, chironomid, malacological, and coleopteran data provide insights in the local micro-climate and vegetation (Rodriguez 1994; Lotter et al. 1997; Meyrick 2001; Ponel et al. 2005). Some of these analyses were

species	pre-GS-2a	GS-2a3	GS-2a2	GS-2a1	GI-1e	GI-1d	GI-1c3	GI-1c2	GI-1c1	GI-1b	GI-1a	GS-1b	GS-1a	GH
Rangifer tarandus	S	^{14}C / S	^{14}C / S	S / nd	nd	nd						nd	nd	nd
Equus sp.	^{14}C / S	^{14}C / S	^{14}C / S	S / nd	^{14}C / nd	^{14}C / nd	^{14}C	^{14}C / S	^{14}C / S	^{14}C / S	^{14}C / S	S / nd	nd	nd
Alces alces				^{14}C / nd	^{14}C / S	^{14}C / nd	S	S	^{14}C / S	^{14}C / S	^{14}C / S	nd	nd	nd
Sus scrofa		S	S	S / nd	nd	nd		S	S	S	S	S / nd	nd	nd
Cervus elaphus	S	^{14}C / S	S	S/ nd	S / nd	^{14}C / nd	^{14}C / S	^{14}C / S	^{14}C / S	^{14}C / S	^{14}C / S	^{14}C / S	nd	nd
Bison sp. / *Bison priscus*	S	^{14}C / S	S	nd	nd	nd						nd	nd	nd
Bison sp. / *Bos* sp.				S / nd	^{14}C / S	^{14}C / S	^{14}C / S	^{14}C / S	^{14}C / S	^{14}C / S	^{14}C / S	nd	nd	nd
Bos sp. / *Bos primigenius*				nd	nd	nd			S	S	S	nd	nd	nd
Ovibos moschatus				nd	nd	nd						nd	nd	nd
Alopex lagopus	S	S	S	S / nd	nd	nd						nd	nd	nd
Lepus timidus	S	S	S	nd	nd	nd						nd	nd	nd
Saiga tatarica		S	S	nd	nd	nd						nd	nd	nd
Capreolus capreolus				nd	S / nd	S / nd	S	S	^{14}C / S	^{14}C / S	^{14}C / S	S / nd	nd	nd
Castor fiber				nd	nd	^{14}C / nd	^{14}C / S	^{14}C / S	S	S	S	nd	nd	nd

Tab. 77 Presence of selected fauna and indicator species in the Central Rhineland. **^{14}C** directly dated specimens; **S** stratigraphic attribution and/or other species in the assemblage were ^{14}C-dated to this period; **nd** no or very sparse data from this period. Dark grey shaded: reliable presence; light grey shaded: possible presence. For Greenland events see **tab. 64**, additional sub-divisions were made: **pre-GS-2a** LGM to onset GS-2a (< 16,230 years cal. b2k/c. 13,400 ^{14}C-BP); **GS-2a$_3$** first part of GS-2a with generally lower oxygen isotope values (16,230 – c. 15,400 years cal. b2k/c. 13,400-12,900 ^{14}C-BP); **GS-2a$_2$** intermediate phase of GS-2a with fluctuating oxygen isotope values (c. 15,400-15,000 years cal. b2k/c. 12,900-12,600 ^{14}C-BP); **GS-2a$_1$** last phase of GS-2a with generally higher oxygen isotope values (c. 15,000 > years cal. b2k/c. 12,600 > ^{14}C-BP); **GS-1b** pre-Vedde Ash or mid-GS-1 (c. 12,770-12,170 years cal. b2k); **GS-1a** post-Vedde Ash or mid-GS-1 (c. 12,170-11,670 years cal. b2k).

regularly carried out on archaeological sites but often suffered from poor conditions of preservation for this type of material. However, due to the very local nature, most results from palaeoecological stratigraphies are not relevant for the present study except for the discussion on the patchiness of the landscape. The mollusc assemblages from the Late Magdalenian assemblages in the Paris Basin suggests partially temperate forested areas (Étiolles) and partially cold steppic environments (Pincevent, Marsangy; Rodriguez 1994). However, the Late Magdalenian occupation was associated at all sites with very low proportions or absence of aquatic species, also in otherwise moist areas. Thus, the malacological data also pointed to the significant aridity of this period. Nevertheless, the diversity of vegetation indicated by the molluscs increased at many of the Late Magdalenian sites suggesting that in the floodplain of sheltered river valleys the first shrub communities or light forests could develop. In the Central Rhineland, a malacological study at Gönnersdorf provided indications for a ground covering vegetation with rare mesophile elements and single species inhabiting steppe, hygrophile, or semi-forested habitats (Puisségur 1978). In general, the results were comparable to those from the French sites but various ground movements such as bioturbation, sediment flows, or hydrological infiltration at the site could have caused an admixture of diachronic elements.
In addition, some mammals and particularly small mammals allowed for further assumptions on the regional development of the climate and the vegetation (Cordy 1991; Rabenstein 1991; cf. Price 2003; Hernández Fernández/Peláez-Campomanes 2005; Fahlke 2009). Comparable to the plant material, these remains suffered in the Lateglacial Interstadial from dating uncertainties. However, the lack of directly dated plant

species	pre-GS-2a	GS-2a$_3$	GS-2a$_2$	GS-2a$_1$	GI-1e	GI-1d	GI-1c$_3$	GI-1c$_2$	GI-1c$_1$	GI-1b	GI-1a	GS-1b	GS-1a	GH
Rangifer tarandus	^{14}C / S	^{14}C / S	^{14}C / S	^{14}C / S	^{14}C / S	^{14}C / S	^{14}C / nd	^{14}C / nd	nd	nd	^{14}C	^{14}C / S	^{14}C / S	^{14}C
Equus sp.	^{14}C / S	^{14}C / S	^{14}C / S	^{14}C / S	^{14}C / S	^{14}C / S	S / nd	S / nd	nd	nd	S	S	S	S
Alces alces			S	S	S		nd	nd	nd	nd		S	S	S
Sus scrofa		S	S	S	S		nd	nd	nd	nd		S	S	^{14}C / S
Cervus elaphus		S	S	S	S	S	nd	nd	nd	^{14}C / nd	^{14}C	^{14}C / S	S	S
Bison sp. / *Bison priscus*		S	S	S	S	S	nd	nd	nd	nd				
Bison sp. / *Bos* sp.			S	S	S	S	S / nd	S / nd	nd	nd				
Bos sp. / *Bos primigenius*		S	S	S	S	S	nd	nd	nd	nd		S	S	S
Ovibos moschatus		^{14}C / S	^{14}C / S	^{14}C / S	^{14}C		nd	nd	nd	nd				
Alopex lagopus		S	S	^{14}C / S	^{14}C / S	^{14}C / S	nd	nd	nd	nd		S	S	S
Lepus timidus							nd	nd	nd	nd	S	S	S	S
Saiga tatarica		S	S	S	S	S	nd	nd	nd	nd				
Capreolus capreolus		S	S	S	S	S	nd	nd	nd	nd		S	S	S
Castor fiber		S	S	S	S	S	nd	nd	nd	nd		S	S	S

Tab. 78 Presence of selected fauna and indicator species in the western uplands. **^{14}C** directly dated specimens; **S** stratigraphic attribution and/or other species in the assemblage were ^{14}C-dated to this period; **nd** no or very sparse data from this period. Dark grey shaded: reliable presence; light grey shaded: possible presence. For Greenland events see **tab. 64** and for the additional sub-divisions see **tab. 77**.

species	pre-GS-2a	GS-2a$_3$	GS-2a$_2$	GS-2a$_1$	GI-1e	GI-1d	GI-1c$_3$	GI-1c$_2$	GI-1c$_1$	GI-1b	GI-1a	GS-1b	GS-1a	GH
Rangifer tarandus	^{14}C / S	^{14}C / S	^{14}C / S	^{14}C / S	^{14}C / S	^{14}C / S	^{14}C / S	S				nd	nd	
Equus sp.	S	S	^{14}C / S	^{14}C / S	^{14}C / S	^{14}C / S	^{14}C	^{14}C / S	^{14}C / S	S	S	^{14}C / S	^{14}C / S	^{14}C / S
Alces alces												nd	nd	
Sus scrofa	S	S	S	^{14}C / S	^{14}C / S	^{14}C / S	S	S		S	S	S / nd	nd	^{14}C / S
Cervus elaphus	S	S	S	^{14}C / S	^{14}C / S	^{14}C / S	^{14}C / S	S	S	^{14}C / S	^{14}C / S	S / nd	S / nd	S
Bison sp. / *Bison priscus*		S	S	S	S	S	S					nd	nd	
Bison sp. / *Bos* sp.	^{14}C / S	S	S	S	S	S	S	S				nd	nd	
Bos sp. / *Bos primigenius*			S	S	S	^{14}C / S	^{14}C / S	^{14}C / S	^{14}C / S	^{14}C / S	^{14}C / S	^{14}C / S	S / nd	S
Ovibos moschatus												nd	nd	
Alopex lagopus												nd	nd	
Lepus timidus			S	S	S	S	S					nd	nd	
Saiga tatarica												nd	nd	
Capreolus capreolus					S	S	S	S	S	S	S	nd	nd	
Castor fiber												nd	nd	

Tab. 79 Presence of selected fauna and indicator species in the Paris Basin. **^{14}C** directly dated specimens; **S** stratigraphic attribution and/or other species in the assemblage were ^{14}C-dated to this period; **nd** no or very sparse data from this period. Dark grey shaded: reliable presence; light grey shaded: possible presence. For Greenland events see **tab. 64** and for the additional sub-divisions see **tab. 77**.

remains and preserved pollen deposits affected in particular the Late Pleniglacial period from which some well dated faunal assemblages were preserved.
Some of the species in these assemblages help to qualify the surrounding landscape such as steppe pika (*Ochotona pusilla*) and saiga antelope (*Saiga tatarica*) which inhabit arid, continental steppe regions or collared lemmings (*Dicrostonyx* sp.) and arctic fox *(Alopex lagopus*) which are circumpolar species indicating glacial habitats (van Kolfschoten 1995). In contrast, beaver (*Castor fiber*) requires significant wetland conditions and roe deer (*Capreolus capreolus*) is a representative of forest communities. Directly dated samples from these specimens are sparse. In particular, the small mammal species were dated only on singular samples in Britain (Hedges et al. 1998a; Bronk Ramsey et al. 2002; cf. Price 2003). These samples produced age ranges from the Lateglacial Stadial to the Holocene and, thus, are not useful in the present study. Some medium to large sized mammals can be used for a comparison between the sub-areas.
However, none of the indicator species detailed above was dated directly in the French sub-area and, in fact, only roe deer was determined within the archaeological assemblages of the Grotte du Gouy and Conty. For many northern French sites, this absence could reflect the poor preservation but also on the sites with relatively good preservation, such as Pincevent, these species are absent and the species lists are in general not short. Thus, the limited diversity of the species in the French sub-area is partially due to a human prey selection bias (cf. Gaudzinski/Street 2003) and partially due to the absence of extreme conditions such as high arctic or very arid climates. The absence of beaver can perhaps be explained by poorer conditions of preservation for this smaller mammal in a more temperate environment with more intense pedogenetic processes and possibly due to the position of most sites along the Seine or Somme and at some distance to more suitable habitats for beavers such as smaller streams, brooks, or limnic environments. In contrast, several specimens were found and dated in the western uplands and the Central Rhineland and can be included in the following outline of the faunal development.

Development of the selected species in the sub-areas

In the following, the presence of the selected fauna species and also the FADs after the LGM and the LADs in the Lateglacial for these species are considered per sub-area. Moreover, directly dated specimens in the surrounding areas of Central Europe are also reckoned to understand the distribution of these species after the LGM.
Thus far, no indications were found for the presence of faunal species in the study area during the LGM and immediately after. However, some evidence for the presence of animal in north-western Europe during this period were found outside the study area (**fig. 62**). For instance, a FAD for horse and reindeer came from Wiesbaden-Igstadt which is located only a few kilometres south of the southern limit of the western upland zone. Horse remains from this assemblage were dated approximately to the end of the LGM with an oldest date of 19,320 ± 240 years ^{14}C-BP (OxA-7502; 23,610-22,450 years cal. b2k; Hedges et al. 1998a; cf. Street/Terberger 2004). Reindeer remains were also dated at the site but yielded younger results (**tab. 73**), even though the material was considered as quasi-contemporaneous with the horse remains according to the spatial distribution (Street/Terberger 1999, 264f.). In fact, the calibrated age ranges overlapped significantly (**fig. 63**). The older dates were congruent with dates on reindeer remains from the middle horizon of the Swiss Kastelhöhle-Nord (Street/Terberger 2004) and dated to the late GS-3 and early GI-2 (approximately 24,000 to 22,350 years cal. b2k). The younger dates made on reindeer from Wiesbaden-Igstadt are congruent with two dates made on human remains from the Mittlere Klause and fall calibrated to GI-2 (c. 23,050 to 21,350 years cal. b2k), a less rigid period (Street/Terberger/Orschiedt 2006; cf. Sánchez

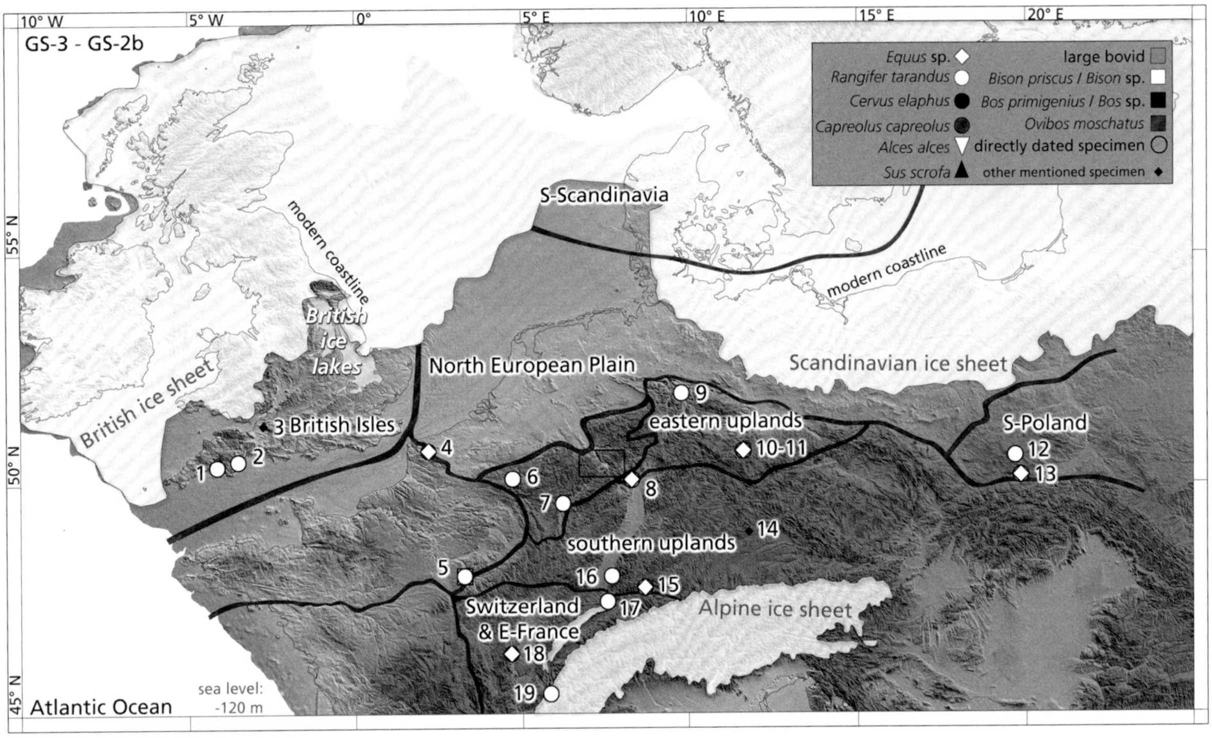

Fig. 62 Map of sites yielding material of selected fauna species during the late LGM and post-LGM to pre-GS-2a and sub-areas outside the study area. **1** Reindeer Rift; **2** Kent's Cavern; **3** Soldier's Hole; **4** Hallines; **5** Thèmes-Ferme de la Bouvière; **6** Trou des Blaireaux; **7** Schlaederbach valley; **8** Wiesbaden-Igstadt; **9** Aschenstein; **10** Ranis-Ilsenhöhle; **11** Kniegrotte; **12** Maszycka Cave; **13** Deszczowa Cave; **14** Mittlere Klause; **15** Kesslerloch; **16** Munzingen; **17** Kastelhöhle-Nord; **18** Solutré; **19** Abri de la Fru. – For further details see text.

Goñi 1991; Delpech 2012). The Mittlere Klause material, the archaeological assemblages from Wiesbaden-Igstadt (Street/Terberger 1999) and Kastelhöhle-Nord (Terberger/Street 2002) provided the evidence of a first return of human hunters into Central Europe after the LGM, presumably during a milder interstadial period. Moreover, these dates indicated that animals such as reindeer and horse could range in western Central Europe around GI-2.

A single date made by the Zürich laboratory on an indeterminate bone from Wiesbaden-Igstadt produced a somewhat younger date (UZ-3768/ETH-13380: 17,210 ± 135 years ^{14}C-BP; Street/Terberger 1999). When calibrated, this date fell to approximately GS-2c (21,090-20,290 years cal. b2k). Though similar to the reindeer dates from the site, the date was younger and not congruent with the group of Oxford dates. Since it is a single result and contamination was considered a possible source of error at the site (Street/Terberger 1999, 267), this date is considered a doubtful outlier in this study.

Besides these archaeological materials, unmodified reindeer remains from the Aschenstein assemblage east of the western upland zone dated to this late LGM and GI-2 period (Terberger et al. 2009). Furthermore, a red deer calcaneum from the French site Solutré (Pestle/Colvard/Pettitt 2006), a cow (*Bos* sp.) tibia from the British Soldier's Hole (Gowlett et al. 1986b), and a reindeer antler from the Polish Deszczowa Cave (Cyrek et al. 2000) were also dated to this period.

In the Aschenstein collection, further reindeer material with traces of human use was dated to the transition from the Late Pleniglacial to the early Lateglacial Interstadial (Terberger et al. 2009). The assemblage originated from a mid-20th century excavation and the spatial and stratigraphic position of the material was not documented. Therefore, the attribution of further material, also from other species such as horse, steppe wisent, musk ox, mammoth, arctic fox, arctic hare, and several bird species found at the site remained un-

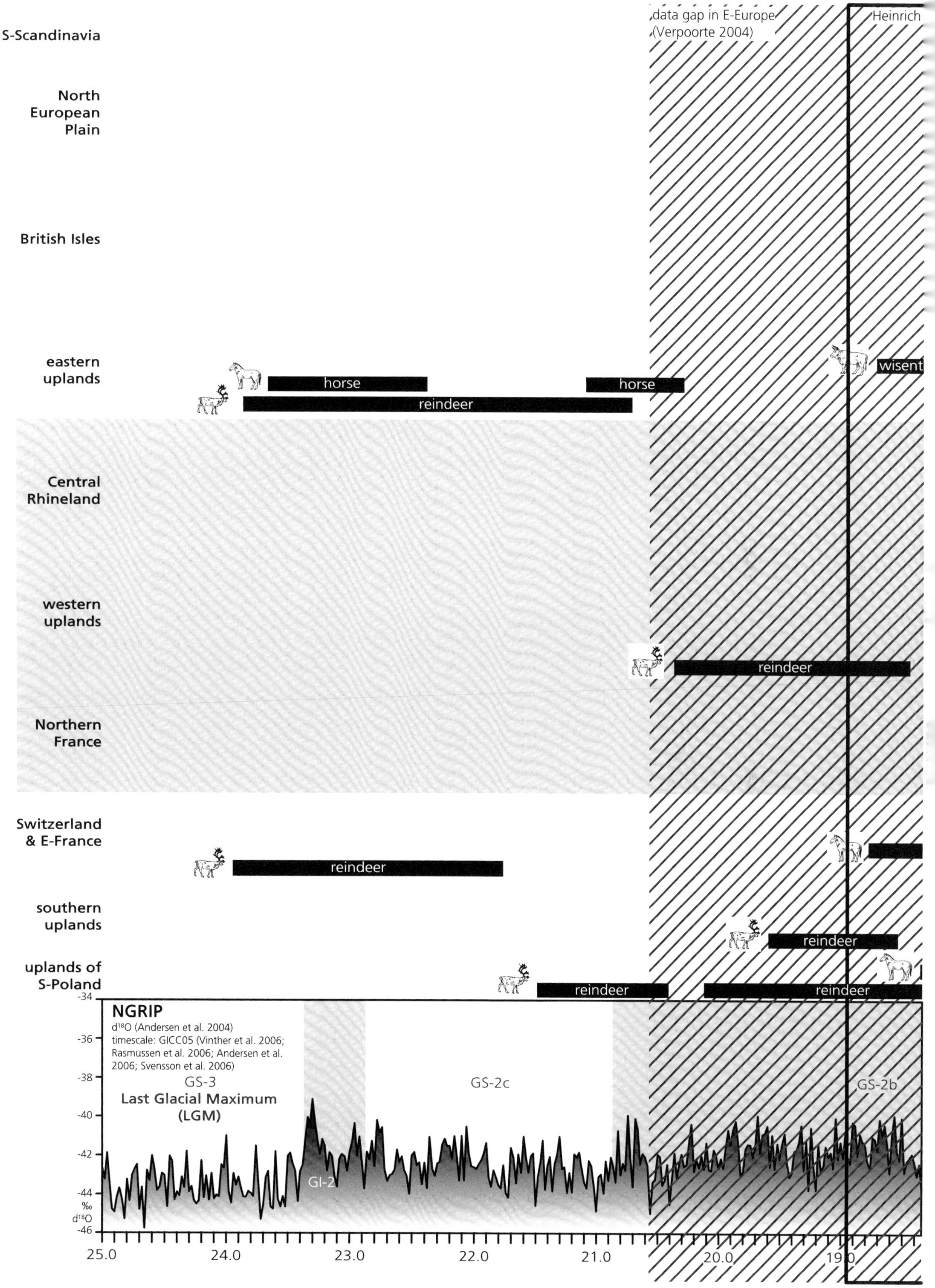

Fig. 63 Presence of selected fauna species in north-western Europe based on calibrated ^{14}C dates (black background) and indirect evidence (grey background; only in study area) in relation to the oxygen isotope record of NGRIP (see **fig. 53**).Grey shaded areas represent periods of more interstadial values than the surrounding values (for instance, the values in GS-2b are more interstadial than in GS-2a but

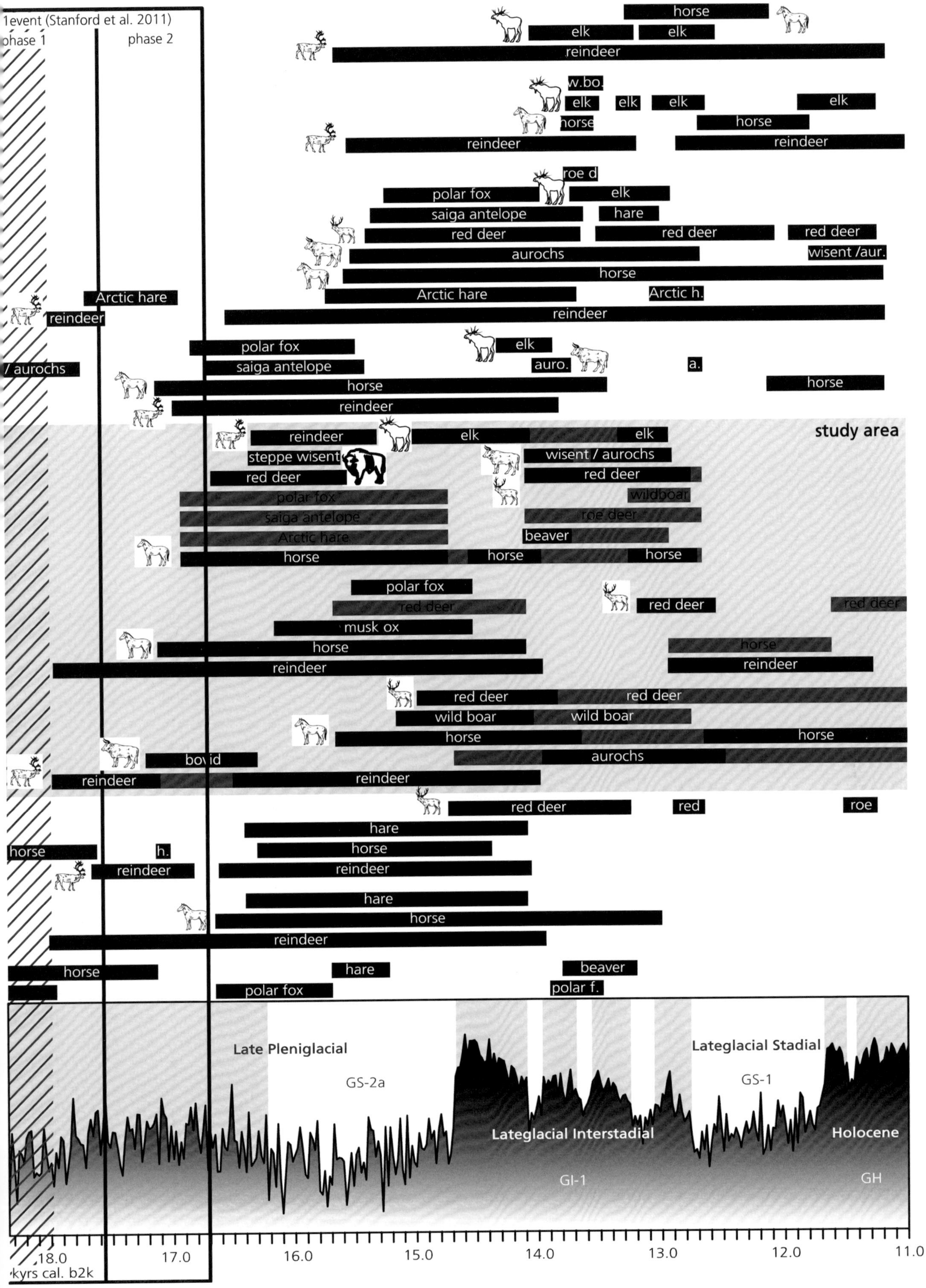

still these values are as stadial as the values in GS-1). The event limits are set according to **fig. 29** and Blockley et al. 2012, fig. 1. Note that some calibrated age ranges were capped due to the additional stratigraphic evidence such as the cover by the LST. – For further details see text.

certain. However, many pieces appeared to represent natural intrusions which were deposited during the LGM to the Late Pleniglacial and some remains such as the ones of roe deer were probably even younger elements (Terberger et al. 2009). Consequently, only reindeer can thus far be ascertained for the late LGM to early post-LGM period at this site.

The reliability of the Solutré specimen is, for various reasons, difficult to estimate: The sample was dated at the Oxford laboratory in the period in which some contamination in the ultra-filtration process still occurred but the date is congruent with previous dates from the Solutrean deposits (Montet-White/Evin/Stafford 2002). However, the piece was found in a collection from the 19th century and a relation to one of the hearths from the reindeer age was noted, although this hearth may not necessarily be equivalent to the Solutrean. Furthermore, the small faunal assemblage from the Solutrean was poorly preserved and in a previous examination, no red deer remains were determined (Olsen 1989). Thus, the sample might represent a single piece which could have reached the site as an element of the provision brought to the site from elsewhere, for example from a south-western French refuge (cf. Sommer et al. 2008). Therefore, the context of this single evidence has to be ascertained before it can be accepted as evidence for the presence of red deer in this area during this period.

The specimen from Soldier's Hole was found in a lower deposit (unit 4, spit 16) at the site and the date was made on amino acids in an early phase of the Oxford accelerator unit (cf. Burleigh 1986). Other samples from this unit produced considerably older ages suggesting a pre-LGM accumulation. An admixture of material from different periods could be a possible explanation but a contamination of the sample cannot be entirely excluded. However, the Cheddar Gorge where the Soldier's Hole was located (**fig. 62**) was only at a distance of some 35 km south-east of the maximum extent of the British glaciers during the LGM. If a fast regression of the ice sheet and the surrounding permafrost zone was considered possible in this area a reestablishment of the environment to provide a suitable habitat for large grazers such as red deer still appears improbable. Therefore, this evidence is doubted in this project.

The reindeer antler from horizon VIII in the southern Polish Deszczowa Cave produced the following date: Gd-10212: 17,480 ± 150 years ^{14}C-BP; 21,490-20,410 years cal. b2k (Cyrek et al. 2000; Wojtal 2007). This result is comparable in age to the doubted date from Wiesbaden-Igstadt. In contrast, a reindeer long bone from the same deposit produced a younger date and increased the age range of this deposit into GS-2b (Gd-9464: 16,150 ± 280 years ^{14}C-BP; 20,120-18,720 years cal. b2k; Cyrek et al. 2000; Wojtal 2007). Furthermore, dates on bones of woolly rhinoceros (*Coelodonta antiquitatis*; Nadachowski et al. 2009) and willow ptarmigan (*Lagopus lagopus*; Lorenc 2006) produced LGM results. Presumably, the position of the samples within the thick horizon as well as the location of the samples on the site (inside or outside the cave) could explain some discrepancy (Lorenc 2006, 47 f.). Even though the antler was found near a structure interpreted as hearth, it was unmodified and not unambiguously related to human activity. Nevertheless, besides the younger horse date from Wiesbaden, this date represents the only possible result determined to species in western Central Europe for the period between 17,500 and 16,500 years ^{14}C-BP (*c.* 21,250-19,500 years cal. b2k). Calibrated, this period relates to the transition from GS-2c to GS-2b. Within GS-2b a rapid rise in sea-level occurred and also the onset of the Heinrich 1 event in the broad sense is related to this event (cf. Stanford et al. 2011b, fig. 5).

In Eastern Europe, a period of particularly sparse evidence for human settlement followed the LGM dating between 17,000 and 15,000 years ^{14}C-BP (Verpoorte 2004). This period did not correlate to the onset of Heinrich 1 event as could be suggested but occurred shortly after the onset of GS-2b. Due to significant loess deposition, the reduced river activities, and the decreasing vegetation cover (cf. Fletcher et al. 2010), Alexander Verpoorte suggested that the decreasing evidence relates to a retreat of humans triggered by an increase in aridity (Verpoorte 2004, 263). Besides the aridity, seasonality and wind intensity could have been

additionally limiting factors (cf. Dietrich/Seelos 2009) that could have had different effects in Western and in Eastern Europe.

However, the most significant aridity in Europe was assumed to relate to the main phase of the Heinrich 1 event dated between 15,000 and 13,500 years ^{14}C-BP (McCabe/Clark/Clark 2005; Stanford et al. 2011b). Thus, from this period no data should be available if the general aridity on the continent was the major limiting factor for the dispersal of vegetation and fauna and, consequently, humans into northern Europe.

In fact, comparable to Eastern Europe only very few dates from the period between approximately 17,000 and 15,000 years ^{14}C-BP have been achieved in north-western Europe. However, the first results from the studied sub-areas of this project appeared after the onset of GS-2b. These were made on several reindeer remains from the horizon III in the Trou des Blaireaux (**tab. 74**) and provided the FAD for this species in the western uplands. Two dates from the base of the stratigraphy provided results from the period between 16,500 and 15,800 years ^{14}C-BP (*c.* 20,200-18,750 years cal. b2k) and three dates from an intermediate concentration and one date from an upper concentration in this horizon III produced results between 14,000 and 13,300 years ^{14}C-BP (*c.* 17,550-15,550 years cal. b2k; Charles 1996). Thus, a gap of approximately 1,200 radiocarbon years occurred in the data series from Trou des Blaireaux between 15,800 and 14,000 years ^{14}C-BP.

The older set of dates was supplemented by a reindeer antler found in the Schlaederbach valley at the foot of the Plateau Haed in Oetrange (Luxembourg) which was dated to the period between 16,500 and 15,600 years ^{14}C-BP (20,370-18,450 years cal. b2k; Gilot 1970). This older series was not related to human occupation (Charles 1996; Housley et al. 1997) and indicated only the presence of reindeer in this area during this part of the Late Pleniglacial. The ^{14}C results from Trou des Blaireaux and the Schlaederbach valley were dated with the conventional method in the 1970s and 1980s and except the Schlaederbach valley specimen, these dates were made on bulked antler samples. Thus, the dates on bulked samples at the base of the Trou Blaireaux sequence could also be the result of material of different ages being dated. A redating of single specimens is desirable to ascertain the presence of reindeer in Central Europe during this period of generally sparse evidence. Nevertheless, since several old dates originated from the site and the general stratigraphic succession seems to confirm the results, the dates from horizon III of the Trou des Blaireaux in combination with the Schlaederbach valley date are considered as the FAD of reindeer in the western uplands.

In addition to the older series from the western uplands, the dated mammoth vertebrae from the upper horizon in Hallines resulted in an age around 16,000 years ^{14}C-BP but the association of this piece with the archaeology remains uncertain (Fagnart 1997, 42). A horse tooth and a red deer rib were also found in this horizon but if the mammoth material represents fossil material the horse and red deer remains as well as the archaeology could be considerably younger. In fact, the small mammal and the mollusc assemblages found in this deposit indicated a Late Pleniglacial age comparable to Gönnersdorf. Moreover, the archaeological material was also similar to the Late Magdalenian and, thus, the mammoth remains were probably fossil. Whether the mammoth died in the vicinity of the site or the material was brought to the site by Lateglacial hunters for further use remains to be examined for example by isotope analysis.

However, a series of dates made on reindeer material from the southern German site Munzingen produced comparable results to the ones from the western uplands (between 16,200 and 15,250 years ^{14}C-BP, approximately 19,600-18,550 years cal. b2k) and was considered among the earliest evidences of Magdalenian remains in Central Europe (Pasda 1998; Kozłowski et al. 2012; Street/Jöris/Turner 2012). The end of the date range from Munzingen is correlated with a period of particularly low values in the oxygen isotope record reflecting the first indications of the Heinrich 1 event (Stanford et al. 2011b). The faunal assemblage recovered in the 1976-1977 excavation of the site was poorly preserved and only reindeer was determined (Pasda 1994). In contrast, several campaigns between 1914 and 1960 yielded beside reindeer material, re-

mains of horse, woolly rhinoceros, arctic hare, and wolverine (*Gulo gulo*) as well as teeth of red fox (*Vulpes vulpes*) and mammoth ivory. Due to patchy documentation, the exact relation of these remains and whether they are representatives of a contemporary biome cannot be evaluated. However, the species reflect a cold environment comparable to the previously described assemblages.

Calibrated ages made on samples of human, horse, and reindeer remains from the Polish Maszycka Cave (approximately 15,800-14,280 years ^{14}C-BP; c. 19,470-17,120 years cal. b2k; Kozłowski et al. 2012) also fell in this part of low values during GS-2b. Although the ^{14}C date on the horse sample appeared a bit younger than the other dates, it was consistent with the rest of the series after the calibration. Besides reindeer and horse, the collection contained mammoth ivory and remains of woolly rhinoceros, saiga antelope, red deer, brown bear as well as cave bear (*Ursus spelaeus*) and remains determined as wisent (*Bison* sp.) as well as some attributed to cattle (*Bos* sp.; Kozłowski et al. 1995). The rich organic artefact inventory from the site was comparable to the Munzingen assemblage attributed to a middle Magdalenian (Kozłowski et al. 1995). The variety of faunal species seemed likewise in Munzingen increased but since some material originated from a late 19th century collection, the integrity of the assemblage remained again uncertain.

Three dates made on horse samples from Solutré also produced results which fell calibrated to the period of lowered values in GS-2b between 15,210 and 14,505 years ^{14}C-BP (Hedges et al. 1997; Pestle/Colvard/Pettitt 2006). These dates were also assumed to relate to a Magdalenian assemblage of the site (Montet-White/Evin/Stafford 2002; Pestle/Colvard/Pettitt 2006).

At the southern limit of the French sub-area, a single direct date on a reindeer bone from Thèmes-Ferme de la Bouvière fell also into the period after 15,000 years ^{14}C-BP (Higham et al. 2007). However, the reliability of this date was questioned because two further dates from this assemblage produced significantly younger results and the archaeological assemblage suggested an early Upper Palaeolithic age.

A humerus from a large bovid (*Bison* sp./*Bos* sp.) found in the Thuringian Ranis-Ilsenhöhle also resulted in a date from this early Heinrich 1 event period (Higham et al. 2007). This sample came from a yellow layer (horizon 4) which was also related to a middle Magdalenian (Grünberg 2006). The δ^{13}C value of the specimen was surprisingly low but, perhaps, this value could be explained by depletion due to the nutritional intake. However, a horse bone from the same stratigraphic unit yielded a pre-LGM date. In fact, this series for the Ilsenhöhle showed some disturbances in the stratigraphic integrity of this cave site. Moreover, this date series was made during a period in which still some problems in the ultra-filtration process occurred. Therefore, this result remains doubtful.

Thus, although no gap can be identified in the date series for north-western Europe between the LGM and the onset of the Heinrich 1 event, the indications for human and faunal presence in this period are very scattered and occasionally still uncertain. The few determined samples indicate only species inhabiting cold to arctic habitats such as reindeer or woolly rhinoceros. These species were supplemented by horse during milder episodes (**fig. 63**). Possibly, the variety of preserved species in the assemblages increased with the first indications of the Heinrich 1 event. After the period of particularly sparse data, an arctic fauna with reindeer was found in most parts of north-western Europe. For example, the first reliable results of reindeer from south-western England (Kent's Cavern, Reindeer Rift) and a possible evidence of arctic hare in western England (Pin Hole) dated to this period. However, during the main phase (phase 2) of the Heinrich 1 event (Stanford et al. 2011b), these northern ranges seemed to be abandoned again.

After the younger set of dates from the Trou des Blaireaux, the next youngest date for reindeer in the western sub-areas is the dated specimen from Saint Mihiel followed by dates from Gönnersdorf. Due to the relatively large standard deviations no gap in the reindeer dates from the western uplands can be identified, although the few finds suggested that during the most intense period during the first part of the main

phase of Heinrich 1 evidence from the western uplands could be lacking. However, since the end of this phase 2 of the Heinrich event, a regular presence of reindeer in this eastern sub-area was attested.

In contrast to the reindeer dates, an AMS date on horse from horizon II of the Trou des Blaireaux resulted in an age of 13,330 ± 160 years ^{14}C-BP (OxA-4200; 17,110-15,710 years cal. b2k; Hedges et al. 1994) and indicate the general reliability of the stratigraphic sequence. Nevertheless, this date was made in the series with the ion-exchange pretreatment (lab. Code: AI) which could cause some contamination and, in fact, a bone of brown bear (*Ursus arctos*) from the same horizon resulted in a considerably younger age (Lv-1386: 12,440 ± 180 years ^{14}C-BP; 15,490-13,850 years cal. b2k; Charles 1996; Lanting/van der Plicht 1996). Moreover, the distribution of lithic material suggested some admixture within this assemblage, probably due to animal burrows (Charles 1996; Housley et al. 1997). However, dates of horse from Gönnersdorf and the lower horizon in Andernach were only slightly younger than this sample from the Trou des Blaireaux. Consequently, a major dispersal of horse into the sub-areas can only be proven for the transition from GS-2b to GS-2a and, thus, the period after the main phase of Heinrich 1. Although horse was present at Wiesbaden-Igstadt in approximately GI-2, no remain of this species was directly dated between 17,000 and 15,000 years ^{14}C-BP and a regular occurrence in north-western Europe was only proven after 14,000 years ^{14}C-BP. Therefore, these later dates represented the FADs of horse in the western upland zone and the Central Rhineland respectively.

With the end of the main phase of the Heinrich 1 event, a spread of the late Magdalenian into the middle-range mountain ranges and a significant increase in preserved faunal material can be noted (**fig. 63**). This increase further intensified with the onset of GS-2a. Besides reindeer and horse, hare and arctic fox were frequently identified and dated. This spread can be observed in the western uplands as well as in the Central Rhineland but northern France appeared comparable to the British Isles and the North European Plain still unaffected by this spread. Only a date of reindeer from Rinxent also indicates the presence of this species in northern France at the onset of GS-2a. The climatic harshness of this period is reflected by the occasional presence of saïga antelope in the eastern upland zone and the Central Rhineland and the occurrence of musk ox at the northern limit of the western upland zone.

Saïga antelope was found in some Magdalenian assemblages such as Gönnersdorf, Saint Mihiel, Trou de Chaleux (Charles 1998), or Kniegrotte (Feustel 1974; Hedges et al. 1998a; Höck 2000) but also in faunal assemblages from Creswellian sites such as Soldier's Hole and Gough's Cave (Jacobi 1980; Currant 1991; Jacobi 2004). Remains from the latter three sites were directly dated and produced calibrated age ranges from the Late Pleniglacial to the mid-Lateglacial Interstadial (see **tab. 73**). Thus, the last appearance date (LAD) for this species in Central Germany was in the Late Pleniglacial and on the British Isles the LAD fell to the early Lateglacial Interstadial. Possibly, this late appearance was related to the return of arid and glacial conditions in Britain during GI-1d. Even though only a few pieces of saïga material were determined at the British sites, the number of sites (n = 5) indicate a relatively frequent occurrence of this species in southern England and, in particular in the Cheddar region during this period (Currant 1987).

Musk ox remains were only identified in cave sediments from the northern limit of the western uplands. Since many of these remains originated from early collections the integrity was questioned and some specimens from the Grotte du Goyet, Trou da Somme, and Trou de Chaleux were directly dated (see **tab. 74**). Two pieces were cut-marked and indicated the human influence (Stevens 2009). The resulting dates were congruent and suggested a time of death during GS-2a.

The harsh climatic conditions of GS-2a were also indicated by the environmental data from some Late Magdalenian Paris Basin sites such as Verberie and most of the horizon IV assemblages at Pincevent (Taborin 1994). However, the radiometric results usually indicated an attribution to the second half of GS-2a or even to the early Lateglacial Interstadial. For example, the horse remains from Le Grand Canton and Le Tureau des

Gardes represented the FAD of this species in northern France and date comparable to Bois Laiterie (Krueger 1997). According to techno-typological analyses, the archaeological material found in these different sites also was comparable (Julien/Rieu 1999; Weber 2006; Valentin 2008a; Sano/Maier/Heidenreich 2011). ^{14}C dates from the Late Magdalenian concentrations of Verberie provided comparable results but many of the results from the Late Magdalenian of Pincevent produced even older dates (Bodu 2004; Débout et al. 2012). Thus, the Late Magdalenian occurred contemporaneously with the so-called faciès Cepoy-Marsangy or the dates from these Late Magdalenian assemblages were too young (see p. 474-481). If these dates were to young and the assemblages were, dated according to the archaeological evidence, comparable to the Central Rhineland and the Belgian sites, horse and, in particular, reindeer were regularly present in northern France at the onset of GS-2a.

In contrast, red deer remains were regularly found in the Belgian cave collections and in Gönnersdorf and the lower horizon of Andernach. From the latter two sites, these remains were identified as special items such as jewellery teeth (Street et al. 2006; Street/Turner 2013) which can be considered as imports from elsewhere. Comparably, in the Belgian assemblages remains of this species represented either special goods or represented Holocene intrusions (Hedges et al. 1994). Thus, red deer was probably still not present in the northern range of the uplands during GS-2a (cf. Sommer et al. 2008).

However, in the second half of GS-2a a gradual amelioration of the climate led to an increasing suitability for faunal expansions into the northern landscapes. Many of the selected species were present on the British Isles during this period (**fig. 63; tab. 80**) and also the first arctic indicatiors from the North European Plain fall into this transitional period between the Late Pleniglacial and the early Lateglacial Interstadial (Mortensen 2007; Grimm/Weber 2008).

The appearance of species indicating a different habitat to the arctic tundra and the grass steppe appeared in the study area at the transition from GS-2a to the early Lateglacial Stadial closely following on the expansion of many arctic species into these northern areas. Thus, a northward shift of the different habitat limits began around the end of the Late Pleniglacial.

For example, remains of wild boar occurred infrequently in assemblages attributed to the Late Pleniglacial such as in the Late Magdalenian horizon V of the Wildscheuer cave at the eastern limit of the Central Rhineland (Terberger 1993), in Rinxent, Abri du Mammouth, Bonnières-sur-Seine, Trou des Nutons, Trou de Frontal, and Grotte de Sy Verlaine (Charles 1996). Probably, these remains represented intrusions and/or admixtures from younger deposits because most of these sites were disturbed during 19th and early 20th century excavations of the sites.

However, directly dated material from the lower horizon of Le Closeau proved the presence of wild boar in the Seine valley during the transition of the Late Pleniglacial to the Lateglacial Interstadial (Bodu/Debout/Bignon 2006). These dates represent the FADs for wild boar in the northern French sub-area. Directly dated specimens from the other sub-areas resulted only in Holocene age and, therefore, a FAD is not given although this species appeared at least in the Central Rhineland during the Lateglacial (see p. 443). Younger material of wild boar found in the *loci* 4 and 46 were considered as Mesolithic intrusions or suggested possible contamination. The former is possible in some parts of the site due to the complex stratigraphy.

In addition, elk also occurred during this first faunal transition period. During the Lateglacial, this species was only found in the eastern sub-area and directly dated specimens come only from the Central Rhineland. Thus, the elk from Gönnerdorf south-west provided the FAD in the Central Rhineland. However, remains of a possible female elk were also determined within the faunal assemblage from the Magdalenian horizons of Bois Laiterie (Gautier 1997). This determination was mainly based on the size of the remains but also on a morphological dental feature. A recent reanalysis of the lithic inventory from the site revealed some technical differences in comparison to typical Late Magdalenian assemblages but also came to the conclusion

of a single episode inventory (Sano/Maier/Heidenreich 2011; cf. Straus/Orphal 1997). Thus, if the elk was related to the archaeology because it was found within the archaeological horizon, it represented the oldest indirectly dated evidence of elk presence in the western uplands. This indirect date would date the expansion of elk into the mid to late GS-2a. However, the remains were not directly dated and, thus, a somewhat younger admixture comparable to the elk in Gönnersdorf cannot be excluded completely. Yet, the Gönnersdorf results proved the presence of elk in this area at the onset of the Lateglacial Interstadial.

Besides these two species, red deer expanded its range into Western Europe during this early phase of the Lateglacial Interstadial. The Late Weichselian expansion of red deer was recently described based on ^{14}C dates, stratigraphic and spatial distributions as well as genetic evidence (Sommer et al. 2008). Its presence in western Europe was proven by several dates from the British Isles (Jacobi 2004; Jacobi/Higham 2009) and a palaeontological sample from Conty produced an age from the Late Pleniglacial to Lateglacial Interstadial transition (OxA-6257: 12,300 ± 120 years ^{14}C-BP, 14,990-13,830 years cal. b2k; Ponel et al. 2005). Even though this date was made in the problematic series with ion-exchange pretreatment (lab. code: AI) which could cause contamination of the sample, this type of problem was not indicated by additional signature values. Furthermore, the stratigraphic position was in accordance. Since this period, red deer had been found regularly in the archaeological assemblages of the northern French sub-area. Two further bones of a cervid were dated from the lower horizon of Le Closeau and resulted in almost identical ages as the Conty specimen. However, these pieces were not determined to species level. Although red deer appeared the more probable species, they might also reflect the presence of reindeer in the assemblage. A date of a red deer calcaneum from the Swiss Kesslerloch equally yielded a date from this transition period (KIA-33351: 12,335 ± 45 years ^{14}C-BP, 14,730-14,050 years cal. b2k; Napierala 2008a) and suggested a reoccupation of this site comparable to the south-western area of Gönnersdorf. Since the position of dated skeletal material from the lower horizon of Andernach remained ambiguous, these dates are not used as FAD for the presence of red deer in the Central Rhineland. Although red deer occurred in several assemblages in the western uplands, the uncertainty of the relation with the Late Pleniglacial material remained comparable to the Andernach material. Furthermore, admixture of this species into the Belgian assemblages was considered possible because several of the faunal collections originated from 19th and early 20th century excavations. This doubt was further sustained when a bone from the Trou des Nutons was dated to the mid-Holocene (Hedges et al. 1994). Thus, the presence of red deer in the western upland zone and particularly in the Central Rhineland remains questionable for the Late Pleniglacial. Moreover, no data of red deer could be attributed to the early Lateglacial Interstadial with some certainty in the eastern sub-areas. Possibly, red deer remains found in a stratigraphic higher position at Gönnersdorf could date comparable to the elk from the south-western area but these pieces were not yet directly dated. In the western upland zone, no assemblage from the mid-Lateglacial Interstadial is known and, thus, the very late appearance of red deer seems unsurprising.

Thus, in the west, wild boar and red deer supplemented the cold and open fauna of, predominantly, horse during the transition from the Late Pleniglacial to the Lateglacial Interstadial. In the eastern sub-areas the reliable assemblages are very few but the directly dated material also suggested the continued presence of horse (Stevens et al. 2009a; Stevens et al. 2009b) but supplemented by elk.

The relatively late reindeer dates in northern France and the LADs of reindeer were already discussed within the probability distributions (see p. 428). However, the question which ^{14}C date is the last reliable could be further discussed but clearly reindeer occurred in the northern French sub-area until the cold event in the early Lateglacial Interstadial as suggested by the evidence from Marsangy. According to this LAD in GI-1d, the data confirmed the previous suggestion that reindeer left France around GI-1d (Lang 1998, 85). This last appearance of reindeer in northern France co-occurred with the disappearance of reindeer in the French

Jura (Drucker/Bridault/Cupillard 2012). In the Central Rhineland, reindeer had already disappeared after the Late Magdalenian settlement of Gönnersdorf and the Wildweiberlei, whereas a few indications for a survival into the early Lateglacial Interstadial comparable to the Northern French evidence also exist in the western uplands.

LADs for horses are not given since single specimen occur throughout the Lateglacial Interstadial and it represented one of the most common species among the material from the late Lateglacial Stadial to early Holocene transition in northern France (Fagnart 1997) and also occurred occasionally in the northern Rhineland (Street 1991). Moreover, the colonisation history of this species seemed related to open landscapes and as a result of the human change of the landscape, the range patterns of horses were complex during the Holocene (Sommer et al. 2011). However, a significant decrease in the presence of horse can be recorded around the disappearance of reindeer in the northern French assemlages around GI-1d. Also in the Central Rhineland sites, horses were only found singularly from the mid-Lateglacial onwards.

In contrast, red deer occurred during this period regularly in the assemblages from the Central Rhineland and it often formed the dominant species in the archaeological assemblages (see **tab. 15**). Furthermore, the faunal composition changed significantly in this period. Besides elk that occurred frequently in the Central Rhineland assemblages, beaver became occasionally attested. A direct date of beaver which indicates a contrasting landscape to the saïga was made on a scapula fragment found in the upper horizon of Andernach 2. The calibrated age range of this sample suggested the first appearance of this species in the Central Rhineland during the mid-Lateglacial Interstadial (GrA-16987: 12,050 ± 70 years ^{14}C-BP, 14,100-13,700 years cal. b2k; Kegler 2002).

In addition, roe deer supplemented the species in all sub-areas. Two ^{14}C dates made on roe deer material from the Central Rhineland have potentially been contaminated (see **tab. 75**). However, roe deer was identified in almost all the FMG assemblages; only in the limited excavation area of Urbar and in the Boppard assemblage it has thus far not been determined. Hence, the presence of roe deer in the Central Rhineland can indirectly be dated to the onset of the mid-Lateglacial Interstadial. Equally, roe deer was found in many Belgian cave assemblages such as Trou des Nutons, Trou de Chaleux (Charles 1998), or Goyet (Germonpré 1997). However, these remains usually originated from old excavations and their association with the cold adapted Magdalenian fauna was questioned. Thus far, roe deer remains from these collections were not directly dated but other questionable elements in these assemblages such as wild boar or red deer were demonstrated to be of Holocene age (Hedges et al. 1994). In northern France, roe deer was determined in the assemblages of Grotte de Gouy (Bordes et al. 1974) and the lower horizon of Conty (Coudret/Fagnart 2006). Even though the remains were not directly dated, the evidence from Conty is reliable and the assemblage was well dated to the transition from the mid- to the late Lateglacial Interstadial (see p. 237-239, p. 479-481, and **figs 69-71, tabs 44. 76**). The material from the Grotte de Gouy was dated to the transition from the early to the mid-Lateglacial Interstadial but again the association of roe deer with the other material inside the cave remained uncertain (Fosse 1997). A direct date from Reichwalde produced an early Lateglacial Interstadial age (GrA-15437: 12,350 ± 50 years ^{14}C-BP, 14,750-14,070 years cal. b2k; Vollbrecht 2005). Even though this date is in accordance with the stratigraphic position in an early Lateglacial Interstadial soil, it remains questionable because the sampled material consisted of calcined bones which produced occasionally unreliable dates in the Lateglacial (Lanting/Niekus/Stapert 2002). Another bone of roe deer was found in the early Lateglacial horizon B of the southern Polish Komarowa Cave (Wojtal 2007). However, within this horizon giant deer (*Megaloceros giganteus*) remains were found which dated to the early Upper Palaeolithic. Using these dates, otherwise undetected movements within the cave sediment were proven. Consequently, the attribution of the roe deer remain to the early Lateglacial Interstadial remains uncertain. Nevertheless, an expansion of roe deer from eastern Europe, possibly from a Carpathian refuge,

after the LGM was previously suggested based on the palaeogeographic and genetic evidence (Sommer et al. 2009).

Comparable to roe deer, large bovids were found in most assemblages from the mid- and late Lateglacial Interstadial in the sub-areas. If these remains could be determined to species they usually originated from aurochs (Coudret/Fagnart 2006; Baales 2002). On the British Isles some specimens were dated already to the transition between the Late Pleniglacial and the early Lateglacial Interstadial (Jacobi 2004). Perhaps, these early occurrence in the north-western corner of Europe can confirm the evidence of aurochs alongside steppe wisent and musk ox in the Belgian assemblages (Germonpré/Sablin 2002), although an admixture cannot be fully excluded in these assemblages.

In northern France, wild boar probably remained present in this period. Among the faunal remains recovered from the Grotte de Gouy, wild boar was determined (Bordes et al. 1974; cf. Bignon/Bodu 2006). Even though the recovery of the material was again doubtful, the assemblage reflected a temperate forest community in which wild boar did not appear intrusive. An undetermined bone from this assemblage was dated to the transition from the early Lateglacial Interstadial to the mid-Lateglacial Interstadial (Fosse 1997, 242). The possibly contaminated wild boar sample from the *locus* 4 of Le Closeau produced a mid-Lateglacial Interstadial age but neither in the intermediate nor the upper horizon from this site were wild boar remains identified. However, this absence could be due to the significant decrease in bone preservation in these sediments.

After the establishment of this forest faunal community around the onset of GI-1c, this composition remained stable until the end of the Lateglacial Interstadial in the three sub-areas. Furthermore, in the northern Europe an expansion of this community can be registered around GI-1c_3 (**fig. 63**). In particular, elk moved to these northern areas but also wild boar (Hanik 2009) and roe deer (Richards et al. 2005) were occasionally found further in the north.

However, elks were still found in the Central Rhineland during the late Lateglacial Interstadial. Elks from Miesenheim 4 and Niederbieber 19 (previously 2; Street/Terberger 2004; cf. Fiedel et al. 2013) yielded the LAD of this in the Central Rhineland.

In contrast to northern France, wild boar was only determined in late Lateglacial Interstadial assemblages of the Central Rhineland (Niederbieber, Boppard). Perhaps, this late appearance in the Central Rhineland was related with the expansion of denser forests at the end of the Lateglacial Interstadial. During this period, wild boar also appeared in several Dutch assemblages such as Doetinchem (Johansen et al. 2000), or Wierden Enterse Akkers (Deeben et al. 2006). In sites further to the north such as Stellmoor (Bratlund 1999) or Bromme (Mathiassen 1948), this species seemed again to represent a Holocene intrusion. Nevertheless, an expansion north-eastwards of this species can be identified at the end of the Lateglacial Interstadial. In northern France, at Saleux 114 which also dated to the late Lateglacial Interstadial wild boar was found again (Fagnart 1997; Coudret/Fagnart 2006) proving its constant presence in this sub-area during the Lateglacial Interstadial.

The second, directly dated piece of red deer from the Paris Basin came from a test trench in Saleux and resulted in a late Lateglacial age (Beta-170494: 11,180 ± 50 years ^{14}C-BP, 13,180-12,940 years cal. b2k; Coudret/Fagnart 2004). However, red deer was determined in all assemblages from Saleux (Coudret/Fagnart 2006). For the western uplands the identification of this species was problematic due to the lack of reliable assemblages. However, an unmodified mandible of red deer from one of the Presle caves produced a calibrated age of late Lateglacial Interstadial to early Lateglacial Stadial indicating the presence of this species in the western uplands.

From the North European Plain and southern Scandinavia, this species is almost absent. A single date on a barbed point from Lemförde am Dümmer must be considered critically regarding the presence of this species

in this region (Veil et al. 1991; Riede et al. 2010). However, near the late Lateglacial Bromme site at Trollesgrave red deer remains were recovered from lake sediments that were considered of the same age as the site (Fischer/Mortensen 1977). Nevertheless, this relation cannot be proven unambiguously and the remains were not dated (cf. Fischer et al. 2013). Therefore, a Holocene age of this material cannot be excluded.
With the return of cold climate conditions during the Lateglacial Stadial the faunal composition again changed in northern Europe. For instance, reindeer returned into the western uplands (Baales 1996). Thus, the LAD of reindeer in the western uplands is of a Holocene age.
Nevertheless, some forested elements could have survived in sheltered areas. The evidence is relatively sparse because in the Central Rhineland as well as in the Seine valley, no sites from this period are preserved. Thus, the absence of wild boar from northern French sites during the Lateglacial Stadial related probably to the absence of reliable assemblages because the species is again known from Holocene assemblages in Le Closeau. The survival of wild boar in the western upland zone throughout the Lateglacial Stadial is poorly understood because the presence of this species in Ahrensburgian assemblages such as Hohle Stein (Kallenhardt) or Fond-de-Forêt was also assumed to represent Holocene intrusions (Baales 1996). Furthermore, elk was found in the Ahrensburgian assemblage of the Hohle Stein (Kallenhardt) but could also represent a Holocene admixture in this case (Baales 1996) or a visitor from the North European Plain where it remained a common species throughout the Lateglacial (Schmölcke/Zachos 2005; Riede et al. 2010).
In the assemblage from the Hohler Stein (Kallenhardt), red deer remains were also found from the Lateglacial Stadial horizon but again a Holocene intrusion cannot be excluded. However, red deer as well as aurochs were still attested in the fauna in a top horizon of Belloy-sur-Somme (Fagnart 1997). This indirect evidence indicates that these species survived in the western sub-area during this cold event. The Holocene record of red deer is again very numerous in the Paris Basin, the Central Rhineland, and the western upland (cf. Sommer et al. 2008). Frequently, aurochs were found in Mesolithic assemblages such as Bedburg-Königshoven (Street 1991).
Furthermore, roe deer and wild boar became also common representatives of the temperate forests in all the sub-areas again and elk was also frequent in most parts of Europe including northern France (Schmölcke/Zachos 2005).

In conclusion, reindeer seemed to appear quasi-contemporaneously as a frequent inhabitant in the sub-areas around the onset of GS-2a. The disappearance at the onset of the Lateglacial Interstadial began in the east and ended in the west. The retreat of the reindeer seemed a gradual process with some specimen occurring until GI-1d in the Paris Basin. However, the numbers of animals changed significantly between some assemblages in Pincevent horizon IV to the singular examples at Marsangy. Possibly, the migration behaviour of this species changed already with the onset of the Lateglacial Interstadial leaving only few individuals to range in the northern French sub-area and the larger herds to establish elsewhere (**fig. 64**). With the onset of GI-1c, reindeer had probably left the study area. Only in the northern part of the western uplands was a return of reindeer documented during the Lateglacial Stadial.
During the Lateglacial Interstadial, the number of determined horses also became significantly smaller but this species was still regularly found in the assemblages of the different sub-areas. These finds showed the continued presence of this species but in how far the group sizes changed from the larger steppe herds to small and solitary forest inhabitants (cf. MacFadden 1992, 263-298) remains a matter of discussion.
Elk occurred in western Central Europe during the transition of the Late Pleniglacial to the Lateglacial Interstadial. The absence of this species from northern France and most assemblages in the western uplands suggested that this large cervid migrated onto the North European Plain and into western Europe through the water rich, northern environments, probably from refugia in eastern Europe. In regard to the relation of

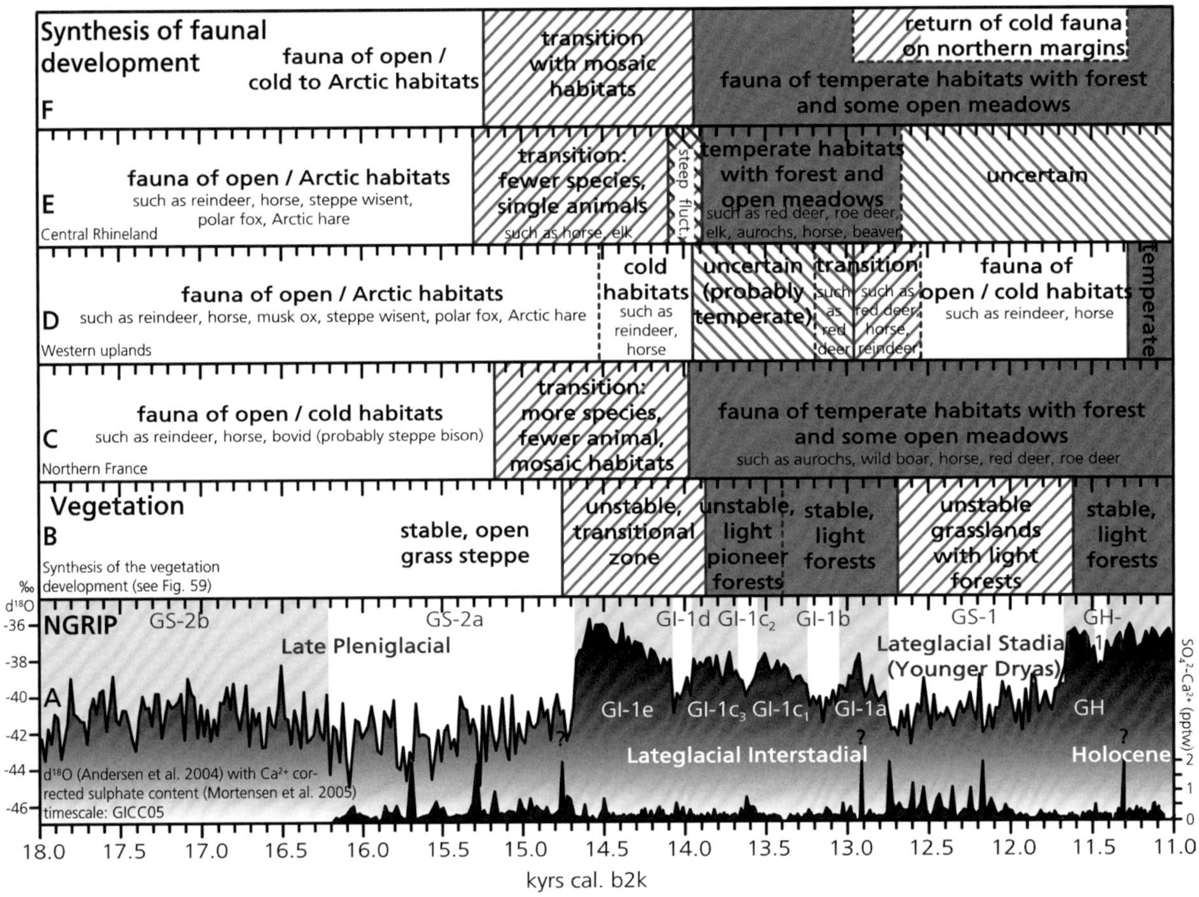

Fig. 64 Synthesis of faunal development in the sub-areas of this study (**C-F**; see **figs 59-60. 62**) in comparison to the synthesis of the vegetation development (**B**; see **fig. 58**) and the oxygen isotope record as well as the gypsum corrected sulphate content of NGRIP (**A**; see **fig. 53**). White backgrounds: cold and open environments; hatched background: transitional environments; grey shaded background: temperate forest environments.

this species to early stages of forest succession (Peterson 1955), the quick development of denser vegetation in the river valleys of the Paris Basin (see p. 388-394) possibly hindered the expansion of elks further to the west. Another possibility is a lack of moist areas providing sufficient nutrition for elks. The later argument could also help to explain the absence of beaver in this western sub-area during the Lateglacial. This species was first documented in the Central Rhineland at the transition from GI-1d to GI-1c.

Also since this period, large bovids, mainly aurochs, and red deer became regular prey of human hunters in the Central Rhineland and the northern French sub-area.

For the distribution of wild boar the different vegetation developments in the study area seemed to have played an important role. In the early forested environments of the French sub-area, wild boar were quickly established as regular inhabitants, whereas this species seemed to have colonised the western uplands including the Central Rhineland only after the local vegetation developed into a denser forest community.

According to these few indicator species, the arctic (arctic fox, arctic hare, musk ox) and arid (saïga antelope) climatic conditions and the grass dominated steppe zone was superseded as the predominant landscape in north-western Europe by forested wetlands (beaver, roe deer) at the onset of the mid-Lateglacial Interstadial (GI-1c_3). Thus, these indicator species further confirm the impression based on the selected species that the major transition in the faunal composition occurred considerably later than the onset of the Lateglacial Interstadial (**fig. 64**) and seemed to rather correlate with the onset of the mid-Lateglacial Interstadial.

site	lab. no.	years ^{14}C-BP	± years	material	species	comment	years cal. b2k	ref.
reindeer (*Rangifer tarandus*), n = 181								
Kastelhöhle-Nord, Solothurn (CH) – SEF	OxA-9738**	19,620	140	bone	*Rangifer tarandus*	modified	23,950-23,190	1-2
Aschenstein, Niedersachsen (D) – EU	KIA-33773	19,570	100	antler	*Rangifer tarandus*	unworked	23,860-22,900	3
Kastelhöhle-Nord, Solothurn (CH) – SEF	OxA-9739**	19,200	150	bone	*Rangifer tarandus*	modified	23,540-22,540	1-2
Aschenstein, Niedersachsen (D) – EU	KN-2712	18,820	180	bone/antler	*Rangifer tarandus*	palaeontological specimens; PROBLEMATIC: bulked sample	23,050-22,370	3
Wiesbaden-Igstadt, Hessen (D) – EU	OxA-6809*	18,670	160	bone	*Rangifer tarandus*		23,040-21,840	2; 4-5
Kastelhöhle-Nord, Solothurn (CH) – SEF	OxA-9737**	18,530	150	bone	*Rangifer tarandus*	modified	22,730-21,770	1-2
Wiesbaden-Igstadt, Hessen (D) – EU	OxA-7501	18,220	180	bone	*Rangifer tarandus*		22,650-21,410	2; 4-5
Wiesbaden-Igstadt, Hessen (D) – EU	OxA-7500	17,820	240	bone	*Rangifer tarandus*		22,110-20,710	2; 4-5
Deszczowa Cave, Śląskie (PL) – SP	Gd-10212	17,480	150	antler	*Rangifer tarandus*		21,490-20,410	6
Deszczowa Cave, Śląskie (PL) – SP	Gd-9464	16,150	280	long bone	*Rangifer tarandus*		20,120-18,720	6
Munzingen, Baden-Württemberg (D) – SU	OxA-4785*	16,060	140	bone	*Rangifer tarandus*	PROBLEMATIC: possibly contaminated by chemical preservatives	19,610-18,890	7-8
Munzingen, Baden-Württemberg (D) – SU	*ETH-7499*	*15,700*	*135*	metatarsus	*Rangifer tarandus*	REJECT: possibly contaminated by chemical preservatives and unusually low δ^{13}C value	×	7-8
Munzingen, Baden-Württemberg (D) – SU	OxA-4786*	15,670	140	bone	*Rangifer tarandus*	PROBLEMATIC: possibly contaminated by chemical preservatives	19,190-18,630	7-8
Maszycka Cave, Małopolskie (PL) – SP	Ly-2454	15,490	310	antler	*Rangifer tarandus*	modified; PROBLEMATIC: large standard deviation	19,470-17,950	9-10
Munzingen, Baden-Württemberg (D) – SU	OxA-4783*	15,400	130	bone	*Rangifer tarandus*	PROBLEMATIC: possibly contaminated by chemical preservatives	18,940-18540	7-8
Maszycka Cave, Małopolskie (PL) – SP	KIA-39226	15,025	50	antler	*Rangifer tarandus*	artefact: point	18,660-17,940	10

Tab. 80 ^{14}C dates from selected faunal samples in north-western Europe outside the study area. If the sub-assemblage is known from which the sample originated, the sub-assemblage is given behind the site name. Otherwise the province and the country is given, as well as the sub-area. Abbreviations: **SEF** Switzerland and eastern France; **SU** southern uplands; **EU** eastern uplands; **SP** southern Polish uplands; **NEP** North European Plain; **SCA** southern Scandinavia; **GB** British Isles; **?** uncertain species determination (behind probable species); ***** dates which were pretreated by the use of ion-exchanged gelatin (Lab.code: AI) in the Oxford series (cf. Jacobi/Higham 2009, 1896); ****** dates which might be contaminated due to the use of a method leaving traces of a humectant in the collagen (Lab.code: AF*) in the Oxford series (cf. Higham et al. 2007, S55 & S2). Doubtful dates are shaded grey. Rejected dates are shaded grey and set in italics and, in addition, the main reason for rejection is given in comment. For further details see p. 259-263, p. 265-269, and p. 412-429. The dates were calibrated with the calibration curve of the present study (see p. 358-364) and the calibration program CalPal (Weninger/Jöris/Danzeglocke 2007). The result range of 95 % confidence is given for the calibrated ages (years cal. b2k). – References (ref.): **1** Bronk Ramsey et al. 2002; **2** erberger/Street 2002; **3** Terberger et al. 2009; **4** Hedges et al. 1998a; **5** Street/Terberger 1999; **6** Cyrek et al. 2000; **7** Pasda 1994; **8** Pasda 1998; **9** Kozłowski et al. 1995; **10** Kozłowski et al. 2012; **11** Jacobi/Higham 2009; **12** Housley et al. 1997; **13** Napierala 2008a; **14** Hedges et al. 1997; **15** Pion 1997; **16** Hedges et al. 1998b; **17** Hedges et al. 1989; **18** Schuler 1994; **19** Bowman/Ambers/Lesse 1990; **20** Albrecht 1979; **21** Fischer/Tauber 1986; **22** Grimm/Weber 2008; **23** Holm/Rieck 1992; **24** Aaris-Sørensen/Mühldorff/Petersen 2007; **25** Aaris-Sørensen 2009; **26** Hedges et al. 1987; **27** Jacobi 2004; **28** Rubin/Suess 1956; **29** Bridault et al. 2000; **30** Drucker et al. 2003; **31** Münnich 1957; **32** Fischer 1996; **33** Petersen/Johansen 1996; **34** Hedges et al. 1994; **35** Riede et al. 2010; **36** Hedges et al. 1991; **37** Hedges et al. 1992; **38** Jacobi/Higham/Lord 2009; **39** Aaris-Sørensen 1995; **40** Clausen 2004; **41** Larsson et al. 2002; **42** Lanting/van der Plicht 1996; **43** Hedges et al. 1996b; **44** Hedges et al. 1995; **45** Stensager 2004; **46** Gowlett et al. 1986a; **47** Barendsen/Deevey/Gralenski 1957; **48** Weber/Grimm/Baales 2011; **49** Hedges et al. 1990b; **50** Gowlett et al. 1986b; **51** Burleigh 1986; **52** Ambers/Matthews/Burleigh 1985; **53** Currant 1991; **54** Mellars 1969; **55** Barker/Burleigh/Meeks 1971; **56** Pestle/Colvard/Pettitt 2006; **57** Stevens/Hedges 2004; **58** Grünberg 2006; **59** Higham et al. 2007; **60** Bodu et al. 2009b; **61** Verpoorte/Šída 2009; **62** Benecke et al. 2006; **63** Richards et al. 2000; **64** Gillespie et al. 1985; **65** Vencl 1995; **66** Hedges et al. 1996a; **67** Kaiser/de Klerk/Terberger 1999; **68** Bronk Ramsey et al. 2000; **69** Burleigh/Hewson/Meeks 1976; **70** Hedges et al. 1990a; **71** Hanik 2009; **72** Fiedorczuk/Schild 2002; **73** Heinemeier/Rud 2000; **74** Kabaciński/Schild 2005; **75** Benecke/Heinrich 2003; **76** Meiri et al. 2013; **77** Drucker et al. 2009; **78** Barton/Roberts 1996; **79** Veil et al. 1991; **80** Richards et al. 2005; **81** Bailey et al. 1996; **82** Hedges et al. 1993a; **83** Hedges et al. 1988a; **84** Tietz 2005; **85** Street 1998a; **86** Irish et al. 2008; **87** Vollbrecht 2005.

site	lab. no.	years ^{14}C-BP	± years	material	species	comment	years cal. b2k	ref.
Maszycka Cave, Małopolskie (PL) – SP	KIA-32225	14,885	60	antler	*Rangifer tarandus*	artefact: navettes	18,720-17,840	10
Reindeer Rift, Devon/England (GB) – GB	OxA-17160**	14,550	55	calcaneum	*Rangifer tarandus*	no association with archaeology	17,980-17,620	11
Munzingen, Baden-Württemberg (D) – SU	OxA-4784*	14,510	110	bone	*Rangifer tarandus*	PROBLEMATIC: possibly contaminated by chemical preservatives	18,010-17,530	7
Kent's Cavern, Devon/England (GB) – GB	OxA-14826	14,395	60	astralagus	*Rangifer tarandus*		17,870-17,510	11
Munzingen, Baden-Württemberg (D) – SU	OxA-4788*	14,270	120	bone	*Rangifer tarandus*	PROBLEMATIC: possibly contaminated by chemical preservatives	17,800-17,080	7-8
Kesslerloch, Schaffhausen (CH) – SEF	OxA-5749*	14,150	100	antler	*Rangifer tarandus*	modified, shed antler, possible collection of fossil material	17,660-17,060	12-13
Abri de la Fru, Savoie/Rhône-Alpes, (F) – SEF	OxA-5260* (Ly-130)	14,060	130	metapodium	*Rangifer tarandus*		17,660-16,980	14-15
Abri de la Fru, Savoie/Rhône-Alpes, (F) – SEF	OxA-4937* (Ly-89)	13,810	110	mandible	*Rangifer tarandus*		17,170-16,930	14-15
Kesslerloch, Schaffhausen (CH) – SEF	OxA-5750*	13,670	100	antler	*Rangifer tarandus*	modified, shed antler, possible collection of fossil material	17,100-16,820	12-13
Munzingen, Baden-Württemberg (D) – SU	ETH-7500	13,560	120	metatarsus	*Rangifer tarandus*	PROBLEMATIC: possibly contaminated by chemical preservatives	17,110-16,550	7-8
Kniegrotte, Thüringen (D) – EU	OxA-4832*	13,310	110	scapula	*Rangifer tarandus*	cutmarks	16,980-15,940	16
Munzingen, Baden-Württemberg (D) – SU	OxA-4820*	13,230	110	bone	*Rangifer tarandus*	PROBLEMATIC: possibly contaminated by chemical preservatives	16,920-15,640	7-8
Kniegrotte, Thüringen (D) – EU	OxA-4845*	13,120	130	tibia	*Rangifer tarandus*	probable marrow extraction	16,720-15,440	16
Teufelsküche, Baden-Württemberg (D) – SU	ETH-7501	13,080	120	metatarsus	*Rangifer tarandus*	PROBLEMATIC: also treated with chemicals? unusually low $\delta^{13}C$ value	16,620-15,420	7
Pin Hole, Derbyshire/England (GB) – GB	OxA-1936	13,050	250	antler	*Rangifer tarandus*	palaeontological specimen; PROBLEMATIC: large standard deviation	16,970-14,970	17
Schussenquelle I, Baden-Württemberg (D) – SU	KN-4251	13,050	120	antler	*Rangifer tarandus*		16,540-15,380	18
Kesslerloch, Schaffhausen (CH) – SEF	B-3329	12,970	180	bone	*Rangifer tarandus*	PROBLEMATIC: bulked sample of mainly reindeer	16.620-15,060	13
Abri Stendel XVIII, Niedersachsen (D) – EU	OxA-10471**	12,860	75	antler	*Rangifer tarandus*	PROBLEMATIC: possibly contaminated series	15,730-15,290	1
Schussenquelle I, Baden-Württemberg (D) – SU	KN-4250	12,860	120	scapula	*Rangifer tarandus*		15,980-15,140	18
Picken's Hole, Somerset/England (GB) – GB	*BM-2118R*	*12,710*	*1,500*	metacarpus	*Rangifer tarandus*	uncertain association to archaeology; revised version not measured date; REJECT: very large standard deviation	×	19
Petersfels 1, AH 4, Baden-Württemberg (D) – SU	H-7145-7303	12,700	100	metatarsus	*Rangifer tarandus*		15,620-14,820	20

Tab. 80 (continued)

site	lab. no.	years ^{14}C-BP	± years	material	species	comment	years cal. b2k	ref.
Petersfels 1, AH 3, Baden-Württemberg (D) – SU	H-7138-7057	12,685	75	metatarsus	*Rangifer tarandus*		15,500-14,940	20
Oelknitz, Thüringen (D) – EU	OxA-5717*	12,670	110	calcaneum	*Rangifer tarandus*		15,640-14,560	16
Petersfels 1, AH 3, Baden-Württemberg (D) – SU	H-7135-6879	12,670	100	long bone	*Rangifer tarandus?*		15,600-14,640	20
Teufelsbrücke, Thüringen (D) – EU	OxA-5726*	12,640	130	humerus	*Rangifer tarandus*	black bone	15,650-14,410	7
Petersfels 1, AH 4, Baden-Württemberg (D) – SU	H-7144-7302	12,630	95	ribs	*Rangifer tarandus*		15,560-14,520	20
Oelknitz, Thüringen (D) – EU	OxA-5714*	12,620	120	maxilla	*Rangifer tarandus*	cutmarks	15,580-14,420	16
Petersfels 1, AH 3, Baden-Württemberg (D) – SE	H-7133-6877	12,580	130	ribs and vertebras	*Rangifer tarandus*		15,570-14,250	20
Poggenwisch, Schleswig-Holstein (D) – NEP	K-4332	12,570	115	bone	*Rangifer tarandus*		15,510-14,310	21-22
Gough's New Cave, Somerset/England (GB) – GB	OxA-18064	12,535	55	antler	*Rangifer tarandus*	artefact: perforated rod; redating of OxA-2797	15,310-14,510	11
Petersfels 1, AH 3, Baden-Württemberg (D) – SU	H-7140-7058	12,530	90	tibia	*Rangifer tarandus*		15,410-14,290	20
Slotseng, kettle hole, Syddanmark (DK) – SCA	AAR-906	12,520	190	antler/bone	*Rangifer tarandus*	modified	15,640-13,920	22; 23-25
Schussenquelle I, Baden-Württemberg (D) – SU	ETH-6155	12,510	130	vertebra	*Rangifer tarandus*		15,470-14,070	18
Aveline's Hole, Somerset/England (GB) – GB	OxA-1122	12,480	130	antler	*Rangifer tarandus*		15,420-14,020	26-27
Castlepook Cave, Cork (IRL) – GB	OxA-3602*	12,480	130	antler	*Rangifer tarandus*	base	15,420-14,020	14
Stellmoor, kettle hole, Schleswig-Holstein (D) – NEP	W-261	12,450	200	antler	*Rangifer tarandus*	uncertain association with archaeology; PROBLEMATIC: pretreatment procedure?	15,550-13,830	22; 28
Poggenwisch, Schleswig-Holstein (D) – NEP	K-4331	12,440	115	bone	*Rangifer tarandus*		15,320-14,000	21-22
Poggenwisch, Schleswig-Holstein (D) – NEP	K-4577	12,440	115	bone	*Rangifer tarandus*	butchering marks; carnivore gnawing; PROBLEMATIC: unusually high δ^{13}C value	15,320-14,000	21-22
Abri de Rochedane, Doubs/Franche-Comté (F) – SEF	OxA-8030 (Ly-709)	12,420	75	bone	*Rangifer tarandus*		15,170-14,050	29-30
Slotseng, kettle hole, Syddanmark (DK) – SCA	AAR-8159	12,410	70	antler	*Rangifer tarandus*		15,150-14,030	22; 24
Munzingen, Baden-Württemberg (D) – SU	OxA-4787*	12,370	100	bone	*Rangifer tarandus*	PROBLEMATIC: possibly contaminated by chemical preservatives	15,140-13,940	7-8
Aschenstein, Niedersachsen (D) – EU	KIA-33772	12,366	61	antler	*Rangifer tarandus*	debris of spall production	14,830-14,070	1
Poggenwisch, Schleswig-Holstein (D) – NEP	KIA-32926	12,365	60	antler	*Rangifer tarandus*	modified	14,820-14,060	22

Tab. 80 (continued)

site	lab. no.	years ^{14}C-BP	± years	material	species	comment	years cal. b2k	ref.
Meiendorf pond, Schleswig-Holstein (D) – NEP	K-4329	12,360	110	antler	*Rangifer tarandus*		15,150-13,910	21-22
Poggenwisch, Schleswig-Holstein (D) – NEP	KIA-32927	12,330	55	antler	*Rangifer tarandus*	with attached cranial bone	14,750-14,030	22
Petersfels 1, AH 3, Baden-Württemberg (D) – SU	H-7136-6890	12,320	90	bone	*Rangifer tarandus*		14,860-13,940	20
Meiendorf pond, Schleswig-Holstein (D) – NEP	*H-38-121B*	*12,300*	*300*	antler	*Rangifer tarandus*	organic fraction; REJECT: without further pre-treatment	×	22; 31
Slotseng, kettle hole, Syddanmark (DK) – SCA	AAR-8157	12,299	41	antler	*Rangifer tarandus*	modified	14,670-13,990	22; 24
Slotseng, kettle hole, Syddanmark (DK) – SCA	AAR-8165	12,290	75	vertebra	*Rangifer tarandus*		14,760-13,920	22; 24
Oelknitz , Thüringen (D) – EU	OxA-5712*	12,270	110	radius	*Rangifer tarandus*		14,850-13,810	16
Poggenwisch, Schleswig-Holstein (D) – NEP	KIA-32925	12,265	55	antler	*Rangifer tarandus*	modified; PROBLEMATIC: unusually low $δ^{13}C$ value	14,570-13,930	22
Slotseng, kettle hole, Syddanmark (DK) – SCA	AAR-8160	12,240	50	vertebra	*Rangifer tarandus*	with embedded flint projectile	14,360-13,960	22; 24
Slotseng, kettle hole, Syddanmark (DK) – SCA	AAR-8162	12,220	100	antler	*Rangifer tarandus*		14,680-13,760	22; 24
Slotseng, kettle hole, Syddanmark (DK) – SCA	AAR-8163	12,205	65	tibia	*Rangifer tarandus*		14,320-13,880	22; 24
Slotseng, kettle hole, Syddanmark (DK) – SCA	AAR-8164	12,190	50	rib	*Rangifer tarandus*		14,220-13,900	22; 24
Stellmoor, kettle hole, Schleswig-Holstein (D) – NEP	K-4261	12,190	125	antler	*Rangifer tarandus*		14,740-13,660	21-22
Stellmoor, kettle hole, Schleswig-Holstein (D) – NEP	K-4328	12,180	130	bone	*Rangifer tarandus*		14,720-13,640	21-22
Slotseng, kettle hole, Syddanmark (DK) – SCA	AAR-8158	12,165	55	antler/bone	*Rangifer tarandus*	modified	14,190-13,870	22; 24
Køge Bugt 1, Sjælland (DK) – SCA	AAR-1036	12,140	110	antler	*Rangifer tarandus*	modified; also known as »offshore Solrød Strand«	14,370-13,690	22; 32-33
Munzingen, Baden-Württemberg (D) – SU	*H4738-4660*	*12,130*	*130*	bone	*Rangifer tarandus*	GH 4; REJECT: poor bone preservation; disturbed site	×	7-8
Villestofte, Syddanmark (DK) – SCA	AAR-4187	12,080	90	bone	*Rangifer tarandus*		14,170-13,690	24
Slotseng, kettle hole, Syddanmark (DK) – SCA	AAR-8161	12,065	80	antler	*Rangifer tarandus*	modified	14,130-13,690	22; 24
Teufelsküche, Baden-Württemberg (D) – SU	*ETH-7503*	*12,040*	*120*	metatarsus	*Rangifer tarandus*	REJECT: unusually low $δ^{13}C$ value	×	7
Church Hole Cave, Nottinghamshire/England (GB) – GB	OxA-3717*	12,020	100	antler	*Rangifer tarandus?*	artefact: point/rod with scooped end	14,120-13,640	34

Tab. 80 (continued)

site	lab. no.	years ^{14}C-BP	± years	material	species	comment	years cal. b2k	ref.
Meiendorf pond, Schleswig-Holstein (D) – NEP	H-38-121A	12,000	200	antler	*Rangifer tarandus*	PROBLEMATIC: large standard deviation	14,500-13,380	22; 31
Fox Hole Cave, Derbyshire/England (GB) – GB	OxA-1493	11,970	120	antler	*Rangifer tarandus*	artefact: point/rod with scooped end	14,110-13,550	17
Teufelsküche, Baden-Württemberg (D) – SU	*ETH-7502*	*11,960*	*120*	metatarsus	*Rangifer tarandus*	REJECT: unusually low δ^{13}C value	×	7
Borneck Kammer III, Schleswig-Holstein (D) – NEP	KIA-33949	11,940	50	humerus	*Rangifer tarandus*	PROBLEMATIC: unusually high δ^{13}C value	13,930-13,650	35
Petersfels 3, AH 4, Baden-Württemberg (D) – SU	H-4743-4137	11,890	130	bone/antler	*Rangifer tarandus*	PROBLEMATIC: unusual dating of reindeer in SU	14,060-13,420	20
Gough's New Cave, Somerset/England (GB) – GB	*OxA-2797**	*11,870*	*110*	antler	*Rangifer tarandus*	artefact: perforated rod; REJECT: redating available (OxA-18064)	×	11; 27; 36
Meiendorf pond, Schleswig-Holstein (D) – NEP	*W-281*	*11,870*	*200*	antler	*Rangifer tarandus*	REJECT: carbonate fraction?	×	22; 28
Meiendorf pond, Schleswig-Holstein (D) – NEP	*W-264*	*11,790*	*200*	antler	*Rangifer tarandus*	REJECT: without pretreatment?	×	22; 28
Victoria Cave, North Yorkshire/England (GB) – GB	OxA-2455*	11,750	120	antler	*Rangifer tarandus?*	artefact: double bevelled point	13,820-13,340	37-38
Poggenwisch, Schleswig-Holstein (D) – NEP	*W-271*	*11,750*	*200*	antler	*Rangifer tarandus*	REJECT: without pretreatment	×	22; 28
Pellegård, Hovedstaden (DK) – SCA	K-4875	11,650	120	antler	*Rangifer tarandus*	PROBLEMATIC: unusually high δ^{13}C value	13,770-13,210	24
Victoria Cave, North Yorkshire/England (GB) – GB	OxA-2457*	11,590	130	radius	*Rangifer tarandus*		13,710-13,150	37
Nørre Lyngby, Nordjylland (DK) – SCA	AAR-1511	11,570	110	rib	*Rangifer tarandus*	cutmarks	13,630-13,190	25; 39
Klappholz LA 63, Schleswig-Holstein (D) – NEP	AAR-2785	11,560	110	antler	*Rangifer tarandus*	artefact: Lyngby axe	13,620-13,180	25; 40
Hässleberga 1, Skåne (S) – SCA	LuA-4489	11,410	130	humerus	*Rangifer tarandus*		13,550-12,990	41
Hässleberga 1, Skåne (S) – SCA	Ua-3296	11,390	90	antler	*Rangifer tarandus*	shed	13,410-13,050	41
Nørre Lyngby, Nordjylland (DK) – SCA	K-6188	11,370	165	bone	*Rangifer tarandus*		13,550-12,910	25; 39
Hässleberga 1, Skåne (S) – SCA	LuA-4492	11,300	140	metatarsus	*Rangifer tarandus*		13,410-12,890	41
Køge Bugt 2, Sjælland (DK) – SCA	K-4321	11,290	160	bone	*Rangifer tarandus*		13,440-12,840	24
Kinsey Cave, North Yorkshire/England (GB) – GB	OxA-2456*	11,270	110	antler	*Rangifer tarandus*	artefact fragment	13,320-12,920	37
Poggenwisch, Schleswig-Holstein (D) – NEP	*GrN-11262*	*11,250*	*50*	antler	*Rangifer tarandus*	association with archaeology uncertain; REJECT: from Preboreal peat, thus, stratigraphic uncertainties	×	22; 42
Nørre Lyngby, Nordjylland (DK) – SCA	AAR-1910	11,190	100	bone	*Rangifer tarandus*		13,270-12,830	25; 39
Nørre Lyngby, Nordjylland (DK) – SCA	K-6189	11,190	135	bone	*Rangifer tarandus*		13,330-12,730	25; 39
Nørre Lyngby, Nordjylland (DK) – SCA	AAR-1909	11,180	130	bone	*Rangifer tarandus*		13,300-12,740	25; 39

Tab. 80 (continued)

site	lab. no.	years ^{14}C-BP	± years	material	species	comment	years cal. b2k	ref.
Torbryan Six Cave, Devon/England (GB) – GB	*OxA-3894**	*11,130*	*100*	tooth	*Rangifer tarandus*	cutmarks; uncertain association; REJECT: unusually low $\delta^{13}C$ value	×	43
Nørre Lyngby, Nordjylland (DK) – SCA	AAR-1908	11,120	160	bone	*Rangifer tarandus*	PROBLEMATIC: unusually high $\delta^{13}C$ value	13,300-12,660	25; 39
Kalø Vig, Sjælland (DK) – SCA	K-7067	10,990	110	antler	*Rangifer tarandus*		13,070-12,670	24
Kilgreany Cave, Waterford (IRL) – GB	OxA-4240*	10,990	120	bone	*Rangifer tarandus*		13,100-12,660	14
Mickelsmosse, Skåne (S) – SCA	OxA-2791*	10,980	110	antler	*Rangifer tarandus*	club	13,070-12,670	44
Victoria Cave, North Yorkshire/England (GB) – GB	OxA-2454*	10,970	120	mandible	*Rangifer tarandus*		13,090-12,650	37
Hässleberga 1, Skåne (S) – SCA	LuA-4491	10,920	140	calcaneum	*Rangifer tarandus*		13,080-12,600	41
Dyndeby, Hovedstaden (DK) – SCA	K-4874	10,880	150	antler	*Rangifer tarandus*		13,080-12,560	24
Edenvale Cave, Clare (IRL) – GB	OxA-3701*	10,850	80	bone	*Rangifer tarandus*		12,970-12,610	14
Odense Kanal, Syddanmark (DK) – SCA	AAR-9298	10,815	65	antler	*Rangifer tarandus*	artefact: Lyngby axe	12,800-12,640	45
Victoria Cave, North Yorkshire/England (GB) – GB	OxA-2607*	10,810	100	antler	*Rangifer tarandus*	artefact: point with two rows of barbs	12,930-12,570	37
Klemensker, Hovedstaden (DK) – SCA	K-4877	10,780	145	antler	*Rangifer tarandus*	PROBLEMATIC: unusually high $\delta^{13}C$ value	13,010-12,410	24
Ossom's Cave, Staffordshire/England (GB) – GB	OxA-631	10,780	160	mandible	*Rangifer tarandus*	PROBLEMATIC: from disturbed context	13,040-12,360	46
Hässleberga 2, Skåne (S) – SCA	LuA-4493	10,770	150	metacarpus	*Rangifer tarandus*		13,010-12,370	41
Meiendorf pond, Schleswig-Holstein (D) – NEP	*Y-158.2*	*10,760*	*250*	antler	*Rangifer tarandus*	REJECT: collagen date with uncertain pretreatment	×	22; 47
Chelm's Combe Shelter, Somerset/England (GB) – GB	OxA-17828	10,655	45	metacarpus	*Rangifer tarandus*	re-dating of OxA-1781	12,710-12,550	11
Hässleberga 1, Skåne (S) – SCA	LuA-4494	10,640	120	humerus	*Rangifer tarandus*		12,790-12,310	42
Nahe LA 11, Schleswig-Holstein (D) – NEP	*KIA-23369*	*10,610*	*80*	humerus	*Rangifer tarandus*	REJECT: humic acid; too little amount of carbon	×	48
Arreskov, Syddanmark (DK) – SCA	OxA-3173*	10,600	100	antler	*Rangifer tarandus*	club	12,750-12,310	32
Ossom's Cave, Staffordshire/England (GB) – GB	OxA-632	10,600	140	antler	*Rangifer tarandus*	spike	12,830-12,110	46
Chelm's Combe Shelter, Somerset/England (GB) – GB	*OxA-1781*	*10,600*	*200*	metacarpus	*Rangifer tarandus*	REJECT: redating available (OxA-17828)	×	11; 49
Hässleberga 1, Skåne (S) – SCA	LuA-4490	10,580	140	radius	*Rangifer tarandus*		12,810-12,090	41
Nahe LA 11, Schleswig-Holstein (D) – NEP	KIA-23372	10,544	49	antler/bone	*Rangifer tarandus*	with attached cranial bone	12,670-12,390	48
Bølling Sø, Midtjylland (DK) – SCA	K-7079	10,540	80	antler	*Rangifer tarandus*		12,740-12,220	24
Chelm's Combe Shelter, Somerset/England (GB) – GB	OxA-17831	10,480	45	mandible with teeth	*Rangifer tarandus*	re-dating of OxA-1784	12,650-12,210	11
Bjerremarks Gård, Sjælland (DK) – SCA	AAR-4210	10,450	70	antler	*Rangifer tarandus*		12,640-12,080	24
Gough's New Cave, Somerset/England (GB) – GB	OxA-1461	10,450	110	maxilla	*Rangifer tarandus*		12,710-11,990	27; 49
Hässleberga 2, Skåne (S) – SCA	LuA-4496	10,450	140	metatarsus	*Rangifer tarandus*		12,750-11,910	41

Tab. 80 (continued)

site	lab. no.	years ^{14}C-BP	± years	material	species	comment	years cal. b2k	ref.
Køge Bugt 3, Sjælland (DK) – SCA	K-4322	10,380	140	bone	*Rangifer tarandus*		12,690-11,770	24
Earl's Barton, Northamptonshire/England (GB) – GB	OxA-803	10,320	160	antler	*Rangifer tarandus*	artefact: Lyngby axe	12,690-11,530	50
Stellmoor, kettle hole, Schleswig-Holstein (D) – NEP	Y-159.2	10,320	250	antler	*Rangifer tarandus*	PROBLEMATIC: large standard deviation	12,860-11,220	47-48
Nørregård, Hovedstaden (DK) – SCA	*K-4876*	*10,280*	*140*	antler	*Rangifer tarandus*	REJECT: unusually high $\delta^{13}C$ value	×	24
Hässleberga 1, Skåne (S) – SCA	Ua-3294	10,265	140	skull	*Rangifer tarandus*		12,610-11,450	41
Chelm's Combe Shelter, Somerset/England (GB) – GB	*OxA-1784*	*10,230*	*110*	mandible with teeth	*Rangifer tarandus*	REJECT: redating available (OxA-17831)	×	11; 49
Victoria Cave, North Yorkshire/England (GB) – GB	OxA-2453*	10,220	110	antler	*Rangifer tarandus*	artefact: double bevelled point	12,410-11,490	37
Chelm's Combe Shelter, Somerset/England (GB) – GB	BM-2431	10,220	130	tibia	*Rangifer tarandus*		12,500-11,380	27; 51
Hässleberga 1, Skåne (S) – SCA	LuA-4495	10,200	130	pelvis	*Rangifer tarandus*		12,450-11,330	41
Gough's Old Cave, Somerset/England (GB) – SCA	OxA-1120	10,190	120	antler	*Rangifer tarandus*		12,370-11,370	25
Chelm's Combe Shelter, Somerset/England (GB) – GB	BM-2318	10,190	130	radius	*Rangifer tarandus*		12,410-11,330	27; 51
Nahe LA 11, Schleswig-Holstein (D) – NEP	KIA-23370	10,172	45	antler/bone	*Rangifer tarandus*		12,120-11,680	48
Horsens, Nordjylland (DK) – SCA	K-7080	10,170	80	antler	*Rangifer tarandus*		12,210-11,490	24
Chelm's Combe Shelter, Somerset/England (GB) – GB	OxA-17829	10,150	40	mandible with teeth	*Rangifer tarandus*	re-dating of OxA-1782	12,110-11,630	11
Sun Hole, Somerset/England (GB) – GB	OxA-14827**	10,145	55	phalange	*Rangifer tarandus*	cutmarks	12,140-11,540	11; 38
Nahe LA 11, Schleswig-Holstein (D) – NEP	*KIA-23371*	*10,142*	*49*	antler/bone	*Rangifer tarandus*	REJECT: unusually high $\delta^{13}C$ value	12,120-11,560	48
Chelm's Combe Shelter, Somerset/England (GB) – GB	*OxA-1782*	*10,140*	*100*	mandible with teeth	*Rangifer tarandus*	REJECT: redating available (OxA-17829)	×	11; 49
Stellmoor, kettle hole, Schleswig-Holstein (D) – NEP	K-4326	10,140	105	bone	*Rangifer tarandus*		12,260-11,300	22; 32
Kjelleklintegård, Sjælland (DK) – SCA	*K-7069*	*10,140*	*120*	antler	*Rangifer tarandus*	REJECT: unusually high $\delta^{13}C$ value	12,280-11,280	24
Stellmoor, kettle hole, Schleswig-Holstein (D) – NEP	K-4327	10,130	105	antler	*Rangifer tarandus*	PROBLEMATIC: uncertain attribution	12,240-11,280	22; 32
Søhjem, Hovedstaden (DK) – SCA	K-4870	10,120	140	antler	*Rangifer tarandus*		12,300-11,220	24
Meiendorf? or Stellmoor, Schleswig-Holstein (D) – NEP	K-4330	10,110	85	bone	*Rangifer tarandus*	PROBLEMATIC: falsely labled?	12,180-11,300	22; 32
Stellmoor, kettle hole, Schleswig-Holstein (D) – NEP	K-4262	10,110	105	antler	*Rangifer tarandus*		12,220-11,260	22; 32
Stellmoor, kettle hole, Schleswig-Holstein (D) – NEP	K-4578	10,100	100	bone	*Rangifer tarandus*		12,180-11,260	22; 32

Tab. 80 (continued)

site	lab. no.	years ^{14}C-BP	± years	material	species	comment	years cal. b2k	ref.
Hässleberga 1, Skåne (S) – SCA	Ua-3295	10,055	80	skull	*Rangifer tarandus*		12,020-11,260	41
Strangegård, Hovedstaden (DK) – SCA	K-4871	10,050	130	antler	*Rangifer tarandus*		12,170-11,170	24
Dead Man's Cave, South Yorkshire/England (GB) – GB	OxA-5804*	10,020	80	phalange	*Rangifer tarandus*	PROBLEMATIC: either not associated with archaeology or too young date	11,940-11,220	43
Stellmoor, kettle hole, Schleswig-Holstein (D) – NEP	K-4325	10,010	100	bone	*Rangifer tarandus*		11,990-11,190	22; 32
Stellmoor, kettle hole, Schleswig-Holstein (D) – NEP	K-4581	9,990	105	antler	*Rangifer tarandus*		11,970-11,170	22: 32
Stellmoor, kettle hole, Schleswig-Holstein (D) – NEP	K-4579	9,980	105	antler	*Rangifer tarandus*		11,960-11,160	22; 32
Grejsdalen, Midtjylland (DK) – SCA	K-7078	9,950	70	antler	*Rangifer tarandus*		11,780-11,180	24
Vedde, Sjælland (DK) – SCA	K-7077	9,940	100	antler	*Rangifer tarandus*		11,870-11,150	24
King Arthur's Cave, Herefordshire/England (GB) – GB	OxA-6839*	9,930	90	bone	*Rangifer tarandus*		11,800-11,160	1
Stellmoor, kettle hole, Schleswig-Holstein (D) – NEP	K-4323	9,930	100	antler	*Rangifer tarandus*		11,860-11,140	22; 32
Soldier's Hole, Somerset/England (GB) – GB	BM-2249	9,930	210	metacarpus	*Rangifer tarandus*	PROBLEMATIC: large standard deviation	12,230-10,870	52
Gough's New Cave, Somerset/England (GB) – GB	Q-1581	9,920	130	antler	*Rangifer tarandus*		11,930-11,090	53
Stellmoor, kettle hole, Schleswig-Holstein (D) – NEP	K-4324	9,900	105	antler	*Rangifer tarandus*		11,830-11,110	22; 32
Stellmoor, kettle hole, Schleswig-Holstein (D) – NEP	K-4580	9,810	100	antler	*Rangifer tarandus*		11,550-11,030	22; 32
Dead Man's Cave, South Yorkshire/England (GB) – GB	BM-440 (b)	9,750	110	antler	*Rangifer tarandus*	PROBLEMATIC: unassociated with archaeology	11,480-10,760	54-55
Aveline's Hole, Somerset/England (GB) – GB	OxA-802	9,670	110	antler	*Rangifer tarandus*		11,380-10,700	50
Meiendorf pond, Schleswig-Holstein (D) – NEP	Y-158	9,540	130	antler	*Rangifer tarandus*	PROBLEMATIC: pretreatment procedure?	11,320-10,520	22; 47
Stellmoor, kettle hole, Schleswig-Holstein (D) – NEP	W-262	9,500	200	bone	*Rangifer tarandus*	PROBLEMATIC: pretreatment procedure?, large standard deviation	11,410-10,290	22; 28
Stellmoor, kettle hole, Schleswig-Holstein (D) – NEP	*Y-159*	*9,310*	*260*	antler	*Rangifer tarandus*	REJECT: insufficient pretreatment	11,380-9,900	47
Nørre Lyngby, Nordjylland (DK) – SCA	AAR-8919	9,110	65	antler	*Rangifer tarandus*	artefact: Lyngby axe	10,510-10,190	25; 45
Munzingen, Baden-Württemberg (D) – SU	*OxA-4789**	*9,080*	*80*	bone	*Rangifer tarandus*	REJECT: preservative glue	×	8
Meiendorf pond, Schleswig-Holstein (D) – NEP	*Y-158.1*	*7,060*	*400*	antler	*Rangifer tarandus*	REJECT: carbonate fraction	×	22; 47

Tab. 80 (continued)

site	lab. no.	years ^{14}C-BP	± years	material	species	comment	years cal. b2k	ref.
Meiendorf pond, Schleswig-Holstein (D) – NEP	*H-38-121C*	*6,150*	*500*	antler	*Rangifer tarandus*	REJECT: carbonate fraction	×	22; 31
Stellmoor, kettle hole, Schleswig-Holstein (D) – NEP	*Y-159.1*	*5,340*	*200*	antler	*Rangifer tarandus*	REJECT: residue	×	47
horse (*Equus* sp.), n=140								
Wiesbaden-Igstadt, Hessen (D) – EU	OxA-7502	19,320	240	phalange 2	*Equus* sp.		23,660-22,580	2; 4
Wiesbaden-Igstadt, Hessen (D) – EU	OxA-7406	19,200	160	phalange 3	*Equus* sp.		23,550-22,510	2; 4
Wiesbaden-Igstadt, Hessen (D) – EU	OxA-6808*	19,080	160	shaft	*Equus* sp.?		23,540-22,380	2; 4
Wiesbaden-Igstadt, Hessen (D) – EU	UZ-3768 / ETH-13380	17,210	135	tibia	*Equus* sp.	PROBLEMATIC: singular date	21,090-20,290	2
Solutré, Saône-et-Loire/Bourgogne (F) – SEF	OxA-6730*	15,080	130	P3/ P4	*Equus* sp.	cutmarks; PROBLEMATIC: singular date	18,780-17,900	14
Solutré, Saône-et-Loire/Bourgogne (F) – SEF	OxA-6731*	14,570	130	calcaneum	*Equus* sp.	cutmarks	18,090-17,530	14
Solutré, Saône-et-Loire/Bourgogne (F) – SEF	OxA-13299**	14,565	60	maxilla	*Equus* sp.		18,010-17,610	56
Maszycka Cave, Małopolskie (PL) – SP	Ly-2453	14,520	240	bone	*Equus* sp.	PROBLEMATIC: large standard deviation	18,360-17,120	9-10
Kesslerloch, Schaffhausen (CH) – SEF	KIA-11828	13,858	55	radius	*Equus* sp.		17,130-17,010	13
Kniegrotte, Thüringen (D) – EU	OxA-4852*	13,520	130	lumbar vertebra	*Equus* sp.	cutmarks and fracture	17,120-16,400	16; 57
King Arthur's Cave, Herefordshire/England (GB) – GB	*OxA-6734**	*13,320*	*130*	M1/ M2	*Equus* sp.	REJECT: unusually low δ^{13}C value	×	11; 57
Kniegrotte, Thüringen (D) – EU	OxA-4846*	13,190	130	femur	*Equus* sp.	cutmarks and marrow fracture	16,880-15,520	16; 57
Nebra, Sachsen-Anhalt (D) – EU	OxA-11893**	13,160	60	humerus	*Equus* sp.		16,610-15,650	58-59
Kniegrotte, Thüringen (D) – EU	OxA-4848*	13,150	130	metatarsus	*Equus* sp.	cutmarks	16,790-15,470	16; 57
Geißenklösterle, Baden-Württemberg (D) – SU	OxA-6254*	13,130	100	bone	*Equus* sp.		16,650-15,530	57
Abri Stendel XVIII, Niedersachsen (D) – EU	OxA-10470	13,105	70	bone	*Equus* sp.		16,480-15,560	1; 57
Nebra, Sachsen-Anhalt (D) – EU	OxA-11892**	13,070	60	humerus	*Equus* sp.		16,320-15,560	58-59
Monruz, Sect. 1, lower layer, Neuchâtel (CH) – SEF	OxA-20699	13,055	60	bone	*Equus* sp.		16,300-15,540	60
Kesslerloch, Schaffhausen (CH) – SEF	KIA-11827	13,052	53	metacarpus 3	*Equus* sp.		16,270-15,550	13
Petersfels 1, AH 4, Baden-Württemberg (D) – SU	H-7142-7348	12,980	90	bone	*Equus* sp.		16,280-15,320	20
Abri Stendel XVIII, Niedersachsen (D) – EU	OxA-10494**	12,970	70	bone	*Equus* sp.		16,160-15,360	57
Saaleck, Sachsen-Anhalt (D) – EU	OxA-11891**	12,945	60	humerus	*Equus* sp.		15,930-15,370	58-59

Tab. 80 (continued)

site	lab. no.	years ^{14}C-BP	± years	material	species	comment	years cal. b2k	ref.
Gough's New Cave, Somerset/England (GB) – GB	*OxA-3413**	*12,940*	*140*	cervical vertebra	*Equus* sp.	cutmarks; REJECT: redating available (OxA-16292); unusually high $\delta^{13}C$ value	×	11; 27; 34
Kesslerloch, Schaffhausen (CH) – SEF	*KIA-11829*	*12,897*	*53*	metacarpus 3	*Equus* sp.	REJECT: unusually high $\delta^{13}C$ value	×	13
Teufelsbrücke, Thüringen (D) – EU	OxA-5722*	12,860	130	phalange 3	*Equus* sp.	white	16,080-15,080	16
Champréveyres, sect. 1, lower layer, Neuchâtel (CH) – SEF	OxA-20700	12,815	65	bone	*Equus* sp.		15,660-15,260	60
Champréveyres, sect. 1, lower layer, Neuchâtel (CH) – SEF	OxA-20701	12,805	75	bone	*Equus* sp.		15,660-15,220	60
Oelknitz, Thüringen (D) – EU	OxA-5716*	12,790	110	metacarpus 3	*Equus* sp.		15,780-15,020	16
Saaleck, Sachsen-Anhalt (D) – EU	OxA-11890**	12,780	60	metatarsus	*Equus* sp.		15,610-15,210	58-59
Kesslerloch, Schaffhausen (CH) – SEF	*KIA-11825*	*12,774*	*54*	phalange 1	*Equus* sp.	REJECT: unusually high $\delta^{13}C$ value	×	13
Oelknitz, Thüringen (D) – EU	OxA-5713*	12,740	120	phalange 2	*Equus* sp.	cutmarks	15,740-14,820	16
Keblice, Ústecký (CZ) – SU	GrA-37169	12,730	60	tooth	*Equus* sp.		15,530-15,090	61
Abri Fuchskirche, Thüringen (D) – EU	KIA-12928	12,721	65	metacarpus	*Equus* sp.		15,530-15,050	62
Gough's New Cave, Somerset/England (GB) – GB	OxA-11241	12,710	90	tooth	*Equus* sp.		15,590-14,910	57
Wallendorf-Weinberg, Sachsen-Anhalt (D) – EU	OxA-13849	12,685	55	femur	*Equus* sp.	formerly known as Friedensdorf and Kriegsdorf	15,450-15,010	58-59
Keblice, Ústecký (CZ) – SU	GrA-37168	12,680	50	tooth	*Equus* sp.		15,450-15,010	62
Petersfels 1, AH 3, Baden-Württemberg (D) – SU	H-7141-6985	12,680	110	teeth or mandible	*Equus* sp.		15,650-14,610	20
Gough's New Cave, Somerset/England (GB) – GB	*OxA-4106**	*12,670*	*120*	cervical vertebra	*Equus* sp.	cutmarks; REJECT: redating available (OxA-18068)	15,670-14,510	11; 27; 34
Oelknitz, Thüringen (D) – EU	OxA-8075	12,660	80	bone	*Equus* sp.		15,490-14,810	57
Oelknitz, Thüringen (D) – EU	OxA-8076	12,630	75	bone	*Equus* sp.		15,510-14,630	57
Ranis-Ilsenhöhle 5, Thüringen (D) EU	OxA-12052**	12,615	50	metatarsus	*Equus* sp.		15,390-14,830	58-59
King Arthur's Cave, Herefordshire/England (GB) – GB	OxA-17725	12,610	55	M1/ M2	*Equus* sp.		15,440-14,680	11
Sun Hole, Somerset/England (GB) – GB	*OxA-14476*	*12,610*	*90*	P4	*Equus* sp.	fractured; same dentary as OxA-14477; REJECT: too low % wt. coll.	×	11
Abri de la Fru, Savoie/Rhône-Alpes (F) – SEF	OxA-4408*	12,600	120	bone	*Equus* sp.		15,570-14,370	14
Gough's New Cave, Somerset/England (GB) – GB	OxA-16292	12,585	55	cervical vertebra	*Equus* sp.	cutmarks; redating of OxA-3413	15,440-14,560	11
Pixie's Hole, Devon/England (GB) – GB	OxA-14068	12,585	60	P3/ P4	*Equus* sp.	fractured	15,430-14,550	11
Gough's New Cave, Somerset/England (GB) – GB	OxA-17833	12,570	45	phalange 2	*Equus* sp.	cutmarks; redating of OxA-465	15,390-14,550	11
Mother Grundy's Parlour, Derbyshire/England (GB) – GB	OxA-19505	12,545	50	bone/tooth	*Equus* sp.		15,330-14,530	11

Tab. 80 (continued)

site	lab. no.	years ^{14}C-BP	± years	material	species	comment	years cal. b2k	ref.
Sun Hole, Somerset/England (GB) – GB	OxA-14438	12,545	55	tibia	*Equus* sp.	fractured	15,330-14,530	11
Mother Grundy's Parlour, Derbyshire/England (GB) – GB	OxA-4102*	12,540	140	premolar	*Equus* sp.		15,560-14,080	27; 34
Sun Hole, Somerset/England (GB) – GB	*OxA-14477*	*12,540*	*75*	M1	*Equus* sp.	same dentary as OxA-14476; high C/N ratio; REJECT: too low % wt. coll.	×	11
Gough's Old Cave, Somerset/England (GB) – GB	OxA-587	12,530	150	phalange 1	*Equus* sp.	cutmarks	15,560-14,040	27; 46; 63
Gough's New Cave, Somerset/England (GB) – GB	OxA-18068	12,520	55	cervical vertebra	*Equus* sp.	cutmarks; redating of OxA-4106	15,250-14,490	11
Kesslerloch, Schaffhausen (CH) – SEF	KIA-11826	12,502	52	metacarpus 3	*Equus* sp.		15,210-14,450	13
Gough's New Cave, Somerset/England (GB) – GB	*OxA-592*	*12,500*	*160*	metapodium	*Equus* sp.	same sample as BM-2187R; REJECT: amino acids of OxA-591	×	27; 46
Gough's Cave, Somerset/England (GB) – GB	OxA-12104	12,495	50	M1/ M2	*Equus* sp.		15,170-14,450	11; 27
Gough's Cave, Somerset/England (GB) – GB	OxA-18065	12,490	55	phalange 1	*Equus* sp.	cutmarks; redating of OxA-3452	15,200-14,360	11
Sun Hole, Somerset/England (GB) – GB	OxA-18705	12,490	45	M1	*Equus* sp.	cutmarks	15,170-14,450	11
Petersfels 1, AH 3, Baden-Württemberg (D) – SU	H-7137-7067	12,470	100	teeth	*Equus* sp.		15,320-14,080	20
Gough's New Cave, Somerset/England (GB) – GB	*OxA-464*	*12,470*	*160*	metacarpus	*Equus* sp.	cutmarks; REJECT: redating available (OxA-17832); unusually low δ^{13}C value	×	27; 64
Gough's New Cave, Somerset/England (GB) – GB	BM-2186R	12,470	240	metapodium	*Equus* sp.	PROBLEMATIC: large standard deviation	15,690-13,770	27; 53
Mother Grundy's Parlour, Derbyshire/England (GB) – GB	OxA-19504	12,455	55	bone/tooth	*Equus* sp.		15,180-14,180	11
Geißenklösterle, Baden-Württemberg (D) – SU	OxA-5158*	12,450	120	bone	*Equus* sp.		15,330-14,010	57
Hostim, Středočeský (CZ) – SU	Ly-1108	12,420	470	scapula and longbone	*Equus* sp.	PROBLEMATIC: large standard deviation	16,440-13,200	65
Gough's New Cave, Somerset/England (GB) – GB	OxA-17832	12,415	50	metacarpus	*Equus* sp.	cutmarks; redating of OxA-464	15,090-14,090	11
Gough's Cave, Somerset/England (GB) – GB	*OxA-3452**	*12,400*	*110*	phalange 1	*Equus* sp.	cutmarks; REJECT: redating available (OxA-18065)	×	11; 27; 34
Gough's New Cave, Somerset/England (GB) – GB	BM-2188R	12,380	230	metapodium	*Equus* sp.	PROBLEMATIC: large standard deviation	15,520-13,680	27; 53
Gough's New Cave, Somerset/England (GB) – GB	*OxA-590*	*12,370*	*150*	atlas	*Equus* sp.	cutmarks; same sample as BM-2183R; REJECT: amino acids of OxA-589	×	27; 46
Gough's New Cave, Somerset/England (GB) – GB	OxA-465	12,360	170	phalange 2	*Equus* sp.	cutmarks	15,340-13,780	11; 27; 63-64
Gough's New Cave, Somerset/England (GB) – GB	*BM-2183R*	*12,350*	*160*	atlas	*Equus* sp.	cutmarks; REJECT: same sample as OxA-589 and OxA-590	×	27; 53

Tab. 80 (continued)

site	lab. no.	years ^{14}C-BP	± years	material	species	comment	years cal. b2k	ref.
Gough's New Cave, Somerset/England (GB) – GB	OxA-589	12,340	150	atlas	*Equus* sp.	cutmarks; same sample as BM-2183R and OxA-590	15,260-13,780	27; 46
Mother Grundy's Parlour, Derbyshire/England (GB) – GB	OxA-3400*	12,340	110	premolar	*Equus* sp.		15,110-13,870	27; 34
Victoria Cave, North Yorkshire/England (GB) – GB	OxA-15078	12,325	50	atlas	Equus ferus	cutmarks	14,740-14,020	38
Mother Grundy's Parlour, Derbyshire/England (GB) – GB	OxA-19503	12,315	55	bone/tooth	*Equus* sp.		14,740-13,980	11
Gough's New Cave, Somerset/England (GB) – GB	*BM-2187R*	*12,300*	*200*	metapodium	*Equus* sp.	cutmarks; REJECT: same sample as OxA-591 and OxA-592	×	27; 53
Mother Grundy's Parlour, Derbyshire/England (GB) – GB	OxA-19507	12,280	50	bone/tooth	*Equus* sp.		14,630-13,950	11
Mother Grundy's Parlour, Derbyshire/England (GB) – GB	OxA-5698*	12,280	110	premolar	*Equus* sp.	incised/cutmarks	14,870-13,830	27; 43
Mother Grundy's Parlour, Derbyshire/England (GB) – GB	OxA-3398*	12,280	110	premolar	*Equus* sp.		14,870-13,830	27; 34
Oelknitz, Thüringen (D) – EU	OxA-5709*	12,270	120	metacarpus 3	*Equus* sp.	cutmarks	14,900-13,780	16
Gough's New Cave, Somerset/England (GB) – GB	OxA-591	12,260	160	metapodium	*Equus* sp.	same sample as BM-2187R and OxA-592	15,100-13,660	27; 46
Gough's New Cave, Somerset/England (GB) – GB	BM-2184R	12,250	160	calcaneum	*Equus* sp.		15,080-13,640	27; 53
Kent's Cavern, Devon/England (GB) – GB	OxA-5692*	12,250	110	metacarpus	*Equus* sp.	cutmarks; PROBLEMATIC: unusually low C/N ratio	14,820-13,780	27; 43
Kent's Cavern, Devon/England (GB) – GB	OxA-8002	12,240	100	tooth	*Equus* sp.	cutmarks	14,770-13,770	1; 27; 57
King Arthur's Cave, Herefordshire/England (GB) – GB	OxA-6733*	12,230	100	molar	*Equus* sp.	intentionally fractured	14,730-13,770	27; 57
Three Holes Cave, lower hearth, Devon/England (GB) – GB	OxA-4494*	12,220	110	P2	*Equus* sp.		14,760-13,720	66
Robin Hood's Cave, Derbyshire/England (GB) – GB	OxA-6324*	12,210	100	bone	*Equus* sp.		14,630-13,750	57
Gough's New Cave, Somerset/England (GB) – GB	BM-2185R	12,200	250	metapodium	*Equus* sp.	PROBLEMATIC: large standard deviation	15,350-13,390	27; 53
Petersfels 1, AH 3, Baden-Württemberg (D) – SU	H-7139-7300	12,180	100	front limb	*Equus* sp.		14,470-13,750	20
Mother Grundy's Parlour, Derbyshire/England (GB) – GB	OxA-8739	12,170	80	premolar	*Equus* sp.	transversely fractured	14,290-13,810	1; 27; 57
Mother Grundy's Parlour, Derbyshire/England (GB) – GB	OxA-19506	12,155	50	bone/tooth	*Equus* sp.		14,180-13,860	11
King Arthur's Cave, Herefordshire/England (GB) – GB	OxA-6732*	12,150	100	tooth	*Equus* sp.		14,330-13,730	57

Tab. 80 (continued)

site	lab. no.	years ^{14}C-BP	± years	material	species	comment	years cal. b2k	ref.
Three Holes Cave, lower hearth, Devon/England (GB) – GB	*OxA-3890**	*12,150*	*110*	tooth	*Equus* sp.	fractured (snapped); REJECT: unusually low δ^{13}C value	14,410-13,690	27; 66
Oelknitz, Thüringen (D) – EU	OxA-5710*	12,080	110	metatarsus 3	*Equus* sp.	cutmarks	14,230-13,670	16
Oelknitz, Thüringen (D) – EU	OxA-5711*	12,050	110	metacarpus 4	*Equus* sp.	cutmarks	14,180-13,660	16
Mother Grundy's Parlour, Derbyshire/England (GB) – GB	OxA-6666*	12,040	80	tooth	*Equus* sp.		14,090-13,690	57
Buttentalhöhle, Baden-Württemberg (D) – SU	OxA-4981*	12,040	120	bone	*Equus* sp.		14,190-13,630	57
Fox Hole Cave, Derbyshire/England (GB) – GB	OxA-6311*	12,030	90	bone	*Equus* sp.		14,110-13,670	57
King Arthur's Cave, Herefordshire/England (GB) – GB	OxA-6631*	12,000	80	tooth	*Equus* sp.		14,060 1 13,660	57
Mother Grundy's Parlour, Derbyshire/England (GB) – GB	OxA-8738	11,970	75	premolar	*Equus* sp.	transversely fractured	14,010-13,630	1; 57
Three Holes Cave, Devon/England (GB) – GB	OxA-1499	11,970	150	tooth	*Equus* sp.	lower hearth?	14,200-13,480	17
Fox Hole Cave, Derbyshire/England (GB) – GB	OxA-6310*	11,920	130	bone	*Equus* sp.		14,090-13,450	57
Endingen, Horst VI, Mecklenburg-Vorpommern (D) – NEP	UtC-5681	11,830	50	rib	*Equus* sp.	artefact: blade-like shaped rib	13,790-13,510	67
Oelknitz, Thüringen (D) – EU	OxA-5715*	11,810	110	metacarpus 3	*Equus* sp.		13,860-13,420	16; 57
Dead Man's Cave, South Yorkshire/England (GB) – GB	OxA-6326*	11,800	100	bone	*Equus* sp.		13,830-13430	57
Kent's Cavern, Devon/England (GB) – GB	OxA-8003	11,800	180	tooth	*Equus* sp.		14,050-13,250	57; 68
Dead Man's Cave, South Yorkshire/England (GB) – GB	OxA-6327*	11,630	100	bone	*Equus* sp.		13,710-13,230	57
Sun Hole, Somerset/England (GB) – GB	OxA-4986*	11,530	120	M1	*Equus* sp.		13,610-13,130	66
Petersfels 3, AH 2/3, Baden-Württemberg (D) – SU	H-4741-4145	11,300	85	vertebra	*Equus* sp.		13,310-12,990	20
Hässleberga 1, Skåne (S) – SCA	Ua-3969	11,190	100	skull	*Equus* sp.		13,270-12,830	41
Hässleberga 1, Skåne (S) – SCA	Ua-3293	11,180	95	mandible	*Equus* sp.		13,240-12,840	41
Moughton Fell Cave, North Yorkshire/England (GB) – GB	OxA-6321*	11,070	100	bone	*Equus* sp.		13,140 12,700	57
Fox Hole Cave, Derbyshire/England (GB) – GB	OxA-6312*	10,980	90	bone	*Equus* sp.		13,040-12,680	57
Ossom's Cave, Staffordshire/England (GB) – GB	OxA-6316*	10,920	90	tooth	*Equus* sp.		13,010-12,650	57
Bridget Pot Shelter, Somerset/England (GB) – GB	OxA-6632*	10,810	75	bone	*Equus* sp.		12,830-12,630	57
Hässleberga 1, Skåne (S) – SCA	Ua-4763	10,725	110	skull	*Equus* sp.		12,810-12,450	41

Tab. 80 (continued)

site	lab. no.	years ^{14}C-BP	± years	material	species	comment	years cal. b2k	ref.
Sewell's Cave, North Yorkshire/England (GB) – GB	OxA-6636*	10,715	75	bone	*Equus* sp.		12,760-12,560	57
Hässleberga 2, Skåne (S) – SCA	LuA-4497	10,610	130	radius	*Equus* sp.		12,810-12,170	41
Robin Hood's Cave, Derbyshire/England (GB) – GB	BM-604	10,590	90	humerus	*Equus* sp.		12,740-12,340	69
Hässleberga 2, Skåne (S) – SCA	Ua-4765	10,510	95	tibia	*Equus* sp.		12,740-12,100	41
Hässleberga 2, Skåne (S) – SCA	Ua-4764	10,495	95	pelvis	*Equus* sp.		12,730-12,090	41
Robin Hood's Cave, Derbyshire/England (GB) – GB	OxA-6323*	10,460	90	bone	*Equus* sp.		12,690-12,050	57
Kaster, Nordrhein-Westfalen (D) – NEP	OxA-1392 (OxA-78?)	10,380	140	bone	*Equus* sp.	PROBLEMATIC: disturbed context	12,690-11,770	17; 57
Chelm's Combe Shelter, Somerset/England (GB) – GB	OxA-1785	10,370	110	metacarpal	*Equus* sp.		12,610-11,890	49
Sproughton, Devil's Wood Pit, Suffolk/England (GB) – GB	OxA-6315*	10,290	100	bone	*Equus* sp.		12,530-11,690	57
Three Ways Wharf, Scatter A, Greater London/England (GB) – GB	OxA-1788	10,270	100	molar	*Equus* sp.		12,490-11,650	70
Aveline's Hole, Somerset/England (GB) – GB	*OxA-6671**	*10,220*	*90*	tooth	*Equus* sp.	REJECT: unusually low δ^{13}C value and high C/N	×	57
Flixton 2, North Yorkshire/England (GB) – GB	OxA-6319*	10,150	80	bone	*Equus* sp.		12,190-11,430	57
Flixton 2, North Yorkshire/England (GB) – GB	OxA-6328*	10,150	90	bone	*Equus* sp.		12,220-11,380	57
Wolf's Den, Somerset/England (GB) – GB	OxA-6633*	10,125	70	bone	*Equus* sp.		12,150-11,390	57
Flixton 2, North Yorkshire/England (GB) – GB	OxA-6318*	10,090	90	bone	*Equus* sp.		12,150-11,270	57
Teufelsbrücke, Thüringen (D) – EU	OxA-5727*	10,040	120	mandible	*Equus* sp.	white; PROBLEMATIC: very young for the context	12,130-11,170	16; 57
Three Ways Wharf, Scatter A, Greater London/England (GB) – GB	OxA-1902	10,010	120	mandible	*Equus* sp.		12,050-11,170	70
Kendrick's Cave, Conwy/Wales (GB) – GB	OxA-111	10,000	200	mandible	*Equus* sp.	artefact: engraved mandible; PROBLEMATIC: very young for context and large standard deviation	12,290-11,010	64
Flixton 2, North Yorkshire/England (GB) – GB	*OxA-6329**	*9,160*	*80*	bone	*Equus* sp.	REJECT: unusually high C/N ratio	×	57
Wustermark 22, Brandenburg (D) – NEP	?	9,135	75	metatarsus	*Equus* sp.	PROBLEMATIC: disturbed context in alluvial deposits	10,560-10,200	71
Wilczyce, Świętokrzyskie (PL) – SP	*Ua-15721*	*8,415*	*100*	*bone*	Equus *sp.*	Artefact; REJECT: stratigraphic position not in accordance	×	72

Tab. 80 (continued)

site	lab. no.	years ^{14}C-BP	± years	material	species	comment	years cal. b2k	ref.
Victoria Cave, North Yorkshire/England (GB) – GB	*OxA-6634**	*1,740*	*40*	*bone*	Equus *sp.*	REJECT: unusually low $\delta^{13}C$ value and high C/N	×	57
Gough's Cave, Somerset/England (GB) – GB	*OxA-6668**	*1,190*	*40*	*astralagus*	Equus *sp.*	cutmarks; REJECT: unusually low $\delta^{13}C$ value	×	27
Bromme, Sjælland (DK) – SCA	*AAR-4540*	*1,040*	*65*	*tooth*	Equus *sp.*	REJECT: unusually low $\delta^{13}C$ value; apparently too young and no association with archaeology	×	73
Mirkowice 33, Wielkopolskie (PL) – NEP	*UtC-8492*	*165*	*32*	*premolar*	Equus *sp.*	REJECT: apparently too young and association uncertain	×	22; 74
elk (*Alces alces*), n=21								
High Furlong, Lancashire/England (GB) – GB	*OxA-X-2100-6*	*24,410*	*100*	*metacarpus*	*Alces alces*	REJECT: extracted humics/preservatives; non-routine chemistry	×	38
High Furlong, Lancashire/England (GB) – GB	*OxA-151*	*21,500*	*250*	*metatarsus*	*Alces alces*	artefact: barbed point; REJECT: extracted humics/preservatives	×	64
High Furlong, Lancashire/England (GB) – GB	*OxA-X-2066-43*	*16,100*	*70*	*metacarpus*	*Alces alces*	REJECT: extracted humics/preservatives; non-routine chemistry	×	38
High Furlong, Lancashire/England (GB) – GB	*OxA-150*	*12,400*	*300*	*metatarsus*	*Alces alces*	REJECT: amino-acids, presumably contaminated	×	64
Arrie, Skåne (S) – SCA	*St-13310*	*12,390*	*150*	*metatarsus*	*Alces alces*	REJECT: redating available (LuS-7685)	×	25
Abri Fuchskirche, Thüringen (D) – EU	KIA-12926	12,232	50	P4	*Alces alces*		14,320-13,960	62
Abri Fuchskirche, Thüringen (D) – EU	KIA-12925	12,158	50	M3	*Alces alces*		14,180-13,860	62
Vonsmose, Midtjylland (DK) – SCA	K-6124	11,770	190	bone	*Alces alces*		14,030-13,190	25
Borneck Kammer III, Schleswig-Holstein (D) – NEP	KIA-33950	11,770	55	tibia	*Alces alces*		13,750-13,470	35
High Furlong, Lancashire/England (GB) – GB	OxA-13075**	11,715	50	metacarpus	*Alces alces*		13,710-13,430	38
High Furlong, Lancashire/England (GB) – GB	OxA-11151	11,660	60	metacarpus	*Alces alces*	PROBLEMATIC: non-routine chemistry	13,680-13,320	38
Bart's Shelter, Cumbria/England (GB) – GB	OxA-11646	11,600	70	P3 / P4	*Alces alces*		13,580-13,260	38
Braunsbedra, Sachsen-Anhalt (D) – EU	OxA-13283**	11,400	45	antler	*Alces alces ?*	artefact: fish hook	13,340-13,140	58-59
Arrie, Skåne (S) – SCA	LuS-7685	11,345	70	metatarsus	*Alces alces*	re-dating of St-13310	13,320-13,040	25
Coniston(e) Dib, North Yorkshire (GB) – GB	OxA-2847*	11,210	90	bone	*Alces alces ?*	artefact: point	13,260-12,900	37-38
Hässleberga 2, Skåne (S) – SCA	LuA-3969	11,040	130	skull	*Alces alces*		13,150-12,670	41
Klein Nordende D, Schleswig-Holstein (D) – NEP	KIA-33951	11,035	50	femur	*Alces alces*		13,050-12,730	35
Berlin-Tiergarten, Berlin (D) – NEP	KIA-4937	10,730	40	bone	*Alces alces*		12,740-12,620	75

Tab. 80 (continued)

site	lab. no.	years ^{14}C-BP	± years	material	species	comment	years cal. b2k	ref.
Bromme, Sjælland (DK) – SCA	*AAR-4539*	*10,720*	*90*	lumbar vertebra	*Alces alces*	REJECT: too small amount of carbon; minimum age	×	73
Wustermark 22, Brandenburg (D) – NEP	?	10,005	70	antler	*Alces alces*	artefact; PROBLEMATIC: disturbed context in alluvial deposits	11,870-11,230	71
Meiendorf pond, Schleswig-Holstein (D) – NEP	*KIA-33952*	*6,900 (c.)*	*200 (c.)*	bone	*Alces alces*	REJECT: too small amount of carbon; unusually low δ^{13}C value	×	35
wild boar (*Sus scrofa*) / domestic pig (*Sus scrofa domesticus*), n=3								
Wustermark 22, Brandenburg (D) – NEP	unknown	11,720	45	femur	*Sus scrofa*	PROBLEMATIC: disturbed context in alluvial deposits	13,720-13,440	71
Pin Hole, Derbyshire/England (GB) – GB	*OxA-1469*	*3,750*	*80*	bone	*Sus scrofa domesticus*	REJECT: Holocene admixture	×	17
Gough's New Cave, Somerset/England (GB) – GB	*OxA-815*	*1,740*	*60*	bone	*Sus scrofa domesticus*	REJECT: Holocene admixture	×	50
red deer (*Cervus elaphus*), n=36								
Solutré, Saône-et-Loire/Bourgogne (F) – SEF	OxA-13298**	19,740	90	calcaneum	*Cervus elaphus*	PROBLEMATIC: only red deer remain in the assemblage	23,950-23,350	56
Gough's New Cave, Somerset/England (GB) – GB	*OxA-466*	*12,800*	*170*	metatarsal	*Cervus elaphus*	cutmarks; REJECT: unusually low δ^{13}C value	×	11; 27; 63-64
Aveline's Hole, Somerset/England (GB) – GB	OxA-17722	12,565	50	phalange 2	*Cervus elaphus*	cutmarks; possibly redating of OxA-1121	15,380-14,540	11
Gough's New Cave, Somerset/England (GB) – GB	OxA-16378	12,515	50	metatarsal	*Cervus elaphus*	cutmarks; redating of OxA-466	15,220-14,500	11
Gough's New Cave, Somerset/England (GB) – GB	OxA-17845	12,500	50	phalange 2	*Cervus elaphus* (formerly bovid)	cutmarks; redating of OxA-1071	15,190-14,470	11
Gough's New Cave, Somerset/England (GB) – GB	*OxA-3412**	*12,490*	*120*	tibia	*Cervus elaphus*	cutmarks; REJECT: redating available (OxA-18067)	15,420-14,060	11; 27; 34
Aveline's Hole, Somerset/England (GB) – GB	OxA-1121	12,380	130	phalange	*Cervus elaphus* (previously: bovid)	cutmarks; possibly redating available (OxA-17722)	15,250-13,890	26-27
Kesslerloch, Schaffhausen (CH) – SEF	KIA-33351	12,335	45	calcaneus	*Cervus elaphus*		14,730-14,050	13
Ossom's Cave, Staffordshire/England (GB) – GB	OxA-15215	12,310	50	bone/antler	*Cervus elaphus*		14,710-13,990	76
Gough's New Cave, Somerset/England (GB) – GB	*OxA-1071*	*12,300*	*180*	phalange 2	*Cervus elaphus* (previously bovid)	cutmarks; REJECT: redating available (OxA-17845)	15,290-13,650	11; 26-27
Kent's Cavern, Devon/England (GB) – GB	OxA-13683**	12,270	45	metacarpus	*Cervus elaphus*	cutmarks	14,500-13,980	38
King Arthur's Cave, Herefordshire/England (GB) – GB	OxA-6844*	12,250	100	bone	*Cervus elaphus*		14,790-13,790	1
Abri de Rochedane, Doubs, Franche-Comté (F) – SEF	GrA-21512	12,250	70	radius	*Cervus elaphus*		14,600-13,880	77

Tab. 80 (continued)

site	lab. no.	years ^{14}C-BP	± years	material	species	comment	years cal. b2k	ref.
Gough's New Cave, Somerset/England (GB) – GB	OxA-18067	12,245	55	tibia	*Cervus elaphus*	cutmarks; redating of OxA-3412	14,420-13,940	11
King Arthur's Cave, Herefordshire/England (GB) – GB	OxA-1563	12,210	120	tooth	*Cervus elaphus*		14,770-13,690	17
King Arthur's Cave, Herefordshire/England (GB) – GB	OxA-1562	12,120	120	tooth	*Cervus elaphus*		14,370-13,650	17
Aveline's Hole, Somerset/England (GB) – GB	OxA-801	12,100	180	antler	*Cervus elaphus*		14,760-13,480	27; 50
Pixie's Hole, Devon/England (GB) – GB	OxA-5796*	12,070	90	pelvis	*Cervus elaphus*	associated with hearth	14,160-13,680	16; 78
Three Holes Cave, lower hearth, Devon/England (GB) – GB	OxA-3891*	11,980	100	metatarsus	*Cervus elaphus*	fractured	14,060-13,620	27; 43
Plunkett Cave, Keshcorran Caves, Sligo (IRL) – GB	OxA-3693*	11,790	120	bone	*Cervus elaphus*		13,840-13,400	14
Abri de Rochedane, Doubs, Franche-Comté (F) – SEF	GrA-21516	11,600	80	metatarsus	*Cervus elaphus*		13,600-13,240	77
Abri de Rochedane, Doubs, Franche-Comté (F) – SEF	GrA-21514	11,570	70	metatarsus	*Cervus elaphus*		13,560-13,240	77
Porth-y-Waen, Shropshire/England (GB) – GB	OxA-1946	11,390	120	bone/antler?	*Cervus elaphus?*	artefact: barbed point, one row	13,500-12,980	49
Hyaena Den, Wookey Hole, Somerset/England (GB) – GB	OxA-5700*	11,320	120	phalange 2	*Cervus elaphus*		13,390-12,950	43
Lemförde am Dümmer, Niedersachsen (D) – NEP	Hv-14972	10,955	315	antler	*Cervus elaphus*	artefact: point; PROBLEMATIC: only red deer remain in N-Germany before the Holocene	13,440-12,200	79
Chelm's Combe Shelter, Somerset/England (GB) – GB	OxA-1783	10,910	110	mandible	*Cervus elaphus*		13,030-12,630	49
Abri de Rochedane, Doubs, Franche-Comté (F) – SEF	GrA-23147	10,880	50	metatarsus	*Cervus elaphus*		12,900-12,660	77
Abri de Rochedane, Doubs, Franche-Comté (F) – SEF	GrA-23150	10,880	50	metatarsus	*Cervus elaphus*		12,900-12,660	77
Abri de Rochedane, Doubs, Franche-Comté (F) – SEF	GrA-21518	10,830	70	metatarsus	*Cervus elaphus*		12,840-12,640	77
Elderbush Cave, Staffordshire (GB) – GB	OxA-811	10,600	110	vertebra and ribs	*Cervus elaphus*	cutmarks; PROBLEMATIC: uncertain conservation history	12,770-12,250	50
Hyaena Den, Wookey Hole, Somerset/England (GB) – GB	OxA-6728*	10,460	90	tooth	*Cervus elaphus*		12,690-12,050	4
Three Holes Cave, upper hearth, Devon/England (GB) – GB	OxA-4478*	10,020	80	humerus	*Cervus elaphus*	PROBLEMATIC: unusually high $\delta^{13}C$ value	11,940-11,220	43
Elderbush Cave, Staffordshire/England (GB) – GB	*OxA-812*	*9,000*	*130*	bone/charcoal?	*Cervus elaphus?*	REJECT: sample material is questionable	10,510-9,750	50
Abri de Rochedane, Doubs, Franche-Comté (F) – SEF	GrA-21519	8,640	60	metatarsus	*Cervus elaphus*	PROBLEMATIC: unusually low $\delta^{13}C$ value	9,810 -9,530	77

Tab. 80 (continued)

site	lab. no.	years ^{14}C-BP	± years	material	species	comment	years cal. b2k	ref.
Three Holes Cave, Devon/England (GB) – GB	OxA-4491*	6,330	75	bone	*Cervus elaphus*	PROBLEMATIC: Neolithic intrusion?	7,490-7,130	43
Three Holes Cave, Devon/England (GB) – GB	OxA-4492*	6,120	75	bone	*Cervus elaphus*	PROBLEMATIC: Neolithic intrusion?	7,300-6,820	43
large bovids (*Bison priscus / Bos* sp. / *Bos primigenius*), n=31								
Soldier's Hole, Somerset/England (GB) – GB	OxA-649	19,300	400	tibia	*Bos* sp.	PROBLEMATIC: singular date	24,100-22,260	50
Ranis-Ilsenhöhle 4, Thüringen (D) – EU	OxA-12049**	14,780	60	humerus	*Bos* sp. / *Bison* sp.	PROBLEMATIC: singular date and disturbed context	18,720 -17,720	58-59
Pin Hole, Derbyshire/England (GB) – GB	OxA-1615	12,480	160	astralagus	*Bos primigenius*		15,510-13,950	17; 27
Kent's Cavern, Devon/England (GB) – GB	OxA-7994	12,430	80	bone/teeth	*Bos primigenius*	cutmarks	15,210-14,050	27; 68
Kendrick's Cave, Conwy/Wales (GB) – GB	OxA-6146*	12,410	100	bone	*Bos* sp. / *Bison* sp.	PROBLEMATIC: uncertain association	15,230-13,990	80
Pin Hole, Derbyshire/England (GB) – GB	OxA-1471	12,400	140	astragalus	*Bos primigenius*		15,300-13,900	17; 27
Bob's Cave, Devon/England (GB) – GB	OxA-5860*	12,380	90	astralagus	*Bos primigenius*	cutmarks	15,130-13,970	27
Bob's Cave, Devon/England (GB) – GB	OxA-5861*	12,290	90	phalange	*Bos primigenius*	cutmarks	14,810-13,890	27; 81
Holywell Coombe, Kent/England (GB) – GB	OxA-1752	12,280	140	scapula	*Bos primigenius*		15,050-13,730	82
Mother Grundy's Parlour, Derbyshire/England (GB) – GB	OxA-734	12,190	140	bone	*Bos* sp. / *Bison* sp.	found in hearth context	14,820-13,620	50
Mother Grundy's Parlour, Derbyshire/England (GB) – GB	OxA-733	12,060	160	bone	*Bos* sp. / *Bison* sp.	found in hearth context	14,420-13,540	50
Abri Fuchskirche, Thüringen (D) – EU	KIA-12927	12,030	52	radius	*Bos primigenius*		14,030-13,710	62
Gough's New Cave, Somerset/England (GB) – GB	OxA-588	12,030	150	phalange	*Bos primigenius*		14,280-13,560	27; 46; 63
Pixie's Hole, Devon/England (GB) – GB	OxA-5795*	11,910	90	metacarpus	*Bos* sp.		13,990-13,510	16
Gough's New Cave, Somerset/England (GB) – GB	OxA-813	11,900	140	astragalus	*Bos primigenius*		14,090-13,410	27; 50
Kent's Cavern, Devon/England (GB) – GB	OxA-1203	11,880	120	mandible	*Bos primigenius*		14,000-13,440	83
Kent's Cavern, Devon/England (GB) – GB	BM-2168R	11,800	420	atlas	*Bos* sp. / *Bison* sp.	PROBLEMATIC: large standard deviation	15,160-12,680	19
Pixie's Hole, Devon/England (GB) – GB	OxA-5794*	11,630	120	scapula	*Bos* sp.	scratch marks, spiral breakage; PROBLEMATIC: uncertain association	13,750-13,190	16; 78
Kent's Cavern, Devon/England (GB) – GB	BM-2168	11,570	410	atlas	*Bos* sp. / *Bison* sp.	PROBLEMATIC: large standard deviation	14,450-12,610	52
Pin Hole, Derbyshire/England (GB) – GB	OxA-1937	10,970	110	tibia	*Bos primigenius*	PROBLEMATIC: younger admixture or contaminated?	13,060-12,660	17

Tab. 80 (continued)

site	lab. no.	years ^{14}C-BP	± years	material	species	comment	years cal. b2k	ref.
Berzdorf, Sachsen (D) – EU	KIA 15252-2	10,810	50	skull	*Bos primigenius*		12,770-12,650	84
Berzdorf, Sachsen (D) – EU	KIA 15252-1	10,750	35	skull	*Bos primigenius*		12,730-12,650	84
Bedburg-Königshoven, Nordrhein-Westfalen (D) – NEP	*KN-4138*	*10,670*	*100*	skull	*Bos primigenius*	REJECT: from Preboreal peat	12,770-12,410	1; 85
Bedburg-Königshoven, Nordrhein-Westfalen (D) – NEP	KN-4137	10,290	100	skull	*Bos primigenius*		12,530-11,690	85
Bedburg-Königshoven, Nordrhein-Westfalen (D) – NEP	KN-4139	10,140	100	skull	*Bos primigenius*		12,240-11,320	85
Bedburg-Königshoven, Nordrhein-Westfalen (D) – NEP	KN-4136	10,020	100	rib	*Bos primigenius*		12,000-11,200	85
Mother Grundy's Parlour, Derbyshire/England (GB) – GB	*OxA-3399**	*9,910*	*90*	premolar	*Bos* sp. / *Bison* sp.	REJECT: unusually high δ^{13}C value	11,780-11,140	34
Bedburg-Königshoven, Nordrhein-Westfalen (D) – NEP	KN-4135	9,740	100	rib	*Bos primigenius*		11,450-10,770	1; 85
Mother Grundy's Parlour, Derbyshire/England (GB) – GB	*OxA-3395*	*8,480*	*95*	bone	*Bos* sp.	PROBLEMATIC: unusually low δ^{13}C value; REJECT: charred	9,680-9,320	34
Three Holes Cave, Devon/England (GB) – GB	OxA-4493*	5,060	70	tooth	*Bos primigenius*		6,010-5,690	66
Broken Cavern, Devon/England (GB) – GB	OxA-3207*	5,015	80	tooth	*Bos taurus*	juvenile	6,010-5,610	66
Three Holes Cave, Devon/England (GB) – GB	OxA-4495*	5,010	70	tooth	*Bos primigenius*		5,990-5,630	66
Ossom's Cave, Staffordshire/England (GB) – GB	OxA-629	2,030	80	tibia	*Bos* sp. / *Bison* sp.	cutmarks	2,250-1,850	46
others, n=16								
Boltinggaards Skov, Sjælland (DK) – SCA	AAR-1977	14,040	200	bone	*Saiga tatarica*	PROBLEMATIC: large standard deviation	17,720-16,920	25
Boltinggaards Skov, Sjælland (DK) – SCA	AAR-1456	13,880	140	bone	*Saiga tatarica*		17,410-16,890	25
Ranis-Ilsenhöhle 2, Thüringen (D) – EU	OxA-12047**	13,450	60	femur	*Capreolus capreolus*	PROBLEMATIC: species determination roe deer or saiga?	16,980-16,540	58-59
Wilczyce, Świętokrzyskie (PL) – SP	Oxa-16728	13,180	60	tooth	*Alopex lagopus*	artefact: drilled; PROBLEMATIC: uncertain association; sunken into ice wedge	16,650-15,690	86
Kniegrotte, Thüringen (D) – EU	OxA-4850*	13,160	140	tibia	*Alopex lagopus*	cutmarks	16,830-15,470	16
Kniegrotte, Thüringen (D) – EU	OxA-4849*	13,130	120	cranium/horn core	*Saiga tatarica*		16,720-15,480	16
Kniegrotte, Thüringen (D) – EU	OxA-4853*	13,090	130	skull/cranium	*Saiga tatarica*		16,680-15,400	16

Tab. 80 (continued)

site	lab. no.	years ^{14}C-BP	± years	material	species	comment	years cal. b2k	ref.
Gough's New Cave, Somerset/England (GB) – GB	OxA-1200	12,400	110	mandible, partial	*Alopex lagopus*		15,230-13,950	27; 63; 83
Gough'sNew Cave, Somerset/England (GB) – GB	OxA-463	12,380	160	calcaneum	*Saiga tatarica*	unmodified	15,340-13,820	27; 64
Reichwalde 5049, Sachsen (D) – NEP	*GrA-15437*	*12,350*	*50*	bone	*Capreolus* sp.	REJECT: calcined	×	87
Soldier'sHole, Somerset/England (GB) – GB	OxA-1464	12,100	140	metacarpal	*Saiga tatarica*		14,440-13,600	17
Wilczyce, Świętokrzyskie (PL) – SP	Ua-15723	11,890	105	bone	*Alopex lagopus*	PROBLEMATIC: uncertain association; sunken into ice wedge	13,990-13,470	72
Kendrick's Cave, Conwy/Wales (GB) – GB	OxA-6116*	11,795	65	bone	*Capreolus capreolus*	artefact: ochre-stained & engraved tally	13,760-13,480	80
Wilczyce, Świętokrzyskie (PL) – SP	Ua-15722	11,665	135	tooth	*Castor fiber*	PROBLEMATIC: uncertain association; sunken into ice wedge	13,800-13,200	72
Kesslerloch, Schaffhausen (CH) – SEF	KIA-33352	9,920	40	metacarpus	*Capreolus capreolus*	PROBLEMATIC: Holocene admixture?	11,520-11,240	13
Gough's Old Cave, Somerset/England (GB) – GB	OxA-1119	9,320	120	bone	*Castor fiber*	PROBLEMATIC: Holocene admixture	10,920 -10,240	26

Tab. 80 (continued)

HUMAN BEHAVIOUR AND CHRONOLOGY

In the following, results of the analysed archaeological material are described. The chronological relation of the assemblages represents an important issue since only a reliable chronological succession of the assemblages allows for considerations about interrelations of the sites and causal connections with the climatic and environmental changes.

Chronologies of the analysed sites

For most of the studied sites, the chronology was clear and outlined in the previous presentation of the material (see Material-Archaeology, p. 75-244). However, for some assemblages the internal chronology required further discussion after the evaluation and calibration of the ^{14}C dates. The chronological order of the sites is of major interest since it is the base for the establishment of sub-steps in the transition from the Late Magdalenian to the FMG.

Central Rhineland

This project started in the Central Rhineland where 12 Lateglacial complexes were recovered from ten sites. These complexes consisted of several concentrations at Andernach, Gönnersdorf, Kettig, and Niederbieber. Problems regarding the internal chronology at these multi-concentration sites were previously discussed. For a comparison within a process of behavioural change, the possibility of different accumulation periods for these complexes ranging between days and centuries must be kept in mind.
The oldest complexes were assigned to the Late Magdalenian at Gönnersdorf and the lower horizon of Andernach. The general chronological position of Gönnersdorf was relatively undisputed, even though the internal chronology remained unclear. A separate phase within concentration I was considered possible due to material refitting, seasonal indicators, and the ^{14}C ages from this area that tend to be younger than in the other concentrations (see **tab. 20**). However, the more recently published faunal analysis qualified the previous seasonal indicators as incomplete and revealed that activities within concentration I were indeed related to the other concentrations at the site (Street/Turner 2013). The calibrated ^{14}C dates from Gönnersdorf form a consistent set (**fig. 65**) with significant overlap of the probability distributions if separated by concentration. Consequently, the four main concentrations of Gönnersdorf are indistinguishable on a general chronostratigraphic level and must be considered as quasi-contemporaneous.
The results from the lower horizon in Andernach are as comparably indistinguishable as the Late Magdalenian at Gönnersdorf. Although the spatial analysis suggested two possible phases, the environmental indicators and the stratigraphic evidence made a chronological distinction impossible. Moreover, the 15 reliable Late Magdalenian dates scatter evenly across the concentrations I-III. In a χ^2-test, fourteen of these dates are consistent with a single event with a weighted mean of 13,080 ± 25 ^{14}C-BP ($p = 18.0\,\%$). The fifteenth date is a recently dated sample of horse recovered from a pit in concentration III (OxA-V-2223-37) and statistically compatible with the next youngest Late Magdalenian date (OxA-1127, **tab. 20**; Stevens et al. 2009b, 140). However, this young date is younger than the majority of the other reliable dates and also significantly older than the next youngest reliable date (OxA-V-2218-39). After calibration this young Late Magdalenian result overlaps with the calibrated age ranges of several other Late Magdalenian dates but is still inconsistent with the five oldest results. These older results were mainly found in concentration II, whereas the younger results

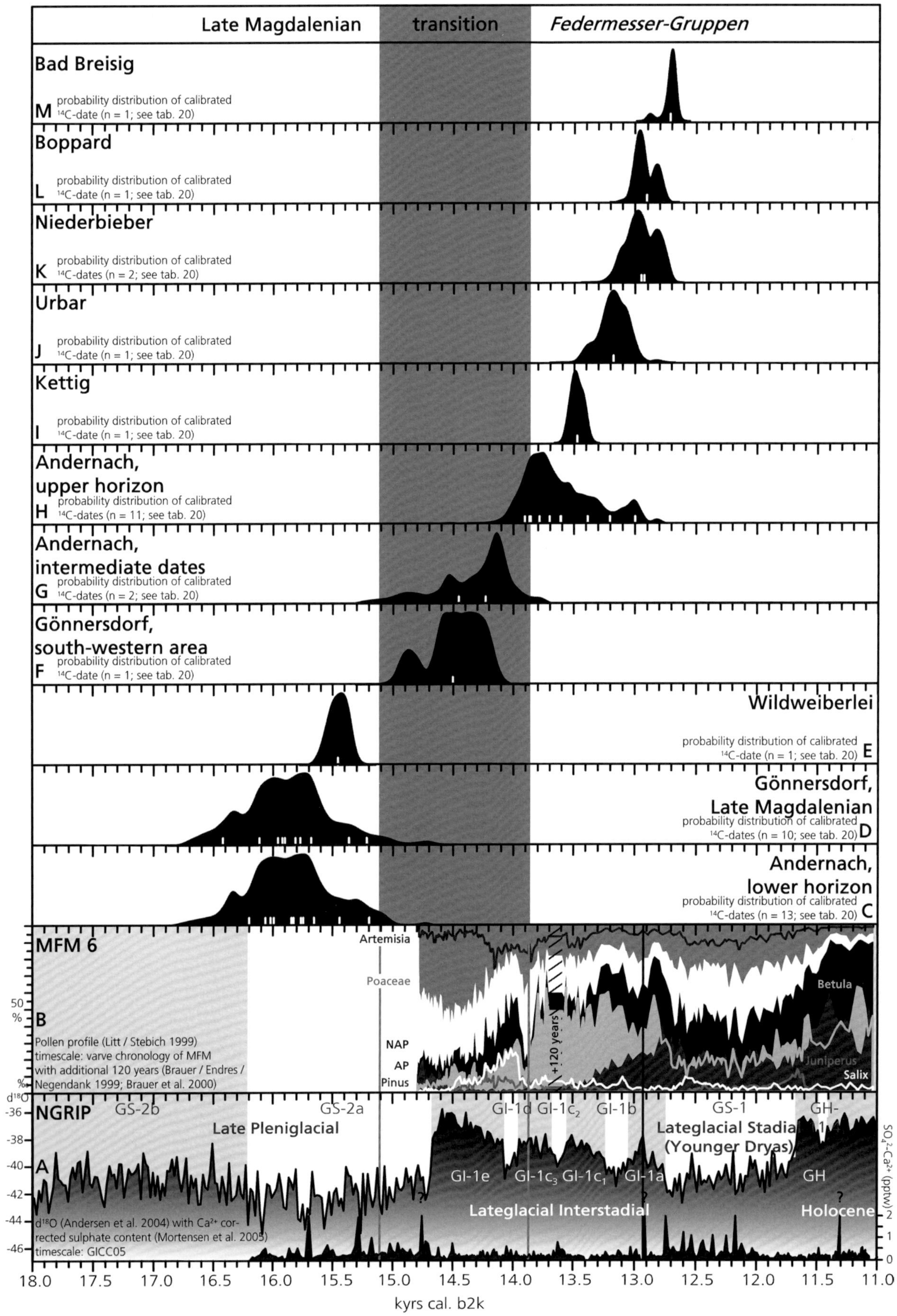

Fig. 65 Probability distributions of the calibrated ^{14}C dates from assemblages from the Central Rhineland (**C-M**) contrasted by the Meerfelder Maar pollen diagram (**B**) and the oxygen isotope record of NGRIP (**A**; see **fig. 53**).

originated mainly from concentration I and III. Thus, two possibly distinct phases within the Late Magdalenian at Andernach are visible in the calibrated age ranges. Nevertheless, this evidence is considered too faint thus far and, therefore, the material from the lower horizon is treated as a single complex until further evidence is presented.

According to the environmental indicators and the radiometric dating, the Late Magdalenian complexes from Gönnersdorf and the lower horizon at Andernach must be considered as quasi-contemporaneous. The amount of archaeological material, its complex spatial distribution, and the radiometric results suggest a use of these sites over a considerable period during the Late Pleniglacial.

In contrast, the small assemblage of the Wildweiberlei yielded only one reliable ^{14}C date and few environmental indicators. The environment appeared comparable to Gönnersdorf and Andernach and the calibrated date fell into the second half of the age range for the Late Magdalenian at these two sites. Consequently, this small assemblage must also be considered quasi-contemporaneous with the last Late Magdalenian occupations at the two Neuwied basin sites. Perhaps, the Wildweiberlei site was used once for a relatively short period and, thus, represents a contrast to the more permanently inhabited open-air sites at Gönnersdorf and Andernach.

A plateau in the calibration curve begins around 12,600 years ^{14}C-BP (c. 15,200 years cal. b2k) and continues until 11,700 years ^{14}C-BP (c. 13,550 years cal. b2k), with an increasing declination of the curve beginning from 12,000 years ^{14}C-BP (c. 13,750 years cal. b2k) onwards. Thus, at the end of the Late Magdalenian age range and at the onset of the FMG dates, the calibration results are less distinct and result occasionally in longer fringes of the probability distributions. Based on the environmental data at the different sites, many of these fringes can be ignored in the Central Rhineland record. Only the elk date from Gönnersdorf and the intermediate dates from Andernach fell completely into the plateau period. Moreover, the results from Bonn-Oberkassel, Irlich, and the main group of dates from the upper horizon in Andernach 2 fell into the final phase of this plateau. Therefore, these dates require some more consideration for a precise attribution. The Gönnersdorf sample was made on elk, which was otherwise unknown from the Late Pleniglacial and found in a stratigraphically higher position. The calibrated age range attributed the date to the last centuries of the Late Pleniglacial and the first temperate event (GI-1e) in the Lateglacial Interstadial. Thus, the radiometric age and the stratigraphic position as well as faunal succession are in accordance with an age at the transition from the Late Pleniglacial to the early Lateglacial Interstadial. However, the sample cannot be related unambiguously with the lithic projectile points. Besides these points, the lithic material recovered in the south-western area appeared to represent a Late Magdalenian assemblage.

Besides the early Late Magdalenian set of dates from Andernach, three further sets of ^{14}C ages can be distinguished in this assemblage (**fig. 66**). An intermediate horizon was identified by two ^{14}C dates made on horse and chamois. Thus far, the archaeological material could not be distinguished unambiguously from the main cluster of archaeological material from the upper horizon. In contrast to the intermediate data set, which falls onto the plateau, the main and late data sets from Andernach fall along a portion of the calibration curve with some steep fluctuations. A first steep decrease (**fig. 66C**, steep fluctuation 1) followed by a comparably steep increase (**fig. 66C**, steep fluctuation 2) occurred approximately between 14,100 and 14,000 years cal. b2k. This period correlated to the first increase of forested vegetation in the Meerfelder Maar record (MFM; **fig. 66B**) and the coldest part of the early Lateglacial Interstadial (GI-1d; **fig. 66A**). This attribution seemed unreliable in comparison with the temperate fauna of this horizon. Furthermore, even though this fluctuation falls beyond the ^{14}C age ranges for the intermediate and a portion of the main data set (**fig. 66**), the majority of dates from the main cluster remain inconsistent with the older dates. The main cluster relates to a small plateau followed by a long-stretched fluctuation in the calibration curve. At the end of this period another steep decline in the curve occurred (**fig. 66C**, steep fluctuation 3). Due to this

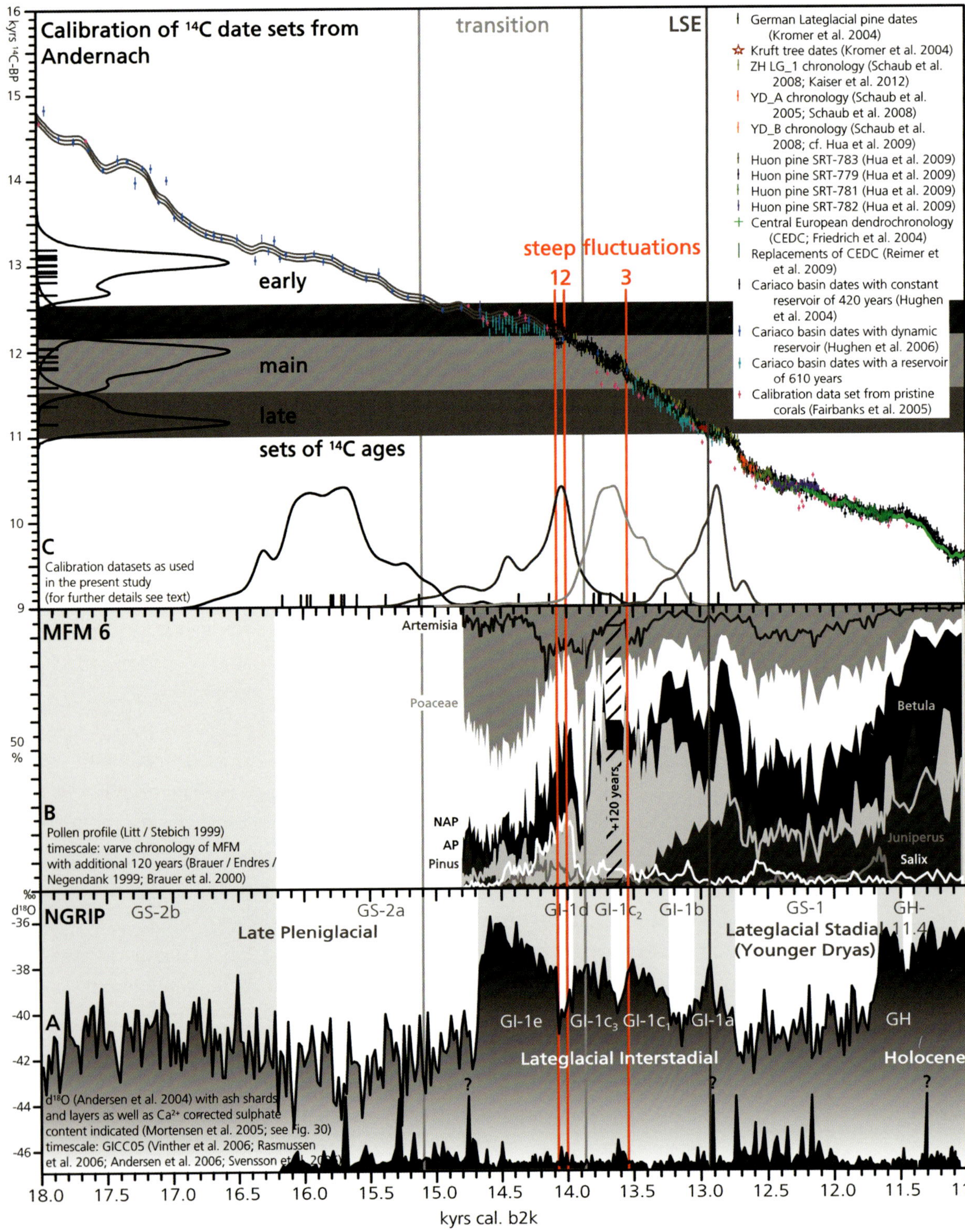

Fig. 66 Calibration of the different data sets of Andernach (**C**; see text and **tab. 20**) contrasted by the Meerfelder Maar pollen diagram (**B**) and the oxygen isotope record of NGRIP (**A**; see **fig. 53**). The limits of the transition between the Late Magdalenian and the FMG are indicated by light grey lines (see **fig. 65**), the Laacher See eruption (LSE) and steep moves in the calibration data which are mentioned in the text are indicated by dark grey lines. – For further details see text.

steep decline, the calibrated age of the older Andernach 3 date overlaps with the calibration results of the main group of dates. However, the remaining two dates of the late phase are not explicable by fluctuations and either reveal a younger period, possibly associated with a phase of chalcedonies in the southern part of the site, or the two dates were contaminated. Consequently, the fluctuations in the calibration curve cannot explain the complete date range from the intermediate to the late data sets and at least two distinct episodes are reflected by the radiometric results. The archaeological remains of the two episodes can be divided into the main accumulation from Andernach 2 and the accumulations from Andernach 3 and the south of Andernach 2. The latter material could not be separated from the main material in Andernach 2 based on the literature (cf. Veil 1982; Bolus 1984; Street 1993; Stapert/Street 1997). Thus, the values given for Andernach 2-FMG in the following material sub-chapters reflect the remains of at least two chronologically distinct episodes. Moreover, if the intermediate dates were considered as indistinguishable from to this main phase, this assemblage accumulated over a considerable period of time. According to the faunal composition, the main FMG phase at Andernach 2 reflected a temperate forest community. This additional information combined with the record from the MFM further narrows the position of the site to the early mid-Lateglacial Interstadial (GI-1c_3; **fig. 66**). The complex spatial patterns and the presence of at least three hearths in this main FMG accumulation could further support the hypothesis of repetitive visits to this site during the mid-Lateglacial Interstadial. The detailed spatial analysis is on-going and could make possible the distinction between different phases within the main concentration and the material associated to these phases. Analysis of the faunal assemblage from Andernach 3 is also still incomplete, but the identified species include red deer and large bovid, presumably aurochs. These species also indicate a temperate forest environment, conditions that prevailed in the late Lateglacial Interstadial to which this phase was dated. Since the horizons were sealed by the LST, this marker sets an upper limit to the age range of the Lateglacial occupation at Andernach.
The seven ^{14}C dates from Bonn-Oberkassel are statistically inconsistent with the hypothesis of a single event. They form two sub-sets, the first comprising the samples from the man (OxA-4790), the dog (KIA-4163), and the penis bone of bear (OxA-4791). This chronologically younger sub-set is similar to the results from Irlich (see p. 472). The second sub-set consists of a sample from the female (OxA-4792) and three remains of dog (KIA-4161, KIA-4163, OxA-4793). However, the oldest and the youngest date from the site were made on left and right ulnae, which were considered as mirror pieces from a sub-adult dog (Street 2002). The presence of two dogs which were almost the same age and physique at the time of death as suggested by the ^{14}C dates appears improbable. Indications for contamination were not apparent although the Oxford dates were made with the ion-exchange pretreatment method, which could result in contamination of the sample, particularly, if the sample yielded only small amounts of carbon (Jacobi/Higham 2009, 1896). Furthermore, preliminary results of a stable isotope analysis on the human remains revealed that their diet was presumably supplemented by freshwater fish (Giemsch/Schmitz in Holzkämper et al. 2014), which has been suggested elsewhere as a potential source of contamination for ^{14}C dates on human remains (Olsen et al. 2010; cf. Nalawade-Chavan et al. 2013); this effect usually resulted in older dates. In addition, if the difference in ages were mainly attributed to nutrituion, the proportion of freshwater fish would have been more significant for the dog and the woman than for the man. Thus, the differences in the radiometric results can probably be attributed to some type of contamination and/or reservoir effect. Therefore, new ^{14}C dates accompanying the stable isotope analysis are desirable. However, assuming that Verworn's conclusion of a double burial is correct (Verworn/Bonnet/Steinmann 1919, 191 f.) and the contamination of the ^{14}C ages is of only minor degree, the comparison of the probability distributions with the calibration curve again indicates that the significant wiggles in the calibration curve could have also contributed to the ambiguous results (**fig. 67**).

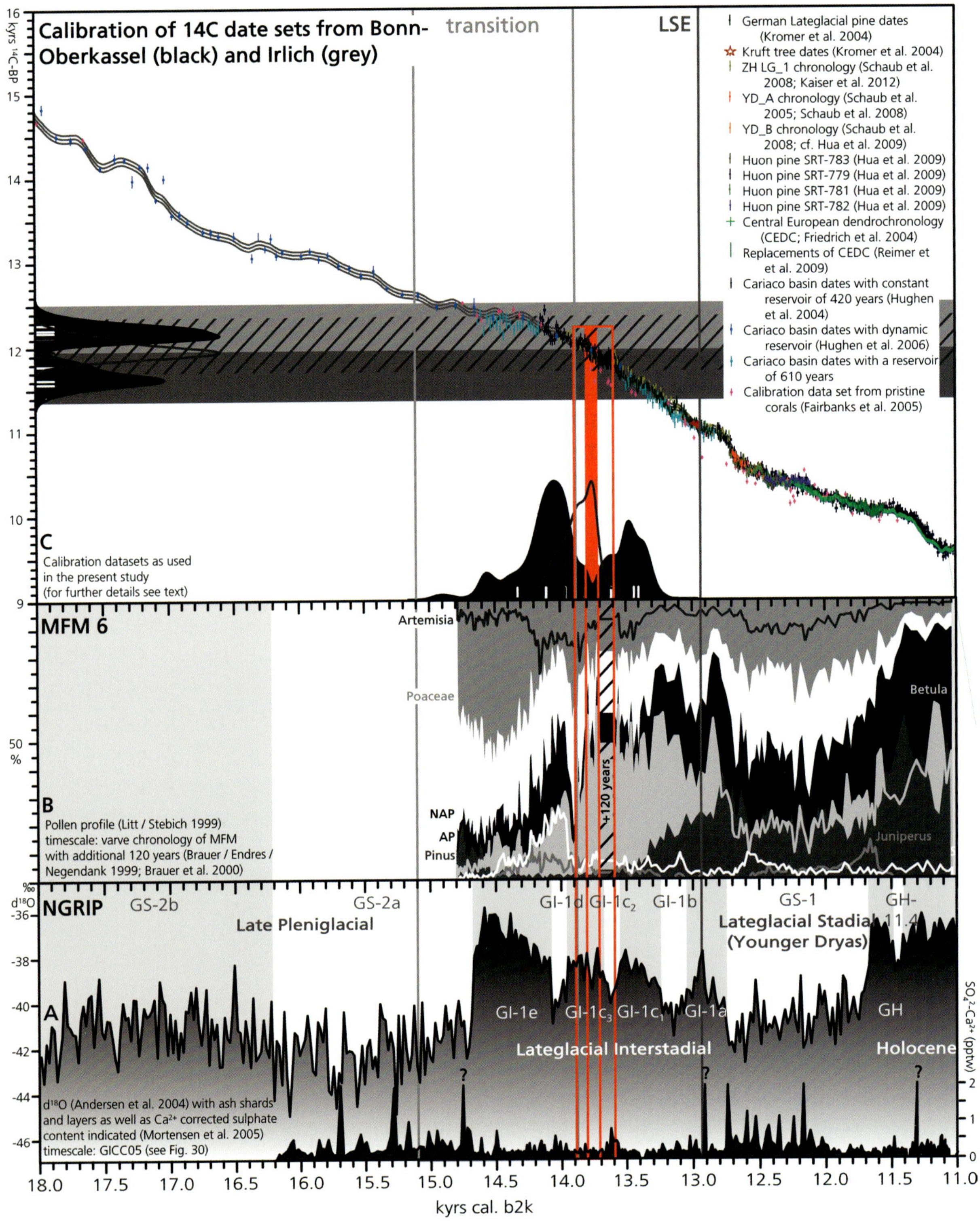

Fig. 67 Calibration of the different date sets of Bonn-Oberkassel (black fields) and Irlich (grey lines; **C**; see text and **tab. 20**) contrasted by the Meerfelder Maar pollen diagram (**B**) and the oxygen isotope record of NGRIP (**A**; see **fig. 53**). The limits of the transition between the Late Magdalenian and the FMG are indicated by light grey lines (see **fig. 65**) and the Laacher See eruption (LSE) are indicated by dark grey lines. – For further details see text.

Five AMS dates were taken on the human material from Irlich and produced only slightly heterogeneous results (**tab. 20**; cf. Bronk Ramsey et al. 2002), which can be evaluated more precisely by the attribution of those remains to the various individuals. Besides the uncoloured, adult skull fragment (OxA-9876) from the Hallstatt period, an uncoloured or brownish coloured bone (OxA-9736) yielded an early Lateglacial Interstadial age that is incompatible with an Oxford date taken on the adult femur (OxA-9847) and the other Oxford dates in a χ^2-test. However, the sample is compatible with another date made in the Utrecht laboratory (UtC-9221). Hence, either the adult bones were from two different individuals that died several centuries apart, or one of the samples was contaminated. Since the reclassification produced no indication of a second adult and attributed further cranial fragments to the young female, the contamination hypothesis appears more probable. In fact, the four Oxford dates were made in the series of Oxford AMS dates (numbers between OxA-9361 and OxA-11851) that might be affected significantly by traces of humectant left in the collagen (Higham et al. 2007, S2 and S55). redating of some of the affected samples resulted in ages that on average were 120 ^{14}C years younger than the previous dates. If a redating of the OxA-9736 date altered the age consistent with these other redatings, OxA-9736 would yield a similar age to the Utrecht date and, thus, be compatible with the other Oxford dates. Nevertheless, although the different colour might indicate different conditions of preservation and, thus, single out the OxA-9736 sample, the other Oxford dates could equally be affected by the contamination. Moreover, the only compatible date from Utrecht was taken on a sample which was considered a juvenile or neonate, possibly from the same individual that supplied a sample for one of the younger Oxford dates (OxA-9848). Perhaps, the samples for OxA-9848 and UtC-9221 originated from the youngest child as the OxA-9848 sample was a rib, and ribs were only attributed to the youngest individual during the reclassification. In addition, the two younger Oxford results (OxA-9847; OxA-9848) and the Utrecht date are congruent with each other in a weighted mean of 11,977 ± 42 ^{14}C-BP (p = 36.67 %). Calibrated, this weighted mean falls along the transition between early to mid-Lateglacial Interstadial; thus, the dates could again be affected by the inconsistency of the carbon isotopes in the atmosphere during the first part of the Lateglacial Interstadial (**fig. 67**). Possibly, the dates can be explained by the same steep decline as the Bonn-Oberkassel dates (see p. 470). In this case, the burial episodes were comparable. In comparison with the MFM, these episodes occurred during a period when light forests became established in the western upland area. According to the calibrated age ranges, this period corresponded to the end of the transition between the Late Magdalenian and the FMG in the Central Rhineland.

The other ^{14}C dates for FMG sites in the Central Rhineland were in accordance with the assumed chronology. However, several dates from these sites for example Niederbieber, had be rejected due to the stratigraphic positions below the LST. In addition, many dates, such as those from Kettig, were rejected due to the uncertainty resulting from dating calcined bone from the Lateglacial (Lanting/Aerts-Bijma/van der Plicht 2001; Lanting/Niekus/Stapert 2002).

According to the calibrated age ranges (**fig. 65**), the transition between the Late Magdalenian and the FMG in the Central Rhineland can be set between 15,100 and 14,000 years cal. b2k. This period corresponds to the end of the Late Pleniglacial and the early Lateglacial Interstadial. However, with the more precisely attributed main group of dates from Andernach 2-FMG, this transitional period could be expanded to approximately 13,900 years cal. b2k. After this transition, a relatively clear succession of sites until shortly after the LSE can be established. Nevertheless, some of the more complex sites such as Kettig and Niederbieber yielded very few dates, therefore, the duration of the occupation of these sites could be underestimated. The single acceptable date from Kettig was made on bulked material and could represent a mean of heterogeneous ages. Two concentrations were observed in Kettig, and with a single possible radiometric date, it remained unclear whether these concentrations were distinct or quasi-contemporaneous episodes. Therefore, new dates from this site are particularly desirable. Comparably, Niederbieber yielded several distinct

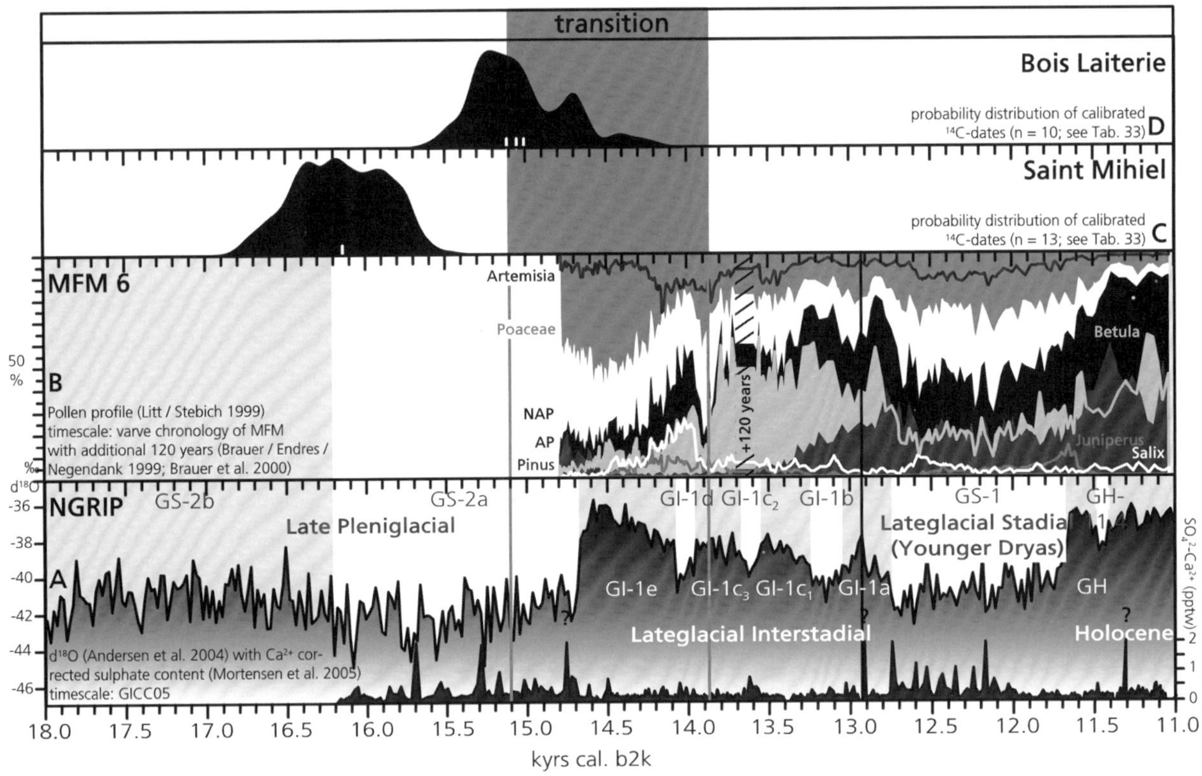

Fig. 68 Probability distributions of the calibrated ^{14}C dates from assemblages from the western uplands (**C-D**) contrasted by the Meerfelder Maar pollen diagram (**B**) and the oxygen isotope record of NGRIP (**A**; see **fig. 53**). The transition period between the Late Magdalenian and the FMG as identified in the Central Rhineland (see **fig. 65**) is indicated in grey. – For further details see text.

concentrations but only two technically acceptable dates. For one of these dates (OxA-1135), the association with the archaeological scatter remained uncertain. Moreover, the samples for both dates originated from concentrations along the periphery, whereas the main area yielded no reliable date. The number of concentrations and their interpretation as repeatedly visited hunting preparation camps (Gelhausen 2011b; Gelhausen 2011a) suggested a longer use of this site that is not reflected in the few radiometric results. Thus, although the dating for many FMG sites suggested very short-lived occupation, this impression resulted from the scarcity of reliable dates. At least for Kettig and Niederbieber, longer periods of use are possible.

Western uplands

The two assemblages from this sub-area are described in relation to the sites from the Central Rhineland to identify their chronological position in the transition process.
The Saint Mihiel assemblage can be attributed to the Late Magdalenian based on the faunal composition and the ^{14}C date (see **tab. 33**). However, the sample for the radiometric measurement was a bulked sample of reindeer remains. The considerable number of unmodified antler remains at the site could indicate that shed and/or fossil material was gathered at the site. A date containing fossil material would result in an older age than the moment of gathering the piece. Thus, even though the ^{14}C date yielded a slightly older result than the data set from Gönnersdorf (**fig. 68**), the assemblage from Saint Mihiel could have formed quasi-contemporaneously with Gönnersdorf and the lower horizon of Andernach.

In contrast, the calibrated age range of Bois Laiterie fell at the onset of the transition period between the Late Magdalenian and the FMG in the Central Rhineland (**fig. 68**) and the onset of the plateau in the calibration curve. The fauna reflected a very heterogeneous habitat and was possibly the result of stratigraphic disturbances, for example caused by a Mesolithic burial. Even though the radiometric dates for this site are very consistent, the direct dating of some unusual species, such as the cave bear, cave hyena, elk, musk ox, steppe bison, and the possible roe deer/saiga antelope material, is desirable to further evaluate their meaning for the Lateglacial occupation. However, the major species which were unambiguously related to human use were still very comparable to the Late Magdalenian assemblages from the Central Rhineland. Thus, according to the faunal evidence and the radiometric results, the Bois Laiterie material can be attributed to the Late Pleniglacial. Nevertheless, the site can be considered as the first assemblage within the transition process. In fact, a reanalysis of the lithic material highlighted differences from the Late Magdalenian (Sano/Maier/Heidenreich 2011), although many similarities in the archaeological assemblages remained consistent with an attribution to the Late Magdalenian such as the stone pavement with occasional engravings and the drilled shells (Miller/López Bayón 1997; Lejeune 1997).

Northern France

Even though the collection of ^{14}C dates from Lateglacial sites has grown to a considerable size in the northern France sub-area (n = 140), many problems and uncertainties remain. For example, several sites produced regularly younger ^{14}C dates than were considered possible based on the stratigraphic and environmental evidence (cf. Bodu 2004; Bodu et al. 2009b; Débout et al. 2012). Several reasons probably caused these arbitrary results.

In the present study, charcoal samples from several of these sites were excluded as possible intrusions from stratigraphically higher positions. From the intermediate and upper horizon of Le Closeau many dates were made on charcoal samples that appeared to yield reliable Lateglacial ages but many of these charcoal patches could have originated from natural wild fires, thus the reliable dates do not necessarily provide information on the occupation period. Besides the probable movement of this material in the light sediment of the river floodplains of various northern French sites, a reanalysis of hearth material from Marsangy indicated that previously assumed wood charcoal samples were heavily contaminated by other materials (Bodu et al. 2009b). Moreover, spatial distributions suggested repeated visits for many of the northern French sites (e.g. Julien et al. 1999; Bodu 2010). Consequently, bulked samples, often necessary for the conventional dating, from these sites were particularly questionable. Nevertheless, several single faunal remains dated in the early days of the AMS method also resulted in unexpected ages, and many of the Oxford dates were made within technically problematic dating series (Jacobi/Higham 2009).

If these ambiguous AMS dates and those from bulked samples were excluded from the record, the number of ^{14}C dates from the Paris Basin decreased significantly (**tab. 44**). However, the problem of considerably younger results would still not be solved. For example, some well-selected samples of charcoal and bone from the Late Magdalenian deposit of Pincevent yielded surprisingly young calibrated ages (**fig. 69**). Furthermore, two of three dates made on horse remains from *locus* 10 at Le Tureau des Gardes (AA-

Fig. 69 Probability distributions of the calibrated ^{14}C dates from assemblages from northern France (**C-U**) contrasted by the synchronised northern French pollen diagram (**B**) and the oxygen isotope record of NGRIP (**A**; see **fig. 53**). * assemblages not analysed in detail in this project. Probability distributions from the Seine-Yonne-confluence basin are set in grey. The transition period between the Late Magdalenian and the FMG as identified in the Central Rhineland (see **fig. 65**) is indicated by a grey background. – For further details see text.

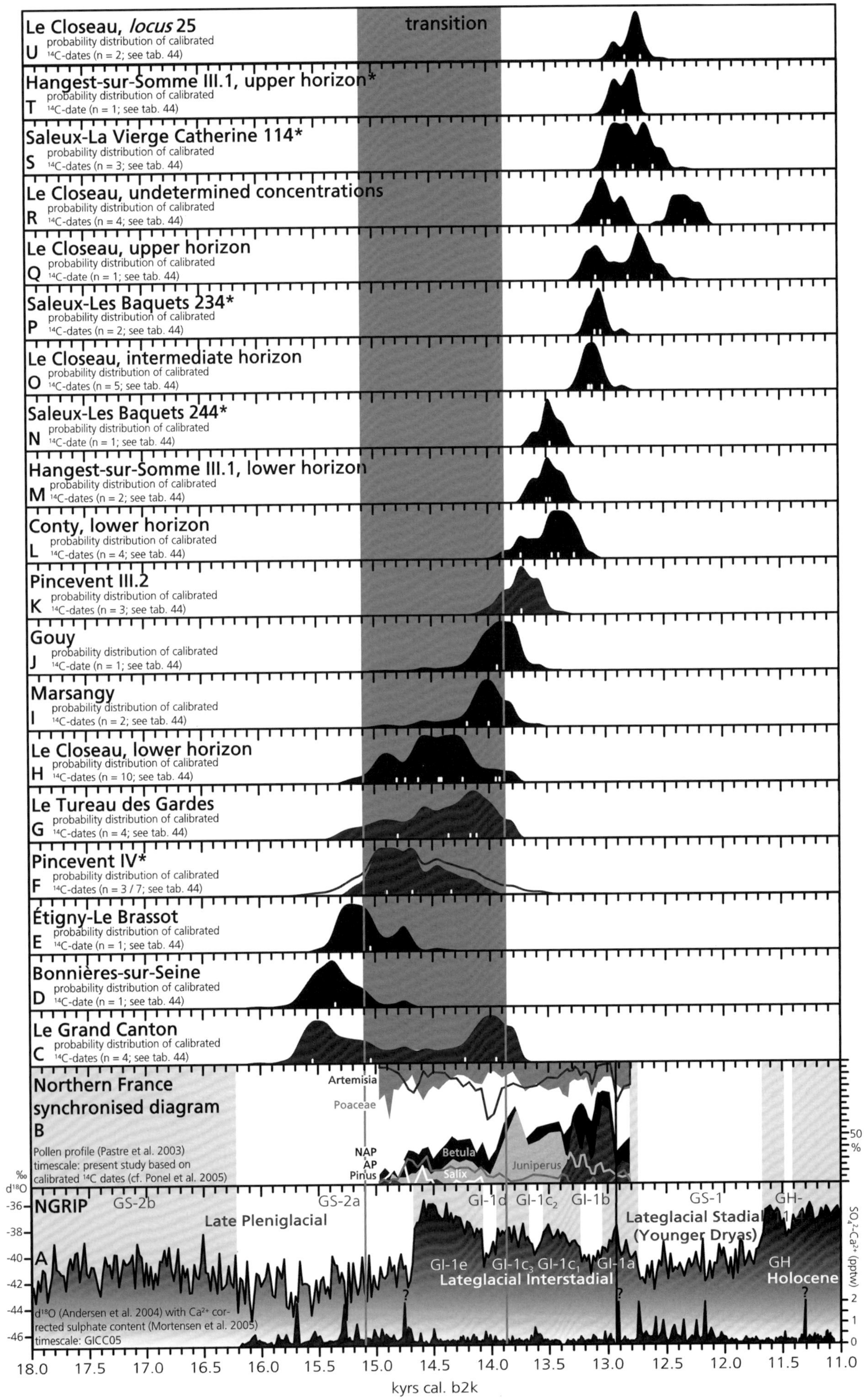

transition
Le Closeau, locus 25
probability distribution of calibrated
U 14C-dates (n = 2; see tab. 44)
Hangest-sur-Somme III.1, upper horizon*
probability distribution of calibrated
T 14C-date (n = 1; see tab. 44)
Saleux-La Vierge Catherine 114*
probability distribution of calibrated
S 14C-dates (n = 3; see tab. 44)
Le Closeau, undetermined concentrations
probability distribution of calibrated
R 14C-dates (n = 4; see tab. 44)
Le Closeau, upper horizon
probability distribution of calibrated
Q 14C-date (n = 1; see tab. 44)
Saleux-Les Baquets 234*
probability distribution of calibrated
P 14C-dates (n = 2; see tab. 44)
Le Closeau, intermediate horizon
probability distribution of calibrated
O 14C-dates (n = 5; see tab. 44)
Saleux-Les Baquets 244*
probability distribution of calibrated
N 14C-date (n = 1; see tab. 44)
Hangest-sur-Somme III.1, lower horizon
probability distribution of calibrated
M 14C-dates (n = 2; see tab. 44)
Conty, lower horizon
probability distribution of calibrated
L 14C-dates (n = 4; see tab. 44)
Pincevent III.2
probability distribution of calibrated
K 14C-dates (n = 3; see tab. 44)
Gouy
probability distribution of calibrated
J 14C-date (n = 1; see tab. 44)
Marsangy
probability distribution of calibrated
I 14C-dates (n = 2; see tab. 44)
Le Closeau, lower horizon
probability distribution of calibrated
H 14C-dates (n = 10; see tab. 44)
Le Tureau des Gardes
probability distribution of calibrated
G 14C-dates (n = 4; see tab. 44)
Pincevent IV*
probability distribution of calibrated
F 14C-dates (n = 3 / 7; see tab. 44)
Étigny-Le Brassot
probability distribution of calibrated
E 14C-date (n = 1; see tab. 44)
Bonnières-sur-Seine
probability distribution of calibrated
D 14C-date (n = 1; see tab. 44)
Le Grand Canton
probability distribution of calibrated
C 14C-dates (n = 4; see tab. 44)
Northern France synchronised diagram
B
Pollen profile (Pastre et al. 2003)
timescale: present study based on calibrated 14C dates (cf. Ponel et al. 2005)
Artemisia
Poaceae
NAP
AP
Pinus
Betula
Salix
Juniperus
50 %
‰ d18O
-36
-38
-40
-42
-44
-46
NGRIP
A
GS-2b
GS-2a
Late Pleniglacial
GI-1d
GI-1c2
GI-1b
GS-1
GH-
Lateglacial Stadial
(Younger Dryas)
GI-1e
GI-1c3
GI-1c1
GI-1a
Lateglacial Interstadial
GH
Holocene
d18O (Andersen et al. 2004) with Ca2+ corrected sulphate content (Mortensen et al. 2005)
timescale: GICC05
SO42--Ca2+ (pptw)
2
1
0
18.0 17.5 17.0 16.5 16.0 15.5 15.0 14.5 14.0 13.5 13.0 12.5 12.0 11.5 11.0
kyrs cal. b2k

44214 and AA-44215) returned relatively young results, whereas the date of AA-44216 was consistent with stratigraphic and archaeological expectations. Despite the stratigraphic and environmental inconsistency a chronological divergence was considered possible for these young Late Magdalenian dates (Débout et al. 2012). In contrast, the date from Étigny-Le-Brassot and numerous dates from the lower horizon in Le Closeau were also made more recently and yielded results older than expected (Bignon/Bodu 2006). Comparably, archaeological material from the lower horizon in Le Closeau and Gouy dated at least 200 ^{14}C years younger (**tab. 44**) than material from horizon III.2 at Pincevent. Only few independent indications for a chronological attribution of this younger horizon at Pincevent were found such as the stratigraphy and the generally temperate fauna but these indicators were in accordance with the calibrated ^{14}C date from this horizon. For the offset between Le Closeau, Gouy, and Pincevent III.2, the wiggling calibration curve could provide a possible explanation, but the Late Magdalenian discrepancies are usually too large to be explicable by fluctuations in the calibration curve. Consequently, these results further affirmed that many previous dates from the Paris Basin were possibly unreliable. According to the climatic and environmental comparisons in the present project, a lag development of the Paris Basin in contrast to the relatively consistent record of the Late Magdalenian in the Central Rhineland, the western uplands, or Switzerland (cf. Débout et al. 2012) appears improbable. However, in comparison with the synchronised and NGRIP-tuned pollen profile, many of the younger dates correlated with a period in the early Lateglacial Interstadial in which open vegetation returned. Nevertheless, the evidence from Marsangy suggested that in this period loamy deposits had already formed, which was not the case in the Late Magdalenian horizons of Pincevent, Le Tureau des Gardes, or Le Grand Canton. The question remains: Why do the samples from these sites produce such young dates?

The dates from the Late Magdalenian concentrations at Étiolles or Verberie are generally consistent with the Late Magdalenian records from the neighbouring areas. Younger dates from these sites could be linked to one of the previously named sources of contamination. However, some young dates, slightly older or comparable to the result from the south-western area of Gönnersdorf, remained unexplained. Perhaps, these samples also related to a last Late Magdalenian visit to these sites. To answer whether these last visits contained material that showed initial changes in the behavioural repertoires, such as found in Bois Laiterie or Gönnersdorf, a comprehensive presentation of these sites, their *loci*, and their material is required due to the complex stratigraphic and spatial distribution of the material (Audouze/Enloe 1997; Olive 2004; Enloe 2006; Olive/Pigeot 2006). Analyses of these sites are still on-going, but clarifications on the assemblages from which the dated material originated may help to evaluate the chronological questions in this area. The TL dates and the reliable dates of Étiolles (Olive 2004, tab. 1) indicate that U5-P15 can unambiguously be attributed to the Late Pleniglacial. Only a small horse concentration (A17) produced a significantly younger date which suggested the comparability to the horse date at Andernach. At Verberie, the oldest date (GifA-95454: 12,950 ± 130 years ^{14}C-BP; 16,380-15,180 years cal. b2k; Bodu 2004) originated from an intermediate horizon (II.2), whereas the youngest date (GifA-99421: 12,300 ± 120 years ^{14}C-BP; 14,990-13,830 years cal. b2k; Bodu 2004) originated from a lower horizon (II.3; Audouze/Enloe 1997). From this lower horizon another older date originated (GifA-99106: 12,520 ± 120 years ^{14}C-BP; 15,470-14,110 years cal. b2k; Bodu 2004) and a further old date came from the upper horizon (II.1; GifA-95453: 12,430 ± 120 years ^{14}C-BP; 15,300-13,980 years cal. b2k; Bodu 2004). Thus, three of the dates fall at the transition between the Late Pleniglacial and the Lateglacial Interstadial. The fauna at Verberie was dominated by reindeer, indicating sufficiently cold and open areas. This assessment was partially supported by the small malacological assemblage from the archaeological horizons (David 1994; Rodriguez 1994). The malacological sample taken on top of the archaeological horizon indicated an amelioration based on increased numbers of forest species and dry indicators. These environmental parameters could also relate to a later period in the

Lateglacial Interstadial. Since this sample overlaid the archaeological horizons, the youngest archaeological horizon should not be younger than the early Lateglacial Interstadial, but more likely dates to the youngest period of the Late Pleniglacial.

In the two sites near Marolles-sur-Seine (Le Tureau des Gardes and Le Grand Canton), the horizontal and vertical stratigraphy was complex and occasionally featured imprecise distinctions between earlier and later episodes of occupation. At both sites, material suggesting an archaeologically younger development was found. Thus, admixture of different events could have resulted in younger dates. Some areas at these sites were reliably attributed to the Late Magdalenian yet still yielded younger ages (e.g. Le Tureau des Gardes 6). However, this argument is not applicable to Pincevent. Although the dated micro-charcoal produced older results, those dates were still too young in comparison with the stratigraphy, the environmental setting, and the results from the neighbouring regions. The recently dated bone was again clearly too young. However, the three problematic sites were located within a basin formed by the confluence of the Seine and the Yonne rivers. Perhaps, some undetected geochemical process contaminated the organic material deposited in the floodplain of this basin. The absorption of some type of reservoir by drinking the waters in this basin during prehistoric times seems an improbable explanation because the few dates from the nearby site Le Tilloy appeared less affected. A reservoir due to dissolved carbon from carboniferous mountains, for example, would result in older ages, not younger ones. Therefore, contamination with younger residues such as humic acids are considered more probable. More detailed geochemical analyses within this basin are suggested to identify potential contaminations. As such, the dates from this area are considered with caution.

Additionally, the relatively old ^{14}C date from Hallines (**tab. 44**) was not unambiguously associated with the actions of humans. Even though the faunal composition appeared as comparably heterogeneous as Bois Laiterie, suggesting a very late Late Pleniglacial age, the questionable relationship with archaeological material limits the usefulness in a chronological argumentation. Thus, this assemblage has to be attributed mainly stratigraphically. According to the stratigraphic position, two episodes must be assumed, both probably during the Late Pleniglacial. A more precise attribution would be desirable; until then, the site is placed at the onset of the succession of sites.

Comparably, the date of Bonnières-sur-Seine probably predates the archaeological material. Based on this *terminus post quem*, the stratigraphy, the composition of the faunal and the lithic assemblage, the archaeological material is considered quasi-contemporaneous with the inventory of Bois Laiterie.

In addition, the material from Cepoy and the lower horizon from Belloy-sur-Somme yielded no ^{14}C date thus far. No organic material was preserved at either site, so these assemblages are only attributed stratigraphically. Cepoy was dated to the early Lateglacial Interstadial due to a change in the sediment regime. The lower horizon of Belloy-sur-Somme was found within a soil from the first part of the Lateglacial Interstadial, probably younger than the Cepoy material. However, the Marsangy material was found in a loamy deposit, perhaps linking the formation of a soil in northern France to this early Lateglacial Interstadial.

In Le Closeau, the probability distributions of calibrated ^{14}C dates from *loci* attributed to the intermediate horizon and those from the *locus* 25 indicated a clear succession. However, the calibrated age range for the upper horizon encompassed both periods. Furthermore, the calibrated ages from undetermined *loci* produced an age range comparable to those from the intermediate and upper horizon, and one younger than the age range of *locus* 25. Thus, the dates from *locus* 45 and possibly the one from structure IV in the greyish deposits originated from the early Younger Dryas.

These assemblages were attributed to different archaeological units in northern France, mainly according to typo-technological observations (Schmider 1971; Fagnart 1997; Valentin 2008a). With this archaeological distinction, the calibration results, and the stratigraphic considerations about the sites, a succession of the

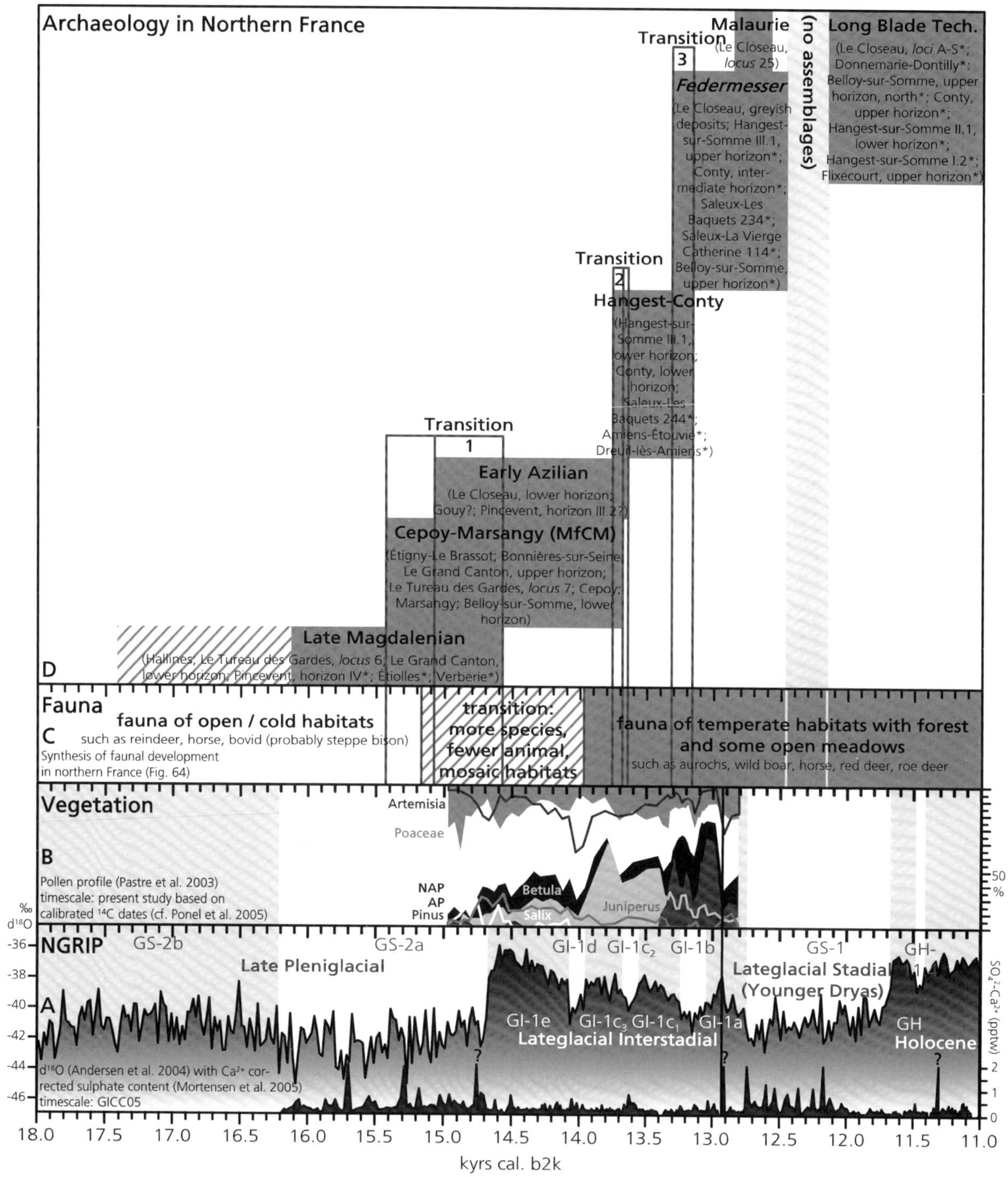

Fig. 70 Succession of archaeological units (bold) and attributed assemblages in northern France (**D**) contrasted by the synthesised faunal (**C**; see **fig. 64**) and vegetation development in northern France (**B**; see **fig. 59**) as well as the oxygen isotope record of NGRIP (**A**; see **fig. 53**). Some similarities can arise from the fact that the vegetation development was synchronised with the oxygen istotpe record and the faunal database relied mainly on the material from the archaeological assemblages. * assemblages not analysed in detail in this project. – For further details see text.

archaeological units can be established for northern France (**fig. 70**). This succession shows that the behavioural development in northern France was only interrupted during the mid-Lateglacial Stadial to which no assemblage was reliably attributed.

Assuming that the different typo-technological units reflected different behavioural recipes (cf. Mesoudi/O'Brien 2008b), a gradual replacement of some behaviours is revealed by the often considerable temporal overlap of these units. Three periods of transition in which three, or at least two, different units occur concomitantly can be identified.

In the first transition, the Late Magdalenian occurs contemporaneously with assemblages attributed to the Final Magdalenian MfCM (MfCM), and both were accompanied by Early Azilian assemblages. The end of this transition is set to the youngest reliable attribution of the Late Magdalenian. This period relates to the end of the Late Pleniglacial and the onset of the Lateglacial Interstadial.

However, after this transition the MfCM and the Early Azilian continue to co-occur. This co-existence can be explained by an insufficient chronological resolution or two different modes of adaptation, where two different groups behave differently or one group behaves differently depending on the seasonal cycle or the function of the site. In the latter case, a supposed ethno-historical differentiation of typo-technological units (Audouze/Valentin 2010) would not be applicable to these groups. In fact, the chronological resolution of these assemblages is problematic due to the plateau in the calibration curve at the transition from the Late Pleniglacial to the Lateglacial Interstadial. However, a co-occurrence cannot be disproven by the chronology. The comparability of the behavioural expressions seen in MfCM and Early Azilian assemblages is discussed in the following chapters.

The co-occurrence of these entities ends with the second transition period when these behavioural repertoires appear for a very short period contemporaneously with inventories from the lower horizons of Hangest-sur-Somme III.1 and Conty. The assemblages from the Somme region were considered as a very early faciès of FMG (Coudret/Fagnart 1997) and comparable to the Hengistbury Head type industries of southern England (Barton et al. 2009). The very short intervals between the appearance of these Hangest-Conty assemblages and the disappearance of the MfCM and the Early Azilian imply a very rapid process. According to the correlation of the present study, this transition relates to the end of a first expansion of trees to northern France and the end of a more unstable period in the NGRIP climate record (**fig. 70**). After the transitional period, these inventories are the only types of assemblages identified in northern France until the third transition.

The third transition in northern France marks the overlap of these Hangest-Conty assemblages with those attributed to the typical FMG, as they were also found in the Central Rhineland. It correlates with the onset of pine-dominated forests in northern France. Although not as long as the overlaps during Transition 1, the third transition gives the impression of a more gradual change from one type of behaviour to another. The identification of a fourth or fifth transition was impossible due to the gap in the record. For instance, according to the calibration results, the typical Malaurie assemblage from Le Closeau, *locus* 25 forms a sub-set of these FMG assemblages. The behavioural recipe associated with Malaurie points (Laborian) possibly superseded the FMG in the mid-Lateglacial Stadial, as was indicated in south-western France (Le Tensorer 1981; Célérier/Chollet/Hantaï 1997). How the Long Blade Technology assemblages were related to these Laborian behaviours remains a matter of debate, but new data can be revealed by on-going research on the assemblages of the Long Blade Technology and their affiliations (personal communication, Mara-Julia Weber).

In conclusion, the varying lengths of overlap imply that the process of change in northern France was a continuous but uneven development.

A comparative and successional representation of all Lateglacial complexes considered in this study (**fig. 71**) reveals that relatively few assemblages fall into the transition period identified in the Central Rhineland (Bois Laiterie; Étigny-Le-Brassot; Le Closeau, lower horizon; Gönnersdorf, south-western area; Cepoy; Marsangy; Gouy; Belloy-sur-Somme, lower horizon). The calibrated age ranges of Le Tureau des Gardes, Le Grand Canton, and Pincevent IV are also attributed to the transition period. However, some sub-assemblages of Le

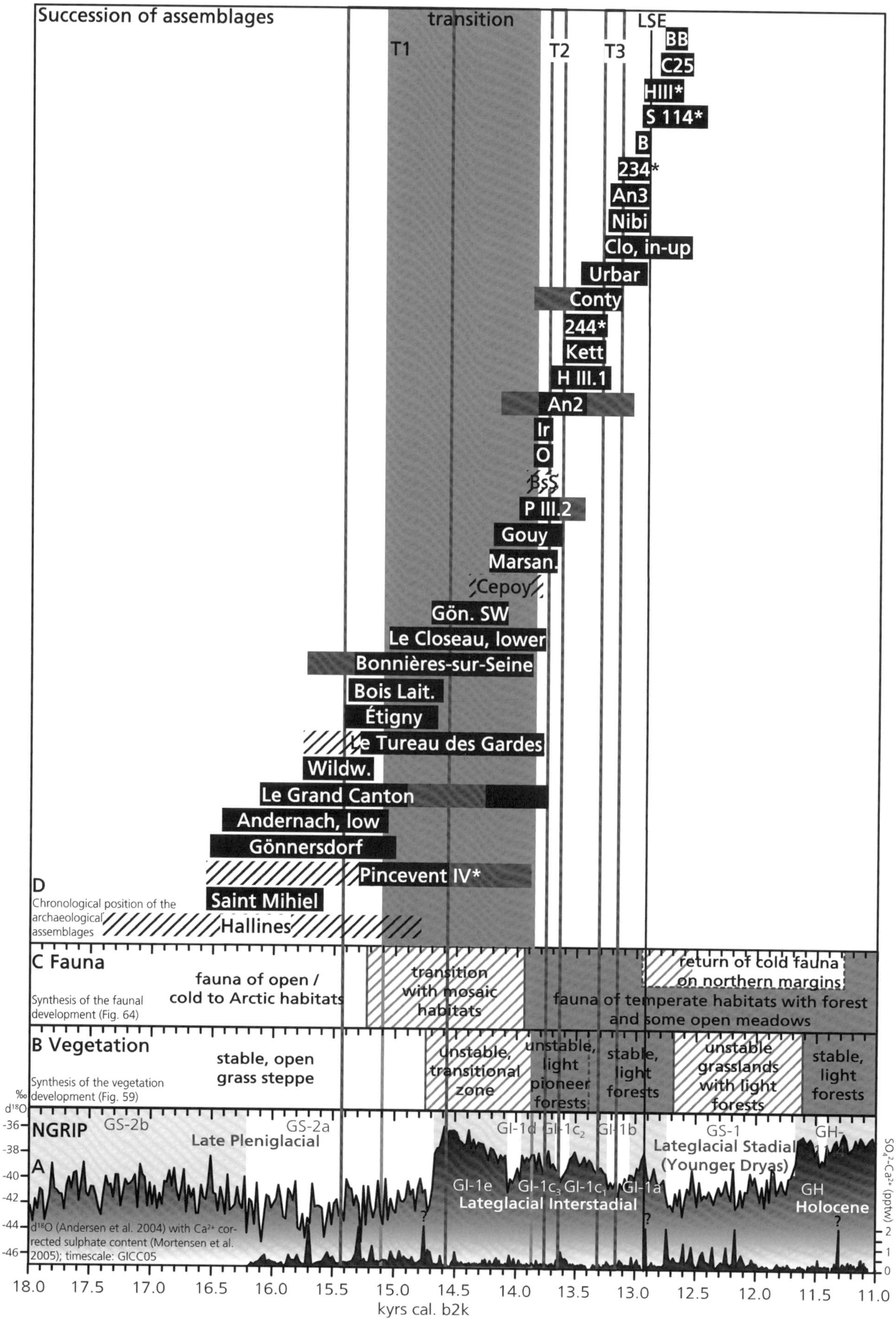
Succession of assemblages
transition
T1
T2
T3
LSE
BB
C25
HIII*
S 114*
B
234*
An3
Nibi
Clo, in-up
Urbar
Conty
244*
Kett
H III.1
An2
Ir
O
BsS
P III.2
Gouy
Marsan.
Cepoy
Gön. SW
Le Closeau, lower
Bonnières-sur-Seine
Bois Lait.
Étigny
Le Tureau des Gardes
Wildw.
Le Grand Canton
Andernach, low
Gönnersdorf
Pincevent IV*
D
Chronological position of the archaeological assemblages
Saint Mihiel
Hallines
C Fauna
Synthesis of the faunal development (Fig. 64)
fauna of open / cold to Arctic habitats
transition with mosaic habitats
fauna of temperate habitats with forest and some open meadows
return of cold fauna on northern margins
B Vegetation
Synthesis of the vegetation development (Fig. 59)
stable, open grass steppe
unstable, transitional zone
unstable, light pioneer forests
stable, light forests
unstable grasslands with light forests
stable, light forests
‰ d18O
NGRIP
GS-2b
GS-2a
Late Pleniglacial
GI-1d
GI-1c2
GI-1b
GS-1
GH-
Lateglacial Stadial (Younger Dryas)
A
GI-1e
GI-1c3
GI-1c1
GI-1a
Lateglacial Interstadial
GH
Holocene
d18O (Andersen et al. 2004) with Ca2+ corrected sulphate content (Mortensen et al. 2005); timescale: GICC05
SO42--Ca2+ (pptw)
kyrs cal. b2k

Tureau des Gardes and Le Grand Canton could possibly originate from this period but the Late Magdalenian assemblages from these sites and from Pincevent IV were certainly older.
In a direct comparison between the transitions in northern France and the transition period in the Central Rhineland (**fig. 71**), Transitions 1 and 2 with the co-occurrence of the MfCM and the Early Azilian, covered the Rhenish transition period. However, according to the archaeological units, Transitions 1 and 3 related to the behavioural change that occurred in the Central Rhineland. In addition, the comparison of the Central Rhineland assemblages with the northern French ones reveals that the faciès represented by the assemblages from the lower horizons at Conty and Hangest-sur-Somme III.1, and possibly by the inventory from Saleux-Les Baquets 244, was quasi-contemporaneous with the main cluster in the upper horizon of Andernach, as well as Kettig and Urbar. Thus, in this mid-Lateglacial Interstadial phase, at least two regional sub-groups can be established for the Somme basin and the Central Rhineland. In contrast, the material from the greyish deposits in Le Closeau as well as the inventories from Saleux-Les Baquets 234 and Saleux-La Vierge Catherine 114 (Fagnart 1997; Coudret/Fagnart 2004; Coudret/Fagnart 2006), dated to the late Lateglacial Interstadial, are similar to the quasi-contemporaneous assemblages in the Central Rhineland.
Fewer changes in the typo-technological units in the Central Rhineland indicate more conservative behaviours in this region than in northern France. Perhaps the richer availability of raw materials in northern France allowed for increased experimentation, which led tp more technical innovations than in the conservative Central Rhineland.

Changes in the Lateglacial exploitation strategies

Acquisition of resources

A first step in the exploitation strategies is the acquisition of the resources. The fossil and mineral resources were presented in the archaeological material chapters according to distance classes (**tabs 12. 25. 36**).
In these tables, a gap can frequently be observed between the very local (5-10 km surrounding of the site) and the next regional raw materials. This gap appears in the Late Magdalenian as well as in the FMG sites and in almost all sites that yielded besides local also a considerable amount of regional raw materials which seemed not to result from exchange but rather from procurement. In the Central Rhineland and the western uplands this gap lies usually around 15-25 km, whereas in the Paris Basin the gap in the few assemblages yielding non local raw materials encompassed 30-40 km. Since the exact acquisition sites of the raw materials are often unknown and, in particular, in the Paris Basin fluvial transport played an important role in diffusing the material, this picture of raw material distances is further indistinct and this diffusion possibly compensates the gap. Therefore, further analyses with more precise localisations of material origins (e.g.

◄

Fig. 71 Succession of assemblages named in this study (**D**) contrasted by the synthesised faunal (**C**; see **fig. 64**) and vegetation development (**B**; see **fig. 59**) as well as the oxygen isotope record of NGRIP (A; see **fig. 53**). Black horizontal bars: probable chronological attribution according to calibrated ^{14}C ages; dark grey horizontal bars: range of unreliable calibrated ^{4}C ages; hatched area: possible age attribution. The transition period between the Late Magdalenian and the FMG as identified in the Central Rhineland (see **fig. 65**) is indicated by a grey background. The limits of the three transition periods in northern France (T1–T3) are indicated by dark grey lines (see **fig. 70**). * assemblages not analysed in detail in this project. Abbreviations: **low** lower horizon; **Wildw.** Wildweiberlei; **Bois Lait.** Bois Laitèrie; **Gön.SW** Gönnersdorf, south-western area; **Marsan.** Marsangy; **P III.2** Pincevent, horizon III.2; **BsS** Belloy-sur-Somme, lower horizon; **O** Bonn-Oberkassel; **Ir** Irlich; **An2** Andernach, upper horizon, main cluster; **H III.1** Hangest-sur-Somme III.1, lower horizon; **Kett** Kettig; **244** Saleux-Les Baquets, *locus* 244; **Clo in-up** Le Closeau, intermediate to upper horizon and undetermined *loci*; **Nibi** Niederbieber; **234** Saleux-Les Baquets, *locus* 234; **B** Boppard; **S 114** Saleux-La Vierge Catherine, *locus* 114; **HIII** Hangest-sur-Somme III.1, upper horizon; **C25** Le Closeau, *locus* 25; **BB** Bad Breisig. – For further details see text.

Pettitt/Rockman/Chenery 2012)[47] are desirable in the future to further locate the source/acquisition areas of raw materials in relation to the site at which the materials were used to allow for further considerations about the Lateglacial settlement systems.

In a more detailed comparison of the tables, a decreasing importance of very long distance sources (250 km >) is apparent between the Late Magdalenian sites and the FMG sites in the Central Rhineland (**tab. 12**). These distances referred particularly to the few numbers of (fossil) shell materials which could be interpreted as special goods. These special good materials were mostly absent from the comparative sites in the western uplands and northern France; only in Bois Laiterie, the upper horizon of Le Grand Canton, and in Marsangy were molluscs found. Probably, these specimens originated from fossil deposits in the Paris Basin and, thus, from a distance of 65-235 km to these sites. Moreover, good quality flint material was available locally in these areas. Therefore, very long distance imports were generally absent from the northern French Late Magdalenian sites[48] and the majority of lithic artefacts was usually made of local raw material. Consequently, these areas are not very useful to exactly identify the end of very far connections. However, since these very long distance imports are already absent in the Wildweiberlei and, presumably, in the younger episode in the south-western area of Gönnersdorf (assuming the jasper artefacts belong to concentration II), the end of these connections can be set to the younger part of the Late Magdalenian occupation (**fig. 72**).

Besides the absence of far distance resources on FMG sites in the Central Rhineland, a tendency towards a major use of local or regional raw materials can be observed on these sites (**tabs 12. 83**; cf. Street et al. 2006, 765-770). In the Late Magdalenian concentrations, local materials such as Tertiary quartzite or indurated slate were usually supplemented by significant amounts of materials such as Western European flint, Baltic flint, or foreign chalcedonies from more distant sources (60 km >). The local to regional Tertiary quartzite clearly dominated in the assemblages attributed to Andernach I and III but from both concentrations the lithic inventory was only partially recovered. This lack of approximately two thirds of the material (Eickhoff-Cziesla 1992, 308) could alter the proportions significantly. Furthermore, according to the spatial analysis and the filling of the pits, Tertiary quartzite was used throughout the occupation of concentration I but it was also used during the last stage of the occupation of this concentration and, thus, other, foreign raw materials could have been cleared out of this concentration (Eickhoff-Cziesla 1992, 309). Therefore, a more numerous supplement to these concentrations by a more distant raw material cannot be excluded. Furthermore, in the concentration III of Gönnersdorf, indurated slate was the most dominant raw material and was not supplemented by an equally numerous distant raw material. Yet, in this concentration the raw material composition was very heterogeneous with several distant raw materials supplementing the inventory (see **tab. 12**, p. 105 f., p. 117, and p. 121). Thus, the Wildweiberlei assemblage, where only the local indurated slate dominated, forms an exception in the Late Magdalenian assemblages. In contrast, the foreign Western European and Baltic flint were only dominant in Kettig and Niederbieber 7 and 20 in FMG assemblages. The other FMG sites and concentrations were dominated by the local indurated slate, Tertiary quartzite, or the regional Muffendorf chalcedony and, thus, during this period people covered the majority of their lithic raw material supply by local to regional sources.

47 The problem of an element analysis in many parts of north-western Europe is that the results can identify the site where the material had been formed but that the raw material has been diffused by glaciers or rivers from this formation site and, thus, the source site where humans gathered the raw material is not identical with the formation site. These source sites remain difficult to identify since the post-formation deposition conditions hardly influenced the material. Often, only the cortex patterns and the shape of complete nodules make some considerations about the depositional processes possible but this information does still not allow the identification of a precise deposition site.

48 Exceptions are jet pearls from habitation 1 of Pincevent which originated from outside the Paris Basin (Leroi-Gourhan/Brézillon 1966, 279) and a mollusc from Étiolles which was of possible Atlantic origin (Alvarez Fernández 2001, 554). In this period, the next Atlantic coast was approximately 435 km west-south-westwards of Étiolles. Known jet sources in southern France are located 430 km south of Pincevent and locations in southern Germany are over 475 km eastwards of Pincevent.

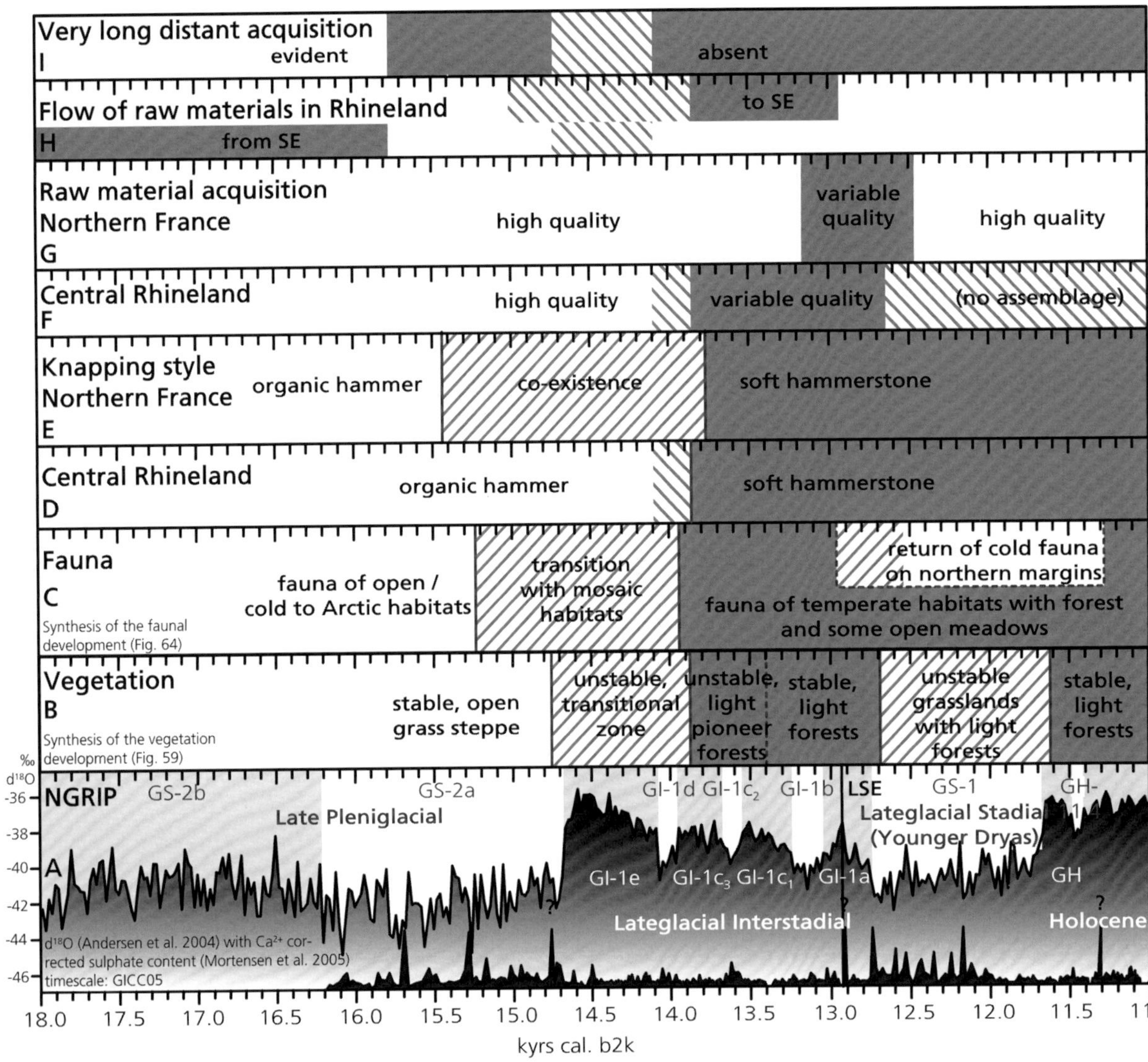

Fig. 72 Behavioural developments related to raw material acquisition (**D-I**) contrasted by the synthesised faunal (**C**; see **fig. 64**) and vegetation development (**B**; see **fig. 59**) as well as the oxygen isotope record of NGRIP (**A**; see **fig. 53**). Hatched areas (low left to up right): transition periods; hatched areas (up left to low right): uncertainties due to the lack of reliable assemblages. – For further details see text.

Nevertheless, except for the chalcedonies and the siliceous oolite from the region around the Rhine-Main basin, foreign raw materials known from the Late Magdalenian assemblages werestill present among the raw materials of FMG sites. Consequently, contacts to these source areas were still evident in the period of the FMG but the relevance of these resources for the everyday supply had decreased.

The chalcedonies used in the Late Magdalenian inventories originated probably from the Main region but no material from this region was identified among the FMG assemblages in the Central Rhineland. However, in the Main region the FMG site Rüsselsheim 122 was located approximately 30 km west of the potential Late Magdalenian source for chalcedonies and the inventory yielded small numbers of Tertiary quartzite which possibly originated from the Taunus and flint varieties which were presumably from the Baltic moraines and/or from the Meuse gravels (Loew 2006). Thus, the acquisition of the latter resource set the Main region in at least a transitional relation to the Central Rhineland during the period before the LSE. In contrast, the raw material from the Late Magdalenian site Dreieich-Götzenhain in the Main region suggested close relations to southern Germany and comprised no northern material (Terberger et al. 2013). Thus, the flow of

raw materials along the Rhine seemed to have changed its direction between the Late Magdalenian when raw materials were taken northwards and the FMG when materials were taken southwards (**fig. 72H**). When exactly this change in direction occurred is difficult to estimate due to the limited evidence. The dating of the sites in the Main region was based on the stratigraphic positions and the techno-typological attribution which made only an approximate chronological position possible. These positions are in accordance with the Late Magdalenian (Dreieich-Götzenhain) and typical FMG sites (Rüsselsheim 122) in the Central Rhineland. In the Central Rhineland, the single artefact from Irlich could possibly be a Muffendorf chalcedony and in the upper horizon of Andernach 2, chalcedonies from Muffendorf were already used. Thus, the change of the raw material flow seemed accomplished then. However, chalcedonies were unrelated to the south-western area of Gönnersdorf and they were also not found among the Wildweiberlei assemblage. Consequently, the transport of materials from the south-east to the Central Rhineland could have stopped as early as the younger period of the Late Magdalenian and, thus, around the same time when long distance relations ceased. Comparable to other distant raw materials such as the Western European flint, the connections between the Rhine-Main region and the Central Rhineland were not permanently capped after the Late Magdalenian but due to the lack of sites with a reliable and detailed chronology from the Main area, it remains a matter of debate when exactly during the transition period the reversed flow began. Moreover, a few artefacts made of a jasper from the Upper Rhine valley were found in the south-western area at Gönnersdorf. These jasper artefacts were spatially related to red deer remains (Buschkämper 1993, Plan 11; Street/Turner 2013, Plan 40) but a connection to activities of a Late Magdalenian occupation of the site, in particular of concentration II cannot be excluded (Buschkämper 1993, 50-52; Street/Turner 2013, 134 f.). Nevertheless, if the jasper was related to the later episode of the south-western area very long distance transport from south-eastern origins would still have occurred within the transition period in the Central Rhineland. Thus, the discontinuation of these far connections had to be placed towards the end of the transition in the Central Rhineland.

An increasing use of the local resources was considered elsewhere as a sign for increasing knowledge about the local resources (Conneller 2007; cf. Rockman 2003a). However, in the Late Magdalenian assemblages from the Central Rhineland a very high quality Tertiary quartzite was already used suggesting that the landscape and its mineral sources were already well explored by the Late Magdalenian hunter-gatherers. Moreover, local sources of fossil material appeared to be also well known by these people because a source for the possibly burnable fossil wood material, which is scarce today, was regularly found in the Late Magdalenian hearths (see **tab. 12** and cf. p. 167). In fact, a difference which is observable between the Late Magdalenian and the FMG assemblages is a decreasing purity and nodule size of the local as well as the distant raw materials and the form in which the distant raw materials were transported (Street et al. 2006; cf. Floss 1994, 332-336). This difference suggests that the size of the raw pieces as well as the purity and, thus, the knapping abilities of a raw material were of a minor importance for the FMG. Thus, the quality was considered sufficient and/or the ability to handle imperfect raw materials which were available at numerous spots in the local and regional surrounding had increased.

This tendency can also be observed in the Paris Basin where FMG sites were no longer set near good quality sources (Valentin 2008b) or nodules of mediocre quality were occasionally selected from these resources for the FMG assemblages (Bodu/Valentin 1997). An attentive choice of the raw materials comparable to the Late Magdalenian acquisition was still observed in assemblages attributed to the Early Azilian such as the lower horizon at Le Closeau or the assemblage from Gouy (Bodu/Valentin 1997). In contrast, the raw material in the upper horizons of Le Closeau appeared more variable in its quality. Moreover, in the Somme valley where raw material of good quality is easily available, the typical FMG inventories such as Saleux 114 or the upper horizon of Hangest-sur-Somme III.1 used a good quality, fine grained Coniacian flint. Nevertheless,

in the assemblages such as the lower horizons of Belloy-sur-Somme, Conty, and Hangest-sur-Somme III.1 an even better, very fine grained, excellent raw material from the Upper Turonian to Lower Coniacian was used (Fagnart 1997). Thus, the decreasing relevance of the raw material properties affect the raw material acquisition in northern France later than in the Central Rhineland (see **fig. 71**).

In both sub-areas, this changed raw material choice (**fig. 72**) occurred in combination with a prevalence in the technical behaviour which was associated with specific characteristics on the raw material blanks attributed to the use of a soft hammerstone (Bodu/Valentin 1997; Valentin 2008a; Ludovic Mevel, written communication)[49]. However, these specific characteristics which distinguish a different knapping style from the typical Late Magdalenian one related to organic hammers occurred already in assemblages of the MfCM (Valentin 2008a) and also in Bois Laiterie (Sano/Maier/Heidenreich 2011). In addition, this knapping style was dominant in the lower horizons of Le Closeau, Hangest-sur-Somme III.1, and Conty (Valentin 2008a). In the assemblages of the MfCM, the relation of the soft hammerstone knapping style with the production of blanks with a straighter longitudinal profile and the use of these blanks in the production of lithic projectile tips was demonstrated (Valentin 1995). The introduction of the technical behaviour associated with the soft hammerstone appeared more flexible and allowed for a wider range of raw materials to be transformed to a designated form. Perhaps, this flexibility turned the balance towards an increasing use of this behavioural recipe. However, the raw material acquisition in the assemblages of the MfCM and the Early Azilian was still comparable to the Late Magdalenian one. In fact, the acquisition of high-quality raw materials did not change until the appearance of typical FMG assemblages in northern France. Thus, the change of acquisition patterns was not instantly connected to the introduction of an alternative technical behaviour but more closely to the discontinued use of the classic knapping style which apparently required a more rigid choice of high-quality raw materials. Nevertheless, after the abandonment of the classic Late Magdalenian blank production process, high-quality raw materials still remained the first choice in the flint-rich area of northern France and raw materials of more variable qualities were only collected during the late Lateglacial Interstadial. Perhaps, the decreasing qualitative choice in northern France was related to an increasing use of harder hammerstones. This tendency was shown for southern Scandinavian assemblages where hard hammerstones became the most important knapping instrument in Brommean inventories during the late Lateglacial Interstadial (Madsen 1992; Madsen 1996). However, on-going technological analyses of northern French inventories and possible comparisons with northern European assemblages can help to evaluate this possibility in the near future.

Exploitation of lithic resources

The size and composition of a lithic assemblage is influenced by raw material types, including the knapping properties, along with the knapping style employed, as well as the function and formation of the site (cf. Löhr 1979; Richter 1990). Understanding these influences, permits a further consideration about the role of a site in the settlement system of a hunter-gatherer group.

49 Even though these characteristics are well described based on material of experienced modern flintknappers (Pelegrin 2000; cf. Madsen 1996), it must remain a matter of debate whether these characteristics were solely related to a different knapping instrument until further experiments with a variety of knapping instruments, raw materials, styles, and larger samples produce a statistically reliable corpus for this correlation. In addition, Ludovic Mevel's on-going technological analysis of the Kettig FMG assemblage might contribute further important observations to this debate because an organic hammer with a zone of scars, presumably from flintknapping, was found within a lithic inventory described as being made by knapping directly with a hard hammerstone (Baales 2002).

assemblage size class	artefacts ≥ 1 cm	artefacts total
5 (large)	**Gönnersdorf**; Le Tureau des Gardes; Étigny-Le Brassot, south; Cepoy; Le Grand Canton, upper horizon; Marsangy; *Le Closeau, greyish layers; Niederbieber*; Le Tureau des Gardes, *locus* 10; Cepoy, sector 1 and sector 2; Le Grand Canton, sector 2, upper horizon; Marsangy N19	**Gönnersdorf**; **Andernach, lower horizon**; *Kettig; Niederbieber; Bad Breisig*
4	**Andernach, lower horizon**; Le Closeau, lower horizon; Belloy-sur-Somme, lower horizon; *Bad Breisig;* **Le Tureau des Gardes, *locus* 6** and *locus* 7	Andernach, upper horizon; **Gönnersdorf II**; **Andernach I and IV**; Le Tureau des Gardes, *locus* 7; *Niederbieber 4+17a, 6+10a, and 18; Andernach 3-FMG*
3	Andernach, upper horizon; Kettig; Le Closeau, *locus* 46 and *loci* 4+50; Gönnersdorf, south-western area; *Niederbieber 1, 4+17a, and 9; Andernach 3-FMG*	Niederbieber 1, 5, 8, 12, 13, and 14
2	Bois Laiterie; Bonnières-sur-Seine; **Andernach IV**; Le Grand Canton, sector 1, upper horizon; *Andernach 2-FMG; Niederbieber 5, 6+10a, 7, and 12*	Bois Laiterie; **Andernach II**; *Niederbieber 9, 11, 16, and 20*
1	**Hallines**; **Saint Mihiel**; **Wildweiberlei**; Hangest-sur-Somme III.1, lower horizon; *Urbar*; **Gönnersdorf IV**; **Le Grand Canton, sector 2, lower horizon**; *Niederbieber 8, 10, 13, 14, 18, and 20*	**Saint Mihiel**; Hangest-sur-Somme III.1, lower horizon; Conty, lower horizon; *Urbar*; **Gönnersdorf IV** and south-western area; *Andernach 2-FMG; Niederbieber 7, 10, and 17*
0 (small)	Gouy; Pincevent III.2; *Niederbieber 3, 11, 15, 16, 17, and 19; Le Closeau, locus 25*	**Wildweiberlei**; Pincevent III.2; **Andernach III**; *Niederbieber 3, 15, and 19*

Tab. 81 Attribution of the studied assemblages to assemblage size classes (see p. 273 f. and **tab. 52**). Late Magdalenian assemblages are set in bold and *Federmesser-Gruppen* assemblages are set in italics.

Firstly, assemblages were differentiated by the number of lithic artefacts to distinguish them by size. Subsequently, the general composition of the lithic inventory was differentiated in relation to the number of cores and retouched artefacts as well as the composition and diversity of the retouched artefacts. A comparison of the assemblages by weight rather than number of the lithic material brought to the site is desirable. For example, the 76 kg of lithic raw material from Gönnersdorf (Floss 1994, 219 f.) provided almost 25 % more artefacts (n ≈ 33,000) than the 334 kg from Le Grand Canton (n = 25,175; Valentin et al. 1999b, tabs 14-15). Certainly in relation to resource exploitation and transport conditions, the weight of lithic raw material imported seemed of a greater importance than the numbers of lithics. However, these values are not regularly presented in the literature and, thus far, further comparisons can therefore not be accomplished.

Generally, comparing assemblages composed of artefacts ≥ 1 cm in size indicates that most of these assemblages are of a small to medium assemblage size (assemblage size classes 1 and 2), whereas larger assemblages (assemblage size classes 4 and 5) were usually found on sites with several concentrations (**tab. 81**). Very small assemblages (assemblage size class 0) were only found in the context of the Early Azilian and the FMG, including the Malaurie *locus* 25 of Le Closeau. However, some of these very small inventories were considered as special task assemblages such as Niederbieber 3, or incomplete excavations such as Gouy. In these inventories, the knapping process and consequential debris was of minor importance or neglected during collection, which can explain the very small numbers (≤ 500 artefacts ≥ 1 cm). The lack of such small, possibly specialised assemblages from the Late Magdalenian could be due to the choice of sites included in this project. However, assemblages considered as special task camps such as Beeck in the northern Rhineland (Jöris/Schmitz/Thissen 1993) or Götzenhain-Dreieich (Terberger et al. 2013) yielded more than 1,000 artefacts ≥ 1 cm. Moreover, lithic workshops and/or lithic procurement sites such as Kanne, Orp, or Mesch (Vermeersch/Lauwers/van Peer 1985; Vermeersch et al. 1987; Rensink 2000) provided large numbers of lithic artefacts, mainly debris. Thus, the absence of very small lithic assemblages from Late Magdalenian

site	no.
Hallines	1
Saint Mihiel	2
Gönnersdorf	3
Andernach, lower horizon	4
Le Grand Canton, sector 2, lower horizon	5
Wildweiberlei	6
Le Tureau des Gardes, *locus* 6	7
Le Tureau des Gardes, *locus* 7	8
Le Tureau des Gardes, *locus* 10	9
Étigny-Le Brassot, south	10
Bois Laiterie	11
Le Grand Canton, sector 1	12
Le Grand Canton, sector 2, upper horizon	13
Bonnières-sur-Seine	14
Le Closeau, *locus* 46	15
Le Closeau, *loci* 4+50	16
Gönnersdorf SW	17
Cepoy, sector 1	18
Cepoy, sector 2	19
Marsangy	20
Gouy	21
Pincevent III.2	22
Belloy-sur-Somme	23
Andernach 2-FMG	24
Hangest-sur-Somme III.1, lower horizon	25
Kettig	26
Conty, lower horizon	27
Urbar	28
Le Closeau, intermediate, upper, and undetermined horizons (greyish deposits)	29
Niederbieber	30
Andernach 3-FMG	31
Le Closeau, *locus* 25	32
Bad Breisig	33

Tab. 82 Numeration of assemblages in the graphs.

contexts[50], perhaps reflects a behavioural habit where lithic material was not discarded unless there was sufficient replacement available. Similarly, a knapping process was not started unless absolutely necessary, or the knapping process created usually more than 500 pieces. Another possible explanation of the lack of these assemblages is that these inventories were not diagnostic enough to be identified as Late Magdalenian remains. However, this overview indicates that the largest variety of assemblage sizes was found in the FMG assemblages.

Even though assemblages of the assemblage size class 0 are absent, Late Magdalenian concentrations are not generally larger than the ones attributed to the FMG. In contrast, some FMG concentrations produced more artefacts than Late Magdalenian ones, in particular in the comparison of total numbers of artefacts (**tab. 81**). In fact, some of the Late Magdalenian inventories fall into the small and very small assemblage size class if the total number of artefacts is considered. This observation suggests that more and tendentially smaller material remained at FMG sites. This increase in discarded material could relate to a change in knapping style, probably using mineral hammers resulting in more splintering of the lithic raw materials during the knapping process. Nevertheless, very small assemblages were all attributed to the use of soft hammerstone but the scarcity of knapping in these assemblages can also explain the lack of numerous splintered material. Furthermore, even though Late Magdalenian concentrations such as Gönnersdorf II were considered to be repeatedly visited or occupied for a longer period, these inventories did not provide more artefacts than Niederbieber 4+17a or Bad Breisig which were usually considered to be short-term camps (Grimm 2004; Gelhausen 2011b). The alternative suggestion that FMG concentrations were also formed by longer occupations or single accumulations formed by repetitive visits is contradicted by the small overall quantities of preserved fauna. The well defined and organised spatial distributions of the lithic artefacts and splinters, the clearly separated concentrations on repetitively visited FMG sites such as Niederbieber (Gelhausen 2011a) or

50 The assemblages of the south-western area from Gönnersdorf and Bois Laiterie were not further attributed to an archaeological unit but summarised in a not further specified Final Magdalenian group as a provisional solution. According to the recent results (Sano/Maier/Heidenreich 2011), an attribution of the Bois Laiterie inventory to the MfCM seems possible but requires further confirmation because it is located outside the main concentration of this archaeological unit (cf. Valentin 2008a, 132-136) and also contained elements of the Late Magdalenian (Lejeune 1997; Straus/Orphal 1997).

site	assemblage size class	density index	% cores	main raw material	exploitation index	% retouched artefacts	function index
Hallines	1 (×)	**158** (×)	**5.82** (×)	Senonian flint	17 (×)	15.4 (×)	2.7
Saint Mihiel	1 (1)	50 (150)	1.00 (0.33)	Western European flint	100 (300)	2.7 (0.9)	2.7
Gönnersdorf	5 (5)	48 (119)	0.94 (0.38)	Tertiary quartzite; Western European flint; Baltic flint	**107** (×)	14.7 (5.9) / 13.1 (5.3)	15.7 / 13.9
Gönnersdorf II	× **(5)**	× (138)	× (0.09)	Western European flint	× **(1,062)**	× (5.2)	55.4
Gönnersdorf IV	1 (1)	7 (22)	**3.28** (1.02)	chalcedony	31 (99)	21.3 (6.6)	6.5
Andernach, lower horizon	4 (5)	27 (179)	0.84 (0.13)	Tertiary quartzite; Western European flint	119 (797)	28.1 (4.2)	33.3
Andernach I	× **(4)**	× **(209)**	× (0.18)	Tertiary quartzite; Baltic flint	× **(543)**	× (3.3)	17.7
Andernach II	× (2)	× **(341)**	× (0.05)	Western European flint	× **(1,930)**	× (8.7)	167.7
Andernach III	× (0)	× (59)	× (0.45)	Tertiary quartzite; Baltic flint; chalcedony	× (223)	× (16.2)	36.2
Andernach IV	2 **(4)**	11 (175)	0.75 (0.05)	Western European flint	133 (2,202)	23.0 **(1.4)**	30.6
Le Grand Canton, sector 2, lower horizon	1 (×)	41 (×)	0.49 (×)	Senonian gravel flint	**206** (×)	**2.6** (×)	5.3
Wildweiberlei	1 (0)	24 (28)	4.83 (4.23)	indurated slate (lydite)	21 (24)	17.1 (15.0)	3.5
Le Tureau des Gardes	5 (×)	**132** (×)	1.55 (×)	local Cretaceous flint	65 (×)	**5.7** (×)	3.7
Le Tureau des Gardes, *locus* 6	**4** (×)	**853** (×)	1.41 (×)	local Cretaceous flint	71 (×)	**5.8** (×)	4.2
Le Tureau des Gardes, *locus* 7	4 (4)	54 (114)	1.11 (0.53)	local Cretaceous flint	90 (190)	10.4 (4.9)	9.3
Le Tureau des Gardes, *locus* 10	5 (×)	**166** (×)	**1.96** (×)	local Cretaceous flint	51 (×)	8.4 (×)	4.3
Étigny-Le Brassot, south	5 (×)	41 (×)	× (×)	Senonian gravel flint	× (×)	**1.2** (×)	×
Bois Laiterie	2 (2)	58 (107)	0.22 (0.12)	Western European flint	454 (842)	14.0 (7.5)	63.5
Le Grand Canton, upper horizon	5 (×)	24 (×)	**4.13** (×)	Senonian gravel flint	24 (×)	**4.2** (×)	1.0
Le Grand Canton, sector 1	2 (×)	17 (×)	**18.41** (×)	Senonian gravel flint	5 (×)	11.1 (×)	0.6
Le Grand Canton, sector 2, upper horizon	5 (×)	47 (×)	**3.51** (×)	Senonian gravel flint	29 (×)	**3.9** (×)	1.1
Bonnières-sur-Seine	2 (×)	44 (×)	0.53 (×)	local flint	188 (×)	**1.0** (×)	1.9
Le Closeau, lower horizon	4 (×)	6 (×)	0.83 (×)	Campanian flint	**120** (×)	8.8 (×)	~10.5
Le Closeau, *locus* 46	3 (×)	12 (×)	1.00 (×)	Campanian flint	**100** (×)	10.6 (×)	8.7
Le Closeau, *loci* 4+50	3 (×)	14 (×)	0.81 (×)	Campanian flint	**124** (×)	8.7 (×)	13.1
Gönnersdorf SW	3 (1)	20 (25)	0.27 (0.21)	Baltic flint; Western European flint	**375** (467)	6.6 (5.3)	24.7
Cepoy	5 (×)	85 (×)	1.18 (×)	local Cretaceous gravel flint	85 (×)	**0.9** (×)	0.8
Cepoy, sector 1	5 (×)	71 (×)	1.74 (×)	local Cretaceous gravel flint	58 (×)	**1.0** (×)	0.6
Cepoy, sector 2	5 (×)	**100** (×)	0.69 (×)	local Cretaceous gravel flint	**146** (×)	**0.8** (×)	~1.2
Marsangy	5 (×)	**102** (×)	1.75 (×)	local Cretaceous flint	57 (×)	**3.0** (×)	1.7
Marsangy N19	5 (×)	**179** (×)	**1.90** (×)	local Cretaceous flint	53 (×)	**3.0** (×)	1.6

Tab. 83 Indices of lithic material referring to the size and use of the studied assemblages. Assemblage size classes refer to numbers of lithic artefacts ≥1 cm (total) found at a site/concentration. The classes are **0** 1-500 (1-1,500); **1** 501-1,000 (1,501-3,000); **2** 1,001-2,000 (3,001-6,000); **3** 2,001-4,000 (6,001-12,000); **4** 4,001-8,000 (12,001-24,000); **5** 8,000> (24,000>). Density index is calculated by dividing the no. of artefacts ≥1 cm by excavated m². Exploitation index is formed by dividing no. of artefacts ≥1 cm by no. of cores. The % cores is given for all cores and core fragments (see **tabs 11. 24. 35**) in relation to lithic artefacts ≥1 cm. In parentheses the indices are given calculated with no. of total artefacts. The same applies to % retouched artefacts. Function index is the no. of formally retouched artefacts divided by the no. of cores and core fragments. In bold, values that are indicative for a wasteful exploitation strategy.

site	assemblage size class	density index	% cores	main raw material	exploitation index	% retouched artefacts	function index
Gouy	0 (×)	× (×)	0.86 (×)	Cretaceous flint	**116** (×)	13.8 (×)	16.0
Pincevent III.2	0 (0)	2 (2)	**2.05** (1.78)	local gravel flint	49 (56)	6.3 (5.5)	3.1
Belloy-sur-Somme	4 (×)	14 (×)	1.68 (×)	local Turonian and Coniacian flint	59 (×)	**2.5** (×)	1.5
Andernach, upper horizon	3 **(4)**	16 (81)	1.74 (0.33)	Tertiary quartzite; chalcedony	58 (×)	7.4 (1.4)	4.2
Andernach 2-FMG	2 (1)	13 (28)	**1.82** (0.86)	Tertiary quartzite	55 (116)	11.3 (5.4)	6.2
Hangest-sur-Somme III.1, lower horizon	1 (1)	6 (13)	**2.50** (1.08)	local Turonian and Coniacian flint	40 (93)	11.9 (5.1)	4.8
Kettig	3 **(5)**	16 (100)	1.59 (0.25)	Tertiary quartzite; Western European flint	63 (395)	9.2 **(1.5)**	5.7
Conty, lower horizon	× (1)	× (20)	× (0.15)	local Turonian flint	× (×)	× (2.1)	14.0
Urbar	1 (1)	30 (97)	1.55 (0.49)	Tertiary quartzite	65 (205)	23.3 (7.3)	15.0
Le Closeau, greyish layers	5 (×)	2 (×)	**2.00** (×)	Campanian flint	50 (×)	**2.0** (×)	1.0
Niederbieber	5 (5)	19 (116)	1.38 (0.22)	Tertiary quartzite; chalcedony	72 (449)	9.0 **(1.4)**	6.3
Niederbieber 1	3 (3)	54 (166)	1.28 (0.41)	chalcedony	78 (242)	10.5 (3.4)	8.2
Niederbieber 3	0 (0)	9 (50)	0.39 (0.07)	chalcedony	258 (1,402)	3.5 (0.6)	9
Niederbieber 4 (+17a)	3 **(4)**	29 **(233)**	1.37 (0.17)	chalcedony	73 **(593)**	12.3 **(1.5)**	9
Niederbieber 5	2 (3)	39 (184)	1.46 (0.31)	Tertiary quartzite	68 (×)	6.7 **(1.4)**	4.6
Niederbieber 6 (+10a)	2 **(4)**	41 **(288)**	0.76 (0.11)	chalcedony; Tertiary quartzite	131 (924)	5.7 (0.8)	7.4
Niederbieber 7	2 (1)	29 (63)	1.54 (0.71)	Baltic flint	65 (141)	12.0 (5.5)	7.8
Niederbieber 8	1 (3)	7 (151)	0.93 (0.04)	Tertiary quartzite	107 (2,267)	6.9 **(0.3)**	7.4
Niederbieber 9	3 (2)	17 (40)	1.49 (0.63)	Tertiary quartzite	67 (158)	8.3 (3.6)	5.6
Niederbieber 10	1 (1)	35 (101)	1.36 (0.48)	Tertiary quartzite	74 (210)	7.2 (2.5)	5.3
Niederbieber 11	0 (2)	9 (85)	**2.20** (0.23)	chalcedony; Tertiary quartzite	45 (430)	13.2 **(1.4)**	6
Niederbieber 12	2 (3)	24 (148)	**1.90** (0.31)	Tertiary quartzite	53 (324)	5.9 (1.0)	3.1
Niederbieber 13	1 (3)	18 (178)	1.29 (0.13)	Tertiary quartzite	77 **(770)**	7.2 **(0.7)**	5.6
Niederbieber 14	1 (3)	18 (157)	**2.93** (0.33)	indurated slate; chalcedony	34 (300)	11.9 **(1.4)**	4.1
Niederbieber 15	0 (0)	10 (18)	**2.47** (1.27)	chalcedony	41 (79)	14.8 (7.6)	1.3
Niederbieber 16	0 (2)	11 (146)	1.68 (0.13)	indurated slate; chalcedony	60 **(769)**	7.4 **(0.6)**	6
Niederbieber 17	0 (1)	3 (40)	1.24 (0.10)	chalcedony	81 **(1,030)**	38.9 (3.1)	15.5
Niederbieber 18	1 **(4)**	19 **(368)**	0.14 (0.01)	indurated slate (greenish; Western European flint; chalcedony	715 (13,991)	1.4 (0.1)	9
Niederbieber 19	0 (0)	6 (10)	**4.40** (2.44)	chalcedony	23 (41)	**5.7** (3.1)	10
Niederbieber 20	1 (2)	16 (57)	1.02 (0.29)	Baltic flint	98 (340)	4.6 (1.3)	4.5
Andernach 3-FMG	3 **(4)**	21 (151)	1.70 (0.24)	chalcedony	59 (×)	5.1 (0.7)	3.0
Le Closeau, *locus* 25	0 (×)	5 (×)	0.79 (×)	Campanian flint	**126** (×)	**3.0** (×)	10
Bad Breisig	4 (5)	119 (910)	**3.09** (0.41)	Tertiary quartzite	32 (247)	5.0 (0.7)	1.6

Tab. 83 (continued)

Rekem (De Bie/Caspar 2000), and the massive structural elements found in the Late Magdalenian concentrations indicate the longer use, whether repetitive or continuous, at these sites. Accordingly, it seems that the careful preparation of the Late Magdalenian cores produced some waste material with a series of good quality blanks that were, in general, further used. In contrast, the FMG knapping process produced a lot of discarded blanks and debris. Consequently, the larger assemblages are probably an expression of a more wasteful handling of lithic raw materials in the FMG than the Late Magdalenian. Furthermore, although the Andernach 2-FMG site was possibly repeatedly visited, the concentrations are not as clearly defined as the Niederbieber concentrations but the site also produced a relatively small assemblage with tendentially bigger pieces. Thus, this increasingly wasteful handling seemed to further increase within the FMG resulting in a high quantity of discard at Bad Breisig (**tab. 81**).
In fact, the only assemblage of a single Late Magdalenian concentration yielding more artefacts ≥ 1 cm than most FMG concentrations (only Bad Breisig was larger) was Le Tureau des Gardes, *locus* 6. *Locus* 6 was only excavated over a very small area, which explains the very high density index (**tab. 83**). However, the assemblages from Le Tureau des Gardes were recovered from depressions in the palaeosurface (Lang 1998, 84). Consequently, the possibility that these concentrations represent sediment traps and, thus, accumulation by natural processes rather than a single anthropogenic concentration can only be excluded though a more comprehensive excavation and detailed spatial and micromorphological analyses that are not yet published (Lang 1998, 104; cf. Julien/Rieu 1999). In addition, several of the *loci* from this site produced material attributed to the MfCM where the knapping style attributed to the use of soft hammerstones was first identified. Inventories of this archaeological unit yielded the most numerous lithic assemblages. Thus, did the use of this knapping style create more discard material and did the change of the knapping style result in a more wasteful handling of the resources? Could this wasteful style only appear in a region such as the Paris Basin with rich lithic resources? To answer these questions further attributes of the inventories must be considered (see below).
However, besides the MfCM, the Early Azilian occurred in the same sub-area and, in these assemblages, the change of knapping style towards the use of soft hammerstone was well established. Since the number of analysed sites varied between the archaeological units, the distribution of sites to the different assemblage size classes only makes it possible to establish general comparative observations about the occurrence of different assemblage sizes in the archaeological units (**fig. 73**). Nevertheless, the assemblage sizes of both, the Early Azilian and the MfCM, were bivariate but the distributions are alternating. The few Early Azilian assemblages were generally small to medium-sized, whereas the MfCM inventories were medium-to very large-sized. In fact, all archaeological units yielded a bivariate distribution of assemblage size classes, except for the many FMG concentrations that produced a monovariate distribution. This difference is possibly due to the number of analysed sites and if equal numbers were studied for the Late Magdalenian, the MfCM, or the Early Azilian a comparable monovariate distribution would appear. However, the peaks of these distributions are comparable for the Early Azilian and the FMG. Perhaps, the Late Magdalenian assemblages were similar to the former but might also be tendentially bigger. It is already apparent that assemblages of the MfCM yielded many large assemblages (**fig. 73**).
The assemblages of the lower horizons of Hangest-sur-Somme III.1 and Conty are rather small and the early FMG assemblages of the Central Rhineland are also small to medium-sized, whereas on younger sites such as Niederbieber, Bad Breisig, and also the greyish deposits of Le Closeau are large and very large assemblages again present. In these FMG horizons, the concentrations were well defined and usually appeared to respect one another (cf. Bodu 1998; Gelhausen 2011a). In contrast, the large assemblages of the MfCM sites were often found in dense but diffuse distributions of material, occasionally around hearths, which seem to be unrelated to one another. Assuming that this pattern was not primarily caused by post-depositional

Fig. 73 Generalised frequency of assemblages according to their assemblage size. Black area: Late Magdalenian; grey hatched area: Final Magdalenian faciès Cepoy-Marsangy (Cepoy-Marsangy); light grey hatched area: Early Azilian; grey area: *Federmesser-Gruppen*.

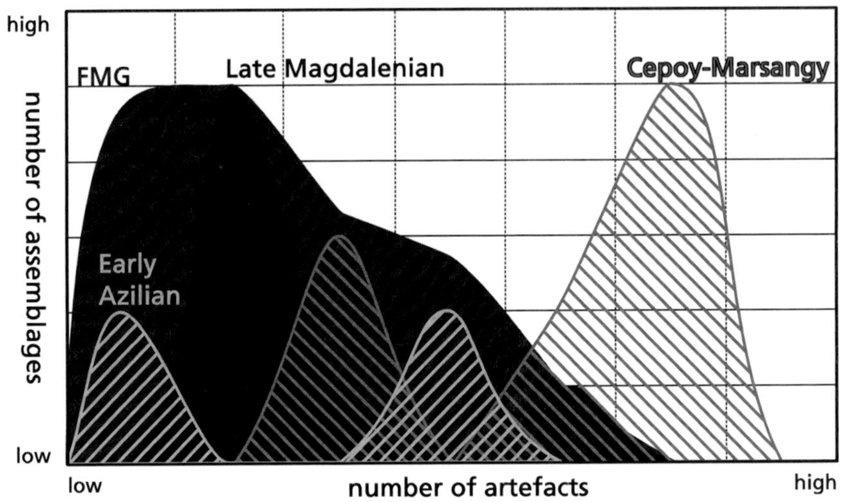

processes (cf. Julien et al. 1999), these latter horizons appear to have accumulated over a longer period of time resulting in a greater diffusion by more intense settlement dynamics and/or partial cover of previous concentrations. However, the undifferentiated presentation of some of these sites such as the sectors 1 of Cepoy or Le Grand Canton and the unclear accumulation history (and still questionable post-depositional influences) of these assemblages can only partially be responsible for this impression. Single concentrations such as the southern concentration of Étigny-Le Brassot or Marsangy N19 yielded more than 8,000 artefacts ≥ 1 cm and, thus, fall into the group of the largest assemblages (assemblage size class 5). The sector 2 of Le Grand Canton was analysed in greater detail (Julien/Rieu 1999) and also yielded some very dense clusters with occasionally 7-10 cores per squaremetre (Julien et al. 1999, fig. 84D). Thus, the impression of very wasteful handling of the raw materials on these sites persists. Nevertheless, to identify a wasteful handling of raw materials further criteria have to be considered.

For this project, some of the criteria identified as indicative of wasteful handling of lithic resources can be further tested. For instance, besides the amounts of material (assemblage size class) brought to a site, the density of this material on the excavated site, the percentage of cores (% cores), and the relation of these cores to the debris material was calculated for the analysed sites (**tab. 83**). In addition, the relation of the blanks produced at a site to those used, with or without formal retouch, and those that were abandoned without signs of use could be an indicator for an efficient exploitation of the lithic resources[51]. However, this efficiency in lithic products remains difficult to evaluate. For example, for some cores the state of exploitation in which they arrived at the site cannot be exactly reconstructed; some blanks could have arrived at the site prepared and were discarded there. Therefore, the efficiency of secondary exploitation cannot be considered further in this project.

Based on the available values, a wasteful handling of material could be characterised by a high assemblage size class for a single concentration with a high artefact density at the site (density index). For example, a high proportion of cores that produced little other material (low exploitation index for artefacts ≥ 1 cm) but many small pieces (high exploitation index for total artefacts). Furthermore, a low number of blanks transformed into formally retouched tools can be considered as a sign of unfocused blank production if the unretouched specimens were not intensely used. However, this later characteristic also depends on the function of a site because large numbers of formally retouched artefacts cannot be expected on a raw material acquisition/exploitation site. In contrast, a more economic use of raw materials would result in a medium to small

51 At this point, the author likes to thank Ludovic Mevel, Nanterre, for the fruitful discussion.

assemblage size class for a single concentration with an intermediate density assuming that the knapping was a more organised activity and, thus, distributed in a more restricted space. These economic assemblages had very few abandoned cores at the site (low % cores) assuming that useful blanks were imported and still usable cores exported. Furthermore, the exploitation index for artefacts ≥1 cm would also be low assuming that in addition to the cores, blanks were exported from the site. In contrast to the wasteful handling, an economic use would not probably tolerate a raw material that splinters a lot and, thus, the exploitation index for the total artefacts would also be low. Moreover, the remaining lithic artefacts should bear traces of intensive use in an efficient exploitation strategy. Since use-wear analyses are infrequently conducted and usually focused on only a small sample of the unretouched material, this indicator cannot be used in the present comparison. However, the percentage of formally retouched artefacts could provide a proxy for the transformation of the blanks into tools and, thus, the secondary use of the raw material.
To visualise the temporal development of these parameters, graphs were created that give the age range, the archaeological attribution, and the value of the analysed parameter for each analysed assemblage (**fig. 74A**). To prevent an overemphasis of outlying values, a complete distribution of values is given in a small overview, whereas in the main graph mostly contains the main cluster of values.
In general, the density index (di; **tab. 83**) appears to develop from very variable values to relatively low values (**fig. 74B**). Thus, the impression that the Early Azilian and FMG sites were rather ephemeral is further confirmed by the di. The Early Azilian as well as the FMG assemblages range below or at the lower limit of Late Magdalenian or MfCM densities. Bad Breisig forms a single, clear outlier with a di of 119 and also the di of the Urbar assemblage is with a value of 30 higher than in the other FMG sites that range between values of 2 and 21. The excavation of Bad Breisig was oriented along the main concentration and so stopped where the main density decreased. Comparably, only some parts of a concentration were excavated at Urbar resulting in a higher di. In contrast, the vast excavation area of Le Closeau comprised several, almost sterile, square-metres resulting in a lower value as a consequence of these sterile areas. Therefore, the values from this site form the lower range of the FMG assemblages. However, if the single concentrations of Niederbieber are considered (**tab. 83**), they reveal a larger range of dis from 3 to 55. Thus, the low values displayed by the Le Closeau assemblages appear not only as a result of the large excavation area and the values fall well within the occasionally, very ephemeral concentrations of the FMG.
The majority of the Late Magdalenian assemblages yielded density values of 24 to 50 and, thus, higher than the range of the main FMG units but within the range of the single concentrations of Niederbieber. Assuming that sites such as Gönnersdorf and possibly also Saint Mihiel were occupied over a longer period or visited repetitively, this longer use resulted in Late Magdalenian sites having a higher density. Nevertheless, the occupation of Niederbieber was also considered to have been repetitively used over a longer period (Gelhausen 2011b; Gelhausen 2011a). At this location, behaviour resulted in contrary values due to the spread of the concentrations over a larger area. Single FMG concentrations that yielded dense clusters were possibly also used over a longer period such as Niederbieber 1 (di = 54). This site is interpreted as being used for several weeks during the late winter/early spring (Bolus 1992, 183). Concentration 4

Fig. 74 **A** Template graph for values given per time and archaeological unit. **B** Values of the density index given per time and archaeological unit. – Circles were set around the main cluster of assemblages attributed to the same archaeological unit. The distribution of all assemblages is given in the overview where also the convex hulls were set per archaeological unit (see p. 281 f.). In the overview, names are only given for sites which are not shown in the main graph. Abbreviations: **LTG** Le Tureau des Gardes; **Cepoy 1/2** Cepoy, sector 1/2; **LGC lower** Le Grand Canton, lower horizon; **LGC 1/2, up** Le Grand Canton, sector 1/2, upper horizon; **An 2/3 up(per)** Andernach 2/3, upper horizon; **Pin. III.2** Pincevent, horizon III.2; **Han III.1** Hangest-sur-Somme III.1, lower horizon; **Conty** Conty, lower horizon; **Le Closeau, in-up** Le Closeau, intermediate to upper horizons (greyish deposits). – For further details see text.

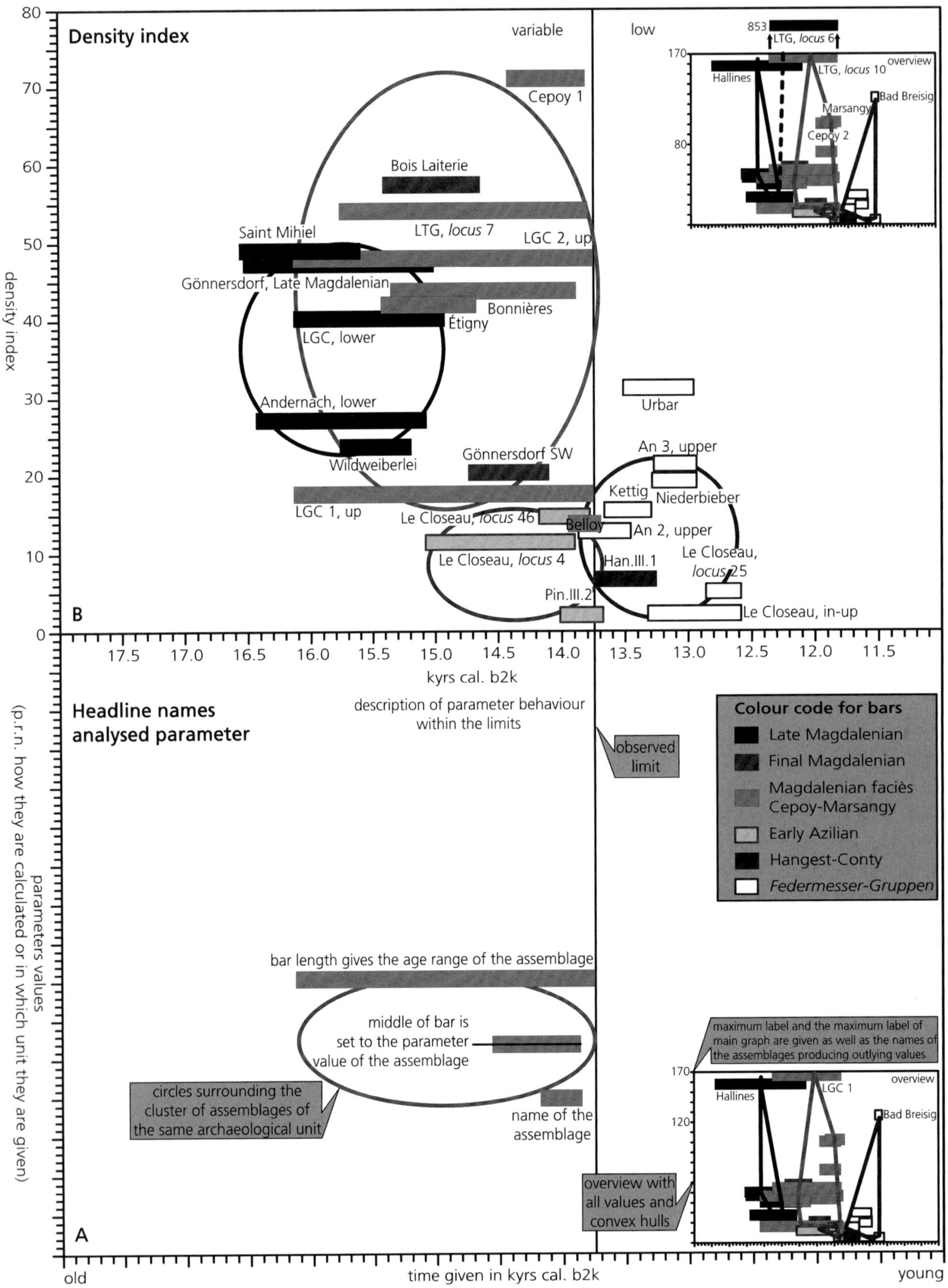

Density index
variable
low
853
LTG, locus 6
170
overview
Hallines
LTG, locus 10
Bad Breisig
Marsangy
Cepoy 2
80
Cepoy 1
Bois Laiterie
Saint Mihiel
LTG, locus 7
LGC 2, up
Gönnersdorf, Late Magdalenian
Bonnières
Étigny
LGC, lower
Urbar
Andernach, lower
An 3, upper
Wildweiberlei
Gönnersdorf SW
Kettig
Niederbieber
LGC 1, up
Le Closeau, locus 46
Belloy
An 2, upper
Le Closeau, locus 4
Han.III.1
Le Closeau, locus 25
Pin.III.2
Le Closeau, in-up
B
density index
($n_{artefacts}$ / $n_{excavated\ sq.m}$)
80
70
60
50
40
30
20
10
0
17.5
17.0
16.5
16.0
15.5
15.0
14.5
14.0
13.5
13.0
12.5
12.0
11.5
kyrs cal. b2k
Headline names analysed parameter
description of parameter behaviour within the limits
observed limit
Colour code for bars
Late Magdalenian
Final Magdalenian
Magdalenian faciès Cepoy-Marsangy
Early Azilian
Hangest-Conty
Federmesser-Gruppen
bar length gives the age range of the assemblage
middle of bar is set to the parameter value of the assemblage
circles surrounding the cluster of assemblages of the same archaeological unit
name of the assemblage
maximum label and the maximum label of main graph are given as well as the names of the assemblages producing outlying values
170
overview
Hallines
LGC 1
120
Bad Breisig
overview with all values and convex hulls
A
parameters values
(p.r.n. how they are calculated or in which unit they are given)
old
time given in kyrs cal. b2k
young

at Niederbieber (di = 29) was assumed to have been used even longer due to the more »spoiled« appearance (Bolus 1992, 183). These more intense settlement dynamics with occasional clearings of the main activity area(s) certainly occurred during the Late Magdalenian (Sensburg 2007; Sensburg/Moseler 2008). These clearings resulted in a widening of the main concentration, whereas in FMG assemblages a new concentration was installed after a certain limit of lithic density was reached. Presumably, this more readily made reinstallation of the FMG concentrations was partially due to the lighter constructions on these sites that were easier to move. Moreover, lighter lithic material was covered more easily underneath the heavy plates or within the pits of Late Magdalenian concentrations allowing for a higher density of this material to accumulate within a concentration. Nevertheless, Late Magdalenian concentrations were generally dense and yielded the most compact cluster, uncovered in the *locus* 6 of Le Tureau des Gardes with 853 artefacts on average per squaremetre. However, the uncertainties of a human accumulation of these concentrations were pointed out previously. Furthermore, the accumulation in Hallines (di = 158) was also very high. This assemblage was recovered from only a very small test pit area and included the material of at least two episodes (see p. 183-189), which could explain the relatively high di. In contrast, the assemblage from the lower horizon of Andernach was not very dense (di = 27) but in this excavation area only three major concentrations were cut and not completely excavated. Furthermore, the wasteful blank production process seems clearly under-represented either because it was of a minor importance or because the knapping workshops or the refuse area were not excavated (pers. comm. Martin Street, Neuwied, and Ludovic Mevel, Nanterre). For the small Wildweiberlei assemblage (di = 24), the question remains, which material was lost due to the position in a cliff (Lateglacial disposal of waste into the valley), the early excavations (neglect of small pieces), and/or the destruction of the cave. Thus, the accumulation of 40 to 50 artefacts per squaremetre seems the more common density on Late Magdalenian sites and on several sites of the MfCM (**fig. 74B**).

The assemblages of the south-western area from Gönnersdorf and Bois Laiterie fall slightly out of the typical range of the Late Magdalenian di but into the very variable range of the MfCM di. In comparison with the Late Magdalenian assemblages, the material on the south-western area in Gönnersdorf was more scattered (di = 20) and the material in the cave of Bois Laiterie was too dense (di = 58). The di of the south-western area of Gönnersdorf is comparable to the one from sector 1 of Le Grand Canton and also to values from the lower horizons of Belloy-sur-Somme as well as Le Closeau and several FMG sites in the Central Rhineland (**fig. 74B**). However, on some of these sites large surfaces containing almost empty spaces were uncovered resulting in a lower di. The factor of used versus excavated spaces could also explain the low value in the south-western area of Gönnersdorf. Moreover, refittings showed that some of the material from the Late Magdalenian areas were recovered from this part of the site suggesting that it was partially a Late Magdalenian refuse area. Thus, without the Late Magdalenian input the assemblage appears even smaller and, thus, the di would be even lower. This assumption makes the south-western assemblage from Gönnersdorf appear as a very ephemeral episode in contrast to the dense Late Magdalenian clusters at the same site.

In Le Closeau, Conty, and Pincevent, horizon III.2, large surfaces were uncovered alongside the main concentrations lowering the di of these assemblages. However, the densities do not increase significantly in Pincevent or Le Closeau if the calculation was only focused on the sections with the Lateglacial concentrations (see **tab. 83**). Thus, these sites seem to represent this first trend towards a different, more ephemeral use of space. Since a typo-technologically comparable assemblage to Conty was found in the lower horizon of Hangest-sur-Somme III.1 and yielded also a cluster that was not very dense, this ephemeral character continued in these as well as the FMG sites.

In contrast, sites from the MfCM frequently yielded high and occasionally very high di values. The uncertainties of the concentrations from Le Tureau des Gardes discussed previously and *locus* 10 from this sites, which

also yielded a very high di (di = 166) has to be considered comparably problematic. *Locus* 7 was presented in more detail (Weber 2006) and this concentration was only slightly denser (di = 54) than Saint Mihiel and a marginally less dense than Bois Laiterie. However, the densities in Marsangy (di = 102) and the sector 2 of Cepoy (di = 100) were very high even for MfCM sites. In Marsangy, the high values were caused by the very dense cluster around the hearth N19 (di = 179; **tab. 83**), which probably represented a knapping workshop (see p. 225 f.). The high densities of the material in sector 2 of Cepoy were probably also formed by several knapping episodes that were evident in this area (Wenzel 2009, 47-61). Nevertheless, other activity areas were identified in this sector which should balance out for the debris-rich knapping areas. Consequently, the very high density value at Cepoy, sector 2 reflects either an intense use of lithic material or a very wasteful knapping process. The latter can be tested by examining the percentage of cores and the exploitation index. In fact, the cores form only a very small portion of the material from sector 2 of Cepoy (**tab. 83**). Moreover, the value for sector 1 of Cepoy is also among the highest (di = 71). The upper horizon of sector 1 of Le Grand Canton yielded a very low value, unlike the comparably low value from Belloy (di = 14), could not be explained by the large unexcavated surface but partially by the destruction of the surface by various processes and the preservation of only a semi-circular concentration (Alix et al. 1993). The composition of this half concentration dominated by large pieces such as cores and core tablets suggests that it was a blank production workshop or a refuse area for the knapping process. The latter appeared more probable because small flakes and splinters were rare.

Thus, during the early period of this study, a very variable range of densities from small to high and occasionally very high values occur. In contrast, whereas in the second half only a limited range could be observed. The variable values point to the methodological problem of differentiating between concentrations, used areas, and excavated areas. This problem also reflects the differentiated use of space which was more focused in the Magdalenian than during the FMG. The impression of a more limited range of results from the disappearence of dense and very dense concentrations that correlates with the disappearance of assemblages attributed to the MfCM in this study. In general, the higher densities on sites of the MfCM seems related to intense knapping processes that resulted in areas densely covered with artefacts. However, to determine whether this knapping process was particularly wasteful, it is important to consider the composition of the assemblages with special focus on cores.

Although the majority of Lateglacial assemblages (artefacts ≥ 1 cm) yielded a very comparable ratio of cores (**tab. 83**), the general temporal trend of this ratio seems to suggests a gradual increase in number of cores during the studied time period (**fig. 75A**). Thus, more cores were used to produce the lithic material. However, a comparison with the assemblage size classes reveals that the general size of the assemblage and the presence of cores was not a one-to-one ratio. This finding emphasises that further factors such as technical abilities, raw material, spatial organisation, site function, and the duration of the occupation played a role in the size and composition of the assemblages.

The proportion of cores lies usually between 0.2 and 2.5 % in the analysed assemblages. The Late Magdalenian assemblages yielded values between 0.5 and 1.5 %. Contrastingly, the values of the Early Azilian inventories ranged between 0.8 and 2.1 % and the ratio of cores in FMG assemblages ranged between 1.4 and 2 %. The assemblages of the MfCM of the Magdalenian seemed to bridge between the Early Azilian and the Late Magdalenian values byranging between 0.5 to 2 %. However, if the horizon III.2 from Pincevent is excluded, the Early Azilian had a more limited range between 0.8 and 1 % cores. The percentage of cores in this upper horizon at Pincevent lies closer to the upper limits of MfCM or FMG assemblages than to the Early Azilian. The high ratio of the cores can be explained possibly, by the general character of this part of the excavation appearing as an area of waste disposal material and no actual knapping spots. The ratio of cores in the assemblages from Bois Laiterie and the south-western area of Gönnersdorf lies with 0.2 and

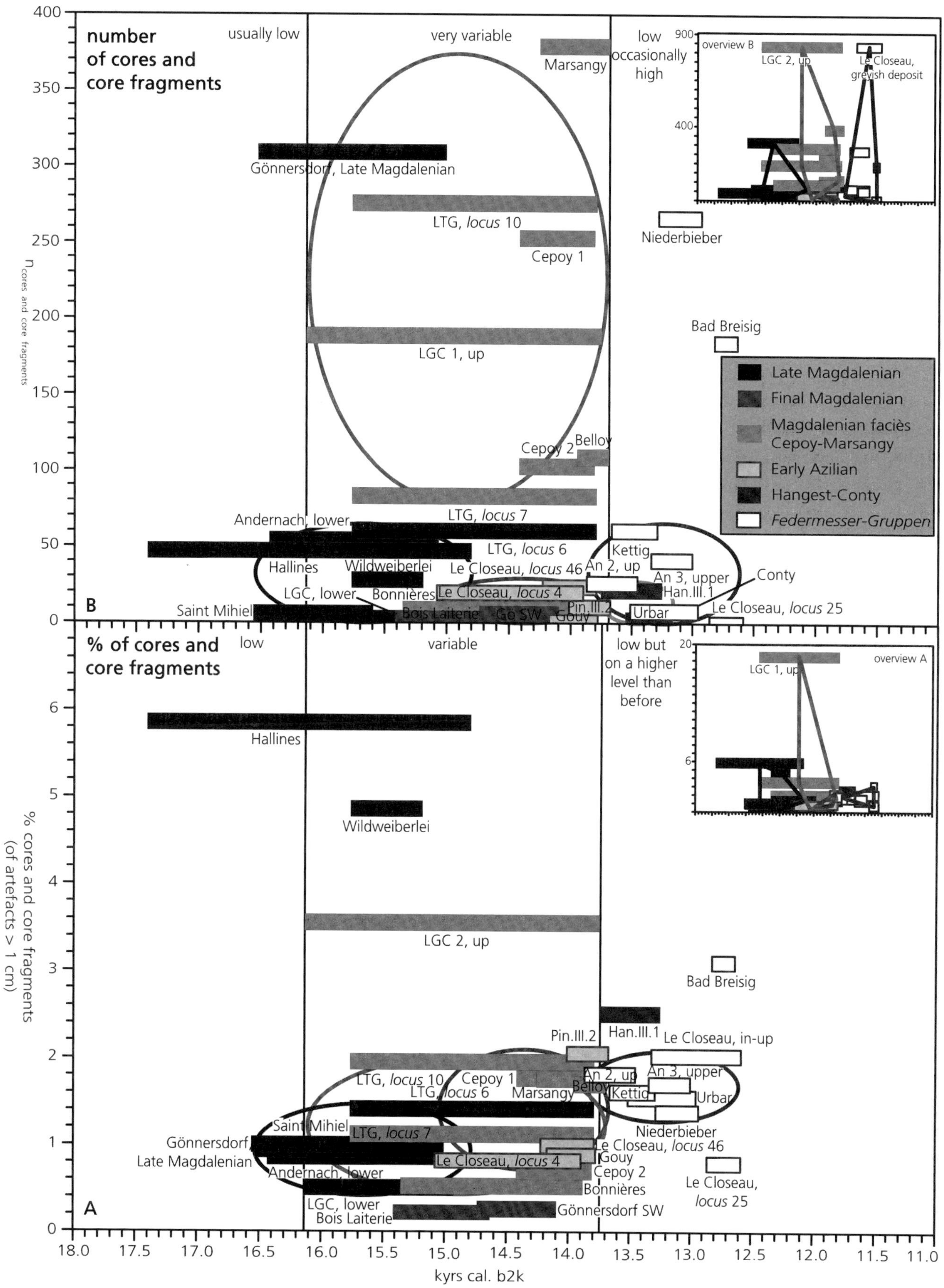

number of cores and core fragments
usually low
very variable
low occasionally high
B
% of cores and core fragments
low
variable
low but on a higher level than before
A
n cores and core fragments
% cores and core fragments (of artefacts > 1 cm)
kyrs cal. b2k
Marsangy
Gönnersdorf, Late Magdalenian
LTG, locus 10
Cepoy 1
Niederbieber
LGC 1, up
Bad Breisig
Late Magdalenian
Final Magdalenian
Magdalenian faciès Cepoy-Marsangy
Early Azilian
Hangest-Conty
Federmesser-Gruppen
Cepoy 2
Belloy
LTG, locus 7
Andernach, lower
LTG, locus 6
Kettig
Hallines
Wildweiberlei
Le Closeau, locus 46
An 2, up
An 3, upper
Conty
LGC, lower
Bonnières
Le Closeau, locus 4
Han.III.1
Saint Mihiel
Bois Laiterie
Go SW
Gouy
Pin.III.2
Urbar
Le Closeau, locus 25
overview B
LGC 2, up
Le Closeau, greyish deposit
overview A
LGC 1, up
Hallines
Wildweiberlei
LGC 2, up
Bad Breisig
Han.III.1
Pin.III.2
Le Closeau, in-up
LTG, locus 10
Cepoy 1
LTG, locus 6
Marsangy
Belloy
An 2, up
An 3, upper
Kettig
Urbar
Niederbieber
Saint Mihiel
Gönnersdorf, Late Magdalenian
LTG, locus 7
Le Closeau, locus 46
Le Closeau, locus 4
Gouy
Andernach, lower
Cepoy 2
Bonnières
LGC, lower
Bois Laiterie
Gönnersdorf SW
Le Closeau, locus 25

0.3 % below the Late Magdalenian range. Perhaps, these particularly low amount of cores was due to the function of the concentrations.

In contrast, these general ranges were exceeded in the assemblages from the upper horizon of Le Grand Canton, in the Late Magdalenian assemblages of Hallines and the Wildweiberlei as well as in Bad Breisig.

In Bad Breisig, as well as in sector 1 of Le Grand Canton, only half a concentration was found. High values at these sites are surprising but could be explained by the discard of cores for Le Grand Canton and the possibility of a nearby source of fragmented raw material at Bad Breisig (see p. 164 f.). In the Wildweiberlei assemblage, excavations during the early 20th century could have influenced the composition of the recovered assemblage with larger pieces such as cores being more readily collected and archived. In Hallines, recovery by an amateur collector could have introduced similar biases into the collection. Moreover, the small area of excavation at Hallines only represents the limits of a much larger concentration. If this excavated area represented a knapping or refuse area that could explain the over-representation of cores. Comparably, if the depressions of Le Grand Canton are sediment traps, lighter lithic material could have been washed out and heavier pieces such as cores would have remained trapped in the depressions. This scenario could explain the high values of these assemblages. In fact, the highest densities in sector 2 of Le Grand Canton, particularly smaller material, accumulated where the terrain gradually slopes into the deepest part of the area (Julien et al. 1999, 144 fig. 66). However, according to the spatial analysis of this sector, the distribution of artefacts was considered to be almost undisturbed by post-depositional processes but clearings of the main activity zones during the occupation were assumed possible (Julien et al. 1999, 152 f.). Accordingly, the large excavation area which also contained the refuse areas would explain the high amount of cores. In this case the di would be higher than on not so extensively excavated sites.

However, the di values are very similar for sector 2 of Le Grand Canton, Gönnersdorf, and Saint Mihiel. Therefore, the question could be posed: Would Late Magdalenian sites such as Gönnersdorf and Saint Mihiel, which were not excavated much beyond the limits of the main concentrations, have revealed a higher amount of cores if more of the surrounding area was excavated? In fact, the numbers of cores are very low on Late Magdalenian sites (**fig. 75B**) ranging between 6 and 60 cores. In contrast, the number of cores found at Gönnersdorf (n = 309) where in the south-western area a scattered, potential, refuse area was excavated, is clearly higher. Nevertheless, the outstanding number of cores from Gönnersdorf was not due to this refuse area but was mainly due to the numerous cores found within concentration III (n ≈ 145; cf. Franken/Veil 1983, Abb. 57. 82. 96). The other Late Magdalenian concentrations of Gönnersdorf yielded values that ranged between 28 and approximately 35 cores (see **tab. 11**) and, thus, well within the range of the other Late Magdalenian assemblages. In analogy to Gönnersdorf, clearings and the extension of the excavation area cannot solely explain the exceptionally high values of Le Grand Canton.

In contrast, for the other Late Magdalenian assemblages yielding exceptionally high ratios of cores (Hallines and Wildweiberlei), the suggested reason was that smaller material was not archived suggesting some bias towards larger specimens. This observation was also made for the assemblage from Le Grand Canton, sector 1 (Alix et al. 1993) where 188 cores were found but only a small assemblage (n = 1,021). This variation within the assemblage was due to either the collection represented a specific refuse area or to the washing out of material and redeposition. However, both explanations do not apply to sector 2 where also a high ratio of cores was found. Hence, was smaller material such as blades possibly also taken away from Le Grand

Fig. 75 Percentage (**A**) and numbers of cores and core fragments (**B**) in relation to the lithic assemblages (≥ 1 cm) given per time and archaeological unit. Circles were set around the main cluster of assemblages attributed to the same archaeological unit. The distribution of all assemblages is given in the overview where also the convex hulls were set per archaeological unit (see p. 281 f.). In the overview, names are only given for sites which are not shown in the main graph. Abbreviations see **fig. 74**. – For further details see text.

Canton by the Lateglacial hunter-gatherers? In this case, the site could have served besides the hunting as a lithic workshop.
From Le Grand Canton 89 kg from a total of 334 kg were attributed to cores, some 155 kg identified as preparation material, and over 48 kg were determined as blanks (Valentin et al. 1999b, Tabl. 14). These vast amounts, along with the incidental preparation of cores due to their natural shape (Valentin et al. 1999b, 81), and limited exploitation of cores, strengthens the idea of a lithic extraction site or blank production workshop near such a site. In contrast, the amount of aimed material remaining at the site suggests that this workshop was not necessarily intended to produce material for export but the intense lithic blank production seemed rather intended for the local use. Since butchering activity played an important role at this site (Bridault/Bemilli 1999), the unretouched blanks could have been used in this process. Therefore, a traceological analysis of the remaining blanks without further modification would be of particular interest (cf. Bodu/Mevel 2008; Sano 2012a 170-217). An additional factor was probably the selected length of the cores that usually clustered around 9 cm in size (Valentin et al. 1999b, 76-79). Moreover, blank production was mainly focused on bladelets (Valentin et al. 1999b, 83) that is also reflected in the dimensions of the projectiles which were particularly small for the northern French assemblages (see p. 529-532). Perhaps, the convenience of a good raw material source in combination with a reliable hunting spot where freshly knapped, sharp edged material was used for butchering as well as reequipping and resulted in an apparently wasteful handling of the lithic resource and the exceptional values for Le Grand Canton.
The number of cores (n = 827) further sustains the exceptional position of the assemblage from sector 2 of Le Grand Canton, even though the general range of cores between 84 and 379 specimens from sites attributed to the MfCM was already very high (**fig. 75B**). However, the intermediate, undetermined, and upper horizons including *locus* 25 from Le Closeau yielded a similar number of cores (n = 827) as sector 2 of Le Grand Canton but the ratio of cores in Le Closeau (% cores = 2.0) was similar to the other FMG assemblages. Thus, these concentrations were clearly different to the exploitation strategy from sector 2 of Le Grand Canton.
The FMG sites yielded usually comparable numbers to the Late Magdalenian assemblages ranging between 1 and 61 cores (**tab. 83**). Besides the numerous concentrations of the greyish deposits at Le Closeau, Niederbieber also has numerous concentrations, whilst Bad Breisig is the exception. Although the total number of cores found at Niederbieber is very high (n = 267), the number of cores found per concentration at this site is within the lower half of the FMG distribution with number between 1 and 33 cores and an average of approximately 14 cores per concentration (see **tab. 11**).
The material from Le Closeau was found in some 52 concentrations and the number of cores ranged further than the usual FMG sites with 0 to 91 cores found in the concentrations. Thus, on average almost 16 cores were found in a concentration in the greyish deposits at Le Closeau. If these numbers and ratios are further separated by the different horizons, the intermediate horizon yielded a range from 1 to 91 cores and an average number of 21 cores per concentration. Yet, these numerous cores formed only 0.8 to 7.3 % of the material found in the concentrations with an average of 1.6 % forming a smaller part of the material than in the other concentrations from the greyish deposits. In contrast, cores usually made up to 2.3 % of the assemblage (range between 0.9 and 12.5 %) in the concentrations attributed to the upper horizon but only 16 cores were found on average per concentration and the actual number of cores found in the concentrations ranged between 2 and 68. Thus, the assemblages attributed to the upper horizon were very similar to the main FMG cluster, composed mainly of assemblages from the Central Rhineland. In comparison with material from the intermediate horizon, the assemblages of the upper horizon yielded in general a smaller number of artefacts. Moreover, the undetermined concentrations were the smallest assemblages and the lowest values: with 0 to 40 cores found in the concentrations and an average of 11 cores per concentration.

Apparently, it is easier to assign material to specific typo-technological attributes with a greater quantity of material. In *locus* 25, only a single core was found. However, in the other excavated areas of Le Closeau (sud RN13) more material was recovered from this top horizon suggesting an increase of cores per concentration (range: 0-142; average: 33 cores/concentration) and also an increase of the ratio of cores with values between 1.3 and 11.6 % and an average share of 3.7 % in the assemblage as a whole. However, since these assemblages are claimed to be partially the result of post-depositional processes, this trend cannot be sustained further without a spatial analysis of this part of the site.
Nevertheless, the relatively similar numbers and ratios in Le Closeau suggest that the number of cores necessary to produce a certain number of blanks and, thus, the exploitation strategy was similar for all these concentrations. However, if the assemblages from the intermediate horizon are tendentially older than the ones from the upper horizon, which were certainly older than the material from the top horizon, a chronological trend was observable within the FMG assemblages. Use of a higher number of cores that produced larger quantities of material leading to the overexploitation of these cores that hence formed a smaller ratio of the assemblages. Subsequently, cores became less numerous but also less productive and, thus, the ratio increased. Finally, the number of cores again increased but their share of the assemblages further increased suggesting that numerous unproductive cores were abandoned at the site (or numerous blanks were exported from the site). This latter argument is most comparable to the assemblages from Le Grand Canton with relatively high ratios of cores and very high numbers of cores.
In contrast to Le Closeau and Niederbieber, Bad Breisig as well as sector 1 from Le Grand Canton were only half of a lithic concentration. However, the number of cores found in Le Grand Canton, sector 1 were within the range of cores found in concentrations of the MfCM, even though the ratio was relatively high. This difference can be explained by a differentiated spatial distribution (cf. Julien et al. 1999) and partial preservation. In contrast, Bad Breisig appears to be a unique FMG assemblage in terms of the ratios of cores (3.1 %) as well as the number of cores (n=184). Hence, the exploitation of the lithic resources, mainly the Tertiary quartzite, at Bad Breisig was more comparable to the handling of the flint at Le Grand Canton or the material in the top horizon of Le Closeau sud RN 13. This behaviour is in contrast to the exploitation scheme before the LSE in the Central Rhineland or the upper horizon of Le Closeau. Presumably, this change was due to the presence of a reliable sources in the immediate vicinity of the site. The cores in Bad Breisig were usually used in short production sequences with little preparations (pers comm. Ludovic Mevel, Nanterre) or the cores were abandoned as soon as preparations became necessary.
A similar case of raw material exploitation can also be seen in the Late Magdalenian assemblage of Gönnersdorf where the majority of cores (n=309) came from concentration III (n ≈ 145). These cores were predominantly made of the small, local indurated slate gravels (n ≈ 122; cf. Franken/Veil 1983, Abb. 57). These small gravels as a major source are comparable to the relatively small dimensions of the gravel flints at Le Grand Canton. In contrast to the flint, it is difficult to obtain a steady sequence of regular bladelets from the fissured indurated slate but this raw material is also available in regular quantities in the nearby gravels of the Rhine. Consequently, instead of receiving a number of blanks from one core, the same number of blanks was obtained from many cores. Thus, for different reasons (poor knapping abilities of the indurated slate and, perhaps, the weathered Tertiary quartzite, need for numerous blanks), the easy accessibility to larger quantities of smaller-size material resulted in a behaviour that appears similarly wasteful in concentration III of Gönnersdorf, the upper horizon of sector 2 in Le Grand Canton, and Bad Breisig. Perhaps, in the sector 2 some admixture of Late Magdalenian material resulted in the similarities observed with the material from Gönnersdorf. However, Bad Breisig was chronologically distinct from the two other assemblages. Thus, the different attribution and chronological position of these concentrations mark this wastefulness as a diachronic phenomenon.

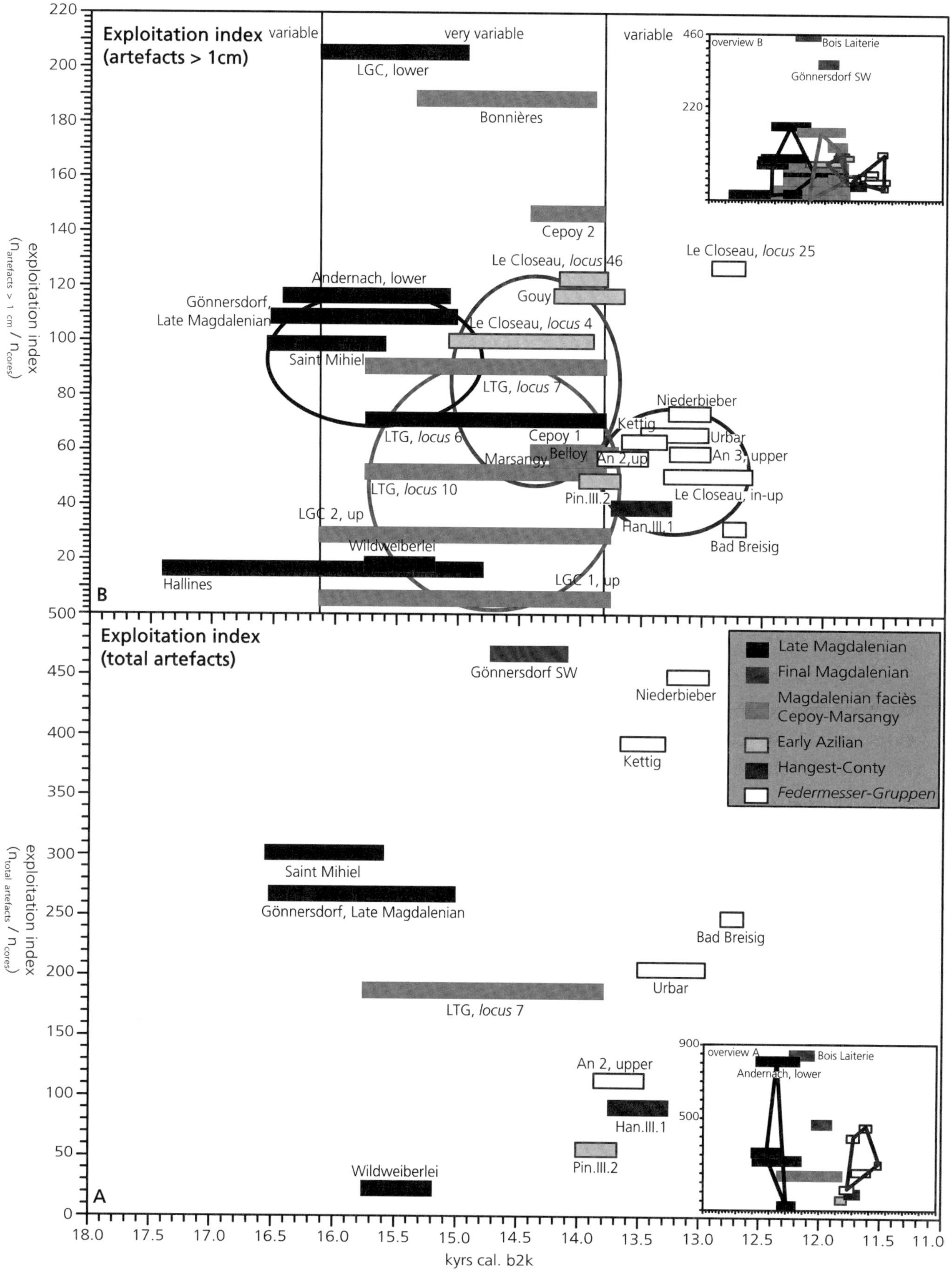
Exploitation index (artefacts > 1cm)
variable
very variable
variable
LGC, lower
Bonnières
Cepoy 2
Le Closeau, locus 46
Le Closeau, locus 25
Andernach, lower
Gönnersdorf, Late Magdalenian
Gouy
Le Closeau, locus 4
Saint Mihiel
LTG, locus 7
Niederbieber
Kettig
Urbar
LTG, locus 6
Cepoy 1
Belloy
An 2,up
An 3, upper
Marsangy
LTG, locus 10
Pin.III.2
Le Closeau, in-up
LGC 2, up
Han.III.1
Bad Breisig
Wildweiberlei
Hallines
LGC 1, up
B
overview B
Bois Laiterie
Gönnersdorf SW
exploitation index ($n_{artefacts > 1 cm} / n_{cores}$)
Exploitation index (total artefacts)
Gönnersdorf SW
Niederbieber
Kettig
Late Magdalenian
Final Magdalenian
Magdalenian faciès Cepoy-Marsangy
Early Azilian
Hangest-Conty
Federmesser-Gruppen
Saint Mihiel
Gönnersdorf, Late Magdalenian
Bad Breisig
Urbar
LTG, locus 7
An 2, upper
Han.III.1
Pin.III.2
Wildweiberlei
overview A
Bois Laiterie
Andernach, lower
A
exploitation index ($n_{total\ artefacts} / n_{cores}$)
kyrs cal. b2k

Likewise the ratio of cores (0.8 %), the number of cores in the lower horizon of Le Closeau (n ≥47) was significantly smaller than the values of the concentrations from the greyish deposits (% cores = 2.0; n = 827). The other assemblages attributed to the Early Azilian yielded comparably low numbers of cores. These numbers clearly distinguish the Early Azilian assemblages from the assemblages of the MfCM except for the site of Bonnières-sur-Seine that yielded only seven cores. This very low number for assemblages of this faciès can be explained by several factors: Firstly, the assemblage comes from a very small area of a partially excavated rockshelter that was first examined in the early 20th century. Moreover, the site was located near an accessible flint source (Valentin 1995, 404) and, thus, some blocks may have only been tested in Bonnières and were then removed to other localities. In addition, the site was considered as a short-term observation point, possibly used several times during the Lateglacial or a once used residential camp (see p. 196-204). Thus, the function of this site appeared more comparable to the function of the sites of the lower horizon from Le Closeau than the usual sites of the MfCM that were frequently associated with hunting and butchering. The assemblages from the south-western area of Gönnersdorf and from Bois Laiterie are also similar to the Early Azilian assemblages but with even lower numbers and ratios of cores. This group of assemblages with low number of cores forms a contrast to the quasi-contemporary assemblages with high numbers of cores, mainly of the MfCM, whereas the ratio of cores was similar in both groups. Consequently, the handling of cores and the raw material was comparable in these assemblages, only the amounts of material used on the sites differed which was already seen in the assemblage size classes.

The exploitation index provides the ratio of artefacts to cores and, thus, should produce a comparable result to the percentage of cores. In addition, the exploitation indices allow for an estimate of how fragmented the material became on site. This degree of fragmentation contributes to an impression of the amount and quality of the raw material brought to a site in order to allow for a successful knapping episode that eventually resulted in an abandoned core. This information relates to transport possibilities and the procurement strategies of hunter-gatherers and, therefore, makes considerations possible about the position of resource economy within the settlement system.

In the present project, a further refined index of the ratio of blanks to cores is not calculated, even though this index could be used to approximate the exploitation potential of a single core allowing for insights in the quality of the raw material, the technical abilities of the knappers, and/or the exploitation strategy (see p. 270-272). However, some of the possible information of such an index were already given for several studied assemblages. For example, technical abilities used in Late Magdalenian assemblages were demonstrated by single expressions from FMG inventories to be comparably elaborated (Gelhausen 2011a, 24). In contrast to the Late Magdalenian, these skills were not an integral part of the lithic production for everyday use. Furthermore, the raw material sources were often shown to be comparable between the Late Magdalenian or the Early Azilian and the FMG (Bodu/Valentin 1997; Floss 2002) but the quality of the raw material was often described as decreasing. Nevertheless, since high-quality material is often still identifiable at the source areas, this mediocre quality seemed to represent a choice, which considered these materials as sufficient for the tasks required. In general, FMG groups selected a lower quality of lithic raw materials than Late Magdalenian groups (Stapert/Street 1997; Barton/Roberts 1996; Floss 2002). However, low quality indurated slate was already used in the Late Magdalenian assemblages In the Central Rhineland

Fig. 76 Values of the exploitation index calculated with the complete lithic assemblage (**A**) and with artefacts ≥ 1 cm (**B**) given per time and archaeological unit. Circles were set around the main cluster of assemblages attributed to the same archaeological unit. The distribution of all assemblages is given in the overview where also the convex hulls were set per archaeological unit (see p. 281 f.). In the overview, names are only given for sites which are not shown in the main graph. Abbreviations see **fig. 74**. – For further details see text.

such as at Gönnersdorf III or Wildweiberlei. In addition, in the Paris Basin and the Somme region, the fine-grained flint material used in Lateglacial Interstadial contexts possessed more consistent knapping properties in comparison to the Rhenish Tertiary quartzite. Thus, if predominant raw materials and their splintering properties had an observable effect on the assemblage compositions and exploitation indices, the northern French sites would regularly form a separate group from the sites in Central Rhineland. The use of flint at these sites would have resulted in greater comparability to the sites from the western uplands and to the Late Magdalenian assemblages of the Central Rhineland than to the FMG sites from the Central Rhineland. As this pattern was not observed consistently within the analysed archaeological record, then the quality of the raw materials appears to be of minor importance in creating similarities and differences between these regions and between the Late Magdalenian and the FMG.

In contrast, the exploitation index (ei; **fig. 76**) reveals a clear differentiation, particularly between the assemblages attributed to the MfCM and those attributed to the inventories of the Early Azilian. The former generally yielded values between 5 and 90, with only Cepoy, sector 2 (ei = 146) and Bonnières-sur-Seine (ei = 188) having exceptional values. For the latter, the explanations have been previously discussed, but a careless excavation and/or documentation and spatial limitations do not apply to the sector 2 of Cepoy. Possibly, the high density of the lithic material in combination with this high ei and the locally accessible gravel flints suggest that this site is comparable to the suggested exploitation character of Bonnières-sur-Seine where blocks and cores were mainly prepared and tested and exported to other localities. The amount of material received from a single core in Le Grand Canton are in general small. Perhaps, this is due to outwashing of smaller material or dumping of cores in sector 1 which in fact yielded the lowest value (ei = 5) for this archaeological unit. However, this explanation was rejected for sector 2 (Julien et al. 1999), which produced the second lowest value with an ei of 29. Thus, again the export of blanks should be considered a possible explanation for the composition of this assemblage and, consequently, the blank production as a major activity at this site. The intentional production of a few, large blanks for butchering can be excluded based on the dimensions of the projectiles (see p. 529-532) and the cores (see p. 495-502). However, to what degree smaller, unretouched blanks were used in the processing of the prey cannot be determined without further use-wear analyses. The number of retouched artefacts in the assemblage of sector 2 were intermediate but the function index was relatively low indicating the larger importance of cores in the lithic material. Nevertheless, the presence of at least some 74 horses and the composition of the lithic assemblage with mainly burins and end-scrapers indicates that the processing of the prey was an important factor at this site.

The ei of the Early Azilian assemblages ranges between values of 49 and 124. If the horizon III.2 of Pincevent is excluded the range is further limited to values of 100 to 124 and, thus, clearly above the majority of the assemblages attributed to the Magdalenian MfCM. The Pincevent, horizon III.2 yielded comparable values for the ei based on artefacts ≥ 1 cm and based on the total number of artefacts which suggests that small splinters were hardly recovered. This observation further sustains the impression of this material representing waste scatters rather than knapping spots.

In contrast, the Late Magdalenian assemblages have a very varied ei of 17 (Hallines) to 206 (Le Grand Canton, sector 2, lower horizon). This very broad range contributes to the picture of very differentiated site function resulting in very different exploitation strategies. However, the second highest value for a Late Magdalenian assemblage comes from the lower horizon of Andernach (ei = 119) that suggest to narrow the range by excluding the lower horizon of Le Grand Canton, sector 2 (a special case due to selective excavation of a special activity zone). Furthermore, the assemblages from Bois Laiterie (ei = 453.5) and the south-western area of Gönnersdorf (ei = 375.2) yielded very small ratios as well as total numbers of cores and stand out with high exploitation values (**fig. 76B**, overview B; **tab. 83**). A use as workshop or exploitation site can be excluded for these concentrations due to the low numbers of cores and the lack of a nearby lithic source.

Instead, the low numbers and ratios indicate that these sites were supplied by mainly imported material and the flintknapping was no particular activity at these sites.

Thus, this index suggests that there was greater variability of assemblages in the Late Magdalenian and the MfCM, whilst the FMG assemblages appear very similar. The latter yielded values ranging from 50 to 72 with the youngest assemblages from Bad Breisig (ei = 32) and *locus* 25 of Le Closeau (ei = 126) forming clear outliers. The assemblage from Hangest-sur-Somme (ei = 40) falls into this larger range of the FMG assemblages. However, if the single assemblages at Niederbieber or Le Closeau are considered, the range of FMG concentrations is considerably enlarged: The values from the concentrations in the greyish deposits of Le Closeau range between 5 and 148 and in Niederbieber the ei of the single concentrations ranges from 23 to 715. The upper limit on the latter site was reached in concentration 18 with the next highest value from the special task camp in concentration 3 (ei = 258) followed by the double concentration 6 and 10a (ei = 131). Thus, these values exceed even the outliers of Bois Laiterie and Gönnersdorf but seem to indicate a special task area (concentration 3; Bolus 1992) or possible knapping spots or refuse areas (concentration 18; Gelhausen 2011c). The adjacent area 19 yielded the lowest value (ei = 23) and, thus, it seems probable that these areas complement one another which can be further sustained by the distribution of Tertiary quartzite and Baltic flint (Gelhausen 2011c, Abb. 11b; 12b) as well as the faunal materials (Gelhausen 2011c, Abb. 14). The second lowest ei comes from Niederbieber 14 (ei = 34) and, thus, excluding the complement areas and the special task camp, the range for the concentrations of Niederbieber are well in accordance with the limits set by the assemblages of Bad Breisig and *locus* 25 of Le Closeau. Values below 25 at Le Closeau were mainly associated with partially excavated concentrations, supplementary concentrations, concentrations from the disturbed channel area, or areas with a thin scatter of material. Values above 125 occurred only twice (*loci* 19 and 34) and were related to dense and larger clusters. This detailed comparison shows that different assemblages were also present within the FMG range. However, the very low values such as found in Hallines, Wildweiberlei (ei = 21), or Le Grand Canton are hardly found in FMG concentrations. Thus, the range of ei in FMG assemblages was compared with the Late Magdalenian range located in the upper part of this range.

The ei of total assemblages (ei_T) can only be calculated for a few sites. The value of the lower horizon of Andernach ($ei_T = 797$) forms, together with Bois Laiterie ($ei_T = 842$), clear outliers to the other assemblages (**fig. 76A**). The other assemblages range between the value of the south-western area of Gönnersdorf ($ei_T = 467$) and the value from the Wildweiberlei ($ei_T = 24$). The latter further sustains the previous consideration that material of smaller dimensions was not archived for this assemblage. The other two Late Magdalenian assemblages, Gönnersdorf ($ei_T = 265$) and Saint Mihiel ($ei_T = 300$), produced very comparable results.

The lower horizon of Hangest-sur-Somme III.1 yielded a comparably low value ($ei_T = 93$) to horizon III.2 of Pincevent ($ei_T = 56$). This observation is perhaps due to the splintering qualities of the raw material used in northern France combined with the aim for particularly large blanks during this time period. However, a higher value was reached in *locus* 7 of Le Tureau des Gardes ($ei_T = 190$), perhaps, indicating the still present Magdalenian trend of producing smaller material such as bladelets. The FMG assemblages vary considerably from the upper horizon of Andernach 2 ($ei_T = 116$) to the example of Niederbieber ($ei_T = 449$). Again the range is significantly enlarged if the single concentrations of Niederbieber are considered (**tab. 83**). Besides the previously described assemblages, which show the same pattern in the ei_T as in the ei, dense clusters (concentrations 8 and 17) yielded very high values marking them as almost undisturbed knapping areas where the small splinters remained *in situ* (cf. Gelhausen 2011a). However, most of the concentrations from Niederbieber that yielded high values were dominated by chalcedony indicating that this raw material had a tendency to produce more small splinters than other raw materials.

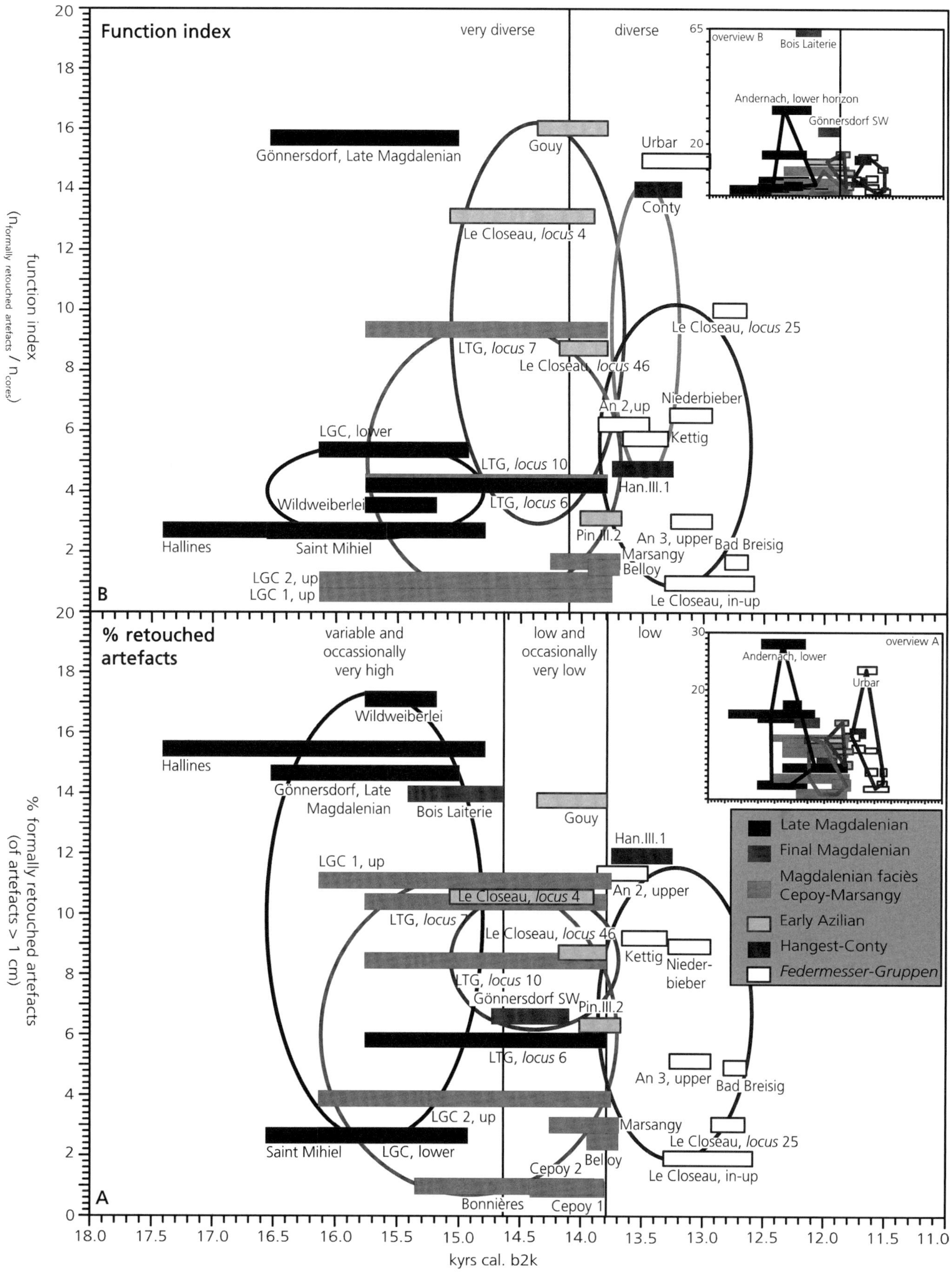
Function index
very diverse
diverse
overview B
Bois Laiterie
Andernach, lower horizon
Gönnersdorf SW
Gönnersdorf, Late Magdalenian
Gouy
Urbar
Conty
Le Closeau, locus 4
Le Closeau, locus 25
LTG, locus 7
Le Closeau, locus 46
Niederbieber
An 2,up
Kettig
LGC, lower
LTG, locus 10
Han.III.1
Wildweiberlei
LTG, locus 6
Hallines
Saint Mihiel
Pin.III.2
An 3, upper
Bad Breisig
Marsangy
Belloy
LGC 2, up
LGC 1, up
Le Closeau, in-up
B
function index
(n formally retouched artefacts / n cores)
% retouched artefacts
variable and occasionally very high
low and occasionally very low
low
overview A
Andernach, lower
Urbar
Wildweiberlei
Hallines
Gönnersdorf, Late Magdalenian
Bois Laiterie
Gouy
Han.III.1
LGC 1, up
An 2, upper
Le Closeau, locus 4
LTG, locus 7
Le Closeau, locus 46
Kettig
Nieder-bieber
LTG, locus 10
Gönnersdorf SW
Pin.III.2
LTG, locus 6
An 3, upper
Bad Breisig
LGC 2, up
Marsangy
Le Closeau, locus 25
Saint Mihiel
LGC, lower
Belloy
Cepoy 2
Le Closeau, in-up
Bonnières
Cepoy 1
A
Late Magdalenian
Final Magdalenian
Magdalenian faciès Cepoy-Marsangy
Early Azilian
Hangest-Conty
Federmesser-Gruppen
% formally retouched artefacts (of artefacts > 1 cm)
kyrs cal. b2k

In general, besides the knapping style and the economy of handling raw materials, the excavated area had some impact on the composition of the assemblage. This impact was particularly the case if the settlement organisation was more complex including sites with special functions as well as specialised areas and clearing episodes on residential sites such as in the Late Magdalenian (cf. Sensburg 2007). Therefore, the function index and the diversity of the material is compared further to take possible site function into consideration.

The function of an assemblage also had some influence on the intensity of the exploitation performed at this site. Besides the blank production process, the production and use of formally retouched artefacts reflected an important field of functions and influenced the composition of the lithic assemblage. In comparison, the ratio of formally retouched artefacts decreased from the Late Magdalenian of 2.6 and 28.1 % to the FMG values of between 2.0 and 11.3 % (**fig. 77A; tab. 83**). The outstanding value from Urbar (% retouched = 23.3) among the FMG variety can be explained by the selective excavation of the site where a single specific activity area was recovered. The relatively high value of the Early Azilian assemblage from Gouy (% retouched = 13.8) could be similarly explained because only a superficial excavation was conducted at this site. Moreover, the value from the lower horizon of Andernach (% retouched = 28.1) appears outstanding in comparison to the following value from Wildweiberlei being only 17.1 %. However, single Late Magdalenian concentrations such as Andernach IV or Gönnersdorf IV also yielded values above 20 %, whereas the single FMG concentrations from Niederbieber provided values just above 15 % (**tab. 83**). Thus, the Late Magdalenian assemblages contained, in general, a higher ratio of formally retouched artefacts than FMG assemblages. Niederbieber 17 is an exception with 38.9 %. According to the spatial analysis, this very small assemblage perhaps originated from a sheltered area, possibly a dwelling structure where a variety of activities were performed (Gelhausen 2011a, 232-237). However, the relatively high ratio in this concentration is related to the total number of artefacts that are particularly small. The percentage of the formally retouched artefacts from Niederbieber 17 is well within the range of the other concentrations from Niederbieber (**tab. 83**). The ratios of the MfCM inventories already fit approximately into the FMG range with values up to 11.1 % (Le Grand Canton, sector 1, upper horizon) but some assemblages contained percentages of formally retouched artefacts as low as 0.8 % (Cepoy, sector 1). In contrast, the Early Azilian assemblages were composed of 6.3 to 13.8 % formally retouched artefacts. The highest value came from Gouy where the inventory was possibly biased against larger and formally retouched artefacts. The remaining three values had up to 10.6 % into the upper range of the MfCM. In the lower horizon from Hangest-sur-Somme III.1 (% retouched = 11.9), some more formally retouched artefacts were found compared to the main FMG assemblages but the ratio was within the usual range of single Niederbieber concentrations. The value of the south-western area of Gönnersdorf (% retouched = 6.6) falls within the range of all archaeological units, whereas the higher ratio from Bois Laiterie (% retouched = 14.0) placed this assemblage into the range of the Late Magdalenian.

To consider the importance of formally retouched artefacts, in contrast to the blank production process, the function index (fi; see p. 272-275) was calculated for all assemblages (**fig. 77B; tab. 83**). This index revealed some clear differentiation within the Late Magdalenian assemblages which ranged between values of 2.7 (Saint Mihiel and Hallines) and 5.3 (lower horizon of Le Grand Canton, sector 2) but also yielded some very

Fig. 77 Percentages of formally retouched artefacts (**A**) and values of the function index (**B**) given per time and archaeological unit. Circles were set around the main cluster of assemblages attributed to the same archaeological unit. The distribution of all assemblages is given in the overview where also the convex hulls were set per archaeological unit (see p. 281 f.). In the overview, names are only given for sites which are not shown in the main graph. Abbreviations see **fig. 74**. – For further details see text.

high values such as the values of Gönnersdorf (fi = 15.7) and the lower horizon of Andernach (fi = 33.3) in general and of Gönnersdorf II (fi = 55.4) and Andernach II (fi = 167.7) in particular. These very high values suggest that in some assemblages the blank production was of very low importance and/or numerous retouched artefacts were used at the site, perhaps due to a higher degree of specialisation. Very low values were probably not present due to the selection of sites in this study.
The wide range remained throughout the transition period. The indices of the south-western area of Gönnersdorf (fi = 24.7) and of Bois Laiterie (fi = 63.5) are exceptionally high but in this period fi values as low as 0.6 (Le Grand Canton, sector 1, upper horizon) were recorded. These very low values reflect the importance of blank production on these sites. The highest values of the MfCM were reached in Le Tureau des Gardes (fi of *locus* 7 = 9.3; fi of *locus* 10 = 4.3) but otherwise the fi was considerably lower and ranged between values of 1.7 and 0.6. This major range points to the importance of blank production of sites attributed to the MfCM. In contrast, the Early Azilian sites yielded values between 3.1 (horizon III.2 of Pincevent) and 16.0 (Gouy). The values were also relatively high for the lower horizons of Conty (fi = 14.0) and Hangest-sur-Somme III.1 (fi = 4.8). Although the range of the FMG assemblages was lower with values between 1.0 and 10, the Hangest-Conty range was still comparable to the FMG range due to Urbar yielding an exceptionally high fi of 15.0. However, this value was possibly due to the selective character of the inventory. In contrast to the separated clusters of the Magdalenian units, the FMG assemblages formed no apparent groups suggesting that the functional variability of these assemblages was of a more gradual nature. Nevertheless, the assemblages of Bad Breisig (fi = 1.6) and the greyish deposits of Le Closeau (fi = 1.0) yielded values that were within the range of the low cluster of the MfCM sites. These unusually consistent values of cores and formally retouched artefacts distinguish the blank production process at these sites as of major importance and, thus, characterise these sites as a type of lithic exploitation sites. Thus, values suggesting an ostensibly wasteful handling of raw materials on these sites could be the result of the function of these sites.
In fact, the only assemblage that fit most of the criteria suggested to identify a wasteful handling of raw material (high % cores, low ei, high ei_T, high di, high assemblage size class) is Bad Breisig. Further assemblages (Hallines, the upper horizon of sector 2 in Le Grand Canton, *locus* 10 of Le Tureau des Gardes, Marsangy, and the greyish deposits of Le Closeau as well as Niederbieber 11 and 14) matched only three of the criteria. However, the differentiation between an ei and ei_T was only possible for a few assemblages. For instance, the ei_T could not be calculated for the upper horizon of Le Grand Canton and most northern French assemblages. The few available double values (**tab. 83**) show that, in general, the two indices developed comparably indicating that more artefacts ≥ 1 cm were found on a Lateglacial site, the more total artefacts were found. Some changes between the two values were observable suggesting that the lithic raw material influenced the number of splinters.
According to the archaeological attribution, the assemblages that matched three or more criteria of wasteful raw material handling were mostly sites of the Magdalenian faciès Cepoy Marsangy and the FMG. Thus, the suggestion of these sites as representing a particular wasteful raw material handling seemed partially supported. However, according to the sub-area, the majority of these sites were located in northern France where a wide range of excellent lithic raw materials was locally available and the large amount of waste could rather relate to a function of the sites than a particularly careless use of the resources.
How far the majority of criteria can be also applied to Gönnersdorf III, remains a matter of reanalysis of this assemblage and a more detailed presentation of the material. However, the handling of particularly numerous cores was already observed in this inventory as well as the upper horizon of Le Grand Canton and Bad Breisig. They indicate that a wasteful handling in the sense of abandoning numerous cores is a diachronic phenomenon and the handling of resources was not in general more wasteful at FMG sites than during the Late Magdalenian. The three mentioned inventories suggest that an abandonment of numerous cores in

the Lateglacial occurred usually at sites where plentiful, small-sized material was available in a very nearby source. Thus, it appears that in these assemblages, quantity was a substitute for larger dimensions.
Since wastefulness of lithic resources was considered as an explanation for the large and very large assemblages, the three small or very small assemblages (Hallines, Niederbieber 11 and 14) reveal that the acquiescence of producing much lithic waste with only few tools does not necessarily create large assemblages. In all three assemblages the ratio of cores was high but the ei was relatively low suggesting that not many larger blanks were abandoned in the assemblage. The two Niederbieber concentrations yielded a relatively high ei_T as a third matching criteria. However, in both concentrations chalcedony was among the dominant raw materials suggesting that the number of splinters was very high due to the choice of raw material. The other Niederbieber concentrations partially matched one or two of the criteria, and, consequently, the Niederbieber assemblage indicates no particularly wasteful handling of raw material. In Hallines, the third matching criteria was the density of the artefacts which can be explained by the restricted excavation area. This explanation can also explain the low ei and high percentage of cores: Assuming that spatial organisation was more complex at Late Magdalenian sites, this pattern could reflect a knapping spot from which the ready made blanks were removed or refuse area where only cores and knapping debris was abandoned. Thus, the values of these three concentration could be explained by other behaviours than wastefulness.
The other assemblages matching three or more criteria (the upper horizon of sector 2 in Le Grand Canton, *locus* 10 of Le Tureau des Gardes, Marsangy, the greyish deposits of Le Closeau, and Bad Breisig) were all large or very large and had a low ei. In all these assemblages, except Le Tureau des Gardes, *locus* 10, very low values of retouched artefacts were found resulting in a low fi. Except Marsangy, these assemblage also yielded high percentages of cores which further sustained the blank production as the major task in these assemblages. This major task attributed these assemblages (the upper horizon of sector 2 in Le Grand Canton, Marsangy, the greyish deposits of Le Closeau, and Bad Breisig) to a type of sites focused on lithic exploitation which explained the very high values. Thus, function rather than a particular careless handling caused these very large assemblages. In Bad Breisig, Marsangy, and Le Tureau des Gardes, *locus* 10, the density of artefacts was very high suggesting that large quantities of raw material were transformed at least at the two latter, widely excavated sites. In Le Tureau des Gardes, *locus* 10, the formally retouched artefacts were numerous and, for a Lateglacial assemblage, formed a relatively typical proportion. In summary, the material abandoned at this site was not very numerous in relation to the cores with a limited number of blanks transformed into formally retouched tools. Thus, no intense lithic exploitation was performed at this site but since the assemblage is nevertheless very large, a relatively wasteful handling and/or a longer use of the site can be assumed. The examples from Hallines and, particularly, Niederbieber 11 and 14 suggest that to create a numerous assemblage as found in *locus* 10 of Le Tureau des Gardes, both factors need to co-occur in the Lateglacial. Nevertheless, large assemblages were more frequently the result of an intense lithic exploitation at a site.
In general, the ei_T was not directly related to the density or the assemblage size class but a regular relation to high function indices could be observed. Thus, the production of formally retouched artefacts seemed to have produced more splinters and where this production was intended it also guided the blank production towards a more numerous blank exploitation of the cores. However, density and assemblage size were closely related, except for Hallines and Bois Laiterie. The former was already explained by the restricted excavation area and the cave of Bois Laiterie can be explained comparably with the restriction of space resulted in a higher density. The restriction or rather the openness of space also explains some very large sites with neither a high amount of blank production nor high densities such as Gönnersdorf, the lower horizon of Le Closeau, Belloy, Niederbieber, and Étigny-Le Brassot. The large open space made a large settlement at favourable spots possible whether these were occupied concurrently or repetitively.

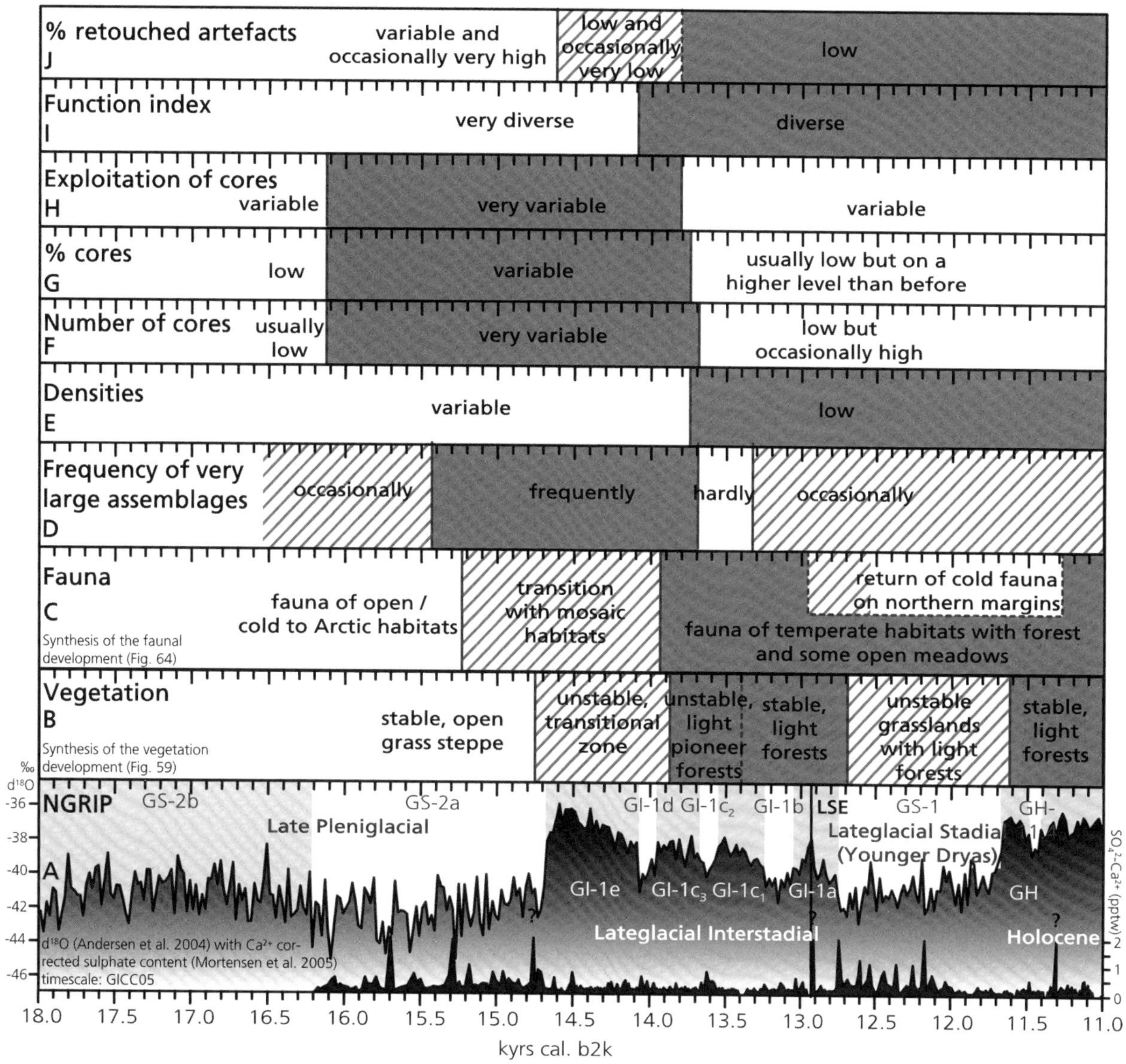

Fig. 78 Developments in lithic assemblages related to lithic raw material exploitation (**D-J**) contrasted by the synthesised faunal (**C**; see **fig. 64**) and vegetation development (**B**; see **fig. 59**) as well as the oxygen isotope record of NGRIP (**A**; see **fig. 53**). Hatched areas: transition periods. – For further details see text.

These examples further sustained that based on the suggested characteristics, a wasteful handling of the lithic resource can be rejected for most sites as an explanation for their large size. Consequently, besides the intense lithic exploitation, accumulation is the most common reason for large and very large assemblages in Lateglacial north-western Europe. This accumulation was either due to a long continued and/or a repetitive use of the site and/or by the presence of numerous inhabitants. These factors possibly explain the differences between the assemblages and can be further refined by consideration of the function of the sites and their place in the settlement system.

In summary, the comparison of the sites according to indicators related to the exploitation of lithic resources reveals some clear changes in the behaviour of Lateglacial hunter-gatherers. The Late Magdalenian assemblages were characterised by a relative wide variability and diversity in the reflected behaviour (**fig. 78**). This behaviour sustains the impression of sites being used for very specific purposes. This specific function resulted in the assemblages forming clusters within the used ranges. In the transition period, this specification

seemed to disappear and the assemblages spread throughout the full range that was occasionally further enlarged. As a result, the assemblages seemed to be more variable, even though the activities conducted at the sites were less differentiated than before. In the process of this transition, the sites of the Early Azilian and those of the MfCM often overlap with a tendency towards the opposing ends of the continuum but only in numerical comparisons such as the number of artefacts or cores, they form distinct clusters. These overlaps and similarities raise the question how different these two archaeological units truly are. Could they represent sites with two, possibly complementary purposes? Was the difference in the function expressed in this period also by a variation in the techno-typological characteristics of the assemblages? Usually, the values of the Late Magdalenian assemblages, in particular the ones from the Central Rhineland and the western uplands, were more similar to the range of the Early Azilian sites than to the results from the MfCM sites. Following the idea of two different functions, this similarity could indicate a more residential character for the older Azilan and a more specialised character of the MfCM sites. However, further details of the functionality of the assemblages are necessary to confirm this conclusion.

At the end of the transition period, the differentiation of sites according to specific purposes seems widely abandoned and, therefore, the variability of the assemblages decreased. Apparent numerical differences seem to only result from differing durations of use and/or the number of concentrations forming the site. Perhaps, the more distinct values of Le Closeau, *locus* 25 and Bad Breisig signal that towards the end of the studied period more specialised assemblages begin reappearing. A more wasteful handling of lithic resources was not apparent. In a direct comparison of the Late Magdalenian and the FMG assemblages, the approximate number of blanks produced per core becomes reduced and instead the proportion of the cores increases. This change could be explained by the decreasing dimensions of the collected raw material nodules. In fact, the numbers of cores found per site remain, in general, in the same relatively narrow range. Only sites that were considered to be used over a longer period, repetitively or continuously, such as Gönnersdorf or Le Closeau were higher numbers of cores recorded. However, in the transition period, higher values occurred more frequently raising the question as to whether these sites were used over a longer time and/or more often?

Besides the faunal material (see p. 534-548), the diversity of the lithic assemblages forms another important line of evidence for the study of the occupation duration and the function of sites as well as the process of change.

Diversity of the lithic assemblages

Diversity of single lithic assemblages was previously introduced as a marker for a diverse set of activities and/or a longer occupation of a site (Löhr 1979; Richter 1990). However, in a model of changing behavioural codices, diversity can also mark a transformation between two phases of stable compositions. In this transformation, the older stable composition (consistent phase) disintegrated (post-consistent and threshold phase) and by testing of innovations and inventions (threshold and formative phase) causing a higher diversity, new characteristics were selected resulting in a new fixed composition (new consistent phase; Clarke 1968, 276-283; cf. Shennan 2008). Besides the diversity of the assemblage composition, this concept of change can also be applied to single sub-sections of this assemblage (Eerkens/Lipo 2005; White 2008) such as projectile implements (Mesoudi/O'Brien 2008a; Hamilton/Buchanan 2009). In all these models, the social transmission of the complex behavioural recipes underlying the production of the assemblages was emphasised as a stabilising factor (cf. Mesoudi/O'Brien 2008b). Assuming that the stabilising function of the transmission systems was formed by a general social consensus to secure survival within unstable environ-

ments (Stein Mandryk 1993), changes in the stability of the transmission may also reflect a destabilisation of the social consensus. In this context, diversity of lithic assemblages becomes an important factor in the study of a transition process in social networks. Consequently, diversity of retouched artefact assemblages in a quasi-contemporaneous comparison shows a functional quality and in a diachronic perspective allows for considerations about the stage of social network stability.

Thus, the composition of complete lithic assemblage, formally retouched artefacts, and, particularly, of the LMP was analysed for this project to reveal changes in different types of diversity that can occur in lithic assemblages.

In regard to the general composition of the lithic assemblages, some observations were already made in the previous section. In a short summary: The Late Magdalenian assemblages differ significantly from one another in the proportion of formally retouched artefacts (**fig. 77A**) as well as of cores and core fragments (**fig. 75A**), in particular on a single site such as the lower horizon of Andernach (**tab. 83**). The proportion of formally retouched artefacts as well as cores and core fragments decreased during the transition period between the Late Magdalenian and the FMG. Nevertheless, the proportion of unretouched blanks and debris did not increase significantly because both equated up to values comparable to the share that was contributed by a single of these categories in a Late Magdalenian assemblage. As a result of the alignment of the values, the assemblage composition of the FMG sites remains relatively stable, in particular in Niederbieber where the assemblages from numerous concentrations are almost identical. This similarity of assemblages in the FMG implies that the share of the main activities was similar for all concentrations (cf. Gelhausen 2011a), whereas in the Late Magdalenian activities were more clearly differentiated in space. During the transition period, sites of the MfCM and the Early Azilian seem to form two extremes within a continuum. The former sites appeared to be more focused on the lithic exploitation and/or the blank production process and yielded significant shares of cores and core fragments. In contrast, the Early Azilian sites seem more dominated by indicators for the use of formally retouched artefacts.

As previously suggested, the relation of cores and core fragments to formally retouched artefacts within an assemblage could reveal the function of a site. This relations permits considerations of occupation duration. Usually, these considerations presume some extreme values. However, on several Lateglacial sites blank production as well as retouching and use of artefacts was performed suggesting a non-specialised activity. Moreover, the limit towards significantly higher values is not fixed but depends on the studied record.

In the present record, several assemblages from the different archaeological units yielded rather inconsistent values. Usually, these sites ranged between 0 and 13 % for formally retouched artefacts and 0 and 4 % for cores and core fragments (**fig. 79A** and detail view A). Very few sites and concentrations yielded extreme values (**tab. 83**) suggesting some specialisation such as the lower horizon of Andernach, Hallines, Wildweiberlei, Urbar, and the upper horizon of sector 1 in Le Grand Canton. However, the former two Late Magdalenian sites and Urbar were not comprehensively excavated and, thus, the specialisation could reflect the function of the excavated activity area/s. In the Wildweiberlei, selective processes before and within the recovery of the material cannot be excluded leading to a disproportional preservation of the lithic material other than cores and formally retouched pieces. Besides the outstanding value for cores and core fragments in Le Grand Canton, sector 1, the values of sector 2 were also clearly dominated by cores and core fragments marking this site as one focussed on specialised blank production. Compared to other sites in this region, the choice of smaller raw material units could explain the extreme values. Moreover, sector 1 perhaps either represented a knapping refuse area and/or was possibly affected by post-depositional processes.

Besides the outstanding values, the values for the upper horizon of sector 2 in Le Grand Canton, Bad Breisig, Gouy, Bois Laiterie, and the Late Magdalenian assemblage of Gönnersdorf were contrasted with the other values. The former two sites were more dominated by cores indicating that the blank production was of greater

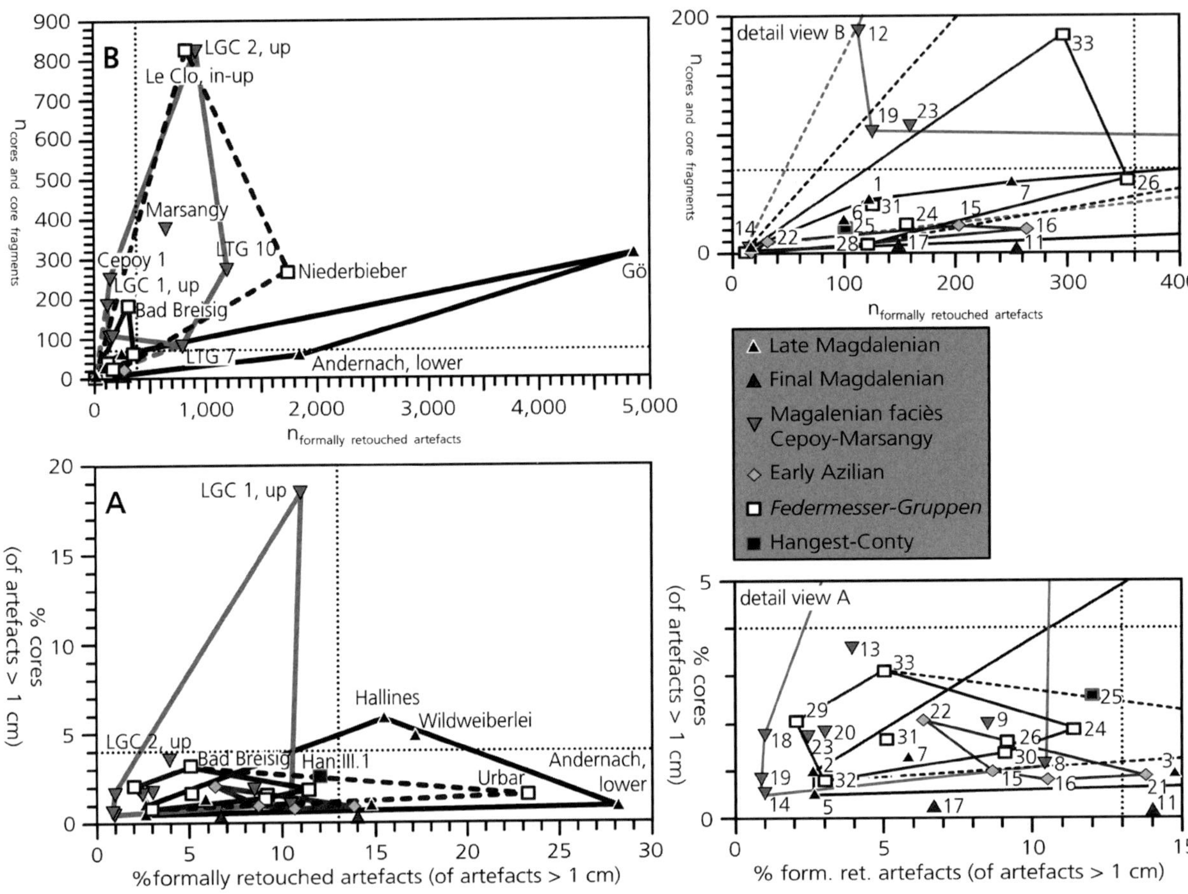

Fig. 79 Percentages (**A**) and numbers of formally retouched artefacts in contrast to cores and core fragments (**B**) given per site and archaeological unit. The distribution of all assemblages is given in the main view where also the convex hulls were set per archaeological unit (see p. 281 f.). Numbers for the sites are used in the detail views according to **tab. 82**. In the detail view B, the unlabled symbols around the 0 points are: Saint Mihiel, the lower horizon of Le Grand Canton, Gouy, and *locus* 25 of Le Closeau. The lines of the convex hulls are also given in the detail views. If an assemblage produced unusually outstanding values for the archaeological unit, the line is dashed. Black dotted lines: the limits of observable breaks in the data beyond which the values can be considered as extreme. For abbreviations see **fig. 74**. – For further details see text.

importance, whereas the latter three assemblages yielded a larger share of formally retouched artefacts suggesting that these sites had a specialised character related to the use of the retouched artefacts.
However, if the inventories were numerically compared instead of using proportional values (**fig. 79B** and detail view B), the results differ, perhaps because the actual numbers are influenced by both the function and by duration of the occupation or the number of inhabitants. In this comparison, the assemblages attributed to the MfCM stand out by their large numbers of cores and core fragments with the exception of Bonnières. Bonnières was previously described as differing by its location from the other sites of this units and, perhaps, this spatial restriction was also the reason for the lower numbers recovered from this site. The Late Magdalenian assemblage of Gönnersdorf, Bad Breisig, Niederbieber, and the intermediate, upper, and undetermined concentrations of Le Closeau also contained significant numbers of cores and core fragments. For the latter two sites, the number of visible concentrations suggests that these sites were used repetitively over a considerable period. Both sites also contained higher numbers of formally retouched artefacts and, therefore, had normal proportional values.
If the single concentrations of these sites, along with the lower horizon of Andernach, Late Magdalenian concentrations of Gönnersdorf, and Marsangy N19 are considered, the tendency does not change only

the amplitude of the outstanding values becomes smaller. For example, Marsangy N19 was more similar to Bad Breisig with a clear dominance of cores. Gönnersdorf III contained the greatest number of cores for the Late Magdalenian assemblages but this number was still lower than those from Marsangy N19 or Bad Breisig. The numbers of formally retouched artefacts from Gönnersdorf I and II were still exceptional. This result further emphasises the functional and spatial differentiation of Magdalenian sites resulting in almost exclusive distributions for the Magdalenian faciès Cepoy Marsangy and the Late Magdalenian assemblages. In contrast, the numerous concentrations from Niederbieber and the greyish deposits of Le Closeau were scattered in a dense cluster near the zero points with some concentrations containing more cores and others clearly more formally retouched artefacts. In general, the Le Closeau concentrations yielded tendentially more cores, whereas the Central Rhineland concentrations were dominated by more formally retouched tools. This tendency hints at a continuity of different raw material behaviours in the Paris Basin and the Central Rhineland. Nevertheless, the numbers were usually below 100 pieces for cores as well as for formally retouched artefacts. Thus, the number of concentrations and, consequently, the duration of the use influenced the high numbers on those sites if compared at a site level. The assemblages from *loci* 7 and 10 of Le Tureau des Gardes, the upper horizon of sector 2 in Le Grand Canton, and Marsangy contained comparably high numbers of cores and formally retouched artefacts. The assumption that these sites were formed by a longer duration of use at the locations appears very probable. Moreover, the Late Magdalenian concentrations of Gönnersdorf and from the lower horizon of Andernach contained outstanding values of formally retouched artefacts and following the assumption based on Niederbieber and the greyish deposits of Le Closeau, a longer duration of the occupation for these concentrations (repetitive or continuous) also seems a possible conclusion. However, the large quantity of formally retouched artefacts singles out these assemblages suggesting an intense and regular use of the formally retouched artefacts on the sites. In the lower horizon of Andernach, the larger values are not accompanied by higher values of cores. Probably, this difference was due to the restriction of the Andernach excavation area which did not encompass a blank production area or a knapping refuse zone. Nevertheless, this comparison further sustains a major difference of the Late Magdalenian and the FMG settlement organisations: In the Late Magdalenian, various activities were performed in a restricted area creating a tell-like accumulation, occasionally cleaned up, whereas this process seemed almost absent on FMG sites. Most activities were still performed in a small, restricted area on those sites with a very similar layout (Gelhausen 2011a) but once the material accumulation became to dense or the activities were finished, this spot was not reused and the occupation was moved to another, often nearby spot. These changes in occupied areas resulted sometimes in a large areals with several, clearly distinct and very similar concentrations reminiscent of planned suburban areas. The Early Azilian concentrations of Le Closeau were previously considered as comparable to the Late Magdalenian layouts of more intensely used and architecturally limited activity areas (Jöris/Terberger 2001). More detailed analyses on this site have shown that special working zones were occasionally performed in spatially distinct satellite spots (Bodu/Debout/Bignon 2006). These distinct areas were also observed in Pincevent, horizon III.2 (Bodu/Orliac/Baffier 1996; Orliac 1996b) but these more ephemeral concentrations already resembled the layout of Niederbieber (Gelhausen 2011a). In contrast, the sites of the MfCM as well as the Magdalenian scatters of Monruz (Bullinger/Leesch/Plumettaz 2006) and Champréveyres (Leesch 1997) appear as a mixture of these two types of spatial behaviour with an accumulation of concentrations in the periphery of one another and partially merging into one another. These sites near an open water source were considered as favourable hunting grounds on which the carcasses could be instantly processed (Müller et al. 2006) in the proximity of one or more hearths (Leesch et al. 2010). The composition of the lithic inventories was comparably heterogeneous in the Swiss examples (Bullinger/Leesch/Plumettaz 2006, 99) as in the FMG sites and clearings, except that maintaining a fire also appeared absent. The dense cluster of stone slab refitting was

suggested as a result of reuse of previous constructions as source for the creation of new hearth constructions (Leesch et al. 2010). Thus, the activity areas remained at approximately the same spot but gradually shifted in the periphery of previous accumulations along the lake shore. Possibly, this shifting was due to dry grounds and the exact position where large prey animals perished. These gradual movements of successive parties resulted in large areals covered by the site without clear limits of single episodes. In contrast, on the northern French sites attributed to the MfCM, clearings, for instance of knapping debris, and more specialised activity areas were still identified (Julien et al. 1999). A comparable gradual movement between the repetitive visits of the sites as seen on the Swiss lake shores had also resulted in a dense scatter of material over a large area but often with still detectable limits.

On sites mainly used as hunting camps, the numbers and proportions of formally retouched artefacts could be influenced by the numerous, mainly broken LMP. However, subtracting the LMP from the formally retouched artefacts in this record only resulted in a compressed version of the previously described distributions and, thus, did not substantially alter the interpretation.

The convex hull (see p. 281 f.) shows that the Late Magdalenian assemblages in this record tend to be dominated by formally retouched artefacts rather than cores and core fragments. This trend could be a result of the choice of the sites. However, the numbers from potential extraction sites such as Eyserheide (Rensink 2012), Orp (Vermeersch et al. 1987; Wenzel 2009), and Kanne (Vermeersch/Lauwers/van Peer 1985) fall within the range of this convex hull. Subtracting the LMP, the hull becomes compressed but remains an elongated shape with fewer cores and more formally retouched artefacts. Furthermore, this trend is also observable for the Early Azilian sites, whereas the convex hulls of the MfCM and the FMG sites clearly differ from this trend and tend to be more dominated by cores. Besides an increasing number of cores, this trend also reflects a decrease in importance of formally retouched artefacts. To find out whether this decrease was due to a wider decline of standardised artefacts or due to a specific part of the inventory, the diversity of the formally retouched inventories is of interest. Moreover, this diversity was also proposed to be useful to further specify the function of site as well as the duration of their use (Löhr 1979).

For a first general impression of the diversity of the formally retouched artefacts, the seven general classes are compared (**tab. 84**). Through this comparison, a clear distinction can be made between the Late Magdalenian assemblages and the preceding period: In Late Magdalenian assemblages, the seven classes are all usually present and this observation also applies to more specialised hunting sites such as Pincevent (Leroi-Gourhan/Brézillon 1966; Bodu et al. 2006b) or the concentrations along the shores of the Lake Neuchâtel (Bullinger/Leesch/Plumettaz 2006; Leesch et al. 2010). In the studied record, some classes were only missing from the assumed hunting camps at Saint Mihiel and in the lower horizon of Le Grand Canton, sector 2. The missing classes in both cases were truncations, composite tools, and other retouched artefacts. The absence of the latter class is surprising but could be due to the lack of activity areas where unstandardised forms were used or discarded, a very high standardisation in the production of retouched implements, the wish of the presenters to attribute any retouched piece to a standardised type, or pieces of uncertain attribution being unmentioned in the presentation of the site. The lack of composite tools is not so surprising if the formation of these tools is considered as a type of reuse for a tool and thereby becomes comparable to the reuse as splintered pieces which was suggested to occur late in the occupation process (see **fig. 27**; Löhr 1979). However, truncations were thought to occur very early within the appearance of formally retouched artefacts on a site. Therefore, the absence of truncations seemed surprising again and could be explained by a transformation of all these artefacts into burins or the absence of the activity in which truncations were used from the excavated areas. The latter is a very probable explanation, in particular, because both assemblages are very small and, in fact, could reflect the very first stage of an inventory where only the material introduced from elsewhere was abandoned at the sites. The next suggested stage in the devel-

site	function index	% ret. artefacts	retouched artefact groups	missing group/s of retouched artefacts	dominant group/s	Simpson index for ret. artefacts	end-scraper-burin index	% borers	% LMP	point-burin index
Hallines	2.7	15.4 (×)	7	0	**burins**, end-scrapers	0.2931	0.5	18.9	2.5	0
Saint Mihiel	2.7	2.7 (0.9)	4	truncations, composite tools, others	LMP, end-scrapers	0.2656	1.7	18.8	31.3	0.3333
Gönnersdorf	15.7 / 13.9	14.7 (5.9) / 13.1 (5.3)	7	0	LMP, burins	0.2467	0.2	9.9 / 11.1	39.7	0
Gönnersdorf I	×	× (×)	7	0	LMP, burins	0.2398	0.1	10.9	33.1	×
Gönnersdorf II	55.4	× (5.2)	7	0	**LMP**, (others), burins	0.2618	0.7	10.8	44.6	×
Gönnersdorf III	×	×	7	0	**LMP**, burins	0.2544	0.3	16.4	41.0	×
Gönnersdorf IV	6.5	21.3 (6.6)	7	0	**LMP**, burins	0.2422	0.2	2.8	47.3	×
Andernach, lower horizon	33.3	28.1 (4.2)	7	0	(others), burins, LMP	0.2224	0.6	5.4	20.0	0.0025
Andernach I	17.7	× (3.3)	7	0	end-scrapers, LMP	0.2248	1.8	7.9	23.0	×
Andernach II	167.7	× (8.7)	7	0	burins, LMP	0.2474	0.1	9.9	30.0	×
Andernach III	36.2	× (16.2)	7	0	LMP, burins	0.2326	0.8	5.0	30.9	×
Andernach IV	30.6	23.0 (1.4)	7	0	**burins**, (others), LMP	0.2874	0.2	3.6	13.5	×
Le Grand Canton, sector 2, lower horizon	5.3	2.6 (×)	4	truncations, composite tools, others	**LMP**, burins	0.3672	0.4	6.3	50.0	0
Wildweiberlei	3.5	17.1 (15.0)	7	0	**(others)**, LMP, burins	0.681	0.5	9.1	20.2	0
Le Tureau des Gardes	3.7	5.7 (×)	7	0	LMP, burins	0.1900	0.9	12.8	30.7	0.0352
Le Tureau des Gardes, *locus* 6	4.2	5.8 (×)	7	0	**LMP**, (others), borers	0.2916	1.0	11.7	49.0	×
Le Tureau des Gardes, *locus* 7	9.3	10.4 (4.9)	7	0	LMP, (others), end-scrapers	0.2335	1.4	10.2	39.9	0.1861
Le Tureau des Gardes, *locus* 10	4.3	8.4 (×)	7	0	LMP, burins	0.1861	0.7	12.1	28.3	×
Étigny-Le Brassot, south	×	1.2 (×)	×	borers, truncations, composite tools, others	end-scrapers, LMP	0.1379	1.7	×	23.0	0.5333
Bois Laiterie	63.5	14.0 (7.5)	7	0	**LMP**, burins	0.2459	0.7	8.3	44.9	0.3871
Le Grand Canton, upper horizon	1.0	4.2 (×)	7	0	burins, end-scrapers	0.2097	0.6	12.0	16.9	0.0172
Le Grand Canton, sector 1	0.6	11.1 (×)	7	0	burins, LMP	0.2070	0.7	6.2	25.7	×
Le Grand Canton, sector 2, upper horizon	1.1	3.9 (×)	7	0	burins, end-scrapers	0.2119	0.6	12.7	15.8	×
Bonnières-sur-Seine	1.9	1.0 (×)	5	truncations, others	LMP, borers	0.2426	1.0	30.8	30.8	1
Le Closeau, lower horizon	~10.5	8.8 (×)	6	borers	**(others)**, *end-scrapers, LMP*	0.3633	3.4	0.0	16.4	0.3793
Le Closeau, *locus* 4+50	13.1	10.6 (×)	6	borers	**(others)**, end-scrapers, LMP	0.3239	2.0 (2.1)	0.0	17.2	0.4783
Le Closeau, *locus* 46	8.7	8.7 (×)	5	borers, truncations	**(others)**, end-scrapers, LMP	0.4183	12.3	0.0	13.9	×
Gönnersdorf SW	24.7	6.6 (5.3)	7	0	LMP, burins	0.2569	0.2	8.1	39.2	0.0882
Cepoy	0.8	0.9 (×)	6-7	composite tools	end-scrapers, borers, LMP	0.1802	2.4	18.7	18.4	0.6207
Cepoy, sector 1	0.6	1.0 (×)	7	0	end-scrapers, borers	0.1962	2.2	23.9	17.6	0.7778

Tab. 84 IIndices referring to the function of the studied assemblages. Function index is the no. of formally retouched artefacts (ret. artefacts) divided by the no. of cores and core fragments. % ret. artefacts is calculated in relation to artefacts ≥ 1 cm and in parentheses to total artefacts. The two (or three, if the »others« group is included) most numerous groups are given and if a single group formed more than 40 % of the formally retouched artefacts, this group is set in bold. Simpson index is the sum of the square-results of dividing the number of implements within a retouched artefact group by the total number of retouched artefacts. End-scraper-burin index is the no. of end-scrapers divided by the number of burins. % values (if not specified) calculated in relation to all retouched artefacts. % borers/LMP is calculated in relation to all formally retouched artefacts. Point-burin index is the no. of points (see **tab. 85**) divided by the number of burins.

site	function index	% ret. artefacts	retouched artefact groups	missing group/s of retouched artefacts	dominant group/s	Simpson index for ret. artefacts	end-scraper-burin index	% borers	% LMP	point-burin index
Cepoy, sector 2	~1.2	0.8 (×)	6	composite tools	(others), end-scrapers, LMP	0.1996	2.6	12.8	13.9	0.2727
Marsangy	1.7	3.0 (×)	7	0	burins, LMP	0.2006	0.4	17.3	26.2	0.1341
Marsangy N19	1.6	3.0 (×)	7	0	LMP, burins	0.2020	0.4	21.4	27.5	0.1216
Gouy	16.0	13.8 (×)	6	others	LMP, borers, burins	0.2344	0.3	18.8	37.5	1.3333
Pincevent III.2	3.1	6.3 (5.5)	5	borers, composite tools	(others), end-scrapers, truncations	0.2570	4.0	0.0	9.7	2
Belloy-sur-Somme	1.5	2.5 (×)	6	composite tools	**(others)**, end-scrapers, burins	0.2984	1.1	6.3	12.7	0.5238
Andernach, upper horizon	4.2	7.4 (1.4)	6	borers	**LMP**, (others), end-scrapers, burins	0.2877	1	0.0	45.7	0.6585
Andernach 2-FMG	6.2	11.3 (5.4)	6	borers	**LMP**, (others), end-scrapers	0.2691	1.1	0.0	42.3	0.8182
Hangest-sur-Somme III.1, lower horizon	4.8	11.9 (5.1)	5	borers, composite tools	**LMP**, burins	0.3599	0.2	0.0	52.0	0.1482
Kettig	5.7	9.2 (1.5)	7	0	end-scrapers, LMP	0.216	3.1	1.4	28.4	0.6111
Conty, lower horizon	14.0	× (2.1)	6	composite tools	**LMP**, burins	0.3708	0.3	4.8	57.1	1.6667
Urbar	15.0	23.3 (7.3)	5	borers, composite tools	**end-scrapers,** LMP	0.245	49	0.0	10.8	3
Le Closeau, greyish deposits	1.0	2.0 (×)	7	0	**LMP**, end-scrapers	0.3723	3.8	2.9	58.3	0.2258
Niederbieber	6.3	9.0 (1.4)	7	0	LMP, (others), end-scrapers	0.2123	1.2	2.1	31.9	0.4025
Niederbieber 1	8.2	10.5 (3.4)	6	composite tools	LMP, (others), end-scrapers	0.681	1.2	1.8	25.4	0.2245
Niederbieber 3	9	3.5 (0.6)	3	all (except LMP, burins, others)	**LMP**, burins	0.4321	×	0.0	55.6	0
Niederbieber 4 (+17a)	9	12.3 (1.5)	6	composite tools	LMP, end-scrapers	0.2329	2.2	0.8	30.8	0.2059
Niederbieber 5	4.6	6.7 (1.4)	6	composite tools	(others), LMP, truncations	0.2582	0.9	1.9	32.4	1.4286
Niederbieber 6 (+10a)	7.4	5.7 (0.8)	6	composite tools	**LMP**, (others), truncations	0.3522	1	1.0	51.0	3
Niederbieber 7	7.8	12.0 (5.5)	7	0	burins, (others), truncations	0.2209	0.4	4.0	18.4	0.2286
Niederbieber 8	7.4	6.9 (0.3)	5	borers, composite tools	burins, LMP	0.2739	0.2	0.0	32.4	0.2308
Niederbieber 9	5.6	8.3 (3.6)	6	composite tools	LMP, (others), end-scrapers	0.2518	2.3	2.3	37.4	0.9286
Niederbieber 10	5.3	7.2 (2.5)	6	composite tools	**LMP**, (others), end-scrapers	0.2699	2	1.7	43.1	1
Niederbieber 11	6	13.2 (1.4)	7	0	**LMP**, end-scrapers	0.2665	2.3	1.5	40.9	0.2857
Niederbieber 12	3.1	5.9 (1.0)	5	borers, composite tools	**LMP**, (others), burins	0.2923	0.4	0.0	40.0	0.3076
Niederbieber 13	5.6	7.2 (0.7)	6	composite tools	**LMP**, end-scrapers, (others)	0.2824	3	2.0	40.0	0.5
Niederbieber 14	4.1	11.9 (1.4)	6	composite tools	LMP, end-scrapers	0.2026	1.4	5.6	28.1	0.1333
Niederbieber 15	1.3	14.8 (7.6)	5	borers, composite tools	LMP, end-scrapers, (others)	0.229	2.8	0.0	26.2	0.5
Niederbieber 16	6	7.4 (0.6)	6	composite tools	**LMP**, burins	0.4131	0.3	2.4	61.3	0.25
Niederbieber 17	15.5	38.9 (3.1)	5	borers, composite tools	LMP, truncations	0.2336	0.6	0.0	34.9	0.4167
Niederbieber 18	9	1.4 (0.1)	2	all (except LMP, burins)	**LMP**, end-scrapers	0.46	×	0.0	60.0	×
Niederbieber 19	10	5.7 (3.1)	4	borers, composite tools, others	**LMP**, end-scrapers, truncations, burins	0.4815	1	0.0	66.7	×
Niederbieber 20	4.5	4.6 (1.3)	6	composite tools	**LMP**, burins	0.3353	0.2	6.7	51.1	×
Andernach 3-FMG	3.0	5.1 (0.7)	6	borers	**LMP**, burins	0.3123	0.8	0.0	50.0	0.4737
Le Closeau, *locus* 25	10	3.0 (×)	1	all (except LMP)	LMP	1.0000	×	0.0	100	×
Bad Breisig	1.6	5.0 (0.7)	6	borers	**LMP**, end-scrapers	0.2844	3.9	0.0	38.5	0.76

Tab. 84 (continued)

opment of a Magdalenian lithic inventory is an initial blank production with the production of burins and first bladelets that were transformed into backed bladelets before truncations were made. The relatively low numbers and percentages of the cores in the assemblages of Saint Mihiel and the lower horizon of Le Grand Canton, sector 2 (**fig. 79**) further support the impression of these assemblages reflecting only a brief occupation. The Late Magdalenian inventories were generally dominated by the LMP (**tab. 84**) confirming Jürgen Richter's suggestion that the LMP reflect a basic group of retouched artefacts on Magdalenian sites (Richter 1990). However, on some sites such as the lower horizon of Andernach, burins occurred in an even greater number than LMP. In Andernach, the different focus of activities becomes apparent particularly in regard to the most numerous group. Although LMP are always numerous, the most numerous group varies in almost each concentration, further suggesting that some concentrations were perhaps supplementary installations.

In general, the appearance of all retouched artefact groups also applies to the sites of the MfCM. Only in sector 2 of Cepoy and in the lower horizon of Belloy-sur-Somme, composite tools were not identified and in Bonnières-sur-Seine, no truncations or other retouched artefacts were found. Besides reasons for the absence of these groups, the limited excavation area at Bonnières could be an additional bias for these groups. Moreover, the rich lithic sources at Belloy-sur-Somme as well as at Cepoy created a surplus of blanks making reuse unnecessary and, perhaps, the use of one instrument for various tasks unwanted. Until the lithic inventory from Étigny-Le Brassotis presented comprehensively, it remains uncertain which groups are missing. According to a short, preliminary presentation (Lhomme et al. 2004), borers seemed to be absent along with the three groups not present at Saint Mihiel and in the lower horizon of Le Grand Canton, sector 2. Based on the evidence from Le Tureau des Gardes, Laurent Lang suggested that besides burins, (heavily retouched) borers could indicate work with antler material and that a decrease in their number could reflect a diminishing importance of antler as a resource (Lang 1998, 99). However, use-wear analysis on *becs* from the Late Magdalenian site Verberie and the perforators of Gönnersdorf revealed that mostly these pieces were in fact used for piercing and boring (Beyries/Janny/Audouze 2005; Sano 2012b). The heavily retouched specimens from Verberie were generally used on hard organic material such as bone or antler but the finer pieces from Gönnersdorf were used on a variety of material such as stone, bone, antler, ivory, tooth, or shell. Thus, the disappearance of this group of artefacts might also indicate a decreasing importance of working hard organic material but the abandonment of drilling activities of hard substances and/or the use of standardised pieces for this activity seems to more generally reflect the decrease in importance of borers. The changes in the blank production process which were widely observed in the assemblages of the MfCM produced more solid and pointed flakes and blades which could perhaps have been substituted for the borers without the need for further retouch. Comparably, no borers were found in the Early Azilian assemblages of Le Closeau and Pincevent III.2 where the knapping technique using soft hammerstone had become exclusive. Besides LMP and burins, end-scrapers become a dominant group in these inventories. The most numerous groups become more heterogeneous in these assemblages and no group reached values of over 40 % as did the LMP and occasionally burins in the Late Magdalenian. Besides a greater spatial specialisation among the Late Magdalenian, the generally lower proportion of formally retouched artefacts in assemblages of the MfCM is probably of less importance regarding assemblage composition.

In contrast to the Magdalenian inventories, Early Azilian and the FMG assemblages infrequently contained the complete set of formally retouched artefacts and usually this set was only found in horizons formed by several concentrations. Moreover, truncations were usually present in these inventories. Besides the lack of borers, no composite tools were found in Pincevent III.2 and the group of other retouched artefacts was absent from Gouy. The absence of these groups is comparable to Saint Mihiel and the lower horizon of Le Grand Canton, sector 2. The most numerous groups of retouched artefacts varied significantly between

these assemblages. End-scrapers became more frequent than burins and besides these groups, other retouched artefacts became of greater importance in the Early Azilian assemblages.
Composite tools were most frequently missing in FMG assemblages followed by borers. In contrast to the MfCM and the Early Azilian assemblages, LMP were the clearly dominant group again which frequently covered more than 40 % of the retouched artefacts. Besides this group, end-scrapers were the second most numerous group. Infrequently, burins formed a more numerous group. Besides these typical groups, truncations and other retouched artefacts were also occasionally dominant groups. In some FMG inventories, only very few formally retouched artefacts were found and comprised usually of pieces identified as LMP and burin. Thus, even though the numbers and shapes of the burins became unstable in the FMG assemblages, they remained one of the more essential lithic artefact types.
According to the missing as well as the dominant groups of retouched artefacts in this project (**tab. 84**), the LMP can be confirmed as a basic group (cf. Richter 1990) throughout the Late Pleniglacial and the Lateglacial Interstadial. In fact, most Lateglacial assemblages contained 30-50 % LMP among the retouched artefacts (**fig. 80A**). Nevertheless, these values differed, sometimes considerably, between the sites and within the assemblages (**tab. 84**). Tendentially, the FMG assemblages contained a larger proportion of LMP than the Late Magdalenian. The FMG assemblages ranged generally between 28.4 and 58.3 % LMP but the value of Urbar (% LMP = 10.8 %) formed a clear outlier. Since the site was only partially excavated, the low proportion could be a result of this bias. In contrast, the Late Magdalenian assemblages contained 20 to 50 % LMP with Hallines (% LMP = 2.5 %) again representing an outlier. Likewise at Urbar, the restriction of the excavation area, which caused high values of intense blank production on this site, probably also caused the low share of LMP. However, the trend from the Late Magdalenian to the FMG assemblages was not a continuous increase because the inventories of the MfCM contained relatively small proportions of LMP (range of % LMP: 12.7-39.9 %) and the ratio further decreased in the Early Azilian assemblages (range of % LMP: 9.7-17.2 %). The assemblage from Gouy formed an exception to the latter with a value of 37.5 %. The superficial excavation at this site could have also biased this value. Since some of the MfCM sites were considered as hunting and/or butchering locales, the low values are surprising. These generally lower values become explicable if the types of LMP are taken into account (**tab. 85**). In the Late Magdalenian assemblages such as Gönnersdorf, small backed bladelets were the aimed shape which often was produced by the intentional breakage of larger backed blades into smaller sections (see **pl. 1, 29-30**). This procedure resulted in numerous pieces including the debris which often bore partial retouch. According to the limited evidence (Leroi-Gourhan 1983), several of these fragments were used as composite tools. In contrast, only a few of these small implements were found on sites of the MfCM or the Early Azilian. On these sites, lithic points were the major LMP. The points were generally considered to be used as single implements glued in or on top of a shaft (cf. Valentin 2008a; figs 31-32). Thus, counted per projectile significantly fewer lithic inserts were necessary and less debris material was produced resulting probably in an overall decrease of LMP numbers and proportions, although the same number of ready made projectiles might be present at the site.
Thus, the number of points per site could help balance for this possible disproportion. However, besides previously mentioned difficulties for estimating the number of points, this number is also dependent on the size of the occupation and/or number of occupants or occupying episodes. To take this assemblage size into account, the points can be set in relation to a group of the formally retouched artefacts attributed to the *fond commun*. Since burins are the second most frequently present type, the point-burin index can be used for this count (**fig. 80B**). Burins were usually considered as antler or bone working tools, in particular for the fabrication of spalls which were often the preform of organic points. These numbers also changed significantly. Therefore, the index was proposed as a possible indicator for the significance of organic points in contrast to lithic projectiles (see p. 275-282). Thus, the suggestion that the increase of lithic points was

site	total	displayed / % of total	ref.	indet./ % of displayed	couteau à dos	backed blade(let)	simple point	straight-backed point	curve-backed point	angle-backed point	tanged point	no. of types	no. of points
Hallines	3	1 / 33.3	1	0 / 0	**1* / 100**	0	0	0	0	0	0	1	0
Saint Mihiel	5	6 / 120.0	2	5 / 83.3	0	0	0	0	0	**1 / 16.7**	0	1	1
Gönnersdorf	1,927	67 / 3.5	5	35 / 52.2	0	**32 / 47.8**	0	0	0	0	0	1	0
Andernach, lower horizon	367	46 / 12.5	3-4	32 / 69.6	0	**13 / 28.3**	0	1	0	0	0	2	1
Le Grand Canton, lower horizon	8	8 / 100	7	8 / 100	0	0	0	0	0	0	0	0	0
Wildweiberlei	20	15 / 75.0	6	9 / 60.0	0	**6 / 40.0**	0	0	0	0	0	1	0
Le Tureau des Gardes	1,057	28 / 2.7	12-13	6 / 21.4	0	1	0	0	3	**18 / 64.3**	0	3	21°
Le Tureau des Gardes, locus 7**	313	16 / 5.1	13	0 / 0 (189 / 60.4)	0 (0)	0 (14)	0	0 (0)	2 (4)	**14 / 87.5 (106 / 33.9)**	0 (0)	2 (3)	16° (124)
Étigny-Le Brassot, south	23	8 / 34.8	11	0 / 0	0	0	0	0	**6 / 75.0**	2	0	2	8
Bois Laiterie	254	47 / 18.5	9-10	27 / 57.5	2	**6 / 12.8**	2	1	5	1	3	**7**	12
Le Grand Canton, upper horizon	174	59 / 33.9	7	45 / 76.3	0	**8 / 13.6**	1	1	1	3	0	**5**	6
Bonnières-sur-Seine	4	4 / 100	8	1 / 25.0	0	1	0	0	0	**2 / 50.0**	0	2	2
Le Closeau, locus 4+50	45	18 / 40.0	14-15	7 / 38.9	0	0	2	0	**7 / 38.9**	2	0	3	11
Gönnersdorf SW	58	34 / 58.6	16	29 / 85.3	0	**2 / 5.9**	0	0	1	**2 / 5.9**	0	3	3
Cepoy	47	17 / 36.2	17-18	0 / 0	0	0	1	0	2	**14 / 82.4**	0	3	17
Cepoy, sector 1	25	14 / 56.0	18	0 / 0	0	0	1	0	2	**11 / 78.6**	0	3	14
Marsangy	168	69 / 41.1	19	39 / 56.5	1	5	2	1	5	**13 / 18.8**	3	**7**	24
Marsangy N19	86	35 / 40.7	19	22 / 62.9	1	2	0	1	0	**7 / 20.0**	1	**5**	9
Gouy	6	6 / 100	20	1 / 16.7	1	0	0	0	**3 / 50.0**	1	0	3	4
Pincevent III.2	5	5 / 100	25-26	1 / 20.0	0	0	0	0	**4 / 80.0**	0	0	1	4
Belloy-sur-Somme	20	18 / 90	1	4 / 22.2	1	3	3	0	2	**6 / 40.0**	0	**5**	11
Irlich	1	1 / 100	21	1 / 100	0	0	0	0	0	0	0	0	0
Andernach, upper horizon	128	128 / 100	22-24	94 / 73.4	3	4	5	1	**20 / 16.4**	1	0	**6**	27
Andernach 2-FMG	66	71 / 107.6	22	52 / 73.2	1*	0	5	0	**13 / 19.7**	0	0	3	18

Tab. 85 Numbers of LMP types in the studied assemblages. The most numerous defined type is given in bold and, in addition, the % of all displayed figures. The number of types refers to the number of determined types and excluding indetermined pieces. * Magdalenian (pointed) blades; ** the numbers set in parantheses refer to numbers given in Lang 1998, 28-36 (the plates from this report were not available to the present author); *** in the original publication the fragments were counted, whereas in the present study the complete pieces are counted; ° these numbers are according to what was found displayed in publications, the given numbers in the publications are considerably higher; × unknown. References (ref.): **1** Fagnart 1997, figs 20, 40, 41, 89, and 166; **2** Stocker et al. 2006, figs 6-7; **3** Floss/Terberger 2002, Abb. 120. 124; **4** Holzkämper 2006, Taf. 16; **5** Franken/Veil 1983, Taf. 22 and 33; **6** Terberger 1993, Taf. 72; **7** Valentin et al. 1999b, figs 22 and 26-27; **8** Barois-Basquin/Charier/Lécolle 1996, fig. 10; **9** Straus/Orphal 1997, figs 10-17; **10** Sano/Maier/Heidenreich 2011, figs 5-6; **11** Lhomme et al. 2004, fig. 32; **12** Alix et al. 1993, fig. 20; **13** Weber 2006, fig. 4; **14** Bodu/Mevel 2008, fig. 6; **15** Bodu 1998, fig. 418; **16** Buschkämper 1993, Abb. 6 Taf. 2 and 3; **17** Valentin 1995, Planche 38; **18** Wenzel 2009, fig. 52; **19** Schmider 1992a, figs 106, 108-109, and 111; **20** Bordes et al. 1974, fig. 1; **21** Baales 2002, Abb. 8, 80-81, and 83-84; **22** Bolus 1984, Abb. 21, 35, 39, 78, 91, and 129; **23** Kegler 1999; **24** Kegler 2002; **25** Bodu/Orliac/Baffier 1996, fig. 72; **26** Orliac 1996b, fig. 78; **27** Baales/Mewis/Street 1998; **28** Bolus 1992; **29** Loftus 1984; **30** Loftus 1985; **31** Gelhausen 2011a; **32** Gelhausen 2011b; **33** Husmann 1988; **34** Husmann 1989; **35** Thomas 1990; **36** Freericks 1989; **37** Wenzel 2004; **38** Baales/Grimm/Jöris 2001; **39** Grimm 2004; **40** Grimm 2003.

site	total	displayed / % of total	ref.	indet./ % of displayed	couteau à dos	backed blade(let)	simple point	straight-backed point	curve-backed point	angle-backed point	tanged point	no. of types	no. of points
Hangest-sur-Somme III.1 lower horizon	51	15 / 29.4	1	1 / 6.7	4	**6 / 40.0**	0	0	2	2	0	4	4
Kettig	100	96 / 96.0	21	63 / 65.6	0	11	7	2	**12 / 12.5**	1	0	**5**	22
Conty, lower horizon	24	19 / 79.2	1	4 / 21.1	**7 / 36.8**	1	0	0	4	3	0	4	7
Urbar	13	13 / 100	27	6 / 46.2	0	1	1	1	**2 / 15.4**	**2 / 15.4**	0	**5**	6
Niederbieber	551	393 / 71.3	28-16	245 / 62.3	8	45	5	12	**60 / 15.3**	18	0	**6**	95
Niederbieber 1	69	75 / 108.7	28-29	54 / 72.0	0	**10 / 13.3**	0	0	**10 / 13.3**	1	0	3	11
Niederbieber 2 (18+19+20)	35	29 / 82.9	30-31	23 / 79.3	0	**2 / 6.9**	0	1	**2 / 6.9**	0	0	3	3
Niederbieber 3	5	5 / 100	28	5 / 100	0	0	0	0	0	0	0	0	0
Niederbieber 4+17a	80	78 / 97.5	28; 32	59 / 75.6	1*	**11 / 14.1**	0	1	5	1	0	**5**	7
Niederbieber 5	34	41 / 120.6	33-34	26 / 63.4	0	5	3	0	**6 / 14.6**	1	0	4	10
Niederbieber 6+10a	53	43 / 81.1	32; 35	26 / 60.5	0	**8 / 18.6**	0	1	6	2	0	4	9
Niederbieber 7	23	12 / 52.2	36	4 / 33.3	0	0	0	1	3	**4 / 33.3**	0	3	8
Niederbieber 8	12	5 / 41.7	32	2 / 40.0	0	0	0	0	**2 / 40.0**	1	0	2	3
Niederbieber 9	65	27 / 41.5	32	5 / 18.5	**7 / 25.9**	2	0	3	**7 / 25.9**	3	0	**5**	13
Niederbieber 10	25	11 / 44.0	32	5 / 45.5	0	1	0	0	**5 / 45.5**	0	0	2	5
Niederbieber 11	27	7 / 25.9	32	3 / 42.9	0	**2 / 28.6**	0	1	1	0	0	3	2
Niederbieber 12	26	9 / 34.6	32	3 / 33.3	0	2	0	1	**3 / 33.3**	0	0	3	4
Niederbieber 13	20	6 / 30.0	32	4 / 66.7	0	0	0	0	**2 / 33.3**	0	0	1	2
Niederbieber 14	25	7 / 28.0	32	5 / 71.4	0	0	0	0	**2 / 28.6**	0	0	1	2
Niederbieber 15	11	5 / 45.5	32	3 / 60.0	0	0	0	0	**1 / 20.0**	**1 / 20.0**	0	2	2
Niederbieber 16	19	7 / 36.8	32	6 / 85.7	0	0	0	0	0	**1 / 14.3**	0	1	1
Niederbieber 17	22	12 / 54.5	32	4 / 33.3	0	**3 / 25.0**	2	0	**3 / 25.0**	0	0	4	5
Andernach 3-FMG	62	57 / 91.9	23-24	42 / 73.7	2	4	0	1	**7 / 12.5**	1	0	**5**	9
Boppard	×	4 / ×	37	0	0	0	0	**1 / 25.0**	**1 / 25.0**	**1 / 25.0**	**1 / 25.0**	4	4
Le Closeau, locus 25	10***	10 / 100	15	4 / 40	0	**1 / 10.0**	0	0	**4 / 40**	**1 / 10**	0	3	5
Bad Breisig	114	83 / 72.8	38-40	57 / 68.7	0	7	0	**10 / 12.1**	6	3	0	4	19

Tab. 85 (continued)

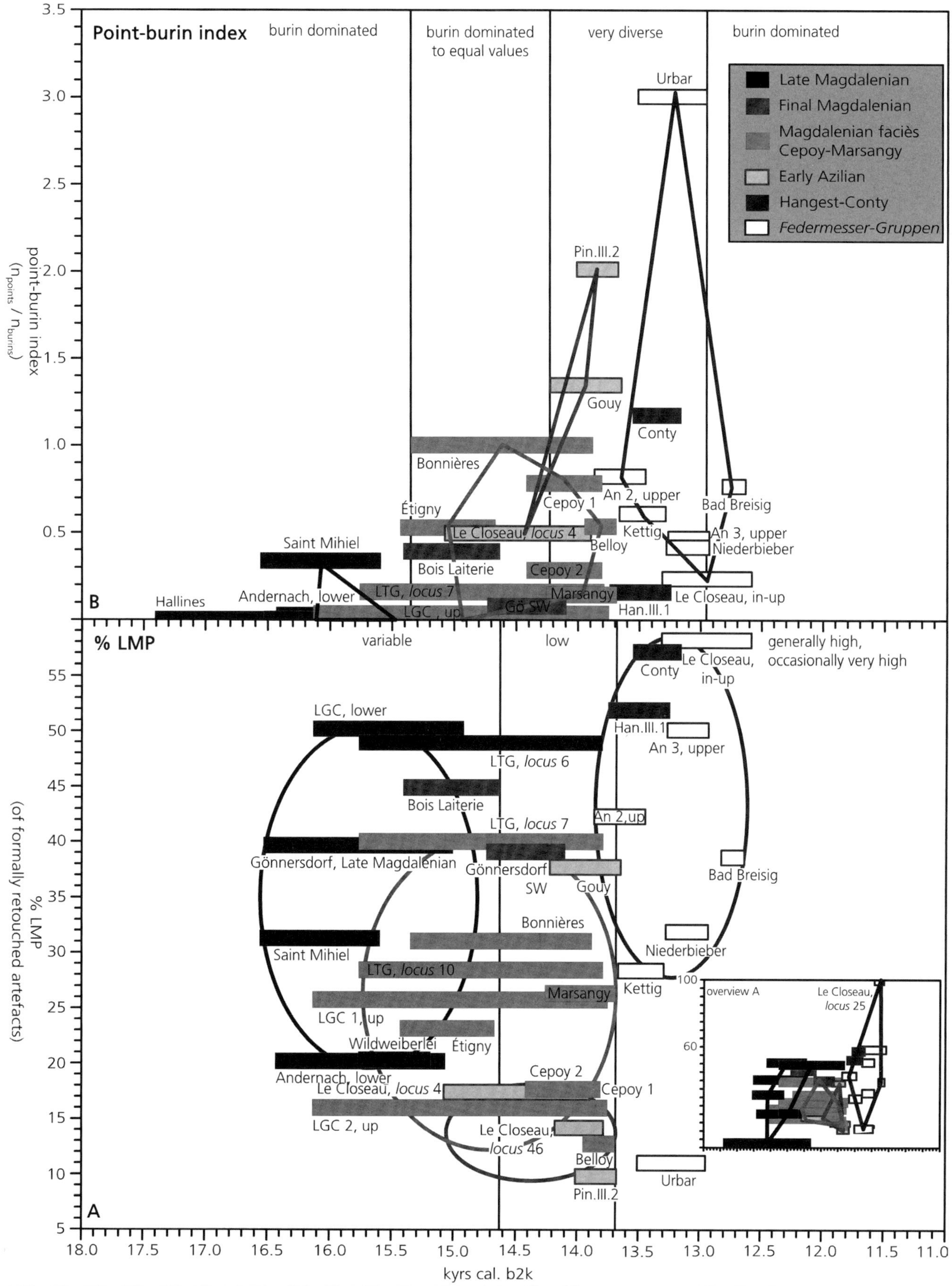

Point-burin index
burin dominated
burin dominated to equal values
very diverse
burin dominated
Late Magdalenian
Final Magdalenian
Magdalenian faciès Cepoy-Marsangy
Early Azilian
Hangest-Conty
Federmesser-Gruppen
point-burin index (n_{points} / n_{burins})
Urbar
Pin.III.2
Gouy
Conty
Bonnières
An 2, upper
Cepoy 1
Bad Breisig
Étigny
Le Closeau, locus 4
Kettig
Belloy
An 3, upper
Saint Mihiel
Niederbieber
Bois Laiterie
Cepoy 2
LTG, locus 7
Marsangy
Le Closeau, in-up
Andernach, lower
Hallines
LGC, up
Gö SW
Han.III.1
B
% LMP
variable
low
generally high, occasionally very high
Le Closeau, in-up
LGC, lower
LTG, locus 6
An 3, upper
Bois Laiterie
An 2,up
Gönnersdorf, Late Magdalenian
Gönnersdorf SW
Saint Mihiel
LTG, locus 10
LGC 1, up
Wildweiberlei
Andernach, lower
Le Closeau, locus 4
LGC 2, up
Le Closeau, locus 46
Pin.III.2
overview A
Le Closeau, locus 25
% LMP (of formally retouched artefacts)
A
kyrs cal. b2k

possibly due to the gradual disappearance of antler as resource for projectiles (Lang 1998, 96-99), perhaps, also explains the decrease of backed bladelets.

However, in the present record, the phases observable in the proportions of the LMP and the point-burin index differ suggesting no direct relation between the values. Nevertheless, the lowest LMP percentages co-occur temporally with the appearance of point-dominated assemblages. In particular, the inventories of Pincevent, horizon III.2 and Cepoy 1 are characterised by a low proportion of LMP and a high (point-dominated) result in the point-burin index. The LMP as well as the burins did not belong to the most numerous groups in both inventories and small changes in the number of points could result in obvious changes in the index value. Nevertheless, Pincevent III.2 is a very small inventory of formally retouched artefacts and drawings of the LMP were completely displayed in publications. Thus, a change in these values is not very probable but this possibility remains possible for other inventories and could shift the limits of the phases in the point-burin index. The most outstanding correlation of low LMP proportion and a high point-burin index occurs in the Urbar assemblage. From this assemblage all LMP were displayed and changing values seem improbable. However, this assemblage came from a restricted excavation and the complete assemblage could again alter the value. In the concentrations from Niederbieber, very high LMP percentages occurred with very low as well as very high point-burin indices but low LMP percentages were only accompanied by burin-dominated index values. Even though lithic points were the usual LMP in the FMG inventories, the ratio of LMP sometimes became very high. Moreover, in these assemblages backed bladelets occurred again and many fragments which could not be further determined were also found on these sites. Thus, the decrease of LMP values due to the introduction of points applied possibly for the initial period of use but certainly not for the continuous use.

In regard to the relationship of organic and lithic projectile points, the point-burin index suggests that the importance of lithic armatures increases, in particular in the Early Azilian, but that working of hard organic tissues remains of some importance. In order to consider the importance of antler and bone working on Lateglacial sites, the burins were also set in relation to end-scrapers and the proportions of borers as alternative tools was used (**fig. 81**). These values further refine the picture of the development of the Lateglacial inventories, specifically in regard to possible indicators for working of hard organic tissues. Thus, for the validity of the results, sites with preserved organic artefacts are also of interest.

The point-burin index is, generally, burin-dominated in Late Magdalenian assemblages because only single pointed implements were found at these sites. The end-scraper-burin index is also dominated by burins and the ratio of borers range between 5.35 and 18.85 %. Only in single assemblages such as Andernach IV or Gönnersdorf IV values as low as 2.75 % occurred. Following the suggestion that burins and, possibly, borers were indicative of working hard organic materials (cf. Lang 1998, 96-99), this activity was of some importance in the Late Magdalenian assemblages. This assumption is further sustained by the numerous organic remains reflecting the production of bone, antler, and ivory implements at sites with a good organic preservation (Tinnes 1994; Bodu et al. 2006b).

The values from Bois Laiterie and the south-western area of Gönnersdorf fall in the range of Late Magdalenian assemblages. In Bois Laiterie, the ratio of the points was marginally higher than in the Late Magdale-

←

Fig. 80 **A** Percentages of LMP given per time and archaeological unit. Circles were set around the main cluster of assemblages attributed to the same archaeological unit. The distribution of all assemblages is given in the overview where also the convex hulls were set per archaeological unit (see p. 281 f.). In the overview, names are only given for sites which are not shown in the main graph. **B** Values of the point-burin index given per time and archaeological unit. The complete distribution is already given and the convex hulls are therefore set in the main view. Besides Hallines and the lower horizon of Andernach, the values of Gönnersdorf, Wildweiberlei, and the lower horizon of Le Grand Canton, sector 2 fall to the bottom of graph B. Abbreviations see **fig. 74**. – For further details see text.

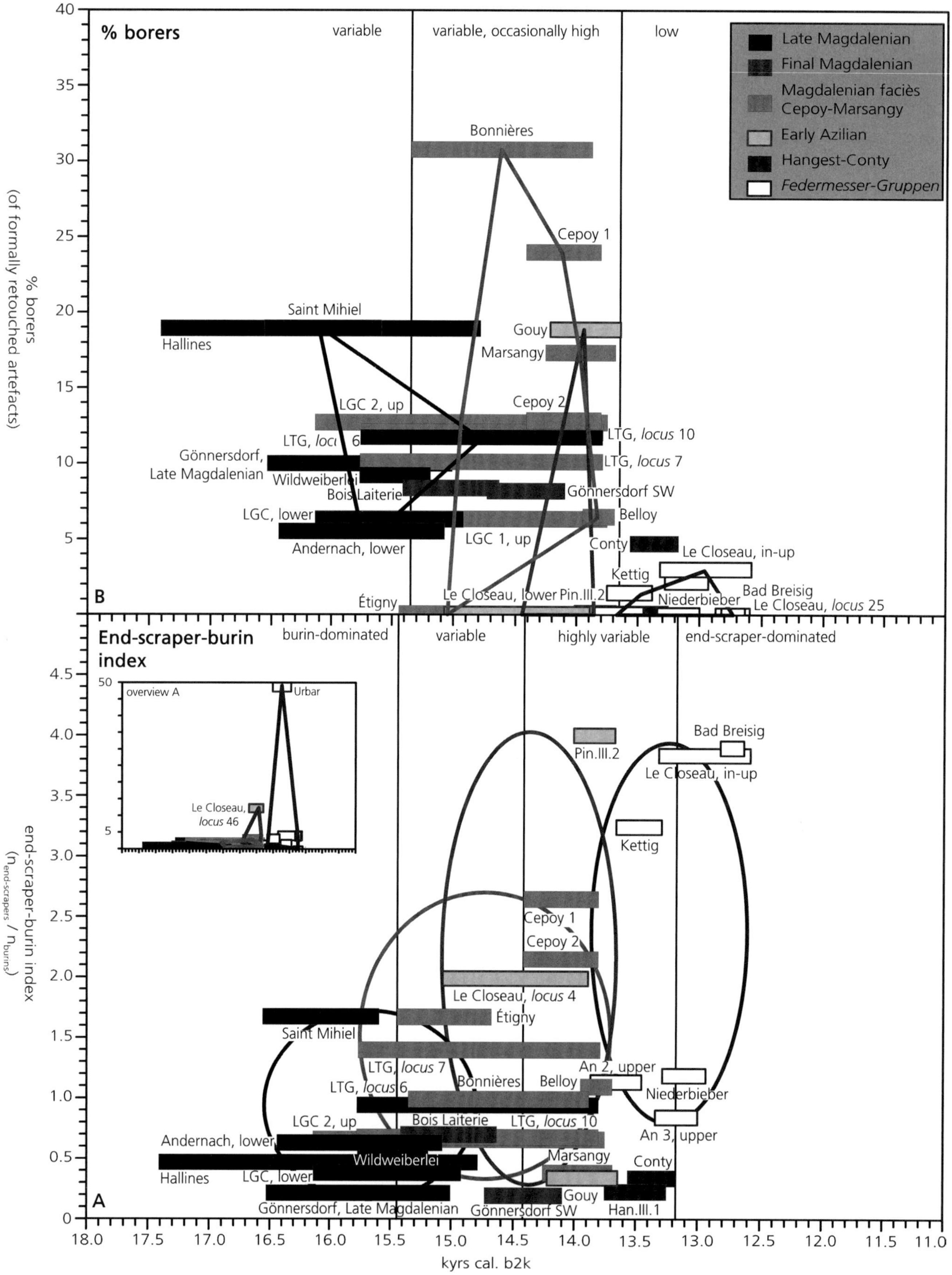
% borers
variable
variable, occasionally high
low
Late Magdalenian
Final Magdalenian
Magdalenian faciès Cepoy-Marsangy
Early Azilian
Hangest-Conty
Federmesser-Gruppen
Bonnières
Cepoy 1
Saint Mihiel
Hallines
Gouy
Marsangy
LGC 2, up
Cepoy 2
LTG, locus 10
LTG, locus 7
Gönnersdorf, Late Magdalenian
Wildweiberlei
Bois Laiterie
Gönnersdorf SW
LGC, lower
Belloy
Andernach, lower
LGC 1, up
Conty
Le Closeau, in-up
Kettig
Étigny
Le Closeau, lower
Pin.III.2
Niederbieber
Bad Breisig
Le Closeau, locus 25
% borers (of formally retouched artefacts)
B
End-scraper-burin index
burin-dominated
variable
highly variable
end-scraper-dominated
overview A
Urbar
Le Closeau, locus 46
Pin.III.2
Bad Breisig
Le Closeau, in-up
Kettig
Cepoy 1
Cepoy 2
Le Closeau, locus 4
Étigny
Saint Mihiel
LTG, locus 7
An 2, upper
LTG, locus 6
Bonnières
Belloy
Niederbieber
LGC 2, up
Bois Laiterie
LTG, locus 10
An 3, upper
Andernach, lower
Wildweiberlei
Marsangy
Conty
Hallines
LGC, lower
Gouy
Gönnersdorf, Late Magdalenian
Gönnersdorf SW
Han.III.1
end-scraper-burin index ($n_{end\text{-}scrapers}$ / n_{burins})
A
kyrs cal. b2k

nian and in the south-western area of Gönnersdorf, the burins were proportionally more frequent than in Late Magdalenian assemblages.

The point-burin index in the inventories of the MfCM is also burin-dominated but to a much lesser extent than within the Late Magdalenian. This greater variability of the index reflects the more regular appearance of pointed implements on these sites. However, the end-scraper-burin index is also higher in this archaeological unit than in the Late Magdalenian, in particular at Cepoy. A high value of this index indicates an end-scraper dominance. However, the proportion of borers was, in general, similar to values from Late Magdalenian. Thus far, no borers were reported from Étigny and the outstanding value from Bonnières-sur-Seine can be explained with the small, selectively excavated assemblage. Nevertheless, Cepoy 1 also yielded a relatively high proportion of borers. Looking at the actual numbers of the formally retouched artefacts (**tabs 14. 27. 38**), the numbers of burins as well as of borers increase occasionally in the MfCM in relation to the Late Magdalenian assemblages. However, the number of end-scrapers usually increases more significantly in the former inventories. In a traceological analysis of selected flints from Gönnersdorf II, Katsuhiro Sano showed that, at least in a longer occupied site such as Gönnersdorf, the use of burins and end-scrapers is relatively similar because, besides hide-working, end-scrapers were also used in working antler or bone. Taking this finding into account, was the increasing number of end-scrapers perhaps also an indicator for an intense working of hard organic material on these sites? In Cepoy 1 were the end-scraper-burin index is high, the number of burins is relatively low as well as the number of LMP in general. Combining these indicators, at least the function of Cepoy 1 can be assumed as focused on the exploitation of animal carcasses, in particular the hard tissue. In general, the inventories of the MfCM tended to produce values that indicate this activity was possibly a major task on these sites. However, the occasionally poor preservation and not yet accomplished analyses prohibited a confirmation of this tendency from the faunal record. In the material from Le Tureau des Gardes as well as in the upper horizon of Le Grand Canton, no organic artefacts were found.

The Early Azilian sites are very heterogeneous. The small inventories of Gouy and Pincevent III.2 were dominated by points rather than burins suggesting that the number of burins was low. Consequently, the high dominance of end-scrapers at Pincevent III.2 is also not surprising but the assemblage of Gouy is burin-dominated which is due to the single end-scraper recovered from the superficial excavation. Moreover, Gouy also provided a high proportion of borers in contrast to the other Early Azilian sites where, usually, no borers were found. Perhaps, this difference can be explained by site specifics and the additional task at this site where the cave walls were engraved (Martin 2007a). Thus, a traceological analysis of this material proposed that besides the processing of hard organic material, also mineral material was scraped and/or engraved (Plisson 2007). That burins were used for engravings can be supported by the results from Gönnersdorf II where a burin facet was found to be used in stone processing (Sano 2012b). However, end-scrapers were more frequently used to process stone material but they were suggested to have been used in the production of stone lamps and mortars (Sano 2012b, 269). According to the same study, all analysed types of artefacts were used to process stone material which could explain the unremarkable values of Pincevent III.2 where an engraved horse head was found (Allain 1976). Besides this single engraving, the structure and hardness of the cortex presumably resulted in relatively few use-wear traces and, possibly, influenced the

Fig. 81 **A** Values of the end-scraper-burin index given per time and archaeological unit. Circles were set around the main cluster of assemblages attributed to the same archaeological unit. The distribution of all assemblages is given in the overview where also the convex hulls were set per archaeological unit (see p. 281 f.). In the overview, names are only given for sites which are not shown in the main graph. **B** Percentages of borers given per time and archaeological unit. The complete distribution is already given and the convex hulls are therefore set in the main view. Besides the *loci* 4, 46, and 25 of Le Closeau, the horizon III.2 of Pincevent, and Bad Breisig, the values of the upper horizon in Andernach 2 and 3, Urbar as well as lower horizon at Hangest-sur-Somme III.1 fall to the bottom of graph B. Abbreviations see **fig. 74**. – For further details see text.

choice of the artefacts used for the engraving. Thus, leaving out only the results from Gouy, the relatively high end-scraper-burin indices and the absence of borers in the Early Azilian assemblages suggest that the processing of hard organic material was of no major importance at these sites, although the point-burin index from *locus* 4 of Le Closeau falls into the upper range of those from MfCM sites. Moreover, organic artefacts were only found in the *locus* 4 of Le Closeau. These pointed rib fragments were suggested to be used in processing animal skins (Bemilli 1998, 402) but again these implements were possibly introduced to the site from elsewhere. However, if the processing of animal skins was an activity performed at the site and the dominant use of end-scrapers was working hides, the relatively high number of end-scrapers and, consequently, the high end-scraper-burin index could be explained.

Even though points were clearly dominant among the LMP in the FMG, these assemblages were generally burin-dominated as highlighted by the point-burin index. The exceptional value of Urbar could be related to a spatial differentiation and a restricted excavation area in which burins were not extensively used and from which borers were absent. A major activity in this area was the use of end-scrapers which reached a high number (n=98) for the small area. Consequently, the end-scraper-burin index is exceptionally high in this assemblage where also remains of a particularly high number of red deer (MNI=7) were found suggesting a specialised working area. The end-scraper-burin index can be bisected in a sub-group with almost equal values of end-scrapers and burins (upper horizon of Andernach 2 and 3, Niederbieber) and a sub-group with a clearly end-scraper-dominated retouched artefact inventory (Bad Breisig, Kettig, Le Closeau, greyish deposits). Borers were only present in small proportions in the greyish deposits of Le Closeau, Kettig, and Niederbieber. Thus, assuming that burins and borers were mainly indicators of antler and bone working, this activity was either of no greater importance in the FMG or performed with other types of artefacts. In particular, the borers almost disappeared from these assemblages.

However, in Kettig where an organic projectile was found in a FMG context, the proportions of the LMP are relatively low and the point-burin index is moderately higher than in the Late Magdalenian assemblages. The comprehensive publication of the lithic material (Baales 2002) made a determination of most LMP possible and, thus, the values of this inventory will not significantly change. Consequently, the use of lithic points in this assemblage appears of greater importance than processing hard organic raw material. Even though the Kettig assemblage contained the organic artefacts and had an exceptional organic preservation for FMG sites, no debris material was found at the site which would indicate a local production of the organic artefacts. Thus, the organic artefacts were possibly discarded at Kettig but produced elsewhere.

Although the values from the lower horizons of Hangest-sur-Somme III.1 and Conty were close to the range of the FMG assemblages, these assemblages contained end-scrapers similar in proportion to burins within Late Magdalenian inventories. From Conty, only a fragment of a Lyngby axe-type artefact has been published thus far (Fagnart 1997, 112-118). Except for the engraving, burins were probably not needed in the production of this piece and, thus, a high point-burin index does not seems surprising. However, burins remained the most numerous artefacts besides the LMP in this inventory and, consequently, this artefact group was also dominant in the end-scraper-burin index. The borers were of a proportion which came close to the lower range of the Late Magdalenian assemblages. In conclusion, this inventory was similar to the typical Late Magdalenian inventories and suggests rather the regular processing of hard organic material.

In summary, the development from the high values of borers in the Late Magdalenian to the low values of the FMG as well as from the clearly burin-dominated Late Magdalenian assemblages to the less burin- and more end-scraper-dominated inventories of the FMG does not appear to be a gradual process. The values from the MfCM sites and the Early Azilian which fall into the transitional period contained occasionally higher proportions of borers than Late Magdalenian inventories and scatter over a wider range than the FMG or the Late Magdalenian sites. However, the three developmental processes resulted in incongruent limits of the sub-periods in these developments. The increasing proportion of end-scrapers was the first registered change followed by a congruent increase of points and borers. The variability of the end-scraper-burin index becomes greater followed by increase in the variability of the point-burin index. The next change is the realtively sudden disappearance of borers followed by the disappearance of the burin-dominated and the establishment of only point-dominated and mainly end-scraper-dominated assemblages. Finally, the ratio of points in relation to burins decreases again. At the onset, an increasing diversity seems to indicate a connected development, this impression is lost within the period of highest diversity. In the FMG assemblages, the diversity decreases and a stabilisation is again observable but the proportion of end-scrapers differentiates two types, with the inventories from the lower horizons of Hangest-sur-Somme III.1 and Conty possibly three types of assemblages that might be a relevant factor in the organisation of the settlement system (see p. 557).

The appearance of large variation between the assemblages raises some questions. For example, why did borers become redundant? Were they no longer of use or were they replaced in their function? This question leads to the challenge of whether the function of artefacts such as burins and/or end-scrapers changed in this period. A comprehensive compilation and evaluation of the already accomplished use-wear studies (Plisson 1985; Moss 1986; Plisson 2002; Vaughan 2002; Beyries/Janny/Audouze 2005; Janny et al. 2006; Plisson 2007; Sano 2009; Sano 2012b) supplemented by further analyses of inventories such as Cepoy 1, Bonnières-sur-Seine, or Conty could provide an answer for some of these questions but this endeavour exceeds the possibilities of this project.

Another set of questions refers to the diversity of the assemblages. One possible explanation for the increased variety is that the single assemblages became more specialised and the range reflected sites with supplementary functions. This development would suggest that a progression of the Late Magdalenian behaviour of spatial differentiation of the activities might have occurred. In contrast, if the diversity of the inventories showed no increasing specialisation several explanations are possible. A general explanation would be that the sites were visited for a major task but that besides this task, further activities were also performed at the site. According to this model, special task camps disappeared from the settlement system and were replaced by sites with a more residential character and a special purpose. Another explanation can be found in the accumulation of material over a longer time at the sites. This long duration of use and changing functions during this time could create diverse assemblages with very different compositions. In this case, the use of a location for special purpose in a seasonal round seemed improbable for these sites because the purpose for their use was unstable.

To identify specialised inventories the Simpson diversity index (Sdi) was suggested as useful (see p. 274 f.). This index was also considered to help evaluating the duration of an occupation with more diverse inventories being occupied, usually, for a longer period (Richter 1990). Therefore, the interpretation must take into account the occupation period as an influencing factor.

In the present record, the high degree of similarity of Late Magdalenian and the FMG assemblages (**fig. 82**) seems surprising after the previous results as well as to classic concepts assuming a clear difference of the two archaeological units (see p. 55-74). In fact, some specialised outliers were recorded in the FMG assemblages. From *locus* 25 of Le Closeau, only LMP were recovered and, thus, this inventory yielded an outstanding value (Sdi = 1.0) but whether this concentration represented a single unit or the activity area of a larger unit remains uncertain. The other outlier came again from Urbar, which was previously considered as the excavation of only a specialised activity area. The more continuously or more often used sites such as Gönnersdorf (Sdi = 0.25), the lower horizon of Andernach (Sdi = 0.22), and Niederbieber (Sdi = 0.22) yielded values that indicate the most diverse inventories (**tab. 84**) and, thus, support the suggestion that a greater diversity develops on sites which were used for a longer period. However, the short-termed assemblages

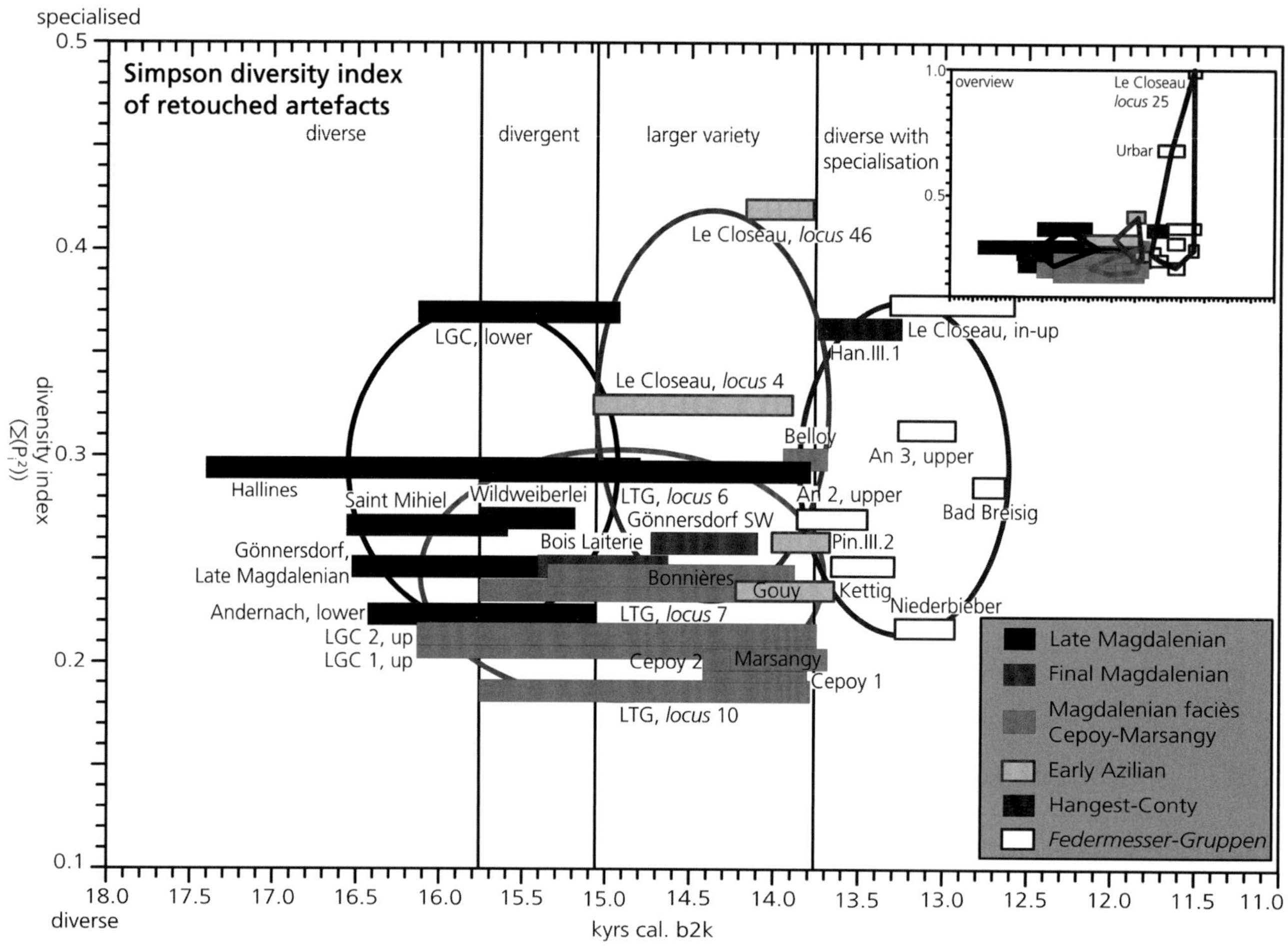

Fig. 82 Values of the Simpson diversity index given per time and archaeological unit. Circles were set around the main cluster of assemblages attributed to the same archaeological unit. The distribution of all assemblages is given in the overview where also the convex hulls were set per archaeological unit (see p. 281 f.). In the overview, names are only given for sites which are not shown in the main graph. Abbreviations see **fig. 74**. – For further details see text.

of the south-western area from Gönnersdorf (Sdi = 0.26) and Bois Laiterie (Sdi = 0.25) also fall in this range, whereas the inventories found in the greyish deposits of Le Closeau (Sdi = 0.37) were assumed to be used over a very long period. These deposits yielded the specialised end of the general FMG range. The LMP in these deposits reached proportions of almost 60 % and indicate a rather special purpose of these concentrations. In particular, the comparison with the Niederbieber concentrations which were interpreted as camps for the preparation and/or postprocessing of a hunting event (Gelhausen 2011b; Gelhausen 2011a) shows the outstanding values of Le Closeau. In Niederbieber, the LMP composed around 30 % of the inventory and the Sdi ranges between a value of 0.20 and 0.48. In the concentrations found in the greyish deposits of Le Closeau, the Sdi ranges between 0.22 (excluding those without any formally retouched artefacts) and 1. Thus, the lower limit of the range is comparable between Niederbieber and Le Closeau but in the latter many more specialised concentrations were found. However, ten concentrations of the greyish deposits yielded a Sdi of 1 but only four of these inventories were dominated by LMP, three by the other artefact group, two by end-scrapers, and one by borers. The suggestion that several concentrations formed supplementary units can thus be considered likely despite the dominance of the LMP in most of the concentrations within these greyish deposits. Nevertheless, the high proportion of the LMP indicates that the occupation at this location had a special purpose. Were these concentrations related to a hunting camp where additional tasks were accomplished? Why were the proportions so high? In comparison, the numbers of LMP ranged between 5 and 80 per concentration in Niederbieber, whereas in the greyish deposits of Le Closeau, the LMP ranged between 0 and 95 pieces per *locus*. Thus, in regard to these similar numbers, an interpretation of the greyish deposits of Le Closeau comparable to Niederbieber seems possible. The different proportion originate in the smaller numbers of the inventories in Le Closeau and, thus, the inventories were frequently formed by only a few or a single retouched artefact groups. However, how far the various *loci* in Le Closeau were related and formed a single occupation episode requires a more comprehensive spatial analyses including inter-concentration refitting attempts. These analyses could further clarify how far the spatial organisations of the greyish deposits in Le Closeau and the one in Niederbieber (Gelhausen 2011c; Gelhausen 2011a) differ from one another.

In comparison to the Late Magdalenian and the FMG assemblages, the inventories of the MfCM produced comparable but tendentially lower values, whereas the values of the Early Azilian vary considerably. The co-occurrence of these two archaeological units results in a large variety of indices.

The values from the MfCM sites range mainly between a Sdi of 0.19 and 0.24, except for the lower horizon of Belloy-sur-Somme which yielded an Sdi of 0.30. For many of the low values such as the ones from Cepoy 1 or the upper horizon of Le Grand Canton, sector 2, the accumulation over a longer period seemed a probable explanation which is in accordance with the archaeological evidence. However, the Sdi of single concentrations of this archaeological unit was also low. For example, the Sdi of Le Tureau des Gardes, *locus* 7 (Sdi = 0.23) fell into the lower range of the Late Magdalenian and Marsangy N19 (Sdi = 0.20) lies below the Late Magdalenian range. If the low values in these single concentrations were to be explained by accumulation over a longer period, a continuous occupation during which diverse activities were performed or repetitive visits to the same spot with different purposes of the visit must be assumed. The quantity of lithic material found in the two concentrations is relatively high: Marsangy N19 falls to assemblage size class 5 and *locus* 7 of Le Tureau des Gardes to class 4 (**tab. 81**). Thus, a longer use of the locale could also be explained the large assemblage size. Moreover, these inventories appear thereby similar to the Late Magdalenian assemblages of the Central Rhineland where many activities were performed in a restricted area leaving a large accumulation of artefacts. However, detailed spatial analyses are necessary to reveal the development of these concentrations and make a distinction between a long continuous or a repetitive occupation possible. The outstanding value of Belloy-sur-Somme is related to an area with an ephemeral lithic scatter and a concentration around a hearth with a possible satellite working area (Fagnart 1997, 65-68). This spatial distribution differs from those observed in Le Tureau des Gardes, *locus* 7 and Marsangy N19 but it is similar to the model of FMG concentrations in Niederbieber (Gelhausen 2011a, 271 f.). Nevertheless, the broad expanse of archaeological material on these sites suggests that the spatial dispersal to the periphery of previous spots of occupation began to occur on these sites.

In Early Azilian inventories, the values from Gouy (Sdi = 0.23) and Pincevent III.2 (Sdi = 0.26) formed a lower group, whereas *loci* 4 and 50 (Sdi = 0.32) and 46 (Sdi = 0.42) of the Le Closeau formed the upper end. Thus, the overlap with the assemblages of the MfCM is minimal. The value for the complete lower horizon of Le Closeau (Sdi = 0.36) is similarly high compared to the greyish deposits. In the lower horizon, the range of the Sdi encompasses values from 0.32 to 0.46. Thus, the range is considerably smaller than in the greyish deposits. In contrast to Niederbieber and the majority of concentrations from the greyish deposits of Le Closeau, the other artefact group was the most numerous group instead of the LMP in the inventories of the lower horizon as well as the areas south of the RN 13. For these concentrations, dominated by the other artefact group, a clear preference of using unstandardised tools can be supported. In addition, several hundred pieces were found with macroscopic use-wear traces showing the intensive use of the lithic artefacts at the site. The differences in the composition of the inventories suggests that the function of the site differed, vertically, from the lowermost to highest horizons and compared to the majority of the greyish deposits. The spatial distribution of the material in the lower horizon with a main activity area where the material was densely scattered and some outlying special activity areas still closely resembled Late Magdalenian spatial organisations (Bodu/Debout/Bignon 2006; Bodu 2010). The single, ephemeral locations in the greyish deposits might also reflect some special working zones but the intensively used main activity areas are absent in these deposits. This different spatial organisation could have further increased the variation between these horizons. A tendency towards more ephemeral activity areas can be detected in the Early Azilian assemblages of Gouy and Pincevent, horizon III.2 which were still relatively diverse. The limited spatial information from Gouy prohibits further consideration about the development of this inventory. However, the restricted area in the small cave, the superficial excavation, and the small assemblage make the accumulation of various episodes improbable as a reason for the documented diversity. Thus, comparable to many Late Magdalenian assemblages, a relatively diverse inventory was already used in a short occupation episode. In Pincevent, horizon III.2, a large, ephemerally scattered area was utilised and, thus, was already more similar to the FMG type of spatial organisation.

Thus, the previously documented difference between the Late Magdalenian, tell-like use of a single restricted area and the suburban-like dispersal of small units over a large space on the FMG sites in the Central Rhineland can also be observed in the Paris Basin. The change between the two different types of using space occurred between Early Azilian assemblages and the FMG occupation of Le Closeau and, more precisely this change occur abruptly within the late phase of the Early Azilian and the MfCM. However, the increased variability of assemblage compositions in these archaeological units can be a first indicator of a less rigid differentiation of site functions that would have had some impact on the organisation of the settlement system (see p. 548-559).

Besides initial variation in the spatial behaviour and the settlement patterns, the increased variability of the retouched artefact inventories, in particular in combination with the dominance of the unspecified others group and artefacts exhibiting macroscopic traces of use, could reflect a less normative production and use of lithic implements. This decreasing conformism to a given standard should be even more apparent on a lower analytical level such as within a single artefact group. Among the LMP, the increased use of points was already mentioned. This group of artefacts was usually attributed to the hunting equipment, in particular the LMP were considered as parts of projectiles. Since these weapons need to be precise in order to be effective and, thus, make hunting a successful subsistence strategy, a relatively high standardisation can be assumed for the different parts of this equipment. Nevertheless, an aim for improvement and a necessity to adapt to fluctuations in the availability and quality of the resources and the prey, leads to a continuous trial-and-error process which causes some diversity in the LMP. Moreover, the choice of the prey can cause some variation in terms of the projectiles used (Ellis 1997). However, when a standard fails to be useful due to a different prey and/or the failure of necessary resources, this process will probably be intensified to prevent alimentary deficiencies. Consequently, the Lateglacial environmental change can be assumed as such an enforcing situation on a wide geographic area. Since the Late Magdalenian societies formed a large information network and the trial-and-error process was probably intensified in several places, testing of new ideas from elsewhere further increased the pool of possibilities. Thus, in these periods high diversity in the LMP can be expected. The diversity of the LMP was specified per assemblage by the number of LMP types present at the site (**tab. 85**). These values were based on a survey of published figures of the LMP and, thus, the result is partially biased by the lack of readable presentations of LMP and occasionally by the lack of spatial information for the LMP making a precise attribution to a sub-assemblage impossible.

According to the current record, the diversity of the LMP is very low in the Late Magdalenian assemblages. Usually, only backed bladelets and blades were identified. In the Early Azilian the values are marginally higher with up to three types occurring in an assemblage. Curve-backed points were always present but angle-backed points were also frequently identified. Usually, the assemblages of the MfCM were similarly diverse with two or three different types of LMP being found in the inventory. In contrast to the Early Azilian, the angle-backed points were the most common type and curve-backed points were more infrequent and less numerous. Backed bladelets or blades occurred in all assemblages except for Cepoy and Étigny in small numbers. However, some of these assemblages (the upper horizon of Le Grand Canton, Marsangy, Belloy-sur-Somme) yielded a greater diversity of LMP types. For the former inventory, the duration of the use and the size of the assemblage could explain some of the diversity. Therefore, this assemblage requires a more detailed spatial analysis. In Marsangy N19, the diversity was also smaller than on the complete site. However, Marsangy N19 was still as diverse as the almost contemporary Belloy-sur-Somme assemblage which was probably represented a single short-termed occupation. The pieces from south-western concentration of Gönnersdorf fall into this usual group of inventories yielding three different types of LMP and likewise the MfCM, these types were backed bladelets, angle-backed points, and a curve-backed point. In contrast, a complete set of LMP types was found in the publications about Bois Laiterie. This diversity indicates that the inhabitants of Bois Laiterie seemed to create or test a range of backed implements. However, in comparison with other inventories from the present record and taking into account the possibility that the site was formed by several hunting episodes, a more detailed spatial analysis of the LMP, in particular of the more recently identified point types (Sano 2009; Sano/Maier/Heidenreich 2011) appears necessary. The FMG assemblages are more diverse usually with three to four and in exceptional cases also five different LMP types occurring on the sites. In larger units such as the upper horizon of Andernach or Niederbieber, the different concentrations contributed to an almost complete set of LMP types. The curve-backed points were the most frequent type followed by backed bladelets and blades and tanged points were the least frequent type followed by simple points. The assemblages from the lower horizon of Conty and Hangest-sur-Somme III.1 produced four types of LMP including curve-backed points as well as angle-backed points but backed blades and couteaux à dos formed the most frequent LMP types.

Besides the diversity of the LMP shapes, the dimensions of the projectile implements play an important role in their use within composite tools requiring some norms to allow for a replacement of defective parts without a discard of the complete projectile. The production of the single LMP is one of the smallest possible analytical levels and very fine differences in the dimensions of the implements are probably beyond the perceptibility of the maker (Eerkens 2000). However, since the function of the complete projectile limits the variability of these implements, the dimensions should spread around an aimed mean. A high standardisation of the equipment would allow for only a small variance, whereas an increased variance reflected a more flexible and possibly different tool kit. Therefore, the variance of the three dimensions of the measured LMP (see p. 275-282) are compared per assemblage and archaeological unit (**fig. 83**).

In general, the length (**fig. 83C**) appeared a more variable dimension than width (**fig. 83B**) or thickness (**fig. 83A**). As long as the implements were inserted into a groove or notch in the shaft, a very restricted standard for the thickness of the implements is necessary. This high standardisation was clearly observable in the assemblages of the Late Magdalenian as well as in the MfCM, Bois Laiterie, and the south-western area of Gönnersdorf. However, in the Early Azilian assemblages of Gouy and Pincevent, horizon III.2 as well as in the lower horizons of Conty and Hangest-sur-Somme III.1, the variance increases considerably. However, the variance of the Gouy assemblage falls as low as that from Étigny if the single *couteau à dos* (**pl. 12, 5**) is excluded. The same applies to Hangest-sur-Somme III.1, lower horizon which produced, without the very big *couteaux à dos* (**pls 13, 7; 14, 2-3. 7**), a variance comparable to Bois Laiterie. In the lower horizon of Conty, the *couteaux à dos* formed a significant part of the record (**pls 12, 10. 12-13; 13, 3-4**) and the variance changes comparable to Hangest-sur-Somme III.1 when these implements were removed. However,

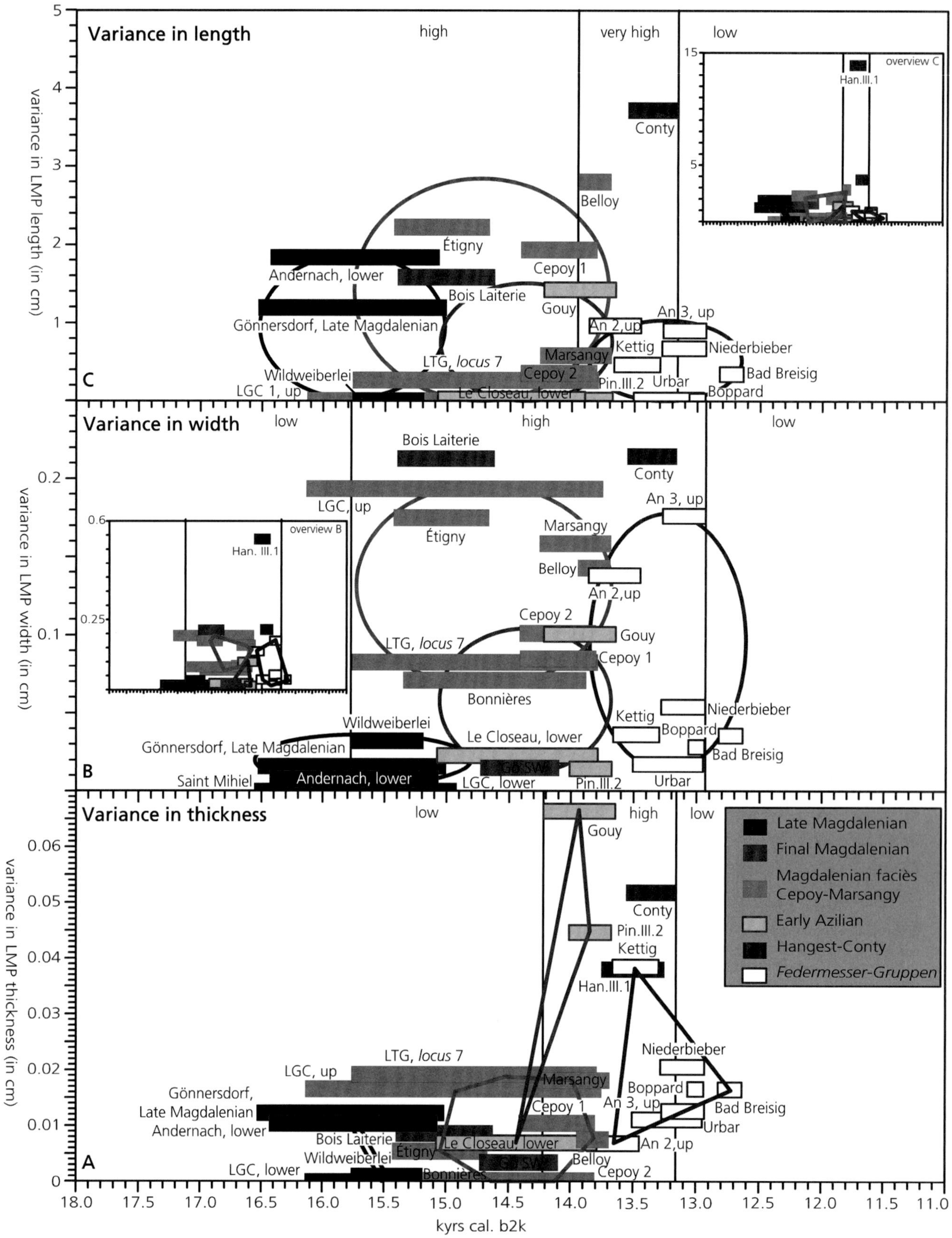

Fig. 83 Variance of the completely preserved thickness (**A**), width (**B**), and length (**C**) of the analysed LMP (see **pls 1-14**) given per time and archaeological unit. Circles were set around the main cluster of assemblages attributed to the same archaeological unit in graphs **B** and **C**. The distribution of all assemblages is given in the overview where also the convex hulls were set per archaeological unit (see p. 281 f.). In the overview, names are only given for sites which are not shown in the main graph. In **A** the complete distribution is already given and the convex hulls are therefore set in the main view. Abbreviations see **fig. 74**. – For further details see text.

in Pincevent, horizon III.2, no *couteau à dos* was found and the outstanding variance remains. The FMG assemblages had generally a higher variance than the Late Magdalenian assemblages but did not exceed the variance of the MfCM inventories by considerable amount. Only the assemblage of Kettig formed an exception with a variance in thickness comparable to the original value from the lower horizon of Hangest-sur-Somme III.1. In Kettig, no *couteau à dos* was found and the microlithic simple points fell into the range of the curve-backed points. However, a single, thick, curve-backed point (**pl. 3, 10**) formed a significant outlier and after removing this piece, the variance of Kettig falls into the usual range of the FMG inventories. Thus, the inventories of the MfCM were the first assemblages in which also wider variances were allowed. In the Early Azilian assemblages, the very strict standardisation did no longer occur and the couteau à dos established a distinct group of implements.

The variance in width was also very standardised in Late Magdalenian assemblages but became considerably more variable with the appearance of the MfCM and, in particular with the co-occurrence of points and backed bladelets. Perhaps, the absence of the narrower bladelets is the reason for the lower variance of width in the Early Azilian assemblages. However, the value of Gouy falls in the lower range of the MfCM but if the *couteau à dos* is excluded again the range is similar to Kettig. In this case the Early Azilian variance in width appeared comparably restricted as the Late Magdalenian one. In the FMG assemblages, the variance of width increases but remains, mainly, in the range of the Early Azilian except for the inventories from the upper horizon of Andernach which were more comparable to the upper range of the MfCM. The LMP from these assemblages formed a very heterogeneous groups of which some very thick pieces were considered as knives (Kegler 2002). Thus, in these assemblages the LMP were possibly used for various tasks which allowed variable standardisations of the width. However, this standardisation does not seem to be related to a specific shape. The values for the lower horizon of Conty and Hangest-sur-Somme III.1 were again very high but decreased to values in the range of the Early Azilian assemblages (Conty) and the lower range of the MfCM (Hangest-sur-Somme III.1) without the *couteaux à dos*.

The variance in length is already greater in the Late Magdalenian. However, this impression can be false since the problem of a distinction between broken and complete backed bladelets was difficult because many pieces were intentionally broken. Thus, the unbroken forms had to be considered as complete, even though they may not represent the pieces that were used. If only the preserved lengths of the Late Magdalenian assemblages were considered they again scattered closely around a value of 1.0 except for Saint Mihiel where a large angle-backed point fragment was found besides a small bladelet-like LMP. The variance of length was in general more restricted in the other archaeological units which appears reasonable if these implements were used as a projectile head and need to be in balance with the shaft. In particular, the Early Azilian and the MfCM sites yielded very low values, whereas the FMG assemblages were again more variable. Perhaps, this increasing variance as well as the frequent occurrence of backed bladelets reveals the return of a laterally inserted type of implements. Nevertheless, the variance of length in some assemblages was very high. The variances of Conty, Hangest-sur-Somme III.1, and Gouy id not change as significantly as in the previous results when the *couteaux à dos* were removed from the dataset. Without these heavy implements, the values ranged around the original value of Gouy. The value of Cepoy 1 sank to a value between Cepoy 2 and Marsangy if a single, small, simple point (**pl. 10, 11**) was excluded. Was this small piece a Mesolithic intrusion or reflected a second type of projectile implement or did it show the possible range of lithic implements on this site? A detailed analysis of the LMP inventory with a special focus of potential use-wear might provide some interesting results for this site. In Belloy, a comparably small angle-backed point increased the variance. Without this piece the value fell to the upper range of the FMG assemblages. In Étigny only a few pieces were completely preserved. A single curve-backed point (**pl. 8, 22**) was considerably smaller than the preserved angle-backed point (**pl. 8, 24-25**). Without this piece the variance dropped to a value below the one of Le Tureau des Gardes, *locus* 7. The high value of Bois Laiterie decreased to a comparably low value if the piece, which appears as a mixture of a tanged point and a Magdalenian blade (**pl. 7, 13**), is rejected. However, this piece was identified by the use-wear analysis and the breakage pattern as a projectile head (Sano 2009). These changes in the variance of the last four sites showed that, besides the *couteaux à dos*, two different types of projectiles, one with a lighter lithic implement and one with a heavier lithic implement, were possibly used in these assemblages.

Thus, the variance of length reveals a general difference between the Late Magdalenian and the other assemblages: The latter were more restrictive about the characteristics than the Late Magdalenians. This greater variability in the length is congruent with a general comparison of the LMP from this study with pieces from Thuringia. On average, the Thuringian pieces were over 1 cm shorter and 0.1 cm thinner but as wide as the implements from this study (Bock et al. 2013, tab. 10). This comparison emphasises the importance of the width in Late Magdalenian LMP inventories. Moreover, in some assemblages of the MfCM as well as in Bois Laiterie indications for two different sets of projectile implements were found. However, the suggested variability of FMG assemblages, particularly, in regard to the backed implements (Barton et al. 2009) cannot be, unambiguously, supported. The present data which is based on FMG sites from the Central Rhineland show that the shape diversity and variance of these implements within single assemblages increased. This increasing variability indicates that standardisation of the hunting equipment was less restrictive and the establishment of *couteaux à dos* as independent group as well as the possible presence of two size classes on some sites such as Kettig also suggested a greater repertoire in the equipment of these hunter-gatherers than in the Late Magdalenian.

Based on ethnographic comparisons, the differences in shape and dimension of lithic projectile implements was considered as being related to the size of the prey (Ellis 1997). In the lower horizons of Hangest-sur-Somme III.1 and Conty, aurochs (*Bos primigenius*) were the main prey. Thus, was the larger equipment, in particular the use of large couteaux à dos, related to the hunting and/or processing of this large herbivore and did this specialisation cause the differences in the inventories of mainly red deer (*Cervus elaphus*) hunting communities in the Paris Basin and the Central Rhineland? Moreover, did the variable faunal spectrum of Bois Laiterie cause the high diversity of LMP types? These questions indicate the importance of contextualising the lithic industries with data from the faunal remains.

In summary, the diversity of the assemblages was influenced by the occupation duration and the function of an assemblage but also by the norms of spatial behaviour and the interrelated presettings of the material equipment. In the equipment represented by formally retouched artefacts, the first changes became apparent with an increasing diversity range (**fig. 84**). In particular, the hunting equipment was subject to variation which is reflect in an instability of normed widths. However, although different LMP types already occurred in this period, they remained singular phenomenons. A more frequent appearance of different LMP types accompanied with a clear differentiation in the overall size of these implements that occurred later concomitantly with an increasing importance of end-scrapers in comparison to burins. Shortly after, LMP point shapes also increased in number compared to burins and the proportion of borers began varying considerably. This increasing differentiation of assemblages was thereafter also observable in the general composition of the assemblages with core-dominated inventories appearing regularly besides assemblages dominated by formally retouched artefacts. Later, the general presence of a complete set of formally retouched artefacts vanished and the diversity of this retouched inventories became very variable forming a large range from more specialised to very diverse assemblages. Besides the continued diversity of LMP width, types, and size groups, the proportion of LMP in the retouched inventories decreased, temporally, to a low that was possibly related to the near disappearance of backed bladelets. In a next step, the differentiation of assemblages which were dominated by end-scraper to those which were burin-dominated increased significantly. This

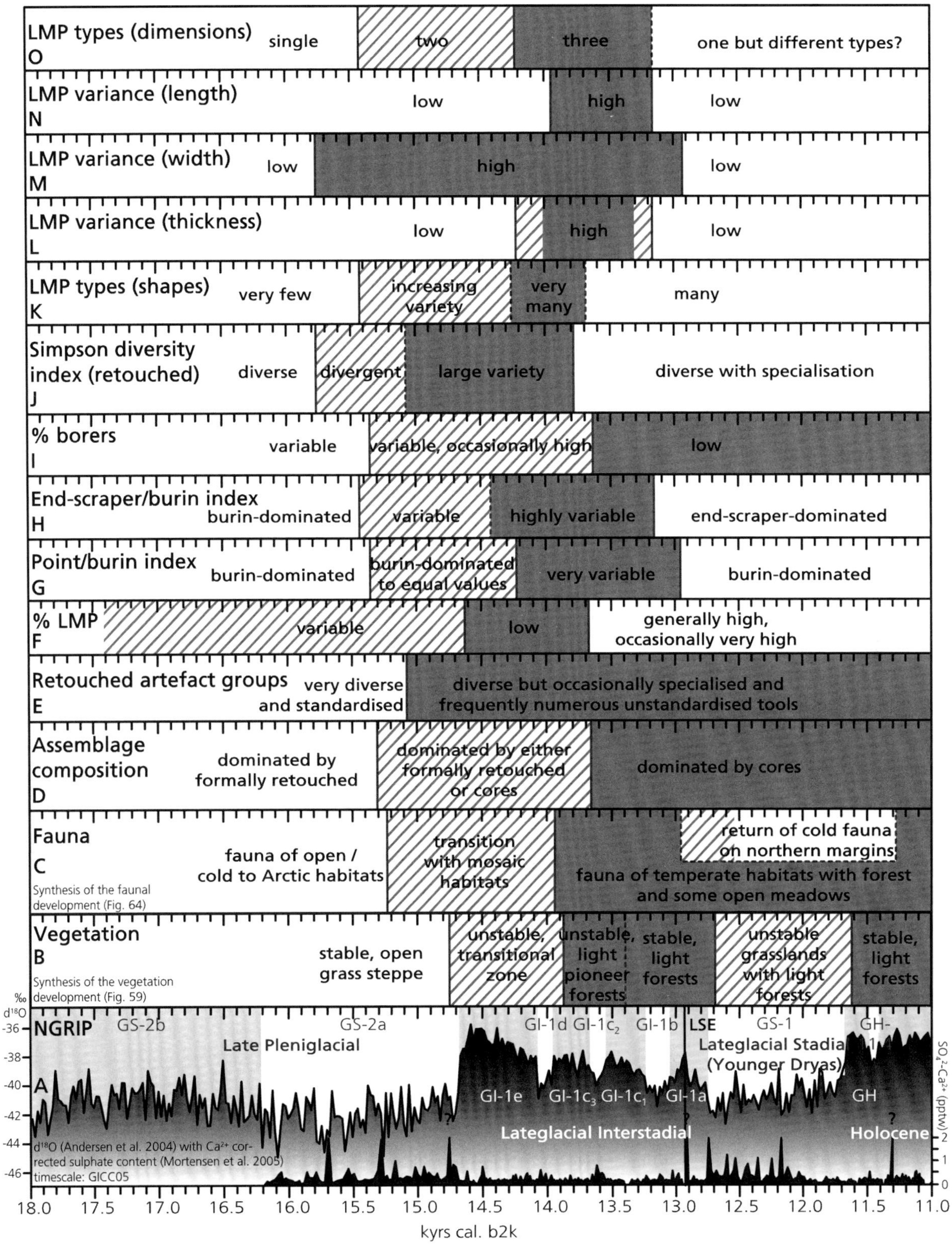

Fig. 84 Developments in lithic assemblages related to the diversity of the lithic inventory (**D-J**) contrasted by the synthesised faunal (**C**; see **fig. 64**) and vegetation development (**B**; see **fig. 59**) as well as the oxygen isotope record of NGRIP (**A**; see **fig. 53**). Hatched areas: transition periods. – For further details see text.

increasing differentiation also affected the relation of points and burins on a site. This increasing variability in the point-burin index was accompanied by the appearance of the highest number of LMP shape types and size types. The thickness of the LMP also began varying and reflected the regular appearance of large *couteaux à dos* in the LMP inventories. At this point, no standard for LMP seemed to exist any more and/or that these implements were used for a range of purposes.
On these almost simultaneously occurring alterations followed the highest variability of thickness as well as length in the LMP, possibly reflecting the search for new standards. In fact, from this point on, the assemblages seem variable in almost all categories and on various levels between the composition of the complete lithic assemblage to the composition of a single LMP. This high variability suggests that no common standard existed in this period.
An establishment of new standards begins with a decrease in the variability of assemblage diversities, followed by a decrease in LMP types and an increase of their proportion followed shortly after by an extreme decrease of borers and the establishment of relatively equal assemblage compositions with only more core-dominated assemblages as the exception. The variance of the thickness in LMP ceases later and the disappearance of exceptionally large *couteaux à dos* follows even later. With the disappearance of these very large LMP, the variance of LMP lengths also ends and, thus, the number of different size classes in the LMP seems to end. Moreover, end-scrapers become more dominant than burins. However, burins remain important and somewhat later they become generally more frequently discarded than points. At this point the variance in the width of LMP also normalises again.

The occasional repetitive appearance of when changes occurred show the interrelation of acquisition, exploitation, and use of lithic resources. Alterations in one of these sectors caused or were influenced by a change in another sector. However, the presentation, in particular of the diversity of the assemblages, also showed the restrictions in the interpretation due to spatial limitations and reveals the importance of spatial organisation on hunter-gatherer sites. Moreover, this finding further emphasises the necessity of spatial analyses for the interpretation of Late Pleistocene assemblages. The results of these analyses make further comparisons and, consequently, considerations about the spatial organisation on a larger scale possible by setting sites into relation to other sites in the same region and allowing for regional models which can be compared with other regions.
However, before the assemblages can be further discussed in regard to their function and, thus, their position in the regional and supra-regional settlement systems, the exploitation of the faunal resources has to be considered. This part of the assemblages is of relevance for considerations about subsistence strategies, duration of the occupation, and the function of the site.

Exploitation of faunal resources

The evidence from faunal resources permits a multitude of analyses to be undertaken, such as hunting specialisation (Gaudzinski/Street 2003), the selection of a specific age structure (Bignon 2006), the seasonality of the hunt (Enloe 1997), hunting methods and equipment (Bratlund 1991; Bratlund 1996b; Leduc 2014a), standardisation in the processing of the animals (Charles 1997b; Bridault/Bignon/Bemilli 2003; Leduc 2014b), and the spatial organisation of these processes (Enloe/David 1989; Street/Turner 2013) as well as the production of artefacts (Tinnes 1994; Álvarez Fernández 1999; Pétillon 2006; Brasser 2012; Leduc 2012). These analyses contribute to the understanding of hunting, subsistence, and exploitation strategies.

However, faunal remains were preserved in only a few Lateglacial assemblages, where the quality of preservation varied considerably. Moreover, the archaeozoological analyses are not accomplished for all of the sites considered in this project. For instance, although species were determined in a preliminary note about the Conty assemblage (Coudret/Fagnart 2006), counts of the skeletal pieces and the individuals was not made available for the present author on time (Auguste 2012). Therefore, the number of the assemblages which can be compared numerically is further limited.

The most apparent difference in regard to the exploitation of faunal resources is the absence of organic tools on most FMG sites. In Late Magdalenian assemblages a variety of organic tools were usually found (Tinnes 1994). In contrast, Early Azilian and FMG assemblages produced only a few fragments of organic implements (**tabs 17. 41**) in well preserved assemblages such as in the lower horizon of Le Closeau (Bodu 1998, 234. 285), Conty (Fagnart 1997, 112-118), and Kettig (Baales 2002). Even though conditions of preservation were considered as a possible reason for this scarcity, the presence of isolated organic implements in well preserved assemblages indicates that this resource decreased in importance and organic implements were no longer produced on a regular basis. Reindeer antler and ivory were substituted as raw material by red deer antler, as shown in Kettig and Conty. Red deer do not usually occur in large herds, such as reported for reindeer. Furthermore, only male red deer grow antlers, whereas antlers are known from both sexes in reindeer. Consequently, a decreasing availability of this resource can be assumed with the disappearance of reindeer herds.

Furthermore, the presence of some of the selected species in the dated assemblages helps to identify changes in the prey selection. Bearing in mind the often unevaluated relation of the faunal remains to the lithic assemblages, these few analysed assemblages can serve as models in the settlement system for other sites with no organic preservation. For a general comparison of the size of the faunal assemblages, a modified version of Gerd-Christian Weniger's classification system (Weniger 1989) was used which took into account the ethologies of the different groups of the hunted mammals (see p. 283 f.). Summing up the classes (**tab. 86**) produced a reference value for the number of hunting episodes and/or the size of the hunting party contributing to the assemblage. This value was of course also influenced by the diversity of the assemblage but, based on the assumption that a coincidental hunt of different large mammals is improbable, animal diversity appears to reflect different hunting episodes, thus justifying a higher value in the sense mentioned above. This value made further considerations about the function of the site and the duration of its use possible and, moreover, allowed a comparison between the very large Late Magdalenian and the smaller FMG assemblages to be undertaken.

The most apparent difference is the choice of prey. Reindeer (*Rangifer tarandus*) was only determined in Late Magdalenian and MfCM inventories, whereas roe deer (*Capreolus capreolus*) was only found in FMG assemblages, at Gouy and Conty. Wildboar (*Sus scrofa*) was typically found in Early Azilian sites as well as Bonnières-sur-Seine and again in late FMG sites of the Central Rhineland. Horse (*Equus* sp.) was the dominant species in the Late Magdalenian assemblages as well as in some MfCM assemblages. Horse also occurs regularly in the Early Azilian and FMG assemblages but are no longer dominant. Besides horses, large bovids (*Bison priscus/Bos primigenius*) were the most constantly hunted larger mammals in the Lateglacial. In the older assemblages, these two species were supplemented by reindeer and in the younger assemblages by red deer (*Cervus elaphus*). In Gönnersdorf II, Andernach II, Bois Laiterie, Étigny-Le Brassot, and Marsangy, reindeer and red deer co-occured suggesting a transitional period in which both species were available. If these sites are considered according to their sub-area, this transitional period ended earlier in the Central Rhineland than in the western uplands and northern France. In the FMG assemblages, red deer was occasionally more intensely hunted. In general, the choice of the prey appeared more heterogeneous in the FMG assemblages than in the Late Magdalenian. However, the average for classes found in an assemblage

site	class *Rangifer tarandus*	class *Equus* sp.	class *Bison priscus / Bos primigenius*	class *Cervus elaphus*	class *Alces alces*	class *Capreolus capreolus*	class *Sus scrofa*	sum of classes	Simpson diversity index	% smaller mammals
Saint Mihiel	1	1	1	0	0	0	0	3	0.4167	23.7 (7.8)
Gönnersdorf	1	4	1	1	0	0	0	7	0.2239 / 0.2295*	40.5 / 41.9* (17.7 / 18.2*)
Gönnersdorf I	1	2	1	0	0	0	0	4	0.2850	65.1 (17.5)
Gönnersdorf II	1	3	1	1	0	0	0	6	0.3263	29.1 (21.8)
Gönnersdorf III	1	1	0	0	0	0	0	2	0.2628	21.7 (17.4)
Gönnersdorf IV	0	1	1	0	0	0	0	2	0.5918	0
Andernach, lower horizon	1	2	1	1	0	0	0	5	0.1786	21.7 (4.4)
Andernach II	1	2	1	1	0	0	0	5	0.2016	22.9 (2.9)
Andernach IV	1	1	1	0	0	0	0	3	0.1405	18.2 (9.1)
Wildweiberlei	1	1	0	0	0	0	0	2	0.2000	26.7 (20.0)
Le Tureau des Gardes	2	4	1	0	0	0	0	7	0.4996	3.8 (3.8)
Le Tureau des Gardes, *locus* 6	1	1	0	0	0	0	0	2	0.4150	10.0 (10.0)
Le Tureau des Gardes, *locus* 7	0	1	1	0	0	0	0	2	0.7813	0
Le Tureau des Gardes, *locus* 10	1	2	1	0	0	0	0	4	0.5022	3.3 (3.3)
Étigny-Le Bras-sot, south	p	0	0	p	0	0	0	×	×	×
Bois Laiterie	1	1	1	1	1	0	0	5	0.1174	63.6 (18.2)
Le Grand Can-ton, sector 1, upper horizon	p	p	0	0	0	0	0	×	×	×
Le Grand Can-ton, sector 2, upper horizon	1	4	1	0	0	0	0	6	0.7371	0
Bonnières-sur-Seine	0	1	0	1	0	0	1	3	0.3878	0
Le Closeau, lower horizon	0	1	1	1	0	0	1	4	0.2769	9.1 (9.1)
Le Closeau, *locus* 4	0	1	1	1	0	0	1	4	0.2188	12.5 (12.5)
Le Closeau, *locus* 46	0	1	0	1	0	0	1	3	0.2857	7.1 (7.1)
Gönnersdorf SW	0	1	0	1	1	0	0	3	0.3600	0
Marsangy	1	1	0	1	0	0	0	3	0.3333	0
Gouy	0	0	0	p	0	p	p	×	×	×
Pincevent III.2	0	0	p	p	0	0	0	×	×	p
Bonn-Oberkas-sel	0	0	p	p	0	0	0	×	×	×
Irlich	0	0	0	p	0	0	0	×	×	×

Tab. 86 Classes and indices of faunal assemblages. For the classes see **tab. 55**. **p** presence of a species at a site meaning that it was already determined in a preliminary study; **×** values that cannot be calculated. The class columns were coloured comparably to the ones of **tab. 63** with the darkest background for the highest number and the lightest for the lowest number. The sum of classes counts the different classes together to give an impression of the size of the assemblage. % smaller mammals based on MNI and in relation to total MNI. In parentheses are values without small carnivores. Simpson index is the sum of the square-results of dividing the MNI of each species by the total MNI. Sub-assemblages in these columns are set in shaded light grey. ***** in parentheses index without Gönnersdorf SW; ****** according to the conditions of preservation and partially the spatial distribution, the smaller mammals found at the site were younger intrusions.

site	class *Rangifer tarandus*	class *Equus* sp.	class *Bison priscus / Bos primigenius*	class *Cervus elaphus*	class *Alces alces*	class *Capreolus capreolus*	class *Sus scrofa*	sum of classes	Simpson diversity index	% smaller mammals
Andernach 2-FMG, upper horizon	0	1	1	1	1	1	0	5	0.2018	9.5 (9.5)
Hangest-sur-Somme III.1 lower horizon	0	p	p	0	0	0	0	×	×	×
Kettig	0	1	1	2	0	0	0	4	0.1720	21.7 (8.7)
Conty, lower horizon	0	0	p	p	0	p	0	×	×	×
Urbar	0	1	1	2	0	0	0	4	0.6296	0
Le Closeau, greyish deposits	0	1	0	1	0	1	0	3	0.2857	14.3 (0)
Niederbieber	0	1	1	3	3	1	1	10	0.1910	12.5 (8.3)
Niederbieber 1	0	1	0	1	1	1	1	5	0.1600	20.0 (0)
Niederbieber 2 (18+19+20)	0	1	0	1	2	0	0	4	0.2778	16.7 (16.7)
Niederbieber 3	0	1	0	0	1	0	0	2	0.3333	33.3 (33.3)
Niederbieber 4+17a	0	1	1	1	0	0	0	3	0.3878	0
Niederbieber 5	0	0	1	1	0	0	0	2	0.3750	0
Niederbieber 6+10a	0	1	0	1	2	0	0	4	0.5000	0
Niederbieber 7	0	0	0	1	0	0	1	2	0.3333	33.3 (33.3)
Niederbieber 10	0	0	0	1	0	0	1	2	0.3750	0
Niederbieber 11	0	0	1	1	0	0	0	2	0.5000	0
Niederbieber 12	0	0	1	0	0	0	0	1	1.0000	0
Niederbieber 13	0	0	1	1	0	0	0	2	0.5000	0
Niederbieber 14	0	1	0	1	0	0	0	2	0.5000	0
Niederbieber 15	0	0	0	1	0	0	0	1	1.0000	0
Niederbieber 17	0	1	1	0	0	0	0	2	0.5000	0
Andernach 3-FMG	0	0	0	p	0	0	0	×	×	×
Boppard	0	0	0	p	0	0	p	×	×	×
Bad Breisig	0	1	0	1	0	1	0	3	0.4400	0**

Tab. 86 (continued)

per archaeological unit consistently produced a value of 3 meaning that, usually, 3 different species of the selected mammals were introduced to a Lateglacial concentration.

The comparison of the summed value of the classes indicated that a general value between 2 and 4 for a single concentration appeared to be typical (**tab. 86**). Thus, a value of 2 to 4 can be considered as the norm of hunting per occupation unit in the Lateglacial. The Late Magdalenian assemblages ranged between a value of 2 and 6 with an average for the sub-assemblages of 3.2. Except for the upper horizon in sector 2 of Le Grand Canton, values between 2 and 4 were found in the MfCM sites resulting in an average of 3 for the sub-assemblages. Similarly, the two Early Azilian assemblages from the lower horizon of Le Closeau yielded values of 3 and 4. The FMG assemblages had again a wider range of values between 1 and 5 with an average of 2.9 for the sub-assemblages.

However, on larger sites such as Niederbieber, Gönnersdorf, Le Tureau des Gardes, and the upper horizon of Le Grand Canton, sector 2, these smaller values accumulated producing higher ones. In comparison, these accumulated values appeared particularly different. The FMG sub-assemblages had relatively even values between 2 and 4, whereas the Late Magdalenian assemblages showed a clearer differentiation between concentrations yielding a value of 2 or 3 and those with a value of 4 to 6. For instance, Gönnersdorf III and IV yielded the typical value 2, whereas concentrations I and II reached values of 4 and 6. Concentration II in the lower horizon at Andernach also produced a value of 5, whereas in Andernach IV only a value of 3 was found. Whether this difference on Late Magdalenian sites was due to a shorter occupation of those parts of the sites with the lower values or due to complementary concentrations remains to be discussed in the context of the settlement system.

In Bois Laiterie, only a single concentration was recovered which yielded a value almost as high as Gönnersdorf II. However, this value was due to five different classes, whereas in Gönnersdorf II only three classes formed this high value. Thus, not only the differences in the sum of the classes, but also the diversity of the assemblages forming this sum is different between the archaeological units. High values in the Late Magdalenian as well as in some of the MfCM sites were attained by the typical three different classes of which one usually yielded a higher value. In contrast, in the FMG assemblages as well as at Bois Laiterie, values of five were only reached when five different classes were present. This difference implies that Late Magdalenian assemblages were possibly more specialised than FMG inventories. This result can be tested by the application of the Simpson diversity index (see p. 284f.) and will be discussed later in more detail. The high value of Bois Laiterie raises the question whether this assemblage was created during a single, longer episode in which different provisions reached the site or whether the paved construction was used during several revisits in which different animals had been hunted.

In contrast to the Late Magdalenian, concentrations with values as high as 5 or higher are almost absent from the sub-assemblages of the FMG, the Early Azilian, and, probably, the MfCM. The upper horizon of Le Grand Canton, sector 2 yielded a value as high as Gönnersdorf II but this material originated from a much larger area and was more dispersed. Nevertheless, without a spatially differentiated presentation of the faunal material, concentrations comparable to Gönnersdorf I or II cannot be entirely excluded for this site. In Niederbieber, only concentrations 1 and 6+10a yielded higher values of 5 and 4. Although the value of area 2 was also higher, it resulted from two, possibly three different clusters (cf. Gelhausen 2011c). Comparably, in Kettig as well as in the upper horizon of Andernach 2, several concentrations formed again the higher values and, thus, each concentration yielded a smaller value. The value of 4 from Urbar can be considered as a higher value because the assemblage came from a very restricted excavation area. Perhaps, this area was the main faunal processing zone of the site. If the complete site had been excavated the value could remain the same but could also possibly increase. Moreover, this assemblage yielded indications for an occupation during the cold season. Niederbieber 1 was also assumed to be an occupation during the cold season based on the close relation to Niederbieber 4 and the seasonal indicators from this concentration. Following this line of evidence, larger assemblages in the FMG concentrations seemed to develop during the cold season. In contrast, the sites with higher values from two or three concentrations such as Kettig and the upper horizon of Andernach 2 yielded indications for a settlement during warmer seasons. In contrast, Niederbieber 2 produced a higher value which was probably formed by two or three concentrations but also contained indications of winter occupation and, even though Niederbieber 4 contained indications for a cold period occupation, the assemblage provided no higher value. Consequently, the results from Niederbieber 2 and 4 contradict a simple seasonal explanation. In Niederbieber, assemblages with higher values were the only ones at the site containing remains of elk, except for the special task camp in concentration 3. In the larger Andernach inventory, elk was also determined. Thus, were assemblages containing elk more probable to also contain material of other hunting episodes? Could this explanation also be applied to the diverse inventory from Bois Laiterie? Thus far no elk remains were found in Urbar nor in Kettig and, perhaps, assemblages containing elk bones were only indicative of better conditions of preservation.

The problem of very poor preservation prohibited the visibility of accumulated faunal remains such as in the greyish deposits of Le Closeau. The very few faunal remains from all these concentrations created only a typical value of 3. Most Niederbieber concentrations which contained faunal remains produced a typical value of 2 or 3, only concentrations 12 and 15 had a very small value of 1. Some of these values have to be seen critically because the conditions of preservation at this site deteriorated drastically after the removal of the protecting cover of pumice (Gelhausen 2011a). Consequently, the accumulated value of 10 for Niederbieber must be regarded as a minimal value. Moreover, the generally lower average value in FMG assemblages than in the other archaeological units must also be seen critically. Thus, bone preservation played an important role in Niederbieber as well as in the greyish deposits of Le Closeau.
As well as poor preservation due to taphonomic processes, the faunal material at many FMG sites often consisted of only small, calcined pieces. An ethnographic example described an ideal of »nothing is wasted« (Pasda/Odgaard 2011) in a successful episode of caribou hunting in the harsh hinterland environments of Western Greenland. In this ideal, the complete animal was exploited and the bones were smashed and cooked to extract the bone grease. This process also meant that very small fragments could be disposed near the dwelling (Pasda/Odgaard 2011, 42) since they were no longer attractive to vermin or other predators. As a result, the faunal material found at these sites was mainly composed of small bone fragments. A possible explanation for the necessity of conforming to this ideal was that »The heavy exploitation of fat resources may be the only means of survival in any community where there is dietary stress.« (Pasda/Odgaard 2011, 37). In fact, the study in Western Greenland also found evidence for deviation from this ideal in the case of sufficient surplus, for instance due to communal hunting episodes and/or modern equipment[52]. Transferring this example to the North-West-European Lateglacial, many very small faunal fragments were found in the pits from Gönnersdorf and Andernach (Street 1993; Bergmann/Holzkämper 2002; Street/Turner 2013). Probably, these pieces are evidence of conforming to a »nothing is wasted« strategy in these assemblages. In the upper horizon of Le Grand Canton, the fragmented state and the under-representation of some skeletal elements were interpreted comparably (Bridault/Bemilli 1999, 55). In the faunal assemblage of Kettig, the under-representation of bone grease rich material in combination with the presence of supposed cooking stones was also suggested as possibly indicative of the cooking of bones (Baales 2002, 204 f.). However, cooking does not usually alter the appearance of the material in the same way as burning and small, burnt or calcined pieces were often found on FMG sites. Burning of small faunal waste also required no additional disposal of the material elsewhere to keep vermin and predators away from the site. An increase of small faunal material being finally discarded in a fire was already observed in the Early Azilian assemblages of Le Closeau (Bignon/Bodu 2006, 408. 414). If these pieces reflect a similar type of exploitation behaviour as do the small pieces in the pits of Gönnersdorf and Andernach, then the method of extracting bone grease had changed between these sites. Perhaps the evidence from the upper horizon of Le Grand Canton, sector 2 was still in accordance with the Late Magdalenian behaviour found at Gönnersdorf and Andernach, shifting the possible alteration in bone grease extraction between the MfCM occupation of this site and the Early Azilian in Le Closeau. However, if burning material was necessary for site hygiene, then the »nothing is wasted« strategy was no longer practised at Early Azilian and FMG sites, since the boiled bones did not require this special treatment. Based on the present evidence, the question whether a change in the bone grease extraction method or an abandonment of grease extraction paired with a persistent unwillingness to »take out the rubbish« led to an increase of burnt bone fragments cannot be answered. A detailed reanalysis of the fragmented faunal assemblages might therefore be of some inter-

52 The study also indicates that only some older hunters in Greenland still conform to this ideal, whereas modern hunters again leave much more waste at the hunting sites.

est as well as, perhaps, the results of the archaeozoological analysis of the well preserved faunal assemblage from Conty (cf. Auguste 2012).

Following the example from Western Greenland, an abandonment of this strategy would show a period of sufficient surplus, whereas conforming reflected a buffering mechanism against nutritional uncertainties. Besides a more intense exploitation of the usual prey, a broadening of the prey spectrum could also safeguard against these uncertainties. At this point the impression that Late Magdalenian assemblages appeared in general more specialised than FMG sites has to be discussed.

A previous study on hunting specialisation of Magdalenian assemblages suggested that the diversity index of Simpson calculated for the faunal assemblages is a useful indicator in the distinction between specialised hunting camps and residential sites (Gaudzinski / Street 2003). Moreover, the relation of this diversity to the total MNI was supposed to be further indicative of the type of site from which the assemblage originated (**tab. 58**). However, in different environments the resulting values of this index diverge probably in relation to the total MNI of the assemblage due to the diversity and density of the available species in general. In small assemblages reflecting possibly only a single hunting episode, this difference is not identifiable due to the small numbers. In assemblages reflecting a longer use such as base camps or repetitively visited sites such as agglomerations of various groups or hunting camps, a difference between the site types as well as between the different environments can generally be expected due to the diversity and the total number of hunted species brought to the site. The characteristics for each site type are likely to alter due to the different function and tasks performed at the site. In different environments, the intensity of these characteristics are assumed to differ due to varying possibilities of exploitation. In particular, these possibilities relate to the necessary adaptations of the subsistence strategy to species with different social structures inhabiting the environments. In a more uniform grassland with a lower species diversity and larger group sizes, a single species, for instance horse, was probably hunted more frequently than any species with a smaller group size in a diverse forest environment. Thus, this index was calculated for the assemblages which were published with sufficient details to base the discussion on a common value (**tab. 86**). A relation of this index to the total MNI of the sites is used later in the discussion about the Lateglacial settlement system.

For some northern French Magdalenian assemblages, a specialisation was already suggested due to the predominance of reindeer or horse in the faunal assemblages, in particular for Le Tureau des Gardes (Gaudzinski / Street 2003). This finding is replicated in the present project. However, the positioning of these assemblages in a chronological succession reveals the Late Magdalenian results clearly contrast with those of the MfCM (**fig. 85**). The Late Magdalenian assemblage form two clearly separated groups. The two large Central Rhineland assemblages as well as the smaller one from the Wildweiberlei form a dense set of diverse faunal inventories, even though horse appeared frequently at the sites. Based on the other archaeological evidence, these values reflect probably some type of base camp or residential site to which prey from various other places was introduced. The northern French assemblage from Le Tureau des Gardes, *locus* 6 and the assemblage from Saint Mihiel in the western uplands resulted in almost identical, more specialised values. A detailed archaeozoological analysis has not been published so far for the former assemblage. In Saint Mihiel, only few traces of human use were identified, including engravings on antler, due to the poor surface preservation of the material. However, almost 500 antler remains were probably accumulated by humans, possibly as a type of raw material cache to be used later for the production of antler spalls (cf. Stocker et al. 2006). Since both assemblages were not recovered from a comprehensive excavation, the attribution to a different group must be considered as provisional and, in the future, should be further tested with additional inventories. In analogy to Lewis Binford's settlement system models (Binford 1980), the clear difference between the two groups (**fig. 85**) support the idea of a clear functional differentiation of at least two types of sites in the Late Magdalenian, the base camp and the hunting camp. A variety of faunal mate-

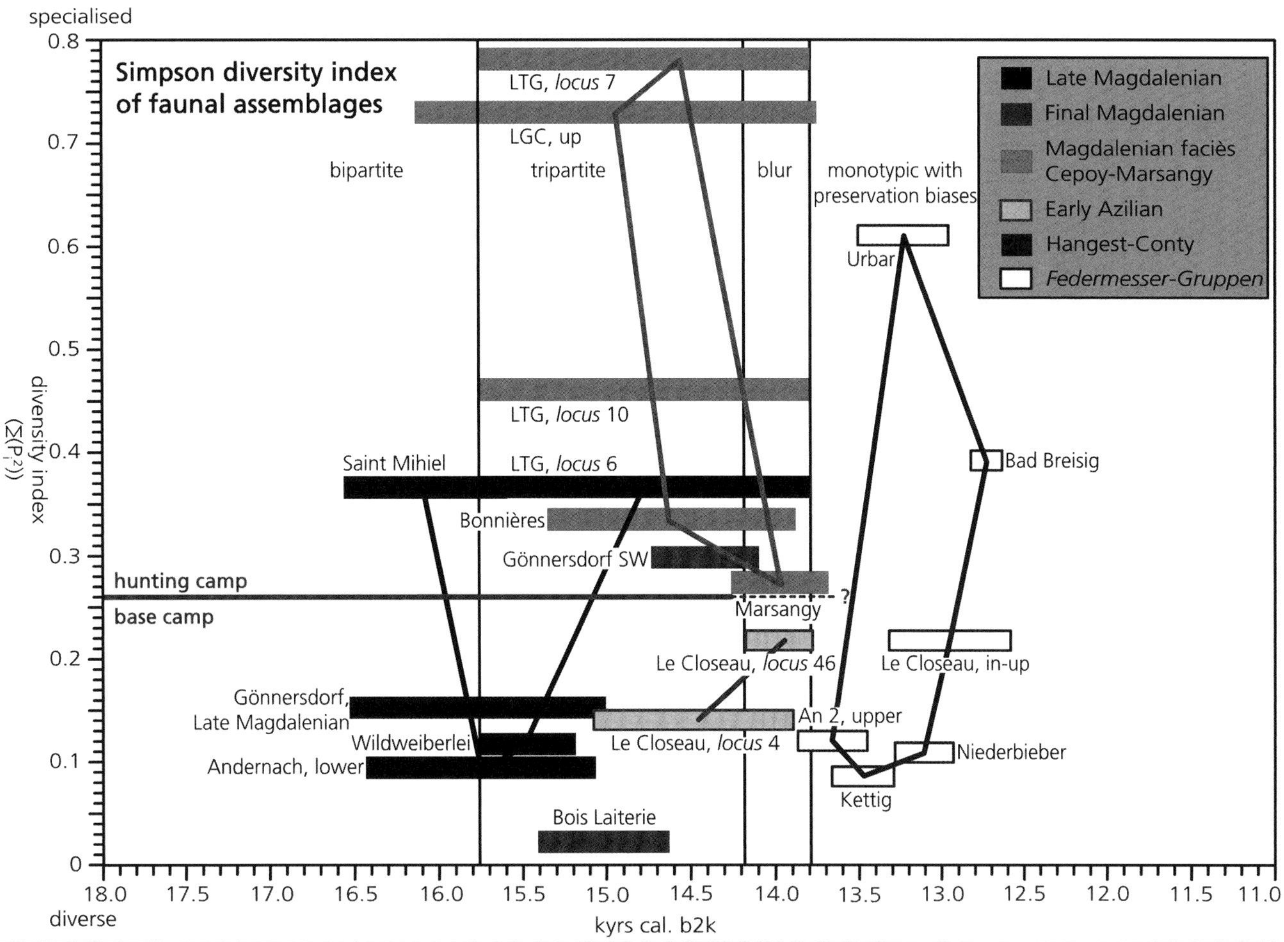

Fig. 85 Values of the Simpson diversity index given per time and archaeological unit. The convex hulls (see p. 281 f.) were set around the main cluster of assemblages attributed to the same archaeological. It was not possible to form a convex hull for the two Early Azilian assemblages, nevertheless, they are connected by a line for better comparability. An approximate limit between more base camp-type assemblages (i. e. diverse) and more hunting-camp-type assemblages (i. e. specialised) was set to the middle between the two Late Magdalenian groups of assemblages. Abbreviations see **fig. 74**. – For further details see text.

rial was brought to base camps and these resources were occasionally intensely processed, clearly shown by the preserved state of the material, modifications as well as the spatial dispersal (Street/Turner 2013). At hunting camps, one species was found almost exclusively. According to the concept of a hunting or butchering site, indications for dismembering and portioning of the prey should be found but the remains should mainly relate to waste material. However, in Saint Mihiel a considerable amount of useful antlers and a large number of long bone fragments as well as some indications of art were found (Stocker et al. 2006). Thus, the site appeared not as a pure hunting or butchering site. Accordingly, Claude Stocker and colleagues interpreted the excavated portion of the site as a consumption zone with a potential butchering area in the unexcavated vicinity (Stocker et al. 2006, 36). Moreover, the difference of the diversity indices of Saint Mihiel and the single Gönnersdorf concentrations is considerably smaller than the difference to the complete Gönnersdorf assemblage. In particular, Gönnersdorf IV yielded a high value indicating a specialisation similar to that found in the MfCM assemblages. Was Gönnersdorf IV a specialised hunting camp or was the specialisation a result of a small assemblage and provision from other areas of the site? According to the spatial analysis (Sensburg/Moseler 2008) and the faunal remains (Street/Turner 2013) in this part of the site, the former can be rejected and the latter considered more probable.

In general, the MfCM assemblages can also be divided in two groups. The more diverse group yielded values comparable to those of the hunting camp-type inventories of the Late Magdalenian. In addition, the values of Le Tureau des Gardes, *locus* 7 and the upper horizon of Le Grand Canton formed a very specialised group (**fig. 85**). However, different conditions of preservation on a site as well as a greater preservation probability of various elements from different species were considered as an important factor influencing the diversity of the faunal assemblages from Le Tureau des Gardes (Bridault 1996; Lang 1998, 89f.). In fact, the differential preservation of faunal elements has been discussed as an important factor in the formation of assemblage compositions for several decades (Poplin 1976). Differentiating conditions of preservation were found at several Lateglacial sites such as Gönnersdorf, Andernach, or Le Closeau but at these sites the composition of the faunal assemblage did not seem to be affected. Moreover, based on an analysis of the faunal material from the upper horizon of Le Grand Canton and the *loci* 5, 6, and 10 of Le Tureau des Gardes, Olivier Bignon doubted that this taphonomic factor alone could have resulted in the apparent under-representation of other mammal species at these sites (Bignon 2006, 187). Nevertheless, differential preservation could be an explanation for the further distinction of these assemblages from the already specialised group of assemblages. Alternatively, these sites were only in use for short periods and/or were located near reliable and productive hunting grounds for which an additional provision was not necessary. The number of artefacts and activities identified on these sites contradict a very short-termed episode and the diversity of the assemblage also makes an accumulation of several short-term episodes an improbable scenario. In contrast, favourable hunting patches which were recurrently visited but also used for a variety of other purposes (cf. Müller et al. 2006), could explain the number and diversity of artefacts and activities found in Le Tureau des Gardes and Le Grand Canton. In combination with differential conditions of preservation, this location near favourable hunting patches could have resulted in the assemblage appearing more specialised. In contrast, the outstanding inventory of Le Tureau des Gardes, *locus* 7 encompassed only a small assemblage (MNI=8). Thus, an interpretation comparable to Gönnersdorf IV (see above) is also possible. Most diverse is the youngest assemblage from Marsangy which is only 0.05 higher on the index than the Early Azilian *locus* 46 from Le Closeau. However, the very poor state of preservation of the faunal assemblage prohibited the calculation of a more precise MNI for the different species. In better preserved assemblages, this value could become more specialised but also more diverse. In the latter case, for example, the Marsangy inventory would resemble the Early Azilian assemblage of Le Closeau, *locus* 4.

The Early Azilian inventories formed in general a third group with a wide range of values. In comparison to the diverse Late Magdalenian assemblages, species diversity increased slightly in *locus* 4 of Le Closeau and decreased in *locus* 46 of the same site. In fact, this difference is inverse to the previously described assumptions of the influence of conditions of preservation on diversity at this site. Conditions of preservation were mediocre in *locus* 4, whereas very good preservation was described for the more specialised *locus* 46 (Bignon/Bodu 2006). Sample size was approximately the same in both assemblages but the proportion of the determined pieces varied. However, the number of individuals per species and the total MNI decreased significantly in both *loci* in comparison to the majority of MfCM assemblages. The larger and possibly younger assemblage of *locus* 46 was comparable to the small assemblage of Marsangy and together these values resulted in an indistinct separation of the more diverse group of the MfCM and the Early Azilian assemblages. Moreover, the assemblage from the south-western area of Gönnersdorf produced a value which was slightly higher than the one from Marsangy, placing this part of the site into the group of more specialised camps of the Late Magdalenian. However, the attribution of animals to the assemblage was difficult due to the diffuse outlines of this younger phase of occupation at the site and, therefore, the MNI for this site is low. This area can possibly be interpreted as comparable to Gönnersdorf IV. In comparison, the spatial structure of the south-western corner was very different from Gönnersdorf IV and since the south-western area was the only evidence of a younger phase, the question arises which other concentration provided for this part. Bois Laiterie yielded an outstandingly diverse value as was already indicated by the distribution of the selected species. In fact, the value falls clearly below those from large camp sites of the Late Magdalenian, interpreted as base or residential sites, and the value of Wildweiberlei assemblage which was also found in a small cave. The most probable explanation is that this relatively large faunal assemblage was formed during several occupation events which fell into the period of a changing environment, rather than a single long-term settlement in a hyper-mosaic landscape.

The Simpson index decreases even further in the majority of FMG assemblages. However, three outliers were observed: Urbar, Bad Breisig, and Le Closeau. The two former assemblages represented partial inventories. For the latter, the calculation of MNI was biased again by the poor preservation and the low numbers of preserved and determinable faunal remains. Therefore, these highly specialised inventories in the FMG are probably the result of poor preservation. Niederbieber 1 yielded a value lower than Kettig and the upper horizon of Andernach 2. The other concentrations at Niederbieber ranged approximately between a value comparable to that from Marsangy to just below the value for Bad Breisig indicating, perhaps, that a hunting camp-type also existed in the FMG. This scatter was in accordance with the hypothesis that a more diverse environment, inhabited by generally smaller, non-migratory groups of prey, also resulted in a more generalised subsistence strategy which was reflected in a more diverse faunal assemblage.

Besides an increasing diversity within the large mammal spectrum, the incorporation of smaller species such as smaller mammals, fish, or birds into the diet was considered as a step to broaden the food spectrum (broad spectrum revolution) in areas experiencing resource imbalance either due to decreased environmental productivity or to increased population demand (Zeder 2012). As in larger mammal communities, the social structure and the territoriality of potential smaller prey can also differ between different environments but most of these communities are fast growing and their group sizes can vary considerably at local levels, buffering and overprinting the environmental fluctuations (Campbell et al. 2005; Barrio et al. 2013). Consequently, smaller mammal communities commonly found in large and fast growing groups was considered important as a supplementary resource in times of environmental instability and crisis (Munro 2003) and population pulses (Stiner et al. 1999). Ruth Charles pointed further to the complete use of fur-bearing mammals and carnivores (Charles 1997b) and Werner Müller emphasised the importance of non-alimentary parts of small mammals besides their nutritional value (Müller 2004). Furthermore, Anne Bridault and Laure Fontana showed the increasing importance of smaller mammal species during 13,000 and 12,000 years ^{14}C-BP (*c.* 15,890-13,730 years cal. b2k) in a study on Lateglacial environmental changes based on the faunal compositions in mountainous regions (Bridault/Fontana 2003).

For these reasons, the proportion of small mammals in the studied assemblages was considered with and without carnivores (**fig. 86; tab. 86**).

In general, a development from high to low values is observable in both sets (**fig. 86**). In detail, decreasing proportions first occurred in the inventory of Le Tureau des Gardes, *locus* 6. At this site, selective preservation as well as the small number of remains in the assemblage could explain the difference to the other Late Magdalenian assemblages. If the carnivores are removed from all assemblages, the value of *locus* 6 is located in an intermediate position. Nevertheless, the Early Azilian values were generally lower than those of the Late Magdalenian except for Le Tureau des Gardes, *locus* 6 which in both calculations was lower than the results from *locus* 4 of Le Closeau. If the carnivores are removed, the Early Azilian values appeared almost unaltered and rank in the lower part of the Late Magdalenian range. In the MfCM assemblages, smaller mammals with and without carnivores were very infrequent. These assemblages usually rank below the Late Magdalenian range and were also distinctly lower than the Early Azilian assemblages. At the majority of FMG sites, no smaller mammals were preserved. Those with smaller mammals fell approximately into

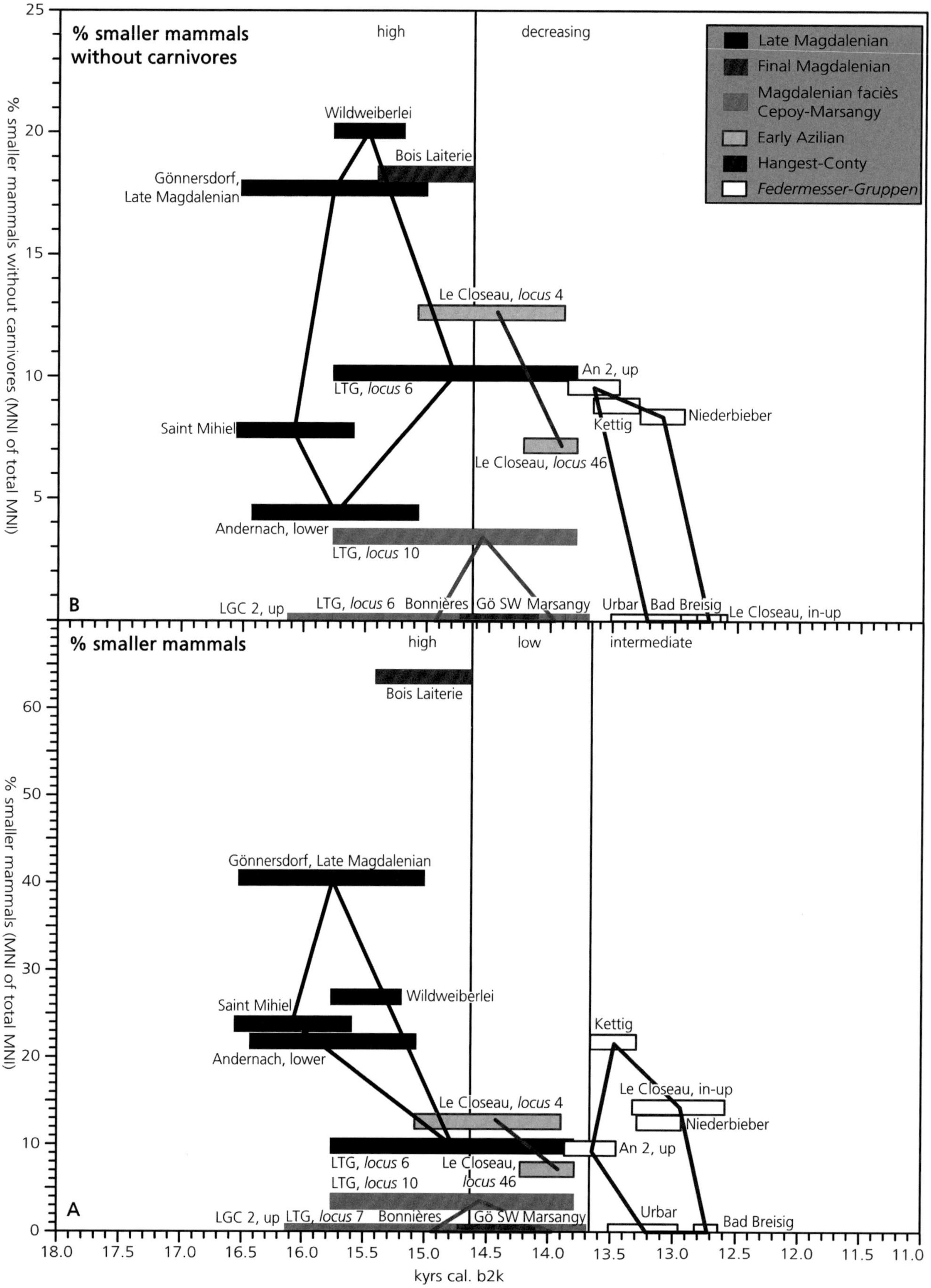

% smaller mammals without carnivores
high
decreasing
Late Magdalenian
Final Magdalenian
Magdalenian faciès Cepoy-Marsangy
Early Azilian
Hangest-Conty
Federmesser-Gruppen
Wildweiberlei
Bois Laiterie
Gönnersdorf, Late Magdalenian
Le Closeau, locus 4
An 2, up
LTG, locus 6
Saint Mihiel
Kettig
Niederbieber
Le Closeau, locus 46
Andernach, lower
LTG, locus 10
B
LGC 2, up
LTG, locus 6
Bonnières
Gö SW
Marsangy
Urbar
Bad Breisig
Le Closeau, in-up
% smaller mammals without carnivores (MNI of total MNI)
% smaller mammals
high
low
intermediate
Bois Laiterie
Gönnersdorf, Late Magdalenian
Wildweiberlei
Saint Mihiel
Kettig
Andernach, lower
Le Closeau, in-up
Le Closeau, locus 4
Niederbieber
An 2, up
LTG, locus 6
Le Closeau, locus 46
LTG, locus 10
A
LGC 2, up
LTG, locus 7
Bonnières
Gö SW
Marsangy
Urbar
Bad Breisig
% smaller mammals (MNI of total MNI)
kyrs cal. b2k

the range of the Early Azilian assemblages from Le Closeau. Only Kettig reached values of the lower range of the Late Magdalenian inventories, but after the carnivores had been removed, all FMG assemblages became comparable to the value from Le Tureau des Gardes, *locus* 6.

Thus, the two graphs are relatively similar suggesting that the impact of carnivores usually changed in a way comparable to that of the smaller mammals. If all these mammals were hunted, this tendency would suggest, as Charles had pointed out (Charles 1997b), carnivores were treated in a similar way to other mammals (cf. Street/Turner 2013, 161-176). However, comparable conditions of preservation would also result in a linked trend. Exceptions to linked percentages hint occasionally at non-taphonomic causes. For example, one exception is the outstanding value of smaller mammals in Bois Laiterie which becomes comparable to Late Magdalenian inventories when the carnivores are removed. Some of these smaller carnivores such as badger (*Meles meles*) can be assumed to have taken shelter in the cave and, perhaps, reached the site without human influence. A comparable strong decline of the values can be seen in Saint Mihiel and the lower horizon of Andernach as well as Kettig and the greyish deposits of Le Closeau. The assemblage from Kettig sank to a value comparable to the other FMG assemblages after the proportion of carnivores was removed (**tab. 86**). In this case, several smaller carnivores such as red fox (*Vulpes vulpes*), weasel (*Mustela* sp.), and marten (*Martes* sp.) had increased the value. An attribution to natural intrusions at the open air site cannot be completely excluded for these remains. In Le Closeau, the strong decrease was due to the conditions of preservation. Since only a small number of individual animals could be determined in the greyish deposits, a single badger had increased the proportion of the smaller mammals significantly. Thus, in both cases natural rather than human agents could be the reason for the varying values. In contrast, in the lower horizon of Andernach and at Saint Mihiel the decrease is due, in particular, to high proportions of arctic fox (*Alopex lagopus*). At Gönnersdorf, this species was heavily exploited, particularly in concentration I (Street/Turner 2013, 161-176). Presumably, this exploitation was also important in the lower horizon of Andernach (cf. Álvarez Fernández 1999). The values of the assemblages without carnivores were lower than the Early Azilian assemblage of Le Closeau, *locus* 4. Thus, the introduction of other small mammals such as hare (*Lepus* sp.) was not as clearly displayed in these assemblages as, for example, in Gönnersdorf and the Wildweiberlei. Nevertheless, the different values of smaller mammals with and without carnivores in these assemblages indicate that humans could have been an important agent in the creation of these values, even though the small size of the excavations at these sites have to be kept in mind.

Consequently, the question must be discussed whether differential preservation as well as restricted excavations were major agents in creating the observed development. In fact, the preservation was usually good at Late Magdalenian sites and these assemblages yielded the largest proportion of smaller mammals. The higher proportions found in poorly preserved inventories such as in the greyish deposits of Le Closeau or the beaver bone in Niederbieber 3 can be considered critically because they were frequently based on a single individual. However, the proportions did not necessarily increase with better conditions of preservation, such as were found in Le Closeau, *locus* 46 and the upper horizon of Andernach 2. Thus, the observed development towards lower proportions of smaller mammals found in the assemblages of the Lateglacial Interstadial seemed not directly related to preservation. However, the impression of slightly increased values towards the late Lateglacial Interstadial could be the result of generally poor conditions of

←

Fig. 86 Percentages of smaller mammal MNI from the total determined MNI given per time and archaeological unit. **A** All smaller mammals. – **B** Smaller mammals without carnivores. The convex hulls (see p. 281 f.) were set around the main cluster of assemblages attributed to the same archaeological. It was not possible to form a convex hull for the two Early Azilian assemblages, nevertheless, they are connected by a line for better comparability. Abbreviations see **fig. 74**. – For further details see text.

preservation and small assemblages of bones identified to species. Moreover, the distribution of the values for smaller mammals (**fig. 86A**) recorded at Late Magdalenian sites seems to decrease in association with the extent of the concentration recovered by excavation. In contrast, the widely excavated south-western area of Gönnersdorf contained no smaller mammals suggesting some spatial differentiation where this type of fauna occurred on Magdalenian sites. In addition, the incompletely excavated FMG sites, Urbar and Bad Breisig, yielded no material attributed to smaller mammals, whereas the more complete FMG sites contained smaller mammals in proportions similar to lower ones of the Late Magdalenian. Perhaps this difference indicates that the distribution of faunal material was spatially more distinct at FMG sites than previously thought possible. Consequently, the excavated portion of the site had some effect on the outcome of faunal analyses.

Furthermore, surfaces of the organic material were frequently affected by post-depositional processes making archaeozoological analyses of human modifications on these remains difficult. Besides restricting the comparability of processing methods and behavioural standards, the lack of traces of butchering produces problems when identifying human introductions in contrast to a natural background fauna. The previously mentioned burnt faunal material can only partially help to solve the problem because depending on the attention paid to the setting of the site and the discard process, carelessly swept floors could contain elements which were disposed of in the fire without being introduced as waste by humans. Perhaps recurrently appearing species can possibly be considered as an indicator of human intervention, assuming that a regular appearance at these sites was not a random event. An example of this is the frequent occurrence of arctic fox at Late Magdalenian sites. However, at several of these sites the exploitation of arctic fox has been clearly attested and based on this knowledge, the assumption that foxes were also exploited as a resource at other sites becomes more credible. On the other hand, several species could regularly occur at human settlements where they scavenge on debris. In these cases, their presence was not part of a random process but also not a result of direct human impact. In particular, badgers fall into this uncertain category on Lateglacial Interstadial sites because these omnivorous animals appear regularly in the faunal lists (**tabs 15. 39**). No modified remains of this species were thus far identified, although their fur could be good alternative to furs of arctic foxes. In contrast, beavers (*Castor fiber*) are more likely to be human introductions based on the general habitat and behaviour of this species (Pinto/Santos/Rosell 2009) as well as the frequent appearance of their remains among burnt bone material. Nevertheless, various difficulties restrict the identification of exploitation of small mammals, in particular at FMG sites. Due to these problems all smaller mammals and carnivores were incorporated in the calculation of the values. Consequently, if some of these remains resulted from natural intrusions which had to be excluded, the proportion of smaller mammals in the potential prey assemblages would decrease further. This phenomenon would result in an even clearer manifestation of the general trend of a decreasing importance of smaller mammals in the Lateglacial Interstadial.

According to the hypotheses related to the broad spectrum revolution, the decreasing proportion of small mammals in the development indicated less alimentary pressure either due to an increasing productivity of the environment or a decrease of the population density or both. However, conditions became more favourable for the Lateglacial hunter-gatherers allowing for this decreasing exploitation intensity. In accordance with the finding that a »nothing is wasted« ideal (cf. Pasda/Odgaard 2011) was probably conformed to in the Late Magdalenian and in some MfCM assemblages but possibly no longer in the Early Azilian and the FMG, an increasing alimentary security can be assumed for the hunter-gatherer communities in the Lateglacial Interstadial.

However, the transition in faunal exploitation strategies began much earlier (**fig. 87**). A first step was the appearance of red deer, elk, and wild boar in the various assemblages, followed afterwards by an increasing variety in the diversity of faunal assemblages, in particular due to the occurrence of some hyper-specialised

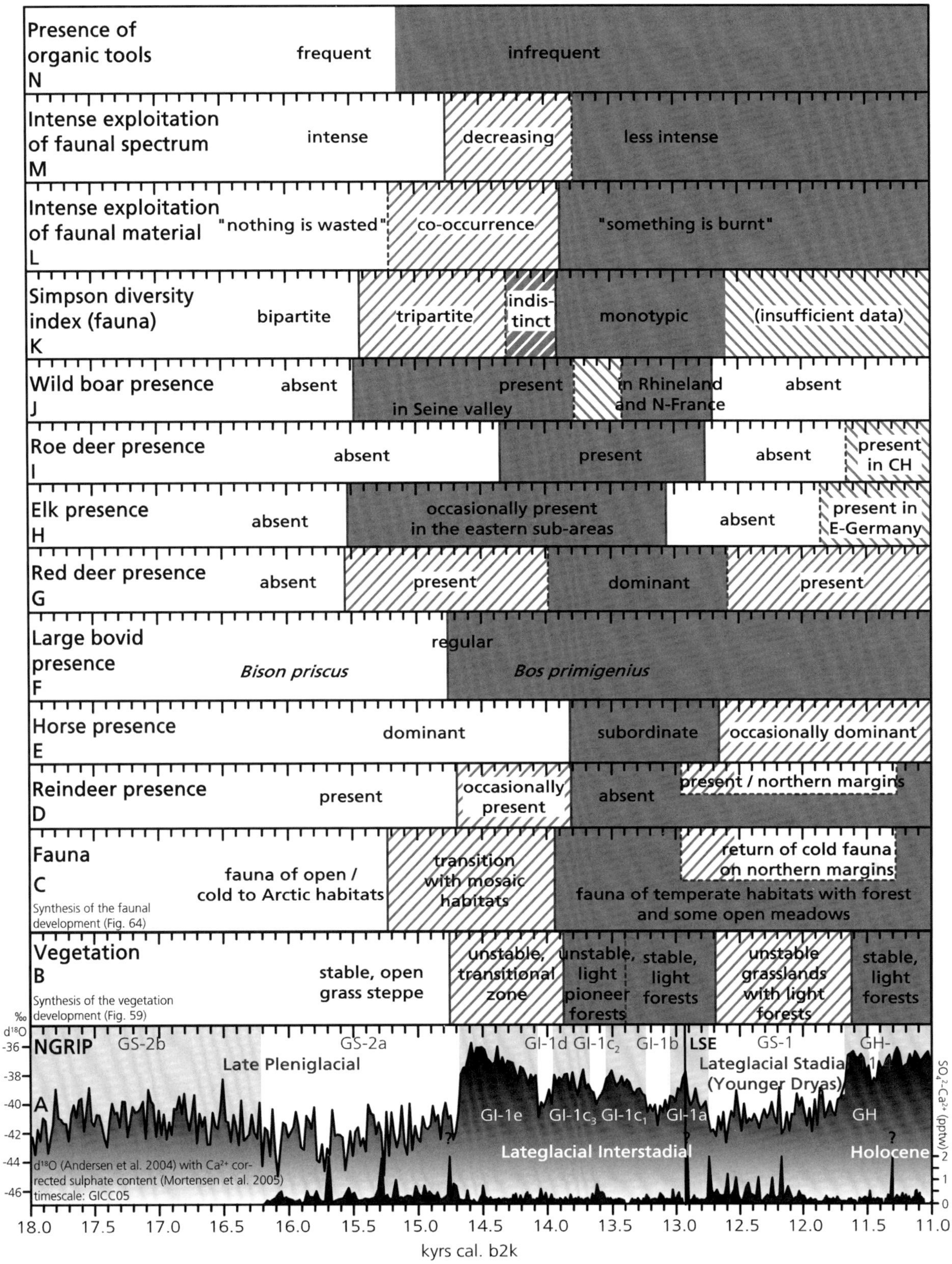

Fig. 87 Developments in faunal assemblages related to faunal exploitation (**D-N**) contrasted by the synthesised faunal (**C**; see **fig. 64**) and vegetation development (**B**; see **fig. 59**) as well as the oxygen isotope record of NGRIP (**A**; see **fig. 53**). Hatched areas indicate transition periods. – For further details see text.

sites. Later the first assemblages with increased numbers of burnt bones appeared almost concomitant with the disappearance of frequently made organic tools. The next wave of changes was the replacement of bison (*Bison* sp., usually *Bison priscus*) by another large bovid (*Bos* sp.), often determined as aurochs (*Bos primigenius*), reindeer becoming only a sporadic visitor in the studied sub-areas, and a decrease of smaller mammal faunas, in particular carnivores in the faunal assemblages. Later roe deer (*Capreolus capreolus*) appears in the archaeological assemblages and, shortly after, the differentiation of diverse and more specialised faunal assemblages becomes indistinct. The next cluster of changes begins with red deer becoming dominant in some assemblages. Thereafter, mainly diverse assemblages are found and the few, very specialised assemblages were found to be incomplete for various reasons. Perhaps, due to preservation, the ideal of »nothing is wasted« could no longer be proven by the evidence of very small, fragmented bones on these sites. However, the trend of burnt material, mentioned previously, could indicate that this ideal was no longer conformed to or that this method of most intense exploitation had changed. At the same time, reindeer disappeared as well as assemblages in which horses were the dominant prey. Wild boar was for a short period not determined in an assemblage of the study area but was possibly present in eastern Germany (see **tab. 80**), appearing again in the assemblages of Niederbieber, Boppard, and Saleux 114 (Fagnart 1997; Coudret/Fagnart 2006). The later disappearance of elk followed by roe deer and wild boar from the archaeological assemblages can only partially be attributed to a change in the faunal composition because the number of assemblages with organic preservation further decreases in this period. Thus, during the Lateglacial Stadial and the early Holocene, the dataset of the study area becomes in some parts insufficient to evaluate further developments in faunal exploitation.

After considering the changes in the exploitation strategies, it becomes apparent that it is necessary to consider these values in the context of the organisation, the length of the occupation, and the function of the sites. However, having presented all this information further assumptions about the settlement system and the changes therein become possible.

Changes in the Lateglacial settlement behaviour

In the previous sub-chapter, the functions of the studied assemblages were already considered in some detail. These results are combined in the following and compared to the idealised models suggested previously (see p. 269f. and p. 286f.).

Different levels in the spatial behaviour of Lateglacial hunter-gatherers can be analysed based on the archaeological material. Besides the position of the items and, hence, the spatial organisation within a concentration and on a site, the overall composition and the amount of material help to suppose the main tasks accomplished at a site and, consequently, the general purpose of a site and the duration of its use. In combination with the spatial organisation and seasonal indicators, the duration can be further differentiated into sites which were used repetitively and those which were used more continuously. In a regional comparison of quasi-contemporaneous sites which were characterised in regard to their function and the duration of their use, the spatial behaviour in a limited territory and, thus, the regional settlement system can be modelled. This model is further contributed by the information about the raw material acquisition which allows for further assumptions on the regional to super-regional spatial organisation of the studied groups. Moreover, detailed comparisons of the archaeological types can help sustain an information exchange between different regions and, thus, the spatial extent of information networks. These information networks were considered as the main fundament of the social organisation in prehistoric groups (Clarke

1968; Gamble 1983). Hence, besides reflecting a change in the material output of human behaviour, the above presented analyses also allow for the identification of social changes.

The importance of the spatial organisation became particularly apparent in the last chapter by several outliers in the record which were best explained with specific task zones being excavated such as at Hallines, the upper horizon of Le Grand Canton, sector 1, or Urbar. Even though zoning of a site was found in all studied archaeological units, the intensity of some outliers suggested that the zoning was usually more restrictive on Late Magdalenian than on FMG sites. However, detailed intra-concentration analyses were not possible in the present project but were in some cases accomplished previously (Julien et al. 1999; Bodu/Debout/Bignon 2006; Sensburg 2007; Sensburg/Moseler 2008; Bodu 2010; Gelhausen 2011a; Jöris/Street/Turner 2011). These concentration units were considered as single occupation episodes and the base for assumptions about the diversification of lithic assemblages by function and duration (cf. Löhr 1979; Richter 1990). This concept appears valid for the ephemeral FMG inventories which were positioned at some distance to one another and, thus, preserved the abandonment of the archaeological material almost unaltered. In one case, the horizontal distribution appeared so well preserved that the last sitting position of a single stone knapper was possibly identifiable (Baales 2003b; Gelhausen 2010). This almost undisturbed pattern can be assumed to result from a relatively short-termed event because in a longer occupation, the position where material was originally dropped would be altered by the intentional (cleaning) or unintentional moving (kicking) or transformation (trampling, burning) by the inhabitants. In contrast, large installations such as the concentrations of Gönnersdorf were often cleaned up (Sensburg 2007) and possibly reused in a tell-like manner during several occupation events (Jöris/Street/Turner 2011). In this case, the question arises which duration of use was reflected by the diversity of the lithic inventory: The number of times in which the installation was revisited or the time of a single visit. According to the original Löhr model (Löhr 1979), the cleaning of the site during a visit was considered and assumed to occur in sites which were settled for a longer period. However, if the purposes of the visits were different, possibly depending on the period of the year when the installation was visited and the cleaning up was related to a certain site hygenie to not attract predators or vermin which could destabilise the installation, the interpretations based on this model could be ambiguous. Thus, the relation between concentrations on a site are an important matter in the characterisation of the sites. Moreover, the possible contemporaneity of two or more concentrations on a site further throw up the discussion of social organisations within groups: Were there larger agglomerations or were two or three parties occasionally joining forces? As previously mentioned, questions relating to the contemporaneity are difficult to answer even with a detailed spatial analysis because the interpretation of the results often allows for conflicting models (cf. Cziesla 1990; Jöris/Street/Turner 2011). Nevertheless, although these questions and interpretation problems cannot be solved in this study, the change in the use of space as an interesting factor of standardisation of life spaces can be recorded independent of whether this material accumulation was the result of a single or a few, longer occupation events or several short events. Comparing the layout of the different sites with multiple concentrations in the present record already reveals some clear differences: The large, tell-like used installations on the here studied Late Magdalenian sites are usually clearly separated although the archaeological material sometimes scatters between these areas. However, some works and special dump zones were spatially restricted in the relatively large areas. Frequently, material was transferred between these installations either as part of the differentiation of working zones or as raw material supply from abandoned structures (Terberger 1997;Street/Turner 2013). The concentrations of the Early Azilian in Le Closeau were comparably organised but revealed some satellite concentrations which were separated by several metres from the main activity area (**fig. 88**; cf. Jöris/Terberger 2001; Bodu/Debout/Bignon 2006). The connection between these concentrations were not as frequent as on Late Magdalenian sites. In contrast, the sites of the MfCM along river banks showed the

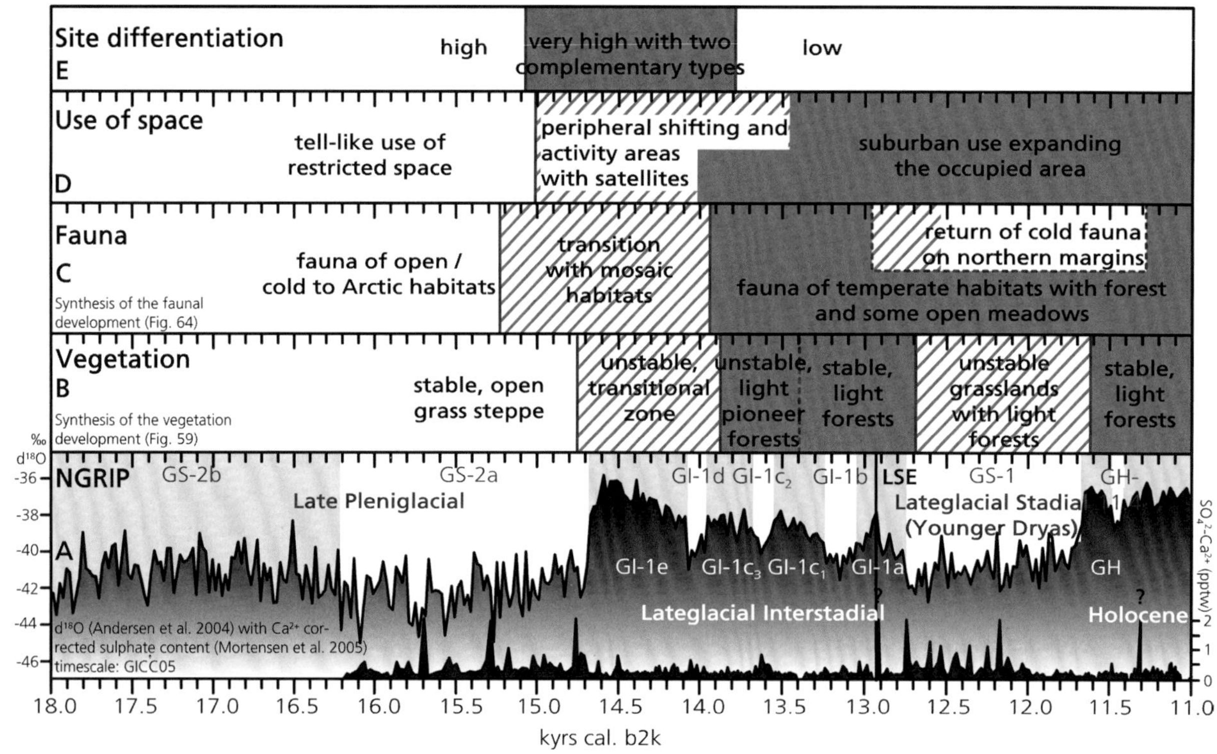

Fig. 88 Developments in the Lateglacial assemblages related to settlement behaviour (**D-E**) contrasted by the synthesised faunal (**C**; see **fig. 64**) and vegetation development (**B**; see **fig. 59**) as well as the oxygen isotope record of NGRIP (**A**; see **fig. 53**). Hatched areas: transition periods. – For further details see text.

gradual shift of much lighter installations to the periphery of a previously used area. This spatial behaviour was also found on Late Magdalenian sites along lake shores in Switzerland where one installation consisted usually of only one or two stone-filled hearths (Leesch/Cattin/Müller 2004; Müller et al. 2006; Leesch et al. 2010). Material from previous installations was occasionally reused on these sites resulting in often complex spatial interrelations on these large and often densely scattered sites. The early FMG assemblage from the upper horizon of Andernach 2 appears as a reminiscence of the peripheral moving of activity areas because in the late MfCM sites such as Belloy-sur-Somme and late Early Azilian sites such as Pincevent III.2, more separate, small concentrations already appear which conform to layout comparable to the single concentrations at the FMG site Niederbieber (Gelhausen 2011a). In Niederbieber, this uniformity of the concentrations reminds of suburban areas planned on a drawing board. Many of the single FMG concentrations also conform to this general layout. In Niederbieber, only very few connections between the various concentrations were found. In combination with the highly similar layout, the concentrations give an independent impression rather than reflecting different spots used in a single, longer-termed occupation. Thus, the number of revisits was perhaps readable from the number of concentrations. In this case, the complete site of Niederbieber could be comparable to a single Late Magdalenian concentration of Gönnersdorf in regard to the number of visits on the site.

Based on the above named differences in the formation of the sites, the diversity of the formally retouched artefact inventories as indicator for the function and duration of the occupation (cf. Löhr 1979; Richter 1990) must be considered in relation to the size of the inventory. In addition, the diversity of the faunal assemblages in relation to the size was considered as an indicator for the function of a site (cf. Gaudzinski/Street 2003) and the duration of its use (**tab. 58**). The distinction between high and low values indicating frequency and diversity remain again a matter of definition and the present database.

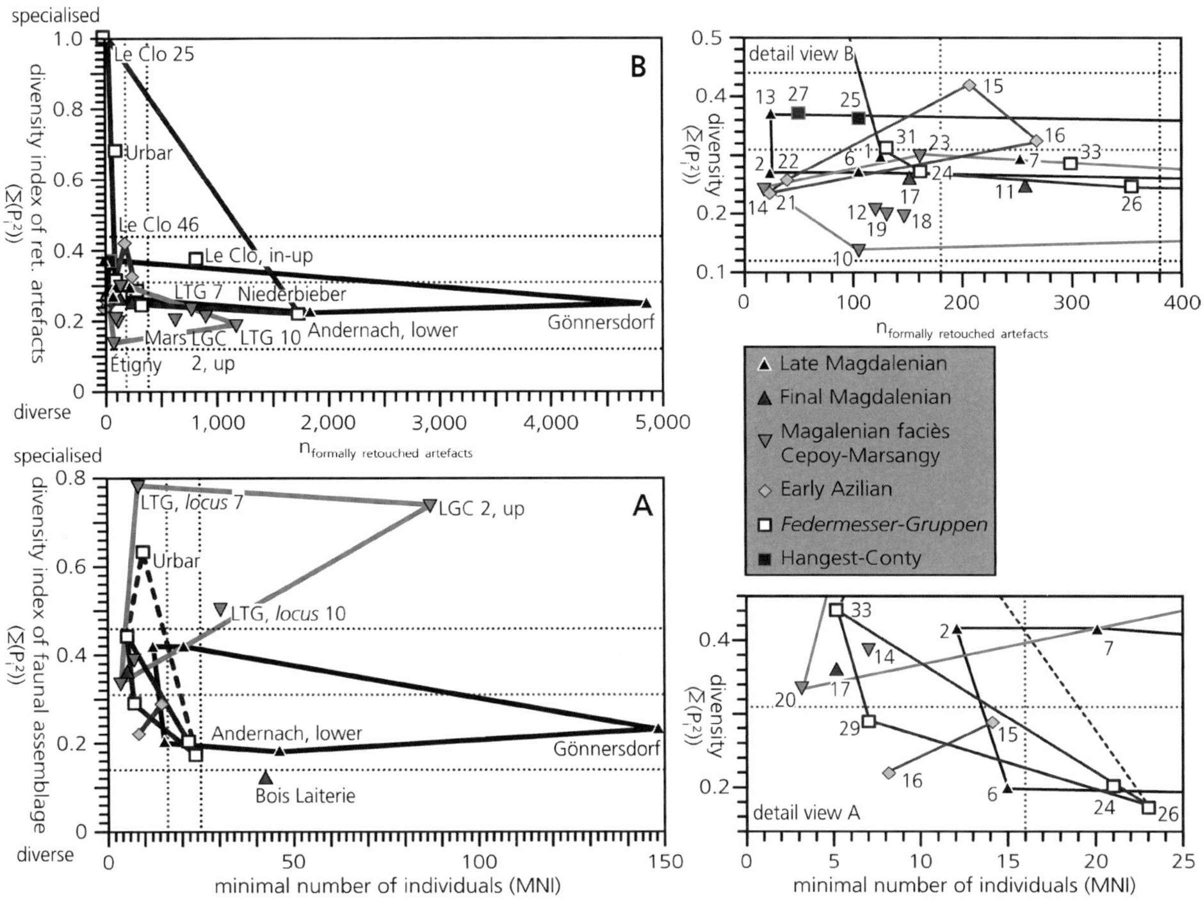

Fig. 89 Diversity of faunal assemblages (**A**) and formally retouched artefacts (**B**) in contrast to the minimal number of individuals determined in the assemblages given per site and archaeological unit. The distribution of all assemblages is given in the main view where also the convex hulls were set per archaeological unit (see p. 281 f.) which are also indicated in the detail view. Numbers for the sites are used in the detail views according to **tab. 82**. Black dotted lines: limits of observable breaks in the data beyond which the values can be considered as extreme; grey dotted lines: sub-division of the usual values into low and high values. For abbreviations see **fig. 74**. – For further details see text.

In the comparison of the faunal assemblages (**fig. 89A**), an important observation is that the convex hulls of Late Magdalenian and MfCM assemblages are almost exclusive and that the convex hull of the FMG overlaps the ends of both types of inventories. The dichotomy of the Late Magdalenian and the MfCM assemblages cannot be explained by a difference in function alone. The Late Magdalenian sample includes Saint Mihiel and *locus* 6 of Le Tureau des Gardes which were considered as specialised hunting sites (Bridault/Lang/Rieu 1997). In fact, these assemblages fall into the lower range of the MfCM hull which is formed by the poorly preserved material of Marsangy and the problematic inventory of Bonnières-sur-Seine. The other assemblages represented some hyper-specialised sites of which the upper horizon in Le Grand Canton, sector 2 and the *locus* 10 of Le Tureau des Gardes were also very large. In these cases, differential conditions of preservation alone appear unreliable as an explanation but in combination with many visits and extremely standardised hunting patterns, they form a probable scenario of the hyper-specialisation. In regard to the very standardised purpose and behaviour of visiting a site, these assemblages appear to be produced by typical Magdalenians. The Late Magdalenian assemblages range between diverse and semi-diverse inventories but except for Gönnersdorf and the lower horizon of Andernach no outstandingly large assemblages were found. For Gönnersdorf as well as for Andernach, this significant amount of diverse faunal material brought to the site seems indicative for a longer use, possibly by several revisits but cer-

tainly also by longer stays on these sites. Larger assemblages were not found in the FMG. They were a bit more specialised and generally smaller than the Late Magdalenian sites. However, the sites of Kettig and the upper horizon of Andernach 2 were comparably diverse as the Wildweiberlei assemblage, whereas the partially excavated assemblage from Bad Breisig was approximately as specialised as Saint Mihiel and Le Tureau des Gardes, *locus* 6. Nevertheless, the value of the greyish deposits in Le Closeau ranged between those two values indicating either a less strict differentiation between the FMG sites or, more probable, the impact of poorer preservation in these deposits. Moreover, if in Bad Breisig, as in Urbar, only a specialised part of the site was excavated, the high value must be questioned. If the single concentrations of Niedebieber are taken into account, they yielded usually smaller assemblages than Bad Breisig but comparable specialised inventories. Perhaps, the clear distinction between specialised and diverse sites became indistinct in the FMG inventories. The Early Azilian assemblages of the lower horizon of Le Closeau range among the more diverse inventories but as with the FMG, these inventories were rather small. In general, the tendency towards smaller inventories becomes apparent in the Early Azilian and, perhaps, later MfCM. In summary, the Late Magdalenian yielded two general groups of data with a more specialised group and a diverse one which was usually formed by very large assemblages. The diverse group is also present among the FMG inventories but the distinction to more specialised inventories became possibly unclear. In contrast, the diverse group was also found in the Early Azilian but the more specialised group was absent, whereas the MfCM assemblages yielded only specialised, occasionally hyper-specialised and no diverse inventory. Supposing that the two types of sites were necessary to form a complete settlement cycle in the Lateglacial, the Early Azilian and the MfCM sites probably formed two complementary sets of sites.
The diversity of the formally retouched inventory (**fig. 89B**) does not reflect the specialisation of the MfCM sites nor the diversity of the Early Azilian sites but rather points to the contrary. The Early Azilian assemblages appear more specialised and the MfCM sites are more diverse. Thus, these inventories appear again as complementary sets. However, the specialisation of the Early Azilian sites results from the undefined other formally retouched artefacts group. In addition to a significant amount of macroscopically splintered material which indicate intense use or trampling, this specialisation shows the decreasing reliance on standardised implements in these inventories. The numbers of formally retouched artefacts found on a small group of MfCM sites (Le Tureau des Gardes, *loci* 7 and 10, the upper horizon of Le Grand Canton 2, and Marsangy) suggested along with the faunal inventories a longer use of the site. In particular, the relation of the formally retouched artefacts to cores (see **fig. 79**) proves the large amount of material used at these sites. This relation also indicated that the exploitation of lithic resources played an important role besides the specialised hunting events. Perhaps, this purpose also contributed to the diversity of the formally retouched inventories with the production of rejected implements rather than used material. Among the FMG inventories, Urbar as well as the *locus* 25 of Le Closeau form highly specialised outliers. Besides sustaining the special task character of these inventories, these high values reflect the small size of the assemblages. Exceptionally large FMG assemblage were produced by the accumulations in Niederbieber and the greyish deposits of Le Closeau. The former yielded approximately as many retouched artefacts as the lower horizon in Andernach. However, the Gönnersdorf assemblage was considerably larger pointing to the exceptional position of Gönnersdorf also among the Late Magdalenian inventories. The comparison of the diversity of the smaller Late Magdalenian assemblages indicated little differences between these assemblages. This lack of difference also between the potential hunting camps and the sites with a more residential character reveals the very standardised tool set of the Late Magdalenian which was brought to all site types. Thus, diversity of the formally retouched artefacts could possibly not serve for a functional or temporal differentiation of these sites. In fact, contrasting the diversity indices of the analysed inventories (**fig. 90**) to understand the use of the sites, the Late Magdalenian assemblages formed again two groups with Gönnersdorf, the lower horizon

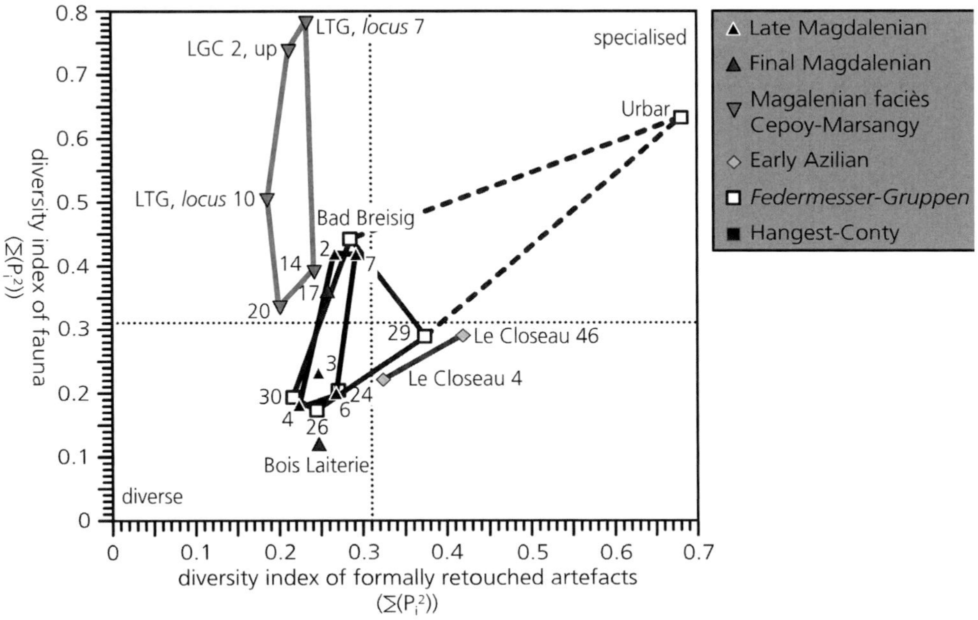

Fig. 90 Diversity of faunal assemblages in contrast to formally retouched artefacts given per site and archaeological unit. The convex hulls were set per archaeological unit (see p. 281 f.). Numbers for the sites are used according to **tab. 82**. Black dotted lines: limits of observable breaks in the data beyond which the values can be considered as extreme (see **fig. 89**). For abbreviations see **fig. 74**. – For further details see text.

of Andernach, and the Wildweiberlei assemblage falling into the diverse sector and the assemblages of Saint Mihiel and *locus* 6 of Le Tureau des Gardes as well as the assemblage from the south-western area of Gönnersdorf falling into the diverse artefact but more specialised faunal assemblages. In this sector, all MfCM assemblages were located, whereas the Early Azilian assemblages of Le Closeau as well as the greyish deposits of Le Closeau fall into the opposite sector with a high diversity in the faunal assemblages but a low diversity in the retouched artefacts. Thus, the site types become further differentiated in this period. The FMG assemblages reveal an almost congruent pattern with the Late Magdalenian inventories. However, the more specialised fauna and diverse artefacts sector is only represented by Bad Breisig. Although this site shows high similarities with the MfCM sites, for instance in regard to the importance of cores, the inventory was not completely excavated which could lead to the increased specialised values. Moreover, the assemblage of Urbar yielded a completely outstanding position in the specialised fauna and specialised artefact assemblages sector showing this assemblage to be biased by the excavated area.

The duration of the use whether recurrent or continuous can also be measured by the evident structures (**tabs 19. 32. 43**) found on the sites. The hearths indicated by stone setting or stone filling as well as those indicated by the colouring or other transformation of the underlying sediment or the accumulation of burnt material prove the existence of fire on a site. However, the former show some effort in the construction of the fireplace, whereas these efforts were not detectable in the other types of hearths. Some efforts in maintaining and use of these latter hearths were visible by the distribution of burnt material showing occasional clearings and allowing for the assumption of a longer use. Nevertheless, in several of the Niederbieber concentrations only a single dense cluster of the burnt material suggested a short-term use of the fireplace. Previously, the use of open fires as places of waste disposal were suggested and, perhaps, explain the increasing proportion of burnt material on Lateglacial sites (see **tabs 11. 35**; cf. Kind 1985). These various types of hearths can be moved easier than large pavements and/or pits. In contrast, the occasionally massive installations of pavements are more sustainable and make reuse of the same installation after some

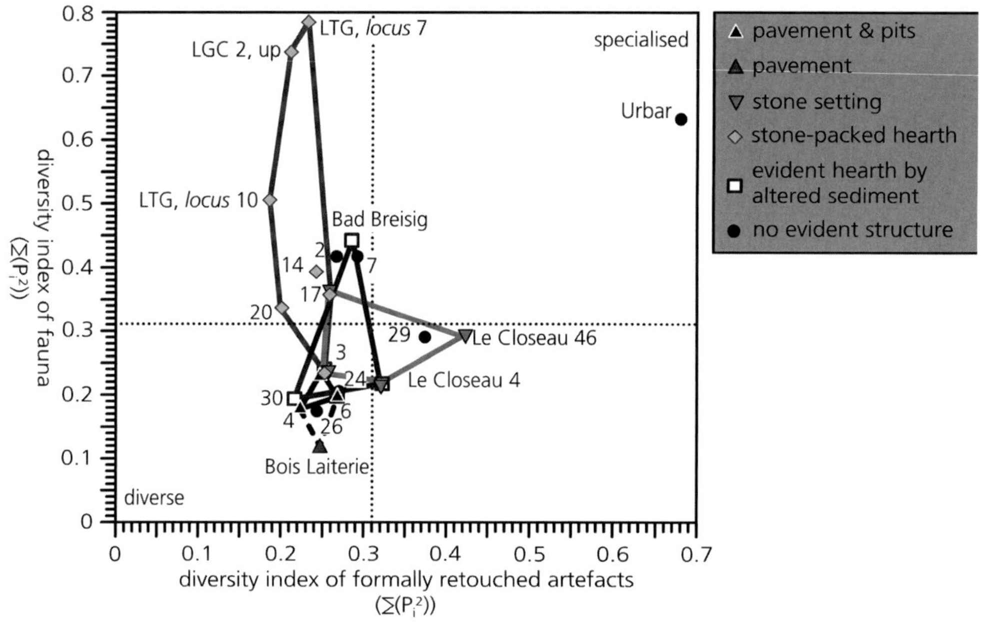

Fig. 91 Diversity of faunal assemblages in contrast to formally retouched artefacts given per site and evident structures. Some sites such as Gönnersdorf yielded several types of evident structures. The convex hulls were set per evident structure type (see p. 281 f.). Numbers for the sites are used according to **tab. 82**. Black dotted lines: limits of observable breaks in the data beyond which the values can be considered as extreme (see **fig. 89**). For abbreviations see **fig. 74**. – For further details see text.

time has passed possible. In fact, pavements often in combination with several pits were only found on very diverse sites (**fig. 91**). Stone-filled hearths were found in very diverse retouched artefact assemblages but seemed independent of the diversity of the fauna. In contrast, stone settings were related to a variety of assemblages, usually of intermediate diversity in regard to the faunal assemblages as well as the retouched inventories. In Gönnersdorf, these structures co-occur but otherwise they are exclusive. Hearths which became evident due to colouration of the sediment appeared also at a variety of sites.

Finally, comparing the seasonal indicators known for the sites (**fig. 92**), the dichotomy of the Early Azilian sites and the MfCM sites is further sustained. Although in all assemblages providing seasonal evidence, this evidence suggested an almost yearly use of the sites, in MfCM inventories the evidence was stronger for the light season, whereas for Early Azilian sites the evidence was more numerous for the dark season. Thus, in the transitional period, the spatial differentiation between hunting and more residential camps appeared stricter than in the Late Magdalenian. In the more solid winter camps, the hunter-gatherers lived on a variety of prey species and the tasks performed necessitated no standardised lithic inventory. During the lighter months, lighter dwellings or no structures were need and a few hearth constructions were sufficient. In this period, regular hunting episodes were performed besides which intense lithic production was performed. In the FMG this seasonal differentiation is no longer visible.

Based on the previous assumption of how the archaeological material would differentiate site types and additional assumptions on the general availability of faunal resources, a hypothetical model can be developed to propose which position sites would take in this model (**fig. 93**) and compare this model with the results of this study (**figs 90-92**).

In general, assuming that the group size of inhabitants of a base camp and their daily tasks did not change, the diversity of the formally retouched inventories should be comparable for all base camps. However, the territorial and reproduction behaviour of the prey influences the duration for which a base camp could be used. Different attributes can be used to describe the faunal compositions (uniform – diverse; migratory –

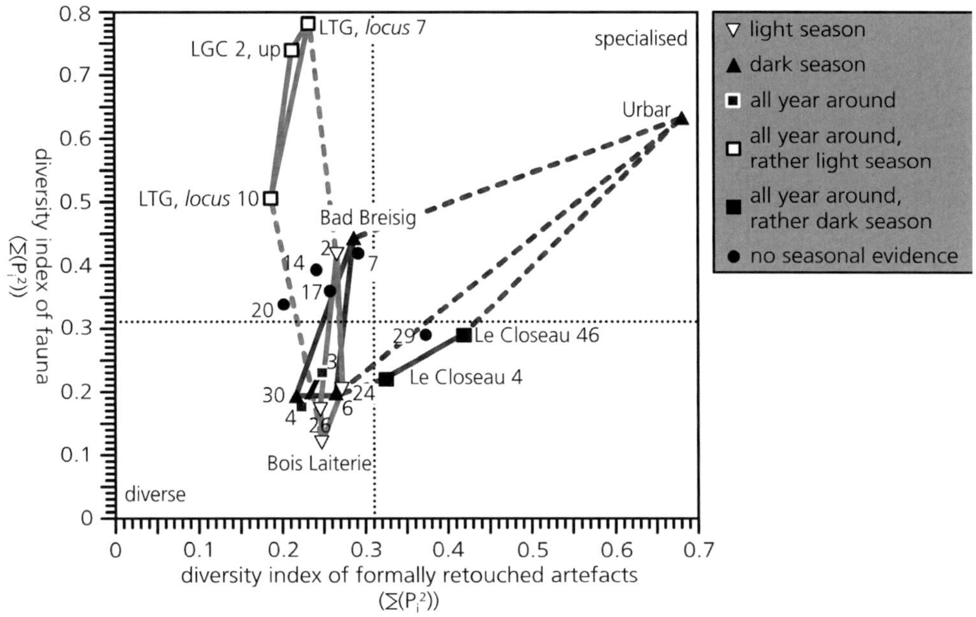

Fig. 92 Diversity of faunal assemblages in contrast to formally retouched artefacts given per site and seasonal indicator. The convex hulls were set per season (see p. 281 f.). Numbers for the sites are used according to **tab. 82**. Black dotted lines: limits of observable breaks in the data beyond which the values can be considered as extreme (see **fig. 89**). For abbreviations see **fig. 74**. – For further details see text.

nonmigratory; large herds – small groups). These attributes are often combined in two groups which are mainly assumed to correspond to an open environment (uniform, migratory, large herds) and a forested environment (diverse, nonmigratory, small groups). Assuming that the intensity of lithic tool use alters with the duration of the site use, base camps in forested areas with more diverse faunas would produce higher diversity values for the faunal assemblage but smaller diversity for the retouched artefact inventories than base camps in more uniform landscapes.

In a landscape characterised by animals which are faithful to their habitat, prey animals can be found there all year round but the supply of prey animals is limited. Thus, the occupation duration for base camps in these areas are limited due to the supply. In contrast, in landscapes with large-scale migration movements, new groups of the same species can replace the hunted ones but often these movements are seasonally restricted. However, if the landscape was characterised by a uniform faunal composition over a large area, the possibilities for the rest of the year were limited to either follow the herds or to supplement the seasonal resources with species available during other seasons. Consequently, if a base camp existed in a uniform environment it could be occupied for a longer period with a succession of seasonally available faunal resources. The seasonal variation between the site types from different environments becomes probably most evident in hunting camps. Seasonally migrating animals are usually caught at a specific place during the migration period, whereas species which are residential in an area can be hunted all year round in a favourable place. Favourable places such as watering places are usually visited by various species. Thus, hunting small numbers of a special prey in favourable place allows for an opportunistic supplement with other species. In contrast, the chance of hunting supplementary species when hunting a migratory herd at a topographically suitable hunting spot is improbable and, in particular, if a larger number of animals is hunted the processing of the wanted species consumes more time making additional work unprofitable. Consequently, a hunting place in a diverse environment can result in slightly higher diversity of the faunal assemblage than hunting camps where the focus was set on a single herd species. The total numbers of determined individual (MNI)

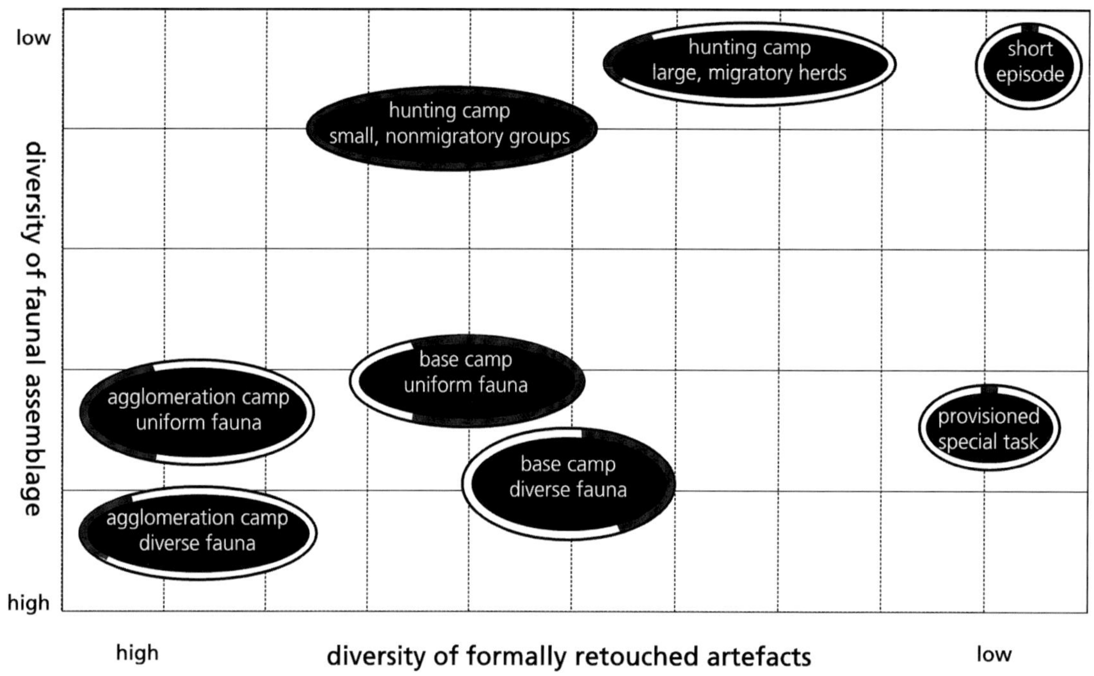

Fig. 93 Hypothetical distribution of different site types in relation to the diversity of the lithic formally retouched artefacts and the faunal assemblage. In addition, the presumable seasonal indicators from these site types are given as grey shaded area in an idealised annual cycle. – For further details see text.

should further differentiate these hunting spots. Places focused on large herds should also produce larger numbers of animals than those focused on smaller groups. Moreover, at mass hunt sites accomplishing tasks unrelated to this event is improbable. However, the time to process numerous animals would also require some space for other activities such as sleeping or eating (cf. Stocker et al. 2006). However, at more opportunistic hunting sites additional tasks could be embedded and, thus, an increased diversity of the lithic inventories seems possible. Nevertheless, repetitive, short stops could also result in more specialised inventories. Besides the base and the hunting camps, agglomeration camps were regularly considered. This type is partially difficult to distinguish from repetitively visited base or hunting camps where also more diverse material could be accumulated (cf. discussions on Niederbieber, Gelhausen 2011a; Gelhausen 2011b; or Rekem, Caspar/De Bie 1996; De Bie/Caspar 2000).

In general, larger agglomerations of groups from different territories should result in a more diverse faunal composition. Depending on the uniformity of the environment between the different territories, this diversity could be only a bit higher (very uniform landscape) to significantly higher (mosaic landscape) than at a single base camp. However, if this agglomeration were continuous for a longer period of time people would have to exploit the local faunal resources. Therefore, the meeting of several groups occurred probably in seasons when large herds were hunted or these meetings were set in particularly favourable places and/or these camps were only sustained for a short duration. In the latter case, sites with a very high diversity but a very limited occupation period as shown by the seasonal indicators could be assumed to represent an agglomeration camp rather than a repetitively visited base camp. In addition, in areas where agglomeration as well as base camps are present, the seasonality of the two types should be exclusive.

Thus, the connection with the seasonality shows which influence the faunal composition in the landscape can have on the human mobility patterns. The more diverse but resident, small group species of the forested environment had a generally smaller faunal biomass per square kilometre than the less diverse, often migratory and larger herd fauna of the open grasslands. This difference required a higher mobility of the

Lateglacial forest hunters to provide the necessary faunal resources. Thus, these hunters had to either move their residential site frequently (high residential mobility) or cover increasing distances to the base camp from hunting grounds (high logistical mobility; cf. Kelly 1991; Kelly 1992).
Comparing this hypothetical model with the results of the present study, the differences between Early Azilian and MfCM sites match the assumptions on base and hunting camps in areas with non-migratory species very well. The inventories of the Late Magdalenian were closer in the faunal diversity and almost identical in the diversity of the lithic inventory. However, the slight decrease in the diversity of the faunal as well as the lithic inventory on the supposed hunting camp sites resembled more the distribution of sites in a uniform landscape. Furthermore, in comparison with the hypothetical model, Gönnersdorf and the lower horizon of Andernach are more similar to the base camps in uniform landscapes with an almost annual occupation than to the agglomeration camps (see **tab. 60**).
According to this model, Bois Latierie was also best explained as a base camp in a mosaic landscape. However, based on the faunal data and also the diversity in the LMP spectrum, this assemblage could also represent a palimpsest of a revisited hunting camp during a period of extreme environmental changes. The south-western area of Gönnersdorf could also represent a site continuity in the period of change but in contrast to the previous occupation of the site, the main purpose could have changed to a short hunting camp. The FMG assemblages were of more interest because in these inventories base and special task sites seemed less distinct.
The assemblages of the lower horizons of Conty and Hangest-sur-Somme III.1 cannot yet be tested against this model. However, due to their lithic inventories (see p. 525), these sites were occasionally separated from the group of FMG values. Since Conty and Hangest-sur-Somme III.1 were possible hunting camps specialised in aurochs, the results promise to be interesting.
Finally, besides these direct evidences from the inventories of the studied sites, the interrelation of various areas can be considered based on the acquired raw materials. Analyses locating the precise source/acquisition area of raw materials were mentioned to be desirable and might become easier to accomplish in the future (Pettitt/Rockman/Chenery 2012; Frahm et al. 2014). They are particularly desirable in regard to this project because a gap in the raw material acquisition distances was observed and if this gap proved valid, it suggested a specific territorial behaviour in which people moved between acquisition sites over a minimal distance. This pattern was independent of the archaeological group and, thus, seemed to sustain the major changes during this period. The possible minimal distance incorporated the local exploitation zones around the sites and a type of untouched transition zone which related approximately to another full exploitation zone. This pattern also occurs when exchange and, thus, the meeting of two groups is excluded due to the amount of the regional or exogenous raw material. Consequently, this pattern can also be related to the settlement behaviour of a single group. Moreover, this pattern seemed independent whether a group was sent to the next mineral source for logisitical reasons such as raw material procurement or the complete group moved their residence. Thus, movements related to the exploitation of a mineral resource seemed optimal within 5-10km around a site. For sources further away logistical or residential moves had to be considered and these moves seemed only reasonable for sources which were at least another exploitation zone away (**fig. 94A**). Perhaps, this distance could be explained by recovery cycles of the exploited vegetation and the fauna around the sites. Nevertheless, since the decision where to move next was made at least with each residential move, a relatively small territory could be completely exploited with a few moves (**fig. 94B**). These distant moves made a recovery of the ecotope in an exploited zone possible before the next human visit. This behaviour would add to a potential seasonal cycle a fluctuation for environmental sustainability. In addition, this behaviour allowed the intense use of a favourable area, for instance with a high resource concentration such as vegetation and fauna diversity due to water availability in river valleys (cf. Butzer

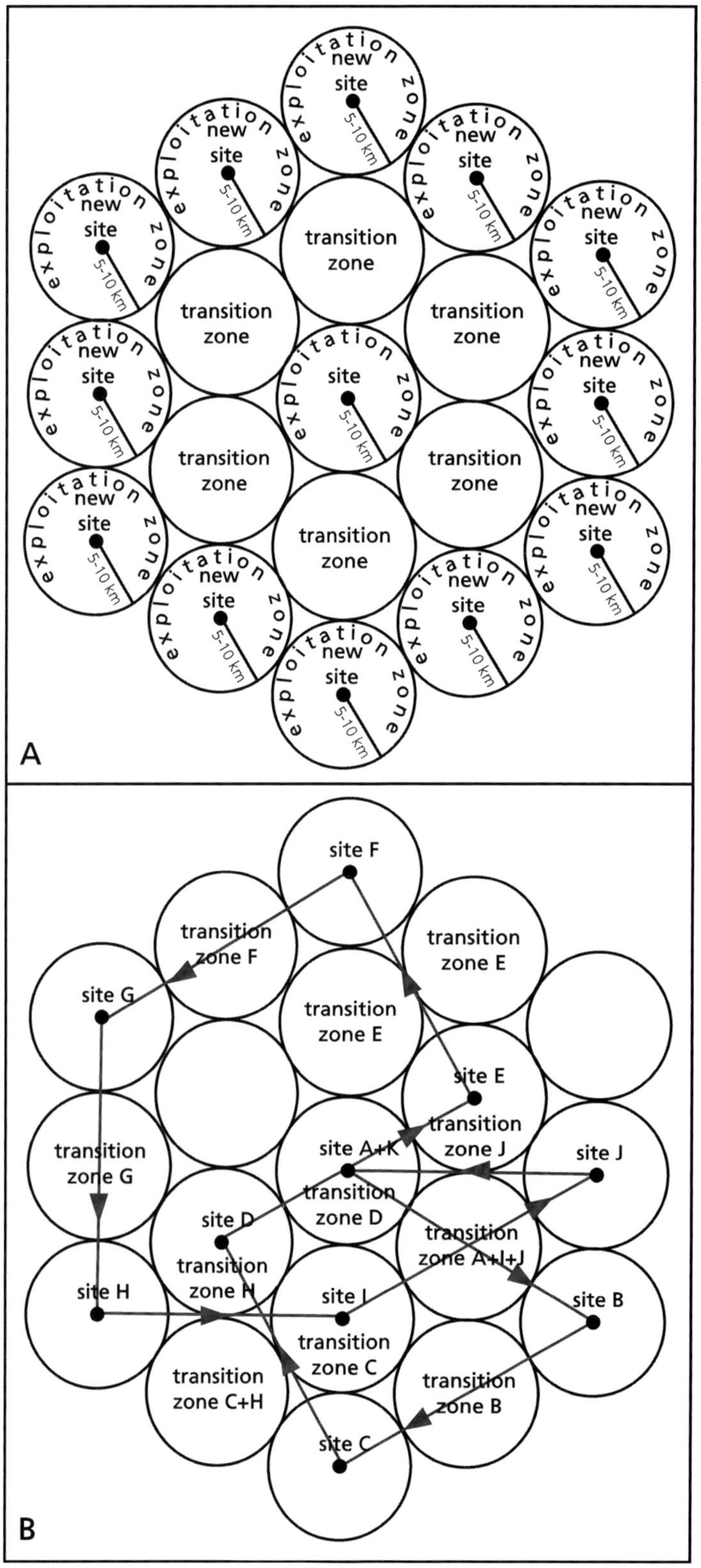

Fig. 94 Hypothetical model of site movements based on a gap in the distances of mineral raw material sources. **A** Possibilities of moves from one site to the next over a transitional zone; **B** use of a territory with the type of moves described in A. Grey line: connection of the successive sites A-K along the straight way between the sites.

1982, figs 12-2 and 14-2). If these observations can be further verified in the future, they provide further implications about the Lateglacial exploitation strategies and the prehistoric understanding of ecological sustainability.

Moreover, the precise spatial location of acquisition sites in relation to the studied sites allows for further suggestions about the territorial and settlement behaviour of Lateglacial groups and, thus, the social organisation of space. In particular, the loss of »long distance« acquisitions indicates that the very large information networks which were sustained during the Late Magdalenian disappeared. Eventhough these connections seemed not very intense, they probably served a wide-span buffering network (cf. Stein Mandryk 1993). Their disappearance can reflect a decreasing need of these security patterns. Nevertheless, distant contacts were still identifiable in the acquisition distances and suggest that contacts at a regional distance were probably sustained.

CHRONOLOGY OF CHANGES IN LATEGLACIAL NORTH-WESTERN EUROPE

The comprehensive presentation of the climatic and environmental developments, in particular in regard to their chronology, was necessary to allow for a synthesis of these results with the developments of the human society in the same temporal framework. Considerations about the impact of climatic and environmental changes on human behaviour are only possible within a consistent and reliable synthetic approach. In the analyses performed at a single site, the indicators of environmental development can be found within the same horizon as the archaeological remains and, thus, a quasi-contemporaneity can generally be assumed after the exclusion of natural intrusions. Comparisons of contemporary or even causally related behaviour across wider geographical areas often lack this direct correlation and other means of chronological correlation are required. As shown in the previous chapters, this correlation on a common (calendar) timescale is necessary to consider two locally distinct developments as quasi-contemporary.

Even though the methodological problems such as the necessity of radiocarbon calibration or the biasing conditions of preservation should be considered as possible reasons for mistrusting the temporal successions in detail, approximate coincidences must be assumed valid based on the coherency of the present chronological framework. However, a problem with evaluating the limits in the archaeological record remained for sites formed by numerous repetitive visits. Limits based on these sites often have a significant standard deviation, even if the limit was only set when at least two sites of overlapping dating showed the changes in the attributes. Nevertheless, the number of analysed attributes (n = 37) helps to further sustain the observations.

To establish the process of change from the Late Magdalenian to the FMG, the previously presented summary graphs were joined to a single graph (**fig. 95**). The analysed archaeological attributes were then ordered chronologically according to the first limit within an attribute. In a second graph, all limits were used to create a cumulative presentation of the appearance of the changes (**fig. 96**).

The chronological order according to the first limit already reveals a generally gradual character of the process of change in the studied period of time. The appearances of first limits in the 37 assembled attributes range over more than 3,500 years. This finding indicates that adapting behavioural patterns was common in Lateglacial hunter-gatherer societies and gradual evolution appeared to have been the norm. Moreover, another important observation of this plot is that many alterations in the archaeological record were not directly related to climatic or environmental changes. Thus, even though the changes appeared gradually, chain reactions were not straightforwardly detectable. A classic causal chain that climate change was promptly followed by environmental variation which again forced humans to react directly was not found.

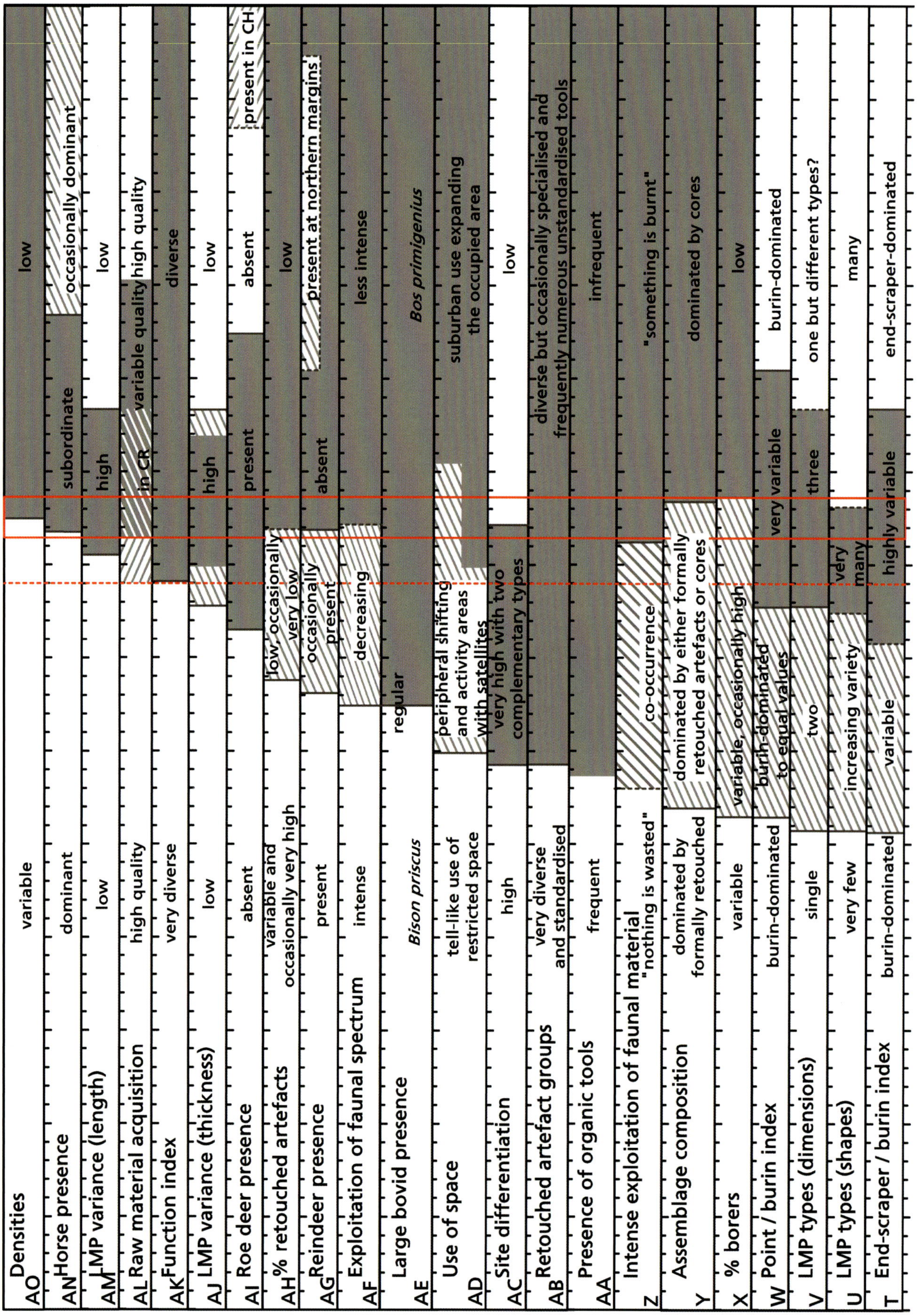
AO Densities
variable
low
AN Horse presence
dominant
subordinate
occasionally dominant
AM LMP variance (length)
low
high
low
AL Raw material acquisition
high quality
variable quality in CR
high quality
AK Function index
very diverse
diverse
AJ LMP variance (thickness)
low
high
low
AI Roe deer presence
absent
present
absent
present in CH
AH % retouched artefacts
variable and occasionally very high
low, occasionally very low
low
AG Reindeer presence
present
occasionally present
absent
present at northern margins
AF Exploitation of faunal spectrum
intense
decreasing
less intense
AE Large bovid presence
Bison priscus
regular
Bos primigenius
AD Use of space
tell-like use of restricted space
peripheral shifting and activity areas with satellites
suburban use expanding the occupied area
AC Site differentiation
high
very high with two complementary types
low
AB Retouched artefact groups
very diverse and standardised
diverse but occasionally specialised and frequently numerous unstandardised tools
AA Presence of organic tools
frequent
infrequent
Z Intense exploitation of faunal material
"nothing is wasted"
co-occurrence
"something is burnt"
Y Assemblage composition
dominated by formally retouched
dominated by either formally retouched artefacts or cores
dominated by cores
X % borers
variable
variable, occasionally high
low
W Point / burin index
burin-dominated
burin-dominated to equal values
very variable
burin-dominated
V LMP types (dimensions)
single
two
three
one but different types?
U LMP types (shapes)
very few
increasing variety
very many
many
T End-scraper / burin index
burin-dominated
variable
highly variable
end-scraper-dominated

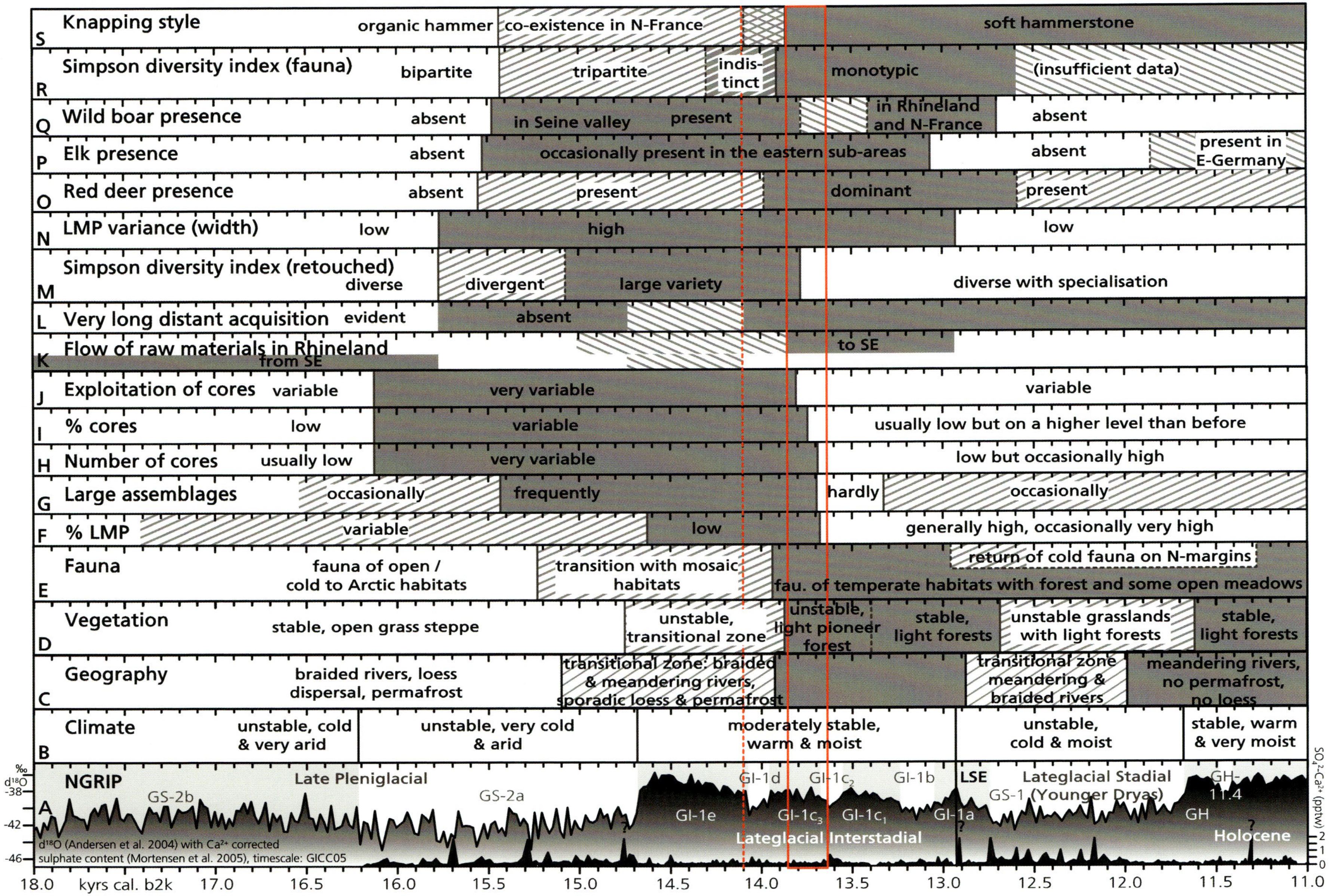

Fig. 95 Synthesis of changes in the archaeological material (**E-AO**) contrasted by the synthesised faunal (**E**; see **fig. 64**) and vegetation development (**D**; see **fig. 59**) as well as the changes in the physical environment (**C**; see p. 365-370), the climatic synthesis and the oxygen isotope record of NGRIP (**A-B**; see **fig. 53**). Hatched areas: transition periods; red lines: phase of rapid and accumulated change; dashed line: beginning of the prelude. – For further details see text.

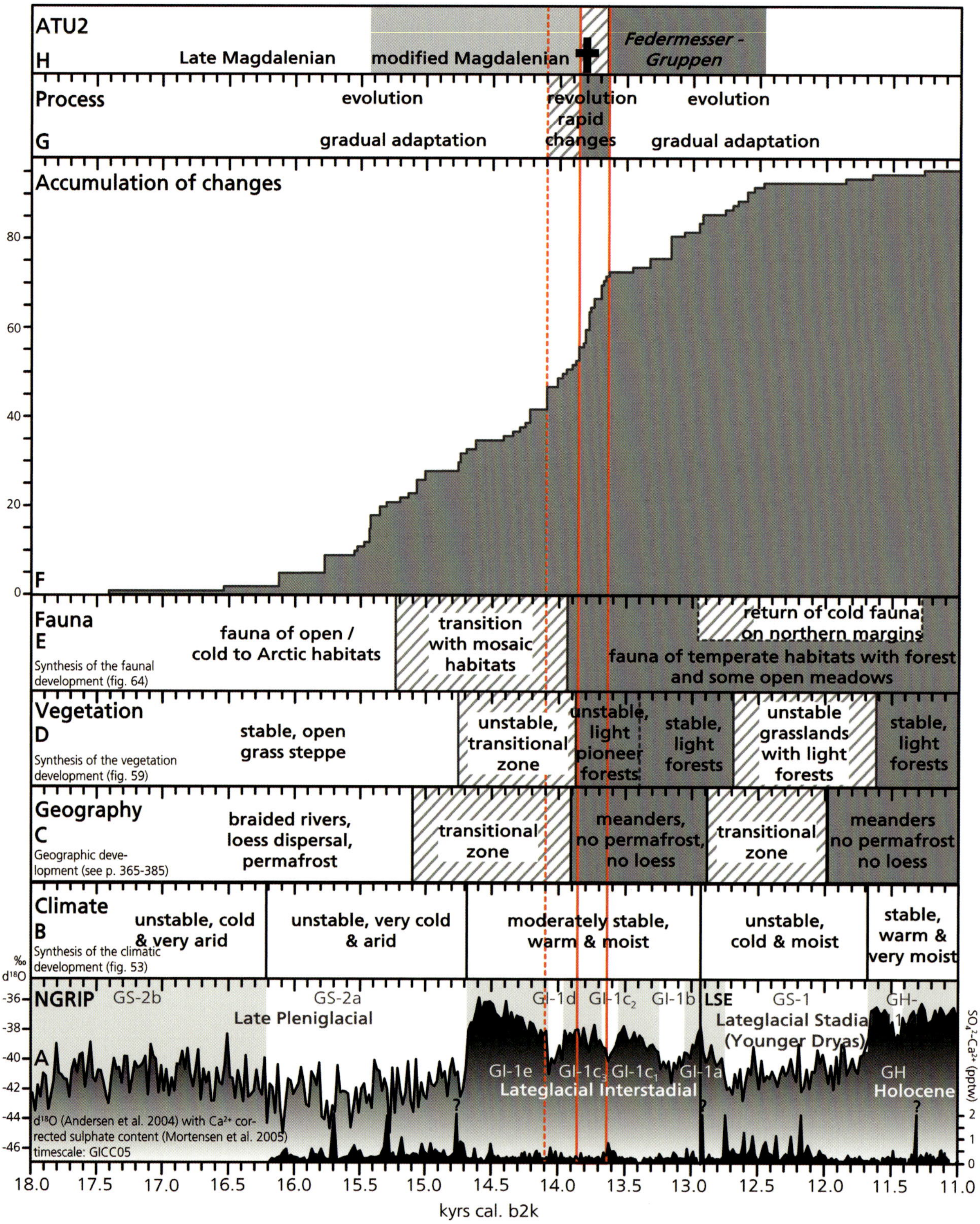

Fig. 96 Quantitative accumulation of changes observed in the archaeological material (**F**; see **fig. 95**), description of the transition process (**G**) and changes of the meso-scale archaeological taxonomic units (ATU2, **G**; see **tab. 62**) contrasted by the synthesised faunal (**E**; see **fig. 64**) and vegetation development (**D**; see **fig. 59**) as well as the changes in the physical environment (**C**; see p. 365-370), the climatic synthesis and the oxygen isotope record of NGRIP (**A-B**; see **fig. 53**). Hatched areas: transition periods; red lines: phase of rapid and accumulated change; dashed line: beginning of the prelude. – For further details see text.

In fact, variation in the archaeological record, in particular in the LMP inventories, occasionally preceded climatic and environmental changes. However, several of the considered lines of evidence were interrelated and, consequently, co-occurring changes in these interrelated characteristics appeared such as the increasing presence of various prey fauna in the inventories followed by a change in the diversity index or the appearance of a new knapping style accompanied by variations in the shape and dimensions of the LMP.

If the plot is examined for the main changes in the climatic record, the most significant change was the onset of the Lateglacial Interstadial. With this change only the first decrease of reindeer in the archaeological record coincided. The counting error of the ice-core record is approximately 200 years at this point. In a range of 200 years around this onset, changes in only two further attributes before and three attributes after the decrease of reindeer are observed. Thus, the onset of the Lateglacial Interstadial was no prominent break in the behavioural evolution. Reindeer, a species adapted rather to arctic environments, can be assumed to have reacted sensitively to the amelioration process and to have followed a more cold climate habitat further north. As reindeer possibly became rarer, it was only hunted in an opportunistic way. Some decades after the decreasing importance of reindeer, the proportion of the formally retouched artefacts dropped to a significant low. Whether this development was related to climate change remains uncertain and even a direct connection to the disappearance of reindeer can be questioned because the decrease of formally retouched artefacts was connected to the appearance of numerous sites with an intense blank production. These sites seem related to the frequent hunting of horses which were also previously hunted but, with the decreasing presence of reindeer, became the main focus prey. The hunt of these smaller, possibly semi-residential groups of animals made a different type of hunting camp possible on which additional tasks such as lithic raw material exploitation could be performed besides the hunting and butchering of the prey. However, shortly before the onset of the Lateglacial Interstadial, the proportion of smaller mammals was already decreasing. This change was possibly due to increasing alimentary security based on the development of more mosaic landscapes.

Some changes in the archaeological record appeared around the onset of GI-1d when cold and dry conditions returned, probably initiating the aeolian dispersal of coversands north of the study area (**fig. 42**). Perhaps due to the more severe impacts occurring northwards, no changes were found in the vegetation or faunal record. In the Central Rhineland assemblages, the quality of the acquired raw materials became more variable as well as the knapping style, whereas the variability in the function index decreased. Furthermore, connections over very long distances and those from the Central Rhineland to the Southeast were no longer identifiable. Together these changes suggest a deterioration in the standards of the well-connected Magdalenian groups. Nevertheless, all these changes are closely related to the end of the age range of the south-western area of Gönnersdorf and the beginning of the age range from the upper horizon of Andernach but no unambiguous equivalence in records from northern France. In the following centuries, changes again appeared more gradual.

A more gradual change in the climate record was the onset of the Lateglacial Stadial. Some changes were possibly related to this climatic fluctuation but the poor database at this transition makes such correlations difficult.

Analysing the plot in regard to the vegetation record, a non-congruent reaction to the climatic developments was highlighted. For instance, the transitional zone from a grass steppe to a light forest environment preceded the onset of the Lateglacial Interstadial in the isotope records of Greenland. The onset of this vegetation zone coincides with the settlement of the south-western area of Gönnersdorf and a possible reestablishment of long distance relations if the jasper pieces from this area were attributed to this later occupation. Moreover, approximately concurrent with this period, steppe bison disappeared from the archaeological inventories and was substituted by aurochs. This correlation of the presence of two differently adapted spe-

cies with changes in the vegetation seems unsurprising and can be interpreted comparably to the decreasing presence of reindeer. In addition, decreasing proportions of smaller mammals in faunal assemblages correlate with this onset and raises the question of whether plants were substituted for the smaller mammals. The preservation of vegetation remains from this period is poor and makes a pursuit with this hypothesis fruitless for the moment. The next change in the vegetation record was the wide-spread establishment of pioneer forests. This change was concomitant with several changes in the archaeological record such as the abandonment of the regular use of organic hammers in the knapping process, the establishment of generally diverse faunal inventories, the disappearance of clear evidences for conforming to a »nothing is wasted« ideal, the appearance of more uniform site types, proportions of smaller mammals becoming almost insignificant, the final disappearance of reindeer from the study area, the dominant acquisition of qualitatively poorer raw materials, and a decreasing importance of horse. However, the burial sites found in the Central Rhineland also originate from the onset of this period. The majority of change to a more boreal fauna occurred shortly before or possibly in many places concomitant with this vegetation change. Changes accomplished shortly before the shift in the vegetation were mainly relate to a new succession in the faunal record. For instance, red deer became a dominant species with the accomplishment of the faunal shift. At approximately the same time, the more suburban use of space on a site appeared and, in particular, the high variances of length and thickness of the LMP are very probably related to this faunal change. The earlier onset of the transitional fauna zone was also closely related with the appearance of burnt faunal material. The later establishment of stable light forests in the study areas was only accompanied by the appearance of wild boar in the assemblages from the Central Rhineland and northern France.
The numerous limits observed around the transition to a generally more forested environment were possibly the result of a temporally close occurrence of changes in the fauna and plant environment as well as the rapidity of this establishment after the cold phase within the early Lateglacial Interstadial. Further changes occurred within the 250 years following the establishment of first pioneer forests with a more boreal fauna and, thus, almost 75 % of the observed attributes are affected by these changes marking this as the period of major change in the archaeological record. Besides the number and tempo of changes, the quality of the changing patterns increases. This higher quality can be established by relating the analysed attributes to the provisional hierarchy of archaeological taxonomic units (**tab. 62**). Moreover, after this period the values typical for FMG are generally established.

In summary, the number of recorded differences made the establishment of a general development in the archaeological record possible and, besides methodological reasons, the behaviour and, in particular, the resilience of the Magdalenian provides a plausible explanation for the observed succession of these not necessarily related developments (see Discussion, p. 565-574).

DISCUSSION

STUDYING CHANGE

An initial aim of this project is to examine the reorganisation of social systems in the context of climatic and environmental change. This aim was frequently approached in archaeological analyses (Weiss/Bradley 2001; Haug et al. 2003; Tainter 2008, 44-45; Bettinger/Richerson/Boyd 2009; Langlais 2011; Bradtmöller et al. 2012) but fragmented data often permitted different, occasionally contrasting interpretations.

For example, in the Late Pleistocene, the transformation from the Late Magdalenian to the Azilian, the so-called Azilianisation, was regularly the subject of conflicting assumptions: The reflections of the abbot Henri Breuil about the French late Upper Palaeolithic record resulted in the statement: »Revolution, the Azilian is one«[53] (Breuil 1913, 216). Based mainly on the same record, Denise de Sonneville-Bordes came to the contrasting conclusion that the replacement of the Magdalenian by the Azilian was a progressive passage (Sonneville-Bordes 1966). These two different interpretations reflect different perceptions of historical developments as either revolutionary or evolutionary processes. The two types differ in tempo and mode in which variations appear: In a revolution, variations occur in a quick, step-wise transition, whereas slow, gradual transformations are considered as evolutionary processes (see Introduction, p. 1-5).

Contrasting perceptions of the process of change were also a subject in biology focusing on the presence of major revolutionary processes in a generally gradual process of evolution (Simpson 1944; Eldredge/Gould 1972; Eldredge et al. 2005). As important factors of identifying and defining these revolutionary periods were named: temporal resolution of the record and taxonomic level on which the research was focused (Thomson 1992). In ecology, different tempos were suggested to be influenced by the different levels of complex adaptive systems and these different tempos drove adaptive cycles but also sustained the complete social-ecological system in making it resilient (Holling 2001; Holling/Gunderson 2002).

Regarded as another social-ecological system, human behaviour should also be studied with varying focus of research to be able to study change. For example, changes in individual behaviour appear constantly (cf. Eerkens 2000; Eerkens/Lipo 2005) and small groups can adapt quickly to short-term local changes such as emigrating from areas affected by volcanic eruptions but return to these areas as soon as the surrounding appears again habitable (cf. Grattan 2006). Besides transformation in a behaviour, also appearance and disappearance can, hence, provide meaningful observations. Nevertheless, a common behavioural repertoire such as expressed for example by Magdalenian art was observed in the archaeological record over several millenia. Consequently, comparable to social-ecological systems, slow and fast processes also occur in human behaviour.

Nevertheless, the time frame has to be chosen deliberately. For example, a study of the period shortly before and after the above exemplified eruption would result in the detection of a significant change in the human behaviour (presence, then absence). In contrast, a prolonged time frame after the eruption would make the observation possible that the emigrants return. Comparably, constant but small variations in knapping techniques result in a long-term change (Hamilton/Buchanan 2009) that is only observable in a sufficiently long time frame. Moreover, these examples lead to another important question: What is perceived as change?

53 Translated by the present author. The original reads: »Révolution, L'Azilien en est une«.

A comprehensive answer cannot be provided by this study and for an often used »end of civilisation« scenario to raise awareness of the finality of civilisation, including the one's own, the Lateglacial record appears not to be a good analogy. To quote a popular song: »It's the end of the world as we know it« (R. E. M. 1987, Document) but life went on. The established FMG behavior of the Lateglacial Interstadial was not the end of behavioural development nor was the Late Magdalenian a beginning of this development. This study comprises only a short extract of human behavioural history assumed to encompass a single adaptive cycle of a social system.

However, the example of different perceptions of this process in the archaeological record from southwestern France helps to provide an interesting point to an answer to the above posed question: The dataset of Denise de Sonneville-Bordes was more fine-grained than that of abbot Breuil. Moreover, his previously mentioned quote continues: »No more figurative animals in art, only paintings on pebbles and walls with geometric or schematic elements. Revolution in the work of bone and deer antler«[54] (Breuil 1913, 216f.). Art can be regarded as a communicative expression and to be able to understand the content correctly, transmission of information is necessary, in particular, when abstraction is involved such as in schematic and geometric signs or letters of the alphabet. Thus, Breuil based his interpretation on a change in the social transmission of information. In contrast, Denise de Sonneville-Bordes based her considerations mainly on statistical variations of the lithic inventories and regarded the typology of organic barbed points also as evidence for the gradual development of the Azilian from the Magdalenian (Sonneville-Bordes 1966). Her analysis became possible by the greater focus on details. Moreover, this focus was set on variations in the individual manufacturing of equipment and the use of this equipment at various sites. As a result, the contrasting perceptions can be explained by different resolutions and a focus on different levels of human groups. These differences revealed fast processes in smaller groups and slow ones in larger entities. Apparently, both analyses provide valid information but only if the results were integrated in a common analytical system. Consequently, a study of change must consider along with the time frame the level at which a social system is studied to distinguish developments that resulted in faster, gradual and slower, more step-wise processes.

These processes forming tempo and mode of the Azilianisation became particularly important when this development was assumed as an adaptive response to the rapidly changing climate and environment of the Lateglacial (Bosinski 1989). Climate as a globally effective trigger has been previously considered as a motor for large-scale changes in biological communities (Firbas 1939; cf. von Koenigswald 2002). Even though triggers of change and the influence of these triggers on the process of change are important subjects of research, also in archaeology, the connection of potential triggers to changing parameters remains difficult to prove in prehistoric records due to the often uncertain interrelation, the inconsistent presence of chronological indicators, and the decreasing temporal resolution.

Therefore, time and the progress of time are important factors in a study of change. Even though methods of chronological attribution and dating were constantly refined in the last century (Weber/Grimm 2009), time remains a difficult topic in the archaeological record (Crema/Bevan/Lake 2010). In particular, preservation of high-resolution data generally decreases with time, whereas susceptibility to contamination increases for this data. The often demanded need for carefully cross-checking between all chronological indicators (Johnson 1952, 101) and of establishing a relation between the dated material to the archaeological record (Dean 1978) became particularly apparent when precision of radiometric dating results was increased. In addition to the contextual imprecisions, radiometric series also demonstrated the general possibility of meth-

54 Translated by the present author. The original reads: »Plus d'art animalier, seulement des peintures, sur galet et sur paroi, d'éléments schématique ou géométrique. Révolution dans le travail de l'os et du bois de Cerf«.

odological failures. Therefore, all methods should be considered critically concerning the potential failure in the application as well as in the results. For example, apparently reliable (calibrated) radiometric dates also require testing against the environmental context from which the samples were taken and the reliability of the environmental context should equally be confirmed. All reliable chronostratigraphical indicators from a site should be consistent in a synthetic chronostratigraphic position. However, besides reliable results from natural sciences, the archaeological record has to be resolved sufficiently to understand site formation processes and the position of dated material (dated event) in relation to human activities which shall be dated (target event). Therefore, spatial distributions are documented in detail and the most meaningful material is usually selected for radiocarbon dating or other types of radiometric dating (cf. Richter et al. 2009). However, often many human activities remained temporally indistinct. For example, even though sub-assemblages were occasionally clearly identifiable, the chronological relation remained uncertain (cf. Gelhausen 2011b; Jöris/Street/Turner 2011). In other cases, a longer chronological use appeared plausible but the material accumulation remained indistinguishable (cf. Julien et al. 1999; Sensburg 2007). Thus, resolutions of site internal chronologies are limited by the spatial differentiation of the material and its attribution to a single process (Cziesla 1990; Sensburg 2007; Sensburg/Moseler 2008; Gelhausen 2011b). The resolution of archaeological material can subsequently range from site level to intra-site and/or intra-concentration level to a level of single activity sequences performed at a site. As indicated above with the different perceptions of the Azilianisation, the different resolutions of the archaeological record can provide various pieces of information but the resolution should be selected appropriate for the aimed project.

Finally, if a very high temporal resolution was created and a close temporal relation of variation in a potential trigger and changes in another parameter was established, this observation does still not prove that the potential trigger has provoked changes in the studied parameter. These changes could also represent coincidental processes or both, trigger and parameter, responded concomitantly to another factor. Therefore, an attempt of plausibly connecting trigger and responder should be made.

A precise terminology is necessary to relate different parts of prehistoric records at variable levels, in varying resolutions, and from different areas and to communicate these relations as a contribution to further refine studies of these records. The use of defined terms should usually help a reader to follow the presented argument, but in the present study definitions became apparent as playing an important role in causing misconceptions. For example, in the Weichselian Lateglacial records, it often remains a matter of preference which Bølling definition is applied, which definition for the onset of the Lateglacial Stadial, or which attributes to distinguish an Azilian point. To prevent these sources of potential misunderstandings, attempts such as the one for pollen-chronological sequences (Litt/Stebich 1999) that defined hierarchically structured stratotypes and gave increasingly detailed indicators of the defined sections are a good beginning to come to a general understanding. However, this consensus was partially neglected due to local traditions, inapplicability in the local archives, or non-transferability to other proxies. Nevertheless, without a more common consensus, further uncertainties and misunderstandings will appear and prevent comparisons at different scales and of different disciplines. Yet, only this broad comparison can help reveal general developments concomitantly with locally differing movements.

This scientific need for a large consensus to simplify communication is not very different from the methodological topic and the subject of the present study. With a further refinement of the archaeological taxonomic units, a methodological attempt was made to systematise the archaeological record to create a common reference frame in which archaeological units can be compared at various levels. The subject of this study, the Lateglacial record, also reflects the effort to keep a wide-spread community united by common expressions and standards. In some cases this effort was successful, in others it was not. In future, the present study will hopefully be considered among the former.

In summary, to produce meaningful results, studying change requires precision. This precision has to be effective in different parts and levels of the analysis. For example, previously defined terms should be used exactly or rejected to prevent unnecessary confusion if they are not useful. Comparably, the resolution of chronology as well as the analysed part of the record should be accurately clarified to build a solid, though doubtlessly improvable fundament for interpretations. This dissected way of building an analytical basis should prevent misconceptions due to foci on different scales and reveal differences and similarities at comparable levels of the records. Moreover, the knowledge formed in this way could also permit identification of interdependence of the various levels and, as with classic and quantum physics, detect different mechanisms. Thus, a more profound understanding of the development of human behaviour becomes possible. In fact, this study of change already revealed the ability of humans to vary their behavioural repertoires. Exposed to a selective process, this variation helps to abandon a previous standard and to manifest another, better suited variation as a new standard. Thus, studying behavioural change reflects human behavioural adaptation and, thus, helps to understand human evolution.

REVOLUTION OR EVOLUTION?

In the previous chapters, several lines of evidence were collected (see Material, p. 7-244) and compared (see p. 293-564) to consider different assessments of the reorganisation of the Lateglacial social-ecological system in north-western Europe. A detailed schedule of the process of change from the Late Magdalenian to the FMG was established based on material from the Central Rhineland, southern Belgium, and northern France. Analytical values of the studied sites were given and positioned chronologically to reveal the timing of changes in the archaeological assemblages. Then, the observed changes were set in a chronological relation to one another and the climatic and environmental developments and some suggestions about functional changes were given (see Results-Chronology of changes, p. 559-564). After this presentation of the chronological succession, identifying possible mechanisms and causes of the Lateglacial social reorganisation is approached and an interpretation of the results is formulated. Therefore, observed changes are differentiated according to the level to which these changes refer and reasons for these change in relation to changing environments are considered.
According to the present analysis, the process of change in human societies during the Lateglacial occurred on many levels such as the molecular level of LMP dimension, the micro-level of formally retouched artefact groups disappearing, the appearance of the MfCM on the meso-scale, and finally the observable variations in settlement behaviour. In general, the tempo of this process was gradual but a relatively short period was observed in which an accumulation of changes occurred within a few centuries. The appearance of the typical FMG mode of life as defined previously with a greater flexibility in the various behavioural elements such as the raw material acquisition, the blank production, the variety of formally retouched artefacts, in particular of the LMP, and the dimensional changes of these implements as well as the more ephemerally appearing settlement organisation was not related with the prominent climatic event at the onset of the Lateglacial Interstadial. Thus, among the possible triggers for the observed behavioural modifications, climate change in the modern sense of global warming does not appear as a directly effective and/or very important factor. Nevertheless, at the onset of a short but severe cold phase (GI-1d) some 600 years after this rapid warming the tempo of changes appeared accelerated but only some 200 years later, during the subsequent warm phase (GI-1c), a cluster of rapid behavioural changes occurred concomitantly with the establishment of forested environments, which were persistent for several centuries afterwards. This delayed reaction and the

ostensive dissent of the process being an evolutionary as well as a revolutionary process can be explained by the human disposition to form dynamic, complex, and adaptive social networks as a survival strategy and by the developmental history of the original network.

During the Magdalenian resettlement phase, Europe was still a sub-marginal landscape which recovered gradually from the LGM. This cold maximum and the main retreat into southern refuges around 28,000 years ago had resulted in a significant decline of human populations in Europe (Torroni et al. 1998; Pereira et al. 2005; Soares et al. 2010). However, some 10,000 years later, the human groups of the Iberian/south-western French refuge had recovered enough to expand during the Late Pleniglacial into the continuously growing territories in Northern Europe. These territories grew with the regression of ice-sheets and the subsequent disappearance of continuous permafrost. A gradual recolonisation by pioneer vegetation followed by first arctic animal species appeared. However, this recolonisation depended on various factors such as soil development and shelter. Consequently, a continuous, wave-like advance seemed improbable. The recolonisation must rather be visualised as a patchy wiggling or weaving movement, most probable along larger water courses which provided sufficient security and stability of water supply in the arid landscape. This uneven movement resulted in a landscape formed by different biotopes in which resources were unevenly distributed. Assuming that some of these resources were necessitated for a human survival, a clustered distribution of sites can be assumed (cf. Butzer 1982) and is, in fact, still identifiable around 16,000 years cal. b2k by the various Late Magdalenian centres in the Thuringian Basin, the Central Rhineland, or the Paris Basin. However, the intensified aridity during the Heinrich 1 event caused possibly a faster ice sheet decline and strengthened the aeolian deposition of fine-grained sediments. Although loess deposition, possibly in the form of loess storms, affected parts of the landscape at the onset of the studied period, the examples from the Late Magdalenian inventories found within these aeolian deposits prove that this climatic and environmental nuisance formed no restriction to the expansion process. In particular, the almost year round occupation of Gönnersdorf (Street/Turner 2013) indicates that this possibly seasonal phenomenon of winds loaded with sediment was no limiting factor. However, a low population density (cf. Bocquet-Appel et al. 2005; Kretschmer 2012) in combination with a persisting climatic and environmental instability required large networks to secure the survival of clustered groups migrating into the newly available landscapes and the stability of the human metapopulation in these insecure surroundings (cf. Hanski 1998). Various strategies to reduce risk and uncertainty and to compensate for stress caused by minor variations of their environments were applied to ensure survival across critical periods (cf. Stein Mandryk 1993). For instance, these groups had to conform to a very conservative lifestyle to remain connected and able to communicate across these large areas to allow for a sustenance of security networks. Thus, survival referred in this period to single individuals as well as the community and, thus, these strategies formed the resilience of the Magdalenian.

In particular, the subsistence strategy was modified during the Late Pleniglacial to provide the largest possible nutritional security which was expressed by a generalist approach (Gaudzinski/Street 2003) and first steps of the broad spectrum revolution in the diet reflected in the incorporation of higher proportions of smaller mammals (see **tab. 86** and p. 543-546; cf. Munro 2003; Zeder 2012), fish, birds (cf. Street/Turner 2013), and, perhaps, marine resources (Pétillon 2008a; Langley/Street 2013). Furthermore, evidence shows that the Late Magdalenian hunters exploited their prey to a degree comparable to an ethnographic »nothing is wasted« ideal (cf. Pasda/Odgaard 2011). However, this intensive alimentary exploitation and the partially seasonal availability of resources required also a strictly organised mobility behaviour in the landscape. Mobility patterns were additionally shaped by finite or limited resources such as wood. Perhaps, the poverty in wood during the arid Heinrich 1 event resulted in a complex composite weapon system which could save this resource. Nevertheless, hunting equipment which was used at a distance requires a relatively high precision and, thus, the organic and lithic production were adapted to serve high quality products of a consistent quality and regularity. To guarantee this precision on which the hunting success and, thus, the alimentary security of the group partially depended, humans have to be trained (cf. Pigeot 1990; Schmider 1992e) and regularly monitored to prevent variation (Eerkens/Lipo 2005; Hamilton/Buchanan 2009).

These compensating behaviours paid tribute to an enormous need for security (Maslow 1943) in these human groups and resulted in a remarkable resilience. For hunter-gatherers, a predictable environment contributes to a feeling of security because the predictability allows for planning and preparation (Rowley-Conwy/Zvelebil 1989). Thus, predictability of the environment is an important factor for hunter-gatherer behaviour (Halstead/O'Shea 1989b). However, as demonstrated by the Late Magdalenian, hunter-gatherers can also cope with variable environments, for instance by allowing for more variation within their behavioural patterns (Jochim 1991). This ability to adapt was also considered as an important factor in the resilience of a social-ecological system (Walker et al. 2006).

This pattern is also seen in the Late Pleniglacial where, besides the high conformity, the ability to adjust to short-term variation was sustained. This ability was reflected by the sustenance of almost continuous variation on molecular and micro-scale levels such as variability in the use of hunting equipments (% LMP), in the lithic equipment (Simpson diversity index of formally retouched artefacts), and the provisioning of lithic resources (% and number of cores, exploitation of cores) as well as a variety of sites (site types, function index). In addition, the quick adjustment was also observable on a meso-scale level when unfamiliar species such as red deer, elk, and wild boar were incorporated in the prey spectrum soon after their arrival in the sub-areas. The material from Bois Laiterie and the south-western part of Gönnersdorf reflect this adaptability of the Final Magdalenian.

The seasonally variable availability of the newly appearing faunal species was low and the increasing availability of a variety of larger mammal species allowed for an occasional neglect of the strict »nothing is wasted« ideal. Moreover, the increasing surplus created wastes which could attract scavengers and vermin. Thus, to keep regularly visited locations free of attractive faunal residues, burning of these wastes was a possible solution which first appeared during the transition towards the Lateglacial Interstadial. However, the occurrence of the burnt material could also indicate the appearance of a different method of extracting grease but since the variety of the burnt faunal remains appears not very selective (Bignon/Bodu 2006), this possibility is considered less probable. Another possible explanation for the appearance of burnt instead of fragmented bones could be provided by the seasonal evidence. According to seasonal indications, the sites where the burning of bones occurred were tendentially used in winter. Thus, the meagre winter/dark season game was possibly no longer used for nutritional provision and instead the summer provision was consumed. Nevertheless, according to this scenario the nutritional stress in winter had certainly decreased. Possibly, this increasing ability of the applied subsistence strategy to provide for the dark season by surplus created in the summer months was also detectable in the appearance of super-specialised sites which often resulted from recurrent horse hunting episodes in favourable valley situations. This type of site also led to a variation on the mega-scale by the creation of a less restrictive organisation of space use at sites around 15,000 years cal. b2k, perhaps, due to the decreasing loess deposition. Combined with shorter intervals of returning to the site, the smaller sediment cover permitted the reuse of material previously transported to the site. This reuse led to a decreasing need to transport additional materials to the site. However, on the tendential winter sites, a classic Late Magdalenian organisation was preserved (Jöris/Terberger 2001). These locations were still clearly structured and, thus, appeared not as intensively used as the tendential summer locations. Although some Late Magdalenian occupations seemed similarly tidy such as Gönnersdorf IV, recurrent settlement dynamics overprinted the originally tidy structure in the majority of larger Late Magdalenian concentrations (Terberger 1997; Sensburg 2007; Sensburg/Moseler 2008).

These indicators mainly from the faunal and the spatial record suggest that the Magdalenian hunter-gatherers, which were adapted to the harsh Late Pleniglacial environments, reacted to an apparently more favourable period beginning in the second half of GS-2a. In this favourable period, non-conformity to an intense exploitation strategy was not punished and/or sufficient surplus could be produced to provide for harsher seasons. Subsequently, the Late Magdalenian was modified by adaptations to a more favourable and less punishing environment, mainly by allowing for a wider flexibility in the behavioural repertoire. Moreover, the MfCM and the Early Azilian were shown to be different on various scales but the remarks on subsistence strategies and seasonality of these sites suggest an interdependence of these two complexes.

Furthermore, with the increase of vegetation productivity and the appearance of the first shrubs during GI-1e, previously limited resources such as plant fibre, plant tars, and wood became available in increasing amounts. This change allowed for a more profound change on the molecular scale in the hunting equipment (LMP types – shapes and dimensions). The effectiveness of these new systems seemed to slowly replace the more complex organic systems. Perhaps, the gradually decreasing availability of antler material further contributed to this decline. With the continued presence of the vegetation resource, the projectile system could be altered which was possibly also detected on the molecular level of the variance of LMP thickness around 14,200 years cal. b2k.

However, the return of harsh climatic conditions during GI-1d, including the deposition of some massive coversands in parts of Northern Europe (Kolstrup 2007), were related in the record with the definitive disappearance of long distance connections, in particular the relations from the Central Rhineland to the areas in the south-east, and a further increase of the LMP variability (LMP types – shapes and dimensions). The former reflected changes in the socio-economic networks between different regions and can therefore be related to the meso-scale. Shortly after the onset of GI-1d, the appearance of small, centrally organised concentrations and the decreasing function variability revealed also changes in the spatial behaviour and, thus, referred to the analytical mega-scale.

Perhaps, the dispersal of more coarse-grained material hindered the travel in some regions and possibly led to the emigration from the most intensely affected areas. In addition, with the previously seen greater flexibility of the behaviour, the conforming to common standards which sustained large social networks must be questioned. With the potential population increase, new ideas about hunting equipment could have been introduced and/or an adaptation to optimise the hunt on returning colder adapted species became useful. Nevertheless, population pressure due to emigrants in the areas unaffected by the deposition of coversand was not observable by a greater density of sites or an increasing exploitation intensity of the prey. In fact, the number of sites decreased and also the intensity of their use indicated no apparent increase of the number of inhabitants. These indicators rather suggested a decreasing population at the end of GI-1d. Occasionally, the diversity of the faunal record was increased but it seemed that the modified Magdalenian hunters were no longer aware of the highly conservative behavioural patterns to compensate for nutritional stress. During this transition from GI-1d to GI-1c_3, the graves in the Central Rhineland and the period with the highest density of changes relating to different scales occurred. The evidence from Irlich (Orschiedt/Berg/Flohr 2011) suggested that malnutrition might have affected these hunter-gatherers and occasionally favoured fatal diseases or resulted in starvation. Nevertheless, these graves show some classic Magdalenian behaviours such as the use of ochres, figurine art, and the deposition of complete bodies in an earthen grave (cf. Wüller 1999). However, these behaviours were previously unknown from this region. Perhaps, a wish to return to old behavioural recipes was thereby reflected that would prove the severity of this situation. The engravings at Gouy were possibly an expression of a comparable wish to return to the »old ways«.

Studies of complex adaptive systems in modern geography made a distinction between planning strategies of just-in-time and just-in-case (Alfasi/Portugali 2004). If this approach is transferred to hunter-gatherer groups, the transmission of various behavioural recipes without direct necessity is a just-in-case strategy and clearly helps to compensate for insecure environments. Among modern Evenk in Siberia, Ole Grøn observed that these hunter-gatherers knew about cyclic developments in their environment and had a concept of the interactions which he assumed comparable to the modern concepts of ecology (Grøn 2012). This knowledge was due to the lack of scientific terms transmitted in a spiritual terminology and the shaman of the Evenk kept this knowledge. A comparable case of a transmission of ecological cycles by the use of myths was previously reported from the sea people (*tareumiut*) and the inland people (*nunamiut*) of the Alaskan Iñupiat Eskimos (Minc/Smith 1989). Images could play an important role in these transmission chains. The very naturalistic engravings of animals on the slate plates of Gönnersdorf and Andernach are, therefore, explicable as relevant indicators for a transmission strategy. In particular, since some of these images showed species that were almost extinct at this time such as woolly rhinoceros or mammoth (cf. Stuart 2005; Bosinski 2008) or species which could only be found at a distance of several hundred kilometres from the sites such as seals (Hansen 2006), these potential instructions were certainly not meant for an instant use near the sites. In this context, the Late Magdalenian seemed to apply a just-in-case strategy that together with the naturalistic images became infrequent at the end of the Late Pleniglacial and during the early Lateglacial Interstadial. Thus, the transmission of behavioural recipes without direct use might not have been sustained in the modified Magdalenian, possibly due to the lack of occasional need and punishment of non-remembering.

At the end of the cold period, light forests rapidly expanded in the study area but the environment and the climate remained unstable for the next centuries. Probably, this instability led to the inability to successfully adapt to this environment because the important factor of predictability (Halstead/O'Shea 1989b) was difficult to establish. A possible way to compensate for changing resource availability even within yearly rounds is the change of group sizes (cf. Stein Mandryk 1993). Thus, were the hunter-gatherer of the FMG perhaps organised in smaller groups that roamed the area? Smaller groups could not split up easily into special task groups and, hence, the differentiation between the sites would decrease. The result would be a more collector-like settlement system (cf. Binford 1980) with some agglomeration camps which allowed for the exchange of information at least with the neighbouring groups. Seasonal revisiting of sites with this undifferentiated site type would produce the sub-urban type of space use.

Numerous changes occurred in the 200 years following the forestation. These changes seem to reflect the collapse of the modified Magdalenian way of life. However, based on the adaptations introduced previously, this collapse gave rise to the fully established FMG in the Central Rhineland and, probably, the Paris Basin. In the north of France, another set of adaptations was chosen on the molecular and micro-scale to create the new standard behaviour. Thus, in addition to the absence of evidence for long distance relations, different behavioural patterns are observed in the study area reflecting the trend towards a regionalisation in this period. Even though the lower horizons at Hangest-sur-Somme III.1 and Conty show differences in the material remains, the faunal assemblages indicate that the regionalisation was rather caused by a different function than a completely different way of life.

However, after this long period of gradual change over at minimum 2,100 years, why and how could a collapse of the Magdalenian and the emergence of the FMG appear so rapidly within less than 500 years? According to the social brain hypothesis, formation of beneficial alliances such as the Late Magdalenian occurred as an answer to social competition and led to the development of cooperative behaviour (Byrne 1996; Charlton 1997; Dunbar 1998). The members of this cooperation agreed on norms and limits in the behaviour to which they conformed for a general benefit. Although some variation within these behavioural limits were allowed and essentially needed to remain flexible concerning the natural demands, some social restriction for complying with these limits are necessary to let the majority consider the alliance or, more precisely, the following of the behavioural recipe as generally beneficial. Such a balance of equality is particularly required in an environment with restricted resources where competition and accordingly beneficial alliances can become essential for survival. In modern politics and environmental studies, the reason for this conservative behaviour would be called risk mitigation or management because these groups had to survive in an almost empty social space that faced a dense and rigid environment. Resilience is the outcome of this flexible conformistic behaviour. Large distances needed to be crossed to sustain securing social networks. Therefore, it is unsurprising that the highly adaptable but also highly conformist social system of the Late Magdalenian was the last almost pan-European hunter-gatherer entity which shared a common set of traditions from Portugal in the west to eastern Poland and from the Mediterranean to the limits of the North European Plain.

Breaking the limits of alliances, for example due to a poverty of resources, this behaviour is either isolated or punished by other allies or causes the collapse of the existing agreement. Thus, single, small infringements to an agreement can perhaps be tolerated by the community that needs to remain particularly flexible in dispersedly settled areas. Therefore, an increasing diversification and, thus, decreasing pan-European conformism revealed by the archaeological material was probably not regarded as dangerous in the subsequent periods of the Lateglacial as well as the Holocene. Nevertheless, alliances break up rapidly after a threshold of behavioural flexibility is surpassed. This threshold is when a new alliance offers to be more convenient or competition promises to be more beneficial than allying (cf. Han/Pereira/Santos 2012). However, a new alliance can only fulfil its promises if it is supported by a substantial part of the community members. Thus, the emergence of new alliances and, consequently, new rules for socially selected behaviour often occur in a phase transition-like manner (cf. Gavrilets/Duenez-Guzman/Vose 2008) meaning that some gradual changes occurred prior to an accelerated collapse of the old alliance or agreement and the emergence of the new alliance. Comparably, resilience theory also suggested changes prior to the collapse phase (Abel/Cumming/Anderies 2006).

Furthermore, the Lateglacial transition was compared to people adopting to sub-marginal landscapes before. Considered from the point of view of a Late Magdalenian hunter-gatherer: Was the environment becoming more sub-marginal or was the transition into the Lateglacial Interstadial rather perceived as increasing prosperity? And what happens to hunter-gatherers confronted with a rich and less punishing environment than before? According to the present results, the transition rather seemed initiated by increasingly decadent behaviour in a landscape producing significant surplus for a society which was highly adapted to intense faunal exploitation. The lithic assemblages of Bad Breisig, sector 2 of Le Grand Canton, and concentration III of Gönnersdorf showed that humans tend to exploit resources that appear to be plentiful and easily accessible without longer transport distances in a wasteful, decadent manner independent of their environmental and behavioural background.

Based on this reasoning, interactions of humans with their surrounding ecosystem can be assumed to shape and limit the resilience of these often inert human systems. The resilience and the ecological tolerance range reflected the ability of past societies to adjust their policies of population growth and economic demands in response to climate and environmental change.

In general, the results of this study are comparable to the findings of an analysis about the impact of volcanic eruptions on past societies (Grattan 2006). Minor impacts were shown to have major effects on societies which were already destabilised, whereas major impacts had only little or no impact on resilient communities. Comparably, at the onset of the Lateglacial Interstadial, the impact of climate and environmental change was compensated by the resilient Magdalenian groups through minor adaptations for almost an-

other millenium. However, continuous uncontrolled variation during this period resulted in a destabilised alliance which could no longer persist during less severe climatic and environmental fluctuations.
In conclusion, climate change, in particular global warming, is no disaster for well connected societies that are flexible enough to adapted to changing conditions. However, if social agreements are already disintegrating the impact of minor changes can have major consequences. Thus, also resilient societies facing a climate change must consider long-term results of their small-scale adaptations and strengthen social cohesion to prevent a later failure due to minor external triggers.

SUMMARY

AIM OF THE PROJECT

The process of reorganisation in a social system is examined in the present study with a particular focus on the impact of climatic and environmental changes on this process.
In social-ecological resilience theory, reorganisation of social-ecological systems was described as one phase in an adaptive cycle which is framed by a proceeding phase of release or decline and a succeeding phase of exploitation or growth and contrasted by a conservation or consolidation phase (Holling 2001; Abel/Cumming/Anderies 2006; Rosen/Rivera-Collazoa 2012). Thus, in this terminology reorganisation equates the collapse of a system. In the present study, this phase is termed collapse, whereas reorganisation is understood as the complete process between two consolidation phases. The transition from an exploitation to a conservation phase was described as a slow progress in contrast to the other transitions which were considered as rather fast changes. A comparable discussion of a variable tempo in the process of evolution was led in biology about macro-evolutions (Simpson 1944) and punctuated equilibriums (Eldredge/Gould 1972; Eldredge et al. 2005). In this discussion, the temporal resolution of the records and the referenced scale of detail were identified as important factors to distinguish between the gradual development of evolution as a standard and »revolutionary periods« in which changes seemed cumulated and appeared rapidly (Thomson 1992; Mayr 1996). Physical adaptation is considered as a rather slow process. In contrast, communication, that maintains social networks of common behavioural patterns, enables these networks to change much faster and, therefore, also to be changed in a revolutionary process.
To be able to distinguish between slow evolutionary and fast revolutionary processes in the reorganisation of social systems (cf. Holling 2001; Walker et al. 2012), human behaviour has to be studied in a sufficiently fine-grained chronological framework. Changes in climate and environment were considered as external drivers of human behavioural change. To understand the impact of these external drivers on human behaviour, true shifts in those external systems such as alterations between stadial and interstadial periods should be selected as case studies. Therefore, only studies of Pleistocene societies make an understanding of the behavioural alterations in times of true climate change possible. Thus far, temporal resolution necessary for analysing an adaptive cycle in relation to such unambiguous climate changes is only reached at the end of the Pleistocene.
In north-western Europe, the rapid climatic and environmental changes during this Weichselian Lateglacial were related to the so-called Azilianisation process (cf. Bosinski 1989). This process refers to the transformation from the Late Pleistocene Magdalenian societies to the Azilian groups of the Lateglacial Interstadial. Based on the south-western French record, the Azilianisation was previously perceived contrastingly either as revolutionary process (Breuil 1913) or as evolutionary process (Sonneville-Bordes 1966). In this project, a detailed chronology of this process is established based on the archaeological record from north-western Europe to examine the progress of change. In particular, the two consolidation phases of the Late Magdalenian and the *Federmesser-Gruppen* (FMG) represent an ideal case study of an *in-situ* adaptive cycle in this area. North-western Europe was deserted during the Last Glacial Maximum (LGM) and afterwards only the south-western European Magdalenian expanded into this unpopulated area. A sustainable settlement was first established there during the Late Magdalenian. In this wide-spread, uniform substratum of the Late Magdalenian, acculturation processes can be excluded as causes of change. Consequently, climatic

and environmental changes were the only major external drivers. Therefore, the archaeological record is contextualised in the climatic and environmental developments of the Weichselian Lateglacial to examine these developments as potential drivers for changes in human behaviour. Further expansion into Northern Europe and increasing regionalisation subsequently led to the development of distinct traditions. Exchange between these traditions represented another potential external stimulus. Hence, the development from the Late Magdalenian to the FMG can be regarded as a last adaptation process undisturbed by acculturation influences.

CASE STUDY

Three major parts of the Weichselian Lateglacial record were examined: The climate record that often also served to establish a reliable chronostratigraphy, the environmental record, and the archaeological record of north-western Europe. A special emphasis was put on the chronology of the appearance of changes in the single domains. These changes were compared to answer the main questions relating to the tempo and mode of the social reorganisation as well as the relation of this process to the climatic and environmental development.

Climate and chronostratigraphy

Climate and chronostratigraphy are global systems and, therefore, the records to reconstruct these systems also need to be selected more globally (see Material-Climate, p. 7-30). In this study, these records are Greenland ice-core data from NGRIP in the GICC05 chronology (Andersen et al. 2004; Rasmussen et al. 2006; Svensson et al. 2008; Steffensen et al. 2008; Wolff et al. 2010), deep sea sediment stratigraphies from the Cariaco basin offshore Venezuela (Hughen et al. 2006) and the Porcupine Seabight offshore Ireland (Peck et al. 2007), and mainly European terrestrial archives. Terrestrial archives from outside Europe came only from Chinese speleothems found in the Hulu Cave (Wang et al. 2001) and the Qingtian Cave (Liu et al. 2008) and from Tasmanian Huon pines which provided a dendrochronological sequence (Hua et al. 2009). Further dendrochronological material incorporated in this project was recovered in Central Europe (Friedrich et al. 2004; Kromer et al. 2004) and, in particular, in Switzerland (Schaub et al. 2008b; Kaiser et al. 2012). In addition, several varved lake sediments were used including the Eifel Maar lakes (Zolitschka et al. 2000; Brauer et al. 2008), the northern German varved lakes (Merkt/Müller 1999), the East German Rehwiese sequence (Neugebauer et al. 2012), and two Polish varved lakes (Ralska-Jasiewiczowa et al. 1998; Goslar et al. 1999).

Since the development of climate is globally effective due to the thermohaline cycle, major changes (»events«) can be registered almost synchronously. These events are individually identifiable and can be used to construct a consistent chronostratigraphy to which other stratigraphies can be correlated on a global scale. Besides the global climate system, sub-systems and specific regions exist in which the records are more similar than on a global scale. In this study, the oxygen isotope record from a Greenland ice-core was used as a proxy to create a template for this climatic eventstratigraphy. The Greenland data reflects the climate in the sub-system of the northern hemisphere and, in particular, the North Atlantic region. This record was selected because of its continuity into the Weichselian Lateglacial and its relevance for north-western Europe that is also influenced by the North Atlantic climate. Previously defined limits of the Weichselian Lateglacial climate events were either made on other ice-core records, in other ice-core chronologies, and/or on other proxy data such as the deuterium excess (d, see p. 592; Björck et al. 1998; Lowe et al. 2008; Blockley et al. 2012). Thus, the limits of the climatic events for a template record of this study were newly defined based on oxygen isotope values (see p. 245-247; **tab. 63**). By this comparison, the onset of the Weichselian Lateglacial Interstadial (GI-1) around 14,690 years before the year 2000 A.D. (cal. b2k; **tab. 64**) was determined as the most intense climatic change in the period between 18,000 and 10,000 years cal. b2k (see p. 293-310). The second most intense climatic change was the onset of the recent warm period, the Holocene (GH), around 11,680 years cal. b2k. The onset of stadial periods such as the Lateglacial Stadial (GS-1) appeared more diffuse in this record. The limit between GI-1 and GS-1 was set to 12,770 years cal. b2k but the transition stretched over 250 years. The onset of the sub-event GS-2a was also set with some caution to 16,230 years cal. b2k because it was previously not detected in other proxies (Lowe et al. 2008). In contrast to the interstadial periods (GI-1; GH), stadials showed a great instability with generally larger amplitudes, in particular during the Late Pleistocene (GS-2). Perhaps, this increasing instability made a definition of an onset for stadials so difficult.

In general, observations and limits of the oxygen isotope eventstratigraphy were further confirmed by the records from Polish varved lakes and Chinese speleothems (see p. 319-334). The former are more similar to the Greenland ice-core records indicating the greater influence of the North Atlantic climate on these records and a decreasing influence on sites that are located farther within the Eurasian landmass. However, the chronologies of the Polish records were anchored to other chronologies and, thus, are not independent and continuous. Nevertheless, some interesting differences to the Greenland oxygen isotope record could be identified in these records. For instance, the onset of the Lateglacial Stadial (GS-1) was identifiable as a sharp and rapid event making a more precise definition of this limit possible (**tab. 66**). The Chinese records were radiometrically dated. The advantages of this type of dating is that it is punctual and, thus, independent of disturbances and that it circumvents the accumulation of counting errors that occurs in annually layered stratigraphies such as the Greenland ice-core records or the laminated sediments from the Polish lakes. However, in addition to the air temperature, the Chinese oxygen isotope records were also influenced by hydrology. This factor can offset the reaction of the record to the changes in air temperature significantly as shown by the duration of the Lateglacial sub-events (**tab. 68**). Therefore, dates for the limits in these records (**tab. 66**) should not be used to set limits of the oxygen isotope eventstratigraphy more precisely.

The isotope data of the Central European dendrochronology (CEDC) was also influenced by the hydrology and the onset of the Holocene appeared comparably delayed in this dataset as in the Hulu Cave. Thus far, the CEDC is continuous into the older part of GS-1 (Friedrich et al. 2004; Schaub et al. 2008b). It was formed by tree-ring growth patterns (Friedrich et al. 2001b). The onset of the Holocene determined by this pattern is similar to the onset of the Holocene in the oxygen isotope record from NGRIP. A connection between the CEDC and the German Lateglacial pine chronology (GLPC; Kromer et al. 2004) and the partially identical Central European Lateglacial Master chronology (CELM; Kaiser et al. 2012) was not yet established dendrochronologically due to insufficient data around the onset of the Lateglacial Stadial. Different attempts to bridge the gap were made using sequences of radiocarbon dates provided by Swiss and Tasmanian pine records (**fig. 71**; Schaub et al. 2008b; Hua et al. 2009; Reimer et al. 2009). These attempts neglected the possibility to test the position of the resulting correlation with the position of the Laacher See eruption (LSE) identified in the GLPC against the position of this tephra in other chronologies. If the GLPC and based on the identical parts the CELM are positioned according to the position of the LSE in the record from the Meerfelder Maar, the CELM record overlaps over a short period of time with the CEDC. The sequences from the Swiss and the Tasmanian records are correlative with the radiocarbon sequences of the GLPC and the CEDC in this position. With this provisional connection the dendrochronology became continuous into the early Lateglacial Interstadial. A comparison of patterns of tree-ring growth from this Lateglacial Interstadial part of the dendrochronology with the same part in the NGRIP isotope record showed significant offsets (**tab. 69**). This difference is probably again due to more numerous influences on tree-ring growth such as hydrology and overprinting of global climate by more local conditions. Consequently, a direct correlation of dendrochronological data with isotope records from Greenland in a high-resolution frame must also be considered critical.

Besides the Polish lakes, sequences with laminated sediments were also found in the Cariaco basin, Eifel maars, north-western German lakes, and at Rehwiese in eastern Germany. The Eifel maars, in particular the correlation of the Meerfelder Maar and the Holzmaar, provided an almost continuous chronology over the past 24,000 years. Tephra from the LSE was found in these sequences. The other records were correlated to the CEDC or the Meerfelder Maar record. In general, patterns observed in the Greenland eventstratigraphy were also identified in these records. However, the limits of the climate events were defined on various proxies in the laminated archives such as sediment composition or pollen frequencies. Again, clear offsets were observed between all records (**tab. 71**). In particular, the duration of the periods assumed to be equivalent to the events in the Greenland isotope record varied considerably (**tab. 70**). Changes in the composition of the sediment were usually more similar to the onset and duration of events in the NGRIP isotope record than vegetation data which yielded very inconsistent results. Thus, depending on the analysed proxy, the limits of the events shifted by several decades within a single record (**fig. 31**). The evaluation of a consistent chronology based on these variable proxies is very difficult because observable differences cannot be attributed with certainty to prolonged response times, false correlation, or chronological disturbances such as a hiatus or non-annual deposition of the laminae. Consequently, these records should only be correlated using independent markers such as tephra layers or radiometric dates.

In summary, tuning of climatic records is useful in a coarse-grained chronostratigraphy but it is not advisable in high-resolution approaches due to different response times. Different response times are due to different, occasionally contrasting impacts of main influencing factors, such as temperature and hydrology, and the varying influence of these factors on different proxies. Sequences based on vegetation data suggested that resilience was a strong factor interfering with the construction of a consistent chronostratigraphy.

Nevertheless, the NGRIP isotope template was also supplemented by a tephrochronological dataset (Mortensen et al. 2005). This additional record allowed for a very precise chronological positioning of marker horizons (Turney et al. 2006; Lowe 2011). Tephras are therefore useful correlation points. Some tephras were also identified in the European terrestrial sequences and, thus, allowed for an evaluation of the precision of the different chronologies. However, the Laacher See eruption (LSE) was not unambiguously identified in the NGRIP record, although it forms the most important volcanic marker horizon in Lateglacial Central Europe.

This marker often appeared in laminated sequences of north-western Europe but it was also integrated in the dendrochronologically record. Having positioned the marker horizon of the LSE in the laminated records of Europe and possibly having identified it in the NGRIP tephrochronological record, the dendrochronological record which comprises data for this event can be correlated in this chronostratigraphy.

With the correlation of the dendrochronological data sets using the LSE, the Lateglacial calibration record was automatically reset. In comparison, the proposed position of the dendrochronological calibration data slightly differs from the two most commonly used calibration curves (IntCal09 and CalPal-2007$_{HULU}$). Therefore, a new Lateglacial calibration curve was constructed for the present study (see p. 358-364). The use of this new calibration curve is more appropriate in this project to establish the complete analyses in a consistent chronostratigraphic system of climate and environmental change to which the human changes were related by ^{14}C-dated archaeological records.

Environmental reconstruction

For the environmental reconstruction a variety of datasets, stratigraphies (see Material-Environment, p. 30-48), and dates (see Material-Databases, p. 49-53) were used. In particular, the creation of maps of Lateglacial north-western Europe to visualise the development of the physical geography during the studied time period required a compilation of various data (see p. 253-259). Regression stages of the European ice sheets were taken from the literature as far as possible (Lundqvist/Wohlfarth 2001; Clark et al. 2004; Ivy-Ochs et al. 2008) and interpolated in analogy to the given information for the areas and stages for which these limits are thus far unknown. Furthermore, the rising sea level was selected according to a global dataset (Weaver et al. 2003) and the mid-point of four determined Lateglacial sub-periods (**tab. 47**). For the physical base maps, the NASA shuttle radar topography mission (SRTM) data combined with bathymetric data by the Scripps Institution of Oceanography, University of California San Diego was used (ftp://topex.ucsd.edu/pub/srtm30_plus/; Becker et al. 2009; cf. Sandwell/Smith 2009). Modern disturbances such as open quarry fields or sediment fans of rivers were graphically revised but this data could not be corrected for isostasy or aeolian deposits. Thus, the base maps can only give an approximation of the past landscape. A digital elevation model (DEM) with a sufficient temporal resolution is yet not available for north-western Europe but could be a fruitful project in the future. In the studied area, the impact of isostasy and aeolian deposition was relatively insignificant since the Lateglacial and such details would be lost anyway due to precision of the dataset. Based on these maps, results from various studies about the physical geography were compiled in a short history of the landscape development in north-western Europe (see p. 370-383). In this compilation, restrictions for occupation by plants and mammals as well as for human mobility such as ice sheets, permafrost, or large water bodies were focussed on but more short-term obstructions such as earth-quakes and aeolian deposition were also included.

The cold landscape of the Late Pleniglacial was characterised by braided rivers, occasional permafrost, and irregular aeolian dust deposition. The fronts of the major ice sheets that had been retreating significantly since the LGM were very inconsistent in this period and some glaciers still formed barriers also in the lower mountain ranges such as the Vosges Mountains. Seasonality had presumably a strong effect on this landscape. The transition to the warm landscape of the Lateglacial Interstadial appeared gradually creating a longer interval in which some rivers still flowed in braided systems, whereas others were already settled in meandering river beds. In some areas, permafrost had already completely disappeared and soil horizons started to develop, whereas sporadic permafrost still persisted in some other locations. Although aeolian dispersal in general ceased, irregular loess deposition continued in some areas. These differences formed a very mosaic landscape. Ice sheets retreated rapidly resulting in structural instability provoking earth-quakes and tsunamis in Northern Europe. Moreover, global meltwaters quickly raised the sea-level and first areas in the north were submerged. This period of transition started during the end of the Late Pleniglacial and continued into the early Lateglacial Interstadial. During the GI-1c event, the warm landscape was completely established with mostly meandering river systems and ubiquitous soil development. Probably, a major ice sheet existed only in Scandinavia, whereas the other ice sheets had shrunken to glacier fields or completely disappeared. In Northern Europe further parts of the landscape were submerged but the rise of the sea-level had slowed down. The developments during this temperate period were so sustainable that in the succeeding stadial period, conditions comparable to the transitional period rather than to the Late Pleniglacial reappeared.

Pollen and other vegetation data from the Eifel maars are available for the Holocene, Lateglacial Stadial, and Lateglacial Interstadial as well as for a short period of the Late Pleistocene. For the Paris Basin, a synchronised pollen sequence was established from the end of the Late Pleniglacial to the onset of the Lateglacial Stadial.

This relative sequence was supplemented by ^{14}C dates of which only four dates were considered reliable in a technical audit. This technical audit was one part of an evaluation accomplished for all ^{14}C dates used in the present study; the other part was a contextual audit which was performed before and after calibration (see p. 259-265). Using the previously constructed radiocarbon calibration curve, the four reliable dates of the French synchronised pollen profile were calibrated and the most reliable position was established by correlating the results along the NGRIP oxygen isotope record (see p. 384-388). The profile was stretched and compressed according to these calibration results and, thus, tuned to the NGRIP eventstratigraphy. A comparison of this tuned French sequence and the pollen profile from the Eifel (**fig. 52**) revealed that the French record begins earlier than the Eifel one, that some general differences in the vegetation development exist between the regions, and that changes usually occurred earlier in the French record. The general differences can partially be explained by different local climates with northern France tending to be drier than the Eifel region. The generally earlier response of the French diagram can be due to the correlation along the isotope eventstratigraphy but the clearly earlier beginning of the record might be related to the earlier disappearance of permafrost from this region and the onset of biome development.

At the end of the Late Pleniglacial, the pollen profiles suggest a sparse vegetation cover with wide-spread grasslands (**fig. 53**). The vegetation begins to gradually become denser around the onset of the Lateglacial Interstadial. During the first warm sub-event (GI-1e), arboreal pollen (AP) increases slowly, perhaps influenced by increased volcanic output in this transitional period. The AP increase was mainly due to the expansion of shrub communities, in particular, by juniper, willow, and birch. However, a first important spread of trees possibly appeared in the study area during the second part of GI-1e. This gradual development was stopped abruptly by a rapid expansion of open herbal steppe vegetation. Whether this sudden decrease of AP was related to returning cold conditions during GI-1d cannot be answered with certainty, changes in other factors such as aridity, wind intensity, or wind tracks could also be possible explanations. These factors could have changed after the onset of the severe cold episode of GI-1d. In fact, although the vegetation, in general, becomes denser and, in particular, light pioneer forests establish, the vegetation development is very unsteady with quickly alternating periods of dense and open vegetation until GI-1c_1. During this second half of the Lateglacial Interstadial, the light forest environments established and gradually turned from pioneer to established forest communities. Perhaps, this stabilisation was related to a significantly decreased volcanic sulphate output. Towards the Lateglacial Stadial, the vegetation became more open again but with values comparable to the early Lateglacial Interstadial suggesting that shrub communities and light forests persisted in sheltered locations.

Preservation of pollen seems to begin at the onset of the first transitional phase in the development of the physical geography. Thus, information about the Late Pleniglacial is small in this record and supplementary data has to be collected. This supplement was mainly provided by directly ^{14}C-dated macro-remains as well as macro-remains from archaeological contexts such as charcoal. The latter indicate the availability of wood, usually willow and pine, during the Late Pleniglacial. However, the exact origin of this wood cannot be determined, although the presence of some trees in sheltered areas appears possible. With the onset of the Lateglacial Interstadial, the preservation of datable material as well as the diversity of determined species increased. In general, these species reflected expanding shrub communities, in particular of birch species, comparable to the pollen profiles. During the unstable period in the pollen profiles, willow became important besides the birches. With the stabilisation of forest communities, pine becomes dominant among the macro-remains and continues to be dominant during the Lateglacial Stadial.

Comparable to the macro-remains, the faunal record was mainly reconstructed by directly ^{14}C-dated and determined specimens as well as species found in the archaeological context. Using probability distributions of calibrated ^{14}C dates (**figs 60-61**) to establish population dynamics of commonly hunted species such as horse or reindeer produces results that are no more meaningful than first and last appearance data (see p. 412-432). This ineffectiveness is revealed by a comparison with the presence of the same species in the archaeological and palaeontological record (see **tabs 77-79**). The probable reason is a considerable bias of the datasets by human subsistence strategies, selective preservation, and modern research interests resulting in unbalanced dating strategies. However, with the record of directly ^{14}C-dated species supplemented by well contextualised determinations of species from reliably dated records, a well funded chronology of presence and absence of some selected species can be established for several regions from the LGM to the onset of the Holocene (**fig. 63**).

This compilation reveals that after the LGM the faunal record still remains small until the end of GS-2b and the main Heinrich 1 event (cf. Stanford et al. 2011b). After the main phase, the fauna reflects an open vegetation community with several larger mammals such as reindeer or horse that are known in modern periods to have occasionally formed large herds. These herds were supplemented by a rich faunal community which contains grass steppe inhabitants such steppe bison but mostly species that are found in modern arctic regions such as arctic hare, arctic fox, and occasionally musk ox. A northward shift of this faunal community began at the end of the Late Pleniglacial. In the studied regions, typical inhabitants of arctic tundra such as musk ox disappeared and other species associated with more temperate environments appeared such as elk, red deer, and wild boar. A regional difference is the absence of wild boar in the western upland zone including the Central Rhineland and the absence of elk in northern France. Possibly, this absence was related to the lower moisture and higher vegetation cover in the Northern French floodplains. However, horses became the dominant species in the assemblages but reindeer still occurred sporadically in northern France. Reindeer disappeared from the study area with the end of the cold sub-event of GI-1d. The importance of horse decreased significantly at the same time but horse was still present in the areas until the Holocene. After GI-1d, species indicating temperate forests such as roe deer appeared. In the Central Rhineland, these were supplemented by wetland species such as beaver. In northern France, aurochs also indicated forest environments. Wild boar was attested only for the end of the Lateglacial Interstadial in the Central Rhineland when the importance of pine also increased in this region. The temperate faunal communities were not replaced completely during the Lateglacial Stadial. In areas adjacent to the North European Plain, more arctic species such as arctic fox and reindeer reappeared but in northern France aurochs remained present in addition to horse. During the Holocene, temperate species expanded rapidly into the northern regions of the study area. This development indicated that in the faunal communities, likewise the physical geography and the vegetation record, a transitional period bridged the period of the Late Magdalenian and the period of the FMG. The beginning of the transitional phase appeared gradual but the end was relatively rapid and, therefore, the latter appeared as more fundamental.

Behavioural patterns based on the archaeological record

The archaeological record (see Material-Archaeology, p. 75-244) examined in the present study is based on well known sites from the Central Rhineland, the Paris Basin and the Somme region, and the uplands between these areas. These sites were chosen because of the comparable archaeological record which was previously interpreted as consolidated information system (the extended Nebra group, Floss/Terberger 2002). In the material chapters, the published information about the sites and their archaeological material were summarised. This information was compiled per sub-area in eleven model tables (**tabs 10-44**). The internal chronology as well as the general chronostratigraphic position of each site was evaluated based on available chronological indicators. In particular, the ^{14}C dates were examined along the same criteria as the environmental database and, in addition, the relation of the radiometric result to the dated archaeological material was considered (see p. 265-269; cf. Dean 1978). Compared within the sub-areas, periods of transition were established for the Central Rhineland (**fig. 65**) and northern France (**fig. 69**). In the Central Rhineland, this transitional period was relatively long due to the relatively sparse archaeological record from this time. In the French record, three periods were identified in which different archaeological units overlap. These periods imply a continuous but uneven process of change. In a next step, a succession of sites from all studied areas was established (**fig. 70**) and suggested that the long transitional period from the Central Rhineland encompassed the two first transitions of northern France. Moreover, the succession of the archaeological sites suggests that during the mid-Lateglacial Interstadial two different varieties existed concomitantly in northern France and the Central Rhineland. After the third transition, a variety more comparable to the one in Central Rhineland was also established in northern France. Thus, according to the archaeological units, only one transition can be identified in the Central Rhineland in contrast to three transitions in France. This observation suggests a more conservative behaviour in the Central Rhineland than in northern France.

Besides the variable tempo, this different perception also arises from the examined part of the archaeological record. To systematise changes in the archaeological record according to their effect, this project uses a refined version of a hierarchical system (**tab. 62**) introduced previously for the study of human material culture and population history (see p. 289-291; Foley/Lahr 1997; Gamble et al. 2005). Archaeological material systematised in this hierarchy reflects an increasing number of individuals which conformed to the same behavioural pattern to create the archaeological observation. For example, a single individual made a retouch on an artefact on a molecular scale, a family group organised the space of a site on a micro-scale, one or several groups formed a settlement system on the meso-scale, and generations formed a tradition of using space on a macro-scale. Variations on the individual scale are probable to occur constantly due to human imperfection in reproducing behaviours and behavioural expressions (Eerkens 2000; Eerkens/Lipo 2005). Correction by others (Hamilton/Buchanan 2009) leads to a decrease of unintentional variation on higher hierarchical levels, whereas the number of people who must agree and/or be able to conform to new behavioural patterns increases. Thus, the higher in this hierarchical system a change is observed, the more effective was it for the Lateglacial society. However, some changes in the archaeological record were probably interrelated because the different parts belonged to the same complex behavioural recipe (cf. Mesoudi/O'Brien 2008b). For example, the production of lithic points at sites attributed to the MfCM possibly represented a complex behavioural recipe: To produce short but straight blades for the transformation into points, the knapping instrument was changed to a soft hammerstone and this change also led to different reduction strategies that subsequently permitted a different raw material economy (Valentin 1995; Valentin 2000; Valentin/Julien/Bodu 2002; Valentin 2008a; Valentin 2008b). A chronological vicinity of changes in such interrelated parts of the record should not be considered as multiple but a single process of decision making that resulted in the fully new complete behavioural recipe.

To establish the progress of variations more precisely and allow for an evaluation of the observed variations, several parts of the archaeological record were analysed on several levels from the molecular scale such as the measures of lithic projectiles to the meso-scale, for example, the long distance connections that are expressed by the acquisition of resources. A particular focus was set on exploitation strategies (see p. 485-509 and p. 534-548) and spatial organisation (see p. 548-559) as behaviours that are interdependent with the natural environment and that are therefore particularly sensitive towards alterations in this environment. The acquisition of resources, in particular, lithic raw materials is often difficult to establish (see p. 569f.) because the exact sources are not known or, in the case of secondary deposits such as river gravels, spread over a wide area. However, a detailed itemization of raw materials found at the analysed sites and the nearest possible origin of these materials (**tabs 12. 25. 36**) confirmed the previously observed disappearance of very long distance transports in the study areas and the increasing use of local resources (cf. Floss 1994; Baales 2006b; Street et al. 2006). Increasing knowledge of the landscape and, in particular, of source qualities seems no probable explanation for this variation because the Late Magdalenian already knew and exploited high-quality local resources. A connection with technical innovations permitting a more flexible use of the lithic raw materials appears a more likely explanation. In addition, the comparison of the distances indicated a wider local zone in northern France than in the Central Rhineland. Moreover, this comparison also revealed a tendency of exploiting the immediate local surrounding and again a more distant regional surrounding. Intermediate distances were hardly identified. This gap in the raw material acquisition distances could be compensated by the diffuse distribution of some raw materials. However, if this gap proved valid in a future, more precise source location survey this behaviour could implicate specific site movement behaviour (**fig. 94**) that, perhaps, was performed to allow the exploited ecology to recover. Furthermore, a comparison with neighbouring areas of the Central Rhineland suggested that the directions of a flow of raw materials along the Rhine might have changed during the transition from the Late Pleniglacial to the Lateglacial Interstadial. A changed direction of this flow could reflect variations in the seasonality and routes of the settlement system in this extended region.

To establish changes in exploitation strategies, numerous parameters were produced for the lithic (see p. 270-272) and faunal assemblages (see p. 282-285) of the analysed sites. In particular, the lithic inventory was studied with some detail in regard to the applied exploitation as well as the observed diversity assuming that the settlement type and the spatial organisation affected the lithic raw material economy at the sites (see p. 286f.). The resulting parameters were directly compared in a temporal succession per site and per attributed archaeological unit (**fig. 74A**).

Firstly, the lithic assemblages were compared by their numerical size identifying the assemblages of FMG sites as tending to be small and MfCM assemblages as particularly numerous (**fig. 73**). This observation gives the impression of a wasteful handling of resources at the latter sites. To further test this impression further criteria indicating a wasteful exploitation were considered. These criteria are the density of lithic material on a site, the number of cores and the portion this group makes up of the complete assemblage, the relation of cores to the debris material of the site termed the exploitation index, and the relation of cores to formally retouched artefacts termed function index. The density was strongly biased by the spatial organisation of the sites and the limits of the archaeological excavations resulting in some extreme outliers. Nevertheless, a clear difference was found between the generally dense Late Magdalenian and MfCM sites on the one hand and the ephemeral Early Azilian and FMG sites on the other hand (**fig. 74B**). In the comparison of the cores, increasing proportions of cores were found in the chronologically younger assemblages, whereas the numbers of cores remained, in general, comparable (**fig. 75**). However, several MfCM and some FMG assemblages yielded clearly higher numbers of cores than the other sites. Usually, the site formation could explain the high numbers at FMG sites with obviously repetitive visits creating higher numbers of cores for the complete site. If these numbers were considered per concentration they fell in the normal range. Bad Breisig formed the only exception. The exploitation index showed that the number of knapped material per core generally decreased from the Late Magdalenian over the Early Azilian to the FMG (**fig. 76**). However, in the Late Magdalenian, two types of assemblages were found which appeared to relate to different types of use of these sites. In the MfCM assemblages, this index varied considerably with the upper horizon of Le Grand Canton producing particularly low values. The function index showed again a difference between the sites of the Late Magdalenian with the large sites of Gönnersdorf and Andernach being clearly dominated by retouched artefacts and smaller sites and possible hunting camps yielding a smaller dominance of retouched artefacts (**fig. 77b**). The values of the Early Azilian assemblages scattered between the two groups of the Late Magdalenian, whereas the values of the MfCM assemblages fell to the lower Late Magdalenian group and even below this one. The FMG assemblages had a comparable range indicating the decreasing numerical importance of formally retouched artefacts and the increasing proportion of the blank production process. For the majority of MfCM sites, the hypothesis of a particularly wasteful handling of raw material could be qualified due to the excavated areas and the high values were explained by an intense blank production performed at these sites. By contrast, for Gönnersdorf III, the upper level of Le Grand Canton, and Bad Breisig the criteria suggested a more wasteful handling of raw materials revealing this behaviour as a diachronic phenomenon. However, in these assemblages small-sized raw material nodules that were available in large numbers in the immediate vicinity were used. The number of these small-sized pieces seemed to have been a substitute for fewer but larger raw material nodules. This trend in connection with the less restrictive acquisition of raw materials resulting also in the use of smaller-sized material during the time of the FMG could explain the impression of an increasingly wasteful exploitation strategy in the Lateglacial. Nevertheless, a greater wastefulness could be rejected as an explanation of large and very large assemblages. Intense lithic exploitation and a frequent use of sites appear better explanations of large and very large assemblages in this period. The comparison per archaeological unit further demonstrated that in the Late Magdalenian sites seemed to be used for special purposes and, therefore, produced sometimes very different values. In contrast, MfCM, Early Azilian, and FMG sites appeared more similar. The latter sites seemed as a combination of the Late Magdalenian site types, whereas the MfCM and the Early Azilian sites appeared to reflect a single, opposing site type. This finding raises the question whether these two archaeological units represented two complementary units.

Diversity of lithic assemblages was considered as indicator for time in various senses: as an indicator for the occupation duration of the site (Löhr 1979; Richter 1990) as well as the position in a typo-technological chronology. The latter further makes an evaluation of the diversity between quasi-contemporary assemblages possible and, thus, allows considerations about the stability of socially given norms. Therefore, diversity of lithic assemblages is important in the study of a transition process. The composition of the complete lithic inventory that had already served in the discussion of the exploitation, the composition of the formally retouched artefacts, and, in particular, of the LMP was therefore analysed (see p. 272-282).

The proportion of the formally retouched artefacts showed a significant variability in Late Magdalenian assemblages (**fig. 77a**). In general, this proportion decreased during the Lateglacial with the disappearance of inventories that were strongly dominated by formally retouched artefacts. The latter becomes apparent when directly compared to the numbers and proportions of cores (**fig. 84**). A more detailed comparison of the structure of the formally retouched artefacts (**tab. 83**) indicated a decreasing conformism to classical tool types. In addition, this comparison showed that the LMP are the most basic group of formally retouched artefacts followed by burins. These two groups were usually found within Lateglacial inventories of formally retouched artefacts. The proportion of the supposed elements of hunting equipment were variable but in general in a similar range but at Late Magdalenian sites smaller proportions were occasionally found, whereas the FMG assemblages sometimes yielded higher proportions (**fig. 80A**). The MfCM and the Early Azilian sites produced generally lower values of LMP. Perhaps, this observation was due to the general absence of backed bladelets in these assemblages. In late Magdalenian sites, intentional breakage of backed bladelets was recorded already resulting in higher numbers of LMP than the use of points. In addition, these bladelets were probably used in higher numbers per projectile, whereas the large lithic points from the MfCM and Early Azilian sites were considered to be used as single lithic implement. Thus, a considerably higher number of LMP was produced for the same number of projectiles when backed bladelets were used. Perhaps, this explained the higher proportions at FMG and some Late Magdalenian sites. To evaluate a suggestion that the use of lithic points and a decrease of burins in Lateglacial assemblages reflected a lower availability of suitable organic sources such as antler for the production of hunting equipment (Lang 1998), the number of points, burins, end-scrapers, and borers were compared. In the context of reindeer gradually disappearing from northern France during the faunal transition period, this suggestion seemed to offer a possible explanation for changes in the archaeological record. However, based on the present results (**figs 80B-81**), this suggestion cannot be confirmed, although it can also not be completely rejected. In fact, the detailed comparison indicated a considerable variety between the assemblages, between the archaeological units as well as within the single archaeological units. This variation raised some questions: Were borers no longer of use or was the function performed with another artefact? Was the function of specific tool types, perhaps, varied within the Lateglacial? A compilation of the existing traceological results and further, diachronic analyses of numerous pieces from different inventories would be desirable in the future to provide further insights in this complex subject. In the context of possible changes in the tasks performed with a specific formally retouched artefact group and the previously indicated differentiation in the use of space in this period, the question must be asked whether the suggested use of the Simpson diversity index to evaluate the duration of Magdalenian occupations (Richter 1990) can be applied likewise for other Lateglacial archaeological units (cf. Gelhausen 2011b). The results for the Late Magdalenian and the other archaeological units are generally similar but the variety increases in these other units (**fig. 82**), in particular, Early Azilian and FMG values cover a greater range and produced more often specialisations in single artefact groups. These specialisations have to be considered in the spatial organisation of sites and the excavation limits. Therefore, the use of this index as indicator for occupation duration must be supplemented by a spatial analysis of the concentration and its position in the site context as well as in the settlement system.

The diversity in the LMP was established by shape (**tab. 85**) and by dimension (**fig. 83**) to document changes probably associated with the hunting equipment that is considered as most sensitive to changes in the natural environment. Furthermore, a greater variation in this record can also be understood as an experimentation phase which is of some interest studying behavioural change. The lowest variability of shapes was found in the Late Magdalenian and the greatest variability was found in the mid-Lateglacial Interstadial from approximately GI-1d to mid-GI-1c but the variability of shapes remained relatively high afterwards. This period equates the end of the transition period in the Central Rhineland and towards the end of this transition also the highest variability in the dimension of the LMP appears suggesting that in some assemblages two or three size classes of LMP existed which were either used in different tasks or in different weapon systems. Besides this indication of variability, the variances of the LMP also revealed important standard measures. For example, width appeared to be a very standardised value in Late Magdalenian backed bladelets, whereas length was the standardised parameter of *Federmesser*. Probably, these standards were closely related to the hafting of the lithic implements into the projectile shaft.

Some sites provided in addition to the lithic assemblage also a faunal record. Although preservation conditions varied, the better preserved FMG assemblages such as Kettig showed that the production and use of organic tools had significantly decreased in comparison to the Late Magdalenian. Possibly, the faunal material was substituted by wood that was not preserved. The choice of prey animals generally follows the appearance and disappearance of the animals in the studied area. However, in the Central Rhineland the transition in the faunal assemblages was shorter indicating the more conservative behaviour. The number of different species is bisected for the Late Magdalenian again according to the different site types, whereas the FMG yielded a small range of numbers (**tab. 86**). The preservation of the bone material suggested that a change in the disposal of faunal remains occurred with material on FMG sites being usually burnt. In Late Magdalenian sites, very small pieces appeared that were considered in an intense exploitation of animal resources comparable to ethnographic examples of a »nothing is wasted« strategy (Pasda/Odgaard 2011).

This cooking of very small faunal fragments allowed a disposal near the site without attracting predators and vermin. Perhaps, for the same hygienic reasons not so intensely exploited material had to be burnt at FMG sites. This difference implies that at FMG sites a sufficient surplus existed that made discard of potential nutrition possible. The exploitation of smaller mammals such as hare but also carnivores was comparably considered as buffering mechanism in insecure environments (Charles 1997b; Stiner et al. 1999; Munro 2003; Zeder 2012). The proportion of these smaller mammals decreased over time (**fig. 86**) sustaining the impression that the intensity of exploitation of alimentary resources decreased from the Late Magdalenian to the FMG. In particular, values were very low at MfCM sites. To what degree selective preservation at these sites was a major agent in the creation of this pattern cannot be fully evaluated. However, this selective preservation had possibly some influence on the specialisation index of the faunal assemblages (**fig. 85**) resulting in some hyper-specialised assemblages. The Late Magdalenian faunal assemblages displayed a bipartition that reflected the often before observed different site types. This differentiation becomes inconsistent during the second transition in northern France. The FMG faunal assemblages appear rather diverse.

Thus, a change in the spatial behaviour during the Lateglacial was identifiable by the archaeological record. In particular, the diversity of the formally retouched inventories in comparison to the number of formally retouched artefacts (**fig. 89b**) and the diversity of the faunal assemblage compared to the minimal number of individuals showed very different functions of the sites (**fig. 89a**). The position in these systems shows the short-lived opportunistic character of most assemblages. These comparisons also revealed some Late Magdalenian, MfCM, and FMG assemblages as exceptional outliers and, again, the almost opposite character of MfCM and Early Azilian assemblages becomes obvious. This opposition becomes further apparent when the sites were displayed in a direct comparison of the diversity of formally retouched artefacts and faunal assemblages (**fig. 90**). According to a predefined connection of this display to the settlement system (**tab. 59**), the MfCM sites would reflect long-term hunting camps or short-term base camps, whereas the Early Azilian sites would represent provisioned special task camps or opportunistic episodes. The Late Magdalenian sites form again two groups between potential base or agglomeration camps and hunting-camps, whereas the FMG sites encompassed the Late Magdalenian sites and ranged between the MfCM and the Early Azilian sites. Providing additionally information about settlement structures in this comparison (**fig. 91**) shows that pavement and pits were only found in large base camp or agglomeration sites. Hearths identified by stone setting and those identified by stone packing seemed to occur, in general, on different types of sites. Alteration of the sediment tended to be more common on site types comparable to those with the stone setting. Putting the seasonal indication for these sites on this display (**fig. 92**), a distinctive trend can be observed with the MfCM being more frequently used during the light season and Early Azilian sites tending to be used during the dark season. This dichotomy provides further arguments for different, possibly complementary, functions of these archaeological units. Moreover, the connection of dark season and almost all year round use of base camps and hunting camps being rather in the light season during the Late Magdalenian is disintegrated at FMG assemblages that show nor clear relation between site type and seasonality. Establishing a hypothetical model of site type distribution in relation to the diversity of the formally retouched artefacts and the faunal assemblages as well as in relation to the faunal environment (**fig. 93**), the dichotomy of the MfCM and the Early Azilian sites matches well to the differences of hunting and base camps in areas with non-migratory species. Besides the different functions, the spatial layout of frequently used sites varied considerably with a tell-like use during the Late Magdalenian and occasionally the Early Azilian, a peripheral shifting at MfCM sites, to a suburban style with clearly separated but almost identical concentrations at FMG sites.

Chronology of changes in the Lateglacial record

The compilation of the various changes reveals several points:
Vegetation and fauna did not react directly to climate changes because further factors such as the soil formation also played an important role in these developments. Human behaviour was similar and reacted to a variety of external stimuli among which climate was not a directly effective and/or important factor; environmental developments which resulted in resource variation were certainly of a greater importance.
The Lateglacial hunter-gatherers gradually changed their behavioural patterns which can be regarded as an evolutionary process (**fig. 95**). When less people were involved in the decision-making process variations occurred more readily and, thus, on lower analytical scales variation appeared more frequently. Sometimes observed variations were interrelated and an accumulation of changes in interrelated parameters must be considered as a change in a complex behavioural recipe. However, in a relatively short period (c. 250 years) in the mid-Lateglacial Interstadial several changes in many analysed parameters, also several unrelated ones, and at different analytical scales were observed. This short-term process can be considered as a revolution or step-wise change (**fig. 96**). After this rapid episode of change, FMG sites were established in the Central Rhineland. Consequently, the transformation from the Late Magdalenian to the FMG can best be described as a phase transition-like process with a gradual beginning and a sudden, almost exponential rise if a critical threshold is passed. The starting point of this accelerated development can be seen in the cold phase of GI-1d and the subsequent establishment of light forests environments. This major change in the vegetation and the faunal record coincided with the onset of the accumulated changes in the archaeological records. In northern France, another faciès was first established after this rapid rise. The number of characteristics that this faciès shared with the FMG of the Central Rhineland was relatively high and, perhaps, the differences can be attributed to a specific environmental adaptation.

EXPLAINING REORGANISATION OF SOCIAL SYSTEMS IN THE CONTEXT OF CLIMATIC AND ENVIRONMENTAL CHANGES

In a study of changes in behavioural patterns, precision in chronology, terminology, and the research focus have to be endeavoured to provide and fill a common research frame. This endeavour was attempted in this analysis for climatic, environmental, and archaeological data from the Weichselian Lateglacial. The records and the combination of these records in the same frame delivered the above presented results.
An interpretation of these results must consider that humans form dynamic, complex, and adaptive social networks as survival strategy. Further survival strategies shape the behavioural standards of these networks. These specific behavioural standards were contextualised in their natural and social surrounding and adapted to changes occurring in these environments. Consequently, behavioural variation is also directed by the developmental history of these standards.
Contextualising the Lateglacial record in the recolonisation process of north-western Europe after the LGM, the still relatively low population density during the Late Magdalenian in combination with a persisting climatic and environmental instability required large networks to secure the survival of the single groups as well as of the metapopulation (cf. Stein Mandryk 1993; Hanski 1998). To remain connected across these wide areas, a very conservative lifestyle had to be sustained. This sustenance of security networks also functioned as distributor of information. The fast exchange of information also worked as a buffer-

ing mechanism against the insecure surroundings. In particular, modification in subsistence and hunting strategies were communicated quickly to provide nutritional security. Smaller variations of the hunting equipment appeared regularly and useful innovations were spread rapidly during this period. However, these high-precision composite weapons had to be geared to the needs of the hunters, the characteristics of the prey, and the availability of the single components. Thus, they formed a complex behavioural recipe (cf. Mesoudi/O'Brien 2008b) in which acquisition, techniques of the production and use, and the application of the instrument had to be balanced. Variation in one of these parts could have resulted in changes in other parts. Subsistence strategies were, in general, modified towards a generalist approach (cf. Gaudzinski/Street 2003) and first steps on a broad spectrum revolution were also detectable in the incorporation of smaller mammals (see p. 543-546; cf. Munro 2003; Zeder 2012), birds (cf. Street/Turner 2013), fish, and, possibly, marine resources in the diet (cf. Pétillon 2008a; Langley/Street 2013). Moreover, the faunal material was intensely exploited similar to a »nothing is wasted« strategy known from ethnographic examples (Pasda/Odgaard 2011). Mobility patterns in this environment of limited resources and important social contacts across longer distances had to be strictly organised. To guarantee the functioning of this complete system, individuals had to be trained (cf. Pigeot 1990) and regularly monitored to prevent variation to be accumulated (Eerkens/Lipo 2005; Hamilton/Buchanan 2009) resulting in a potential non-functioning that could have fatal consequence for the individual as well as a complete group. Thus, conservative behaviours compensated for an enormous need for security (Maslow 1943) in these human groups and resulted in a remarkable resilience. This resilience of the Late Magdalenian was sustained by the ability to adapt to unstable environments by small-scale variations in the behavioural patterns (cf. Rowley-Conwy/Zvelebil 1989; Jochim 1991; Walker et al. 2006). Furthermore, this resilience formed the developmental history that directed the following period of change.

During the transition towards the Lateglacial Interstadial and the gradual changes in the climatic, physical, and environmental conditions, hunter-gatherer groups in north-western Europe remained in this Magdalenian modus and reacted to these changes with minor modification of their behaviour. The partial coexistence of old, often migratory and new, often non-migratory prey animals, the increasing availability of resources such as wood along with antler, and the temperate climatic conditions with milder winters formed a favourable and prosperous landscape for highly adapted hunter-gatherers. In this surrounding, sufficient surplus could be created without intense exploitation and non-conformity was not necessarily punished by harsh climatic and environmental conditions. With the continuity of this more favourable context, safeguarding mechanisms such as the intense faunal exploitation were gradually neglected, abandoned, and finally forgotten. In this period, behaviours became more flexible and did not sustain connectivity.

The return of harsher climatic conditions combined with the return of aeolian sediment transport during the cold sub-event GI-1d resulted in nutritional stress that could no longer be compensated by buffering mechanisms such as increased exploitation intensity or long-distance security networks. This seems to have resulted in a decreasing population and the collapse of the Magdalenian. Thus, uncontrolled variation in combination with neglect of safeguarding strategies led finally to the collapse of Magdalenian lifestyles.

The collapse of the Magdalenian was the beginning of the Azilian. With the return of temperate but unstable environments, previous behavioural patterns were rapidly replaced. Within 250 years, a FMG lifestyle was established in the Central Rhineland and another Azilian faciès in northern France. Although some relevant changes in the behavioural repertoire were observed, the Azilian was clearly descended from the transitional phase when uncontrolled variation had produced behaviours adapted to temperate environments with denser vegetation. In particular, Early Azilian assemblages indicate these variations in lithic techniques and prey choice but the spatial behaviour detected at these sites clearly places this archaeological unit among the Magdalenian groups.

An explanation for this sudden replacement after the long period of gradual variation can be found in the formation of beneficial alliances to answer, besides insecure environments, also to social competition that led to the development of cooperative behaviour (Byrne 1996; Charlton 1997; Dunbar 1998). The highly adaptable but also highly conformist social system of the Late Magdalenian was a beneficial alliance and formed the last almost pan-European hunter-gatherer entity which shared a common set of traditions from Portugal in the west to eastern Poland and from the Mediterranean to the limits of the North European Plain. However, alliances can break up rapidly when a new alliance offers to be more convenient or competition promises to be more beneficial than allying (cf. Han/Pereira/Santos 2012). The emergence of new alliances has to be supported by a substantial part of the community members to be beneficial and, consequently, new rules for socially selected behaviour often occur in a phase transition-like manner (cf. Gavrilets/Duenez-Guzman/Vose 2008; Abel/Cumming/Anderies 2006). This manner was observed in the archaeological record from Lateglacial north-western Europe.

Furthermore, in an analysis about the impact of volcanic eruptions on past societies, minor impacts were shown to have had major effects on societies which were already destabilised, whereas major impacts had only little or no impact on resilient communities (Grattan 2006). This result is comparable to findings of this study.

In conclusion, climate change is no disaster for well connected societies that are flexible enough to adapted to changing conditions. However, if social agreements are already disintegrating the impact of minor changes can have major consequences. Thus, also resilient societies facing a climate change must consider long-term results of their small-scale adaptations and strengthen social cohesion to prevent a later failure due to minor external triggers.

APPENDIX

ABBREVIATIONS AND GLOSSARY

A.D. — *anno domini.* Referring to »after year 1 of Christian (Gregorian) calendar«.

AMS — accelerator mass spectrometry. By the use of this instrument, carbon ions can be counted (Fifield 1999). This precise counting allows precise calculations of the isotopic composition of a sample and, hence, its age. The AMS is often set equivalent with this ^{14}C-dating method. The AMS is able to produce quicker results than the classic β-counting (Libby 1952) and the precision of AMS dating was improved in the last decade and can meanwhile also reach down to 0.2 % (Bronk Ramsey/Higham/Leach 2004; Higham/Jacobi/Bronk Ramsey 2006, 181). Moreover, for AMS dating a small sample size is often sufficient (Trumbore 2000, 48 tab. 1). Nevertheless, the small size of the samples may also cause problems because the effect of small contaminations rises with the decreasing size of the sampled material (Wohlfarth et al. 1998, 144; Bronk Ramsey 2008, 259). Therefore, minimum amounts of datable carbon are also required for AMS dating. At the moment this minimum is usually ranging around 1 mg C (Higham/Jacobi/Bronk Ramsey 2006; Bronk Ramsey 2008).

AP — arboreal pollen. Pollen from trees and shrubs such as *Betula* sp., *Salix* sp., or *Pinus* sp.

a.s.l. — above (modern) sea level.

^{10}Be — beryllium isotope 10. This isotope is a cosmogenic radionuclides comparable to ^{14}C, meaning it is »produced in the atmosphere by the cascade of nuclear reactions induced by the high-energy galactic cosmic rays« (Muscheler et al. 2008, 2). It is measured in the ice-core records (Muscheler et al. 2000; Muscheler et al. 2004) as well as in moraines (Ivy-Ochs et al. 2008). The latter allows for the calculation of exposure ages of previously glaciated grounds but requires supplementary information in advance. This supplementary information includes for example the production rate of the cosmogenic radionuclides being effective at the specific location, the snow cover as well as the erosion rate (Ivy-Ochs et al. 2006, 121 f.). Even with this information the results are still given with large standard deviations (cf. Ivy-Ochs et al. 2009, 2141 f. tabs 1-2 and fig. 4). Based on the similar production, the ^{10}Be flux record of the Greenland ice-cores should reflect the same major changes as the $\Delta^{14}C$ record (Muscheler et al. 2008). However, dating uncertainties, climatic influence on the ^{10}Be deposition, and the influence of the carbon cycle on the distribution of ^{14}C between the major carbon reservoirs could cause offsets between the two records (Muscheler et al. 2008, 2). Therefore, some objections to an exact correlation were particularly discussed for the Lateglacial period and a critical use of the correlative data was proposed (Hughen et al. 2000).

b2k — before 2,000. Referring to »before 2000 A.D.«. This reference was introduced as correlation point, i.e. zero year in the construction of a stratigraphic timescale based on a multi-parameter analysis of the Greenland ice-core records (Rasmussen et al. 2006, X-2). If BC ages are converted to b2k ages it needs to be noted that the year zero does not exist in the Christian calendar (i.e. b2k age = BC age + 2000 – 1; cf. Spurk et al. 1998, 1107 note 4).

β-counting — β-counting or »conventional dating« is the classic method in radiocarbon dating (Libby 1952). The remaining radioactivity of the sample is measured/counted in this method (Trumbore 2000). Thus, with increasing age of the sample the measuring time also increased to produce a reliable statistical counting. Furthermore, the amount of sample material necessary for receiving reliable results is relatively high (10 g>; Trumbore 2000). In the past, this requirement often resulted in the dating of bulked material. These bulked samples only yielded a reliable date if the material originated from a single event such as a hunting episode. In the other cases, the dates were more or less arbitrary admixtures with frequently low precision and high standard deviations. Basically, due to the necessary amount of sample material, Pleistocene archaeological samples are currently mainly dated by the use of AMS.

BC — before Christ. Referring to »before year 1 of Christian (Gregorian) calendar«.

BP — before present/physics. Referring to »before 1950 A.D.«; originally used for the timescale of radiocarbon measurements (Godwin 1962) but later also adopted for other scientific timescales. If BC ages are

	converted to BP ages it needs to be noted that the year zero does not exist in the Christian calendar (i. e. BP age = BC age + 1950 – 1; Spurk et al. 1998, 1107 note 4).
b. s. l.	below (modern) sea level.
^{14}C	carbon isotope 14; also radiocarbon. Willard F. Libby introduced this instable natural carbon isotope as a means for dating organic samples from the past (Libby/Anderson/Arnold 1949; Libby 1952).
c.	*circa*. approximately.
Ca^{2+}	calcium ions. These ions are the soluble part of the mineral aerosols deposited in ice layers (Ruth et al. 2007). The Ca^{2+} content is one of the parameters frequently analysed in ice-core sequences because it can help to identify annual layers in the ice-core record (Rasmussen et al. 2006). Furthermore, these calcium ions, particularly, in combination with the insoluble dust concentration are assumed to »reflect both source strength and transport conditions from terrestrial sources« (Steffensen et al. 2008, 681). These conditions allow further assumptions on the organisation of the atmospheric dust circulation and, thus, on the climate system.
CBP groups	Curve-Backed Point groups. Assemblages in which curve-backed points (Kozłowski 1987) are more abundant than other point types. Thus, the term is identical with the term arch-backed point techno-complex (Schild 1984). Most common types of Lateglacial curve-backed points are bipoints, *Federmesser sensu stricto*, penknife points, and Malaurie points. Thus, the CBP groups can also be regarded as equivalent with the Azilian *sensu lato* (see p. 65-74). However, assemblages with dominantly straight-backed points or angle-backed points are excluded by this definition. Assemblages where mainly these types of points were found formed further sub-groups such as the Angle-Backed or Shouldered Point groups which occurred partially contemporaneous with CBP groups.
cal.	calendar. This addition to reference points means that the datum or age is expressed on a sun year (calendar/solar) timescale. Generally, cal. is referring to the Gregorian calendar.
CELM	Central European Lateglacial Master Chronology. Chronology based on cross-dated raw data of Swiss and southern German tree-ring sequences building an almost continuous connection to the Central European dendrochronology spanning the Holocene into the Mid-Younger Dryas (Kaiser et al. 2012).
CEDC	Central European Dendrochronology. A continuous tree-ring record back to the mid-Lateglacial Stadial based on German and Swiss oak and pine material (Friedrich et al. 2004; Schaub et al. 2008b; Hua et al. 2009). Besides the comprehensive ^{14}C data set, the record also provided insights into climatic and environmental developments by means of tree-ring widths (Friedrich et al. 1999; Friedrich et al. 2001b).
CFA	continuous flow analysis (Rasmussen et al. 2006). This term refers to a broad range of techniques which are automated procedures allowing the simultaneous analysis of several chemical elements from a liquid sample such as ice water (Spolaor et al. 2013).
cf.	*confer*. compare further.
D	deuterium (^2H). Deuterium is the stable isotope 2 of hydrogen. In palaeoclimatology, D usually refers to δD which is the ratio of ^2H to ^1H expressed in ‰. In ice-core records and tree-rings, it is considered as a proxy for the local temperature change (Petit et al. 1999; McCarroll/Loader 2004). Since in both types of records the hydrogen isotopes originate from precipitation, the hydrological cycle also has some influence on this proxy.
d	deuterium excess. The deuterium excess is the difference of eight times the ^{18}O content from the deuterium ratio ($d = \delta D - 8\delta^{18}O$). It is assumed to be a proxy for past ocean surface temperatures at the moisture-source region (Steffensen et al. 2008). Besides the temperature at the source region, d is influenced by the position of this source region and, thus, by the seasonal precipitation patterns in the North Atlantic (Masson-Delmotte et al. 2005; Sodemann et al. 2008).
$\Delta^{14}C$	level of atmospheric ^{14}C. This level is calculated by the comparison of ^{14}C with ^{12}C isotopes and the modern standard (Stuiver/Polach 1977).
di	density index. In this study, the di is used to give the density of artefacts on an excavated site. It is calculated by dividing the number of artefacts by the number of excavated square-metres (see p. 270, **tab. 83**, and p. 491-495).
ECM	electrical conductivity measurement. Electrical conductivity in ice-cores depends on the acidity of the ice. This acidity changes in measurable degrees with the input of alkaline dust (Taylor et al. 1993a) or

volcanic aerosols (Moore et al. 1992). Therefore, the method is used as a proxy for the moisture regime at the source region of the dust or as marker for volcanic activities.

e.g. — *exempli gratia.* for example.

ei — exploitation index of artefacts ≥1 cm. In the present project, the ei is calculated by dividing the number of artefacts ≥1 cm by the number of cores (see p. 271 f., **tab. 83**, p. 491 f., and p. 500-503). The index is used to give an approximation of artefacts produced by a single core.

ei_T — exploitation index of total numbers of artefacts. In contrast to the ei above, this index is calculated by dividing the total number of artefacts by the number of cores (see p. 271 f., **tab. 83**, p. 491 f., and p. 503). Thereby also the splinters are encountered and, thus, allowing for some assumptions on the splintering of the raw material either due to its properties or the knapping performance.

ELA — equilibrium line altitude. In glacier studies, the term refers to an altitude where a stability of total net ablation and accumulation of the glacier is reached within a year (e.g. Benn/Lehmkuhl 2000; Osmaston 2005; Zemp/Hoelzle/Haeberli 2007; Federici et al. 2008). Thus, the variation of this theoretical line indicates climatic fluctuations.

ELSA — Eifel laminated sediment archive (Sirocko et al. 2005). Within this project sediment cores from dry volcanic lakes in the Eifel region are taken. These cores produced high-resolution (annually layered) litho- and biostratigraphic sequences spanning together a period from the middle Pleistocene to the Holocene. In the present study ELSA refers in particular to a combined sequence of undisturbed sections from four ELSA cores, which form the »ELSA greyscale stack 2005« (Sirocko et al. 2005, 834). The greyscale results from the sediment being darker in warmer periods due to higher organic carbon content and lighter in colder periods due to higher amounts of silt-sized quartz, i.e. it can be taken as an indicator of increasing/decreasing organic material in the catchment area of the lakes.

esp. — especially.

eventstratigraphy — succession of identifiable sections in which distinguishable, widely occurring, short-term phenomena (<100 kyrs) form the geochronological system such as geomagnetic reversals (Kauffman 1988). – In the present study, eventstratigraphy usually refers to the Greenland oxygen isotope eventstratigraphy (Björck et al. 1998; Walker et al. 1999; Lowe et al. 2008). The Greenland oxygen isotope record is largely influenced by the climate regime of the North Atlantic. In particular, changes between periods of cold and dry climate (Greenland stadials, GS) and periods of warm and moist climate (Greenland interstadials, GI) resulted in very different values. The developments of these periods form distinguishable sections which are often separated by very short-term shifts. Thus, these periods can be described as events. These events are counted parallelly (GS-1, GI-1, GS-2, GI-2, GS-3 etc.) top-down in the stratigraphy of the ice-cores, meaning from young to old. The Holocene is in fact the current Greenland interstadial but counting only began after the first significant shift in the record, i.e. the onset of the Holocene. Therefore, the Holocene is not numbered and abbreviated as GH. Significant shifts towards generally lower or higher values than in the period before are used as limits of the events. Due to the annually formed ice layers in the Greenland ice-cores, these limits can be dated precisely. Since the significant shifts in the Greenland records reflect major changes in the North Atlantic climate regime, the onsets of these Greenland events were assumed as markers which should also affect the marine and terrestrial environments. Therefore, these climate events should be detectable in other records and the onsets of these marine and terrestrial changes were often tuned to the Greenland chronology.

etc. — *et cetera.* and the rest/and so forth.

FAD — first appearance datum. Commonly used in palaeontology, it gives the oldest dating for a specific phylogenetic order; usually it is used on the species level (Fahlke 2009, 13). In the present project, the FAD is given according to directly ^{14}C-dated remains of a specific species in the sub-areas of this study. These dates are further limited between the LGM and the onset of the Holocene (cf. Aaris-Sørensen 2009).

fi — function index. This index is calculated by dividing the number of formally retouched artefacts by the number of cores (see p. 273, **tab. 83**, and p. 504-506). The fi is considered to reflect the relation of the major functions within the lithic assemblages in this study.

FMG — *Federmesser-Gruppen/Federmessergruppen/groupe à Federmesser*. This term was introduced by Hermann Schwabedissen (Schwabedissen 1944b; Schwabedissen 1944a, Schwabedissen 1954) and can refer to a (chronological significant) sub-group of the Azilian (cf. Baales 2002). However, occasionally the term is also used synonymous to the CBP groups and/or the Azilian. The main projectiles within the assemblages attributed to the FMG are small lithic points with a curved, blunting retouch of one lateral edge ending in a single tip, the so called *Federmesser*. These points represent one of the most common types of the curve-backed points and are chronologically insignificant (Célérier/Nisole/Beaune 1993, 89-92; cf. Ikinger 1998). Furthermore, in the lithic inventories of the FMG small end-scrapers, unstandardised burins, and some various truncations occur. Little is known about the organic tool inventory (e.g. Baales 2002; cf. Clausen 2004) as well as the personal ornamentation or art of the FMG. Evident settlement structures are rare and only the distribution of the archaeological material allows some latent structures to be identified. The concentrations are generally small dense clusters of lithic artefacts often distributed around an area with burnt artefacts and/or burnt bones which are interpreted as latent hearths. In general, FMG characterised in this way were found in a light forest environment. Unambiguous dates for the FMG place them in the second half of the Lateglacial Interstadial (i.e. GI-1c_1 – GI-1a).

GH — Greenland Holocene. See eventstratigraphy, p. 593.

GI — Greenland Interstadial. See eventstratigraphy, p. 593.

GICC05 — Greenland ice-core chronology 2005 (Vinther et al. 2006; Rasmussen et al. 2006; Andersen et al. 2006; Svensson et al. 2006; Rasmussen et al. 2008; Svensson et al. 2008; Wolff et al. 2010). The currently (effective: mid-2011) best available, annually resolved ice-core chronology for the last 60,000 years based on the analysis of seven chemical parameters added up by the visible stratigraphy from the records of the Greenland ice-cores GRIP, NGRIP, DYE-3 and GISP2. The resolution of the single parameters differed between 3 and 50 years (Rasmussen et al. 2006, X-3 tab. 1). In the Holocene and below the beginning of the Lateglacial Interstadial the records were correlated by volcanic marker events as documented in the ECM (electrical conductivity measurement) and furthermore the NH_4^+ record (Rasmussen et al. 2008). The chronology was counted back 60,200 years (Svensson et al. 2008) and an age model was applied to it further back to 123,000 years (Wolff et al. 2010).

GISP2 — Greenland Ice Sheet Project 2 (e.g. Grootes et al. 1993; Zielinski et al. 1996; Stuiver/Grootes 2000). This over 3,050 m long Greenland ice-core was drilled at 72.58 N and 38.47 W (3,208 m a.s.l.), some 28 km westwards of the GRIP ice-core (Taylor et al. 1993b). Visible layers were counted in the Pleistocene part of this ice-core (Alley et al. 1997; Meese et al. 1997).

GLPC — German Lateglacial pine chronology. Abbreviation used in the present work for the thus far floating data set of ^{14}C-dated Lateglacial tree-rings of mainly pine (Kromer et al. 2004) but also some poplars which were recovered in Kruft within the deposits from the LSE (Baales/Bittmann/Kromer 1998; Friedrich et al. 1999; Kromer et al. 2004).

GRIP — Greenland Ice Core Project (e.g. Johnsen et al. 1992; Taylor et al. 1993a; Johnsen et al. 1995; Johnsen et al. 2001; Masson-Delmotte et al. 2005). This Greenland ice-core was drilled 3,028.8 m deep at 72.58 N and 37.64 W (3,238 m a.s.l.), i.e. 28 km east of the GISP2 ice-core, and covers approximately the last 250,000 years, although beyond 160,000 years the lamination becomes indistinct (Dansgaard et al. 1993). It was originally defined as the stratotype for the Greenland isotope eventstratigraphy (Björck et al. 1998; Walker et al. 1999).

GS — Greenland Stadial. See eventstratigraphy.

Heinrich event — A Heinrich event refers to a period of massive iceberg discharges in the North Atlantic (Heinrich 1988). This discharge is identified by the deposition of notable amounts of ice-rafted debris (IRD) and occurred periodically in the North Atlantic with frequencies of 5,000-14,000 years (Stanford et al. 2011b). Heinrich events are usually assumed to result from massive collapses of ice sheets in the northern hemisphere (Peck et al. 2007). The resulting iceberg flux led to a halt in the North Atlantic thermo-haline conveyor belt (Rahmstorf 2002). These processes are assumed to have caused exceptionally cold and arid climates in Europe (Bartov et al. 2003; McManus et al. 2004; Wohlfarth et al. 2008; Sima et al. 2009).

HULU — speleothem record from five cave stalagmites in the Hulu Cave encompassing the time period between c. 11,000-75,000 years cal. b2k (Wang et al. 2001). The record yielded a calcite $\delta^{18}O$ sequence which is closely connected to the Asian monsoon. The stalagmites have been partially band counted and additionally $^{230}Th/^{234}U$ dated providing a precise timescale for the encompassed time period (cf. Weninger/Jöris 2008).

IACP — Inner or Intra-Allerød Cold Period (Lehman/Keigwin 1992). This period occurred within the generally temperate phase of the Allerød in the North Atlantic records. The IACP is assumed to be equivalent to the isotopic event GI-1b in the Greenland ice-core archives, the Gerzensee oscillation in the Swiss lake environments (Lotter et al. 1992), and the Killarney oscillation in North American lake records (Levesque/Cwynar/Walker 1997). Since the impact of this oscillation was recognisable around the North Atlantic seaboard this period was also named »Amphi-Atlantic Oscillation« (Levesque et al. 1993).

i.e. — *id est*: this (what is said before) is/means

IRD — ice-rafted debris. Layers of this material in sediment cores from the North Atlantic indicate Heinrich events. The debris originated from icebergs which melted and thereby set indissoluble material free which was enclosed in the ice during its formation and movement processes. Thus, an analysis of the petrological and geochemical composition of the debris provides indications for the origin of the icebergs (Peck et al. 2007; Bigg et al. 2010). The layer thickness indicates the intensity of the iceberg discharge. In addition, the latitudinal position at which these layers occur provides an indication of the magnitude of the iceberg discharge.

ITCZ — Inter-Tropical Convergence Zone. The ITCZ is an atmospheric zone located around the equator. It is strongly influenced by the sun and, consequently, the position of this zone shifts seasonally with the angle of the sun. The intense heating of the air in the equatorial area by the sun causes the air to rise. Subsequently, the cooler air masses of the northern and the southern hemisphere push into this zone leading to cloud formation and heavy rains in the zone and wind formation around this zone (e.g. Hasenrath 2002). Furthermore, in the Atlantic the ITCZ is also closely connected to the sea-surface temperatures (e.g. Marshall et al. 2001).

kyrs — kilo years. 1,000 years.

LAD — last appearance datum. Commonly used in palaeontology, it gives the youngest dating for a specific phylogenetic order; usually it is used on the species level (Fahlke 2009, 13). In the present project, the LAD is given according to directly ^{14}C-dated remains of a specific species in the sub-areas of this study. These dates are further limited between the LGM and the onset of the Holocene (cf. Aaris-Sørensen 2009).

LGM — Last Glacial Maximum. Depending on the definition the LGM refers to the time of the maximum low of the global sea level or the last maximal ice sheet advance. In general, the latter is in Europe correlated with GS-3 (*c.* 27,500-23,340 ± 298 years cal. b2k, Andersen et al. 2006) or, precisely, late GS-3 (cf. Weninger/Jöris 2008). However, the marine records and the global ice sheet developments as well as terrestrial indicators suggested a more complex development resulting in a range of times given for the LGM from GS-3 to approximately GS-2c (Clark et al. 2009; Starnberger/Rodnight/Spötl 2011). Also in the archaeological record, a period with scarce archaeological evidences is mainly related to the LGM and dated around 23,000-16,000 years ^{14}C-BP (*c.* 28,000-19,000 years cal. b2k; Terberger/Street 2002; cf. Verpoorte 2004). In particular, archaeological material is almost absent from Central Europe between 22,000-19,000 years ^{14}C-BP (*c.* 27,000-22,500 years cal.b2k).

LMP — laterally (abruptly) modified (laminar) pieces. A general term for »backed pieces« introduced by Jean-Paul Caspar and Marc de Bie (Caspar/De Bie 1996). A comparable term »abruptly modified pieces« was used by Roger Jacobi (Jacobi 2004, 34-45). While LMP refers to the position of the retouch, Jacobi's term refers to the retouch itself. Under both terms almost all Upper and Final Palaeolithic projectile implements and their fragments as well as various »knives« can be understood. However, LMP can also refer to not abruptly retouched pieces, which were possibly not used as projectiles (laterally retouched pieces), whereas abruptly modified pieces could also refer to various types of truncations. However, in the present work, LMP refers to laminar pieces with abrupt modification along at least one lateral edge.

LSE — Laacher See (volcano) eruption. The LSE is a major volcanic marker event in the late Lateglacial Interstadial, which spread visible ashes (see below) across most parts of western Central Europe (Schmincke/Park/Harms 1999, Baales et al. 2002).

LST — Laacher See Tephra. These are the ashes of the LSE, which have been identified as visible ashes as well as microtephras across most parts of Central Europe (e.g. Baales et al. 2002; Davies et al. 2002; Turney et al. 2006). These are chemically and mineralogically well defined (Wörner/Schmincke 1984) and therefore could e.g. be excluded from volcanic ashes preserved in the northern Greenland ice-core records (Mortensen et al. 2005, 214).

MfCM — (Final) Magdalenian faciès Cepoy-Marsangy. This term was introduced by Béatrice Schmider (Schmider 1982) and further defined by Boris Valentin (Valentin 1995; Valentin 2008a) based on the inventories of the eponymous sites Cepoy and Marsangy. The MfCM was considered as a special variety of the Late Magdalenian in the Paris Basin which contained lithic points, usually of an angle-backed shape, and heavy borers but almost no backed bladelets similar to Final Magdalenian assemblages in northern Europe (Schmider 1982). Moreover, the regular use of soft hammerstones in the production of blanks for points was suggested to result in different, more suitable characteristics from the classic Late Magdalenian blank production process (Valentin 1995). This process was meanwhile also observed in northern European assemblages attributed to the classic Hamburgian (Weber 2012). Thus, some kind of connection between northern Europe and northern France were considered a probable explanation for the related development (cf. Valentin 2008a). The lithic inventory was generally similar to a Late Magdalenian assemblage, besides the lithic points, the heavy *bec,* and the absence of backed bladelets. Organic tools or special goods were not found on these sites, although the organic preservation was occasionally good. The sites were usually set along river banks and contained a more or less dense scatters of stone-filled hearths surrounded by large amounts of lithic and faunal material. The dates for these sites appeared surprisingly young but are comparable to the dating of the similar northern European development (Grimm/Weber 2008).

MFM — Meerfelder Maar. This abbreviation is only used in the present study since the chronostratigraphic (e.g. Brauer/Endres/Negendank 1999) and environmental record (e.g. Litt/Stebich 1999) from the MFM represent one of the fundamental records to which human behaviour is compared.

MONREPOS — MONREPOS Archaeological Research Centre and Museum for Human Behavioural Evolution, part of the Römisch-Germanisches Zentralmuseum, Forschungsinstitut für Archäologie which is a member of the Leibniz Gemeinschaft. MONREPOS is located at Schloss Monrepos in Neuwied/Rhine (Germany). Research at MONREPOS is focused on the process of becoming human. This research focus is generally borne by the themes of chronologies, strategies, and social networks. Analyses of these themes are based particularly on hunter-gatherer societies, mainly from the Pleistocene period.

NAP — non-arboreal pollen. These are all herbaceous pollen such as *Artemisia* sp., Poaceae, or Cyperaceae.

NGRIP — North Greenland ice core project (Andersen et al. 2004). At this Greenland ice-core position located at 75.10 N and 42.32W (2,917m a.s.l.) an approximately 3,085m deep ice-core sequence was drilled (Andersen et al. 2004). The sequence comprises approximately the last 123,000 years and was mainly used for the construction of the GICC05 replacing the GRIP ice-core partially as stratotype (Rasmussen et al. 2006; Vinther et al. 2006).

NH_4^+ — ammonium cation. These atoms are used as a seasonal tracer in the formation of the ice-core chronology (Rasmussen et al. 2006). NH_4^+ concentration is particularly high during the summer season but during cold periods the annual signal is not very clear.

^{18}O — oxygen isotope 18, but correctly: »$\delta^{18}O$ which is defined as the‰ deviation in $^{18}O/^{16}O$ ratio relative to the SMOW (Standard Mean Ocean Water)« (Hammer/Clausen/Tauber 1986, 284 note 1). This ratio is linked generally to the climate system by temperature (Johnsen et al. 1992, 312) and precipitation (Mattey et al. 2008; Vinther et al. 2009; cf. Landais et al. 2010). Thus, it is commonly used as a palaeoproxy for air temperature at the coring site (Steffensen et al. 2008), although modern comparisons have shown a constant underestimation, presumably due to the additional dependency on the hydrological cycle (Masson-Delmotte et al. 2005).

p — (statistical) probability

PAZ — pollen assemblage zone. The term is used in palynology to group characteristic pollen communities of a defined stratigraphic unit.

p.r.n. — *pro re nata*. as the occasion arises; if necessary.

RGZM — Römisch-Germanisches Zentralmuseum, Leibniz-Forschungsinstitut für Archäologie. The RGZM is the parent research institution of MONREPOS and located in Mainz (Germany). Beside MONREPOS, it encompasses further five research fields and departments (see VAT, p. 598) studying the time period from the Lower Palaeolithic to the Middle Ages. In addition, the RGZM runs a publishing house and several laboratories and workshops specialised in conservation. The RGZM is a foundation under German law and a member of the Leibniz Gemeinschaft. Thus, the RGZM is a research institution of national importance and federal scientific interest. Therefore, it is financed by the federal and the individual German States with substantial involvement of the state of Rhineland-Palatinate and a contribution by the town of Mainz.

SRTM Shuttle Radar Topography Mission. This mission of the US-American NASA was accomplished in 2000 and aimed »[...] to generate the most complete high-resolution digital topographic database of Earth.« (http://www2.jpl.nasa.gov/srtm/, verified 23-Feb-2012). Therefore, land surface of Earth were scanned. This elevation data contains (modern) noise such as anthropogenic disturbances created by roads, buildings, or extraction pits and environmental variations in the form of trees or aeolian sand or loess dunes. In particular, the latter developments have changed the landscape of Northern Europe significantly since the LGM. Moreover, interesting parts for Late Weichselian research are data of landscapes which are submerged today. These data were usually obtained by depth-sounding on ships that were subsequently combined with the SRTM-data to provide also a dataset containing the bathymetric data. This combined dataset is used in this project (SRTM30_Plus, version 6.0; available for non-profit use from the Scripps Institution of Oceanography, University of California San Diego at: ftp://topex.ucsd.edu/pub/srtm30_plus/; Becker et al. 2009; cf. Sandwell/Smith 2009). These combined data reach a precision of c. 900 m (30 × 30 arc-seconds; for further information and references read the Readme.V6.0 and References_SRTM30_Plus text files provided at the homepage mentioned above). Further detailed data of the complete surface of Earth in a single data set can be expected in the near future by the European TanDEM-X project (www.dlr.de/hr/en/desktopdefault.aspx/tabid-2317 verified 23-Feb-2012).

ss09 An age model formed mainly by data from the DYE-3 and GRIP ice-cores. The model was based on layer counting (mainly in DYE-3), correlation of DYE-3 and GRIP along volcanic markers in this upper parts, and an age estimate for the lower part of GRIP according to an ice flow model in which accumulation was considered as a function of ^{18}O values (Johnsen et al. 1992; Dansgaard et al. 1993; Johnsen et al. 1995; Johnsen et al. 1997). This age model was usually used as the standard Greenland ice-core chronology until the updated version (ss09sea) was published.

ss09sea An updated version of the age model ss09 which became also known as GRIP2001 (Johnsen et al. 2001). In contrast to ss09, the improved ss09sea used more precise seawater-corrected isotope values for the ice flow model. From 2001 to 2006 (publication of GICC05), this age model was the most commonly used ice-core chronology.

SST sea-surface temperature. The temperature below the water surface of oceans, i. e. in the top layer of the ocean water column. These temperatures reflect major changes in the global heat conveyor and also the hydrological transport system because evaporation and sea ice formation depend on the SST. Thus, near continental shores these temperatures are important indicators for the terrestrial climate (Palmer/Zhaobo 1985; Hurrell 1995; Rodwell/Rowell/Folland 1999; Sutton/Hodson 2005). In palaeoclimatology, various proxies are used to estimate the SST such as strontium/calcium (Sr/Ca) ratios in shallow water corals (Alibert/McCulloch 1997), magnesium/calcium (Mg/Ca) ratios in fossil material (Mashiotta/Lea/Spero 1999), or frequencies of foraminifera species in sediment cores (Kucera et al. 2005).

$^{230}Th/^{234}U$ relation of thorium isotope 230 to uranium isotope 234. Also known as uran-thorium dating. Equally to radiocarbon, this dating method is based on a decay process including radioactive elements (e. g. Schwarcz 1980). However, the half life age of uranium is 245,500 years and, thus, significantly longer than in the radiocarbon cycle, but the thorium isotope 230 also decays with a half life of 75,000 years. Consequently, the limit of this method is set by the half life of thorium. Nevertheless, approximately ten times older material than in ^{14}C dating can be dated by $^{230}Th/^{234}U$, thus, approximately 0.5 million years old material (Lawrence Edwards/Chen/Wasserburg 1987) if the material contains a minimum of c. 0.01 g (Hellstrom 2012).

TIMS thermal ionization mass spectrometry. With this instrument a high precision isotope abundance measurement is performed on environmental samples (Heumann et al. 1995). In particular, ratios of elements used in **geochronology** and **tracer** studies such as $^{230}Th/^{234}U$ are measured. Meanwhile, further improvements in this field allowed for a decrease of the necessary sample material and an increase of the accuracy and the precision of the results compared with other dating techniques (Luo et al. 1997).

TL thermoluminescence. TL is a radiometric dating method (Daniels/Boyd/Saunders 1953). It is commonly used in the dating of Pleistocene sites, especially, for sites which produced no material for radiocarbon dating (Richter 2007). In the TL method radiation damage such as burning is identified, for instance, on flints by measuring of the radiation dose in relation to the »undamage« radiation dose. On the temporal scale the measured radiation dose accumulates in a stable and linear process (Richter 2007) and, consequently, the results can be calculated in calendar years. In general, the dates resulting from TL measurements represent good indications for the actual age of human activity because the burning events, particularly of lithic material, can often be closely connected to hearths. Furthermore, an increasing number of assemblages can be dated with this method due to the decreasing amount of required material for the dating procedure (Richter/Krbetschek 2006).

UMT — Ulmener Maar Tephra. Volcanic deposits of the youngest volcanic eruption in the Eifel field forming the Ulmener Maar (Zolitschka/Negendank/Lottermoser 1995). The eruption occurred in the early Holocene (c. 10,945 varve-years b2k) and the deposits can be found in the surrounding Maar sequences such as the Schalkenmehrener Maar, Holzmaar, and Meerfelder Maar.

VAT — Department of volcanology, archaeology and history of technology studies. VAT is a diachronic focused department of the RGZM located within the Eastern Eifel volcanic field at Mayen (Germany). The research focus is set on the emergence of the industrial landscape within which the department is located.

Weichselian — last major glacial cycle as identified in the stratigraphies of Northern Europe. According to the stratigraphic nomenclature, the Weichselian encompasses the period from GS-5 to the onset of the Holocene (Litt et al. 2007, 46). The Weichselian ice sheet formation is generally correlated to the Würmian from the Alpine region and the Devensian from the British Isles.

WGS 84 — World Geodetic System 1984. WGS 84 is a standard coordinate system to which also the GPS (global positioning system) today is referenced.

REFERENCES

Aaris-Sørensen 1995: K. Aaris-Sørensen, Palaeoecology of a Late Weichselian vertebrate fauna from Nørre Lyngby, Denmark. Boreas 24, 1995, 355-365.

2009: K. Aaris-Sørensen, Diversity and dynamics of the mammalian fauna in Denmark throughout the last glacial-interglacial cycle, 115-0 kyr BP. Fossils and Strata 57 (Chichester 2009).

Aaris-Sørensen/Mühldorff/Petersen 2007: K. Aaris-Sørensen / R. Mühldorff / E. B. Petersen, The Scandinavian reindeer (*Rangifer tarandus* L.) after the last glacial maximum: time, seasonality and human exploitation. Journal of Archaeological Science 34, 2007, 914-923.

Abel/Cumming/Anderies 2006: N. Abel / D. H. M. Cumming / J. M. Anderies, Collapse and reorganization in social-ecological systems: Questions, some ideas, and policy implications. Ecology and Society 11 (1) Art. 17, 2006, 1-25.

Agren/Zackrisson 1990: J. Agren / O. Zackrisson, Age and size structure of *Pinus sylvestris* populations on mires in central and northern Sweden. Journal of Ecology 78, 1990, 1049-1062.

Albrecht 1979: G. Albrecht, Magdalénien-Inventare vom Petersfels. Siedlungsarchäologische Ergebnisse der Ausgrabungen 1974-1976. Tübinger Monographien zur Urgeschichte 6 (Tübingen 1979).

Alessio/Bella/Cortesi 1964: M. Alessio / F. Bella / C. Cortesi, University of Rome Carbon-14 dates II. Radiocarbon 6, 1964, 77-90.

Alfasi/Portugali 2004: N. Alfasi / J. Portugali, Planning Just-in-Time versus planning Just-in-Case. Cities 21, 2004, 29-39.

Alibert/McCulloch 1997: C. Alibert / M. T. McCulloch, Strontium/calcium ratios in modern porites corals from the Great Barrier Reef as a proxy for sea surface temperature: Calibration of the thermometer and monitoring of ENSO. Paleoceanography 12, 1997, 345-363.

Alix/Rieu 1999: P. Alix / J.-L. Rieu, La découverte et la fouille. In: Julien/Rieu 1999, 19-23.

Alix et al. 1993: P. Alix / A. Averbouh / L. Binter / P. Bodu / A. Boguszewski / C. Cochin / V. Deloze / P. Gouge / V. Krier / C. Leroyer / D. Mordant / M. Philippe / J.-L. Rieu / P. Rodriguez / B. Valentin, Nouvelles recherches sur le peuplement magdalénien de l'interfluve Seine-Yonne: Le Grand Canton et Le Tureau des Gardes à Marolle-sur-Seine (Seine-et-Marne). Bulletin de la Société préhistorique française 90, 1993, 196-218.

Allain 1975: J. Allain, Circonscription du Centre. Eure-et-Loire, Loiret, Loir-et-Cher, Indre-et-Loir, Indre, Cher. Gallia Préhistoire 17, 1974 (1975), 465-489.

1976: J. Allain, Cepoy (Loiret). Le site paléolithique de la Pierre aux Fées. In: A. Leroi-Gourhan / J. Allain / M. Brézillon (eds), Livret-guide de l'excursion A1. Sud du Bassin Parisien. IX Congrès U.I.S.P.P., Nice (Nice 1976) 86-89.

1979: J. Allain, L'industrie lithique et osseuse de Lascaux. In: A. Leroi-Gourhan / J. Allain (eds), Lascaux inconnu (Paris 1979) 87-120.

Allain et al. 1978: J. Allain / A. Aubourg / F. Guillon / D. Jagu, Le site préhistorique de La Pierre aux Fées. Bulletin de la Société d'Emulation de Montargis numéro spécial, 1978, 3-20.

1985: J. Allain / R. Desbrosse / J. K. Kozłowski / A. Rigaud, Le Magdalénien à navettes. Gallia Préhistoire 28, 1985, 37-124.

Allason-Jones/Jones 2001: L. Allason-Jones / J. M. Jones, Identification of »Jet« artefacts by reflected light microscopy. European Journal of Archaeology 4, 2001, 233-251.

Alley et al. 1997: R. B. Alley / C. A. Shuman / D. A. Meese / A. J. Gow / K. C. Taylor / K. M. Cuffey / J. J. Fitzpatrick / P. M. Grootes / G. A. Zielinski / M. Ram / G. Spinelli / B. Elder, Visual-stratigraphic dating of the GISP2 ice core: basic, reproducibility, and application. Journal of Geophysical Research 102 (C12), 1997, 26367-26381.

Alvarez Fernández 2001: E. Alvarez Fernández, L'axe Rhin-Rhône au Paléolithique supérieur récent: l'exemple des mollusques utilisés comme objets de parure. L'Anthropologie 105, 2001, 547-564.

1999: E. Álvarez Fernández, Las perlas de madera fósil del terciario y los objetos de adorno-colgantes sobre dientes de zorro y ciervo del Magdaleniense de Gönnersdorf y de Andernach-Martinsberg-2 (Neuwied, Rheinland Pfalz, Alemania). Zephyrus 52, 1999, 79-106.

2009: E. Álvarez Fernández, Magdalenian personal ornaments on the move: A review of the current evidence in Central Europe. Zephyrus 63, 2009, 45-59.

Amantov/Fjeldskaar/Cathles 2011: A. Amantov / W. Fjeldskaar / L. Cathles, Glacial erosion/sedimentation of the Baltic region and the effect of the postglacial uplift. In: Harff/Björck/Hoth 2011, 53-74.

Ambers/Matthews/Burleigh 1985: J. Ambers / K. Matthews / R. Burleigh, British Museum natural radiocarbon measurements XVIII. Radiocarbon 27, 1985, 508-524.

Amick 1996: D. S. Amick, Regional patterns of Folsom mobility and land use in the American Southwest. World Archaeology 27, 1996, 411-426.

Ammann et al. 1994: B. Ammann / A. F. Lotter / U. Eicher / M. J. Gaillard / B. Wohlfarth / W. Haeberli / G. Lister / M. Maisch / F. Niessen / C. Schlüchter, The Würmian Late-glacial in lowland Switzerland. Journal of Quaternary Science 9, 1994, 119-125.

2000: B. Ammann / H. J. B. Birks / S. J. Brooks / U. Eicher / U. von Grafenstein / W. Hofmann / G. Lemdahl / J. Schwander / K. Tobolski / L. Wick, Quantification of biotic responses to rapid climatic changes around the Younger Dryas – a synthesis. Palaeogeography, Palaeoclimatology, Palaeoecology 159, 2000, 313-347.

Anchukaitis et al. 2010: K. J. Anchukaitis / B. M. Buckley / E. R. Cook / B. I. Cook / R. D. D'Arrigo / C. M. Ammann, Influence of volcanic eruptions on the climate of the Asian monsoon region. Journal of Geophysical Research – Atmospheres 37, 2010, L22703-1-5.

Andersen et al. 2004: K. K. Andersen / N. Azuma / J. M. Barnola / M. Bigler / P. Biscaye / N. Caillon / J. Chappellaz / H. B. Clausen / D. Dahl-Jensen / H. Fischer / J. Flückiger / D. Fritzsche / Y. Fujii / K. Goto-Azuma / K. Grønvold / N. S. Gundestrup / M. Hansson / C. Huber / C. S. Hvidberg / S. J. Johnsen / U. Jonsell / J. Jouzel / S. Kipfstuhl / A. Landais / M. Leuenberger / R. Lorrain / V. Masson-Delmotte / H. Miller / H. Motoyama / H. Narita / T. Popp / S. O. Rasmussen / D. Raynaud / R. Rothlisberger / U. Ruth /

D. Samyn / J. Schwander / H. Shoji / M. L. Siggard-Andersen / J. P. Steffensen / T. Stocker / A. E. Sveinbjörnsdóttir / A. Svensson / M. Takata / J. L. Tison / T. Thorsteinsson / O. Watanabe / F. Wilhelms / J. W. C. White, High-resolution record of Northern Hemisphere climate extending into the last interglacial period. Nature 431, 2004, 147-151.

2006: K. K. Andersen / A. Svensson / S. J. Johnsen / S. O. Rasmussen / M. Bigler / R. Röthlisberger / U. Ruth / M.-L. Siggaard-Andersen / J. P. Steffensen / D. Dahl-Jensen / B. M. Vinther / H. B. Clausen, The Greenland Ice Core Chronology 2005, 15-42 ka. Part 1: Constructing the time scale. Quaternary Science Reviews 25, 2006, 3246-3257.

Andrén/Björck/Johnsen 1999: T. Andrén / J. Björck / S. Johnsen, Correlation of Swedish glacial varves with the Greenland (GRIP) oxygen isotope record. Journal of Quaternary Science 14, 1999, 361-371.

Andreoli et al. 2006: R. Andreoli / T. Rosique / M. Schmidt / J.-M. Carozza, La dernière phase glaciaire du haut bassin de la Fecht (Vosges, France): dynamique de l'englacement et chronologie relative de la déglaciation. Géomorphologie: relief, processus, environnement 2006, 23-36.

Andrews et al. 1995: J. T. Andrews / A. E. Jennings / M. Kerwin / M. Kirby / W. Manley / G. H. Miller / G. Bond / B. MacLean, A Heinrich-like event, H-0 (DC-0): Source(s) for detrital carbonate in the North Atlantic during the Younger Dryas Chronozone. Paleoceanography 10, 1995, 943-952.

Antoine 1997: P. Antoine, Modifications des systèmes fluviatiles à la transition Pléniglaciaire – Tardiglaciaire et à l'Holocène: L'example du bassin de la Somme (nord de la France). Géographie physique et Quaternaire 51, 1997, 1-14.

Antoine et al. 2001: P. Antoine / D.-D. Rousseau / L. Zöller / A. Lang / A.-V. Munaut / C. Hatté / M. Fontugne, High-resolution record of the last Interglacial-glacial cycle in the Nussloch loess-palaeosol sequences, Upper Rhine Area, Germany. Quaternary International 76-77, 2001, 211-229.

2003a: P. Antoine / J.-P. Coutard / P. Gibbard / B. Hallegouet / J.-P. Lautridou / J.-C. Ozouf, The Pleistocene rivers of the English Channel region. Journal of Quaternary Science 18, 2003, 227-243.

2003b: P. Antoine / A.-V. Munaut / N. Limondin-Lozouet / P. Ponel / J. Dupéron / M. Dupéron, Response of the Selle River to climatic modifications during the Lateglacial and Early Holocene (Somme Basin-Northern France). Quaternary Science Reviews 22, 2003, 2061-2076.

2003c: P. Antoine / J. Catt / J.-P. Lautridou / J. Sommé, The loess and coversands of northern France and southern England. Journal of Quaternary Science 18, 2003, 309-318.

2009: P. Antoine / D.-D. Rousseau / O. Moine / S. Kunesch / C. Hatté / A. Lang / H. Tissoux / L. Zöller, Rapid and cyclic aeolian deposition during the Last Glacial in European loess: a high-resolution record from Nussloch, Germany. Quaternary Science Reviews 28, 2009, 2955-2973.

2012: P. Antoine / J. P. Fagnart / P. Auguste / P. Coudret / N. Limondin-Lozouet / P. Ponel / A. V. Munaut / A. Defgnée / A. Gauthier / C. Fritz (eds), Conty, Vallée de la Selle (Somme, France): séquence tardiglaciaire de référence et occupations préhistoriques (Paris 2012).

2013: P. Antoine / D.-D. Rousseau / J.-P. Degeai / O. Moine / F. Lagroix / S. Kreutzer / M. Fuchs / C. Hatté / C. Gauthier / J. Svoboda / L. Lisá, High-resolution record of the environmental response to climatic variations during the Last Interglacial-Glacial cycle in Central Europe: the loess-palaeosol sequence of Dolní Věstonice (Czech Republic). Quaternary Science Reviews 67, 2013, 17-38.

Ardalan et al. 2011: A. Ardalan / C. F. C. Kluetsch / A.-B. Zhang / M. Erdogan / M. Uhlén / M. Houshmand / C. Tepeli / S. R. M. Ashtiani / P. Savolainen, Comprehensive study of mtDNA among Southwest Asian dogs contradicts independent domestication of wolf, but implies dog-wolf hybridization. Ecology and Evolution 1, 2011, 373-385.

Arts 1988: N. Arts, A survey of Final Palaeolithic archaeology in the southern Netherlands. In: Otte 1988, 287-356.

2012: N. Arts, Milheeze – A very large Federmesser site on the northern edge of the Peel peat bog (Southern Netherlands) and its environmental and social context. In: Niekus et al. 2012, 267-279.

Arts/Deeben 1987: N. Arts / J. Deeben, On the northwestern border of Late Magdalenian territory: Ecology and archaeology of early Late Glacial band societies in northwestern Europe. In: J. M. Burdukiewicz / M. Kobusiewicz (eds), Late glacial in Central Europe. Culture and environment (Wrocław 1987) 25-66.

Asch 2005: K. Asch, IGME 5000: 1:5 Million International Geological Map of Europe and Adjacent Areas – final version for the internet. BGR (Hannover 2005; www.bgr.de/karten/igme5000/igme5000_finalversion2005_g.jpg; last accessed: 05.02.2010).

Atkinson/Briffa/Coope 1987: T. C. Atkinson / K. R. Briffa / G. R. Coope, Seasonal temperatures in Britain during the past 22,000 years, reconstructed using beetle remains. Nature 325, 1987, 587-591.

Audouze 1999: F. Audouze, New advances in French prehistory. Antiquity 73, 1999, 167-175.

Audouze/Beyries 2007: F. Audouze / S. Beyries, Chasseurs de renne d'hier et d'aujourd'hui. In: S. Beyries / V. Vaté (eds), Les civilisations du renne d'hier et d'aujourd'hui. Approches ethnohistoriques, archéologiques et anthropologiques. XXVIIe rencontres internationales d'archéologie et d'histoire d'Antibes (Antibes 2007) 185-208.

Audouze/Cattin 2011: F. Audouze / M.-I. Cattin, Flint wealth versus scarcity. Consequences for Magdalenian apprenticeship. Lithic technology 36, 2011, 109-126.

Audouze/Enloe 1997: F. Audouze / J. G. Enloe, High resolution archaeology at Verberie: limits and interpretations. World Archaeology 29, 1997, 195-207.

Audouze/Leroi-Gourhan 1981: F. Audouze / A. Leroi-Gourhan, France: a continental insularity. World Archaeology 13, 1981, 170-189.

Audouze/Valentin 2010: F. Audouze / B. Valentin, A paleohistorical approach to Upper Paleolithic structural change. In: D. J. Bolender (ed.), Eventful archaeologies. New approaches to social transformation in the archaeological record (Albany 2010) 29-47.

Auguste 2012: P. Auguste, La grande faune de Conty: Taxonomie, écologie et palethnographie. In: Antoine et al. 2012, 95-124.

Baales 1996: M. Baales, Umwelt und Jagdökonomie der Ahrensburger Rentierjäger im Mittelgebirge. Monographien des RGZM 38 (Bonn 1996).

1998: M. Baales, Zur Fortführung der Ausgrabungen auf dem spätpaläolithischen Siedlungsareal von Niederbieber (Stadt Neuwied). Ein Arbeitsbericht. Archäologisches Korrespondenzblatt 28, 1998, 339-356.

1999: M. Baales, Neue Untersuchungen zum Spätpaläolithikum des Neuwieder Beckens: Einige Aspekte des Federmesser-Fundplatzes Kettig, Kr. Mayen-Koblenz. In: E. Cziesla / T. Kersting / S. Pratsch (eds), Den Bogen spannen… Festschrift für Bernhard Gramsch zum 65. Geburtstag (Weissbach 1999) 55-66.

2001: M. Baales, From lithics to spatial and social organization: Interpreting the lithic distribution and raw material composition at the Final Palaeolithic site of Kettig (Central Rhineland, Germany). Journal of Archaeological Science 28, 2001, 127-141.

2002: M. Baales, Der spätpaläolithische Fundplatz Kettig. Untersuchungen zur Siedlungsarchäologie der Federmesser-Gruppen am Mittelrhein. Monographien des RGZM 51 (Mainz 2002).

2003a: M. Baales, Vulkanismus und Archäologie des Eiszeitalters am Mittelrhein. Die Forschungsergebnisse der letzten dreißig Jahre. Jahrbuch des RGZM 49, 2002 (2003), 43-80.

2003b: M. Baales, Zur Fortführung der Ausgrabungen auf dem spätpaläolithischen Siedlungsareal von Niederbieber (Stadt Neuwied). Erste Ergebnisse der siedlungsarchäologischen Analyse. Jahrbuch des RGZM 48, 2001 (2003), 163-198.

2006a: M. Baales, Environnement et archéologie durant le Paléolithique final dans la région du Rhin moyen (Rhénanie, Allemagne): conclusions des 15 dernières années de recherches. L'Anthropologie 110, 2006, 418-444.

2006b: M. Baales, Some special aspects of Final Palaeolithic silex economy in the Central Rhineland (western Germany). Der Anschnitt Beiheft 19, 2006, 239-245.

Baales/Bittmann/Kromer 1998: M. Baales / F. Bittmann / B. Kromer, Verkohlte Bäume im Trass der Laacher See-Tephra bei Kruft (Neuwieder Becken). Ein Beitrag zur Datierung des Laacher See-Ereignisses und zur Vegetation der Allerød-Zeit am Mittelrhein. Archäologisches Korrespondenzblatt 28, 1998, 191-204.

Baales/Grimm/Jöris 2001: M. Baales / S. Grimm / O. Jöris, Hunters of the »Golden Mile«. The late Allerød Federmessergruppen site at Bad Breisig, Central Rhineland, Germany. Notae Praehistoricae 21, 2001, 67-72.

Baales/Jöris 2001: M. Baales / O. Jöris, Zwischen Nord und Süd. Ein spätallerødzeitlicher Rückenspitzen-Fundplatz bei Bad Breisig, Kr. Ahrweiler (Mittelrhein, Rheinland-Pfalz). Die Kunde N.F. 52, 2001, 275-292.

Baales/Mewis/Street 1998: M. Baales / S. U. Mewis / M. Street, Der Federmesser-Fundplatz Urbar bei Koblenz (Kreis Mayen-Koblenz). Jahrbuch des RGZM 43, 1996 (1998), 241-279.

Baales/Street 1996: M. Baales / M. Street, Hunter-gatherer behavior in a changing late glacial landscape: Allerød archaeology in the Central Rhineland, Germany. Journal of Anthropological Research 52, 1996, 281-316.

1998: M. Baales / M. Street, Late Palaeolithic Backed-Point assemblages in the northern Rhineland: current research and changing views. Notae Praehistoricae 18, 1998, 77-92.

Baales et al. 2002: M. Baales / O. Jöris / M. Street / F. Bittmann / B. Weninger / J. Wiethold, Impact of the Late Glacial Eruption of the Laacher See Volcano, Central Rhineland, Germany. Quaternary Research 58, 2002, 273-288.

Badura/Jary/Smalley 2013: J. Badura / Z. Jary / I. Smalley, Sources of loess material for deposits in Poland and parts of Central Europe: The lost Big River. Quaternary International 296, 2013, 15-22.

Baede et al. 2001: A. P. M. Baede / E. Ahlonsou / Y. Ding / D. Schimel, The climate system: an overview. In: J. T. Houghton / Y. Ding / D. J. Griggs / M. Noguer / P. J. van der Linden / X. Dai / K. Maskell / C. A. Johnson (eds), Climate Change 2001: The Scientific Basis. Contribution of working group I to the third assessment report of the Intergovernmental Panel on Climate Change (IPCC) (Cambridge 2001) 85-98.

Bailey et al. 1996: J. F. Bailey / M. B. Richards / V. A. Macaulay / I. B. Colson / I. T. James / D. G. Bradley / R. E. M. Hedges / B. C. Sykes, Ancient DNA suggests a recent expansion of European cattle from a diverse wild progenitor species. Proceedings of the Royal Society of London. Series B: Biological Sciences 263, 1996, 1467-1473.

Bakke et al. 2009: J. Bakke / Ø. Lie / E. Heegaard / T. Dokken / G. H. Haug / H. H. Birks / P. Dulski / T. Nilsen, Rapid oceanic and atmospheric changes during the Younger Dryas cold period. Nature Geoscience 2, 2009, 202-205.

Bałaga/Goslar/Kuc 1998: K. Bałaga / T. Goslar / T. Kuc, A comparative study on the Late-Glacial/Holocene climatic changes recorded in the laminated sediments of Lake Perespilno – introductory data. In: Ralska-Jasiewiczowa et al. 1998, 175-180.

Ball/Nordengren/Wallin 2001: J. P. Ball / C. Nordengren / K. Wallin, Partial migration by large ungulates: characteristics of seasonal moose *Alces alces* ranges in northern Sweden. Wildlife Biology 7, 2001, 39-47.

Ballin et al. 2010: T. B. Ballin / A. Saville / R. Tipping / T. Ward, An Upper Palaeolithic flint and chert assemblage from Howburn Farm, South Lanarkshire, Scotland: First results. Oxford Journal of Archaeology 29, 2010, 323-360.

Banks et al. 2006: W. E. Banks / F. d'Errico / H. L. Dibble / L. Krishtalka / D. West / D. I. Olszewski / A. T. Peterson / D. G. Anderson / J. C. Gillam / A. Montet-White / M. Crucifix / C. W. Marean / M.-F. Sánchez-Goñi / B. Wohlfarth / M. Vanhaeren, Eco-cultural niche modeling: New tools for reconstructing the geography and ecology of past human populations. PaleoAnthropology 2006, 68-83.

2008: W. E. Banks / F. d'Errico / A. T. Peterson / M. Vanhaeren / M. Kageyama / P. Sepulchre / G. Ramstein / A. Jost / D. Lunt, Human ecological niches and ranges during the LGM in Europe derived from an application of eco-cultural niche modeling. Journal of Archaeological Science 35, 2008, 481-491.

Barbetti et al. 2004: M. Barbetti / Q. Hua / U. Zoppi / D. Fink / Y. Zhao / B. Thomson, Radiocarbon variations from the Southern Hemisphere, 10,350-9700 cal BP. Nuclear Instruments and Methods in Physics Research B 223-224, 2004, 366-370.

Bard/Hamelin/Delanghe-Sabatier 2010: E. Bard / B. Hamelin / D. Delanghe-Sabatier, Deglacial Meltwater Pulse 1B and Younger Dryas Sea Levels Revisited with Boreholes at Tahiti. Science 327, 2010, 1235-1237.

Bard et al. 2000: E. Bard / F. Rostek / J.-L. Turon / S. Gendreau, Hydrologial impact of Heinrich events in the subtropical Northeast Atlantic. Science 289, 2000, 1321-1324.

Barendsen/Deevey/Gralenski 1957: G. W. Barendsen / E. S. Deevey / L. J. Gralenski, Yale Natural Radiocarbon Measurements III. Science 126, 1957, 908-919.

Barker/Burleigh/Meeks 1971: H. Barker / R. Burleigh / N. Meeks, British Museum natural radiocarbon measurements VII. Radiocarbon 13, 1971, 157-188.

Barois-Basquin/Charier/Lécolle 1996: B. Barois-Basquin / M.-A. Charier / F. Lécolle, Un abri de la fin du Paléolithique supérieur à Bonnières-sur-Seine (Yvelines). Bulletin de la Société préhistorique française 93, 1996, 33-42.

Barr 1975: J. H. Barr, The Late Upper Paleolithic station of Moosbühl – A geological study and dating of the Magdalenian occupation. Mitteilungen der Naturforschenden Gesellschaft in Bern N.F. 32, 1975, 3-14.

Barrio et al. 2013: I. C. Barrio / J. Herrero / C. G. Bueno / B. C. López / A. Aldezabal / A. Campos-Arceiz / R. García-González, The successful introduction of the alpine marmot Marmota marmota in the Pyrenees, Iberian Peninsula, Western Europe. Mammal Review 43, 2013, 142-155.

Barton 1997: C. M. Barton, Stone tools, style, and social Identity: an evolutionary perspective on the archaeological record. Archeological Papers of the American Anthropological Association 7, 1997, 141-156.

Barton 1992: R. N. E. Barton, Hengistbury Head, Dorset. 2: The Late Upper Palaeolithic & Early Mesolithic sites. Oxford University Committee for Archaeology Monograph 34 (Oxford 1992).

Barton/Dumont 2000: N. Barton / S. Dumont, Recolonisation and settlement of Britain at the end of the Last Glaciation. In: Valentin/Bodu/Christensen 2000, 151-162.

Barton/Roberts 1996: R. N. E. Barton / A. J. Roberts, Reviewing the British Late Upper Palaeolithic: New evidence for chronological patterning in the Lateglacial record. Oxford Journal of Archaeology 15, 1996, 245-265.

1997: R. N. E. Barton / A. J. Roberts, Systèmes économiques et modalités technique dans l'Ouest de la Grande-Bretagne au cours du Tardiglaciaire. In: Fagnart/Thévenin 1997, 507-516.

Barton/Roberts/Roe 1991: N. Barton / A. J. Roberts / D. A. Roe (eds), The Late Glacial in north-west Europe: Human adaptation and environmental change at the end of the Pleistocene (London 1991).

Barton et al. 2003: R. N. E. Barton / R. M. Jacobi / D. Stapert / M. J. Street, The Late-glacial reoccupation of the British Isles and the Creswellian. Journal of Quaternary Science 18, 2003, 631-643.

2009: N. Barton / S. Ford / S. Collcutt / J. Crowther / R. Macphail / E. Rhodes / A. van Gijn, A Final Upper Palaeolithic site at Nea Farm, Somerley, Hampshire and some reflections on the occupation of Britain in the Lateglacial Interstadial. Quartär 56, 2009, 7-35.

Bartov et al. 2003: Y. Bartov / S. L. Goldstein / M. Stein / Y. Enzel, Catastrophic arid episodes in the Eastern Mediterranean linked with the North Atlantic Heinrich events. Geology 31, 2003, 439-442.

Bar-Yosef 1998: O. Bar-Yosef, The Natufian culture in the Levant, threshold to the origins of agriculture. Evolutionary Anthropology 6, 1998, 159-177.

Baskin 1990: L. M. Baskin, Population dynamics of reindeer. Rangifer 3, 1990, 151-156.

Bassett et al. 2005: S. E. Bassett / G. A. Milne / J. X. Mitrovica / P. U. Clark, Ice sheet and solid earth influences on far-field sea-level histories. Science 309, 2005, 925-928.

Beaulieu/Reille 1992: J. L. de Beaulieu / M. Reille, The last climatic cycle at La Grande Pile (Vosges, France). A new profile. Quaternary Science Reviews 11, 1992, 431-438.

Becker et al. 2009: J. J. Becker / D. T. Sandwell / W. H. F. Smith / J. Braud / B. Binder / J. Depner / D. Fabre / J. Factor / S. Ingalls / S. H. Kim / R. Ladner / K. Marks / S. Nelson / A. Pharaoh / R. Trimmer / J. von Rosenberg / G. Wallace / P. Weatherall, Global bathymetry and elevation data at 30 arc seconds resolution: SRTM30_PLUS. Marine Geodesy 32, 2009, 355-371.

Bellec/Diesing/Schwarzer 2010: V. Bellec / M. Diesing / K. Schwarzer, Late Quaternary evolution of gravel deposits in Tromper Wiek, south-western Baltic Sea. Journal of Coastal Research 51, 2010, 173-186.

Bemilli 1998: C. Bemilli, Le Closeau. Etude de la faune du niveau inférieur. In: Bodu 1998, 380-410.

Benecke/Heinrich 2003: N. Benecke / D. Heinrich, Neue Daten zur Entwicklung der Huftierfauna im Tiefland zwischen Elbe und Oder im Spätglazial und Altholozän. Archeozoologia 21, 2003, 19-36.

Benecke et al. 2006: N. Benecke / R. Bollongino / M. Küßner / C. Weber, Zur Datierung und Fauna des spätglazialen Schichtkomplexes am Abri Fuchskirche I bei Allendorf, Lkr. Saalfeld-Rudolstadt. Alt-Thüringen 39, 2006, 121-130.

Benn/Lehmkuhl 2000: D. I. Benn / F. Lehmkuhl, Mass balance and equilibrium-line altitudes of glaciers in high-mountain environments. Quaternary International 65/66, 2000, 15-29.

Bennett/Willis 2002: K. D. Bennett / K. J. Willis, Pollen. In: J. P. Smol / H. J. B. Birks / W. M. Last / R. S. Bradley / K. Alverson (eds), Tracking environmental change using lake sediments (Dordrecht 2002) 5-32.

Bennike/Jensen 2011: O. Bennike / J. B. Jensen, Postglacial, relative shore-level changes in Lillebælt, Denmark. Geological Survey of Denmark and Greenland Bulletin 23, 2011, 37-40.

Bennike/Seppänen 2004: O. S.-K. K. Bennike / A. Seppänen, Reinvestigation of the classic late-glacial Bølling Sø sequence, Denmark: chronology, macrofossils, Cladocera and chydorid ephippia. Journal of Quaternary Science 19, 2004, 465-478.

Berg 1994: A. von Berg, Allerödzeitliche Feuerstellen unter dem Bims im Neuwieder Becken (Rheinland-Pfalz). Archäologisches Korrespondenzblatt 24, 1994, 355-365.

Bergmann/Holzkämper 2002: S. Bergmann / J. Holzkämper, Die Konzentration IV des Magdalénien von Andernach-Martinsberg Grabung 1994-1996. Erste Ergebnisse. Archäologisches Korrespondenzblatt 32, 2002, 471-486.

Bettinger/Richerson/Boyd 2009: R. Bettinger / P. Richerson / R. Boyd, Rethinking the origins of agriculture. Constraints on the development of agriculture. Current Anthropology 50, 2009, 627-631.

Beyries/Janny/Audouze 2005: S. Beyries / F. Janny / F. Audouze, Débitage, matière première et utilisation des becs sur le site de Verberie »Le buisson campin« (Oise) dans le nord de la France. Revue Archéologique de Picardie spécial 22, 2005, 15-24.

Bianchi/Allison 2009: T. S. Bianchi / M. A. Allison, Large-river delta-front estuaries as natural »recorders« of global environmental change. Proceedings of the National Academy of Sciences of the United States of America 106, 2009, 8085-8092.

Bigg et al. 2010: G. R. Bigg / R. C. Levine / C. D. Clark / S. L. Greenwood / H. Haflídason / A. L. C. Hughes / A. Nygård / H. P. Sejrup, Last glacial ice-rafted debris off southwestern Europe: the role of the British-Irish Ice Sheet. Journal of Quaternary Science 25, 2010, 689-699.

Bignon 2006: O. Bignon, De l'exploitation des chevaux aux stratégies de subsistance des Magdaléniens du Bassin parisien. Gallia Préhistoire 48, 2006, 181-206.

2009: O. Bignon, Comparaisons taphonomiques de sites magdaléniens du Bassin parisien: les faunes de Ville-Saint-Jacques (Seine-et-Marne) et Étiolles (Essonne). Annales de Paléontologie 95, 2009, 97-116.

Bignon/Bodu 2006: O. Bignon / P. Bodu, Stratégie cynégétique et mode de vie à l'Azilien ancien dans le Bassin parisien: les apports de l'exploitation des chevaux du Closeau (niveau inférieur; Rueil-Malmaison, Hauts-de-Seine). L'Anthropologie 110, 2006, 401-417.

Binford 1980: L. R. Binford, Willow smoke and dogs' tails: Hunter-gatherer settlement systems and archaeological site formation. American Antiquity 45, 1980, 4-20.

Birks/Willis 2008: H. J. B. Birks / K. J. Willis, Alpines, trees, and refugia in Europe. Plant Ecology & Diversity 1, 2008, 147-160.

Birks 2003: H. H. Birks, The importance of plant macrofossils in the reconstruction of Lateglacial vegetation and climate: examples from Scotland, western Norway, and Minnesota, USA. Quaternary Science Reviews 22, 2003, 453-473.

Bittmann/Pasda 1999: F. Bittmann / C. Pasda, Die Entwicklung einer Düne während der letzten 12000 Jahre. Untersuchungsergebnisse von Groß Lieskow (Stadt Cottbus) in der Niederlausitz. Quartär 49-50, 1999, 39-54.

Björck 1995a: S. Björck, A review of the history of the Baltic Sea. 13.0-8.0 ka BP. Quaternary International 27, 1995, 19-40.

1995b: S. Björck, Late Weichselian to early Holocene development of the Baltic Sea – with implications for coastal settlements in the southern Baltic region. In: A. Fischer (ed.), Man and sea in the Mesolithic. Coastal settlement above and below present sea level. Proceedings of the international symposium, Kalundborg, Denmark 1993 (Oxford 1995) 23-34.

Björck/Koç/Skog 2003: S. Björck / N. Koç / G. Skog, Consistently large marine reservoir ages in the Norwegian Sea during the Last Deglaciation. Quaternary Science Reviews 22, 2003, 429-435.

Björck et al. 1998: S. Björck / M. J. C. Walker / L. C. Cwynar / S. Johnsen / K.-L. Knudsen / J. J. Lowe / B. Wohlfarth / INTIMATE members, An event stratigraphy for the Last Termination in the North Atlantic region based on the Greenland ice-core record: a proposal by the INTIMATE group. Journal of Quaternary Science 13, 1998, 283-292.

Blaauw et al. 2007: M. Blaauw / J. A. Christen / D. Mauquoy / J. van der Plicht / K. D. Bennett, Testing the timing of radiocarbon-dated events between proxy archives. The Holocene 17, 2007, 283-288.

2010: M. Blaauw / B. Wohlfarth / J. A. Christen / L. Ampel / D. Veres / K. A. Hughen / F. Preusser / A. Svensson, Were last glacial climate events simultaneous between Greenland and France? A quantitative comparison using non-tuned chronologies. Journal of Quaternary Science 26, 2010, 387-394.

Blackwell/Buck 2003: P. G. Blackwell / C. E. Buck, The Late Glacial human reoccupation of north-western Europe: new approaches to space-time modelling. Antiquity 77, 2003, 232-240.

Blockley/Donahue/Pollard 2000a: S. P. E. Blockley / R. E. Donahue / A. M. Pollard, Radiocarbon calibration and Late Glacial occupation in northwest Europe. Antiquity 74, 2000, 112-121.

2000b: S. P. E. Blockley / R. E. Donahue / A. M. Pollard, Rapid human response to Late Glacial climatic change: a reply to Housley et al. (2000). Antiquity 74, 2000, 427-428.

Blockley/Pinhasi 2011: S. P. E. Blockley / R. Pinhasi, A revised chronology for the adoption of agriculture in the Southern Levant and the role of Lateglacial climatic change. Quaternary Science Reviews 30, 2011, 98-108.

Blockley et al. 2006: S. P. E. Blockley / S. M. Blockley / R. E. Donahue / C. S. Lane / J. J. Lowe / A. M. Pollard, The chronology of abrupt climate change and Late Upper Palaeolithic human adaptation in Europe. Journal of Quaternary Science 21, 2006, 575-584.

2007: S. P. E. Blockley / C. S. Lane / A. F. Lotter / A. M. Pollard, Evidence for the presence of the Vedde Ash in Central Europe. Quaternary Science Reviews 26, 2007, 3030-3036.

2008: S. P. E. Blockley / C. Bronk Ramsey / C. S. Lane / A. F. Lotter, Improved age modelling approaches as exemplified by the revised chronology for the Central European varved lake Soppensee. Quaternary Science Reviews 27, 2008, 61-71.

2012: S. P. E. Blockley / C. S. Lane / M. Hardiman / S. O. Rasmussen / I. K. Seierstad / J. P. Steffensen / A. Svensson / A. F. Lotter / C. S. M. Turney / C. Bronk Ramsey, Synchronisation of palaeoenvironmental records over the last 60,000 years, and an extended INTIMATE event stratigraphy to 48,000 b2k. Quaternary Science Reviews 36 (The INTegration of Ice core, Marine and Terrestrial records of the last termination [INTIMATE] 60,000 to 8000 BP), 2012, 2-10.

Boaretto et al. 2002: E. Boaretto / C. Bryant / I. Carmi / G. Cook / S. Gulliksen / D. Harkness / J. Heinemeier / J. McClure / E. McGee / P. Naysmith, Summary findings of the fourth international radiocarbon intercomparison (FIRI) (1998-2001). Journal of Quaternary Science 17, 2002, 633-637.

Bock et al. 2013: C. Bock / M. Genschow / M. Hellmich / S. Köhler / M. Mewes / G. Otto / C. Pasda / M. Pollock / R. Roa Romero / C. Rüdel / J. Weiß / D. Zeiß, Steinartefakte aus Saaleck und ihre Stellung im späten Jungpaläolithikum zwischen Weißer Elster, Saale und Unstrut. Archäologisches Korrespondenzblatt 43, 2013, 141-158.

Bock et al. 1985: W. Bock / B. Menke / E. Strehle / H. Ziemus, Neuere Funde des Weichselspätglazials in Schleswig-Holstein. Eiszeitalter und Gegenwart 35, 1985, 161-180.

Bocquet-Appel et al. 2005: J.-P. Bocquet-Appel / P.-Y. Demars / L. Noiret / D. Dobrowsky, Estimates of Upper Palaeolithic metapopulation size in Europe from archaeological data. Journal of Archaeological Science 32, 2005, 1656-1668.

Bodu 1992: P. Bodu, Galets et percuteurs. In: Schmider 1992f, 116-127.

1993: P. Bodu, Analyse typo-technologique du materiel lithique de quelques unites du site magdalénien de Pincevent (Seine-et-Marne). Applications spatiales, economiques et sociales [dissertation, Paris 1993].

1998: P. Bodu (ed.), »Le Closeau«. Deux années de fouille sur un gisement azilien et belloisien en bord de Seine, Rueil-Malmaison (Hauts-de-Seine). Document final de synthèse de sauvetage urgent (Saint-Denis, Paris 1998).

2000a: P. Bodu, Les faciès tardiglaciaires à grandes lames rectlinignes et les ensembles à pointes de Malaurie dans le sud du Basin parisien: quelques réflexions à partir de l'example du gisement du Closeau (Hauts-de-Seine). In: P. Crotti (ed.), Meso '97. Actes de la table ronde »Épipaléolithique et Mésolithique«, Lausanne, 21-23 novembre 1997 (Lausanne 2000) 9-28.

2000b: P. Bodu, Que sont devenus les Magdaléniens du Bassin parisien? Quelques éléments de réponse sur le gisement azilien du Closeau (Rueil-Malmaison, France). In: Valentin/Bodu/Christensen 2000, 315-339.

2004: P. Bodu, Données réunies: Datations absolues obtenues sur les séquences archéologiques tardiglaciaires du sud du Bassin parisien. In: Valentin/Bodu/Julien 2004, 175-177.

2010: P. Bodu, Espaces et habitats au Tardiglaciaire dans le Bassin Parisien. In: E. B. W. Zubrow / F. Audouze / J. G. Enloe (eds), The Magdalenian household: Unraveling domesticity (Albany, N.Y 2010) 176-197.

Bodu/Cary 1998: P. Bodu / S. Cary, Le niveau ancien. In: Bodu 1998, 322-343.

Bodu/Chehmana/Debout 2007: P. Bodu / L. Chehmana / G. Debout, Le Badegoulien de la moitié nord de la France. Un état des connaissances. Bulletin de la Société préhistorique française 104, 2007, 661-679.

Bodu/Debout/Bignon 2006: P. Bodu / G. Debout / O. Bignon, Variabilité des habitudes tardiglaciaires dans le Bassin parisien: l'organisation spatiale et sociale de l'Azilien ancien du Closeau. Bulletin de la Société préhistorique française 103, 2006, 711-728.

Bodu/Mevel 2008: P. Bodu / L. Mevel, Enquête autour des lames tranchantes de l'Azilien ancien. Le cas du niveau inférieur du Closeau (Rueil-Malmaison, Hauts-de-Seine, France). L'Anthropologie 112, 2008, 509-543.

Bodu/Orliac/Baffier 1996: P. Bodu / M. Orliac / D. Baffier, L'unité d'occupation de la section 27. In: Gaucher 1996, 68-85.

Bodu/Valentin 1997: P. Bodu / B. Valentin, Groupes à Federmesser ou Aziliens dans le sud et l'ouest du Bassin parisien. Propositions pour un nouveau modèle d'évolution. Bulletin de la Société préhistorique française 94, 1997, 341-348.

Bodu et al. 2006a: P. Bodu / A. Averbouh / O. Bignon / G. Dumarçay, Une autre unité sur le niveau IV0? L'unité Y127 de la section 44. Gallia Préhistoire 48, 2006, 140-148.

2006b: P. Bodu / M. Julien / B. Valentin / G. Debout / A. Averbouh / C. Bemilli / S. Beyries / O. Bignon / G. Dumarçay / J. G. Enloe / D. Joly / A. Lucquin / R. J. March / M. Orliac / M. Vanhaeren, Un dernier hiver à Pincevent: les Magdaléniens du niveau IV 0 (Pincevent, La Grande-Paroisse, Seine-et-Marne). Gallia Préhistoire 48, 2006, 1-180.

2009a: P. Bodu / M. Olive / B. Valentin / O. Bignon / G. Debout, Où sont les haltes de chasse? Discussion à partir des sites tardiglaciaires du Bassin Parisien. In: Valentin 2009, 191-202.

2009b: P. Bodu / G. Debout / G. Dumarçay / D. Leesch / B. Valentin, Révision de la chronologie magdalénienne dans le Bassin Parisien et alentours: Nouveaux résultats. In: Valentin 2009, 91-107.

2011: P. Bodu / M. Olive / B. Valentin / O. Bignon / G. Débout, Where are the hunting camps? A discussion based on Lateglacial sites in the Paris Basin. Palethnologie 3, 2011, 231-250.

Boekschoten/Stapert 1993: G. R. Boekschoten / D. Stapert, Rings and sectors: a computer package for spatial analysis. With examples from Oldeholtwolde and Gönnersdorf. Helinium 33, 1993, 20-35.

Bohmers 1947: A. Bohmers, Jong-Palaeolithicum en Vroeg-Mesolithicum. In: H. E. van Gelder / P. Glazema / G. A. Bontekoe / H. Halbertsma / W. Glasbergen (eds), Een kwart Eeuw Oudheidkundig Bodemonderzoek in Nederland. Gedenkboek A. E. Van Giffen (Meppel 1947) 129-201.

1956: A. Bohmers, Statistics and graphs in the study of flint assemblages. II. A preliminary report on the statistical analysis of the Younger Palaeolithic in northwestern Europe. Palaeohistoria 5, 1956, 7-25.

1961: A. Bohmers, Statistiques et graphiques dans l'étude des industries lithiques préhistoriques. V. Considérations générales aus sujet du Hambourgien, du Tjongerien, du Magdalénien et de l'Azilien. Palaeohistoria 8, 1961, 15-37.

Bohncke/Wijmstra 1988: S. Bohncke / L. Wijmstra, Reconstruction of Late-Glacial lake-level fluctuation in The Netherlands based on palaeobotanical analyses, geochemical results and pollen-density data. Boreas 17, 1988, 403-425.

Bokelmann 1979: K. Bokelmann, Rentierjäger am Gletscherrand in Schleswig-Holstein? Ein Diskussionsbeitrag zur Erforschung der Hamburger Kultur. Offa 36, 1979, 12-22.

1991: K. Bokelmann, Some new thoughts on old data on human and reindeer in the Ahrensburgian tunnel valley in Schleswig-Holstein, Germany. In: Barton/Roberts/Roe 1991, 72-81.

Bokelmann/Heinrich/Menke 1983: K. Bokelmann / D. Heinrich / B. Menke, Fundplätze des Spätglazials am Hainholz-Esinger Moor, Kreis Pinneberg. Offa 40, 1983, 199-239.

Bolus 1984: M. Bolus, Die spätpaläolithischen Steinartefakte vom Martinsberg in Andernach [unpublished M.A. thesis, Köln 1984].

1991: M. Bolus, Niederbieber and Andernach. Examples for the Final Palaeolithic settlement patterns in the Neuwied Basin (Central Rhineland). In: A. Montet-White (ed.), Les bassins du Rhin et du Danube au Paléolithique supérieur: environnement, habitat et systèmes d'échange. Colloque de la commission 8 de l'Union internationale des sciences préhistoriques et protohistoriques, Mayence 1986 (Liège 1991) 116-133.

1992: M. Bolus, Die Siedlungsbefunde des späteiszeitlichen Fundplatzes Niederbieber (Stadt Neuwied). Ausgrabungen 1981-1988. Monographien des RGZM 22 (Bonn 1992).

2002: M. Bolus, Gerölle zur Silexbearbeitung aus dem Magdalénien-Horizont vom Martinsberg in Andernach. In: Floss/Terberger 2002, 175-183.

Bolus/Street 1985: M. Bolus / M. Street, 100 Jahre Eiszeitforschung am Martinsberg in Andernach. Archäologisches Korrespondenzblatt 15, 1985, 1-9.

Bond et al. 1992: G. C. Bond / H. B. W. Heinrich / L. Labeyrie / J. McManus / J. Andrews / S. Huon / R. Jantschik / S. Clasen / C. Simet / K. Tedesco / M. Klas / G. Bonani / S. Ivy, Evidence for massive discharge of icebergs into the North Atlantic ocean during the last glacial. Nature 360, 1992, 245-249.

1999: G. C. Bond / W. Showers / M. Elliot / M. Evans / R. Lotti / I. Hajdas / G. Bonani / S. Johnson, The North Atlantic's 1-2 kyr Climate Rhythm: Relation to Heinrich Events, Dansgaard / Oeschger Cycles and the Little Ice Age. In: P. U. Clark / R. S. Webb / L. D. Keigwin (eds), Mechanisms of global climate change at millennial time scales (Washington, D.C. 1999) 35-58.

Bondevik et al. 2005: S. Bondevik / J. Mangerud / S. Dawson / A. Dawson / Ø. Lohne, Evidence for three North Sea tsunamis at the Shetland Islands between 8000 and 1500 years ago. Quaternary Science Reviews 24, 2005, 1757-1775.

Bonenfant et al. 2004: C. L. L. E. Bonenfant / A. Mysterud / R. Langvatn / N. C. Stenseth / J.-M. Gaillard / F. Klein, Multiple causes of sexual segregation in European red deer: enlightenments from varying breeding phenology at high and low latitude. Proceedings of the Royal Society B: Biological Sciences 271, 2004, 883-892.

Bordes 1958: F. Bordes, Nouvelles fouilles à Laugerie-Haute Est. Premiers résultats. L'Anthropologie 62, 1958, 205-244.

Bordes/Sonneville-Bordes 1979: F. Bordes / D. de Sonneville-Bordes, L'Azilianisation dans la vallée de la Dordogne: les données de la Gare de Couze (Dordogne) et de l'abri Morin (Gironde). In: de Sonneville-Bordes 1979, 449-459.

Bordes et al. 1974: F. Bordes / M. J. Graindor / P. Martin / Y. Martin, L'industrie de la grotte ornée de Gouy (Seine-Maritime). Bulletin de la Société préhistorique française 71, 1974, 115-118.

Bos 2001: J. A. A. Bos, Lateglacial and Early Holocene vegetation history of the northern Wetterau and the Amöneburger Basin (Hessen), central-west Germany. Review of Palaeobotany and Palynology 115, 2001, 177-212.

Bosinski 1977: H. Bosinski, Die Rondelle des Magdalénien-Fundplatzes Gönnersdorf. Ein Arbeitsbericht. Quartär 27/28, 1977, 153-160.

Bosinski 1978: G. Bosinski, Der Poggenwischstab. Bonner Jahrbücher 178, 1978, 83-92.

1979: G. Bosinski, Die Ausgrabungen in Gönnersdorf 1968-1976 und die Siedlungsbefunde der Grabung 1968. Der Magdalénien-Fundplatz Gönnersdorf 3 (Stuttgart 1979).

1987: G. Bosinski, Die Kunst des Magdalénien im Rheinland. In: H. Müller-Beck / C. Albarrán (eds), Die Anfänge der Kunst vor 30.000 Jahren [exhibition catalogue Tübingen] (Stuttgart 1987) 52-59.

1989: G. Bosinski, Die große Zeit der Eiszeitjäger. Europa zwischen 40000 und 10000 v. Chr. Theodor-Mommsen-Vorlesung von November 1987. Jahrbuch des RGZM 34, 1987 (1989), 3-139.

1992: G. Bosinski, The representation of female figures in the Rhineland Magdalenian. Proceedings of the Prehistoric Society 57, 1992, 51-64.

1996: G. Bosinski, Die Gravierungen des Magdalénien-Fundplatzes Andernach-Martinsberg. Jahrbuch des RGZM 41, 1994 (1996), 19-58.

2008: G. Bosinski, Tierdarstellungen von Gönnersdorf. Nachträge zu Mammut und Pferd sowie die übrigen Tierdarstellungen. Der Magdalénien-Fundplatz Gönnersdorf 9 (Regensburg 2008).

Bosinski/d'Errico/Schiller 2001: G. Bosinski / F. d'Errico / P. Schiller, Die gravierten Frauendarstellungen von Gönnersdorf. Der Magdalénien-Fundplatz Gönnersdorf 8 (Stuttgart 2001).

Bosinski/Fischer 1974: G. Bosinski / G. Fischer, Die Menschendarstellungen von Gönnersdorf der Ausgrabung von 1968. Der Magdalénien-Fundplatz Gönnersdorf 1 (Wiesbaden 1974).

1980: G. Bosinski / G. Fischer, Mammut- und Pferdedarstellungen von Gönnersdorf. Der Magdalénien-Fundplatz Gönnersdorf 5 (Stuttgart 1980).

Bosinski/Hahn 1972: G. Bosinski / J. Hahn, Der Magdalénienfundplatz Andernach-Martinsberg. In: H. Löhr (ed.), Beiträge zum Paläolithikum im Rheinland (Bonn 1972) 81-257.

Bosinski/Richter 1997: G. Bosinski / J. Richter, Paläolithikum und Mesolithikum. Geschichtlicher Atlas der Rheinlande. Beiheft II / I. Publikationen der Gesellschaft für Rheinische Geschichtskunde 12 N. F., Abt. 1b (Köln 1997).

Bosinski et al. 1982: G. Bosinski / R. Braun / E. Turner / P. Vaughan, Ein spätpaläolithisches Retuscheurdepot von Niederbieber/Neuwieder Becken. Archäologisches Korrespondenzblatt 12, 1982, 295-312.

Bosselin/Djindjian 1988: B. Bosselin / F. Djindjian, Un essai de structuration du Magdalénien français à partir de l'outillage lithique. Bulletin de la Société préhistorique française 85, 1988, 304-331.

Boulton et al. 2001: G. S. Boulton / P. Dongelmans / M. Punkari / M. Broadgate, Palaeoglaciology of an ice sheet through a glacial cycle: the European ice sheet through the Weichselian. Quaternary Science Reviews 20, 2001, 591-625.

Bourillet et al. 2003: J.-F. Bourillet / J.-Y. Reynaud / A. Baltzer / S. Zaragosi, The »Fleuve Manche«: the submarine sedimentary features from the outer shelf to the deep-sea fans. Journal of Quaternary Science 18, 2003, 261-282.

Bouvier 1977: J.-M. Bouvier, Un gisement préhistorique: La Madeleine (Périgueux 1977).

Bowman/Ambers/Lesse 1990: S. G. E. Bowman / J. C. Ambers / M. N. Lesse, Re-evaluation of British Museum radiocarbon dates issued between 1980 and 1984. Radiocarbon 32, 1990, 59-79.

Bradley/Stanford 2004: B. Bradley / D. Stanford, The North Atlantic ice-edge corridor: a possible Palaeolithic route to the New World. World Archaeology 36, 2004, 459-478.

Bradtmöller et al. 2012: M. Bradtmöller / A. Pastoors / B. Weninger / G.-C. Weniger, The repeated replacement model – Rapid climate change and population dynamics in Late Pleistocene Europe. Quaternary International 247, 2012, 38-49.

Brasser 2012: M. Brasser, Horse exploitation at the Late Upper Palaeolithic site of Oelknitz (Thuringia, Germany) with special reference to canine modifications. Quaternary International 252, 2012, 175-183.

Bratlund 1991: B. Bratlund, A study of hunting lesions containing flint fragments on reindeer bones at Stellmoor, Schleswig-Holstein, Germany. In: Barton/Roberts/Roe 1991, 193-207.

1996a: B. Bratlund, A survey of the subsistence and settlement pattern of the Hamburgian Culture in Schleswig-Holstein. Jahrbuch des RGZM 41, 1994 (1996), 59-93.

1996b: B. Bratlund, Hunting Strategies in the Late Glacial of Northern Europe: A Survey of the Faunal Evidence. Journal of World Prehistory 10, 1996, 1-48.

1999: B. Bratlund, A survey of the Ahrensburgian faunal assemblage of Stellmoor. In: S. K. Kozłowski / J. Gurba / L. L. Zaliznyak (eds), Tanged points cultures in Europe. Read at the International

Archaeological Symposium, Lublin, September, 13-16, 1993 (Lublin 1999) 47-59.

2002: B. Bratlund, The faunal remains of Wilczyce. In: Eriksen/Bratlund 2002, 101-107.

Brauer/Endres/Negendank 1999: A. Brauer / C. Endres / J. F. W. Negendank, Lateglacial calendar year chronology based on annually laminated sediments from Lake Meerfelder Maar, Germany. Quaternary International 61, 1999, 17-25.

Brauer et al. 1994: A. Brauer / I. Hajdas / J. F. W. Negendank / B. Rein / H. Vos / B. Zolitschka, Warvenchronologie. Eine Methode zur absoluten Datierung und Rekonstruktion kurzer und mittlerer solarer Periodizitäten. Geowissenschaften 12, 1994, 325-332.

1999: A. Brauer / C. Endres / C. Günter / T. Litt / M. Stebich / J. F. W. Negendank, High resolution sediment and vegetation responses to Younger Dryas climate change in varved lake sediments from Meerfelder Maar, Germany. Quaternary Science Reviews 18, 1999, 321-329.

2000a: A. Brauer / C. Günter / S. J. Johnsen / J. F. W. Negendank, Land-ice teleconnections of cold climatic periods during the last Glacial/Interglacial transition. Climate Dynamics 16, 2000, 229-239.

2000b: A. Brauer / C. Endres / B. Zolitschka / J. F. W. Negendank, AMS radiocarbon and varve chronology from the annually laminated sediments record of the Lake Meerfelder Maar, Germany. Radiocarbon 42, 2000, 355-368.

2008: A. Brauer / G. H. Haug / P. Dulski / D. M. Sigman / J. F. W. Negendank, An abrupt wind shift in western Europe at the onset of the Younger Dryas cold period. Nature Geoscience 1, 2008, 520-523.

Breuil 1913: H. Breuil, Les subdivisions du Paléolithique supérieur et leur signification. Congrès international d'anthropologie et d'archéologie préhistoriques. 14e session, Genève 1912 (Genève 1913) 165-238.

1954: H. Breuil, Le Magdalénien. Bulletin de la Société préhistorique française 51, 1954, 59-66.

Breuil/Graindor 1959: H. Breuil / M.-J. Graindor, Découverte de la grotte inscrite des Formures près de Rouen (Normandie). Comptes-rendus des séances de l'Académie des Inscriptions et Belles-Lettres 103, 1959, 19.

Brézillon 1968: M. N. Brézillon, La dénomination des objets de pierre taillée. Matériaux pour un vocabulaire des préhistoriens de langue française. Supplément à Gallia Préhistoire 4 (Paris 1968).

Bridault 1996: A. Bridault, Le problème de l'exploitation du gibier au Tureau-des-Gardes (Marolles-sur-Seine, Seine-et-Marne). In: Y. Pautrat (ed.), Paléolithique supérieur et Épipaléolithique dans le Nord-Est de la France. Actes de la table ronde de Dijon, 7 et 8 octobre 1995 (Dijon 1996) 141-151.

Bridault/Bemilli 1999: A. Bridault / C. Bemilli, La chasse et le traitement des animaux. In: Julien/Rieu 1999, 50-64.

Bridault/Bignon/Bemilli 2003: A. Bridault / O. Bignon / C. Bemilli, L'éxploitation du cheval à la fin du Tardiglaciaire dans le Bassin Parisien. In: S. Costamagno / V. Laroulandie (eds), Mode de vie au Magdalénien. Apports de l'archéozoologie / Zooarchaeological insights into Magdalenian lifeways. Actes du XIVème congrès UISPP, Université de Liège, Belgique, 2-8 septembre 2001, section 6: paleolithique superieur, colloque 6 (Oxford 2003) 33-45.

Bridault/Fontana 2003: A. Bridault / L. Fontana, Enregistrement des variations environnementales par les faunes chassées, dans les zones de moyenne montagne d'Europe occidentale, au tardiglaciaire et au début de l'holocène. In: M. Patou-Mathis / H. Bocherens (eds), Le rôle de l'environnement dans les comportements des chasseurs-cueilleurs préhistoriques. Actes du XIVème Congrès UISPP, Université de Liège, Belgique, 2-8 septembre 2001 (Oxford 2003) 55-65.

Bridault/Lang/Rieu 1997: A. Bridault / L. Lang / J. L. Rieu, Les sites magdaléniens du Grand Canton et du Tureau-des-Gardes à Marolles-sur-Seine (Seine-et-Marne). In: A. Thévenin / A. Villes (eds), Le paléolithique supérieur de l'Est de la France: de l'Aurignacien à l'Ahrensbourgien. Colloque de Chamont 1994 (Reims 1997) 119-128.

Bridault et al. 2000: A. Bridault / L. Chaix / G. Pion / C. Oberlin / S. Thiébault / J. Argant, Position chronologique du renne (Rangifer tarandus L.) à la fin du Tardiglaciaire dans les Alpes du Nord françaises et le Jura méridional. In: G. Pion (ed.), Le Paléolithique supérieur récent: nouvelles données sur le peuplement et l'environnement. Actes du table ronde de Chambéry, Mars 1999 (Paris 2000) 47-57.

Brock et al. 2011: F. Brock / S. Lee / R. A. Housley / C. Bronk Ramsey, Variation in the radiocarbon age of different fractions of peat: A case study from Ahrenshöft, northern Germany. Quaternary Geochronology 6, 2011, 550-555.

Broecker 2003: W. S. Broecker, Does the trigger for abrupt climate change reside in the ocean or in the atmosphere? Science 300, 2003, 1519-1522.

Broecker et al. 2010: W. S. Broecker / G. H. Denton / R. L. Edwards / H. Cheng / R. B. Alley / A. E. Putnam, Putting the Younger Dryas cold event into context. Quaternary Science Reviews 29, 2010, 1078-1081.

Bronk Ramsey 2008: C. Bronk Ramsey, Radiocarbon dating: revolutions in understanding. Archaeometry 50, 2008, 249-275.

Bronk Ramsey/Higham/Leach 2004: C. Bronk Ramsey / T. Higham / Ph. Leach, Towards high-precision AMS: progress and limitations. Radiocarbon 46, 2004, 17-24.

Bronk Ramsey/van der Plicht/Weninger 2001: C. Bronk Ramsey / J. van der Plicht / B. Weninger, »Wiggle matching« radiocarbon dates. Radiocarbon 43, 2001, 381-389.

Bronk Ramsey et al. 2000: C. Bronk Ramsey / P. B. Pettitt / R. E. M. Hedges / G. W. L. Hodgins / D. C. Owen, Radiocarbon dates from the Oxford AMS system: Archaeometry datelist 29. Archaeometry 42, 2000, 243-254.

2002: C. Bronk Ramsey / T. F. G. Higham / D. C. Owen / A. W. G. Pike / R. E. M. Hedges, Radiocarbon dates from the Oxford AMS system: Archaeometry datelist 31. Archaeometry 44, 2002, 1-149.

2012: C. Bronk Ramsey / R. A. Staff / C. L. Bryant / F. Brock / H. Kitagawa / J. van der Plicht / G. Schlolaut / M. H. Marshall / A. Brauer / H. F. Lamb / R. L. Payne / P. E. Tarasov / T. Haraguchi / K. Gotanda / H. Yonenobu / Y. Yokoyama / R. Tada / T. Nakagawa, A complete terrestrial radiocarbon record for 11.2 to 52.8 kyr B.P. Science 338, 2012, 370-374.

Brooks/Mayle/Lowe 1997: S. J. Brooks / F. E. Mayle / J. J. Lowe, Chironomid-based Lateglacial climatic reconstruction for southeast Scotland. Journal of Quaternary Science 12, 1997, 161-167.

Brooks et al. 2011: A. J. Brooks / S. L. Bradley / R. J. Edwards / N. Goodwyn, The palaeogeography of Northwest Europe during the last 20,000 years. Journal of Maps 7, 2011, 573-587.

Brunnacker 1978a: K. Brunnacker (ed.), Geowissenschaftliche Untersuchungen in Gönnersdorf. Der Magdalénien-Fundplatz Gönnersdorf 4 (Wiesbaden 1978).

1978b: K. Brunnacker, Geologischer Überblick. In: Brunnacker 1978a, 5-34.

1978c: K. Brunnacker, Stratigraphie im Bereich der Gönnersdorfer Siedlung: Profilaufbau. In: Brunnacker 1978a, 35-44.

1978d: K. Brunnacker, Stratigraphie im Bereich der Gönnersdorfer Siedlung: C14-Datierung. In: Brunnacker 1978a, 44-45.

1978e: K. Brunnacker, Stratigraphie im Bereich der Gönnersdorfer Siedlung: Geologisch-pedologische Befunde. In: Brunnacker 1978a, 65-79.

Buck/Litton/Scott 1994: C. E. Buck / C. D. Litton / E. M. Scott, Making the most of radiocarbon dating: some statistical considerations. Antiquity 68, 1994, 252-263.

Bullinger/Lämmli/Leuzinger-Piccand 1998: J. Bullinger / M. Lämmli / C. Leuzinger-Piccand, Le site magdalénien de plein air de Moosbühl: nouveaux éléments de datation et essai d'interprétation des données spatiales. Annuaire de la Société de Préhistoire et d'Archéologie 80, 1997 (1998), 7-26.

Bullinger/Leesch/Plumettaz 2006: J. Bullinger / D. Leesch / N. Plumettaz, Le site magdalénien de Monruz, 1. Premiers éléments pour l'analyse d'un habitat de plein air. Archéologie neuchâteloise 33 (Hauterive 2006).

Buoncristiani/Campy 2011: J.-F. Buoncristiani / M. Campy, Quaternary glaciations in the French Alps and Jura. In: Ehlers/Gibbard/Hughes 2011, 117-126.

Burdukiewicz 1981: J. M. Burdukiewicz, Creswellian and Hamburgian. (The Shouldered Point Technocomplex). In: J. K. Kozłowski / S. K. Kozłowski (eds), Prehistorie de la Grande Plaine de l'Europe. Actes du colloque international organisé dans le cadre de X^{e} congrès U.I.S.P.P. Mexico (commissions 10e et 14e) (Kraków, Warszawa 1981) 44-56.

1986: J. M. Burdukiewicz, The late Pleistocene shouldered point assemblages in western Europe (Leiden 1986).

Burdukiewicz/Schmider 2000: J. M. Burdukiewicz / B. Schmider, Analyse comparative des pointes à cran hambourgiennes du Bassin de l'Oder et des pointes à cran magdaléniennes du Bassin parisien. In: Valentin/Bodu/Christensen 2000, 97-108.

Burky et al. 1998: R. R. Burky / D. L. Kirner / R. E. Taylor / P. E. Hare / J. R. Southon, ^{14}C dating of bone using γ-carboxyglutamic acid and α-carboxyglycine (aminomalonate). Radiocarbon 40, 1998, 11-20.

Burleigh 1986: R. Burleigh, Radiocarbon dates for human and animal bones from Mendip caves. Proceedings of the University of Bristol Spelaeological Society 17, 1986, 267-274.

Burleigh/Hewson/Meeks 1976: R. Burleigh / A. Hewson / N. Meeks, British Museum natural radiocarbon measurements VIII. Radiocarbon 18, 1976, 16-42.

Buschkämper 1993: T. Buschkämper, Die Befunde im Südwestteil der Gönnersdorfer Grabungsfläche [unpublished M.A. thesis, Köln 1993].

Busschers et al. 2005: F. S. Busschers / H. J. T. Weerts / J. Wallinga / P. Cleveringa / C. Kasse / H. de Wolf / K. M. Cohen, Sedimentary architecture and optical dating of Middle and Late Pleistocene Rhine-Meuse deposits – fluvial response to climate change, sea-level fluctuation and glaciation. Netherlands Journal of Geosciences (Geologie en Mijnbouw) 84, 2005, 25-41.

2007: F. S. Busschers / C. Kasse / R. T. van Balen / J. Vandenberghe / K. M. Cohen / H. J. T. Weerts / J. Wallinga / C. Johns / P. Cleveringa / F. P. M. Bunnik, Late Pleistocene evolution of the Rhine-Meuse system in the southern North Sea basin: imprints of climate change, sea-level oscillation and glacio-isostacy. Quaternary Science Reviews 26, 2007, 3216-3248.

Butzer 1982: K. W. Butzer, Archaeology as human ecology (Cambridge 1982).

Byrne 1996: R. W. Byrne, Machiavellian intelligence. Evolutionary Anthropology 5, 1996, 172-180.

Campbell 1977: J. B. Campbell, The Upper Palaeolithic of Britain. A study of man and nature in the late Ice Age (Oxford 1977).

Campbell et al. 2005: R. D. Campbell / F. Rosell / B. A. Nolet / V. A. A. Dijkstra, Territory and group sizes in Eurasian beavers (*Castor fiber*): echoes of settlement and reproduction? Behavioral Ecology and Sociobiology 58, 2005, 597-607.

Campos et al. 2010: P. F. Campos / E. Willerslev / A. Sher / L. Orlando / E. Axelsson / A. Tikhonov / K. Aaris-Sørensen / A. D. Greenwood / R.-D. Kahlke / P. Kosintsev / T. Krakhmalnaya / T. Kuznetsova / P. Lemey / R. MacPhee / C. A. Norris / K. Shepherd / M. A. Suchard / G. D. Zazula / B. Shapiro / M. T. P. Gilbert, Ancient DNA analyses exclude humans as the driving force behind late Pleistocene musk ox (*Ovibos moschatus*) population dynamics. Proceedings of the National Academy of Sciences of the United States of America 107, 2010, 5675-5680.

Capitan/Peyrony 1928: L. Capitan / D. Peyrony, La Madeleine: son gisement, son industrie, ses oeuvres d'art. Institut international d'anthropologie (Paris 1928).

Carr et al. 2006: S. J. Carr / R. Holmes / J. J. M. van der Meer / J. Rose, The Last Glacial Maximum in the North Sea Basin: micromorphological evidence of extensive glaciation. Journal of Quaternary Science 21, 2006, 131-153.

Caspar/De Bie 1996: J.-P. Caspar / M. De Bie, Preparing for the hunt in the Late Paleolithic camp at Rekem, Belgium. Journal of Field Archaeology 23, 1996, 437-460.

Cattelain 1997: P. Cattelain, Hunting during the Upper Paleolithic: bow, spearthrower, or both. In: H. Knecht (ed.), Projectile technology (New York NY 1997) 213-240.

Célérier 1979: G. Célérier, Inventaire morphologique de pointes aziliennes en Périgord. Un projet de rationalisation. In: de Sonneville-Bordes 1979, 461-466.

Célérier/Chollet/Hantaï 1997: G. Célérier / A. Chollet / A. Hantaï, Nouvelles observations sur l'évolution de l'Azilien dans les gisements du Bois-Ragot (Vienne) et de Pont d'Ambon (Dordogne). Bulletin de la Société préhistorique française 94, 1997, 331-336.

Célérier/Nisole/Beaune 1993: G. Célérier / S. Nisole / S. de Beaune, L'abri sous roche de Pont d'Ambon à Bourdeilles (Dordogne). Gallia Préhistoire 35, 1993, 1-137.

Charles 1996: R. Charles, Back into the North: the radiocarbon evidence for human recolonisation of the North-Western Ardennes after the Last Glacial Maximum. Proceedings of the Prehistoric Society 62, 1996, 1-17.

1997a: R. Charles, Does the Belgian Creswellian exist? Lithics: The Journal of the Lithic Studies Society 17/18, 1997, 25-45.

1997b: R. Charles, The exploitation of carnivores and other furbearing mammals during the North-western European Late and Upper Paleolithic and Mesolithic. Oxford Journal of Archaeology 16, 1997, 253-277.

1998: R. Charles, Late Magdalenian chronology and faunal exploitation in the north-western Ardennes. British Archaeological Reports, International Series 737 (Oxford 1998).

1999: R. Charles, Garrod and the Belgian Creswellian. In: W. Davies / R. Charles (eds), Dorothy Garrod and the progress of the Palaeolithic. Studies in the prehistoric archaeology of the Near East and Europe (Oxford 1999) 57-75.

Charlton 1997: B. G. Charlton, Injustice, inequality and evolutionary psychology. Journal of Health Psychology 2, 1997, 413-425.

Chaussé 2003: C. Chaussé, Les nappes alluvialles de la basse vallée de l'Yonne, approches géométriques et chronostratigraphiques et l'apport de l'étude de la Nappe de Soucy à la compréhension des occupations du Paléolithique inférieur de Soucy (Lille 2003).

2005: C. Chaussé, Les horizons pédologiques tardiglaciaires du Closeau à Rueil-Malmaison (92). Premiers éléments de diagnose. In: B. Valentin / P. Bodu / M. Julien (eds), Habitats et peuplements tardiglaciaires du Bassin parisien. Bilan des activités de 2003 à 2005 (Paris 2005) 87-93.

Cheddadi/Bar-Hen 2009: R. Cheddadi / A. Bar-Hen, Spatial gradient of temperature and potential vegetation feedback across Europe during the late Quaternary. Climate Dynamics 32, 2009, 371-379.

Cheddadi et al. 2006: R. Cheddadi / G. G. Vendramin / T. Litt / L. François / M. Kageyama / S. Lorentz / J.-M. Laurent / J.-L. de Beaulieu / L. Sadori / A. Jost / D. Lunt, Imprints of glacial refugia in the modern genetic diversity of Pinus sylvestris. Global Ecology and Biogeography 15, 2006, 271-282.

Cheynier 1966: A. Cheynier, Considérations sur le passage du Magdalénien à l'Azilien. Bulletin de la Société préhistorique française 63, 1966, CXXXV-CXXXVIII.

Chmielewska 1961: M. Chmielewska, Huttes d'habitation épipaléolithiques de Witów, district de Leczyca. Acta Archaeologica Universitatis Lodziensis 10, 1961, 1-137.

Chmielewska/Chmielewski 1960: M. Chmielewska / W. Chmielewski, Stratigraphie et chronologie de la dune de Witów, distr. de Leczyca. Biuletyn Periglacjalny 8, 1960, 132-141.

Chmielewski 1961: W. Chmielewski, Civilisations épipaléolithiques en Pologne centrale. Bulletin de la Société des Sciences et des Lettres de Lódz 12, 1961, 1-passim.

Christensen 1998: M. Christensen, Analyse fonctionnelle des outils lithiques du niveau à Federmesser ancien du Closeau. In: Bodu 1998, 363-384.

Churski 1998: Z. Churski, Location of Lake Gościąż and the Gostynińskie Lake District. In: Ralska-Jasiewiczowa et al. 1998, 17-18.

Clark et al. 2001: P. U. Clark / S. J. Marshall / G. K. C. Clarke / S. W. Hostetler / J. M. Licciardi / J. T. Teller, Freshwater forcing of abrupt climate change during the Last Glaciation. Science 293, 2001, 283-287.

2002: P. U. Clark / N. G. Pisias / T. F. Stocker / A. J. Weaver, The role of the thermohaline circulation in abrupt climate change. Nature 415, 2002, 863-869.

2004: C. D. Clark / D. J. A. Evans / A. Khatwa / T. Bradwell / C. J. Jordan / S. H. Marsh / W. A. Mitchell / M. D. Bateman, Map and GIS database of glacial landforms and features related to the last British Ice Sheet. Boreas 33, 2004, 359-375.

2009: P. U. Clark / A. S. Dyke / J. D. Shakun / A. E. Carlson / J. Clark / B. Wohlfarth / J. X. Mitrovica / S. W. Hostetler / A. M. McCabe, The Last Glacial Maximum. Science 325, 2009, 710-714.

2012: C. D. Clark / A. L. C. Hughes / S. L. Greenwood / C. Jordan / H. P. Sejrup, Pattern and timing of retreat of the last British-Irish Ice Sheet. Quaternary Glaciation History of Northern Europe. Quaternary Science Reviews 44, 2012, 112-146.

Clarke 1968: D. L. Clarke, Analytical Archaeology (London 1968).

Clausen 1998: I. Clausen, Neue Untersuchungen an späteiszeitlichen Fundplätzen der Hamburger Kultur bei Ahrenshöft, Kr. Nordfriesland. Ein Vorbericht. Archäologische Nachrichten aus Schleswig-Holstein 8, 1998, 8-49.

1999: I. Clausen, Alt Duvenstedt, Kr. Rendsburg-Eckernförde. Jungpaläolithische Stationen LA 120A und LA 121. Offa 53, 1999, 372-373.

2004: I. Clausen, The Reindeer antler axe of the Allerød period from Klappholz LA 63, Kreis Schleswig-Flensburg/Germany. Is it a relict of the Federmesser, Bromme or Ahrensburg Culture? In: Terberger/Eriksen 2004, 141-164.

2010: I. Clausen, Auf den Spuren von Alfred Rust und seinen Rentierjägern. Aktuelle Untersuchungen am Stellmoorhügel bei Ahrensburg, Kr. Stormarn. Archäologische Nachrichten aus Schleswig-Holstein 2010, 15-19.

Clottes/Arnold 2001: J. Clottes / M. Arnold (eds), La grotte Chauvet. L'art des origines. Collection »Arts rupestres« (Paris 2001).

Coles 1998: B. J. Coles, Doggerland: a speculative survey. Proceedings of the Prehistoric Society 64, 1998, 45-81.

2000: B. J. Coles, Doggerland: the cultural dynamics of a shifting coastline. In: K. Pye / J. R. L. Allen (eds), Coastal and estuarine environments. Sedimentology, geomorphology and geoarchaeology (Bath 2000) 393-401.

Conard/Bolus 2003: N. J. Conard / M. Bolus, Radiocarbon dating the appearance of modern humans and timing of cultural innovations in Europe: new results and new challenges. Journal of Human Evolution 44, 2003, 331-371.

Conard/Floss 1999: N. J. Conard / H. Floss, Une pierre peinte du Hohle Fels (Baden-Württemberg, Allemagne) et la question de l'art pariétal paléolithique en Europe centrale. Paléo 11, 1999, 167-176.

Conard/Malina 2011: N. J. Conard / M. Malina, Neue Eiszeitkunst und weitere Erkenntnisse über das Magdalénien vom Hohle Fels bei Schelklingen. Archäologische Ausgrabungen Baden-Württemberg 2010 (2011), 56-60.

Conneller 2007: C. Conneller, Inhabiting new landscapes: Settlement and mobility in Britain after the Last Glacial Maximum. Oxford Journal of Archaeology 26, 2007, 215-237.

Conneller/Ellis 2007: C. Conneller / C. Ellis, A Final Upper Palaeolithic Site at La Sagesse Convent, Romsey, Hampshire. Proceedings of the Prehistoric Society 73, 2007, 191-228.

Coope/Lemdahl 1995: G. R. Coope / G. Lemdahl, Regional differences in the Lateglacial climate of northern Europe based on coleopteran analysis. Journal of Quaternary Science 10, 1995, 391-395.

Coope et al. 1998: G. R. Coope / G. Lemdahl / J. J. Lowe / A. Walkling, Temperature gradients in northern Europe during the last glacial-Holocene transition (14-9 ^{14}C kyr BP) interpreted from coleopteran assemblages. Journal of Quaternary Science 13, 1998, 419-433.

Cooper/Sheets 2012: J. Cooper / P. D. Sheets (eds), Surviving sudden environmental change. Answers from archaeology (Boulder Colo. 2012).

Cordain et al. 2000: L. Cordain / J. Brand Miller / S. B. Eaton / N. Mann / S. H. A. Holt / J. D. Speth, Plant-animal subsistence ratios and macronutrient energy estimations in worldwide hunter-gatherer diets. American Journal of Clinical Nutrition 71, 2000, 682-692.

Cordy/Lacroix 1997: J.-M. Cordy / P. Lacroix, Bio- et chronostratigraphie de la grotte due Bois Laiterie (Profondeville, Namur) à partir des microvértébres. In: Otte/Straus 1997, 161-176.

Cordy 1990: J.-M. Cordy, La faune du tardiglaciaire en Haute-Normandie. In: G. Fosse / A. Ropars / G. Habasque (eds), Derniers chasseurs, premiers agriculteurs (Rouen 1990) 13-14.

1991: J.-M. Cordy, Palaeoecology of the Late Glacial and early Postglacial of Belgium and neighbouring areas. In: Barton/Roberts/Roe 1991, 40-47.

Coudret/Fagnart 1997: P. Coudret / J.-P. Fagnart, Les industries à *Federmesser* dans le bassin de la Somme: Chronologie et identité des groupes culturels. Bulletin de la Société préhistorique française 94, 1997, 349-359.

2004: P. Coudret / J.-P. Fagnart, Les fouilles du gisement paléolithique final de Saleux (Somme). Revue Archéologique de Picardie 1/2, 2004, 3-17.

2006: P. Coudret / J.-P. Fagnart, Données préliminaires sur les habitats des groupes de la tradition à *Federmesser* du bassin de la Somme. Bulletin de la Société préhistorique française 103, 2006, 729-740.

2012: P. Coudret / J.-P. Fagnart, Les occupations préhistoriques du »Marais de Conty« (Somme). In: Antoine et al. 2012, 61-88.

Courty 1997: M. A. Courty, Étude micro-stratigraphique de la Grotte du Bois Laiterie (Provinve de Namur, Belgique). In: Otte/Straus 1997, 113-140.

Crema/Bevan/Lake 2010: E. R. Crema / A. Bevan / M. W. Lake, A probabilistic framework for assessing spatio-temporal point patterns in the archaeological record. Journal of Archaeological Science 37, 2010, 1118-1130.

Cremadès 1992: M. Cremadès, Les cortex gravés. In: Schmider 1992f, 232-243.

Croisset/Schmider 1992: E. de Croisset / B. Schmider, Les produits de débitge. In: Schmider 1992f, 87-107.

Currant 1987: A. P. Currant, Late Pleistocene saiga antelope *Saiga tatarica* on Mendip. Proceedings of the University of Bristol Spelaeological Society 18, 1987, 74-80.

1991: A. P. Currant, A Late Glacial Interstadial mammal fauna from Gough's Cave, Somerset, England. In: Barton/Roberts/Roe 1991, 48-50.

Cyrek et al. 2000: K. Cyrek / A. Nadachowski / T. Madeyska / Z. Bocheński / T. Tomek / P. Wojtal / B. Miękina / G. Lipecki / A. Garapich / B. Rzebik-Kowalska / E. Stworzewicz / M. Wolsan / J. Godawa / R. Kościów / Ł. Fostowicz-Frelik / Z. Szyndlar, Excavation in the Deszczowa Cave (Kroczyckie Rocks, Częstochowa Upland, Central Poland). Folia Quaternaria 71, 2000, 5-84.

Cziesla 1990: E. Cziesla, Siedlungsdynamik auf steinzeitlichen Fundplätzen. Methodische Aspekte zur Analyse latenter Strukturen. Studies in Modern Archaeology 2 (Bonn 1990).

2001: E. Cziesla, Zur Besiedlungsgeschichte von Berlin-Brandenburg: Die Anfänge. In: Gehlen/Heinen/Tillmann 2001, 381-396.

2007a: E. Cziesla, Einige Hypothesen zur Verwendung zweireihiger Widerhakenspitzen des nordeuropäischen Flachlandes. In: M. Masojć / T. Płonka / B. Ginter / S. K. Kozłowski (eds), Contributions to the Central European Stone Age. Papers dedicated to the late Professor Zbigniew Bagniewski (Wrocław 2007) 19-32.

2007b: E. Cziesla, Robbenjagd in Brandenburg? Gedanken zur Verwendung großer Widerhakenspitzen. Ethnographisch-Archäologische Zeitschrift 48, 2007, 1-48.

Cziesla/Masojć 2007: E. Cziesla / M. Masojć, Die einreihige Widerhakenspitze aus Węgliny (polnische Niederlausitz) und ihre kulturgeschichtliche Deutung. Archäologisches Korrespondenzblatt 37, 2007, 457-469.

Cziesla/Pettitt 2003: E. Cziesla / P. Pettitt, AMS-^{14}C-Datierungen von spätpaläolithischen und mesolithischen Funden aus dem Bützsee (Brandenburg). Archäologisches Korrespondenzblatt 33, 2003, 21-38.

Dahl-Jensen et al. 2002: D. Dahl-Jensen / N. S. Gundestrup / H. Miller / O. Watanabe / S. J. Johnsen / J. P. Steffensen / H. B. Clausen / A. Svensson / L. B. Larsen, The NorthGRIP deep drilling programme. Annales of Glaciology 35, 2002, 1-4.

Daniels/Boyd/Saunders 1953: F. Daniels / C. A. Boyd / D. F. Saunders, Thermoluminescence as a research tool. Science 117, 1953, 343-349.

Dansgaard et al. 1993: W. Dansgaard / S. J. Johnsen / H. B. Clausen / D. Dahl-Jensen / N. S. Gundestrup / C. U. Hammer / C. S. Hvidberg / J. P. Steffensen / A. E. Sveinbjörnsdottir / J. Jouzel / G. Bond, Evidence for general instability in past climate from 250 kyr ice-core record. Nature 364, 1993, 218-220.

David 1994: F. David, La faune de Pincevent et Verberie. In: Y. Taborin (ed.), Environnements et habitats magdaléniens dans le centre du Bassin parisien (Paris 1994) 105-110.

David/d'Errico/Thévenin 1998: S. David / F. d'Errico / A. Thévenin, L'art mobilier de Ranchot (Jura) et de Rochedane (Doubs). In: C. Cupillard (ed.), Les derniers chasseurs-cueilleurs du Massif jurassien et de ses marges. (13000-5500 avant Jésus-Christ) [exhibition catalogue Besançon] (Lons-le-Saunier 1998) 192-200.

Davies et al. 2002: S. M. Davies / N. P. Branch / J. J. Lowe / C. S. M. Turney, Towards a European tephrochronological framework for Termination 1 and the Early Holocene. Philosophical Transactions of the Royal Society of London A 360, 2002, 767-802.

De Angelis et al. 2003: M. De Angelis / J. Simões / H. Bonnaveira / J. D. Taupin / R. J. Delmas, Volcanic eruptions recorded in the Illimani ice core (Bolivia): 1918-1998 and Tambora periods. Atmospheric Chemistry and Physics 3, 2003, 1725-1741.

De Bie/Caspar 2000: M. De Bie / J.-P. Caspar, Rekem. A Federmesser camp on the Meuse River bank. Acta Archaeologica Lovaniensia Monographiae 10 (Leuven, Asse-Zellik 2000).

De Bie/Van Gils 2006: M. De Bie / M. Van Gils, Les habitats des groupes à *Federmesser* (aziliens) dans le Nord de la Belgique. Bulletin de la Société préhistorique française 103, 2006, 781-790.

De Bie/Van Gils/Deforce 2009: M. De Bie / M. Van Gils / K. Deforce, Human occupation in a Late Glacial landscape: The *Federmessergruppen* site complex at Lommel Maatheide (Belgium). In: Street/Barton/Terberger 2009, 77-87.

De Bie/Vermeersch 1998: M. De Bie / P. M. Vermeersch, Pleistocene-Holocene transition in Benelux. Quaternary International 49/50, 1998, 29-43.

d'Errico 1994: F. d'Errico, L'art gravé azilien. De la technique à la signification. Supplément à Gallia Préhistoire 31 (Paris 1994).

d'Errico/Ucelli Gnesutta 1999: F. d'Errico / P. Ucelli Gnesutta, L'art mobilier épigravettien de la grotte de Settecannelle (Viterbo, Italie). Contexte archéologique, analyse technique et stylistique. L'Anthropologie 103, 1999, 121-160.

d'Errico et al. 2011: F. d'Errico / W. E. Banks / M. Vanhaeren / V. Laroulandie / M. Langlais, PACEA geo-referenced radiocarbon database. PaleoAnthropology 2011, 1-12.

De Smedt et al. 2013: P. De Smedt / M. Van Meirvenne / N. S. Davies / M. Bats / T. Saey / J. De Reu / E. Meerschman / V. Gelorini / A. Zwertvaegher / M. Antrop / J. Bourgeois / P. De Maeyer / P. A. Finke / J. Verniers / P. Crombé, A multidisciplinary approach to reconstructing Late Glacial and Early Holocene landscapes. Journal of Archaeological Science 40, 2013, 1260-1267.

Dean 1978: J. S. Dean, Independent dating in archaeological analysis. Advances in Archaeological Method and Theory 1, 1978, 223-255.

Débout et al. 2012: G. Débout / M. Olive / O. Bignon / P. Bodu / L. Chehmana / B. Valentin, The Magdalenian in the Paris Basin: New results. Quaternary International 272-273, 2012, 176-190.

Deeben 1988: J. Deeben, The Geldrop sites and the *Federmesser* occupation of the Southern Netherlands. In: Otte 1988, 357-398.

Deeben et al. 2006: J. Deeben / O. Brinkkemper / B. Groenewoudt / R. Lauwerier, Een Federmesser-site van de Enterse Akkers (gemeente Wierden, Overijssel). In: B. J. Groenewoudt / R. M. van Heeringen / G. H. Scheepstra (eds), Het zandeilandenrijk van Overijssel. Bundel verschenen ter gelegenheid van de pensionering van A. D. Verlinde als archeoloog in, vor en van Overijssel (Amersfoort 2006) 49-82.

Deevey et al. 1954: E. S. Deevey, Jr. / M. S. Gross / G. E. Hutchinson / H. L. Kraybill, The natural C14 contents of material from hard-water lakes. Proceedings of the National Academy of Sciences of the United States of America 40, 1954, 285-288.

Delibrias et al. 1976: G. Delibrias / M. T. Guillier / J. Evin / J. Thommeret / Y. Thommeret, Datations absolues des depots quaternaires et des sites prehistoriques par la methode du Carbone 14. In: H. de Lumley (ed.), La préhistoire française. Les civilisations paléolithiques et mésolithiques de la France (Paris 1976) 1499-1514.

Delibrias/Guillier/Labeyrie 1970: G. Delibrias / M. T. Guillier / J. Labeyrie, Gif natural radiocarbon measurements V. Radiocarbon 12, 1970, 421-443.

Delisle et al. 2009: G. Delisle / S. Grassmann / B. Cramer / J. Messner / J. Winsemann, Estimating episodic permafrost development in northern Germany during the Pleistocene. In: M. J. Hambrey / P. Christoffersen / N. F. Glasser / B. Hubbard (eds), Glacial Sedimentary Processes and Products (Hoboken 2009) 109-119.

Deloze et al. 1999: V. Deloze / V. Krier / C. Leroyer / P. Rodriguez / M. Fontugne / M. Julien, Approches chronostratigraphique et paléoenvironnementale. In: Julien/Rieu 1999, 26-39.

Delpech 2012: F. Delpech, Biostratigraphy of the Solutrean layers of Laugerie-Haute (Les Eyzies, Dordogne). Archaeological implications. Paléo 23, 2012, 105-116.

Delporte 1966: H. Delporte, Le Paléolithique dans le Massif central: I. Le Magdalénien des vallées supérieures de la Loire et de l'Allier. Bulletin de la Société préhistorique française 63, 1966, 181-207.

Deplazes et al. 2013: G. Deplazes / A. Luckge / L. C. Peterson / A. Timmermann / Y. Hamann / K. A. Hughen / U. Rohl / C. Laj / M. A. Cane / D. M. Sigman / G. H. Haug, Links between tropical rainfall and North Atlantic climate during the last glacial period. Nature Geoscience 6, 2013, 213-217.

Desbrosse/Kozłowski 1989: R. Desbrosse / J. K. Kozłowski, Les origines du »Creswello-Hambourgien«. L'Anthropologie 93, 1989, 183-188.

Deschamps et al. 2012: P. Deschamps / N. Durand / E. Bard / B. Hamelin / G. Camoin / A. L. Thomas / G. M. Henderson / J. Okuno / Y. Yokoyama, Ice-sheet collapse and sea-level rise at the Bølling warming 14,600 years ago. Nature 483, 2012, 559-564.

Deville/Gautier 1997: J. Deville / A. Gautier, The avifauna of la grotte du Bois Laiterie. In: Otte/Straus 1997, 215-218.

Dewez 1987: M. Dewez, Le Paléolithique supérieur récent dans les grottes de Belgique. Publications d'Histoire de l'Art et de l'Archéologie de l'Université catholique de Louvain 57 (Leuven 1987).

Dietrich/Seelos 2009: S. Dietrich / K. Seelos, The reconstruction of paleo wind directions for the Eifel region (Central Europe) during the period 40.3-12.9 kaBP. Climate of the Past 5, 2009, 2157-2176.

Dobres/Hoffman 1994: M.-A. Dobres / C. R. Hoffman, Social agency and the dynamics of prehistoric technology. Journal of Archaeological Method and Theory 1, 1994, 211-258.

Dokken/Jansen 1999: T. M. Dokken / E. Jansen, Rapid changes in the mechanism of ocean convection during the last glacial period. Nature 401, 1999, 458-461.

Dörfler et al. 2012: W. Dörfler / I. Feeser / C. van den Bogaard / S. Dreibrodt / H. Erlenkeuser / A. Kleinmann / J. Merkt / J. Wiethold, A high-quality annually laminated sequence from Lake Belau, Northern Germany: Revised chronology and its implications for palynological and tephrochronological studies. The Holocene 22, 2012, 1413-1426.

Drucker/Bridault/Cupillard 2012: D. G. Drucker / A. Bridault / C. Cupillard, Environmental context of the Magdalenian settlement in the Jura Mountains using stable isotope tracking (^{13}C, ^{15}N, ^{34}S) of bone collagen from reindeer (Rangifer tarandus). Quaternary International 272-273, 2012, 322-332.

Drucker et al. 2003: D. Drucker / H. Bocherens / A. Bridault / D. Billiou, Carbon and nitrogen isotopic composition of red deer (Cervus elaphus) collagen as a tool for tracking palaeoenvironmental change during the Late-Glacial and Early Holocene in the northern Jura (France). Palaeogeography, Palaeoclimatology, Palaeoecology 195, 2003, 375-388.

2009: D. G. Drucker / A. Bridault / P. Iacumin / H. Bocherens, Bone stable isotopic signatures (^{15}N, ^{18}O) as tracers of temperature variation during the Late-glacial and early Holocene: case study on red deer Cervus elaphus from Rochedane (Jura, France). Geological Journal 44, 2009, 593-604.

Ducasse 2012: S. Ducasse, What is left of the Badegoulian »interlude«? New data on cultural evolution in southern France between 23,500 and 20,500 cal. BP. Quaternary International 272-273, 2012, 150-165.

Ducasse/Langlais 2007: S. Ducasse / M. Langlais, Entre Badegoulien et Magdalénien, nos coeurs balancent... Approche critique des industries lithiques du Sud de la France et du Nord-Est espagnol entre 19000 et 16500 BP. Bulletin de la Société préhistorique française 104, 2007, 771-785.

Dumarçay 2001: G. Dumarçay, Étude méthodologique des roches chauffées: approche classique, pétrologique et archéométrique des roches du gisement magdalénien final d'Ètigny le Brassot (Yonne) [unpublished M.A. thesis, Paris 2001].

Dunbar 1998: R. I. M. Dunbar, The social brain hypothesis. Evolutionary Anthropology 6, 1998, 178-190.

Ebert et al. 2010: C. Ebert / D. Huckschlag / H. K. Schulz / U. Hohmann, Can hair traps sample wild boar (*Sus scrofa*) randomly for the purpose of non-invasive population estimation? European Journal of Wildlife Research 56, 2010, 583-590.

Edwards/Brooks 2008: R. Edwards / A. Brooks, The island of Ireland: Drowning the myth of an Irish land-bridge? The Irish Naturalists' Journal 29, 2008, 19-34.

Eerkens 2000: J. W. Eerkens, Practice makes within 5 % of perfect: visual perception, motor skills, and memory in artifact variation. Current Anthropology 41, 2000, 663-667.

Eerkens/Lipo 2005: J. W. Eerkens / C. P. Lipo, Cultural transmission, copying errors, and the generation of variation in material culture and the archaeological record. Journal of Anthropological Archaeology 24, 2005, 316-334.

Ehlers/Gibbard/Hughes 2011: J. Ehlers / P. L. Gibbard / P. D. Hughes (eds), Quaternary Glaciations – Extent and Chronology. A closer look (Amsterdam 2011).

Eickhoff 1988: S. Eickhoff, Die Artefakte aus westeuropäischem Feuerstein des Magdalenien-Fundplatzes Gönnersdorf [unpublished M.A. thesis, Köln 1988].

1989: S. Eickhoff, Zur Analyse der latenten Strukturen und der Siedlungsdynamik der Konzentration II von Gönnersdorf. Jahrbuch des RGZM 36, 1989, 117-158.

Eickhoff-Cziesla 1992: S. Eickhoff-Cziesla, Wohnplatzstrukturen – Strukturierung eines Wohnplatzes. Zur Besiedlungsgeschichte des Magdalénien-Fundplatzes Andernach [unpublished dissertation, Köln 1992].

Eldredge/Gould 1972: N. Eldredge / S. J. Gould, Punctuated equilibria: an alternative to phyletic gradualism. In: T. J. M. Schopf (ed.), Models in paleobiology (San Francisco 1972) 82-115.

Eldredge et al. 2005: N. Eldredge / J. N. Thompson / P. M. Brakefield / S. Gavrilets / D. Jablonski / J. B. C. Jackson / R. E. Lenski / B. S. Lieberman / M. A. McPeek / W. Miller, III, The dynamics of evolutionary stasis. Paleobiology 31, 2005, 133-145.

Ellenberg et al. 2001: H. Ellenberg / H. E. Weber / R. Düll / V. Wirth / W. Werner / D. Paulißen, Zeigerwerte von Pflanzen in Mitteleuropa. Scripta geobotanica 18 (Göttingen 2001).

Ellis 1997: C. J. Ellis, Factors influencing the use of stone projectile tips. An ethnographic perspective. In: H. Knecht (ed.), Projectile technology (New York NY 1997) 37-74.

Embry-Barbier 1997: A. Embry-Barbier, Analyse palynologique de la grotte du Bois Laiterie. In: Otte/Straus 1997, 141-142.

Emery-Barbier/Rodriguez 1996: A. Emery-Barbier / P. Rodriguez, Pollen et coquilles. In: Gaucher 1996, 53-64.

Enloe 1997: J. G. Enloe, Seasonality and age structure in remains of Rangifer tarandus: Magdalenian hunting strategy at Verberie. Anthropozoologica 25-26, 1997, 95-102.

2000: J. G. Enloe, Readaptation: changes in Magdalenian subsistence and social organization. In: Peterkin/Price 2000, 115-120.

2006: J. G. Enloe, Geological processes and site structure: assessing integrity at a Late Paleolithic open-air site in northern France. Geoarchaeology 21, 2006, 523-540.

Enloe/David 1989: J. G. Enloe / F. David, Le remontage des os par individus: le partage du renne chez les Magdaléniens de Pincevent (La Grande Paroisse, Seine-et-Marne). Bulletin de la Société préhistorique française 86, 1989, 275-281.

Eriksen 1991: B. V. Eriksen, Change and continuity in a prehistoric hunter-gatherer society: a study of cultural adaptation in late glacial – early postglacial southwestern Germany. Archaeologica venatoria 12 (Tübingen 1991).

1996: B. V. Eriksen, Resource exploitation, subsistence strategies, and adaptiveness in Late Pleistocene – Early Holocene Northwest Europe. In: L. G. Straus / B. V. Eriksen / J. M. Erlandson / D. R. Yesner (eds), Humans at the end of the Ice Age. The archaeology of the Pleistocene-Holocene transition (New York 1996) 101-128.

2002: B. V. Eriksen, Fossil mollusks and exotic raw materials in Late Glacial and Early Postglacial find contexts: A complement to lithic studies. In: Fisher/Eriksen 2002, 27-52.

Eriksen/Bratlund 2002: B. V. Eriksen / B. Bratlund (eds), Recent studies in the Final Palaeolithic of the European Plain. Proceedings of a U.I.S.P.P. symposium, Stockholm, 14.-17. October 1999 (Aarhus 2002).

European Commission/Eurostat 2002: European Commission / Eurostat (eds), Water Sheds Pan Europe, scale 1 Million. GISCO reference database (Bruxelles 2002).

Eusterhues et al. 2002: K. Eusterhues / J. Lechterbeck / J. Schneider / U. Wolf-Brozio, Late- and Post-Glacial evolution of Lake Steisslingen (I). Sedimentary history, palynological record and inorganic geochemical indicators. Palaeogeography, Palaeoclimatology, Palaeoecology 187, 2002, 341-371.

Evan 2012: A. T. Evan, Atlantic hurricane activity following two major volcanic eruptions. Journal of Geophysical Research – Atmospheres 117, 2012, D06101.

Evin/Marien/Pachiaudi 1975: J. Evin / G. Marien / C. Pachiaudi, Lyon natural radiocarbon measurements V. Radiocarbon 17, 1975, 4-34.

1978: J. Evin / G. Marien / C. Pachiaudi, Lyon natural radiocarbon measurements VII. Radiocarbon 20, 1978, 19-57.

Faegri/Iversen 1950: K. Faegri / J. Iversen, Text-book of modern pollen analysis (Copenhagen 1950).

Fagnart 1982: J.-P. Fagnart, Le gisement Paléolithique Final du Hameau d'Étouvie à Amiens (Somme). Revue Archéologique de Picardie 4, 1982, 36-48.

1997: J.-P. Fagnart, La fin des temps glaciaires dans le nord de la France. Approches archéologiques et environnementales des occupations humaines du Tardiglaciaire. Mémoires de la Société préhistorique française 24 (Rivery 1997).

Fagnart/Coudret 2001: J.-P. Fagnart / P. Coudret, Fouille Programmée de Saleux – Les Baquets (Somme). Autorisation de fouille pluriannuelle (1999-2001) rapport de synthèse: année 2001. Site no. 80 724 005 AP. Conseil Général de la Somme et ESA 8018 du Centre National de la Recherche Scientifique (C.N.R.S.) (Amiens 2001).

Fagnart/Thévenin 1997: J.-P. Fagnart / A. Thévenin (eds), Le Tardiglaciaire en Europe du Nord-Ouest. Chronostratigraphie et environnement des occupations humaines du Tardiglaciaire et du début de l'Holocène en Europe du Nord-Ouest. Actes du 119e congrès national des sociétés historiques et scientifiques à Amiens, 26-30 octobre 1994 (Paris 1997).

Fahlke 2009: J. M. Fahlke, Der Austausch der terrestrischen Säugetierfauna an der Pleistozän/Holozän-Grenze in Mitteleuropa [dissertation, Bonn 2009].

Fairbanks 1989: R. G. Fairbanks, A 17,000-year glacio-eustatic sea level record: influence of glacial melting rates on the Younger Dryas event and deep-ocean circulation. Nature 342, 1989, 637-642.

Fairbanks et al. 2005: R. G. Fairbanks / R. A. Mortlock / T. C. Chiu / L. Cao / A. Kaplan / T. P. Guilderson / T. W. Fairbanks / A. L. Bloom / P. M. Grootes / M. J. Nadeau, Radiocarbon calibration curve spanning 10,000 to 50,000 years BP based on paired $^{230}Th/^{234}U/^{238}U$ and ^{14}C dates on pristine corals. Quaternary Science Reviews 25, 2005, 1781-1796.

Farmer 1994: M. F. Farmer, The origins of weapon systems. Current Anthropology 35, 1994, 679-681.

Farr et al. 2007: T. G. Farr / P. A. Rosen / E. Caro / R. Crippen / R. Duren / S. Hensley / M. Kobrick / M. Paller / E. Rodriguez / L. Roth / D. Seal / S. Shaffer / J. Shimada / J. Umland / M. Werner / M. Oskin / D. Burbank / D. Alsdorf, The Shuttle Radar Topography Mission. Reviews of Geophysics 45, 2007, RG2004.

Federici et al. 2008: P. R. Federici / D. E. Granger / M. Pappalardo / A. Ribolini / M. Spagnolo / A. J. Cyr, Exposure age dating and Equilibrium Line Altitude reconstruction of an Egesen moraine in the Maritime Alps, Italy. Boreas 37, 2008, 245-253.

Feustel 1956: R. Feustel, Vier jungpaläolithische Freilandstationen in Ostthüringen. Alt-Thüringen 2, 1955/1956 (1956), 1-26.

1974: R. Feustel, Die Kniegrotte. Eine Magdalénien-Station in Thüringen. Veröffentlichungen des Museums für Ur- und Frühgeschichte Thüringens 5 (Weimar 1974).

1979: R. Feustel, Le Magdalénien final en Thuringe (R.D.A.). In: de Sonneville-Bordes 1979, 877-887.

1980: R. Feustel (ed.), Magdalénienstation Teufelsbrücke. II: Paläontologischer Teil. Weimarer Monographien zur Ur- und Frühgeschichte 3 (Weimar 1980).

Feustel/Teichert/Unger 1963: R. Feustel / M. Teichert / K. P. Unger, Die Magdalénienstation Lausnitz in der Orlasenke. Alt-Thüringen 6, 1962/1963 (1963), 57-103.

Fiebig/Ellwanger/Doppler 2011: M. Fiebig / D. Ellwanger / G. Doppler, Pleistocene glaciations of southern Germany. In: Ehlers/Gibbard/Hughes 2011, 163-173.

Fiedel 2011: S. J. Fiedel, The mysterious onset of the Younger Dryas. Quaternary International 242, 2011, 262-266.

Fiedel et al. 2013: S. J. Fiedel / J. R. Southon / R. E. Taylor / Y. V. Kuzmin / M. Street / T. F. G. Higham / J. van der Plicht / M.-J. Nadeau / S. Nawalade-Chavan, Assessment of interlaboratory pretreatment protocols by radiocarbon dating an elk bone found below Laacher See Tephra at Miesenheim IV (Rhineland, Germany). Radiocarbon 55, 2013, 1443-1453.

Fiedorczuk/Schild 2002: J. Fiedorczuk / R. Schild, Wilczyce – a new late Magdalenian site in Poland. In: Eriksen/Bratlund 2002, 91-100.

Fiedorczuk et al. 2007: J. Fiedorczuk / B. Bratlund / E. Kolstrup / R. Schild, Late Magdalenian feminine flint plaquettes from Poland. Antiquity 81, 2007, 97-105.

Fifield 1999: L. K. Fifield, Accelerator mass spectrometry and its applications. Reports on Progress in Physics 62, 1999, 1223-1274.

Firbas 1939: F. Firbas, Vegetationsentwicklung und Klimawandel in der mitteleuropäischen Spät- und Nacheiszeit. Die Naturwissenschaften 27, 1939, 81-89.

Fischer 1990: A. Fischer, On being a pupil of a Flintknapper of 11,000 years ago. A preliminary analysis of settlement organization and flint technology based on conjoined flint artefacts from Trollesgave site. In: E. Cziesla / S. Eickhoff / N. Arts / D. Winter (eds), The big puzzle. International Symposium on Refitting Stone Artefacts. Monrepos 1987 (Bonn 1990) 447-464.

1991: A. Fischer, Pioneers in deglaciated landscapes: The expansion and adaptation of Late Palaelithic societies in southern Scandinavia. In: Barton/Roberts/Roe 1991, 100-121.

1996: A. Fischer, At the border of human habitat. The Late Palaeolithic and Early Mesolithic in Scandinavia. In: Larsson 1996b, 157-176.

Fischer/Heinemeier 2003: A. Fischer / J. Heinemeier, Freshwater reservoir effect in ^{14}C dates of food residue on pottery. Radiocarbon 45, 2003, 449-466.

Fischer/Mortensen 1977: A. Fischer / B. N. Mortensen, Trollesgavebopladsen. Et eksempel på anvendelse of EDB inden for arkeologien. Nationalmuseets Arbejdsmark 1977, 90-95.

Fischer/Tauber 1986: A. Fischer / H. Tauber, New C-14 datings of Late Palaeolithic cultures from northwestern Europe. Journal of Danish Archaeology 5, 1986, 7-13.

Fischer et al. 2013: A. Fischer / M. F. Mortensen / P. S. Henriksen / D. R. Mathiassen / J. Olsen, Dating the Trollesgave site and the Bromme culture – chronological fix-points for the Lateglacial settlement of southern Scandinavia. Journal of Archaeological Science 40, 2013, 4663-4674.

Fisher/Eriksen 2002: L. E. Fisher / B. V. Eriksen (eds), Lithic raw material economies in Late Glacial and early Postglacial Europe (Oxford 2002).

Fitch/Thomson/Gaffney 2005: S. Fitch / K. Thomson / V. Gaffney, Late Pleistocene and Holocene depositional systems and the palaeogeography of the Dogger Bank, North Sea. Quaternary Research 64, 2005, 185-196.

Flenniken/Ozbun 1988: J. J. Flenniken / T. L. Ozbun, Experimental analysis of Plains grooved abraders. Plains Anthropologist 33, 1988, 37-52.

Fletcher et al. 2010: W. J. Fletcher / M. F. Sánchez Goñi / J. R. M. Allen / R. Cheddadi / N. Combourieu-Nebout / B. Huntley / I. Lawson / L. Londeix / D. Magri / V. Margari / U. C. Müller / F. Naughton / E. Novenko / K. Roucoux / P. C. Tzedakis, Millennial-scale variability during the last glacial in vegetation records from Europe. Quaternary Science Reviews 29, 2010, 2839-2864.

Floss 1994: H. Floss, Rohmaterialversorgung im Paläolithikum des Mittelrheingebietes. Monographien des RGZM 21 (Bonn 1994).

2002: H. Floss, Climate and raw material behavior: A case study from late Pleistocene hunter-gatherers in the Middle Rhine area of Germany. In: Fisher/Eriksen 2002, 79-88.

Floss/Terberger 1987: H. Floss / Th. Terberger, Die Konzentration II von Andernach. Ein Beitrag zur Kenntnis der Variationsbreite spätjungpaläolithischer Steinartefaktensembles. Archäologisches Korrespondenzblatt 17, 1987, 287-294.

2002: H. Floss / Th. Terberger (eds), Die Steinartefakte des Magdalénien von Andernach (Mittelrhein). Die Grabungen 1979-1983. Tübinger Arbeiten zur Urgeschichte 1 (Rahden/Westf. 2002).

Foley/Lahr 1997: R. Foley / M. M. Lahr, Mode 3 technologies and the evolution of modern humans. Cambridge Archaeological Journal 7, 1997, 3-36.

Folz et al. 1998: E. Folz / N. Mercier / L. Froget / H. Valladas, Datations par luminescence stimulée optiquement. In: Bodu 1998, 60-61.

2001: E. Folz / P. Bodu / P. Bonte / J.-L. Joron / N. Mercier / J.-L. Reyss, OSL dating of fluvial quartz from Le Closeau, a Late Paleolithic site near Paris – Comparison with ^{14}C chronology. Quaternary Science Reviews 20, 2001, 927-933.

Fort/Pujol/Cavalli-Sforza 2004: J. Fort / T. Pujol / L. L. Cavalli-Sforza, Palaeolithic populations and waves of advance. Cambridge Archaeological Journal 14, 2004, 53-61.

Fortin/Fortin 2009: D. Fortin / M.-E. Fortin, Group-size-dependent association between food profitability, predation risk and distribution of free-ranging bison. Animal Behaviour 78, 2009, 887-892.

Fosse 1997: G. Fosse, Le Paléolithique récent et final du Bassin de la Basse-Seine (Yvelines-Eure – Seine-Maritime). In: Fagnart/Thévenin 1997, 233-244.

Fougère 2011: F. Fougère, From the ethnographic modelling of nomadic behaviours to archaeological site functions: Determining attribution critera. Palethnologie 3, 2011, 41-60.

Frahm et al. 2014: E. Frahm / B. A. Schmidt / B. Gasparyan / B. Yeritsyan / S. Karapetian / K. Meliksetian / D. S. Adler, Ten seconds in the field: rapid Armenian obsidian sourcing with portable XRF to inform excavations and surveys. Journal of Archaeological Science 41, 2014, 333-348.

Franken/Veil 1983: E. Franken / S. Veil, Die Steinartefakte von Gönnersdorf. Der Magdalénien-Fundplatz Gönnersdorf 7 (Wiesbaden 1983).

Frechen/Oches/Kohfeld 2003: M. Frechen / E. A. Oches / K. E. Kohfeld, Loess in Europe – mass accumulation rates during the Last Glacial Period. Quaternary Science Reviews 22, 2003, 1835-1857.

Freericks 1989: M. Freericks, Der Fundplatz Niederbieber. Fläche 17/46-22/52 [unpublished M.A. thesis, Köln 1989].

Friedrich et al. 2001a: M. Friedrich / M. Knipping / P. van der Kroft / A. Renno / S. Schmidt / O. Ullrich / J. Vollbrecht, Ein Wald am Ende der Eiszeit. Untersuchungen zur Besiedlungs-, Landschafts- und Vegetationsentwicklung an einem verlandeten See im Tagebau Reichswalde, Niederschlesischer Oberlausitzkreis. Arbeits- und Forschungsberichte zur Sächsischen Bodendenkmalpflege 43, 2001, 21-94.

2001b: M. Friedrich / B. Kromer / K. F. Kaiser / M. Spurk / K. A. Hughen / S. J. Johnsen, High-resolution climate signals in the Bølling-Allerød Interstadial (Greenland Interstadial 1) as reflected in European tree-ring chronologies compared to marine varves and ice-core records. Quaternary Science Reviews 20, 2001, 1223-1232.

1999: M. Friedrich / B. Kromer / M. Spurk / J. Hofmann / K. F. Kaiser, Paleo-environment and radiocarbon calibration as derived from Lateglacial / Early Holocene tree-ring chronologies. Quaternary International 61, 1999, 27-39.

2004: M. Friedrich / S. Remmele / B. Kromer / J. Hofmann / M. Spurk / K. F. Kaiser / C. Orcel / M. Küppers, The 12,460-year Hohenheim oak and pine tree-ring chronology from Central Europe – A unique annual record for radiocarbon calibration and paleoenvironment reconstructions. Radiocarbon 46, 2004, 1111-1122.

Frisch 2010: W. Frisch, Der Auerochs. Das europäische Rind. Vom größten, durch den Menschen verdrängten und ausgerotteten Landsäugetier, seinen Ahnen und Nachkommen (Starnberg 2010).

Fritz 2012: C. Fritz, Le bois de cerf gravé et les cortex incisés de Conty. In: Antoine et al. 2012, 89-90.

Fu et al. 2016: Q. Fu / C. Posth / M. Hajdinjak / M. Petr / S. Mallick / D. Fernandes / A. Furtwängler / W. Haak / M. Meyer / A. Mittnik / B. Nickel / A. Peltzer / N. Rohland / V. Slon / S. Talamo / I. Lazaridis / M. Lipson / I. Mathieson / S. Schiffels / P. Skoglund / A. P. Derevianko / N. Drozdov / V. Slavinsky / A. Tsybankov / R. Grifoni Cremonesi / F. Mallegni / B. Gély / E. Vacca / M. R. González Morales / L. G. Straus / C. Neugebauer-Maresch / M. Teschler-Nicola / S. Constantin / O. T. Moldovan /S. Benazzi / M. Peresani / D. Coppola / M. Lari / S. Ricci / A. Ronchitelli / F. Valentin / C. Thevenet / K. Wehrberger / D. Grigorescu / H. Rougier / I. Crevecoeur / D. Flas / P. Semal / M. A. Mannino / C. Cupillard / H. Bocherens, N. J. Conard / K. Harvati / V. Moiseyev / D. G. Drucker / J. Svoboda / M. P. Richards / D. Caramelli / R. Pinhasi / J. Kelso / N. Patterson / J. Krause / S. Pääbo / D. Reich, The genetic history of Ice Age Europe. Nature 534, 2016, 200-205.

Gaffney/Thomson/Fitch 2007: V. Gaffney / K. Thomson / S. Fitch (eds), Mapping Doggerland. The Mesolithic landscapes of the Southern North Sea (Oxford 2007).

Gambier et al. 2000: D. Gambier / H. Valladas / N. Tisnérat-Laborde / M. Arnold / F. Bresson, Datation de vestiges humains présumés du Paléolithique supérieur par la méthode du Carbone 14 en spectrométrie de masse par accélérateur. Paléo 12, 2000, 201-212.

Gamble 1983: C. Gamble, Culture and society in the Upper Palaeolithic of Europe. In: G. Bailey (ed.), Hunter-gatherer economy in prehistory. A European perspective (Cambridge 1983) 201-211.

Gamble et al. 2004: C. Gamble / W. Davies / P. Pettitt / M. Richards, Climate change and evolving human diversity in Europe during the last glacial. Philosophical Transactions of the Royal Society of London B 359, 2004, 243-254.

2005: C. Gamble / W. Davies / P. Pettitt / L. Hazelwood / M. Richards, The archaeological and genetic foundations of the European population during the Late Glacial: Implications for »agricultural thinking«. Cambridge Archaeological Journal 15, 2005, 193-223.

Garrod 1926: D. A. E. Garrod, The Upper Palaeolithic age in Britain (Oxford 1926).

Gaucher 1996: G. Gaucher (ed.), Fouilles de Pincevent II. Le site et ses occupations récentes. L'environnement tardi- et post-glaciaire et les témoins postérieurs au Magdalénien. Mémoires de la Société préhistorique française 23 (Saint-Etienne 1996).

Gaudzinski-Windheuser 2011: S. Gaudzinski-Windheuser, An introduction to living structures and the history of occupation at the Late Upper Palaeolithic site of Oelknitz (Thuringia, Germany). In: Gaudzinski-Windheuser et al. 2011b, 127-139.

2012: S. Gaudzinski-Windheuser, Indication for social interaction during the Central European Late Upper Palaeolithic: Evidence from the Magdalenian site of Oelknitz, Structure 1 (Thuringia, Germany). Quaternary International 252, 2012, 165-174.

Gaudzinski-Windheuser/Jöris 2006: S. Gaudzinski-Windheuser / O. Jöris (eds), 600 000 Jahre Menschheitsgeschichte in der Mitte Europas. Begleitbuch zur Ausstellung im Museum für die Archäologie des Eiszeitalters, Schloss Monrepos, Neuwied (Mainz 2006).

Gaudzinski/Street 2003: S. Gaudzinski / M. Street, Reconsidering hunting specialisation in the German Magdalenian faunal record. In: S. Costamagno / V. Laroulandie (eds), Mode de vie au Magdalénien. Apports de l'archéozoologie / Zooarchaeological insights into Magdalenian lifeways. Actes du XIVème congrès UISPP, Université de Liège, Belgique, 2-8 septembre 2001, section 6: paleolithique superieur, colloque 6 (Oxford 2003) 11-21.

Gaudzinski-Windheuser et al. 2011a: S. Gaudzinski-Windheuser / O. Jöris / M. Sensburg / M. Street / E. Turner, Foreword. In: Gaudzinski-Windheuser et al. 2011b, 1-9.

2011b: S. Gaudzinski-Windheuser / O. Jöris / M. Sensburg / M. Street / E. Turner (eds), Site-internal spatial organization of hunter-gatherer societies: Case studies from the European Palaeolithic and Mesolithic. RGZM – Tagungen 12 (Mainz 2011).

Gautier 1997: A. Gautier, The macromammal remains of la grotte du Bois Laiterie. In: Otte/Straus 1997, 177-196.

Gavrilets/Duenez-Guzman/Vose 2008: S. Gavrilets / E. A. Duenez-Guzman / M. D. Vose, Dynamics of alliance formation and the egalitarian revolution. PLoS One 3, 2008, e3293.

Gearey/Marshall/Hamilton 2009: B. R. Gearey / P. Marshall / D. Hamilton, Correlating archaeological and palaeoenvironmental records using a Bayesian approach: a case study from Sutton Common, South Yorkshire, England. Journal of Archaeological Science 36, 2009, 1477-1487.

Gebhardt 1998: A. Gebhardt, Le contexte géomorphologique du Closeau. In: Bodu 1998, 62-86.

Gehlen/Heinen/Tillmann 2001: B. Gehlen / M. Heinen / A. Tillmann (eds), Zeit-Räume: Gedenkschrift für Wolfgang Taute (Bonn 2001).

Gelhausen 2010: F. Gelhausen, Verteilungsmuster ausgewählter Fundkonzentrationen des allerødzeitlichen Fundplatzes Niederbieber, Stadt Neuwied (Rheinland-Pfalz) – Grabung 1996-1999. Jahrbuch des RGZM 54, 2007 (2010), 1-23.

2011a: F. Gelhausen, Siedlungsmuster der allerødzeitlichen Federmesser-Gruppen in Niederbieber, Stadt Neuwied. Monographien des RGZM 90 (Mainz 2011).

2011b: F. Gelhausen, Subsistence strategies and settlement systems at the Federmessergruppen site of Niederbieber (Central Rhineland, Germany). In: Gaudzinski-Windheuser et al. 2011b, 159-173.

2011c: F. Gelhausen, Die Fundkonzentrationen der Fläche II des allerødzeitlichen Fundplatzes Niederbieber, Stadt Neuwied (Rheinland-Pfalz). Jahrbuch des RGZM 56, 2009 (2011), 1-38.

Gelhausen / Kegler / Wenzel 2004: F. Gelhausen / J. F. Kegler / S. Wenzel, Latent dwelling structures in the Final Palaeolithic: Niederbieber IV, Andernach-Martinsberg 3, Berlin-Tegel IX. Notae Praehistoricae 24, 2004, 69-79.

Gelhausen/Kegler/Wenzel 2005: F. Gelhausen / J. Kegler / S. Wenzel, Hütten oder Himmel? Latente Behausungsstrukturen im Spätpaläolithikum Mitteleuropas. Jahrbuch des RGZM 51, 2004 (2005), 1-22.

Germonpré 1997: M. Germonpré, The Magdalenian upper horizon of Goyet and the late Upper Palaeolithic recolonisation of the Belgian Ardennes. Bulletin de l'Institut royal des sciences naturelles de Belgique (Sciences de la terre) 67, 1997, 167-182.

Germonpré/Lázničková-Galetová/Sablin 2012: M. Germonpré / M. Lázničková-Galetová / M. V. Sablin, Palaeolithic dog skulls at the Gravettian Předmostí site, the Czech Republic. Journal of Archaeological Science 39, 2012, 184-202.

Germonpré/Sablin 2002: M. Germonpré / M. Sablin, Preliminary results on the large bovids of the Belgian Magdalenian. Notae Praehistoricae 22, 2002, 71-73.

Germonpré et al. 2009: M. Germonpré / M. V. Sablin / R. E. Stevens / R. E. M. Hedges / M. Hofreiter / M. Stiller / V. Després, Fossil dogs and wolves from Palaeolithic sites in Belgium, the Ukraine and Russia: osteometry, ancient DNA and stable isotopes. Journal of Archaeological Science 36, 2009, 473-490.

Geyh 2005: M. A. Geyh, Handbuch der physikalischen und chemischen Altersbestimmung (Darmstadt 2005).

Geyh et al. 1983: M. A. Geyh / G. Roeschmann / T. A. Wijmstra / A. A. Middledorp, The unreliability of ^{14}C dates obtained from buried sandy podzols. Radiocarbon 25, 1983, 409-416.

Gibbard 1988: P. L. Gibbard, The history of the great Northwest European rivers during the past three million years. Philosophical Transactions of the Royal Society of London B (Biological Sciences) 318, 1988, 559-602.

Gilg et al. 2000: O. Gilg / R. Sané / D. V. Solovieva / V. I. Pozdnyakov / B. Sabard / D. Tsanos / C. Zöckler / E. G. Lappo / E. E. Syroechkovski, Jr. / G. Eichhorn, Birds and mammals of the Lena Delta Nature Reserve, Siberia. Arctic 53, 2000, 118-133.

Gillespie et al. 1985: R. Gillespie / J. A. J. Gowlett / E. T. Hall / R. E. M. Hedges / C. Perry, Radiocarbon dates from the Oxford AMS System: Archaeometry Datelist 2. Archaeometry 27, 1985, 237-246.

Gillespie/Hedges 1983: R. Gillespie / R. E. M. Hedges, Sample chemistry for the Oxford High Energy Mass Spectrometer. Radiocarbon 25, 1983, 771-7741.

Gilot 1970: E. Gilot, Louvain natrual radiocarbon measurements IX. Radiocarbon 12, 1970, 553-558.

1997: É. Gilot, Index général des dates Lv-: Laboratoire du carbone 14 de Louvain/Louvain-la-Neuve. Studia praehistorica Belgica 7 (Liège 1997).

Ginter/Połtowicz 2007: B. Ginter / M. Połtowicz, Magdalenian settlement in Poland before the Bølling oscillation. In: M. Kobusiewicz / J. Kabaciński (eds), Studies in the Final Palaeolithic settlement of the Great European Plain (Poznań 2007) 7-19.

Ginter et al. 2005: B. Ginter / M. Połtowicz / M. Pawlikowski / S. Skiba / J. Trąbska / A. Wacnik / M. Winiarska-Kabacińska / P. Wojtal, Dzierżysław 35 – ein neuer Fundplatz des Magdalénien in Oberschlesien. Archäologisches Korrespondenzblatt 35, 2005, 431-446.

Girard/Bintz/Bocquet 1981: M. Girard / P. Bintz / A. Bocquet, La végétation et les climats au Tardiglaciaire et à l'Holocène en Savoie d'après l'étude pollinique des grottes de Saint-Thibaud-de-Couz. Bulletin de l'Association française pour l'étude du quaternaire 18, 1981, 89-106.

Gob 1990: A. Gob, Chronologie du Mésolithique en Europe. Atlas des dates ^{14}C. Histoire de l'art et archéologie 1 (Liège 1990).

Godwin 1962: H. Godwin, Half-life of radiocarbon. Nature 195, 1962, 984.

Goodwin 1999: D. Goodwin, The importance of ethology in understanding the behaviour of the horse. Equine Veterinary Journal supplement 28, 1999, 15-19.

Goslar 1998a: T. Goslar, Absolute age of floating varve chronology of Lake Gościąż. In: M. Ralska-Jasiewiczowa / T. Goslar / T. Madeyska / L. Starkel (eds), Lake Gościąż, Central Poland. A monographic study. Part 1 (Krakow 1998) 110-111.

1998b: T. Goslar, Floating varve chronology of Lake Gościąż. In: M. Ralska-Jasiewiczowa / T. Goslar / T. Madeyska / L. Starkel (eds), Lake Gościąż, Central Poland. A monographic study. Part 1 (Krakow 1998) 97-99.

1998c: T. Goslar, Late-Glacial sediments of Lake Gościąż – chronological background. In: M. Ralska-Jasiewiczowa / T. Goslar / T. Madeyska / L. Starkel (eds), Lake Gościąż, Central Poland. A monographic study. Part 1 (Krakow 1998) 119-124.

1998d: T. Goslar, Record of laminae thickness of the Lake Gościąż sediments, and its correlation with absolutely dated tree-ring width sequences. In: M. Ralska-Jasiewiczowa / T. Goslar / T. Madeyska / L. Starkel (eds), Lake Gościąż, Central Poland. A monographic study. Part 1 (Krakow 1998) 104-110.

Goślar/Arnold/Pazdur 1995: T. Goślar / M. Arnold / M. F. Pazdur, The Younger Dryas cold event – was it synchronous over the North Atlantic region? Radiocarbon 37, 1995, 63-70.

1998: T. Goslar / M. Arnold / M. F. Pazdur, Variations of atmospheric ^{14}C concentrations at the Pleistocene/Holocene transition, reconstructed from the Lake Gościąż sediments. In: Ralska-Jasiewiczowa et al. 1998, 162-171.

Goslar et al. 1998a: T. Goslar / A. Pazdur / M. F. Pazdur / M. Arnold / I. Hajdas, Correlation of radiocarbon and varve chronologies of Lake Gościąż. In: Ralska-Jasiewiczowa et al. 1998, 100-104.

1998b: T. Goslar / M. Ralska-Jasiewiczowa / L. Starkel / D. Demske / T. Kuc / B. Łącka / K. Szeroczyńska / W. Bodumił / K. Więckowski, Discussion of the Late-Glacial recorded in the Lake Gościąż sediments. In: Ralska-Jasiewiczowa et al. 1998, 171-175.

1999: T. Goslar / K. Bałaga / M. Arnold / N. Tisnerat / E. Starnawska / M. Kuźniarski / L. Chróst / A. Walanus / K. Więckowski, Climate-related variations in the composition of the Lateglacial and Early Holocene sediments of Lake Perespilno (eastern Poland). Quaternary Science Reviews 18, 1999, 899-911.

2000: T. Goslar / M. Arnold / N. Tisnerat-Laborde / C. Hatté / M. Paterne / M. Ralska-Jasiewiczowa, Raciocarbon calibration by means of varves versus ^{14}C ages of terrestrial macrofossils from Lake Gościąż and Lake Perespilno, Poland. Radiocarbon 42, 2000, 335-348.

Gould 1994: S. J. Gould, Tempo and mode in the macroevolutionary reconstruction of Darwinism. Proceedings of the National Academy of Sciences of the United States of America 91, 1994, 6764-6771.

Gouveia et al. 2008: C. Gouveia / R. M. Trigo / C. C. DaCamara / R. Libonati / J. M. C. Pereira, The North Atlantic Oscillation and European vegetation dynamics. International Journal of Climatology 28, 2008, 1835-1847.

Gowlett et al. 1986a: J. A. J. Gowlett / E. T. Hall / R. E. M. Hedges / C. Perry, Radiocarbon dates from the Oxford AMS system: Archaeometry Datelist 3. Archaeometry 28, 1986, 116-125.

1986b: J. A. J. Gowlett / R. E. M. Hedges / I. A. Law / C. Perry, Radiocarbon dates from the Oxford AMS system: Archaeometry Datelist 4. Archaeometry 28, 1986, 206-221.

1987: J. A. J. Gowlett / R. E. M. Hedges / I. A. Law / C. Perry, Radiocarbon dates from the Oxford AMS system: Archaeometry Datelist 5. Archaeometry 29, 1987, 125-155.

Gradstein/Ogg/Smith 2004: F. Gradstein / J. Ogg / A. Smith (eds), A Geological Time Scale 2004 (Cambridge 2004).

Graf/Timmreck 2001: H. F. Graf / C. Timmreck, A general climate model simulation of the aerosol radiative effects of the Laacher See eruption (10,900 B.C.). Journal of Geophysical Research – Atmospheres 106, 2001, 14747-14756.

Grafenstein et al. 1992: U. von Grafenstein / H. Erlenkeuser / J. Müller / A. Kleinmann-Eisenmann, Oxygen isotope records of benthic ostracods in Bavarian lake sediments. Reconstruction of Late and Post Glacial air temperatures. Die Naturwissenschaften 79, 1992, 145-152.

1999: U. von Grafenstein / H. Erlenkeuser / A. Brauer / J. Jouzel / S. J. Johnsen, A Mid-European decadal isotope-climate record from 15,500 to 5000 years B.P. Science 284, 1999, 1654-1657.

Graindor 1959: M.-J. Graindor, Circonscription de Caen. Gallia Préhistoire 2, 1959, 82-90.

1965: M.-J. Graindor, Informations archéologiques. Circonscription de Caen. Gallia Préhistoire 8, 1965, 21-31.

Graindor/Martin 1972: M.-J. Graindor / Y. Martin, L'art préhistorique de Gouy (Paris 1972).

Grattan 2006: J. Grattan, Aspects of Armageddon: An exploration of the role of volcanic eruptions in human history and civilization. Quaternary International 151, 2006, 10-18.

Grimm 2003: S. B. Grimm, Der spätpaläolithische Fundplatz bei Bad Breisig/Kreis Ahrweiler [unpublished M.A. thesis, Köln 2003].

2004: S. B. Grimm, Ein spätallerødzeitlicher Fundplatz bei Bad Breisig, Kreis Ahrweiler. Berichte zur Archäologie an Mittelrhein und Mosel 9, 2004, 11-32.

2007: S. B. Grimm, Maps of late glacial NW-Europe. NW-EU 10W-20E 45-60N. available since 2007-09-19. Römisch-Germanisches Zentralmuseum; Forschungsbereich Altsteinzeit (Schloss Monrepos, Neuwied 2007; http://web.rgzm.de/Late-glacial-nw-europe.html [This webpage is meanwhile shut down. Newer versions are available if you search for »Change (Karten)« at: http://monrepos-rgzm.de/tl_files/monrepos/content/projektarchiv].

Grimm/Jensen/Weber 2012: S. Grimm/D. S. Jensen/M.-J. Weber, A lot of good points – Havelte points in the context of Lateglacial tanged points in north-western Europe. In: Niekus et al. 2012, 251-266.

Grimm/Weber 2008: S. B. Grimm/M.-J. Weber, The chronological framework of the Hamburgian in the light of old and new ^{14}C dates. Quartär 55, 2008, 17-40.

Grøn 1990: O. Grøn, Studies in settlement patterns and submarine bogs: results and strategy for further research. In: P. M. Vermeersch / P. van Peer (eds), Contributions to the Mesolithic in Europe. Papers presented at the Fourth International Symposium »The Mesolithic in Europe«, Leuven 1990 (Leuven 1990) 81-86.

2012: O. Grøn, Our grandfather sent the elk – some problems for hunter-gatherer predictive modelling. Quartär 59, 2012, 175-188.

Grønnow 1987: B. Grønnow, Meiendorf und Stellmoor revisited. An analysis of Late Palaeolithic reindeer exploitation. Acta Archaeologica (København) 56, 1985 (1987), 131-166.

Grootes et al. 1993: P. M. Grootes / M. Stuiver / J. W. White / S. Johnsen / J. Jouzel, Comparison of oxygen isotope records from the GISP2 and GRIP Greenland ice cores. Nature 366, 1993, 552-554.

Groves 1994: C. P. Groves, Morphology, habitat, and taxonomy. In: L. Boyd / K. A. Houpt (eds), Przewalski's horse. The history and biology of an endangered species (Albany NY 1994) 39-60.

Grünberg 2006: J. M. Grünberg, New AMS dates of Palaeolithic and Mesolithic camp sites and single finds in Saxony-Anhalt and Thuringia (Germany). Proceedings of the Prehistoric Society 72, 2006, 95-112.

Guillot/Guillot/Thévenin 1995: G. Guillot/O. Guillot/A. Thévenin, Nouvelles stations à Federmesser de la moyenne et basse Saulx, départment de la Meuse. In: Thévenin 1995, 25-34.

2000: G. Guillot/O. Guillot/A. Thévenin, Le gisement à pointes à dos courbe de Louppy-le Château, canton de Vaubécourt (Meuse). Bulletin de la Société Préhistorique Luxembourgeoise 19, 1997 (2000), 101-111.

Guiter et al. 2003: F. Guiter / V. Andrieu-Ponel / J.-L. de Beaulieu / R. Cheddadi / M. Calvez / P. Ponel / M. Reille / T. Keller / C. Goeury, The last climatic cycles in Western Europe: a comparison between long continuous lacustrine sequences from France and other terrestrial records. Quaternary International 111, 2003, 59-74.

Gupta et al. 2007: S. Gupta / J. S. Collier / A. Palmer-Felgate / G. Potter, Catastrophic flooding origin of shelf valley systems in the English Channel. Nature 448, 2007, 342-345.

Haase et al. 2007: D. Haase/J. Fink/G. Haase/R. Ruske/M. Pécsi/H. Richter/M. Altermann/K. D. Jäger, Loess in Europe – its spatial distribution based on a European Loess Map, scale 1:2,500,000. Quaternary Science Reviews 26, 2007, 1301-1312.

Habel/Assmann 2010: J. C. Habel / T. Assmann (eds), Relict Species. Phylogeography and Conservation Biology (Berlin, Heidelberg 2010).

Haeberli et al. 2010: W. Haeberli / J. Noetzli / L. Arenson / R. Delaloye / I. Gärtner-Roer / S. Gruber / K. Isaksen / C. Kneisel / M. Krautblatter / M. Phillips, Mountain permafrost: development and challenges of a young research field. Journal of Glaciology 56, 2010, 1043-1058.

Hajdas 2008: I. Hajdas, Radiocarbon dating and its applications in Quternary studies. Eiszeitalter und Gegenwart 57, 2008, 2-24.

Hajdas/Bonani/Zolitschka 2000: I. Hajdas/G. Bonani/B. Zolitschka, Radiocarbon dating of varve chronologies: Soppensee and Holzmaar lakes after ten years. Radiocarbon 42, 2000, 349-353.

Hajdas et al. 1993: I. Hajdas / S. D. Ivy / J. Beer / G. Bonani / D. Imboden / A. F. Lotter / M. Sturm / M. Suter, AMS radiocarbon dating and varve chronology of Lake Soppensee: 6000 to 12000 ^{14}C years BP. Climate Dynamics 9, 1993, 107-116.

1995: I. Hajdas / B. Zolitschka / S. D. Ivy-Ochs / J. Beer / G. Bonani / S. A. G. Leroy / J. F. W. Negendank / M. Ramrath / M. Suter, AMS radiocarbon dating of annually laminated sediments from Lake Holzmaar, Germany. Quaternary Science Reviews 14, 1995, 137-143.

Hall/Meiklejohn/Bumby 2011: K. Hall / I. Meiklejohn / A. Bumby, Marion Island volcanism and glaciation. Antarctic Science 23, 2011, 155-163.

Halperin 1980: R. H. Halperin, Ecology and mode of production: Seasonal variation and the division of labor by sex among hunter-gatherers. Journal of Anthropological Research 36, 1980, 379-399.

Halstead/O'Shea 1989a: P. Halstead / J. O'Shea (eds), Bad year economics. Cultural responses to risk and uncertainty. New directions in archaeology (Cambridge 1989).

1989b: P. Halstead / J. O'Shea, Cultural responses to risk and uncertainty. Introduction. In: Halstead/O'Shea 1989a, 1-7.

Hamilton/Buchanan 2007: M. J. Hamilton / B. Buchanan, Spatial gradients in Clovis-age radiocarbon dates across North America suggest rapid colonization from the north. Proceedings of the National Academy of Sciences of the United States of America 104, 2007, 15625-15630.

2009: M. J. Hamilton / B. Buchanan, The accumulation of stochastic copying errors causes drift in culturally transmitted technologies: Quantifying Clovis evolutionary dynamics. Journal of Anthropological Archaeology 28, 2009, 55-69.

Hammer/Clausen/Tauber 1986: C. U. Hammer / H. B. Clausen / H. Tauber, Ice-core dating of the Pleistocene/Holocene boundary applied to a calibration of the ^{14}C time scale. Radiocarbon 28, 1986, 284-291.

Hammer/Harper/Ryan 2001: Ø. Hammer / D. A. T. Harper / P. D. Ryan, PAST: Paleontological Statistics Software Package for Education and Data Analysis. Palaeontologia Electronica 4, 2001, 1-9.

Han/Pereira/Santos 2012: T. A. Han/L. M. Pereira/F. C. Santos, The emergence of commitments and cooperation. In: V. Conitzer / M. Winikoff / L. Padgham / W. van der Hoek (eds), Proceedings of the 11th international conference on autonomous agents and multi-agent systems (AAMAS 2012). June, 4-8, 2012, Valencia, Spain 2012. Volume 1 (Richland/South Carolina) 559-566.

Hanebuth/Stattegger/Bojanowski 2009: T. J. J. Hanebuth / K. Stattegger / A. Bojanowski, Termination of the Last Glacial Maximum sea-level lowstand: The Sunda-Shelf data revisited. Global and Planetary Change 66, 2009, 76-84.

Hanik 2009: S. Hanik, Schweine im Alleröd. Archäozoologische Neuigkeiten aus Wustermark, Lkr. Havelland. Archäologie in Berlin und Brandenburg 2007 (2009), 45.

Hansen 2006: M. K. Hansen, Beyond seals: The representation of seals on engraved slate plaquettes from the Magdalenian site Gönnersdorf (Central Rhineland, Germany) [M.A. thesis, Tromsø 2006].

Hanski 1998: I. Hanski, Metapopulation dynamics. Nature 396, 1998, 41-49.

Hantke 1980: R. Hantke, Eiszeitalter 2. Die jüngste Erdgeschichte der Schweiz und ihrer Nachbargebiete: letzte Warmzeiten, Würm-Eiszeit, Eisabbau und Nacheiszeit der Alpen-Nordseite vom Rhein- zum Rhone-System (Thun 1980).

1983: R. Hantke, Eiszeitalter 3. Die jüngste Erdgeschichte der Schweiz und ihrer Nachbargebiete: westliche Ostalpen mit ihrem bayerischen Vorland bis zum Inn-Durchbruch und Südalpen zwischen Dolomiten und Mont Blanc (Thun 1983).

Harff et al. 2007: J. Harff / W. Lemke / R. Lampe / F. Lüth / H. Lübke / M. Meyer / F. Tauber / U. Schmölcke, The Baltic Sea Coast – A model of interrelations among geosphere, climate, and anthroposphere. The Geological Society of America 426, 2007, 133-142.

Harff/Björck/Hoth 2011: J. Harff / S. Björck / P. Hoth (eds), The Baltic Sea Basin. Central and Eastern European Development Studies (CEEDES) (Berlin, Heidelberg 2011).

Hartley 2010: M. Hartley, Qualitative risk assessment of the role of the feral wild boar (*Sus scrofa*) in the likelihood of incursion and the impacts on effective disease control of selected exotic diseases in England. European Journal of Wildlife Research 56, 2010, 401-410.

Hasenrath 2002: S. Hasenrath, The intertropical convergence zone of the eastern Pacific revisted. International Journal of Climatology 22, 2002, 347-356.

Hatté/Guiot 2005: C. Hatté / J. Guiot, Palaeoprecipitation reconstruction by inverse modelling using the isotopic signal of loess organic matter: application to the Nußloch loess sequence (Rhine Valley, Germany). Climate Dynamics 25, 2005, 315-327.

Haug et al. 2003: G. H. Haug / D. Günther / L. C. Peterson / D. M. Sigman / K. A. Hughen / B. Aeschlimann, Climate and the collapse of Maya civilization. Science 299, 2003, 1731-1735.

Hayden 1981: B. Hayden, Research and development in the Stone Age: Technological transitions among hunter-gatherers [and comments and reply]. Current Anthropology 22, 1981, 519-548.

Haynes 1991: G. Haynes, Mammoths, mastodonts, and elephants. Biology, behavior, and the fossil record (Cambridge 1991).

Heaton/Blackwell/Buck 2009: T. J. Heaton / P. G. Blackwell / C. E. Buck, A Bayesian approach to the estimation of radicarbon calibration curves: The IntCal09 methodology. Radiocarbon 51, 2009, 1151-1164.

Hebblewhite/Pletscher 2002: M. Hebblewhite / D. H. Pletscher, Effects of elk group size on predation by wolves. Canadian Journal of Zoology 80, 2002, 800-809.

Heckel 2009: C. Heckel, Physical characteristics of mammoth Ivory and their implications for ivory work in the Upper Paleolithic. Mitteilungen der Gesellschaft für Urgeschichte 18, 2009, 71-91.

Hedges 2002: R. E. M. Hedges, Bone diagenesis: An overview of processes. Archaeometry 44, 2002, 319-328.

Hedges/Millard 1995: R. E. M. Hedges / A. R. Millard, Bones and groundwater: towards the modelling of diagenetic processes. Journal of Archaeological Science 22, 1995, 155-164.

Hedges/Millard/Pike 1995: R. E. M. Hedges / A. R. Millard / A. W. G. Pike, Measurements and relationships of diagenetic alteration of bone from three archaeological sites. Journal of Archaeological Science 22, 1995, 201-209.

Hedges et al. 1987: R. E. M. Hedges / R. A. Housley / I. A. Law / C. Perry / J. A. J. Gowlett, Radiocarbon dates from the Oxford AMS system: Archaeometry datelist 6. Archaeometry 29, 1987, 289-306.

1988a: R. E. M. Hedges / R. A. Housley / I. A. Law / C. Perry, Radiocarbon dates from the Oxford AMS system: Archaeometry datelist 7. Archaeometry 30, 1988, 155-164.

1988b: R. E. M. Hedges / R. A. Housley / I. A. Law / C. Perry / E. Hendy, Radiocarbon dates from the Oxford AMS system: Archaeometry datelist 8. Archaeometry 30, 1988, 291-305.

1989: R. E. M. Hedges / R. A. Housley / I. A. Law / C. Bronk Ramsey, Radiocarbon dates from the Oxford AMS system: Archaeometry datelist 9. Archaeometry 31, 1989, 207-234.

1990a: R. E. M. Hedges / R. A. Housley / C. Bronk Ramsey / G. J. van Klinken, Radiocarbon dates from the Oxford AMS system: Archaeometry datelist 11. Archaeometry 32, 1990, 211-237.

1990b: R. E. M. Hedges / R. A. Housley / I. A. Law / C. Bronk Ramsey, Radiocarbon dates from the Oxford AMS system: Archaeometry datelist 10. Archaeometry 32, 1990, 101-108.

1991: R. E. M. Hedges / R. A. Housley / C. Bronk Ramsey / G. J. van Klinken, Radiocarbon dates from the Oxford AMS system: Archaeometry datelist 13. Archaeometry 33, 1991, 279-296.

1992: R. E. M. Hedges / R. A. Housley / C. Bronk Ramsey / G. J. van Klinken, Radiocarbon dates from the Oxford AMS system: Archaeometry datelist 14. Archaeometry 34, 1992, 141-159.

1993a: R. E. M. Hedges / R. A. Housley / C. Bronk Ramsey / G. J. van Klinken, Radiocarbon dates from the Oxford AMS system. Archaeometry datelist 17. Archaeometry 35, 1993, 305-326.

1993b: R. E. M. Hedges / R. A. Housley / C. Bronk Ramsey / G. J. van Klinken, Radiocarbon dates from the Oxford AMS system: Archaeometry datelist 16. Archaeometry 35, 1993, 147-167.

1994: R. E. M. Hedges / R. A. Housley / C. Bronk Ramsey / G. J. van Klinken, Radiocarbon dates from the Oxford AMS system: Archaeometry datelist 18. Archaeometry 36, 1994, 337-374.

1995: R. E. M. Hedges / R. A. Housley / C. Bronk Ramsey / G. J. van Klinken, Radiocarbon dates from the Oxford AMS system: Archaeometry datelist 19. Archaeometry 37, 1995, 195-214.

1996a: R. E. M. Hedges / P. B. Pettitt / C. Bronk Ramsey / G. J. van Klinken, Radiocarbon dates from the Oxford AMS system: Archaeometry datelist 22. Archaeometry 38, 1996, 391-415.

1996b: R. E. M. Hedges / P. B. Pettitt / C. Bronk Ramsey / G. J. van Klinken, Radiocarbon dates from the Oxford AMS system: Archaeometry datelist 22. Archaeometry 38, 1996, 391-415.

1997: R. E. M. Hedges / P. B. Pettitt / C. Bronk Ramsey / G. J. van Klinken, Radiocarbon dates from the Oxford AMS system: Archaeometry datelist 24. Archaeometry 39, 1997, 445-471.

1998a: R. E. M. Hedges / P. B. Pettitt / C. Bronk Ramsey / G. J. van Klinken, Radiocarbon dates from the Oxford AMS system: Archaeometry datelist 26. Archaeometry 40, 1998, 437-455.

1998b: R. E. M. Hedges / P. B. Pettitt / C. Bronk Ramsey / G. J. van Klinken, Radiocarbon dates from the Oxford AMS system: Archaeometry datelist 25. Archaeometry 40, 1998, 227-239.

Heikkilä/Fontana/Seppa 2009: M. Heikkilä / S. L. Fontana / H. Seppa, Rapid Lateglacial tree population dynamics and ecosystem changes in the eastern Baltic region. Journal of Quaternary Science 24, 2009, 802-815.

Heinemeier/Rud 2000: J. Heinemeier / N. Rud, AMS ^{14}C dateringer, Århus 1999. Arkæologiske udgravninger i Danmark 1999 (2000), 296-313.

Heinen 2005: M. Heinen, Sarching '83 und '89/90. Untersuchungen zum Spätpaläolithikum und Frühmesolithikum in Südost-Deutschland. Edition Mesolithikum 1 (Kerpen-Loogh 2005).

2010: M. Heinen, Der Federmesser-Fundplatz von Wesseling im Rheinland. Presentation at annual meeting of Arbeitsgemeinschaft Mesolithikum (Schleswig 20.2.2010).

Heinrich 1988: H. Heinrich, Origin and consequences of cyclic ice rafting in the Northeast Atlantic Ocean during the past 130,000 years. Quaternary Research 29, 1988, 142-152.

Heiri/Lotter 2001: O. Heiri / A. F. Lotter, Effect of low count sums on quantitative environmental reconstructions: an example using subfossil chironomids. Journal of Paleolimnology 26, 2001, 343-350.

Hellstrom 2012: J. Hellstrom, Absolute dating of cave art. Science 336, 2012, 1387-1388.

Henke 1986: W. Henke, Die magdalénienzeitlichen Menschenfunde von Oberkassel bei Bonn. Bonner Jahrbücher 186, 1986, 317-366.

Henke/Schmitz/Street 2006: W. Henke / R. W. Schmitz / M. Street, Die späteiszeitlichen Funde von Bonn-Oberkassel. In: Rheinisches Landesmuseum 2006, 243-255.

Hernández Fernández/Peláez-Campomanes 2005: M. Hernández Fernández / P. Peláez-Campomanes, Quantitative palaeoclimatic inference based on terrestrial mammal faunas. Global Ecology and Biogeography 14, 2005, 39-56.

Heumann et al. 1995: K. G. Heumann / S. Eisenhut / S. Gallus / E. H. Hebeda / R. Nusko / A. Vengosh / T. Walczyk, Recent developments in thermal ionization mass spectrometric techniques for isotope analysis. A review. Analyst 120, 1995, 1291-1299.

Heuschen 2007: W. Heuschen, Altsteinzeitlichen Elchjägern auf der Spur – erste Untersuchungen am Oberflächenfundplatz Windeck. Archäologie im Rheinland 2006 (2007), 50-52.

Heuschen et al. 2006: W. Heuschen / F. Gelhausen / S. B. Grimm / M. Street, Ein verzierter Retuscheur aus dem mittleren Siegtal (Nordrhein-Westfalen). Archäologisches Korrespondenzblatt 36, 2006, 17-28.

Hewitt 2000: G. Hewitt, The genetic legacy of the Quaternary ice ages. Nature 405, 2000, 907-913.

Heyman et al. 2013: B. M. Heyman / J. Heyman / T. Fickert / J. M. Harbor, Paleo-climate of the central European uplands during the last glacial maximum based on glacier mass-balance modeling. Quaternary Research 79, 2013, 49-54.

Higham/Bronk Ramsey/Owen 2004: T. Higham / C. Bronk Ramsey / C. Owen (eds), Radiocarbon and archaeology. Fourth international symposium, St Catherine's College, Oxford, 9-14 April 2002 (Oxford 2004).

Higham/Jacobi/Bronk Ramsey 2006: T. F. G. Higham / R. M. Jacobi / C. Bronk Ramsey, AMS radiocarbon dating of ancient bone using ultrafiltration. Radiocarbon 48, 2006, 179-195.

Higham et al. 2007: T. F. G. Higham / C. Bronk Ramsey / F. Brock / D. Baker / P. Ditchfield, Radiocarbon dates from the Oxford AMS System: Archaeometry datelist 32. Archaeometry 49, 2007, supplement 1.

2011: T. F. G. Higham / C. Bronk Ramsey / F. Brock / D. Baker / P. Ditchfield, Radiocarbon dates from the Oxford AMS System: Archaeometry datelist 34. Archaeometry 53, 2011, 1067-1084.

Hijma et al. 2012: M. P. Hijma / K. M. Cohen / W. Roebroeks / W. E. Westerhoff / F. S. Busschers, Pleistocene Rhine-Thames landscapes: geological background for hominin occupation of the southern North Sea region. Journal of Quaternary Science 27, 2012, 17-39.

Hilgers et al. 2001: A. Hilgers / E. Gehrt / A. Janotta / U. Radtke, A contribution to the dating of the northern boundary of the Weichselian Loess Belt in Northern Germany by luminescence dating and pedological analysis. Quaternary International 76-77, 2001, 191-200.

Hiller et al. 2003: A. Hiller / C. Tinapp / P. M. Grootes / M.-J. Nadeau, Ungewöhnliche Probleme bei der ^{14}C-Datierung organischer Komponenten und Fraktionen fluviatiler Sedimente aus der Aue der Weißen Elster bei Leipzig. Eiszeitalter und Gegenwart 52, 2003, 4-12.

Hiscock 1981: P. Hiscock, Comments on the use of chipped stone artefacts as a measure of »intensity of site usage«. Australian Archaeology 13, 1981, 30-34.

2002: P. Hiscock, Quantifying the size of artefact assemblages. Journal of Archaeological Science 29, 2002, 251-258.

Höck 1995: C. Höck, Die Frauenstatuetten von Gönnersdorf und Andernach. Jahrbuch des RGZM 40, 1993 (1995), 253-316.

2000: C. Höck, Das Magdalénien der Kniegrotte. Ein Höhlenfundplatz bei Döbritz, Saale-Orla-Kreis. Weimarer Monographien zur Ur- und Frühgeschichte 35 (Stuttgart 2000).

Hoek 1997: W. Z. Hoek, Late-glacial and early Holocene climatic events and chronology of vegetation development in the Netherlands. Vegetation History and Archaeobotany 6, 1997, 197-213.

2000: W. Z. Hoek, Abiotic landscape and vegetation patterns in the Netherlands during the Weichselian Late Glacial. Netherlands Journal of Geosciences (Geologie en Mijnbouw) 79, 2000, 497-509.

2001: W. Z. Hoek, Vegetation response to the ~14.7 and ~11.5 ka cal. BP climate transitions: is vegetation lagging climate? Global and Planetary Change 30, 2001, 103-115.

Hoek/Bohncke 2001: W. Z. Hoek / S. J. P. Bohncke, Oxygen-isotope wiggle matching as a tool for synchronising ice-core and terrestrial records over Termination 1. Quaternary Science Reviews 20, 2001, 1251-1264.

Hoek et al. 1999: W. Z. Hoek / S. J. P. Bohncke / G. M. Ganssen / T. Meijer, Lateglacial environmental changes recorded in calcareous gyttja deposits at Gulickshof, southern Netherlands. Boreas 28, 1999, 416-432.

Hoffmann/Reicherter 2012: G. Hoffmann / K. Reicherter, Soft-sediment deformation of Late Pleistocene sediments along the southwestern coast of the Baltic Sea (NE Germany). International Journal of Earth Sciences (Geologische Rundschau) 101, 2012, 351-363.

Holling 2001: C. S. Holling, Understanding the complexity of economic, ecological, and social systems. Ecosystems 4, 2001, 390-405.

Holling/Gunderson 2002: C. S. Holling / L. H. Gunderson, Resilience and adaptive cycles. In: L. H. Gunderson / C. S. Holling (eds), Panarchy. Understanding transformations in human and natural systems (Washington, D.C. 2002) 25-62.

Hollister/Webber/Bay 2005: R. D. Hollister / P. J. Webber / C. Bay, Plant response to temperature in northern Alaska: Implications for predicting vegetation change. Ecology 86, 2005, 1562-1570.

Holm 2003: J. Holm, Rentierjäger im Norden. Archäologie in Deutschland 3, 2003, 54-56.

Holm/Rieck 1992: J. Holm / F. Rieck, Istidsjægere ved Jelssøerne. Hamburgkulturen i Danmark. Skrifter fra Museumsrådet for Sønderjyllands Amt 5 (Haderslev 1992).

Holst 2010: D. Holst, Hazelnut economy of early Holocene hunter-gatherers: a case study from Mesolithic Duvensee, northern Germany. Journal of Archaeological Science 37, 2010, 2871-2880.

2014: D. Holst, Subsistenz und Landschaftnutzung im Frühmesolithikum: Nussröstplätze am Duvensee. Monographien des RGZM 120 (Mainz 2014).

Holzkämper 2006: J. Holzkämper, Die Konzentration IV des Magdalénien von Andernach-Martinsberg, 1994-1996 [dissertation, Köln 2006].

Holzkämper et al. 2014: J. Holzkämper / I. Kretschmer / A. Maier / M. Baales / A. von Berg / J. A. A. Bos / M. Bradtmöller / K. Edinborough / S. Flohr / L. Giemsch / S. B. Grimm / J. Hilpert / A. J. Kalis / T. Kerig / D. Leesch / J. Meurers-Balke / L. Mevel / J. Orschiedt / M. Otte / A. Pastoors / P. Pettitt / E. Rensink / J. Richter / F. Riede / I. Schmidt / R. W. Schmitz / S. Shennan / M. Street / Y. Tafelmaier / M.-J. Weber / K. P. Wendt / G.-C. Weniger / A. Zimmermann, The Upper-Late Palaeolithic Transition in Western Central Europe. Typology, Technology, Environment and Demography. Report on the workshop held in Rösrath, 21st-24th June 2012. Archäologische Informationen 36, 2014, 161-186.

Höpken 1995: C. Höpken, Die Steinartefakte von Kamphausen und Galgenberg bei Möchengladbach und deren Stellung im späten Jungpaläolithikum Nordwesteuropas. Archäologische Informationen 18, 1995, 99-103.

Houmark-Nielsen/Henrik Kjær 2003: M. Houmark-Nielsen / K. Henrik Kjær, Southwest Scandinavia, 40-15 kyr BP: palaeogeography and environmental change. Journal of Quaternary Science 18, 2003, 769-786.

Houpt/Boyd 1994: K. A. Houpt / L. Boyd, Social behaviour. In: L. Boyd / K. A. Houpt (eds), Przewalski's horse. The history and biology of an endangered species (Albany NY 1994) 229-254.

Housley et al. 1997: R. A. Housley / C. S. Gamble / M. Street / P. Pettitt, Radiocarbon evidence for the Lateglacial human recolonisation of northern Europe. Proceedings of the Prehistoric Society 63, 1997, 25-54.

2012: R. A. Housley / C. S. Lane / V. L. Cullen / M. J. Weber / F. Riede / C. S. Gamble / F. Brock, Icelandic volcanic ash from the Late-glacial open-air archaeological site of Ahrenshöft LA 58 D, North Germany. Journal of Archaeological Science 39, 2012, 708-716.

Hua et al. 2009: Q. Hua / M. Barbetti / D. Fink / K. F. Kaiser / M. Friedrich / B. Kromer / V. A. Levchenko / U. Zoppi / A. M. Smith / F. Bertuch, Atmospheric ^{14}C variations derived from tree rings during the early Younger Dryas. Quaternary Science Reviews 28, 2009, 2982-2990.

Hubbard et al. 2009: A. Hubbard / T. Bradwell / N. Golledge / A. Hall / H. Patton / D. Sugden / R. Cooper / M. Stoker, Dynamic cycles, ice streams and their impact on the extent, chronology and deglaciation of the British-Irish ice sheet. Quaternary Science Reviews 28, 2009, 758-776.

Huber et al. 2010: K. Huber / K. Weckström / R. Drescher-Schneider / J. Knoll / J. Schmidt / R. Schmidt, Climate changes during the last glacial termination inferred from diatom-based temperatures and pollen in a sediment core from Längsee (Austria). Journal of Paleolimnology 43, 2010, 131-147.

Huet/Thévenin 1995: F. Huet / A. Thévenin, Les gisements à Federmesser d'Éclaron, canton de Saint-Dizier (Haute-Marne). In: Thévenin 1995, 13-22.

Huet et al. 1995: F. Huet / G. Guillot / O. Guillot / J. Krzyzanowski / A. Thévenin, Les occupations à Federmesser entre Meuse et Marne dans les vallées de l'Ornain et de la Saulx. In: Thévenin 1995, 43-48.

Hughen et al. 1996a: K. A. Hughen / J. T. Overpeck / L. C. Peterson / R. F. Anderson, The nature of varved sedimentation in the Cariaco Basin, Venezuela, and its palaeoclimatic significance. Geological Society, London, Special Publications 116, 1996, 171-183.

1996b: K. A. Hughen / J. T. Overpeck / L. C. Peterson / S. Trumbore, Rapid climate changes in the tropical Atlantic region during the last deglaciation. Nature 380, 1996, 51-54.

1998a: K. A. Hughen / J. T. Overpeck / S. J. Lehman / M. Kashgarian / J. Southon / L. C. Petersonk / R. Alley / D. M. Sigman, Deglacial changes in ocean circulation from an extended radiocarbon calibration. Nature 391, 1998, 65-68.

1998b: K. A. Hughen / J. T. Overpeck / S. J. Lehman / M. Kashgarian / J. R. Southon / L. C. Peterson, A new ^{14}C calibration data set for the last deglaciation based on marine varves. Radiocarbon 40, 1998, 483-494.

2000: K. A. Hughen / J. R. Southon / S. J. Lehman / J. T. Overpeck, Synchronous radiocarbon and climate shifts during the last deglaciation. Science 290, 2000, 1951-1954.

2004a: K. Hughen / S. Lehman / J. Southon / J. Overpeck / O. Marchal / C. Herring / J. Turnbull, ^{14}C activity and global carbon cycle changes over the past 50,000 years. Science 303, 2004, 202-207.

2004b: K. A. Hughen / T. I. Eglinton / L. Xu / M. Makou, Abrupt tropical vegetation response to rapid climate changes. Science 304, 2004, 1955-1959.

2004c: K. A. Hughen / J. R. Southon / C. J. H. Bertrand / B. Frantz / P. Zermeño, Cariaco Basin calibration update: revisions to calendar and ^{14}C chronologies for core PL07-58PC. Radiocarbon 46, 2004, 1161-1187.

2006: K. Hughen / J. Southon / S. Lehman / C. Bertrand / J. Turnbull, Marine-derived ^{14}C calibration and activity record for the past 50,000 years updated from the Cariaco Basin. Quaternary Science Reviews 25, 2006, 3216-3227.

Hughes 1998: S. S. Hughes, Getting to the point: Evolutionary change in prehistoric weaponry. Journal of Archaeological Method and Theory 5, 1998, 345-408.

Huijzer/Isarin 1997: A. S. Huijzer / R. F. B. Isarin, The reconstruction of past climates using multi-proxy evidence: an example of the Weichselian pleniglacial in Northwest and Central Europe. Quaternary Science Reviews 16, 1997, 513-533.

Huijzer/Vandenberghe 1998: B. Huijzer / J. Vandenberghe, Climatic reconstruction of the Weichselian Pleniglacial in northwestern and Central Europe. Journal of Quaternary Science 13, 1998, 391-417.

Hurrell 1995: J. W. Hurrell, Decadal trends in the North Atlantic Oscillation: regional temperatures and precipitation. Science 269, 1995, 676-679.

Husmann 1988: H. Husmann, Die Fläche 37/40-42/45 des späteiszeitlichen Fundplatzes Niederbieber/Neuwieder Becken [unpublished M.A. thesis, Köln 1988].

1989: H. Husmann, Die Konzentration V (Fläche 37/40-42/45) des späteiszeitlichen Fundplatzes Niederbieber. Neue Erkenntnisse zur internen Chronologie. Archäologisches Korrespondenzblatt 19, 1989, 117-127.

Ikeda/Tajika 2002: T. Ikeda / E. Tajika, Carbon cycling and climate change during the last glacial cycle inferred from the isotope records using an ocean biogeochemical carbon cycle model. Global and Planetary Change 35, 2002, 131-141.

Ikinger 1998: E.-M. Ikinger, Der endeiszeitliche Rückenspitzen-Kreis Mitteleuropas. GeoArchaeoRhein 1 (Münster 1998).

Ikinger/Ikinger 1998: A. Ikinger / E.-M. Ikinger, Erd- und Siedlungsgeschichte des Sporns von Niederbieber am Mittelrhein. In: A. Ikinger / J. Bloemendal (eds), Festschrift Wolfgang Schirmer. Geschichte aus der Erde (Münster 1998) 195-212.

Ioviță 2009: R. Ioviță, Ontogenetic scaling and lithic systematics: method and application. Journal of Archaeological Science 36, 2009, 1447-1457.

2011: R. Ioviță, Shape variation in Aterian tanged tools and the origins of projectile technology: A morphometric perspective on stone tool function. PLoS One 6, 2011, e29029-1-14.

Irish et al. 2008: J. D. Irish / B. Bratlund / R. Schild / E. Kolstrup / H. Królik / D. Mańka / T. Boroń, A late Magdalenian perinatal human skeleton from Wilczyce, Poland. Journal of Human Evolution 55, 2008, 736-740.

Isarin 1997: R. F. B. Isarin, Permafrost distribution and temperatures in Europe during the Younger Dryas. Permafrost and Periglacial Processes 8, 1997, 313-333.

Isarin/Bohncke 1999: R. F. B. Isarin / S. J. P. Bohncke, Mean July temperatures during the Younger Dryas in Northwestern and Central Europe as inferred from climate indicator plant species. Quaternary Research 51, 1999, 158-173.

IUGG (CCS) – UNEP – UNESCO 2005: IUGG (CCS) – UNEP – UNESCO (eds), Fluctuations of Glaciers. 1995-2000 (Volume VIII) (Zürich 2005).

Iversen 1942: J. Iversen, En pollenanalytisk tidsfæstelse af ferskvandslagene ved Nørre Lyngby. Med Bemærkninger om de senglaciale Naturforhold i Danmark. Bulletin of the Geological Society of Denmark (Meddelelser fra Dansk Geologisk Forening) 10, 1942, 130-151.

1947: J. Iversen, Plantevækst, Dyreliv og Klima i det senglaciale Danmark. Geologiska Föreningens i Stockholm Förhandlingar (GFF) 69, 1947, 67-78.

Ivy-Ochs et al. 2006: S. Ivy-Ochs / H. Kerschner / P. W. Kubik / C. Schlüchter, Glacier response in the European Alps to Heinrich Event 1 cooling: the Gschnitz stadial. Journal of Quaternary Science 21, 2006, 115-130.

2008: S. Ivy-Ochs / H. Kerschner / A. Reuther / F. Preusser / K. Heine / M. Maisch / P. W. Kubik / C. Schlüchter, Chronology of the last glacial cycle in the European Alps. Journal of Quaternary Science 23, 2008, 559-573.

2009: S. Ivy-Ochs / H. Kerschner / M. Maisch / M. Christl / P. W. Kubik / C. Schlüchter, Latest Pleistocene and Holocene glacier variations in the European Alps. Quaternary Science Reviews 28, 2009, 2137-2149.

Jacobi 1980: R. M. Jacobi, The Upper Palaeolithic of Britain with special reference to Wales. In: J. A. Taylor (ed.), Culture and environment in prehistoric Wales. Selected essays (Oxford 1980) 15-100.

1988: R. M. Jacobi, Towards a British lateglacial archaeology. In: Otte 1988, 427-446.

2004: R. Jacobi, The Late Upper Palaeolithic Collection from Gough's Cave, Cheddar, Somerset and Human Use of the cave. Proceedings of the Prehistoric Society 70, 2004, 1-92.

Jacobi/Higham 2009: R. M. Jacobi / T. F. G. Higham, The early Lateglacial re-colonization of Britain: new radiocarbon evidence from Gough's Cave, southwest England. Quaternary Science Reviews 28, 2009, 1895-1913.

2010: R. M. Jacobi / T. F. G. Higham, The Later Upper Palaeolithic recolonisation of Britain: New results from AMS radiocarbon dating. In: N. Ashton / S. G. Lewis / C. Stringer (eds), The ancient human occupation of Britain (Amsterdam, Boston, London 2010) 223-248.

Jacobi/Higham/Lord 2009: R. M. Jacobi / T. F. G. Higham / T. C. Lord, Improving the chronology of the human occupation of Britain during the Late Glacial. In: Street/Barton/Terberger 2009, 7-25.

Jacobi/Roberts 1992: R. Jacobi / A. Roberts, A new variant on the Creswellian angle-backed blade. Lithics: The Newsletter of the Lithic Studies Society 13, 1992, 33-39.

Janes 1989: R. R. Janes, A comment on microdebitage analyses and cultural site-formation processes among tipi dwellers. American Antiquity 54, 1989, 851-855.

Janny et al. 2006: F. Janny / F. Audouze / S. Beyries / D. Keeler, Les burins du niveau supérieur du site de Verberie – Le Buisson Campin (France). De la gestion des supports à l'utilisation des outils: Un pragmatisme bien tempéré. ArchéoLogiques 2, 2006, 255-275.

Jardón Giner 1997: P. Jardón Giner, Analyse tracéologique du materiel lithique de la grotte du Bois Laiterie. In: Otte/Straus 1997, 325-327.

Jary/Ciszek 2013: Z. Jary / D. Ciszek, Late Pleistocene loess-palaeosol sequences in Poland and western Ukraine. Quaternary International 296, 2013, 37-50.

Jensen et al. 2005: J. B. Jensen / O. Bennike / W. Lemke / A. Kuijpers, The Storebælt gateway to the Baltic. Geological Survey of Denmark and Greenland Bulletin 7, 2005, 45-48.

2002: J. B. Jensen / K. S. Petersen / P. Konradi / A. Kuijpers / O. Bennike / W. Lemke / R. Endler, Neotectonics, sea-level changes and biological evolution in the Fennoscandian Border Zone of the southern Kattegat Sea. Boreas 31, 2002, 133-150.

Jessen 1935: K. Jessen, Archaeological dating in the history of North Jutland's vegetation. Acta Archaeologica (København) 5, 1934 (1935), 185-214.

Jochim 1991: M. A. Jochim, Archeology as long-term ethnography. American Anthropologist (New Series) 93, 1991, 308-321.

Johansen/Stapert 2004: L. Johansen / D. Stapert, Oldeholtwolde. A Hamburgian family encampment around a hearth (Lisse 2004).

Johansen et al. 2000: L. Johansen / J. N. Lanting / R. C. G. M. Lauwerier / M. J. L. T. Niekus / D. Stapert / I. L. M. Stuijts, De Federmesser-vindplaats bij Doetinchem (Gld.): natuurwetenschappelijk onderzoek. Paleo-Aktueel 11, 2000, 9-14.

Johnsen et al. 1992: S. J. Johnsen / H. B. Clausen / W. Dansgaard / K. Fuhrer / N. Gundestrup / C. U. Hammer / P. Iversen / J. Jouzel / B. Stauffer / J. P. Steffensen, Irregular glacial interstadials recorded in a new Greenland ice core. Nature 359, 1992, 311-313.

1995: S. Johnsen / D. Dahl-Jensen / W. Dansgaard / N. Gundestrup, Greenland palaeotemperatures derived from GRIP bore hole temperature and ice core isotope profiles. Tellus 47B, 1995, 624-629.

1997: S. J. Johnsen / H. B. Clausen / W. Dansgaard / N. S. Gundestrup / C. U. Hammer / U. Andersen / K. K. Andersen / C. S. Hvidberg / D. Dahl-Jensen / J. P. Steffensen / H. Shoji / Á. E. Sveinbjörnsdóttir / J. White / J. Jouzel / D. Fisher, The $\delta^{18}O$ record along the Greenland Ice Core Project deep ice core and the problem of possible Eemian climatic instability. Journal of Geophysical Research – Oceans 102, 1997, 26397-26410.

2001: S. J. Johnsen / D. Dahl-Jensen / N. Gundestrup / J. P. Steffensen / H. B. Clausen / H. Miller / V. Masson-Delmotte / A. E. Sveinbjörnsdottir / J. White, Oxygen isotope and palaeotemperature records from six Greenland ice-core stations: Camp Century, Dye-3, GRIP, GISP2, Renland and NorthGRIP. Journal of Quaternary Science 16, 2001, 299-307.

Johnson 1952: F. Johnson, The significance of the dates for archaeology and geology. In: W. F. Libby, Radiocarbon dating (Chicago 1952) 97-111.

Jones/Montanarella/Jones 2005: A. Jones / L. Montanarella / R. Jones (eds), Soil atlas of Europe. EUR 21676 EN (Luxembourg 2005).

Jones et al. 2002: R. T. Jones / J. D. Marshall / S. F. Crowley / A. Bedford / N. Richardson / J. Bloemendal / F. Oldfield, A high resolution, multiproxy Late-glacial record of climate change and intrasystem responses in northwest England. Journal of Quaternary Science 17, 2002, 329-340.

Jöris/Schmitz/Thissen 1993: O. Jöris / R.-W. Schmitz / J. Thissen, Beeck: Ein »special-task-camp« des Magdalénien. Neue Aspekte zum späten Jungpaläolithikum im Rheinland. Archäologisches Korrespondenzblatt 23, 1993, 259-273.

Jöris/Street/Turner 2011: O. Jöris / M. Street / E. Turner, Spatial analysis at the Magdalenian site of Gönnersdorf (Central Rhineland, Germany) – an introduction. In: Gaudzinski-Windheuser et al. 2011b, 53-80.

Jöris/Terberger 2001: O. Jöris / Th. Terberger, Zur Rekonstruktion eines Zeltes mit trapezförmigem Grundriß am Magdalénien-Fundplatz Gönnersdorf/Mittelrhein – eine »Quadratur des Kreises?«. Archäologisches Korrespondenzblatt 31, 2001, 163-172.

Jöris/Weninger 1998: O. Jöris / B. Weninger, Extension of the ^{14}C calibration curve to ca. 40,000 cal BC by synchronizing Greenland $^{18}O/^{16}O$ ice core records and North Atlantic foraminifera profiles: a comparison with U/Th coral data. Radiocarbon 40, 1998, 495-504.

2000: O. Jöris / B. Weninger, Radiocarbon calibration and the absolute chronology of the Late Glacial. In: Valentin/Bodu/Christensen 2000, 19-54.

Jöris et al. 2011: O. Jöris / M. Street / Th. Terberger / B. Weninger, Radiocarbon dating the Middle to Upper Palaeolithic transition: The demise of the last Neanderthals and the first appearance of anatomically modern humans in Europe. In: S. Condemi / G.-C. Weniger (eds), Continuity and discontinuity in the peopling of Europe. One hundred fifty years of Neanderthal study (Dordrecht 2011) 239-298.

Jouzel et al. 1997: J. Jouzel / R. B. Alley / K. M. Cuffey / W. Dansgaard / P. Grootes / G. Hoffmann / S. J. Johnsen / R. D. Koster / D. Peel / C. A. Shuman / M. Stievenard / M. Stuiver / J. White, Validity of the temperature reconstruction from water isotopes in ice cores. Journal of Geophysical Research – Oceans 102, 1997, 26471-26487.

Julien 1982: M. Julien, Les harpons magdaléniens. Supplément à Gallia Préhistoire 17 (Paris 1982).

Julien/Karlin/Bodu 1987: M. Julien / C. Karlin / P. Bodu, Pincevent: où en est le modèle théorique aujourd'hui? Bulletin de la Société préhistorique française 84, 1987, 335-342.

Julien/Rieu 1999: M. Julien / J.-L. Rieu (eds), Occupations du Paléolithique supérieur dans le sud-est du Bassin parisien. Documents d'archéologie française; Série archéologie préventive – Autoroutes A 5/A 160 78 (Paris 1999).

Julien et al. 1999: M. Julien / P. Alix / A. Bridault / B. Valentin, Analyse de la répartition spatiale. In: Julien/Rieu 1999, 132-155.

Kabaciński/Schild 2005: J. Kabaciński / R. Schild, The Hamburgian site at Mirkowice: a chronological framework. Fontes Archaeologici Posnanienses 41, 2005, 15-18.

Kabaciński/Sobkowiak-Tabaka 2010: J. Kabaciński / I. Sobkowiak-Tabaka, Between East and West – a new site of the Federmessergruppen in Poland. Quartär 57, 2010, 139-154.

Kageyama et al. 2004: M. Kageyama / S. Charbit / C. Ritz / M. Khodri / G. Ramstein, Quantifying ice-sheet feedbacks during the last glacial inception. Geophysical Research Letters 31, 2004, L24203.1-4.

Kaiser 1993: K. F. Kaiser, Beiträge zur Klimageschichte vom Hochglazial bis ins frühe Holozän, rekonstruiert mit Jahrringen und Molluskenschalen aus verschiedenen Vereisungsgebieten (Winterthur 1993).

2002: K. Kaiser (ed.), Die jungquartäre Fluß- und Seegenese in Nordostdeutschland. Beiträge zur Tagung in Hohenzieritz (Mecklenburg) vom 26.-28. Februar 2002 (Greifswald 2002).

2004: K. Kaiser, Geomorphic characterization of the Pleistocene – Holocene transition in Northeast Germany. In: Terberger/Eriksen 2004, 53-73.

Kaiser/Clausen 2005: K. Kaiser / I. Clausen, Palaeopedology and stratigraphy of the Late Palaeolithic Alt Duvenstedt site, Schleswig-Holstein (Northwest Germany). Archäologisches Korrespondenzblatt 35, 2005, 447-466.

Kaiser/de Klerk/Terberger 1999: K. Kaiser / P. de Klerk / Th. Terberger, Die »Riesenhirschfundstelle« von Endingen: geowissenschaftliche und archäologische Untersuchungen an einem spätglazialen Fundplatz in Vorpommern. Eiszeitalter und Gegenwart 49, 1999, 102-123.

Kaiser et al. 2009: K. Kaiser / A. Hilgers / N. Schlaak / M. Jankowski / P. Kühn / S. Bussemer / K. Przegiętka, Palaeopedological marker horizons in northern central Europe: characteristics of Lateglacial Usselo and Finow soils. Boreas 38, 2009, 591-609.

2012: K. F. Kaiser / M. Friedrich / C. Miramont / B. Kromer / M. Sgier / M. Schaub / I. Boeren / S. Remmele / S. Talamo / F. Guibald / O. Sivan, Challenging process to make the Lateglacial tree-ring chronologies from Europe absolute – an inventory. Quaternary Science Reviews 36, 2012, 78-90.

Karlin/Julien 1994: C. Karlin / M. Julien, Prehistoric technology: a cognitive science? In: C. Renfrew / E. B. W. Zubrow (eds), The ancient mind. Elements of cognitive archaeology (Cambridge 1994) 152-164.

Kars/Wallinga/Cohen 2008: R. H. Kars / J. Wallinga / K. M. Cohen, A new approach towards anomalous fading correction for feldspar IRSL dating – tests on samples in field saturation. Radiation Measurements 43, 2008, 786-790.

Kasse 2002: C. Kasse, Sandy aeolian deposits and environments and their relation to climate during the Last Glacial Maximum and Lateglacial in northwest and central Europe. Progress in Physical Geography 26, 2002, 507-532.

Kasse et al. 2005: C. Kasse / W. Z. Hoek / S. J. P. Bohncke / M. Konert / J. W. H. Weijers / M. L. Cassee / R. M. van der Zee, Late Glacial fluvial response of the Niers-Rhine (western Germany) to climate and vegetation change. Journal of Quaternary Science 20, 2005, 377-394.

Kauffman 1988: E. G. Kauffman, Concepts and methods of high-resolution event stratigraphy. Annual Review of Earth and Planetary Sciences 16, 1988, 605-654.

Keeling 1995: R. F. Keeling, The atmospheric oxygen cycle. The oxygen isotopes of CO_2 and O_2 and the O_2/N_2 ratio. Reviews of Geophysics Supplement 1995, 1253-1262.

Kegler 1999: J. Kegler, Die retuschierten Artefakte der oberen Fundschicht von Andernach-Martinsberg, Grabung 1994-1996 [unpublished M.A. thesis, Köln 1999].

2002: J. Kegler, Die federmesserzeitliche Fundschicht des paläolithischen Siedlungsplatzes Andernach-Martinsberg, Grabung 1994-1996. Archäologisches Korrespondenzblatt 32, 2002, 501-516.

2007: J. Kegler, Das Azilien von Mas d'Azil. Der chronologische und kulturelle Kontext der Rückenspitzengruppen in Südwesteuropa [dissertation, Köln 2007].

Kelly 1991: R. L. Kelly, Sedentism, sociopolitical inequality, and resource fluctuations. In: S. A. Gregg (ed.), Between bands and states (Carbondale 1991) 135-158.

1992: R. L. Kelly, Mobility/Sedentism: concepts, archaeological measures, and effects. Annual Review of Anthropology 21, 1992, 43-66.

Kempe et al. 2002: S. Kempe / W. Roesendahl / B. Wiegand / A. Eisenhauer, New speleothem dates from caves in Germany and their importance for the Middle and Upper Pleistocene climate reconstruction. Acta Geologica Polonica 52, 2002, 55-61.

Kerschner/Ivy-Ochs 2008: H. Kerschner / S. Ivy-Ochs, Palaeoclimate from glaciers: Examples from the Eastern Alps during the Alpine Lateglacial and early Holocene. Global and Planetary Change 60, 2008, 58-71.

Kiden/Denys/Johnston 2002: P. Kiden / L. Denys / P. Johnston, Late Quaternary sea-level change and isostatic and tectonic land movements along the Belgian-Dutch North Sea coast: geological data and model results. Journal of Quaternary Science 17, 2002, 535-546.

Kind 1985: C.-J. Kind, Die Verteilung von Steinartefakten in Grabungsflächen. Ein Modell zur Organisation alt- und mittelpaläolithischer Siedlungsplätze. Urgeschichtliche Materialhefte 7 (Tübingen 1985).

Kitagawa/van der Plicht 1998: H. Kitagawa / J. van der Plicht, Atmospheric radiocarbon calibration to 45,000 yr B.P: Late Glacial fluctuations and cosmogenic isotope production. Science 279, 1998, 1187-1190.

Kleinmann/Merkt/Müller 2002: A. Kleinmann / J. Merkt / H. Müller, Sedimentologische und palynologische Untersuchungen an Ablagerungen des Siethener Sees und Blankensees (Brandenburg). – erste Ergebnisse. In: Kaiser 2002, 59-62.

de Klerk 2004: P. de Klerk, Confusing concepts in Lateglacial stratigraphy and geochronology: origin, consequences, conclusions (with special emphasis on the type locality Bøllingsø). Review of Palaeobotany and Palynology 129, 2004, 265-298.

2008: P. de Klerk, Patterns in vegetation and sedimentation during the Weichselian Late-glacial in north-eastern Germany. Journal of Biogeography 35, 2008, 1308-1322.

de Klerk et al. 2008: P. de Klerk / W. Janke / P. Kühn / M. Theuerkauf, Environmental impact of the Laacher See eruption at a large distance from the volcano: Integrated palaeoecological studies from Vorpommern (NE Germany). Palaeogeography, Palaeoclimatology, Palaeoecology 270, 2008, 196-214.

Klijn/de Haes 1994: F. Klijn / H. A. U. de Haes, A hierarchical approach to ecosystems and its implications for ecological land classification. Landscape Ecology 9, 1994, 89-104.

Knapp et al. 1999: A. K. Knapp / J. M. Blair / J. M. Briggs / S. L. Collins / D. C. Hartnett / L. C. Johnson / E. G. Towne, The keystone role of bison in North American tallgrass prairie. Bison increase habitat heterogeneity and alter a broad array of plant, community, and ecosystem processes. BioScience 49, 1999, 39-50.

Kobusiewicz 1999: M. Kobusiewicz, Ludy łowiecko-zbierackie północno-zachodniej Polski. Prace Komisji Archeologicznej / Polska Akademia Nauk 19 (Poznań 1999).

Koç/Jansen/Haflídason 1993: N. Koç / E. Jansen / H. Haflídason, Paleoceanographic reconstructions of surface ocean conditions in the Greenland, Iceland and Norwegian Seas through the last 14 ka based on diatoms. Quaternary Science Reviews 12, 1993, 115-140.

Koči 1978: A. Koči, Stratigraphie im Bereich der Gönnersdorfer Siedlung: Paläomagnetik. In: Brunnacker 1978a, 45.

von Koenigswald 1999: W. von Koenigswald, Paläoökologie und Vorkommen des pleistozänen Auerochsen (*Bos primigenius* BOJANUS 1827) im Vergleich zu den großen Rindern des Pleistozäns. In: G.-C. Weniger (ed.), Archäologie und Biologie des Auerochsen. Wissenschaftliche Schriften des Neanderthal Museums 1 (Mettmann 1999) 23-33.

2002: W. von Koenigswald, Lebendige Eiszeit. Klima und Tierwelt im Wandel (Darmstadt 2002).

2004: W. von Koenigswald, Das Quartär: Klima und Tierwelt im Eiszeitalter Mitteleuropas. Biologie in unserer Zeit 34, 2004, 151-158.

Kolstrup 2007: E. Kolstrup, Lateglacial older and younger coversand in northwest Europe: chronology and relation to climate and vegetation. Boreas 36, 2007, 65-75.

Konradi 2000: P. B. Konradi, Biostratigraphy and environment of the Holocene marine transgression in the Heligoland Channel, North Sea. Bulletin of the Geological Society of Denmark 47, 2000, 71-79.

Korn 1993: E.-M. Korn, Siedlungsmuster der Allerödzeit am Beispiel des Fundplatzes Niederbieber [unpublished M.A. thesis, Köln 1993].

Körner 2007: C. Körner, The use of »altitude« in ecological research. Trends in Ecology and Evolution 22, 2007, 569-574.

Kossinna 1911: G. Kossinna, Die Herkunft der Germanen. Zur Methode der Siedlungsarchäologie. Mannus-Bibliothek 6 (Würzburg 1911).

Kossler et al. 2011: A. Kossler / P. Tarasov / G. Schlolaut / T. Nakagawa / M. Marshall / A. Brauer / R. Staff / C. Bronk Ramsey / C. Bryant / H. Lamb / D. Demske / K. Gotanda / T. Haraguchi / Y. Yokoyama / H. Yonenobu / R. Tada / Suigetsu 2006 project members, Onset and termination of the late-glacial climate reversal in the high-resolution diatom and sedimentary records from the annually laminated SG06 core from Lake Suigetsu, Japan. Palaeogeography, Palaeoclimatology, Palaeoecology 306, 2011, 103-115.

Koster 2009: E. A. Koster, The »European Aeolian Sand Belt«: Geoconservation of Drift Sand Landscapes. Geoheritage 1, 2009, 93-110.

1988: E. A. Koster, Ancient and modern cold-climate aeolian sand deposition: A review. Journal of Quaternary Science 3, 1988, 69-83.

Kozłowski/Kozłowski 1996: J. K. Kozłowski / S. K. Kozłowski, Le paléolithique en Pologne. Collection l'homme des origines série préhistoire d'Europe 2 (Grenoble 1996).

Kozłowski 1987: S. K. Kozłowski, Remarks on the origins of the Polish curved backed point assemblages. In: J. M. Burdukiewicz / M. Kobusiewicz (eds), Late glacial in Central Europe. Culture and environment (Wrocław 1987) 241-253.

Kozłowski et al. 1995: S. K. Kozłowski / E. Sachse-Kozłowska / A. Marshack / T. Madeyska / H. Kierdorf / A. Lasota-Moskalewska / G. Jakubowski / M. Winiarska-Kabacińska / Z. Kapica / A. Wierciński, Maszycka Cave. A Magdalenian site in southern Poland. Jahrbuch des RGZM 40, 1993 (1995), 115-252.

2012: S. K. Kozłowski / M. Połtowicz-Bobak / D. Bobak / Th. Terberger, New information from Maszycka Cave and the Late Glacial recolonisation of Central Europe. Quaternary International 272-273, 2012, 288-296.

Kra 1986: R. Kra, Standardizing procedures for collecting, submitting, recording, and reporting radiocarbon samples. Radiocarbon 28, 1986, 765-775.

Kramer 2005: A. E. Kramer, Trade in mammoth ivory, helped by global thaw, flourishes in Russia. The New York Times, 25th March 2005.

Kretschmer 2012: I. Kretschmer, Palaeodemography of the Magdalenian – Estimating population density of hunter-gatherer during the Late Upper Palaeolithic in Europe. Quaternary International 279-280, 2012, 256.

Krienke 2002: K. Krienke, Sedimentation in Eisstauseen und lokalen Becken des Weichsel-Hochglazials innerhalb der Küstenregion Vorpommerns (NE-Deutschland). In: Kaiser 2002, 63-66.

Krinner et al. 2004: G. Krinner / J. Mangerud / M. Jakobsson / M. Crucifix / C. Ritz / J. I. Svendsen, Enhanced ice sheet growth in Eurasia owing to adjacent ice-dammed lakes. Nature 427, 2004, 429-432.

Kristensen et al. 2007: T. B. Kristensen / M. Huuse / J. A. Piotrowski / O. R. Clausen, A morphometric analysis of tunnel valleys in the eastern North Sea based on 3D seismic data. Journal of Quaternary Science 22, 2007, 801-815.

2008: T. B. Kristensen / J. A. Piotrowski / M. Huuse / O. R. Clausen / L. Hamberg, Time-transgressive tunnel valley formation indicated by infill sediment structure, North Sea. – the role of glaciohydraulic supercooling. Journal of Quaternary Science 33, 2008, 546-559.

Kromer et al. 2001: B. Kromer / S. W. Manning / P. I. Kuniholm / M. W. Newton / M. Spurk / I. Levin, Regional $^{14}CO_2$ offsets in the troposphere: magnitude, mechanisms, and consequences. Science 294, 2001, 2529-2532.

Kromer et al. 2004: B. Kromer / M. Friedrich / K. A. Hughen / F. Kaiser / S. Remmele / M. Schaub / S. Talamo, Late Glacial ^{14}C ages from a floating, 1382-ring pine chronology. Radiocarbon 46, 2004, 1203-1209.

Krueger 1997: H. W. Krueger, Radiocarbon dating and isotopic analysis of the human remains. In: Otte/Straus 1997, 365-367.

Krukowski 1939: S. Krukowski, Paleolit. In: S. Krukowski / J. Kostrzewski / R. Jakimowicz, Prahistoria ziem polskich (Kraków 1939-1948) 1-117.

Kuc/Różański/Dubliński 1998: T. Kuc / K. Różański / M. Dubliński, Isotopic indicators of the Late-Glacial/Holocene transition recorded in the sediments of Lake Gościąż. In: Ralska-Jasiewiczowa et al. 1998, 158-162.

Kucera et al. 2005: M. Kucera / M. Weinelt / T. Kiefer / U. Pflaumann / A. Hayes / M. Weinelt / M.-T. Chen / A. C. Mix / T. T. Barrows / E. Cortijo / J. Duprat / S. Juggins / C. Waelbroeck, Reconstruction of sea-surface temperatures from assemblages of planktonic foraminifera: multi-technique approach based on geographically constrained calibration data sets and its application to glacial Atlantic and Pacific Oceans. Quaternary Science Reviews 24, 2005, 951-998.

Kutterolf et al. 2013: S. Kutterolf / M. Jegen / J. X. Mitrovica / T. Kwasnitschka / A. Freundt / P. J. Huybers, A detection of Milankovitch frequencies in global volcanic activity. Geology 41, 2013, 227-230.

Łącka/Starnawska/Kuźniarski 1998: B. Łącka / E. Starnawska / M. Kuźniarski, Mineralogy and geochemistry of the Younger Dryas sediments from Lake Gościąż. In: Ralska-Jasiewiczowa et al. 1998, 124-128.

Laj et al. 2002: C. Laj / C. Kissel / A. Mazaud / E. Michel / R. Muscheler / J. Beer, Geomagnetic field intensity, North Atlantic Deep Water circulation and atmospheric $\Delta^{14}C$ during the last 50 kyr. Earth and Planetary Science Letters 200, 2002, 177-190.

Landais et al. 2010: A. Landais / G. Dreyfus / E. Capron / V. Masson-Delmotte / M. F. Sánchez-Goñi / S. Desprat / G. Hoffmann / J. Jouzel / M. Leutenberger / S. Johnsen, What drives the millenial and orbital variations of the $\delta^{18}O_{atm}$? Quaternary Science Reviews 29, 2010, 235-246.

Landmann et al. 1996: G. Landmann / A. Reimer / G. Lemcke / S. Kempe, Dating Late Glacial abrupt climate change in the 14,570 long continuous varve record of Lake Van, Turkey. Palaeogeography, Palaeoclimatology, Palaeoecology 122, 1996, 107-118.

Lane et al. 2011: C. S. Lane / S. P. E. Blockley / C. Bronk Ramsey / A. F. Lotter, Tephrochronology and absolute centennial scale synchronisation of European and Greenland records for the last glacial to interglacial transition: A case study of Soppensee and NGRIP. Quaternary International 246, 2011, 145-156.

2012a: C. S. Lane / S. P. E. Blockley / A. F. Lotter / W. Finsinger / M. L. Filippi / I. P. Matthews, A regional tephrostratigraphic framework for central and southern European climate archives during the Last Glacial to Interglacial transition: comparisons north and south of the Alps. Quaternary Science Reviews 36, 2012, 50-58.

2012b: C. S. Lane / S. P. E. Blockley / J. Mangerud / V. C. Smith / Ø. S. Lohne / E. L. Tomlinson / I. P. Matthews / A. F. Lotter, Was the 12.1 ka Icelandic Vedde Ash one of a kind? Quaternary Science Reviews 33, 2012, 87-99.

Lang 1998: L. Lang (ed.), Marolles-sur-Seine »Le Tureau des Gardes« (Seine-et-Marne). Vestiges d'implantations du Paléolithique supérieur. Fouilles de sauvetage urgent. Document final de synthèse (Bazoches-lès-Bray, Saint-Denis 1998).

Langlais 2007: M. Langlais, Des identités qui se cherchent... Apports des industries lithiques à la question de l'origine du Magdalénien moyen dans le Sud-Ouest européen. Bulletin de la Société préhistorique française 104, 2007, 759-770.

2008: M. Langlais, Chronology and territories in the Magdalenian between the Rhône and Ebro rivers: the exemple of lithics points. Palethnologie 1, 2008, 211-240.

2011: M. Langlais, Processes of change in Magdalenian societies in the Pyrenean isthmus (20-16 ky cal BP). Antiquity 85, 2011, 715-728.

Langlais et al. 2012: M. Langlais / S. Costamagno / V. Laroulandie / J.-M. Pétillon / E. Discamps / J.-B. Mallye / D. Cochard / D. Kuntz, The evolution of Magdalenian societies in South-West France between 18,000 and 14,000 calBP: Changing environments, changing tool kits. Quaternary International 272-273, 2012, 138-149.

Langley/Street 2013: M. C. Langley / M. Street, Long range inland-coastal networks during the Late Magdalenian: Evidence for individual acquisition of marine resources at Andernach-Martinsberg, German Central Rhineland. Journal of Human Evolution 64, 2013, 457-465.

Lanting/Aerts-Bijma/van der Plicht 2001: J. N. Lanting / A. T. Aerts-Bijma / J. van der Plicht, Dating of cremated bones. Radiocarbon 43, 2001, 249-254.

Lanting/van der Plicht 1996: J. N. Lanting / J. van der Plicht, De ^{14}C-chronologie van de Nederlandse Pre-en Protohistorie. I: Laat-Paleolithicum. Palaeohistoria 37/38, 1996, 71-125.

Lanting/Niekus/Stapert 2002: J. Lanting / M. Niekus / D. Stapert, Dateringen aan gecremeerd bot uit het Paleolithicum en Vroeg-Mesolithicum: Een overzicht van de stand van zaken. Paleo-Aktueel 13, 2002, 30-36.

Larsson 1996a: L. Larsson, The colonization of south Sweden during the deglaciation. In: Larsson 1996b, 141-155.

1996b: L. Larsson (ed.), The earliest settlement of Scandinavia and its relationship with neighbouring areas. Symposium Ystad, 10th-13th November 1994 (Lund 1996).

1999: L. Larsson, Perspectives on the colonisation of the Scandinavian Peninsula. In: M. Kobusiewicz / J. K. Kozłowski (eds), Post-Pleniglacial re-colonisation of the Great European Lowland. Papers presented at the conference organised by the International Union of Prehistoric and Protohistoric Sciences, Commission 8, held at the Jagellonian University, Kraków, in June 1998 (Kraków 1999) 175-196.

2009: L. Larsson, After the cold: Mammal finds, hunting and cultural affinities during the Preboreal. In: Street/Barton/Terberger 2009, 131-140.

Larsson et al. 2002: L. Larsson / R. Liljegren / O. Magnell / J. Ekström, Archaeo-faunal aspects of bog finds from Hässleberga, southern Scania, Sweden. In: Eriksen/Bratlund 2002, 61-74.

Lartet/Christy 1875: E. Lartet / H. Christy, Reliquiae Aquitanicae. Being contributions to the archaeology and palaeontology of Périgord and adjoining provinces of southern France (London, Paris 1875).

Lawrence Edwards/Chen/Wasserburg 1987: R. Lawrence Edwards / J. H. Chen / G. J. Wasserburg, ^{238}U-^{234}U-^{230}Th-^{232}Th systematics and the precise measurement of time over the past 500,000 years. Earth and Planetary Science Letters 81, 1987, 175-192.

Lawson/Bonsall 1986: T. Lawson / C. Bonsall, The Palaeolithic in Scotland: a reconsideration of evidence from Reindeer Cave, Assynt. In: S. N. Collcutt (ed.), The Palaeolithic of Britain and its nearest neighbours. Recent trends (Sheffield 1986) 85-89.

Le Tensorer 1981: J.-M. Le Tensorer, Le Paléolithique de l'Agenais. Cahiers du Quaternaire 3 (Paris 1981).

1998: J.-M. Le Tensorer, Le Paléolithique en Suisse. Préhistoire d'Europe 5 (Grenoble 1998).

Lebreton et al. 2010: V. Lebreton / E. Messager / L. Marquer / J. Renault-Miskovsky, A neotaphonomic experiment in pollen oxidation and its implications for archaeopalynology. Review of Palaeobotany and Palynology 162, 2010, 29-38.

Leduc 2012: C. Leduc, Ungulates exploitation for subsistence and raw material, during the Maglemose culture in Denmark: the example of Mullerup site (Sarauw's Island) in Sjælland. Danish Journal of Archaeology 1, 2012, 62-81.

2014a: C. Leduc, New Mesolithic hunting evidence from bone injuries at Danish Maglemosian sites: Lundby Mose and Mullerup (Sjælland). International Journal of Osteoarchaeology 24, 2012, 476-491.

2014b: C. Leduc, A specialized Early Maglemosian site at Lundby Mose (Zealand, Denmark): a contribution to the understanding of Maglemosian patterns of animal resource exploitation. Journal of Archaeological Science 41, 2014, 199-213.

Leesch 1997: D. Leesch, Hauterive-Champréveyres. Un campement magdalénien au bord du lac de Neuchâtel Neuchâtel. Cadre chronologique et culturel, mobilier et structures, analyse spatiale (secteur 1). Archéologie neuchâteloise 19 (Neuchâtel 1997).

Leesch/Bullinger 2012: D. Leesch / J. Bullinger, Identifying dwellings in Upper Palaeolithic open-air sites – The Magdalenian site at Monruz and its contribution to analysing palimpsests. In: Niekus et al. 2012, 165-181.

Leesch/Cattin/Müller 2004: D. Leesch / M.-I. Cattin / W. Müller, Hauterive-Champréveyres et Neuchâtel-Monruz. Témoins d'implantations magdaléniennes et aziliennes sur la rive nord du lac de Neuchâtel. Archéologie neuchâteloise 31 (Hauterive 2004).

Leesch et al. 2010: D. Leesch / J. Bullinger / M.-I. Cattin / W. Müller / N. Plumettaz, Hearths and hearth-related activities in Magdalenian open-air sites: The case studies of Champréveyres and Monruz (Switzerland) and their relevance to an understanding of Upper Paleolithic site structure. In: M. Połtowicz-Bobak / D. Bobak (eds), The Magdalenian in Central Europe. New finds and concepts. Conference devoted to the studies on the Magdalenian culture in Central Europe, held in September 2005 in Rzeszów (Rzeszów 2010) 53-69.

2012: D. Leesch / W. Müller / E. Nielsen / J. Bullinger, The Magdalenian in Switzerland: Re-colonization of a newly accessible landscape. Quaternary International 272-273, 2012, 191-208.

Lehman/Keigwin 1992: S. J. Lehman / L. D. Keigwin, Sudden changes in North Atlantic circulation during the last deglaciation. Nature 356, 1992, 757-762.

Lejeune 1997: M. Lejeune, L'art mobilier magdalénien final de la grotte du Bois Laiterie. In: Otte/Straus 1997, 293-317.

Lemke et al. 2002: W. Lemke / J. B. Jensen / O. Bennike / R. Endler, Spät- und postglaziale Flüsse und Seen in der heutigen westlichen Ostsee. In: Kaiser 2002, 73-77.

Lemonnier 1986: P. Lemonnier, The study of material culture today: Toward an anthropology of technical systems. Journal of Anthropological Archaeology 5, 1986, 147-186.

Leotard/Otte 1988: J.-M. Leotard / M. Otte, Occupation Paléolithique final aux grottes de Presle fouilles de 1983-1984 (Aiseau – Belgique). In: Otte 1988, 189-215.

Léotard/Lacroix 1997: J. M. Léotard / P. Lacroix, Historique de la decouverte. In: Otte/Straus 1997, 7-15.

Léotard 1993: J.-M. Léotard, Tourisme magdalénien dans la région Dinantaise. Notae Praehistoricae 12, 1993, 63-64.

Lericolais/Auffret/Bourillet 2003: G. Lericolais / J.-P. Auffret / J.-F. Bourillet, The Quaternary Channel River: seismic stratigraphy of its palaeo-valleys and deeps. Journal of Quaternary Science 18, 2003, 245-260.

Leroi-Gourhan 1943: An. Leroi-Gourhan, Evolution et techniques 1: L'Homme et la matiere (Paris 1943).

1945: An. Leroi-Gourhan, Evolution et technique 2: Milieu et Technique (Paris 1945).

1964: An. Leroi-Gourhan, Le Geste et la Parole 1: Technique et langage (Paris 1964).

1965: An. Leroi-Gourhan, Le Geste et la Parole 2: la Memoire et les Rythmes (Paris 1965).

1983: An. Leroi-Gourhan, Une tête de sagaie à armature de lamelles de silex à Pincevent (Seine-et-Marne). Bulletin de la Société préhistorique française 80, 1983, 154-156.

1993: An. Leroi-Gourhan, Gesture and Speech. October Books (Cambridge / Massachusetts 1993).

Leroi-Gourhan/Brézillon 1966: An. Leroi-Gourhan / M. Brézillon, L'habitation magdalénienne no. 1 de Pincevent près Montreau (Seine-et-Marne). Gallia Préhistoire 9, 1966, 263-385.

1972: An. Leroi-Gourhan / M. Brézillon, Fouilles de Pincevent. Essai d'analyse ethnographique d'un habitat magdalénien (la section 36). Supplément à Gallia Préhistoire 7 (Paris 1972).

Leroi-Gourhan 1978: Ar. Leroi-Gourhan, Analyse pollinique. In: Brunnacker 1978a, 46-55.

Leroy et al. 2000: S. A. G. Leroy / B. Zolitschka / J. F. W. Negendank / G. Seret, Palynological analyses in the laminated sediment of Lake Holzmaar (Eifel, Germany): duration of Lateglacial and Preboreal biozones. Boreas 29, 2000, 52-71.

Leroyer 1994: C. Leroyer, Le paysage végétal au Tardiglaciaire: apport de la palynologie. In: Taborin 1994, 59-64.

Leroyer/Allenet 1998: C. Leroyer / G. Allenet, Analyse pollinique des sites archéologiques de Rueil-Malmaison »Le et Les Closeaux«. In: Bodu 1998, 99-112.

Leroyer/Allenet Ribemont 2011: C. Leroyer / G. de Allenet Ribemont, Novelles données polliniques sue les séquences des Bazoches-lès-Bray »Le tureau à la caille« (Nov. 2010). In: B. Valentin (ed.), Paléolithique final et Mésolithique dans le Bassin parisien et ses marges. Habitats, sociétés et environnements. Rapport d'activités pour 2010 (Nanterre 2011) 43-47.

Letocart 1970: L. Letocart, Un gisement du paléolithique final à Obourg »St. Macaire« (Hainaut). In: K. Gripp / R. Schütrumpf / H. Schwabedissen (eds), Frühe Menschheit und Umwelt. Alfred Rust zum 65. Geburtstag (Köln 1970) 352-361.

Letourneux/Pétillon 2008: C. Letourneux / J. M. Pétillon, Hunting lesions caused by osseous projectile points: experimental results and archaeological implications. Journal of Archaeological Science 35, 2008, 2849-2862.

Levesque/Cwynar/Walker 1997: A. J. Levesque / L. C. Cwynar / I. R. Walker, Exceptionally steep north-south gradients in lake temperatures during the last deglaciation. Nature 385, 1997, 423-426.

Levesque et al. 1993: A. J. Levesque / F. E. Mayle / I. R. Walker / L. C. Cwynar, The Amphi-Atlantic Oscillation: A proposed late-glacial climatic event: A contribution to the »North Atlantic seaboard programme« of IGCP-253 »Termination of the Pleistocene«. Quaternary Science Reviews 12, 1993, 629-643.

Levy 1983: T. E. Levy, The emergence of specialized pastoralism in the southern Levant. World Archaeology 15, 1983, 15-36.

Lewis/Maddy/Scaife 2001: S. G. Lewis / D. Maddy / R. G. Scaife, The fluvial system response to abrupt climate change during the last cold stage: the Upper Pleistocene River Thames fluvial succession at Ashton Keynes, UK. Global and Planetary Change 28, 2001, 341-359.

Lewontin 1970: R. C. Lewontin, The units of selection. Annual Review of Ecology and Systematics 1, 1970, 1-18.

Lhomme et al. 2004: V. Lhomme / N. Connet / C. Chaussé / C. Bemilli / J.-J. Bahain / P. Voinchet, Les sites et les industries lithiques du Paléolithique inférieur, moyen et supérieur de la basse vallée de l'Yonne dans leur contextes chronostratigraphiques. Bilan de dix ans d'activité archéologique pluridisciplinaire dans le sud-est du bassin parisien. Bulletin de la Société préhistorique française 101, 2004, 701-739.

Libby 1952: W. F. Libby, Radiocarbon dating (Chicago 1952).

Libby/Anderson/Arnold 1949: W. F. Libby / E. C. Anderson / J. R. Arnold, Age determination by radiocarbon content: world-wide assay of natural radiocarbon. Science 109, 1949, 227-228.

Limondin 1998: N. Limondin, Analyses malacologiques sur le gisement du Closeau. In: Bodu 1998, 113-132.

Limondin-Lozouet et al. 2002: N. Limondin-Lozouet / A. Bridault / C. Leroyer / P. Ponel / P. Antoine / C. Chaussé / A.-V. Munaut / J.-F. Pastre, Evolution des écosystèmes de fond de vallée en France septentrionale au cours du Tardiglaciaire: l'apport des indicateurs biologiques. In: J.-P. Bravard / M. Magny (eds), Les fleuves ont une histoire. Paléo-environnement des riviéres et des lacs français depuis 15.000 ans. Table ronde, 21-23 juin 2001, Motz, Savoie (Paris 2002) 45-62.

Litt/Schmincke/Kromer 2003: T. Litt / H.-U. Schmincke / B. Kromer, Environmental response to climatic and volcanic events in central Europe during the Weichselian Lateglacial. Quaternary Science Reviews 22, 2003, 7-32.

Litt/Stebich 1999: T. Litt / M. Stebich, Bio- and chronostratigraphy of the lateglacial in the Eifel region, Germany. Quaternary International 61, 1999, 5-16.

Litt et al. 2001: T. Litt / A. Brauer / T. Goslar / J. Merkt / K. Bałaga / H. Müller / M. Ralska-Jasiewiczowa / M. Stebich / J. F. W. Negendank, Correlation and synchronisation of Lateglacial continental sequences in northern central Europe based on annually laminated lacustrine sediments. Quaternary Science Reviews 20, 2001, 1233-1249.

2007: T. Litt / K.-E. Behre / K.-D. Meyer / H.-J. Stephan / S. Wansa, Stratigraphische Begriffe für das Quartär des norddeutschen Vereisungsgebietes. Eiszeitalter und Gegenwart 56, 2007, 7-65.

Liu et al. 2008: D. Liu / Y. Wang / H. Cheng / R. L. Edwards / X. Kong / X. Wang / J. Wu / S. Chen, A detailed comparison of Asian Monsoon intensity and Greenland temperature during the Allerød and Younger Dryas events. Earth and Planetary Science Letters 272, 2008, 691-697.

Loew 2006: S. Loew, Rüsselsheim 122 und die Federmessergruppen am Unteren Main [dissertation, Köln 2006].

Loftus 1982: J. Loftus, Ein verzierter Pfeilschaftglätter von Fläche 64/70-73/78 des spätpaläolithischen Fundplatzes Niederbieber/ Neuwieder Becken. Archäologisches Korrespondenzblatt 12, 1982, 313-316.

1985: J. Loftus, Der spätpaläolithische Fundplatz Niederbieber. Fläche 64/70-81/77 [unpublished M.A. thesis, Köln 1985].

Loftus 1984: R. Loftus, Der spätpaläolithische Fundplatz Niederbieber. Fläche 33/32-38/39 [unpublished M.A. thesis, Köln 1984].

Löhr 1995a: H. Löhr, Alsdorf. In: G. Bosinski / M. Street / M. Baales (eds), The Palaeolithic and Mesolithic of the Rhineland. 14th congress of the International Union for Quaternary Research, August 3-10, 1995, Berlin, Germany (München 1995) 951-954.

1995b: H. Löhr, Katzensteine. In: G. Bosinski / M. Street / M. Baales (eds), The Palaeolithic and Mesolithic of the Rhineland. 14th congress of the International Union for Quaternary Research, August 3-10, 1995, Berlin, Germany (München 1995) 947-949.

1979: H. Löhr, Der Magdalénien-Fundplatz Alsdorf, Kreis Aachen-Land. Ein Beitrag zur Kenntnis der funktionalen Variabilität jungpaläolithischer Stationen [unpublished dissertation, Tübingen 1979].

Lombard/Haidle 2012: M. Lombard / M. N. Haidle, Thinking a bow-and-arrow set: Cognitive implications of Middle Stone Age bow and stone-tipped arrow technology. Cambridge Archaeological Journal 22, 2012, 237-264.

Lompré 2003: A. Lompré, Une nouvelle étude techno-fonctionnelle appliquée à un ensemble magdalénien de bâtons percés. Préhistoire Anthropologie méditerranéennes 12, 2003, 147-163.

Lonergan/Maidment/Collier 2006: L. Lonergan / S. C. R. Maidment / J. S. Collier, Pleistocene subglacial tunnel valleys in the central North Sea basin: 3-D morphology and evolution. Journal of Quaternary Science 21, 2006, 891-903.

López Bayón/Lacroix/Léotard 1997: I. López Bayón / P. Lacroix / J.-M. Léotard, Étude des restes malacologique de la grotte du Bois Laiterie. In: Otte/Straus 1997, 145-160.

López Bayón et al. 1997: I. López Bayón / L. G. Straus / J.-M. Léotard / P. Lacroix / E. Teheux, L'industrie osseuse du Magdalénien du Bois Laiterie. In: Otte/Straus 1997, 257-277.

Lorblanchet 1973: M. Lorblanchet, L'art préhistorique de Gouy. Notes bibliographiques. Bulletin de la Société préhistorique française 70, 1973, 200-204.

Lorenc 2006: M. Lorenc, Radiocarbon dating of some Late Pleistocene faunal assemblages in caves in Poland. Acta zoologica cracoviensia 49A, 2006, 41-61.

Lotter 1991a: A. F. Lotter, Absolute dating of the Late-Glacial period in Switzerland using annually laminated sediments. Quaternary Research 35, 1991, 321-330.

1991b: A. F. Lotter, How long was the Younger Dryas? Preliminary evidence from annually laminated sediments of Soppensee (Switzerland). Hydrobiologia 214, 1991, 53-51.

1999: A. F. Lotter, Late-glacial and Holocene vegetation history and dynamics as shown by pollen and plant macrofossil analyses in annually laminated sediments from Soppensee, central Switzerland. Vegetation History and Archaeobotany 8, 1999, 165-184.

Lotter/Birks/Zolitschka 1995: A. F. Lotter / H. J. B. Birks / B. Zolitschka, Late-glacial pollen and diatom changes in response to two different environmental perturbations: volcanic eruption and Younger Dryas cooling. Journal of Paleolimnology 14, 1995, 23-47.

Lotter et al. 1992: A. F. Lotter / U. Eicher / U. Siegenthaler / H. J. B. Birks, Late-glacial climatic oscillations as recorded in Swiss lake sediments. Journal of Quaternary Science 7, 1992, 187-204.

1997: A. F. Lotter / H. J. B. Birks / W. Hofmann / A. Marchetto, Modern diatom, cladocera, chironomid, and chrysophyte cyst assemblages as quantitative indicators for the reconstruction of past environmental conditions in the Alps. I. Climate. Journal of Paleolimnology 18, 1997, 395-420.

2012: A. F. Lotter / O. Heiri / S. Brooks / J. F. N. van Leeuwen / U. Eicher / B. Ammann, Rapid summer temperature changes during Termination 1a: high-resolution multi-proxy climate reconstructions from Gerzensee (Switzerland). Quaternary Science Reviews 36, 2012, 103-113.

Lowe 2011: D. J. Lowe, Tephrochronology and its application: A review. Quaternary Geochronology 6, 2011, 107-153.

Lowe 2001: J. J. Lowe, Abrupt climatic changes in Europe during the last glacial-interglacial transition: the potential for testing hypotheses on the synchroneity of climatic events using tephrochronology. Global and Planetary Change 30, 2001, 73-84.

Lowe/Walker 2000: J. J. Lowe / M. J. C. Walker, Radiocarbon dating the Last Glacial-Interglacial transition (ca. 14-9 ^{14}C ka BP) in terrestrial and marine records: The need for new quality assurance protocols. Radiocarbon 42, 2000, 53-68.

Lowe et al. 1999: J. J. Lowe / H. H. Birks / S. J. Brooks / G. R. Coope / D. D. Harkness / F. E. Mayle / C. Sheldrick / C. S. M. Turney / M. J. C. Walker, The chronology of palaeoenvironmental changes during the Last Glacial-Holocene transition: towards an event stratigraphy for the British Isles. Journal of the Geological Society, London 156, 1999, 397-410.

2008: J. J. Lowe / S. O. Rasmussen / S. Björck / W. Z. Hoek / J. P. Steffensen / M. J. C. Walker / Z. C. Yu / INTIMATE group, Synchronisation of palaeoenvironmental events in the North Atlantic region during the Last Termination: a revised protocol recommended by the INTIMATE group. Quaternary Science Reviews 27, 2008, 6-17.

Lücke/Brauer 2004: A. Lücke / A. Brauer, Biogeochemical and mirco-facial fingerprints of ecosystem response to rapid Late Glacial climatic changes in varved sediments of Meerfelder Maar (Germany). Palaeogeography, Palaeoclimatology, Palaeoecology 211, 2004, 139-155.

Lücke et al. 2003: A. Lücke / G. H. Schleser / B. Zolitschka / J. F. W. Negendank, A Lateglacial and Holocene organic carbon isotope record of lacustrine palaeoproductivity and climatic change derived from varved lake sediments of Lake Holzmaar, Germany. Quaternary Science Reviews 22, 2003, 569-580.

Lundqvist/Wohlfarth 2001: J. Lundqvist / B. Wohlfarth, Timing and east-west correlation of south Swedish ice marginal lines during the Late Weichselian. Quaternary Science Reviews 20, 2001, 1127-1148.

Luo et al. 1997: X. Luo / M. Rehkämper / D.-C. Lee / A. N. Halliday, High precision $^{230}Th/^{232}Th$ and $2^{34}U/^{238}U$ measurements using energyfiltered ICP magnetic sector multiple collector mass spectrometry. International Journal of Mass Spectrometry and Ion Processes 171, 1997, 105-117.

Lyman/van Pool/O'Brien 2009: R. L. Lyman / T. L. van Pool / M. J. O'Brien, The diversity of North American projectile-point types, before and after the bow and arrow. Journal of Anthropological Archaeology 28, 2009, 1-13.

MacFadden 1992: B. J. MacFadden, Fossil horses. Systematics, paleobiology and evolution of the family equidae (Cambridge 1992).

Madeyska 2002: T. Madeyska, Evidence of climatic variations in loess and cave Palaeolithic sites of southern Poland and western Ukraine. Quaternary International 91, 2002, 65-73.

Madsen 1992: B. Madsen, Hamburgkulturens flintteknologi i Jels. In: Holm/Rieck 1992, 93-131.

1996: B. Madsen, Late Palaeolithic cultures of south Scandinavia – tools, traditions and technology. In: Larsson 1996b, 61-73.

Magny 2001: M. Magny, Palaeohydrological changes as reflected by lake-level fluctuations in the Swiss Plateau, the Jura Mountains and the northern French Pre-Alps during the Last Glacial-Holocene transition: a regional synthesis. Global and Planetary Change 30, 2001, 85-101.

Magny/Bégeot 2004: M. Magny / C. Bégeot, Hydrological changes in the European midlatitudes associated with freshwater outbursts from Lake Agassiz during the Younger Dryas event and the early Holocene. Quaternary Research 61, 2004, 181-192.

Magny et al. 2006: M. Magny / G. Aalbersberg / C. Bégeot / P. Benoit-Ruffaldi / G. Bossuet / J.-R. Disnar / O. Heiri / F. Laggoun-Defarge / F. Mazier / L. Millet / O. Peyron / B. Vannière / A.-V. Walter-Simonnet, Environmental and climatic changes in the Jura mountains (eastern France) during the Lateglacial-Holocene transition: A multi-proxy record from Lake Lautrey. Quaternary Science Reviews 25, 2006, 414-445.

Maher/Banning/Chazan 2011: L. A. Maher / E. B. Banning / M. Chazan, Oasis or Mirage? Assessing the role of abrupt climate change in the prehistory of the southern Levant. Cambridge Archaeological Journal 21, 2011, 1-30.

Maher/Thompson 2012: B. A. Maher / R. Thompson, Oxygen isotopes from Chinese caves: records not of monsoon rainfall but of circulation regime. Journal of Quaternary Science 27, 2012, 615-624.

Makaske/Maas/van Smeerdijk 2008: B. Makaske / G. J. Maas / D. G. van Smeerdijk, The age and origin of the Gelderse IJssel. Netherlands Journal of Geosciences (Geologie en Mijnbouw) 87, 2008, 323-337.

Makos/Nitychoruk/Zreda 2013: M. Makos / J. Nitychoruk / M. Zreda, Deglaciation chronology and paleoclimate of the Pięciu Stawów Polskich/Roztoki Valley, high Tatra Mountains, Western Carpathians, since the Last Glacial Maximum, inferred from ^{36}Cl exposure dating and glacier-climate modelling. Quaternary International 293, 2013, 63-78.

Malec 1978: F. Malec, Umweltverhältnisse: Kleinsäugerfauna. In: Brunnacker 1978a, 105-227.

Mangerud et al. 1974: J. Mangerud / S. T. Andersen / B. E. Berglund / J. J. Donner, Quaternary stratigraphy of Norden, a proposal for terminology and classification. Boreas 3, 1974, 109-128.

Marchand et al. 2004: G. Marchand / S. Blanchet / G. Chevalier / J. Y. Gallais / M. Le Goffic / N. Naudinot / E. Yven, La fin du Tardiglaciaire sur le Massif Armoricain: territoires et cultures matérielles. Paléo 16, 2004, 137-170.

2008: G. Marchand / S. Sicard / P. Forré / J.-F. Nauleau, De la pelle mécanique aux remontages lithiques: espace habité et techniques au Tardiglaciaire sur l'habitat des Chaloignes (Mozé-sur-Louet, Maine-et- Loire). Revue archéologique de l'Ouest 25, 2008, 7-52.

Marciniak 1998: B. Marciniak, Younger Dryas diatom assemblages of Lake Gościąż. In: Ralska-Jasiewiczowa et al. 1998, 143-148.

Marshall 2007: D. M. Marshall, Ethnopalynological applications in land and water based archaeology [dissertation, College Station TX 2007).

Marshall et al. 2001: J. Marshall / Y. Kuschnir / D. Battisti / P. Chang / A. Czaja / R. Dickson / J. Hurrell / M. McCartney / R. Saravanan / M. Visbeck, North Atlantic climate variability: phenomena, impacts and mechanisms. International Journal of Climatology 21, 2001, 1863-1898.

Martin 2004: Y. Martin, Découvertes de peintures dans la grotte paléolithique de Gouy (Seine-Maritime): apport d'un éclairage inhabituel dans l'étude de l'art pariétal. Comptes Rendus Palevol 3, 2004, 143-156.

2007a: Y. Martin, The engravings of Gouy: France's northernmost decorated cave. In: Pettitt/Bahn/Ripoll 2007, 140-193.

2007b: Y. Martin, Une sculpture paléolithique inédite: la silhouette féminine en bas-relief de Gouy (Seine-Maritime, France). Comptes Rendus Palevol 6, 2007, 345-358.

2010: Y. Martin, La grotte de Gouy (Seine-Maritime). Préhistoire du Sud-Ouest 18, 2010, 9-25.

Mashiotta/Lea/Spero 1999: T. A. Mashiotta / D. W. Lea / H. J. Spero, Glacial-interglacial changes in Subantarctic sea surface temperature and $\delta^{18}O$-water using foraminiferal Mg. Earth and Planetary Science Letters 170, 1999, 417-432.

Maslow 1943: A. H. Maslow, Theory of Human Motivation. Psychological Review 50, 1943, 370-396.

1955: A. Maslow, Deficiency motivation and growth motivation. In: M. R. Jones (ed.), Nebraska symposium on motivation (Lincoln, NE 1955) 1-30.

Masson-Delmotte et al. 2005: V. Masson-Delmotte / J. Jouzel / A. Landais / M. Stievenard / S. J. Johnsen / J. W. C. White / M. Werner / A. Sveinbjornsdottir / K. Fuhrer, GRIP deuterium excess reveals rapid and orbital-scale changes in Greenland moisture origin. Science 309, 2005, 118-121.

Mathiassen 1948: T. Mathiassen, En senglacial boplads ved Bromme. Aarbøger for nordisk Oldkyndighed og Historie 1946 (1948), 121-197.

Mattey et al. 2008: D. Mattey / D. Lowry / J. Duffet / R. Fisher / E. Hodge / S. Frisia, A 53 year seasonally resolved oxygen and carbon isotope record from a modern Gibraltar speleothem: Reconstructed drip water and relationship to local precipitation. Earth and Planetary Science Letters 269, 2008, 80-95.

Mayr 1996: E. Mayr, The modern evolutionary theory. Capstone address, presented at the 75th annual meeting of the American Society of Mammalogists, Burlington, VT, June 1995. Journal of Mammalogy 77, 1996, 1-7.

McCabe/Clark/Clark 2005: A. M. McCabe / P. U. Clark / J. Clark, AMS ^{14}C dating of deglacial events in the Irish Sea Basin and other sectors of the British-Irish ice sheet. Quaternary Science Reviews 24, 2005, 1673-1690.

McCabe/Cooper/Kelley 2007: A. M. McCabe / J. A. G. Cooper / J. T. Kelley, Relative sea-level changes from NE Ireland during the last glacial termination. Journal of the Geological Society 164, 2007, 1059-1063.

McCarroll/Loader 2004: D. McCarroll / N. J. Loader, Stable isotopes in tree rings. Quaternary Science Reviews 23, 2004, 771-801.

McCormac et al. 2004: F. G. McCormac / A. G. Hogg / P. G. Blackwell / C. E. Buck / T. F. G. Higham / P. J. Reimer, SHCal04 Southern hemisphere calibration, 0-11.0 cal kyr BP. Radiocarbon 46, 2004, 1087-1092.

McManus et al. 2004: J. F. McManus / R. Francois / J. M. Gherardi / L. D. Keigwin / S. Brown-Leger, Collapse and rapid resumption of Atlantic meridional circulation linked to deglacial climate changes. Nature 428, 2004, 834-837.

Meadows et al. 1972: D. H. Meadows / D. L. Meadows / J. Randers / W. W. Behrens III, The limits to growth. A report for the Club of Rome's project on the predicament of mankind (New York 1972).

Meese et al. 1997: D. A. Meese / A. J. Gow / R. B. Alley / G. A. Zielinski / P. M. Grootes / M. Ram / K. C. Taylor / P. A. Mayewski / J. F. Bolzan, The Greenland Ice Sheet Project 2 depth-age scale: methods and results. Journal of Geophysical Research 102 (C12), 1997, 26411-26423.

Meiri et al. 2013: M. Meiri / A. M. Lister / T. F. G. Higham / J. R. Stewart / L. G. Straus / H. Obermaier / M. R. González Morales / A. B. Marín-Arroyo / I. Barnes, Late-glacial recolonization and phylogeography of European red deer (*Cervus elaphus* L.). Molecular Ecology 22, 2013, 4711-4722.

Mellars 1969: P. Mellars, News and Notes. Radiocarbon dates for a new Creswellian site. Antiquity 43, 1969, 308-310.

Ménot et al. 2006: G. Ménot / E. Bard / F. Rostek / J. W. H. Weijers / E. C. Hopmans / S. Schouten / J. S. Sinninghe Damsté, Early reactivation of European rivers during the last deglaciation. Science 313, 2006, 1623-1625.

Merkt 1994: J. Merkt, The Alleroed – duration and climate as derived from laminated lake sediments. Terra Nostra 1994/1, 59-63.

Merkt/Kleinmann 1998: J. Merkt / A. Kleinmann, Die Entstehung und Entwicklung des Wollingster Sees und seiner Ablagerungen. Mitteilungen der Arbeitsgemeinschaft Geobotanik in Schleswig-Holstein und Hamburg 57, 1998, 17-27.

Merkt/Müller 1999: J. Merkt / H. Müller, Varve chronology and palynology of the Lateglacial in Northwest Germany from lacustrine sediments of Hämelsee in Lower Saxony. Quaternary International 61, 1999, 41-59.

Merton 1949: R. K. Merton, Social theory and social structure. Toward the codification of theory and research (Glencoe, Ill. 1949).

Mesoudi/O'Brien 2008a: A. Mesoudi / M. J. O'Brien, The cultural transmission of Great Basin projectile point technology I: an experimental simulation. American Antiquity 73, 2008, 3-28.

2008b: A. Mesoudi / M. J. O'Brien, The learning and transmission of hierarchical cultural recipes. Biological Theory 3, 2008, 63-72.

Metzger et al. 2005: M. J. Metzger / R. G. H. Bunce / R. H. G. Jongman / C. A. Mücher / J. W. Watkins, A climatic stratification of the environment of Europe. Global Ecology and Biogeography 14, 2005, 549-563.

Mevel 2004: L. Mevel, Des lames pourquoi faire? L'exemple de le niveau inférieur du Closeau à Rueil-Malmaison. In: Valentin/Bodu/Julien 2004, 103-115.

Mevel/Grimm 2019: L. Mevel / S. B. Grimm, From one camp to another. In: B. V. Eriksen / E. Rensink / S. K. Harris (eds), The Final Palaeolithic of Northern Eurasia. Proceedings of the Amersfoort, Schleswig and Burgos UISPP Commission Meetings. Schriften des Museums für Archäologie Schloss Gottorf: Ergänzungsreihe 13 (Kiel, Schleswig 2019) 105-136.

Mewis 1993: S. U. Mewis, Der späteiszeitliche Fundplatz Urbar bei Koblenz [unpublished M.A. thesis, Köln 1993].

Meyer/Harff 2005: M. Meyer / J. Harff, Modelling palaeo coastline changes of the Baltic Sea. Journal of Coastal Research 21, 2005, 598-609.

Meyrick 2001: R. A. Meyrick, The development of terrestrial mollusc faunas in the »Rheinland region« (western Germany and Luxembourg) during the Lateglacial and Holocene. Quaternary Science Reviews 20, 2001, 1667-1675.

Michalczyk et al. 2010: I. M. Michalczyk / L. Opgenoorth / Y. Luecke / S. Huck / B. Ziegenhagen, Genetic support for per-glacial survival of *Juniperus communis* L. in Central Europe. The Holocene 20, 2010, 887-894.

Miller/López Bayón 1997: R. Miller / I. López Bayón, Refit analysis of the Bois Laiterie plaquettes. In: Otte/Straus 1997, 279-291.

Miller 2012: R. Miller, Mapping the expansion of the Northwest Magdalenian. Quaternary International 272-273, 2012, 209-230.

Miller/Noiret 2009: R. Miller / P. Noiret, Recent results for the Belgian Magdalenian. In: Street/Barton/Terberger 2009, 39-53.

Minc/Smith 1989: L. Minc / K. Smith, The spirit of survival: cultural responses to resource variability in North Alaska. In: Halstead/O'Shea 1989a, 8-39.

Mistrot 1998: V. Mistrot, Les micromammifères du Closeau (Tr IV). In: Bodu 1998, 344-346.

Mol/Vandenberghe/Kasse 2000: J. Mol / J. Vandenberghe / C. Kasse, River response to variations of periglacial climate in mid-latitude Europe. Geomorphology 33, 2000, 131-148.

Mollison 1928: T. Mollison, Die Deutung zweier Fundstücke von Oberkassel. Anthropologischer Anzeiger 5, 1928, 156-160.

Molodkov/Bolikhovskaya 2006: A. N. Molodkov / N. Bolikhovskaya, Long-term palaeoenvironmental changes recorded in palynologically studied loess-palaeosol and ESR-dated marine deposits of Northern Eurasia: Implications for sea-land correlation. Quaternary International 152-153, 2006, 48-58.

Montet-White/Evin/Stafford 2002: A. Montet-White / J. Evin / T. Stafford, Les datations radiocarbone des amas osseux de Solutré. In: J. Combier / A. Montet-White (eds), Solutré 1968-1998 (Paris 2002) 181-189.

Moore et al. 1992: J. C. Moore / E. W. Wolff / H. B. Clausen / C. U. Hammer, The chemical basis for the electrical stratigraphy of ice. Journal of Geophysical Research – Solid Earth 97, 1992, 1887-1896.

Moreau et al. 2012: J. Moreau / M. Huuse / A. Janszen / P. van der Vegt / P. L. Gibbard / A. Moscariello, The glaciogenic unconformity of the southern North Sea. Geological Society, London, Special Publications 368, 2012, 99-110.

Morel 1993: P. Morel, Une chasse à l'ours brun il y a 12.000 ans: nouvelle découverte à la grotte du Bichon (La Chaux-de-Fonds). Archäologie in der Schweiz 16, 1993, 110-117.

Moreno et al. 2010: A. Moreno / H. Stoll / M. Jiménez-Sánchez / B. Valero-Gracés / E. Ito / R. L. Edwards, A speleothem record of glacial (25-11.6 kyr BP) rapid climatic changes from northern Iberian Peninsula. Global and Planetary Change 71, 2010, 218-231.

Mörner 1999: N. A. Mörner, Paleo-tsunamis in Sweden. Physics and Chemistry of the Earth, Part B: Hydrology, Oceans and Atmosphere 24, 1999, 443-448.

Mörner 2005: N.-A. Mörner, An interpretation and catalogue of paleoseismicity in Sweden. Tectonophysics 408, 2005, 265-307.

2008: N.-A. Mörner, Tsunami events within the Baltic. Polish Geological Institute Special Papers 23, 2008, 71-76.

Moros et al. 2004: M. Moros / K. Emeis / B. Risebrobakken / I. Snowball / A. Kuijpers / J. McManus / E. Jansen, Sea surface temperatures and ice rafting in the Holocene North Atlantic: climate influences on northern Europe and Greenland. Quaternary Science Reviews 23, 2004, 2113-2126.

Mortensen et al. 2005: A. K. Mortensen / M. Bigler / K. Grönvold / J. P. Steffensen / S. J. Johnsen, Volcanic ash layers from the Last Glacial Termination in the NGRIP ice core. Journal of Quaternary Science 20, 2005, 209-219.

Mortensen 2007: M. F. Mortensen, Biostratigraphy and chronology of the terrestrial late Weichselian in Denmark. – New investigations of the vegetation development based on pollen and plant macrofossils in the Slotseng basin [dissertation, Aarhus 2007].

Mortensen et al. 2011: M. F. Mortensen / H. H. Birks / C. Christensen / J. Holm / N. Noe-Nygaard / B. V. Odgaard / J. Olsen / K. L. Rasmussen, Lateglacial vegetation development in Denmark – New evidence based on macrofossils and pollen from Slotseng, a small-scale site in southern Jutland. Quaternary Science Reviews 30, 2011, 2534-2550.

de Mortillet 1869: G. de Mortillet, Essai d'une classification des cavernes et des stations sous abri, fondée sur les produits de l'industrie humaine. Matériaux pour l'Histoire primitive et naturelle de l'Homme 5, 1869, 172-179.

1872: G. de Mortillet, Classification des diverses périodes de l'âge de la pierre. Revue d'Anthropologie 1, 1872, 432-459.

Moseler 2014: F. Moseler, Feuerstelle oder »Heizofen«? – Eine funktionale Analyse von Brandstrukturen aus dem späten Magdalénien [dissertation, Mainz 2014].

Moss 1986: E. H. Moss, Further work on the functions of flint tools at Pincevent (Seine-et-Marne), France: sections 36 and 27. In: D. A. Roe (ed.), Studies in the Upper Palaeolithic of Britain and Northwestern Europe (Oxford 1986) 175-185.

Müller 2004: W. Müller, One horse or a hundred hares? Small game exploitation in an Upper Palaeolithic context. In: J. P. Brugal / J. Desse (eds), Petits animaux et sociétés humaines. Du complément alimentaire aux ressources utilitaires. XXIVe rencontres internationales d'archéologie et d'histoire d'Antibes (Antibes 2004) 231-240.

Müller/Kleinmann 1998: H. Müller / A. Kleinmann, Palynologische Untersuchung eines Sedimentprofiles aus dem Wollingster See. Mitteilungen der Arbeitsgemeinschaft Geobotanik in Schleswig-Holstein und Hamburg 57, 1998, 44-52.

Müller et al. 2006: W. Müller / D. Leesch / J. Bullinger / M.-I. Cattin / N. Plumettaz, Chasse, habitats et rythme des déplacements: réflexions à partir des campements magdaléniens de Champréveyres et Monruz (Neuchâtel, Suisse). Bulletin de la Société préhistorique française 103, 2006, 741-752.

Münnich 1957: K. O. Münnich, Heidelberg Natural Radiocarbon Measurements I. Science 126, 1957, 194-199.

Munro 2003: N. D. Munro, Small game, the younger dryas, and the transition to agriculture in the southern levant. Mitteilungen der Gesellschaft für Urgeschichte 12, 2003, 47-71.

Murdock/Wilson 1972: G. P. Murdock / S. F. Wilson, Settlement patterns and community organization. Ethnology 11, 1972, 254-295.

Muscheler et al. 2000: R. Muscheler / J. Beer / G. Wagner / R. C. Finkel, Changes in deep-water formation during the Younger Dryas event inferred from ^{10}Be and ^{14}C records. Nature 408, 2000, 567-570.

2004: R. Muscheler / J. Beer / G. Wagner / C. Laj / C. Kissel / G. M. Raisbeck / F. Yiou / P. W. Kubik, Changes in the carbon cycle during the last deglaciation as indicated by the comparison of ^{10}Be and ^{14}C records. Earth and Planetary Science Letters 219, 2004, 325-340.

2005: R. Muscheler / J. Beer / P. W. Kubik / H. A. Synal, Geomagnetic field intensity during the last 60,000 years based on ^{10}Be and ^{36}Cl from the Summit ice cores and ^{14}C. Quaternary Science Reviews 24, 2005, 1849-1860.

2008: R. Muscheler / B. Kromer / S. Björck / A. Svensson / M. Friedrich / K. F. Kaiser / J. Southon, Tree rings and ice cores reveal ^{14}C calibration uncertainties during the Younger Dryas. Nature Geoscience 1, 2008, 263-267.

Musil 2000: R. Musil, Domestication of wolves in Central European Magdalenian sites. In: S. J. Crockford (ed.), Dogs through time. An archaeological perspective. Proceedings of the 1st ICAZ symposium on the history of the domestic dog at the 8th congress of the International Council for Archaeozoology (ICAZ98), August 23-29, 1998 in Victoria, B.C., Canada (Oxford 2000) 21-28.

2010: R. Musil, Palaeoenvironment at Gravettian sites in Central Europe with emphasis on Moravia (Czech Republic). Quartär 57, 2010, 95-123.

Nadachowski et al. 2009: A. Nadachowski / P. Wojtal / B. Miękina / G. Lipecki / M. Żarski / M. Urabanowski / K. Ochman / M. Krawczyk / G. Jakubowski / T. Tomek, Late Pleistocene environment of the Częstochowa Upland (Poland) reconstructed on the basis of faunistic evidence from archaeological cave sites (Kraków 2009).

Nakoinz 2009: O. Nakoinz, Zentralortforschung und zentralörtliche Theorie. Archäologisches Korrespondenzblatt 39, 2009, 361-380.

Nalawade-Chavan et al. 2013: S. Nalawade-Chavan / J. McCullagh / R. Hedges / C. Bonsall / A. Boronean / C. Bronk Ramsey / T. Higham, Compound specific radiocarbon dating of essential and non-essential amino acids: Towards determination of dietary reservoir effects in humans. Radiocarbon 55, 2013, 709-719.

Napierala 2008a: H. Napierala, Die Tierknochen aus dem Kesslerloch. Neubearbeitung der paläolithischen Fauna. Beiträge zur Schaffhauser Archäologie 2 (Schaffhausen 2008).

2008b: H. Napierala, Viele Jäger sind des Hasen Tod? Magdalénienzeitliche Hasenjagd am Kesslerloch. Mitteilungen der Gesellschaft für Urgeschichte 17, 2008, 55-67.

Napierala/Uerpmann 2010: H. Napierala / H.-P. Uerpmann, A »new« Palaeolithic dog from Central Europe. International Journal of Osteoarchaeology 22, 2010, 127-137.

Naudinot 2008: N. Naudinot, Lithic weapon elements in Western France (Brittany and Pays de la Loire) during the late glacial period: a proposed chrono-cultural organization. Palethnologie 1, 2008, 241-268.

2010: N. Naudinot, Dynamiques techno-économiques et de peuplement au Tardiglaciaire dans le Grand-Ouest de la France [dissertation, Rennes 2010].

Nehren 2001: R. Nehren, Zur Urgeschichte einer niederrheinischen Kleinlandschaft. Ergebnisse eines archäologischen Prospektionsprojekts am Unterlauf der Inde, Kreis Düren. In: Gehlen/Heinen/Tillmann 2001, 55-65.

Neugebauer et al. 2012: I. Neugebauer / A. Brauer / N. Dräger / P. Dulski / S. Wulf / B. Plessen / J. Mingram / U. Herzschuh / A. Brande, A Younger Dryas varve chronology from the Rehwiese palaeolake record in NE-Germany. The INTegration of Ice core, Marine and TErrestrial records of the last termination (INTIMATE) 60,000 to 8000 BP. Quaternary Science Reviews 36, 2012, 91-102.

Nicoud et al. 2011: E. Nicoud / P. Bodu / C. Chaussé / L. Chesnaux, Un gisement du Tardiglaiciaire dans le sud de la Nièvre (Decize, Zac du Four à Chaux, diagnostic INRAP, Février 2010). In: B. Valentin (ed.), Paléolithique final et Mésolithique dans le Bassin parisien et ses marges. Habitats, sociétés et environnements. Rapport d'activités pour 2010 (Nanterre 2011) 239-249.

Niekus 2006: M. J. L. T. Niekus, A geographically referenced ^{14}C database for the Mesolithic and the early phase of the Swifterbant culture in the northern Netherlands. Palaeohistoria 47/48, 2006, 41-99.

Niekus/Stapert/Johansen 1998: M. J. L. T. Niekus / D. Stapert / L. Johansen, Een Federmesser-Vindplaats bij Doetinchem (Gld.), met vijf retouchoirs rond een haard. Paleo-Aktueel 9, 1998, 18-22.

Niekus et al. 2012: M. J. L. T. Niekus / R. N. E. Barton / M. Street / Th. Terberger (eds), A mind set on flint. Studies in the honour of Dick Stapert (Groningen 2012).

Nielsen-Marsh et al. 2006: C. Nielsen-Marsh / A. Gernaey / G. Turner-Walker / R. Hedges / A. Pike / M. Collins, The chemical degradation of bone. In: M. Cox / S. Mays (eds), Human osteology. In archaeology and forensic science (London 2006) 439-454.

Niemann/Behling 2008: H. Niemann / H. Behling, Late Quaternary vegetation, climate and fire dynamics inferred from the El Tiro record in the southeastern Ecuadorian Andes. Journal of Quaternary Science 23, 2008, 203-212.

Nilsson 2006: A. Nilsson, Palaeomagnetic secular variations in the varved sediments of Lake Gosciaz, Poland: testing the stability of the natural remanent magnetization and validity of relative palaeointensity estimates. Examensarbeten i geologi vid Lunds universitet 193 (Lund 2006).

Nixon et al. 2003: S. Nixon / Z. Trent / C. Marcuello / C. Lallana, Europe's water. An indicator-based assessment. Topic report / European Environment Agency 1 (Copenhagen 2003).

Nobis 1986: G. Nobis, Die Wildsäugetiere in der Umwelt des Menschen von Oberkassel bei Bonn und das Domestikationsproblem von Wölfen im Jungpaläolithikum. Bonner Jahrbücher 186, 1986, 368-376.

Norton/Hampel 2010: K. P. Norton / A. Hampel, Postglacial rebound promotes glacial re-advances – a case study from the European Alps. Terra Nova 22, 2010, 297-302.

Nowak 2005: R. M. Nowak, Walker's carnivores of the world (Baltimore 2005).

Nývlt/Engel/Tyráèek 2011: D. Nývlt / Z. Engel / J. Tyráèek, Pleistocene glaciations of Czechia. In: Ehlers/Gibbard/Hughes 2011, 37-46.

Obbink/Carlson/Klinkhammer 2010: E. A. Obbink / A. E. Carlson / G. P. Klinkhammer, Eastern North American freshwater discharge during the Bølling-Allerød warm periods. Geology 38, 2010, 171-174.

O'Brien/Lyman 2002: M. J. O'Brien / R. L. Lyman, Evolutionary Archeology: Current status and future prospects. Evolutionary Anthropology 11, 2002, 26-36.

Odgaard 1999: B. V. Odgaard, Fossil pollen as a record of past biodiversity. Journal of Biogeography 26, 1999, 7-17.

Odum 1953: E. P. Odum, Fundamentals of ecology (Philadelphia, London 1953).

Ojala et al. 2012: A. E. K. Ojala / P. Francus / B. Zolitschka / M. Besonen / S. F. Lamoureux, Characteristics of sedimentary varve chronologies – A review. Quaternary Science Reviews 43, 2012, 45-60.

Olausson 2010: D. Olausson, Experimental flintknapping replication – A valuable method of archaeological analysis. In: H. G. Nami (ed.), Experiments and interpretation of traditional technologies: Essays in honor of Errett Callahan (Buenos Aires 2010) 37-55.

Olive 2004: M. Olive, À propos du gisement magdalénien d'Étiolles (Essonne): réflexion sur la fonction d'un site paléolithique. Bulletin de la Société préhistorique française 101, 2004, 797-813.

Olive/Pigeot 2006: M. Olive / N. Pigeot, Réflexions sur le temps d'un séjour à Étiolles (Essonne). Bulletin de la Société préhistorique française 103, 2006, 673-682.

Olsen 1989: S. L. Olsen, Solutré: A theoretical approach to the reconstruction of Upper Palaeolithic hunting strategies. Journal of Human Evolution 18, 1989, 295-327.

Olsen et al. 2008: J. Olsen / J. Heinemeier / P. Bennike / C. Krause / K. M. Hornstrup / H. Thrane, Characterisation and blind testing of radiocarbon dating of cremated bone. Journal of Archaeological Science 35, 2008, 791-800.

2010: J. Olsen / J. Heinemeier / H. Lübke / F. Lüth / Th. Terberger, Dietary habits and freshwater reservoir effects in bones from a Neolithic NE German cemetery. Radiocarbon 52, 2010, 635-644.

Onoratini/Defleur/Joris 1996: G. Onoratini / A. Defleur / C. Joris, Mise en évidence du Magdalénien ancien II dans les gorges de l'Ardèche. Bulletin de la Société préhistorique française 93, 1996, 25-32.

Orliac 1996a: M. Orliac, Chronologie et topographie des sols d'habitat. In: Gaucher 1996, 35-51.

1996b: M. Orliac, L'occupation de la section 17. In: Gaucher 1996, 87-92.

Orschiedt/Berg/Flohr 2011: J. Orschiedt / A. von Berg / S. Flohr, Human remains from the Late Upper Paleolithic of Irlich, Germany – a preliminary report Poster presented at 80th Annual Meeting of the American Association of Physical Anthropologists (Minneapolis, MN April, 12th-16th 2011).

Osmaston 2005: H. Osmaston, Estimates of glacier equilibrium line altitudes by the area x altitude, the area x altitude balance ratio and the area x altitude balance index methods and their validation. Quaternary International 138-139, 2005, 22-31.

Osterkamp/Romanovsky 1999: T. E. Osterkamp / V. E. Romanovsky, Evidence for warming and thawing of discontinuous permafrost in Alaska. Permafrost and Periglacial Processes 10, 1999, 17-37.

Otlet/Slade 1974: R. L. Otlet / B. S. Slade, Harwell Radiocarbon Measurements I. Radiocarbon 16, 1974, 178-191.

Otte 1988: M. Otte (ed.), De la Loire à l'Oder. Les civilisations du Paléolithique final dans le Nord-Ouest européen. Actes du colloque de Liège, décembre 1985 (Oxford 1988).

2000: M. Otte, Final Palaeolithic in the Northwest: migrations and seasons. In: Peterkin/Price 2000, 169-177.

Otte/Straus 1997: M. Otte / L. G. Straus (eds), La grotte du Bois Laiterie. Recolonisation magdalénienne de la Belgique (Liège 1997).

Ovdov et al. 2011: N. D. Ovdov / S. J. Crockford / Y. V. Kuzmin / T. F. G. Higham / G. W. L. Hodgins / J. van der Plicht, A 33,000-year-old incipient dog from the Altai Mountains of Siberia: Evidence of the earliest domestication disrupted by the Last Glacial Maximum. PLoS One 6, 2011, e22821-1-7.

Palmer/Zhaobo 1985: T. N. Palmer / S. Zhaobo, A modelling and observational study of the relationship between sea surface temperature in the North-West Atlantic and the atmospheric general circulation. Quarterly Journal of the Royal Meteorological Society 111, 1985, 947-975.

Park/Schmincke 1997: C. Park / H.-U. Schmincke, Lake formation and catastrophic dam burst during the Late Pleistocene Laacher See eruption (Germany). Naturwissenschaften 84, 1997, 521-525.

Pasda 1994: C. Pasda, Das Magdalénien in der Freiburger Bucht. Materialhefte zur Archäologie in Baden-Württemberg 25 (Stuttgart 1994).

1998: C. Pasda, Der Beginn des Magdaléniens in Mitteleuropa. Archäologisches Korrespondenzblatt 28, 1998, 175-190.

2001: C. Pasda, Das Knochengerät vom spätpaläolithischen Fundplatz Kleinlieskow in der Niederlausitz. Ein Essay zum steinzeitlichen Angelhaken. In: Gehlen/Heinen/Tillmann 2001, 397-408.

2002: C. Pasda, A short note on man in the Allerød/Younger Dryas environment of Lower Lusatia (Brandenburg, Germany). In: Eriksen/Bratlund 2002, 123-128.

2012: C. Pasda, Kulturentwicklung oder kulturspezifische Lebensweise? Ein Beitrag zur Ethnographie des Paläolithikums. Archäologisches Korrespondenzblatt 42, 2012, 1-14.

Pasda/Odgaard 2011: K. Pasda / U. Odgaard, Nothing is wasted: The ideal »nothing is wasted« and divergence in past and present among caribou hunters in Greenland. Quaternary International 238, 2011, 35-43.

Passchier et al. 2010: S. Passchier / C. Laban / C. S. Mesdag / K. F. Rijsdijk, Subglacial bed conditions during Late Pleistocene glaciations and their impact on ice dynamics in the southern North Sea. Boreas 39, 2010, 633-647.

Pastre 1998: J.-F. Pastre, Recherche de la téphra du Laacher See dans le sol Alleröd de la séquence du Closeau. In: Bodu 1998, 98.

Pastre et al. 1997: J.-F. Pastre / M. Fontugne / C. Kuzucuoglu / C. Leroyer / N. Limondin-Lozouet / M. Talon / N. Tisnérat, L'évolution tardi et postglaciaire des lits fluviaux au nord-est de Paris (France). Relations avec les données paléoenvironnementales et l'impact anthropique sur les versants. Géomorphologie 3, 1997, 291-312.

2000: J.-F. Pastre / C. Leroyer / N. Limondin-Lozouet / C. Chaussé / M. Fontugne / A. Gebhardt / C. Hatté / V. Krier, Le Tardiglaciaire des fonds de vallée du Bassin Parisien (France). Quaternaire 11, 2000, 107-122.

2003: J.-F. Pastre / N. Limondin-Lozouet / C. Leroyer / P. Ponel / M. Fontugne, River system evolution and environmental changes during the Lateglacial in the Paris Basin (France). Quaternary Science Reviews 22, 2003, 2177-2188.

Pays et al. 2012: O. Pays / D. Fortin / J. Gassani / J. Duchesne, Group dynamics and landscape features constrain the exploration of herds in fusion-fission societies: The case of European roe deer. PLoS One 7, 2012, e34678-1-8.

Pearson 1987: G. W. Pearson, How to cope with calibration. Antiquity 61, 1987, 98-103.

Peck et al. 2007: V. L. Peck / I. R. Hall / R. Zahn / F. Grousset / S. R. Hemming / J. D. Scourse, The relationship of Heinrich events and their European precursors over the past 60 ka BP: a multi-proxy ice-rafted debris provenance study in the North East Atlantic. Quaternary Science Reviews 26, 2007, 862-875.

Pelegrin 1992: J. Pelegrin, Remarques sur le débitage laminaire. In: Schmider 1992f, 108-115.

2000: J. Pelegrin, Les techniques de débitage laminaire au Tardiglaciaire: critères de diagnose et quelques réflexions. In: Valentin/Bodu/Christensen 2000, 73-86.

2009: J. Pelegrin, Cognition and the emergence of language: A contribution from lithic technology. In: S. A. de Beaune / F. L. Coolidge / T. G. Wynn (eds), Cognitive archaeology and human evolution (Cambridge 2009) 95-108.

Peltier/Fairbanks 2006: W. R. Peltier / R. G. Fairbanks, Global glacial ice volume and Last Glacial Maximum duration from an extended Barbados sea level record. Quaternary Science Reviews 25, 2006, 3322-3337.

Pennington 1981: W. Pennington, Records of a lake's life in time: the sediments. Hydrobiologia 79, 1981, 197-219.

Pereira et al. 2005: L. Pereira / M. Richards / A. Goios / A. Alonso / C. Albarrán / O. Garcia / D. M. Behar / M. Gölge / J. Hatina / L. Al-Gazali / D. G. Bradley / V. Macaulay / A. Amorim, High-resolution mtDNA evidence for the late-glacial resettlement of Europe from an Iberian refugium. Genome Research 15, 2005, 19-24.

Pernaud 1997: J.-M. Pernaud, Le site du Bois-Laiterie (Province de Namur, Belgique) rapport de l'analyse anthracologique des niveaux du Magdalénien supérieur (12,600 B.P.). In: Otte/Straus 1997, 143-144.

1998: J.-M. Pernaud, Rapport de l'analyse anthracologique. In: Bodu 1998, 133-139.

Pestle/Colvard/Pettitt 2006: W. Pestle / M. Colvard / P. Pettitt, AMS dating of a recently rediscovered juvenile human mandible from Solutre (Saone-et-Loire, France). Paléo 8, 2006, 285-292.

Peterkin/Price 2000: G. L. Peterkin / H. A. Price (eds), Regional approaches to adaptation in late Pleistocene western Europe. Colloquium Minneapolis 1995 and New Orleans 1996 (Oxford 2000).

Peters 1969: I. Peters, Botanische Untersuchungen in Gönnersdorf. Voläufige Ergebnisse. Germania 47, 1969, 47-52.

Peters et al. 2010: C. Peters / W. E. N. Austin / J. Walden / F. D. Hibbert, Magnetic characterisation and correlation of a Younger Dryas tephra in North Atlantic marine sediments. Journal of Quaternary Science 25, 2010, 339-347.

Petersen/Johansen 1996: P. V. Petersen / L. Johansen, Tracking Late Glacial reindeer hunters in eastern Denmark. In: Larsson 1996b, 75-88.

Peterson 1955: R. L. Peterson, North American moose (Toronto 1955).

Pétillon 2006: J.-M. Pétillon, Des Magdaléniens en armes. Technologie des armatures de projectile en bois de Cervidé du Magdalénien supérieur de la grotte d'Isturitz (Pyrénées-Atlantiques) (Treignes 2006).

2008a: J.-M. Pétillon, First evidence of a whale-bone industry in the western European Upper Paleolithic: Magdalenian artifacts from Isturitz (Pyrénées-Atlantiques, France). Journal of Human Evolution 54, 2008, 720-726.

2008b: J.-M. Pétillon, What are these barbs for? Preliminary study on the function of the Upper Magdalenian barbed weapon tips. Palethnologie 1, 2008, 66-97.

Pétillon et al. 2011: J.-M. Pétillon / O. Bignon / P. Bodu / P. Cattelain / G. Debout / M. Langlais / V. Laroulandie / H. Plisson / B. Valentin, Hard core and cutting edge: experimental manufacture and use of Magdalenian composite projectile tips. Journal of Archaeological Science 38, 2011, 1266-1283.

Petit et al. 1999: J. R. Petit / J. Jouzel / D. Raynaud / N. I. Barkov / J. M. Barnola / I. Basile / M. Bender / J. Chappellaz / M. Davis / G. Delaygue / M. Delmotte / V. M. Kotlyakov / M. Legrand / V. Y. Lipenkov / C. Lorius / L. Pepin / C. Ritz / E. Saltzman / M. Stievenard, Climate and atmospheric history of the past 420,000 years from the Vostok ice core, Antarctica. Nature 399, 1999, 429-436.

Pettitt 2011: P. Pettitt, The palaeolithic origins of human burial (London 2011).

Pettitt/Bahn/Ripoll 2007: P. Pettitt / P. Bahn / S. Ripoll (eds), Palaeolithic cave art at Creswell Crags in European context (Oxford 2007).

Pettitt/Rockman/Chenery 2012: P. Pettitt / M. Rockman / S. Chenery, The British Final Magdalenian: Society, settlement and raw material movements revealed through LA-ICP-MS trace element analysis of diagnostic artefacts. Quaternary International 272-273, 2012, 275-287.

Pettitt/White 2012: P. Pettitt / M. White (eds), The British Palaeolithic. Hominin societies at the edge of the Pleistocene world. Routledge archaeology of Northern Europe (London, New York 2012).

Pettitt et al. 2003: P. B. Pettitt / W. Davies / C. S. Gamble / M. B. Richards, Palaeolithic radiocarbon chronology: quantifying our confidence beyond two half-lives. Journal of Archaeological Science 30, 2003, 1685-1693.

Peyrony 1936: D. Peyrony, L'Abri de Villepin (Dordogne). Magdalénien supérieur et Azilien. Bulletin de la Société préhistorique française 33, 1936, 253-272.

Peyrony/Peyrony 1938: D. Peyrony / É. Peyrony, Laugerie-Haute près des Eyzies (Dordogne). Archives de l'Institut de paléontologie humaine 19 (Paris 1938).

Pidoplichko 1998: I. H. Pidoplichko, Upper Palaeolithic dwellings of mammoth bones in the Ukraine: Kiev-Kirillovskiī, Gontsy, Dobranichevka, Mezin and Mezhirich. British Archaeological Reports, International Series 712 (Oxford 1998).

Piette 1889: E. Piette, Un groupe d'assises représentant l'époque de transition entre les temps quaternaires et les temps modernes. Compte rendu de l'Academie des Sciences 108, 1889, 422-424.

1895: E. Piette, Hiatus & Lacune. Vestiges de la période de transition dans la grotte du Mas d'Azil. Bulletin de la Société d'Anthropologie de Paris 1895, 1-36.

Pigeaud 2007: R. Pigeaud, Determining style in Palaeolithic cave art: a new method derived from horse images. Antiquity 81, 2007, 409-422.

Pigeaud et al. 2010: R. Pigeaud / V. Plagnes / M. Bouchard / J.-J. Bahain / C. Causse / S. Demailly / C. Falguères / É. Laval / F. Noël / Rodet / Joël / H. Valladas / P. Walter, Analyses archéométriques dans la grotte ornée Mayenne-Sciences (Thorigné-en-Charnie, Mayenne). L'Anthropologie 114, 2010, 97-112.

Pigeot 1987: N. Pigeot, Magdaléniens d'Étiolles. Économie de débitage et organisation sociale (l'unité d'habitation U5). Supplément à Gallia Préhistoire 25 (Paris 1987).

1990: N. Pigeot, Technical and social actors. Flintknapping specialists and apprentices at Magdalenian Etiolles. Archaeological Review from Cambridge 9, 1990, 126-141.

Pinto/Santos/Rosell 2009: B. Pinto / M. J. Santos / F. Rosell, Habitat selection of the Eurasian beaver (*Castor fiber*) near its carrying capacity: an example from Norway. Canadian Journal of Zoology 87, 2009, 317-325.

Pion 1990: G. Pion, L'abri de La Fru à Saint-Christophe (Savoie). Gallia Préhistoire 32, 1990, 65-123.

1997: G. Pion, L'abri de la Fru à Saint-Christophe-la-Grotte (Savoie): L'Azilien ancien du début de l'Alleröd. Bulletin de la Société préhistorique française 94, 1997, 319-326.

Pittman/Zeigler 2007: T. S. Pittman / K. R. Zeigler, Basic human needs. In: A. W. Kruglanski / E. T. Higgins (eds), Social psychology. Handbook of basic principles (New York 2007) 473-489.

Plisson 1985: H. Plisson, Etude fontionnelle d'outillages lithiques préhistoriques par l'analyse des micro-usures: Recherche méthodologique et archéologique [unpublished dissertation, Paris 1985].

2002: H. Plisson, Étude fonctionnelle de l'outillage en quartzite des concentrations magdaléniennes I et III d'Andernach. In: Floss/Terberger 2002, 139-157.

2007: H. Plisson, La fonction des outils de silex dans les grottes ornées paléolithiques. In: J. Evin (ed.), Un siècle de construction du discours scientifique en préhistoire: congrès du centenaire de la Société préhistorique française. XXVIe Congrès préhistorique de France, Avignon, 21-25 septembre 2004 (Paris 2007) 125-132.

2009: H. Plisson, Bilan tracéologique préliminaire sur les lamelles à dos du Magdalénien inférieur. In: Valentin 2009, 45-48.

Polly/Eronen 2010: P. D. Polly / J. T. Eronen, Mammal associations in the Pleistocene of Britain: implications of ecological niche modelling and a method for reconstrucing palaeoclimate. In: N. Ashton / S. G. Lewis / C. Stringer (eds), The ancient human occupation of Britain (Amsterdam, Boston, London 2010) 279-304.

Połtowicz 2007: M. Połtowicz, The Magdalenian period in Poland and neighbouring areas. Archaeologia Baltica 7, 2007, 21-28.

Połtowicz-Bobak 2009: M. Połtowicz-Bobak, Magdalenian settlement in Poland in the light of recent research. In: Street/Barton/Terberger 2009, 55-66.

2012: M. Połtowicz-Bobak, Observations on the late Magdalenian in Poland. Quaternary International 272-273, 2012, 297-307.

Ponel et al. 2005: P. Ponel / R. Coope / P. Antoine / N. Limondin-Lozouet / C. Leroyer / A.-V. Munaut / J.-F. Pastre / F. Guiter, Lateglacial palaeoenvironments and palaeoclimates from Conty and Houdancourt, northern France, reconstructed from beetle remains. Quaternary Science Reviews 24, 2005, 2449-2465.

Poplin 1976: F. Poplin, Les grands vertébrés de Gönnersdorf fouilles 1968. Der Magdalénien-Fundplatz Gönnersdorf 2 (Wiesbaden 1976).

1978: F. Poplin, Großtierfauna: Données de la grande faune sur le climat et l'environnement. In: Brunnacker 1978a, 98-104.

1992: F. Poplin, Les restes osseux animaux. In: Schmider 1992f, 37-44.

Popping/Beijerinck 1933: H. J. Popping / W. Beijerinck, Een palaeolithische nederzetting aan het Kuinderdal nabij Oosterwolde. Tijdschrift van het Aardrijkskundig Genootschap 50, 1933, 178-219.

Posth et al. 2016: C. Posth / G. Renaud / A. Mittnik / D. G. Drucker / H. Rougier / C. Cupillard / F. Valentin / C. Thevenet / A. Furtwängler / C. Wißing / M. Francken / M. Malina / M. Bolus / M. Lari / E. Gigli / G. Capecchi / I. Crevecoeur / C. Beauval / D. Flas / M. Germonpré / J. van der Plicht / R. Cottiaux / B. Gély / A. Ronchitelli / K. Wehrberger / D. Grigourescu / J. Svoboda / P. Semal / D. Caramelli / H. Bocherens / K. Harvati / N. J. Conard / W. Haak / A. Powell / J. Krause, Pleistocene mitochondrial genomes suggest a single major dispersal of Non-Africans and a Late Glacial population turnover in Europe. Current Biology 26, 2016, 1-7.

Pouclet/Juvigné 2009: A. Pouclet / E. Juvigné, The Eltville tephra, a late Pleistocene widespread tephra layer in Germany, Belgium and The Netherlands; symptomatic compositions of the minerals. Geologica Belgica 12, 2009, 93-103.

Power et al. 2008: M. J. Power / J. Marlon / N. Ortiz / P. J. Bartlein / S. P. Harrison / F. E. Mayle / A. Ballouche / R. Bradshaw / C. Carcaillet / C. Cordova / S. Mooney / P. Moreno / I. C. Prentice / K. Thonicke / W. Tinner / C. Whitlock / Y. Zhang / Y. Zhao / R. S. Anderson / R. Beer / H. Behling / C. Briles / K. J. Brown / A. Brunelle / M. Bush / P. Camill / G. Q. Chu / J. Clark / D. Colombaroli / M. Daniels / J. Dodson / E. Doughty / M. E. Edwards / W. Fisinger / D. Foster / M. J. Gaillard / D. G. Gavin / E. Gobet / S. Haberle / D. J. Hallett / P. Higuera / G. Hope / S. Horn / J. Inoue / P. Kaltenrieder / L. Kennedy / Z. C. Kong / C. Larsen / C. J. Long / J. Lynch / B. Lynch / M. McGlone / S. Meeks / S. Mensing / G. Meyer / T. Minckley / J. Mohr / R. Newnham / R. Noti / W. Oswald / J. Pierce / P. J. H. Richard / B. J. Shuman / H. Takahara / J. Toney / C. Turney / C. Umbanhower / M. Vandergoes / B. Vanniere / E. Vescovi / M. Walsh / X. Wang / N. Williams / J. Wilmshurst / J. H. Zhang, Changes in fire activity since the Last Glacial Maximum: an assessment based on a global synthesis and analysis of charcoal data. Climate Dynamics 30, 2008, 887-907.

Praeg 2003: D. Praeg, Seismic imaging of mid-Pleistocene tunnel-valleys in the North Sea Basin-high resolution from low frequencies. Journal of Applied Geophysics 53, 2003, 273-298.

Price 2003: C. R. Price, Late Pleistocene and Early Holocene small mammals in south west Britain. British Archaeological Reports, British Series 347 (Oxford 2003).

Pucek/Belousova 2004: Z. Pucek / I. P. Belousova (eds), European bison. Status survey and conservation action plan (Gland, Cambridge 2004).

Puisségur 1978: J. J. Puisségur, Stratigraphie im Bereich der Gönnersdorfer Sieldung: Faune malacologique. In: Brunnacker 1978a, 56-59.

Pushkina/Raia 2008: D. Pushkina / P. Raia, Human influence on distribution and extinctions of the late Pleistocene Eurasian megafauna. Journal of Human Evolution 54, 2008, 769-782.

Pye 1995: K. Pye, The nature, origin and accumulation of loess. Quaternary Science Reviews 14, 1995, 653-667.

Rabenstein 1991: R. Rabenstein, Klimaökologische Untersuchungen in der Nordeifel – Die kleinen Säugetiere der Ahrensburger Schichten aus der Grabung am Kartstein. Landschaftsverband Rheinland, Das Rheinische Landesmuseum Bonn, Berichte aus der Arbeit des Museums 1991, 50-53.

Rahmstorf 2002: S. Rahmstorf, Ocean circulation and climate during the past 120,000 years. Nature 419, 2002, 207-214.

Raisor 2004: M. J. Raisor, Determining the antiquity of dog origins: Canine domestication as a model for the consilience between molecular genetics and archaeology [dissertation, College Station TX 2004).

Ralska-Jasiewiczowa/Demske/van Geel 1998: M. Ralska-Jasiewiczowa / D. Demske / B. van Geel, Late-Glacial vegetation history recorded in the Lake Gościąż sediments. In: M. Ralska-Jasiewiczowa / T. Goslar / T. Madeyska / L. Starkel (eds), Lake Gościąż, Central Poland. A monographic study. Part 1 (Krakow 1998) 128-143.

Ralska-Jasiewiczowa et al. 1998: M. Ralska-Jasiewiczowa / T. Goslar / T. Madeyska / L. Starkel (eds), Lake Gościąż, Central Poland. A monographic study. Part 1 (Krakow 1998).

Rasmussen et al. 2006: S. O. Rasmussen / K. K. Andersen / A. M. Svensson / J. P. Steffensen / B. M. Vinther / H. B. Clausen / M. L. Siggaard-Andersen / S. J. Johnsen / L. B. Larsen / D. Dahl-Jensen / M. Bigler / R. Röthlisberger / H. Fischer / K. Goto-Azuma / M. E. Hansson / U. Ruth, A new Greenland ice core chronology for the last glacial termination. Journal of Geophysical Research 111, 2006, D06102.

2008: S. O. Rasmussen / I. K. Seierstad / K. K. Andersen / M. Bigler / D. Dahl-Jensen / S. J. Johnsen, Synchronization of the NGRIP, GRIP, and GISP2 ice cores across MIS 2 and palaeoclimatic implications. Quaternary Science Reviews 27, 2008, 18-28.

Redman 2005: C. L. Redman, Resilience theory in archaeology. American Anthropologist 107, 2005, 70-77.

Redman/Kinzig 2003: C. L. Redman / A. P. Kinzig, Resilience of past landscapes: resilience theory, society, and the longue durée. Conservation Ecology 7, 2003, 14.1-19.

Reicherter/Kaiser/Stackebrandt 2005: K. Reicherter / A. Kaiser / W. Stackebrandt, The post-glacial landscape evolution of the North German Basin: morphology, neotectonics and crustal deformation. International Journal of Earth Sciences (Geologische Rundschau) 94, 2005, 1083-1093.

Reimer et al. 2004: P. J. Reimer / M. G. L. Baillie / E. Bard / A. Bayliss / J. W. Beck / C. J. H. Bertrand / P. G. Blackwell / C. E. Buck / G. S. Burr / K. B. Cutler / P. E. Damon / R. L. Edwards / R. G. Fairbanks / M. Friedrich / T. P. Guilderson / A. G. Hogg / K. A. Hughen / B. Kromer / G. McCormac / S. Manning / C. Bronk Ramsey / R. W. Reimer / S. Remmele / J. R. Southon / M. Stuiver / S. Talamo / F. W. Taylor / J. van der Plicht / C. E. Weyhenmeyer, IntCal04 terrestrial radiocarbon age calibration, 26-0 ka BP. Radiocarbon 46, 2004, 1029-1058.

2009: P. J. Reimer / M. G. L. Baillie / E. Bard / A. Bayliss / J. W. Beck / P. G. Blackwell / C. Bronk Ramsey / C. E. Buck / G. S. Burr / R. L. Edwards / M. Friedrich / P. M. Grootes / T. P. Guilderson / I. Hajdas / T. J. Heaton / A. G. Hogg / K. A. Hughen / K. F. Kaiser / B. Kromer / F. G. McCormac / S. W. Manning / R. W. Reimer / D. A. Richards / J. R. Southon / S. Talamo / C. S. M. Turney / J. van der Plicht / C. E. Weyhenmeyer, IntCal09 and Marine09 radiocarbon age calibration curves, 0-50,000 years cal BP. Radiocarbon 51, 2009, 1111-1150.

Reimers/Colman 2006: E. Reimers / J. E. Colman, Reindeer and caribou (*Rangifer tarandus*) response towards human activities. Rangifer 26, 2006, 55-71.

Reitner 2007: J. M. Reitner, Glacial dynamics at the beginning of Termination I in the Eastern Alps and their stratigraphic implications. Quaternary International 164-165, 2007, 64-84.

Rensink 1995: E. Rensink, On Magdalenian mobility and land use in north-west Europe. Some methodological considerations. Archaeological Dialogues 2, 1995, 85-119.

2000: E. Rensink, Regional-scale variation and the Magdalenian record of Northwestern Europe. In: Peterkin/Price 2000, 133-146.

2002: E. Rensink, Late Palaeolithic sites in the Maas valley of southern Netherlands: prospects, surveys and results. In: Eriksen/Bratlund 2002, 181-188.

2012: E. Rensink, Magdalenian hunter-gatherers in the northern loess area between the Meuse and Rhine – New insights from the excavation at Eyserheide (SE Netherlands). Quaternary International 272-273, 2012, 251-263.

Renssen/Isarin 2001: H. Renssen / R. F. B. Isarin, The two major warming phases of the last deglaciation at ~14.7 and ~11.5 ka cal BP in Europe: climate reconstructions and AGCM experiments. Global and Planetary Change 30, 2001, 117-153.

Renssen/Vandenberghe 2003: H. Renssen / J. Vandenberghe, Investigation of the relationship between permafrost distribution in NW Europe and extensive winter sea-ice cover in the North Atlantic Ocean during the cold phases of the Last Glaciation. Quaternary Science Reviews 22, 2003, 209-223.

Rheinisches Landesmuseum Bonn 2006: Rheinisches Landesmuseum Bonn (ed.), Roots. Wurzeln der Menschheit (Mainz, Bonn 2006).

Rhodes et al. 2003: E. J. Rhodes / C. Bronk Ramsey / Z. Outram / C. Batt / L. Willis / S. Dockrill / J. Bond, Bayesian methods applied to the interpretation of multiple OSL dates: high precision sediment ages from Old Scatness Broch excavations, Shetland Isles. Quaternary Science Reviews 22, 2003, 1231-1244.

Richards et al. 2000: M. P. Richards / R. E. M. Hedges / R. Jacobi / A. Currant / C. Stringer, Gough's Cave and Sun Hole Cave human stable isotope values indicate a high animal protein diet in the British Upper Palaeolithic. Journal of Archaeological Science 27, 2000, 1-3.

2005: M. P. Richards / R. Jacobi / J. Cook / P. B. Pettitt / C. B. Stringer, Isotope evidence for the intensive use of marine foods by Late Upper Palaeolithic humans. Journal of Human Evolution 49, 2005, 390-394.

Richter 2007: D. Richter, Advantages and limitations of thermoluminescence dating of heated flint from Paleolithic sites. Geoarchaeology 22, 2007, 671-683.

Richter/Krbetschek 2006: D. Richter / M. Krbetschek, A new thermoluminescence dating technique for heated flint. Archaeometry 48, 2006, 695-705.

Richter et al. 2009: D. Richter / G. Tostevin / P. Škrdla / W. Davies, New radiometric ages for the Early Upper Palaeolithic type locality of Brno-Bohunice (Czech Republic): comparison of OSL, IRSL, TL and ^{14}C dating results. Journal of Archaeological Science 36, 2009, 708-720.

Richter 1990: J. Richter, Diversität als Zeitmaß im Spätmagdalénien. Archäologisches Korrespondenzblatt 20, 1990, 249-257.

Richter 1992: P. B. Richter, Ein spätglazialer Fundplatz auf dem Friedhof in Melbeck, Ldkr. Lüneburg. Nachrichten aus Niedersachsens Urgeschichte 61, 1992, 3-32.

Rick 1987: J. W. Rick, Dates as data: An examination of the Peruvian Preceramic radiocarbon record. American Antiquity 52, 1987, 55-73.

Rick/Vellanoweth/Erlandson 2005: T. C. Rick / R. L. Vellanoweth / J. M. Erlandson, Radiocarbon dating and the »old shell« problem: direct dating of artifacts and cultural chronologies in coastal and other aquatic regions. Journal of Archaeological Science 32, 2005, 1641-1648.

Riede 2008: F. Riede, The Laacher See-eruption (12,920 BP) and material culture change at the end of the Allerød in Northern Europe. Journal of Archaeological Science 35, 2008, 591-599.

2009: F. Riede, The loss and re-introduction of bow-and-arrow-technology: A case study from the northern European Late Palaeolithic. Lithic Technology 34, 2009, 27-46.

2010: F. Riede, Hamburgian weapon delivery technology: a quantitative comparative approach. Before Farming 2010/1, 1-18.

2011: F. Riede, Steps towards operationalizing an evolutionary archaeological definition of culture. In: B. W. Roberts / M. Vander Linden (eds), Investigating archaeological cultures: material culture, variability, and transmission (New York et al. 2011) 245-270.

Riede/Edinborough/Thomas 2009: F. Riede / K. Edinborough / M. Thomas, Tracking Mesolithic demography in time and space and its implications for explanations of culture change. In: P. Crombé / M. van Strydonck / J. Sergant / M. Boudin / M. Bats (eds), Chronology and evolution within the Mesolithic of North-West Europe. Proceedings of an international meeting, Brussels, May 30th-June 1st 2007 (Newcastle 2009) 177-194.

Riede/Kristensen 2010: F. Riede / I. K. Kristensen, Skaftglatter. Skalk 6, 2010, 3-6.

Riede/Laursen/Hertz 2011: F. Riede / S. T. Laursen / E. Hertz, Federmesserkulturen i Danmark. Belyst med udgangspunkt i en amatørarkæologs flintsamling. Kuml – Årbog for Jysk Arkæologisk Selskab 2011, 9-38.

Riede et al. 2010: F. Riede / S. B. Grimm / M.-J. Weber / J. M. Fahlke, Neue Daten für alte Grabungen. Ein Beitrag zur spätglazialen Archäologie und Faunengeschichte Norddeutschlands. Archäologisches Korrespondenzblatt 40, 2010, 297-316.

Rieu/March/Soler-Mayor 1999: J.-L. Rieu / R. J. March / B. Soler-Mayor, Les témoins de combustion. In: Julien/Rieu 1999, 96-129.

Ritz/Stocker/Müller 2008: S. P. Ritz / T. F. Stocker / S. A. Müller, Modeling the effect of abrupt ocean circulation change on marine reservoir age. Earth and Planetary Science Letters 268, 2008, 202-211.

Robinson 1988: S. Robinson, Statave (Statistic Average). BASIC program (Menlo Park CA 1988 [unpubl.]).

Roblin-Jouve 1992: A. Roblin-Jouve, Etude géomorphologique. In: Schmider 1992f, 19-30.

1994: A. Roblin-Jouve, Le milieu physique. In: Taborin 1994, 12-35.

1996: A. Roblin-Jouve, Pincevent et ses environs: essai de paléogéographie. In: Gaucher 1996, 15-33.

Rockman 2003a: M. H. Rockman, Landscape learning in the Late Glacial recolonization of Britain [unpublished dissertation, Tucson 2003].

2003b: M. Rockman, Knowledge and learning in the archaeology of colonization. In: M. Rockman / J. Steele (eds), Colonization of unfamiliar landscapes. The archaeology of adaptation (London 2003) 3-24.

Rodrigues et al. 2010: T. Rodrigues / J. O. Grimalt / F. Abrantes / F. Naughton / J.-A. Flores, The last glacial-interglacial transition (LGIT) in the western [eastern] mid-latitudes of the North Atlantic: Abrupt sea surface temperature change and sea level implications. Quaternary Science Reviews 29, 2010, 1853-1862.

Rodriguez 1994: P. Rodriguez, La malacologie: contribution à la paléocologie et la chronologie des habitats magdaléniens. In: Taborin 1994, 39-58.

Rodriguez-Iturbe 2000: I. Rodriguez-Iturbe, Ecohydrology: A hydrologic perspective of climate-soil-vegetation dynamics. Water Resource Research 36, 2000, 3-9.

Rodwell/Rowell/Folland 1999: M. J. Rodwell / D. P. Rowell / C. K. Folland, Oceanic forcing of the wintertime North Atlantic Oscillation and European climate. Nature 398, 1999, 320-323.

Rosen/Rivera-Collazoa 2012: A. M. Rosen / I. Rivera-Collazoa, Climate change, adaptive cycles, and the persistence of foraging economies during the late Pleistocene/Holocene transition in the Levant. Proceedings of the National Academy of Sciences of the United States of America 109, 2012, 3640-3645.

Rosendahl et al. 2006: G. Rosendahl / K.-W. Beinhauer / M. Löscher / K. Kreipl / R. Walter / W. Rosendahl, Le plus vieil arc du monde? Une pièce intéressante en provenance de Mannheim, Allemagne. L'Anthropologie 110, 2006, 371-382.

Rots 2005: V. Rots, Wear traces and the interpretation of stone tools. Journal of Field Archaeology 30, 2005, 61-73.

Rots/Stapert/Johansen 2002: V. Rots / D. Stapert / L. Johansen, De Cheddar- en Creswell-Spitsen van Zeijen (Dr.): »Projectielen« of Messen? Paleo-Aktueel 13, 2002, 24-29.

Rowley-Conwy/Zvelebil 1989: P. Rowley-Conwy / M. Zvelebil, Saving it for later: Storage by prehistoric hunter-gatherers in Europe. In: Halstead/O'Shea 1989a, 40-56.

Rozoy 1978: J. G. Rozoy, Les derniers chasseurs. L'Epipaléolithique en France et en Belgique. Essai de synthèse (Charleville 1978).

1988: J.-G. Rozoy, Le Magdalénien supérieur de Roc-la-Tour I dans le contexte franco-belgo-rhénan. In: Otte 1988, 137-162.

1990: J.-G. Rozoy, Les plaquettes gravées magdaléniennes de Roc-la-Tour I. In: J. Clottes (ed.), L'art des objets au Paléolithique. Colloque International à Foix-Le Mas-d'Azil, Novembre 1987. 1: L'Art Mobilier et son Contexte (Paris 1990) 261-277.

Rubin/Suess 1956: M. Rubin / H. E. Suess, U.S. Geological Survey Radiocarbon Dates III. Science 123, 1956, 442-448.

Rundgren 1995: M. Rundgren, Biostratigraphic evidence of the Allerød-Younger Dryas-Preboreal oscillation in northern Iceland. Quaternary Research 44, 1995, 405-416.

Rust 1943: A. Rust, Die alt- und mittelsteinzeitlichen Funde von Stellmoor (Neumünster 1943).

1958: A. Rust, Die jungpaläolithischen Zeltanlagen von Ahrensburg. Offa-Bücher 15 (Neumünster 1958).

Ruth et al. 2007: U. Ruth / M. Bigler / R. Röthlisberger / M.-L. Siggaard-Andersen / S. Kipfstuhl / K. Goto-Azuma / M. E. Hansson / S. J. Johnsen / H. Lu / J. P. Steffensen, Ice core evidence for a very tight link between North Atlantic and east Asian glacial climate. Geophysical Research Letters 34, 2007, L03706, 1-5.

SADL 2005: SADL (Spatial Applications Division Leuven), GISCO DataBase Manual. Research and Development at the Katholieke Universiteit Leuven (Leuven 2005).

Sagan 1995: C. Sagan, The demon-haunted world. Science as a candle in the dark (New York 1995).

Sánchez Goñi 1991: M. F. Sánchez Goñi, On the Last Glacial Maximum and interstadials during the Solutrean: a contradiction? Current Anthropology 32, 1991, 573-575.

Sandersen/Jørgensen 2003: P. B. E. Sandersen / F. Jørgensen, Buried Quaternary valleys in western Denmark – occurrence and inferred implications for groundwater resources and vulnerability. Journal of Applied Geophysics 53, 2003, 229-248.

Sandwell/Becker 2009: D. T. Sandwell / J. J. Becker, SRTM30_plus: Data fusion of SRTM land topography with measured and estimated seafloor topography. University of California (San Diego, CA 2009; ftp://topex.ucsd.edu/pub/srtm30_plus/README.V6.0.txt).

Sandwell/Smith 2009: D. T. Sandwell / W. H. F. Smith, Global marine gravity from retracked Geosat and ERS-1 altimetry: Ridge segmentation versus spreading rate. Journal of Geophysical Research 114, 2009, B01411.

Sano 2009: K. Sano, Hunting evidence from stone artefacts from the Magdalenian cave site Bois Laiterie, Belgium: a fracture analysis. Quartär 56, 2009, 67-86.

2012a: K. Sano, Functional variability in the Late Upper Palaeolithic of north-western Europe. A traceological approach. Universitätsforschungen zur Prähistorischen Archäologie 219 (Bonn 2012).

2012b: K. Sano, Functional variability in the Magdalenian of north-western Europe: A lithic microwear analysis of the Gönnersdorf K-II assemblage. Quaternary International 272-273, 2012, 264-274.

Sano/Maier/Heidenreich 2011: K. Sano / A. Maier / S. M. Heidenreich, Bois Laiterie revisited: functional, morphological and technological analysis of a Late Glacial hunting camp in north-western Europe. Journal of Archaeological Science 38, 2011, 1468-1484.

Saville/Ballin 2009: A. Saville / T. B. Ballin, Upper Palaeolithic evidence from Kilmelfort Cave, Argyll: a re-evaluation of the lithic assemblage. Proceedings of the Society of Antiquaries of Scotland 139, 2009, 9-45.

Schaub et al. 2005: M. Schaub / K. F. Kaiser / B. Kromer / S. Talamo, Extension of the Swiss Lateglacial tree-ring chronologies. Dendrochronologia 23, 2005, 11-18.

2008a: M. Schaub / U. Büntgen / K. F. Kaiser / B. Kromer / S. Talamo / K. K. Andersen / S. O. Rasmussen, Lateglacial environmental variability from Swiss tree rings. Quaternary Science Reviews 27, 2008, 29-41.

2008b: M. Schaub / K. F. Kaiser / D. C. Frank / U. Büntgen / B. Kromer / S. Talamo, Environmental change during the Allerød and Younger Dryas reconstructed from Swiss tree-ring data. Boreas 37, 2008, 74-86.

Schettler/Rein/Negendank 1999: G. Schettler / B. Rein / J. F. W. Negendank, Geochemical evidence for the Holocene palaeodischarge variations in lacustrine records from the Westeifel Volcanic Field, Germany: Schalkenmehrener Maar and Meerfelder Maar. The Holocene 9, 1999, 381-400.

Schiffer 1986: M. B. Schiffer, Radiocarbon dating and the »old wood« problem: The case of the Hohokam chronology. Journal of Archaeological Science 13, 1986, 13-30.

Schild 1984: R. Schild, Terminal Paleolithic of the North European Plain: A Review of Lost Chances, Potential, and Hopes. Advances in World Archaeology 3, 1984, 193-274.

1996: R. Schild, The North European Plain and eastern Sub-Balticum between 12,700 and 8,000 BP. In: L. G. Straus / B. V. Eriksen / J. M. Erlandson / D. R. Yesner (eds), Humans at the end of the Ice Age. The archaeology of the Pleistocene-Holocene transition (New York 1996) 129-157.

Schirmer 1990: W. Schirmer, Die Goldene Meile. In: W. Schirmer (ed.), Rheingeschichte zwischen Mosel und Maas (Hannover 1990) 94-98.

Schmider 1971: B. Schmider, Les industries lithiques du Paléolithique supérieur en Ile-de-France. Supplément à Gallia Préhistoire 6 (Paris 1971).

1982: B. Schmider, The Magdalenian culture of the Paris river-basin and its relationship with the Nordic cultures of the late Old Stone Age. World Archaeology 14, 1982, 259-269.

1984: B. Schmider, Les habitations magdaléniennes de Marsagny (Vallée de l'Yonne). In: H. Berke / J. Hahn / C.-J. Kind (eds), Jungpaläolithische Siedlungsstrukturen in Europa. Kolloquium 8-14. Mai 1983 Reisensburg/Günzburg (Tübingen 1984) 169-180.

1992a: B. Schmider, Le travail du silex: L'outillage. In: Schmider 1992f, 129-223.

1992b: B. Schmider, Les structures d'habitation. In: Schmider 1992f, 49-86.

1992c: B. Schmider, Les témoins esthétiques. In: Schmider 1992f, 228-231.

1992d: B. Schmider, Présentation du gisement du Pré-des-Forges à Marsangy. In: Schmider 1992f, 9-18.

1992e: B. Schmider, Synthèse des données sur le Magdalénien de Marsangy. In: Schmider 1992f, 245-261.

1992f: B. Schmider (ed.), Marsangy. Un campement des derniers chasseurs magdalénien, sur le bords de l'Yonne (Liège 1992).

Schmidt 1912: R. R. Schmidt, Die diluviale Vorzeit Deutschlands (Stuttgart 1912).

Schmincke/Park/Harms 1999: H.-U. Schmincke / C. Park / E. Harms, Evolution and environmental impacts of the eruption of Laacher See Volcano (Germany) 12,900 a BP. Quaternary International 61, 1999, 61-72.

Schmitz 2009: R. W. Schmitz, Hier liegt der Hund begraben. Die späteiszeitliche Doppelbestattung von Bonn-Oberkassel. In:

D. Planck / J. Heiligmann / N. J. Conard (eds), Eiszeit. Kunst und Kultur [exhibition catalogue Stuttgart] (Ostfildern 2009) 70.

Schmitz/Thissen 1997: R.-W. Schmitz / J. Thissen, Aktuelle Untersuchungen zum endpleistozänen/frühholozänen Fundplatz Bonn-Oberkassel. Ein Vorbericht. Archäologische Informationen 19, 1997, 197-203.

Schmölcke/Zachos 2005: U. Schmölcke / F. E. Zachos, Holocene distribution and extinction of the moose (*Alces alces*, Cervidae) in Central Europe. Mammalian Biology – Zeitschrift für Säugetierkunde 70, 2005, 329-344.

Schmölcke et al. 2006: U. Schmölcke / E. Endtmann / S. Klooss / M. Meyer / D. Michaelis / B.-H. Rickert / D. Rößler, Changes of sea level, landscape and culture: A review of the south-western Baltic area between 8800 and 4000 BC. Palaeogeography, Palaeoclimatology, Palaeoecology 240, 2006, 423-438.

Schrag et al. 2002: D. P. Schrag / J. F. Adkins / K. McIntyre / J. L. Alexander / D. A. Hodell / C. D. Charles / J. F. McManus, The oxygen isotopic composition of seawater during the Last Glacial Maximum. Quaternary Science Reviews 21, 2002, 331-342.

Schuler 1994: A. Schuler, Die Schussenquelle. Eine Freilandstation des Magdalénien in Oberschwaben. Materialhefte zur Archäologie in Baden-Württemberg 27 (Stuttgart 1994).

Schulte-Dornberg 2000: G. Schulte-Dornberg, Benutzte Felsgesteine und Gerölle im Magdalénien von Andernach-Martinsberg und ihre mögliche Funktion [unpublished M.A. thesis, Köln 2000].

Schulte-Dornberg 2002: G. Schulte-Dornberg, Benutzte Gerölle im Magdalénien von Andernach-Martinsberg und ihre mögliche Funktion. Archäologisches Korrespondenzblatt 32, 2002, 487-499.

Schütrumpf 1937: R. Schütrumpf, Die paläobotanisch-pollenanalytische Untersuchung. In: A. Rust, Das altsteinzeitliche Rentierjägerlager Meiendorf (Neumünster 1937) 10-47.

Schuur et al. 2009: E. A. G. Schuur / J. G. Vogel / K. G. Crummer / H. Lee / J. O. Sickman / T. E. Osterkamp, The effect of permafrost thaw on old carbon release and net carbon exchange from tundra. Nature 459, 2009, 556-559.

Schwab 1985: H. Schwab, Gagat und Bernstein auf dem Rentierjägerhalt Moosbühl bei Moosseedorf (Kanton Bern). Mit einem Beitrag von Curt W. Beck. In: R. Fellmann / G. Germann / K. Zimmermann (eds), Jagen und Sammeln. Festschrift für Hans-Georg Bandi zum 65. Geburtstag (3. September 1985) (Bern 1985) 259-266.

Schwabedissen 1939: H. Schwabedissen, Zur Entstehung des Nordischen Kreises. Forschungen und Fortschritte 15, 1939, 142-144.

1944a: H. Schwabedissen, Die mittlere Steinzeit im westlichen Norddeutschland. unter besonderer Berücksichtigung der Feuersteinwerkzeuge. Offa-Bücher 7 (Neumünster 1944).

1944b: H. Schwabedissen, Eine neue magdalénienartige Kulturgruppe im nordwesteuropäischen Flachland und deren Bedeutung für die Entstehung des Nordischen Kreises. Forschungen und Fortschritte 20, 1944, 51-52.

1954: H. Schwabedissen, Die Federmesser-Gruppen des nordwesteuropäischen Flachlandes. Zur Ausbreitung des Spät-Magdalénien. Offa-Bücher 9 (Neumünster 1954).

1955: H. Schwabedissen, Zur Auswertung steinzeitlicher Oberflächenfundplätze. Federmesser-Zivilisation und Mesolithikum. Eiszeitalter und Gegenwart 6, 1955, 159-169.

1957: H. Schwabedissen, Das Alter der Federmesser-Zivilisation auf Grund neuer naturwissenschaftlicher Untersuchungen. Eiszeitalter und Gegenwart 8, 1957, 200-209.

Schwantes 1933: G. Schwantes, Die ältesten Bewohner des mittleren Norddeutschland. Forschungen und Fortschritte 9, 1933, 261-262.

Schwarcz 1980: H. P. Schwarcz, Absolute age determination of archaeological sites by uranium series dating of travertines. Archaeometry 22, 1980, 3-24.

Schwark/Zink/Lechterbeck 2002: L. Schwark / K. Zink / J. Lechterbeck, Reconstruction of postglacial to early Holocene vegetation history in terrestrial Central Europe via cuticular lipid biomarkers and pollen records from lake sediments. Geology 30, 2002, 463-466.

Schweingruber 1978: F. Schweingruber, Umweltverhältnisse: Flora. Holzanalytische Untersuchungen. In: Brunnacker 1978a, 82-97.

Schwendler 2012: R. H. Schwendler, Diversity in social organization across Magdalenian Western Europe ca. 17-12,000 BP. Quaternary International 272-273, 2012, 333-353.

Scott et al. 2004: E. M. Scott / C. Bryant / I. Carmi / G. Cook / S. Gulliksen / D. Harkness / J. Heinemeier / E. McGee / P. Naysmith / G. Possnert / H. van der Plicht / M. van Strydonck, Precision and accuracy in applied ^{14}C dating: some findings from the Fourth International Radiocarbon Inter-comparison. Journal of Archaeological Science 31, 2004, 1209-1213.

Seelos/Sirocko/Dietrich 2009: K. Seelos / F. Sirocko / S. Dietrich, A continuous high-resolution dust record for the reconstruction of wind systems in central Europe (Eifel, Western Germany) over the past 133 ka. Geophysical Research Letters 36, 2009, L20712-1-6.

Sejrup et al. 2000: H. P. Sejrup / E. Larsen / J. Landvik / E. L. King / H. Haflidason / A. Nesje, Quaternary glaciations in southern Fennoscandia: evidence from southwestern Norway and the northern North Sea region. Quaternary Science Reviews 19, 2000, 667-685.

2009: H. P. Sejrup / A. Nygård / A. M. Hall / H. Haflidason, Middle and Late Weichselian (Devensian) glaciation history of southwestern Norway, North Sea and eastern UK. Quaternary Science Reviews 28, 2009, 370-380.

Sensburg 2007: M. Sensburg, Die räumliche Organisation der Konzentration IIa von Gönnersdorf. Struktur und Dynamik eines magdalénienzeitlichen Siedlungsbefundes am Mittelrhein. Monographien des RGZM 69 (Mainz 2007).

Sensburg/Moseler 2008: M. Sensburg / F. Moseler, Die Konzentrationen IIb und IV des Magdalénien-Fundplatzes Gönnersdorf (Mittelrhein). Monographien des RGZM 73 (Mainz 2008).

Serangeli/Terberger 2006: J. Serangeli / Th. Terberger, Sondierungen an der Magdalénien-Fundstelle von Götzenhain-Ost, Kreis Offenbach. Mitteilungen der Gesellschaft für Urgeschichte 15, 2006, 49-61.

Shennan 2008: S. Shennan, Analytical archaeology. In: J. Bintliff (ed.), A companion to archaeology (Oxford 2008) 3-20.

Shennan/Edinborough 2007: S. Shennan / K. Edinborough, Prehistoric population history: from the Late Glacial to the Late Neolithic in Central and Northern Europe. Journal of Archaeological Science 34, 2007, 1339-1345.

Shennan et al. 2000: I. Shennan / K. Lambeck / R. Flather / B. Horton / J. McArthur / J. Innes / J. Lloyd / M. Rutherford / R. Wingfield, Modelling western North Sea palaeogeographies and tidal changes during the Holocene. In: I. Shennan / J. Andrews (eds), Holocene land-ocean interaction and environmental change around the North Sea (London 2000) 299-319.

2006: I. Shennan / S. Bradley / G. Milne / A. Brooks / S. Bassett / S. Hamilton, Relative sea-level changes, glacial isostatic modelling and ice-sheet reconstructions from the British Isles since the Last Glacial Maximum. Journal of Quaternary Science 21, 2006, 585-599.

Siegenthaler/Oeschger 1980: U. Siegenthaler / H. Oeschger, Correlation of ^{18}O in precipitation with temperature and altitude. Nature 285, 1980, 314-317.

Sima et al. 2009: A. Sima / D.-D. Rousseau / M. Kageyama / G. Ramstein / M. Schulz / Y. Balkanski / P. Antoine / F. Dulac / C. Hatte, Imprint of North-Atlantic abrupt climate changes on western European loess deposits as viewed in a dust emission model. Quaternary Science Reviews 28, 2009, 2851-2866.

Simmons 2007: A. H. Simmons, The neolithic revolution in the Near East. Transforming the human landscape (Tucson 2007).

Simpson 1944: G. G. Simpson, Tempo and mode in evolution (New York 1944).

Sirocko et al. 2005: F. Sirocko / K. Seelos / K. Schaber / B. Rein / F. Dreher / M. Diehl / R. Lehne / K. Jäger / M. Krbetschek / D. Degering, A late Eemian aridity pulse in central Europe during the last glacial inception. Nature 436, 2005, 833-836.

Smalley et al. 2009: I. Smalley / K. O'Hara-Dhand / J. Wint / B. Machalett / Z. Jary / I. Jefferson, Rivers and loess: The significance of long river transportation in the complex event-sequence approach to loess deposit formation. Quaternary International 198, 2009, 7-18.

Smith/Bonsall 1991: C. Smith / C. Bonsall, Late Upper Palaeolithic and Mesolithic chronology: points of interest from recent research. In: Barton/Roberts/Roe 1991, 208-212.

Smith/Sandwell 2003: B. Smith / D. Sandwell, Accuracy and resolution of shuttle radar topography mission data. Geophysical Research Letters 30, 2003, 1467-1470.

Soares et al. 2010: P. Soares / A. Achilli / O. Semino / W. Davies / V. Macaulay / H.-J. Bandelt / A. Torroni / M. B. Richards, The Archaeogenetics of Europe. Current Biology 20, 2010, R174-R183.

Sodemann et al. 2008: H. Sodemann / V. Masson-Delmotte / C. Schwierz / B. M. Vinther / H. Wernli, Interannual variability of Greenland winter precipitation sources: 2. Effects of North Atlantic Oscillation variability on stable isotopes in precipitation. Journal of Geophysical Research – Atmospheres 113, 2008, D12111.

Sokal/Rohlf 1962: R. R. Sokal / F. J. Rohlf, The comparison of dendrograms by objective methods. Taxon 11, 1962, 33-40.

Sommer et al. 2009: R. S. Sommer / J. M. Fahlke / U. Schmölcke / N. Benecke / F. E. Zachos, Quaternary history of the European roe deer *Capreolus capreolus*. Mammal Review 39, 2009, 1-16.

2008: R. S. Sommer / F. E. Zachos / M. Street / O. Jöris / A. Skog / N. Benecke, Late Quaternary distribution dynamics and phylogeography of the red deer (*Cervus elaphus*) in Europe. Quaternary Science Reviews 27, 2008, 715-733.

2011: R. S. Sommer / N. Benecke / L. Lõugas / O. Nelle / U. Schmölcke, Holocene survival of the wild horse in Europe: a matter of open landscape? Journal of Quaternary Science 26, 2011, 805-812.

Sonneville-Bordes 1960: D. de Sonneville-Bordes, Le Paléolithique Supérieur en Périgord (Bordeaux 1960).

1966: D. de Sonneville-Bordes, L'évolution du Paléolithique supérieur en Europe occidentale et sa signification. Bulletin de la Société préhistorique française 63, 1966, 3-34.

1979: D. de Sonneville-Bordes (ed.), La fin des temps glaciaires en Europe. Chronostratigraphie et écologie des cultures du Paléolithique final. Actes du colloque international du Centre National de la recherche scientifique 271. Talence, 24-28 mai 1977 (Paris 1979).

Sonneville-Bordes/Perrot 1956: D. de Sonneville-Bordes / J. Perrot, Lexique typologique du Paléolithique supérieur. Outillage lithique (suite et fin). Bulletin de la Société préhistorique française 53, 1956, 547-559.

Soula 2000: L. Soula, Contribution à l'étude des groupes magdaléniens supérieur et final du Bassin parisien: le gisement du Brassot à Étigny dans l'Yonne [unpublished M.A. thesis (Maîtrise), Paris 2000].

Southon 2002: J. Southon, A first step to reconciling the GRIP and GISP2 ice-core chronologies, 0-14,500 yr B.P. Quaternary Research 57, 2002, 32-37.

Southon et al. 2012: J. Southon / A. L. Noronha / H. Cheng / R. L. Edwards / Y. Wang, A high-resolution record of atmospheric ^{14}C based on Hulu Cave speleothem H82. Quaternary Science Reviews 33, 2012, 32-41.

Spier 1997: F. Spier, Les occupations humaines de la fin du Tardiglaciaire et du début de l'holocène entre Ardennes et Vosges. In: Fagnart/Thévenin 1997, 409-427.

Spier 2000: F. Spier, L'Épipaléolithique de Luxembourg: quelques aspects. Bulletin de la Société préhistorique luxembourgeoise 19, 1997 (2000), 91-99.

Spolaor et al. 2013: A. Spolaor / P. Vallelonga / J. Gabrieli / M. Roman / C. Barbante, Continuous flow analysis method for determination of soluble iron and aluminium in ice cores. Analytical and Bioanalytical Chemistry 405, 2013, 767-774.

Spötl/Mangini/Richards 2006: C. Spötl / A. Mangini / D. A. Richards, Chronology and paleoenvironment of Marine Isotope Stage 3 from two high-elevation speleothems, Austrian Alps. Quaternary Science Reviews 25, 2006, 1127-1136.

Spurk et al. 1998: M. Spurk / M. Friedrich / J. Hofmann / S. Remmele / B. Frenzel / H. H. Leuschner / B. Kromer, Revisions and extension of the Hohenheim oak and pine chronologies: new evidence about the timing of the Younger Dryas/Preboreal transition. Radiocarbon 40, 1998, 1107-1116.

Stampfuss/Schütrumpf 1970: R. Stampfuss / R. Schütrumpf, Harpunen der Allerödzeit aus Dinslaken, Niederrhein. Bonner Jahrbücher 170, 1970, 19-35.

Stanford et al. 2006: J. D. Stanford / E. J. Rohling / S. E. Hunter / A. P. Roberts / S. O. Rasmussen / E. Bard / J. McManus / R. G. Fairbanks, Timing of meltwater pulse 1a and climate responses to meltwater injections. Paleoceanography 21, 2006, PA4103.1-9.

2011a: J. D. Stanford / R. Hemingway / E. J. Rohling / P. G. Challenor / M. Medina-Elizalde / A. J. Lester, Sea-level probability for the last deglaciation: A statistical analysis of far-field records. Global and Planetary Change 79, 2011, 193-203.

2011b: J. D. Stanford / E. J. Rohling / S. Bacon / A. P. Roberts / F. E. Grousset / M. Bolshaw, A new concept for the paleoaceanographic evolution of Heinrich event 1 in the North Atlantic. Quaternary Science Reviews 30, 2011, 1047-1066.

Stapert 1990: D. Stapert, Within the tent or outside? Spatial patterns in Late Palaeolithic sites. Helinium 30, 1990, 14-35.

1991: D. Stapert, The ring and sector method: Intrasite spatial analysis of Stone Age sites, with special reference to Pincevent. Palaeohistoria 31, 1989 (1991), 1-57.

1992: D. Stapert, Rings and sectors: Intrasite spatial analysis of stone age sites [unpublished dissertation, Groningen 1992].

2003: D. Stapert, Towards dynamic models of Stone Age settlements. In: S. A. Vasil'ev / O. Soffer / J. Kozłowski (eds), Perceived landscapes and built environments. The cultural geography of Late Palaeolithic Eurasia. Actes du XIVème congrès UISPP, Université de Liège, Belgique, 2-8 septembre 2001, section 6: paléolithique supérieur, colloques 6.2 & 6.5 (Oxford 2003) 5-15.

2005: D. Stapert, Het Laat-Paleolithicum in Noord-Nederland. In: J. Deeben / E. Drenth / M. F. van Oorsouw / L. Verhart (eds), De steentijd van Nederland (Meppel 2005) 143-169.

Stapert/Johansen 1997: D. Stapert / L. Johansen, Ring & sector analysis, and site »IT« on Greenland. Palaeohistoria 37/38, 1995/1996 (1997), 29-69.

Stapert/Street 1997: D. Stapert / M. Street, High resolution or optimum resolution? Spatial analysis of the Federmesser site at Andernach, Germany. World Archaeology 29, 1997, 172-194.

Stapert/Terberger 1991: D. Stapert / Th. Terberger, Gönnersdorf concentration III: Investigating the possibility of multiple occupations. Palaeohistoria 31, 1989 (1991), 59-95.

Stapert/Veenstra 1988: D. Stapert / H. J. Veenstra, The section at Usselo: Brief description, grain-size distributions, and some remarks on the archaeology. Palaeohistoria 30, 1988, 1-28.

Starkel 2002: L. Starkel, Younger Dryas-Preboreal transition documented in the fluvial environment of Polish rivers. Global and Planetary Change 35, 2002, 157-167.

Starkel/Gębica/Superson 2007: L. Starkel / P. Gębica / J. Superson, Last Glacial-Interglacial cycle in the evolution of river valleys in southern and central Poland. Quaternary Science Reviews 26, 2007, 2924-2936.

Starnberger/Rodnight/Spötl 2011: R. Starnberger / H. Rodnight / C. Spötl, Chronology of the Last Glacial Maximum in the Salzach palaeoglacier area (Eastern Alps). Journal of Quaternary Science 26, 2011, 502-510.

Steele 2010: J. Steele, Radiocarbon dates as data: quantitative strategies for estimating colonization front speeds and event densities. Journal of Archaeological Science 37, 2010, 2017-2030.

Steelman/Rowe 2004: K. L. Steelman / M. W. Rowe, Non-destructive plasma-chemical extraction of carbon from organic artefacts. In: Higham/Bronk Ramsey/Owen 2004, 35-42.

Steffensen et al. 2008: J. P. Steffensen / K. K. Andersen / M. Bigler / H. B. Clausen / D. Dahl-Jensen / H. Fischer / K. Goto-Azuma / M. Hansson / S. J. Johnsen / J. Jouzel / V. Masson-Delmotte / T. Popp / S. O. Rasmussen / R. Röthlisberger / U. Ruth / B. Stauffer / M.-L. Siggaard-Andersen / Á. E. Sveinbjörnsdóttir / A. Svensson / J. W. C. White, High-resolution Greenland ice core data show abrupt climate change happens in few years. Science 321, 2008, 680-684.

Stein Mandryk 1993: C. A. Stein Mandryk, Hunter-gatherer social costs and the nonviability of submarginal environments. Journal of Anthropological Research 49, 1993, 39-71.

Stensager 2004: A. O. Stensager, Lyngby-rentakken er fra Maglemosetid. Aktuel Arkæologi 6, 2004, 32-33.

Stephan 1999: E. Stephan, Sauerstoffisotopenverhältnisse im Knochengewebe großer terrestrischer Säugetiere. Klimaproxies für das Quartär in Mittel- und Westeuropa. Tübinger Geowissenschaftliche Arbeiten (TGA), Reihe E: Mineralogie, Petrologie und Geochemie 6 (Aachen 1999).

Stevens/Hedges 2004: R. E. Stevens / R. E. M. Hedges, Carbon and nitrogen stable isotope analysis of northwest European horse bone and tooth collagen, 40,000 BP-present: Palaeoclimatic interpretations. Quaternary Science Reviews 23, 2004, 977-991.

Stevens et al. 2009a: R. E. Stevens / M. Germonpré / C. A. Petrie / T. C. O'Connell, Palaeoenvironmental and chronological investigations of the Magdalenian sites of Goyet Cave and Trou de Chaleux (Belgium), via stable isotope and radiocarbon analyses of horse skeletal remains. Journal of Archaeological Science 36, 2009, 653-662.

2009b: R. E. Stevens / T. C. O'Connell / R. E. M. Hedges / M. Street, Radiocarbon and stable isotope investigations at the Central Rhineland sites of Gönnersdorf and Andernach-Martinsberg, Germany. Journal of Human Evolution 57, 2009, 131-148.

Stewart/Lister 2001: J. R. Stewart / A. M. Lister, Cryptic northern refugia and the origins of modern biota. Trends in Ecology and Evolution 16, 2001, 608-613.

Stewart/Lonergan/Hampson 2012: M. Stewart / L. Lonergan / G. Hampson, 3D seismic analysis of buried tunnel valleys in the Central North Sea: tunnel valley fill sedimentary architecture. Geological Society, London, Special Publications 368, 2012, 173-184.

Stiner et al. 1999: M. C. Stiner / N. D. Munro / T. A. Surovell / E. Tchernov / O. Bar-Yosef, Paleolithic population growth pulses evidenced by small animal exploitation. Science 283, 1999, 190-194.

Stocker et al. 2006: C. Stocker / J.-M. Cordy / M. Patou-Mathis / A. Thévenin, Le gisement magdalénien de la Roche Plate à Saint-Mihiel (Meuse, France). Bulletin de la Société préhistorique luxembourgeoise 25, 2003 (2006), 23-41.

Stodiek 1987: U. Stodiek, Fussgönheim – zwei spätjungpaläolithische Fundplätze in der Vorderpfalz. Archäologisches Korrespondenzblatt 17, 1987, 31-41.

Stokes/Smiley 1968: M. A. Stokes / T. L. Smiley, An introduction to tree-ring dating (Chicago 1968).

Strauch/Tembrock 1978: F. Strauch / M. L. Tembrock, Umweltverhältnisse: Beschaffung von Werkstoffen. Marine Molluskenreste. In: Brunnacker 1978a, 231-233.

Straus 1997: L. G. Straus, La Grotte du Bois Laiterie: The site, stratigraphy, chronology and excavation. In: Otte/Straus 1997, 25-63.

Straus/Leesch/Terberger 2012: L. Straus / D. Leesch / Th. Terberger, The Magdalenian settlement of Europe: An introduction. Quaternary International 272-273, 2012, 1-5.

Straus/Martinez 1997: L. G. Straus / A. Martinez, Site formation/disturbance processes, spatial distributions, site structure and activity areas. In: Otte/Straus 1997, 65-112.

Straus/Meltzer/Goebel 2005: L. G. Straus / D. J. Meltzer / T. Goebel, Ice Age Atlantis? Exploring the Solutrean-Clovis »connection«. World Archaeology 37, 2005, 507-532.

Straus/Orphal 1997: L. G. Straus / J. Orphal, The Bois Laiterie Magdalenian lithic industry. In: Otte/Straus 1997, 219-255.

Street 1986: M. Street, Ein Wald der Allerödzeit bei Miesenheim, Stadt Andernach (Neuwieder Becken). Archäologisches Korrespondenzblatt 16, 1986, 13-22.

1991: M. Street, Bedburg-Königshoven: A Pre-Boreal Mesolithic site in the Lower Rhineland (Germany). In: Barton/Roberts/Roe 1991, 256-270.

1993: M. J. Street, Analysis of late palaeolithic and mesolithic faunal assemblages in the Northern Rhineland, Germany [unpublished dissertation, Birmingham 1993].

1996: M. Street, The Late Glacial faunal assemblage from Endingen, Lkr. Nordvorpommern. Archäologisches Korrespondenzblatt 26, 1996, 33-42.

1998a: M. Street, A Preboreal lithic assemblage from the Lower Rhineland site of Bedburg-Königshoven. In: N. Ashton / F. Healy / P. Pettitt (eds), Stone age archaeology. Essays in honour of John Wymer (Oxford 1998) 165-173.

1998b: M. Street, The archaeology of Pleistocene-Holocene transition in the northern Rhineland, Germany. Quaternary International 49/50, 1998, 45-67.

2002: M. Street, Ein Wiedersehen mit dem Hund von Bonn-Oberkassel. Festschrift für Günter Nobis. Bonner Zoologische Beiträge 50, 2002, 269-290.

Street/Baales/Weninger 1994: M. Street / M. Baales / B. Weninger, Absolute Chronologie des späten Paläolithikums und Frühmesolithikums im nördlichen Rheinland. Archäologisches Korrespondenzblatt 24, 1994, 1-28.

Street/Barton/Terberger 2009: M. Street / N. Barton / Th. Terberger (eds), Humans, Environment and Chronology of the Late Glacial of the North European Plain. Proceedings of Workshop 14, 15th U.I.S.P.P Congress, Lisbon, September 2006 (Mainz 2009).

Street/Jöris/Turner 2012: M. Street / O. Jöris / E. Turner, Magdalenian settlement in the German Rhineland – An update. Quaternary International 272-273, 2012, 231-250.

Street/Terberger 1999: M. Street / Th. Terberger, The last Pleniglacial and the human settlement of Central Europe: new information from the Rhineland site of Wiesbaden-Igstadt. Antiquity 73, 1999, 259-272.

2004: M. Street / Th. Terberger, The radiocarbon chronology of the German Upper Palaeolithic: Fifteen years of cooperation with ORAU. In: Higham/Bronk Ramsey/Owen 2004, 281-302.

Street/Terberger/Orschiedt 2006: M. Street / Th. Terberger / J. Orschiedt, A critical review of the German Paleolithic hominin record. Journal of Human Evolution 51, 2006, 551-579.

Street/Turner 2013: M. Street / E. Turner, The faunal remains from Gönnersdorf. Monographien des RGZM 104 (Mainz 2013).

Street et al. 2006: M. Street / F. Gelhausen / S. Grimm / F. Moseler / L. Niven / M. Sensburg / E. Turner / S. Wenzel / O. Jöris, L'occupation du bassin de Neuwied (Rhénanie centrale, Allemagne) par les Magdaléniens et les groupes à Federmesser (aziliens). Bulletin de la Société préhistorique française 103, 2006, 753-780.

Streif 2004: H. Streif, Sedimentary record of Pleistocene and Holocene marine inundations along the North Sea coast of Lower Saxony, Germany. Quaternary International 112, 2004, 3-28.

Stuart 2005: A. J. Stuart, The extinction of woolly mammoth (*Mammuthus primigenius*) and straight-tusked elephant (*Palaeoloxodon antiquus*) in Europe. Quaternary International 126-128, 2005, 171-177.

Stuart et al. 2002: A. J. Stuart / L. D. Sulerzhitsky / L. A. Orlova / Y. V. Kuzmin / A. M. Lister, The latest woolly mammoths (*Mammuthus primigenius* Blumenbach) in Europe and Asia: a review of the current evidence. Quaternary Science Reviews 21, 2002, 1559-1569.

2004: A. J. Stuart / P. A. Kosintsev / T. F. G. Higham / A. M. Lister, Pleistocene to Holocene extinction dynamics in giant deer and woolly mammoth. Nature 431, 2004, 684-689.

Stuiver 1982: M. Stuiver, A high-precision calibration of the AD radiocarbon time scale. Radiocarbon 24, 1982, 1-26.

Stuiver/Grootes 2000: M. Stuiver / P. M. Grootes, GISP2 oxygen isotope ratios. Quaternary Research 53, 2000, 277-284.

Stuiver/Kra 1986: M. Stuiver / R. S. Kra, Notice to readers and contributers. Radiocarbon 28, 1986, I.

Stuiver/Polach 1977: M. Stuiver / H. A. Polach, Reporting of ^{14}C data. Radiocarbon 19, 1977, 355-363.

Stutz 1997: A. J. Stutz, Seasonality of Magdalenian cave occupations in the Mosan Basin: Cementum increment data from Bois Laiterie, Chaleux, and the Trou da Somme. In: Otte/Straus 1997, 197-204.

Surovell/Brantingham 2007: T. A. Surovell / P. J. Brantingham, A note on the use of temporal frequency distributions in studies of prehistoric demography. Journal of Archaeological Science 34, 2007, 1868-1877.

Sutton/Hodson 2005: R. T. Sutton / D. L. R. Hodson, Atlantic ocean forcing of North American and European summer climate. Science 309, 2005, 115-118.

Svendsen et al. 2004: J. I. Svendsen / H. Alexanderson / V. I. Astakhov / I. Demidov / J. A. Dowdeswell / S. Funder / V. Gataullin / M. Henriksen / C. Hjort / M. Houmark-Nielsen / H. W. Hubberten / Ó. Ingólfsson / M. Jakobsson / K. H. Kjær / E. Larsen / H. Lokrantz / J. P. Lunkka / A. Lyså / J. Mangerud / A. Matiouchkov / A. Murray / P. Möller / F. Niessen / O. Nikolskaya / L. Polyak / M. Saarnisto / C. Siegert / M. J. Siegert / R. F. Spielhagen / R. Stein, Late Quaternary ice sheet history of northern Eurasia. Quaternary Science Reviews 23, 2004, 1229-1271.

Svensson et al. 2006: A. Svensson / K. K. Andersen / M. Bigler / H. B. Clausen / D. Dahl-Jensen / S. M. Davies / S. J. Johnsen / R. Muscheler / S. O. Rasmussen / R. Röthlisberger / J. P. Steffensen / B. M. Vinther, The Greenland Ice Core Chronology 2005, 15-42 ka. Part 2: Comparison to other records. Quaternary Science Reviews 25, 2006, 3258-3267.

2008: A. Svensson / K. K. Andersen / M. Bigler / H. B. Clausen / D. Dahl-Jensen / S. M. Davies / S. J. Johnsen / R. Muscheler / F. Parrenin / S. O. Rasmussen / R. Röthlisberger / I. Seierstad / J. P. Steffensen / B. M. Vinther, A 60 000 year Greenland stratigraphic ice core chronology. Climate of the Past 4, 2008, 47-57.

Szeroczyńska 1998: K. Szeroczyńska, Cladocera analysis in the Late-Glacial sediments of Lake Gościąż. In: Ralska-Jasiewiczowa et al. 1998, 148-158.

Szmidt et al. 2009: C. Szmidt / J.-M. Pétillon / P. Cattelain / C. Normand / C. Schwab, Premières dates radiocarbone pour le Magdalénien d'Isturitz (Pyrénées-Atlantiques). Bulletin de la Société préhistorique française 106, 2009, 588-592.

Taborin 1994: Y. Taborin (ed.), Environnements et habitats magdaléniens dans le centre du Bassin parisien. Documents d'archéologie française 43 (Paris 1994).

Tainter 2008: J. A. Tainter, The collapse of complex societies. New studies in archaeology (Cambridge 2008).

Tauber 1966: H. Tauber, Copenhagen Radiocarbon Dates VII. Radiocarbon 8, 1966, 213-234.

Taute 1963: W. Taute, Funde der spätpaläolithischen »Federmesser-Gruppen« aus dem Raum zwischen mittlerer Elbe und Weichsel. Berliner Jahrbuch für Vor- und Frühgeschichte 3, 1963, 62-107.

1968: W. Taute, Die Stielspitzen-Gruppen im nördlichen Europa. Ein Beitrag zur Kenntnis der späten Altsteinzeit. Fundamenta (Monographien zur Urgeschichte), Reihe A 5 (Köln, Graz 1968).

Taylor et al. 1993a: K. C. Taylor / C. U. Hammer / R. B. Alley / H. B. Clausen / D. Dahl-Jensen / A. J. Gow / N. S. Gundestrup / J. Kipfstuhl / J. C. Moore / E. D. Waddington, Electrical conductivity measurements from GISP2 and GRIP Greenland ice cores. Nature 366, 1993, 549-552.

1993b: K. C. Taylor / G. W. Lamorey / G. A. Doyle / R. B. Alley / P. M. Grootes / P. A. Mayewski / J. W. C. White / L. K. Barlow, The »flickering switch« of late Pleistocene climate change. Nature 361, 1993, 432-436.

Teller/Leverington 2004: J. T. Teller / D. W. Leverington, Glacial Lake Agassiz: A 5000 yr history of change and its relationship to the $\delta^{18}O$ record of Greenland. Geological Society of America Bulletin 116, 2004, 729-742.

Teller/Leverington/Mann 2002: J. T. Teller / D. W. Leverington / J. D. Mann, Freshwater outbursts to the oceans from glacial Lake Agassiz and their role in climate change during the last deglaciation. Quaternary Science Reviews 21, 2002, 879-887.

Terberger 1993: K. Terberger, Das Lahntal-Paläolithikum. Materialien zur Vor- und Frühgeschichte von Hessen 11 (Wiesbaden 1993).

1996: Th. Terberger, Die »Riesenhirschfundstelle« von Endingen, Lkr. Nordvorpommern. Spätglaziale Besiedlungsspuren in Nordostdeutschland. Archäologisches Korrespondenzblatt 26, 1996, 13-32.

1997: Th. Terberger, Die Siedlungsbefunde des Magdalénien-Fundplatzes Gönnersdorf. Konzentrationen III und IV. Der Magdalénien-Fundplatz Gönnersdorf 6 (Stuttgart 1997).

Terberger/Eriksen 2004: Th. Terberger / B. V. Eriksen (eds), Hunters in a changing world. Environment and archaeology of the Pleistocene-Holocene transition (ca. 11 000-9000 B.C.) in Northern Central Europe. Workshop of the U.I.S.P.P.-Commission XXXII at Greifswald in September 2002 (Rahden/Westf. 2004).

Terberger/Street 2002: Th. Terberger / M. Street, Hiatus or continuity? New results for the question of Pleniglacial settlement in Central Europe. Antiquity 76, 2002, 691-698.

Terberger et al. 2009: Th. Terberger / G. Tromnau / M. Street / G.-C. Weniger, Die jungpaläolithische Fundstelle Aschenstein bei Freden an der Leine, Kr. Hildesheim (Niedersachsen). Quartär 56, 2009, 87-103.

Terberger et al. 2013: Th. Terberger / H. Floss / P. Heinzelmann / A. Kotula / J. Serangeli, Down the Rhine river ca. 16,000 years ago: new evidence from the site Dreieich-Götzenhain, Hesse. In: A. Pastoors / B. Auffermann (eds), Pleistocene foragers: Their culture and environment. Festschrift in honour of Gerd-Christian Weniger for his sixtieth birthday (Mettmann 2013) 101-115.

Thévenin 1990: A. Thévenin, Du Dryas III au début de l'Atlantique: pour une approche méthodologique des industries et des territoires das l'est de la France. Revue archéologique de l'Est 42, 1990, 177-212.

1995: A. Thévenin (ed.), Épipaléolithique et Mésolithique entre Seine et Rhin. Table ronde d'Ancerville 1989 (Paris 1995).

2000: A. Thévenin, Le gisement magdalénien à scalènes de la Guillotine à Chariez (Haute-Saône). Actes de la Table ronde de Metz les 23 et 24 novembre 1996. Bulletin de la Société préhistorique luxembourgeoise 19, 1997 (2000), 67-84.

2003: A. Thévenin, Les pointes à cran de la fin du Tardiglaciaire: Chronologie et contexte culturel. In: E. Ladier (ed.), Les pointes à cran dans les industries lithiques du Paléolithique supérieur récent. de l'oscillation de Lascaux à l'oscillation de Bölling. Actes de la table ronde de Montauban (Tarn-et-Garonne), 26-27 avril 2002 (Montauban 2003).

Thiébault 1994: S. Thiébault, Analyse anthracologique. In: Taborin 1994, 118-119.

Thissen/Krull/Weiner 1997: J. Thissen / H. P. Krull / J. Weiner, Eine Station des Creswellian im Rheinland? Der spätpaläolithische Oberflächenfundplatz Kleinenbroich. Bonner Jahrbücher 196, 1996 (1997), 373-396.

Thomas 1990: R. Thomas, Der spätpaläolithische Fundplatz Niederbieber/Stadt Neuwied. Fläche 7/30-11/34 [unpublished M.A. thesis, Köln 1990].

Thomsen et al. 2008: K. J. Thomsen / A. S. Murray / M. Jain / L. Bøtter-Jensen, Laboratory fading rates of various luminescence signals from feldspar-rich sediment extracts. Radiation Measurements 43, 2008, 1474-1486.

Thomson 1992: K. S. Thomson, Macroevolution: The morphological problem. American Zoologist 32, 1992, 106-112.

Thordarson/Self 2003: T. Thordarson / S. Self, Atmospheric and environmental effects of the 1783-1784 Laki eruption: A review and reassessment. Journal of Geophysical Research – Atmospheres 108, 2003, 4011.

Thornalley et al. 2011: D. J. R. Thornalley / S. Barker / W. S. Broecker / H. Elderfield / I. N. McCave, The deglacial evolution of North Atlantic deep convection. Science 331, 2011, 202-205.

Tietz 2005: O. Tietz, The first certain discovery of an aurochs (*Bos primigenius* BOJANUS 1827) from the Weichselian Lateglacial in Germany. Zeitschrift für Geologische Wissenschaften, Berlin 33, 2005, 273-282.

Timmreck/Graf 2006: C. Timmreck / H.-F. Graf, The initial dispersal and radiative forcing of a Northern Hemisphere mid-latitude super volcano: a model study. Atmospheric Chemistry and Physics 6, 2006, 35-49.

Tinnes 1994: J. Tinnes, Die Geweih-, Elfenbein- und Knochenartefakte der Magdalénienfundplätze Gönnersdorf und Andernach [unpublished dissertation, Köln 1994].

2002: J. Tinnes, Ein Schlaginstrument aus Rengeweih. In: Floss/Terberger 2002, 185-187.

Tixier 1968: J. Tixier, Informations archéologiques. Circonscription de Lorraine. Gallia Préhistoire 11, 1968, 343-347.

1973: J. Tixier, Circonscription de Lorraine. Gallia Préhistoire 16, 1973, 439-461.

Tolksdorf et al. 2013: J. F. Tolksdorf / F. Turner / K. Kaiser / E. Eckmeier / M. Stahlschmidt / R. A. Housley / K. Breest / S. Veil, Multiproxy analyses of stratigraphy and palaeoenvironment of the Late Palaeolithic Grabow floodplain site, Northern Germany. Geoarchaeology: An International Journal 28, 2013, 50-65.

Torroni et al. 1998: A. Torroni / H.-J. Bandelt / L. D'Urbano / P. Lahermo / P. Moral / D. Sellitto / C. Rengo / P. Forster / M.-L. Savontaus / B. Bonné-Tamir / R. Scozzari, mtDNA analysis reveals a major Late Paleolithic population expansion from southwestern to northeastern Europe. American Journal of Human Genetics 62, 1998, 1137-1152.

Toucanne et al. 2009: S. Toucanne / S. Zaragosi / J. F. Bourillet / P. L. Gibbard / F. Eynaud / J. Giraudeau / J. L. Turon / M. Cremer / E. Cortijo / P. Martinez / L. Rossignol, A 1.2 Ma record of glaciation and fluvial discharge from the West European Atlantic margin. Quaternary Science Reviews 28, 2009, 2974-2981.

2010: S. Toucanne / S. Zaragosi / J.-F. Bourillet / V. Marieu / M. Cremer / M. Kageyama / B. van Vliet-Lanoë / F. Eynaud / J.-L. Turon / P. L. Gibbard, The first estimation of Fleuve Manche palaeoriver discharge during the last deglaciation: Evidence for Fennoscandian ice sheet meltwater flow in the English Channel ca 20-18 ka ago. Earth and Planetary Science Letters 290, 2010, 459-473.

Toussaint/Becker/Lacroix 1998: M. Toussaint / A. Becker / P. Lacroix, Recherches 1997-1998 aux grottes de Goyet, à Gesves, province de Namur. Notae Praehistoricae 18, 1998, 33-44.

Toussaint et al. 1993: M. Toussaint / A. Becker / M. Burhenne / J. M. Cordy / E. Gilot / J. Heim / E. Juvigné, Le Trou Jadot à Comblain-au-Pont (Province de Liège, Belgique). Paléoécologie et archéologie d'un site du Paléolithique supérieur récent. Etudes et recherches archéologiques de l'Université de Liège 58 (Liège 1993).

Tromnau 1992: G. Tromnau, Anmerkungen zur Rengeweih-Harpune von Meiendorf. In: E.-B. Krause / B. Mecke (eds), Ur-Geschichte im Ruhrgebiet. Festschrift Arno Heinrich (Gelsenkirchen 1992) 79-83.

Trumbore 2000: S. E. Trumbore, Radiocarbon geochronology. In: J. S. Noller / J. M. Sowers / W. R. Lettis (eds), Quaternary geochronology: methods and application (Washington, D.C. 2000) 41-60.

Turk 2012: J. Turk, Last interglacial-glacial cycle and an attempt to reconstruct Würmian paleohumidity of the European Alps. Journal of Mountain Science 9, 2012, 322-330.

Turner/Harris/Highwood 2012: A. G. Turner / B. M. Harris / E. J. Highwood, Monsoon collapse following super-volcanic eruption? EGU General Assembly Conference Abstracts 14, 2012, 8256.

Turner et al. 2013: F. Turner / J. F. Tolksdorf / F. Viehberg / A. Schwalb / K. Kaiser / F. Bittmann / U. von Bramann / R. Pott / U. Staesche / K. Breest / S. Veil, Lateglacial/early Holocene fluvial reactions of the Jeetzel river (Elbe valley, northern Germany) to abrupt climatic and environmental changes. Quaternary Science Reviews 60, 2013, 91-109.

Turner-Walker 2008: G. Turner-Walker, The chemical and microbial degradation of bones and teeth. In: R. Pinhasi / S. Mays (eds), Advances in human palaeopathology (Chichester/England, Hoboken/NJ 2008) 3-29.

Turney et al. 2000: C. S. M. Turney / G. R. Coope / D. D. Harkness / J. J. Lowe / M. J. C. Walker, Implications for the dating of Wisconsinan (Weichselian) Late-Glacial events of systematic radiocarbon age differences between terrestrial plant macrofossils from a site in SW Ireland. Quaternary Research 53, 2000, 114-121.

Turney et al. 2006: C. S. M. Turney / K. van den Burg / S. Wastegård / S. M. Davies / N. J. Whitehouse / J. R. Pilcher / C. Callaghan, North European last glacial-interglacial transition (LGIT; 15-9 ka) tephrochronology: extended limits and new events. Journal of Quaternary Science 21, 2006, 335-345.

Tweddle/Edwards 2010: J. C. Tweddle / K. J. Edwards, Pollen preservation zones as an interpretative tool in Holocene palynology. Review of Palaeobotany and Palynology 161, 2010, 59-76.

Uehara et al. 2006: K. Uehara / J. D. Scourse / K. J. Horsburgh / K. Lambeck / A. P. Purcell, Tidal evolution of the northwest European shelf seas from the Last Glacial Maximum to the present. Journal of Geophysical Research – Oceans 111, 2006, C09025.

UNEP/DEWA~Europe 2004: UNEP/DEWA~Europe (eds), Freshwater in Europe – Facts, Figures and Maps. An environmental atlas (Geneva 2004).

Urdea/Reuther 2009: P. Urdea / A. U. Reuther, Some new data concerning the Quaternary glaciation in the Romanian Carpathians. Geographica Pannonica 13, 2009, 41-52.

Usinger 1981: H. Usinger, Zur spät- und frühen postglazialen Vegetationsgeschichte der Schleswig-Holsteinischen Geest nach einem Pollen- und Pollendichtediagramm aus dem Esinger Moor. Pollen et Spores 23, 1981, 389-432.

1985: H. Usinger, Pollenstratigraphische, vegetations- und klimageschichtliche Gliederung des »Bölling-Alleröd-Komplexes« in Schleswig-Holstein und ihre Bedeutung für die Spätglazial-Stratigraphie in benachbarten Gebieten. Flora 177, 1985, 1-43.

1998: H. Usinger, Pollenanalytische Datierung spätpaläolithischer Fundschichten bei Ahrenshöft, Kr. Nordfriesland. Archäologische Nachrichten aus Schleswig-Holstein 8, 1998, 50-73.

Valentin 1995: B. Valentin, Les groupes humains et leurs traditions au Tardiglaciaire dans le Bassin Parisien. Apports de la technologie lithique comparée [unpublished dissertation, Paris 1995].

2000: B. Valentin, L'usage des percuteurs en pierre tendre pour le débitage des lames. Circonstances de quelques innovations au cours du Tardiglaciaire dans le Bassin parisien. In: G. Pion (ed.), Le Paléolithique supérieur récent: nouvelles données sur le peuplement et l'environnement. Actes du table ronde de Chambéry, Mars 1999 (Paris 2000) 253-260.

2008a: B. Valentin, Jalons pour une paléohistoire des derniers chasseurs (XIVe-VIe millénaire avant J.-C.). Cahiers Archéologiques de Paris 1 (Paris 2008).

2008b: B. Valentin, Productions lithiques magdaléniennes et aziliennes dans le Bassin parisien: disparition d'une économie programmée. The Arkeotek Journal 2, 2008, 1-54.

2009: B. Valentin (ed.), Paléolithique final et Mésolithique dans le Bassin parisien et ses marges. Habitats, sociétés et environnements. Rapport d'activités pour 2009 (Nanterre 2009).

Valentin/Bodu/Christensen 2000: B. Valentin / P. Bodu / M. Christensen (eds), L'Europe centrale et septentrionale au Tardiglaciaire. Actes de la Table-ronde internationale de Nemours, 14-16 mai 1997 (Nemours 2000).

Valentin/Bodu/Julien 2004: B. Valentin / P. Bodu / M. Julien (eds), Habitats et peuplements tardiglaciaires du Bassin parisien. Rapport d'activité pour 2004 (Nanterre 2004).

Valentin/Fosse/Billard 2004: B. Valentin / G. Fosse / C. Billard, Aspects et rythmes de l'azilianisation dans le Bassin parisien. Caractérisation de l'industrie lithique recueille au Cornet (locus 33) à Ambenay, Eure. Gallia Préhistoire 46, 2004, 171-209.

Valentin/Julien/Bodu 2002: B. Valentin / M. Julien / P. Bodu, Lithic raw material economy in the Late Glacial of the Paris Basin: Case studies from Magdalenian, Federmesser and Long Blade technology hunter-gatherers. In: Fisher/Eriksen 2002, 117-132.

Valentin/Pigeot 2000: B. Valentin / N. Pigeot, Éléments pour une chronologie des occupations magdaléniennes dans le Bassin parisien. In: Valentin/Bodu/Christensen 2000, 129-138.

Valentin et al. 1999a: B. Valentin / P. Bodu / S. de Beaune / J.-L. Rieu / A. Bridault, Le terroir et ses ressources. In: Julien/Rieu 1999, 42-47.

1999b: B. Valentin / P. Bodu / A. Hantaï / M. Philippe, L'industrie lithique. In: Julien/Rieu 1999, 66-94.

Valladas 1994: H. Valladas, Chronologie des sites du Magdalénien final du Bassin parisien. In: Taborin 1994, 65-68.

van Asch et al. 2013: N. van Asch / O. Heiri / S. J. P. Bohncke / W. Z. Hoek, Climatic and environmental changes during the Weichselian Lateglacial Interstadial in the Weerterbos region, the Netherlands. Boreas 42, 2013, 123-139.

van der Hammen 1957: T. van der Hammen, The stratigraphy of the Late-Glacial. Geologie en Mijnbouw 19, 1957, 250-254.

van der Hammen/van Geel 2008: T. van der Hammen / B. van Geel, Charcoal in soils of the Allerød-Younger Dryas transition were the result of natural fires and not necessarily the effect of an extra-terrestrial impact. Netherlands Journal of Geosciences (Geologie en Mijnbouw) 87, 2008, 359-361.

van der Plicht et al. 2004: J. van der Plicht / J. W. Beck / E. Bard / M. G. L. Baillie / P. G. Blackwell / C. E. Buck / M. Friedrich / T. P. Guilderson / K. A. Hughen / B. Kromer / F. G. McCormac / C. Bronk Ramsey / P. J. Reimer / R. W. Reimer / S. Remmele / D. A. Richards / J. R. Southon / M. Stuiver / C. E. Weyhenmeyer, NotCal04 – comparison/calibration ^{14}C records 26-50 cal kyr BP. Radiocarbon 46, 2004, 1225-1238.

van der Plicht/Hogg 2006: J. van der Plicht / A. Hogg, A note on reporting radiocarbon. Quaternary Geochronology 1, 2006, 237-240.

van Geel/Coope/van der Hammen 1989: B. van Geel / G. R. Coope / T. van der Hammen, Palaeoecology and stratigraphy of the Lateglacial type section at Usselo (The Netherlands). Review of Palaeobotany and Palynology 60, 1989, 25-129.

van Huissteden/Kasse 2001: J. van Huissteden / C. Kasse, Detection of rapid climate change in Last Glacial fluvial successions in The Netherlands. Global and Planetary Change 28, 2001, 319-339.

van Kolfschoten 1995: T. van Kolfschoten, On the application of fossil mammals to the reconstruction of the palaeoenvironment of northwestern Europe. Acta zoologica cracoviensia 38, 1995, 73-84.

van Leeuwaarden/Janssen 1987: W. van Leeuwaarden / C. R. Janssen, Differences between valley and upland vegetation development in eastern Noord-Brabant, The Netherlands, during the Late Glacial and Early Holocene. Review of Palaeobotany and Palynology 52, 1987, 179-204.

van Liedekerke/Jones/Panagos 2006: M. van Liedekerke / A. Jones / P. Panagos, 1km Raster version of the European Soil Database (v. 2.0). European Soil Bureau Network & European Commission. EUR 19945 EN. European Soil Bureau Network; European Commission (Bruxelles 2006; http://eusoils.jrc.it/ESDB_Archive/ESDBv2/index.htm).

van Neer 1997: W. van Neer, Fish remains from the Upper Magdalenian in the grotte du Bois Laiterie. In: Otte/Straus 1997, 205-213.

van Vliet-Lanoë et al. 1992: B. van Vliet-Lanoë / M. Helluin / J. Pellerin / B. Valadas, Soil erosion in Western Europe: from the last interglacial to the present. In: M. Bell / J. Boardman (eds), Past and present soil erosion. Archaeological and geographical perspectives (Oxford 1992) 104-114.

Vandenberghe 2001: J. Vandenberghe, Permafrost during the Pleistocene in North West and Central Europe. In: R. Paepe / V. Melnikov (eds), Permafrost response on economic development, environmental security and natural resources. Proceedings of the NATO Advanced Research Workshop on Permafrost Response on Economic Development, Environmental Security and Natural Resources; Novosibirsk, Russia, 12-16 November 1998 (Dordrecht 2001) 185-194.

2003: J. Vandenberghe, Climate forcing of fluvial system development: an evolution of ideas. Quaternary Science Reviews 22, 2003, 2053-2060.

Vandenberghe et al. 2012: J. Vandenberghe / H. Renssen / D. M. Roche / H. Goosse / A. A. Velichko / A. Gorbunov / G. Levavasseur, Eurasian permafrost instability constrained by reduced sea-ice cover. Quaternary Science Reviews 34, 2012, 16-23.

Vardy et al. 2010: M. E. Vardy / L. J. W. Pinson / J. M. Bull / J. K. Dix / T. J. Henstock / J. W. Davis / M. Gutowski, 3D seismic imaging of buried Younger Dryas mass movement flows: Lake Windermere, UK. Geomorphology 118, 2010, 176-187.

Vaughan 2002: P. C. Vaughan, Gebrauchsspurenanalyse der Feuersteinartefakte aus Konzentration II. In: Floss/Terberger 2002, 159-172.

Veil 1982: S. Veil, Der späteiszeitliche Fundplatz Andernach-Martinsberg. Germania 60, 1982, 391-424.

Veil/Terberger 2009: S. Veil / Th. Terberger, Kunst und Umwelt im Wandel. Was wurde aus der Eiszeitkunst? In: D. Planck / J. Heiligmann / N. J. Conard (eds), Eiszeit. Kunst und Kultur [exhibition catalogue Stuttgart] (Ostfildern 2009) 347-351.

Veil et al. 1991: S. Veil / M. A. Geyh / J. Merkt / U. Müller / U. Staesche, Eine Widerhakenspitze aus Lemförde am Dümmer, Landkreis Diepholz. Überlegungen zu ihrer Einordnung in das Spätglazial. Neue Ausgrabungen und Forschungen in Niedersachsen 19, 1991, 1-19.

2012: S. Veil / K. Breest / P. Grootes / M.-J. Nadeau / M. Hüls, A 14 000-year-old amber elk and the origins of northern European art. Antiquity 86, 2012, 660-673.

Velichko et al. 2002: A. A. Velichko / N. Catto / A. N. Drenova / V. A. Klimanov / K. V. Kremenetski / V. P. Nechaev, Climate changes in East Europe and Siberia at the Late glacial-holocene transition. Quaternary International 91, 2002, 75-99.

Vencl 1970a: S. Vencl, Das Spätpaläolithikum in Böhmen. Anthropologie (Brno) 8, 1970, 3-68.

1970b: S. Vencl, Die böhmische Fazies der Federmesser-Gruppen. In: K. Gripp / R. Schütrumpf / H. Schwabedissen (eds), Frühe Menschheit und Umwelt. Alfred Rust zum 65. Geburtstag (Köln 1970) 375-381 Taf. 119-124.

1995: S. Vencl, Hostim: Magdalenian in Bohemia. Pamétky Archeologické Supplementum 4 (Praha 1995).

Vermeersch/Lauwers/van Peer 1985: P. M. Vermeersch / R. Lauwers / P. van Peer, Un site magdalénien à Kanne (Limbourg). Archaeologica Belgica 1, 1985, 17-54.

Vermeersch et al. 1987: P. M. Vermeersch / N. Symens / P. Vyncker / G. Gijselings / R. Lauwers, Orp, site Magdalénien de plein air (comm. de Orp-Jauche). Archaeologica Belgica 3, 1987, 7-56.

Verpoorte 2004: A. Verpoorte, Eastern Central Europe during the Pleniglacial. Antiquity 78, 2004, 257-266.

2009: A. Verpoorte, Limiting factors on early modern human dispersals: The human biogeography of late Pleniglacial Europe. Quaternary International 201, 2009, 77-85.

Verpoorte/Šída 2009: A. Verpoorte / P. Šída, The Magdalenian colonisation of Bohemia (Czech Republic). Archäologisches Korrespondenzblatt 39, 2009, 325-332.

Verworn/Bonnet/Steinmann 1919: M. Verworn / R. Bonnet / G. Steinmann, Der diluviale Menschenfund von Obercassel bei Bonn (Wiesbaden 1919).

Vinther et al. 2006: B. M. Vinther / H. B. Clausen / S. J. Johnsen / S. O. Rasmussen / K. K. Andersen / S. L. Buchardt / D. Dahl-Jensen / I. K. Seierstad / M. L. Siggaard-Andersen / J. P. Steffensen / A. M. Svensson / J. Olsen / J. Heinemeier, A synchronized dating of three Greenland ice cores throughout the Holocene. Journal of Geophysical Research 111, 2006, D13102.

2009: B. M. Vinther / S. L. Buchardt / H. B. Clausen / D. Dahl-Jensen / S. J. Johnsen / D. A. Fisher / R. M. Koerner / D. Raynaud / V. Lipenkov / K. K. Andersen / T. Blunier / S. O. Rasmussen / J. P. Steffensen / A. M. Svensson, Holocene thinning of the Greenland ice sheet. Nature 461, 2009, 385-388.

Vollbrecht 2005: J. Vollbrecht, Spätpaläolithische Besiedlungsspuren aus Reichwalde. Reichwalde 1 (Dresden 2005).

Vos et al. 1997: H. Vos / A. Sanchez / B. Zolitschka / A. Brauer / J. F. W. Negendank, Solar activity variations recorded in varved sediments from the crater lake of Holzmaar – a maar lake in the Westeifel Volcanic Field, Germany. Surveys in Geophysics 18, 1997, 163-182.

Waldmann/Jöris/Baales 2001: G. Waldmann / O. Jöris / M. Baales, Nach der Flut. Ein spätallerødzeitlicher Rückenspitzen-Fundplatz bei Bad Breisig. Archäologisches Korrespondenzblatt 31, 2001, 173-184.

Walker 1995: M. J. C. Walker, Climatic changes on Europe during the last glacial/interglacial transition. Quaternary International 28, 1995, 63-76.

Walker/Coope/Lowe 1993: M. J. C. Walker / G. R. Coope / J. J. Lowe, The Devensian (Weichselian) Lateglacial palaeoenvironmental record from Gransmoor, East Yorkshire, England. Quaternary Science Reviews 12, 1993, 659-680.

Walker et al. 1999: M. J. C. Walker / S. Björck / J. J. Lowe / L. C. Cwynar / S. Johnsen / K. L. Knudsen / B. Wohlfarth / INTIMATE group, Isotopic »events« in the GRIP ice core: a stratotype for the Late Pleistocene. Quaternary Science Reviews 18, 1999, 1143-1150.

2003: M. J. C. Walker / G. R. Coope / C. Sheldrick / C. S. M. Turney / J. J. Lowe / S. P. E. Blockley / D. D. Harkness, Devensian Lateglacial environmental changes in Britain: a multi-proxy environmental record from Llanilid, South Wales, UK. Quaternary Science Reviews 22, 2003, 475-520.

2006: B. Walker / L. Gunderson / A. Kinzig / C. Folke / S. Carpenter / L. Schultz, A handful of heuristics and some propositions for understanding resilience in social-ecological systems. Ecology and Society 11/1, 2006, Art. 13, 1-15.

2009: M. Walker / S. Johnsen / S. O. Rasmussen / T. Popp / J.-P. Steffensen / P. Gibbard / W. Hoek / J. Lowe / J. Andrews / S. Björck / L. C. Cwynar / K. Hughen / P. Kershaw / B. Kromer / T. Litt / D. J. Lowe / T. Nakagawa / R. Newnham / J. Schwander, Formal definition and dating of the GSSP (Global Stratotype Section and Point) for the base of the Holocene using the Greenland NGRIP ice core, and selected auxiliary records. Journal of Quaternary Science 24, 2009, 3-17.

2012: B. H. Walker / S. R. Carpenter / J. Rockstrom / A.-S. Crépin / G. D. Peterson, Drivers, »slow« variables, »fast« variables, shocks, and resilience. Ecology and Society 17/3, 2012, Art. 30, 1-4.

Wallinga et al. 2004: J. Wallinga / T. E. Törnqvist / F. S. Busschers / H. J. T. Weerts, Allogenic forcing of the late Quaternary Rhine-Meuse fluvial record: the interplay of sea-level change, climate change and crustal movements. Basin Research 16, 2004, 535-547.

Walsh 1998: S. L. Walsh, Fossil datum and paleobiological event terms, paleontostratigraphy, chronostratigraphy, and the definition of land mammal »age« boundaries. Journal of Vertebrate Paleontology 18, 1998, 150-179.

Wang et al. 2001: Y. J. Wang / H. Cheng / R. L. Edwards / Z. S. An / J. Y. Wu / C. C. Shen / J. A. Dorale, A high-resolution absolute-dated Late Pleistocene monsoon record from Hulu Cave, China. Science 294, 2001, 2345-2348.

2005: P. Wang / S. Clemens / L. Beaufort / P. Braconnot / G. Ganssen / Z. Jian / P. Kershaw / M. Sarnthein, Evolution and variability of the Asian monsoon system: state of the art and outstanding issues. Quaternary Science Reviews 24, 2005, 595-629.

Ward/Larcombe 2008: I. Ward / P. Larcombe, Determining the preservation rating of submerged archaeology in the post-glacial southern North Sea: a first-order geomorphological approach. Environmental Archaeology 13, 2008, 59-83.

Ward/Larcombe/Lillie 2006: I. Ward / P. Larcombe / M. Lillie, The dating of Doggerland – post-glacial geochronology of the southern North Sea. Environmental Archaeology 11, 2006, 207-218.

Wastegård 2005: S. Wastegård, Late Quaternary tephrochronology of Sweden: a review. Quaternary International 130, 2005, 49-62.

Weaver et al. 2003: A. J. Weaver / O. A. Saenko / P. U. Clark / J. X. Mitrovica, Meltwater pulse 1A from Antarctica as a trigger of the Bølling-Allerød warm interval. Science 299, 2003, 1709-1713.

Weber 2006: M.-J. Weber, Typologische und technologische Aspekte des Fundplatzes Le Tureau des Gardes 7 (Seine-et-Marne, Frankreich). – ein Beitrag zur Erforschung des Magdalénien im Pariser Becken. Archäologisches Korrespondenzblatt 36, 2006, 159-178.

2012: M.-J. Weber, From technology to tradition – Re-evaluationg the Hamburgian-Magdalenian relationship. Untersuchungen und Materialien zur Steinzeit in Schleswig-Holstein und im Ostseeraum 5 (Neumünster 2012).

Weber/Grimm 2009: M.-J. Weber / S. B. Grimm, Dating the Hamburgian in the context of Lateglacial chronology. In: P. Crombé / M. van Strydonck / J. Sergant / M. Boudin / M. Bats (eds), Chronology and evolution within the Mesolithic of North-West Europe. Proceedings of an international meeting, Brussels, May 30th-June 1st 2007 (Newcastle 2009) 3-21.

Weber/Grimm/Baales 2011: M.-J. Weber / S. B. Grimm / M. Baales, Between warm and cold – Impact of Dryas III on human behavior in Central Europe. Quaternary International 242, 2011, 277-301.

Weber et al. 2010: M.-J. Weber / I. Clausen / R. A. Housley / C. E. Miller / F. Riede, New information on the Havelte Group site Ahrenshöft LA 58 D (Nordfriesland, Germany) – Preliminary results of the 2008 fieldwork. Quartär 57, 2010, 7-24.

Weiss/Bradley 2001: H. Weiss / R. S. Bradley, What drives societal collapse? Science 291, 2001, 609-610.

Welker 2004: W. Welker, Späteiszeitliche Jäger in Boppard. Zur Entdeckung eines einzigartigen Fundplatzes im oberen Mittelrheintal. Heimatkundlicher Arbeitskreis des Verkehrs- und Verschönerungs-Verein Boppard (eds), Rund um Boppard – Beiträge zur Geschichte der Stadt Boppard (Boppard 2004).

Weniger 1989: G.-C. Weniger, The Magdalenian in western Central Europe: settlement pattern and regionality. Journal of World Prehistory 3, 1989, 323-372.

2000: G.-C. Weniger, Magdalenian barbed points: Harpoons, spears and arrowheads. In: C. Bellier / P. Cattelain / M. Otte (eds), La chasse dans la préhistoire. Actes du colloque international de Treignes, 3-7 octobre 1990 (Bruxelles 2000) 79-87.

1986: B. Weninger, High-precision calibration of archaeological radiocarbon dates. In: C. Ambros / M. Žilková / M. Líšková (eds), Papers of the symposium held at the Institute of Archaeology of the Slovak Academy of Sciences Nové Vozokany October 28-32, 1985 (Nitra 1986) 11-53.

Weninger/Jöris 2004: B. Weninger / O. Jöris, Glacial radiocarbon age calibration: The CalPal program. In: Higham/Bronk Ramsey/ Owen 2004, 9-15.

2008: B. Weninger / O. Jöris, A ^{14}C age calibration curve for the last 60 ka: the Greenland-Hulu U/Th timescale and its impact on understanding the Middle to Upper Paleolithic transition in Western Eurasia. Journal of Human Evolution 55, 2008, 772-781.

Weninger/Jöris/Danzeglocke 2003: B. Weninger / O. Jöris / U. Danzeglocke, Climate Archaeology with Fortran. Fortran Source 19, 2003, 3.

2007: B. Weninger / O. Jöris / U. Danzeglocke, CalPal-2007. Cologne Radiocarbon Calibration & Palaeoclimate Research Package (Köln 2007; www.calpal.de: last accessed: 04.06.2007).

Weninger et al. 2008: B. Weninger / R. Schulting / M. Bradtmöller / L. Clare / M. Collard / K. Edinborough / J. Hilpert / O. Jöris / M. Niekus / E. J. Rohling / B. Wagner, The catastrophic final flooding of Doggerland by the Storegga Slide tsunami. Documenta Praehistorica 35, 2008, 1-24.

2009: B. Weninger / L. Clare / E. J. Rohling / O. Bar-Yosef / U. Böhner / M. Budja / M. Bundschuh / A. Feurdean / H. G. Gebel / O. Jöris / J. Linstädter / P. A. Mayewski / T. Mühlenbruch / A. Reigruber / G. Rollefson / D. Schyle / L. Thissen / H. Todorova / C. Zielhofer, The impact of rapid climate change on prehistoric societies during the Holocene in the Eastern Mediterranean. Documenta Praehistorica 902, 2009, 551-583.

2011: B. Weninger / K. Edinborough / L. Clare / O. Jöris, Concepts of probability in radiocarbon analysis. Documenta Praehistorica 38, 2011, 1-20.

Wennrich et al. 2005: V. Wennrich / B. Wagner / M. Melles / P. Morgenstern, Late Glacial and Holocene history of former Salziger See, Central Germany, and its climatic and environmental implications. International Journal of Earth Sciences (Geologische Rundschau) 94, 2005, 275-284.

Wenzel 2004: S. Wenzel, Ein verziertes Knochengerät spätaltsteinzeitlicher Hirschjäger aus Boppard. Archäologie in Rheinland-Pfalz 2003 (2004), 13-15.

2009: S. Wenzel, Behausungen im Späten Jungpaläolithikum und im Mesolithikum in Nord-, Mittel- und Westeuropa. Monographien des RGZM 81 (Mainz 2009).

Wenzel/Álvarez Fernández 2004: S. Wenzel / E. Álvarez Fernández, La espátula de Boppard (Boppard, Renania-Palatinado, Alemania) y sus paralelos en Europa a finales del Paleolítico superior y en el Mesolítico. Zephyrus 57, 2004, 137-151.

Werts/Jahren 2007: S. P. Werts / A. H. Jahren, Estimation of temperatures beneath archaeological campfires using carbon stable isotope composition of soil organic matter. Journal of Archaeological Science 34, 2007, 850-857.

Whallon 2006: R. Whallon, Social networks and information: non-»utilitarian« mobility among hunter-gatherers. Journal of Anthropological Archaeology 25, 2006, 259-270.

White 2008: A. A. White, A developmental perspective on technological change. World Archaeology 40, 2008, 597-608.

White 1993: R. White, Introduction. In: Leroi-Gourhan 1993, xiii-xxii.

Whittaker/Nogués-Bravo/Araújo 2007: R. J. Whittaker / D. Nogués-Bravo / M. B. Araújo, Geographical gradients of species richness: a test of the water-energy conjecture of Hawkins et al. (2003) using European data for five taxa. Global Ecology and Biogeography 16, 2007, 76-89.

Whittington/Fallick/Edwards 1996: G. Whittington / A. E. Fallick / K. J. Edwards, Stable oxygen isotope and pollen records from eastern Scotland and a consideration of Late-glacial and early Holocene climate change for Europe. Journal of Quaternary Science 11, 1996, 327-340.

Wick/Lemcke/Sturm 2003: L. Wick / G. Lemcke / M. Sturm, Evidence of Lateglacial and Holocene climatic change and human impact in eastern Anatolia: high-resolution pollen, charcoal, isotopic and geochemical records from the laminated sediments of Lake Van, Turkey. The Holocene 13, 2003, 665-675.

Wickham-Jones/Woodman 1998: C. R. Wickham-Jones / P. C. Woodman, Studies on the early settlement of Scotland and Ireland. Quaternary International 49/50, 1998, 13-20.

Widlok et al. 2012: T. Widlok / A. Aufgebauer / M. Bradtmöller / R. Dikau / T. Hoffmann / I. Kretschmer / K. Panagiotopoulos / A. Pastoors / R. Peters / F. Schäbitz / M. Schlummer / M. Solich / B. Wagner / G.-C. Weniger / A. Zimmermann, Towards a theoretical framework for analyzing integrated socio-environmental systems. Quaternary International 274, 2012, 259-272.

Wiessner 1983: P. Wiessner, Style and social information in Kalahari San projectile points. American Antiquity 48, 1983, 253-276.

Willis/van Andel 2004: K. J. Willis / T. H. van Andel, Trees or no trees? The environments of central and eastern Europe during the Last Glaciation. Quaternary Science Reviews 23, 2004, 2369-2387.

Winterhalder/Leslie 2002: B. Winterhalder / P. Leslie, Risk-sensitive fertility. The variance compensation hypothesis. Evolution and Human Behavior 23, 2002, 59-82.

Wintle/Lancastert/Edwards 1994: A. G. Wintle / N. Lancastert / S. R. Edwards, Infrared stimulated luminescence (IRSL) dating of late-Holocene aeolian sands in the Mojave Desert, California, USA. The Holocene 4, 1994, 74-78.

Wohlfarth et al. 1998: B. Wohlfarth / G. Skog / G. Possnert / B. Holmquist, Pitfalls in the AMS radiocarbon-dating of terrestrial macrofossils. Journal of Quaternary Science 13, 1998, 137-145.

2006: B. Wohlfarth / M. Blaauw / S. M. Davies / M. Andersson / S. Wastegård / A. Hormes / G. Possnert, Constraining the age of Lateglacial and early Holocene pollen zones and tephra horizons in southern Sweden with Bayesian probability methods. Journal of Quaternary Science 21, 2006, 321-334.

2008: B. Wohlfarth / D. Veres / L. Ampel / T. Lacourse / M. Blaauw / F. Preusser / V. Andrieu-Ponel / D. Kéravis / E. Lallier-Vergès / S. Björck / S. M. Davies / J.-L. de Beaulieu / J. Risberg / A. Hormes / H. U. Kasper / G. Possnert / M. Reille / N. Thouveny / A. Zander, Rapid ecosystem response to abrupt climate changes during the last glacial period in western Europe, 40-16 ka. Geology 36, 2008, 407-410.

Wojtal 2007: P. Wojtal, Zooarchaeological studies of the Late Pleistocene sites in Poland (Kraków 2007).

Wolff et al. 2010: E. W. Wolff / J. Chappellaz / T. Blunier / S. O. Rasmussen / A. Svensson, Millennial-scale variability during the last glacial: The ice core record. Quaternary Science Reviews 29, 2010, 2828-2838.

Wolstedt 1956: P. Wolstedt, Die Geschichte des Flußnetzes in Norddeutschland und angrenzenden Gebieten. Eiszeitalter und Gegenwart 7, 1956, 5-12.

Wörner/Schmincke 1984: G. Wörner / H. U. Schmincke, Mineralogical and Chemical Zonation of the Laacher See Tephra Sequence (East Eifel, W. Germany). Journal of Petrology 25, 1984, 805-835.

Wüller 1993: B. Wüller, Die chronologische Stellung des »contour découpé« aus dem Magdalénien-Grab von Oberkassel bei Bonn. Archäologische Informationen 16, 1993, 144-146.

1999: B. Wüller, Die Ganzkörperbestattungen des Magdalénien. Universitätsforschungen zur Prähistorischen Archäologie 57 (Bonn 1999).

Wurth et al. 2000: G. Wurth / S. Niggemann / N. Frank / A. Mangini / D. Richter, Jungquartäre Stalagmiten aus Höhlen des Gottesackergebietes (Allgäuer Alpen) als Archiv für Paläoumweltbedingungen. Karst und Höhle 2000, 183-197.

Zazzo/Saliège 2011: A. Zazzo / J. F. Saliège, Radiocarbon dating of biological apatites: A review. Palaeogeography, Palaeoclimatology, Palaeoecology 310, 2011, 52-61.

Zeder 2012: M. A. Zeder, The Broad Spectrum Revolution at 40: Resource diversity, intensification, and an alternative to optimal foraging explanations. Journal of Anthropological Archaeology 31, 2012, 241-264.

Zeeberg 1998: J. Zeeberg, The European sand belt in eastern Europe – and comparison of Late Glacial dune orientation with GCM simulation results. Boreas 27, 1998, 127-139.

Zemp/Hoelzle/Haeberli 2007: M. Zemp / M. Hoelzle / W. Haeberli, Distributed modelling of the regional climatic equilibrium line altitude of glaciers in the European Alps. Global and Planetary Change 56, 2007, 83-100.

Zhilin 1996: M. G. Zhilin, The western part of Russia in the Late Palaeolithic – Early Mesolithic. In: Larsson 1996b, 273-284.

Zielinski 2000: G. A. Zielinski, Use of paleo-records in determining variability within the volcanism – climate system. Quaternary Science Reviews 19, 2000, 417-438.

Zielinski et al. 1996: G. A. Zielinski / P. A. Mayewski / L. D. Meeker / S. Whitlow / M. S. Twickler, A 110,000-yr record of explosive volcanism from the GISP2 (Greenland) ice core. Quaternary Research 45, 1996, 109-118.

Zimmermann et al. 2004: A. Zimmermann / J. Richter / T. Frank / P. Wendt, Landschaftsarchäologie II: Überlegungen zu Prinzipien einer Landschaftsarchäologie. Berichte der Römisch-Germanischen Kommission 85, 2004, 37-95.

2009: A. Zimmermann / K. P. Wendt / T. Frank / J. Hilpert, Landscape Archaeology in Central Europe. Proceedings of the Prehistoric Society 75, 2009, 1-53.

Ziswiler 1965: V. Ziswiler, Bedrohte und ausgerottete Tiere. Eine Biologie des Aussterbens und des Überlebens. Verständliche Wissenschaft 86 (Berlin, Heidelberg 1965).

Zolitschka 1998: B. Zolitschka, Paläoklimatische Bedeutung laminierter Sedimente. Holzmaar (Eifel, Deutschland), Lake C2 (Nordwest-Territorien, Kanada) und Lago Grande di Monticchio (Basilicata, Italien). Relief, Boden, Paläoklima 13 (Berlin 1998).

Zolitschka/Negendank/Lottermoser 1995: B. Zolitschka / J. F. W. Negendank / B. G. Lottermoser, Sedimentological proof and dating of the Early Holocene volcanic eruption of Ulmener Maar (Vulkaneifel, Germany). Geologische Rundschau 84, 1995, 213-219.

Zolitschka et al. 2000: B. Zolitschka / A. Brauer / J. F. W. Negendank / H. Stockhausen / A. Lang, Annually dated late Weichselian continental paleoclimate record from the Eifel, Germany. Geology 28, 2000, 783-786.

PLATES 1-14

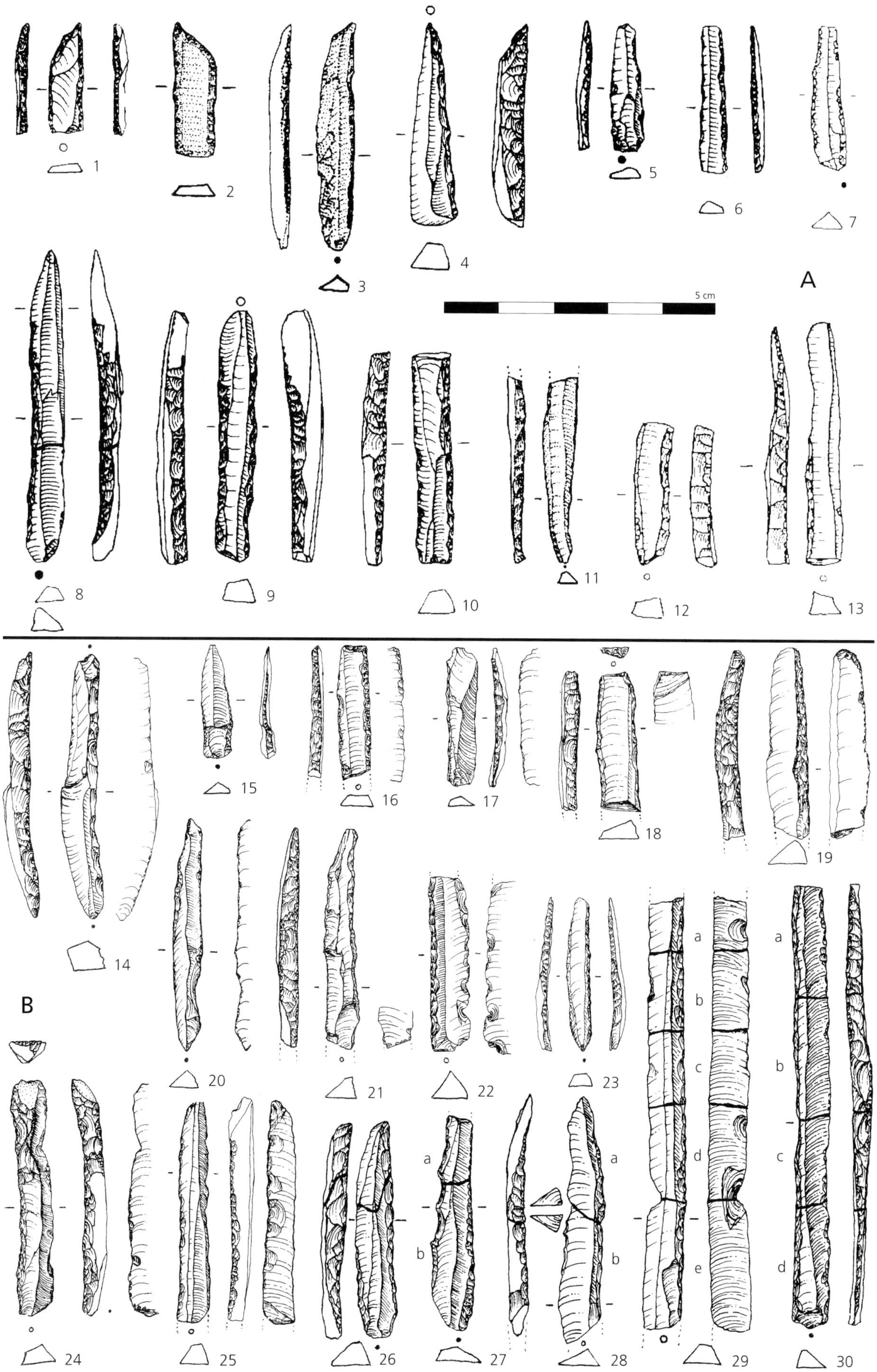

LMP. **A** Andernach, lower horizon: **1-6. 8-11** Floss/Terberger 2002, Abb. 120. 124; **7. 12-13** Holzkämper 2006, Taf. 16. – **B** Gönnersdorf, concentrations I-IV: **14-30** Franken/Veil 1983, Taf. 22. 33.

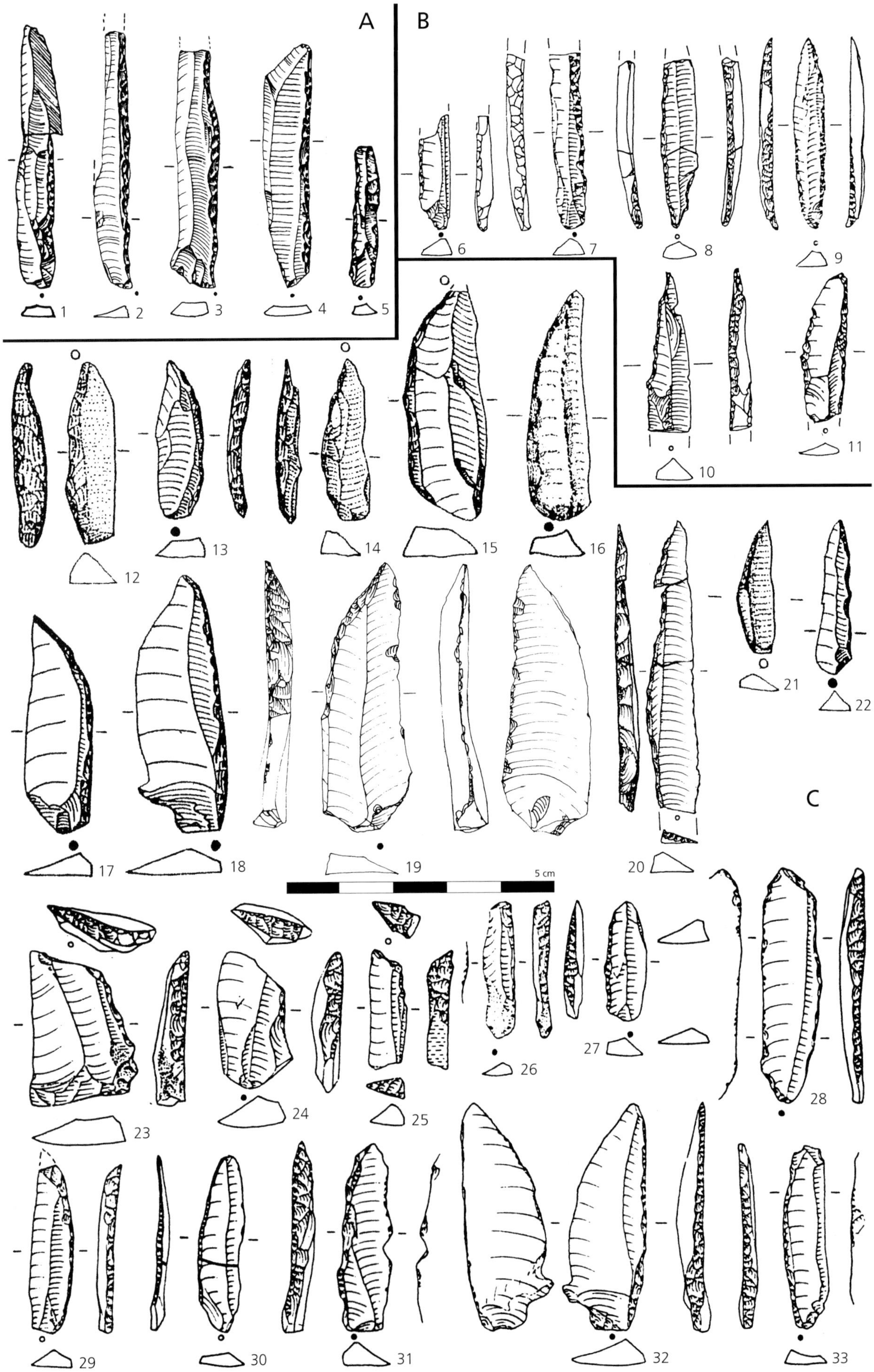

LMP. **A** Wildweiberlei: **1-5** Terberger 1993, Taf. 72. – **B** Gönnersdorf, south-western area: **6-11** Buschkämper 1993, Abb. 6 Taf. 3. – **C** Andernach, upper horizon: **12-18. 21-22** Bolus 1984, Taf. 2; **19-20** Veil 1982, Abb. 3; **23-33** Kegler 1999, Taf. 1-2. 4 (**12-22** Andernach 2. – **23-33** Andernach 3).

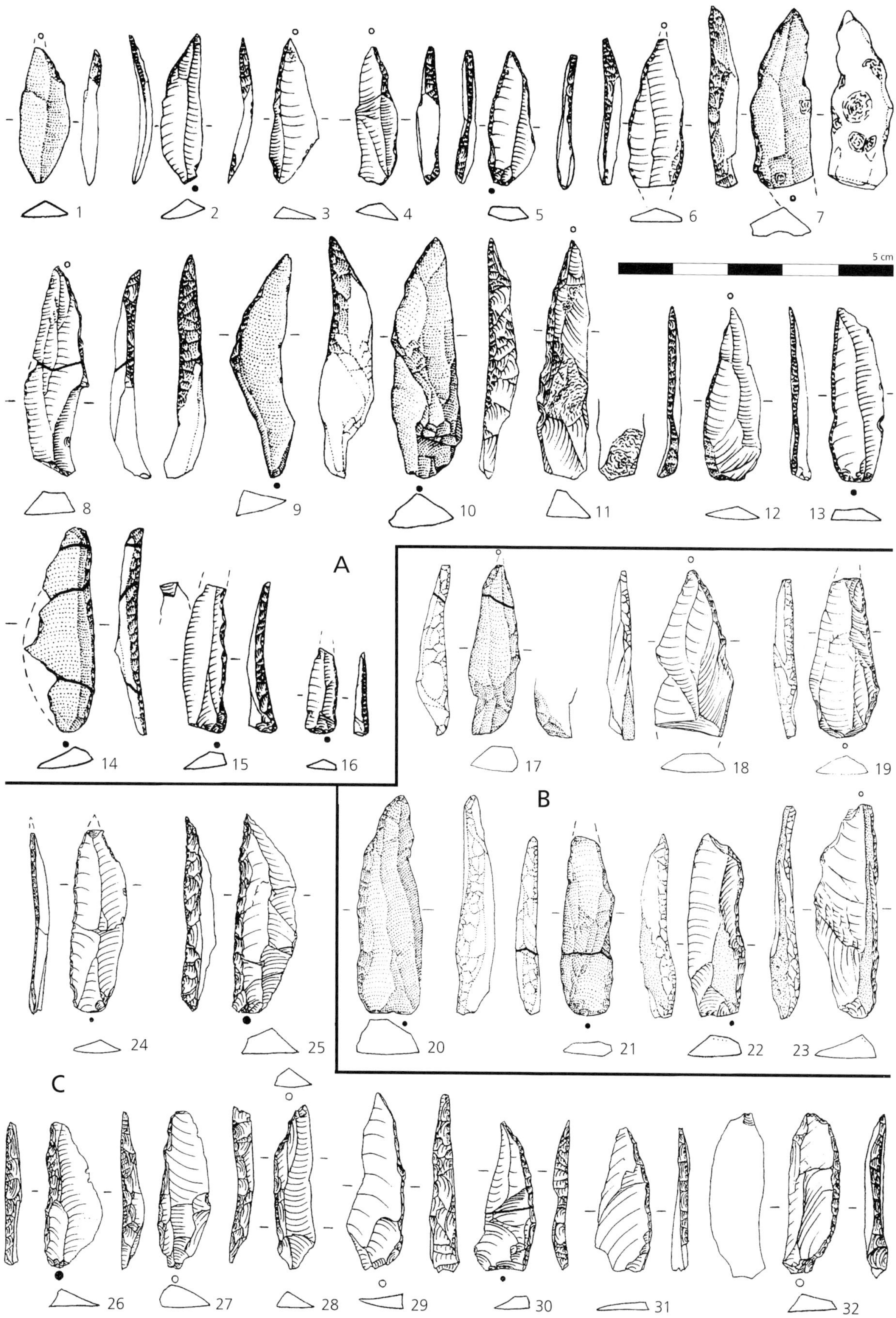

LMP. **A** Kettig: **1-16** Baales 2002, Abb. 80-81. 84. – **B** Urbar: **17-23** Baales/Mewis/Street 1998, Abb. 15. – **C** Niederbieber 1: **24-32** Bolus 1992, Abb. 35.

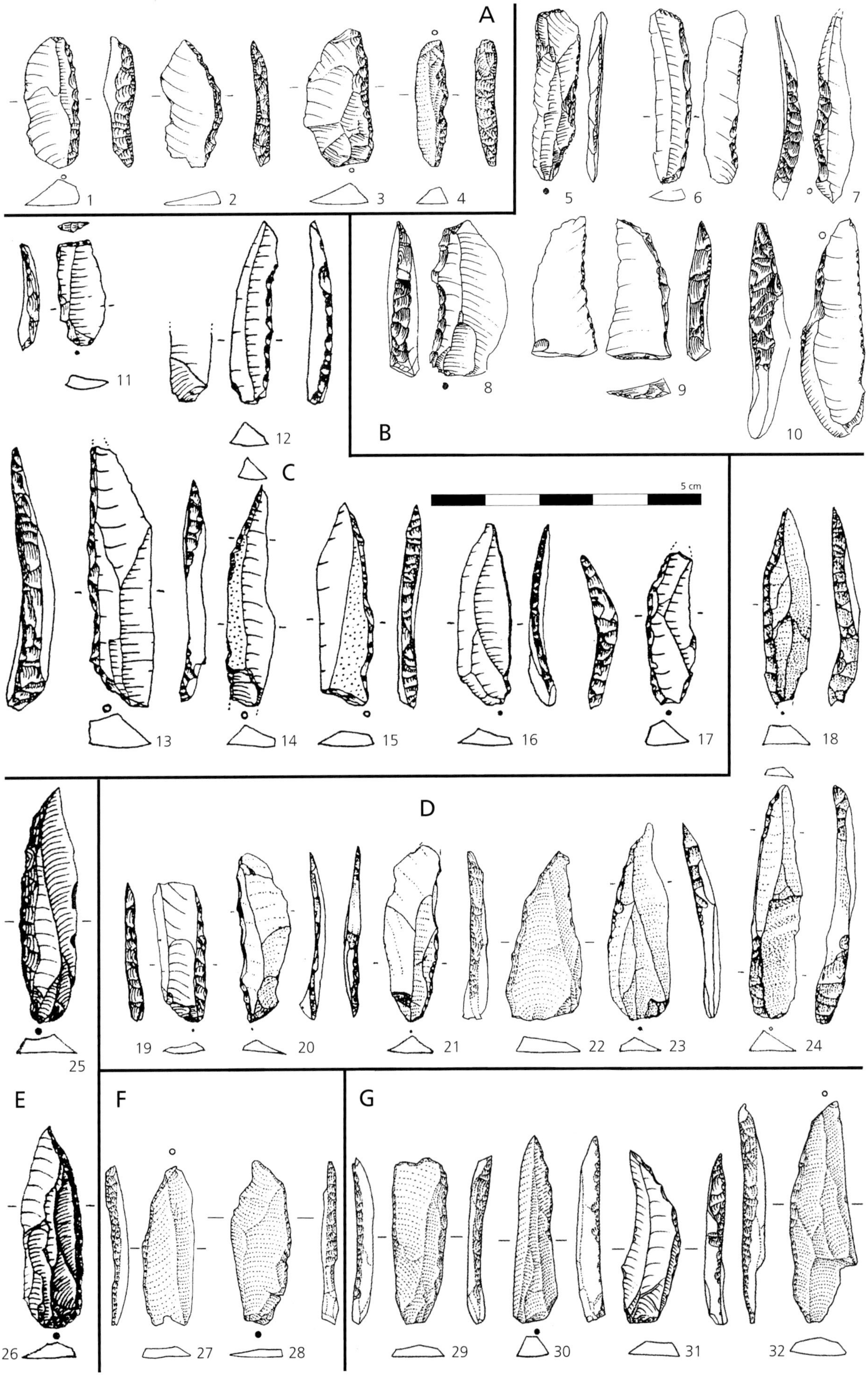

LMP. **A** Niederbieber 2: **1-4** Gelhausen 2011a, Abb. 18. – **B** Niederbieber 4: **5-10** Bolus 1992, Abb. 91. – **C** Niederbieber 5: **11-15** Husmann 1989, Abb. 3; **16-17** Husmann 1988, Taf. 2-3. – **D** Niederbieber 6: **18-21. 23-24** Thomas 1990, Taf. 1-2; **22** Gelhausen 2011b, Taf. 14. – **E** Niederbieber 7: **25-26** Freericks 1989, Taf. 1. – **F** Niederbieber 8: **27-28** Gelhausen 2011b, Taf. 8. – **G** Niederbieber 10: **29-32** Gelhausen 2011b, Taf. 13.

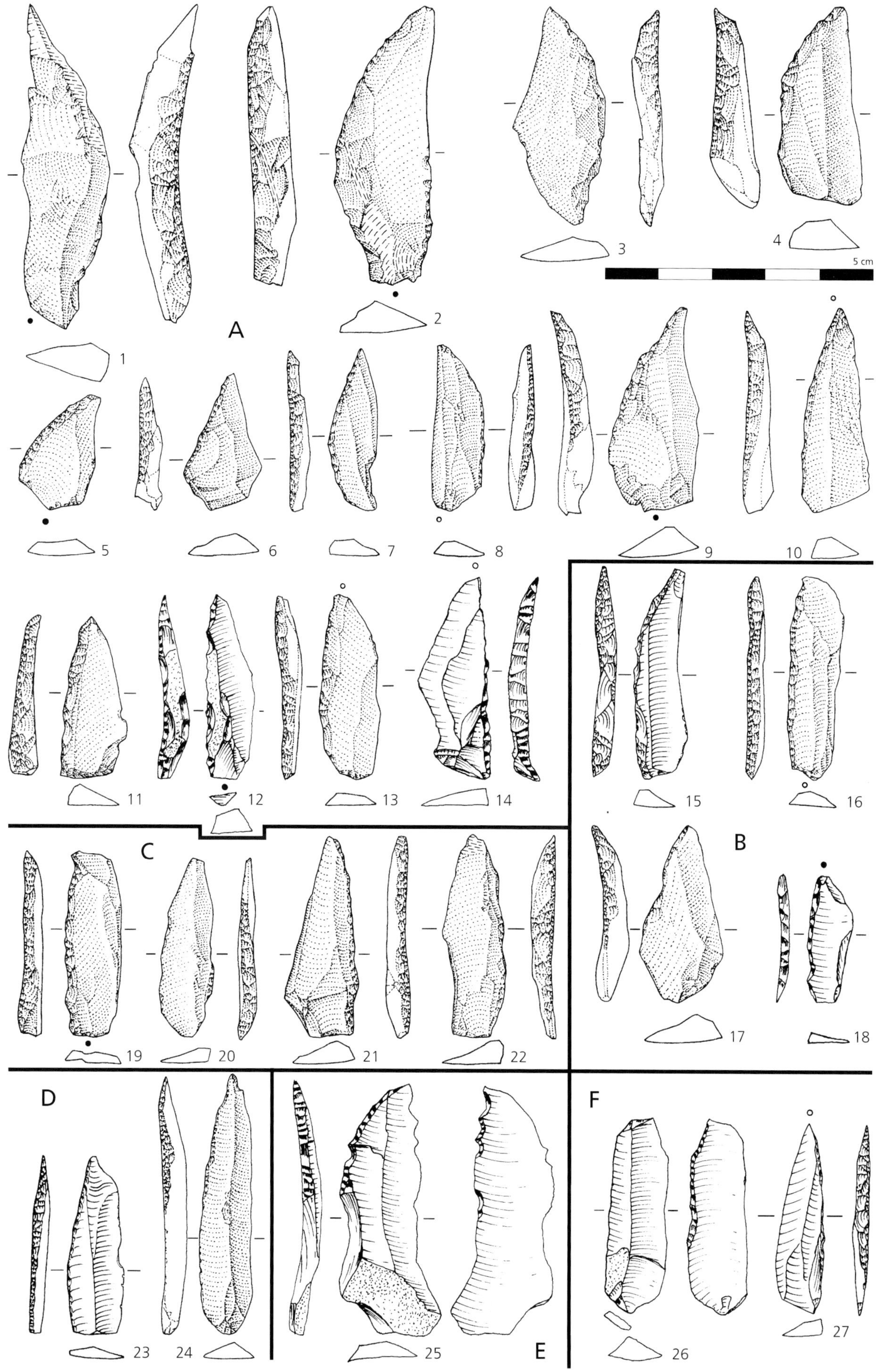

LMP. **A** Niederbieber 9: **1-14** Gelhausen 2011b, Taf. 9-10. – **B** Niederbieber 11: **15-18** Gelhausen 2011b, Taf. 15. – **C** Niederbieber 12: **19-22** Gelhausen 2011b, Taf. 16. – **D** Niederbieber 13: **23-24** Gelhausen 2011b, Taf. 17. – **E** Niederbieber 14: **25** Gelhausen 2011b, Taf. 19.– **F** Niederbieber 15: **26-27** Gelhausen 2011b, Taf. 20.

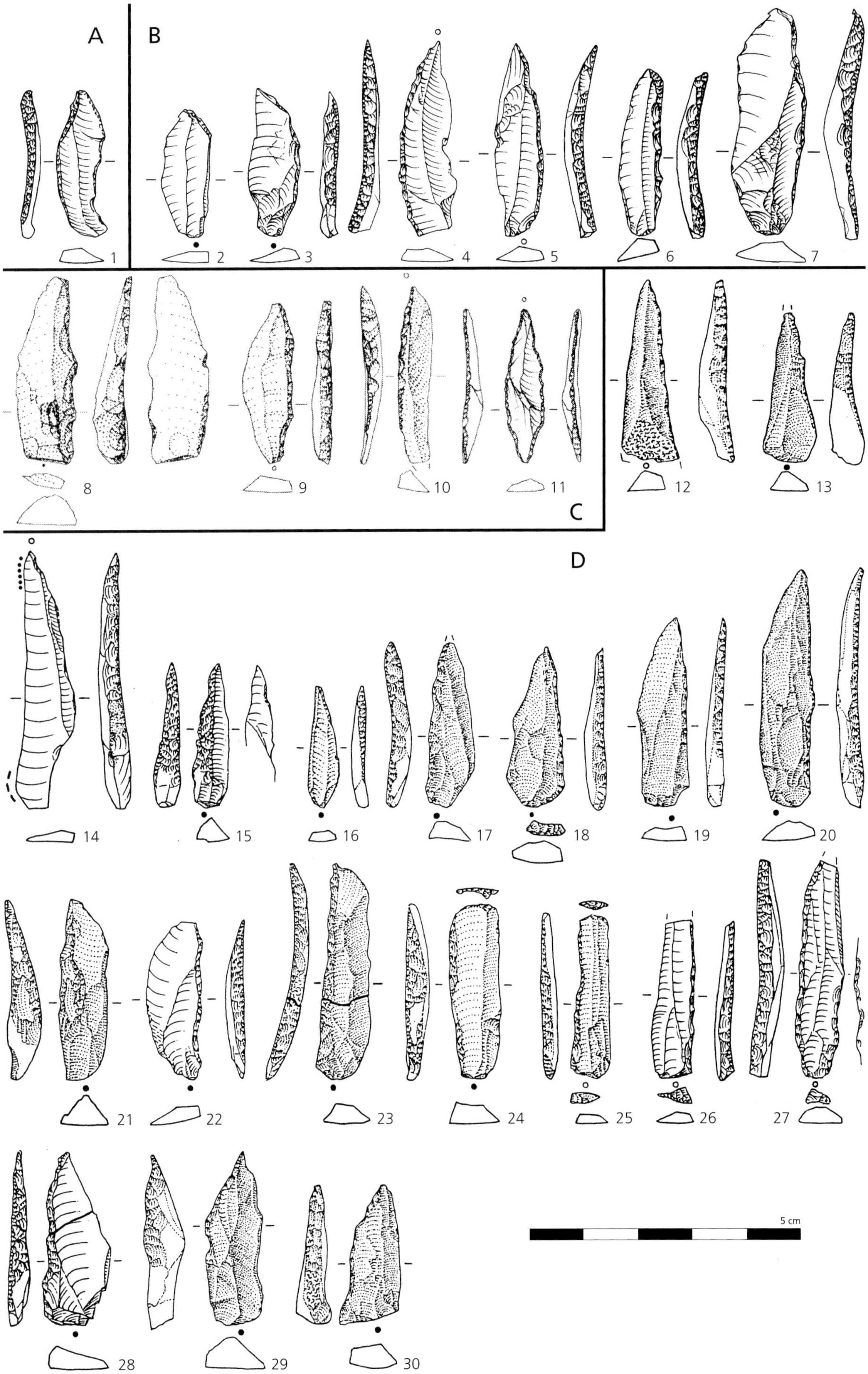

LMP. **A** Niederbieber 16: **1** Gelhausen 2011b, Taf. 21. – **B** Niederbieber 17: **2-7** Gelhausen 2011b, Taf. 22. – **C** Boppard: **8-11** Wenzel 2004, Abb. 1. – **D** Bad Breisig: **12-13. 17. 19. 21-23. 30** Grimm 2003, Taf. 9. 11-12; **14-16. 18. 20. 24-29** Grimm 2004, Abb. 9.

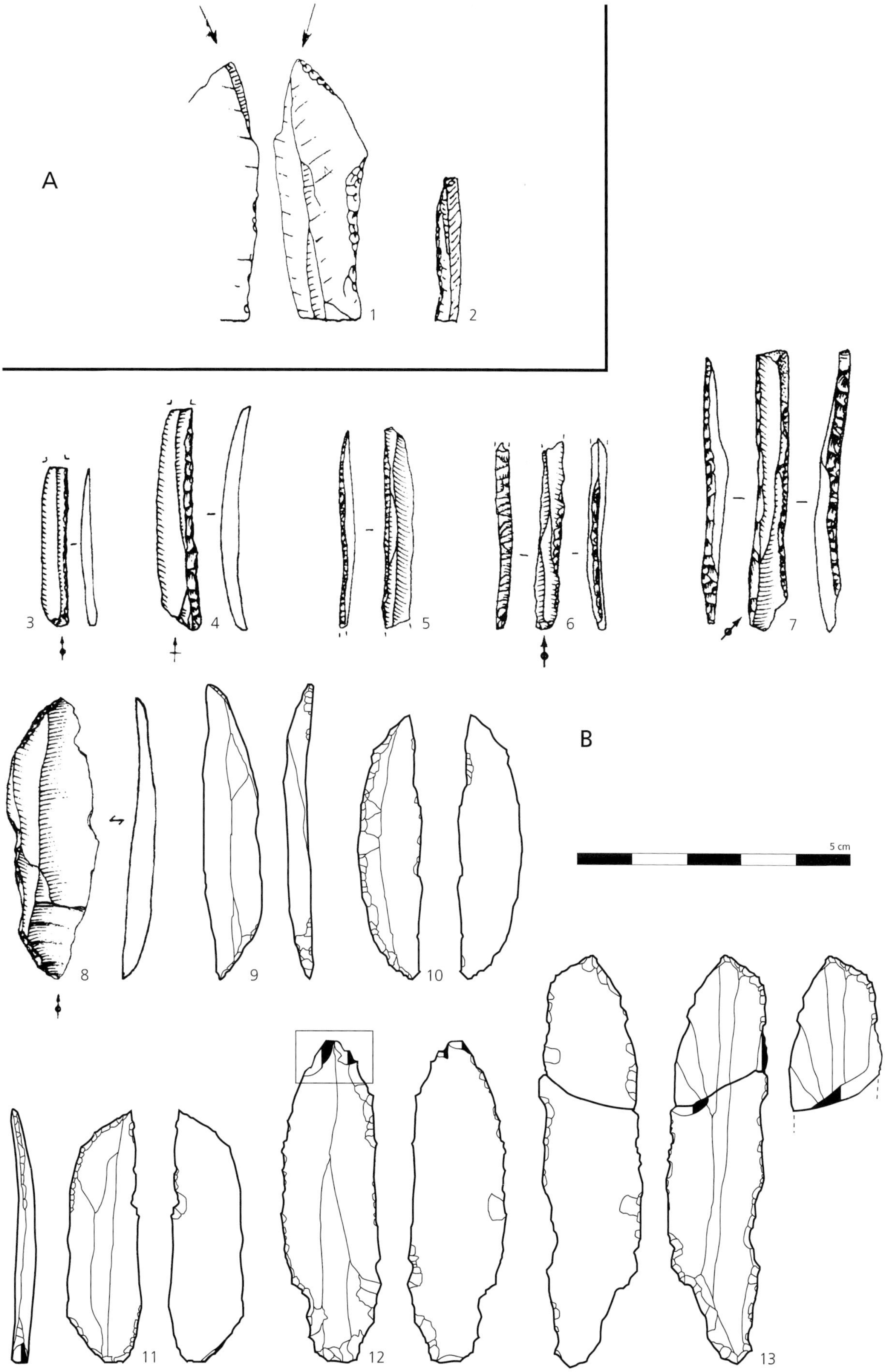

LMP. **A** Saint Mihiel: **1-2** Stocker et al. 2006, figs 6-7. – **B** Bois Laiterie: **3-8** Straus/Orphal 1997, figs 12. 17; **9-13** Sano 2009, figs 16. 18-19.

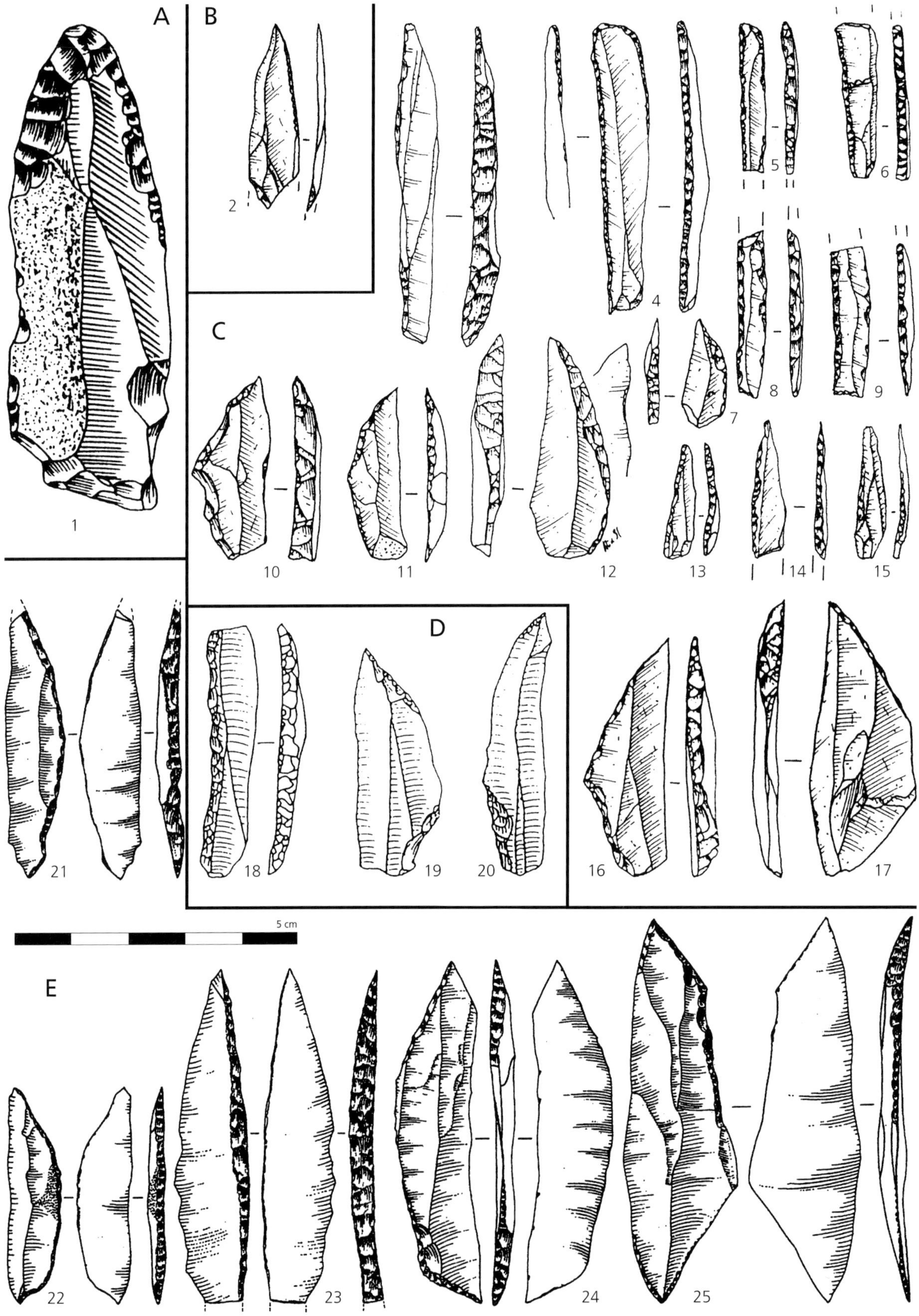

LMP. **A** Hallines: **1** Fagnart 1997, fig. 20. – **B** Le Grand Canton, lower horizon: **2** Valentin et al. 1999b, fig. 22. – **C** Le Grand Canton, upper horizon: **3-17** Valentin et al. 1999b, figs 26-27. – **D** Bonnières-sur-Seine: **18-20** Barois-Basquin/Charier/Lécolle 1996, fig. 10. – **E** Étigny Le Brassot, south: **21-25** Lhomme et al. 2004, fig. 32.

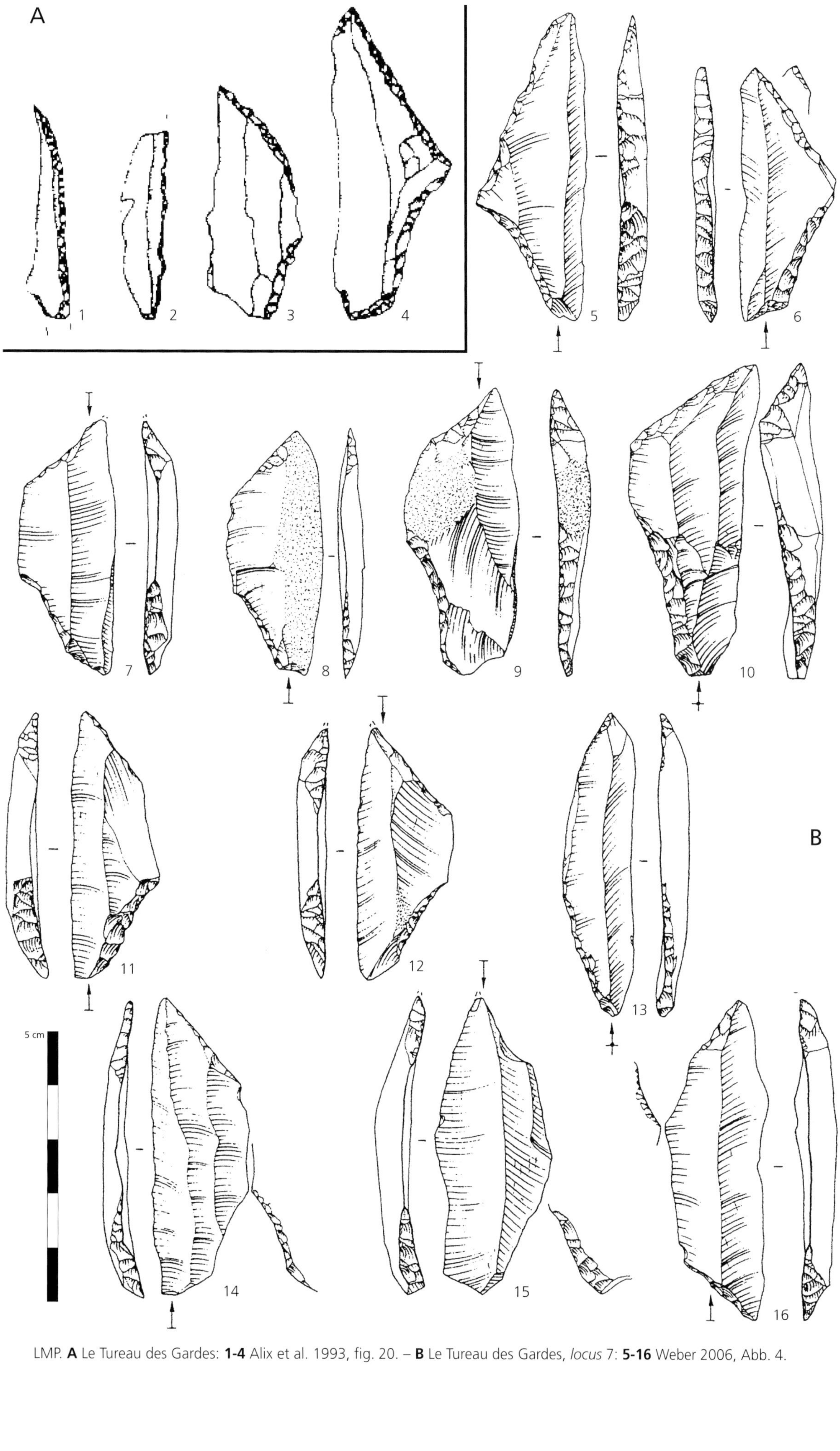

LMP. **A** Le Tureau des Gardes: **1-4** Alix et al. 1993, fig. 20. – **B** Le Tureau des Gardes, *locus* 7: **5-16** Weber 2006, Abb. 4.

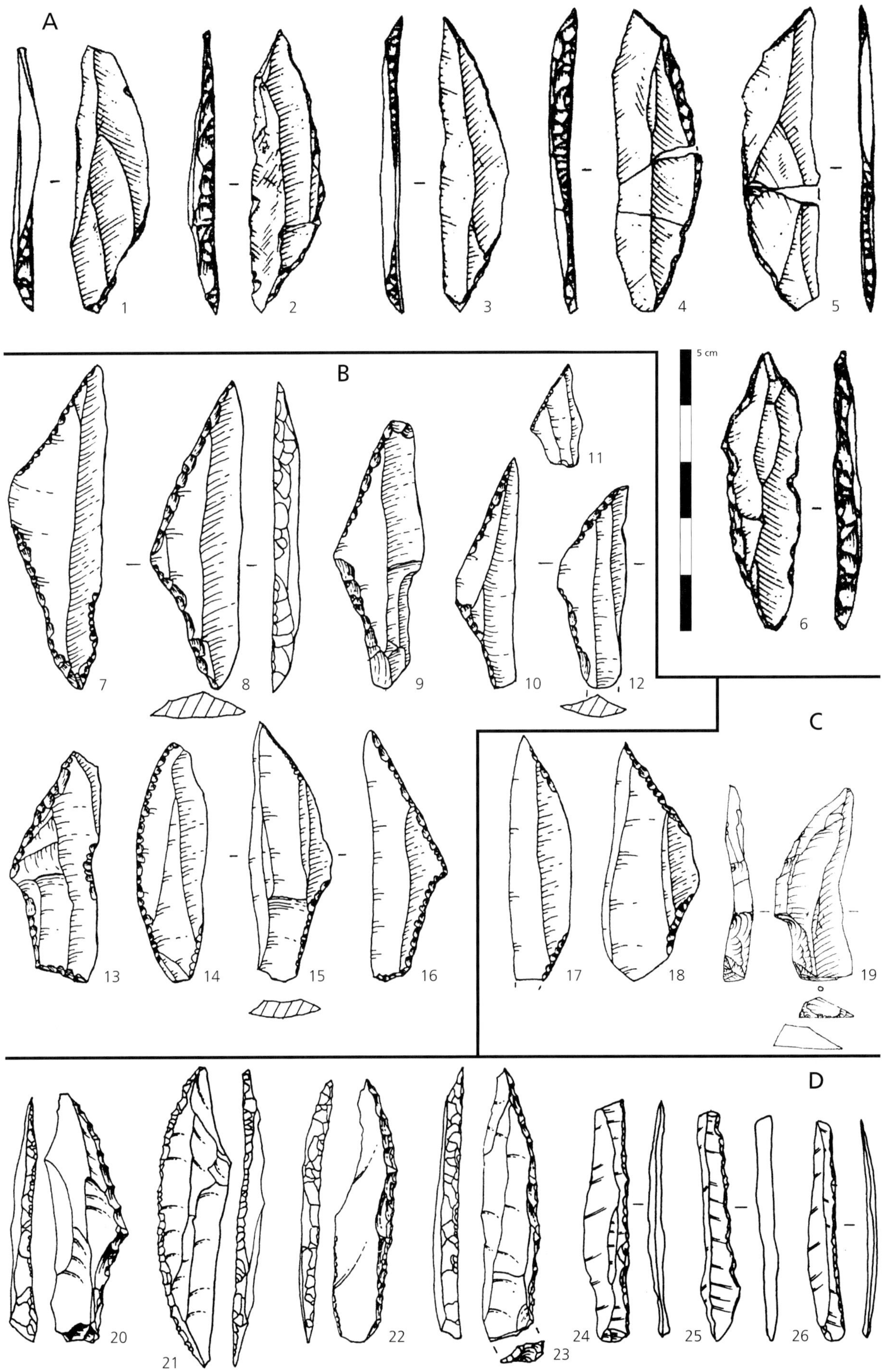

LMP. **A** Le Closeau, lower horizon, *locus* 4: **1-6** Bodu 1998, fig. 418. – **B** Cepoy, horizon IV, sector 1: **7-16** Valentin 1995, Planche 38. – **C** Cepoy, horizon IV, sector 2: **17-19** Wenzel 2009, fig. 52. – **D** Marsangy: **20-26** Schmider 1992a, figs 106. 108. 111 (**20-22** D14; **23** H17-D14; **24-26** H17).

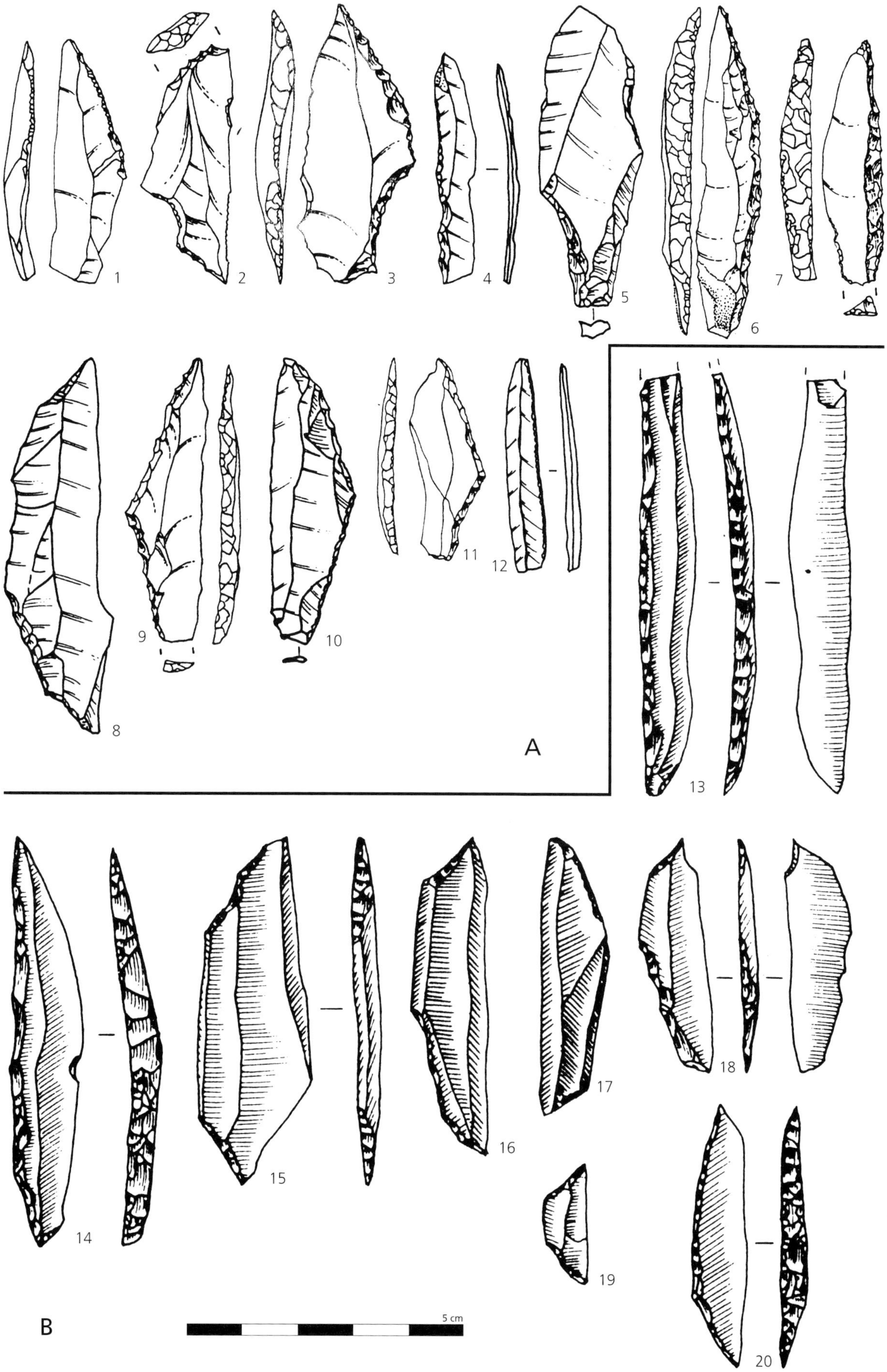

LMP. **A** Marsangy: **1-12** Schmider 1992a, figs 106. 108-109. 111 (**1-2** H17; **3-5** H17-N19; **6-12** N19). – **B** Belloy-sur-Somme, lower horizon: **13-20** Fagnart 1997, figs 40-41.

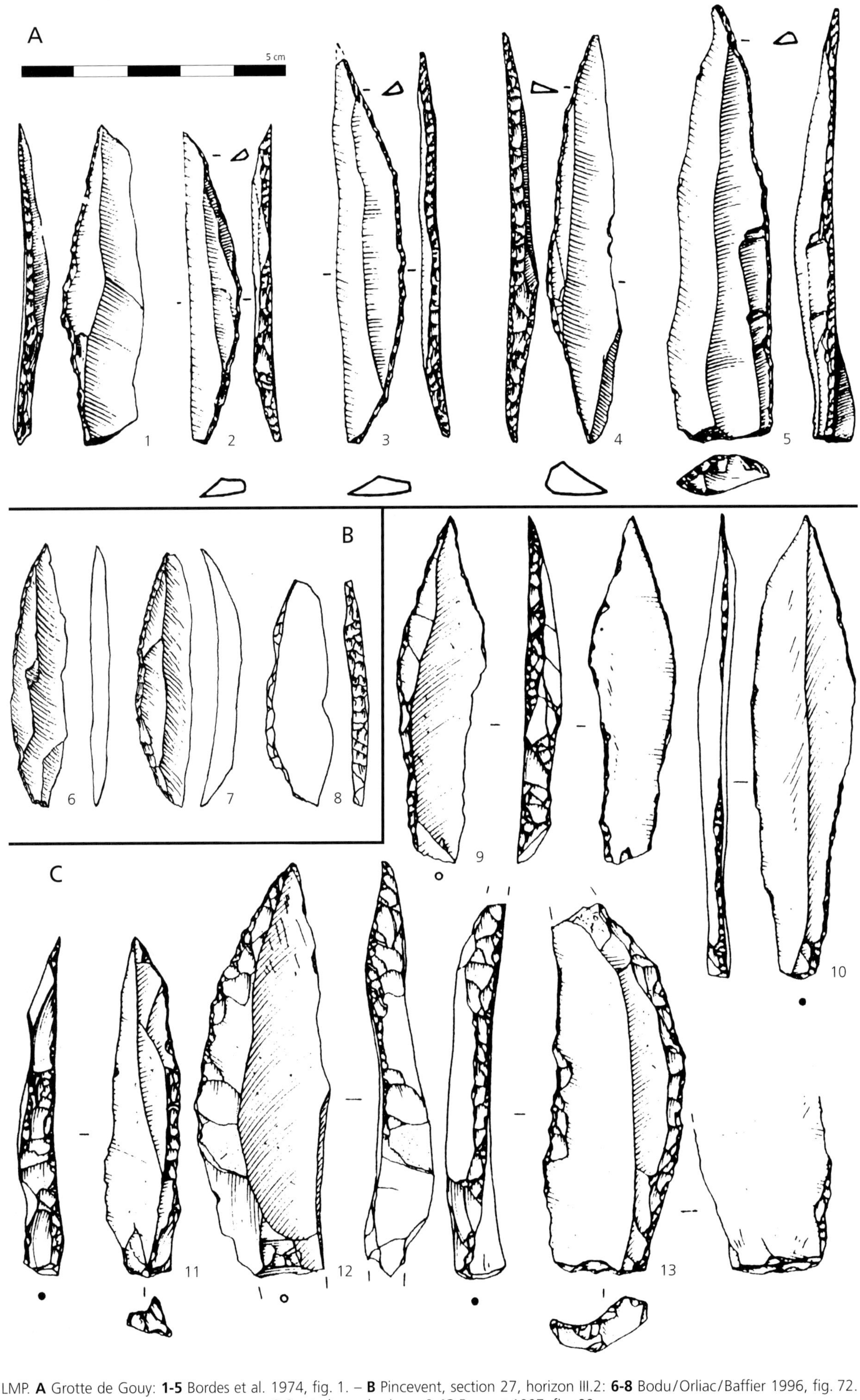

LMP. **A** Grotte de Gouy: **1-5** Bordes et al. 1974, fig. 1. – **B** Pincevent, section 27, horizon III.2: **6-8** Bodu/Orliac/Baffier 1996, fig. 72. – **C** Conty, lower horizon: **9-13** Fagnart 1997, fig. 89.

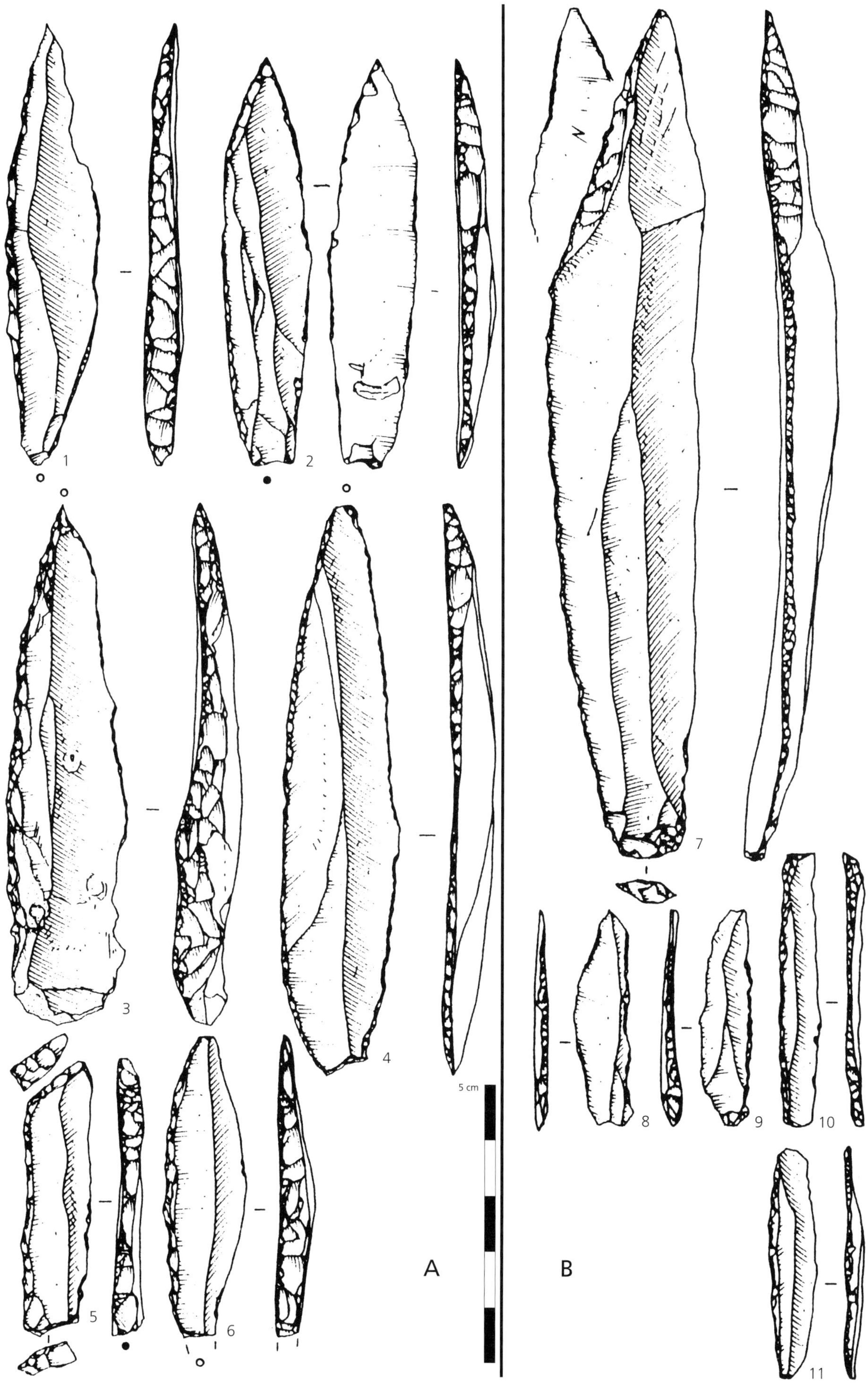

LMP. **A** Conty, lower horizon: **1-6** Fagnart 1997, fig. 89. – **B** Hangest-sur-Somme III.1, lower horizon: **7-11** Fagnart 1997, fig. 166.

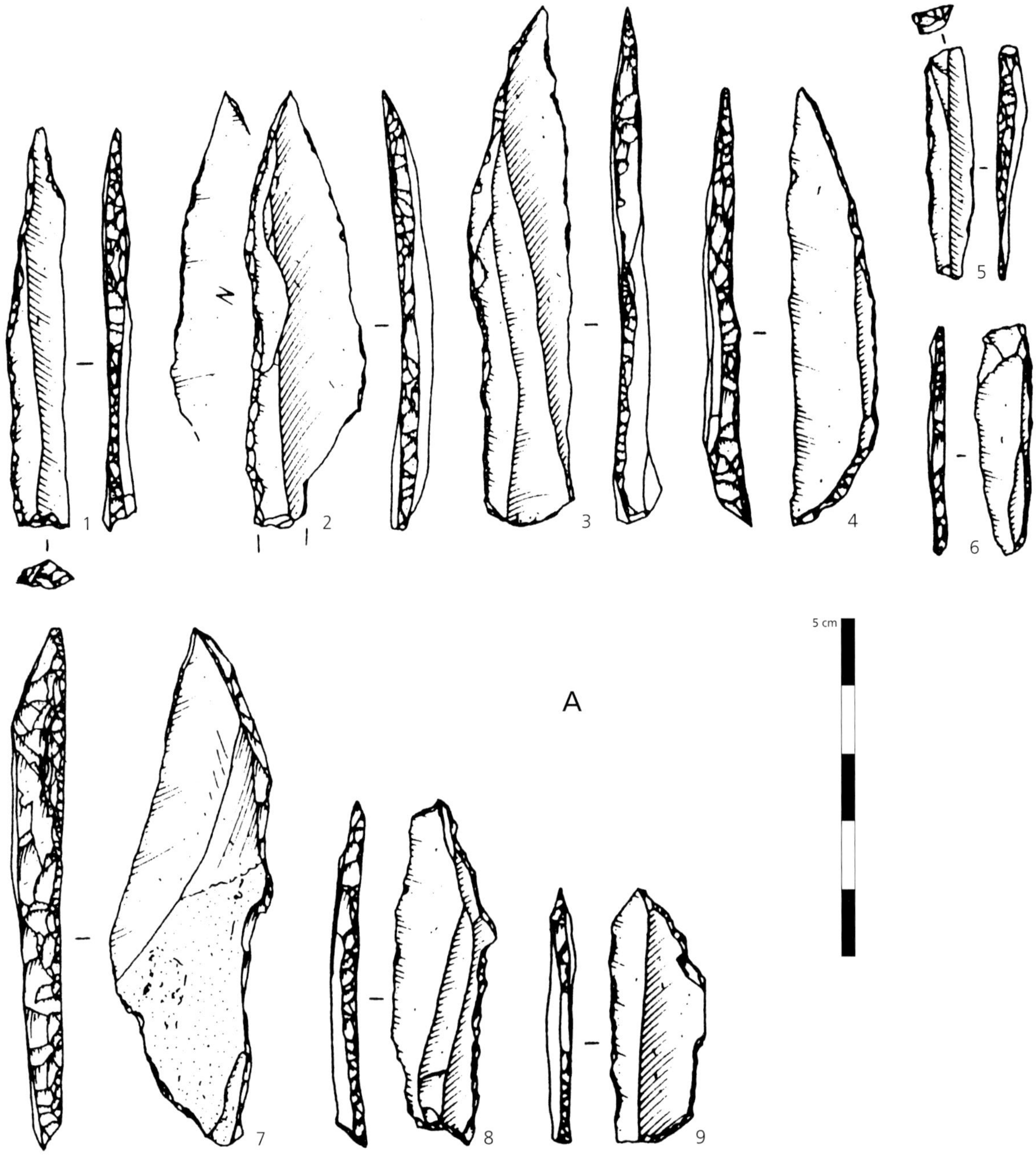

Plate 14 LMP. **A** Hangest-sur-Somme III.1, lower horizon: **1-9** Fagnart 1997, fig. 166.